uth   *P a c i f i c*

Wuvulu

Manus

New
Hanover

145° east

150° east

B i s m a r c k   S e a

New

I r e l a n d

Rabaul

5° south

Wewak

pik   River

Ambunti

Adelbert Mts.

Karkar

Madang
Astrolabe Bay
Bongu

Kimbe

Annanberg

Ramu River

Usino

New   B r i t a i n

ntral   Range

orgera
ende Highlands

Mount
Hagen

Mt Wilhelm
4509 m

Kubor Ra.

Goroka

YUS   Huan
Peninsula

Tari

Mt Giluwe
4368 m

Lae   Finschhafen

ake Kutubu

Purari River

Okapa

Huon Gulf

t Bosavi
2397 m

Crater Mt

Kratke
Range

Markham River

A P U A   N E W   G U I N E A

Aseki

Wau

Solomon   Sea

Morobe

Lakekamu Basin

Kerema

Owen   Stanley

Mt Albert
Edward
3993 m

Iloma

Popondetta

Trobriand
Islands

Woodlark

G u l f   o f

Mt Victoria
4073 m

Tufi   D'Entrecasteaux

Goodenough

Fergusson   Islands

riomo

Daru

P a p u a

Port Moresby

Range

Mt Suckling
3676 m

Normanby

Louisiade   Archipelago

Strait

nks

145° east

150° east

Cape Rodney

Mt. Simpson
3039 m

Milne Bay

Misima

Amazon Bay

Suau

Tagula   Rossel

10° south

pe

rk

nsula

C o r a l   S e a

A

BIRDS OF NEW GUINEA

DISTRIBUTION, TAXONOMY, AND SYSTEMATICS

Birds of New Guinea

Distribution, Taxonomy, and Systematics

Bruce M. Beehler *and* Thane K. Pratt

Mary LeCroy, *Technical Editor*

Princeton University Press, Princeton *and* Oxford

Published by Princeton University Press
41 William Street, Princeton, New Jersey 08540

In the United Kingdom:
Princeton University Press
6 Oxford Street, Woodstock, Oxfordshire OX20 1TW

press.princeton.edu

Cover: Raggiana Bird of Paradise,
photograph by Bruce Beehler

Frontispiece: Rainbow Lorikeets,
illustration by John Anderton

ISBN 978-0-691-16424-3

Library of Congress Control Number: 2015939779

British Library Cataloging-in-Publication
Data is available

Edited by Amy Hughes
Designed by Carol Beehler
Typeset in Adobe Minion
Cover designed by Amanda Weiss

Printed in the United States of America
on acid-free paper. ∞

Recommended citation:
Beehler, Bruce M., and Thane K. Pratt. 2016. *Birds of
New Guinea: Distribution, Taxonomy, and Systematics.*
Princeton University Press, Princeton, NJ, USA

10 9 8 7 6 5 4 3 2 1

Dedicated to the memory of our parents, who fostered our love of birds and nature

Contents

Preface · 8

PART ONE

Introduction · 11
  The New Guinea Region · 11
  New Guinea in Context · 12
  New Guinea's Bird Regions · 13
  References and Data Sources · 16
  Systematics · 18
  Taxonomy · 21
  Species Concepts · 22
  Treatment of Subspecies · 23
  The Avifauna · 25
  Historical Biogeography · 25
  The Checklist and Nature Conservation · 26
  The Future · 27
Layout of the Accounts · 28
  Terms, Usage, and Abbreviations · 34
Figures · 37

PART TWO

Casuariiformes · 40
  Casuariidae · 40
Galliformes · 44
  Megapodiidae · 44
  Phasianidae · 50
Anseriformes · 53
  Anseranatidae · 53
  Anatidae · 54
Phoenicopteriformes · 61

Podicipedidae · 61
Columbiformes · 63
  Columbidae · 63
Phaethontiformes · 92
  Phaethontidae · 92
Procellariiformes · 93
  Oceanitidae · 94
  Hydrobatidae · 96
  Procellariidae · 97
Ciconiiformes · 103
  Ciconiidae · 103
  Pelecanidae · 104
  Threskiornithidae · 105
  Ardeidae · 108
  Fregatidae · 116
  Sulidae · 117
  Phalacrocoracidae · 120
  Anhingidae · 122
Otidiformes · 123
  Otididae · 123
Gruiformes · 123
  Rallidae · 124
  Gruidae · 135
Cuculiformes · 136
  Centropodidae · 136
  Cuculidae · 138
Caprimulgiformes · 148
  Podargidae · 149
  Caprimulgidae · 150
  Aegothelidae · 153
  Hemiprocnidae · 157

Apodidae · 157
Charadriiformes · 163
   Burhinidae · 163
   Haematopodidae · 164
   Recurvirostridae · 165
   Charadriidae · 166
   Jacanidae · 171
   Scolopacidae · 171
   Turnicidae · 184
   Glareolidae · 186
   Laridae · 187
   Stercorariidae · 197
Accipitriformes · 198
   Pandionidae · 199
   Accipitridae · 200
Strigiformes · 213
   Tytonidae · 213
   Strigidae · 216
Bucerotiformes · 219
   Bucerotidae · 219
Coraciiformes · 220
   Meropidae · 220
   Coraciidae · 221
   Halcyonidae · 222
   Alcedinidae · 233
Falconiformes · 235
   Falconidae · 235
Psittaciformes · 238
   Cacatuidae · 238
   Psittrichasidae · 240
   Psittaculidae · 241
Passeriformes · 271
   Pittidae · 271
   Ptilonorhynchidae · 274
   Climacteridae · 282
   Maluridae · 284
   Meliphagidae · 288
   Acanthizidae · 325
   Pomatostomidae · 340
   Orthonychidae · 341
   Cnemophilidae · 342
   Melanocharitidae · 345
   Paramythiidae · 352
   Psophodidae · 354

Cinclosomatidae · 355
Machaerirhynchidae · 360
Cracticidae · 362
Artamidae · 365
Rhagologidae · 367
Campephagidae · 368
Neosittidae · 379
Oreoicidae · 381
Eulacestomatidae · 383
Pachycephalidae · 384
Oriolidae · 400
Rhipiduridae · 406
Dicruridae · 413
Ifritidae · 415
Paradisaeidae · 416
Melampittidae · 438
Monarchidae · 440
Corvidae · 453
Laniidae · 454
Petroicidae · 455
Alaudidae · 473
Hirundinidae · 474
Pycnonotidae · 478
Phylloscopidae · 479
Zosteropidae · 481
Acrocephalidae · 487
Locustellidae · 489
Cisticolidae · 492
Sturnidae · 494
Turdidae · 498
Muscicapidae · 500
Dicaeidae · 503
Nectariniidae · 506
Passeridae · 510
Estrildidae · 511
Motacillidae · 520

PART THREE

Bibliography · 525
Geographic Gazetteer · 560
   J. L. Mandeville and W. S. Peckover
Index of English Bird Names and Topics · 633
Index of Scientific Names · 647

Preface

ERNST MAYR, the paterfamilias of New Guinea ornithology, published his *List of New Guinea Birds* in 1941. It serves as the model for this publication. Mayr's seminal treatise to this day remains a cherished collectible as well as an informative reference for students of the avifauna of New Guinea. Offering this update, some seventy-five years later, acknowledges that the job Mayr did back before World War II was remarkably thorough and incisive. Some of the troublesome systematic and taxonomic issues that Mayr identified remain only partially resolved today, but molecular systematics has provided a major step forward, especially at the higher taxonomic levels. Exciting challenges await whomever takes up the avifauna of New Guinea as a focus of study, and we hope our publication will encourage a new generation of field, museum, and molecular-systematic ornithologists to work to resolve the persistent enigmas highlighted in Mayr's *List* and revisited here.

We started this book in the year 2000, at the same time that we started work on our field guide, *Birds of New Guinea, Second Edition* (Pratt & Beehler 2014). Both projects progressed in lockstep. Just as the *Species-Checklist of the Birds of New Guinea* (Beehler & Finch 1985) provided the taxonomic foundation of the first edition of *Birds of New Guinea*, so the present new book (which we informally call "the Checklist") delivers the scientific documentation for the second edition of the field guide. This new Checklist vastly expands the taxonomic scope of the 1985 work, containing very detailed accounts of distribution, systematics, plus treatment to the subspecies level. This expansion was necessary because of a wealth of new information on bird distribution within the region, wholesale changes to the classification of New Guinea birds, and the need to sort out the long-neglected treatment of subspecies. Because of this flood of new information, the Checklist's purpose outgrew the limits of the field guide, and this book became a truly independent and comprehensive avifaunal treatise. We have tracked down thousands of bits of information new and old pertaining to the distribution of the New Guinea birds and have traced the revolution in systematics for the avifauna. In this Checklist we have delved into the whys and wherefores of the classification of this great avifauna.

Towards the end of the field guide project, the burden of producing two lengthy and highly-detailed books proved too much. It was then decided to complete the field guide first, with the Checklist to follow. As a result, the distributional and taxonomic information of the two books matches closely, but not perfectly, owing to newly-published studies and some post-poned systematic decisions that are presented now for the first time.

For both authors this project has been a labor of friendship and collaboration. Many have generously lent a helping hand. We start with Mary LeCroy of the American Museum of Natural History, New York (AMNH). Mary is, in many ways, the keeper of the archive for New Guinea and its birdlife. We have gone to her to resolve the countless details surrounding the collectors, places, expeditions, type descriptions, and study skins that can be amassed only at a great natural history museum. She has worked for five decades in the American Museum's Department of Ornithology. No one knows more. We salute her dedication, friendship, and generosity. She was the natural candidate for Technical Editor.

Our respective home institutions have each offered a productive and supportive platform for long-term research. For BMB this is the Division of Birds of the Smithsonian's National Museum of Natural History in Washington, DC (here abbreviated as USNM). For TKP it is the Department of Vertebrate Zoology of the Bernice P. Bishop Museum, Honolulu (BPBM). We thank both institutions for their help.

Others have aided in various ways. Peter Mumford spent a summer assembling data for an early version of the text. Jennifer L. Mandeville completed the late Bill Peckover's New Guinea geographic gazetteer, which appears here as an appendix. Fellow scientists and scholars have shared their knowledge with us. Jared Diamond has provided advice on our work at various stages as well as important information on distributions of many species inhabiting New Guinea. David Bishop reviewed our manuscript at several stages and provided many insightful corrections and additions. The manuscript in draft was reviewed and critiqued by Sebastian van Balen, K. David Bishop, Walter Boles, Brian Coates, Jack Dumbacher, Guy Dutson, Chris Filardi, Ben Freeman, David Gibbs, Frank Gill, Leo Joseph, Mary LeCroy, Roger Pasquier, Richard Schodde, Christophe Thébaud, and Iain Woxvold. Valuable technical advice was offered by Allen Allison, Michael Andersen, Richard Banks, Brett Benz, Enrico Borgo, Les Christidis, Nigel Cleere, John Cox, Joel Cracraft, Normand David, Mark Denil, Edward C. Dickinson, Giuliano Doria, Carla Dove, Scott Edwards, Clifford Frith, Francesco Germi, Brian Gill, Gary Graves, Philippa Horton, Helen James, Knud Jønsson, Leo Joseph, Allan Keith, Henny van de Kerkhof, Kellee Koenig, Andy Mack, Steven van der Mije, Borja Mila, Storrs Olson, Dewi Prawiradilaga, Paul Scofield, Neil Stronach, Paul Sweet, Carlo Violani, and Bret Whitney. Also we thank the excellent technical staff of the ornithology units of the Museum Zoologicum Bogoriense, Cibinong, Indonesia; the Natural History Museum, Tring, England; the National Museum of Natural History, Leiden, Netherlands; Museo Civico di Storia Naturale G. Doria, Genoa, Italy; the American Museum of Natural History, New York; the Bernice Bishop Museum, Honolulu; and the National Museum of Natural History, Smithsonian Institution, Washington, DC. Jacob Saucier provided a critical reading of the text in the proof stage. Robert Kirk and Princeton University Press made publication possible; Ellen Foos expertly managed the complicated publication process; Amy K. Hughes provided detailed and insightful copyediting; and Carol Beehler designed the book to make it compact, handsome, and readable. Finally, John Anderton provided the lovely bird drawings that grace the text.

PART I

Southern Crowned Pigeons, illustration by John Anderton

Introduction

The New Guinea Region

Our region of coverage follows Mayr (1941: vi), who defined the natural region that encompasses the avifauna of New Guinea, naming it the "New Guinea Region." It comprises the great tropical island of New Guinea as well as an array of islands lying on its continental shelf or immediately offshore. This region extends from the equator to latitude 12° south and from longitude 129° east to 155° east; it is 2,800 km long by 750 km wide and supports the largest remaining contiguous tract of old-growth humid tropical forest in the Asia-Pacific (Beehler 1993a). The Region includes the Northwestern Islands (Raja Ampat group) of the far west—Waigeo, Batanta, Salawati, Misool, Kofiau, Gam, Gebe, and Gag; the Aru Islands of the southwest—Wokam, Kobroor, Trangan, and others; the Bay Islands of Geelvink/Cenderawasih Bay—Biak-Supiori, Numfor, Mios Num, and Yapen; Dolak Island of south-central New Guinea (also known as Dolok, Kimaam, Kolepom, Yos Sudarso, or Frederik Hendrik); Daru and Kiwai Islands of eastern south-central New Guinea; islands of the north coast of Papua New Guinea (PNG)—Kairiru, Muschu, Manam, Bagabag, and Karkar; and the Southeastern (Milne Bay) Islands of the far southeast—Goodenough, Fergusson, Normanby, Kiriwina, Kaileuna, Woodlark, Misima, Tagula/Sudest, and Rossel, plus many groups of smaller islands (see the endpapers for a graphic delimitation of the Region).

Politically, the island of New Guinea is bisected at longitude 141° east. To the west is Indonesian New Guinea (comprising Papua and Papua Barat Provinces). To the east of the line is the mainland portion of Papua New Guinea. Although this abrupt north-south boundary line is an artificial product of colonial-era claims, today this line, in effect, separates Asia (to the west) from the Pacific (to the east).

Indonesian New Guinea includes the western half of mainland New Guinea plus the islands of Geelvink (Cenderawasih) Bay, the Aru Islands, and the Raja Ampat Islands—all territory covered in this book. Papua New Guinea encompasses territory in the New Guinea region—the eastern half of the island of New Guinea and the islands of Milne Bay Province, as well as territory outside the Region and not covered in this book—the Bismarck and Admiralty Islands and the northernmost of the Solomon Islands. Thus, much of insular Papua New Guinea is *not* included in this treatment.

The New Guinea region *does not* include Seram, the Southeast Islands of Indonesia, the Kai Islands, the Torres Strait Islands, Long and Umboi Islands, New Britain, New Ireland, Manus, or the Solomon Islands. The northern Melanesian avifauna inhabiting these last six entities is admirably treated in Mayr & Diamond (2001) and Dutson (2011).

The postmodern political side of geographic names is problematic—is it Maluku or Moluccas? Nusa Tenggara or Lesser Sundas? Tagula or Sudest? Rossel or Yela? We have tended toward conservatism here (particularly because the main users of this type of work are people interested in the history of ornithology), as the older names have received more use in the literature and are, quite simply, better known and more widely used in the science. Our geographic gazetteer in the back of the book (Appendix) makes an initial attempt to present *all* the geographic names, so one can locate and identify both the new and old name here, even if only the "old" name appears in the text accounts. Some of the more prominent choices appear in our usage chart at the end of this introductory section.

With regard to seabirds, our treatment includes records within *ca.* 50 km of the Mainland coastline and *ca.* 25 km of any fringing New Guinea island. Also included are waters encompassed by embayments (the limit is a straight line between major projecting points on the Mainland). That said, we do not include any territorial waters of Australia (which in the Torres Strait approaches northward to the shores of the New Guinea mainland) or the Solomon Islands. We strongly encourage much additional seabird-watching in New Guinea's waters. These efforts should be timed to coincide with the annual spring and fall movements of these long-distance migrants and mainly should focus on the western and eastern extremities of the Region, where north-south water passages encourage concentrations of the birds where they can bypass the substantial east-west land barrier posed by mountainous mainland New Guinea.

New Guinea in Context

Aside from continental Australia, the only island larger than New Guinea is ice-capped Greenland. Among tropical islands, New Guinea is the largest and highest (it is larger and substantially higher than either Madagascar or Borneo) and still supports tropical glaciers in the far west of its high Central Ranges.

New Guinea is the geographic hub of the southwest Pacific—situated at the heart of an array of tropical island arcs that are home to a wonderful assemblage of bird species featured in this book. Australia lies just to the south. New Guinea and Australia share the Australian plate and thus the same tectonic history—New Guinea is the high, wet, and equatorial sector, whereas continental Australia is the low, dry, and temperate sector. To the west lie the Moluccas (Maluku) and Lesser Sundas (Nusa Tenggara) of Indonesia. To the north and northwest lie the Philippines, Palau, and the Mariana Islands. The Bismarck, Admiralty, Caroline, Marshall, and Gilbert Islands lie to the northeast and east, and the Solomons, Vanuatu, New Caledonia, and Fiji to the southeast.

New Guinea supports the Pacific's richest humid-forest avifauna. By contrast, Australia hosts the Pacific's richest savanna and dry-zone avifauna. Both rest atop the Australian continental craton, isolated by deepwater barriers from Sundaland to the west and from the Melanesian islands to the northeast and southeast. Whereas the differences between the avifaunas of New Guinea and Australia are mainly products of their distinct environments, the differences distinguishing New Guinea's avifauna from that of southeast Asia are biogeographic in origin. Wallace's Line, an ancient deepwater barrier, marks the eastern limit of many continental Asian bird lineages, separating the continental avifaunas of southeast Asia from those of Australia–New Guinea. In a similar manner, moving from New Guinea eastward into island Melanesia, one suddenly encounters oceanic avifaunas poor in many of the lineages that are widespread in the Australia–New Guinea region (*e.g.*, Australasian robins, Australasian warblers, bowerbirds, and birds of paradise) yet distinctively rich in a subset of Australasian lineages that have exploded across this insular geography (*e.g.*, monarchs and whistlers). Many bird groups that evolved on the Australian continent have not been very successful in colonizing oceanic island archipelagoes.

In sum, New Guinea is the Pacific's version of Andean South America—wet, tropical, cordilleran, and species-rich. And New Guinea's vast rain forests and montane cloud forests, like those of the Andean region, remain a treasure trove for research ornithologists. Here, future students of biogeography will come to address the distributional and systematic mysteries that are embedded in the species accounts that constitute the bulk of this book.

One thing that makes New Guinea different from cordilleran South America is its incredible cultural and linguistic diversity. Its many indigenous peoples live customary lifestyles and speak more than a thousand distinct languages (not dialects). These rural-dwelling people have had a long and close relationship with the birdlife, and village naturalists are often incredibly knowledgeable about the habits of these birds (Diamond 1966). In many instances, the astute guidance and cheerful assistance of local New Guinean naturalists have ensured the success of Western field ornithologists, especially in the period after World War II. Both authors can attest that their doctoral and postdoctoral field researches were immeasurably aided by the contribution of New Guinean field collaborators.

New Guinea's Bird Regions

New Guinea is geographically complex, and poses a challenge to the novice seeking to study its avifauna. It is hard enough learning the scientific and English names of nearly 800 bird species. Add to that the difficulty of learning where these species live in a mountainous world twice the size of California but with far fewer cities, roads, and other identifying features. Here, we employ a set of 15 standardized names for New Guinea's bird regions. These New Guinean bird regions articulate ornithogeographic zones, defined by areas of species or subspecies endemism and bounded by physiographic barriers that separate abutting ranges of sister forms. These are adapted from the bird areas highlighted in the first edition of the *Birds of New Guinea* field guide (Beehler *et al.* 1986) and influenced by Birdlife International's endemic bird areas as well as the World Wildlife Fund's Pacific ecoregions. They appear in the accompanying map (fig. 1), and are briefly described below, from northwest to southeast. Note that these updated bird regions are also used in the revised *Birds of New Guinea* field guide (Pratt & Beehler 2014).

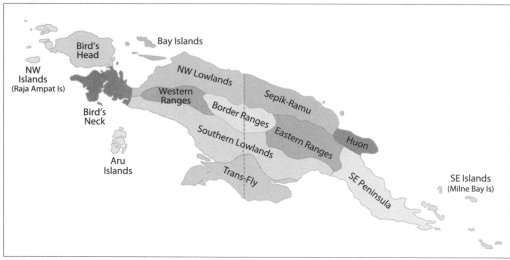

Figure 1. New Guinea Bird Regions

Northwestern Islands/NW Islands (Raja Ampat or Western Papuan Islands). The islands of this region include Waigeo, Batanta, Salawati, Misool, Kofiau, Gam, Gag, and Gebe (plus other smaller islands). They are home to an array of Moluccan species found nowhere else in the New Guinea region, plus six endemics: *Aepypodius bruijnii, Tanysiptera elliotii, Pitohui cerviniventris, Symposiachrus julianae, Cicinnurus respublica,* and *Paradisaea rubra.*

Bird's Head (Vogelkop, Berau or Doberai Peninsula). This area features the Arfak and Tamrau Mountains, which host a number of montane specialties (some shared with the Bird's Neck): *Charmosyna papou, Amblyornis inornata, Melipotes gymnops, Melidectes leucostephes, Sericornis rufescens, Parotia sefilata, Paradigalla carunculata, Astrapia nigra,* and *Lonchura vana.* The region also includes substantial lowlands in its southern sector.

Bird's Neck. An expanse of rugged and isolated low mountains featuring fjords on the southern coast, this is a little-studied and physiographically spectacular region. Specialties shared with the Bird's Head include *Amblyornis inornata* (or an undescribed relative), *Melipotes gymnops, Melidectes leucostephes, Parotia sefilata,* and a *Paradigalla* not yet identified to species. Note that the Bird's Neck includes the Onin Peninsula (home of the Fakfak Mountains); the Bomberai Peninsula (home of the Kumawa Mountains); and the Wandammen Peninsula (home of the Wandammen or Wondiwoi Mountains), all of which are important areas of montane endemism, mainly at the subspecies level.

Bay Islands/Bay Is (islands of Geelvink Bay/Teluk Cenderawasih). This region features Biak and Supiori Islands (a matched pair) and Numfor, Mios Num, and Yapen Islands. Biak-Supiori and Numfor are oceanic islands and support endemic species of birds: *Centropus chalybeus, Otus beccarii, Tanysiptera riedelii, Seicercus misoriensis, Symposiachrus brehmii,* and others on Biak; *Tanysiptera carolinae* and *Seicercus maforensis* on Numfor; *Micropsitta geelvin-kiana* and *Myiagra atra* on Biak and Numfor. Some of these endemic species also range out to small islands in the bay: *Megapodius geelvinkianus*, *Ducula geelvinkiana*, and *Eos cyanogenia*. Substantial and mountainous Yapen Island is a land-bridge island with some endemism at the subspecies level.

Northwestern Lowlands/NW Lowlands. Essentially the vast drainage of the Mamberamo basin, which includes the Mamberamo, Tariku, Taritatu/Idenburg, and Van Daalen Rivers, it combines lowland forest, swamps, and small but important north coastal ranges (Foja, Van Rees, and Cyclops). Specialties include *Rallicula mayri*, *Psittaculirostris salvadorii*, *Amblyornis flavifrons*, *Ptiloprora mayri*, *Philemon brassi*, *Drepanornis bruijnii*, and *Parotia berlepschi*.

Aru Islands/Aru Is. Composed of islands of uplifted coral with extensive mangrove channels, this region shares avian specialties with the Southern Lowlands. *Eulabeornis castaneoventris* and *Dicaeum hirundinaceum* are recorded in the New Guinea region only from the Aru Islands. *Ptilinopus wallacii*, a Moluccan specialty, is also found in these islands.

Western Ranges/Western Ra. This is the great high sector of New Guinea's main cordillera, with a number of summits exceeding 4,500 m, and several small glaciers on Carstensz Massif/Puncak Jaya (4,884 m). It includes (historically) the Charles Louis, Weyland, Nassau, and Oranje Mountains, now called the Sudirman (western) and Jayawijaya (eastern) ranges. Specialties include *Anurophasis monorthonyx*, *Melionyx nouhuysi*, *Oreornis chrysogenys*, *Astrapia splendidis-sima*, *Lonchura teerinki*, and *Lonchura montana*.

Border Ranges/Border Ra. Only marginally less impressive than the Western Ranges, the Border Ranges include great summits such as Mount Mandala and Mount Capella and share some montane specialties with the Western Ranges such as *Melionyx nouhuysi*, *Astrapia splen-didissima*, and *Lonchura montana*.

Southern Lowlands/S Lowlands. This is a vast expanse of lowland rain forest that transitions to swamp forest and mangrove in the west and east, and seasonally flooded savanna in the central sector. Specialty species include *Sericulus ardens*, *Chenorhamphus campbelli*, *Pseudo-rectes incertus*, and *Paradisaea apoda*. In the far northwest, where the Southern Lowlands meet the Bird's Neck and the Northwestern Lowlands, one finds an ornithogeographic mixing zone where the ranges of many sister species and subspecies meet. That area would bear additional field study. Numerous species that seem to have originated in the Southern Lowlands have spilled eastward into the southern watershed of the Southeastern Peninsula, for instance, *Casu-arius casuarius*, *Talegalla fuscirostris*, and *Psittaculirostris desmarestii*.

Trans-Fly. The southern bulge of New Guinea, this area's large expanses of monsoon woodland and savanna share many species with Australia. Endemics include *Megalurus albolim-batus*, *Lonchura nevermanni*, and *Lonchura stygia*. It is an important area for migratory waders as well as resident waterbirds and migratory waterbirds from Australia.

Sepik-Ramu. The eastern counterpart to the Northwestern Lowlands, this is the interior basin of two rivers—the Sepik and the Ramu—isolated from the Northwestern Lowlands by a series of low ranges near the Papua New Guinea–Papua border. It includes much lowland rain forest plus some fire-generated grassland patches and grassy marshlands of the Sepik. It also encompasses Papua New Guinea's North Coastal Ranges and the Adelbert Mountains. Its sole endemic is *Sericulus bakeri*.

Eastern Ranges/Eastern Ra. This bird region includes the central highlands of Papua New Guinea (Kaijende Highlands, Mount Giluwe, Mount Hagen, Kubor Mountains, Schrader Range, Bismarck Range, and Kratke Mountains). It extends westward to the Strickland River gorge and eastward to the Kratke Mountains, beyond which lies the Watut-Tauri Gap, which marks the northwestern terminus of the mountains of the Southeastern Peninsula. The Eastern Ranges are home to two endemics: *Melionyx princeps* and *Astrapia mayeri*.

Huon Peninsula/Huon Penin. This bird region contains a compact collection of high ranges (Finisterre, Saruwaged/Sarawaget, Cromwell, and Rawlinson) isolated from the Eastern Ranges by the broad lowland Markham and Ramu valleys. It is home to six endemic or near-endemic species: *Amblyornis germana, Melipotes ater, Melidectes foersteri, Parotia wahnesi, Astrapia rothschildi,* and *Paradisaea guilielmi*.

Southeastern Peninsula/SE Peninsula. The region comprises the Herzog and Kuper Mountains as well as the Owen Stanley and Wharton Ranges, the latter two substantial mountain chains with peaks to 4,400 m. The peninsula also includes coastal lowlands and some river basins. It is home to the endemics *Tanysiptera danae, Amblyornis subalaris, Parotia helenae, Lonchura caniceps,* and *Lonchura monticola*.

Southeastern Islands/SE Islands. This is the southeastern counterpart to the Northwestern Islands at the opposite end of New Guinea. Major islands include Goodenough, Fergusson, and Normanby (constituting the D'Entrecasteaux Archipelago); the Trobriand Islands; Misima, Rossel, and Tagula Islands (the Louisiade Archipelago); and Woodlark Island. Endemics include *Myzomela albigula, Meliphaga vicina, Manucodia comrii,* and *Paradisaea decora,* among others.

References and Data Sources

Regional taxonomic checklists are historical in nature and reflect the cumulative efforts of many dedicated individuals over long periods. This checklist is built largely on previously published checklists, field guides, handbooks, journals, and newsletters, which will continue to serve as the core of knowledge about the nomenclature, taxonomy, systematics, distribution, and natural history of the birds of New Guinea for decades to come. They should be studied by anyone working on ornithology of the Region. We list the most important of these here, by date of publication.

Salvadori, Tommaso. 1880–1882. *Ornitologia della Papuasia e delle Molucche.* 3 volumes. Stamperia reale G.B. Paravia e co. di I. Vigliardi, Torino, Italy.

Peters, James L. (and subsequent editors). 1931–1987. *Check-list of Birds of the World.* Harvard University Press and Museum of Comparative Zoology, Cambridge, Massachusetts.

Mayr, Ernst. 1941. *List of New Guinea Birds.* American Museum of Natural History, New York, New York. [Now online at: http://dx.doi.org/10.5962/bhl.title.68262.]

Rand, Austin L., & E. Thomas Gilliard. 1967. *Handbook of New Guinea Birds.* Wiedenfeld and Nicolson, London/The Natural History Press, Garden City, New York.

Coates, Brian J. 1985, 1990. *The Birds of Papua New Guinea.* 2 volumes. Dove Publications, Alderley, Queensland, Australia.

Beehler, Bruce M., T. K. Pratt, & D. A. Zimmerman. 1986. Birds of New Guinea. Princeton University Press, Princeton, New Jersey.

Marchant, S., P. J. Higgins, *et al.* (eds.). 1990–2006. *Handbook of Australian, New Zealand and Antarctic Birds.* 7 volumes. Oxford University Press, Melbourne, Australia.

del Hoyo, Josep, and collaborators (eds.). 1992–2012. *Handbook of Birds of the World.* 16 volumes. Lynx Edicions, Barcelona, Spain.

Schodde, Richard, & Ian J. Mason. 1999. *The Directory of Australian Birds.* Passerines. CSIRO Publishing, Collingwood, Victoria, Australia.

Mayr, Ernst, & Jared M. Diamond. 2001. *The Birds of Northern Melanesia: Speciation, Ecology, and Biogeography.* Oxford University Press, Oxford, UK.

Coates, Brian J., and William S. Peckover. 2001. *Birds of New Guinea and the Bismark Archipelago.* Dove Publications, Alderley, Queensland, Australia.

Christidis, Les, & Walter E. Boles. 2008. *Systematics and Taxonomy of Australian Birds.* CSIRO Publishing, Collingwood, Victoria, Australia.

Dutson, Guy. 2011. *Birds of Melanesia: Bismarcks, Solomons, Vanuatu, and New Caledonia.* Princeton University Press, Princeton, New Jersey.

Dickinson, Edward C., & J. V. Remsen (eds.). 2013. *The Howard and Moore Complete Checklist of the Birds of the World.* 4th edition. Volume 1: Non-Passerines. Aves Press, Eastbourne, UK.

Dickinson, Edward C., & L. Christidis (eds.). 2014. *The Howard and Moore Complete Checklist of the Birds of the World.* 4th edition. Volume 2: Passerines. Aves Press, Eastbourne, UK.

Pratt, Thane K., & Bruce M. Beehler. 2014. *Birds of New Guinea.* 2nd edition. Princeton University Press, Princeton, New Jersey.

del Hoyo, J., & N. J. Collar. 2014. *Illustrated Checklist of the Birds of the World.* Volume 1: Non-passerines. Lynx Edicions, Barcelona, Spain.

Gill, Frank, & David Donsker (eds.). 2015. *IOC World Bird Names* (version 5.1). Available at http://www.worldbirdnames.org [dx.doi.org/10.14344/IOC.ML.5.1].

In addition to the above references, this checklist synthesizes a wide assortment of primary sources, which we cite within the text and list in the bibliography. We note in particular the following serial publications that over the decades have featured important publications on the birds of New Guinea: the *Emu,* the *Muruk,* and the *Papua New Guinea Bird Society Newsletter.* For every substantive decision or new field record (post Mayr 1941), we have attempted to cite an authoritative source. In the instances where we ourselves have been the source of novel treatments or distributional records, we provide the distributional details and a reasoned argument in support of a taxonomic or systematic decision.

Systematics

Systematics is the science of delineating the pattern and history of evolutionary diversification of life on earth. The last decade has seen explosive growth in the publication of systematic revisions of avian groups based on the computational analysis of base-pair sequences of mitochondrial and nuclear DNA and, most recently, sampling of genomes. This has been a boon to the understanding of the composition and placement of orders, families, and genera. It has also led to major reorganization of the list as we have come to know it (*e.g.*, Cracraft 2013, 2014, del Hoyo & Collar 2014, Gill & Donsker 2015).

We entirely accept the practice of phylogenetic systematics and branching trees. The field of systematics has left the realm of phenetics and similarity and is firmly camped in the notion of monophyly and phylogeny. We note that DNA sequence analysis has been revolutionary but may not yet offer the magic bullet. Many issues confound the resolution of phylogenies using molecular data, and in not a few instances, morphological characters and molecules do not agree (Sorenson *et al.* 2003, Joseph 2008, Losos *et al.* 2012). Apparently the science of molecular systematics has some way to go before we fully access the secrets of DNA and properly understand all aspects of the relationship between genotype and phenotype.

For this checklist, our task has been to sort through the array of technical systematic publications and generate a synthesis of the results that best captures current understanding of the history of diversification of the birds of the New Guinea region. The task is not always straightforward, for two reasons. First, new studies produce novel results that often do no align with previously supported hypotheses. Second, it is not uncommon for different research teams employing varying techniques to produce disparate results for the same avian lineage—often using essentially the same molecular data set. Furthermore, the quality and rigor of analysis varies. At this time, the analysis of avian evolution using DNA sequence data remains in its early stages. We tend to side with the studies that employ the larger and more diverse (especially nuclear) molecular data sets and which are also informed by additional biological knowledge of the lineage being studied.

Every phylogeny, whether based on morphology or molecules, is no more than an evolutionary hypothesis, subject to future testing, revision, and refinement. However, there is a widespread perception that gives unwarranted "power" to molecules over morphology in systematic analysis. The unspoken belief that molecules don't mislead, whereas morphology does (because of homoplasy) is in no way resolved (see Losos *et al.* 2012, among others). It is essential to resist the temptation to see a molecular phylogeny as a magical revelation of evolutionary truth. Yet not only is it difficult to envision how DNA can mislead, the computer algorithms deployed to generate phylogenetic trees operate behind a veil, and do indeed appear magical.

In fact, the inability of molecular systematists to generate consensus or replicability, especially where sequence evolution has been sparse and/or random, points clearly to the need for general practitioners to proceed with caution. The generation of multiple systematic hypotheses, even when employing the same sequence data uploaded from GenBank, is the enemy of taxonomic and checklist stability. Systematists generate abundant publications with novel results, but to what effect? Practicing ecologists and conservationists wanting nomenclatural stability find frustration in this continual change in the details of systematic relationships and the need for repeated checklist revision. The end users of the new systematics await improved algorithms, better understanding of the inner workings of the DNA molecule, a capability to combine molecular and nonmolecular character data in tree generation, and the humility needed to generate good science, which is all about the replicability of results.

The accounts treating orders, families, genera, and species form the bulk of this work. In each, we provide minimal discussion of historical systematic treatment, and instead focus on the most recent informative work. Because of the comprehensive coverage provided in Schodde & Mason (1999), Christidis & Boles (2008), B. J. Gill *et al.* (2010), Dickinson & Remsen (2013), and Dickinson & Christidis (2014) among others, we feel it is not necessary to tread that ground in detail once again. Moreover, the discussion of all past systematic treatments, with their attendant conflicting systematic hypotheses, provides historical background but little in the way of ontological synthesis that leads to a higher degree of resulting systematic insight.

Our most challenging decisions have been at the ordinal and familial level, where there has been wholesale resetting of the sequence of nonpasserine orders and passerine families, as well as the breaking apart of lineages long treated as close relatives (Barker *et al.* 2004, Hackett *et mult.* 2008, Cracraft 2013, 2014, Jarvis *et al.* 2014). For purposes of this checklist, we have chosen to build our list from a combination of original research (*e.g.*, Hackett *et mult.* 2008, Jønsson *et al.* 2011b, Jarvis *et al.* 2014, Aggerbeck *et al.* 2014) as well as synthetic analyses (Dickinson & Remsen 2013, Dickinson & Christidis 2014, del Hoyo & Collar 2014, Gill & Donsker 2015). Our objective has been to provide a well-researched and well-referenced platform from which field and museum biologists may continue the study of the birds of New Guinea productively. The most recent whole-genome analysis of avian orders (*e.g.*, Jarvis *et al.* 2014) has, we think, provided a major step forward in strengthening the understanding of the avian tree at that higher level. Moreover, we hope the existence of our Checklist will, itself, provide some stability of usage over the years to come.

With few exceptions (noted in the text), we follow the latest evidence on systematics and relationships, so long as supporting data and analysis are deemed substantive and consonant with what we know of the biology of the lineage in question. On occasion, analyses may suggest the entire reordering of an existing species or generic sequence, yet are at odds with the known biology and morphology of the taxon in question. In these instances, we have proceeded with caution, accepting only components of the overall analysis supported with additional evidence. Thus, we overrule traditional names and analyses cautiously, and only when novel data provides compelling reason to do so. That said, we see our list as the "2015 model" not the "1985 model" (*e.g.*, Beehler & Finch 1985). We have been advised by some that the 2015 model is unproven and for the benefit of users, our new checklist should stick with the familiar 1985 sequence of orders and families. We are unable to do that in good conscience. Although this new sequence will prove difficult initially for fresh users, it reflects today's truth, which in years to come may indeed come to be tried and true. We certainly hope so.

The systematists' job would be easier if it were not for two issues: a shortage of reliable tissue samples from the birds of New Guinea and the limitations of the cutting-edge algorithms and statistics being applied to complex systematic problems. The issue of inadequate molecular data on the species of New Guinea's birds can only be remedied by the work of future fieldworkers in tandem with the effort of collaborating laboratory scientists. There are many instances where we have been forced to make checklist decisions based on field observations or laboratory results of uncertain provenance. Moreover, we have had to decide which molecular systematic analyses may or may not be couched in good scientific technique or statistical method. There is also the issue of generating properly identified and clean tissue samples (Moyle *et al.* 2013). Moreover, dealing with species groups that may exhibit introgressive hybridization creates problems for generating stable trees (Lavretsky *et al.* 2014).

Whereas 20th-century scholarship in avian systematics was largely based in building a case for a systematic decision through the marshaling of data related to plumage, behavior, and morphology, today's molecular systematic treatments are opaque to most nonspecialist readers. To demonstrate the instability of current molecular analyses, we review the phylogenetic placement of six important New Guinean honeyeater genera in publications from 2004 to 2014. The phylogenetic results for *Myzomela*, *Philemon*, *Melidectes*, *Meliphaga*, and *Ptiloprora* are as follows:

(*Philemon* (*Myzomela*+*Ptiloprora*) (*Melidectes*+*Meliphaga*))	(Driskell & Christidis 2004)
((*Myzomela*+*Philemon*) *Meliphaga*)	(Gardner *et al.* 2010; not included: *Ptiloprora* and *Melidectes*)
(*Myzomela*+*Ptiloprora*)(*Philemon* (*Melidectes*+*Meliphaga*))	(Nyári & Joseph 2011)
(*Myzomela*+*Philemon*)(*Ptiloprora* (*Melidectes*+*Meliphaga*))	(Joseph *et al.* 2014b)
(*Philemon* (*Myzomela*+*Ptiloprora*))(*Melidectes*+*Meliphaga*)	(Andersen *et al.* 2014a)

Of the possible generic pairings above, only the *Melidectes*+*Meliphaga* pairing is stable, appearing in four phylogenies (and not evaluated in the fifth). The overall instability makes it difficult for researchers generating regional lists to know how to proceed. Is it simply a matter of selecting the most recent publication? Or is it a matter of selecting the publication with the most base pairs or the most genes? Or the one with the best taxonomic sampling?

A more sinister problem is when lab teams do not know the biology of the birds they study and thus cannot tell when a computer-generated tree is uninformative. A published molecular analysis for bowerbirds using mitochondrial DNA (mtDNA) provided an example of this. The study aimed to trace the evolution of bower design across a molecular phylogeny. In this instance, the resulting tree was as follows:

(((((*Amblyornis inornatus*/Fakfak Mts+*Amblyornis inornatus*/Arfak Mts) *Archboldia papuensis*) *Amblyornis macgregoriae*) *Prionodura newtoniana*) *Amblyornis subalaris*)

Here, the mtDNA provided an uninformative tree, probably through no fault of the practitioners. The failing in the bowerbird analysis was to ignore a wide array of morphological and behavioral characters that could aid the fine-tuning of the published phylogeny or at least alert the team that their molecular phylogeny was flawed. The unambiguously monophyletic genus *Amblyornis*—supported by an array of morphological and behavioral synapomorphies—is scattered across the tree, and obvious monotypic lineages (*Archboldia, Prionodura*) are interdigitated among the *Amblyornis* taxa. There are simply too many nonconforming characters to list in detail here. But for example, *Archboldia*'s mat bower is unique among the bowerbirds and shares essentially no characters with the bower of *Amblyornis*. *Prionodura* exhibits a unique plumage and a unique two-pole bower design. In the same vein, the *Amblyornis* species build bowers that are unique in being based around a single pole, indicating that all members of the genus should be clustered. Moreover *Amblyornis* plumage is as uniform as the plumage of any genus in New Guinea, and clearly distinct from the plumages of *Archboldia* and *Prionodura*. And yet the publication of this phylogeny—with its beguiling molecular backing—presumably influences systematic and taxonomic thinking in spite of what, to us, are irreconcilable flaws.

As part of our research for this book, we have examined various museum collections in order to study and diagnose species and subspecies. Both authors have worked on the collection at the AMNH on multiple visits; BMB has visited the collections at Tring on two occasions and Leiden once; BMB is a research associate at the USNM, and TKP is a research associate at the BPBM. Both of those two institutions hold substantial reference collections of birds from New Guinea. We have, in addition, requested and received color photographs of specimens of interest from the museums at Genoa, Leiden, and Cibinong.

Taxonomy

Taxonomy is the science of naming and classifying the earth's life-forms according to natural evolutionary relationships. Taxonomic decisions are made based on systematic relationships in combination with the history of name creation and use over time. In its bookkeeping and historical dimensions, taxonomic nomenclature is driven by the International Code of Zoological Nomenclature, or ICZN (International Commission on Zoological Nomenclature 1999), which can be both complex and subject to interpretation. In this realm we have chosen to follow the authorities and their published determinations. We cite sources of taxonomic decisions in all instances. Here, the world lists of Dickinson & Remsen (2013), Dickinson & Christidis (2014), their predecessor Dickinson (2003), del Hoyo & coeditors (1992–2012), and Gill & Donsker (2015) have been of great value in drawing together the latest taxonomic judgments in the scientific literature. Dickinson *et al.* (2011) highlighted the degree to which history and precedence play a role in taxonomy. Research into the original sources of bird names is not for the faint of heart. With respect to authority and original date of publication of type descriptions, we have generally followed the decisions presented by Dickinson & Remsen (2013) and Dickinson & Christidis (2014), as informed by Dickinson *et al.* (2011). In many instances, Mary LeCroy checked the original publication.

Species Concepts

There are many avian species concepts, and the thinking on what constitutes a species changes with the decades. Our view is founded upon the classic Mayrian Biological Species Concept (BSC; see Mayr 1963, 1970), but also incorporates a more contemporary vision of species delimitations (Sangster 2014). Most ornithologists today, ourselves included, hold to a much narrower species delimitation than did Mayr. We still appreciate Mayr's concept of "superspecies," yet in virtually all instances we treat the component taxa as full species rather than well-marked subspecies (see Mayr & Short 1970), while at the same time lowering the bar for full-species recognition of Mayr's "semi-species."

We illustrate our differing viewpoint with the following example. Mayr (1941, 1962) saw New Guinea's golden-plumaged, lowland-dwelling *ardens* population of *Sericulus* bowerbird as a subspecies of the masked form *aureus* from New Guinea's northern foothills. He treated the two as a single species—*Sericulus aureus*. By contrast, we recognize the two forms as full species—well delimited by unique plumage characters and exhibiting clear differences in habitat preference and geography. Moreover, three recent treatments of this group (Frith & Frith 2004, Frith & Frith 2008, and Dickinson & Christidis 2014) support our more narrow species concept, and Zwiers *et al.* (2008) provided a molecular phylogeny indicating the two are not even sister forms. We thus do not hesitate to treat the two as distinct species—*Sericulus aureus* and *Sericulus ardens*. This is but one of many examples in which we treat as two species an assemblage of phenotypically varying populations that Mayr treated as one.

Some authorities argue that the splitting of a single species into two should be accompanied by a freestanding analytical paper published in the scientific literature. It is our opinion that this bar for splitting species is unnecessarily high, especially given the narrowing of the species concept over the last several decades. We assert that the bar for splitting should be as low as the bar for not splitting. In other words, the decision to not split up a species in the face of supporting evidence is as substantive as the decision to split (Gill 2014). Thus, we have split up a number of traditional species herein, and have felt no compunction toward doing so. For example, we recognize here the following novel species pairs: *Charmosyna papou* and *stellae*, *Heteromyias albispecularis* and *armiti*, and *Rhamphocharis crassirostris* and *piperata*. We perhaps err on the side of liberality in our desire to tell the world about full species hidden within New Guinea's avifauna.

We prefer the BSC over the phylogenetic species concept (PSC), in part because the BSC recognizes subspecies, species, and genus, whereas the PSC acknowledges only species and genus. Thus we highlight three levels of relatedness in our treatment, and subspecies form an important component of this checklist that provides additional power of resolution (see below).

However, we do recognize the challenge of distinguishing between species-level taxa and sub-species-level taxa, and turn to guidance provided by Winker & Haig (2010), Tobias *et al.* (2010), and del Hoyo & Collar (2014). Notably, we see the phenomenon of hybridization as informative but not decisive when distinguishing species from subspecies. So long as each lineage maintains a stable phenotype over much of its range, we do not see presence of a hybrid zone as proof that the two focal lineages are conspecific. In New Guinea, we treat the hybridizing *Melidectes rufocrissalis* and *belfordi* as distinct species but, at the same time, we treat the distinct but hybridizing subspecies populations of *Paradisaea raggiana* as conspecific. We do not doubt the correctness of our decision with these two examples. That said, we have less confidence regarding our treatment of "*Sericornis virgatus*" as nothing more than as series of stable hybrid crosses between *S. beccarii* and *S. nouhuysi.*

We cannot entirely embrace the PSC for one substantial reason—the PSC considers as a species any population that can be defined by one or more characters. In our experience in New Guinea, there are so many situations in which two populations differ but only in one or two minor ways. We see these as obvious subspecies, not species.

Treatment of Subspecies

A subspecies is a geographically defined allopatric population that exhibits some measurable and readily detectable morphological traits distinguishing it from an adjacent population—another subspecies of that species. Subspecies are identified and named to help *us*—museum ornithologists and fieldworkers, who attempt to define and analyze the diversity of life. We know from many molecular studies that some subspecies represent populations that exhibit a distinct molecular fingerprint, whereas other subspecies do not (*e.g.*, Ross & Bouzat 2014). That said, practicing ornithologists can see plumage, size, and morphology but cannot see molecules, so we believe describing and delineating morphologically distinct populations as subspecies has some merit even today, especially for fieldworkers.

Despite the utility of delimiting subspecies, we guess that as many as half the currently accepted subspecies on earth listed in contemporary world checklists are poorly or very thinly defined and without true merit. Why are there so many "bad" subspecies in circulation? It often is related to inadequate sample sizes and large gaps in geographic sampling. It also traditionally came from the earnest desire to describe all variation of life on earth—even minor variation. Moreover, in the culture of expeditionary field ornithology, there has been an incentive for the field ornithologist to come home with new forms to describe. After the expedition has spent large sums of donor money in some far away jungle in New Guinea, the subsequent naming of novel taxa confirms its worthiness and provides the fodder for creating new names that honor generous sponsors. Even the best ornithologists (*e.g.*, Mayr, Rand, Gilliard) succumbed to temptation and named populations that they believed were slightly browner, slightly larger, or slightly shorter-billed. But this all gets back to inadequate samples early on. With this unchecked enthusiasm, the true purpose of new taxa—to aid researchers in understanding the process of geographic differentiation and the evolution of biotic complexity—can be lost in a welter of names. Our job today is to weed out the least informative of these. We have tried to do that here. Mayr (1941) dismissed (rendered into synonymy) 331 subspecies in his monumental list. In our checklist, we dismiss (or synonymize) 400+ more subspecies. We believe at least an additional 50 to 100 probably merit removal from the list and placement in synonymy.

To clarify: Our treatment of subspecies is conservative. We have looked at every named subspecies from the Region with a critical eye, recognizing only well-marked subspecies while relegating many poorly defined forms to synonymy. Our "ideal" subspecies exhibit distinct characters rather than grades: we do not support subspecies based on slight size increments or subspecies that capture endpoints in clinal variation. Thus we do not look with favor on a named subspecies that is defined by being "slightly browner ventrally, on average," but we do favor a form that is "blue-capped, distinct from the black-capped form." Similarly, we find subspecies based on mensural characters particularly unsatisfying unless the size disparity is *substantial*. We here rather arbitrarily select 10 percent as the minimum difference worthy of recognition. This we calculate as follows: $L - S / S > 0.10$, where L is the mean measurement for the larger form and S the mean measurement for the smaller form.

Contrary to tradition, we feel no special need to examine holotypes, unless topotypical material is not available. We note that in many instances a holotype has been chosen to best exhibit the putative characters of the new subspecies, and thus is *unlikely* to be "typical" but rather more likely to be an extreme individual. We prefer to look at a range of topotypical material to determine whether there is some recurrent set of characters that unambiguously distinguishes the population in question.

Finally, we do not believe all "distinct" forms merit recognition, and thus we have synonymized some described subspecies that *do* exhibit some objective but rather minor distinction from the other described forms. These subspecies, in our opinion, do not reach a threshold of distinction worthy of recognition. Our threshold is higher than for past treatments. That said, we have allowed a large number of thinly defined subspecies to squeak through our filter, mainly due to the inability to examine museum material or inadequate sample sizes. The next authority to review this avifauna can, we hope, carry the process to the next stage of refinement. With that in mind, we have noted those forms that we believe need additional scrutiny, and we strongly encourage all taxonomic practitioners to get to work on this underappreciated housecleaning task—not just in the New Guinea region, but worldwide.

We offer an adaptation of Mayr's (1942) definition of subspecies: a geographically defined subdivision of a species that is diagnosable in measurable ways from other subdivisions of that species. We stress diagnosability and nonclinal traits as the building blocks of strong subspecies. Mayr in 1942 linked genetic and phenotypic characters in underpinning subspecies, but today we better realize that in many instances phenotype does not align well with genotype—that is the main reason subspecies are viewed by evolutionary biologists with less favor today than in Mayr's day.

A final note on subspecies: In some subspecies accounts, the subspecific epithet is preceded by a question mark to indicate our uncertainty as to diagnosis. These, of course, need future research.

The Avifauna

The avifauna of New Guinea is rich and taxonomically diverse. The complete bird list for the Region as of February 2015 comprises 101 families, 330 genera, 769 species, and 1,331 forms.

We boil it down to a few salient species numbers below:

	Mayr (1941)	Beehler & Finch (1985)	This Checklist
Breeding land bird species	568	578	630
Seabirds	22	40	46
Palearctic migrants	37	56	60
Australian migrants	22	34	33
The avifauna	649	708	769

The reader will find that the major changes in the list since the publication of Mayr (1941) relate to the whittling down of marginal subspecies and the splitting up of polytypic species—leading to an increase in recognized species and a decrease in recognized subspecies.

Note that this checklist was a work in progress when the field guide *Birds of New Guinea*, second edition, was published in October 2014 (the printed year of publication in that work is 2015 because of a publisher's convention). Thus the reader should be warned that while the two taxonomies have a similar origin and are nearly congruent, they do differ, mainly because of the appearance of new analyses in the scientific literature since the publication of the field guide. A more substantive difference is the delimitation of geographic coverage for seabirds, which in the field guide includes the full extent of surrounding seas but in this checklist is more limited by set distances from the coast.

Historical Biogeography of the New Guinea Region

The avifauna of New Guinea has featured in the formulation of speciation theory and the development of models for historical biogeography (*e.g.*, Mayr 1942, 1963, Diamond 1972). This is, in large part, because the distribution of bird species is so well known and can serve as a model. In the last several decades, weighty multiauthored tomes have sought to provide comprehensive overviews of the various major components of the biota of the New Guinea region (Gressitt 1982, Marshall & Beehler 2007). These include detailed descriptions of plate tectonic scenarios for the Region (Polhemus 2007) as well as summaries of the avifauna and its affinities (Pratt 1982, Mack & Dumbacher 2007). More recently, large-scale work on molecular systematics has been combined with biogeographic analyses (*e.g.*, Deiner *et al.* 2011, Jønsson *et al.* 2010b, 2011b, 2014, Aggerbeck *et al.* 2014).

In brief, the latest research on historical ornithogeography of the Region has produced several take-home points. First, the remarkable corvoid assemblage evolved in Australo-Papua and dispersed outward from this isolated source region (Jønsson *et al.* 2011b). Second, local lineages have evolved in place both on the Australian craton and in island Melanesia (Filardi &

Moyle 2005). At a fine scale, the process of speciation on New Guinea and its fringing islands is complex and not easily summarized (Deiner *et al.* 2011). At best, we find clear repeated patterns of species distribution relating to visible biogeographic barriers, which are highlighted in the New Guinea Bird Regions discussed earlier. The Strickland River gorge appears to be an important species barrier to upper montane forest birds of the Central Ranges. In a similar manner, the meeting place of the eastern terminus of the Eastern Ranges and the mountains of the Huon Peninsula is an effective barrier for lowland-dwelling birds in the northern watershed. In the southern watershed, the Purari River basin serves as an important barrier for many lowland species and subspecies. The eastern sector of the Bird's Neck is also a place where many species' ranges start or end ("Zoogeographer's Gap"). Most of these distributional barriers are formed by the interplay of mountains and basins. Given the estimated youthful ages of most avian species in the Region, we think that the movement of tectonic plates has little to do with current species ranges (*pace* Heads 2001), though it may have influenced the current distribution of higher lineages (subfamilies and families).

The first step towards devising compelling theories of historical biogeography is the delineation of species and their distribution. This is what our Checklist seeks to achieve. We encourage biogeographers to deploy the results found here to take the next step forward toward explaining the evolution of the great New Guinea biota in time and space.

The Checklist and Nature Conservation

Both authors have spent decades actively pursuing nature conservation in the Pacific (Pratt *et al.* 2009, Beehler & Kirkman 2013). What does this checklist have to do with nature conservation? At the most basic level, one must give a name to a thing before society will move to conserve it. The history of ornithology in New Guinea has been: discover, describe, study, and conserve (Beehler & Mandeville 2016). We believe knowing the taxonomy, distribution, and evolutionary history of the birds of New Guinea is the first step toward conserving this rich avifauna. Virtually every recent field ornithologist who has studied New Guinea birds in the rain forest has also invested effort in promoting conservation of these birds and their forests.

This is not the place to provide conservation guidance for New Guinea's birds and their habitats. But we think everyone using this book should think about conservation of the birds they study. Although the forests of New Guinea have been more resistant to wholesale conversion than those of adjacent southeast Asia, future decades will bring new threats. It is better to act sooner rather than later on behalf of this natural treasure. We advocate for (1) locally driven conservation of natural habitats; (2) government-led survey, study, and management of natural resources; and (3) international advocacy for Western investment in the protection and study of the biotas and rain forests of the developing world. Allied to this is the need to conserve cultures and traditional languages in the New Guinea region. Strong local cultures can generate the political will to act to conserve local natural wealth.

The Future

Over the long term, the future of the study and conservation of New Guinea's birdlife is in the hands of local naturalists and scientists. It is the mandate of international scientists to reach out and assist the growth of cohorts of local practitioners whenever possible. In both eastern and western New Guinea, we believe local and national governments should be actively training, equipping, and funding local naturalists to better survey, document, publish, and curate knowledge about collections of birds of the Region. This knowledge base should include photographs, digital sound files, field notes, and yes, museum specimens (study skins, anatomical specimens, skeletal specimens, and DNA samples; see Peterson 2014). We are confident that local museums and research institutions can easily attract the partnership and collaboration of a whole range of international institutions—and with respect to biodiversity survey and study, the more the merrier! International partners can provide training, intellectual input, equipment, supplies, and other resources. Local partners can provide special knowledge, access, linguistic tools, and other intangibles. The mix is usually productive for all concerned. Formal, government-sponsored, and permanent biological surveys for Tanah Papua (western New Guinea) and for Papua New Guinea are institutions whose time has come. These need to be established, staffed, and fully funded.

Layout of the Accounts

The accounts follow a standard format, with brief descriptive narratives for each order, family, genus, species, and subspecies of bird known from the Region.

We provide summary counts of species in each family and genus. These are derived from Dickinson & Remsen (2013) and Dickinson & Christidis (2014). Note that this source checklist includes recently extinct forms, which we thus include in our counts.

The species accounts provide the current English and scientific name; status in the Region; original type citation with type locality; geographic range of the species (including extralimital range); subspecies accounts for all recognized races (including description, synonyms, range, and diagnosis); and, finally, a Notes section that addresses taxonomic and nomenclatural decisions, plus priorities for future research and a list of alternative English and scientific names.

A note on species and generic sequences: At all levels, phylogeny is a branching of lineages that in most instances does not directly translate into a unique linear list. This is because branching is three-dimensional, and branches can be rotated without any loss of systematic information. With multiple opportunities to rotate multiple branches, there can be many linear solutions even to a rather simple phylogeny. Linear lists thus are often only weak approximations of complex historical phenomena. Given this problem, we tend to stick with convention, where possible keeping the species and generic order that are found in world lists or well-accepted familiar treatments. Thus the user of this list should not read too much from the presented sequence of genera and species. Where useful, we discuss these issues in the familial and generic accounts, saying, for instance, "genus x is the sister to genus y and that pairing (genera $x + y$) is, in turn, sister to genus z."

Ordinal Accounts

Each ordinal account focuses on which families are included in the order and the sister relationships this lineage shows to other orders. Our ordinal treatment derives mainly from Hackett *et mult.* (2008), Cracraft (2013, 2014), and Jarvis *et al.* (2014).

Family Accounts

Each family account discusses where the family is situated within the parent order, the number of recent species in the family, its current geographic distribution, and the sources of systematic authority for the treatment of the genera and species within the family.

Genus Accounts

Each genus account lists the number of species in the genus, provides the citation for the generic description and type species, and discusses current distribution of the genus and key characters defining the genus.

Species Accounts

The species accounts constitute the heart of this work. Each species account includes the following:

English Names. Each species account is headed by our selection of the best English name for the species. Names that have been passed over and old names appear at the end of each species account under "Other names". Our English names are derived mainly from past and present checklists (cited earlier in the list of important publications). Our English name selection is an attempt to refine the best of what is already in print, most of which is excellent. Below we focus on the problems with the small percentage of names that suffer a lack of widespread support.

The construction of English names for bird species around the world is fraught with challenges (Gill *et al.* 2009). We are well versed in these challenges (Beehler & Finch 1985, Beehler *et al.* 1986). In this checklist, English names of species adhere to accepted rules of grammatical construction (*e.g., The Chicago Manual of Style*), and take into consideration the following factors:

- The New Guinea region is in Australasia, and hence English name usage should recognize the primacy of regional usage for species that are native to the Region. Thus for species that range into Australia we generally follow the English treatment recommended by Christidis & Boles (2008) for Australian birds. In addition, we use British/Australian rather than American spelling of names (*e.g.,* "Grey Thornbill" rather than "Gray Thornbill").
- We prefer names that have long been in use over the construction of new names. Thus we believe in historical precedent (*e.g.,* preferring Nankeen Night-Heron over the "more descriptive" Rufous Night-Heron). A few problematic names have had such long usage and are so obviously descriptive and preferable that we have not abandoned them just because a bird somewhere else in the world is similarly named; thus, we retain New Guinea Harpy-Eagle and New Guinea Vulturine Parrot *vs.* the recently introduced names Papuan Eagle and Pesquet's Parrot.
- We prefer simpler and shorter names. Thus many of the names we use today are considerably shorter and simpler than those used in Rand & Gilliard (1967). For instance, we have evolved from Mountain Yellow-billed Kingfisher to Mountain Kingfisher, and from Western Black-

capped Lory to Black-capped Lory. More recently, some new longer names have been introduced, and we reject these. We think Long-tailed Buzzard and Doria's Hawk are preferable to the uninformative and overly complicated Long-tailed Honey-Buzzard and Doria's Goshawk.

- We do not believe in the absolute necessity of modifiers. Thus we prefer Crested Tern to the more descriptive Greater Crested Tern. Also, we find acceptable the "nonparallel" construction of the names for two related species, "Crested Tern" and "Lesser Crested Tern."
- We prefer to present two-noun compound names by employing a hyphen when a noun is serving as an adjective. Thus the hyphen in "night-heron" (two nouns), as opposed to "imperial pigeon" (an adjective modifying a noun).
- We follow Parkes (1978) in the capitalization of a group name that is taxonomically informative. Thus we use "Nankeen Night-Heron" and capitalize "Heron," because the bird is, in fact, a heron. By contrast, we leave in lower case a component of a bird name that is not taxonomically informative, such as "Spotted Jewel-babbler," in which we have "babbler" uncapitalized because this bird is not a true babbler.
- We do not use hyphens when a compound name is properly constructed with appropriate modifiers (adjectives or prepositions). For example, we prefer the long-used "bird of paradise" versus the more novel "bird-of-paradise." The argument for the use of hyphens in this example is to quickly indicate to the reader the extent of the group name and keep it distinct from the species name. Supposedly, "Red Bird of Paradise" is more difficult for the reader to comprehend than "Red Bird-of-paradise." We disagree. Another example of a properly constructed name not needing hyphens would be "Eastern Alpine Mannikin"; the group name "mannikin" is modified by two adjectives—"eastern" and "alpine"—and thus there is no need for hyphens.
- We use hyphens mainly for orthographic purposes but also appreciate that the hyphen is useful when the group name imparts information regarding systematic relationship (e.g. Night-Heron, Storm-Petrel; *pace* Gill *et al.* 2009).
- In regard to the trend to move from hyphenated to combined compound words, we do agree that, when possible, hyphenated names should be replaced with single words, when the combination permits. Thus we have evolved from "scrub-wren" to "scrubwren" (the latter is felicitous), but we have not evolved from "bee-eater" to "beeeater," because the triple-*e* construction is difficult to read.
- Possessive is created by adding an apostrophe and an *s*. For example, for a bird named for a person named Edwards, the possessive is "Edwards's."

Scientific Names. Following the English name, we provide the currently valid scientific binomen. In addition, we provide full citations of names in synonymy for all forms that have been synonymized since Mayr's 1941 checklist. We encourage the reader to refer to Mayr (1941) and Peters's *Check-list* (1931–1986) for an initial engagement with the vast listings of additional synonymous names that have been generated over the decades. Additional sources of earlier usage and treatments of the names of the birds of New Guinea can be found in Sharpe (1874–1898), Salvadori (1880–1882), and Mathews (1927, 1930).

Species Status Definitions. On the first line of the account we provide a status assessment for the species, as follows:

endemic	restricted to the New Guinea region
near-endemic	essentially restricted to the New Guinea region, but with a range that marginally extends to small areas adjacent to the Region.
resident	a native breeder to the New Guinea region that also breeds in areas beyond the boundaries of the New Guinea region
visitor	a species that periodically visits the New Guinea region as a nonbreeder
introduced	a nonnative species transported to the Region by human agency (or self-introduced from adjacent areas to which populations were introduced by human agency)
vagrant	a rare visitor recorded in the Region fewer than five times.
visitor and occasional breeder	a species that visits New Guinea from time to time and that occasionally breeds in the Region.
hypothetical	known from only one or few unconfirmed records; or pertaining to an introduced population that may have died out after introduction. Hypothetical species are not included in the New Guinea list counts.
resident and visitor	a species that includes both nonbreeding visitors and a breeding resident population.

Type Description. The second line of the account presents the original type description for the species that appeared in the historical scientific literature.

Dates of Authority. In those instances where there is a discrepancy, we present the currently accepted "true" date of type publication up front with the author's name—"Jones, 1888"—and the printed (visible) date of publication with the volume and page (when it differs)—"*J. of Science* (1887) 55: 223." We do this because we believe this is the easiest to interpret for the reader. This reduces confusion for the bibliographer looking for the work in print, and for the taxonomist looking for the earliest date of publication (put up front). Hence:

Otus hypotheticus Jones, 1888. *Journ. Hypothet. Sci.* (1887) 55: 998

can be clearly interpreted as having as its correct (actual) year of publication as 1888, and having the year of publication printed in the publication as 1887. Knowing the exact date of publication is important for determining seniority of names and helps determine which name has priority over others for proper use today (see Ride *et al.* 1999).

Type Locality. The type locality is the place of collection of the name-bearing type of a nominal species or subspecies (ICZN 1999, 86, Art. 71.1). Our type locality descriptions provide brevity, clarity, and utility by presenting succinct locality data using modern names of places and provinces, and by situating the locality with reference to easily understood regions in modern world geography. Thus, for instance, the original type locality for *Ardea ibis coromanda* was "Coromandel Coast." We have revised this to "southeastern coast of India" to assist the reader who might be unaware of the location of the Coromandel Coast.

Description of Geographic Range. We attempt to provide a clear overview of the species range and, when appropriate, then follow this with a detailed listing of particular sites where the species occurs. We also highlight notable absences. For montane species we focus in particular on presence or absence from the various central and outlying mountain ranges (see Diamond 1985). Locality names used in the account text appear in the geographic gazetteer at the end of the book (but not all of the original type locality names, many of which are archaic as well as outside of the Region). We attempt to characterize the elevational range as well, noting both core range and extreme occurrences. We note here that because of the impact of climate change, elevational ranges have been shifting recently—mainly uphill (see Freeman & Freeman 2014). For species range descriptions, we range from northwest to southeast but nearly always start with mainland New Guinea and end with a separate section for the range outside of the Region ("extralimital").

With monotypic species and subspecies of polytypic species, we have provided superscript letter codes to help identify sources of information. We have done this to save space and to provide a more readable range description. The coding comes in two forms. Two-letter codes provide an indication of a major source reference, as follows: MA = Mayr (1941), CO = Coates (1985, 1990), and DI = Diamond (1972). Single letter codes allow the reader to match a locality record with a specific source that is cited at the end of the sentence. A clarifying example appears below:

Range: Karkar I[MA], Goodenough Is[X], and Smith[Y], Normanby[W], and Batanta I[Z] (Diamond 1986[Z], Jones 1999[Y], Pratt field notes[W], Beehler coll. specimens USNM[X]).

In the hypothetical example above, the record for Karkar I comes from Mayr (1941); that from the Goodenough Island comes from Beehler's specimens at the Smithsonian; that from Smith Island from Jones (1999); that from Normanby Island from the unpublished field notes of Thane Pratt; and that from Batanta I from Diamond (1986).

Subspecies accounts. The species range description is followed by subspecies accounts, when the species is polytypic. Each subspecies account includes the current trinomial, its author, the reference to the original description, synonyms if present, range, and diagnosis—how the subspecies can be distinguished from other subspecies. The subspecies range description includes its entire global range, presented from northwest to southeast, and is not confined to the range in New Guinea.

Notes. The Notes section includes any necessary discussion related to taxonomy, nomenclature, and distribution and ends with a listing of alternative names that have appeared in the recent ornithological literature.

A note on specimen records, field sightings, and uncertainty: In the species accounts, we discuss distributional records of all sorts. Historically, these were mainly generated by the collection of a voucher specimen deposited in a museum collection and often written about in one or more technical publications. Over the last several decades, the collection of new museum voucher specimens has much declined—primarily to be replaced by sight records from the field, which have been described mainly in bird-watching journals. A sight record of a bird does not have the same permanence or weight of a museum specimen. For one, the museum specimen can be repeatedly studied by a wide array of scholars, and it also can be identified to subspecies long after it was collected (which cannot be done with a sight record unless an unambiguous photographic series accompanies it). We encourage future fieldworkers to collect museum specimens of novel records whenever feasible—especially through collaborations between institutions and nations (Peterson 2014). The representation of New Guinea's bird fauna in the museums of the world remains woefully inadequate. For example, this checklist discusses a number of undescribed species and subspecies that have not yet been collected and deposited in museum collections. Because of this, our work remains incomplete, and these novel taxa remain unnamed. That said, we also encourage more collecting and publishing of novel sight records, supported when possible by digital images as well as digital sound recordings. These do not take the place of museum specimens but have considerable value to future researchers, especially when stored on readily accessible websites or other compendia.

Websites such as www.xeno-canto.org and that of the Macaulay Library of the Cornell Lab of Ornithology (macaulaylibrary.org) are important repositories of sound files of birds from New Guinea. Although there is a certain level of uncertainty surrounding sound files because they typically are not associated with a collected voucher museum specimen, they serve an important archival function and more fieldworkers should deposit their sound files in accessible archives. There are also web-based archives of avian imagery that constitute useful references (*e.g.*, the Internet Bird Collection of Lynx Edicions), and we encourage the expansion and greater use of these resources, in spite of the uncertainty surrounding them. Because of time constraints, we have been unable to tap into the wealth of Cornell's eBird field list data for New Guinea. We hope to do this in a next edition.

We attempt to acknowledge uncertainty in this checklist by applying a question mark to indicate the lack of surety surrounding a record. We think imperfect and uncertain records need to be included in checklists, and that uncertainty will always be a part of these sorts of works. Our preference is to not sweep them under the rug, but to highlight them as a way of guiding future workers to areas needing clarification or additional attention.

Abbreviations and Latin Terms

[ ]	hypothetical: these brackets enclose the entire species account when there is no confirmed record for the Region; also applies to introduced species for which there is no proof of establishment.
?	uncertainty: "?" precedes a subspecies or species name in an account to acknowledge uncertainty of diagnosis (usually because of lack of museum specimens)
AMNH	American Museum of Natural History, New York
Arch	Archipelago
BMNH	The Natural History Museum, Tring, England
BPBM	Bernice P. Bishop Museum, Honolulu
ca.	circa (approximately)
cf.	confer or compare
CSIRO	Commonwealth Scientific and Industrial Research Organisation, Australia
distr	district
et al.	*et alia*; and others
et mult.	and multiple others
e.g.	for example
fide	refers to the source of the information discussed
I	Island (e.g., High, Fisher's, and Black I); used for single island or series of single islands
ICZN	International Commission on Zoological Nomenclature
in litt.	*in litteris*; in correspondence
Is	Islands (e.g., D'Entrecasteaux and Trobriand Is); used for island groups or series of island groups
IUCN	International Union for Conservation of Nature, which manages the Red List
i.e.	that is
mm	millimeter: all biometric measurements of birds (wing, tail, bill, etc.) are in millimeters (but "mm" is typically not included in measurements).
ms	manuscript (unpublished)
Mt	Mount, Mountain (*e.g.*, Mt Wilhelm, Smith Mt)
mtDNA	mitochondrial DNA
Mts	Mountains (*e.g.*, Kubor Mts)
n=	"n" is the sample size reported on; thus "n = 2" means two birds were measured.
nec	not elsewhere classified

NG	mainland of NG
NP	National Park
(ns)	not significant
NSW	New South Wales, Australia
NT	Northern Territory, Australia
NZ	New Zealand
pace	peace; used to indicate we respectfully disagree with the cited authority
Penin	Peninsula
PNG	Papua New Guinea
Prov	Province
QLD	Queensland, Australia
R	River
Ra	Range, Ranges
USNM	US National Museum of Natural History, Smithsonian Institution
WMA	Wildlife Management Area
YUS	YUS Conservation Area on the Huon Peninsula

Geographic Terms

See also terms in "New Guinea Bird Regions" (p. 14), and preceding list of abbreviations.

Australia	includes mainland Australia and Tasmania
Bismarck Arch	includes both the main Bismarck islands and the Admiralty group
Central Ra	the central cordillera from the Weyland Mts east to the Owen Stanley Ra of the SE Peninsula (to Milne Bay Prov)
Mainland	the island of New Guinea
PNG	Papua New Guinea
Region	New Guinea region: mainland plus fringing islands (capitalized when freestanding, lower case when accompanying New Guinea)
watershed	The land basin formed by a river system. The English term for watershed as used here is "catchment" (not used).

Conventions

1. Ranges of species and listings of subspecies start in the northwest and end in the southeast.
2. *Pycnonotus aurigaster ?aurigaster*: question mark indicates there is uncertainty regarding the designation of subspecies.
3. Wing measurements: Many wing measurements (provided in millimeters) are extracted from other publications, making it impossible to know whether the measurements are of wing arc (flattened wing) or wing chord (the straight line measurement across the bowed wing). Thus caution is warranted when using these measurements—wing arc in most cases provides a larger measurement than wing chord.
4. Gazetteer coordinates are in decimal degrees, widespread in digital mapping. In these, a complex of three numbers (degrees, minutes, seconds) is simplified to a single number. For latitude, a negative indicates south latitude and a positive indicates north latitude. For longitude, negative is west longitude and positive is east longitude. Thus latitude -5.5 = 5° 30' south latitude. To convert from decimal degrees to traditional degrees+minutes, simply split up the integer from the decimal; the integer converts one-to-one (5 degrees decimal is 5 degrees traditional), whereas the decimal converts to minutes by multiplying by 60. Thus in this example 0.50 × 60 = 30, hence 30 minutes.
5. The *Other names* sections presents other name usages in their original published orthographies, to show the range of treatments in the literature. Thus for the currently-named Purple-tailed Imperial Pigeon, the *Other names* section lists published usages: Rufous-bellied Fruit-pigeon and Rufous-breasted Imperial-pigeon. Note distinct spellings of the generic name.

A request to readers

The authors would appreciate receiving corrections of fact as well as new records, novel systematic analysis, and comments pertaining to material contained here. Please forward these to: brucembeehler@gmail.com or by post to Bruce Beehler, Division of Birds, NHB MRC 116, PO Box 37012, Smithsonian Institution, Washington, DC 20013.

Figures

Figure 2a. *Pachyptila belcheri* specimen from southern New Guinea (Museo Civico di Storia Naturale G. Doria)

Figure 2b. *Neopsittacus pullicauda alpinus* (top) and *N. p. pullicauda*

Figure 2c. *Pachycare flavogriseum lecroyae* (left) and *P. f. subaurantium*

Figure 2d. *Cnemophilus macgregorii* described forms *macgregorii*, *kuboriensis*, and *sanguineus* (bottom to top)

Figure 2e. *Loboparadisea sericea sericea* (left) and *L. s. aurorae*

Figure 2f. *Melanocharis longicauda longicauda* (left) and *M. l. chloris*

Figure 3a. *Rhamphocharis piperata* (top) and *R. crassirostris*, female plumage

Figure 3b. *Rhamphocharis crassirostris*, female plumage, from type series

Figure 3d. *Manucodia chalybatus* (left) and *M. jobiensis*

Figure 3c. *Colluricincla tenebrosa tenebrosa* (left) and *C. t. atra*

Figure 3e. *Parotia lawesii* (left) and *P. helenae*

Figure 3f. *Ptiloris magnificus* (left) and *P. intercedens*

Figure 3h. *Heteromyias armiti, albispecularis,* and *cinereifrons* (top to bottom)

Figure 3g. *Epimachus meyeri albicans, E. m. bloodi,* and *E. m. meyeri* (bottom to top)

PART II

Starry Owlet-nightjar, illustration by John Anderton

ORDER CASUARIIFORMES

We follow the treatment of Christidis & Boles (2008) and Cracraft (2013) in recognizing this order, which encompasses two sister lineages—the cassowaries and the emus. The kiwis and extinct elephant birds comprise a clade that is the sister group to the Casuariiformes (Mitchell *et al.* 2014). The Casuariiformes are huge and flightless terrestrial birds laying giant eggs and inhabiting Australia and New Guinea. Harshman *et mult.* (2008) have provided evidence that the ratite birds (kiwis, rheas, emus, cassowaries, ostriches, *etc.*) are polyphyletic and that there are multiple instances of the evolution of the loss of flight among the ratites. This opens the possibility that the ancestors of the Casuariiformes could have reached Australasia by wing.

Most recently, the composition and sequence of orders within the nonpasserines has been delineated by the molecular analysis of Hackett *et al.* (2008), Cracraft (2013), and Jarvis *et al.* (2014). Differences in sequence among the publications in many instances are the product of a trivial difference in selecting the listing of sister clades. As mentioned in the Introduction, there are usually multiple equally correct ways of presenting the results of a tree when reducing it to a linear sequence.

Family Casuariidae

Cassowaries   [Family 3 spp/Region 3 spp]

The native range of the family extends from northeastern Queensland to New Guinea, Salawati, Batanta, and Yapen Island. In addition, *Casuarius casuarius* has been introduced to Seram and *C. bennetti* to New Britain. The two larger species inhabit lowland forest, whereas the smallest species is confined to upland forest up to the subalpine tree line (Mayr 1940, 1979b, Folch 1992, Christidis & Boles 2008). This Australo-Papuan group is an ancient, compact, and well-delineated family most recently revised by Mayr (1979b), with closest affinities to the Emu, family Dromaiidae (see Christidis & Boles 2008, Hackett *et mult.* 2008, Harshman *et mult.* 2008). It is distinguished from other ratites by the uniform all-black plumage, the multicolored bare neck skin, the bare and ridged crown casque, and the daggerlike inner toenail. Cassowaries are solitary and retiring forest dwellers, especially where they are heavily hunted. Their diet is principally fruit.

GENUS *CASUARIUS*   [Genus 3 spp/Region 3 spp]

Casuarius Brisson, 1760. *Orn.* 1: 46, and 5: 10. Type, by tautonymy, "Casuarius" Brisson = *Struthio casuarius* Linnaeus.

All three species of the genus inhabit the Region. The genus range matches that for the family. Characteristics of the family apply to the genus, which is morphologically homogeneous. The two larger species (*C. casuarius* and *unappendiculatus*) are presumed to be sister species and are largely or wholly allopatric. The considerably smaller and smaller-casqued *C. bennetti* inhabits the various mountain ranges of New Guinea, and in the lower foothills its range may meet those of the larger species. D. Gibbs (*in litt.*) reported closely observing both *C. bennetti* and *C. unappendiculatus* in the coastal foothills of the Tamrau Mountains near Jamursba Medi Beach on the northern coast of the Bird's Head. The smaller species was found slightly inland and marginally higher (200 m) than the larger species, which was observed nearer to the coastline. We suggest that *bennetti* is the oldest lineage, given its montane Mainland distribution and its less well-developed features (casque, wattles, size). Species sequence below follows Mayr (1941) and Beehler & Finch (1985) in placing the smallest form at the head of list. Taking a distinctly different position than earlier in his career (Mayr 1940, 1941), Mayr (1979b: 7–9) suggested that subspecific designations for the three species lacked scientific merit, not because of the absence of geographic variation but because of (1) the current available sample sizes being inadequate to properly distinguish true geographic variation, (2) the inadequate knowledge of plumage and skin coloration changes related to age and sex, and (3) the problem of human-caused geographic mixing brought about by widespread trade-related transport of chicks. We present one exception to this, because of the compelling case offered by Perron (2011). It is imperative that a full study of geographic variation of the three species in New Guinea be undertaken on a priority basis.

Dwarf Cassowary *Casuarius bennetti* ENDEMIC
Casuarius Bennetti Gould, 1858. *Proc. Zool. Soc. London* (1857) 269: pl. 129.—New Britain.

A very shy inhabitant of upland forest interior throughout NG, from near sea level (where mountainous) to 3,620 m (Coates & Peckover 2001). Also Yapen I (Mayr 1941). The species is strictly forest dwelling and is most often encountered on footpaths in the forest.

Extralimital—occurs on New Britain, where presumably introduced from Mainland for food or bride-price. All species of cassowaries are hunted, and the young, when captured live, are raised in villages, so the transport of this species to New Britain is not surprising, given its importance as a source of dietary protein.

SUBSPECIES

a. *Casuarius bennetti westermanni* P. L. Sclater
Casuarius westermanni P. L. Sclater, 1874. *Proc. Zool. Soc. London*: 248.— Arfak Mts, Bird's Head.
RANGE: Arfak Mts and Tamrau Mts.
DIAGNOSIS: Exhibits a large whitish ear patch that extends to nape (Perron 2011). As Perron (2011) explained in detail, the original birds described as *papuanus* and putatively from the Arfak Mts did not exhibit the diagnostic whitish ear patch. Perron's explanation for this anomaly is that the original *papuanus* specimens were probably birds that arrived in Dorey (now Manokwari) via traders who obtained them from the Central Ra (probably the nearby Weyland Mts). M. LeCroy (*in litt.*) noted that many specimens provided by Bruijn and van Renesse van Duivenbode that came from the Weyland Mts were mistakenly attributed to Dorey and the Bird's Head. Dickinson & Remsen (2013) recognized *papuanus*. Davies (2002) treated the Bird's Head form as a distinct species, which he called *C. papuanus*. Documenting photographs of this white-eared form, of live birds procured by S. D. Ripley in the Tamrau Mts, were recently provided to the authors (Mayr & Meyer de Schauensee 1939b, images from the archives of the Philadelphia Zoo courtesy of R. M. Peck and A. Baker).

b. *Casuarius bennetti* subspecies?
Following Mayr (1979b), no subspecies are recognized here but for *westermanni*, because of the work of Perron (2011). But a number of named taxa may, with adequate additional study, be proven to merit recognition. At the very least, the population from NG east of the Bird's Head requires a name, presumably taken from the list below, once characters for each are properly delineated. We list the subspecies recognized by (Mayr 1941) here:

> *Casuarius bennetti papuanus* Schlegel, 1871. *Ned. Tijdschr. Dierk.* (1873) 4: 54.—"Andai, Bird's Head," but apparently from some other unknown part of New Guinea (see Perron 2011).
> *Casuarius bennetti picticollis* P. L. Sclater, 1874. *Rep. Brit. Assoc.*: 138.—Discovery Bay, SE Peninsula.
> *Casuarius bennetti hecki* Rothschild, 1899. *Bull. Brit. Orn. Club* 8: 49.—German NG [probably from the Huon Penin].
> *Casuarius bennetti claudii* Ogilvie-Grant, 1911. *Bull. Brit. Orn. Club* 29: 25.—Iwaka R, 1,220–1,525 m, Western Ra.
> *Casuarius bennetti goodfellowi* Rothschild, 1914. *Bull. Brit. Orn. Club* 35: 7.—Yapen I, Bay Islands.
> *Casuarius bennetti shawmayeri* Rothschild 1937. *Bull. Brit. Orn. Club* 57: 120.—Arau distr of Kratke Mts, Eastern Ra.

D. Gibbs (*in litt.*) noted that after many weeks in the field on Batanta he came away with no records for *C. bennetti*. We encourage research to clarify subspecific geographic variation of all three species of cassowaries in the Region. It would be interesting to know whether the species are interspecifically territorial, given that the two larger lowland species may share habitat in certain localities (*e.g.*, in the sw Bird's Head lowlands). No evidence exists of hybridization between the species. *Other names*—Little, Mountain, or Bennett's Cassowary, *Casuarius papuanus*.

Southern Cassowary *Casuarius casuarius* RESIDENT | MONOTYPIC

Struthio Casuarius Linnaeus, 1758. *Syst. Nat.*, ed. 10, 1: 155.—Seram, s Moluccas.

Inhabits forest and wooded savanna of the lowlands of the Bird's Head and Neck, S Lowlands, Trans-Fly (including Dolak I), SE Peninsula, and Aru Is. Mainland range includes lowlands of the n watershed of the SE Peninsula extending northwest from Milne Bay at least to Cape Ward Hunt. Mainly inhabits lowlands below 300 m (but note that human-transported birds can appear anywhere).

The presence of the species from the NW Lowlands, Sepik-Ramu, Huon Penin, and nw sector of the n watershed of the SE Peninsula requires additional confirmation (see Mayr 1941, Berggy 1978). Extensive field surveys along n coast of Huon Penin (at YUS) provided no indication of any cassowary but *bennetti* (*pace* Berggy 1978). Distribution on the Bird's Head needs clarification with respect to the range of *unappendiculatus*. Gibbs (1994) found *C. casuarius* in the foothills of the Fakfak Mts of the westernmost Bird's Neck, ranging upward into the mts.

Usually found only in the lowlands, though Schodde & Hitchcock (1968) stated the species probably occurred at 750 m near Lake Kutubu, Southern Highlands Prov, and Diamond (1972) stated it probably occurred at 817 m on the floor of the Karimui basin, Chimbu Prov. These two records merit corroboration from modern field sightings, because local reports of cassowaries at the species level tend to be very confusing, given that the concept of "large" and "small" cassowaries is confounded by issues of age and plumage (not to mention human-transport of captive birds). Schodde & Hitchcock's (1968) statements on interspecific differences of egg color and texture also require confirmation. Note also that B. G. Freeman (*in litt.*) reported that cassowaries are essentially extirpated from the Karimui basin as of 2013.

Extralimital—Seram I (where evidently introduced) and ne QLD, Australia, from e Cape York south to the Burdekin drainage, Herberton Ra, and Hinchinbrook I (Mayr 1979b, Marchant & Higgins 1990).

SUBSPECIES

Following Mayr (1979b), no subspecies are recognized. But some named taxa, may, with adequate additional study, merit recognition. We list the subspecies recognized by (Mayr 1941) here:

Casuarius casuarius bicarunculatus P. L. Sclater, 1860. *Proc. Zool. Soc. London*: 211.—Type locality unknown [probably w part of Bird's Head]. Exhibits a wattle on each side of the neck (Rand & Gilliard 1967).

Casuarius casuarius aruensis Schlegel, 1866. *Ned. Tijdschr. Dierk.* 3: 347.—Wammer, Kobror, Aru Is. Exhibits wattles similar to those of *sclaterii* (?) or a pair on the foreneck (Rand & Gilliard 1967).

Casuarius casuarius tricarunculatus Beccari, 1875. *Ann. Mus. Civ. Genova* (1876) 7: 717.—Warbusi, head of Geelvink Bay. Exhibits a pair of wattles on the front of the neck (Rand & Gilliard 1967).

Casuarius casuarius sclaterii Salvadori, 1878. *Ann. Mus. Civ. Genova* 12: 422.—Coast opposite Cornwallis I, SE Peninsula. Exhibits a single median wattle, forked at the tip (Rand & Gilliard 1967).

Casuarius casuarius bistriatus van Oort, 1907. *Notes Leyden Mus.* 29: 205.—Northern coast of the NW Lowlands. Apparently inhabits the n coast of the NW Lowlands (but relative distribution of this and *C. unappendiculatus* is problematic). Exhibits a single median wattle, forked for half its length (Rand & Gilliard 1967).

Casuarius casuarius lateralis Rothschild, 1925. *Bull. Brit. Orn. Club* 46: 30.—Supposedly from Gaza R, ne NG.

NOTES

With this, as with all species of this genus, geographic variation is apparently substantial but inadequately documented through systematic collection of appropriate sample sizes from key localities (controlling for variation related to sex and age). There is need to document the extent of sympatry between *C. unappendiculatus* and *C. casuarius* in the sw Bird's Head lowlands, the Bird's Neck, and the w verge of the Mamberamo basin.

Currently classified by IUCN as Vulnerable. We propose the species might better be classified as Near Threatened at this time, given the abundance of undisturbed forest available for the species in lowland NG. *Other names*—Double-wattled, Two-wattled, or Australian Cassowary.

Northern Cassowary *Casuarius unappendiculatus* ENDEMIC | MONOTYPIC

Casuarius unappendiculatus Blyth, 1860. *Journ. Asiat. Soc. Bengal* 29: 112.—Salawati I, NW Islands (*fide* Gyldenstolpe 1955b).

Restricted to lowland forest of the Bird's Head, NW Lowlands, and Sepik-Ramu; also Batanta, Salawati and Yapen I. Ranges into the hills to *ca.* 700 m (Coates & Peckover 2001). Little known and rarely seen but not uncommon, especially where not hunted.

Localities—w sector of the Bird's Head[MA], Batanta I[K,L], Salawati I[MA], Sorong[Z], Konda on Yapen I[MA], Mamberamo basin[MA], Bernhard Camp on the Taritatu R[Y], Jayapura[Y], s flank Foja Mts[G], Bodim[T], upper Sepik[J], Sepik basin (Gaikarobi)[A], Hunstein Mts[X], Oronga in the Adelbert Mts[Q], Astrolabe Bay[MA], Naru Hills[W] (Rand 1942b[Y], Gyldenstolpe 1955b[Z], Ripley 1964a[T], Gilliard & LeCroy 1966[A], Gilliard & LeCroy 1967[Q], Beehler field notes[G], A. Allison *in litt.*[X], S. F. Bailey 1992a[W], I. Woxvold *in litt.*[J], K. D. Bishop *in litt.*[K], D. Gibbs 2009[L]).

SUBSPECIES

Following Mayr (1979b), no subspecies are recognized here. But a number of named taxa may, with adequate additional study, be proven to merit recognition. We list the subspecies recognized by Mayr (1941) here:

> *Casuarius unappendiculatus unappendiculatus* Blyth, 1860.—Type locality unknown. Known from Salawati I and the w Bird's Head (Rand & Gilliard 1967). D. Gibbs (*in litt.*) noted that birds from the n coast of the Bird's Head and Salawati I exhibit orange rather than red wattles.
>
> *Casuarius unappendiculatus occipitalis* Salvadori, 1875. *Ann. Mus. Civ. Genova* (1876) 7: 718.—Yapen I, Bay Islands. Known from Yapen I and nearby n coastal lowlands (?) (Rand & Gilliard 1967). [Poggi 1996 provided corrected year of publication.]
>
> *Casuarius unappendiculatus philipi* Rothschild, 1898. *Novit. Zool.* 5: 418.—Probably e German NG. Known from the Sepik R east to Astrolabe Bay.
>
> *Casuarius unappendiculatus aurantiacus* Rothschild, 1899. *Bull. Brit. Orn. Club* 8:50.—German NG. Known from the Mamberamo and Sepik basins; perhaps a synonym of the preceding form (Rand & Gilliard 1967).

NOTES

C. unappendiculatus is less well known than *C. casuarius* and merits study. We also recommend attempts to document geographic variation of the species as well as interactions between *C. unappendiculatus* and *C. casuarius* in zones of contact. Field surveys of this species and *C. casuarius* should concentrate on the area west of the Gogol R trending east onto the n scarp of the Huon Penin, as well as on the Bird's Neck and Bird's Head regions. On 19-Feb-1986, K. D. Bishop (*in litt.*) observed an adult *C. unappendiculatus* in the Batanta I lowlands. It exhibited a blue throat, yellow foreneck, and red-yellow side of neck. Gibbs (2009) also reported observing an adult of this species on Batanta.

Currently classified by IUCN as Vulnerable. We propose the species might better be classified as Near Threatened at this time, given the abundance of undisturbed forest available for the species. *Other names*—Single-wattled or One-wattled Cassowary.

ORDER GALLIFORMES

According to the phylogenomic work conducted by Hackett *et mult.* (2008), the Galliformes constitute a monophyletic clade that includes the megapodes (Megapodiidae), quail (Odontophoridae), pheasants (Phasianidae), guineafowl (Numididae), and currasows (Cracidae). This order thus includes most of the ground-foraging fowl-like birds. For details, see discussions of Christidis & Boles (2008), Gill *et al.* (2010), and Cracraft (2013).

Family Megapodiidae

Megapodes or Mound-builders [Family 22 spp/Region 11 spp]

The species of the family range from the Nicobar Islands (north of westernmost Sumatra) and central Indonesia to the Philippines, Moluccas, Micronesia, Lesser Sundas, New Guinea, Australia, Melanesia, and Polynesia. Within the Region the family can be found in virtually all forested environments and from the lowlands into the mountains. This is an Australasian family of darkly plumaged, medium to large, fowl-like birds characterized by the small head, plump body, strong legs, very large feet, and rounded wings. Species are mainly terrestrial, but roost at night in trees. All have some bare skin on the face or facial or neck wattles, which in many cases are colorful. Tails of the brushturkeys are large and tentlike (folded in an A shape), whereas those of the scrubfowl are abbreviated and inconspicuous. We prefer using the Australasian name "scrubfowl" for *Eulipoa* and all species of *Megapodius*, which together constitute a homogeneous assemblage. All species are distinctive in allowing their oversize eggs to be incubated by some natural source of heat (decaying leaves, sun-baked sand, or geothermally warmed soil). A typical species lays its eggs into a constructed mound of leaves in the forest. The hatchlings, the most precocial of birds, dig themselves out of the nest and disperse into the forest apparently without ever interacting with their parents. The megapodes are considered to be the sister group to all other Galliformes (Jones *et al.* 1995, Hackett *et mult.* 2008). Generic assignments and sequence follow the results of Birks & Edwards (2002), further refined by Harris *et al.* (2014). Harris *et al.* showed that there is a brushturkey clade (*Aepypodius*, *Alectura*, *Leipoa*, and *Talegalla*) and a scrubfowl clade (*Megapodius* and *Eulipoa*).

GENUS *AEPYPODIUS* ENDEMIC [Genus 2 spp/Region 2 spp]

Aepypodius Oustalet, 1880. *Compt. Rend. Acad. Sci. Paris* 90: 906. Type, by subsequent designation (Salvadori, 1882. *Orn. Pap. Moluc.* 3: 251), *Talegallus bruijnii* Oustalet.

    The species of this genus are restricted to upland forests of the Region. The range of the genus includes Waigeo, Misool, and Yapen Island of western New Guinea, as well as nearly all mountain ranges of mainland New Guinea (Jones *et al.* 1995). The genus is characterized by the heavy wattling on crown and throat, maroon feathering on rump, large size, blackish plumage, the medium-size tail that is cocked tentlike, and sexual dimorphism. Harris *et al.* (2014) showed that *Aepypodius* is sister to *Alectura*, that this generic pair is sister to *Leipoa*, and that that threesome is sister to *Talegalla*.

Wattled Brushturkey *Aepypodius arfakianus* ENDEMIC

Talegallus arfakianus Salvadori, 1877. *Ann. Mus. Civ. Genova* 9: 333.—Arfak Mts, Bird's Head.

    Inhabits upland interior forest of the mts of NG; also Misool and Yapen I (Diamond 1985, Jones *et al.* 1995), 540[J]–2,800[K] m (I. Woxvold *in litt.*[J], Coates & Peckover 2001[K]). Recorded as low as 300 m on Misool I

(Jones *et al.* 1995). Should be looked for on Batanta I (J. M. Diamond *in litt.*), Salawati I, and the Adelbert Mts (Jones *et al.* 1995).

SUBSPECIES

a. *Aepypodius arfakianus misoliensis* Ripley
Aepypodius arfakianus misoliensis Ripley, 1957. *Postilla* Yale Peabody Mus. 31: 1–2.—Misool I, NW Islands.
RANGE: Known only from interior hill forest of Misool I. Taken at 300 m elevation on Misool (Ripley 1964a).
DIAGNOSIS: Color of comb wattle of adults greenish blue rather than red (Ripley 1964a). Mees (1965) recommended sinking *misoliensis* into the nominate form, but apparently was not aware of the diagnostic distinction in wattle color between the populations.

b. *Aepypodius arfakianus arfakianus* (Salvadori)
RANGE: Mountains of NG and uplands of Yapen I (Diamond 1985), including the Fakfak and Kumawa Mts (Diamond & Bishop 2015).
DIAGNOSIS: Color of comb wattle of adults red, not blue (Jones *et al.* 1995).

NOTES

Jones *et al.* (1995) noted that the population from Yapen I may warrant recognition as a novel subspecies, based on small size. More material is needed to confirm this. An Adelbert Mts sight record from Mackay (1991) requires confirmation, as this species was not recorded in the Adelberts by Gilliard & LeCroy (1967) or Pratt (1983), J. M. Diamond (*in litt.*), or K. D. Bishop (*in litt.*). *Other names*—Wattled Scrubturkey.

Waigeo Brushturkey *Aepypodius bruijnii*   ENDEMIC | MONOTYPIC
Talegallus Bruijnii Oustalet, 1880. *Compt. Rend. Acad. Sci. Paris* 90: 906.—Waigeo I, NW Islands.
    Endemic to the interior upland forest of Waigeo I of the NW Islands (Jones *et al.* 1995, Mauro 2005).

NOTES

This species was long a bird of great mystery until Iwein Mauro tracked it down and documented its life history (Mauro 2005, 2006). Two nest mounds were discovered on 14-May-2002 on Mt Nok on Waigeo I. Given that Wilson's Bird of Paradise (*Cicinnurus respublica*) inhabits the forested uplands of Waigeo and Batanta I, it is interesting to note that the brushturkey apparently does not exhibit the same two-island distribution. Still, it is recommended that effort be invested in further search of the uplands of Batanta for this species, given just how elusive the species has proven to be on Waigeo. The species is listed as Endangered by IUCN. We suggest Threatened is a more realistic classification, given the predominant forest cover on Waigeo. Our choice of English name reflects the fact that this species is a Waigeo I endemic. *Other names*—Bruijn's Brush Turkey, Bruijn's Scrubturkey.

GENUS *TALEGALLA*   ENDEMIC [Genus 3 spp/Region 3 spp]
Talegalla Lesson, 1828. *Man. Orn.* 2: 185. Type, by original designation, *Talegalla cuvieri* Lesson.
    The genus is restricted to mainland New Guinea, Salawati, Misool, and Yapen Island (Jones *et al.* 1995). The genus is characterized by the all-blackish plumage, black rump, and long tent-shaped tail (longer than in *Aepypodius*). Harris *et al.* (2014) provided a tree indicating that *Talegalla* is sister to the three other brushturkey genera.

Red-billed Brushturkey *Talegalla cuvieri*   ENDEMIC | MONOTYPIC
Talegalla Cuvieri Lesson, 1828. *Man. Orn.* 2: 186.—Manokwari, Bird's Head.
    Synonym: *Talegalla cuvieri granti* Roselaar, 1994. *Bull. Zool. Mus. Univ. Amsterdam* 14: 11.—Upper Iwaka R, Bird's Neck.
Inhabits lowland and hill forests of the westernmost sectors of the Region: Salawati[MA] and Misool I[MA], the Bird's Head, the n sector of the Bird's Neck (Warbusi[MA]), and patchily eastward in the w sector of the S Lowlands to the Otakwa R[MA] (see Jones *et al.* 1995); sea level–1,500 m—mainly below 1,000 m (Diamond in Jones *et al.* 1995).

Named subspecies *granti* exhibits very slight size differentiation, not worthy of recognition (compare wing lengths (in mm): nominate Mainland birds at 279–291 versus *granti* at 285–293 [Jones *et al.* 1995: 108]). Additional fieldwork is needed on interaction between this species and *T. fuscirostris* in areas of sympatry. *Other names*—Red-billed Scrubturkey.

Yellow-legged Brushturkey *Talegalla fuscirostris*   ENDEMIC | MONOTYPIC

Talegallus fuscirostris Salvadori, 1877. *Ann. Mus. Civ. Genova* 9: 332.—Mount Epa, Hall Sound, SE Peninsula.

    Synonym: *Talegalla fuscirostris meyeri* Roselaar, 1994. *Bull. Zool. Mus. Univ. Amsterdam* 14: 16.—Southern shore of Geelvink Bay, Bird's Neck.

    Synonym: *Talegalla fuscirostris occidentis* C. M. N. White, 1938. *Ibis*: 763.—Setekwa R, w sector of the S Lowlands.

    Synonym: *Talegalla fuscirostris aruensis* Roselaar, 1994. *Bull. Zool. Mus. Univ. Amsterdam* 14: 15.—Wanumbai, Aru Is.

Inhabits lowland and hill forest of the Bird's Neck, S Lowlands, Trans-Fly, and s watershed of the SE Peninsula, ranging from the Wandammen Penin[J] to Port Moresby[MA]; also the Aru Is[MA] (Jones *et al.* 1995[J]). Sea level–900 m (at Varirata NP; Beehler field notes). Easternmost record from Rigo[MA].

NOTES

Described races based on trivial size variation (see Mees 1982b); the two most distant populations exhibit no difference in size. No differences in plumage or bare-parts coloration. Given the probable youth of the three allospecies of *Talegalla*, one would not expect subspecific variation. The described form *meyeri* may be of hybrid origin (*fuscirostris* × *cuvieri*; see Jones et al. 1995: 113–114). Moreover, the geographic ranges of the described subspecies in several instances bear no relation to known biogeographic patterns displayed by lowland species (especially *meyeri*). The range of this species overlaps with that of *T. cuvieri* (which extends east to the Otakwa R) in the southwesternmost sector of the main body of NG and *T. jobiensis* in the SE Peninsula. In areas of overlap the species segregate elevationally, with *T. jobiensis* at the higher elevation (Diamond 1972, B. G. Freeman *in litt.*). The English name highlights the most prominent field character for the species. *Other names*—Black-billed Brush-turkey, Black-billed Scrubturkey.

Red-legged Brushturkey *Talegalla jobiensis*   ENDEMIC | MONOTYPIC

Talegallus jobiensis A. B. Meyer, 1874. *Sitzungsber. Akad. Wiss. Wien* 69: 74.—Yapen I, Bay Islands.

    Synonym: *Talegalla jobiensis longicauda* A. B. Meyer, 1891. *Abh. Ber. Zool. Mus. Dresden* (1890–91) 3(4): 15.—Stephansort, near Madang, e verge Sepik-Ramu.

Inhabits Yapen I, the NW Lowlands, Sepik-Ramu, and the n watershed of the SE Peninsula (Jones *et al.* 1995); lowland and hill forest up to 1,950 m[J] (Jones *et al.* 1995, I. Woxvold *in litt.*[J]). Status in the s watershed of e NG is uncertain. There is a single specimen record from the Aroa R, nw of Port Moresby (Rothschild & Hartert 1901). Also there are sight records in the s watershed in Crater Mt[J] and camera-trap documentation from 1,500 m on Mt Karimui[K] of the Eastern Ra (Mack & Wright 1996[J], B. G. Freeman *in litt.*[K]). Limits of range uncertain where it apparently overlaps with *T. fuscirostris*. B. G. Freeman (*in litt.*) noted that the two apparently sort elevationally in the Karimui area. Camera-trap and in-the-hand records from four additional sites in s watershed: from upper Fly R east to Purari basin, and from 145 to 1,950 m (I. Woxvold *in litt.*). Apparently widespread in s watershed of the Border Ra and Eastern Ra, where it likely replaces *T. fuscirostris* on areas of steep terrain; both species recorded at the same site in Gulf Prov lowlands, in steep hill *vs.* flat alluvial terrain respectively (Jones *et al.* 1995, I. Woxvold *in litt.*).

NOTES

Putative subspecies based on minor size differences do not merit recognition. Measurements of wing, tarsus, tail, and bill all show overlap, and mean differences in measures are no more than 3% (Jones *et al.* 1995). The three species of *Talegalla* constitute a tightly knit species group, and the individual species exhibit minimal overlap where ranges meet (Jones *et al.* 1995). We recommend studying the vocal, nesting, and mating behavior of the species in areas of interspecific overlap to determine extent of hybridization or species-specific mate selection. *Talegalla* brushturkeys have been reported from 750 to 1,500 m on Mt Sisa and between 900 and 1,200 m, northeast of Lake Kutubu, in the s watershed of the w sector of the Eastern

Ra (Dwyer 1981, Jones *et al.* 1995). Today's knowledge of the vocalizations of these two species should allow easy identification for future field workers. The English name highlights the only distinctive feature of this rather drab species. The three *Talegalla* species exhibit orange, yellow, and red legs. *Other names—* Brown-collared Talegalla, Maroon-throated Brush-turkey, Collared Scrubturkey.

GENUS *EULIPOA* MONOTYPIC

Eulipoa Ogilvie-Grant, 1893. *Cat. Birds Brit. Mus.* 22: 445 (in key), 462. Type, by original designation and monotypy, *Megapodius wallacei* G. R. Gray.

The single species of this genus inhabits the northern and southern Moluccas and one island in the Northwestern Islands (Jones *et al.* 1995). The genus is distinct from *Megapodius* in the following characters: bold patterning on the mantle and wing coverts, distal part of outer web of outer primaries substantially paler than remainder of feathers, whitish vent, and white patch on underwing. Many prior authorities subsumed this form into *Megapodius* (*e.g.*, Beehler & Finch 1985), but more recently Elliott (1994), Jones *et al.* (1995), and Harris *et al.* (2014) recognized this monotypic genus, sister to *Megapodius*.

Moluccan Scrubfowl *Eulipoa wallacei* RESIDENT | MONOTYPIC

Megapodius wallacei G. R. Gray, 1861. *Proc. Zool. Soc. London* (1860): 362, pl. 171.—Halmahera, n Moluccas.

A rare hill-forest resident of Misool I (NW Islands). Ripley (1964a) took his two Misool I specimens from 90 m and 304 m elevation. Ripley (1960) reported that the species is known to lay its eggs in beach sand, although apparently he had no first-hand experience of the Misool birds nesting on the beach.

Extralimital—Halmahera, Ternate, Bacan, Haruku, Buru, Ambon, Seram, ranging up to 1,650 m on Bacan (Jones *et al.* 1995).

NOTES

Status on Misool I needs checking. Apparently the species is in decline in at least some parts of its range, impacted by heavy egg harvesting (Coates & Bishop 1997). Known from Misool I from the two specimens taken by Ripley in 1954 (Ripley 1960). Diamond & Bishop did not record the species on a visit to Misool in Feb-1986. IUCN Red List status Vulnerable. *Other names—*Moluccan Megapode, Wallace's Scrubfowl, *Megapodius wallacei*.

GENUS *MEGAPODIUS* [Genus 13 spp/Region 5 spp]

Megapodius Gaimard, 1823. *Bull. Gén. Univ. Annon. Nouv. Sci.* 2: 450. Type, by subsequent designation (Selby, 1840. *Cat. Gen. Subgen. Types Aves*: 40), *Megapodius freycinet* Gaimard.

The species of this genus range from the Nicobar Islands (and formerly the Andamans) discontinuously to the Philippines, Moluccas, New Guinea, Solomon Islands, Australia, Vanuatu, Micronesia, and Polynesia (Jones *et al.* 1995). The genus formerly inhabited New Caledonia, Fiji, and Samoa, but has been wiped out by humans (Steadman 2006). It is characterized by the drab and sooty plumage, small nuchal crest, and very abbreviated tail. Our species-level treatment follows Jones *et al.* (1995). Sequence of the species is derived from Harris *et al.* (2014: fig. 3). We think the generic term "scrubfowl" is more useful than the name "megapode," which could be applied to any members of the Megapodiidae.

Biak Scrubfowl *Megapodius geelvinkianus* ENDEMIC | MONOTYPIC

Megapodius geelvinkianus A. B. Meyer, 1874. *Sitzungsber. Akad. Wiss. Wien* 69, Abt. 1: 88.—Numfor and Biak I, Bay Islands.

Inhabits the Bay Is of Biak, Owi, Mios Korwar, Numfor, Manim, and Mios Num (Jones *et al.* 1995). Inhabits mainly forest but rarely coastal island scrub.

Our treatment follows Jones *et al.* (1995). A little-studied form that merits additional attention—which applies to many Bay Is allospecific forms. D. Gibbs (*in litt.*) noted that individuals on Biak-Supiori were blackish gray above, with a black pointed crest, reddish face, and obviously fleshy-red legs. By contrast, the Owi I birds were very dark/blackish, with their underparts washed gray, bill tipped pale yellow, bluish facial skin, and legs blackish. Individuals of this genus are known to disperse well across salt water. Perhaps the Owi birds are a mix of *geelvinkianus* and *decollatus*, the latter arriving from Yapen I to the east. *Other names*—Biak Megapode; formerly treated as an insular population of the polytypic *M. freycinet* (*cf.* Mayr 1941).

Orange-footed Scrubfowl *Megapodius reinwardt*   RESIDENT

Megapodius reinwardt Dumont, 1823. *Dict. Sci. Nat.* 29: 416.—Lombok, Lesser Sundas.

Bird's Head and Neck, S Lowlands, s watershed of SE Peninsula, n watershed of the SE Peninsula from East Cape to the Kumusi R, ?NW Islands (?Salawati, ?Batanta)^J, Aru Is, and SE Islands; inhabits forest and scrub from lowlands to 1,800 m (Diamond 1972, Jones *et al.* 1995^J). Some upland birds on the n watershed of mts between the Kokoda Track and Wau are apparently referable to *reinwardt* and seemingly introgressed with *decollatus* (Jones *et al.* 1995: map on p. 215), but see Greenway (1935: 26) for clear indication that birds from the Wau valley (Mt Missim) are dark-legged *decollatus*. Cooper (1988) reported a sighting of the species at 2,600 m above Myola (immediately east of the Kokoda Track).

Extralimital—Lombok and Nusa Penida east through the Lesser Sundas to the Kai and Tanimbar Is to n and ne Australia and islands in the Torres Strait (Jones *et al.* 1995).

SUBSPECIES

a. *Megapodius reinwardt reinwardt* Dumont

Synonym: *Megapodius affinis* A. B. Meyer, 1874. *Sitzungsber. Akad. Wiss. Wien* 69, Abt. 1: 215.—Rubi, Bird's Neck.
Synonym: *Megapodius duperreyii* Lesson & Garnot, 1826. *Bull. Sci. Nat. Férussac* 8: 113.—Manokwari, Bird's Head.
  [For corrected spelling of species epithet see Mees 1982b: 38, Salvadori 1882: 219, M. LeCroy *in litt.*]
Synonym: *Megapodius freycinet aruensis* Mayr, 1938. *Amer. Mus. Novit.* 1006: 7.—Trangan I, Aru Is.

RANGE: From Lombok east to the Aru Is, the Bird's Head and Neck, S Lowlands, s watershed of SE Peninsula, and the n watershed of the SE Peninsula from East Cape to the Kumusi R (Diamond 1972, Jones *et al.* 1995). Mounds attributable to this species were observed near the DINAS field station, Karawawi, S Kumawa Mts in Nov-2014 (C. Thébaud & B. Mila *in litt.*).

DIAGNOSIS: Iris dark brown; bill dark with orange highlights. Feathers of forehead, crown, and nape dark cinnamon brown; crest pointed; cheek feathering dark gray or dark gray-brown; upper mantle medium gray tinged plumbeous; lower mantle bright cinnamon brown; underparts medium gray tinged plumbeous; legs dull orange (Jones *et al.* 1995).

b. *Megapodius reinwardt macgillivrayi* Gray

Megapodius macgillivrayi G. R. Gray, 1862. *Proc. Zool. Soc. London* (1861): 289.—Duchateau and Pig I, Louisiade Arch, SE Islands.

RANGE: Probably all SE Islands, including Kiriwina (Trobriand Is); Goodenough, Fergusson, Dobu, and Normanby^L (D'Entrecasteaux Arch); Conflict Is^K; Misima, Tagula, Duchateau, Rossel, and Pig I (Louisiade Arch); Woodlark, Dugumenu, and Marshall Bennett Is (Woodlark group); and East I in the Bonvouloir Is (Mayr 1941, Jones *et al.* 1995, LeCroy & Peckover 1998, Beehler field notes 29-Sep-2002^K, T. Pratt field notes^L). The Normanby I record is new: Pratt collected a specimen (now at BPBM) and observed it at mounds in 2003.

DIAGNOSIS: Distinct from the nominate form in having much darker, slate-gray plumage generally; mantle dull brownish olive; uppertail coverts dark brown rather than bright chestnut or maroon; crest shorter and broader; forehead partly bare; lesser upperwing coverts partly gray; overall similar to *M. eremita* but for the pale orange-red (not blackish) legs (Jones *et al.* 1995: 221). Dickinson & Remsen (2013) noted this form may be a stabilized hybrid population of *eremita* × *reinwardt*. That seems unlikely, given the current range of *eremita* does not approach the SE Islands.

The subspecific status of the birds from the uplands of the n watershed of the SE Peninsula remains to be determined (Jones *et al.* 1995). Mees (1982b) noted that the original spelling of the form *duperreyii* was corrected by Lesson two years after original publication of the name. Lesson thus becomes the first reviser (ICZN, 1999, Art. 24.2.4). Captain L. I. Duperrey was commander of the corvette *La Coquille* (M. LeCroy *in litt.*). *Other names*—Orange-footed Megapode; formerly treated as a population of *M. freycinet*, as the Common Scrubfowl.

Melanesian Scrubfowl *Megapodius eremita*   RESIDENT | MONOTYPIC
Megapodius eremita Hartlaub, 1868. *Proc. Zool. Soc. London* (1867): 830.—Ninigo Is, Bismarck Arch.

In the NG region, known only from Karkar I; inhabits lowland forest, ranging sparingly to 700 m. On Karkar, *M. eremita* apparently intergrades with *M. decollatus* (Diamond & LeCroy 1979, Jones *et al.* 1995).

Extralimital—Wuvulu I (in the extreme nw of PNG, due north of the NG mainland near the border between PNG and Indonesian NG) east to the large and small islands of the Bismarcks, and the Solomon Is southeast to Makira (Coates 1985, Jones *et al.* 1995).

NOTES
It would be worthwhile to conduct a molecular study of the apparently mixing populations of *Megapodius eremita* and *decollatus* inhabiting the islands north of NG, from Karkar in the west to Umboi in the east. Treated as monotypic following Dickinson & Remsen (2013). *Other names*—Melanesian Megapode; formerly considered a form of *M. freycinet*.

New Guinea Scrubfowl *Megapodius decollatus*   ENDEMIC | MONOTYPIC
Megapodius decollatus Oustalet, 1878. *Bull. Assoc. Sci. France* 31(553): 248.—Kairiru I, off n coast of Sepik-Ramu (*fide* Jones *et al.* 1995).

Inhabits lowland and hill forest of Yapen I[MA], Manam I[MA], Tarawai I[MA], the NW Lowlands, Sepik-Ramu, Huon Penin, and n watershed of the SE Peninsula southeast to the mouth of the Mambare R[MA], ranging as high as 2,950 m in the Saruwaged Ra[J] (Greenway 1935, Ripley 1964a, Gilliard & LeCroy 1966, Freeman *et al.* 2013[J]). Also a few scattered records in the S Lowlands (Jones et al. 1995), from the Setekwa R east to Moroka distr and from the foothills of the upper Purari R (see range map, Jones *et al.* 1995: 209). Birds on Karkar I are hybrids of *M. decollatus* × *M. eremita*. Diamond & LeCroy (1979) noted that there is evidence that *eremita* genes are now dominating this population. In addition, I. Woxvold (*in litt.*) has collected camera-trap records for this northern species in the s watershed of the Western Prov uplands, at Ok Tedi headwaters at the base of the Hindenburg Wall (1,820 m); the upper Fly R; and the Carrington R headwaters in the upper Strickland basin, ca. 950 m.

NOTES
Jones *et al.* (1995) showed that the name *Megapodius affinis* is based on an immature specimen of *M. reinwardt* and hence is not available as a name for this species. Greenway (1935) noted that the specimen collected by Stevens on Mt Missim (in Wau valley) has dark legs and feet, diagnostic of *decollatus*. Birds from Wau exhibit a darker mantle than specimens to the west. *Other names*—formerly treated as a subspecies of *M. freycinet*; *Megapodius affinis*.

Dusky Scrubfowl *Megapodius freycinet*   RESIDENT | MONOTYPIC
Megapodius freycinet Gaimard, 1823 (June). *Bull. Gén. Univ. Annon. Nouv. Sci.* 2: 451.—Waigeo I, NW Islands.
> Synonym: *Megapodius freycinet oustaleti* Roselaar, 1994. *Bull. Zool. Mus. Universit. Amsterdam* 14: 27.—Sorong I, w Bird's Head.
> Synonym: *Megapodius freycinet quoyii* G. R. Gray, 1861. *Proc. Zool. Soc. London* 1861: 289.—Halmahera, Moluccas.

Inhabits coastal forest, mangroves, and edges of swampy woodland through the NW Islands and westernmost Bird's Head; range includes Kofiau[MA], Misool[Q], Waigeo[J], Batanta[MA], Salawati[MA], Arar[A], Boni[B], Saonek[B], Gam[B], Gebe[MA], Gag I[K], Kamuai Is[B], Sorong I[J], several coastal locations north of Sorong on the w end of the

Bird's Head[MA], and small islands (including Hum I[MA]) off the coast of the nw Bird's Head (Gyldenstolpe 1955b[J], Ripley 1964a[A], Mees 1965[Q], Jones *et al.* 1995[B], Johnstone 2006[K]).

Extralimital—n Moluccas (Jones *et al.* 1995).

NOTES

What was formerly considered a single polytypic and widespread species, *M. freycinet* has been broken up into a number of distinct allospecies (see Jones *et al.* 1995 and Harris *et al.* 2014), ranging throughout the Region—*freycinet, geelvinkianus, eremita, decollatus,* and *reinwardt.* Here, the remnant *freycinet* is considered monotypic (see Elliott 1994). The forms *oustaleti* and *quoyii* are insufficiently distinct and are probably points on a weak cline in size of this monotypic form (Jones *et al.* 1995: see table at bottom of p. 185). By contrast, Dickinson & Remsen (2013) recognized four races and state that *quoyi* may deserve species status, based on the work of Birks & Edwards (2002). As recognized here, the species *freycinet* is characterized by the following: pointed hind-crest, plumage overall fuscous, washed with plumbeous; face and side-throat bare; exposed skin dull reddish; iris dark; legs dark (Jones *et al.* 1995). *Other names*—Common Scrubfowl, Dusky Megapode, Incubator Bird.

Family Phasianidae

Pheasants, Old World Quail, and Grouse [Family 178 spp/Region 3 spp]

The species of the family range from North America, Central America, Europe, and Africa to all of mainland and insular Asia, New Guinea, Australia, and New Zealand (McGowan 1994). The family includes several lineages variously treated as either subfamilies or tribes (Kriegs *et al.* 2007, Cracraft 2013). Dickinson & Remsen (2013) placed New Guinea's three quail in the tribe Coturnicini. These are small to medium-large species that inhabit grasslands, ranging to above 3,000 m in the alpine zone of the central cordillera. The species are compact, ground-dwelling game birds with short legs and necks and globular bodies, mainly patterned in browns, grays, and blacks, often with disruptive coloration serving as camouflage in their favored habitat. We follow Dickinson (2003) for generic limits. Generic sequence follows Beehler & Finch (1985).

GENUS *ANUROPHASIS*   ENDEMIC | MONOTYPIC

Anurophasis van Oort, 1910. *Notes Leyden Mus.* 32: 211. Type, by monotypy, *Anurophasis monorthonyx* van Oort.

The single species of the genus is endemic to the Region, restricted to the high alpine grasslands of the Western Ranges. The genus is characterized by the large size, abbreviated tail, strong sexual dichromatism, heavy black barring above and below, and deep rufescent plumage of the adult male. It is presumably closely allied to *Coturnix.* Female plumage exhibits a buff background to underparts rather than rich rufous.

Snow Mountain Quail *Anurophasis monorthonyx*   ENDEMIC | MONOTYPIC

Anurophasis monorthonyx van Oort, 1910. *Notes Leyden Mus.* 32: 212.—Oranje Mts, 3,800 m, c sector of the Western Ra.

Inhabits alpine grasslands and shrublands of the high plateaus north of the highest summits of the Western Ra and the w sector of the Border Ra. Known from the grasslands surrounding Carstensz meadow[J], Yellow and Discovery valleys[J], the Kemabu Plateau[K], Lake Habbema[L], 7 km northeast of Mt Wilhelmina[L], and Mt Mandala[W] (Rand 1942b[L], Ripley 1964a[K], Schodde *et al.* 1975[J], J. M. Diamond *in litt.*[W]). Recorded only from 3,200–4,200 m.

NOTES

We believe the proper adjectival form for the English name is "Snow Mountain." *Other names*—Snow Mountains Quail, *Coturnix monorthonyx.*

Coturnix Bonnaterre, 1791. *Tabl. Encyc. Méth., Ornith.* 1(47): lxxxvii. Type, by tautonymy, "Caille" = *Tetrao coturnix* Linnaeus.

The species of this genus range from Europe and Africa east through Asia, New Guinea, the Bismarck Archipelago, and Australia (McGowan 1994). The genus is characterized by the small size, very short neck and tail, sexual dichromatism, and very short legs. We follow the treatment of Dickinson (2003), who combined *Synoicus, Coturnix,* and *Excalfactoria.* Dickinson & Remsen (2013) split off the species *ypsilophora* and *chinensis* into the genus *Synoicus,* following Kimball *et al.* (2011). We do not follow this change because of the uncertainty of Kimball's results. Bonilla *et al.* (2010) indicated *Coturnix* is a close relative of *Gallus,* although their generic sampling was limited.

Brown Quail *Coturnix ypsilophora*   RESIDENT
Coturnix ypsilophorus Bosc, 1792. *Journ. Hist. Nat. Paris* 2: 297, pl. 39.—Tasmania.

Inhabits lowland grasslands throughout NG but for the Bird's Head; also in the mid-mts of the Eastern Ra and SE Peninsula and in the alpine grasslands in mts of the SE Peninsula. Absent from the Bird's Head and the Western and Border Ra (Mayr 1941, Gyldenstolpe 1955a, Gilliard & LeCroy 1961, Gregory & Johnston 1993, Beehler & Sine 2007). In w NG confined to the lowlands; in Eastern Ra found up to 2,600 m^{CO}; in SE Peninsula occurs to 3,600 m^{MA,CO}.

Extralimital—Lesser Sundas (Flores, Alor, Wetar, Timor, Kisar, Letti, Moa, Luang, Sumba, and Savu) and Australia. Introduced to Fiji and NZ (McGowan 1994).

SUBSPECIES

a. *Coturnix ypsilophora saturatior* (Hartert)
Synoicus ypsilophorus saturatior Hartert, 1930. *Novit. Zool.* 36: 125.—Ifar, Lake Sentani, ne sector of the NW Lowlands.
RANGE: Northwestern Lowlands and Sepik-Ramu. An unconfirmed sight record from Batanta I by S. Hogberg is listed on M. Tarburton's island website.
DIAGNOSIS: The three lowland forms exhibit distinctive gray male plumage, with minor plumage differences. This form is similar to *plumbea,* but the male has dull brown streaking above and below. The females of these three lowland forms are alike.

b. *Coturnix ypsilophora dogwa* (Mayr & Rand)
Synoicus ypsilophorus dogwa Mayr & Rand, 1935. *Amer. Mus. Novit.* 814: 3.—Dogwa, Oriomo R, Trans-Fly.
RANGE: Southern Lowlands (restricted to the Trans-Fly?). Records include Wasur NP (K. D. Bishop *in litt.*).
DIAGNOSIS: Male plumage is dark gray and mainly streaked rather than barred. Underparts pale gray, nearly uniform, or with pale feather centers evident. Dorsal plumage as for *saturatior.*

c. *Coturnix ypsilophora plumbea* (Salvadori)
Synoecus plumbeus Salvadori, 1894. *Ann. Mus. Civ. Genova* 34: 152.—Vakena, SE Peninsula.
RANGE: Lowlands of the SE Peninsula, west in the s watershed to Yule I, and west in the n watershed to the Huon Penin (Mayr 1941); includes the upper Watut/Bulolo to Wau (Greenway 1935).
DIAGNOSIS: As for *dogwa,* but male darker and more sooty gray on underparts; some males exhibit blackish streaking on dorsal surface, others brown streaking (perhaps an age-related difference).

d. *Coturnix ypsilophora mafulu* (Mayr & Rand)
Synoicus ypsilophorus mafulu Mayr & Rand, 1935. *Amer. Mus. Novit.* 814: 1.—Mafulu, s watershed of the SE Peninsula.
> Synonym: *Synoicus ypsilophorus lamonti* Mayr & Gilliard, 1951. *Amer. Mus. Novit.* 1524: 1.—Tomba, s slope of Mt Hagen, c sector of the Eastern Ra.
RANGE: Eastern Ra and s slopes of mts of the SE Peninsula. Mid-montane grasslands from 1,000 to 2,000 m; the population observed in Tari gap presumably also refers to this form (Pratt field notes). Presumably in certain places this subspecies occurs just upslope from populations of *plumbea* and just downslope from *monticola*—a rare example of a triplet of "elevational subspecies," which may perhaps be unique to NG. We follow Dickinson & Remsen (2013) in placing *lamonti* in synonymy with *mafulu.*

DIAGNOSIS: Both males and females are hen-feathered, and their ventral plumage is indistinguishable; male's mantle has some gray mixed in with the browns; female lacks the gray.

e. *Coturnix ypsilophora monticola* (Mayr & Rand)
Synoicus ypsilophorus monticola Mayr & Rand, 1935. *Amer. Mus. Novit.* 814: 2.—Mount Albert Edward, c sector of the SE Peninsula.
RANGE: Known only from alpine shrublands and grasslands above 2,800 m in the Wharton Ra of the SE Peninsula, presumably including the highlands of Mt Albert Edward, Mt Scratchley, and Mt Victoria (Mayr & Rand 1935).
DIAGNOSIS: Like *mafulu* but larger, and dorsal plumage darker and more coarsely patterned; wing (male) 104–109 (Greenway 1973). This is an elevational subspecies and perhaps could be lumped with *mafulu*, which is in the same geography. We keep this example of an elevational subspecies because it is not based on a continuous distribution—the mid-montane grasslands are isolated from the alpine shrublands by a substantial unbroken band of montane forest.

NOTES
We follow Christidis & Boles (2008) in combining *australis* into *ypsilophora* and in placing this species into *Coturnix*, a treatment distinct from that of Beehler & Finch (1985). Dickinson & Remsen (2013) placed the widespread species *ypsilophora* into *Synoicus* along with *chinensis*. The very distinctive population *C. y. dogwa* merits additional study. We follow David & Gosselin (2002a) in considering *Coturnix* as feminine, hence: *C. y. plumbea* (not *plumbeus*). *Other names*—Swamp Quail, *Synoicus ypsilophorus, Coturnix australis*.

King Quail *Coturnix chinensis* RESIDENT
Tetrao chinensis Linnaeus, 1766. *Syst. Nat.*, ed. 12, 1: 277.—Nanjing, China.
Locally distributed through NG grasslands, sea level–2,300 m.
Extralimital—tropical Asia east to the Bismarck Arch and n and e Australia (Mayr & Diamond 2001, Dutson 2011). Introduced to Réunion and Guam (Dickinson & Remsen 2013).

SUBSPECIES
a. *Coturnix chinensis lepida* Hartlaub
Excalfactoria chinensis lepida Hartlaub, 1879. *Sitzungsber. Verh. Naturwiss. Unterh. Hamburg* 7: 3.—Duke of York Islands, off s New Ireland.
> Synonym: *Excalfactoria chinensis papuensis* Mayr & Rand, 1936. *Amer. Mus. Novit.* 868: 1.—Mafulu, s watershed of SE Peninsula.
> Synonym: *Coturnix chinensis novaeguineae* Rand, 1941. *Amer. Mus. Novit.* 1122: 1.—Baliem R, 1,600 m, e sector of the Western Ra.

RANGE: All NG east of the Bird's Neck; also Bismarck Is; sea level–1,220 m (Mayr & Rand 1936, Rand 1942a, Coates 1985). Unconfirmed sight records from Batanta and Salawati I (S. Hogberg in M. Tarburton's island website).
DIAGNOSIS: Considerable variation in ventral plumage pattern of the male in NG and Bismarck Is; we lump *novaeguineae* into *papuensis* because we found specimens exhibiting both plumage types from Lake Daviumbu (AMNH no. 4251521, 425152). Furthermore, we combine this consolidated *papuensis* into the form *lepida*, because differences between the New Guinean and Bismarck birds are marginal (see Peters 1934 and Mayr & Rand 1936).

NOTES
We follow Beehler & Finch (1985), Mayr & Diamond (2001), and Dickinson (2003) and place this species in *Coturnix*. Dickinson & Remsen (2013) placed this species into *Synoicus*. *Other names*—Blue-breasted, Painted, Asian Blue, or Chinese Quail, *Synoicus* or *Excalfactoria chinensis*.

ORDER ANSERIFORMES

The phylogenomic study of Hackett *et mult.* (2008) indicated that the Anseriformes lineage is monophyletic. This clade is sister to the Galliformes. The order comprises three sublineages—the screamers, Anhimidae; the Magpie-Goose, Anseranatidae; and the ducks, geese, and swans, Anatidae (see Christidis & Boles 2008). The sequence of families follows Cracraft (2013). For additional details of historical treatment of the order, see Christidis & Boles (2008) and Gill *et al.* (2010).

Family Anseranatidae

Magpie Goose   MONOTYPIC

A distinct lineage of a single relictual species that ranges from northern and eastern Australia to southern New Guinea. It is the earliest diverging lineage of the Anseriformes (Hackett *et mult.* 2008), and fossils attributed to the Anseranatidae date back to the middle Tertiary in Australia (Worthy & Scanlon 2009). We follow Christidis & Boles (2008) and Dickinson & Remsen (2013) in treating this lineage as a full-fledged family. The family is characterized by the bare facial skin, crown protuberance, long legs, partially webbed feet, and hooked beak with a horny terminal sheath of the maxilla.

GENUS *ANSERANAS*   MONOTYPIC

Anseranas Lesson, 1828. *Man. Orn.* 2: 418. Type, by monotypy, *Anas melanoleuca* Latham = *Anas semipalmata* Latham.

   The single species of the genus inhabits Australia and New Guinea. Generic characters are as for the family.

Magpie Goose *Anseranas semipalmata*   RESIDENT AND VISITOR | MONOTYPIC

Anas semipalmata Latham, 1798. *Trans. Linn. Soc. London* 4: 103.—Near Hawkesbury R, NSW, Australia.

   Breeds in the extensive marshy floodplains of the Trans-Fly and adjacent S Lowlands, from Dolak I east to the Oriomo R. Occurrence and nesting apparently are irregular, depending on rainfall and water level (Hoogerwerf 1959).

   Localities—Dolak I[K], Wasur NP[L], near Merauke[L], Bian Lakes[L], Bensbach R[L], Bula Plains[L], Lake Murray[CO], Lake Pangua[J], Lake Owa[J], near the Pahotouri R[CO], Fly R delta[CO], near Balimo[CO], the Mainland opposite Daru I[CO], and near Port Moresby[W] (Gregory 1997[J], Bishop 1984[K], 2005a[L], K. D. Bishop *in litt.*[W]).

   Bishop (1984) found tens of thousands during surveys of Dolak I, with evidence of breeding. Pale gray young were photographed in Bensbach in Apr-2014; range may also include the Aramia R wetlands east of the Fly R (K. D. Bishop *in litt.*).

   Extralimital—n and e Australia, some dispersing in the postbreeding season to s NG. There is a record for Saibai I in the n Torres Strait Is.

   Other names—Pied or Semipalmated Goose.

Family Anatidae

Ducks, Geese, and Swans   [Family 159 spp/Region 15 spp]

The family exhibits a worldwide distribution. Of the species recorded from New Guinea, eight are resident breeders, three are Palearctic vagrants, three are Australian vagrants, and one is endemic at the generic level; the majority are shared with Australia, and there is at least some seasonal movement between the two landmasses. These are small to large waterbirds with webbed feet, broad and flattened bills, stout bodies, long to very long necks, and waterproof plumage. New Guinean species prefer freshwater swamps, highland lakes, and rivers (Pratt & Beehler 2014). Christidis & Boles (2008) provided a comprehensive review of the systematic work on this family, which is complex. Our generic and species treatment follows Dickinson & Remsen (2013), with some exceptions, in which we follow Christidis & Boles (2008).

GENUS *DENDROCYGNA*   [Genus 8 spp/Region 3 spp]

Dendrocygna Swainson, 1837. *Nat. Hist. Class. Birds* 2: 365. Type, by subsequent designation (Eyton, 1838. *Monogr. Anatidae*: 28), *Anas arcuata* Horsfield.

The species of this genus range from the southern United States, the Caribbean, and Central and South America to Africa, tropical Asia, New Guinea, and Australia (Carboneras 1992d). The genus is characterized by relatively long legs, long neck, and arboreal perching and roosting.

Spotted Whistling Duck *Dendrocygna guttata*   RESIDENT | MONOTYPIC
Dendrocygna guttata Schlegel, 1866. *Mus. Hist. Nat. Pays-Bas, Rev. Méthod. Crit. Coll.* 6 (Anseres): 85.—Sulawesi.

A common inhabitant of freshwater wetlands, lakes, and rivers of lowland NG, Batanta I[J], Salawati I[J,L], Kofiau I[K], and Aru Is[MA], (Johnsgard 1979, Coates 1985, Dutson & Tindige 2006[J], Diamond *et al.* 2009[K], Hornbuckle & Merrill 2004[L]).

Extralimital—s Philippines to Sulawesi, Moluccas, Tanimbar, Bismarck Arch, and Bougainville I (Mayr & Diamond 2001). First recorded in Australia in 1996; it has now bred on Cape York Penin, QLD, and it recently appeared on Melville I, NT (P. Gregory *in litt.*, W. E. Boles *in litt.*).

Other names—Spotted Tree Duck, Spotted Whistle-duck.

Plumed Whistling Duck *Dendrocygna eytoni*   VISITOR | MONOTYPIC
Leptotarsis eytoni (ex Gould) Eyton, 1838. *Monogr. Anatidae*: 111.—Northwestern Australia.

A rare visitor, in small flocks, to the NG region from Australia (Mees 1982b, Coates 1985).

Localities—Daru in Jun-1896[CO], Kurik in Sep-1959[J], Balimo in Oct-1965[M], Moitaka settling ponds in Aug–Sep-1968[K] , Ilimo in Jul-2005[L] (Hoogerwerf 1962[J], Bell 1967a[M], Wolfe 1968[K], P. Gregory *in litt.*[L]). Between 2009 and 2013, Gregory and Vandergaast recorded a flock of as many as 60 spending the nonbreeding season (March–September) near Ilimo in the Port Moresby area (K. D. Bishop *in litt.*).

Extralimital—breeds in n and e Australia (Carboneras 1992d). Vagrant to Makira and Rennell I and New Caledonia (Mayr & Diamond 2001, Dutson 2011).

Other names—Plumed Tree-duck, Grass Whistling-duck.

Wandering Whistling Duck *Dendrocygna arcuata*   RESIDENT AND PROBABLE VISITOR
Anas arcuata Horsfield, 1824. *Zool. Researches Java* 8: pl. [65].—Java. There is some disagreement regarding the plate number; the plates were unnumbered in the original publication (M. LeCroy *in litt.*).

Resident of freshwater wetlands in the NG lowlands. Nomadic populations may also visit from Australia, as suggested by frequent sightings over the Torres Strait Is (Coates 1985). Vagrant in the Wahgi valley (Frith 1982a).

Extralimital—breeds from the n Philippines, Borneo, and Java east to New Britain, n and e Australia, and formerly Fiji and New Caledonia (Carboneras 1992d, Mayr & Diamond 2001, Dutson 2011).

SUBSPECIES

a. *Dendrocygna arcuata australis* Reichenbach

Dendrocygna arcuata (australis) Reichenbach, 1850. *Vollst. Naturgesch., 2 Vögel, 3 Syn. Avium 1 Natatores* 4: col. 7.—Port Essington, NT, Australia.

RANGE: Southern NG and n Australia from the Kimberley Ra to Rockhampton, QLD; occasionally south to NSW and South Australia. Formerly recorded from New Caledonia and Fiji (Johnsgard 1979). In NG, records that may pertain to this race include: Lake Kurumoi[M] (Bird's Neck), Dolak I[J], Kurik[K], Bian Lakes[J], Balimo[CO], Daru[MA], Lake Daviumbu[J], the middle Fly R[MA], Sepik R[MA], Ramu R[MA], Wahgi valley[CO], Huon Gulf[MA], Hall Sound[MA], near Ilimo[L], and Laloki R[MA] (Mees 1982b[K], Erftemeijer & Allen 1989[M], K. D. Bishop *in litt.*[J], P. Gregory *in litt.*[L]).

DIAGNOSIS: Larger than the nominate form (Mees 1982b, Carboneras 1992d). Wing: male 211–227, female 204–219 (Marchant & Higgins 1990).

b. *Dendrocygna arcuata ?pygmaea* Mayr

Dendrocygna arcuata pygmaea Mayr, 1945. *Amer. Mus. Novit.* 1294: 3.—Manlo, Wide Bay, s coast of e New Britain.

RANGE: Johnsgard (1979), Carboneras (1992d), and Dutson (2011) noted the range of this form was restricted to New Britain, possibly Umboi, and perhaps formerly Fiji. Dickinson & Remsen (2013) stated *pygmaea* also occurs in n NG. Carboneras (1992d) noted that individuals from n NG are intermediate between the nominate form and *australis*. Localities that may pertain to this race include: Ramu R[MA], Sepik region[J], and Huon Gulf[MA] (Gilliard & LeCroy 1966[J]).

DIAGNOSIS: A subspecies determined by size. Male wing length in this form is 10% smaller than in the nominate form, and 20% smaller than in *D. arcuata australis* (Mayr 1945). Wing 173–183 (Marchant & Higgins 1990).

NOTES

Additional work is needed to clarify the status of birds from n NG (especially in w NG and the Bird's Head). Presumably size is clinal between the moderate-size birds in se Asia and the largest birds in Australia and s NG. *Other names*—Water Tree-duck, Water Whistle-duck, Diving Whistling Duck.

GENUS *CYGNUS* [Genus 6 spp/Region 1 sp]

Cygnus Bechstein, 1803. *Orn. Taschenb. Deutschl.* 2: 404, note. Type, by monotypy, *Anas olor* Gmelin.

The species of this genus range from North America to southern South America, northern Europe, northern Asia, Australia, and New Zealand. No native populations of this genus inhabit Africa (Carboneras 1992d). The genus is characterized by the very large size, very long neck, very short legs, webbed feet, and white or black-and-white plumage.

Black Swan *Cygnus atratus* VAGRANT | MONOTYPIC

Anas atrata Latham, 1790. *Index Orn.* 2: 834.— New South Wales, Australia.

Two records from NG: a single bird captured and photographed by H. van de Kerkhof near Merauke in 1980 (Beehler 1980a), and two birds at Merauke on 15-Jul-1985 (Parry 1989).

Extralimital—a native breeder in Australia; now the species is known to breed in NZ and Britain (Gill *et al.* 2010, Marchant & Higgins 1990).

Tadorna Boie, 1822. *Tagebuch Reise Norwegen*: 140, 351. *Anas familiaris* Boie, 1822; type, by tautonymy = *Anas tadorna* Linnaeus.

The species of this genus range from Europe to Africa, western Asia, New Guinea, western Australia, and New Zealand (Carboneras 1992d). The genus is characterized by medium-large size and boldly patterned plumage (always a combination of two or more of the following colors: brown, white, black, or green). Dickinson & Remsen (2013) placed *T. radjah* in a monotypic genus, *Radjah*, based on the work of Gonzalez *et al.* (2009) and others, who suggested that *radjah* has been found paraphyletic in *Tadorna* with respect to *Alopochen*, the Egyptian Goose. An alternative may be to include *Alopochen* in an enlarged *Tadorna*. *T. radjah* shares distinctive plumage characters with other members of core *Tadorna*, including bold patterning, large patches of white, and presence of a breast/neck bar. See Kear (2005) and Gill *et al.* (2010) for additional reasons for keeping all the shelducks in this genus.

Raja Shelduck *Tadorna radjah*   RESIDENT

Anas radjah "Garnot" Lesson, 1828. *Man. Orn.* 2: 417.—Buru I, s Moluccas.

Occurs singly or in small groups in freshwater swamps, coasts, and estuaries; mainly coastal, with most records from s NG, particularly the Trans-Fly and Port Moresby region; also from NW Islands, Aru Is, Yapen, Goodenough, and Fergusson I (Mayr 1941, Bell 1970e, Pratt & Beehler 2014).

Extralimital—Moluccas (Buru, Seram) and n fringe of Australia (Carboneras 1992d).

SUBSPECIES

a. *Tadorna radjah radjah* (Lesson)

RANGE: Moluccas; NW Islands but for Gebe, and all NG except perhaps the Trans-Fly (Ripley 1964a, Johnsgard 1979, Carboneras 1992d, but see Mees 1982b). No records from Misool I?
DIAGNOSIS: Distinguished from *rufitergum* (Mees 1982b: 20) by the blacker color of the mantle (Carboneras 1992d).

b. *Tadorna radjah ?rufitergum* Hartert

Tadorna radjah rufitergum Hartert, 1905. *Novit. Zool.* 12: 205.—South Alligator R, NT, Australia.
RANGE: The n tier of Australia to s NG; populations apparently intergrade with the nominate form in s NG (Johnsgard 1979).
DIAGNOSIS: Mantle color richer and more chestnut brown than blackish brown.

NOTES

We follow Gill & Donsker (2015) in spelling the species epithet "Raja," which agrees with the current Indonesian spelling for the word *king*—as in "Raja Ampat Is." *Other names*—White-headed or Black-backed Shelduck, Burdekin Duck or Shelduck, *Radjah radjah*.

GENUS *SALVADORINA*   ENDEMIC | MONOTYPIC

Salvadorina Rothschild & Hartert, 1894. *Novit. Zool.* 1: 683. Type, by monotypy, *Salvadorina waigiuensis* Rothschild & Hartert.

One species in the genus, restricted to the interior uplands of New Guinea. The genus is characterized by the uptilted and pointed tail, the dark back barred narrowly with white, the finely barred sides, the uniformly pale yellow-orange bill, and the unmarked brown head. We follow Mayr (1941), Rand & Gilliard (1967), Mlíkovský (1989), and Dickinson & Remsen (2013) in retaining the monotypic genus *Salvadorina*. It is evidently an isolated form with no close relatives (Kear 2005). The species is typically observed in pairs and most readily observed in alpine lakes, where hunting is minimal. It is behaviorally distinct from typical *Anas* (M. LeCroy *in litt.*).

Salvadori's Teal *Salvadorina waigiuensis*   ENDEMIC | MONOTYPIC

Salvadorina waigiuensis Rothschild & Hartert, 1894. *Novit. Zool.* 1: 683.—Mountains of the NG mainland.
[The type probably originated from the Arfak Mts of the Bird's Head; not known from Waigeo I, despite
its name.]

Widespread in small numbers throughout NG's uplands. Mainly inhabits alpine tarns and torrential
rivers in upland habitats throughout mainland NG (including the Foja and Adelbert Mts and mts of the
Bird's Head). Found from above treeline (4,300 m[J]) to as low as 120 m[K] where the Avi Avi R drains out of
the Chapman Ra into the lowlands of the Lakekamu basin (Beehler *et al.* 1994[K], Sam & Koane 2014[J]). Van
Balen & Rombang (1998) found congregations of the species on alpine lakes above the Freeport Mine in
Jan-1997 and Aug-2000.

NOTES

Currently classified by IUCN as Vulnerable. We propose the species might better be classified as Near
Threatened at this time, given the abundance of undisturbed high-elevation habitat available for the species. *Other names*—Salvadori's Duck, *Anas waigiuensis*.

GENUS *CHENONETTA*   MONOTYPIC

Chenonetta von Brandt, 1836. *Descr. Icon. Anim. Ross. Nov., Aves*, fasc. 1: 5. Type, by monotypy, *Anser
lophotis* von Brandt = *Anas jubata* Latham.

The single living species of the genus inhabits Australia (Johnsgard 1979). A vagrant to New Zealand
and New Guinea (Gill *et al.* 2010). It is rather goose-like in form, sexually dichromatic, small-headed,
short-billed, and large-bodied. See Gill *et al.* (2010) regarding placement of genus within the family.

Australian Wood-Duck *Chenonetta jubata*   VAGRANT | MONOTYPIC

Anas jubata Latham, 1801. *Index Orn.*, Suppl.: lxix.—New South Wales, Australia.

A single record from Ilimo, north of Port Moresby, between 13-Feb-1994 and 20-Nov-1994
(Coates 1995).

Extralimital—Australia.

Other names—Maned Duck.

GENUS *NETTAPUS*   [Genus 3 spp/Region 2 spp]

Nettapus von Brandt, 1836. *Descr. Icon. Anim. Ross. Nov., Aves*, fasc. 1: 5. Type, by monotypy, *Anas madagascariensis* Gmelin = *Anas aurita* Boddaert.

The species of this genus range from sub-Saharan Africa to tropical Asia, New Guinea, and northern
and eastern Australia. The genus is characterized by its diminutive size, very short neck, and proportionally
small head.

Cotton Pygmy Goose *Nettapus coromandelianus*   RESIDENT

Anas coromandeliana Gmelin, 1789. *Syst. Nat.* 1(2): 522.—Southwestern coast of India.

Recorded from quiet freshwater environments of the Sepik[J] and Ramu R[CO], the Eastern Ra to 2,255 m
at Lake Kandep[W], and eastward to Lake Wanum[W], near Lae (Gilliard & LeCroy 1966[J], Bell 1984[W]). Recorded
in 2013 near Timika in the w sector of the S Lowlands (S. van Balen *in litt.*). Its status in the Sepik-Ramu
should be reexamined in light of the devastating effects that introduced plant-eating fish have had on
its environment. Status at Lake Kandep not known; numbers were seen there by L.W.C. Filewood and
the species was collected there in Jul-1966 (an immature specimen in PNG Nat. Mus., collected by W. R.
Wheeler, reported by Bell 1984). Note also record from Aroa Lagoon northwest of Port Moresby, the only s
watershed record (Hicks *et al.* 1988).

Extralimital—s Asia and s China to se Asia and e Australia. Records from se Asia may all be non-breeders (Carboneras 1992d).

a. *Nettapus coromandelianus coromandelianus* (Gmelin)
RANGE: Lowlands of tropical Asia to n NG.
DIAGNOSIS: This form is smaller than *albipennis* from Australia (Carboneras 1992d). Wing 150–167 (Marchant & Higgins 1990).

b. *Nettapus coromandelianus ?albipennis* Gould
Nettapus coromandelianus albipennis Gould, 1842. *Birds Austral.* 6: text to plate 5, labeled "Nettapus coromandelianus?"—Moreton Bay, QLD, Australia.
RANGE: Breeds from n QLD to n NSW. An Australian subspecies, *albipennis* can be expected to occur as a vagrant to s NG (Carboneras 1992d, Coates 1985). Record from Aroa Lagoon (Hicks *et al.* 1988) may be of this form, dispersing from QLD.
DIAGNOSIS: Larger than nominate form (Carboneras 1992d). Wing: male 181–191, female 176–183 (Marchant & Higgins 1990).

NOTES
I. Woxvold (*in litt.*) reported: "Sadly, evidence from local residents in the upper Sepik basin suggests that this and other waterbird species once common locally have become much rarer in recent decades, coinciding with the introduction of a suite of exotic fish species into the Sepik between 1987 and 1997, one or more of which may be responsible for major reductions in native aquatic plant populations." *Other names*—Cotton Teal; White-quilled, Asian, or Indian Pygmy-goose, White Pygmy-Goose.

Green Pygmy Goose *Nettapus pulchellus*   RESIDENT | MONOTYPIC
Nettapus pulchellus Gould, 1842. *Birds Austral.* 6: pl. 4.—Port Essington, NT, Australia.
Inhabits freshwater lakes, swamps, and waterways in the s Bird's Head[A] and Neck, and s NG lowlands to 500 m (Gyldenstolpe 1955b[A], Coates 1985, Coates & Peckover 2001).
Localities—Aimaru Lake[L] in the Bird's Head[CO], Lake Yamur[Q] in the Bird's Neck, near Timika[Z], Dolak I[J], Kurik[J], Merauke distr[MA], Bensbach R[CO], the middle Fly R[J], Lake Daviumbu[X], Lake Teberu in the Purari basin[Y], and near Port Moresby[K] (Beaufort 1909[Q], Rand 1942a[X], Gyldenstolpe 1955b[L], Mees 1982b[J], Mack & Wright 1993[Y], D. Hobcroft *in litt.*[K], S. van Balen *in litt.*[Z]).
Extralimital—Buru and Seram (as a vagrant), and n and ne Australia (Johnsgard 1979, Carboneras 1992d).

NOTES
A sight record from Lake Kopiago requires confirmation, given records of the occurrence of its sister species nearby (Coates 1985). The only confirmation of the breeding status for this species in NG is observation of birds in postbreeding molt in the Bensbach area during early April (Bishop 2005a). *Other names*—Green Dwarf-Goose.

GENUS *ANAS*   [Genus 49 spp/Region 5 spp]

Anas Linnaeus, 1758. *Syst. Nat.*, ed. 10, 1: 122. Type, by subsequent designation (Lesson, 1828. *Man. Orn.* 2: 417), *Anas Boschas* Linnaeus = *Anas platyrhynchos* Linnaeus.
The species of this genus range virtually worldwide (Carboneras 1992d). The genus is characterized by the medium size, short neck, short tail, and diverse plumage dominated by brown and gray.
The Northern Shoveler (*Anas clypeata*) has been recorded as a vagrant to Micronesia, Australia, New Zealand, and New Britain (Johnsgard 1979, Coates 1985, Carboneras 1992d, Mayr & Diamond 2001, Gill *et al.* 2010) and is to be expected in the New Guinea region during the northern winter season. Dickinson & Remsen (2013) placed into the genus *Mareca* a species that we treat as *Anas penelope*.

Eurasian Wigeon *Anas penelope*

VAGRANT | MONOTYPIC

Anas Penelope Linnaeus, 1758. *Syst. Nat.*, ed. 10, 1: 126.—Coasts and swamps of Europe.

Two records from the NG region: a female collected at Sorong on 23-Jan-1949 (Gyldenstolpe 1955b) and a male in eclipse plumage observed near Bereina in Nov-1977 (Heron 1978a).

Extralimital—a Palearctic breeder (Iceland to e Siberia), migrating south in winter to Africa, n India, China, and Japan. Scattered records of wintering birds range from Luzon, Philippines, to New Caledonia (Johnsgard 1979).

NOTES

Dickinson & Remsen (2013) placed the species in the genus *Mareca*. *Other names*—Wigeon, European Wigeon, *Mareca penelope*.

Pacific Black Duck *Anas superciliosa*

RESIDENT (AND VISITOR?)

Anas superciliosa Gmelin, 1789. *Syst. Nat.* 1(2): 537.—Dusky Sound, Fiordland, NZ.

Inhabits freshwater environments from the coast to alpine lakes throughout the Region, including Misima and Woodlark I[J], from sea level to 3,660 m[CO] (Rand & Gilliard 1967, Pratt & Beehler 2014, Pratt field notes[J]). Perhaps also a seasonal visitor to s NG from Australia (Coates 1985).

Extralimital—Sumatra and Sulawesi east through the Moluccas, Palau, Bismarcks, Solomons, Vanuatu, Fiji, Tonga, Samoa, Cook, Society, and Austral Is, thence to Australia and NZ (Johnsgard 1979, Carboneras 1992d, Dutson 2011).

SUBSPECIES

a. *Anas superciliosa superciliosa* Gmelin

Synonym: *Anas superciliosa rogersi* Mathews, 1912. *Austral. Av. Rec.* 1: 33.—Augusta, sw Australia.

RANGE: Southern NG, Australia, and NZ (Marchant & Higgins 1990).
DIAGNOSIS: Substantially larger than *pelewensis*. Wing: male 250–285, female 230–265 (Marchant & Higgins 1990). Includes *rogersi* (see Dickinson & Remsen 2013).

b. *Anas superciliosa pelewensis* Hartlaub & Finsch

Anas superciliosa var. *pelewensis* Hartlaub & Finsch, 1872. *Proc. Zool. Soc. London*: 108.—Palau Is.
RANGE: Micronesia south to n NG, much of Melanesia and Polynesia (Mayr & Diamond 2001).
DIAGNOSIS: Size smaller; wing (male) 224–250.

NOTES

Treated by some authorities as a race of *A. poecilorhyncha*, but recent molecular systematic analysis using mitochondrial and nuclear DNA indicated *A. superciliosa* is sister to *A. luzonica* (Lavretsky *et al.* 2014). Rand & Gilliard (1967) reported that they did not find this duck on the extensive marshes of the Fly R, nor on those of the Taritatu R. However, in the Western Ra it was fairly common along certain mid-montane rivers and on lakes and ponds in the alpine zone. On the middle Sepik R it is thinly distributed in marshes and lagoons, and not uncommon in the Wahgi valley (Coates 1985). *Other names*—Black or Spot-billed Duck, New Zealand Grey Duck.

Grey Teal *Anas gracilis*

RESIDENT AND VISITOR | MONOTYPIC

Anas gracilis Buller, 1869. *Ibis* (ns)5: 41.—Manawatu R area, near Wellington, NZ.

Recorded from many Mainland localities and the Aru Is[MA]. Common along the coast of s NG; a few records from highland lakes of the Central Ra to 3,000 m (Coates 1985).

Localities—Kurik[K], Baliem valley[CO], Lake Habbema[J], Digul R[M], Wahgi R[L], and Lea Lea[CO] (Bangs & Peters 1926[M], Rand 1942b[J], Gyldenstolpe 1955a[L], Mees 1982b[K]). Hornbuckle & Merrill (2004) reported three birds on Salawati I on 11-Sep-2004.

Extralimital—Australia and NZ; also scattered records from New Britain, Vanuatu, New Caledonia, Lord Howe, and Macquarie I (Coates 1985, Mayr & Diamond 2001). Banded Australian birds have been found in NG (Bishop 2005a), and this dispersive species may be a regular migrant to the Region (Marchant & Higgins 1990).

We follow Christidis & Boles (2008) in treating *gracilis* as specifically distinct from *gibberifrons*. Mayr & Diamond (2001) and Dickinson & Remsen (2013) treated *gracilis* as a subspecies of *gibberifrons*. *Other names*—Slender or Sunda Teal, Australian Grey Teal, *Anas gibberifrons*.

[Chestnut Teal *Anas castanea*   HYPOTHETICAL | MONOTYPIC

Mareca castanea Eyton, 1838. *Monogr. Anatidae*: 119, pl. 19.—New South Wales, Australia.

Stronach (1980) reported a single bird in 1980, followed by a single bird collected from a flock of eight birds from near the Bensbach R of the Trans-Fly; and apparently an additional bird was recorded in 1981 (Coates 1985: 101). The collected bird was apparently never prepared as a museum specimen, so it is not available for study. Given the extreme rarity of this species in the n half of Australia, we take a conservative course and treat these records as hypothetical, as they lack the details necessary to evaluate and confirm identification. Breeds in southernmost Australia and winters north to s QLD.]

Northern Pintail *Anas acuta*   VAGRANT

Anas acuta Linnaeus, 1758. *Syst. Nat.*, ed. 10, 1: 126.—Sweden.

One record: two birds at the Moitaka settling ponds near Port Moresby, remaining from late Nov-1978 to Jan-1979 (Finch 1979c).

Extralimital—breeds in n North America and n Eurasia and winters south to Central America, Africa, tropical Asia, Micronesia, Hawai`i, and (rarely) NZ and Polynesia (Johnsgard 1979).

SUBSPECIES

a. *Anas acuta acuta* Linnaeus

RANGE: A Holarctic breeder that winters to Central America, Africa, and tropical Asia, straggling to the w Pacific; records from Umboi and Bougainville I (Carboneras 1992d, Mayr & Diamond 2001, Gill *et al.* 2010, Dutson 2011).

DIAGNOSIS: Wing: male 267–282, female 254–267 (Cramp 1977). Larger than the isolated subspecies from the Kerguelen and Crozet Is (Carboneras 1992d).

Garganey *Anas querquedula*   VISITOR | MONOTYPIC

Anas Querquedula Linnaeus, 1758. *Syst. Nat.*, ed. 10, 1: 126.—Europe.

A scarce northern migrant, arriving in the Region in October and departing by late May. Frequents tidal mudflats, rice fields, lakes, swamps, and settling ponds (Coates 1985).

Localities—Bird's Head[J], Kapare R[K], Merauke R[K], Kumbe R[K], Bian R[K], near Merauke[K], Oriomo R[CO]; Lakes Pangua, Owa, and Abuve[L]; the NG mainland near Daru I[MA]; and Port Moresby[CO] (Bangs & Peters 1926[K], Hoogerwerf 1964[J], Gregory 1997[L]).

Extralimital—breeds in temperate Eurasia, wintering to Africa and tropical Asia, east to New Ireland, New Britain, and Australia (Carboneras 1992d, Mayr & Diamond 2001).

NOTES

A 1966 report of this species from Lake Kandep, Enga Prov, was, in fact, a *Nettapus coromandelianus* (specimen in PNG Nat. Mus.; Coates 1985). Dickinson & Remsen (2013) placed this species in the genus *Spatula*, based on the molecular work of Gonzalez *et al.* (2009). *Other names*—Garganey Teal.

GENUS *AYTHYA*   [Genus 12 spp/Region 1 sp]

Aythya Boie, 1822. *Tagebuch Reise Norwegen*: 308, 351. Type, by monotypy, *Anas marila* Linnaeus.

The species of this genus range from the Americas, Europe, and Africa, east to China and Australia. Largely absent from southeast Asia and the western Pacific (Carboneras 1992d). The genus is characterized by the compact body plan, large head, and simple plumage pattern combining brown, black, and white or gray. All species of the genus exhibit sexual dichromatism.

The Tufted Duck (*Aythya fuligula*) has been recorded from New Britain (Mayr & Diamond 2001) and may be expected as a vagrant to the Region.

Hardhead *Aythya australis*   VISITOR AND OCCASIONAL BREEDER | MONOTYPIC

Nyroca australis Eyton, 1838. *Monogr. Anatidae*: 160.—New South Wales, Australia.
    Synonym: *Aythya australis papuana*, Ripley, 1964. *Yale Peabody Mus. Bull.* 19: 16.—
      Wamena, Baliem valley, w NG.

Scattered records from freshwater locales in NG. Breeding recorded from lagoons of the Aroa R in 1982 (Finch 1982c). In May-1978 a small influx occurred in the Port Moresby region. Observed in small numbers along the Fly R and Elevala R (K. D. Bishop *in litt.*).

    Localities—Waigeo I[K], Anggi Lakes in Arfak Mts[Q], Lake Habbema[K], Wamena[J], Bian R[CO], Kurik[L], Kumbe R[L], Bensbach R[CO], the Trans-Fly[M], Chambri Lakes[CO], Aroa R[CO], and Port Moresby[CO] (Rand 1942b[K], Ripley 1964a[J], Mees 1982b[L], Beehler *et al.* 1986[M], S. F. Bailey *in litt.*[Q]). Recently recorded near Port Moresby (D. Hobcroft *in litt.*).

    Extralimital—breeds in sw and e Australia, with isolated breeding records from Banks I (Vanuatu) and New Caledonia (Dutson 2011). Old records from Tasmania, New Caledonia and NZ; also known from Sulawesi and e Java (Johnsgard 1979, Carboneras 1992d).

NOTES
We follow Mees (1982b) in sinking *papuana*. We follow Marchant & Higgins (1990) in treating the species as monotypic (but see Dickinson & Remsen 2013). On Lake Habbema in the Western Ra, Rand (1942b) found several hundred of these ducks; none of the collected birds showed signs of breeding. Most birds in NG are presumably wintering visitors from Australia. *Other names*—White-eyed Duck, Australian White-eyed Duck, *Nyroca australis*.

ORDER PHOENICOPTERIFORMES

Phylogenomic studies by Hackett *et mult.* (2008) indicated that the grebes and flamingos are sister lineages. These two are sister to the group including the tropicbirds (Phaethontidae), the sandgrouse (Pteroclididae), the pigeons (Columbidae), and the mesites (Mesitornithidae). The order includes a single family in the Region.

Family Podicipedidae

Grebes and Dabchicks   [Family 23 spp/Region 2 spp]

The species of the family range through the Americas, Europe, sub-Saharan Africa, Madagascar, and much of Asia east to arctic Russia, Borneo, Sumatra, the Lesser Sundas, Australia, New Zealand, the Solomon Islands, and Vanuatu (Llimona & del Hoyo 1992). Fossil record of the lineage reaches back to the early Tertiary (W. E. Boles *in litt.*). These are small to medium-size diving birds with sharp bills, rudimentary tails, and lobed toes; legs are set well back on the body. They breed on fresh water and winter on fresh and salt water.

Tachybaptus Reichenbach, 1853. *Avium Syst. Nat.* (1852): iii, pl. 2. Type, by monotypy, *Columbus minor* Gmelin = *Columbus ruficollis* Pallas.

The species of this genus range from New Zealand and Tasmania through Melanesia, parts of tropical Asia, Africa, Madagascar, Europe, the Caribbean, Central America, and South America (Llimona & del Hoyo 1992). The genus is characterized by its small, compact body, small bill, dark plumage, and lack of a crest.

Tricoloured Grebe *Tachybaptus tricolor* RESIDENT

Podiceps (Sylbeocyclus) tricolor G. R. Gray, 1861. *Proc. Zool. Soc. London*: 366.—Ternate, n Moluccas.

Inhabits lakes, slow-moving rivers, and wetlands of the Bird's Head, NW Islands, and patchily through the NW Lowlands, Sepik-Ramu, and Huon Penin; lowlands to 1,500 m.

Extralimital—Long, Umboi, New Britain, Witu, and Lolobau (Dutson 2011); also west to Wallacea (Sulawesi to Seram).

SUBSPECIES

a. *Tachybaptus tricolor tricolor* (G. R. Gray)

RANGE: Moluccas, Lesser Sundas (Lombok to Timor), and Sulawesi, east to nw and nc NG, where it meets and intergrades with *T. t. collaris* on the Huon Penin (Mees 1965). Localities—Kofiau I[M], Misool I[L], ?Salawati[Q], Atinju Lake on the Bird's Head[J], Arfak Mts[MA], Anggi Lakes[W], Baliem valley[K], Lake Sentani[K], Mosso, Yimas Lakes, Baiyer R valley[CO], near Ubaigubi, Alexishafen[W], Madang[MA], Komiminung Swamp of the lower Ramu[CO], (Gyldenstolpe 1955b[J], Ripley 1964a[K], Mees 1965[L], Diamond *et al.* 2009[M], S. F. Bailey *in litt.*[W], S. Hogberg in M. Tarburton's island website[Q]). Sighting from the Kumawa Mts (Diamond & Bishop 2015). DIAGNOSIS: Wing 104 (Rand & Gilliard 1967). See diagnosis for *collaris*.

b. *Tachybaptus tricolor collaris* (Mayr)

Podiceps ruficollis collaris Mayr, 1945. *Amer. Mus. Novit.* 1294: 1.—Bougainville I.

RANGE: Huon Penin (where it intergrades with *T. t. tricolor*), Long, Umboi, New Britain, Lolobau, Witu, New Ireland, and Bougainville I (Mayr & Diamond 2001). Localities—Lae[CO], Finschhafen area[CO], and Kulungtufu[J] (Mayr 1931[J]). DIAGNOSIS: Similar to *T. t. tricolor* but darker throughout; extent of white on inner webs of secondaries much reduced; collar better defined; and black of upper throat restricted to chin, but extending up to the eye (Mayr 1945).

NOTES

We follow Mlíkovský (2010) in recognizing the species *tricolor* as distinct from *ruficollis*. Dickinson & Remsen (2013) and del Hoyo & Collar (2014) treated *tricolor* as a subspecies of *ruficollis*, following Storer (1979). Characters distinguishing *tricolor* include the dark belly plumage and the longer and more massive bill (Mlíkovský 2010). Apparently both *tricolor* and *novaehollandiae* co-occur and probably breed on Lake Sentani and Komiminung Swamp, lower Ramu (Coates 1985). *Other names*—Red-throated or Eurasian Little Grebe, Dabchick; *Podiceps* or *Tachybaptus ruficollis*.

Australasian Grebe *Tachybaptus novaehollandiae* RESIDENT (AND VISITOR?)

Podiceps Novae Hollandiae Stephens, 1826. In Shaw, *General Zool.* 13(1): 18.—New South Wales, Australia.

Inhabits lakes and wetlands throughout NG, mainly in lowlands but with records to 3,225 m (Rand 1942b); found locally in the n watershed, where it coexists with *T. tricolor* (Coates 1985); also Woodlark I (Vang 1991).

Extralimital—Java to Rennell I (Coates & Peckover 2001); range includes NZ (where a recent colonist and breeder), much of Australia, as well as New Britain, Manus I, the Solomon Is, Vanuatu, and New Caledonia (Gill *et al.* 2010, Dutson 2011).

SUBSPECIES

a. *Tachybaptus novaehollandiae novaehollandiae* (Stephens)

RANGE: New Zealand and Australia; also resident, with breeding reported from Port Moresby (Coates

1985); recorded widely from the S Lowlands and alpine lakes in the Western and Border Ra (Rand & Gilliard 1967, Llimona & del Hoyo 1992). Some birds likely are seasonal visitors from Australia. Localities—Lake Habbema[L], Sibil[Q], Merauke[J], Lake Daviumbu[K], and Wahgi R[J] (Rand 1942a[K], 1942b[L], Gyldenstolpe 1955a[J], Mees 1964b[Q]).

DIAGNOSIS: Larger and longer-winged than *incola*; wing: male 105–114, female 105–107 (Rand & Gilliard 1967, Marchant & Higgins 1990).

b. *Tachybaptus novaehollandiae incola* (Mayr)

Podiceps novaehollandiae incola Mayr, 1943. *Emu* 43: 5.—Ifar, Lake Sentani, ne sector of the NW Lowlands.

RANGE: Presumably an endemic breeding subspecies patchily distributed in appropriate habitat in the n watershed of NG. Localities—Lake Sentani[K,Q], Chambri Lake[J] on the Sepik R, Alexishafen[R], Komiminung Swamp in the lower Ramu[R], Bulolo[S], and Popondetta[CO] (Rand 1942b[K], Mayr 1943, Ripley 1964a[Q], Gilliard & LeCroy 1966[J], Storer 1979[S], S. F. Bailey *in litt.*[R]).

DIAGNOSIS: Shorter winged than nominate form: 95–104 (Rand & Gilliard 1967). From the rather brief type description, this form is very small and has a larger bill:wing ratio (Mayr 1943).

NOTES

Records from Lake Habbema, Wamena, and the Baliem R may refer to the southern (nominate) form, but here we consider those birds undiagnosed because their size appears intermediate (Rand 1942b, Ripley 1964a). Lake Habbema and the Baliem valley lie north of the high ranges yet drain southward, making the area one of those that is neither wholly northern nor southern. *Other names*—Australian Little Grebe, Australian or Black-throated Dabchick, *Podiceps novaehollandiae*.

ORDER COLUMBIFORMES

Hackett *et mult.* (2008) reported that this order is monophyletic and possibly sister to the mesites (Mesitornithidae) and the sandgrouse (Pteroclididae). Columbiformes includes a single family, the Columbidae (Cracraft 2013).

Family Columbidae

Pigeons and Doves   [Family 315 spp/Region 52 spp]

The species of the family range worldwide but for northern Canada, Greenland, the central Sahara, and northern Siberia (Baptista *et al.* 1997). The pigeons and doves attain their greatest diversity of size, form, and ecology in New Guinea. Pigeons are plump birds with small heads and short, blunt bills. Arboreal forms have very short legs; most are swift, strong fliers. Pigeons in New Guinea vary greatly in size, from the turkey-size crowned pigeons (75 cm) to the tiny Dwarf Fruit-Dove (14 cm). In the Region, pigeons utilize all forested and woodland habitats. Lowland rain forests support the greatest diversity; some localities are inhabited by more than 20 species. For placement of the family, we follow Christidis & Boles (2008), Hackett *et mult.* (2008), and Jarvis *et mult.* (2014). We follow the molecular phylogeny of Pereira *et al.* (2007) for the generic sequence that follows. See also the tree generated by Shapiro *et al.* (2002) showing how the columbids encompass the Dodo (*Raphus cucullatus*) and Solitaire (*Pezophaps solitaria*).

GENUS *COLUMBA* [Genus 34 spp/Region 2 spp]

Columba Linnaeus, 1758. *Syst. Nat.*, ed. 10, 1: 162. Type, by subsequent designation (Vigors, 1825), *Columba oenas* Linnaeus.

The species of this genus range from Eurasia, Africa, Asia, and Australasia to the western and central Pacific; one species (*C. livia*) has been introduced to urban areas around the world (Baptista *et al.* 1997). The genus no longer includes strictly New World species; these are now assigned to the genus *Patagioenas* (del Hoyo & Collar 2014). Diverse in plumage, the genus *Columba* is characterized by the typically dark or patterned plumage, often with iridescence on the neck, and the standard "Rock Dove" shape.

Rock Dove (Feral Pigeon) *Columba livia*   INTRODUCED

Columba (*domestica*) β *livia* Gmelin, 1789. *Syst. Nat.* 1(2): 769.—Southern Europe.

Naturalized urban populations of the domestic form known from Sorong, Numfor I[W], Aru Is, Fakfak, Vanimo, Madang, Mt Hagen[X], Goroka, Lae, Wau, Port Moresby, Tufi, and probably elsewhere (Beehler *et al.* 1986, Pratt & Beehler 2014, P. Gregory *in litt.*[X], G. Dutson *in litt.*[W]).

Extralimital—current human-assisted distribution of the species extends from North America and South America to much of Europe and Africa, c and tropical Asia, and parts of Australia and NZ (Baptista *et al.* 1997).

Other names—Rock Pigeon.

White-throated Pigeon *Columba vitiensis*   RESIDENT

Columba vitiensis Quoy & Gaimard, 1830. *Voy. Astrolabe, Zool.* 1: 246; *Atlas*, pl. 28.—Fiji.

Widely but thinly distributed in lowland and upland forests of NG, the NW Islands, and the SE Islands, sea level–2,750 m[J] (Stresemann 1923, Ripley 1964a[J]). No records from the Bay Is or Aru Is. Generally scarce or absent, occurring singly and in small flocks. Likely overlooked because it calls so rarely and is not very demonstrative.

Extralimital—islands off n Borneo, the Philippines, Lesser Sundas, Moluccas, Bismarck Arch, Solomons, Vanuatu, New Caledonia, Fiji, and Samoa (Dickinson & Remsen 2013). Extinct on Lord Howe I (Higgins & Davies 1996).

SUBSPECIES

a. *Columba vitiensis halmaheira* (Bonaparte)

Janthaenas halmaheira Bonaparte, 1855. *Consp. Gen. Av.* 2: 44.—Halmahera and Seram, Moluccas.

RANGE: New Guinea, Misool[MA], Salawati[MA], Waigeo[MA], ?Batanta I[J]; Goodenough and Fergusson I[K]; Misima[MA], Tagula[MA], and Rossel I[MA] (LeCroy & Peckover 1998, S. Hogberg in M. Tarburton's island website[J], Pratt field notes 2003[K]). Also Banggai Is, n Sulawesi, Sula Is, Moluccas, Kai Is, Bismarck Arch, and Solomon Is (Coates & Bishop 1997, Mayr & Diamond 2001). Apparently absent from the Aru Is and islands of Geelvink Bay (Mayr 1941, Coates 1985). Our records show the species is currently known from only two outlying mt ranges, those of the Bird's Head[J] and the Huon Penin[K] (Mayr 1931[K], Gyldenstolpe 1955b[J]). The range map in Coates (1985) also indicates the species is absent from the e sector of the S Lowlands.

DIAGNOSIS: Heavily iridescent; striking white throat bordered below by green then purple; broad purple neck band and purple crown (Baptista *et al.* 1997).

NOTES

Other names—Metallic or Wood Pigeon.

GENUS *STREPTOPELIA* [Genus 10 spp/Region 1 introduced sp]

Streptopelia Bonaparte, 1855. *Compt. Rend. Acad. Sci. Paris* 40: 17. Type by subsequent designation (G. R. Gray, 1855. *Cat. Gen. Subgen. Birds*: 150), *Columba risoria* Linnaeus.

The turtle-doves are a widespread and well-known group that ranges through Europe, Africa, and Asia. There are no native species in the Region. One species has been recorded, though it is not certain that it has become established yet. The genus is characterized by the terrestrial foraging habit, the dark neck-patch, and the brown/buff plumage. We retain the species *chinensis* is this genus rather than *Spilopelia* following the explicit recommendation of Johnson *et al.* (2001), which seems counter to their presented phylogenies (*pace* Dickinson & Remson 2014, Gill & Donsker 2015, del Hoyo & Collar 2014).

Spotted Dove *Streptopelia chinensis* INTRODUCED

Columba chinensis Scopoli, 1786. *Delic. Flor. Faun. Insubr.* 2: 94.—Guangzhou, China.

Recently introduced to w NG. A bird of towns and roadsides in agricultural and disturbed habitats. Reported from Sentani (NW Lowlands) by D. Price (*in litt.*) and Biak (Bay Is) by G. Dutson (*in litt.*); expected to spread.

Extralimital—native to se Asia; introduced widely in Australia and across the Pacific (including New Caledonia and NZ; Gill *et al.* 2010, Dutson 2011). Range expected to expand.

SUBSPECIES

a. *Streptopelia chinensis tigrina* (Temminck)

Columba Tigrina Temminck, 1809. In Knip, *Les Pigeons, les Colombes*: 94, pl. 43.—Java and Timor.

RANGE: Eastern India east to the Philippines and Timor (Baptista *et al.* 1997). Presumably birds from w NG derive from Indonesia, home of this subspecies. Photographs of birds from Sentani support this possibility.

DIAGNOSIS: Distinct from the nominate form as follows: pale gray feathering on leading edge of wing; whitish undertail coverts; and wing coverts streaked with black (Baptista *et al.* 1997). Wing 147–150 (Gibbs *et al.* 2001).

NOTES

Other names—Spotted-necked Dove, Spotted Turtle-Dove, *Spilopelia chinensis*.

GENUS *REINWARDTOENA* [Genus 3 spp/Region 1 sp]

Reinwardtoena Bonaparte, 1854. *Compt. Rend. Acad. Sci. Paris* 39: 1112. Type, by monotypy, *Columba reinwardtii* Temminck.

The species of this genus range through Buru, Seram, New Guinea, the Bismarcks, and the Solomons (Baptista *et al.* 1997). Two sister species to the New Guinean species, *R. reinwardtii*, inhabit the Bismarck and Solomon Islands (Mayr & Diamond 2001). The genus is characterized by the very long, rounded tail, the plain all-dark wing and mantle, the large size, and the orange circumorbital skin.

Great Cuckoo-Dove *Reinwardtoena reinwardtii* RESIDENT

Columba reinwardtsi Temminck, 1824. *Planch. Col. d'Ois.*, livr. 42: 248; emended to *reinwardtii* by Temminck, 1839, *Tabl. Méth.* (1838): 80.—Ambon, s Moluccas.

New Guinea, the NW Islands, Bay Is, islands off the n coast, and the D'Entrecasteaux Arch. No records from the Aru Is. Inhabits lowland and montane forest and edge up to 3,380 m (Mayr 1941, Coates 1985). Generally uncommon. Absent from the Trans-Fly.

Extralimital—Moluccas (Coates & Bishop 1997).

SUBSPECIES

a. *Reinwardtoena reinwardtii griseotincta* Hartert

Reinwardtoenas reinwardti griseotincta Hartert, 1896. *Novit. Zool.* 3: 18.—Mailu, Orangerie Bay, SE Peninsula.

RANGE: New Guinea[MA]; NW Islands[MA] (plus Gam I[MA]); Yapen[MA], Kumamba[MA], Kairiru[CO], Manam[MA], and Karkar I[K]; Fergusson and Goodenough I[MA] (Diamond & LeCroy 1979[K]).

DIAGNOSIS: Forehead, face, and upper throat creamy white (Gibbs *et al.* 2001). Head, neck, and lower throat pale gray rather than white or off-white. Abdomen and undertail gray to dark blue-gray; upperwing coverts chiefly dark reddish maroon; wing 230–252. On average, larger than nominate form.

b. *Reinwardtoena reinwardtii brevis* Peters

Reinwardtoena reinwardtsi brevis Peters, 1937. *Check-list Birds World* 3: 82.—Biak I, Bay Islands. Replacement name for *Macropygia Reinwardtii minor* Schlegel, 1873, *Mus. Pays-Bas* 4 (*Columbae*): 106 (not *Macropygia unchall minor* Swinhoe, 1870).

RANGE: Biak I.

DIAGNOSIS: Small size; wing 210–215; neck and underparts whitish not grayish; upperwing coverts mainly blackish (Rand & Gilliard 1967, Gibbs *et al.* 2001).

NOTES

Spelling of the corrected name varies (see Dickinson 2003 *vs.* Dickinson & Remsen 2013 and footnotes therein). In his *Table Méthodique*, Temminck spelled the corrected named *reinwardtii*, which we choose to follow here (M. LeCroy *in litt.*). *Other names*—Giant or Long-tailed Cuckoo-dove.

GENUS *MACROPYGIA* [Genus 9 spp/Region 3 spp]

Macropygia Swainson, 1837. *Classif. Birds* 2: 348. Type, by subsequent designation (Selby, 1840. *Cat. Gen. Sub-gen. Types Aves*), *Columba phasianella* Temminck, May-1821 = *Macropygia amboinensis phasianella* (Temminck, May-1821). This treatment follows the historical nomenclatural interpretation of Schodde (2009).

The species of this genus range from the eastern Himalayas and Tibet, central and southern China, to the Moluccas, Philippines, New Guinea, Bismarck Archipelago, and eastern Australia (Baptista *et al.* 1997). The genus is characterized by the entirely rufous plumage, the long broad tail, the small bill, and the small head.

Brown Cuckoo-Dove *Macropygia amboinensis* RESIDENT

Columba amboinensis Linnaeus, 1766. *Syst. Nat.*, ed. 12, 1: 286.—Ambon, s Moluccas.

All NG, the NW Islands, Aru Is, Bay Is, Manam I, D'Entrecasteaux Arch, and Louisiade Arch, sea level–1,800 m, with records to 2,300 m (Mayr 1941, Mees 1972). Absent from Karkar I, the Trobriand Is, and Woodlark I. Frequents mostly forest-edge habitats and garden areas but also found in forest, especially at lower elevations. On mainland NG can be found in the same habitat as the ecologically similar *M. nigrirostris* (Coates 1985).

Extralimital—Sulawesi, the Moluccas (including Buru, Seram, Seram Laut), the Bismarck Arch, and e Australia (Mayr & Diamond 2001, Dutson 2011).

SUBSPECIES

a. *Macropygia amboinensis doreya* Bonaparte

Macropygia doreya Bonaparte, 1854. *Compt. Rend. Acad. Sci. Paris* 39: 1111.—Manokwari, Bird's Head.

RANGE: Kofiau[MA], Gebe[J], Gag[K], Misool[MA], Salawati[MA], Batanta[MA], and Waigeo I[MA]; Aru Is[MA]; and nw NG, west to the head of Geelvink Bay and Etna Bay (Mees 1972[J], Johnstone 2006[K]). Birds from Gag I presumably of this race.

DIAGNOSIS: Neck and upper breast of male pale pinkish brown, with barring. Birds from Gebe may be referable to the form *M. a. albiceps* (see Mees 1972 and Mees 1982c).

b. *Macropygia amboinensis maforensis* Salvadori

Macropygia maforensis Salvadori, 1878. *Ann. Mus. Civ. Genova* 12: 429, 432.—Numfor I, Bay Islands.

RANGE: Numfor I.

DIAGNOSIS: Male exhibits medium-gray crown and nape, with greenish gloss to the hindneck (Baptista *et al.* 1997).

c. *Macropygia amboinensis griseinucha* Salvadori

Macropygia griseinucha Salvadori, 1876. *Ann. Mus. Civ. Genova* 9: 204.—Mios Num I, Bay Islands.

RANGE: Mios Num I.

DIAGNOSIS: Male of this form differs from *doreya* in the lack of dark barring of underparts (Gibbs *et al.* 2001).

d. *Macropygia amboinensis cinereiceps* Tristram

Macropygia cinereiceps Tristram, 1889. *Ibis*: 558.—Fergusson I, D'Entrecasteaux Arch, SE Islands.

Synonym: *Macropygia goldiei* Salvadori, 1893. *Cat. Birds Brit. Mus.* 21: 338 (in key), 358.—Southeastern Peninsula.
Synonym: *Macropygia kerstingi* Reichenow, 1897. *Orn. Monatsber.* 5: 25.—Ramu R, ne NG.
Synonym: *Macropygia amboinensis balim* Rand, 1941. *Amer. Mus. Novit.* 1102: 5.—Baliem R, 1,600 m, Western Ra.
RANGE: All mainland NG east of the Bird's Neck; also Biak[MA], Yapen[MA], Fergusson[MA] and Goodenough I[MA]; Normanby I record presumably of this race (Pratt field notes 2003). Also islands off sw New Britain: Long, Umboi, Tolokiwa, and Sakar (Mayr & Diamond 2001).
DIAGNOSIS: Male without any dark barring; has pale throat and abdomen, gray crown with whitish forehead, broad maroon-buff breast band. We follow suggestions of Diamond (1972), Mees (1982b), and Dickinson (2003) in subsuming the races *kerstengi* and *balim* into *cinereiceps*. We also follow Mayr (1941) in subsuming the form *goldiei* into *cinereiceps*.

e. *Macropygia amboinensis meeki* Rothschild & Hartert

Macropygia amboinensis meeki Rothschild & Hartert, 1915. *Novit. Zool.* 22: 39.—Manam I, ne NG.
RANGE: Manam I (Mayr 1941).
DIAGNOSIS: Male has abundant fine black barring ventrally, from throat to belly, with throat and breast very pale; broad maroon breast band; prominent broad iridescent band on hindneck.

f. *Macropygia amboinensis cunctata* Hartert

Macropygia doreya cunctata Hartert, 1899. *Novit. Zool.* 6: 214.—Rossel I, Louisiade Arch, SE Islands.
RANGE: Misima, Tagula, and Rossel I (Mayr 1941).
DIAGNOSIS: Male has obsolete barring on breast and belly, with breast and belly buffy tan; overall plumage with a strong brownish wash; throat dull buff.

NOTES

The Australian populations were treated by Baptista *et al.* (1997) as a distinct species, *M. phasianella*. We follow Schodde & Mason (1997: 12–13) and Christidis & Boles (2008) in considering both the NG and Australian forms as *amboinensis*, hence the name Brown Cuckoo-Dove—which would be unavailable if the Australian form were distinct and took that English name. Use of the name "Slender-billed Cuckoo-Dove" is inappropriate, given there is an extant species *Macropygia tenuirostris*. One can read at length the tortuous trail of treatments of this complex lineage in Christidis & Boles (2008: 71–72). The limits of the various geographic forms may be properly resolved by the appropriate application of molecular systematics. *Other names*—Slender-billed, Amboina, or Pink-breasted Cuckoo-dove, Brown Pigeon, *Macropygia phasianella*.

Black-billed Cuckoo-Dove *Macropygia nigrirostris*   RESIDENT | MONOTYPIC

Macropygia nigrirostris Salvadori, 1876. *Ann. Mus. Civ. Genova* 7: 972.—Arfak and Warbusi, Bird's Head.
Widely distributed in NG; also present on Yapen I[MA], Karkar I[K], Goodenough[MA] and Fergusson I[CO], sea level–1,450 m, occasionally to 2,600[CO] m (Diamond & LeCroy 1979[K]). Mainly inhabits upland forest; uncommon elsewhere; found in forest and edge (Coates 1985). Absent from the Trans-Fly.
Extralimital—Bismarck Arch (Mayr & Diamond 2001, Coates & Peckover 2001).

NOTES

We treat the species as monotypic, following Dickinson & Remsen (2013). Gibbs *et al.* (2001) recognized the subspecies *major* from the Bismarck Arch. *Other names*—Rusty or Bar-tailed Cuckoo-dove.

Mackinlay's Cuckoo-Dove *Macropygia mackinlayi*   RESIDENT

Macropygia mackinlayi Ramsay, 1878. *Proc. Linn. Soc. NSW* 2: 286.—Tanna I, Vanuatu.
An insular species reaching the NG region only on Karkar I, sea level–1,000 m (Diamond & LeCroy 1979). Frequents woodland edges of gardens and disturbed areas as well as forest (Coates 1985).
Extralimital—Bismarcks, Solomons, and Vanuatu (Coates 1985).

SUBSPECIES

a. *Macropygia mackinlayi arossi* Tristram, 1879. *Ibis* Oct-1879: 443.—Makira I, Solomon Is.

Synonym: *Macropygia rufa krakari* Rothschild & Hartert, 1915. *Novit. Zool.* 22: 28.—Karkar I, ne NG.
Synonym: *Macropygia rufa goodsoni* Hartert, 1924. *Novit. Zool.* 31: 266.—Saint Matthias Is, north of New Hanover, Bismarck Arch.

RANGE: Karkar I[K]; also apparently ranges though small islands of the Bismarck Arch and the central Solomon Is[L] (Diamond & LeCroy 1979[K], Mayr & Diamond 2001[L]).

DIAGNOSIS: Plumage a rich chestnut red with purplish tinge to upperparts; female slightly paler than the male, and breast more golden rufous and more evenly spotted (Baptista *et al.* 1997). Diamond & LeCroy (1979) subsumed *krakari* into *arossi*; by implication, the intervening race *goodsoni* of the Bismarck Arch must be synonymized also (though not so stated in their paper; see Mayr & Diamond 2001: 377).

NOTES

There is an unconfirmed sighting of a single individual from the coast at Bogia, Madang Prov, on 2-Jun-1973 (Berggy 1978). Coates (1985) noted that no species of *Macropygia* inhabits Bagabag I, just se of Karkar I. *Other names*—Spot-breasted or Rufous Cuckoo-dove.

GENUS *HENICOPHAPS* [Genus 2 spp/Region 1 sp]

Henicophaps G. R. Gray, 1862. *Proc. Zool. Soc. London* (1861): 432. Type, by monotypy, *Henicophaps albifrons* G. R. Gray.

The species of this genus range through the Northwestern Islands, Aru Islands, mainland New Guinea, and New Britain (Baptista *et al.* 1997). The genus is characterized by the purplish-brown dorsal surface, with metallic highlights on the wing, the longish tail and bill, and the white forehead. The phylogeny of Pereira *et al.* (2007) showed this genus to be sister to the ground-doves (*Gallicolumba, Geopelia,* and others). *H. foersteri,* the sister form to New Guinea's *H. albifrons,* inhabits New Britain, Umboi, and Lolobau (Mayr & Diamond 2001).

New Guinea Bronzewing *Henicophaps albifrons* ENDEMIC

Henicophaps albifrons G. R. Gray, 1862. *Proc. Zool. Soc. London* (1861): 432, pl. 44.—Waigeo I, NW Islands.

Endemic to NG, some NW Islands, Yapen I, and Aru Is, lowlands to 2,150 m[CO]. Frequents rain forest and monsoon forest and sometimes tree plantations.

SUBSPECIES

a. *Henicophaps albifrons albifrons* G. R. Gray

RANGE: Waigeo, Gam, ?Misool, Salawati, and Yapen I, and all NG (Mayr 1941).

DIAGNOSIS: Forecrown white; neck purplish brown; underparts pinkish gray (Rand & Gilliard 1967). See photos in Coates (1985: 289–92). Wing 193–214 (Gibbs *et al.* 2001).

b. *Henicophaps albifrons schlegeli* (von Rosenberg)

Rynchaenas Schlegeli von Rosenberg, 1866. *Natuurk. Tijdschr. Nederl. Indië* 29: 143.—Aru Is.

RANGE: Aru Is.

DIAGNOSIS: Forecrown dirty off-white to brownish gray; neck paler than in nominate; underparts much darker, deep wine-colored. Size slightly larger than the nominate form; wing 218 (Gibbs *et al.* 2001).

NOTES

Record from Misool I (Peters 1937, Mayr 1941) is doubtful (Mees 1965). *Other names*—White-capped Ground Pigeon, Jungle Bronzewing Pigeon.

GENUS *GALLICOLUMBA* [Genus 7 spp/Region 1 sp]

Gallicolumba Heck, 1849. *Bilder-Atlas zum Conversations-Lexikon* 1: 434. Type, by monotypy, *Gallicolumba cruenta* (Gmelin) = *Columba luzonica* Scopoli.

The species of this genus range from the Philippines and Sulawesi east to New Guinea (Baptista *et al.* 1997). The genus is characterized by compact form, ground-dwelling habit, and plumage dominated by gray, brown, maroon, and green. All have a pale belly, contrasting with the dark mantle, and a breast spot of red, orange, or yellow. Substantial sexual dichromatism is absent (Jønsson *et al.* 2011a, Moyle *et al.* 2013).

Cinnamon Ground-Dove *Gallicolumba rufigula*　　　　ENDEMIC

Peristera rufigula Pucheran, 1853. *Voy. Pôle Sud, Zool.* 3: 118.—Triton Bay, Bird's Neck.

　　All NG as well as the NW Islands, Aru Is, and Yapen I, sea level–1,700 m.

SUBSPECIES

a. *Gallicolumba rufigula rufigula* (Pucheran)

　　Synonym: *Gallicolumba rufigula orientalis* Rand, 1941. *Amer. Mus. Novit.* 1102: 6.—Kubuna, 100 m elevation, SE
　　Peninsula.

　　Synonym: *Gallicolumba rufigula septentrionalis* Rand, 1941. *Amer. Mus. Novit.* 1102: 6.—Bernhard Camp, 50 m
　　elevation, Taritatu R, Mamberamo basin, s sector of the NW Lowlands.

RANGE: Main NW Islands[X,MA], the Bird's Head, Bird's Neck, Yapen I[J,L], NW Lowlands, Sepik-Ramu, and Huon and SE Peninsulas (Rothschild *et al.* 1932[J], Matheve *et al.* 2009[X] for Batanta, D. Price *in litt.*[L]).

DIAGNOSIS: Side of head has a distinct gray patch; upperwing coverts show broad, clear gray edging contrasting with maroon brown of wing and back (Rand & Gilliard 1967). See photos in Coates (1985: 283–85) and in Coates & Peckover (2001: 53).

b. *Gallicolumba rufigula helviventris* (von Rosenberg)

Ptilopus helviventris von Rosenberg, 1866. *Natuurk. Tijdschr. Nederl. Indië* 29: 144.—Aru Is.

RANGE: Aru Is.

DIAGNOSIS: Similar to *alaris*, but edging of upperwing coverts gray, strongly tinged with vinaceous, and head darker, with buff lores (Rand & Gilliard 1967, Baptista *et al.* 1997). Date of original publication of name follows Mees (1980).

c. *Gallicolumba rufigula alaris* Rand

Gallicolumba rufigula alaris Rand, 1941. *Amer. Mus. Novit.* 1102: 7.—Lake Daviumbu, middle Fly R, Trans-Fly.

RANGE: Southern Lowlands and s foothills of Western, Border, and Eastern Ra east to Karimui (Diamond 1972).

DIAGNOSIS: Similar to nominate form, but side of head pale vinaceous brown without gray; wing coverts edged pale gray (Rand & Gilliard 1967).

NOTES

Matheve *et al.* (2009) observed two individuals on Batanta I at a display area of Wilson's Bird of Paradise on 3-Aug-2009. *Other names*—Goldenheart, Red-throated, or Rufous-throated Ground Dove, Chestnut Quail-dove.

GENUS *ALOPECOENAS*　　　　　　　　　　　　[Genus 11 spp/Region 2 spp]

Alopecoenas "Finsch" Sharpe, 1899. *Hand-list Birds* 1: 90. Type, by monotypy, *Leptotila hoedtii* Schlegel. See Finsch, 1901. *Notes Leyden Mus.* 22: 228.

　　This genus was recently split from *Gallicolumba* based on molecular studies (Jønsson *et al.* 2011a, Moyle *et al.* 2013). The species of the genus range from Seram to New Guinea, Micronesia, and Polynesia. They are similar to those of the preceding genus. Most plumages exhibit a pale breast contrasting with a dark belly. Substantial sexual dichromatism is present. None of the species of this genus has a breast spot (Moyle *et al.* 2013). *Alopecoenas* is masculine.

White-bibbed Ground-Dove *Alopecoenas jobiensis*　　　　RESIDENT

Phlegoenas jobiensis A. B. Meyer, 1875. *Mitt. Zool. Mus. Dresden* 1: 10.—Yapen I, Geelvink Bay.

　　New Guinea and selected n and se islands; sea level–1,600 m, max 2,400 m. Usually rare or absent, but sometimes locally abundant.

　　Extralimital—Bismarck and Solomon Is (Baptista *et al.* 1997, Mayr & Diamond 2001, Dutson 2011).

a. *Alopecoenas jobiensis jobiensis* (A. B. Meyer)

RANGE: New Guinea (except c and w Bird's Head), Yapen[MA], Karkar[MA], Bagabag[K], Manam[MA], Fergusson[MA], and Goodenough I[MA], and the Bismarck Arch[L] (Rand 1942a, 1942b, Diamond 1972, Majnep & Bulmer 1977, Diamond & LeCroy 1979[K], Dutson 2011[L]).

DIAGNOSIS: This form is distinguished from *chalconotus* of the Solomon Is by the male's slighter bill and more expansive patch of purple on mantle (*chalconotus* is mainly black on the mantle); female of *chalconotus* unique in uniformly dark brown plumage (Baptista *et al.* 1997).

NOTES

This species exhibits a patchy geographic distribution, being absent from much of the Bird's Head as well as the central islands of the Solomons chain. Few recent records. Iain Woxvold (*in litt.*) reported: "After tens of thousands of camera trap hours from lowlands to highlands of southern PNG and upper Sepik, not a single picture of this species, though there were Bronze and Cinnamon Ground-Doves aplenty." By contrast, in 1998–99 there was an influx of this species to Varirata NP, and observers recorded as many as 15 birds a day during that unusual period, which may have been related to a bamboo-seeding event (P. Gregory *in litt.*) *Other names*—White-throated or White-breasted Ground-dove, *Gallicolumba jobiensis.*

Bronze Ground-Dove *Alopecoenas beccarii* RESIDENT

Chalcophaps beccarii Salvadori, 1876. *Ann. Mus. Civ. Genova* (1875) 7: 974.—Hatam, Arfak Mts, Bird's Head.

Frequents floor of forest interior of most mts of NG, plus Karkar and Kairiru I; on Mainland from 1,400 m–2,900 m. Mist-netted at 390 m on the Huon Penin. (Freeman *et al.* 2013). On Karkar I recorded from the lowlands to 1,800 m. (Diamond & LeCroy 1979).

Extralimital—Bismarck and Solomon Is (Mayr & Diamond 2001).

SUBSPECIES

a. *Alopecoenas beccarii beccarii* (Salvadori)

RANGE: Mountains of the Bird's Head[MA], Central Ra[MA], Foja Mts[J], PNG North Coastal Ra[K], and Huon Penin[MA] (Beehler *et al.* 2012[J], J. M. Diamond specimens at AMNH[K]).

DIAGNOSIS: Upperparts dark oil green; underwing coverts and wing quills below uniformly sooty gray; lower breast and abdomen sooty gray to black; tail 54 (Rand & Gilliard 1967).

b. *Alopecoenas beccarii johannae* (P. L. Sclater)

Phlogoenas johannae P. L. Sclater, 1877. *Proc. Zool. Soc. London*: 112, pl. 16.—Duke of York Is, New Ireland.

RANGE: Karkar and Kairiru I[MA], as well as the Bismarck Arch; lowland forest, particularly on small coral islands (Diamond & LeCroy 1979, Mayr & Diamond 2001). Also possibly Fergusson I.

DIAGNOSIS: Note bold white orbital ring (absent from Mainland populations). Upperparts chiefly greenish rufous-brown; underwing surfaces not uniform in color, but coverts and inner edges of flight quills pale chestnut; lower breast and belly brown, with some purple tipping on sides of breast; tail 63 (Rand & Gilliard 1967).

NOTES

Subspecific assignment of the population on Kairiru (*cf.* Coates 1985) remains to be determined. The races *johannae* and *beccarii* may constitute two distinct species. The Yule I record (Coates 1985) may be a specimen-labeling error (collectors based on Yule I sent fieldworkers into the Mainland uplands from there). *Other names*—Beccari's Ground Dove, Grey-bibbed Ground-dove, *Gallicolumba beccarii.*

GENUS *GEOPELIA* [Genus 5 spp/Region 2 spp]

Geopelia Swainson, 1837. *Classif. Birds* 2: 348. Type, by monotypy, *Geopelia lineata* Swainson = *Columba striata* Linnaeus.

The species of this genus range from the Malay Peninsula, Sumatra, Java, and the Lesser Sundas to New Guinea and Australia (Baptista *et al.* 1997). The genus is characterized by the slim body, long tail, heavily scaled upper surface, and plain abdomen.

Peaceful Dove *Geopelia placida*

RESIDENT

Geopelia placida Gould, 1844. *Proc. Zool. Soc. London*: 55.—Port Essington, NT, Australia.

Local in c and e NG; also the Aru Is; confined mainly to open woodlands, to 550 m (Coates & Peck-over 2001). Diamond & Bishop (1994) provided the first record for the Aru Is.

Extralimital—widespread in Australia.

SUBSPECIES

a. *Geopelia placida placida* Gould

Synonym: *Geopelia striata papua* Rand, 1938. *Amer. Mus. Novit.* 990: 6.—Wuroi, Oriomo R, Trans-Fly.

RANGE: Inhabits open country and savannas of the Trans-Fly, Madang, Saidor, lower Markham valley, Lae, savanna southeast of Popondetta, and Port Moresby; near sea level. Localities—Merauke[MA], Wasur NP[L], Tabubil[J], Daru[CO], Port Moresby[MA], Aru Is[K], Madang[CO], Dumpu[CO], Saidor[CO], Lae[CO], lower Markham valley[CO], Musa valley[CO], plus n and e Australia (Diamond & Bishop 1994[K], Gregory 1997[J], K. D. Bishop *in litt.*[L]). DIAGNOSIS: Darker than *clelandi* of Western Australia (Baptista *et al.* 1997). The described race *papua* is not distinct from the nominate form.

NOTES

We follow Schodde & Mason (1997), who treated the NG form as *G. placida* and dissected the tradi-tional *G. striata* into three species. By contrast, Christidis & Boles (2008) maintained a single widespread species—*G. striata*. *Other names*—Peaceful Ground Dove; treated as Zebra Dove, *G. striata*, by some authorities.

Bar-shouldered Dove *Geopelia humeralis*

RESIDENT | MONOTYPIC

Columba humeralis Temminck, 1821. *Trans. Linn. Soc. London* 13: 128.—Broad Sound, QLD, Australia.

Synonym: *Geopelia humeralis gregalis* Bangs & Peters, 1926. *Bull. Mus. Comp. Zool. Cambridge* 67: 423.—Wendoe Mer R, w sector of the Trans-Fly.

This species inhabits the lowlands of NG's two main rain shadows—the Port Moresby area and the Trans-Fly, west to Dolak I[X] (Pratt & Beehler 2014, K. D. Bishop *in litt.*). Most records are coastal, but the species is found inland in the Trans-Fly at Lake Daviumbu[CO] and Lake Murray[J] (B. J. Coates *in litt.*[J]). Frequents mangroves, monsoon scrub, and woodland savanna (Coates 1985).

Extralimital—w, n, and e Australia (Baptista *et al.* 1997).

NOTES

There are no substantive plumage distinctions between the NG and Australian birds. Gibbs *et al.* (2001) treated all Australian forms as a single form and also noted the lack of distinction for the NG birds. Mees (1982b) supported the sinking of NG's *gregalis* (*pace* Schodde & Mason 1997).

GENUS *TRUGON*

ENDEMIC | MONOTYPIC

Trugon G. R. Gray, 1849. *List Gen. Birds* 3, app.: 24. Type, by monotypy, *Trugon terrestris* Hombron & Jacquinot = G. R. Gray.

The single species of the genus is restricted to the New Guinea region. The genus is characterized by the large size, peculiar peaked nuchal crest, whitish flash on the face, medium-length tail, and terrestrial rain-forest habit.

Thick-billed Ground-Pigeon *Trugon terrestris*

ENDEMIC

Trugon terrestris G. R. Gray, 1849. *List Gen. Birds* 3, app.: 24.—Triton Bay, Bird's Neck.

Restricted to the NG lowlands and Salawati I, sea level–640 m. Frequents rain forest and monsoon forest (Coates 1985). Apparently no records from the n watershed of the Huon and SE Peninsulas. A sight-ing at 900 m at Varirata NP seems exceptional.

a. *Trugon terrestris terrestris* G. R. Gray
RANGE: Salawati I and the Bird's Head and Bird's Neck (Mayr 1941).
DIAGNOSIS: Face has white patch (not gray); forehead patch pale gray; shoulder patch dull olive brown.

b. *Trugon terrestris mayri* Rothschild
Trugon terrestris mayri Rothschild, 1931. *Bull. Brit. Orn. Club* 51: 69.—Jayapura, ne sector of the NW Lowlands.
RANGE: Northwestern Lowlands and Sepik-Ramu (Mayr 1941).
DIAGNOSIS: Gray rather than white on face; wing coverts only slightly washed with chestnut; flanks paler than in nominate (Baptista *et al.* 1997). Type description noted that this form exhibits an olive-brown mantle, differing from olive-greenish-black mantle of the nominate form; abdomen much whiter; flanks paler rufous cinnamon. It differs from *leucopareia* in absence of chestnut-chocolate color on back and rump.

c. *Trugon terrestris leucopareia* (A. B. Meyer)
Eutrygon leucopareia A. B. Meyer, 1886. *Zeitschr. Ges. Orn.* 3: 29.—Astrolabe Ra, SE Peninsula.
RANGE: Southern NG (Mayr 1941).
DIAGNOSIS: Face has white patch; forehead patch buff whitish; innermost wing coverts are dark purplish brown, distinct from dull gray-brown secondaries and primaries. See photos in Coates (1985: 295–300).

NOTES
Three very thinly drawn subspecies. *Other names*—Grey or Slaty Ground-pigeon, Thick-billed Jungle-pigeon.

GENUS *OTIDIPHAPS* ENDEMIC | MONOTYPIC

Otidiphaps Gould, 1870. *Ann. Mag. Nat. Hist.* (4)5: 62. Type, by monotypy, *Otidiphaps nobilis* Gould.
    The single polytypic species of the genus inhabits the New Guinea mainland, Aru Islands, the Northwestern Islands, Bay Islands, and D'Entrecasteaux Archipelago (Mayr 1941). The genus is characterized by the blackish, tentlike tail, reminiscent of that of a *Talegalla* brushturkey; the distinctive rich-chestnut mantle and wings; blackish underparts; and red bill and iris. Del Hoyo & Collar (2014) recognized four species.

Pheasant Pigeon *Otidiphaps nobilis* ENDEMIC
Otidiphaps nobilis Gould, 1870. *Ann. Mag. Nat. Hist.* (4)5: 62.—Bird's Head.
    Widely distributed in hills and lower mts through the Mainland, Batanta and Waigeo I, Aru Is, and Yapen and Fergusson I; sea level (in hilly country) to 1,900 m, extreme max 2,900 m[J] (Diamond 1985, Pratt & Beehler 2014, B. G. Freeman & A. M. Freeman *in litt.*[J]). No records from the Cyclops Mts.

SUBSPECIES
a. *Otidiphaps nobilis nobilis* Gould
RANGE: Batanta[MA], Waigeo[MA], and Yapen I[J], and mts of w NG: Tamrau[MA], Arfak[MA], Wandammen[J], Kumawa[J], and Foja Mts[J]; also Western[MA] and Border Ra[MA] (presumably in both n and s watersheds) and PNG North Coastal Ra[K] (Diamond 1985[J], specimens at AMNH and BPBM[K]). Eastern extent of range not determined. Presumably this form meets *cervicalis* in w PNG. Said to be absent from Salawati I (Diamond 1985). No specimen exists for the Yapen I population; it is placed here based on geography.
DIAGNOSIS: Distinct black crest on hind-crown; nuchal patch is iridescent green and purple, offset from bronze hind-collar; underparts bluish purple (Rand & Gilliard 1967, Baptista *et al.* 1997). Iris dark red. Wing: male 184–212, female 181–195.

b. *Otidiphaps nobilis aruensis* Rothschild
Otidiphaps nobilis aruensis Rothschild, 1928. *Bull. Brit. Orn. Club* 48: 88.—Aru Is.
RANGE: Aru Is.
DIAGNOSIS: No crest; nuchal patch large and whitish gray; neck and breast deep greenish blue; abdomen blackish purple (Rand & Gilliard 1967, Baptista *et al.* 1997). Wing 183 (Gibbs *et al.* 2001).

c. *Otidiphaps nobilis cervicalis* Ramsay
Otidiphaps nobilis var. *cervicalis* Ramsay, 1880. *Proc. Linn. Soc. NSW* 4: 470.—Goldie R, s watershed of SE Peninsula.
RANGE: Eastern Ra, Sepik-Ramu, and mts of Huon and SE Peninsulas (Diamond 1972, 1985).
DIAGNOSIS: No crest; nuchal patch dove gray; underparts oil green; mantle rich maroon brown. Wing 214–219 (Gibbs *et al.* 2001). See photo in Coates (1985: 302–3).

d. *Otidiphaps nobilis insularis* Salvin & Godman
Otidiphaps insularis Salvin & Godman, 1883. *Proc. Zool. Soc. London*: 33.—Fergusson I, D'Entrecasteaux Arch, SE Islands.
RANGE: Fergusson I^MA. Has not been encountered in recent times.
DIAGNOSIS: No crest; entire head and nape black; back pale (Rand & Gilliard 1967, Baptista *et al.* 1997). This is the most distinctive of the four races; note lack of a pale nuchal patch; wing 198 (Gibbs *et al.* 2001).

NOTES
The races are distinct and have been split off as four separate species by del Hoyo & Collar (2014). *Other names*—Magnificent or Noble Ground Pigeon, Green-collared or Grey-collared Pigeon; the subspecies as Green-naped, Grey-naped, White-naped, and Black-naped Pheasant-pigeons.

GENUS *GOURA*   ENDEMIC [Genus 3 spp/Region 3 spp]

Goüra Stephens, 1819. In Shaw, *Gen. Zool.* 11, pt. 1: 119. Type, by subsequent designation (G. R. Gray, 1840. *List. Gen. Birds*: 59), *G. coronata* (Lath.) = *Columba coronata* Linnaeus = *Columba cristata* Pallas. The species of this genus inhabit the New Guinea mainland, the four main Northwestern Islands, the Aru Islands, and Biak and Yapen Islands (Mayr 1941). The genus is characterized by the very large size, predominantly powdery-blue plumage, and prominent filamentous sagittal crest. The population on Biak Island is presumed to have been transported there by humans.

Western Crowned Pigeon *Goura cristata*   ENDEMIC | MONOTYPIC
Columba cristata Pallas, 1764. In Vroeg, *Cat. Ois., Adumbr.*: 2.—Fakfak Mts, Bird's Neck.
   Synonym: *Goura coronata minor* Schlegel, 1864. *De Dierentuin*: 208.—Waigeo I, NW Islands.
   Synonym: *Goura cristata pygmaea* Mees, 1965. *Nova Guinea, Zool.* 31: 160.—Waigama, Misool I, NW Islands.
Inhabits lowland forest and mangroves of the NW Islands, Bird's Head, and Bird's Neck, up to 150 m (Hoek Ostende *et al.* 1997, Pratt & Beehler 2014).
   Extralimital—also found on Seram, where it was almost certainly introduced (Baptista *et al.* 1997).

NOTES
We treat the species as monotypic (Gibbs *et al.* 2001). For another view see Mees (1965). Using Mees's data, we find the means for wing length as follows: Misool (n=4) 318; Waigeo (n=3) 343; Salawati (n=2) 359; and Bird's Head (n=8) 370. These data can be viewed as clinal: the largest birds are on the Mainland; the next largest on Salawati, the island with the closest connection to the Mainland; and the smallest sizes on Waigeo and Misool, both distant from the Mainland. Misool is farthest from the Mainland and has the smallest birds. The size cline can be ordered as follows: Misool birds are 8% smaller than Waigeo birds; Waigeo birds 5% smaller than Salawati birds; Salawati birds 4% smaller than Mainland birds. In no case is the size ratio greater than 10% (our arbitrary minimum for size-related subspecies).
   Presumably the e edge of the geographic range of this western species contacts the w edge of the ranges of the two eastern species (*victoria* in n NG and *scheepmakeri* in s NG). It would be useful to study the nature of species interactions at these meeting points. *Other names*—Blue, Common, Grey, or Masked Crowned-pigeon or Goura.

Southern Crowned Pigeon *Goura scheepmakeri* ENDEMIC

Goura scheepmakeri Finsch, 1876. *Proc. Zool. Soc. London* (1875): 631, pl. 68.—Probably mainland NG opposite Yule I, SE Peninsula.

Inhabits lowland forest and mangroves of the S Lowlands from the Mimika R east at least to Orangerie Bay of the s watershed of the SE Peninsula (Mayr 1941). Sea level–500 m (Coates & Peckover 2001).

SUBSPECIES

a. *Goura scheepmakeri sclaterii* Salvadori
Goura sclaterii Salvadori, 1876. *Ann. Mus. Civ. Genova* 9: 45.—Fly R.
Synonym: *Goura scheepmakeri wadai* Yamashina, 1944. *Bull. Biogeog. Soc. Japan* 14: 1–2.—Bian R, s NG.
RANGE: Southern NG from the Mimika R to the Fly R and presumably the Strickland basin to the w edge of the range of the nominate form. Reported east to the Hegigio R (I. Woxvold *in litt.*).
DIAGNOSIS: The deep reddish-maroon breast patch extends up toward chin; shoulder maroon; wing speculum ivory white (Rand & Gilliard 1967). Wing 341–365 (Gibbs *et al.* 2001).

b. *Goura scheepmakeri scheepmakeri* Finsch
RANGE: Southeastern Peninsula from Orangerie Bay northwest to the Purari R (Purari specimens at Austral. Nat. Wildl. Coll.).
DIAGNOSIS: Throat blue-gray; bend of wing dark blue-gray to blackish; wing speculum pale gray (Rand & Gilliard 1967). Wing 370–380 (Gibbs *et al.* 2001).

NOTES
It will be useful to determine where the two subspecies' ranges meet—somewhere between the Fly and Purari R. The species is certainly continuously distributed in this intervening zone. The two forms were raised to species status by del Hoyo & Collar (2014). *Other names*—Great Goura or Guria; the subspecies as Scheepmaker's and Sclater's Crowned-pigeons.

Victoria Crowned Pigeon *Goura victoria* ENDEMIC

Lophyrus Victoria Fraser, 1844. *Proc. Zool. Soc. London*: 136.—Islands in Geelvink Bay.

Inhabits lowland forest and mangroves of Yapen and Biak I, the NW Lowlands, Sepik-Ramu, and a few places in the n lowlands of the SE Peninsula (Pratt & Beehler 2014). Absent from the n coastal lowlands between Astrolabe Bay and Morobe. Frequents the interior of alluvial forest, mainly in the lowlands, but in places such as the Jimi valley it extends upward to as high as 600 m (Gyldenstolpe 1955a, Majnep & Bulmer 1977, Coates & Peckover 2001, K. D. Bishop *in litt.*).

SUBSPECIES

a. *Goura victoria victoria* (Fraser)
RANGE: Yapen I[MA]; also Biak I[MA] (where introduced?).
DIAGNOSIS: Small; wing 316–332 (Gibbs *et al.* 2001). Crest rather sparse.

b. *Goura victoria beccarii* Salvadori
Goura beccarii Salvadori 1876, *Ann. Mus. Civ. Genova* 8: 406.—Humboldt Bay, ne sector of the NW Lowlands.
RANGE: Northern NG from the Siriwo R to Astrolabe Bay; also the n watershed of the SE Peninsula (from Morobe village to Collingwood Bay[CO]).
DIAGNOSIS: Large; wing 360–390 (Gibbs *et al.* 2001).

NOTES
Along the coast of ne NG the species is apparently absent from Madang southeast to the mouth of the Waria R. Much of this coastal zone lacks expanses of alluvial lowland forest so perhaps has been a natural barrier to colonization. This natural gap in distribution on the Mainland is evidence of the poor colonizing ability of this species and makes it most likely that it must have been historically transported to Biak (an oceanic island) by humans seeking to populate Biak's forest with this valuable game species. *Other names*—Victoria or White-tipped Goura, White-tipped Crowned-pigeon.

GENUS *CALOENAS*

Caloenas G. R. Gray, 1840. *List Gen. Birds*: 59. Type, by original designation, *Caloenas nicobarica* (Gmelin) = *Columba nicobarica* Linnaeus.

The single species of the genus ranges from the Andaman and Nicobar Islands, Indonesia, and the Philippines to Palau, many small islands in the New Guinea region, and the Solomon Islands (Baptista *et al.* 1997). The genus is characterized by the profuse long, dangling neck hackles, all-white tail, and small-island habit.

Nicobar Pigeon *Caloenas nicobarica*   RESIDENT
Columba nicobarica Linnaeus, 1758. *Syst. Nat.*, ed. 10, 1: 164.—Nicobar Is.

A small-island "tramp" species that frequents small wooded islands and islets mainly in w, n, and far se NG; visits forests on larger islands and sometimes the Mainland.

Extralimital—Andaman and Philippine Is to the Bismarck and Solomon Is, including Rennell (Mayr & Diamond 2001); has been recorded to 500 m on New Britain (Coates 1985). Mainly a small-island form; found on the following larger islands: Mindanao, Negros, the Lesser Sundas, Seram, New Britain, New Ireland, and many of the main Solomon Is (Gibbs *et al.* 2001).

SUBSPECIES

a. *Caloenas nicobarica nicobarica* (Linnaeus)
RANGE: Andaman and Nicobar Is, Indonesia, and the Philippines to small islands in the NG region and the Solomon Is (Baptista *et al.* 1997). Localities—(mainly from Mayr 1941) Pecan Is near Misool[Q], Gag[L], Kofiau[MA], Salawati[MA], Batanta[J], Waigeo[MA], Sorong[MA], islands near Manokwari[MA], Adi[CO], Numfor[MA], Ayawi[MA], Biak[MA], Mios Num[MA], Yapen[MA], islands off the n coast (Seleo, Tarawai, Schouten, Karkar)[MA], Pig I off Madang[K], Huon Gulf, and SE Islands: Trobriand Is, Woodlark, Egum Atoll, Normanby, Hastings, Deboyne, Misima, Renard, Duchateau, Tagula, and Rossel I[MA] (Mayr & Meyer de Schauensee 1939a, Mees 1965[Q], Rand & Gilliard 1967, Johnstone 2006[L], D. Gibbs *in litt.*[J], P. Gregory *in litt.*[K]). Few or no records from most of the s coast, and no documented reports from the Aru Is, Gebe, Manam, Goodenough, or Fergusson I (Pratt & Beehler 2014).
DIAGNOSIS: Differs from *pelewensis* of Palau by slightly larger size and greener, longer neck hackles (Baptista *et al.* 1997). Wing: male 255–264, female 243–254 (Gibbs *et al.* (2001).

NOTES
Other names—Hackled, White-tailed, or Vulturine Pigeon.

GENUS *CHALCOPHAPS*   [Genus 3 spp/Region 3 spp]

Chalcophaps Gould, 1843. *Birds Austral.* pt. 13: pl. [8] (= vol. 5: text to pl. 62 of bound vol.). Type, by monotypy, *Columba chrysochlora* Wagler.

The species of this genus range from tropical Asia and Taiwan to the Moluccas, Lesser Sundas, eastern and northern Australia, New Guinea, Bismarcks, Solomons, Vanuatu, and New Caledonia (Baptista *et al.* 1997). This small genus is characterized by the compact shape, medium-length brown tail, green wings, brown body plumage, short vinaceous legs, red bill, and terrestrial habit.

Common Emerald Dove *Chalcophaps indica*   RESIDENT
Columba indica Linnaeus, 1758. *Syst. Nat.*, ed. 10, 1: 164.—Ambon, s Moluccas.

Inhabits Gebe, Gag, and Kofiau in the NW Islands and Numfor, Biak, and Mios Num in Geelvink Bay; near sea level.

Extralimital—tropical Asia to the Moluccas and Lesser Sundas (Gibbs *et al.* 2001).

SUBSPECIES

a. *Chalcophaps indica indica* (Linnaeus)
RANGE: From Kashmir, Hainan, and the Ryukyu Is, south across the Malay Penin, Indochina, Philippines,

and the Greater and Lesser Sunda Is, east to Alor and Sumba, the Moluccas, and the westernmost of the NW Islands: Gebe[K], Gag[J], Kofiau[MA], and ?Batanta[L] (Mees 1972[K], Johnstone 2006[J], unconfirmed record from M. Tarburton's island website[L]).

DIAGNOSIS: Male exhibits paler crown, mantle, and breast feathering than *minima*, less dark vinaceous brown, with a more buffy wash.

b. *Chalcophaps indica minima* Hartert
Chalcophaps indica minima Hartert, 1931. *Orn. Monatsber.* 39: 144.—Numfor I, Bay Islands.
RANGE: Numfor, Biak, and Mios Num I (Mayr 1941).
DIAGNOSIS: Crown, mantle, and abdomen dark and rich vinaceous brown; male has prominent pale gray forehead and eyebrow and minimal pale gray crescent at bend of wing. Wing 133.5–150 (Gibbs *et al.* 2001).

NOTES
The division of the two species within the lineage (see below) has been long recognized (Goodwin 1983, Schodde & Mason 1997: 12; Gibbs *et al.* 2001). We follow Rasmussen & Anderton (2005) in treating these two as full species, *indica* and *longirostris* (not *indica* and *chrysochlora* as suggested by Christidis & Boles 2008; see explanation in Dickinson & Remsen 2013: 72 footnote 4). The two sibling species can be diagnosed as follows: *C. indica* male exhibits prominent whitish-gray forehead and eyebrow; small whitish crescent at bend of wing. *C. longirostris* male has brown forecrown and eyebrow; large whitish patch at bend of wing. *Other names*—Green-winged Pigeon, Green-winged Ground Dove, Emerald Dove.

Pacific Emerald Dove *Chalcophaps longirostris* RESIDENT
Chalcophaps longirostris Gould, 1848. *Introd. Birds Austral.*: 78.—Port Essington, NT, Australia.
    Synonym: *Columba chrysochlora* Wagler, 1827. *Syst. Av., Columba*, sp. 79.—New South Wales, Australia. Dickinson & Remsen (2013: 72, footnote 5) noted that this name was based on a series of specimens representing several forms and has been excluded from application through lectotypification (Schodde & Mason 1997).
Locally distributed in e NG, including most SE Islands, as far west as the Sepik R (August R) in n NG, and to Kurik in s NG, sea level–1,300[CO] m (Mees 1982b, Pratt & Beehler 2014). In NG, where *C. stephani* inhabits lowland rain forest, this species frequents more open habitats including forest edge and second growth (Coates 1985).
    Extralimital—nw and e Australia, Temotu Is, Vanuatu, and New Caledonia (Dutson 2011).

SUBSPECIES
a. *Chalcophaps longirostris rogersi* Mathews, 1912. *Novit. Zool.* 18: 187.—Cairns, QLD, Australia.
RANGE: From the Trans-Fly[L] east in the s watershed to Milne Bay[M]; in the n watershed from the Sepik basin east to Milne Bay[M]; also Manam I[MA], Bagabag I[K], the D'Entrecasteaux Arch (Goodenough[MA], Fergusson[MA], Normanby[J], Dobu[MA]), Trobriand Is[MA], Woodlark group (Woodlark, Dugumenu, Gawa)[MA]; Bonvouloir Is (East and Hastings Is)[MA]; and Misima, Tagula, and Rossel of the Louisiade Arch[MA] (Diamond & LeCroy 1979[K], Mees 1982b[L], Dickinson & Remsen 2013[M], Pratt specimens at BPBM[J]). Also Lord Howe I, e Australia, Vanuatu, and New Caledonia (Dutson 2011).
DIAGNOSIS: When compared to the nominate form, male crown and mantle less gray; also a slightly slaty-blue cast to nape, and underparts paler and buffier (Gibbs *et al.* 2001). See photos in Coates (1985: 276–77).

NOTES
The male of this species lacks the white forehead; exhibits a prominent white shoulder patch. We do not hyphenate the two-word group name, as we consider "emerald" to be an adjective. Through lectotypification, the earlier name *chrysochlora* Wagler, 1827, no longer applies to this species; see Schodde & Mason (1997: 27). *Other names*—Green-winged Pigeon or Ground Dove, Emerald Dove; formerly considered conspecific with *C. indica*.

Stephan's Emerald Dove *Chalcophaps stephani* RESIDENT
Chalcophaps stephani Pucheran, 1853. *Voy. Pôle Sud, Zool.* 3: 119; *Atlas*, pl. 28, fig. 2.—Triton Bay, Bird's Neck.
    Inhabits forest and edge through NG, the NW Islands, Aru Is, Yapen, Tarawai, Karkar, Bagabag, and Manam I; and the D'Entrecasteaux Arch (Mayr 1941); sea level–700 m but in places to 1,420 m[K] (Coates

& Peckover 2001, Freeman & Freeman 2014[K]). Note the record of a bird found dead at 4,350 m on the Carstensz Massif of the Western Ra (Coates & Peckover 2001).

Extralimital—Sulawesi, Sula Is, Kai Is, Bismarck Arch, and Solomon Is (Mayr & Diamond 2001, Coates 1985).

SUBSPECIES

a. *Chalcophaps stephani stephani* Pucheran
RANGE: Kai, Aru, the four main NW Islands plus Gag I[K], NG, Yapen I, Tarawai, Karkar[J], Manam, and Bagabag I[J], Goodenough, Fergusson, and Normanby I[A], and the Bismarck Arch (Mayr 1941, Diamond & LeCroy 1979[J], Johnstone 2006[K], Pratt specimens at BPBM[A]).
DIAGNOSIS: Male—compact white forehead spot; crown dark maroon brown; mantle rich maroon chestnut; throat paler than breast, which is paler than mantle; broad dark bar on rump; female—forehead spot gray with a touch of white above base of bill; crown rich brown like mantle and breast; throat paler than breast; broad dark bar on rump.

NOTES

Its presence on Manam I is uncertain; listed for the island by Mayr (1941) but not by Diamond & LeCroy (1979). Formation of the English group name for the birds of this genus follows that of the preceding species, contra Baptista *et al.* (1997). *Other names*—Stephan's Green-winged Pigeon, Stephan's Dove or Ground-dove, Brown-backed Emerald Dove.

GENUS *MEGALOPREPIA* [Genus 2 spp/Region 1 sp]

Megaloprepia Reichenbach, 1853. *Nat. Syst. Vogel* (1852): xxvi. Type, by original designation, *Columba magnifica* Temminck.

The two species of the genus range from the northern Moluccas east through New Guinea to eastern Australia (Baptista *et al.* 1997). The genus is characterized by the large body size, long tail, and pale crown. We resurrect this small genus following the molecular analysis and taxonomic recommendations of Cibois *et al.* (2014).

Wompoo Fruit-Dove *Megaloprepia magnifica* RESIDENT
Columba magnifica Temminck, 1821. *Trans. Linn. Soc. London* 13: 125.—New South Wales, Australia.

Frequents rain forest, monsoon forest, and gallery forest throughout NG, the NW Islands, Yapen, Manam, Kairiru, Karkar, Bagabag, and Normanby I, sea level–1,450 m (Stresemann 1923, Mayr 1941). Also possibly the Aru Is (Dickinson & Remsen 2013).

Extralimital—e Australia (Baptista *et al.* 1997).

SUBSPECIES

a. *Megaloprepia magnifica puella* (Lesson & Garnot)
Columba puella Lesson & Garnot, 1827. *Bull. Sci. Nat. Geol.* 10: 400.—Manokwari, Bird's Head.
   Synonym: *Megaloprepia magnifica alaris* Stresemann & Paludan, 1932. *Novit. Zool.* 38: 183.—Waigeo I, NW Islands.
RANGE: Northwestern Islands (but for Kofiau and Gebe) and the Bird's Head (Mayr 1941).
DIAGNOSIS: Small; whitish spotting on wing coverts; crissum rich yellow; undertail coverts greenish. Wing 155–168 (Gibbs *et al.* 2001). The form *alaris* was synonymized by Mayr (1941) and Gyldenstolpe (1955b).

b. *Megaloprepia magnifica poliura* (Salvadori)
Megaloprepia poliura Salvadori, 1878. *Ann. Mus. Civ. Genova* 12: 426, 427.—Mount Epa, near Hall Sound, SE Peninsula.
   Synonym: *Megaloprepia poliura septentrionalis* A. B. Meyer, 1893. *Abh. Ber. Zool. Mus. Dresden* 4(3): 25.—Northern and e NG and Yapen I.
   Synonym: *Megaloprepia magnifica interposita* Hartert, 1930. *Novit. Zool.* 36: 114.—Wandammen Penin, Bird's Neck.
RANGE: All NG east of the Bird's Head. Also Yapen, Manam, Bagabag[K], Karkar, and Normanby I[J] (Mayr 1941, Diamond & LeCroy 1979[K], Pratt field notes 2003[J]).

DIAGNOSIS: Crissum dull olive yellow; undertail paler; wing spotting yellowish. Wing 164–179 (Gibbs *et al.* 2001). It is more likely the Aru Is population (if confirmed) would be allied to this subspecies, not *puella* (*pace* Dickinson & Remsen 2013). We follow Diamond & LeCroy (1979) and Mees (1982b) in subsuming *septentrionalis* and *interposita* into this form. See photos in Coates (1985: 247–50) and Coates & Peckover (2001: 63).

NOTES

R. Schodde (*in litt.*) has confirmed that there is a single specimen in Leiden, collected by Schluter and Maso on 25-Feb-1926 (no. 5554), whose museum label states "Aroe Is." Diamond & Bishop (1994) did not record this species in their work there. It is surprising there are not additional records from this major historical collecting locus. For the time being, we consider the locality for the Schluter and Maso specimen in doubt, and the species' status on the Aru Is needs confirmation. *Other names*—Magnificent Fruitdove, Wompoo Pigeon, *Ptilinopus magnificus*.

GENUS *PTILINOPUS* [Genus 49 spp/Region 15 spp]

Ptilinopus Swainson, 1825. *Zool. Journ.* 1: 473. Type, by monotypy, *Ptilinopus purpuratus* var. *regina* Swainson.

The species of this genus range from Sumatra and the Philippines south and east to Sulawesi, the Lesser Sundas, the New Guinea region, eastern Australia, the Bismarcks, the Solomon Islands, Vanuatu, New Caledonia, Fiji, and many islands in Micronesia and Polynesia (Baptista *et al.* 1997). The genus is characterized by the compact size, sexual dichromatism, soft gut, arboreal habit, and multicolor plumage featuring green and gray but also one or more bright colors. A recent molecular study by Cibois *et al.* (2014) clarified relationships within the genus. We follow their taxonomic recommendations, including splitting off the genera *Megaloprepia*, *Ramphiculus*, and *Chrysoena* from *Ptilinopus*.

Dwarf Fruit-Dove *Ptilinopus nainus* ENDEMIC | MONOTYPIC
Columba naina Temminck, 1835. *Planch. Col. d'Ois.*, livr. 95: pl. 565.—Triton Bay, Bird's Neck.
Synonym: *Ptilinopus nanus minimus* Stresemann & Paludan, 1932. *Novit. Zool.* 38: 182.—Waigeo I, NW Islands.
Widespread through the NG lowlands but apparently absent from much of the NW Lowlands and the n coast of the SE Peninsula between Milne Bay and the Kumusi R (Mayr 1937c, Mees 1965); also found in the NW Islands[MA] (Misool[Q], Salawati[MA], Waigeo[MA], Gam[P]), to 1,250 m[J] (Mees 1965[Q], B. G. Freeman & A. M. Freeman *in litt.*[J], Beehler field notes 12-Mar-2005[P]). S. van Balen (*in litt.*) provided sight records for the w and ne sectors of the NW Lowlands (see below).

Localities—Sorong[MA], Teminaboean in the Bird's Head[X], Triton Bay[Z], Waropen Atas[M], Bernhard Camp on the Taritatu R[Q], Nimbokrang[M,J], upper Sepik[K], Anguganak in West Sepik Prov[L], Pagwi[W], near Madang[K], Markham valley[O], Hydrographer Mts[MA], Varirata NP[CO] (Rand 1942b[Q], Gyldenstolpe 1955b[X], Greig-Smith 1978[L], Hoek Ostende *et al.* 1997[Z], N. Brickle & R. Tizard 2009 field observation[J], I. Woxvold *in litt.*[K], F. Crome *in litt.*[W], S. van Balen *in litt.*[M], C. Eastwood in *Muruk* 6(3): 19[O].

NOTES

Generic assignment was questioned by Cibois *et al.* (2014). For the present we prefer to leave this species in *Ptilinopus*. We follow Mayr (1937c) and Dickinson & Remsen (2013) on the spelling of the species epithet (but see Mayr 1941, Rand & Gilliard 1967, Beehler & Finch 1985). We follow the recommendation of Mees (1965) in treating the species as monotypic. Variation is clinal and best not memorialized in formal subspecies names (see notes in Gibbs *et al.* 2001: 515). Rand & Gilliard (1967) questioned the Misool I record, but Mees (1965) examined museum specimens attributed to Misool. It is now important to obtain field observations from Misool confirming its presence. There are few records from much of the NW Lowlands and the n sector of the SE Peninsula. Geographic ranges of certain species in the genus *Ptilinopus* exhibit weird patchiness of distribution in mainland NG, in which large blocks of suitable habitat are seemingly unoccupied, exemplified by this species, *P. wallacii*, *P. aurantiifrons*, and *P. viridis*. *Other names*—Least, Little, or Small Green Fruit Dove, *Ptilinopus nanus*.

Superb Fruit-Dove *Ptilinopus superbus*   RESIDENT

Columba Superba Temminck, 1809. In Knip, *Les Pigeons, les Colombes*: 75, pl. 33.—Halmahera, n Moluccas.
   All NG and most satellite islands, sea level–1,800 m (Pratt & Beehler 2014).
   Extralimital—Sulawesi and the Moluccas to the Bismarck Arch, Solomon Is, and e Australia (Baptista *et al.* 1997, Mayr & Diamond 2001).

SUBSPECIES

a. *Ptilinopus superbus superbus* (Temminck)
RANGE: All NG; NW Islands[MA] (but absent from Gag, Gebe, Kofiau); also Biak[J,K], Numfor[MA], and Yapen I[MA]; Aru Is[MA]; Schouten Is[CO]; Kairiru[CO], Bagabag[CO], Karkar[MA], and Manam I[MA]; D'Entrecasteaux Arch[MA], Trobriand Is[CO], and Louisiade Arch[MA] (including Rossel I); also Moluccas, Bismarck and Solomon Is, and Australia (King 1979[K], Baptista *et al.* 1997, Bishop ms[J]). J. P. Dumbacher (*in litt.*) found the species on Nurata I off Kitava I in the Trobriands.
DIAGNOSIS: Distinguished from *temminckii* of Sulawesi by lack of purple wash on breast above the black band (in male); and (in female) reduced dark patch on crown (green forecrown, dark hind-crown).

NOTES

Dickinson & Remsen (2013) split off the Sulawesi population as *P. temminckii*. *Other names*—Purple-capped or Purple-crowned Fruit-dove, Purple-crowned Pigeon, Superb Fruit-pigeon.

Mountain Fruit-Dove *Ptilinopus bellus*   ENDEMIC | MONOTYPIC

Ptilonopus bellus P. L. Sclater, 1874. *Proc. Zool. Soc. London* (1873): 696, pl. 57.—Hatam, Arfak Mts, Bird's Head.
   All NG mt ranges, plus mts of Karkar[J], Fergusson[K], and Goodenough I[MA], 200–3,260 m[CO] (Diamond & LeCroy 1979[J], Pratt specimen at BPBM[K]). This upland forest specialist is rare in flat lowlands and the lowest hills of the Mainland.

NOTES

We split this species from *P. rivoli*. Inhabits species-rich upland forest. Male has yellow center to whitish breast crescent; dark green trailing edge to breast crescent; magenta forehead spot; red-orange iris; gray chin; dark blue spots on scapulars; large purple patch on belly; undertail with green feathering and pale yellow feather edges; bluish cast to feathering around eye; female has blue-green crown; gray chin; green dorsal and ventral plumage; undertail with pale yellow feather edgings trending up onto lower breast; no dark blue spotting on scapulars; reddish-yellow iris; bill yellow with a greenish tint. See photo in Coates & Peckover (2001: 62). *Other names*—High Mountain Fruit Dove.

White-bibbed Fruit-Dove *Ptilinopus rivoli*   RESIDENT

Columba Rivoli Prévost, 1843. In Knip, *Les Pigeons* (1839–43), ed. 2, 2: 107, pl. 57.—Duke of York Is, New Ireland.
   A small-island tramp species inhabiting Mios Num and Yapen I (Bay Is) and widespread in the SE Islands, where can be found inhabiting smaller and more isolated islands and islets, to 900 m.
   Extralimital—Bismarck Is (Mayr & Diamond 2001).

SUBSPECIES

a. *Ptilinopus rivoli miquelii* (Schlegel)
Ptilopus Miquelii Schlegel (ex von Rosenberg ms), 1871. *Ned. Tijdschr. Dierk.* (1873) 4: 22.—Mios Num and Yapen I, Bay Islands.
RANGE: Mios Num[MA], Woendi Atoll (east of Biak)[J], and Yapen I[MA], Geelvink Bay[K] (Hoek Ostende *et al.* 1997[K], Junge 1956[J]). Lowland forest up to 450 m.
DIAGNOSIS: Smaller than the nominate form, and male's breast band is pure white; wing 126–134.

b. *Ptilinopus rivoli strophium* Gould
Ptilinopus strophium Gould, 1850. In Jardine, *Contr. Orn.*: 105, note.—Duchateau Is, Louisiade Arch, SE Islands.

RANGE: Primarily inhabits islets of the SE Islands: Trobriand Is[CO]; Tobweyama I[J] (off Normanby I); Bonvouloir Is (East and Hastings I)[MA]; Aion[K], Reu[K], and Waviai I[K] (all off Woodlark); Egum[MA], Yanaba, and Alcester I[MA]; and Duchateau, Misima, Tagula, and Rossel I[MA] of the Louisiade Arch[MA]; also a doubtful record from the islands off the coast between Port Moresby and Milne Bay[MA] (Pratt specimens in BPBM[J], Pratt field notes 2005[K]).

DIAGNOSIS: Male has yellow undertail, iris orange, white breast band, purple trailing edge to breast band, purple belly patch absent, blue spots on scapulars, magenta forehead patch, gray chin, minimal yellow edging to secondaries; female has yellow undertail plumage extending up as yellow scaling onto belly and lower breast, gray chin, rich green forehead, no blue spots on scapulars, fine yellow edging to secondaries (Rand & Gilliard 1967). Larger than *miquelii*, and male's breast band tinged yellow, wing 133–142.

NOTES

We treat the *Ptilinopus rivoli* species group as comprising three species in the Region. There are two distinct white-breasted lowland insular forms (*prasinorrhous* and *rivoli*) and one yellow-breasted upland form (*bellus*). The lowland forms are distinct mainly because of their small-island lowland habit and the white breast band; they occur allopatrically in Geelvink Bay. Our current "reduced" *P. rivoli* is adapted to small islands in the Region with impoverished avifaunas. In the D'Entrecasteaux Arch, it occurs in neighboring sympatry with *P. bellus*, which is in the mts of the two highest islands whereas *P. rivoli* is on the small, low islands, without evidence of population mixing or interbreeding. In addition to plumage color and structure, the two species differ in their short contact calls. We suggest that the specimens supposedly collected from Yapen I were actually taken from a nearby small island, as Yapen is a large and high land-bridge island, where the species *bellus* is more likely to occur. That said, the nominate subspecies of *P. rivoli* does inhabit montane forest on New Britain and New Ireland (Dutson 2011). Splitting subspecies *strophium* and *miquelii* as distinct species separate from nominate *rivoli* (which has a purple belly patch, lacking in *strophium*) is an alternative arrangement that should be explored. *Other names*—White-breasted Fruit-Dove.

Moluccan Fruit-Dove *Ptilinopus prasinorrhous*   RESIDENT | MONOTYPIC

Ptilonopus prasinorrhous G. R. Gray, 1858. *Proc. Zool. Soc. London*: 185.—Kai Is.
    Synonym: *Ptilinopus rivolii buruanus* Hartert & Goodson, 1918. *Novit. Zool.* 25: 265.—Gunung Fogha, Buru I,
        Moluccas.

A small-island specialist of the west. The NW Islands (Schildpad Is and Kofiau, Misool, Salawati, Batanta, Wai[P], Waigeo, Gag, and Gebe I)[MA] and small islands and islets in Geelvink Bay (Padaido Is, Roon, Numfor, Manim, and Mios Korwar I)[MA], and possibly some small islets off Yapen I (Mayr 1937c, Beehler field notes 13-Mar-2005[P]). No reliable records from Yapen or Biak I. Coral islands and coastal woods of larger islands. Diamond & Bishop (1994) reported the first record for the Aru Is region, from nearby Pulau Babi (Pig I).
    Extralimital—s Moluccas (including Buru), Kur Is, Goramlaut Is, and Kai Is (Coates & Bishop 1997).

NOTES

Male has white breast crescent; iris red-orange; medium-size purple breast patch extending posteriorly from white breast crescent; magenta crown patch; dark blue spots on scapulars; no yellow edging to secondaries; green undertail with yellow edges; female has all underparts green; some indication of yellow edges to feathers of undertail; rich green forecrown; all-green upperparts; no dark blue spotting on scapulars; minimal yellow edging of secondaries; wing 130 (Rand & Gilliard 1967). *Other names*—formerly treated as a race of *P. rivoli*.

Yellow-bibbed Fruit-Dove *Ptilinopus solomonensis*   RESIDENT

Ptilonopus solomonensis G. R. Gray, 1870. *Ann. Mag. Nat. Hist.* (4)5: 328.—Makira I, Solomon Is.
    New Guinea range includes various Bay Is (details below). Frequents lowland forest and fruiting trees and shrubs in gardens.
    Extralimital—islands just west of New Britain, small islands off New Britain, New Hanover, various northern offshore islands; also throughout the Solomon Is (Mayr & Diamond 2001, Dutson 2011).

a. *Ptilinopus solomonensis speciosus* (Schlegel)
Ptilopus speciosus Schlegel, 1871. *Ned. Tijdschr. Dierk.* (1873) 4: 23.—Numfor and Biak I, Bay Islands.
RANGE: Numfor, Biak, Owi[X], Traitor's I (Padaido Is), and Marai I, near Yapen I (Mayr 1941, Gibbs 1993[X]).
DIAGNOSIS: Male has a green cap, purple eyespot, and a broad violet belly patch.

NOTES
A tramp species that favors species-poor islands, absent from mainland NG and the larger islands of the Bismarck Arch (Dutson 2011). *P. s. speciosus* was treated as a distinct species, *P. speciosus*, by del Hoyo & Collar (2014). *Other names*—Yellow-breasted or Splendid Fruit-dove.

Wallace's Fruit-Dove *Ptilinopus wallacii*   RESIDENT | MONOTYPIC
Ptilonopus Wallacii G. R. Gray, 1858. *Proc. Zool. Soc. London*: 185, pl. 136.—Aru Is.
Southwestern NG, from Kaimana[J] and the Mimika R[MA] east to Noord R[MA] (S. van Balen *in litt.*—bird hit window and studied in hand[J]), also the Aru Is[MA]; lowland forest near coast and rivers (Mayr 1941). Most recently observed on the Timika Rd and Timika environs (J. Pap *in litt.*, 2014 observation).
Extralimital—Indonesian islands of Babar, Tanimbar, Banda, Kur, Manggur, Taam, Komeer, Bacar, Tual, and Kai Kecil (Gibbs *et al.* 2001).
Other names—Golden-fronted, Yellow-fronted, Golden-shouldered, or Crimson-capped Fruit-dove; Wallace's Green Fruit-dove; Wallace's Fruit-pigeon.

Pink-spotted Fruit-Dove *Ptilinopus perlatus*   ENDEMIC
Columba perlata Temminck, 1835. *Planch. Col. d'Ois.*, livr. 94: pl. 559.—Triton Bay, Bird's Neck.
Frequents rain forest, gallery forest, and edge through NG, the NW Islands, Yapen I, Aru Is, and D'Entrecasteaux Arch, mainly in lowlands, but records to 1,220 m[CO] (Mayr 1941). Diamond (1972) noted this species and *ornatus* tend to occupy mutually exclusive ranges, *perlatus* in the lowlands and *ornatus* in the hills and mts.

SUBSPECIES
a. *Ptilinopus perlatus perlatus* (Temminck)
RANGE: Batanta[J,K,L], Waigeo[MA], Salawati[MA], and Yapen I[MA]; also the Bird's Head east in the n watershed to the Sepik-Ramu lowlands[CO], and east in the s watershed to Triton Bay in the s Bird's Neck[CO] (Beckers *et al.* in M. Tarburton's island website[J], Matheve *et al.* 2009[K], S. Hogberg in M. Tarburton's island website[L]). The exact location where the nominate form gives over to gray-capped *plumbeicollis* in the Sepik-Ramu has not been determined.
DIAGNOSIS: Crown dull greenish yellow; tail without pale terminal band on dorsal surface.

b. *Ptilinopus perlatus zonurus* (Salvadori)
Ptilopus zonurus Salvadori, 1876. *Ann. Mus. Civ. Genova* 9: 197.—Aru Is.
RANGE: Aru Is[MA], S Lowlands, SE Peninsula, and D'Entrecasteaux Arch[CO].
DIAGNOSIS: Similar to *perlatus*, but has tail tip with pale terminal band on dorsal surface.

c. *Ptilinopus perlatus plumbeicollis* (A. B. Meyer)
Ptilopus plumbeicollis A. B. Meyer, 1890. *Ibis*: 422.—Near Madang, e sector of the Sepik-Ramu.
RANGE: Eastern sector of the Sepik-Ramu and the Huon Penin[CO].
DIAGNOSIS: Head is all gray, without the dull yellow-green cap (Baptista *et al.* 1997).

NOTES
Other names—Pink-spotted Fruit-pigeon.

Ornate Fruit-Dove *Ptilinopus ornatus*   ENDEMIC
Ptilopus ornatus Schlegel (ex von Rosenberg ms), 1871. *Ned. Tijdschr. Dierk.* (1873) 4: 52.—Interior of the Bird's Head.
Frequents rain forest and edge across NG, sea level–2,500 m; less common in lowlands (Pratt &

Beehler 2014); records for all outlying ranges but for the Wandammen Mts and apparently absent from much of the S Lowlands and Trans-Fly. Absent from the islands. Found to nest frequently above 2,000 m in montane-forest interior. It appears that this hill-forest species migrates up into the mid-montane forest to breed.

SUBSPECIES

a. *Ptilinopus ornatus ornatus* (Schlegel)
RANGE: Bird's Head.
DIAGNOSIS: Cap is wine red, very distinctive (Rand & Gilliard 1967). Wing 152–153 (Gibbs *et al.* 2001).

b. *Ptilinopus ornatus gestroi* D'Albertis & Salvadori
Ptilonopus gestroi D'Albertis & Salvadori, 1876. *Ann. Mus. Civ. Genova* 7: 834.—Yule I, SE Peninsula.
    Synonym: *Ptilinopus ornatus kaporensis* Rothschild & Hartert, 1901. *Novit. Zool.* 8: 105.—Kapaur, Onin Penin,
    Bird's Neck.
RANGE: From the Bird's Neck (Fakfak and Kumawa Mts on the Onin Penin; Gibbs 1994, Diamond & Bishop 2015) east to Milne Bay in both the n and s watersheds. Includes the Cyclops Mts[MA].
DIAGNOSIS: Cap is mustard yellow (Rand & Gilliard 1967). Characters defining *gestroi* and *kaporensis* are clinal and combined here (Baptista *et al.* 1997, Diamond 1972). Wing: male 150–154, female 145–156 (Gibbs *et al.* 2001). Treated as a full species by del Hoyo & Collar (2014).

NOTES

It is necessary to confirm that the yellow-capped *gestroi* inhabits the w Mamberamo. The Bird's Head form (*ornatus*) is very distinct and perhaps could be considered a sibling species, as was done by del Hoyo & Collar (2014): the characters distinguishing the nominate form are the wine-colored cap and greenish breast band. Dickinson & Remsen (2013) combined *kaporensis* into the nominate form, which is apparently a *lapsus*, as Rand & Gilliard (1967) made clear that *kaporensis* and *gestroi* both exhibit the mustard-yellow crown, unlike the maroon-capped nominate form. *Other names*—Gestroi's Fruit-dove.

Orange-fronted Fruit-Dove *Ptilinopus aurantiifrons*   ENDEMIC | MONOTYPIC
Ptilonopus aurantiifrons G. R. Gray, 1858. *Proc. Zool. Soc. London*: 185, pl. 137.—Aru Is.
    Frequents mangroves, gallery forest, edge, and, less commonly, rain forest, across NG, and Sala-wati[MA], ?Misool[MA], Batanta[MA], the Aru Is[MA], Yapen[MA], Daru I[CO], Killerton Is[CO], and Sariba[CO], Fergusson[MA] and Normanby I[K], sea level–300 m (Rand 1942b, Ripley 1964a, LeCroy *et al.* 1984[K]). Absent from the Huon Penin; but present at Morobe station (Mayr 1941, Pratt & Beehler 2014). Record from Misool I is doubtful (Mees 1965).

Orange-bellied Fruit-Dove *Ptilinopus iozonus*   ENDEMIC
Ptilonopus iozonus G. R. Gray, 1858. *Proc. Zool. Soc. London*: 186.—Aru Is.
    All NG, the NW Islands, Aru Is, and various n islands, sea level–800 m (Mayr 1941, Pratt & Beehler 2014). Frequents rain forest, edge, openings, and mangroves (Coates 1985). Occurs mostly in the lowlands, locally in the hills to *ca.* 800 m (Pratt & Beehler 2014).

SUBSPECIES

a. *Ptilinopus iozonus humeralis* Wallace
Ptilonopus humeralis Wallace, 1862. *Proc. Zool. Soc. London*: 166: pl. 21.—Salawati I, NW Islands.
    Synonym: *Ptilinopus iozonus pseudohumeralis* Rand, 1938. *Amer. Mus. Novit.* 990: 5.—Palmer R, 3 km below its
    junction with the Black R, 100 m elevation, upper Fly R, S Lowlands.
RANGE: ?Batanta[A], Salawati[MA], and Waigeo I[MA]; also the Bird's Head and Neck and S Lowlands as far east as the upper Fly[MA] (M. Tarburton's island website[A]). Apparently this form ranges eastward to the north of the nominate form, which is found only in the Trans-Fly and the Aru Is.
DIAGNOSIS: Shoulder patch gray and wine-colored. Extensive spotting on wings resulting from gray bases to scapulars and median and greater wing coverts. Tail tip green. The description of the subsumed form *pseudohumeralis* was based on specimens with plumage colors that had faded with time; fresh material that was brighter than the older material of "*pseudohumeralis*" is today as dull or duller.

b. *Ptilinopus iozonus iozonus* (G. R. Gray)

RANGE: Aru Is[MA] and Trans-Fly (Mees 1982b).

DIAGNOSIS: Small and indistinct lilac-gray patch on shoulder; much reduced gray spotting on wings. Bangs & Peters (1926) and Mees (1982b) found the Trans-Fly specimens identical to those from the Aru Is (the nominate form).

c. *Ptilinopus iozonus iobiensis* (Schlegel)

Ptilopus humeralis iobiensis Schlegel, 1871. *Mus. Pays-Bas* 4 (*Columbae*): 16.—Yapen I, Bay Islands. [For correct date of publication, see Dickinson & Pieters (2011), *pace* Dickinson & Remsen (2013).]

RANGE: Northern NG: the NW Lowlands and Sepik-Ramu; also Yapen[MA,K], Kairiru[CO], Karkar[J], Manam[MA], and Tarawai I[CO] (Diamond & LeCroy 1979[J], Hoek Ostende *et al.* 1997[K]).

DIAGNOSIS: Throat rich green and obsolete terminal tail band dark green (not gray); substantial gray-violet shoulder patch. We follow Dickinson & Remsen (2013) in using the original subspecies spelling based on the island name Jobi (there is no *j* in the Latin alphabet).

d. *Ptilinopus iozonus finschi* Mayr

Ptilinopus iozonus finschi Mayr, 1931. *Mitt. Zool. Mus. Berlin* 17: 705.—Finschhafen, Huon Penin.

RANGE: Southeastern Peninsula, west to the Huon Penin in the north and west to the Fly R in s NG (Mayr 1941), where it meets *humeralis* (*humeralis* ranges to the north of the nominate form, which apparently is confined to the Trans-Fly and the Aru Is).

DIAGNOSIS: Gray terminal band on tail; large gray patch on secondary coverts; orange belly patch extends up onto lower breast.

[Knob-billed Fruit Dove *Ptilinopus insolitus*   HYPOTHETICAL

Ptilopus insolitus Schlegel, 1863. *Ned. Tijdschr. Dierk.* 1: 61, *Vogels*, pl. 3, f. 3.—New Ireland.

Berggy (1978) reported having observed the species on the n coast at Bogia, Madang Prov, PNG, in 1973. Frequents forest, edge, and disturbed habitats; also visits fruiting trees in cultivated areas. Presumably the birds observed in the NG region would be the nominate form. If the Bogia sight record is valid, one would expect the species to occur on adjacent Manam I. It should be looked for there by future fieldworkers.

Extralimital—throughout the Bismarck Arch. Occurs from sea level up to 1,200 m on New Britain and up to at least 700 m on New Ireland.

Other names—Red-knobbed Fruit Dove.]

Claret-breasted Fruit-Dove *Ptilinopus viridis*   RESIDENT

Columba viridis Linnaeus, 1766. *Syst. Nat.*, ed. 12, 1: 283.—Ambon, s Moluccas.

Note distinctively patchy and marginal Mainland distribution. Inhabits the Bird's Head region, Kumawa and Fakfak Mts, Foja Mts, PNG North Coastal Ra, all the NW Islands, Bay Is, D'Entrecasteaux Arch, and Trobriand Is, 300–1,000 m, locally at sea level (Diamond 1985). This species is absent from the interior of the Mainland and the Huon Penin, but present on islands at both ends of NG.

Extralimital—s Moluccas and Manus I, Lihir I, Tanga Is, and the Solomon Is; absent from New Ireland and New Britain (Baptista *et al.* 1997, Mayr & Diamond 2001, Gibbs *et al.* 2001, Dutson 2011).

SUBSPECIES

a. *Ptilinopus viridis pectoralis* (Wagler)

Columba pectoralis Wagler, 1829. *Isis von Oken* 22: col. 740.—Manokwari, ne sector of the Bird's Head.

RANGE: Kofiau, Gebe, Gag, Waigeo, Salawati[J], Batanta[J], and Misool I[K]; also Bird's Head, east to Geelvink Bay (Mansinam and Andai), and the Bird's Neck, including Triton Bay (Lobo), the Fakfak Mts[J], Kumawa Mts[J], and Wandammen Mts[J] (Mayr 1941, Ripley 1964a[K], Diamond 1985[J]). Mayr (1941) noted that records by von Rosenberg from Numfor and Biak I are erroneous.

DIAGNOSIS: Male exhibits a small, round, maroon patch on breast (lacking in female); a single combined patch of gray on the secondary coverts; and no gray shoulder patch.

b. *Ptilinopus viridis geelvinkianus* (Schlegel)

Ptilopus viridis stirps *Geelvinkiana* Schlegel, 1873. *Ned. Tijdschr. Dierk.* 4: 23.—Numfor, Mios Num, and Biak I, Bay Islands.

> Synonym: *Ptilinopus viridis pseudogeelvinkianus* Junge, 1952. *Zool. Meded. Leiden* 31(22): 247.—Mios Num I, Bay Is.

RANGE: Numfor, Manim, Biak, Mios Num, and Marai I, near Yapen (Mayr 1941, Hoek Ostende *et al.* 1997); the old record from Manokwari is discounted (Mayr 1941).

DIAGNOSIS: Male exhibits a medium-size teardrop-shaped maroon bib (lacking in female); multiple gray spots on the secondary coverts; and a pronounced gray shoulder patch.

c. *Ptilinopus viridis salvadorii* (Rothschild)

Ptilopus salvadorii Rothschild, 1892. *Bull. Brit. Orn. Club* 1: 10.—Yapen I, Bay Islands.

RANGE: Yapen I and nw NG from the Foja and Cyclops Mts east to the Sepik-Ramu in the foothills of the PNG North Coastal Ra (Diamond 1985). Lowland forest up to 600 m. Easternmost record from Mt Turu, 30 km from Wewak (Diamond 1985).

DIAGNOSIS: Male exhibits a medium-size crescent-shaped maroon patch on breast; small gray shoulder spots; and relatively few gray spots on the secondary coverts. Female similar, but breast patch smaller. Type description noted similarity to *pectoralis* but for the paler green of the plumage and larger maroon breast patch.

d. *Ptilinopus viridis vicinus* (Hartert)

Ptilopus lewisii vicinus Hartert, 1895. *Novit. Zool.* 2: 62.—Fergusson I, D'Entrecasteaux Arch, SE Islands.

RANGE: Fergusson[MA], Goodenough[MA] and Normanby I[J,K], and Trobriand Is[CO] (LeCroy *et al.* 1984[J], Pratt field notes 2003[K]).

DIAGNOSIS: Medium teardrop-shaped maroon breast patch of male (smaller in female); multiple gray spots on the secondary coverts; and pronounced gray shoulder patch. Male not distinguishable from the distant form *geelvinkianus*, but *vicinus* is almost sexually monochromatic.

NOTES

Mainland NG distribution is fragmentary, confined to the Bird's Head and Neck, nw NG, and the Sepik-Ramu. The species occurs in six geographic races spread patchily over a far-flung geographic range; whether they should all be retained in a single species beckons investigation, as does their relationship to *P. eugeniae* of the s Solomon Is, which appears to be embedded within *P. viridis* in the phylogeny produced by Cibois *et al.* (2014). David & Gosselin (2002a) highlighted that the subspecies name is correctly spelled *geelvinkianus*. We follow the inclination of Rand & Gilliard (1967) in subsuming *pseudogeelvinkianus* into *geelvinkianus*. Berggy (1978) reported the species from coastal Madang Prov; this record needs corroboration. *Other names*—Red-bibbed, Red-breasted, or Red-throated Fruit-dove.

Beautiful Fruit-Dove *Ptilinopus pulchellus*   ENDEMIC | MONOTYPIC

Columba pulchella Temminck, 1835. *Planch. Col. d'Ois.*, livr. 95: 564.—Triton Bay, Bird's Neck.

> Synonym: *Ptilopus decorus* Madarász, 1910. *Ann. Mus. Nat. Hung.* 8: 173, pl. 2, left-hand fig.—Czinyagi, Astrolabe Bay, e sector of the Sepik-Ramu.

Wet hill forests of NG and the main NW Islands, with records to 1,370 m[CO]. Also can be found in lowland forest adjacent to hills.

NOTES

We follow Goodwin (1983) and the suggestion of Baptista *et al.* (1997) in treating the species as monotypic. Diamond (1972) stated this species and *P. coronulatus* tend to inhabit complementary ranges—not sharing the same forest tract in many instances. *Other names*—Crimson-crowned, Crimson-capped, or Grey-breasted Fruit-dove; Rose-fronted Pigeon.

Coroneted Fruit-Dove *Ptilinopus coronulatus*   ENDEMIC

Ptilonopus coronulatus G. R. Gray, 1858. *Proc. Zool. Soc. London*: 185, pl. 138.—Aru Is.

    Inhabits lowland rain forest and hill forest throughout NG and Salawati, Yapen, Daru, Kairiru, Takar, and Manam I; also Aru Is; sea level–1,200 m (Mayr 1941, Coates 1985). It is also noted that *P. coronulatus* is more common in lowland forest than *P. pulchellus*. Bell (1982a), on his Brown R lowland forest study site, made >1,500 observations of *P. coronulatus* and only five observations of *P. pulchellus*. Both species have been recorded from the Hunstein Mts (A. Allison *in litt.*), at Varirata NP (P. Gregory *in litt.*), at Lake Murray (B. J. Coates *in litt.*), and in Gulf Prov, PNG (I. Woxvold *in litt.*).

SUBSPECIES

a. *Ptilinopus coronulatus geminus* Salvadori
Ptilonopus geminus Salvadori, 1875. *Ann. Mus. Civ. Genova* 7: 786.—Ansus, Yapen I, Bay Islands.
> Synonym: *Ptilonopus trigeminus* Salvadori, 1875. *Ann. Mus. Civ. Genova* 7: 787 (in key).—Sorong and Salawati I, NW Islands. [Date of pub. confirmed by Poggi (1996: 103).]
> Synonym: *Ptilopus quadrigemmus* A. B. Meyer, 1890. *Ibis*: 421.—Near Madang, Astrolabe Bay, e sector of the Sepik-Ramu.
> Synonym: *Ptilopus coronulatus huonensis* A. B. Meyer, 1892. *J. Orn.* 40: 263.—Butaueng, Huon Gulf, s coast of Huon Penin.

RANGE: Salawati, Yapen, and Kurudu I; also the Bird's Head and n NG from Manokwari east to Milne Bay, including all of the Huon Penin (Mayr 1941).
DIAGNOSIS: A clinal race that varies in crown-patch coloration from west to east; breast spot mauve violet rather than claret.

b. *Ptilinopus coronulatus coronulatus* G. R. Gray
RANGE: Aru Is, S Lowlands, and s watershed of the SE Peninsula, from the Mimika R to Milne Bay (Mayr 1941).
DIAGNOSIS: Male has reduced orange on lower belly; purple of belly patch claret, not mauve. See photos in Coates (1985: 235–39) and Coates & Peckover (2001: 60).

NOTES
Other names—Lilac-capped, Lilac-crowned, Coronated, Little Coroneted, or Diadem Fruit-dove, Lilac-crowned Fruit-pigeon.

Rose-crowned Fruit-Dove *Ptilinopus regina* VAGRANT

Ptilinopus purpuratus var. *regina* Swainson, 1825. *Zool. Journ.* 1: 474.—New South Wales, Australia.
> Apparently a vagrant to s NG, with two records known, one of uncertain provenience.
> Extralimital   n and e Australia, and Kai, Tanimbar, and Lesser Sunda Is. Widespread in the Torres Strait Is (Draffan *et al.* 1983, Gibbs *et al.* 2001).

SUBSPECIES

a. *Ptilinopus regina regina* Swainson
RANGE: Eastern Australia from NSW to Cape York and Daru I; vagrant to s NG. Two records: an adult male and a young bird collected on Daru I, in the Trans-Fly, on 9-Jun-1936 (Rand 1942a); and one collected by A. Goldie in 1880 (Frith 1982b). The latter specimen (in the British Museum) is said to have come from near Port Moresby, but the locality is doubtful (Coates 1985).
DIAGNOSIS: Cap rich vinaceous red; breast band dark green; upper breast and face dark greenish gray (Baptista *et al.* 1997).

NOTES
The status of this species on the NG mainland merits assessment. The unconfirmed report from Port Moresby is probably in error. More than 40 years of birding in that region has produced no new sightings. In addition, reports of the species from the Aru Is (Beehler *et al.* 1986, Baptista *et al.* 1997) were apparently in error (J. M. Diamond *in litt.*, M. LeCroy *in litt.*). Not recorded from the Aru Is by Diamond & Bishop (1994), Gibbs *et al.* (2001), or Dickinson & Remsen (2013). *Other names*—Pink-capped, Grey-capped, Blue-spotted, or Swainson's Fruit-dove; Red-crowned or Pink-crowned Pigeon.

Ducula Hodgson, 1836. *Asian Res.* 19: 160. Type, by monotypy, *Ducula insignis* Hodgson.

The species of this genus range from tropical Asia to Christmas Island, the Philippines, Indonesia, New Guinea, Australia, and the islands of the southwestern Pacific east to the Tuamotu, Society, and Marquesas Islands (Baptista *et al.* 1997). The genus is characterized by the fruit-eating habit and the soft gut adapted to digest fruit (not seeds), the large size, small head, medium-length squared-off tail, and proportionately long bills framing a wide mouth; these last two features permit grasping large fruit.

Spectacled Imperial Pigeon *Ducula perspicillata* RESIDENT

Columba perspicillata Temminck, 1824. *Planch. Col. d'Ois.*, livr. 42: 246.—Halmahera I, Moluccas.
Kofiau I, of the NW Islands.
Extralimital—Moluccas.

SUBSPECIES

a. *Ducula perspicillata ?neglecta* (Schlegel)
Carpophaga neglecta Schlegel, 1866. *Ned. Tijdschr. Dierk.* (1865) 3: 195, 344.—Seram, Ambon, Buano, s Moluccas.
RANGE: Seram, Ambon, Saparua, Buano, and Kofiau I.
DIAGNOSIS: Diamond *et al.* (2009) made field observations of this species in Kofiau and noted that the plumage was closest to this form. However, they noted that the birds they observed had pale grayish legs, rather than the red-purple legs reported for this form. Baptista *et al.* (1997) reported that this subspecies can be identified by its paler plumage, with head pale gray, iridescence on upperparts golden-green, and with a silvery bloom to the remiges; while the nominate form is darker, with green of mantle edging up onto nape. However, it is the nominate subspecies that is geographically closest to Kofiau, and the taxonomic status of the Kofiau birds remains undetermined. Specimens need to be collected on Kofiau and analyzed against comparative museum material.

NOTES

Rheindt & Hutchinson (2007) suggested *neglecta* should be treated as a distinct species. Del Hoyo & Collar (2014) followed through on this suggestion (Seram Imperial-pigeon, *D. neglecta*) based on plumage characters and voice. The proposed English name "White-eyed Imperial Pigeon" seems inappropriate for this species, as its iris is dark.

Elegant Imperial Pigeon *Ducula concinna* RESIDENT | MONOTYPIC

Carpophaga concinna Wallace, 1865. *Ibis* (ns)1: 383.—Watubela I, se Moluccas.
Synonym: *Carpophaga concinna separata* Hartert, 1896. *Novit. Zool.* 3: 180.—Kai Is.
Synonym: *Ducula concinna aru* Salomonsen, 1934. *Bull. Brit. Orn. Club* 54: 87.—Aru Is.
Islets fringing the Aru Is (Babi, Karang, and Enu I)[J], the Aru Is proper[L], islands off the Bomberai Penin (Adi I)[K,L], and the coast of the Bomberai Penin[K] (Gyldenstolpe 1955b[L], Diamond 1985[K], Diamond & Bishop 1994[J]).

Extralimital—Moluccas and e Lesser Sundas. Vagrant to Australia. Not resident on Seram (Baptista *et al.* 1997).

NOTES

We follow Baptista *et al.* (1997) in treating this small-island tramp species as monotypic. The record from Biak I (Tindige 2003b) is in error, based on misidentification of *D. geelvinkiana*, which, like *concinna*, lacks a bill knob and has a pale iris. Mayr (1941) sunk the race *aru*. Dickinson & Remsen (2013) did not recognize the form *separata*. See Gibbs *et al.* (2001) for a detailed discussion of this latter form and its variation in e Indonesia. *Other names*—Collared, Blue-tailed, Yellow-eyed, or Gold-eyed Imperial-pigeon.

Pacific Imperial Pigeon *Ducula pacifica* RESIDENT

Columba pacifica Gmelin, 1789. *Syst. Nat.* 1(2): 777.—Tonga.

Recorded from small islands off coastal Sepik-Ramu, and certain SE Islands: the Conflict Is, selected Louisiade Is, and Suau I near se tip of PNG. An island tramp species, usually found only on small islands and atolls. Frequents forest on small islands; possibly also visits littoral forest on the Mainland (Coates 1985).

Extralimital—some outer islands of PNG (Wuvulu, Ninigo Is, Hermit Is, Anchorite Is, Mussau, Tench), small islands in the Solomons, including the Temotu Is, and widespread in e Melanesia and Polynesia (Coates 1985, Baptista *et al.* 1997, Mayr & Diamond 2001, Dutson 2011).

SUBSPECIES

a. *Ducula pacifica sejuncta* Amadon

Ducula pacifica sejuncta Amadon, 1943. *Amer. Mus. Novit.* 1237: 7.—Loof I, Hermit Is, Bismarck Arch.
RANGE: Islands off the coast of nc NG—Seleo and Tarawai (Mayr 1941) and Ninigo and Hermit Is of the w Bismarck Arch (Baptista *et al.* 1997). Also perhaps on islands off the coast of Vanimo[CO] (nw sector of the Sepik-Ramu).
DIAGNOSIS: Paler gray on head, neck, and underparts; smaller than nominate; wings more rounded (Amadon 1943, Baptista *et al.* 1997). Wing 218–230 (Gibbs *et al.* 2001).

b. *Ducula pacifica pacifica* (Gmelin)

Synonym: *Globicera tarrali* Bonaparte, 1854. *Compt. Rend. Acad. Sci. Paris* 39: 1073.—Vanikoro I, Temotu Is, se Solomon Is.
RANGE: Conflict[Q] and Budibudi[K] Is, Anagusa[J], Duchateau[MA], and Teste I[MA] of the Louisiade Arch[MA]; and Suau I[MA] near the se tip of PNG (Dumbacher *et al.* 2010[J], Hobcroft & Bishop ms[K], Beehler field notes 29-Sep-2002[Q]). Also ne and se Solomon Is (Ndai, Rennell); Stewart I of the Temotu Is; and many islands in Polyncsia (Baptista *et al.* 1997).
DIAGNOSIS: Larger than *sejuncta* and darker gray on head and breast. Wing: male 222–256, female 217–244 (Gibbs *et al.* 2001). We follow Baptista *et al.* (1997) in submersion of *tarrali* and the extralimital *intensitincta* and *microcera* into the nominate form.

NOTES

A widely scattered small-island tramp species, not to be found on the large species-rich islands. *Other names*—Pacific Pigeon, Pacific Fruitpigeon.

Spice Imperial Pigeon *Ducula myristicivora* RESIDENT | MONOTYPIC

Columba (myristicivora) Scopoli, 1786. *Delic. Flor. Faun. Insubr.* 2: 94.—Gebe I, NW Islands.

The NW Islands: Kofiau[J], Gebe[Y], Gag[K], Gam[MA], Misool[L], Waigeo[W], Batanta[MA], Salawati[MA], and Arar I[X]; Schildpad Is[MA] (Gyldenstolpe 1955b[W], Ripley 1964a[X], Mees 1965[L], 1972[Y], Johnstone 2006[K], Diamond *et al.* 2009[J]). Inhabits all wooded island habitats, including mangroves.

Extralimital—Widi Is, off e side of Halmahera I, e Moluccas.

NOTES

Readily distinguished from its putative sister species, *D. geelvinkiana*, which has been split off here, by the following characters: larger size, head nearly white, black knob atop bill, narrow white eye-ring, dark red iris, rosy nape, mantle without bronzy sheen, dark underwing, no white band across base of upper mandible. This split is also recognized by del Hoyo & Collar (2014). Wing 257–260 (Gibbs *et al.* 2001). *Other names*—Black-knobbed Fruit-Pigeon, Spice Pigeon.

Geelvink Imperial Pigeon *Ducula geelvinkiana* ENDEMIC | MONOTYPIC

Carpophaga geelvinkiana Schlegel, 1873. *Mus. Pays-Bas* 4 (*Columbae*): 86.—Islands in Geelvink Bay.

A restricted-range island endemic inhabiting Numfor, Biak, and Mios Num I of Geelvink Bay[MA].

NOTES

Easily distinguished from *D. myristicivora* by the following characters: smaller size, no knob atop bill, no white eye-ring, head darker gray, iris pale yellow, mantle with bronzy sheen, rose collar lacking, dull gray underwing, and narrow white band across base of upper mandible; wing 227–232 (Gibbs *et al.* 2001). There is a single hill-forest record from the Arfak Mts, Bird's Head, that may refer to this form (Gibbs *et al.* 2001). *Other names*—formerly considered a race of *D. myristicivora*.

Purple-tailed Imperial Pigeon *Ducula rufigaster* ENDEMIC

Columba rufigaster Quoy & Gaimard, 1830. *Voy. Astrolabe, Zool.* 1: 245, pl. 27.—Manokwari, Bird's Head.
   Widespread in lowland forest throughout NG, the NW Islands, and Yapen I. Inhabits mainly forest interior of lowlands and lower hills, locally to 1,200 m (Coates & Peckover 2001).

SUBSPECIES

a. ***Ducula rufigaster rufigaster*** (Quoy & Gaimard)
   Synonym: *Ducula rufigaster pallida* Junge, 1952. *Zool. Meded. Leiden* 31: 248.—Bivak I, Noord R, s NG.
RANGE: The four main NW Islands, the Bird's Head, Bird's Neck, S Lowlands, and s coast of SE Peninsula (Mayr 1941, Coates 1985, Hoek Ostende *et al.* 1997).
DIAGNOSIS: Mantle and rump plumage has a purplish-red cast; uppertail coverts dark purple (Rand & Gilliard 1967). We follow Diamond (1972) in subsuming *pallida* into this form. See photo in Coates & Peckover (2001: 64, upper-right image).

b. ***Ducula rufigaster uropygialis*** Stresemann & Paludan
Ducula rufigaster uropygialis Stresemann & Paludan, 1932. *Novit. Zool.* 38: 243.—Ramu R, ne NG.
RANGE: Yapen I, the NW Lowlands, Sepik-Ramu, and Huon Gulf area; and presumably the n watershed of the SE Peninsula (Mayr 1941, Coates 1985).
DIAGNOSIS: This form has more uniformly green upperparts and a paler rump and underparts than the nominate (Baptista *et al.* 1997: 233). The birds east of Huon Gulf and the Fly R are intermediate between *rufigaster* and *uropygialis* (Mayr 1941).

NOTES

Some authorities (*cf.* Peters 1937) have included in *D. rufigaster* the Moluccan populations that Baptista *et al.* (1997) placed in *D. basilica*. We follow the treatment of Baptista *et al.* (1997), who postulated that the species *rufigaster*, *basilica*, and *finschii* constitute distinct sister species. *Other names*—Rufous-bellied, Cinnamon, or Purple-tailed Fruit-pigeon, Rufous-breasted Imperial-pigeon.

Rufescent Imperial Pigeon *Ducula chalconota* ENDEMIC

Carpophaga chalconota Salvadori, 1874. *Ann. Mus. Civ. Genova* 6: 87.—Hatam, Arfak Mts, Bird's Head.
   Frequents canopy of montane forest interior on the Bird's Head, Wandammen Penin, Central Ra, and Huon Penin (Mayr 1941); 1,000^M–2,650^X m (Mack & Wright 1996^M, Pratt & Beehler 2014^X). Endemic to mts of the Mainland.

SUBSPECIES

a. ***Ducula chalconota chalconota*** (Salvadori)
RANGE: Arfak and Tamrau Mts of the Bird's Head and Wandammen Mts of the Bird's Neck (Mayr 1941).
DIAGNOSIS: Reddish-purple iridescence on mantle.

b. ***Ducula chalconota smaragdina*** Mayr
Ducula chalconota smaragdina Mayr, 1931. *Mitt. Zool. Mus. Berlin* 17: 706.—Ogeramnang, Huon Penin.
RANGE: Central Ra and mts of Huon Penin (Mayr 1941).
DIAGNOSIS: Compared to the nominate form, this race is slightly larger, greener dorsally, and largely lacking the reddish-purple iridescence on mantle, back, and rump (Baptista *et al.* 1997).

c. ***Ducula chalconota*** undescribed form
RANGE: Upland forest of the Foja Mts (Beehler *et al.* 2012).
DIAGNOSIS: A distinct, undescribed population of this lineage inhabits the uplands of the Foja Mts. It should be treated either as a subspecies of *D. chalconota* or as an allospecies within this lineage (Beehler *et al.* 2012). Three photos of the new form taken by N. Kemp, the bird's discoverer, have been made avail-

able online and show the bird to have the following distinctive features: white eye-ring, whitish chin, gray throat, and pale breast band meeting the rufescent belly. It has not yet been described because no voucher specimen has been secured.

NOTES

Other names—Red-breasted Imperial Pigeon, Rufous-breasted or Mountain Red-breasted Fruitpigeon, Shining Imperial-pigeon.

Floury Imperial Pigeon *Ducula pistrinaria*   RESIDENT

Ducula pistrinaria Bonaparte, 1857. *Consp. Gen. Av.* 2: 34.—Saint George, Solomon Is.

A small-island specialist, with NG records from islands off the coast of the Sepik-Ramu and small islands in the SE Islands bird region. Frequents forest, edge, and clearings.

Extralimital—small islands and wooded coastal habitats of larger islands throughout the Bismarck and Solomon Is (Mayr & Diamond 2001, Dutson 2011).

SUBSPECIES

a. *Ducula pistrinaria rhodinolaema* (Sclater)

Carpophaga rhodinolaema Sclater, 1877. *Proc. Zool. Soc. London*: 555.—Admiralty Is.

RANGE: Small islands off the n coast of NG in Astrolabe Bay: Schouten, Manam, Karkar, and Madang islets (Mayr 1941). Range includes Crown, Long, Tolokiwa, Umboi, Malai; Admiralty Is; and St Matthias Is (Mussau, Emira; Dutson 2011).

DIAGNOSIS: Plumage of face and throat has a pinkish wash; upper mantle and nape greenish, rather than grayish as in *vanwyckii* of New Britain (Baptista *et al.* 1997). Wing 249 (Gibbs *et al.* 2001).

b. *Ducula pistrinaria postrema* Hartert

Ducula pistrinaria postrema Hartert, 1926. *Novit. Zool.* 33: 35.—Egum Atoll, east of D'Entrecasteaux Arch, SE Islands.

RANGE: Southeastern Islands: Deboyne Is[MA], Normanby I[X], Amphlett Is[MA], Misima I[MA], Woodlark group (Woodlark, Egum, Alcester)[MA], Bramble Haven[MA], and Tagula I[X] (Pratt field notes[X]). Also purportedly on small islands off the East Cape[MA]. More recent records include birds seen at Normanby, Rossel, and Reu I (Pratt field notes 2003, 2004) and numerous small islands in the Louisiade Arch (Dumbacher *et al.* 2010).

DIAGNOSIS: Very similar to *rhodinolaema* but apparently has a shorter wing. Wing 222–240 (Gibbs *et al.* 2001). Possibly the subspecies could be subsumed into a single clinal form (Gibbs *et al.* 2001).

NOTES

Hopkins (1988) reported observing a single bird of this species on the Karawari R of the Sepik on 22-Jun-1987. *Other names*—Grey or Island Imperial-pigeon, Grey Fruitpigeon.

Pinon's Imperial Pigeon *Ducula pinon*   ENDEMIC

Columba Pinon Quoy & Gaimard, 1824. *Voy. Uranie, Zool.*: 118, *Atlas*: pl. 28.—Rawak I, off Waigeo I, NW Islands.

All NG and most fringing islands; sea level–500 m, max 1,150 m at Baiyer R Sanctuary in 2013 (I. Woxvold *in litt.*), and 950 m at Crater Mt (Mack & Wright 1996). Frequents forest and edge.

SUBSPECIES

a. *Ducula pinon pinon* (Quoy & Gaimard)

  Synonym: *Carpophaga pinon* var. *rubiensis* A. B. Meyer, 1884. *Sitzungsber. Abh. Nat. Ges. Isis Dresden Abh.* 1: 51.—
  Rubi, Bird's Neck.

RANGE: The four main NW Islands, the Bird's Head, the Aru Is, and s NG east to Milne Bay, and on the n watershed from East Cape to the Kumusi R (Mayr 1941, Gyldenstolpe 1955b). Range includes the Trans-Fly (Mees 1982b).

DIAGNOSIS: Crown pale gray; white feathers fringe red eye skin; scapulars black; upper wing and uppertail coverts uniformly deep gray (Rand & Gilliard 1967).

b. *Ducula pinon jobiensis* (Schlegel)
Carpophaga pinon jobiensis Schlegel, 1871. *Ned. Tijdschr. Dierk.* (1873) 4: 26.—Yapen I, Bay Islands.
RANGE: Yapen I and n NG from the Mamberamo R to the Huon Gulf; also Manam, Bagabag, and Karkar I (Mayr 1941, Diamond & LeCroy 1979). Bulmer collected a specimen assigned to this form in the Baiyer R gorge of the e highlands of PNG (Diamond 1972).
DIAGNOSIS: Compared to the nominate form, the gray plumage is paler and the wing and uppertail coverts are black fringed silvery gray, with a scaly effect (Baptista *et al.* 1997). See photo in Coates & Peckover (2001: 65, upper-left image).

c. *Ducula pinon salvadorii* (Tristram)
Carpophaga salvadorii Tristram, 1882. *Proc. Zool. Soc. London* (1881): 996.—Misima I, Louisiade Arch, SE Islands.
RANGE: Main islands of the D'Entrecasteaux Arch[K]; also Misima, Tagula, and Rossel I (Mayr 1941, Pratt specimen from Normanby I at BPBM[K]).
DIAGNOSIS: Entire head pale pinkish, becoming deeper on upper mantle and back, which are entirely pink except for the black scapulars; underparts are grayer, but still brighter and paler than on other forms, and sharply demarcated from the chestnut of lower breast (Baptista *et al.* 1997); lacks white eye-ring (Rand & Gilliard 1967). This distinctive form was recently raised to species status by del Hoyo & Collar (2014), as the Louisiade Imperial-pigeon.

NOTES
The form *rubiensis* is an intergradient between *jobiensis* and the nominate form and thus not recognized here (Schodde 2006, Gibbs *et al.* 2001, Dickinson & Remsen 2013). *Other names*—Black-shouldered Fruit-pigeon, Pinon or Bare-eyed Imperial-pigeon.

Collared Imperial Pigeon *Ducula mullerii*   NEAR-ENDEMIC | MONOTYPIC
Columba mullerii Temminck, 1835. *Planch. Col. d'Ois.*, livr. 96: pl. 566.—Dourga R, Princess Marianne Strait, Trans-Fly. [No umlaut in *mullerii* in type description.]
> Synonym: *Carpophaga mülleri aurantia* A. B. Meyer, 1893. *Abh. Ber. Zool. Mus. Dresden* (1892–93) 4(3): 25.—
> Geelvink Bay and Astrolabe Bay.
Frequents forest bordering rivers, swamp forest, and mangrove forest of the NW Lowlands, Aru Is, S Lowlands, Trans-Fly, Sepik-Ramu, and s watershed of the SE Peninsula to Port Moresby; lowlands to 200 m (Rand 1942a, Mayr 1941, Bangs & Peters 1926, Rand 1942a, 1942b, Gilliard & LeCroy 1966, Bell 1967a, 1969, Hoek Ostende *et al.* 1997, Mees 1982b). Absent from the Bird's Head.
> Extralimital—vagrant to Saibai[J] and Boigu I, Australia (B. J. Coates *in litt.*[J]).

NOTES
We follow Dickinson & Remsen (2013) in orthography of species epithet *mullerii*. *Other names*—Black-collared or Pink-capped Fruit Pigeon, Müller's Imperial Pigeon, *Ducula muellerii*.

Zoe's Imperial Pigeon *Ducula zoeae*   ENDEMIC | MONOTYPIC
Columba Zoeæ Desmarest, 1826. *Dict. Sci. Nat.*, éd. Levrault 40: 314.—Manokwari, Bird's Head.
> Frequents lowland forest throughout NG, Salawati I, Aru Is, Yapen, Karkar[K], Goodenough and Fergusson[CO], Normanby[X], and Basilaki I (Mayr 1941, Diamond & LeCroy 1979[K], Pratt field notes[X]); locally up to 1,500 m (Coates & Peckover 2001), but up to 2,520 m on Mt Karimui (Freeman & Freeman 2014). The Misima I record in the literature is apparently an error, based on a bird from Basilaki I (LeCroy & Peckover 1998).
> *Other names*—Black-belted or Bar-breasted Fruitpigeon, Banded Imperial Pigeon, Zoe Imperial-pigeon.

Pied Imperial Pigeon *Ducula bicolor* RESIDENT | MONOTYPIC

Columba bicolor Scopoli, 1786. *Delic. Flor. Faun. Insubr.* fasc. 2: 94.—New Guinea.
> Synonym: *Carpophaga (Myristicivora) melanura* G. R. Gray, 1861. *Proc. Zool. Soc. London* (1860): 361.—Bacan I, n Moluccas.

Inhabits coastal forest, mangroves, and coconut plantations of small fringing islands through the NW Islands of Salawati[MA], Misool[MA], and Schildpads[J]; islands near Manokwari[MA]; also possibly Aru Is, where the following species occurs (Ripley 1964a[J]). Range includes a sand islet 2 km north of Batanta I (Beehler field notes 12-Dec-2011); and Mios Kon I, near Kri I (Beehler field notes 14-Dec-2011). Record from Aru Is (Mayr 1941) appears to have been based on a specimen of suspect provenience. Reported to be observed in flight off the coast of the Bomberai Penin near the s Kumawa Mts (C. Thébaud & B. Mila *in litt.*)

Extralimital—mainly on small islands[MA], from the Andaman and Nicobar Is, coastal regions of Cambodia, Thailand, Myanmar, Sumatra, and c and s Philippines east to the Moluccas (Mayr 1941).

NOTES
This species formerly included *D. spilorrhoa* as suggested by Johnstone (1981). We follow Baptista *et al.* (1997), Mayr & Diamond 2001, Dickinson & Remsen (2013), and del Hoyo & Collar (2014) in treating *bicolor* and *spilorrhoa* as specifically distinct. We do not recognize the race *melanura*. Baptista *et al.* (1997) noted that the distribution of the diagnostic plumage characters of *melanura* is apparently the product of nongeographic polymorphism. In addition, the range of *melanura* is entirely encompassed by the very widespread nominate form, which occurs west, north, east, and south of the supposed range of *melanura*—biogeographically improbable for subspecies. We are unable to find records of this species for the mainland Bird's Head (see maps in Goodwin 1983, Gibbs *et al.* 2001). *Other names*—White Nutmeg Imperial Pigeon, Nutmeg Pigeon, White Fruit-pigeon.

Torresian Imperial Pigeon *Ducula spilorrhoa* RESIDENT AND VISITOR | MONOTYPIC

Carpophaga spilorrhoa G. R. Gray, 1858. *Proc. Zool. Soc. London*, 186, 196.—Aru Is.
> Synonym: *Myristicivora bicolor melvillensis* Mathews, 1912. *Austral. Av. Record* 1: 27.—Melville I, NT, Australia.
> Synonym: *Ducula spilorrhoa tarara* Rand, 1941. *Amer. Mus. Novit.* 1102: 5.—Tarara, Wassi Kussa R, Trans-Fly, sc NG.

Small islands and much of coastal NG east of the Bird's Head; also many of the SE Islands, as well as the Aru Is; penetrates the Mainland interior along larger rivers (*e.g.*, Lake Murray; B. J. Coates *in litt.*); sea level–300 m (Pratt & Beehler 2014). Frequents mangroves, mangrove forest, savanna, gallery forest, and rain-forest edge; also low scrub on small islands and islets (Coates 1985). Range includes islands in Geelvink Bay (Biak, Mios Num, Yapen); islands off the n coast (Kumamba, Yamna, Tarawai, Kairiru, Schouten, Manam, Karkar, and Bagabag I); many islands in Milne Bay Prov, including Lusancay and Trobriand Is, Woodlark, Goodenough, Fergusson, and Normanby I; Amphlett Is, Bentley, Misima, Tagula, and Rossel I (Mayr 1941, Diamond & LeCroy 1979, Coates 1985).

Extralimital—n and e Australia, including islands of Torres Strait, with some Australian breeding birds wintering to NG; one specimen from Umboi, and a possible record from New Britain (Dutson 2011).

NOTES
Baptista *et al.* (1997) lumped *melvillensis* and *tarara* into this form. We consider *spilorrhoa* to be monotypic, and follow Dutson (2011) in treating *subflavescens* as specifically distinct (*pace* Mayr & Diamond 2001). *D. spilorrhoa* is distinguished from *subflavescens* by its lack of creamy-white plumage (Baptista *et al.* 1997). See photo in Coates & Peckover (2001: 67). Note that some *spilorrhoa* seen in the SE Islands looked creamy white (Pratt field notes). Christidis & Boles (2008) highlighted the complexity of issues related to the various populations in this species group (*bicolor*, *spilorrhoa*, *luctuosa*, *melanura*, *constans*, and *subflavescens*), which clearly merits a reanalysis using molecular methods. *Other names*—Torres Strait Imperial Pigeon, Eastern Nutmeg Pigeon.

Gymnophaps Salvadori, 1874. *Ann. Mus. Civ. Genova* 6: 86. Type, by original designation, *Gymnophaps albertisii* Salvadori.

The species of this genus range from the Moluccas (Bacan, Buru, Seram) to all of New Guinea, the Bismarck Archipelago, and the Solomon Islands (Baptista *et al.* 1997). Members of the genus are medium-large, flocking, *Columba*-like forest pigeons, characterized by a substantial patch of bare red orbital skin and slaty mantle and wings with a scaly pattern. Its systematic position within the family requires further study (Baptista *et al.* 1997).

Papuan Mountain-Pigeon *Gymnophaps albertisii*   RESIDENT
Gymnophaps albertisii Salvadori, 1874. *Ann. Mus. Civ. Genova* 6: 86.—Andai, Bird's Head.

Virtually all NG mts, Yapen I, Fergusson and Goodenough I, sea level–timberline. Hill and montane forest, most common from 800 m to 2,500 m, less common up to 3,700 m and down to sea level. Frequents forest; sometimes visits disturbed or partially cleared areas. Occurs mostly in hills and mts but is commonly seen in flocks flying over lowland forest in the vicinity of hills. One was found dead on glacial ice at 4,450 m on the Carstensz Massif (Schodde *et al.* 1975). No records yet from the Kumawa Mts.

Extralimital—range extends west to Bacan I and east to New Britain and New Ireland (Mayr & Diamond 2001, Dutson 2011).

SUBSPECIES

a. *Gymnophaps albertisii albertisii* Salvadori
RANGE: Mountains of NG, Fergusson[K], Goodenough[MA], Yapen I[MA], and the Bismarck Is (Pratt field notes 2003[K]).
DIAGNOSIS: Distinguished from *exsul* of Bacan by smaller size, paler plumage, and maroon feathering below eye (Baptista *et al.* 1997). Wing 193–219 (Gibbs *et al.* 2001).

NOTES
Other names—Mountain Pigeon; Papuan, Bare-eyed, or D'Albertis Mountain-pigeon.

ORDER PHAETHONTIFORMES

This is a distinct lineage, not related to the pelicans, where it was traditionally placed, but instead part of a clade that includes the pigeons, mesites, and sandgrouse (Hackett *et mult.* 2008). The tropicbirds occupy their own superorder and have no close relatives (Cracraft 2013).

Family Phaethontidae

Tropicbirds   [Family 3 spp/Region 2 spp]

The species of the family range through tropical seas around the world but are largely absent from much of the eastern Atlantic as well as the eastern Pacific (except near Baja California and the Galapagos). This is a small sea-loving group of medium-size white birds that dive for fish and crustaceans found in salt water. The species exhibit long, pointed wings, a long streamer tail, a substantial pointed beak, and webbed feet. Truly oceanic, they never approach land except when nesting or storm-blown. The species nest in remote colonies on islands (Orta 1992a). Relationships for the family are as for the order.

Phaëthon Linnaeus, 1758. *Syst. Nat.*, ed. 10, 1: 134. Type, by subsequent designation (G. R. Gray, 1840. *List. Gen. Birds*: 80), *Phaethon aethereus* Linnaeus.

The geographic range and generic characters are as for the family.

Red-tailed Tropicbird *Phaethon rubricauda*   VAGRANT | MONOTYPIC

Phaeton rubricauda Boddaert, 1783. *Tabl. Planch. Enlum.*: 57; based on "Paille-en queue de l'Isle de France" of Daubenton: pl. 979.—Mauritius.

Rare in NG waters, with records from the seas south of Port Moresby (Coates 1985). First recorded in the NG region by Tubb (1945), observed in the waters off the s coast of the SE Peninsula.

Extralimital—found through the tropical Pacific and Indian Oceans. Breeds on islands in the tropical Indian Ocean and c Pacific, west to Micronesia and New Caledonia, Vanuatu, and islands of the Great Barrier Reef (*e.g.*, Raine I); sight records from New Britain and the Solomons (Dutson 2011).

NOTES

Orta (1992a) concluded the species should be treated as monotypic. Specimens lacking for Region. W. Goulding (*in litt.* 2015) reported observing an individual collecting nesting material in the forest on Sudest I, indicating the species may breed in the Region.

White-tailed Tropicbird *Phaethon lepturus*   VISITOR | MONOTYPIC

Phaëton lepturus Daudin, 1802. In Buffon, *Hist. Nat.* [ed. Didot], *Quadr.* 14: 319.—Mauritius.

Synonym: *Phaethon lepturus dorotheae* Mathews, 1913. *Austral Avian Rec.* 2: 7.—Near Cairns, QLD, Australia.

An uncommon pelagic visitor to the Region, with records north, south, and east of the Mainland. Records come from the Gulf of Papua[CO], the Trobriand Is[CO], and Woodlark I (Meek collection), as well as islands in Australian territory just south of the Fly R delta[CO]. Records cover different times of the year; nearest recorded nesting is from Tench I off New Ireland (Coates 1985).

Extralimital—Caribbean, Indian Ocean, tropical Atlantic, and w and c tropical Pacific. There is no evidence that it breeds in the Torres Strait Is.

NOTES

Historically, subspecific designation has apparently been based on range combined with passive acceptance of named forms. We follow the inference in Dickinson & Remsen (2013) that this can be properly treated as a monotypic form with color morphs. This treatment has been used by Dutson (2011). *Other names*— Yellow-billed or Long-tailed Tropicbird.

ORDER PROCELLARIIFORMES

The most current higher-level study (Hackett *et mult.* 2008) postulated that this order is monophyletic, sister to the penguins, and that the tubenoses+penguins clade is sister to the pelicans, storks, herons, frigatebirds, boobies, cormorants, and anhingas. Currently, there are no records for the albatrosses (Diomedeidae) from the Region, but the Black-footed Albatross (*Phoebastria nigripes*) has been observed to the north and should be looked for in the Region's waters (see Cheshire 2010: 13).

Family Oceanitidae

Southern Storm-Petrels [Family 8 spp/Region 3 spp]

The species of the family breed mainly in the Southern Hemisphere, but some forms range throughout the world's seas as nonbreeders (Carboneras 1992b, Onley & Scofield 2007, Christidis & Boles 2008). Southern storm-petrels are rare vagrants to New Guinea's pelagic zone. These are small to very small seabirds, with relatively short and somewhat rounded wings, squared-off tails, and long legs that project beyond the tail in flight.

Traditionally, all storm-petrels have been classified in a single family, but recently they have been split into two—the southern and the northern storm-petrels (Nunn & Stanley 1998). The two families differ structurally and are not sister groups. Hackett *et mult.* (2008) showed this lineage to be sister to a clade that includes the diving-petrels (Pelecanoididae), albatrosses, shearwaters, and northern storm-petrels.

GENUS *OCEANITES* [Genus 2 spp/Region 1 sp]

Oceanites Keyserling & Blasius, 1840. *Wirbelthiere Europa's* 1: xciii, 131, 238. Type, by subsequent designation (G. R. Gray, 1841. *List Gen. Birds* 2: 99), *Procellaria wilsonii* Bonaparte = *Procellaria oceanica* Kuhl.

The species of this genus breed in the Antarctic, subantarctic, and the eastern Pacific, and range through the oceans of the world. This small genus is characterized by small size, chocolate plumage, white rump, short wings, square tail, and long legs (Onley & Scofield 2007).

Wilson's Storm-Petrel *Oceanites oceanicus* VISITOR

Procellaria oceanica Kuhl, 1820. *Beitr. Zool. Vergl. Anat., Abth.* 1: 136, pl. 10, fig. 1.—South Georgia, s Atlantic Ocean.

Probably a regular passage migrant through waters south of NG. Cadée (1985) recorded more than 1,200 individuals between the s coast of the Bird's Neck and the Aru Is on 20-Aug-1984. Hoogerwerf (1964) observed the species off the s coast of w NG in September and November; E. Lindgren reported a loose group of five at close range within the reef at Bootless Inlet near Port Moresby on 12-Aug-1980 (Coates 1985); and Cheshire (2010) observed one in Milne Bay Prov *ca.* 50 km southeast of Samarai on 5-Oct-1985 and another *ca.* 40 km north of Saidor, Morobe Prov, on 14-Aug-2007. Warne (1989) reported a single individual 32 km east-southeast of the mouth of the Fly R on 25-Jul-1988. C. Thébaud (*in litt.*) observed the species off the s coast of the Bomberai Penin in Oct-2010.

Extralimital—breeds on islands of the far s oceans; migrates to the Northern Hemisphere (Onley & Scofield 2007). Fairly common during the austral winter in coastal waters of n Australia; also the Coral Sea (Coates 1985).

NOTES

Subspecies not determined. *Other names*—Yellow-webbed Storm-petrel.

GENUS *PELAGODROMA* MONOTYPIC

Pelagodroma Reichenbach, 1853. *Avium Syst. Nat.* (1852): iv. Type, by original designation, *Procellaria marina* Latham.

The single species of the genus breeds in the northern and southern Atlantic, southern verges of Australia, and New Zealand environs. The genus is characterized by the relatively large size, gray rump, and white eyebrow.

White-faced Storm-Petrel *Pelagodroma marina*   VAGRANT
Procellaria marina Latham, 1790. *Index Orn.* 2: 826; based on "Frigate Petrel" of Latham, 1785, *Gen. Synops. Birds* 3: 410.—Off mouth of the Río de la Plata, Argentina.

One record from the Solomon Sea, 125 km northeast of Cape Ward Hunt (Cheshire 2010).

Extralimital—breeds off w and se Australia and NZ, as well as in the Atlantic Ocean, off w Africa and at Tristan da Cunha. A widespread wanderer when not breeding (Onley & Scofield 2007).

NOTES
Subspecies not determined; probably *dulciae* or *maoriana*. *Other names*—Frigate Petrel.

GENUS *FREGETTA*   [Genus 3 spp/Region 1 sp]

Fregetta Bonaparte, 1855. *Compt. Rend. Acad. Sci. Paris*: 41: 1113. Type, by original designation, *Thalassidroma leucogaster* Gould.

The species of this genus range throughout the southern seas (Carboneras 1992b). The genus is characterized by the small size and the distinctive pied plumage pattern with the all-dark hood.

[White-bellied Storm-Petrel *Fregetta grallaria*   HYPOTHETICAL
Procellaria grallaria Vieillot, 1818. *Nouv. Dict. Hist. Nat.*, nouv. éd. 25: 418.—Australia.

Two sight records from the Region: an old record from Orangerie Bay^MA, Central Prov (Macgillivray^MA); and one circling a ship for an extended period 25 km south of Redscar Bay, Central Prov, on 12-Aug-1943 (Tubb 1945). The specific and subspecific determinations of these records are problematic. Not observed by Cheshire (2010) in his extensive oceanic surveys of the Region.

Extralimital—breeds on Lord Howe I and other islands in the s Pacific, Indian, and Atlantic Oceans; in winter typically remains well south of the tropics in the Australasian region (Marchant & Higgins 1990), unlike *F. tropica*. Reported as fairly plentiful in the Coral Sea during the austral winter by Coates (1985), although this record more likely refers instead to the following species.

Other names—Vieillot's Storm-petrel.]

Black-bellied Storm-Petrel *Fregetta tropica*   VISITOR
Thalassidroma tropica Gould, 1844. *Ann. Mag. Nat. Hist.* 13: 366.—Equatorial Atlantic, at 6.55°N, 16.1°W.

Uncommon migrant mainly to Coral Sea during austral winter and spring. Cheshire (2010) found the species widely distributed in small numbers in PNG waters of the n Coral Sea. The most northerly observation was of a bird in the Louisiade Arch on 8-Jul-1988 (11 km from Lunn I, north of the Conflict Is).

Extralimital—breeds on islands of the Southern Ocean, Tristan da Cunha, and possibly Gough I. Nonbreeding birds range widely through the oceans south of the equator; also waters off the s and e coast of Australia to Cape York (Marchant & Higgins 1990); and also to the Arabian Sea and Bay of Bengal (Onley & Scofield 2007).

NOTES
Subspecies not determined. *Other names*—Gould's Storm-petrel

Family Hydrobatidae

Northern Storm-Petrels [Family 16 spp/Region 1 sp]

The birds of this family are often combined with the southern storm-petrels (Carboneras 1992b). The work of Hackett *et mult.* (2008) indicated the two storm-petrel families are not sister groups. The Hydrobatidae have a longer forearm, resulting in a longer, narrower wing that is held with a noticeable bend at the elbow, whereas the Oceanitidae have a shorter, broader, straighter wing, are given to more fluttering flight, and their legs are longer and used for pattering on the surface of the water. This lineage is sister to a clade that includes the shearwaters and diving-petrels.

GENUS *OCEANODROMA* [Genus 15 spp/Region 1 sp]

Oceanodroma Reichenbach, 1853. *Avium Syst. Nat.* (1852): iv. Type, by original designation, *Procellaria furcata* Vieillot.

The species of the genus are long-winged and fork-tailed and either all-dark, blue-gray, or pied. They mainly breed in the Northern Hemisphere or the tropics and winter in the tropics and subtropics (Onley & Scofield 2007).

Additional forms that might be expected to occur in the Region include Swinhoe's Storm-Petrel (*Oceanodroma monorhis*) and Leach's Storm-Petrel (*Oceanodroma leucorhoa*).

Matsudaira's Storm-Petrel *Oceanodroma matsudairae* VAGRANT | MONOTYPIC

Oceanodroma melania matsudariae Nagamichi Kuroda, 1922. *Ibis* (11)4: 311.—Sagami Bay, Honshu, Japan.

Apparently a passage migrant north of the Mainland during austral winter. Cheshire (2010) reported observing three birds *ca.* 20 km north of the mouth of the Sepik on 19-Aug-1985 (apparently near Vokeo and Blupblup I). Photographed out of NG waters, off s New Ireland, in Aug 2007 by Shirihai (2008), who cited Bourne's (1998) mention of a sighting of two individuals thought to be this species on 6-Aug-1997, between Manus and the NG mainland. Hornbuckle & Merrill (2004) reported groups of three and four between Batanta and Salawati I on 12-Sep-2004. Matheve *et al.* (2009) reported one off Salawati I on 5-Aug-2009.

Extralimital—breeds on islands south of mainland Japan (Bonin and Iwo Is) and nonbreeding birds migrate south and west as far as the coast of e Africa. Photographed off Kimbe, w New Britain (D. Hobcroft *in litt.*).

Other names—Sooty Storm-petrel, *Halocyptena* or *Hydrobates matsudairae*.

Family Procellariidae

Petrels and Shearwaters

[Family 84 spp/Region 11 spp]

The species of the family forage across the oceans of the world and breed mainly on islands (Carboneras 1992a, Penhallurick & Wink 2004, Onley & Scofield 2007, Christidis & Boles 2008, Gill *et al.* 2010). At least one species breeds in the Region, whereas the remainder are seasonal visitors to the pelagic zone. These tube-nosed, web-footed seabirds are oceanic, nesting mainly on uninhabited islands free of predators. For decades, our sparse knowledge of this family in the Region reflected the near-absence of fieldwork conducted in the pelagic zone. Cheshire (2010, 2011), however, has laid a foundation to our knowledge at long last, gathering data during 20 research voyages between 1985 and 2007. Scant information exists for seabird colonies in the New Guinea region. There seem to be few, apparently because of depredations by local communities.

This family clusters with the diving-petrels, albatrosses, and storm-petrels (Hackett *et mult.* 2008). Our systematic treatment of the genera follows Penhallurick & Wink (2004) as interpreted by Christidis & Boles (2008), but see also Rheindt & Austin (2005) and Gill *et al.* (2010) for differing views. Systematics of the genera within the family remains contentious, and there is little agreement between major summary treatments. In spite of the new milestones reached by Cheshire's excellent work, we believe there remains abundant opportunity to add substantially to our knowledge of seabirds of the Region. This group might be considered the Region's final frontier, ornithologically.

GENUS *MACRONECTES*

[Genus 2 spp/Region 1 sp]

Macronectes Richmond, 1905. *Proc. Biol. Soc. Wash.* 18: 76. Replacement name for *Ossifraga* Hombron & Jacquinot, 1844, preoccupied. Type, by original designation, *Procellaria gigantea* Gmelin.

The species of this genus range through the cool Southern Ocean to the coastline of Antarctica. The genus includes the largest members of the family, and the species are very long-winged and large-billed (Carboneras 1992a). The Northern Giant Petrel (*Macronectes halli*) should be looked for in the Region as well.

Southern Giant Petrel *Macronectes giganteus*

VAGRANT | MONOTYPIC

Procellaria gigantea Gmelin, 1789. *Syst. Nat.* 1(2): 563.—Admiralty Bay, King George I, South Shetland Is.

One record from the Region: a live banded bird at Kamfo Village, near Kikori, Gulf Prov, S Lowlands, in Jul-1973 (Hudson & Conroy 1975); banded as a nestling on Bird I, South Georgia, in Mar-1973.

Extralimital—breeds on the coast of Antarctica, on subantarctic islands (South Georgia, Macquarie, Heard), and southernmost South America; disperses widely across the Southern Ocean, records to Vanuatu and New Caledonia (Dutson 2011).

NOTES

The form *halli* was once considered conspecific with *giganteus* but now is treated as specifically distinct (Carboneras 1992a). Confidence that the banded individual encountered in NG was indeed *M. giganteus* comes from the fact that the banders were working on both species, which both bred at South Georgia, and thus were expert on identification of birds in the hand (Hudson & Conroy 1975). Apparently the recovered individual was released alive. *Other names*—Antarctic Giant-Petrel, Giant Fulmar.

GENUS *PACHYPTILA* [Genus 6 spp/Region 1 sp]

Pachyptila Illiger, 1811. *Prodromus Syst. Mamm. Avium*: 274. Type, by subsequent designation (Selby, 1840. *Cat. Gen. Sub-gen. Types Class Aves*: 49), *Procellaria Forsteri* Latham = *Procellaria vittata* G. Forster.

The species of this genus range through the southern seas to Antarctica and only rarely into the subtropics. Members of the genus, fiendishly difficult to distinguish from one another, are characterized by small size, white underparts, a blunt tail with a terminal bar, blue-gray and black dorsal patterning, and a characteristic bill shape (Onley & Scofield 2007).

Slender-billed Prion *Pachyptila belcheri* VAGRANT

Heteroprion belcheri Mathews, 1912. *Birds Austral.* 2: 215 (text fig.), 224.—Geelong, Victoria, Australia.

Known from the Region from a specimen collected from Kiwai I, mouth of the Fly R, in Jul-1876 by Luigi D'Albertis. Reidentified from photos (see fig. 2a) of the specimen examined by several experts in the light of our present knowledge of the genus (Penhallurick & Wink 2004, Onley & Scofield 2007).

Extralimital—subantarctic seas, ranging north in austral winter. Breeds on various subantarctic islands and also temperate-zone islands of the southern seas. The population *belcheri* occasionally irrupts into the tropics; recent records from Java and Brazil (P. Scofield *in litt.*).

NOTES

P. belcheri has been treated as a full species by Onley & Scofield (2007) and Dickinson & Remsen (2013). Penhallurick & Wink (2004) consolidated the various prion forms into two species, *P. vittata* and *P. turtur*, based on their molecular systematic work using the mitochondrial gene cytochrome *b*. They embedded *belcheri* into their species *P. vittata*. Regarding that work, we point the reader to the critique of Rheindt & Austin (2005), who offered a differing viewpoint. The safest thing to say is that there is a single specimen record of a whalebird referable to the form *belcheri* from the NG region. *Other names*—the specimen was originally identified as a Fairy Prion, *Pachyptila turtur*.

GENUS *PTERODROMA* [Genus 30 spp/Region 2 spp]

Pterodroma Bonaparte, 1856. *Compt. Rend. Acad. Sci. Paris* 42: 768. Type, by subsequent designation (Coues, 1866. *Proc. Acad. Nat. Sci. Phila.*: 137), *Procellaria macroptera* A. Smith.

The species of this genus range through the oceans of the world (Onley & Scofield 2007). The group is so difficult and fieldwork so fragmentary from New Guinea's waters that little definitive can be said about its occurrence in the Region (but see Cheshire 2010). The species of the genus continue to be split and re-split, making identification of old records difficult. Species are medium-size, compact and lightly built, and mainly dark above and paler below (though some are all dark); note the short bill, small head, and narrow tail.

The following species should be looked for in the Region: Providence Petrel (*Pterodroma solandri*; one record from the Solomons "off Buka" from Shirihai 2008, and one from New Caledonia from Dutson 2011); Kermadec Petrel (*Pterodroma neglecta*; a few records from the Bismarck Sea in Papua New Guinea waters but not near the coast of New Guinea); Cook's Petrel (*Pterodroma cooki*); and Phoenix Petrel (*Pterodroma alba*).

Herald Petrel *Pterodroma heraldica* VAGRANT | MONOTYPIC

Oestrelata heraldica Salvin, 1888. *Ibis*: 357.—Chesterfield Reefs, w Pacific.

A few sight records in the Region from the Coral Sea and Gulf of Papua. Cheshire (2010) documented a sight record from *ca.* 75 km south-southwest of Kerema on 12-Aug-1990 (8.75°S, 145.58°E) and another from *ca.* 40 km south of Abau on 4-Oct-1985 (10.65°S, 149.30°E). These were pale morph individuals.

Extralimital—Australia east to Easter I. Breeds on Raine I, Hunter I, Chesterfield Reefs, Samoa, the Marquesas, and the Pitcairn Is (Onley & Scofield 2007, Dutson 2011). The Raine I breeding site is in the Great Barrier Reef, *ca.* 300 km south-southeast of the Trans-Fly and Daru I, PNG.

NOTES

Jouanin & Mougin (1979) treated this form as a subspecies of *P. arminjoniana*, a treatment followed by Carboneras (1992a). We follow here the species-level treatment of Christidis & Boles (2008). *Other names*—formerly considered a race of Trinidade Petrel, *Pterodroma arminjoniana*.

[Collared Petrel *Pterodroma brevipes*   HYPOTHETICAL | MONOTYPIC

Procellaria brevipes Peale, 1848. *U.S. Explor. Exped.* 8: 294.—Samoa Is.

One sighting near the Budibudi Is in the Woodlark group, within Milne Bay Prov (Hobcroft & Bishop ms).

Extralimital—a poorly known species; breeds in Fiji, Vanuatu, and perhaps elsewhere in Polynesia. Records from the Solomons to New Caledonia (Dutson 2011).

Other names—formerly considered a race of Gould's Petrel, *Pterodroma leucoptera*.]

Pycroft's Petrel *Pterodroma pycrofti*   VAGRANT | MONOTYPIC

Pterodroma pycrofti Falla, 1933. *Rec. Auckland Inst. Mus.* 1: 176.—Taranga, Hen I, NZ.

A single record of an adult with a numbered leg band recovered alive at Lelehudi, East Cape, Milne Bay Prov, SE Peninsula, on 5-May-2005 (Pierce 2009).

Extralimital—breeds on small islands off the North I of NZ. Nonbreeding range unknown, but there are sight records from the n Pacific (Onley & Scofield 2007).

Other names—formerly treated as a race of Stejneger's Petrel, *Pterodroma longirostris*.

GENUS *PSEUDOBULWERIA*   [Genus 4 spp/Region 1 sp]

Pseudobulweria Mathews, 1936. *Ibis* (13)6: 309. Type, by original designation, *Thalassidroma* (*Bulweria*) *macgillivrayi* G. R. Gray.

The species of this genus range through the tropical Indian and Pacific Oceans, but their movements are little known. They are medium-size, long-winged petrels, with a robust black bill, slim body, and a tapering tail. *Pseudobulweria becki* is known from the Bismarcks and should be looked for in the Region.

Tahiti Petrel *Pseudobulweria rostrata*   VISITOR

Procellaria rostrata Peale, 1848. *U.S. Explor. Exped.* 8: 296.—Mountains, at 1,830 m, Tahiti.

In NG waters in small numbers between January and November, where found in the Coral, Solomon, and Bismarck Seas and equatorial Pacific regions of PNG (Cheshire 2010). Bailey (1992b) reported seeing 18 individuals far offshore of Madang on pelagic boat trips in late October and early November. Hicks (1988a) reported six from a boat 24 km off Port Moresby on 5-Apr-1987. Coates (1985: 63) photographed an individual at Port Moresby in May-1970. Wardill & Nando (2000) reported a single bird off the n coast of the Bird's Head on 30-Dec-1999. Cadée (1989) reported the species in low numbers southwest of the s coast of the Bird's Neck. Richards & Rowland (1995) reported 250 in late Nov-1992 off Wasu, Huon Penin, but without supporting description. Hobcroft & Bishop (ms) reported two off the Budibudi Is (in the SE Islands).

Extralimital—breeds in New Caledonia, Fiji, Society, Gambier, and Marquesas Is (Onley & Scofield 2007). Nonbreeders apparently range through the tropical s Pacific.

SUBSPECIES

Subspecies not determined. Expected subspecies, based on geography, is *trouessarti*, which breeds on New Caledonia, the closest breeding area to NG.

NOTES

Confusion with *P. becki* renders earlier records problematic (*e.g.*, Coates 1985). Ripley (in Mayr & Meyer de Schauensee 1939a) reported *P. rostrata* from off Biak I, but the birds could have been *P. becki*. Bourne & Dixon (1971) noted that birds observed by D. M. Simpson off Wuvulu I on 12-Apr-1969 may have been *P. becki*. Apparently *becki* nests high in montane forest atop s New Ireland (Dutson 2011), although two field trips to these high forests by Beehler (in 1976 and 1992) produced no evidence of these breeders (see Beehler 1978d, Beehler & Alonso 2001). Originally collected east of New Ireland (by Rollo Beck) and northwest of Rendova I, Solomon Is (Jouanin & Mougin 1979, Shirihai 2008).

Carboneras (1992a) treated the population *becki* as a subspecies of *rostrata*; however, Onley & Scofield (2007) treated it as a full species. Shirihai (2008) provided evidence that *becki* should be treated as a full species. Field identification of *becki* is problematic, as indicated by the discussion of Shirihai (2008), and should be made with extreme caution. Photographs or specimens are perhaps required. Carboneras (1992a) placed *rostrata* in the genus *Pterodroma*. *Other names—Pterodroma rostrata*.

GENUS *ARDENNA* [Genus 7 spp/Region 3 spp]

Ardenna Reichenbach, 1853. Type, by original designation and monotypy, *Procellaria gravis* O'Reilly, 1818.

The species of this genus range through the seas of the world from arctic to antarctic reaches (Carboneras 1992a). They are medium-size to large procellariids, medium- to long-winged, and mainly white below and dark brown dorsally, though some forms are entirely dark. They are rather similar in plumage to *Puffinus* but are mainly larger. This lineage has been shown to be distinct from *Puffinus* and *Calonectris* by Austin *et al.* (2004) and Penhallurick & Wink (2004). This treatment was followed by Christidis & Boles (2008), Pyle *et al.* (2011), Dickinson & Remsen (2013), and del Hoyo & Collar (2014).

Wedge-tailed Shearwater *Ardenna pacifica* RESIDENT AND VISITOR | MONOTYPIC
Procellaria pacifica Gmelin, 1789. *Syst. Nat.* 1(2): 560.—Kermadec Is.
   Synonym: *Puffinus pacificus chlororhynchus* Lesson, 1831. *Traité Orn.* 8: 613.—Shark Bay, Western Australia.
Breeds in the Region (Bavo I, Port Moresby) and probably winters in the Region as well. Recorded in all months off the s coast, from the mouth of the Katau R^{MA} (west of Daru) to Milne Bay Prov, and off the n coast from Wewak to the east of New Ireland (Coates 1985). Sometimes seen in congregations of hundreds off the s coast. Pale morph is rare in the south but more numerous in the north. Bailey (1992b) reported this species as common offshore from Madang, recording 461 individuals on seven boat trips. Cheshire (2010) reported *ca.* 1,000 birds south of Karkar I on 20-Aug-1985. Cheshire posited that there might be a major migration of birds that breed in NSW, Australia, passing northward through Milne Bay and then westward toward the Philippines through the waters north of NG.

Extralimital—breeds from the Indian Ocean to the c Pacific in various localities; ranges across the tropical and subtropical Pacific and Indian Oceans (Jouanin & Mougin 1979, Carboneras 1992a).

NOTES
We follow Christidis & Boles (2008) and Dickinson & Remsen (2013) in placing this species in *Ardenna*. We follow Carboneras (1992a) in treating the species as monotypic. *Other names—Puffinus pacificus*.

[Sooty Shearwater *Ardenna grisea* HYPOTHETICAL | MONOTYPIC
Procellaria grisea Gmelin, 1789. *Syst. Nat.* 1(2): 564.—New Zealand.
Possibly a rare passage migrant in NG waters during its migration in the austral winter from NZ and e Australia to the n Pacific, but presumably most birds pass far to the east of the Region (Marchant & Higgins 1990). Reported in the Region by Greensmith (1975), but Coates (1985) noted these birds might have been *Puffinus heinrothi*—another dark shearwater. *A. grisea* has been reported several times in the area of the Bismarck Sea west to the Ninigo Is and east to the Solomon Is (Dutson 2011).

Extralimital—breeds in Australia, NZ, and South America, migrating to the n Pacific in May–June and returning in September–October.

NOTES
We follow Christidis & Boles (2008) and Dickinson & Remsen (2013) in placing this form into *Ardenna*. *Other names*—Sombre Shearwater, King Muttonbird, *Puffinus griseus*.]

Short-tailed Shearwater *Ardenna tenuirostris* VISITOR | MONOTYPIC

Procellaria tenuirostris Temminck, 1836. *Planch. Col. d'Ois.*, livr. 99: text to pl. 587.—Seas north of Japan and shores of Korea.

Probably a regular passage migrant through the waters of the SE Islands (Cheshire 2010). There have been sight records, a beach-washed specimen (H. L. Bell *in litt.*), and a live-recovered bird (photos examined by the authors) in 2013 in the interior at Oksibil in the Border Ra (H. van de Kerkhof *in litt.*). Numbers of shearwaters considered to be this species were observed passing Cape Suckling, Central Prov, in a west-to-east direction on 7-May-1983 during a period of strong southeasterly winds (Finch 1983e). Many observed migrating beyond the reef off Port Moresby (Hicks 1988a). The quantitative observations of Cheshire (2010) of major movements of birds north and east of the PNG mainland indicate this species probably is an annual migrant through Regional waters.

Extralimital—breeds in s Australia, migrating to the n Pacific in austral winter.

NOTES

We follow Christidis & Boles (2008) and Dickinson & Remsen (2013) in placing this species in *Ardenna*. *Other names*—Bonaparte's or Slender-billed Shearwater, *Puffinus tenuirostris*.

Flesh-footed Shearwater *Ardenna carneipes* VISITOR | MONOTYPIC

Puffinus carneipes Gould, 1844. *Ann. Mag. Nat. Hist.* 13: 365.—Seas bounding w coast of Australia; breeding on small islands off Cape Leeuwin, Western Australia.

Probably a rare but regular passage migrant in NG waters. Coates (1970a) observed birds in the Region fitting the description of this species in a large feeding flock off Port Moresby on 31-Jan-1970; in Oct-1981, one was seen between Madang and Karkar I (Finch 1981e, 1981g). Cheshire (2010) observed a single bird between East Cape and Normanby I on 14-Jan-1988.

Extralimital—breeds in the s Indian and Pacific Oceans, migrating north after nesting. Breeds on Lord Howe I and islands off sw Australia and NZ; migrates or disperses in austral winter to the n Indian and Pacific Oceans.

NOTES

Some authors considered *A. carneipes* conspecific with *A. creatopus*. We follow Christidis & Boles (2008) and Dickinson & Remsen (2013) in placing this species into the genus *Ardenna*. *Other names*—Pale-footed, Fleshy-footed, or Pink-footed Shearwater, *Puffinus carneipes*.

GENUS *CALONECTRIS* [Genus 4 spp/Region 1 sp]

Calonectris Mathews & Iredale, 1915. *Ibis* (10)3: 590, 592. Type, by original designation, *Procellaria leucomelas* Temminck.

The species of this genus range through the western Pacific, much of the Atlantic, and the southwestern Indian Ocean. This small genus is superficially similar to *Procellaria* but exhibits white linings of the undersurface of the wings.

Streaked Shearwater *Calonectris leucomelas* VISITOR | MONOTYPIC

Procellaria leucomelas Temminck, 1836. *Planch. Col. d'Ois.*, livr. 99: pl. 587.—Seas of Japan and Nagasaki Bay.

A regular and common northern migrant to the Region during the austral summer (mainly October to March; see Yamamoto *et al.* 2010). Seasonally abundant off n NG, the Bismarck Arch, and Bougainville I, extending south and west into the n Coral and Arafura Seas (Coates 1985, Cheshire 2010, Yamamoto *et al.* 2010). K. D. Bishop (*in litt.*) observed small numbers in Jan–Feb-1986 in the NW Islands. Wardill & Nando (2000) reported a westerly movement of 16,000 birds off the n coast of the Bird's Head on 30-Dec-1999. Bailey (1992b) reported the species to be uncommon offshore from Madang, recording 28 individuals in four boat trips. It was seen in numbers in the Banda Sea off the s coast of the Bomberai Penin in Nov-2014 (C. Thébaud & B. Mila *in litt.*).

Extralimital—breeds on islands around Japan, Korea, and Taiwan; winters south to e Australia (Carboneras 1992a, Coates 1985).

Other names—White-faced or Streak-headed Shearwater, *Puffinus leucomelas.*

GENUS *PUFFINUS* [Genus 12 spp/Region 1 sp]

Puffinus Brisson, 1760. *Orn.* 1: 56; and 6: 129. Type, by tautonymy, "Puffinus" Brisson = *Procellaria puffinus* Brünnich.

The species of this genus range through the seas of the world from subarctic to subantarctic reaches. They are small procellariids, mainly white below and blackish dorsally, medium- to long-winged. One form is mainly dark. We split the genus from *Ardenna*, following Austin *et al.* (2004), Penhallurick & Wink (2004), Christidis & Boles (2008), and Pyle *et al.* (2011).

[**Christmas Shearwater** *Puffinus nativitatis*   HYPOTHETICAL | MONOTYPIC

Puffinus (Nectris) nativitatis Streets, 1877. *Bull. U.S. Nat. Mus.* 7: 29.—Kiritimati (Christmas I), Pacific Ocean.

Possible vagrant to the Region. Birds observed in 1978 by Finch (1981d) and Coates (1985) attributed to this species have since been discounted by Cheshire (2010), and B. J. Coates (*in litt.*) has withdrawn his observation. These were more likely *P. heinrothi.* Shirihai (2008) observed *P. nativitatis* on 1-Aug-2007 off the Feni Is, north of NG waters in the Bismarcks. In addition, Cheshire (2010) found *P. nativitatis* north of the Bismarck Is, outside the NG region.

Extralimital—breeds on tropical Pacific islands to the north and east of the Region; ranges throughout the tropical Pacific.

Other names—Black or Christmas Island Shearwater]

[**Hutton's Shearwater** *Puffinus huttoni*   HYPOTHETICAL | MONOTYPIC

Puffinus reinholdi huttoni Mathews, 1912. *Birds Austral.* 2: 47 (key), 77.—"Snares Island," possibly in error (see Gill *et al.* 2010; but Tennyson *et al.* 2014 provided details that indicate Snares I, NZ, may be the correct type locality).

On 7-May-1983, during a period of strong se winds, numbers of small shearwaters identified as *Puffinus huttoni* were seen flying in a west-to-east direction off Cape Suckling, Central Prov, along with *Ardenna pacifica* and small numbers of *A. tenuirostris* (Finch 1983e). Might be expected in the Coral Sea and Torres Strait off the s shores of PNG.

Extralimital—breeds in the mts of the n section of the South I of NZ; nonbreeding range restricted to the waters off the n and e coasts of NZ and the w, s, and e coasts of Australia (Onley & Scofield 2007). It has been collected in the sw Torres Strait in April and May and is expected to occur in NG waters (Draffan *et al.* 1983).]

[**Tropical Shearwater** *Puffinus bailloni*   HYPOTHETICAL

Procellaria nugax a. bailloni Bonaparte, 1857. *Consp. Gen. Avium* 2: 205.—Mauritius.

Expected in the ne sector of the Region. Cheshire (2010) recorded two birds on 7-Oct-1985, *ca.* 85 km east-southeast of Pocklington Reef, in Solomon Is territory.

Extralimital—breeds from the Maldives to Mascarene Is of the tropical Indian Ocean and the n, w, and c Pacific (Onley & Scofield 2007). The form *P. bailloni dichrous* is to be expected in the Region, as it breeds in various sites in the Pacific: Palau, Tahiti, Nauru, Samoa, Line Is, Marquesas, Gambier Is, and Vanuatu; extinct in the Mariana Is (Onley & Scofield 2007).

Other names—split from Audubon's Shearwater, *P. lherminieri.*]

Heinroth's Shearwater *Puffinus heinrothi*   VAGRANT | MONOTYPIC
Pufinns heinrothi Reichenow, 1919. *Journ. Orn.* 67: 225.—Blanche Bay, New Britain.

Cheshire (2010) noted this to be a rare species off the n coast of the PNG mainland. Bailey (1992b) recorded seven birds at sea on three boat trips made off Madang in Sep–Oct-1989 and provided a detailed description of the sightings. Cheshire (2010) encountered individuals in the Bismarck and Solomon Seas outside the Region. Probably breeds in the mts of Bougainville I (Hadden 1981). Coates (1985) surmised it seems possible that at least some of the "Sooty Shearwaters" seen by Greensmith (1975) in 1974–75 were *P. heinrothi*. Greensmith's records included one bird south of Karkar I in October.

Extralimital—A restricted-range species known only from the waters around the Bismarck Arch and w Solomon Is (Onley & Scofield 2007, Dutson 2011).

Other names—some authorities treat this form as conspecific with *P. lherminieri*.

GENUS *BULWERIA*   [Genus 2 spp/Region 1 sp]

Bulweria Bonaparte, 1843. *Nuov. Ann. Sci. Nat. R. Accad. Sci. Instit. Bologna* (1842) 8: 426. Type, by monotypy and virtual tautonymy, *Procellaria bulwerii* Jardine & Selby.

One species has been recorded from the Region as an uncommon visitor. The species of the genus range through the tropical and subtropical oceans of the world. They are long-winged and long-bodied, dark birds, with a prominent dark bill and a long, tapering tail (Onley & Scofield 2007).

Bulwer's Petrel *Bulweria bulwerii*   VISITOR | MONOTYPIC
Procellaria Bulwerii Jardine & Selby, 1828. *Illus. Orn.* 2(4): pl. 65 and text.—Madeira.

Two individuals were observed *ca.* 100 km south-southwest of Kerema, in the Gulf of Papua, on 12-Jul-1990 (Cheshire 2010); 14 individuals were sighted between Sorong and Waigeo I by Argeloo & Dekker (1996) on 9 Oct-1993.

Extralimital—breeds in the Atlantic and Pacific, including Hawai`i, the Marquesas, Phoenix Is, and Johnston Is. Now seen regularly off nw Australia between September and April (Cheshire 2010). Probably migrates from its breeding grounds in the w Pacific to nw Australia by passing through the NW Islands, and thus should be looked for there (Onley & Scofield 2007).

ORDER CICONIIFORMES

We follow Cracraft (2013) in considering the Ciconiiformes to comprise the huge assemblage of waterbirds that includes storks, pelicans, herons, ibises, cormorants, gannets, anhingas, and frigatebirds. This is based on the genomic work of Hackett *et mult.* (2008).

Family Ciconiidae

Storks   [Family 19 spp/Region 1 sp]

The species of the family range from southern North America and Central and South America to Africa, Eurasia, the Philippines, Indonesia, southern New Guinea, and northern Australia. They are very large, long-legged wading birds with huge bills and omnivorous foraging habits (Elliott 1992b). According to Hackett *et mult.* (2008) and Cracraft (2013) the storks are an isolated lineage, sister to the group that includes the pelicans, herons, ibises, anhingas, cormorants, boobies, and frigatebirds.

GENUS *EPHIPPIORHYNCHUS* [Genus 2 spp/Region 1 sp]

Ephippiorhynchus Bonaparte, 1855. *Consp. Gen. Avium* 2: 106. Type, by monotypy and virtual tautonymy, *Ciconia ephippiorhyncha* = *Mycteria senegalensis* Shaw.

The species of this genus range through Africa, India, Australia, and southern New Guinea. They are characterized by the huge, pointed bill, long neck, very long legs, very large size, and pied body plumage.

Black-necked Stork *Ephippiorhynchus asiaticus* RESIDENT
Mycteria asiatica Latham, 1790. *Index Orn.* 2: 670.—India.

In NG breeds in the flooded grassy plains of the Trans-Fly, dispersing elsewhere; seen infrequently from Dolak I east to the Strickland R and the Aramia R wetlands east of the Fly R (Coates 1985, K. D. Bishop *in litt.*); one record from the Port Moresby region (Gilliard 1950).

Extralimital—India and Nepal, s Indochina southeast to n and e Australia (Elliott 1992b).

SUBSPECIES

a. *Ephippiorhynchus asiaticus australis* (Shaw)
Mycteria australis Shaw, 1800. *Trans. Linn. Soc. London* 5: 33.—New South Wales, Australia.
RANGE: Trans-Fly, S Lowlands, islands of the Torres Strait, and nw, n, and e Australia (Coates 1985).
DIAGNOSIS: Exhibits greener sheen on head and neck than nominate form (Elliott 1992b), which exhibits a bluish sheen on head (Marchant & Higgins 1990).

NOTES
Other names—Jabiru, Green-necked Stork, *Xenorhynchus asiaticus*.

Family Pelecanidae

Pelicans [Family 8 spp/Region 1 sp]

The species of the family inhabit wetlands, lakes, estuaries, and seacoasts, from southern and western North America, the Caribbean, the coasts of western and northern South America, much of Africa, Russia, and the Middle East, tropical Asia, and Australia (Elliott 1992a). In New Guinea, the single species representing the family is a regular nonbreeding visitor to the Trans-Fly, and periodically appears throughout the Region during irruption years (*e.g.*, 1977–78). The phylogenomic work of Hackett *et mult.* (2008) suggested that this lineage is sister to the Shoebill (*Balaeniceps rex*, Balaenicipitidae) and Hammerkop (*Scopus umbretta*, Scopidae), and that the clade including those three lineages shows close affinities to the herons. The family is characterized by the large head and huge pouched bill, massive wings, and compact body.

GENUS *PELECANUS* [Genus 8 spp/Region 1 sp]

Pelecanus Linnaeus, 1758. *Syst. Nat.*, ed. 10, 1: 132. Type, by subsequent designation (G. R. Gray, 1840. *List Gen. Birds*: 80), *Pelecanus onocrotalus* Linnaeus.

Species of the genus inhabit all continents but for Antarctica. Defining generic characters are as for the family.

Australian Pelican *Pelecanus conspicillatus* VISITOR AND POSSIBLE BREEDER | MONOTYPIC
Pelecanus conspicillatus Temminck, 1824. *Planch. Col. d'Ois.*, livr. 47: pl. 276.—New South Wales, Australia.

Mainly a nonbreeding visitor to the Region, usually found in wetland habitats, most regularly in the Trans-Fly (Rand 1942a). Bishop (2005a) reported that three of his colleagues had received convincing reports from local informants of former breeding in Wasur NP in the w Trans-Fly. During the spectacular irruption of 1977–78, flocks of varying sizes were observed on the coast, the upper Sepik R[J], and in the highlands (I. Woxvold *in litt.*[1]). Photographs were obtained of a bird on the Ilaga R at 1,950 m in Aug-1959 (Ripley 1964a). Diamond & Bishop (1994) reported a first record of the species for the Aru Is. LeCroy & Peckover (2000) reported a record from Goodenough I on 9-Aug-1988; D. Mitchell (*in litt.*) reported a flock of six from Normanby I on 9-May-2012. Also a record from the Trobriand Is[CO].

Extralimital—breeds in Australia. Nonbreeders irruptively range to NZ, the Bismarck and Solomon Is, Vanuatu, Palau, the Moluccas, Sulawesi and Lesser Sundas, and west apparently to Java (Blakers *et al.* 1984, Elliott 1992a, Mayr & Diamond 2001). Reported to be present in the Torres Strait year-round (Draffan *et al.* 1983). M. Bowe in 1989 (Bishop 2005a) observed nests in the northernmost islands of the Torres Strait (Kerr Islet, just south of Deliverance I)—this is the proven breeding record nearest to the Region.

NOTES

Irruptions into the NG region appear to be a result of a pulse of very productive breeding followed by drought. The mention in Blakers *et al.* (1984: 34) that post-1977 this species began to breed in NG requires further documentation. *Other names*—Spectacled Pelican.

Family Threskiornithidae

Ibises and Spoonbills   [Family 34 spp/Region 5 spp]

The species of this family range throughout the tropical and temperate areas of the world, from North America to Australia (Matheu & del Hoyo 1992). Within the family, De Pietri (2013) recovered *Platalea* as sister to *Threskiornis* and these two as sister to the remaining ibises. A molecular study by Ramirez *et al.* (2013) also found *Platalea* and *Threskiornis* to be sister groups. Hackett *et mult.* (2008) placed the Threskiornithidae as sister to the clade including the herons and the Boat-billed Heron (*Cochlearius cochlearius*).

GENUS *THRESKIORNIS*   [Genus 5 spp/Region 2 spp]

Threskiornis G. R. Gray, 1842. *List Gen. Birds* 2, app.: 13. Type, by original designation, *Tantalus aethiopicus* Latham.

The species of this genus range from Africa and tropical Asia to New Guinea and Australia (Matheu & del Hoyo 1992). This compact genus is typified by the bald head of black skin, matching decurved black bill, and mainly white ventral plumage.

Australian White Ibis *Threskiornis molucca*   RESIDENT AND VISITOR

Ibis moluccca Cuvier, 1829. *Règne Anim.*, éd. nouv. 1: 520 (note).—Moluccas.

Probably mainly a visitor to s NG from Australia, but records exist from the NW Islands and the n watershed.

Localities—Salawati and Waigeo I, the Trans-Fly, the Fly R, Mamberamo, Sepik, and Ramu basins, and the Kemp Welch R/Port Moresby area (Mayr 1941, Rand 1942a, Coates 1985, Bishop 2005a). Nesting colonies have been found in the Trans-Fly on Dolak I (Bishop 1984) and in the Memunggal/Arerr mixed-waterbird colony (Silvius *et al.* 1989); and on Arar I in Sele Strait, southwest of Sorong[CO]. Other breeding evidence from the lower Sepik R[CO]. Some 37,000 birds were noted in Bensbach in late May-1980 (Stronach 1981b). Uncommon in Port Moresby, where apparently only a visitor during the dry season, March–October (K. D. Bishop *in litt.*). Frequents grasslands, open swamps, and other open habitats.

Extralimital—Seram and Great Kai I to Australia, Rennell, and Bellona I of the Solomon Is (Steinbacher 1979, Coates 1985, Mayr & Diamond 2001).

SUBSPECIES

a. *Threskiornis molucca molucca* (Cuvier)

Synonym: *Ibis strictipennis* Gould, 1838 (April). *Synops. Birds Austral.* 4, app.: 7.—New South Wales, Australia.

RANGE: Lesser Sundas to NG and Australia.

DIAGNOSIS: Substantially larger than the se Solomon Is population *pygmaeus* from Rennell and Bellona I. Wing: male 355–398, female 355–372 (Marchant & Higgins 1990). We do not recognize the Australian form *strictipennis*, following Mees (1982c), who showed that size and feather shaft coloration are not reliable for separating this form from the nominate form.

NOTES

It is well established from banding recoveries that at least some of the birds in s NG are visitors from Australia (several banded as young in Victoria have been recovered in the Western and Central Prov, PNG), while sight observations in Torres Strait revealed that annual migration of this species between NG and Australia is quite pronounced. Migratory flocks range in size from *ca.* 20 to 500 birds (Draffan *et al.* 1983). Some authorities (*e.g.*, Steinbacher 1979) have combined *molucca* and *aethiopicus* into a single widespread species. Condon (1975), Matheu & del Hoyo (1992), and Christidis & Boles (2008) asserted that the Australasian form merits specific status. *Threskiornis* is masculine (David & Gosselin 2002b), but the species epithet, *molucca*, is a noun in apposition (the adjectival form would be *moluccana/us*; *fide* R. Schodde *in litt.*). *Other names*—Sacred Ibis, *Threskiornis moluccus* and *T. aethiopicus*.

Straw-necked Ibis *Threskiornis spinicollis*   VISITOR AND RESIDENT? | MONOTYPIC

Ibis spinicollis Jameson, 1835. *Edinburgh New Philos. Journ.* 19: 213.—Murray R, NSW, Australia.

A common dry-season visitor from Australia to the savanna country of the Trans-Fly, where it prefers dry grasslands; a vagrant elsewhere in NG. Suspected of breeding in the Trans-Fly.

D'Albertis recorded the species at Attack I, Fly R delta, on 20-Dec-1875, when large flocks were seen "flying at a great elevation in a northwesterly direction" (cited in Coates 1985). According to Stronach (1981b) the species was present throughout the year at Bensbach, with peak numbers at the beginning of the wet season, when several thousand were present; these subsided to hundreds at the height of the dry season (Coates 1985). Seen fairly regularly by Hoogerwerf (1964) at Kurik on the nearby s coast of Indonesian NG, most often in May and between September and November but, unlike at Bensbach, seldom in the rainy season (Mees 1982b). It was reported by Coates (1985) to be a passage migrant in the Torres Strait during the dry season in flocks of 50–100 birds (Draffan *et al.* 1983). A rare vagrant to Ajkwa I near Timika[K], Bereina[CO], Kanosia Lagoon[M], Port Moresby[L], and Alotau[N] (Wahlberg 1990[M], Pratt field notes[N], K. D. Bishop *in litt.*[L], S. van Balen *in litt.*[K]).

Extralimital—breeds on the Australian mainland and has wandered to Tasmania, Lord Howe I, Norfolk I, and New Ireland (Steinbacher 1979, Mayr & Diamond 2001, Coates 1985).

NOTES

This species was said to nest at the mixed-waterbird colony at Embukam (PNG Trans-Fly; M. Bowe *in litt.* to K. D. Bishop). First-hand confirmation of nesting is needed (Bishop 2005a).

GENUS *PLATALEA*   [Genus 6 spp/Region 2 spp]

Platalea Linnaeus, 1758. *Syst. Nat.*, ed. 10, 1: 139. Type, by subsequent designation (G. R. Gray, 1840. *List Gen. Birds*: 67), *Platalea leucorodia* Linnaeus.

The species of this genus range from the southern United States and Central and South America east to sub-Saharan and eastern Africa, tropical Asia, southern New Guinea, and Australia (Matheu & del Hoyo 1992). Species of this small and compact genus are characterized by the white (or pink and white) plumage, long legs, and long spatulate bill.

Royal Spoonbill *Platalea regia*   VISITOR AND RARE BREEDER | MONOTYPIC

Platalea regia Gould, 1838. *Synops. Birds Austral.* 4, app.: 7.—East coast of NSW, Australia.

A nonbreeding visitor mainly to the savanna country of the Trans-Fly, where it frequents freshwater swamps, lagoons, and estuaries (Coates 1985). Also recorded to breed in the Trans-Fly (Bishop 2005a). Breeding sites include possibly Dolak I (K. D. Bishop *in litt.*); Memunggal/Arerr and Ukra in Wasur NP

(M. Bowe pers. comm. to N. Stronach *in litt.*); and Bondobol at the n edge of the Bula Plains, PNG (M. Bowe pers. comm. to Bishop 2005a). Also a dry-season visitor to the Port Moresby region; and records from the Bird's Head and Geelvink Bay region[CO]. A vagrant to other parts of NG: Ok Tedi[L], Mt Bosavi Mission[CO], and Nondugl[J] (Gyldenstolpe 1955a[J], Coates 1990: 566[L]). Reported to migrate across Torres Strait in large flocks (Coates 1985). Diamond & Bishop (1994) reported a first record for the Aru Is.

Extralimital—breeds mainly in Australia and NZ, but small current breeding populations may exist in Sulawesi, Moluccas, and Lesser Sundas; formerly bred in Java. Old records from Rennell I (Solomons) may have been of breeding birds; visitor to New Caledonia (Dutson 2011).

NOTES
We follow Condon (1975), Matheu & del Hoyo (1992), Christidis & Boles (2008), and Dickinson & Remsen (2013) in considering *regia* as specifically distinct from *leucorodia*, which lacks the distinctive black forehead and face. *Other names*—Black-billed Spoonbill; treated by some authorities as the Eurasian Spoonbill *Platalea leucorodia*.

Yellow-billed Spoonbill *Platalea flavipes*   VISITOR | MONOTYPIC
Platalea flavipes Gould, 1838. *Synops. Birds Austral.* 4, app.: 7.—New South Wales, Australia.

First recorded in New Guinea in the Trans-Fly during Aug-1992 when a flock of 17 were observed at a small pool adjacent to the lower section of the Bensbach R (Bishop 1995). During subsequent years this species was regularly observed by Bishop and several other observers at the same or nearby sites. In addition, R. Edwards reported seeing this species in 1991. Also recorded from the Trans-Fly in Jun-1993 by P. Gregory (Gregory 1994a).

Extralimital—Australia.

GENUS *PLEGADIS*   [Genus 3 spp/Region 1 sp]
Plegadis Kaup, 1829. *Skizz. Entw.-Gesch. Europ. Thierwelt:* 82. Type, by monotypy, *Tantalus falcinellus* Linnaeus.

The species of this genus range from the Americas, Africa, and tropical Asia to New Guinea and Australia (Matheu & del Hoyo 1992). The genus is characterized by the long decurved bill, uniformly dark iridescent plumage, and bare skin in front of the eye.

Glossy Ibis *Plegadis falcinellus*   VISITOR AND RESIDENT? | MONOTYPIC
Tantalus Falcinellus Linnaeus, 1766. *Syst. Nat.*, ed. 12, 1: 241.—Neusiedler See, lower Austria.
     Synonym: *Ibis peregrina* Bonaparte, 1855. *Consp. Gen. Av.* 2: 159.—Java and Sulawesi.

A seasonally common visitor to coastal areas of s NG—Timika area[L,M], Merauke distr[MA], Fly[MA], Kiunga[Z], Katau,[MA] and Laloki R[MA]—particularly the Trans-Fly[K] (Bishop 2005a[K], S. van Balen *in litt.*[L], J. Pap *in litt.*[M], D. Hobcroft *in litt.*[Z]). There are isolated records from the n watershed and highlands. In general, a winter visitor from Australia (March–October), with small numbers remaining until January, if conditions remain suitable (Coates 1985). Possibly sometimes breeds (Bishop 2005a). Gregory (1997) reported hundreds at Lake Daviumbu in Dec-1994.

Extralimital—nearly worldwide distribution in tropics and temperate zone: e US, Caribbean, Africa, s Europe, and tropical Asia east to Australia (Matheu & del Hoyo 1992). Records from Bougainville and Kolombangara (Mayr & Diamond 2001).

NOTES
We follow Steinbacher (1979) and Matheu & del Hoyo (1992) in treating *P. falcinellus* as monotypic. Matheu & del Hoyo (1992) reported the species to breed in the Trans-Fly but remained mute on whose authority. Local informants within the Tonda WMA (PNG) advised M. Bowe (*in litt.* to K. D. Bishop) that this species had bred in a mixed-waterbird nesting colony at Embukam. This requires confirmation, as it would constitute a first breeding record of this species from NG (Bishop 2005a).

Family Ardeidae

Herons, Egrets, and Bitterns
[Family 64 spp/Region 16 spp]

The species of the family range nearly worldwide. In New Guinea, members of the family inhabit wetlands, lakes, rivers, rain-forest streams, and interior highland valleys. They are rarely seen above 2,000 m and are common only in the lowlands. These are medium to very large wading birds, with long pointed beaks, long crooked necks, long legs, and short tails, that consume mainly fish and aquatic invertebrates. Hackett *et mult.* (2008) showed the family to be sister to the Boat-billed Heron (*Cochlearius cochlearius*). The ibises are sister to the heron+boatbill clade, and this clade of three is sister to a clade that includes the Shoebill (*Balaeniceps*), Hammerkop (*Scopus*), and pelicans (*Pelecanus*). Our generic and specific treatment follows Kushlan & Hancock's (2005) monograph of the family. Support for parts of this treatment was recently provided by the molecular phylogeny of Zhou *et al.* (2014), albeit with a limited set of species sampled. See also McCracken & Sheldon (1998) and Sheldon *et al.* (2000). Note that the systematic treatment of the genera *Butorides*, *Ardea*, *Ardeola*, and *Egretta* has been very unstable, with never-ending rearrangements (see summary of these changes in Christidis & Boles 2008).

GENUS *ZONERODIUS*
ENDEMIC | MONOTYPIC

Zonerodius Salvadori, 1882. *Ann. Mus. Civ. Genova* 18: 336. Type, by monotypy, *Ardea heliosyla* Lesson.

The single species of the genus is endemic to the Region. It is a member of the subfamily Tigrisomatinae (Kushlan & Hancock 2005). A sister form, *Tigriornis leucolophus*, inhabits western and central Africa. Apparently this genus is part of an ancient lineage, perhaps analogous to *Harpyopsis*, as both show small or monotypic lineages confined to an old area of tropical humid forest. This genus is characterized by bold barring of the neck, mantle, rump, and tail; the green eye skin; and the forest-interior habitat.

Forest Bittern *Zonerodius heliosylus*
ENDEMIC | MONOTYPIC

Ardea Heliosyla Lesson & Garnot, 1828. *Voy. Coquille, Atlas* 1(7): pl. 44; 1830. *Zool.* 1(16): 722.—Manokwari, Bird's Head.

A solitary inhabitant of small lowland forest-interior streams through NG, Salawati I, and the Aru Is (Mayr 1941); also there are records from interior highland valleys of the Lai and Baiyer R[CO], to 1,650 m[J] (Sam & Koane 2014[J]).

Localities—Salawati I[B], Sorong[K], Manokwari[MA], Mimika R, Setekwa R, Otakwa R, Bodim[B], Bernhard Camp on Taritatu R[Q], Tarara[P], Gaima[P], Oroville Camp[P], Oriomo R[Q], Tarara[P], Baia R[Y], Elevala R[Z], Vanimo[J], Hunstein Mts[W], Sepik R region including the upper Sepik basin[C], Lai R[CO], Baiyer R[CO], near Bundi[L], Astrolabe Bay[B], Pawaia in the Purari basin[A], lower Waria R[X], Hall Sound[CO], Varirata NP[J], Port Moresby area[CO], Vanumai[Q], and Mt Suckling[CO] (PNG Nat. Mus. specimen[A], Stresemann 1923[C], Mayr & Rand 1937[P], Rand 1942a[P], Rand 1942b[Q], Ripley 1964a[B], Dawson *et al.* 2011[X], A. Allison *in litt.*[W], Sam & Koane 2014[L], I. Woxvold *in litt.*[Y], D. Gibbs *in litt.*[K], D. Hobcroft *in litt.*[Z], Beehler field notes 13-Jul-1990[J]).

NOTES

The IUCN Red List status of the species should probably be Least Concern rather than Near Threatened. The species is very difficult to locate and thus is most likely overlooked over most of its range. There is no reason to think this elusive form is suffering a threat to its existence. *Other names*—Zebra Bittern, Zebra Heron, New Guinea Tiger-heron.

Ixobrychus Billberg, 1828. *Synop. Faun. Scand.*, ed. 2, 1(2): 166. Type, by subsequent designation (Stone, 1907. *Auk* 24: 192), *Ardea minuta* Linnaeus.

The species of this genus range from North America to Europe, Africa, tropical Asia, eastern Russia, Japan, Australia, and formerly New Zealand (Martínez-Vilalta & Motis 1992, Gill *et al.* 2010). The genus is characterized by the small size (smallest in the Ardeidae), cryptic habits, and the pale throat patch. Following Christidis & Boles (2008) and Dickinson & Remsen (2013), we here include the slightly larger and aberrant *I. flavicollis*, which is sometimes placed in its own genus, *Dupetor*.

Australian Little Bittern *Ixobrychus dubius*   VISITOR (AND RESIDENT?) | MONOTYPIC
Ixobrychus minutus dubius Mathews, 1912. *Novit. Zool.* 18: 234.—Herdman's Lake, sw Australia.

Recorded in marshes and reedbeds of s NG, probably as a breeder (Rand 1942a, Coates 1985, Pratt & Beehler 2014, K. D. Bishop *in litt.*). Dates of observation suggest it is also a wintering migrant from Australia. A specimen from Lake Daviumbu was ready to lay (Rand 1942a). Jaensch (1995, 1996) reported the species from n NG at Lake Pangua in Jun-1994 and Chambri Lake of the middle Sepik in Jun-1996. Reported from Obo in the middle Fly wetlands (Gregory 1997). Occurs annually at Port Moresby, where seasonally common, but there are few records from elsewhere in NG.

Extralimital—the breeding range of the species as currently circumscribed is sw and e Australia; one recent record from New Caledonia (Martínez-Vilalta & Motis 1992, Dickinson 2003, Dutson 2011).

NOTES

We follow Christidis & Boles (2008) in treating *dubius* as specifically distinct from the widespread *minutus* (*pace* Payne 1979). We consider the species monotypic, following Martínez-Vilalta & Motis (1992). *Other names*—by some authorities treated as *I. minutus*, Little Bittern, Black-breasted Least Bittern, or Black-backed Bittern.

Yellow Bittern *Ixobrychus sinensis*   VISITOR | MONOTYPIC
Ardea Sinensis Gmelin, 1789. *Syst. Nat.* 1(2): 642.—China.

A locally common nonbreeding northern visitor to reedy marshes in the n watershed of NG[CO]; also Waigeo[MA] and Biak I[MA]; few records in s NG[CO]. Apparently an annual visitor to Waigani swamp near Port Moresby from January to early April (Coates 1985); common along the Taritatu R, Mamberamo basin, in March (Rand 1942b). Doubtless overlooked or missed in many places of suitable habitat on the Mainland because of its secretive behavior.

Localities—Anggi Lakes[J], Paniai Lake[K], Taritatu R[A], Baiyer R[L], Astrolabe Bay[MA], Port Moresby area[CO] (Rand 1942b[A], Junge 1953[K], George 1973[L], S. F. Bailey *in litt.*[J]). It is probably this species rather than *I. dubius* that was recorded at Komimnung Swamp, lower Ramu, Madang Prov (Berggy 1978, Coates 1985).

Extralimital—tropical Asia to Japan, Micronesia, and thence to the Bismarck Arch and Bougainville I (Mayr 1941, Martínez-Vilalta & Motis 1992, Mayr & Diamond 2001, Dutson 2011). Occurs in Australia as a vagrant. Nesting records from Micronesia and Bougainville I indicate that local breeding in NG is a possibility.

NOTES

The von Rosenberg specimen supposedly from Misool I is probably from the Aru Is (Mees 1965, 1980). *Other names*—Chinese Little Bittern.

Cinnamon Bittern *Ixobrychus cinnamomeus*   PROBABLE RESIDENT | MONOTYPIC
Ardea cinnamomeus Gmelin, 1789. *Syst. Nat.* 1(2): 643; based on "Cinnamon Heron" of Latham, 1785. *Gen. Synops. Birds* 3: 77.—China.

This bird, which inhabits flooded ricefields and grassy wetlands, is known from a few sight records from the Sorong area of the w Bird's Head (Tindige 2003a, K. D. Bishop *in litt.*).

NOTES

P. Gregory has warned there is a rufous morph of the Black Bittern (*I. flavicollis*) that might be mistaken for this species at times (see notes for that species). However, the flight behavior is very different when the birds flush up from rank, overgrown ricefields (K. D. Bishop *in litt.*).

Black Bittern *Ixobrychus flavicollis* — RESIDENT

Ardea flavicollis Latham, 1790. *Index Orn.* 2: 701.—India.

A widespread breeding resident of wetlands in the NG lowlands, NW Islands, Aru Is, and SE Islands; rarely to 2,200 m (Sims 1956). There is at least one specimen from Biak I but no recent records (Bishop ms).

Extralimital—tropical Asia, Timor, Moluccas, Australia, and the Bismarck and Solomon Is (Mayr & Diamond 2001).

SUBSPECIES

a. *Ixobrychus flavicollis australis* (Lesson)

Ardea australis Lesson, 1831. *Traité Orn.* 8: 572.—Timor, Lesser Sundas.

Synonym: *Ixobrychus flavicollis gouldi* Bonaparte, 1855. *Consp. Gen. Av.* 2: 132.—New South Wales, Australia.

RANGE: Moluccas, Australia, NG; Gebe[Z], Misool[J], and ?Salawati I[K]; Aru Is[MA], and SE Islands (Normanby[L], Misima[N], Woodlark[CO,M], and Tagula I[CO]), plus Bismarck Arch (Rand 1942a, Mees 1965[J], 1972[Z], M. Tarburton's island website[K], LeCroy *et al.* 1984[L], LeCroy & Peckover 1998[N], Pratt field notes 2005[M]) Shaw Mayer collected one on Mt Giluwe at 2,200 m (Sims 1956). Mees (1964b) reported a specimen from Sibil at *ca.* 1,250 m in the Star Mts.

DIAGNOSIS: Given the extreme plumage variation of the form in Australasia, it is difficult to provide characters distinguishing it from the nominate form of se Asia (Marchant & Higgins 1990).

NOTES

Race *gouldi* lumped with *australis* by Payne (1979). P. Gregory (*in litt.*) has noted that there is a rufous morph that may be confused with *I. cinnamomeus*. This is confirmed by Marchant & Higgins (1990: 1055). *Other names*—Yellow-necked or Mangrove Bittern, *Dupetor flavicollis*.

GENUS *NYCTICORAX* [Genus 2 spp/Region 1 sp]

Nycticorax T. Forster, 1817. *Synop. Cat. Brit. Birds*: 59. Type, by tautonymy and monotypy, *Nycticorax infaustus* Forster = *Ardea nycticorax* Linnaeus.

The two species of this genus range through the temperate and tropical New World and Old World to New Guinea, the Solomons, and Australia (Martínez-Vilalta & Motis 1992). The genus is characterized by compact proportions, large head, short neck, short, rather broad bill, and crepuscular habits (Kushlan & Hancock 2005, Martínez-Vilalta & Motis 1992).

Nankeen Night-Heron *Nycticorax caledonicus* — RESIDENT AND VISITOR

Ardea caledonica Gmelin, 1789. *Syst. Nat.* 1(2): 626; based on the "Caledonian Night Heron" of Latham, 1785. *Gen. Synops. Birds* 3: 55.—New Caledonia.

A rather common inhabitant of lakes, wetlands, and river edges, mainly in the NG lowlands but also in the interior highlands to 1,600 m (Gyldenstolpe 1955a); one record to 2,200 m (H. Wild *in litt.*); also found on Kofiau[L], Misool[J], Batanta[K], Salawati[MA], Waigeo[MA], Biak[MA], and ?Yapen I[K], and Aru Is[MA] (Mees 1965[J], Diamond *et al.* 2009[L], M. Tarburton's island website[K]). Large breeding colonies have been observed along the Bensbach R by M. Bowe (K. D. Bishop *in litt.*). Also a visitor to s NG from Australia (Coates & Peckover 2001).

Extralimital—e Java, Borneo, and s Philippines through Indonesia to Micronesia and the Bonin Is, south to the Bismarck Arch, Solomon Is, New Caledonia, and much of Australia (Payne 1979, Martínez-Vilalta & Motis 1992, Mayr & Diamond 2001).

a. *Nycticorax caledonicus australasiae* Vieillot

Ardea australasiae Vieillot, 1823. *Tabl. Encycl. Méthod., Orn.* 3: 1130.—New South Wales, Australia.
    Synonym: *Nycticorax caledonicus hilli* Mathews, 1912. *Novit. Zool.* 18: 233.—Parry Creek, nw Australia.
RANGE: Lesser Sundas, Moluccas, NG, w Bismarck Arch, Australia, and NZ. Straggler to Lord Howe I
(Payne 1979).
DIAGNOSIS: Smaller and paler dorsally than *manillensis* (Marchant & Higgins 1990). Dorsal surface pale rufous
chestnut, lacking the dull undertone of the nominate; presence of a superciliary stripe (Amadon 1942). Much
white on abdomen and flanks and at bend of wing (Martínez-Vilalta & Motis 1992: pl. 29: sp. 37).

NOTES
We follow Gill *et al.* (2010) and Dickinson & Remsen (2013) in the use of the older subspecies name. *Other
names*—Rufous Night-Heron.

GENUS *BUTORIDES*   [Genus 2 spp/Region 1 sp]

Butorides Blyth, 1852. *Cat. Birds, Mus. Asiat. Soc. Bengal* (1849): 281. Type, by monotypy, *Ardea javanica*
Horsfield.
    The range of the species of this genus includes the New World, Africa, tropical Asia, eastern Asia,
southern Japan, New Guinea, the Solomon Islands, Vanuatu, Tahiti, and northern and eastern Australia
(Kushlan & Hancock 2005). *Butorides* is feminine (David & Gosselin 2002b). These are small and compact
herons, similar to *Ixobrychus*.

Striated Heron *Butorides striata*   RESIDENT

Ardea striata Linnaeus, 1758. *Syst. Nat.*, ed. 10, 1: 144.—Suriname.
    A solitary inhabitant of mangroves and swamps of the NG lowlands, NW Islands, Aru Is, and Bay
Is. In NG, mainly coastal and along major rivers, with one record from the Eastern Ra; more local and
uncommon in the n watershed.
    Extralimital—Panama, South America, and Africa; ne, s, and sc Asia, patchily through the Bismarck
and Solomon Is, and Australia (Mayr & Diamond 2001, Kushlan & Hancock 2005, Dutson 2011).

SUBSPECIES
a. *Butorides striata papuensis* Mayr
Butorides striatus papuensis Mayr, 1940. *Amer. Mus. Novit.* 1056: 6.—Numfor I, Bay Islands.
RANGE: Northwestern Islands, the Bird's Head and Neck, Bay Is, and Aru Is (Mayr 1941). Localities—
Koruo Is[J], Kofiau[J], ?Salawati[X], Misool[Q], Pelee/Yetpelle[W], Batanta[K,U], Numfor[Q], Waigeo[Z], Gam[J], Biak[MA], and
Yapen I[MA]; Kambala on the Bomberai Penin[Z], e coast of Geelvink Bay[MA] (Gyldenstolpe 1955b[Z], Mees 1965[Q],
Gibbs 2009[K], Dickinson & Remsen 2013, Beehler field notes 28-Feb-2004[U], Beehler field notes 12- &
16-Dec-2011[J], Beehler field notes 19-Dec-2011[W], S. Hogberg record in Tarburton's island website[X]).
DIAGNOSIS: A dark population (Rand & Gilliard 1967). Greenway (1973) noted Mayr wrote: "Birds of
Numfor, Waigeo, and Yapen are paler than *solomonensis* and darker than *moluccarum.*"

b. *Butorides striata idenburgi* Rand
Butorides striatus idenburgi Rand, 1941. *Amer. Mus. Novit.* 1102: 1.—Bernhard Camp, 50 m elevation,
Taritatu R, Mamberamo basin, s sector of the NW Lowlands.
RANGE: Northwestern Lowlands and Sepik-Ramu (Dickinson & Remsen 2013).
DIAGNOSIS: A pale form, with little rufous color on the underparts and whitish edgings to the upperwing
coverts (Rand 1941).

c. *Butorides striata flyensis* Salomonsen
Butorides striatus flyensis Salomonsen, 1966. *Vidensk. Medd. Dansk Nat. Foren. København* 129: 283.—Lake
Daviumbu, middle Fly R, Trans-Fly.
RANGE: Presumably the entire S Lowlands and the Trans-Fly, east to Port Moresby (Rand 1942a, Coates

1985). Presumably the Aru Is birds are best assigned here on biogeographic grounds (see Mayr 1941). A single bird was observed on the road near Goroka in the Eastern Ra on 31-May-1979 (Beehler field notes). An active nest was found near Port Moresby in April (Mackay & Bell 1968). Records from the Killerton Is[CO] of the n watershed of the SE Peninsula refer to this form.

DIAGNOSIS: Generally paler than *littleri* of ne QLD; underparts pale gray with buff tinge; throat pale, almost white (Salomonsen 1966b).

NOTES

A small, compact, dark-capped heron included in *Ardeola* by some authorities (*e.g.*, Payne 1979). We follow Schodde *et al.* (1980) and Dickinson & Remsen (2013) in circumscription of the species as distinct from *virescens*. Delimiting the geographic extent of the various races needs additional work (see Salomonsen 1966b, Schodde *et al.* 1980). The birds from ne NG and much of SE Peninsula (all n watershed and s watershed east of Port Moresby) have not been diagnosed. Schodde *et al.* (1980) and Dickinson & Remsen (2013) recognize *flyensis*, but Payne (1979) does not. *Other names*—Mangrove or Little Heron, *Ardeola striata, Ardea striata, Butorides striatus*.

GENUS *ARDEA*   [Genus 11 spp/Region 5 spp]

Ardea Linnaeus, 1758. *Syst. Nat.*, ed. 10, 1: 141. Type, by subsequent designation (G. R. Gray, 1840. *List Gen. Birds*: 66), *Ardea cinerea* Linnaeus.

The species of this genus range from the Americas, Eurasia, and Africa east to New Guinea and northern Australia (Kushlan & Hancock 2005). Inclusion of the species *alba* and *ibis* in *Ardea* was supported by Sheldon *et al.* (2000) and Zhou *et al.* (2014). The genus, which includes the drably plumaged and gigantic *Ardea sumatrana* as well as the small and tawny-plumed *Ardea ibis*, encompasses considerable morphological and plumage variation.

Cattle Egret *Ardea ibis*   VISITOR (AND RESIDENT?)

Ardea Ibis Linnaeus, 1758. *Syst. Nat.*, ed. 10, 1: 144.—Egypt.

A 1960s immigrant to NG's open country; increasing but still patchily distributed; most common in Port Moresby and the Trans-Fly (Coates 1985, Beehler *et al.* 1986, K. D. Bishop *in litt.*). Many (or all?) NG records may be migrants or vagrants. No breeding records proven for Region. Prior to 1962, known only from two records in the far west. In 1962–63 occasional flocks of up to 40 birds were seen in the ne sector of the Bird's Head (Hoogerwerf 1971). The first PNG record was of a group of three at the Bensbach R in the Trans-Fly on 19-Oct-1969 (Lindgren 1971). Since then the species has been seen regularly in the Bensbach area, where it may be only a nonbreeding visitor from Australia (Coates 1985). Coates (1990: 566) reported a sighting of birds in breeding plumage in the middle Sepik in May-1987. Regularly observed around Port Moresby with groups as large as 50+; regularly encountered at the Port Moresby airport; often seen in breeding plumage.

Localities—?Batanta I[J], Waigeo I[MA], Kaimana[X], Yamna I[MA], Biak I[K], Manokwari[CO], Bensbach[CO], Trans-Fly[W], middle Sepik[CO], Port Moresby[CO], Misima I[L] (Bishop 2005a[W], Bishop ms[K], Pratt field notes 2004[L], C. Thébaud & B. Mila *in litt.*[X], S. Hogberg in M. Tarburton's island website[J]).

Extralimital—the Americas, Africa, tropical Asia, and Australia; also records from New Britain, New Ireland, Manus I, and New Caledonia (Mayr & Diamond 2001, Dutson 2011).

SUBSPECIES

a. *Ardea ibis coromanda* (Boddaert)

Cancroma Coromanda Boddaert, 1783. *Tabl. Planch. Enlum.*: 54.—Southeastern coast of India.

RANGE: Breeds from s Asia east to Ussuriland, Korea, Japan, Indochina, Philippines, Malaysia, Australia, and NZ. Also ranges to Sumatra, Borneo, NG, Palau, Chuuk, Marianas, and New Caledonia (Payne 1979).

DIAGNOSIS: Larger, with a longer and heavier bill and longer legs than the nominate form; breeding adult exhibits an entirely golden neck, face, and chin. Has been treated as a distinct allospecies by some authorities (*fide* Martínez-Vilalta & Motis 1992).

We follow Kushlan & Hancock (2005) and Christidis & Boles (2008) in generic assignment to *Ardea*. Placed in *Bubulcus* by Martínez-Vilalta & Motis (1992) and Dickinson & Remsen (2013). Generic assignment of this bird over the last several decades has been unstable. *Other names*—Buff-backed Heron, *Bubulcus coromandus*, *Ardeola ibis*.

White-necked Heron *Ardea pacifica*
VISITOR | MONOTYPIC

Ardea pacifica Latham, 1801. *Index Orn.*, Suppl. 1: lxv.—New South Wales, Australia.

Nonbreeding visitor to s NG (the Trans-Fly, Port Moresby), but scattered records along coasts and through the interior of Mainland (Coates 1985, Beehler *et al.* 1986). First recorded by Hoogerwerf (1964) in May-1961 on ricefields at Kurik in the Trans-Fly. Vagrant elsewhere in NG when there is a drought in Australia (Coates 1985). During 1978–79 the species was recorded at a number of localities in the lowlands and highlands: Port Moresby, Finschhafen, Tari, Mt Hagen, and Aiyura (Coates 1985).

Extralimital—breeds in Australia and wanders to NZ and Norfolk I (Martínez-Vilalta & Motis 1992).

NOTES

The name White-necked Heron is more useful than the undeserved epithet "Pacific Heron," which could just as well belong to *Egretta novaehollandiae*. *Other names*—Pacific Heron.

Great-billed Heron *Ardea sumatrana*
RESIDENT | MONOTYPIC

Ardea sumatrana Raffles, 1822. *Trans. Lin. Soc. London* 13(2): 325.—Sumatra.

An uncommon and solitary resident of lowland swamps and waterways of the Region. Occurs mostly in coastal zones, but recorded at 550 m (Kebar valley) and 600 m (Jimi valley).

Localities—Kofiau[B], Waigeo[T], Batanta[H], Misool[Z], Wagamab[H], and Salawati I[Z]; Jeflio[T], Kebar valley[U], Aru Is[MA], Biak I[MA], Yapen I[J], Sorong[MA], Faur[MA], Danau Bira[Q], Bernhard Camp on the Taritatu R[P], Dolak I[K], Princess Marianne Strait[K], Merauke[K], Bensbach R[K], Elevala R[X], upper Sepik[L], Wawoi R[Y], Rentoul R[L], Aure R[L], Purari R[CO], Jimi valley[I], Brahmin station[W], Gogol valley[CO], Huon Gulf[MA], Astrolabe Bay[MA], and Cape Nelson on the n coast of the SE Peninsula[A], Normanby I[M] (Rand 1942b[P], Gyldenstolpe 1955b[T], Mees 1965[Z], Bell 1967a[Y], Gilliard & LeCroy 1970[U], King 1979[J], Diamond 1985[Q], Diamond *et al.* 2009[B], K. D. Bishop *in litt.*[K], Sam & Koane 2014[W], I. Woxvold *in litt.*[L], D. Hobcroft *in litt.*[X], Beehler field notes Dec-2011[H], Beehler field notes 29-Jul-1980[A], Pratt field notes 2003[M]).

Extralimital—breeds from sw Burma, Indochina, and the Philippines to Sumatra, Borneo, e Indonesia, and n Australia.

NOTES

Treated as monotypic by Payne (1979). It is apparently the sister form to *A. insignis* of the e Himalayas and Burma (Martínez-Vilalta & Motis 1992). Presumably this species-pair is, in turn, sister to *A. goliath* of Africa. All possess the distinctive oversized bill. *Other names*—Dusky Grey, Sumatran, or Giant Heron.

Great Egret *Ardea alba*
RESIDENT AND VISITOR

Ardea alba Linnaeus, 1758. *Syst. Nat.*, ed. 10, 1: 144.—Sweden.

A common inhabitant of wetlands, swamps, and mudflats, ranging from the coast up to 2,255 m at Kandep[CO]. Perhaps the most widespread ardeid in NG. Nesting has been recorded in a lagoon near the Aroa R[CO] in the period from February to August and in the w sector of the Trans-Fly (Bishop 2005a).

Extralimital—worldwide in temperate and tropical regions (Condon 1975). Records also patchily through Melanesia (Mayr & Diamond 2001, Dutson 2011).

SUBSPECIES

a. *Ardea alba modesta* J. E. Gray

Ardea modesta J. E. Gray, 1831. *Zool. Misc.*: 19.—India.

RANGE: Tropical Asia, Japan, east to Australia and NZ.

DIAGNOSIS: Subspecies *modesta* is smaller than other races, although populations in Australia and NZ are larger than those in Indian and China. Tibia of breeding bird is pink, like that of European *alba*, but unlike black tibia of African *melanorhynchus* and American *egretta*. Bill of breeding bird black, as in all other races except *egretta*.

NOTES

We follow Johnsgard (1979) and Dickinson & Remsen (2013) in treating the worldwide populations of the Great Egret as a single species. By contrast, Kushlan & Hancock (2005) and Christidis & Boles (2008) treated this eastern form as a full species, *A. modesta*. See detailed discussion in Pratt (2011). Other names—Great White Egret, Large Egret, White Heron, Eastern Great Egret, *Casmerodius albus*, *Ardea modesta*, *Egretta alba*.

Intermediate Egret *Ardea intermedia*   VISITOR AND RESIDENT
Ardea intermedia Wagler, 1829. *Isis von Oken* 6: col. 659.—Java.

A common year-round inhabitant of wetlands in the Region; less common in highland habitats; records to 2,200 m (Mayr 1941, Gilliard & LeCroy 1961, Coates 1985). Known to breed in large numbers in the w Trans-Fly (Bishop 2005a), but mainly a migrant from Australia. Found preserved in glacial ice on the Carstensz Massif in w NG at 4,420 m (Schodde *et al.* 1975).

Extralimital—w Africa through tropical Asia to Sumatra, Philippines, Umboi, New Britain, Manus, Bougainville, Makira, and Australia (Martínez-Vilalta & Motis 1992, Mayr & Diamond 2001, Dutson 2011).

SUBSPECIES

a. *Ardea intermedia plumifera* (Gould)
Herodias plumiferus Gould, 1848. *Proc. Zool. Soc. London* (1847): 221.—New South Wales, Australia.
RANGE: Eastern Indonesia, NG, and Australia. Vagrant to n Melanesia (Dutson 2011).
DIAGNOSIS: Breeding adult has a red bill and tarsus (B. J. Coates *in litt.*). We follow Payne (1979), Martínez-Vilalta & Motis (1992), Kushlan & Hancock (2005), and Dickinson & Remsen (2013) in recognizing the subspecies *plumifera*. This form is treated as a distinct species by del Hoyo & Collar (2014).

NOTES

Other names—Lesser, Plumed, Yellow-billed, Smaller, or Short-billed Egret; *Mesophoyx* or *Egretta intermedia*.

GENUS *EGRETTA*   [Genus 14 spp/Region 4 spp]

Egretta T. Forster, 1817. *Synop. Cat. Brit. Birds*: 59. Type, by monotypy, *Ardea garzetta* Linnaeus.

The species of this genus range from the Americas, Eurasia, and Africa east to New Guinea, Australia, New Zealand, New Caledonia, and the central Pacific (Kushlan & Hancock 2005). The rather diverse genus is characterized by the presence of neck and mantle plumes, medium size, and moderate-length legs.

Pied Heron *Egretta picata*   VISITOR AND RARE BREEDER | MONOTYPIC
Ardea (Herodias) picata Gould, 1845. *Proc. Zool. Soc. London*: 62.—Port Essington, NT, Australia.

Mainly a common nonbreeding inhabitant of marshes and sandbars in lowland NG and the Aru Is (Mayr 1941).

Localities—?Salawati I[J], Timika[M], Baliem valley[CO] (1,650 m[CO]), Bernhard Camp on the Taritatu R[K], Kurik[L], Merauke[MA], Fly R[MA], Balimo[CO], Sepik R[MA], Wahgi and Baiyer valleys[CO], Ramu R[MA], Karimui[DI], and Port Moresby[CO] (Rand 1942b[K], Mees 1982b[L], S. Hogberg in M. Tarburton's island website[J], S. van Balen *in litt.*[M]). Bishop (2005a) reported that M. Bowe observed this species breeding in small groups in somewhat denser stands of *Melaleuca* generally scattered throughout the s parts of Wasur NP (Indonesian NG) and Tonda WMA (PNG).

Extralimital—breeds in Sulawesi and n Australia, and there are nonbreeding records from various sites in between these two endpoints; also New Caledonia (Martínez-Vilalta & Motis 1992, Dutson 2011).

NOTES

The generic disposition of this and most other medium- to small-size Ardeinae has been in dispute (see discussion in Christidis & Boles 2008). This species has been placed in *Ardea*, *Egretta*, *Hydranassa*, and *Notophoyx*. More and better-documented breeding information is needed. *Other names*—Little Pied Heron, Pied Egret, *Notophoyx*, *Hydranassa*, or *Ardea picata*.

White-faced Heron *Egretta novaehollandiae* VISITOR (AND RESIDENT?) | MONOTYPIC

Ardea Novae Hollandiae Latham, 1790. *Index Orn.* 2: 701.—New South Wales, Australia.
> Synonym: *Notophoyx novaehollandiae austera* Ripley, 1964. *Bull. Peabody Mus. Nat. Hist. Yale Univ.* 19: 13.— Wamena, 1,525 m, Baliem valley, Indonesian NG.

A generally uncommon Australian visitor to NG, found foraging in ditches, ponds, and ricefields (Coates 1985). Most frequent in the S Lowlands, but also uncommonly encountered in the highlands (to 1,700 m[CO]) and the Aru Is[MA]; three records of vagrants from the SE Islands: Misima, Woodlark, and Rossel[MA].

Localities—Wamena[K,X], Princess Marianne Strait[MA], Mabaduan[L], Soabi base airstrip in the Western Prov lowlands[J], Sibil[Q], Tabubil[CO], Tari[CO], Koroba[CO], Daru[MA], Baiyer R valley[CO], Wahgi valley[CO], Woitape[CO], Port Moresby[CO], the foot of the Hydrographer Mts[CO], Woodlark[N] and Rossel I[MA] (Rand 1942a[L], 1942b[K], Ripley 1964a[X], Mees 1964b[Q], G. E. Clapp sight record[N], I. Woxvold *in litt.*[J]). No details of breeding in Region located (see Coates 1985).

Extralimital—proven breeding apparently confined to se and sw Australia, NZ, Fiji, and New Caledonia. A nonbreeding visitor to Sulawesi and Lesser Sunda Is; Torres Strait Is; also Bougainville, Guadalcanal, and New Caledonia (Martínez-Vilalta & Motis 1992, Mayr & Diamond 2001, Dutson 2011).

NOTES

Payne (1979) treats the species as monotypic, synonymizing *austera* and *parryi* with the nominate form. Given the species is apparently only a nonbreeding visitor to NG, the endemic upland race *austera* is untenable; the Archbold Expedition collected the *austera* population in the Baliem in 1938 but did not describe it. *Other names*—*Notophoyx novaehollandiae*.

Little Egret *Egretta garzetta* VISITOR

Ardea Garzetta Linnaeus, 1766. *Syst. Nat.*, ed. 12, 1: 237.—Malalbergo, River Reno, south of Ferrara, ne Italy.

A fairly common visitor to freshwater and saltwater wetlands mainly in s NG, to 1,740 m[CO]; also NW Islands, Biak I, and Aru Is (K. D. Bishop *in litt.*). All are apparently migrants from Australia (Coates 1985). No nesting records for NG to date. Birds banded in Australia have been recovered in various places in NG (Coates 1985).

Extralimital—s Europe through Africa and tropical Asia to e Asia, se Asia, Australia, and NZ. Vagrant records extend to New Britain, Solomon Is, New Caledonia, and Palau (Payne 1979, Mayr & Diamond 2001).

SUBSPECIES

a. *Egretta garzetta nigripes* (Temminck)
Ardea nigripes Temminck, 1840. *Man. Orn.*, ed. 2, 4: 376.—Sunda Is.
RANGE: Sumatra and Borneo to NG, Batanta, Kofiau, Misool[J], Waigeo, Salawati, Aru Is, Solomon Is, Palau, Australia, and NZ (Mayr 1941, Mees 1965[J]). Upland localities include the Baiyer valley (1,200 m)[CO], Wahgi valley (1,600 m)[CO], and Paniai Lake[X] at 1,740 m (Gyldenstolpe 1955a[X]). Noted to occur in the Louisiade Arch by Payne (1979), which is not in accordance with other authorities (Peters 1931, Mayr 1941, Coates 1985, Pratt & Beehler 2014). Dutson (2011) noted this species breeds in NG, but we are unaware of documentation of this occurrence.
DIAGNOSIS: Smaller than nominate form; wing 258–272; feet black with yellow soles (Marchant & Higgins 1990).

NOTES
Other names—Snowy Egret.

Eastern Reef-Egret *Egretta sacra*   RESIDENT

Ardea sacra Gmelin, 1789. *Syst. Nat.* 1(2): 640; based on "Sacred Heron" of Latham, 1785. *Gen. Synops. Birds* 3: 93.—Tahiti.

This heron of the seashore and reef is a common breeding resident throughout Region. Present in two color morphs.

Extralimital—breeds from Burma to s Japan, southeast to Australia, New Caledonia, the Loyalty Is, NZ, and east to Polynesia (Payne 1979, Mayr & Diamond 2001).

SUBSPECIES

a. *Egretta sacra sacra* (Gmelin)

RANGE: As for species range but absent from New Caledonia and the Loyalty Is, the range of *E. s. albolineata*.

DIAGNOSIS: Nominate form is smaller than *albolineata* and lacks black marking on the bill (Martínez-Vilalta & Motis 1992). Wing 252–284 (Marchant & Higgins 1990).

NOTES

Our choice of English name is based on its counterpart reference to its sister species, the Western Reef-Egret. *Other names*—Reef Heron, Eastern or Pacific Reef Heron, Pacific Reef-Egret, *Demigretta sacra*.

Family Fregatidae

Frigatebirds   [Family 5 spp/Region 2 spp]

The species of the family range throughout tropical seas (Orta 1992d). Large, soaring seabirds with superb powers of flight, they share the following characters: bill long and hooked at tip; wings very long and pointed (with proportionally the largest surface area, relative to weight, of any bird; Orta 1992d); tail long and deeply forked; legs short; feet small, webbed basally; males have a red gular pouch, which is inflated, balloon-like, during courtship. Frigatebirds are most often seen soaring high on motionless wings. They never dive and rarely float on sea surface; they skim food off the surface or force other seabirds to give up their prey. They are commonly seen over coastal New Guinea and in nearshore waters, but there are no records of breeding from the Region (Coates 1985, Pratt & Beehler 2014, Orta 1992d). Frigatebirds are sister to a clade that encompasses the gannets, cormorants, and anhingas (Hackett *et mult.* 2008).

GENUS *FREGATA*   [Genus 5 spp/Region 2 spp]

Fregata Lacépède, 1799. *Tabl. Méthod. Mamm. Oiseaux*: 15. Type, by subsequent designation (Daudin, 1802. In Buffon, *Hist. Nat.* [ed. Didot], *Quadr.* 14: 317), *Pelecanus aquilus* Linnaeus.

Range, characteristics, and habits are as for the family.

[Christmas Island Frigatebird *Fregata andrewsi*   HYPOTHETICAL | MONOTYPIC

Fregata andrewsi Mathews, 1914. *Austral. Avian Rec.* 2: 120.—Christmas I, e Indian Ocean.

A possible vagrant to waters off s NG, but there are no authenticated records. Simpson (1990) reported sighting the species from Port Moresby on 8-Nov-1987, a bird with an all-white underbody, "possibly a female of this species." In addition, Simpson (1990) reported that a colleague (JB) stated that he had seen this species off Umuda I at the mouth of the Fly R (no details). Future records should provide full documentation, including photographs.

Extralimital—breeds on Christmas I and ranges through the e Indian Ocean and South China Sea. One record from NT, Australia (McKean *et al.* 1975). Other reports for Australia are under review. Dutson (2011) noted observations of birds from the Solomons that meet the description.]

Great Frigatebird *Fregata minor* VISITOR | MONOTYPIC

Pelecanus minor Gmelin, 1789. *Syst. Nat.* 1(2): 572.—Christmas I, e Indian Ocean.
> Synonym: *Pelecanus Palmerstoni* Gmelin, 1789. *Syst. Nat.* 1(2): 573; based on "Palmerston Frigate Pelican" of
> Latham, 1785. *Gen. Synops. Birds* 3: 592.—Palmerston I, Pacific Ocean.

Widespread in NG waters but outnumbered by *F. ariel*. This species and *F. ariel* can be found traveling and loafing together, so care must be taken in identification. There is an inland record from Ok Tedi (Coates 1990: 566).

Extralimital—breeds on small islands across tropical seas of the sw Atlantic, Indian, and Pacific Oceans. Nonbreeders range through the tropics (Dorst & Mougin 1979, Orta 1992d).

NOTES
We follow the suggestion of Orta (1992d) and Dutson (2011) that the species be treated as monotypic. *Other names*—Greater, Pacific, Lesser Frigatebird.

Lesser Frigatebird *Fregata ariel* VISITOR

Atagen ariel G. R. Gray, 1845. *List Gen. Birds* 3: 669, col. pl. [185].—Raine I, QLD, Australia.

A widespread and common year-round visitor to NG waters and coastlines, usually outnumbering *F. minor*. Occasionally blown inland; there are records from Kiunga and Tari gap at 2,750^K m (Coates & Peckover 2001^K, *Muruk* field reports, K. D. Bishop *in litt.*).

Extralimital—w Indian Ocean east to Indonesia, the Coral Sea, and c Pacific (Orta 1992d).

SUBSPECIES

a. *Fregata ariel ariel* (G. R. Gray)
RANGE: Breeds in c and e Indian Ocean and through the c Pacific. Ranges to e Siberia, Japan, Philippines, and Australia to Polynesia (Dorst & Mougin 1979).
DIAGNOSIS: Three doubtful races described based on measurements of bill and wing (Marchant & Higgins 1990). This form currently attributed solely based on distribution. Wing: male 518–550, female 534–562 (Marchant & Higgins 1990).

NOTES
Other names—Least Frigatebird.

Family Sulidae

Boobies and Gannets [Family 10 spp/Region 4 spp]

The species of the family inhabit the oceans of the world (though not the northern Pacific nor much of the Southern Ocean). In the New Guinea region, birds are found in sparse numbers over all oceanic waters, occasionally within sight of land. They are goose-size, with a long, tapered, thick-based bill; long neck; heavy, tapering body; and long, pointed wings and tail. Flight is an alternating series of wingbeats and short glides. These birds plunge-dive into the sea from a height when fishing. Boobies frequent seas near land, usually outside the reefs, rarely along coasts except at breeding islands (Coates 1985, Carboneras 1992c). The phylogenomic work of Hackett *et mult.* (2008) indicated that the family is sister to the cormorants and anhingas, and that the anhinga+cormorant+booby clade is sister to the frigatebirds.

GENUS *PAPASULA*   MONOTYPIC

Papasula Olson & Warheit, 1988. *Bull. Brit. Orn. Club* 108: 10. Type, by monotypy, *Sula abbotti* Ridgway, 1893.

Olson & Warheit (1988) provided osteological evidence supporting recognition of this monotypic genus, sister to *Sula*. Characters defining the genus include the ossified sclerotic ossicles, the distinctive tibiotarsus and tarsometatarsus, the long and narrow wings, and the lack of a distinct juvenile plumage.

Abbott's Booby *Papasula abbotti*   VAGRANT | MONOTYPIC

Sula abbotti Ridgway, 1893. *Proc. U.S. Nat. Mus.* 16: 599.—Assumption I, Seychelles, w Indian Ocean.

Sightings of three birds from open seas between the Bird's Neck and Aru Is were made in February–March (Cadée 1985).

Extralimital—sole remaining population of a few thousand birds nests on Christmas I, Indian Ocean, and disperses as far as the Banda Sea. Originally much more widespread in the Indian and Pacific Oceans. IUCN Red List status: Endangered.

Other names—Sula abbotti.

GENUS *SULA*   [Genus 6 spp/Region 3 spp]

Sula Brisson, 1760. *Orn.* 1: 60; 6: 494. Type by tautonymy, "Sula" Brisson = *Pelecanus leucogaster* Boddaert.

The species of this compact and uniform genus range throughout tropical (and some subtropical) seas of the world. They possess the following characters: long, pointed, conical beak; long and pointed wings; very strong flight; short, pointed tail; and propensity to dive from heights for fish visible below the surface of the sea. Members of the genus exhibit a distinct juvenile plumage.

Masked Booby *Sula dactylatra*   VISITOR AND RESIDENT

Sula dactylatra Lesson, 1831. *Traité Orn.* 8: 601.—Ascension I, Atlantic Ocean.

Rarely recorded in NG waters, but fairly common in open sea. The species may occur irregularly in all NG offshore waters but with greatest frequency in the southeast (Coates 1985). The Region's single known breeding colony is on Pocklington Reef in the SE Islands (2014 photo of adults with nestlings; D. Mitchell *in litt.*). Nearest additional breeding colonies are in the vicinity of the Great Barrier Reef (Raine I, Pandora Cay, and Swain Reefs) in n QLD (Blakers *et al.* 1984). May also breed on isolated Bismarck atolls (Dutson 2011).

Extralimital—widely distributed in tropical seas around the world (Carboneras 1992c).

SUBSPECIES

a. *Sula dactylatra personata* Gould

Sula personata Gould, 1846. *Proc. Zool. Soc. London*: 21.—Raine I, ne QLD, Australia.

RANGE: Breeds on various small islands in the e Indian Ocean and sw and c Pacific, including ne Australia. Apparently a rare breeder in the SE Islands (Pocklington Reef, ne of Rossel I; see species range above); uncommon visitor to seas off the s side of NG from the Great Barrier Reef to East Cape[MA]; birds banded on Raine I have been recovered at various points around the Gulf of Papua[CO]. Localities—Huon Gulf[CO]; Milne Bay, off Port Moresby, and Gulf of Papua[CO]; and Pocklington Reef (D. Mitchell *in litt.*).

DIAGNOSIS: Iris yellow; wing short. Wing (of male Raine I birds) 418–425 (Marchant & Higgins 1990).

NOTES

An effort needs to be made to document further the breeding colonies of this and other booby species within the Region. *Other names*—Blue-faced or White Booby.

Red-footed Booby *Sula sula*

Pelecanus sula Linnaeus, 1766. *Syst. Nat.*, ed. 12, 1: 218.—Barbados, West Indies.

Few records from coastal NG, but fairly common at sea. Nearest known breeding site is Raine I, n QLD, Australia (breeding season June to January).

Extralimital—throughout tropical seas (Dorst & Mougin 1979). Breeds on islands of the n Bismarcks (Dutson 2011).

SUBSPECIES

a. *Sula sula rubripes* Gould

Sula rubripes Gould, 1838. *Synops. Birds Austral.* 4 (app.): 7.—Raine I, QLD, Australia.

RANGE: Breeds in the Indian Ocean and w and c Pacific, including the islands of ne QLD. Locally numerous at sea off Port Moresby in late November and in small numbers between Wuvulu I and Wewak from March to May (Coates 1985); other records are from the vicinity of the Aru Is, Amazon Bay, and Woodlark I (Mayr 1941, Coates 1985).

DIAGNOSIS: Presumably the subspecific designation is by range only. Examination of material obtained in the Region is needed. Marchant & Higgins (1990) considered the named subspecies of doubtful validity.

NOTES

Future researchers should make an effort to confirm whether this species nests on any of NG's offshore seabird islands. Coates (1985: 71) noted that G. Carson reported to E. Lindgren that this species and possibly *S. dactylatra* were seen nesting on the Nuguria Is east of New Ireland (extralimital).

Brown Booby *Sula leucogaster*

Pelecanus Leucogaster Boddaert, 1783. *Tabl. Planch. Enlum.*: 57, pl. 973.—Cayenne, French Guiana.

New Guinea's common booby, seen in virtually all NG waters, usually outside of the reefs. Breeds on islands in Torres Strait, Milne Bay, and probably in the n Bismarcks.

Extralimital—throughout the world's tropical seas.

SUBSPECIES

a. *Sula leucogaster plotus* (Forster)

Pelecanus plotus J. R. Forster, 1844. In Lichtenstein, *Descr. Animalium*: 278.—Near New Caledonia.

RANGE: Recorded from the NW Islands, Aru Is, Yapen I, waters off n and s NG, and the SE Islands. Breeds from the Red Sea and Indian Ocean to islands of w Australia and QLD (including Bramble Cay in n Torres Strait) to the c Pacific. Presumably nonbreeders wander widely. D. Mitchell (*in litt.*) reported evidence of breeding from Budibudi Is (east of Woodlark I), as well as Pocklington Reef (ne of Rossel I) in the SE Islands. Pratt (field notes) recorded breeding on a sand island at Manuga Reefs, between Rossel and Misima I, on 21-Nov-2004; at the time *ca.* 400–500 of these boobies were present, nearly all adult, and there were few nests, with chicks of all ages. J. P. Dumbacher (*in litt.*) reported observing breeding on a tiny rocky outcrop (Oiaveta I) in the Amphlett Is and also in similar habitat between Budibudi and Woodlark (Cannac I). Present in the Region throughout the year; birds banded as young on Raine I, n QLD, in December have been recovered within a few months near Kerema, Gulf Prov, PNG (Coates 1985: 70).

DIAGNOSIS: Larger and uniformly dark dorsally when compared to nominate form (Carboneras 1992c).

NOTES

J. P. Dumbacher (*in litt.*) reported that in the SE Islands, many likely breeding islets are empty because of egg harvesting by local fishermen. *Other names*—White-bellied Booby.

Family Phalacrocoracidae

Cormorants [Family 31 spp/Region 4 spp]

The species of the family range worldwide where there are coastlines and wetlands. They are heavy-bodied, blackish or black-and-white birds, with a long neck and tail, a hooked bill, and webbed feet. Medium to large waterbirds, they swim low in the water and dive from the surface for food; they are often seen perching conspicuously in the open with wings spread while plumage dries. Cormorants are sister to the darters; the clade formed by those two lineages is sister to the boobies and that triplet is sister to the frigatebirds (Hackett *et mult.* 2008). In New Guinea these birds are found on lakes, lagoons, and quiet stretches of rivers (Coates 1985, Orta 1992b). Our treatment follows Christidis & Boles (2008).

GENUS *MICROCARBO* [Genus 5 spp/Region 1 sp]

Microcarbo Bonaparte, 1856. *Compt. Rend. Acad. Sci. Paris* 43: 577. Type, by original designation, *Pelecanus pygmaeus* Pallas.

The five species of this genus range from southern Africa and Europe to Australia and New Zealand. Recognition of the genus follows Siegel-Causey (1988) and Kennedy *et al.* (2000). The lineage is defined by osteological characters and molecular affinities.

Little Pied Cormorant *Microcarbo melanoleucos* RESIDENT AND VISITOR

Hydrocorax melanoleucos Vieillot, 1817. *Nouv. Dict. Hist. Nat.*, nouv. èd. 8: 88.—New South Wales, Australia.

Inhabits various freshwater habitats through NG, the NW Islands, Aru Is, and Numfor in the Bay Is, sea level–3,400 m (Orta 1992b, Coates 1985). The most widespread cormorant in NG. Rare above 1,500 m. The sole e NG breeding record is the report "found breeding at Baiyer R by Bulmer" (Diamond 1972). Bishop (2005a) reported several large breeding sites in the w sector of the Trans-Fly (Dolak I, Wasur NP). One individual, banded as a nestling in Victoria, Australia, on 14-Nov-1974, was shot at Arufi near Morehead, Western Prov, PNG, on 1-Nov-1976, indicating that at least some are visitors from Australia (Coates 1985).

Extralimital—Java and Sulawesi east through Melanesia, Australia, and NZ (Mayr & Diamond 2001, Orta 1992b).

SUBSPECIES

a. *Microcarbo melanoleucos melanoleucos* (Vieillot)
RANGE: Java, Bali, Sulawesi, Lesser Sundas, Moluccas, NW Islands (Kofiau[J], Misool, Salawati, Waigeo)[MA], Aru Is[MA], Numfor I[MA], Biak I, NG, Palau, much of Melanesia, and Australia (Rand 1942b, Ripley 1964a, Dorst & Mougin 1979, Diamond *et al.* 2009[J], K. D. Bishop *in litt.*).
DIAGNOSIS: Size large; underparts white in all individuals (Amadon 1942). Wing 223–240 (Marchant & Higgins 1990).

NOTES
Other names—Little Black and White Shag, *Halietor* or *Phalacrocorax melanoleucos*.

GENUS *PHALACROCORAX* [Genus 26 spp/Region 3 spp]

Phalacrocorax Brisson, 1760. *Orn.* 1: 60; 6: 511. Type, by tautonymy, "Phalacrocorax" Brisson = *Pelecanus carbo* Linnaeus.

The species of this genus range worldwide. The genus is characterized by the distinctive slim shape, long neck, narrow head, long hooked bill, webbed feet, mainly black or black-and-white plumage, distinctive head-up profile when floating in water, and the spread-wing posture when drying plumage. *Phalacrocorax* is separated from *Microcarbo* by osteological and molecular characters.

Little Black Cormorant *Phalacrocorax sulcirostris*   RESIDENT AND VISITOR | MONOTYPIC

Carbo sulcirostris von Brandt, 1837. *Bull. Sci. Acad. Imp. Sci. St. Pétersbourg* 3: col. 56.—New South Wales, Australia.

Locally abundant on freshwater swamps, lakes, and streams on the Mainland. Also recorded from ?Batanta[K] and Fergusson I[MA]; also the Aru Is[CO] (M. Tarburton's island website[K]). A sporadic breeding resident and apparently a common visitor to lowlands (especially to the Trans-Fly) from Australia; local above 500 m; a few records to 2,300 m[CO] in open mid-mt valleys. Fledglings have been found in nesting colonies with *Anhinga novaehollandiae* in swamps northwest of Port Moresby (east of Lea Lea) in July (Wolfe 1969). Bishop (2005a) reported large breeding colonies in the Trans-Fly: Memunggal/Arerr and Kiterr (Indonesian NG), and Embukam (PNG).

Extralimital—Sumatra to Australia and NZ; breeding restricted to Australia, New Caledonia, NZ, Java, and Lesser Sundas. Vagrant to Manus, New Britain, and Bougainville (Orta 1992b, Mayr & Diamond 2001, Hadden 2004, Dutson 2011).

NOTES

The species is best treated as monotypic (Mees 1982b). *Other names*—Little Black Shag.

Australian Pied Cormorant *Phalacrocorax varius*   VAGRANT

Pelecanus varius Gmelin, 1789. *Syst. Nat.* 1(2): 576.—Queen Charlotte Sound, NZ.

One observed at the Moitaka settling ponds in late Jul-1987 (Hicks 1988a, R. Schodde in Coates 1990: 565). The bird was observed on several different days by R. Schodde, R. Hicks, I. Burrows, P. Gregory, and others.

Extralimital—breeds in Australia and NZ. A vagrant to the sw Torres Strait Is (Draffan *et al.* 1983).

NOTES

Expected subspecies is *hypoleucos*, based on distribution of the races. *Other names*—Pied or Great Pied Cormorant.

Great Cormorant *Phalacrocorax carbo*   VISITOR

Pelecanus Carbo Linnaeus, 1758. *Syst. Nat.*, ed. 10, 1: 133.—Rock-nesting form of the n Atlantic Ocean.

Rare in the Region; records from the Mainland and ?Batanta I[J] (M. Tarburton's island website[J]). Probably only a nonbreeding visitor from Australia (Coates 1985).

Extralimital—breeds in e Canada and Greenland east through Europe and many parts of Asia, parts of Africa, Australia, and NZ. Record from Rennell I (Mayr & Diamond 2001).

SUBSPECIES

a. *Phalacrocorax carbo novaehollandiae* (Stephens)

Phalacrocorax Novae Hollandiae Stephens, 1826. In Shaw, *General Zool.* 13(1): 93.—New South Wales, Australia.

RANGE: Breeds in New Caledonia, Rennell, Australia, NZ, Lord Howe, Norfolk, Macquarie, Snares, and Campbell Is. Localities—Batanta I[N], Paniai Lake[O], Alkmaar/Noord R[MA], Danau Bira[K], Dolak I[J], Bensbach R[J], Daru, Balimo[CO], near Tamaria R/Komo[CO], Tari[L], Wapenamanda[L], Tonga[M], Baiyer R[CO], Ramu R, ear Port Moresby[CO] (Melville 1980[O], Bishop 1987[J], Beehler field notes 10-Jul-1987[K], D. Hobcroft *in litt.*[L], P. Gregory *in litt.*[M], M. Tarburton's island website[N]).

DIAGNOSIS: Presumably all birds in NG arrive from the south and refer to the form *novaehollandiae*. A problem in racial designation is created by the fact that birds in NG will mainly be in nonbreeding plumage. Wing: male 332–356, female 324–345 Marchant & Higgins (1990).

NOTES

Other names—Cormorant; Big Black, Large, Large Black, Black, or European Cormorant.

Family Anhingidae

Darters [Family 4 spp/Region 1 sp]

The species of the family range from the southeastern United States, Central America, and northern South America to sub-Saharan Africa, tropical Asia, and Australia. They inhabit wetlands, including lagoons, swamps, shallow bays, and estuaries. Birds of this tiny family are somewhat cormorant-like but are distinguished by the small head; long, straight, and pointed beak (unhooked); the long, squared-off tail; and long, snakelike neck (Orta 1992c). The darter family is a sister form to the cormorants, and the darter+cormorant clade is sister to the boobies (Hackett *et mult.* 2008).

GENUS *ANHINGA* [Genus 4 spp/Region 1 sp]

Anhinga Brisson, 1760. *Orn.* 1: 60; 6: 476. Type, by tautonymy and monotypy, "Anhinga" Brisson = *Plotus anhinga* Linnaeus.

The description for the family applies to the genus. Sangster *et al.* (2013) summarized the argument for treating the four forms of *Anhinga* as full species (*pace* Orta 1992c).

Australasian Darter *Anhinga novaehollandiae*   RESIDENT AND VISITOR | MONOTYPIC

Plotus Novae-Hollandiae Gould, 1847. *Proc. Zool. Soc. London*: 34.—New South Wales, Australia.
    Synonym: *Anhinga rufa papua* Rand, 1938. *Amer. Mus. Novit.* 990: 1.—Lake Daviumbu, middle Fly R, Trans-Fly.
Generally a scarce resident of lowland rivers and lakes of NG, though large concentrations recorded on occasion; also possibly a winter visitor from Australia, straying widely. Found at 1,300 m at Lake Kopiago[CO]. The species has been found breeding in the Trans-Fly in April, the lower Fly in August and September, and near Port Moresby in July and November (Coates 1985).
    Localities—?Salawati I[Z], Wasur NP[J], Kurik[L], Wasur NP[Y], Bensbach R[CO], Trans-Fly[CO], Lake Daviumbu[J], Sepik R[MA], Astrolabe Bay[MA], Hall Sound[MA], near the Aroa R[L], Port Moresby[CO], and Fergusson I[MA] (Rand 1942a[J], Mees 1982b[L], Halse *et al.* 1996[Y], S. Hogberg in Mar-2006 from M. Tarburton's island website[Z]).
    Extralimital—Australia to the Moluccas and Lesser Sundas; also records from Long I and New Britain (Dutson 2011).

NOTES
We follow Schodde *et al.* (2012) and Christidis & Boles (2008) in treating the Australasian form as specifically distinct (*pace* Orta 1992c and Dorst & Mougin 1979). The male of *novaehollandiae* can be distinguished from the geographically nearest sister form *melanogaster* by the pale iris, the crisp pied facial pattern, the short cheek stripe, and the brown spot at the ventral bend of the neck. Following Orta (1992c), we recognize a single form in Australasia. We thus treat *novaehollandiae* as monotypic, in spite of the minor plumage distinction identified by Schodde *et al.* (2012), who showed that the dorsal plumage of female specimens from NG is "dusky" rather than "mid to deep brownish gray," and that "the rich gray centers of the wing coverts are often reduced." As stated elsewhere, we are resistant to recognizing subspecies based on minor shades of gray or brown, which may related to specimen preparation, age, or storage. Even if recognized, the form *papua* would be considered a thinly defined subspecies. Moreover, given the vast twice-annual movement of waterbirds from Australia to the Trans-Fly of NG, the expectation of differentiation between the two populations is low. *Other names*— Darter, Australian or Oriental Darter, *Anhinga rufa*, *A. melanogaster*.

ORDER OTIDIFORMES

Hackett *et mult.* (2008) identified this ordinal-level clade that circumscribes the bustards, which constitute a distinct lineage sister to the cranes, cuckoos, trumpeters (Psophididae), and rails (Cracraft 2013).

Family Otididae

Bustards [Family 26 spp/Region 1 sp]

The species of the family range from Europe and Africa through Asia to Indochina, southern New Guinea, and Australia (Collar 1996). The family comprises large terrestrial birds of open grassy plains and savannas. Defining characters include the short and stout bill; long neck; large and heavy body; long, broad wings; short to medium-length tail; moderately long, stout legs; and cryptically colored upperparts. Males perform remarkable courtship displays. The single New Guinean species inhabits savanna of the Trans-Fly.

GENUS *ARDEOTIS* [Genus 4 spp/Region 1 sp]

Ardeotis Le Maout, 1853. *Règnes Nat. Hist. Ois.*: 339–40. Type, by monotypy, *Otis arabs* Linnaeus.

The species of this genus range from western, central, and southern Africa, southern Asia, southern New Guinea, and Australia (Collar 1996). The genus is characterized by a peculiar flat crown that is black-capped, an unpatterned neck, and a plain brown mantle. The head is held slightly bill-up. These are large and impressive birds of open plains and light woodland. Their flight is strong and steady, on huge, rounded wings. Our treatment follows Collar (1996) and Christidis & Boles (2008).

Australian Bustard *Ardeotis australis*   RESIDENT | MONOTYPIC

Otis Australis J. E. Gray, 1829. In Griffith, *Anim. Kingdom* 8 (3: Aves): 305.—New South Wales, Australia.

Local resident of lowland savanna grasslands of the s Trans-Fly.

Localities—Kurik[J], Wasur NP[J], Bensbach[CO], Tarara[K], and Wuroi[MA] (Rand 1942a[K], Bishop 2005a[J]). Stronach (1981) provided aerial counts of 340 individuals in the Trans-Fly on 24–25-May-1980, of 165 on 8–9-Nov-1980, and of 235 on 17–18-Jan-1981. Record from Dolak I (Mees 1982b) is doubtful (K. D. Bishop *in litt.*).

Extralimital—drier portions of the n two-thirds of Australia (Collar 1996).

NOTES

Coates (1985) summarized breeding evidence for New Guinea. Reports of movement between Australia and s NG require confirmation. *Other names*—Bustard, *Choriotis* or *Eupodotis australis*.

ORDER GRUIFORMES

The Gruiformes appear to constitute a monophyletic assemblage that now includes the cranes, Limpkin (Aramidae), trumpeters (Psophodidae), flufftails (Sarothruridae), finfoots (Heliornithidae), and rails (Hackett *et mult.* 2008, Cracraft 2013).

Family Rallidae

Rails, Crakes, Moorhens, and Coots [Family 138 spp/Region 18 spp]

The species of the family range widely and are absent only from the high Arctic, the Sahara, and the dry interior of Asia (Taylor 1996, 1998). In New Guinea, rails inhabit wetlands, grasslands, scrub, and lowland and montane forest interior. They range from the lowlands to 3,600 m (Pratt & Beehler 2014). The family is a member of the core Gruiformes and is a sister to the finfoots and flufftails (Hackett *et mult.* 2008). New Guinea's species are quite variable in plumage and size—ranging from large to very small, from small-billed to large-billed—but all have strong legs and large feet. The group includes the water rails, forest-rails, bush-hens, swamphens, gallinules, and coots. Many flightless Pacific island forms have become extinct in the last several centuries from human activities on these islands, first by indigenous colonizers, then by western settlers. This one-two punch has decimated the rail fauna of the smaller oceanic islands of the Pacific. Our sequence is informed by the treatments of Christidis & Boles (2008), Kirchman (2012), and Dickinson & Remsen (2013). We also note the morphological work of Livezey (1998) on this lineage. A comprehensive molecular phylogeny of the rails is much needed.

GENUS *RALLICULA* ENDEMIC [Genus 4 spp/Region 4 spp]

Rallicula Schlegel, 1871. *Ned. Tijdschr. Dierk.* (1873) 4: 55. Type, by monotypy, *Rallicula rubra* Schlegel.

The species of this genus range throughout the montane forests of New Guinea (Pratt & Beehler 2014). They are characterized by small size; the forest-dwelling habit; deep chestnut head, neck, and breast; short black bill; and black legs. Ripley (1977) and Taylor (1996) placed this clade into the Asian *Rallina*, which is distinct in being larger, more boldly patterned, and in all instances plain-patterned dorsally. We follow the recommendation of Christidis & Boles (2008) in recognizing *Rallicula*. Frith & Frith (1988a, 1992) enumerated the considerable differences between *Rallicula* and *Rallina* in nesting biology and chick coloration.

Chestnut Forest-Rail *Rallicula rubra* ENDEMIC

Rallicula rubra Schlegel, 1871. *Ned. Tijdschr. Dierk.* (1873) 4: 55.—Arfak Mts, Bird's Head.

Inhabits the floor of montane forest of the Bird's Head and the Western Ra, Border Ra, and w sector of the Eastern Ra, 1,500–3,050 m[CO].

SUBSPECIES

a. *Rallicula rubra rubra* Schlegel
RANGE: Arfak Mts, Bird's Head.
DIAGNOSIS: Male dorsal surface deep vinaceous chestnut with traces of black feather shafting and with nape tinged black (Rand & Gilliard 1967).

b. *Rallicula rubra klossi* Ogilvie-Grant
Rallicula klossi Ogilvie-Grant, 1913. *Bull. Brit. Orn. Club* 31: 104.—Upper Otakwa R, Western Ra.
    Synonym: *Rallicula rubra subrubra* Rand, 1940. *Amer. Mus. Novit.* 1071: 3.—Lake Habbema, Western Ra.
    Synonym: *Rallicula rubra telefolminensis* Gilliard, 1961. *Amer. Mus. Novit.* 2031: 1.—Mt Ifal, 2,195 m, Victor Emanuel Mts, Border Ra.
RANGE: Western, Border, and Eastern Ra (Junge 1953) east to the sw flank of Mount Hagen (photo from Kumul Lodge).
DIAGNOSIS: As for nominate, but male paler, without black nape or black dorsal feather shafting; smaller (Rand & Gilliard 1967). The race *subrubra* is subsumed into *klossi* following Taylor (1996). We also subsume *telefolminensis*, described from a single adult specimen, into *klossi*.

In all likelihood this species will be found in the Tamrau Mts, which has been shown to share virtually all Arfak endemics. *Other names*—Chestnut Rail, New Guinea Chestnut Rail, *Rallina rubra*.

White-striped Forest-Rail *Rallicula leucospila* ENDEMIC | MONOTYPIC
Corethrura? leucospila Salvadori, 1876. *Ann. Mus. Civ. Genova* 7: 975.—Arfak Mts, Bird's Head.

Inhabits montane forests of the Tamrau, Arfak, Kumawa, and Wandammen Mts of the Bird's Head and Neck region, 1,350–1,850 m (Diamond & Bishop 2015).

NOTES
Presumably this species and *Rallicula rubra* share the montane forests of the Arfak Mts. In these situations, *R. rubra* will probably be found at higher elevations. *Other names*—White-striped Chestnut Rail, White-striped Rail, *Rallina leucospila*. The Kumawa Mts population appears distinct (Diamond & Bishop 2015).

Forbes's Forest-Rail *Rallicula forbesi* ENDEMIC
Rallicula forbesi Sharpe, 1887. In Gould, *Birds New Guinea* 23: pl. [70] and text [= 5 of bound vol.].—Owen Stanley Ra, SE Peninsula.

Inhabits montane forests of the Central Ra, Adelbert Mts, and mts of Huon Penin, 1,100–3,000 m^{CO}. Few records west of the PNG Border Ra. Where this species shares the forest with *R. rubra*, the latter occupies mainly higher elevations (Diamond 1969).

SUBSPECIES
a. *Rallicula forbesi forbesi* Sharpe
Synonym: *Rallicula leucospila steini* Rothschild, 1934. *Orn. Monatsber.* 42: 46.—Kunupi 1,500 m, Weyland Mts, Western Ra.
Synonym: *Rallicula rubra dryas* Mayr, 1931. *Mitt. Zool. Mus. Berlin* 17: 709.—Kulung-tufu, Saruwaged Ra, Huon Penin.
RANGE: Central Ra and mts of the Huon Penin (Rand 1942b, Gyldenstolpe 1955a).
DIAGNOSIS: Lower back and wings of male blackish brown; tail (male) 64–79 (Rand & Gilliard 1967). Wing 106.5–117 (n=26; Diamond 1969).

b. *Rallicula forbesi parva* Pratt
Rallicula forbesi parva Pratt, 1983. *Emu* (1982) 82. 119.—Mount Mengam, *ca.* 1,500 m, Adelbert Mts, Madang Prov, PNG.
RANGE: Montane forests of the Adelbert Mts.
DIAGNOSIS: Smaller than the nominate form; wing 100–103 (n=5); mantle color of male dark olive brown, finely vermiculated (Pratt 1983).

NOTES
The races *steini* and *dryas* are here subsumed into the nominate form, and the combined population can be considered clinal. The analyses of Diamond (1969) and Pratt (1983) and a close examination of the material in the AMNH indicated considerable variation in size and plumage coloration, in many cases a product of age- and sex-related condition. We predict most of the described races for the genus will disappear when given further comprehensive scrutiny of additional material from the field. *Other names*—Forbes's Chestnut Rail, Forbes' Rail, *Rallina forbesi*.

Mayr's Forest-Rail *Rallicula mayri* ENDEMIC
Rallicula rubra mayri Hartert, 1930. *Novit. Zool.* 36: 124.—Cyclops Mts, ne sector of the NW Lowlands.

Inhabits montane forest in the Foja Mts, Cyclops Mts, and PNG North Coastal Ra, 1,160–2,070 m^{CO}. For a detailed discussion of the plumages and populations of this species and its close relatives see Diamond (1969) and Pratt (1983).

a. *Rallicula mayri mayri* Hartert

RANGE: Cyclops Mts.

DIAGNOSIS: Upperparts of male uniformly dark reddish brown; underparts chestnut brown; tail like upperparts but with indistinct narrow black bars (Rand & Gilliard 1967).

b. *Rallicula mayri carmichaeli* Diamond

Rallicula mayri carmichaeli Diamond, 1969. *Amer. Mus. Novit.* 2362: 3.—Mount Nibo, Torricelli Mts, n NG.

RANGE: Foja[J,K], Bewani[L], and Torricelli[L] Mts (Diamond 1969[L], Diamond 1985[K], Beehler *et al.* 2012[J]).

DIAGNOSIS: Plumage dull red-brown overall, darker dorsally, paler and richer ventrally; female mantle marked with dark spots that have buff centers (Beehler & Prawiradilaga 2010).

NOTES

The relationships among the populations of *R. forbesi*, *R. leucospila*, *R. mayri*, and *R. rubra* need additional study, especially at the molecular level. *R. mayri* is most similar in plumage to *forbesi* and perhaps should be combined with that species. Hartert originally described *mayri* as a subspecies of *R. rubra*. *Other names*—Mayr's Rail, Mayr's Chestnut Rail, *Rallina mayri*.

GENUS *RALLINA*   [Genus 4 spp/Region 1 sp]

Rallina "Reichenbach" G. R. Gray, 1846. *List Gen. Birds* 3: 595. Type, by original designation, *Rallus fasciatus* Raffles.

The species of this genus range from India and Sri Lanka east through Indochina, Philippines, Indonesia, the New Guinea region, the Bismarck Archipelago, and northeastern Australia (Taylor 1996). The genus is characterized by the bold combination of the red-brown forebody and the barred flanks. Species are larger than those of *Rallicula*.

Red-necked Crake *Rallina tricolor*   RESIDENT | MONOTYPIC

Rallina tricolor G. R. Gray, 1858. *Proc. Zool. Soc. London*: 188.—Aru Is.

Synonym: *Rallina tricolor maxima* Mayr, 1949. *Amer. Mus. Novit.* 1417: 13.—Waigeo I, NW Islands.

Inhabits swampy forest throughout NG; also Waigeo[MA], Batanta[J,Q], Salawati[A], Misool[A], Gag[L], and Kofiau I[B]; Aru Is[MA]; Yapen[MA] and Karkar I[MA,W]; and Normanby I[Z]; lowlands, with records to 1,200 m[R] (Gyldenstolpe 1955b: 380[J], Mees 1965[A], Greenway 1966[Q], Diamond & LeCroy 1979[W], Coates & Peckover 2001[R], Johnstone 2006[L], Diamond et al. 2009[B], K. D. Bishop *in litt.*; Normanby specimen, AMNH no. 784689, adult male, collected on Mt Pabinama, at 820 m, Normanby I, on 5-May- 1956, by Russell F. Peterson[Z]).

Extralimital range—Moluccas, Lesser Sundas, Bismarck Is, and ne QLD, Australia (Coates & Bishop 1997, Mayr & Diamond 2001, Dutson 2011).

NOTES

We follow Taylor (1996: 155) in treating the species as monotypic. For discussion of the synonymized race *maxima* see Mees (1965) and Greenway (1973: 308). *Other names*—Red-necked Rail.

GENUS *LEWINIA*   [Genus 4 spp/Region 1 sp]

Lewinia G. R. Gray, 1855. *Cat. Gen. Subgen. Birds*: 120. Type, by original designation and virtual tautonomy, *Rallus lewinii* Swainson = *Rallus pectoralis* Temminck.

We follow Taylor (1996) and Christidis & Boles (2008) in recognizing *Lewinia*. The genus is characterized by prominently barred flanks contrasting with the plain gray or tan breast, the longish bill with reddish base, the patterned brown mantle, and the cocked tail. *Lewinia* species are difficult to distinguish from certain species of *Rallus*.

Lewin's Rail *Lewinia pectoralis*

Rallus pectoralis Temmick, 1831. *Planch. Col. d'Ois.*, livr. 88: text to pl. 523, part.—New South Wales, Australia.

A scarce and poorly known inhabitant of mid-montane grasslands of the Bird's Head and Central Ra, 1,040–2,600 m (Coates 1985); no records from the uplands of the Huon Penin, but should be looked for there.

Extralimital—Flores I, e and se Australia (formerly also sw Australia; Taylor 1996).

SUBSPECIES

a. *Lewinia pectoralis mayri* (Hartert)
Rallus pectoralis mayri Hartert, 1930. *Novit. Zool.* 36: 121.—Kofo, Anggi Giji Lake, Arfak Mts, Bird's Head.
    Synonym: *Rallus pectoralis connectens* Junge, 1952. *Zool. Meded. Leiden* 31:247.—Paniai Lake, Western Ra.
    Synonym: *Rallus pectoralis captus* Mayr & Gilliard, 1951. *Amer. Mus. Novit.* 1524: 2.—Tomba, near Mt Hagen, Eastern Ra.
RANGE: Mountains of the Bird's Head and the Western, Border, and Eastern Ra, perhaps east to the Kratke Mts[CO]. Localities—Anggi Giji Lake[J], Paniai Lake[P], Araboebivak[M], Bele R[Q], Baliem valley[X], Telefomin[T], Mt Giluwe[W], Ambua Lodge[K], Mt Hagen[Z], Nondugl[L], Weiga[L], Awande[DI] (Hartert 1930[J], Rand 1942b[Q], Junge 1953[P], Mayr & Gilliard 1954[Z], Gilliard & LeCroy 1961[T], Gyldenstolpe 1955b[L], Sims 1956[W], Ripley 1964a[X], Hoek Ostende *et al.* 1997[M], D. Hobcroft *in litt.*[K]).
DIAGNOSIS: Plumage variable. Back feathers edged with olive (Rand & Gilliard 1967). Large; wing (male) 105–109 (Greenway 1973).

b. *Lewinia pectoralis alberti* (Rothschild and Hartert)
Hypotaenidia brachypus alberti Rothschild and Hartert, 1907. *Novit. Zool.* 14: 451.—Owgarra, Angabanga R, s watershed of SE Peninsula.
    Synomyn: *Rallus striatus insulsus* Greenway, 1935. *Proc. New Engl. Zool. Club* 14: 28.—Wau, 1,160 m, upper Watut region, SE Peninsula.
RANGE: Mountains of the SE Peninsula.
DIAGNOSIS: Plumage variable. Like *mayri* but small; wing (male) 93–94 (see Greenway 1973).

NOTES

Ripley (1977) subsumed *R. p. connectens* into *mayri*. We subsume *L. p. insulsus* into *alberti* at the recommendation of Rand & Gilliard (1967) and because of the biogeographically improbable range of this form. We subsume *R. p. captus* into *mayri* based on the reviews of Rand & Gilliard (1967) and Ripley (1977). *Other names*—Water or Slate-breasted Rail, *Rallus pectoralis*.

GENUS *HYPOTAENIDIA*   [Genus 12 spp/Region 2 spp]

Hypotaenidia Reichenbach, 1853. Type, by original designation, *Rallus pectoralis* Reichenbach (not *Rallus pectoralis* Temminck, 1831) = *Rallus philippensis* Linnaeus.

The species of this genus range from Okinawa, the Philippines, Guam, and Sulawesi, east to New Guinea, the Solomon Islands, Vanuatu, Fiji, New Caledonia, Lord Howe Island, eastern Australia, New Zealand, and scattered islands in the central Pacific (Taylor 1996). The genus is characterized by a short bill, a superciliary or malar stripe in most forms, and the presence of a shortish tail (Kirchman 2012).

Barred Rail *Hypotaenidia torquata*

Rallus torquatus Linnaeus, 1766. *Syst. Nat.*, ed. 12, 1: 262.—Philippines.

Known from a few records from the far west of the Region. Apparently inhabits patches of rank grass in the lowlands, coastal swamplands and marshes, and ricefields of the Bird's Head and Salawati I (see White & Bruce 1986).

Extralimital—the Moluccas, Sulawesi, and the Philippines.

a. *Hypotaenidia torquata limaria* Peters
Rallus torquatus limarius Peters, 1934. *Check-list Birds World* 2: 166.—Salawati and Dorey Hum I, NW Islands. Replacement name for *Hypotaenidia saturata* P. L. Sclater (ex Salvadori ms), 1880 (4-August), *Ibis*: 310, note; preoccupied by *Rallus longirostris saturatus* Ridgway, 1880 (6-July).
RANGE: Salawati I[MA] and adjacent w coast of Bird's Head (Sorong[J]), north to Dorey Hum I[MA] (Gyldenstolpe 1955b[J]). Heard near Kawawawi, s Kumawa Mts in Nov-2014 (C. Thébaud *in litt.*)
DIAGNOSIS: Bill greater than 45; similar to *celebensis*, but throat black without white crossbars (Ripley 1977).

NOTES
Apparently a rare resident of the far west of the Region. The range map in Ripley (1977) showed the species range extending east to the mouth of the Mamberamo R. This is presumably in error, and it was probably replicated by Taylor (1996). *Other names—Rallus* or *Gallirallus torquatus.*

Buff-banded Rail *Hypotaenidia philippensis* RESIDENT
Rallus philippensis Linnaeus, 1766. *Syst. Nat.*, ed. 12, 1: 263.—Philippines.
 Inhabits grasslands and garden scrub throughout the Mainland, from sea level to 3,600 m (Coates 1985).
 Extralimital—se Asia, Wallacea, e Australia, NZ, Bismarck and Solomon Is, New Caledonia, Vanuatu, Fiji, Tonga, Samoa, and Niue (Taylor 1996, Coates & Bishop 1997, Mayr & Diamond 2001).

SUBSPECIES
a. *Hypotaenidia philippensis mellori* (Mathews)
Eulabeornis philippensis mellori Mathews, 1912. *Novit. Zool.* 18: 192.—Sandy Hook I, sw Australia.
 Synonym: *Rallus philippensis randi* Mayr & Gilliard, 1951. *Amer. Mus. Novit.* 1524: 4.—11 km ne of Mt Wilhelmina, Western Ra.
 Synonym: *Eulabeornis philippensis yorki* Mathews, 1913. *Austral. Av. Rec.* 2: 6.—Cape York Penin, QLD, Australia.
RANGE: Bird's Head and Neck, Western Ra, Trans-Fly, and S Lowlands; also Australia and Norfolk I.
DIAGNOSIS: Note rufous pectoral band on breast quite dark, crown quite rufous, back heavily spotted, and black bars on flanks conspicuous (Ripley 1977). Rare in the Trans-Fly (K. D. Bishop *in litt.*)

b. *Hypotaenidia philippensis lacustris* (Mayr)
Rallus philippensis lacustris Mayr, 1938. *Amer. Mus. Novit.* 1007: 6.—Ifar, Lake Sentani, n NG.
 Synonym: *Rallus philippensis reductus* Mayr, 1938. *Amer. Mus. Novit.* 1007: 6.—Long I, off ne NG.
 Synonym: *Rallus philippensis wahgiensis* Mayr & Gilliard, 1951. *Amer. Mus. Novit.* 1524: 3.—Nondugl, Wahgi valley, Eastern Ra.
RANGE: Northwestern Lowlands, Sepik-Ramu, Border Ra, Eastern Ra, and apparently SE Peninsula; also Long I. Interior highland records from Mt Giluwe at 2,300 m[K] and Kyaka Territory[L] (Sims 1956[K], Diamond 1972[L]).
DIAGNOSIS: A large and dark form with the pectoral band well developed and complete, forehead and crown blackish, dark rufous ocular stripe, and black flank bars much broader than the intervening white bars (Ripley 1977). There is need to study further the subspecies limits of all birds in the NG region.

NOTES
Junge (1953) provided a detailed discussion on size and plumage variation. Subspecific treatment follows Ripley (1977) and Taylor (1996), with updates based on our own museum review, which led to the sinking of *randi* from high elevations and *reductus* from Long I. Additional work is needed to improve understanding of individual and age-related variation. This species, which inhabits a range of small islands in the Bismarck Arch, remarkably is absent from all of NG's fringing islands (NW Islands, SE Islands, Karkar I, and Manam I). *Other names—*Banded Land Rail, Land Rail, *Rallus* or *Gallirallus philippensis.*

GENUS *EULABEORNIS*

Eulabeornis Gould, 1844. *Proc. Zool. Soc. London*: 56. Type, by monotypy, *Eulabeornis castaneoventris* Gould.

The single species of the genus inhabits small patches of northern Australia and the Aru Islands (Taylor 1996). This monotypic genus is characterized by large size, large dull yellow bill and legs, and rather prominent tail. Molecular evidence indicates this species is part of a Pacific clade that includes the previous three species (Kirchman 2012).

Chestnut Rail *Eulabeornis castaneoventris*   RESIDENT

Eulabeornis castaneoventris Gould, 1844. *Proc. Zool. Soc. London*: 56.—Gulf of Carpenteria, n Australia.

Within the NG region, restricted to the Aru Is. Inhabits mangrove swamps and tropical estuaries. Extralimital—nw QLD and NT, Australia (Taylor 1996).

SUBSPECIES

a. *Eulabeornis castaneoventris sharpei* Rothschild

Eulabeornis castaneiventris sharpei Rothschild, 1906. *Bull. Brit. Orn. Club* 16: 81.—Wokan, Aru Is.

RANGE: Aru Is.

DIAGNOSIS: Differs from the nominate form by deep and rich vinaceous-brown breast and belly; mantle olive brown instead of gray brown. Taylor (1998) noted "bill heavier and deeper, green with a yellow tip and red around nostrils."

NOTES

Other names—Chestnut-bellied Rail, *Gallirallus castaneoventris*.

GENUS *GYMNOCREX*   [Genus 3 spp/Region 1 sp]

Gymnocrex Salvadori, 1875. *Ann. Mus. Civ. Genova* 7: 678. Type, by original designation, *Rallina rosenbergii* Schlegel.

The three species of the genus range from Sulawesi and the Talaud Islands of northeastern Indonesia east to New Guinea, the Aru Islands, and New Ireland. They are forest-dwelling rails, with a diagnostic patch of colored bare skin forming something of a mask; note also the lack of any barring on plumage, and the chestnut mantle. *Gymnocrex* appears to have affinities with the wood rails of the genus *Aramides*.

Bare-eyed Rail *Gymnocrex plumbeiventris*   RESIDENT | MONOTYPIC

Rallus plumbeiventris G. R. Gray, 1862. *Proc. Zool. Soc. London* (1861): 432.—Misool I, NW Islands.

Synonym: *Rallus hoeveni* von Rosenberg, 1866. *Ned. Tijdschr. Dierk.* 3: 349.—Aru Is.

Inhabits lowland forest, wet forest edges, and upland interior marshes of NG, Aru Is[MA], and Misool[Q], Biak[J], and Karkar I[MA] (Mees 1965[Q], Bishop ms[J]); to 2,100 m in Tabubil region. Some populations may be migratory or nomadic (Bell 1984). Most records come from the NW Lowlands, Sepik-Ramu, and w sector of the S Lowlands (*e.g.*, Gilliard & LeCroy 1966), but there are scattered records elsewhere.

Localities—Misool[D], Sorong[Z], Biak I[W], Jayapura[C], Aru Is[MA], Noord R, Fly R[U], Setekwa R[MA], Tabubil[B], Sepik R[L], Wahgi valley[CO], n slope Mt Wilhelm[K], Purari basin[J], Karkar I[MA], and Brown R[Q] (Rand 1942a[U], 1942b[C], Mees 1965[D], Gilliard & LeCroy 1966[L], Bell 1984[Q], Gregory 1995b[B], Hornbuckle & Merrill 2004[Z], Bishop ms[W], Sam & Koane 2014[K], I. Woxvold *in litt.*[J]). The Biak I record is supported by photographs.

Extralimital—n Moluccas; also a single specimen (described as race *intactus* Sclater 1869) is from an unknown locality, either on New Ireland or the Solomon Is (Mayr & Diamond 2001, Dutson 2011).

NOTES

Treated as monotypic following recommendation of Taylor (1996) and the plumage review of Ripley (1977). *Other names—Eulabeornis plumbeiventris*.

Zapornia Stephens, 1824. In Shaw's *Gen. Zool.* 12 (1): 230. Type, by monotypy, *Z. pusilla* Stephens = *Rallus parvus* Scopoli.

The species of this genus range from Africa and Eurasia east to Japan, New Guinea, Australia, and the central Pacific. The diverse genus is characterized by the small and slight body and the small head and bill. Birds are difficult to distinguish from some species of *Laterallus* and *Porzana*.

Baillon's Crake *Zapornia pusilla* RESIDENT (AND VISITOR?)

Rallus pusillus Pallas, 1776. *Reise Versch. Prov. Russ. Reichs.* 3: 700.—Southeastern Siberia.

Scattered records from grassy wetlands in the lowlands and highlands of mainland NG and the NW Islands; lowlands to 2,500 m (Pratt and Beehler 2014).

Extralimital—Europe and s Africa, east to e Asia, Australia, New Caledonia, and NZ. Also winters in India and Indochina (Taylor 1996, Dutson 2011).

SUBSPECIES

a. *Zapornia pusilla mayri* (Junge)

Porzana pusilla mayri Junge, 1952. *Zool. Meded. Leiden* 31: 247.—Paniai Lake, Western Ra.

RANGE: Western, Border, and Eastern Ra to Mt Giluwe (Ripley 1977). Localities—Misool and ?Salawati I[J], the Bird's Head, Weyland Mts[CO], Paniai Lake[L], Mamberamo R, Lake Sentani, Tari[CO], Mt Giluwe[K] (Sims 1956[K], Hoek Ostende *et al.* 1997[L], S. Hogberg in M. Tarburton's island website[J]).

DIAGNOSIS: Differs from *palustris* by smaller size and upper surface of wing and sides of neck warmer brown, less tawny-colored; wing 76–79 (Ripley 1977).

b. *Zapornia pusilla palustris* (Gould)

Porzana palustris Gould, 1843. *Proc. Zool. Soc. London* (1842): 139.—Tasmania.

RANGE: Eastern Ra east of range of *mayri*, and SE Peninsula; also Australia (Dickinson 2003). Localities—Oriomo R, Daru, and the Port Moresby environs[CO].

DIAGNOSIS: Pale, its breast more of a French gray than a dark leaden gray (Ripley 1977). Throat whitish and face medium gray. Wing 77–88 (Ripley 1977).

NOTES

Represented by very few specimens in museum collections, so the validity of these races is uncertain. Diamond (1972) stated that the Mt Giluwe specimens belong to the race *palustris*. It is unclear which subspecies pertains to the records from the Ramu R and Astrolabe Bay. Some Australian birds may winter in NG (Coates 1985). *Other names*—Marsh, Tiny, or Lesser Spotted Crake, *Porzana pusilla*.

Spotless Crake *Zapornia tabuensis* RESIDENT | MONOTYPIC

Rallus tabuensis Gmelin, 1789. *Syst. Nat.* 1(2): 717, no. 20.—Tongatapu group, Tonga.

Synonym: *Porzana tabuensis richardsoni* Rand, 1940. *Amer. Mus. Novit.* 1072: 3.—Lake Habbema, 3,225 m, Western Ra.

Synonym: *Porzana tabuensis edwardi* Gyldenstolpe, 1955. *Arkiv f. Zool.* (2)8(1): 34.—Nondugl, Wahgi valley, Eastern Ra.

Widespread through grassy wetlands in the lowlands and mts of the Bird's Head and Central Ra, sea level–3,150 m[CO] (Beehler *et al.* 1986).

Localities—?Salawati[A], Kebar valley[Q], Anggi Lakes in Arfak Mts[MA], Paniai Lake[CO], Ilaga valley[K], Baliem R[L], Lake Habbema[L], near Dolak I[C], Kurik[W], Tabubil (1997)[X], Tari[CO], Mt Giluwe[B], Nondugl[J], Goroka[CO], Awande[DI], Astrolabe Bay[MA], Wau[CO], near Popondetta[CO], near Bereina[CO], Brown R[CO], Kanudi[CO], Varirata NP[CO], and Port Moresby[CO] (Rand 1942b[L], Gyldenstolpe 1955a[J], Sims 1956[B], Ripley 1964a[K], Hoogerwerf 1964[W], Gilliard & LeCroy 1970[Q], P. Gregory *in litt.*[X], Bishop 2005[C], S. Hogberg in M. Tarburton's island website[A]).

Extralimital—Philippines and Micronesia, south to Australia, NZ, east through New Caledonia to Polynesia (Taylor 1996). Also scattered records from the Bismarck and Solomon Is (Dutson 2011). Possibly a migrant across Torres Strait (Draffan *et al.* 1983).

We treat the species as monotypic. Described subspecies from NG have biogeographically implausible ranges and minimal plumage and size variation. This all can be subsumed into clinal and individual variation, which does not merit the naming of subspecies. See comments by Greenway (1973: 312) and Taylor (1996: 189). *Other names*—Sooty Rail or Crake, *Porzana tabuensis*.

GENUS *AMAURORNIS* [Genus 7 spp/Region 2 spp]

Amaurornis Reichenbach, 1853. *Avium Syst. Nat.* (1852): xxi. Type, by original designation, *Gallinula olivacea* Meyen.

The species of this genus range from sub-Saharan Africa and tropical Asia east to New Guinea, the Bismarck and Solomon Islands, and Australia. This rather diverse genus is characterized by the medium-length neck; shortish bill that is either greenish or horn; the plain unbarred plumage; and the cocked tail (Slikas *et al.* 2002).

White-browed Crake *Amaurornis cinerea* RESIDENT | MONOTYPIC

Porphyrio cinereus Vieillot, 1819. *Nouv. Dict. Hist. Nat.*, nouv. éd. 28: 29.—Java.

Synonym: *Porzana leucophrys* Gould, 1847. *Proc. Zool. Soc. London*: 33.—Port Essington, NT, Australia.

Inhabits marshy grasslands throughout the lowlands of NG, Kofiau[Q], ?Misool[K], and Fergusson I[MA]; exceptionally to 1,830 m[CO] (Mees 1980[K], Taylor 1996, Diamond *et al.* 2009[Q]).

Localities—Utanata R[MA], Taritatu R[K], Humboldt Bay[CO], Sepik R[CO], Kurik[L], Fly R[MA], middle Sepik[J], Awande[DI], Astrolabe Bay[MA], Varirata NP[CO], Popondetta[CO], and Fergusson I[CO] (Rand 1942b[K], Gilliard & LeCroy 1966[J], Mees 1982b[L]).

Extralimital—se Asia, n and ne Australia, Micronesia, the Bismarcks, Solomons, Vanuatu, New Caledonia, Fiji, and Samoa (Taylor 1996, Mayr & Diamond 2001, Dutson 2011).

We follow Dickinson & Remsen (2013) in placement of this species in *Amaurornis* rather than *Poliolimnas* or *Porzana* (Rand & Gilliard 1967, Beehler & Finch 1985), based largely on the molecular phylogeny of Slikas *et al.* (2002). We follow Taylor (1996) in treating the species as monotypic. Mees (1980) noted that the record from Misool I is based on a 1954 sighting by S. D. Ripley. *Other names*—Ashy or Grey-bellied Crake, White-browed Rail, *Poliolimnas* or *Porzana cinerea*.

Rufous-tailed Bush-hen *Amaurornis moluccana* RESIDENT

Porzana moluccana Wallace, 1865. *Proc. Zool. Soc. London*: 480.—Ambon and Ternate, Moluccas.

Widespread in scrub, thick grassland, and forest edge through NG, the NW Islands, Biak, Karkar, Normanby, and Misima I; sea level–1,500 m (Coates 1985).

Extralimital—Moluccas and Sangihe Is, east to the Bismarck and Solomon Is, and south to n and e Australia.

a. *Amaurornis moluccana moluccana* (Wallace)

Synonym: *Gallinula ruficrissa* Gould, 1869. *Birds Austral.*, Suppl., pt. 5, pl. [79] and text.—Cape R, QLD, Australia.

RANGE: All mainland NG; also Misool[MA], Batanta[J,K], Biak[MA], Karkar[L], Normanby[CO], and Misima I[M] (Diamond & LeCroy 1979[L], LeCroy & Peckover 1998[M], Dutson and Tindige 2006[J], S. Hogberg in M. Tarburton's island website[K]).

Extralimital—the Moluccas and Sangihe Is, and n and e Australia.

DIAGNOSIS: Feathers of lower abdomen dull pinkish fawn, forming an indistinct patch; undertail dark olive brown.

David & Gosselin (2002a) noted that the genus name *Amaurornis* should be treated as feminine, and that the species epithet should agree, hence *moluccana*. Examination of specimens from the Moluccas and n Australia show complete overlap of plumage characters by the nominate form and *ruficrissa*. There is also considerable individual variation within populations based on age and specimen preparation. *Other names*—Bush-hen, Pale-vented Moorhen or Waterhen, Rufous-tailed Gallinule, *Amaurornis olivaceus*, *Gallinula olivacea*.

GENUS *MEGACREX*   ENDEMIC | MONOTYPIC

Megacrex D'Albertis & Salvadori, 1879. *Ann. Mus. Civ. Genova* 14: 129. Type, by original designation and monotypy, *Megacrex inepta* D'Albertis & Salvadori.

The single species of the genus is restricted to lowland swamp forest on the New Guinea mainland (Mayr 1941). The genus is characterized by large size, large heavy bill, flightlessness, very short wings, very large legs and feet, abbreviated tail, and dull plumage (Mees 1982b).

New Guinea Flightless Rail *Megacrex inepta*   ENDEMIC

Megacrex inepta D'Albertis & Salvadori, 1879. *Ann. Mus. Civ. Genova* 14: 130.—Upper Fly R.

Inhabits swampy forests, wet thickets, and mangroves of the major river basins of lowland NG. Absent from the SE Peninsula, Bird's Head, and Bird's Neck.

SUBSPECIES

a. *Megacrex inepta inepta* D'Albertis & Salvadori

RANGE: Southern watershed of the NG mainland. Localities—Setekwa R[MA], Noord R[MA], lower Digul R[A], Kumbe[Q], Sturt I[X], Lake Daviumbu[X], Fly R[MA], Elevala R[J], Ketu Creek[W], Kikori R (Gobe, Kopi, Veiru Creek)[Z], and Purari R delta[Z] (Bangs & Peters 1926[A], Rand 1942a[X], Mees 1982b[Q], D. Hobcroft *in litt.*[J], I. Woxvold *in litt.*[Z], P. Gregory *in litt.*[W]).

DIAGNOSIS: Brownish or vinaceous-brown sides of neck and body; rump dark brown (Rand & Gilliard 1967).

b. *Megacrex inepta pallida* Rand

Megacrex inepta pallida Rand, 1938. *Amer. Mus. Novit.* 990: 4.—Hol, Humboldt Bay, n NG.

RANGE: Northern NG. Localities—Mamberamo basin (Bernhard Camp)[K], Humboldt Bay (Hol or Holtekong)[J], and Sepik R[MA] (Rand 1942b[K], Ripley 1964a[J]).

DIAGNOSIS: Exhibits paler, buffier sides of lower neck and body (Taylor 1996), and much paler flanks than the nominate form (Rand 1938a, Ripley 1977).

NOTES

Based on the single n watershed and three s watershed birds that were examined at the AMNH, the two subspecies are very thinly defined, and the species probably merits being treated as monotypic with some slight regional differentiation. *Other names*—Papuan Flightless Rail, Grey-faced Rail, *Habroptila inepta*, *Amaurornis ineptus*.

GENUS *PORPHYRIO*   [Genus 7 spp/Region 1 sp]

Porphyrio Brisson, 1760. *Orn.* 1: 48 and 5: 522. Type, by tautonymy, "Porphyrio" Brisson = *Fulica porphyrio* Linnaeus.

The species of this genus range from eastern North America, Central America, northern South America, and sub-Saharan Africa to tropical Asia, New Guinea, the Solomon Islands, western Pacific Islands, Australia, and New Zealand (Taylor 1996). The genus is characterized by the presence of blue plumage on the head and breast; a frontal shield; a prominent red bill; and a distinctive white undertail patch. The molecular study of Garcia-R. & Trewick (2015) provided support for splitting up the genus.

Purple Swamphen *Porphyrio porphyrio*

Fulica porphyrio Linnaeus, 1758. *Syst. Nat.*, ed. 10, 1: 152.—Lands bordering the w Mediterranean Sea.

Widespread in swamps and wetlands through the Region; local in highlands, to 1,800 m[CO]. Southern areas (*e.g.*, Kurik) receive migrants from Australia (Mees 1982b), but swamphens are apparently absent from some sites in Trans-Fly (Coates 1985). Bishop (2005a) found it breeding at Princess Marianne Strait. Migrates through the Torres Strait[CO].

Extralimital—Spain, w and s Africa, tropical Asia, the Philippines, Indonesia, the Bismarck and Solomon Is, Vanuatu, New Caledonia, Australia, NZ, Palau, and Samoa (Taylor 1996, Mayr & Diamond 2001, Dutson 2011).

SUBSPECIES

a. *Porphyrio porphyrio melanotus* Temminck

Porphyrio melanotus Temminck, 1820. *Man. Orn.*, ed. 2, 2: 701.—New South Wales, Australia.

RANGE: Kai Is, NG, Aru Is, D'Entrecasteaux Arch, Louisiade Arch; also n and e Australia, Lord Howe I, Norfolk I, NZ, and Kermadec Is (Dickinson & Remsen 2013). Localities—?Salawati I[J], Aru Is[MA], Paniai Lake[K], Baliem valley[CO], Kurik[W], Wanggati[W], Merauke[CO], Bensbach[CO], Wabag[CO], Wahgi valley[K], Baiyer valley[CO], Tebi R[CO], middle Sepik[L], Bulolo valley[CO], Port Moresby[CO], Trobriand Is[MA], and Goodenough[CO], Woodlark[MA], Misima[X], and Tagula I[CO] (Gyldenstolpe 1955a[K], Gilliard & LeCroy 1966[L], Mees 1982b[W], LeCroy & Peckover 1998[X], S. Hogberg in M. Tarburton's island website[J]). One bird banded at Townsville, Australia, in 1958, was shot at Kurik, Indonesian NG, in 1961 (Coates 1985).

DIAGNOSIS: Underparts uniformly violet purple; wing (male) 246–275 (Rand & Gilliard 1967). We follow subspecific treatment of Dickinson & Remsen (2013), who simplified local subspecific distributions. For different views, see Ripley (1977) and Taylor (1996).

NOTES

The molecular study of Garcia-R. & Trewick (2015) indicated that the New Guinea population is molecularly distinct from the Australian population. Plumage traits, on the other hand, support keeping the Australian and New Guinea forms as one (Dickinson & Remsen 2013). *Other names*—Swamphen.

GENUS *GALLINULA* [Genus 6 spp/Region 1 sp]

Gallinula Brisson, 1760. *Orn.* 1: 50 and 6. 2. Type, by tautonymy, "Gallinula" Brisson = *Fulica chloropus* Linnaeus.

The species of this genus range through the Americas, Africa, some South Atlantic islands, Europe, Asia, Oceania, and Australia (Taylor 1996). The genus is characterized by its sooty plumage, the bill and frontal shield with red, yellow, or greenish in combination, and the white undertail in most species. The genera *Porphyrio* and *Gallinula* show considerable overlap, and the genera *Fulica* and *Gallinula* exhibit similar intergradation of characters among marginal species.

Dusky Moorhen *Gallinula tenebrosa* RESIDENT

Gallinula tenebrosa Gould, 1846. *Birds Austral.*, pt. 22: pl. 14 (= 6, pl. 73 of bound vol.).—New South Wales and South Australia.

Locally distributed in wetlands and swamps of the lowlands of NG. Surprisingly scarce in the Trans-Fly wetlands (see Bishop 2005a). Some populations are apparently nomadic (Coates 1985). No records from any of NG's fringing islands.

Extralimital—se Borneo, Sulawesi, Lesser Sundas, s Moluccas, New Britain, New Caledonia, and Australia (Taylor 1996, Dutson 2011).

SUBSPECIES

a. *Gallinula tenebrosa frontata* Wallace

Gallinula frontata Wallace, 1863. *Proc. Zool. Soc. London*: 35.—Buru I, s Moluccas.

RANGE: Borneo, Lesser Sunda Is, Sulawesi, s Moluccas, and w and se NG. Localities—Atinju and Aimaru Lakes in the Bird's Head[J], Cyclops Mts[A], near Vanimo[CO], Kikori R[CO], the middle Sepik[W], the Wahgi R at

1,580 m[L], Lake Teberu in the Purari basin[B], Hall Sound[MA], the Laloki R[MA], the Sogeri Plateau[CO], Moitaka settling ponds[CO], Popondetta[CO] (Hartert 1930[A], Gyldenstolpe 1955a[L], 1955b[J], Gilliard & LeCroy 1966[W], Mack & Wright 1993[B]).

DIAGNOSIS: Compared to *neumanni*, darker above and below; mantle washed with olive; larger, wing 198–201 (Rand & Gilliard 1967, Ripley 1977).

b. *Gallinula tenebrosa neumanni* Hartert

Gallinula tenebrosa neumanni Hartert, 1930. *Novit. Zool.* 36: 123.—Ifar, Lake Sentani, n NG.

RANGE: Northwestern Lowlands and Sepik-Ramu.

DIAGNOSIS: Small; wing 170–182; mantle blackish (Rand & Gilliard 1967, Ripley 1977).

NOTES

Other names—Dusky Gallinule.

GENUS *FULICA* [Genus 11 spp/Region 1 sp]

Fulica Linnaeus, 1758. *Syst. Nat.*, ed. 10, 1: 152. Type, by tautonymy, "Fulica" = *Fulica atra* Linnaeus.

The species of this genus exhibit a nearly cosmopolitan distribution, though absent from the Arctic and much of northern South America (Taylor 1996). This very uniform genus is characterized by its entirely plain sooty-black body plumage, blackish head and neck, abbreviated tail, the flanging to the toes, and a white (or yellow/red) bill and frontal shield.

Eurasian Coot *Fulica atra* RESIDENT

Fulica atra Linnaeus, 1758. *Syst. Nat.*, ed. 10, 1: 152.—Sweden.

An uncommon breeding resident on montane lakes of the Bird's Head and the Western and Eastern Ra, to 3,700 m[K] (Schodde *et al.* 1975[K]). Birds that are seen irregularly in the lowlands are perhaps vagrants from Australia (Coates 1985).

Extralimital—w Europe east to Japan, s Asia, Australia, and NZ. Wintering birds range to n and w Africa, se Asia, and Philippines (Taylor 1996).

SUBSPECIES

a. *Fulica atra lugubris* S. Müller

Fulica lugubris S. Müller, 1847. *Verh. Nat. Gesch. Ned., Land-en Volkenk.* 2: 454.—Taman Ideop, 1,968 m, e Java.

> Synonym: *Fulica atra anggiensis* Thompson and Temple 1964. *Proc. Biol. Soc. Wash.* 77: 251.—Anggi Lakes, Arfak Mts, Bird's Head.
> Synonym: *Fulica atra novaeguineae* Rand, 1940. *Amer. Mus. Novit.* 1072: 4.—Lake Habbema, 3,225 m, Western Ra.

RANGE: Uplands of Java, Bali, and w and e NG. Localities—Anggi Lakes[J], Lake Discovery[K], Lake Habbema[MA], Lake Wangbin[CO], Ok Tedi[CO], Lakes Vivien and Louise in the Star Mts of PNG[CO], Lake Iviva[CO], and a montane lake in Crater Mt WMA near Ubaigubi[L] (Gyldenstolpe 1955b[J], Schodde *et al.* 1975[K], Finch & Gillison 1988[L]).

DIAGNOSIS: Plumage dark, often very blackish; usually but not always with prominent white tips to the secondaries; size small; wing 179–186 (Ripley 1977, Taylor 1996).

b. *Fulica atra ?australis* Gould

Fulica australis Gould, 1845. *Proc. Zool. Soc. London*: 2.—Western Australia.

RANGE: Australia. Migrants of this form may be wintering in se NG (Coates 1985). Records that presumably refer to this race include Hisiu, Aroa Lagoon, Moitaka settling ponds, and Tanubada ponds near Port Moresby[CO] (Finch & Gillison 1988).

DIAGNOSIS: Paler, grayer slate-black than *lugubris* on average, with white tips to outer secondaries reduced or absent; wing 180–201 (Ripley 1977).

We follow Dickinson & Remsen (2013) in subspecific treatment of this widespread species, though we note these forms are thinly defined and probably merit revisiting. *Other names*—Coot, Common Coot, Black Coot.

Family Gruidae

Cranes
[Family 15 spp/Region 1 sp]

The species of the family exhibit a patchy yet nearly cosmopolitan range, though absent from South America, insular southeast Asia, and the drier interiors of Africa and Asia (Archibald & Meine 1996). This is a morphologically uniform family of tall and long-necked wading birds plumaged mainly in gray, black, and white, and often with reddish skin on the head or face. Most species exhibit a distinctive "bustle" of elaborated tail coverts that sit atop the tail. Species produce loud bugling calls and perform dancing courtship displays. Most apparently mate for life. Hackett *et mult.* (2008) showed that the cranes are sister to the Limpkin (*Aramidae*), and that this pair of lineages is sister to the trumpeters (*Psophodidae*).

GENUS *GRUS*
[Genus 12 spp/Region 1 sp]

Grus Brisson, 1760. *Orn.* 5: 374–91. Type by absolute tautonomy, *Ardea grus* Linnaeus. [This corrected reference authorized by Opinion 103 of the ICZN.]

The species of this genus range from Alaska to Florida and northern Europe to northern India, Bhutan, Tibet, Japan, northeastern Asia, New Guinea, and Australia (Archibald & Meine 1996). Generic characters include patches of bare red skin on the face, neck, throat, or head; and plumage dominated by expanses of gray, white, and black. All exhibit a thick bustle above the tail, colored gray, white, or black. In New Guinea, the single species inhabits grassy wetlands of the Trans-Fly. Dickinson & Remsen (2013) placed the species *rubicunda* into the genus *Antigone*.

It is surprising that the Sarus Crane (*Grus antigone*) occurs in northern Queensland and in tropical Asia and yet has never been recorded in the Region (Pratt & Beehler 2014: 308).

Brolga *Grus rubicunda*
RESIDENT | MONOTYPIC

Ardea rubicunda Perry, 1810. *Arcana* 6: pl. 22.—Botany Bay, NSW, Australia.

Common inhabitant of grassy wetlands in the s Trans-Fly; often in large flocks, with dry-season congregrations of as many as 600 birds (Stronach 1981b).

Localities—Dolak I[MA], Wasur NP, Bian Lakes, Bensbach, Bula Plains, Morehead R, and Dogwa (Bishop 2005a).

Extralimital—nw, n, e, and se Australia (Archibald & Meine 1996).

NOTES

An old record from the Sepik R (north of Timbunke at 4°S; Kirschbaum) reported by Mayr (1941) has never been corroborated, and thus should be treated with suspicion. The record proabably refers to the White-faced Heron (*Egretta novaehollandiae*), which is known as Blue Crane in rural areas of Australia (P. Gregory *in litt.*). The specific epithet *rubicunda* is adjectival and must agree in gender with its feminine genus. *Other names*—Australian Crane, *Antigone rubicunda*.

ORDER CUCULIFORMES

The Cuculiformes appear to be a well-defined and well-corroborated lineage that includes the coucals (here treated as the Centropodidae), ground cuckoos (Neomorphini), true cuckoos (Cuculinae), malkohas (Phaenicophaeini), and anis (Crotophaginae). Hackett *et mult.* (2008) placed the cuckoos into a clade that includes the rails, finfoots, bustards, cranes, trumpeters, and Limpkin. This clade is sister to a huge clade that includes turacos (Musophagidae), loons (Gaviidae), and many other waterbirds.

Family Centropodidae

Coucals and Couas [Family 39 spp/Region 4 spp]

The species of the family range from Madagascar and southern Africa east through tropical Asia to New Guinea, the Bismarck and Solomon Is, and Australia (Erritzøe *et al.* 2012). We follow Schodde & Mason (1997) and Higgins (1999) in treatment of this lineage as a full family. The lineage is sister to a clade that includes the core cuckoos (Cuculinae; Hackett *et mult.* 2008). The coucals are very distinctive, characterized by the large size; massive tail; thick bill; predominantly black (or black and brown) plumage coloration; coarse body plumage; operculate nostril; zygodactyly; especially long hallux; and prominent, low *hoo*-ing vocalization (Higgins 1999). The couas (*Coua*) are large ground-dwelling cuckoos with distinctive blue facial skin that inhabit Madagascar. Most authorities treat the coucal/coua lineage as a subfamily of Cuculidae (*e.g.*, Cracraft 2013).

GENUS *CENTROPUS* [Genus 28 spp/Region 4 spp]

Centropus Illiger, 1811. *Prodromus Syst. Mamm. Avium*: 205. Type, by subsequent designation (G. R. Gray, 1840, *List Gen. Birds*: 56), *Cuculus aegyptius* Gmelin.

The species of this genus range from Africa and Madagascar to tropical Asia and Australia (Payne 1997). The genus is characterized by nonparasitic nesting habits; very large size; long, straight hallux; long, floppy tail; coarse feathering; short, robust, decurved bill; and crow-like appearance.

Greater Black Coucal *Centropus menbeki* ENDEMIC

Centropus menbeki Lesson & Garnot, 1828. *Voy. Coquille, Atlas*: pl. 33; 1829. *Zool.* 1(13): 600.—Manokwari, Bird's Head.

Frequents forest, edge, and tangles through NG, the NW Islands, Aru Is, and Yapen I. Occurs mostly in the lowlands and hills, max 1,275 m.

SUBSPECIES

a. *Centropus menbeki menbeki* Lesson & Garnot

    Synonym: *Centropus menbeki jobiensis* Stresemann & Paludan, 1932. *Novit. Zool.* 38: 236.—Yapen I, Bay Islands.

RANGE: New Guinea, Misool, Salawati, Batanta, and Yapen I (Mayr 1941).

DIAGNOSIS: Similar to *aruensis* but dorsally darker, and plumage more greenish (Rand & Gilliard 1967). Includes large and small forms. We found specimens of the named form *jobiensis* from Yapen I not diagnosable from Mainland birds from the n coast (*e.g.*, Ifar and Bernhard Camp). This is in agreement with Rand & Gilliard (1967), Payne (2005), and Erritzøe *et al.* (2012).

b. *Centropus menbeki aruensis* (Salvadori)
Nesocentor aruensis Salvadori, 1878. *Ann. Mus. Civ. Genova*, 12: 317.—Aru Is.
RANGE: Aru Is.
DIAGNOSIS: Upperparts dark purplish black (Rand & Gilliard 1967).

NOTES
Dickinson & Remsen (2013: 138) noted that the spelling of *menbeki* is based on the plate published in Mar-1828. *Other names*—Greater, Menbek's, or Ivory-billed Coucal; Black Jungle Coucal.

Biak Coucal *Centropus chalybeus* ENDEMIC | MONOTYPIC
Nesocentor chalybeus Salvadori, 1876. *Ann. Mus. Civ. Genova* 7: 915.—Biak I, Bay Islands.
    Endemic to Biak I. Recent observations indicate that it is probably not uncommon in forest and secondary forest, but it is elusive and more often heard than seen.

NOTES
Analysis using mtDNA suggested *C. chalybeus* is the sister to *C. menbeki* (Sorenson & Payne 2005). *Other names*—Biak Island Coucal.

Lesser Black Coucal *Centropus bernsteini* ENDEMIC | MONOTYPIC
Centropus Bernsteini Schlegel, 1866. *Ned. Tijdschr. Dierk.* 3: 251.—New Guinea.
    Synonym: *Centropus bernsteinii manam* Mayr, 1937. *Amer. Mus. Novit.* 939: 3.—Manam I, ne NG.
?Salawati and Batanta I[J,K], the Bird's Head and Neck, NW Lowlands, S Lowlands, Huon Penin; also Manam I; east in n watershed to Zenag in the Herzog Mts; east in the s watershed to the Turama R[A] (Mayr 1941, Burrows 1993[A], Matheve *et al.* 2009[J], Beehler field notes 19-Dec-2011[K]). Sea level–500 m, max 1,200 m at Zenag (A. H. Miller specimen in Museum of Vertebrate Zoology [MVZ], Berkeley, CA). Frequents tall grass and cane that fringes forest and shoreline.
    Localities—Jeflio[J], Selekobo[J], Bernhard Camp on the Taritatu R[Q], Ifar[K], Gaima[X], Tabubil[A], middle Sepik[L], Turama R[W], Kikori[W], Zenag[B], Sunshine in the lower Bulolo valley[Z] (Rand 1942a[X], 1942b[Q], Gyldenstolpe 1955b[J], Ripley 1964a[K], Gilliard & LeCroy 1966[L], Gregory 1995b[A], Burrows 1993[W], Beehler field notes 31-Aug-1978[Z], A. H. Miller specimen, MVZ[B]).

NOTES
We follow Payne (2005) in treating the species as monotypic. Size differences among putative races are less than 10%. Both this species and *C. phasianinus* were collected at Gaima in the Trans-Fly (Rand 1942a), and they now can be found in sympatry at Kiunga as well (P. Gregory *in litt.*). The map in Erritzøe *et al.* (2012) showing this species inhabiting the SE Peninsula is in error. *Other names*—Bernstein's or Black-billed Coucal.

Pheasant Coucal *Centropus phasianinus* RESIDENT
Cuculus phasianinus Latham, 1801. *Index Orn.*, Suppl.: xxx.—New South Wales, Australia.
    Central and e NG and the D'Entrecasteaux Arch; few records west of the PNG border; sea level–500 m, max 1,800 m in open highland valleys of the Eastern Ra[CO]. Frequents savanna, grassland mixed with scrub, and disturbed habitats. Apparently absent from most of Gulf Prov, PNG (Coates 1985, B. J. Coates *in litt.*).
    Extralimital—nw, n, and e Australia, islands of the Torres Strait, and Timor (Payne 1997).

SUBSPECIES
a. *Centropus phasianinus thierfelderi* Stresemann
Centropus phasianus thierfelderi Stresemann, 1927. *Orn. Monatsber.* 35: 111.—Merauke, Trans-Fly.
RANGE: Trans-Fly, including Dolak I and Merauke distr, Lake Murray[J], and the Fly R, east to the head of the Gulf of Papua, where it presumably meets with *nigricans*; also nw Torres Strait Is (Mees 1982b, Payne 1997, B. J. Coates *in litt.*[J]).
DIAGNOSIS: Wing quills broadly barred and shafted with reddish brown; head and underparts blackish (Rand & Gilliard 1967).

b. *Centropus phasianinus propinquus* Mayr

Centropus phasianinus propinquus Mayr, 1937. *Amer. Mus. Novit.* 939: 4.—Ifar, Lake Sentani, ne sector of the NW Lowlands.

RANGE: Northern NG, from Lake Sentani east to Astrolabe Bay (Mayr 1941). Also Cyclops Mts[A], Baiyer R[B], and Kaironk valley[DI] (Rand 1942b[A], Mayr & Gilliard 1954[B]). Records west of Lake Sentani appear to be absent.

DIAGNOSIS: Like *thierfelderi* but barring of wing quills much reduced and paler, more buff, less reddish brown. Very similar to *nigricans* but smaller (Greenway 1978).

c. *Centropus phasianinus nigricans* (Salvadori)

Polophilus nigricans Salvadori, 1876. *Ann. Mus. Civ. Genova* 9: 17.—Naiabui, Hall Sound, and Yule I, SE Peninsula.

> Synonym: *Centropus phasianinus obscuratus* Mayr, 1937. *Amer. Mus. Novit.* 939: 4.—Fergusson I, D'Entrecasteaux Arch, SE Islands.

RANGE: Southeastern Peninsula and the D'Entrecasteaux Arch. Previously reported from Goodenough and Fergusson I (Mayr 1941); more recently recorded from Normanby I by LeCroy *et al.* (1984) and Pratt (field notes 2003). Transition point of nw verge of range of *nigricans* and se verge of range of *propinquus* needs to be determined. It is some place east of Astrolabe Bay—perhaps the upper Ramu in the south and the n coast of the Huon in the north. *C. p. nigricans* also may range patchily westward in s watershed to range of *thierfelderi.* Not recorded from the Lakekamu basin, which is just east of the Purari (Beehler *et al.* 1994).

DIAGNOSIS: Like *thierfelderi*, but markings on wing quills are broadly barred with reddish brown, and head and underparts are blackish (Greenway 1978). See photo in Coates & Peckover (2001: 93). We follow Mason *et al.* (1984) and Payne (2005) in subsuming *obscuratus* into *nigricans.*

NOTES

Currently the subspecies ranges exhibit geographic gaps; given that the species is a grassland specialist, it is not surprising that in areas of expansive lowland forest there are few records. Questions remain: What race occupies the Huon Penin? What race occupies the e sector of the S Lowlands? These gaps need to be filled by future researchers. Earlier authorities intimated that the species ranges west to the Mamberamo, but various field surveys in the Mamberamo failed to locate it (*e.g.*, Rand 1942b, van Balen *et al.* 2003, Beehler *et al.* 2012). The map of Erritzøe *et al.* (2012) showing the species to be absent from the n watershed of the SE Peninsula is not correct. *Other names*—Common Coucal.

Family Cuculidae

Old World Parasitic Cuckoos   [Family 57 spp/Region 18 spp]

We follow Schodde & Mason (1997: 219) in restricting the family to the Old World true cuckoos and koels, "excluding phaenicophaeids (malkohas) and New World coccyzids, crotophagids and neomorphids." Treatment of the family also excludes the coucals and couas, here placed in a separate family, the Centropodidae. The species of the Cuculidae range from Eurasia and Africa through Australasia and into Oceania (but for northernmost Eurasia and Saharan Africa). This cuckoo family is most distinctive in its nest parasitism of passerines but also is characterized by the following: small and weak legs, zygodactyl feet, fine and decurved bill, eye-ring of colored bare skin, plumage patterning with profuse barring in some plumages, and musical whistled song. New Guinea is especially rich in genera, presumably indicating a long presence and evolutionary development in the region.

Microdynamis Salvadori, 1878. *Ann. Mus. Civ. Genova* 13: 461. Type, by monotypy, *Eudynamis parva* Salvadori.

The single species inhabits the lowland and hill forests of the Mainland and the D'Entrecasteaux Archipelago. The genus is characterized by the hook-tipped bill, slit-like nostril, sexual dichromatism, small size, and ecological restriction to the high forest canopy (Payne 2005). A molecular analysis employing mitochondrial DNA demonstrated that this genus is sister to *Eudynamys* (Sorenson & Payne 2005).

Dwarf Koel *Microdynamis parva*   ENDEMIC

Eudynamis parva Salvadori, 1876. *Ann. Mus. Civ. Genova* 7: 986.—Northwestern NG.

Inhabits canopy of forest and edge through NG and the D'Entrecasteaux Arch, sea level–1,450 m. Known from the three main D'Entrecasteaux islands by records of vocal birds (Beehler & Pratt field notes).

SUBSPECIES

a. *Microdynamis parva parva* (Salvadori)
RANGE: Bird's Head[J], the NW Lowlands[MA], S Lowlands[K], s watershed of SE Peninsula[L] (Mayr & Rand 1936[L], Rand 1942a[K], Gyldenstolpe 1955b[J]). Presumably the birds from the D'Entrecasteaux Arch are allied to this form.
DIAGNOSIS: Upperparts of female russet brown; adult male with paler, buffier belly.

b. *Microdynamis parva grisescens* Mayr & Rand
Microdynamis parva grisescens Mayr & Rand, 1936. *Amer. Mus. Novit.* 868: 1.—Madang, e verge Sepik-Ramu.
RANGE: Jayapura[J], Sepik-Ramu[MA], Huon Penin[K], and n watershed of SE Peninsula east to the Kumusi R[MA] (Rand 1942b[J], Beehler field notes[K]).
DIAGNOSIS: Upperparts of female show a grayish cast; adult male exhibits a dull clay-brown belly (not pale).

NOTES
Microdynamis is feminine (David & Gosselin 2002b). Photograph (see Coates & Peckover 2001) of a putative adult female (red-eyed) from near Wau exhibits a rufous-brown mantle, in spite of being within the range of *grisescens*. The validity of the two subspecies needs further assessment. *Other names*—Black-capped Koel.

GENUS *EUDYNAMYS*   [Genus 3 spp/Region 1 sp]

Eudynamys Vigors & Horsfield, 1827. *Trans. Linn. Soc. London* (1826) 15: 303. Type, by subsequent designation (G. R. Gray, 1840. *List. Gen. Birds*: 57), *Cuculus orientalis* Linnaeus.

The species of this genus range from tropical Asia to Australia (Payne 1997). The genus is characterized by strong plumage dichromatism, the all-black male plumage, the long tail, profusely barred in all female plumages, and the red iris. The nostril is slit-like (Payne 2005).

Eastern Koel *Eudynamys orientalis*   RESIDENT AND VISITOR

Cuculus orientalis Linnaeus, 1766. *Syst. Nat.*, ed. 12, 1: 168.—Ambon, s Moluccas.

Inhabits forest, edge, and gardens through NG, the NW Islands, Aru Is, Padaido Is, Daru I, Manam I, and Karkar I; sea level–1,500 m[CO]. Includes both resident birds and nonbreeding migrants from Australia.

Extralimital—s Moluccas and Lesser Sundas; Kai, Bismarck, and Solomon Is; and n and e Australia. Some Australian populations migrate to e Indonesian islands.

SUBSPECIES

a. *Eudynamys orientalis rufiventer* (Lesson)
Cuculus rufiventer Lesson, 1830. *Voy. Coquille, Zool.*: 622.—Manokwari, Bird's Head.

Synonym: *Eudynamis minima* van Oort, 1911. *Notes Leyden Mus.* 34: 54.—Lorentz [Noord] R, sw NG.

RANGE: Resident in the NW Islands, Aru Is, Karkar I, Bagabag I[K], Manam I[J], and all NG except perhaps the Trans-Fly and the lowlands of the SE Peninsula (Mayr 1941, Rothschild & Hartert 1915a[J], Diamond & LeCroy 1979[K]). Collected on the lower Digul R (Bangs & Peters 1926).

DIAGNOSIS: Female's head rufous with fine blackish markings; back heavily marked rufous and black; breast washed with cinnamon and barred. Payne (2005) recognized the race *minimus*, known only from the Noord R; this form is defined by a size difference of only 10% by wing length, thus of marginal difference. Such a subspecific distribution would be an anomaly.

b. *Eudynamys orientalis subcyanocephalus* Mathews

Eudynamis orientalis subcyanocephalus Mathews, 1912. *Austral. Av. Record* 1(1): 21.—Parry Creek, nw Australia.

RANGE: Breeds in tropical n Australia and winters to s NG from the Merauke distr east to Port Moresby; also recorded from the Aru Is, Padaido Is, Daru I[CO], Port Moresby area[CO], and Kumusi R (Mayr 1941). Coates (1985) stated "possibly resident in Trans-Fly"; this should be explored.

DIAGNOSIS: Female has blackish crown and cheek, with white facial stripe; no rufous spotting on crown; throat very pale buff with heavy dark barring; back blackish with whitish spotting; breast and belly off-white barred with dark sooty-gray. Given the variation found in female and immature plumages of this species, we are uncertain of the substance of the distinctions highlighted by Erritzøe *et al.* (2012).

NOTES

Eudynamys is masculine (David & Gosselin 2002b). We follow the species-level treatment of Schodde & Mason (1997) and Dickinson & Remsen (2013). For another view, see Mayr & Diamond (2001), Payne (2005), and Erritzøe *et al.* (2012). *Other names*—Koel; Indian, Common, Asian, or Pacific Koel; *Eudynamys scolopacea* or *scolopaceus*.

GENUS *URODYNAMIS* MONOTYPIC

Urodynamis Salvadori, 1880. *Orn. Pap. Moluc.* 1: xv and 370. Type, by original designation and monotypy, *Cuculus taitensis* Sparmann.

The single species of the genus breeds in New Zealand and associated islands, and migrates in non-breeding season to the tropical Pacific—mainly central Polynesia. The genus is characterized by the lack of an all-black male plumage, the nostril an oval slit, the wedge-shaped tail, the prominent whitish eyebrow, the absence of barring ventrally, and the yellow or pale brown (not red) iris. Payne (2005) placed this genus sister to *Scythrops*.

Long-tailed Cuckoo *Urodynamis taitensis* VAGRANT | MONOTYPIC

Cuculus taitensis Sparrman, 1787. *Mus. Carlson.* 2: pl. 32.—Tahiti.

Migrant from NZ to small islands of Micronesia, Polynesia, and Melanesia. No records from the Mainland. In the Region, sole record is from Kimuta I, adjacent to Misima I (LeCroy & Peckover 1998).

Extralimital—breeds in NZ; winters mainly from Fiji to the Tuamotu Is; winter range extends from Palau through Micronesia, Melanesia, and Polynesia to Pitcairn I (Gill & Hauber 2012).

NOTES

Should be looked for in the smaller of the SE Islands. Perhaps might also be expected on the small islands off the n coast of the SE Peninsula. Misima I record (Mayr 1941) is in error (LeCroy & Peckover 1998). Higgins (1999) cited a record from n NG and the accompanying map shows a record from near Madang. We have been unable to track down this record, and it is apparently in error (*fide* P. Higgins *in litt.*, and B. J. Gill *in litt.*). *Other names*—Long-tailed Koel, Pacific Long-tailed Cuckoo, *Eudynamys taitensis*.

Scythrops Latham, 1790. *Index Orn.* 1: 141. Type, by monotypy, *Scythrops novaehollandiae* Latham.

The one species in the genus ranges from southeastern Australia to New Guinea, the Bismarck Archipelago, Sulawesi, and the Lesser Sundas (Payne 1997). The genus is characterized by the oversize toucan-like bill, large body, long tail, red eye skin, and predominantly gray plumage.

Channel-billed Cuckoo *Scythrops novaehollandiae*   VISITOR (AND RESIDENT?) | MONOTYPIC

Scythrops novæ Hollandiae Latham, 1790. *Index Orn.*, 1: 141.—New South Wales, Australia.
> Synonym: *Scythrops novaehollandiae fordi* Mason, 1996. *Emu* 96: 225.—Buka, Sulawesi.
> Synonym: *Scythrops novaehollandiae schoddei* Mason, 1996. *Emu* 96: 226.—Talasea, New Britain.

An austral migrant to the lowlands of most or all of the NG region; records to 2,800 m (P. Gregory *in litt.* reported a sighting from Kumul Lodge, Enga Prov). Breeds in Australia, Wallacea, and possibly New Britain. Breeding also possible in NG but not yet proven (Bell 1984). Regional insular records include Gag[J], Kofiau[K], Misool[L], Batanta and Salawati[W], and Waigeo I[MA]; Biak I[X,V]; Aru Is[MA]; Manam and Karkar I[MA]; D'Entrecasteaux[MA,Z,T] and Trobriand Is[MA], and Tagula I[U] (Johnstone 2006[J], Diamond *et al.* 2009[K], Mees 1965[L], Beckers et al. in M. Tarburton's island website[W], K. D. Bishop *in litt.*[X], LeCroy *et al.* 1984[Z], Pratt field notes 2003[T], Pratt field notes 2004[U], S. van Balen observation 1997[V]). No records yet from Gebe, Numfor, Rossel, or Woodlark I. Frequents forest edge, partly cleared areas, and savanna, but can be observed flying overhead just about anywhere. Arrives back in Australia to breed as early as August and departs as late as April (Coates 1985). The fact that the species has been recorded in the Region in all months of the year and is sometimes seen pursued by the Torresian Crow (*Corvus orru*), a confirmed host species in Australia, suggests that there may be a breeding population here (Bell 1970a, Coates 1985). Observed in Kofiau I in late Dec-2011 (Beehler field notes). Observed on 28-Feb-2004 on Batanta (Beehler field notes). S. van Balen recorded it on Biak I in Oct-1997 and Beehler and van Balen recorded it on Biak I on three days in Jan-1997. Possibly breeds in the D'Entrecasteaux Arch, where the species is common, noisy, and in conflict with Torresian Crows during its breeding season (January, February, and November), a time when an Australian migrant would be expected to be absent (Pratt field notes 2003). A pair was observed on Goodenough I on 18-Mar-1976 (Beehler field notes).

Extralimital—breeds in n and e Australia, as well as Sulawesi, Buru, and Flores; wintering birds found in the Moluccas and Bismarck Is (Payne 1997). Rare in the Solomons (Dutson 2011). There may be breeding birds in the Bismarck Arch (Mason & Forrester 1996). Recently recorded in Palawan, Philippines (D. Hobcroft *in litt.*)

NOTES

We treat the species as as monotypic following Payne's (2005) clear statement: "no consistent differences in wing or bill size or plumage pattern are apparent in specimens in all museums." For another view, see Mason & Forrester (1996), who recognized three races and postulated breeding in the Bismarck Arch.

GENUS *RHAMPHOMANTIS*   ENDEMIC | MONOTYPIC

Rhamphomantis Salvadori, 1878. *Ann. Mus. Civ. Genova* 13: 459. Type, by monotypy, *Cuculus megarhynchus* G. R. Gray.

The one species in the genus is restricted to mainland New Guinea, the Northwestern Islands, and the Aru Islands (Payne 1997). The genus is characterized by the small size, plain dull brown plumage, and the narrow, long, pointed, and slightly decurved black bill. The molecular systematic analysis of Sorenson & Payne (2005), using mitochondrial DNA, placed the species as sister to *Chrysococcyx*, into which genus it was subsequently included by Payne (2005). We keep it separate, following Erritzøe *et al.* (2012). This placement is supported by the very distinctive bill and the male plumage.

Long-billed Cuckoo *Rhamphomantis megarhynchus*   ENDEMIC | MONOTYPIC

Cuculus megarhynchus G. R. Gray, 1858. *Proc. Zool. Soc. London*: 184.—Aru Is.

> Synonym: *Rhamphomantis megarhynchus sanfordi* Stresemann & Paludan, 1932. *Orn. Monatsber.* 40: 17.—Waigeo I, NW Islands.

An elusive and infrequently seen inhabitant of lowland forest of the Mainland, Waigeo[MA], ?Salawati[K], Misool I[J], and Aru Is[MA] (Ripley 1964a[J], M. Tarburton's island website[K]). Probably thinly distributed in suitable habitat throughout much of the Mainland lowlands and foothills. Frequents rain forest, monsoon forest, and edge.

Localities—Tamulol on Misool I[X], Waigeo I[CO], Atinju on the Bird's Head[L], Mamberamo R[MA], Passim near Humboldt Bay[CO], Sepik R[MA], upper Fly R[K], Ketu Creek[Y], Strickland R[W], upper Sepik[W], Karawari R[A], Kikori R (Gobe, Kantobo, Kopi)[W], Ramu R[B], Purari basin[W], Lakekamu basin[Z], Kumusi R[MA], Veimauri[U], and Brown R[J], and Sogeri Plateau[Y] (Gyldenstolpe 1955b[L], Ripley 1964a[X], Beehler *et al.* 1994[Z], Gregory 1997[K], B. J. Coates *in litt.*[J], Beehler field notes[U], I. Woxvold *in litt.*[W], P. Gregory *in litt.*[Y], M. LeCroy *in litt.*[A], P. Z. Marki *in litt.*[B]).

NOTES

Treated as monotypic following Payne (1997, 2005). Plumages of this species remain inadequately documented by the limited museum specimen material and understudied in the field. More study skins, photographs, and sound files are needed for a better understanding of this little-known species. Probably more common but overlooked because of its retiring, forest-interior habits. *Other names*—Little Long-billed Cuckoo, *Chalcites megarhynchus*.

GENUS *CHALCITES*   [Genus 8 spp/Region 6 spp]

Chalcites Lesson, 1830. *Traité d'Orn.*, livr. 2: 152. Type by tautonymy, *Cuculus chalcites* "Illiger" = *Cuculus plagosus* Latham, 1801.

The species of the genus range from Malaysia to Australia and New Zealand. We treat the Australasian *Chalcites* as distinct from *Chrysococcyx*, following Schodde & Mason (1997) and Christidis & Boles (2008), but see also Payne (1997) and Sorenson & Payne (2005) for another view. The genus as circumscribed is characterized by the very small size, the metallic (often greenish) dorsal surface, and presence of barring in throat and on breast. Clearly a sister to *Chrysococcyx*, into which Payne (2005) placed all of these species.

Black-eared Cuckoo *Chalcites osculans*   VISITOR | MONOTYPIC

Chalcites osculans Gould, 1847. *Proc. Zool. Soc. London*: 32.—New South Wales, Australia.

A widespread Australian breeder that migrates to the Aru Is[MA]; perhaps migrants also reach s or sw NG (Rand & Gilliard 1967, but see Mees 1982b: 10). Draffan *et al.* (1983) noted that Ashford reported this species to be an uncommon migrant across the Torres Strait.

Extralimital—breeds widely in Australia, and winters to Bacan I, the Kai Is, and Luang, Babar, and Romang I (of e Indonesia; Coates & Bishop 1997).

NOTES

Apparently the report from the w uplands of the Mainland (Coates 1985) is in error (B. J. Coates *in litt.*). Should be looked for in the Trans-Fly. We place the species at the base of the genus (Condon 1975, Schodde & Mason 1997, and see also Higgins 1999 for basal treatment). Several plumage characters distinguish this species from all others of the genus. *Other names*—*Misocalius* or *Chrysococcyx osculans*.

Horsfield's Bronze Cuckoo *Chalcites basalis*   VISITOR | MONOTYPIC

Cuculus basalis Horsfield, 1821. *Trans. Linn. Soc. London* 13: 179.—Java.

An Australian migrant to the savanna and scrub of the Trans-Fly, Aru Is, with records east to near Port Moresby. Coates (1985) reported that its status in s NG is uncertain because virtually all records are undocumented (allowing confusion with *C. minutillus russatus* and subadults of other species). Gregory (*in litt.*) saw it below Bensbach Lodge in Aug-2010. The single most reliable record is of a bird seen and heard near Bereina, Central Prov, PNG, on 13-Jun-1977 (Heron 1977c). The Finschhafen record (Isles & Menkhorst 1976) is unsubstantiated (Coates 1985).

Extralimital—breeds throughout Australia; partly migratory. During austral winter, ranges north to Malay Penin and Sunda Is (Coates & Bishop 1997). Also a migrant to Christmas I in the Indian Ocean (Higgins 1999).

NOTES
Presumably best identified by vocalization, and all future records should discuss voice if possible. Field identification should be made with caution (photographs would be preferred). *Other names*—Narrow-billed, Australian, or Rufous-tailed Bronze Cuckoo, *Chrysococcyx basalis*.

Rufous-throated Bronze Cuckoo *Chalcites ruficollis* ENDEMIC | MONOTYPIC
Lamprococcyx ruficollis Salvadori, 1876. *Ann. Mus. Civ. Genova* (1875) 7: 913.—Hatam, Arfak Mts, Bird's Head.

Mountains of the Bird's Head and Neck (Kumawa Mts[L] and Wandammen Penin), Central Ra, Foja Mts[J], and mts of Huon[K]; 1,800–2,600 m, extremes 1,130–3,230 m[CO] (Mayr 1941, Beehler *et al.* 2012[J], Freeman *et al.* 2013[K], Diamond & Bishop 2015[L]). Frequents forest and edge. *Other names: Chrysococcyx ruficollis*.

Shining Bronze Cuckoo *Chalcites lucidus* VISITOR
Cuculus lucidus Gmelin, 1788. *Syst. Nat.* 1(1): 421.—Queen Charlotte Sound, NZ.

Nonbreeding migrant (March–October) to nonforest and edge habitats in NG region; sea level–1,900 m. So far reported from the Mainland, Karkar I, and the Louisiade Is.

Extralimital—breeds in Australia, NZ, Norfolk I, New Caledonia, Vanuatu, Temotu Is, and s Solomon Is (Rennell, Bellona). Winters to the Lesser Sundas, s Moluccas, Bismarcks, and Solomon Is (Higgins 1999, Mayr & Diamond 2001, Dutson 2011).

SUBSPECIES

a. *Chalcites lucidus plagosus* (Latham)
Cuculus plagosus Latham 1801, *Index Orn.*, Suppl.: xxxi.—New South Wales, Australia.
RANGE: Winter visitor to NG. Breeding range includes e and sw Australia; winter range includes NG, e Indonesia, the Bismarcks, and ?Solomons (Mayr & Diamond 2001, Dutson 2011). Localities—Kurik[L], Star Mts at 1,500 m[CO], Wahgi valley at 1,580 m[J], Kubor Mts[DI], Kyaka territory[DI], Lufa[DI], Wau[M], and Tarawai[MA], Karkar[K], and Bagabag I[K]; (Mayr & Gilliard 1954[J], Diamond & LeCroy 1979[K], Mees 1982b[L], Pratt pers. obs.[M]).
DIAGNOSIS: Male plumage as for female of nominate form, but crown more brown than black, and mantle slightly less green. Female lacks white spots on forehead and over eye; crown bronze; mantle bronze (Payne 2005). Also bill narrow in *plagosus* (Higgins 1999).

b. *Chalcites lucidus lucidus* (Gmelin)
RANGE: Breeds in NZ, Norfolk I, and Chatham Is; winters to the Bismarck and Solomon Is, with a few records from the SE Islands (Woodlark and Misima[MA]).
DIAGNOSIS: Bill broad and relatively large; more speckling on forehead; green sheen on forehead; green mantle; much white in face (Higgins 1999). D. Hobcroft (*in litt.*) recommends distinguishing the nominate form by large white ear patch, white barred forehead, emerald-green crown, and green chest bars.

NOTES
Other names—Golden Bronze-Cuckoo, Golden or Shining Cuckoo, *Chrysococcyx lucidus*.

White-eared Bronze Cuckoo *Chalcites meyerii* ENDEMIC | MONOTYPIC
Chrysococcyx meyerii Salvadori, 1874. *Ann. Mus. Civ. Genova* 6: 82.—Hatam, Arfak Mts, Bird's Head. Replacement name for *Chrysococcyx splendidus* A. B. Meyer, 1874 (not *Cuculus splendidus* G. R. Gray).

All mts of NG, plus Batanta[J] and Goodenough I[K] (Mayr 1941, Greenway 1966[J], Beehler field observation Mar-1976[K]); 500–1,500 m, extremes sea level–2,000 m. No records from the Kumawa or Wandammen Mts. Also apparently absent from the Trans-Fly and other large areas of flat lowlands (Coates 1985). Frequents forest and edge.

Correct species epithet spelling is *meyerii* (Dickinson & Remsen 2013: 145, note 8). *Other names—Chryso-coccyx meyeri.*

Little Bronze Cuckoo *Chalcites minutillus*   RESIDENT AND VISITOR

Chrysococcyx minutillus Gould, 1859. *Proc. Zool. Soc. London*: 128.—Port Essington, NT, Australia.

Resident through NG, the NW Islands (except Gebe and Salawati), Biak I, Aru Is, Tarawai, Manam, and Karkar I, and the D'Entrecasteaux and Trobriand Is (Mayr 1941); 0–500 m, max 1,400 m (Coates 1985). Frequents rain forest, edge, and gardens. Fairly common in the lowlands, where it is the most common *Chalcites.*

Extralimital—se Asia from the Malay Penin through the Sunda Is, Philippines, and Wallacea to n Australia. Occurs in various forms, sometimes regarded as separate species, from se Asia and the Philippines to Australia. In Australia *minutillus* occurs in the north and east, south to ne NSW; *russatus* occurs in n QLD. Southern birds (race *barnardi*) are migratory, wintering north to n QLD, some apparently reaching NG (Ford 1981c). Coates (1985) noted possible occurrence in New Britain, but this is not supported by Dutson (2011)

SUBSPECIES

a. *Chalcites minutillus poecilurus* (G. R. Gray)

Chrysococcyx poecilurus G. R. Gray, 1862. *Proc. Zool. Soc. London* (1861): 431.—Misool I, NW Islands.

RANGE: all NG, and Batanta[K], Gag[L], Kofiau[W], Misool, Waigeo I, Aru Is, Tarawai, Manam, Karkar, Fergusson, Goodenough, Normanby I[J], and Trobriand Is (Mayr 1941, Greenway 1966[K], Mees 1972[L], Diamond *et al.* 2009[W], Dickinson & Remsen 2013, Pratt field notes 2003, heard[J]).

DIAGNOSIS: Forehead and crown brown; mantle bronze to greenish bronze; underparts barred; breast often with a rufous wash (Payne 2005).

b. *Chalcites minutillus misoriensis* (Salvadori)

Lamprococcyx misoriensis Salvadori, 1876. *Ann. Mus. Civ. Genova* (1875) 7: 914.—Biak I, Bay Islands.

RANGE: Biak I. No records subsequent to collection of type (K. D. Bishop *in litt.*).

DIAGNOSIS: Mantle bronze green; no white on wing; underparts barred (Payne 2005).

c. *Chalcites minutillus barnardi* (Mathews)

Chrysococcyx barnardi Mathews, 1912. *Austral. Av. Rec.* 1: 20.—Coomooboolaroo, Dawson R, QLD, Australia.

RANGE: East-central Australia, migrating north to s NG for the nonbreeding season.

DIAGNOSIS: As for the nominate form, but with a dull gloss on mantle and larger size, with wing >100 (Higgins 1999, Payne 2005). Mayr (1941) and Coates (1985) called this form *russatus.*

NOTES

Parker (1981) determined that the name *malayanus* Raffles (1822) is a junior synonym of *Cuculus xantho-rhynchus* Horsfield 1821, and that the next available name for the species is *minutillus.* The mix of names and the complex plumage variation have created a range of systematic solutions by researchers. At this time we follow Payne (2005) at the species and subspecies level but Dickinson & Remsen (2013) at the generic level. See also Parker (1981), Ford (1981c), Mees (2006), and Joseph *et al.* (2011b) in order to get a sense of the history of conflicting assessments. Erritzøe *et al.* (2012) clustered the forms *russatus, misorien-sis,* and *poecilurus* (plus additional Malaysian forms) into a species distinct from *minutillus.* Review of the songs available on the xeno-canto website indicates there are two distinct vocalizations, both of which can be heard in NG and SE Asia: (1) the brief descending series of six or seven notes (a contact call?), and (2) the very rapid, buzzing trill of many notes (the song?). Whether these represent two distinct species, as indicated by Erritzøe, or are simply two differing vocalizations given by a single species (as in some other cuckoo species) should be tested. Del Hoyo & Collar (2014) recognized a single species among the various forms. No doubt, this complex needs a comprehensive molecular systematic analysis to sort out species limits. *Other names*—Malay Bronze-Cuckoo, *Chrysococcyx malayanus.*

Heteroscenes Cabanis and Heine, 1863. *Mus. Hein.* 4(1): 26. Type, by monotypy, *Columba pallida* Latham.

The single species of the genus breeds in Australia and Tasmania and migrates or wanders to New Guinea and the Moluccas (Dickinson & Remsen 2013). *H. pallidus* is almost perfectly intermediate between *Cacomantis* and *Cuculus*. Adult plumage is distinguished by a dark eyeline, plain breast, and barred undertail. Note also the rounded nostril. *Heteroscenes* was shown to be a thinly defined genus sister to *Caliechthrus* in a phylogenetic study comparing mitochondrial DNA (Sorenson & Payne 2005). This single species has bounced from genus to genus over the last several decades. Payne (1997) and Higgins (1999) placed it into *Cuculus*, Christidis & Boles (2008) into *Cacomantis*, and Dickinson & Remsen 2013 into *Heteroscenes*.

Pallid Cuckoo *Heteroscenes pallidus*   VAGRANT | MONOTYPIC

Columba pallida Latham, 1801. *Index Orn.*, Suppl.: lx.—New South Wales, Australia.

A vagrant from Australia (Payne 1997). Records in the NG region from Momi at the head of Geelvink Bay[MA] on 25-Jun-1928; Kurik in the Trans-Fly on 4-Aug-1960 (Mees 1982b); and possibly Bensbach in the Trans-Fly on 1-Jul-1981 (see Stronach 1981c and discussion of this record in Coates 1985: 352).

Extralimital—s and e Australia, migrating to n Australia and e Wallacea (Dickinson & Remsen 2013). Also a straggler to New Zealand (Gill *et al.* 2010).

NOTES
Should be looked for in southernmost NG. Perhaps a regular but overlooked migrant, as the species does winter to the Moluccas (Mayr 1941). *Other names—Cuculus* or *Cacomantis pallidus*.

GENUS *CALIECHTHRUS*   ENDEMIC | MONOTYPIC

Caliechthrus Cabanis & Heine, 1863. *Mus. Hein.* 4(1): 31, note. Replacement name for *Simotes* Blyth 1846 (not Fischer 1829, Mammalia). Type, by monotypy, *Cuculus leucolophus* S. Müller.

The single species in the genus is endemic to the Region (Payne 1997). The genus is characterized by the lack of sexual dichromatism, rounded nostril, and white sagittal stripe on crown. It was presumed by Mayr (1941) to be close to the genus *Eudynamys*, but Payne (2005) postulated a close relationship to *Heteroscenes* and *Cacomantis* based on mitochondrial DNA comparisons (Sorenson & Payne 2005). Payne (2005) placed the following species in *Cacomantis*.

White-crowned Cuckoo *Caliechthrus leucolophus*   ENDEMIC | MONOTYPIC

Cuculus leucolophus S. Müller, 1840. *Verh. Nat. Gesch. Ned., Land-en Volkenk.* 1: 22.—Triton Bay, Bird's Neck.

Frequents the canopy of forest throughout the lowlands and hills of NG and Salawati[MA] and ?Waigeo I[J] (M. Tarburton's island website[J]); sea level–1,740 m[CO], max *ca.* 2,520 m (Freeman & Freeman 2014). Apparently absent from much of the S Lowlands and the Trans-Fly.

NOTES
Distribution in the NW Islands needs to be clarified further in the field. Record from Misool (Mayr 1941) should be dismissed (Mees 1965). In addition, Mayr (1941) mentioned Waigeo I with a question mark. *Other names*—White-crowned Koel, *Cacomantis leucolophus*.

GENUS *CACOMANTIS*   [Genus 7 spp/Region 3 spp]

Cacomantis S. Müller, 1843. *Verh. Nat. Gesch. Ned., Land-en Volkenk.* 6: 177, note. Type, by subsequent designation (Salvadori, 1880, *Orn. Pap. Moluc.* 1: 331), *Cuculus flavus* Gmelin = *Cuculus merulinus* Scopoli.

The species of this genus range from tropical Asia to New Guinea, the Solomon Islands, Vanuatu, Fiji, New Caledonia, and northern, eastern, and southwestern Australia (Payne 1997). The genus is characterized by smallish size, light build, rounded nostril, prominent song, and narrow rounded tail. It is distinguished from *Cuculus* by size and presence of a pale underwing stripe.

Chestnut-breasted Cuckoo *Cacomantis castaneiventris* RESIDENT | MONOTYPIC

Cuculus (Cacomantis) castaneiventris Gould, 1867. *Ann. Mag. Nat. Hist.* (3)20: 269.—Cape York Penin, QLD, Australia.

> Synonym: *Cacomantis arfakianus* Salvadori, 1889. *Orn. Pap. Moluc.*, Agg.: 49.—Arfak Mts, Bird's Head.
> Synonym: *Cacomantis weiskei* Reichenow, 1900. *Orn. Monatsber.* 8: 186.—Aroa R, SE Peninsula.

Mainland NG, the Aru Is[MA,X], the NW Islands (Batanta[L], Misool[Q,X], Salawati[MA,X]), and Yapen I[MA] (Mees 1965[Q], Greenway 1966[L], Hoek Ostende *et al.* 1997[X]). Inhabits forest and edge, especially in hills, from the base of the mts to 1,800 m[CO], max 2,500 m (Coates & Peckover 2001). Resident but probably locally nomadic (Coates 1985). Found in lowland rain forest of the Trans-Fly (Rand 1942a). Occurs mostly at a lower elevation than the resident population of the closely related and ecologically similar *C. flabelliformis*.

Extralimital—ne Australia.

NOTES

Specimens from QLD not separable from specimens from w NG (compare AMNH no. 626184 with 626193). There is some variability in populations, with dark birds and pale birds being found in the same locality, but overall this species is very uniform from one end of the range to the other. We do not hesitate to treat it as monotypic. Dickinson & Remsen (2013) recognized three subspecies. *Other names*—Chestnut-breasted Brush Cuckoo, *Cuculus castaneiventris*.

Fan-tailed Cuckoo *Cacomantis flabelliformis* RESIDENT AND VISITOR

Cuculus flabelliformis Latham, 1801. *Index Orn.*, Suppl.: xxx.—Sydney, NSW, Australia.

Resident race inhabits the mts of the Bird's Head, Central Ra, and Huon Penin. The Australian race winters to the S Lowlands, Aru Is, and interior upland valleys of the Central Ra.

Extralimital—breeds in sw, e, and se Australia, New Caledonia, Vanuatu, and Fiji. Australian and New Caledonian races are partially migratory, the first occasionally visiting the Aru Is and NG during the austral winter, the second occasionally visiting the e Solomon Is (Mayr & Diamond 2001, Dutson 2011).

SUBSPECIES

a. *Cacomantis flabelliformis excitus* Rothschild and Hartert

Cacomantis excitus Rothschild and Hartert, 1907. *Novit. Zool.* 14: 436.—Owgarra, Angabanga R, SE Peninsula.

RANGE: Central Ra and mts of the Bird's Head and Huon Penin (Mayr 1941, Gyldenstolpe 1955a); 1,500–3,000 m[CO], extremes 1,200–3,900 m (Coates & Peckover 2001). Inhabits montane forest and edge, mostly above the range of the ecologically similar *C. castaneiventris*.

DIAGNOSIS: Much darker than the nominate form; grayish black above, chin and throat gray, dusky chestnut on breast and belly (Payne 1997).

b. *Cacomantis flabelliformis flabelliformis* (Latham)

> Synonym: *Cuculus prionurus* Lichtenstein, 1823. *Verz. Doubl. Mus. Berlin*: 9.—New South Wales, Australia.

RANGE: Breeds in Australia. On migration occurs in the Aru Is[MA] and NG (Payne 1997). Migrant localities include: Kurik[J], Nondugl[DI], Awande[DI], and Vanapa R[CO] (Mees 1982b[J]). Abundance and distribution of winter migrants on NG mainland needs study.

DIAGNOSIS: Paler than preceding form; pallid cinnamon below, palest on abdomen; eye-ring yellow (Rand & Gilliard 1967, Payne 1997). See photo in Coates & Peckover (2001: 88). Race *prionurus* subsumed into the nominate race by Payne (1997).

NOTES

We use the species name *flabelliformis* rather than *pyrrhophanus* following Mason (1982). *Other names*—Fan-tailed Brush Cuckoo, Ash-coloured Cuckoo, *Cacomantis* or *Cuculus pyrrhophanus*.

Brush Cuckoo *Cacomantis variolosus*

Cuculus variolosus Vigors & Horsfield, 1827. *Trans. Linn. Soc. London* (1826) 15: 300.—Parramatta, NSW, Australia.

A common species of forest, edge, and gardens through the NG region; sea level–1,300 m, max 1,800 m[CO]. Inhabits many satellite islands, but no records from Mios Num, the Trobriand Is, Woodlark, Misima, Tagula, or Rossel I. Some Australian-breeding birds migrate to NG (Payne 1997, Coates & Peckover 2001).

Extralimital—from peninsular Malaysia, Sumatra, and the Philippines, south and east to the Bismarcks, the Solomons, and n and e Australia (Payne 1997).

SUBSPECIES

a. *Cacomantis variolosus infaustus* Cabanis and Heine

Cacomantis infaustus Cabanis and Heine, 1863. *Mus. Hein.* 4(1): 23.—Misool I, NW Islands.

> Synonym: *Cacomantis assimilis fortior* Rothschild & Hartert, 1914. *Novit. Zool.* 21: 4.—Goodenough I, D'Entrecasteaux Arch, SE Islands.
> Synonym: *Cacomantis variolosus obscuratus* Stresemann & Paludan, 1932. *Novit. Zool.* 38: 202.—Numfor I, Bay Islands.
> Synonym: *Cacomantis variolosus chivae* Mayr & Meyer de Schauensee, 1939. *Proc. Acad. Nat. Sci. Phila.* 91: 23.—Korrido, Biak I, Bay Islands.

RANGE: Morotai, Tidore, Ternate, Halmahera, Bacan, Obi, Seram Laut, Watubela, and Kai Is; Kofiau[L], Gebe[K], and Gag I[J]; Misool, Salawati, Batanta, and Waigeo I; Bird's Head and Neck, and n NG east to the Sepik R; Aru Is; Biak, Numfor, and Yapen I; also Karkar and Manam I (Mayr 1941, Mees 1972[K], Johnstone 2006[J], Diamond *et al.* 2009[L]).

DIAGNOSIS: Dark greenish brown with some iridescence above, crown marginally grayer; ventrally dirty gray and russet mixed, with gray predominating in throat, and russet predominating around abdomen. Feet olive or brownish (Payne 1997). Wing: male 115–121, female 118 (Rand & Gilliard 1967).

b. *Cacomantis variolosus oreophilus* Hartert

Cacomantis variolosus oreophilus Hartert, 1925. *Novit. Zool.* 32: 168.—Hydrographer Mts, SE Peninsula.

RANGE: Southern Lowlands, Huon Penin, SE Peninsula, SE Islands including Goodenough[MA], Fergusson[MA], and Normanby[J,M]; also Karkar[MA] and Manam I[K] (Gyldenstolpe 1955a, Diamond & LeCroy 1979[K], LeCroy *et al.* 1984[J], and Pratt field notes 2003, heard[M]). From the lowlands up to 1,250 m (Diamond 1972).

DIAGNOSIS: Eye-ring dark; short-winged and long-billed (Payne 1997).

c. *Cacomantis variolosus variolosus* (Vigors & Horsfield)

> Synonym: *Cuculus tymbonomus* S. Müller, 1843. *Verh. Nat. Gesch. Ned., Land-en Volkenk.* 6: 177, note 3.—Timor, Lesser Sundas.

RANGE: Breeds in n and e Australia; winters to the Aru, Kai, and Moluccan Is, NG (Geelvink Bay, Otakwa R, Fly R, Astrolabe Bay), NW Islands, and Lesser Sundas (Mayr 1941, Payne 1997).

DIAGNOSIS: Overall paler. Brownish gray above; head pale gray; breast and belly buff, with obsolete barring on lower breast; eye-ring usually gray; iris dark brown; bill black; pale and washed-out looking compared with breeding NG forms.

NOTES

We do not recognize the races *fortior*, *chivae*, *obscuratus*, and *tymbonomus*, based in part on the work of Payne (1997, 2005), as well as our own museum research. *Other names*—Grey-breasted Brush Cuckoo, Square-tailed or Grey-headed Cuckoo, *Cuculus variolosus*.

GENUS *CUCULUS*   [Genus 10 spp/Region 2 spp]

Cuculus Linnaeus, 1758. *Syst. Nat.*, ed. 10, 1: 110. Type, by tautonymy, "Cuculus" = *Cuculus canorus* Linnaeus.

The species of this genus range from northern and western Europe to Africa, Asia, New Guinea, New Britain, and Australia (Payne 1997). The genus is characterized by the long tail; long, pointed wings; profusely barred or streaked underparts; the brown, gray, or black upperparts of the adult male, often with an accipitrine plumage pattern; typically a yellow eye-ring; a fine bill; and small feet and legs.

Himalayan Cuckoo *Cuculus saturatus*   VAGRANT | MONOTYPIC

Cuculus saturatus 'Hodgson', Blyth, 1843. *Journ. Asiat. Soc. Bengal*, 12 (2): 942.—Nepal.

Apparently a vagrant to the NW Islands and Bird's Head (Payne 2005). This species and the following (*C. optatus*) can be distinguished only by vocalization and the shorter wing length of *saturatus*, thus field identification of wintering birds in NG is problematic.

Extralimital—breeds in the Himalayas, s China, and Taiwan; winters to Malaysia and e Indonesia (Payne 2005).

NOTES

Following Payne (1977, 2005) and Christidis & Boles (2008), we recognize three species-level forms of this lineage: *C. saturatus* of the Himalayas and s China, wintering to westernmost NG; *C. lepidus*, the disjunct and sedentary population that breeds in peninsular Malaysia, Sumatra, w Borneo, Java, and the Lesser Sundas (not recorded from NG; see Christidis & Boles 2008: 161); and *C. optatus*, the northeastern form, breeding from w Russia east to Siberia and ne China, and wintering from n peninsular Malaysia and the Philippines southeast to NG and n and e Australia. *C. saturatus* is smaller than *optatus* (wing less than 192), with iris usually brown, and male song distinct, a high note followed by two lower flat notes; female call, *kwik-kwik-kwik*, is higher-pitched than call of *optatus* (Payne 2005). This is yet another species complex for which there is no agreement across authorities (for one recent view distinct from that of Payne, see Erritzøe *et al.* 2012). Molecular analysis is clearly needed here to resolve the impasse. *Other names*—Blyth's Cuckoo, Southern Muted Cuckoo; formerly treated as the Oriental Cuckoo, *C. optatus*.

Oriental Cuckoo *Cuculus optatus*   VISITOR | MONOTYPIC

Cuculus optatus Gould, 1845. *Proc. Zool. Soc. London*: 18.—Port Essington, NT, Australia.

A fairly common nonbreeding visitor to the NG region, arriving from its northern breeding grounds during the northern winter. It is difficult to determine the species of many of the old records, making the status of *optatus* vs. *saturatus* problematic. Records from throughout the Mainland[Q]; also Gebe[J], Misool[K], Salawati[L], Biak[W], Manam[CO], Karkar[Z], and Goodenough I to 1,600 m[T] (Rothschild & Hartert 1915a[Z], Mees 1965[K], 1972[J], Hoogerwerf 1971[Q], Beehler field notes 18-Mar-1976[T], Hornbuckle & Merrill 2004[L], K. D. Bishop *in litt.*[W]). A specimen was recovered from glacial ice at 4,400 m in the Carstensz Massif (Schodde *et al.* 1975).

Extralimital—breeds in w Russia to e Siberia, Korea, and Japan; winters from se Asia and the Philippines through the Bismarcks, the Solomons, n and e Australia, and NZ (Payne 2005, Gill *et al.* 2010).

NOTES

Difficult to distinguish from *saturatus* and treated by some authors (*e.g.*, del Hoyo & Collar 2014) as conspecific. Note larger size of *optatus* (wing >192; Gilliard & LeCroy 1966); typically yellow iris; male song four low *hoop* whistles, the first one soft (Payne 1997). Song lower-pitched than that of *saturatus* (Payne 2005).

Records of *Cuculus canorus* in the NG region (*e.g.*, Mayr 1941: 70; Rand & Gilliard 1967: 235) were in error. Establishing the wintering ranges of *optatus* and *saturatus* in NG will require considerable effort, mainly examining museum material and perhaps molecular methods. It is doubtful these birds vocalize while in NG. *Other names*—Horsfield's Cuckoo, *Cuculus horsfieldi*, *C. saturatus*.

ORDER CAPRIMULGIFORMES

The order Caprimulgiformes encompasses the Oilbird (Steatornithidae), frogmouths, potoos (Nyctibiidae), nightjars, hummingbirds (Trochilidae), swifts, and treeswifts (Hackett *et mult.* 2008, Cracraft 2013). This large and diverse order of land birds, which is nearly cosmopolitan, is characterized by the short and weak legs and tiny feet.

Family Podargidae

Frogmouths [Family 13 spp/Region 2 spp]

The species of the family range through tropical Asia and thence southeastward to New Guinea, Australia, and the Solomon Islands. Frogmouths are medium to large nocturnal birds with a large, flattened head, heavy bill, wide gape, short weak legs, and a long pointed tail (Holyoak 1999b). They occur in forest or partly wooded areas to elevations of more than 1,500 m. Frogmouths typically choose exposed perches from which, at night, they swoop down on large insects, small mammals, birds, lizards, or frogs. The members of the family are distinguished by the very broad, heavy, strong, and hooked bill. The bill of the potoos is less massive, more laterally compressed distally, and its hook is even more pronounced. The Podargidae constitutes a distinct and isolated, deep lineage within the Caprimulgiformes (Braun & Huddleston 2009).

GENUS *PODARGUS* [Genus 3 spp/Region 2 spp]

Podargus Vieillot, 1818, *Nouv. Dict. Hist. Nat.* 27: 151. Type, by monotypy, *Podargus griseus* Vieillot = *Caprimulgus strigoides* Latham.

The species of this genus are confined to Australia and New Guinea; there is no record of the genus from the Bismarck Archipelago (Holyoak 1999b). *Podargus* is characterized by the massive broad bill, cryptic plumage, plumage polymorphism, short legs, large flat head, and long pointed tail.

There is a single unconfirmed sight record of the Tawny Frogmouth (*Podargus strigoides*) from near Morehead, Western Province, Papua New Guinea (Mackay 1976).

Marbled Frogmouth *Podargus ocellatus*   RESIDENT

Podargus ocellatus Quoy & Gaimard, 1830. *Voy. Astrolabe, Zool.* 1: 208; *Atlas*: pl. 14.—Manokwari, Bird's Head.

Inhabits forest and scrub of all NG and the NW Islands; Bay Is; Aru, Trobriand, and D'Entrecasteaux Is; and Tagula I; sea level–1,500 m.

Extralimital—isolated populations in Australia confined to Cape York and the QLD-NSW border region (Higgins 1999). Populations formerly allied with this species in the Solomon Is have been subsequently split off into a distinct genus and species, *Rigidipenna inexpectata* (see Cleere *et al.* 2007).

SUBSPECIES

a. *Podargus ocellatus ocellatus* Quoy & Gaimard

RANGE: All NG, Aru Is, and Misool, Salawati, Waigeo, Batanta[M,W], Yapen, Biak[L], and Mios Num I (Mayr 1941, S. F. Bailey field notes 13-Aug-1994[L], S. F. Bailey sighting 17-Nov-1995[W], Beehler field notes 27-Feb-2004[M]). Biak I sighting should be corroborated.

DIAGNOSIS: Wing 181–193; obscurely sexually dichromatic; female usually rufous with minimal white blotching on ventral surface (Rand & Gilliard 1967).

b. *Podargus ocellatus intermedius* Hartert

Podargus intermedius Hartert, 1895. *Bull. Brit. Orn. Club* 5: x.—Kiriwina, Trobriand Is.

RANGE: Trobriand and D'Entrecasteaux Is (Mayr 1941, Hoek Ostende *et al.* 1997). Also Slade and Hummock I of the Engineer Is (J. Dumbacher field notes).

DIAGNOSIS: Wing 192–243; minimal sexual dichromatism; male with marginally more white blotching; females not very rufous. Some specimens are very large.

c. *Podargus ocellatus meeki* Hartert

Podargus meeki Hartert, 1898. *Bull. Brit. Orn. Club* 8: 8.—Tagula I, Louisiade Arch.

RANGE: Tagula I.

DIAGNOSIS: Wing 190–195; female not more rufous than male but darker, more blotched with blackish brown (Rand & Gilliard 1967). Type description noted female never being rufous, and ventrally much more heavily marked with black when compared to the nominate form and *intermedius*.

NOTES

Status of the described subspecies might be suspect because of the plumage polymorphism combined with sexual dichromatism; however, the insular races seem distinct and share their geographic ranges with numerous other endemic subspecies and species, suggesting they too may be valid. All subspecies need more study, and perhaps fresh specimen material. Voice recordings are available for all taxa, as are tissue samples for DNA comparison. The single sight record from Biak I is curious and should be confirmed with additional sightings or specimens to prove it was not the following species, well known from Biak. *Other names*—Little Papuan Frogmouth.

Papuan Frogmouth *Podargus papuensis*   RESIDENT | MONOTYPIC

Podargus papuensis Quoy & Gaimard, 1830. *Voy. Astrolabe, Zool.* 1: 207; *Atlas*: pl. 13.—Manokwari, Bird's Head.

Forest, edge, and gardens throughout NG, the four main NW Islands, Bay Is (Numfor, Biak, and Yapen), and Aru Is; sea level–1,000 m, max 2,200 m[J] (Mayr 1941, D. Hobcroft *in litt.* at Ambua Lodge[J]). Absent from the SE Islands.

Extralimital—n QLD, Australia.

NOTES

Monotypic, following Holyoak (1999b). *Other names*—Giant or Great Papuan Frogmouth.

Family Caprimulgidae

Nightjars   [Family 92 spp/Region 6 spp]

The species of the family range from North and South America to Europe, Africa, tropical Asia, Melanesia, and Australia (Cleere 1999). Nightjars are cryptic, nocturnal birds with long wings and tail, short and weak legs, and often with a white patch on the outer wing and on the throat. They are most often encountered resting in leaves on the ground or on a tree limb, with the body held in a characteristic horizontal position. Flight is silent and characterized by a peculiar floppy or jerky wingbeat. Christidis & Boles (2008) followed the work of Larsen *et al.* (2007) in treating the species of *Eurostopodus* as distinct at the family level, sister to the Caprimulgidae. We retain the traditional inclusive family treatment, following Dickinson & Remsen (2013).

GENUS *EUROSTOPODUS*   [Genus 6 spp/Region 4 spp]

Eurostopodus Gould, 1838. *Synops. Birds Austral.* 4 (Apr-1838), app.: 1. Type, by subsequent designation (G. R. Gray, 1840. *List Gen. Birds*: 7), *Caprimulgus guttatus* Vigors & Horsfield.

The species of this genus, as now delimited, are restricted to Australasia, ranging from Sulawesi through New Guinea and Australia to New Caledonia and the Solomon Islands (Dickinson & Remsen 2013, del Hoyo & Collar 2014). Formerly included were the Asian species *macrotis* and *temminckii*, now assigned to the genus *Lyncornis* by Cleere (2010). The genus *Eurostopodus* is characterized by the relatively large size, lack of prominent rictal bristles, and absence of white in tail (Cleere 2010). Braun & Huddleston (2009) provided evidence this genus belongs with the Caprimulgidae and should not be split off into its own family. Han *et al.* (2010) demonstrated that *E. papuensis* is an outgroup of the species pair *argus* and *mystacalis*.

Spotted Nightjar *Eurostopodus argus*   VAGRANT | MONOTYPIC

Caprimulgus argus Hartert, 1892. *Cat. Birds Brit. Mus.* 16: 607 (in key), 608.—South Australia (*fide* Schodde & Mason 1997).

Wintering birds recorded from the Aru Is (Mayr 1941, Hoek Ostende *et al.* 1997). Recently a sight record was reported from Obo in the Fly R region by Johnson & Richards (1994). They described the flying bird well, noting the large white wing patches and all-dark tail.

Extralimital—breeds through all Australia but e coast. Winters to the Lesser Sundas (Cleere 1999).

NOTES

Nomenclature follows Schodde & Mason (1980) and Cleere (1999). The latter noted that the named form *guttatus* actually refers to the species *mystacalis*, not *argus*. *Other names*—Spotted Eared-nightjar, *Eurostopodus guttatus*.

White-throated Nightjar *Eurostopodus mystacalis*   VISITOR | MONOTYPIC

Caprimulgus mystacalis Temminck, 1826. *Planch. Col. d'Ois.*, livr. 69: pl. 410.—New South Wales, Australia.
   Synonym: *Caprimulgus guttatus* Vigors & Horsfield, 1826. *Trans. Linn. Soc. London* 15: 192.—Parramatta, NSW, Australia.

Presumably a nonbreeding Australian visitor to grasslands and edge in e and c NG; in the s watershed occurs west to the Fly R; in the n watershed west to Astrolabe Bay (including the Eastern Ra); also one record from the Taritatu R. Sea level–1,500 m.

Localities—Bernhard Camp on the Taritatu R[W], Bensbach[B], Fly R[MA], Kiunga[K], Tabubil[A], Bogia[CO], Ramu R[MA], Astrolabe Bay[MA], Huon Penin[CO], Wahgi valley[CO], Soliabeda[DI], Mt Bosavi[CO], Lake Daviumbu[CO], and Dimisisi in the lower Fly R[CO], and 10 km east of Port Moresby[L] (Rand 1942b[W], Hicks 1988b[L], Gregory 1995b[A], D. Hobcroft *in litt.*[K], K. D. Bishop[B]).

Extralimital—breeds in e Australia; unconfirmed report from Rennell I (Dutson 2011). Reported to be a passage migrant through the Torres Strait in September and October (Draffan *et al.* 1983).

NOTES

Treatment as monotypic follows Cleere (2010) and Dutson (2011). *Other names*—White-throated Eared-nightjar, *Eurostopodus albogularis*.

Papuan Nightjar *Eurostopodus papuensis*   ENDEMIC | MONOTYPIC

Caprimulgus papuensis Schlegel, 1866. *Ned. Tijdschr. Dierk.* 3: 340.—Salawati and the adjacent coast of the w Bird's Head.
   Synonym: *Eurystopodus astrolabae* Ramsay, 1883. *Proc. Linn. Soc. NSW* 8: 20.—Astrolabe Ra, SE Peninsula.

All NG and Salawati I[MA]; sea level–400 m (Beehler *et al.* 1986, Hoek Ostende *et al.* 1997). Frequents glades within the forest and edge. Generally scarce (Rand 1942b, Beehler & Finch 1981, Coates 1985).

NOTES

We follow Cleere (2010) in treating the species as monotypic. *Other names*—Papuan Eared-nightjar.

Archbold's Nightjar *Eurostopodus archboldi*   ENDEMIC | MONOTYPIC

Lyncornis archboldi Mayr & Rand, 1935. *Amer. Mus. Novit.* 814: 4.—Western slope of Mt Tafa, 2,400 m, SE Peninsula.

Frequents montane forest of the Bird's Head, Central Ra, and Huon Penin, from 2,400 m to timberline (records as low as 2,200 m; Coates & Peckover 2001). Distribution apparently patchy. Records include: Arfak Mts[Q], Ilaga[K], Lake Habbema[W], Bele R Camp[W], Hindenburg Ra[A], Tomba on Mt Hagen[L], Mt Giluwe[J], Kumul Lodge[Z], Tari gap[O], Huon Penin[CO], Wharton Ra (Mt Tafa)[CO], Mt Albert Edward[CO], Lake Omha[X] (Rand 1942b[W], Mayr & Gilliard 1954[L], Sims 1956[J], Ripley 1964a[K], Beehler field notes 12-Aug-1987[X], Rowland 1994b[O], Rowland 1995[A], Gibbs 1996[Q], Beehler field notes, multiple dates[Z]). The Huon Penin specimen was collected by the Archbold Expedition on 20-Jun-1964 at their Mt Ulus camp in the Cromwell Mts (AMNH no. 823591; B. J. Coates *in litt.*).

The published record from the Tamrau Mts (map in Cleere 1999) is apparently in error. An occupied nest was photographed on Gunung Indon, Arfak Mts, in 1994 (Gibbs 1996). *Other names*—Mountain Nightjar.

GENUS *CAPRIMULGUS* [Genus 36 spp/Region 2 spp]

Caprimulgus Linnaeus, 1758. *Syst. Nat.*, ed. 10, 1: 193. Type, by tautonymy, *Caprimulgus europaeus* Linnaeus.

The species of this genus range through Eurasia, Africa, Micronesia, Melanesia, and Australia (Barrowclough *et al.* 2006, Larsen *et al.* 2007, Cleere 2010). The genus is characterized by the white patches on wing and tail (at least in males), long rictal bristles around the gape, and cryptic plumage. As traditionally construed, *Caprimulgus* was shown to be polyphyletic (Han *et al.* 2010).

Grey Nightjar *Caprimulgus jotaka*   VAGRANT
Caprimulgus Jotaka Temminck & Schlegel, 1844. In Siebold, *Fauna Japonica, Aves*: 37: pl. 12 (male), pl. 13 (female).—Japan.

Vagrant from Asia. There is one specimen from the e Bird's Head (Mansinam I, near Manokwari), and one field record from the n Bird's Neck (S. van Balen *in litt.*, documented with a sound recording).

Extralimital—breeds from the Himalayas and India to Korea, Manchuria, se Siberia, and Japan, wintering south to Indochina, the Greater Sundas, and the Philippines.

SUBSPECIES

a. *Caprimulgus jotaka jotaka* Temminck & Schlegel
RANGE: Breeds in the Russian far east, Japan, and c and e China, migrating to se Asia; presumably a vagrant to the Bird's Head and Neck.
DIAGNOSIS: Nominate is slightly smaller and paler brown than *hazarae*, less heavily marked on the back, barring on the tail is narrower, and the throat patches are less tawny (Cleere 1998).

NOTES
We follow Rasmussen & Anderton (2005) and Cleere (2010) in treating *jotaka* and *hazarae* as forms of a distinct species, *C. jotaka. Other names*—Jungle or Indian Jungle Nightjar, Japanese Nightjar, *Caprimulgus indicus*.

Large-tailed Nightjar *Caprimulgus macrurus*   RESIDENT
Caprimulgus macrurus Horsfield, 1821. *Trans. Linn. Soc. London* 13: 142.—Java.

Throughout NG and most adjacent large islands; sea level–2,000 m (exceptionally to higher elevations). Frequents forest edge, village environs, mangrove verges, mid-montane grasslands, and riparian sandbanks.

Extralimital—tropical Asia to n and e Australia and the Bismarck Arch (Mayr & Diamond 2001, Dutson 2011).

SUBSPECIES

a. *Caprimulgus macrurus schlegelii* A. B. Meyer, 1874. *Sitzungsber. Akad. Wiss. Wien* 69: 210.—Port Essington, NT, Australia.
    Synonym: *Caprimulgus macrurus yorki* Mathews, 1912. *Novit. Zool.* (1911) 18: 291.—Cape York, QLD, Australia.
    Synonym: *Caprimulgus macrurus meeki* Rothschild and Hartert, 1918. *Novit. Zool.* 25: 321.—Tagula I, Louisiade Arch, SE Islands.
    Synonym: *Caprimulgus macrurus schillmölleri* Stresemann 1931. *Orn. Monatsber.* 39: 170.—Halmahera, Moluccas.
RANGE: All NG; Batanta[MA], Gag[J], Kofiau[K], Misool[L], Salawati[L], and Waigeo I[MA]; Aru Is[MA]; Biak[MA] and Yapen I[MA]; Karkar[P] and Manam I[MA]; Engineer Is[W]; Goodenough[CO], Fergusson[MA], and Normanby I[X,Y]; and Tagula I[MA] (Mees 1965[L], Diamond & LeCroy 1979[P], LeCroy *et al.* 1984[Y], Pratt field notes 2003[X], Johnstone 2006[J], Diamond *et al.* 2009[K], J. Dumbacher *in litt.*[W]). Also Lesser Sundas and s Maluku, Long and Umboi I, New Britain, and coastal n and ne Australia (Cleere 1999, Mayr & Diamond 2001).
DIAGNOSIS: Differs from the nominate form of c Indonesia in having a dark brown breast with fine pale

barring and distinctly buffy tips to wing coverts (more whitish in nominate form). No difference in size (Mees 1977a). Mees (1965, 1977a) synonymized the races *yorki* and *schillmölleri*. Cleere (1999) subsumed *meeki* into this form.

NOTES
We follow the treatment of Mees (1965: 171–72, 1977a) and Cleere (1999). *Other names*—Long-tailed or White-tailed Nightjar, Coffinbird.

Family Aegothelidae

Owlet-nightjars [Family 11 spp/Region 7 spp]

The species of the family are confined to Australia, New Guinea, New Caledonia, and the Moluccas. Similar to but considerably smaller than the frogmouths, the owlet-nightjars are cryptic nocturnal birds with a large gape, a weak bill, small weak legs, and a long tail. Unlike frogmouths, the owlet-nightjars have small facial disks, the bill is framed with long, branched bristles, and the tail is not tapered. The cryptic plumage is rufous, brown, or gray, mottled or barred, and the plumage of some species is highly variable, and thus identifying some birds in the field may prove impossible. Owlet-nightjars inhabit forest and wooded areas throughout New Guinea. They are most easily observed during the day, when flushed from a roost in a tree cavity or tangle of foliage. The owlet-nightjars are a sister lineage to the swifts and hummingbirds (G. Mayr *et al.* 2003, Hackett *et mult.* 2008, Cracraft 2013).

GENUS AEGOTHELES [Genus 10 spp/Region 7 spp]

Aegotheles Vigors & Horsfield, 1827. *Trans. Linn. Soc. London* 15: 194. Type, by monotypy, *Caprimulgus novaehollandiae* Latham = *Caprimulgus cristatus* Shaw.

The species of this genus range from the Moluccas to Australia and New Caledonia; the species in New Zealand became extinct in the Holocene (Gill *et al.* 2010). The center of species richness for the genus is New Guinea. Characters for the genus are as for the family. The gender of the genus is masculine (Peters 1940b: 181, footnote). For generic circumscription we follow Dumbacher *et al.* (2003). We do not recognize *Euaegotheles* at the generic level (see discussion in Schodde & Mason 1997: 314, and Dumbacher *et al.* 2003).

Feline Owlet-nightjar *Aegotheles insignis* ENDEMIC | MONOTYPIC
Aegotheles insignis Salvadori, 1876. *Ann. Mus. Civ. Genova* (1875) 7: 916.—Hatam, Arfak Mts, Bird's Head.
Synonym: *Aegotheles pulcher* Hartert, 1898. *Bull. Brit. Orn. Club* 8: 8.—Mountains of se NG (probably upper Aroa R).
Forests of most or all mts of the Mainland, 1,150–2,800 m (Diamond 1985, Pratt 2000, Beehler *et al.* 2012). Although there are no published records for the Fakfak, Kumawa, or Cyclops Mts, the species should be looked for in all three. Widespread and common though easily overlooked.

NOTES
The form *pulcher* is treated as a synonym of the nominate form following Rand (1942b). The lowland form *tatei* is now treated as a full species, following Pratt (2000). *Other names*—Large Owlet-nightjar.

Starry Owlet-nightjar *Aegotheles tatei* ENDEMIC | MONOTYPIC
Aegotheles insignis tatei Rand, 1941. *Amer. Mus. Novit.* 1102: 10.—8 km below Palmer Junction, 80 m elevation, Fly R.

The species is historically known from the upper Fly R (near Kiunga) and the far SE Peninsula at Amazon Bay (Rand 1941, 1942a, Pratt 2000). It is very poorly known or perhaps very patchily distributed.

The first record for Indonesian NG was made on the Kali Muyu R (40 m elevation) near Kanggup village on 6-Oct-2012, *ca.* 30 km due west of Kiunga (Verbelen 2014); there is a sound recording on the xeno-canto website from this work. I. Woxvold (*in litt.*) observed the species in the e Gulf Prov lowlands.

NOTES

Treatment follows Pratt (2000) and Dumbacher *et al.* (2003). A slender owlet-nightjar, with short, tight plumage. Known to exhibit only a bright rufous plumage. Feather "horns" curve above the eye and taper to a point. Smaller than *A. insignis. Other names*—Rand's or Spangled Owlet-nightjar.

Wallace's Owlet-nightjar *Aegotheles wallacii*   ENDEMIC | MONOTYPIC

Ægotheles wallacii G. R. Gray, 1859. *Proc. Zool. Soc. London:* 154.—Manokwari, Bird's Head.
> Synonym: *Aegotheles wallacei gigas* Rothschild, 1931. *Novit. Zool.* 36: 268.—Mount Derimapa, Gebroeders Mts, Western Ra.
> Synonym: *Aegotheles wallacii manni* Diamond, 1969. *Amer. Mus. Novit.* 2362: 12.—Mount Menawa, Bewani Mts, north-central NG.

Sparsely distributed in the Aru Is, Waigeo I, and through the mts of the Bird's Head, w sector of the Western Ra, the Border Ra, Eastern Ra, and PNG North Coastal Ra; 600–1,500 m, near sea level in sw NG and in the upper Fly (Hartert *et al.* 1936). No records from the Bird's Neck, the NW Lowlands, e Sepik-Ramu, or Huon and SE Peninsulas.

Localities—Tamrau Mts[CO], Arfak Mts[CO], Mt Derimapa of the Gebroeders Mts[K], Kunupi in the Weyland Mts[K], Wanumbai in the Aru Is[MA], Wataikwa R[MA], Bewani and Prince Alexander Mts[D] (970 to 1,100 m), Telefomin[N], Palmer Junction on the upper Fly R at 100 m[MA], Elevala R at 100 m[L], upper May R[J], and Karimui at 1,100 m[DI] (Hartert *et al.* 1936[K], Diamond 1969[D], specimen at BPBM[N], I. Woxvold *in litt.*[J], D. Hobcroft *in litt.*[L]). The records suggest that this is mainly a hill-forest species (Diamond 1969), occurring at elevations above *A. bennettii* and below *A. albertisi.*

NOTES

We treat the species as monotypic. Races have been described based on minor size differences (Holyoak 1999a). A photo of an owlet-nightjar from Waigeo I identified as belonging to this species has been posted online by I. Mauro. *Other names*—White-spotted Owlet-nightjar.

Mountain Owlet-nightjar *Aegotheles albertisi*   ENDEMIC | MONOTYPIC

Ægotheles albertisi P. L. Sclater, 1874. *Proc. Zool. Soc. London* (1873): 696.—Hatam, Arfak Mts, Bird's Head.
> Synonym: *Aegotheles salvadorii* Hartert, 1892. *Cat. Birds Brit. Mus.* 16: 649.—Astrolabe Ra, SE Peninsula.
> Synonym: *Aegotheles albertisi wondiwoi* Mayr & Rand, 1936. *Mitt. Zool. Mus. Berlin* 21: 242.—Mount Wondiwoi, Wandammen Penin, Bird's Neck.
> Synonym: *Aegotheles albertisi archboldi* Rand, 1941. *Amer. Mus. Novit.* 1102: 10.—9 km ne of Lake Habbema, 2,800 m, Western Ra.

Mountains of the Bird's Head and Wandammen Penin, the Central Ra, Kumawa Mts[J], Foja Mts[K], and mts of the Huon Penin; 800–2,900 m[CO] (to 3,700 m on Huon[L]). Frequents forest and edge (Mayr 1941, Diamond 1985[J], Coates & Peckover 2001[L], Beehler *et al.* 2012[K], Pratt & Beehler 2014).

NOTES

The forms *albertisi* and *archboldi* have had a long co-history, initially combined into a single species (Mayr 1941), and then divided (Junge 1953), and now united here once again as a single variable species. Pratt has studied the museum specimens, and Dumbacher *et al.* 2003 have conducted molecular studies that underpin this latest treatment.

These owlet-nightjars exhibit perhaps the most individual variability of any NG bird species. Color morphs in a given population range across a gradient from dark brown to reddish brown with many intermediate shades. Apart from color, individuals also vary in their degree and pattern of barred, chevron, and streaked markings. All characters used to define either *albertisi* or *archboldi* can be found in the populations of both forms, with varying frequency. This variation thwarts any attempt to classify populations into discrete geographic taxa. Individual plumage variation trumps geographic variation, rendering most described subspecies differences as insubstantial. Thus we treat the small montane owlet-nightjars as a single species with substantial nongeographic variation best not memorialized in recognized subspecies.

The molecular results of Dumbacher *et al.* (2003) have been used by Cleere (2010) and Dickinson & Remson (2013) to split this lineage into three species—*albertisi* (far west), *archboldi* (west), and *salvadorii* (east)—and to group them with *wallacii*, owing to strong support for the individual branches representing these taxa and seemingly deep branching among them. However, geographic sampling for *albertisi* and *archboldi* by Dumbacher *et al.* (2003) was severely limited for each population and might be expected to yield strong branch support relative to distant *salvadorii*. Furthermore, the nodes among the *albertisi*, *archboldi*, *salvadorii*, and *wallacii* branches were not well supported and resulted in essentially a four-branch polytomy. To sum up, Dumbacher *et al.* (2003: 545) concluded "We remain cautious about the phylogenetic and taxonomic implication of these results."

An alternative to splitting these taxa into four species is to lump the three taxa that cannot be consistently diagnosed—*albertisi*, *archboldi*, and *salvadorii*. The fourth taxon, *wallacii*, is diagnosable and shares its geographic range with the other three, inhabiting foothills rather than high mountains, and so we maintain its status as a species. A minority opinion might posit that *wallacii* too belongs in the larger combined species as a hill-forest plumage type, were it not that its plumage really is quite distinct and the few recordings of its voice reveal a different song. The relationships of these four populations should be further explored with molecular techniques relying on broad geographic sampling and including nuclear gene data. *Other names:* Archbold's Owlet-nightjar

Barred Owlet-nightjar *Aegotheles bennettii*   ENDEMIC

Aegotheles bennettii Salvadori & D'Albertis, 1876. *Ann. Mus. Civ. Genova* (1875) 7: 816.—Hall Sound, SE Peninsula.

Central and e NG, Aru Is, and D'Entrecasteaux Arch; sea level–800 m, max 1,100 m (Coates & Peckover 2001). Widely distributed in the north and east; absent from w NG. Ranges west in the n watershed to the Taritatu R, and west in the s watershed to the Kumbe R (Trans-Fly; Mees 1982b). Frequents rain forest and monsoon forest.

SUBSPECIES

a. *Aegotheles bennettii wiedenfeldi* Laubmann
Aegotheles bennetti wiedenfeldi Laubmann, 1914. *Orn. Monatsber.* 22: 7.—Sattelberg, Huon Penin.
RANGE: Northern NG from the Taritatu R east to Holnicote Bay (n coast of the SE Peninsula) and including the Huon Penin.
DIAGNOSIS: Large; wing 126–142 (Rand & Gilliard 1967).

b. *Aegotheles bennettii bennettii* Salvadori & D'Albertis
RANGE: Southern NG from the Kumbe R (near Merauke)[J] east to Milne Bay (Mees 1982b[J]).
DIAGNOSIS: Small; wing 121–128 (Rand & Gilliard 1967). See photo in Coates & Peckover (2001: 98).

c. *Aegotheles bennettii terborghi* Diamond
Aegotheles bennettii terborghi Diamond, 1967. *Amer. Mus. Novit.* 2284: 5.—Karimui, Eastern Ra.
RANGE: Karimui basin[DI] (and presumably Crater Mt), se sector of the Eastern Ra.
DIAGNOSIS: Substantially the largest form, with blackish upperparts (Holyoak 1999a). Described from a unique type (Diamond 1972). Wing 154; tail 142 (Diamond 1967).

d. *Aegotheles bennettii plumifer* Ramsay
Aegotheles ? plumifera Ramsay, 1883. *Proc. Linn. Soc.* 8: 21.—Fergusson I, D'Entrecasteaux Arch, SE Islands.
RANGE: D'Entrecasteaux Arch. Orginally reported from Fergusson and Goodenough I (Mayr 1941); more recently collected on Normanby I (Pratt specimens at BPBM). Elevational records indicate this may be an upland form, restricted to forest above *ca.* 500 m.
DIAGNOSIS: Differs from the nominate form as follows: breast with barring reduced and with shaft-streaking and chevron markings present, and plumage overall buffier. See photo in Coates & Peckover (2001: 98). Material at AMNH and specimens recently collected on Fergusson and Normanby I match well with the type description for *plumifer*, so this is undoubtedly a valid insular race. Strangely, the type series

for *plumifer* in the Australian Museum in Sydney, examined by TKP, best aligns with the nominate form, which is inexplicable.

NOTES

The population *affinis* of the Bird's Head, traditionally placed in this species (Mayr 1941) or with *A. cristatus* (Rand & Gilliard 1967) has been given full species status here. The possible species status of *terborghi* and *plumifer* should be investigated. *Other names*—Bennett's or Collared Owlet-nightjar.

Allied Owlet-nightjar *Aegotheles affinis* ENDEMIC | MONOTYPIC
Aegotheles affinis Salvadori, 1876. *Ann. Mus. Civ. Genova* (1875) 7: 917.—Warmendi, Arfak Mts, Bird's Head.
    Arfak Mts of the Bird's Head, 80–1,500 m (Pratt & Beehler 2014). Should be looked for in the Tamrau Mts (and perhaps also the mts of the Bird's Neck, especially the Wandammen Mts).

NOTES

We here treat *affinis* as a distinct species, following the recommendation of Dumbacher *et al.* (2003). Diagnosis as follows: barred pattern like *A. bennettii* but with a brownish cast; thickly feathered, producing a plump shape with a large head; washed buffy brown, with buff cheeks and whitish belly; head coarsely patterned; pale barring on underside of tail feathers narrow and continuous; most birds show a distinct pale collar. Breast usually completely and evenly barred, but belly whitish, more so than in *bennettii* and approaching *cristatus* (*affinis* has been classified as a race of *cristatus*); wing 130–139 mm (Rand & Gilliard 1967). Dickinson & Remsen (2013) placed the form *terborghi* into this species rather than into *bennettii*, based on the molecular work of Dumbacher *et al.* (2003). Plumage characters and geography do not support this association, and as the sampling of *terborghi* by Dumbacher *et al.* was limited to the toe pad of a single specimen, we hold off endorsing the combination of *terborghi* with *affinis* pending future study. Our English name is derived from the species epithet. *Other names*—Vogelkop Owlet-nightjar.

Australian Owlet-nightjar *Aegotheles cristatus* RESIDENT
Caprimulgus cristatus Shaw (in White), 1790. *Journ. Voy. to New South Wales*: 241 and plate.—Sydney, Australia.
    Inhabits savanna woodlands of the Trans-Fly and the dry zone of the s watershed of the SE Peninsula. Extralimital—Australia.

SUBSPECIES

a. *Aegotheles cristatus cristatus* (Shaw)
    Synonym: *Aegotheles cristatus major* Mayr & Rand, 1935. *Amer. Mus. Novit.* 814: 4.—Dogwa, Oriomo R, Trans-Fly, s NG.
RANGE: Australia and the Trans-Fly; presumably the birds collected in the the Port Moresby region refer to this form. Localities—Tarara[MA], upper Wassi Kussa R[CO], Oriomo R[MA], and Port Moresby[M] (3 specimens at the PNG Natl. Mus.[M]).
DIAGNOSIS: Distinguished by larger size and comparatively longer tail than *tasmanicus* of Tasmania; dark gray upperparts (Holyoak 1999a). We follow Holyoak (1999a) in subsuming *major* into the nominate form. Birds from the Trans-Fly are larger than those from the Port Moresby area.

NOTES

The population *affinis* of the Bird's Head, placed in this species by Rand & Gilliard (1967), has been given full species status here. *Other names*—Crested Owlet-nightjar.

Family Hemiprocnidae

Treeswifts [Family 4 spp/Region 1 sp]

The four species of this tropical family range from tropical Asia east to the Solomon Islands (Wells 1999). The family is absent from Australia. These are medium-size aerial birds characterized by a small body, long slim wings, a broad head, and a long and deeply forked tail. They are larger than the swiftlets or swallows. The treeswifts were treated as a subfamily of the Apodidae by Cracraft (2013). We follow Wells (1999) and del Hoyo & Collar (2014) in treating the group as a full family. It is a sister lineage to the swifts (Braun & Huddleston 2009).

GENUS *HEMIPROCNE* [Genus 4 spp/Region 1 sp]

Hemiprocne Nitzsch, 1829. *Obs. Av. Arter. Carot. Comm.*: 15 and note. Type, by subsequent designation (Oberholser, 1906. *Proc. Biol. Soc. Wash.* 19: 68), *Cypselus longipennis* Temminck = *Hirundo longipennis* Rafinesque.

Genus characters are as for the family.

Moustached Treeswift *Hemiprocne mystacea* RESIDENT

Cypselus mystaceus Lesson & Garnot 1827. *Voy. Coquille, Atlas*: pl. 22; 1830. *Zool.*: 647.—Manokwari, Bird's Head.

Throughout NG, the NW Islands, Aru Is, and Bay Is. Lowlands to 1,400 m, max 1,700 m (Mayr 1941, Coates & Peckover 2001). Frequents forest edge, disturbed habitats, and clearings fringed by tall trees for perches. One was recovered from glacial ice at 4,400 m on the Carstensz Massif (Schodde *et al.* 1975).

Extralimital—Moluccas, Bismarcks, and Solomons (Mayr & Diamond 2001, Dutson 2011).

SUBSPECIES

a. *Hemiprocne mystacea mystacea* (Lesson & Garnot)

Synonym: *Hemiprocne mystacea confirmata* Stresemann, 1914. *Novit. Zool.* 21: 110.—Seram, s Moluccas.

RANGE: Misool[O], Salawati[Z], Batanta, Gebe[J], and Waigeo I; Aru Is; Yapen, Numfor, and Biak I; and all NG (Mayr 1941, Mees 1965[Q], 1972[J], Beckers *et al.* in M. Tarburton's island website[Z]).

DIAGNOSIS: Wing 226–248 (Rand & Gilliard 1967). Mees (1965) did not recognize the race *confirmata* and noted considerable nongeographic variation in size and plumage. Perhaps the species should be treated as monotypic.

NOTES

Other names—Whiskered Treeswift, Moustached Swift.

Family Apodidae

Swifts and Swiftlets [Family 95 spp/Region 8 spp]

The species of the family range from North America, South America, Africa, and Eurasia, to Australia and Melanesia (Chantler 1999). These, the most specialized of aerial birds, have cylindrical bodies and long, rather straight wings, with the bend of the wing very close to the body. Diet is exclusively insects caught on the wing; the birds spend most time in the air and rarely are seen at rest, usually only at the roosting or nesting site. This family is sister to the treeswifts. The swift+treeswift lineage is sister to the hummingbirds (Braun & Huddleston 2009). We use the work of Price *et al.* (2004, 2005) and Thomassen *et al.* (2005) to inform our systematic treatment.

Collocalia G. R. Gray, 1840. *List Gen. Birds*: 8. Type, by original designation, *Hirundo esculenta* Linnaeus.

The species of this genus range from the Andaman Islands and the Malay Peninsula east to the Philippines, New Guinea, and Melanesia (Chantler 1999). The genus is characterized by the very small size, light frame, and dorsal plumage blackish, bluish, or greenish. The ability to echolocate is now known from both *Aerodramus* and *Collocalia* (Price *et al.* 2004). A single species (*C. troglodytes*) of the latter genus possesses this ability. The molecular systematic analysis of Price *et al.* (2004), in spite of this discovery, supported the retention of both *Collocalia* and *Aerodramus*.

Glossy Swiftlet *Collocalia esculenta* RESIDENT

Hirundo esculenta Linnaeus, 1758. *Syst. Nat.*, ed. 10, 1: 191.—Ambon, s Moluccas.

All NG and most islands, in hilly country at all elevations. Forages in small or large openings in forest, edge, and along streams and lakes. Typically not found high over closed forest. No records from the Trans-Fly, much of sc NG, Gebe I, or Manam I (Pratt & Beehler 2014).

Extralimital—e Indonesia, Bismarcks, Solomons, New Caledonia, Vanuatu, and ne Australia (Mayr & Diamond 2001, Dutson 2011).

SUBSPECIES

a. *Collocalia esculenta esculenta* (Linnaeus)

Synonym: *Collocalia nitens* Ogilvie-Grant, 1914. *Bull. Brit. Orn. Club* 35: 35.—Otakwa R, 880 m, Western Ra.

Synonym: *Collocalia esculenta erwini* Collin & Hartert, 1927. *Novit. Zool.* 34: 50.—Upper Otakwa R, 2,450 m, Western Ra.

Synonym: *Collocalia esculenta amethystina* Salomonsen, 1983. *Kon. Danske Vidensk. Selsk. Biol. Skrift.* 23(5): 44.—Waigeo I, NW Islands.

Synonym: *Collocalia esculenta numforensis* Salomonsen, 1983. *Kon. Danske Vidensk. Selsk. Biol. Skrift.* 23(5): 44.—Numfor I, Bay Islands.

Synonym: *Collocalia esculenta misimae* Salomonsen, 1983. *Kon. Danske Vidensk. Selsk. Biol. Skrift.* 23(5): 46.—Misima I, Louisiade Arch, SE Islands.

RANGE: Sulawesi, s and n Moluccas, all NG but rare in the Trans-Fly and vicinity; also found in most of the NW Islands[MA], including Batanta[J] and Gag[K] but not Gebe; the Aru Is[MA], all Bay Is[MA], Karkar and Bagabag I[L], all SE Islands[MA,CO] (Greenway 1966[J], Diamond & LeCroy 1979[L], Johnstone 2006[K]).

DIAGNOSIS: A variable form, with smaller birds in lowlands and larger birds in the highlands (Rand 1936b, Rand & Gilliard 1967, Chantler 1999). Bell (1976a) reported it from the Aramia R and K. D. Bishop observed it from the Bian Lakes.

NOTES

Treatment follows the conservative analysis of Mayr (1941) rather than the very liberal analysis of Salomonsen (1983), whose several new taxa are placed in synonymy, as they characterize thinly differentiated forms within a regime of rather substantial variation produced, largely, from different methods deployed in preparing study skins of these very small birds. Moreover, we place *erwini* in synonymy, as it is an "elevational subspecies" that is based on clinal variation and is nongeographic, but that also apparently includes some stochastic variation (most of the birds considered *erwini* are above 2,000 m, and yet some are below 1,000 m). Material from the Aru Is, Waigeo I, Numfor I, and mainland NG was examined at Leiden. The Waigeo material is ventrally very darkly mottled, and that population perhaps merits recognition (*pace* Mayr 1941). Across its range, the subspecies taxonomy of *Collocalia esculenta* could benefit from a careful review. *Other names*—White-bellied Swiftlet.

Aerodramus Oberholser, 1906. *Proc. Acad. Nat. Sci. Phila.* 58: 179 (in key), and 182. Type, by original designation, *Collocalia innominata* Hume.

Species of the genus range from islands near Madagascar and tropical Asia east to New Guinea and Oceania (as far as the Marquesas; Chantler 1999). We follow Chantler (1999) and Christidis & Boles (2008) in recognizing this genus as distinct from *Collocalia*. This exceedingly uniform genus is characterized by the dull gray-brown plumage (darker dorsally, paler ventrally). The ability to echolocate is now known to be found in both this lineage and *Collocalia* (Price *et al.* 2004). That said, the systematic analyses of Price *et al.* (2004, 2005) and Thomassen *et al.* (2005) supported the retention of both *Collocalia* and *Aerodramus*, based on analyses of sequence data from both mitochondrial and nuclear genes. Regarding *Aerodramus*, their analyses included only two New Guinea taxa, *papuensis* and *vanikorensis*, thus our sequence relies partially on traditional placement of the species.

Note that the former inclusion of *A. spodiopygius* in the Region (*cf.* Coates 1985: 394) was an error, based on confusion of a record from Menam Island (north of New Ireland) with Manam Island. Note also that the range map in Higgins (1999: 1059) showing this species widespread in New Guinea is in error.

Three-toed Swiftlet *Aerodramus papuensis*   ENDEMIC | MONOTYPIC

Collocalia whiteheadi papuensis Rand, 1941. *Amer. Mus. Novit.* 1102: 10.—15 km sw of Bernhard Camp, 1,800 m, ne sector of the Western Ra.

The range of this species is poorly known and poorly understood. Apparently widespread but rare on the Mainland. Primarily known from the NG uplands, but there are records from the lowlands to 2,400 m. Specimens from the mts south of the Taritatu R[L], Jayapura[L], s slopes of the Hindenburg Ra[J], Chuave[N], Herowana[O], Wengomanga near Aseki[K], Baroka on Hall Sound[MA], and Kapa-Kapa[M] (Rand 1942b[L], Somadikarta 1967[M], Somadikarta 1975, Rowland 1994a[J], Hamilton et al. 2001[O], BPBM collection[N], Austral. Nat. Wildl. Coll.[K]). Dickinson & Remsen (2013) noted a record from the mts of the Huon Penin.

This species is not separable in the field from *A. nuditarsus*. For what they are worth, sight reports of this species include the upper Fly R[J], Madang foothills[L], the lower Markham valley (Lae and Yalu R)[K], the Watut and Bulolo valleys[K], and near Port Moresby (Watson *et al.* 1962[K], Bailey 1992a[L], see Coates 1985, Gregory 1997[J], Chantler 1999). A flock of large swiftlets with white throats and much longer tails than *Collocalia esculenta* seen at Manam I (Berggy 1978) may also have been this species. The large swiftlet species seen in the Bulolo valley keeps mainly to the gullies. Credibility of any sight records is an issue. Some from the above list may actually refer to *A. nuditarsus* or *hirundinaceus*.

Locally common on the Taritatu R in feeding parties of 20 to 30 birds; these were usually high in the air but occasionally fed low over the marshes, often in association with *Mearnsia novaeguineae*, occasionally with the *Aerodramus vanikorensis* (Rand 1941, 1942b).

NOTES

In hand, note the densely feathered tarsus and presence of three toes (not four). Originally this population was included in *whiteheadi*, a four-toed form (Somadikarta 1967). Shown to be sister to the core lineage by Price *et al.* (2005).

Observations of large swiftlets with a pale throat may be this species, but this plumage character is weak at best even with the bird in the hand and is further subject to interpretation because generally for swiftlets in flight the breast is shaded by the wings and body and so the throat appears paler. *Other names*—Papuan Swiftlet.

Bare-legged Swiftlet *Aerodramus nuditarsus*   ENDEMIC | MONOTYPIC

Collocalia nuditarsus Salomonsen, 1963. *Vidensk. Medd. Dansk. Naturhist. Foren.* (1962) 125: 510.— Baroka, Bioto Creek, near Hall Sound, SE Peninsula.

Range poorly understood. Known from the Central Ra, mainly in the s watershed, from the Mimika R east to the Port Moresby region; 1,600–2,300 m.

Localities—Parimau on the Mimika R[CO], Mt Goliath above 1,500 m[CO], Kubor Mts[DI], upper Kaironk valley in the Schrader Ra[CO], Hamo near Aseki[J], Mindik on the Huon Penin[J], Baroka on Bioto Creek[CO], and Efogi (sight record)[CO] (Austral. Nat. Wildl. Coll.[J]).

NOTES

It has been variously placed in the species *whiteheadi* and *orientalis* by previous revisers (Somadikarta 1967, Salomonsen 1983). This species can be identified in the hand by the presence of four toes (not three) and an unfeathered tarsus (Chantler 1999); also, extensive pale fringes on the eyebrow are distinctive and unique (Pratt museum notes). Treatment as a full species follows Chantler (1999) and Dickinson & Remsen (2013). More distributional information needed, based on birds in the hand and perhaps verified by molecular fingerprinting. *Other names*—New Guinea Swiftlet; formerly treated as Whitehead's Swiftlet, *Collocalia whiteheadi*.

Uniform Swiftlet *Aerodramus vanikorensis*   RESIDENT

Hirundo vanikorensis Quoy & Gaimard, 1830. *Voy. Astrolabe, Zool.* 1: 206; *Atlas*: pl. 12, fig. 3.—Vanikoro I, Temotu Is, se Solomon Is.

All NG lowlands and a scattering of islands (Mayr 1941), sea level–500 m, max 1,600 m on Goodenough I[CO]. No records from Gag, Gebe, Mios Num, Karkar, Manam, or Rossel I.

Extralimital—Sulawesi to Vanuatu; two records from n QLD, Australia; range includes most of the Bismarck and Solomon Is (Mayr & Diamond 2001, Dutson 2011).

SUBSPECIES

a. *Aerodramus vanikorensis yorki* (Mathews)
Collocalia francica yorki Mathews, 1916. *Bull. Brit. Orn. Club* 36: 77.—Peak Point, Cape York, QLD, Australia.
   Synonym: *Collocalia vanikorensis waigeuensis* Stresemann & Paludan, 1932. *Novit. Zool.* 38: 164, 168.—Waigeo I, NW Islands.
   Synonym: *Collocalia vanikorensis steini* Stresemann & Paludan, 1932. *Novit. Zool.* 38: 167.—Numfor I, Bay Islands.
   Synonym: *Collocalia vanikorensis granti* Mayr, 1937. *Amer. Mus. Novit.* 915: 8.—Setekwa R, Nassau Mts, Western Ra.
RANGE: All NG lowlands and Batanta[A], Kofiau[B], Misool[C], possible sightings on Salawati[D], Aru Is, Bay Is (Mayr 1941, Gyldenstolpe 1955b[A], Diamond *et al.* 2009[B], Mees 1965[C], Beckers *et al.* in M. Tarburton's island website[D]). Two records from n Australia. Rarely encountered in the Trans-Fly (K. D. Bishop *in litt.*)
DIAGNOSIS: Blacker above than the following form, with a paler throat and relatively short tail (Chantler 1999).

b. *Aerodramus vanikorensis tagulae* (Mayr)
Collocalia vanikorensis tagulae Mayr, 1937. *Amer. Mus. Novit.* 915: 7.—Tagula I, Louisiade Arch, SE Islands.
RANGE: Goodenough[CO] and Normanby I[E,F], Trobriand Is, and Woodlark, Tagula, and Misima I (Mayr 1941, LeCroy *et al.* 1984[E] and Pratt field notes 2003[F]).
DIAGNOSIS: When compared to the nominate form, note paler upperparts and darker throat (Chantler 1999).

NOTES

Tarsus unfeathered. The two subspecies listed above are probably of doubtful utility and merit a second review in light of Salomonsen's (1983) confused analysis (also see Mees 1965). Bell reported both *C. vanikorensis* and *C. spodiopygia* from Menam I n of New Ireland. Coates transcribed these as Manam I. Neither species occurs on Manam I. Material from sw NG and Misool examined at Leiden. See Mees (1965) for comments on subspecies. *Other names*—Lowland Swiftlet.

Mountain Swiftlet *Aerodramus hirundinaceus*   ENDEMIC

Collocalia fuciphaga hirundinacea Stresemann, 1914. *Verh. Orn. Ges. Bayern* 12: 7.—Setekwa R, Western Ra.

Forages over forest or grasslands through the uplands of NG and Yapen, Karkar, and Goodenough I. Absent from the flat lowlands, and rare below 500 m; common in highlands to above the tree line. Occurs rarely in the lowlands in the vicinity of hills, but exact distribution made difficult by the fact that *hirundinaceus* and *vanikorensis* cannot be distinguished on the wing. No records from the Trans-Fly or from much of the S Lowlands. On Karkar I has not been recorded below 240 m and on Goodenough I apparently restricted to the mts.

SUBSPECIES

a. *Aerodramus hirundinaceus hirundinaceus* (Stresemann)

  Synonym: *Collocalia hirundinacea excelsa* Ogilvie-Grant, 1914. *Bull. Brit. Orn. Club* 35: 34.—Otakwa R, 2,440 m, Western Ra.

RANGE: All uplands of NG; also Karkar[K] and Goodenough I (Mayr 1941, Diamond & LeCroy 1979[K]). Should also be looked for on Fergusson and Normanby I. Swiftlets observed in the mts of Fergusson I could be this species (Pratt field notes 2003).

DIAGNOSIS: Rather pale dorsal surface but bluish cast is visible in flight when viewed from above in sunlight (B. J. Coates *in litt.*); brown cast of ventral plumage reduced (Chantler 1999). The described form *excelsus* is based on the size of high-elevation specimens rather than geography. There is a strong size gradient with elevation in this form.

b. *Aerodramus hirundinaceus baru* (Stresemann & Paludan)

Collocalia vanikorensis baru Stresemann & Paludan, 1932. *Novit. Zool.* 38: 167.—Kampong Baru, Yapen I, Bay Islands.

RANGE: Yapen I.

DIAGNOSIS: Very dark dorsally and with a bluish cast; browner ventrally; heavy tarsal feathering. Wing 111 (Rand & Gilliard 1967, Chantler 1999).

NOTES

Tarsus feathered (Mayr 1937b). Variation in this species probably nongeographic; perhaps should be treated as monotypic.

GENUS *MEARNSIA*   [Genus 2 spp/Region 1 sp]

Mearnsia Ridgway, 1911. *Bull. U.S. Nat. Mus.* 50(5): 686 (in key and note e). Type, by original designation, *Chaetura picina* Tweeddale.

  The species of this genus inhabit the southern Philippines and New Guinea (Chantler 1999). The genus is characterized by the spined tail feathers, glossy plumage, hooked wing shape in flight, broad head, and robust body.

Papuan Spinetailed Swift *Mearnsia novaeguineae*   NEAR-ENDEMIC

Chaetura novaeguineae D'Albertis & Salvadori, 1879. *Ann. Mus. Civ. Genova* 14: 55.—Upper Fly R, S Lowlands.

  Throughout the Mainland in the lowlands and hills up to 550 m. Field sightings from the Bird's Head (J. Hornbuckle & I. Mauro *in litt.* Sep-2004; S. van Balen *in litt.* 28-Oct-1997 near Sorong), Salawati (J. Hornbuckle & I. Mauro *in litt.* Sep-2004; photo from G. Dutson), and Batanta I (Beehler field notes 27-Feb-2004). Frequents forest edge, openings in forest, and clearings along rivers.

  Extralimital—Boigu and Saibai I of the Torres Strait (record accepted by Birds Australia Rarities Committee, D. Hobcroft *in litt.*)

SUBSPECIES

a. *Mearnsia novaeguineae novaeguineae* (D'Albertis & Salvadori)

RANGE: Southern NG from the Mimika R east to the Port Moresby area (Galley Reach), and perhaps to Milne Bay. Birds inhabiting the Bird's Head and the NW Islands remain undiagnosed but may refer to this form.

DIAGNOSIS: Throat gray (Rand & Gilliard 1967).

b. *Mearnsia novaeguineae buergersi* (Reichenow)

Chaetura bürgersi Reichenow, 1917 (October). *J. Orn.* 65: 514.—Malu, Sepik R.

> Synonym: *Chaetura novae-guineae mamberana* Neumann, 1917. *Orn. Montsber.* 25: 153.—Teba, lower Mamberamo R, NW Lowlands.

RANGE: Northwestern Lowlands, Sepik-Ramu, and perhaps east to the Huon Gulf. It is uncertain which subspecies inhabits the n watershed of the SE Peninsula.

DIAGNOSIS: Throat darker, more like back (Rand & Gilliard 1967).

NOTES

Other names—New Guinea Spine-tailed Swift, New Guinea Needletail, *Chaetura novaeguineae.*

GENUS *HIRUNDAPUS*   [Genus 4 spp/Region 1 sp]

Hirundapus Hodgson, 1837. *Journ. Asiat. Soc. Bengal* (1836) 5: 780. Type, by original designation and monotypy, *Cypselus (Chaetura) nudipes* Hodgson.

The species of this genus breed from northeast Asia south to the Philippines and Sulawesi. Birds winter in New Guinea and Australia as well as southeast Asia (Chantler 1999). The genus is characterized by very large size, rapid flight, presence of a bright white flank stripe, and dark breast.

White-throated Needletail *Hirundapus caudacutus*   VISITOR

Hirundo caudacuta Latham, 1801. *Index Orn.*, Suppl.: lvii.—New South Wales, Australia.

In NG, this northern migrant is found mainly in the S Lowlands and SE Peninsula during northern winter. Common in Port Moresby and the Fly R regions from October to April, with the earliest birds arriving at the end of September. Occurs mostly in the lowlands and hills, occasionally up to 1,500 m (Coates & Peckover 2001). Flies over all kinds of habitat. Usually seen in flocks, large and small; nomadic within NG. Movements within NG little known, especially as many sightings are weather-influenced (Coates 1985). Diamond & Bishop (1994) provided a first record for the Aru Is, in early April.

Extralimital—breeds in Siberia, Japan, Korea, Taiwan, and the Himalayas to s China; the northeastern (nominate) form winters in e Australia (Chantler 1999). Stragglers have been reported from Umboi, Lihir, Fiji, and Macquarie I (Coates 1985, Dutson 2011).

SUBSPECIES

a. *Hirundapus caudacutus caudacutus* (Latham)

RANGE: Breeds from e Siberia to Mongolia, Sakhalin, Manchuria and the Japanese islands. Migrates through China to Australia. Records mainly from s and e NG (see above).

DIAGNOSIS: White forehead and lores (Chantler 1999).

NOTES

For generic treatment, we follow Chantler (1999) and Christidis & Boles (2008). *Other names*—White-throated Spinetail, *Chaetura caudacuta.*

GENUS *APUS*   [Genus 15 spp/Region 1 sp]

Apus Scopoli, 1777. *Intro. Hist. Nat.*: 483. Replacement name for *Cypselus* Illiger, 1811, in Sharpe, *Hand-list*; and *Micropus* A. B. Meyer and Wolf, 1810, of many recent authors. Type, by tautonymy, *Hirundo apus* Linnaeus (not preoccupied by *Apos* Scopoli, 1777, Crustacea).

The species of this genus range from temperate Eurasia to Africa and tropical and temperate Asia. Migrants winter to Africa, southeast Asia, New Guinea, and Australia (Chantler 1999). The genus is characterized by the pointed and forked tail, dark belly feathers, and pale throat spot.

Fork-tailed Swift *Apus pacificus*

Hirundo pacifica Latham, 1801. *Index Orn.*, Suppl.: lviii.—New South Wales, Australia.

Migrants from Asia winter in NG, with most records from the s watershed. Also Aru Is. Some birds presumably are passage migrants, moving on to winter in Australia (Mees 1982b, Coates 1985). In NG, of regular occurrence only in the Trans-Fly. Occurs over savanna and open country in the lowlands; probably largely avoids forested terrain.

Extralimital—Breeds from the Himalayas to e Siberia, China, Japan, Korea, and Taiwan; nominate race winters south to Australia. One record from New Georgia I (Dutson 2011).

SUBSPECIES

a. *Apus pacificus pacificus* (Latham)

RANGE: Breeds in e Siberia from the Lena R to Kamchatka, n China, and the Japanese islands. Migrates through s China, Burma, and Indochina to NG and Australia. Localities—Sorong, Aru Is[B], Merauke[A], Bensbach R[CO], Oriomo R[MA], Fly R[CO], Wasu[CO] on the Huon Penin, Port Moresby area[MA], and Hula[CO], where recorded from mid-October to early March (Mees 1982b[A], Diamond & Bishop 1994[B]).

DIAGNOSIS: In comparison to race *kanoi*, overall paler, with larger throat patch and rump patch (Chantler 1999).

NOTES

Other names—Pacific Swift.

ORDER CHARADRIIFORMES

The phylogenomic study of Hackett *et mult.* (2008) indicated that the Charadriiformes include three distinct subclades: a clade with *Haematopus*, *Phegornis*, *Charadrius*, and *Burhinus*; a clade with *Turnix*, *Larus*, and *Dromas*; and a clade with *Thinocorus*, *Pedionomus*, *Rostratula*, *Jacana*, and *Arenaria*. This compares well with the results of Thomas *et al.* (2004: fig. 2) and moderately well with the results of Ericson *et al.* (2003).

Family Burhinidae

Stone-curlews and Thick-knees   [Family 9 spp/Region 2 spp]

The species of the family range from Central and South America, Africa, Europe, and tropical Asia to New Guinea and Australia (Hume 1996). In New Guinea, the species inhabit beaches and coastal flats as well as savanna woodlands in the Trans-Fly. This small and distinctive family is characterized by the crepuscular and nocturnal habit, large size, the peculiar start-and-stop gait, the staring yellow eye, the patterned plumage, and the very long yellow legs. Hackett *et mult.* (2008) identified this lineage as sister to a clade that includes the plovers, the oystercatchers, and the Diademed Sandpiper Plover (*Phegornis mitchellii*).

GENUS *BURHINUS*   [Genus 7 spp/Region 1 sp]

Burhinus Illiger, 1811. *Prodromus Syst. Mamm. Avium*: 250. Type, by monotypy, *Charadrius magnirostris* Latham.

The species of this genus range through Central and South America, Africa, Europe, western, southern, and southeast Asia, New Guinea, and Australia (Hume 1996). The very compact genus is characterized by patterned brown-and-white plumage, long legs, staring yellow eye, the peculiar gait, and the rather short, straight bill.

Bush Stone-curlew *Burhinus grallarius*   RESIDENT | MONOTYPIC

Charadrius grallarius Latham, 1801. *Index Orn.*, Suppl.: 66.—New South Wales, Australia.

In NG, only known from a small area in the s Trans-Fly between Wasur NP[J] and the Bensbach R[K], where it is uncommon in open savanna and patches of *Melaleuca* (Hornbuckle 1991[J], Bishop 2006[K]). First recorded in NG by Lindgren (1971), who collected a specimen of a juvenile.

Extralimital—widespread in Australia.

NOTES

We follow Hume (1996) in the nomenclatural treatment of this species and in considering this species monotypic. *Other names*—Bush Curlew, Southern Stone-curlew, Bush Thick-knee, *Burhinus magnirostris*.

GENUS *ESACUS*   [Genus 2 spp/Region 1 sp]

Esacus Lesson, 1831. *Traité d'Orn.* 7: 547. Type, by monotypy, *Oedicnemus recurvirostris* Cuvier.

The two species of this genus range from tropical Asia to the Philippines, Indonesia, southern New Guinea, and the northern and eastern coasts of Australia (Hume 1996). This small genus is characterized by the large and weirdly shaped bill, the long and sturdy yellow legs, and the strong patterning of the face and of the wings in flight.

Beach Stone-curlew *Esacus magnirostris*   RESIDENT | MONOTYPIC

Œdicnemus magnirostris Vieillot, 1818. *Nouv. Dict. Hist. Nat.*, nouv. éd. 23: 231.—Depuch I, Western Australia.

Sparingly inhabits reef flats, isolated beaches, and uninhabited small fringing islets throughout the Region. To be expected from any uninhabited islet or isolated beach or flat, and will move from island to island (see details in Mayr 1941 and Coates 1985).

Extralimital—Andaman Is, Malay Penin, Philippines, Indonesia, Bismarck and Solomon Is, n and e Australia, New Caledonia, and Vanuatu (Dickinson & Remsen 2013).

NOTES

We follow Hume (1996) in nomenclatural treatment of this species. *Other names*—Beach Curlew, Great or Beach Thick-knee, Reef Stone-curlew, Great Stone Plover, *Burhinus neglectus*.

Family Haematopodidae

Oystercatchers   [Family 13 spp/Region 1 sp]

The species of the family inhabit coastlines around the world, though absent from the Canadian and Russian Arctic, the coast of east Africa, all of southeast Asia, and the Pacific Islands (Hockey 1996). In New Guinea, a single species of the family can be found on coastal mudflats. The members of the family are characterized by the laterally compressed long red or orange bill, the pink legs, the all-black or pied black-and-white plumage, and the brightly colored iris and eye-ring (red or yellow). The family is sister to the avocets and stilts (Hackett *et mult.* 2008).

GENUS *HAEMATOPUS*   [Genus 12 spp/Region 1 sp]

Haematopus Linnaeus, 1758. *Syst. Nat.*, ed. 10, 1: 152. Type, by monotypy, *Haematopus ostralegus* Linnaeus.

Range and characters are as for the family. The Sooty Oystercatcher (*Haematopus fuliginosus*) might be expected as a vagrant to the southern New Guinea coast, as it occurs in northern Australia (Marchant & Higgins 1993).

Australian Pied Oystercatcher *Haematopus longirostris*   VISITOR (AND RESIDENT?) | MONOTYPIC

Haematopus longirostris Vieillot, 1817, *Nouv. Dict. Hist. Nat.*, nouv. éd. 15: 410.—Bernier I, Western Australia.

Possibly a resident in the Aru Is[MA]. Otherwise apparently a rare nonbreeding visitor to the s coast of the Mainland, ranging from the Bird's Neck east to the SE Islands.

Localities—sw Bird's Head[K], Utanata R[MA], Dolak I[K], Komolom I[K], mouth of the Bian R[J], Lampusatu Beach near Merauke[K], mouth of the Bensbach R[K], Katau R[J], Aru Is[CO], Oro Bay, Orangerie Bay[MA], Fergusson[CO] and Kimuta I[Q] (Mees 1982b[J], LeCroy & Peckover 1998[Q], Bishop 2006[K]).

Extralimital—the Kai and Southeast Is of Indonesia and all of Australia's coastlines (Hockey 1996). Reported to breed in the Torres Strait during the dry season (Draffan *et al.* 1983).

NOTES

Treatment follows Baker (1975) and McKean (1978), but also see Mees (1982b), who treated this form as a subspecies of *Haematopus ostralegus*. No breeding records known for the Region, but breeding to be expected (Bishop 2006). *Other names*—Pied Oystercatcher.

Family Recurvirostridae

Avocets and Stilts   [Family 7 spp/Region 2 spp]

The species of the family inhabit temperate North America, Central America, South America (but not the Southern Cone), sub-Saharan Africa, Europe, temperate and tropical Asia, New Guinea, and Australia (Pierce 1996). They occupy wetlands and shallow lagoons. The family is characterized by the long very narrow bill, long neck, very long legs, and boldly patterned all-black, black-and-white, or black-white-and-chestnut plumage. The lineage is sister to the oystercatchers. The avocet+oystercatcher clade is sister to the plovers (Cracraft 2013).

GENUS *HIMANTOPUS*   [Genus 2 spp/Region 1 sp]

Himantopus Brisson, 1760. *Orn.* 1: 46; 5: 33. Type, by tautonymy, "Himantopus" Brisson = *Charadrius himantopus* Linnaeus.

The species of this genus range from the Americas and Europe to Africa, temperate and tropical Asia, New Guinea, Australia, and New Zealand (Pierce 1996). The genus is characterized by the combination of slight build, long neck, strikingly long brightly colored legs, and long, straight, needle-shaped bill.

Black-winged Stilt *Himantopus himantopus*   VISITOR AND RESIDENT

Charadrius Himantopus Linnaeus, 1758. *Syst. Nat.*, ed. 10, 1: 151.—Southern Europe.

A common nonbreeding visitor to the Trans-Fly and Port Moresby regions; elsewhere uncommon or rare. Highest record 1,670 m[CO].

Extralimital—the Americas, Europe, w and c Asia, sub-Saharan Africa, tropical Asia, Long and Umboi I, New Britain, Australia, and NZ (Pierce 1996, Mayr & Diamond 2001).

SUBSPECIES

a. *Himantopus himantopus leucocephalus* Gould

Himantopus leucocephalus Gould, 1837. *Synops. Birds Austral.* 2: pl. 34.—New South Wales, Australia.

RANGE: Java east to NG, thence south to Australia and NZ (Pierce 1996). Bred successfully near Port Moresby in Jul–Aug-1982 (Coates 1985). Localities—Misool[MA], Sorong[J], Aimaru Lake[J], Triton Bay[MA], Taritatu R[L], Dolak I[K], Kurik[Q], Lampusatu Beach[K], Merauke[MA], Ndalir[K], Bensbach R[K], Tonda WMA[K], Tabubil[CO], Balimo[K], Aramia R[K], Tari[CO], Sepik R[MA], Astrolabe Bay[MA], Ramu R[MA], Popondetta, Hall Sound[MA], Kanosia[K], Fergusson[MA], and Misima I[MA], and Port Moresby[CO] (Rand 1942b[L], Gyldenstolpe 1955b[J], Mees 1982b[Q], Bishop 2006[K]).

DIAGNOSIS: White head and a narrow black nape line (down to upper back); white collar separates black mantle from black nape line (Pierce 1996).

NOTES
Mayr & Short (1970) and Mayr & Diamond (2001) treated *leucocephalus* as a distinct species. In contrast, Condon (1975), Pierce (1996), and Christidis & Boles (2008) treated the form as a subspecies of the widespread *H. himantopus*. We follow this latter treatment here. *Other names*—White-headed Stilt, *Himantopus leucocephalus*.

GENUS *RECURVIROSTRA* [Genus 4 spp/Region 1 sp]

Recurvirostra Linnaeus, 1758. *Syst. Nat.*, ed. 10, 1: 151. Type, by monotypy, *Recurvirostra avosetta* Linnaeus.
    The species of this genus range from the Americas and Europe to Africa, temperate Asia, and Australia (Pierce 1996). The genus is characterized by the combination of the bold plumage patterning, long legs, and long, slim, and recurved bill.

Red-necked Avocet *Recurvirostra novaehollandiae* VAGRANT | MONOTYPIC
Recurvirostra Novae-Hollandiae Vieillot, 1816. *Nouv. Dict. Hist. Nat.*, nouv. éd. 3: 103.—Victoria, Australia.
    A single individual, apparently in adult plumage, was observed in Wasur NP in the w Trans-Fly by Bostock (2000).
    Extralimital—Australia. Vagrant to NZ (Gill *et al.* 2010).

Family Charadriidae

Plovers and Dotterels [Family 67 spp/Region 8 spp]

The species of the family range throughout the world but for the Sahara, parts of interior Siberia, and much of Greenland. Three subfamilies—the Pluvialinae, Charadriinae, and Vanellinae—are recorded for the Region (Piersma 1996, Dickinson & Christidis 2014). Only two species are known to be resident. The majority of species are migratory, breeding in the northern latitudes and wintering in the tropics. Migratory species recorded in New Guinea include six Palearctic species and one vagrant from Australia. Other species are to be expected as rare strays. Family members are small to medium-size birds, generally gray, white, and black, with a small bill, longish legs, and distinctive gait, running quickly a few paces, and then halting abruptly. Species are typically found on mudflats, beaches, and other favored wader haunts (Pratt & Beehler 2014). Lapwings (*Vanellus*) are relatively large and have broad, rounded wings; many have wattles about the face and a spur on the wing; some have a long crest. Cracraft (2013) showed the family to cluster with the oystercatchers and the avocets.

GENUS *PLUVIALIS* [Genus 4 spp/Region 2 spp]

Pluvialis Brisson, 1760. *Orn.* 1: 46; 5: 42. Type, by tautonymy, "Pluvialis" Brisson = *Charadrius pluvialis* = *Charadrius apricarius* Linnaeus.
    The species of this genus breed throughout the high Arctic and winter south to Europe, north Africa, the Middle East, tropical Asia, Oceania, Australia, and New Zealand (Wiersma 1996). The genus is characterized by the heavily vermiculated dorsal surface contrasting the white-fringed black ventral surface of the breeding adult. This genus is distinct at the subfamilial level (Baker *et al.* 2012). A possible American Golden Plover (*Pluvialis dominica*) was reported from the Moitaka settling ponds (Finch & Kaestner 1990) but the sighting lacked adequate supporting documentation.

Pacific Golden Plover *Pluvialis fulva*   VISITOR | MONOTYPIC

Charadrius fulvus Gmelin, 1789. *Syst. Nat.* 1(2): 687.—Tahiti.

Widespread in passage and wintering throughout NG, Aru Is, Waigeo I, Biak I, and the D'Entrecasteaux and Louisiade Arch (Mayr 1941). No nonbreeders remain during the northern summer. An abundant visitor from August through May, sea level–2,000 m^{CO}. For the well-studied Port Moresby region, the earliest arrival date is 19-July and the latest departure date is 13-May (Hicks, in Bishop 2006). Singly or in small parties, this species feeds on beaches and mudflats and on short-mown grass at airports; it has been found in interior upland valleys of the Baliem, Wahgi, Baiyer, and Kyaka territory. On the middle Fly R, when the savanna grass was burned in September, this species appeared in small numbers and fed on the bare ridges (Rand & Gilliard 1967).

Extralimital—breeds from n and e Siberia to w Alaska. Winters from the Horn of Africa and India to se Asia, the Pacific, Australia, and NZ (Wiersma 1996).

NOTES

Formerly lumped with *P. dominica*, but all authorities today split these two populations (Wiersma 1996, Christidis & Boles 2008). *Other names*—Eastern or Lesser Golden Plover, *Pluvialis dominica*.

Grey Plover *Pluvialis squatarola*   VISITOR | MONOTYPIC

Tringa Squatarola Linnaeus, 1758. *Syst. Nat.*, ed. 10, 1: 149.—Sweden.

Uncommon on passage and wintering along the Region's coastlines (Coates 1985, Bishop 2006). Some individuals remain throughout the year.

Extralimital—breeds in the far north of Europe, Asia, and North America, migrating to Africa, Australia, and South America (Wiersma 1996).

NOTES

We follow Wiersma (1996) in treating the species as monotypic. *Other names*—Black-bellied Plover, Silver Plover, *Squatarola* or *Charadrius squatarola*.

GENUS *VANELLUS*   [Genus 24 spp/Region 1 sp]

Vanellus Brisson, 1760. *Orn.* 1: 48; 5: 94. Type, by tautonymy, "Vanellus" Brisson = *Tringa vanellus* Linnaeus.

The species of this genus range from South America, Europe, and Africa to tropical Asia, New Guinea, and Australia (Wiersma 1996). The highly varied genus is characterized by the strikingly patterned plumage (especially the bold black-and-white patterns), medium-length to long legs, and a short, straight bill. Some forms include crests and/or wattles and wing spurs.

Masked Lapwing *Vanellus miles*   RESIDENT AND VISITOR

Tringa miles Boddaert, 1783. *Tabl. Planch. Enlum.*: 51.—Timor, Lesser Sundas.

Widespread and locally common in fields, pastures, and marsh edges of s NG from Dolak I east to Port Moresby; also Aru Is (Bishop 2006). Both a local breeder and migrant from Australia. No records from nw NG. Ranges locally up to 500 m; also once at 1,700 m^{CO}. Most records appear to be of winter visitors from Australia, despite the assertion by Marchant & Higgins (1993) that the Australian population does not undergo large scale movements (Bishop 2006). A female in breeding condition was collected at Daru (Rand & Gilliard 1967). Breeding recorded from the Trans-Fly at Dolak I^L, Wasur NP^L, Bensbach R^L, Lake Daviumbu^L, the Sepik R^K, Madang^K, and Port Moresby^K (Bishop 2006^L, B. J. Coates *in litt.*^{K}).

Extralimital—the e half of Australia and NZ. Wintering birds are known from Christmas I, Moluccas, Lesser Sundas, and the New Georgia Is in the Solomons (Wiersma 1996, Mayr 1941, Coates & Bishop 1997, Mayr & Diamond 2001).

SUBSPECIES

a. *Vanellus miles miles* (Boddaert)
RANGE: Mainly e NG, the Aru Is, and n and ne Australia (Dickinson & Remsen 2013).

DIAGNOSIS: The black of the cap does not extend down onto nape or to the side of the breast as it does in *V. m. novaehollandiae*. Additionally, the yellow facial wattle extends above and behind the eye (Wiersma 1996).

b. *V. m. miles* × *V. m. novaehollandiae*
Apparently individuals that show hybrid plumage characters have been seen in Port Moresby (Coates 1985: 178, P. Gregory *in litt.* 15-Apr-2014).

NOTES
Cramp (1983) placed this group into the genus *Hoplopterus*. *Other names*—Masked or Spur-winged Plover, *Lobibyx miles*.

GENUS *ERYTHROGONYS*   MONOTYPIC

Erythrogonys Gould, 1838. *Synops. Birds Austral.* 4: pl. 73 and text. Type, by monotypy, *Erythrogonys cinctus* Gould.
    The single species of the genus ranges from Australia to southern New Guinea (Wiersma 1996). The monotypic genus is characterized by vanelline plumage pattern of bold black, white, and brown, the black-tipped red bill, small size, and the presence of a hind toe (Wiersma 1996). This lineage is sister to the lapwings.

Red-kneed Dotterel *Erythrogonys cinctus*   VISITOR | MONOTYPIC
Erythrogonys cinctus Gould, 1838. *Synops. Birds Austral.* 4: pl. 73 and text.—New South Wales, Australia.
    Presumably a nonbreeding visitor from Australia to muddy edges of wetlands in the Trans-Fly and SE Peninsula. Possibly a regular but scarce visitor to the Bensbach R area in the austral spring (Stronach, in Coates 1985; Bishop 2006). However, not regular at Kurik in nearby se Indonesian NG, where seen only in Apr-1961, when the species was first reported from NG (Hoogerwerf 1964). Two birds were observed at Kanosia Lagoon (nw of Port Moresby) from July to Nov-1982 (Finch 1982e). A record of two adults with an immature bird at Bensbach R in Oct-1980 led Finch (1980c) to speculate that the species may breed in NG. This requires confirmation (Bishop 2006).
    Extralimital—breeds in Australia; nomadic.

NOTES
The range map in Weirsma (1996: 420) showed this species breeding in the Trans-Fly of NG; this treatment is premature. *Other names*—*Charadrius cinctus*.

GENUS *CHARADRIUS*   [Genus 30 spp/Region 4 spp]

Charadrius Linnaeus, 1758. *Syst. Nat.*, ed. 10, 1: 150. Type, by tautonymy, *Charadrius hiaticula* Linnaeus.
    The species of this genus range throughout the world. Some are boreal breeders and highly migratory. Others are sedentary (Wiersma 1996). This large and diverse genus is characterized by the lack of a hind toe (or a vestigial hind toe), the short blunt bill, short neck, dull brown dorsal plumage, and the patterned plumage of the face and throat in most instances. Hornbuckle & Merrill (2004) reported four small plovers on a sandbar on Salawati Island on 12-Sep-2004 that they attributed to the Malay Plover (*C. peronii*). Future fieldworkers should attempt to document this bird's occurrence in the Northwestern Islands in a future autumn migration.

Little Ringed Plover *Charadrius dubius*   RESIDENT AND VAGRANT
Charadrius dubius Scopoli, 1786. *Delic. Flor. Faun. Insubr.* 2: 93.—Luzon, Philippines.
    An uncommon breeding resident on mainland NG, sea level–1,500 m[CO]. The breeding population inhabits gravelly riverbanks. Also encountered in a range of barren or short-grass habitats: airfields, paddocks, playing fields, salt pans, and mangrove flats (Beehler *et al.* 1986). In addition, there are several records of the northern migrant race, *curonicus*, from coastal mudflats (Bishop 2006).

Extralimital—the species has two breeding populations: one in Europe and n Asia, wintering to Africa, Indonesia, and Philippines to NG; the other breeding from the Philippines to the Bismarck Arch (Wiersma 1996, Mayr & Diamond 2001).

a. *Charadrius dubius curonicus* Gmelin
Charadrius curonicus Gmelin, 1789. *Syst. Nat.* 1(2): 692.—Kurland, Latvia.
RANGE: Breeds from Sweden and North Africa across Europe and n Asia to e Siberia, n China, and Japan. Winters throughout the e tropics from Africa to Asia, straggling to NG, Australia, and the Bismarck Arch, where rare (Bishop 2006). Localities—Yapen I[MA], Dolak I, Kurik[J], Wewak, Aroa Lagoon[CO], and Port Moresby[K] (Mees 1982b[J], Bishop 2006[K]). The Port Moresby records include one at the Moitaka settling ponds on 4-Mar-1983 and one at Aroa Lagoon on 19-Mar-1983 (Finch 1983b). Possibly it is a rare, irregular visitor to the subregion.
DIAGNOSIS: *C. d. curonicus* differs from *C. d. dubius* by presence of the small patch of yellow at the base of the lower mandible, narrower yellow eye-ring, distinct black bar in the outer tail feathers (reduced or absent in *dubius*), larger size (wing 109–120 in *curonicus*, 105–114.5 in *dubius*), and very different immature and winter plumage—breast band brownish in *curonicus*, blackish in *dubius*; no black in winter plumage of *curonicus* (Mees 1982b).

b. *Charadrius dubius dubius* Scopoli
Synonym: *Charadrius dubius papuanus* Mayr, 1938. *Amer. Mus. Novit.* 1007: 13.—Upper Setekwa R, southern foot of the Western Ra.
RANGE: Breeding from the Philippines south to NG and the Bismarck Arch (Mayr 1941, Coates 1985, Wiersma 1996). Known from NG's satellite islands from a single record on Fergusson I in Apr-2004 (P. Gregory *in litt.*).
DIAGNOSIS: White line behind black on forehead conspicuous; little or no black in outer tail feathers (Rand & Gilliard 1967: 128). The resident form known from Port Moresby and the Ok Tedi R at km 120 (P. Gregory *in litt.*) has the base of the lower mandible deep pink at all times and is unique in having the eye-ring deep red when breeding (rich yellow when nonbreeding).

NOTES
The race *papuanus* was submerged into the nominate form by Mees (1982b). See Wiersma (1996: 426) for another interpretation. B. J. Coates (*in litt.*) suggested the breeding population from NG should be treated as distinct.

[Rufous-capped Plover *Charadrius ruficapillus*   HYPOTHETICAL | MONOTYPIC

Charadrius ruficapillus Temminck, 1821. *Planch. Col. d'Ois.*, livr. 8: pl. 47, fig. 2 and 5: pl. 68.—New South Wales, Australia.

Expected as a vagrant to beaches of the s coast of the Mainland. Mayr (1941) noted unconfirmed reports from s NG but doubted their validity. Occurs in the Torres Strait Is. S. van Balen has written: "a group of 9 small plovers, resembling Kentish plovers, but without the white collar and thought to be this species, were seen on 13-Feb-1997, out on the mudflat off Pasir Hitam [south of Timika]" (van Balen & Rombang 1998).

Extralimital—breeds in Australia.

Other names—Red-capped Plover.]

Lesser Sand-Plover *Charadrius mongolus*   VISITOR

Charadrius mongolus Pallas, 1776. *Reise Versch. Prov. Russ. Reich.* 3: 700.—Kulussutai, s Transbaikalia.

A northern nonbreeding migrant, common throughout the Region, with some birds remaining in NG year-round (Bishop 2006).

Extralimital—breeds from e Siberia and the Commander Is to Alaska (rarely), and south to Mongolia. Migrants winter to Japan, e China, the Philippines, Caroline and Mariana Is, Lesser Sunda Is, Sulawesi, Moluccas, Australia, Bismarck Arch, and Solomon Is (Wiersma 1996).

a. *Charadrius mongolus mongolus* Pallas

RANGE: Breeds in e Siberia and the Russian far east, wintering to Taiwan and Australia. Presumably a passage migrant in NG, as this is the most common form recorded in nw Australia (Marchant & Higgins 1993).

DIAGNOSIS: Because of the existence of five races in three subspecies groups, and the presence of male, female, juvenile, and breeding and nonbreeding plumages, the identification of specimens or observed birds to subspecies is very difficult, especially in wintering birds (Marchant & Higgins 1993).

b. *Charadrius mongolus ?stegmanni* Portenko

Charadrius mongolus stegmanni Portenko, 1939. *Fauna Anadyr. Kraya*: 159. Replacement name for *C. m. litoralis* Stegmann, 1937. *Orn. Monatsber.* 45: 25.—Bering Is.

RANGE: Kolymskiy, Kamchatka, n Kuril Is, and the Commander Is north to Chukotskiy Penin, wintering to s Ryuku Is, Taiwan, and Australia. Possibly to be expected in NG region as a migrant (Coates 1985).

DIAGNOSIS: See preceding diagnosis. This form may be the dominant form wintering in e Australia, and hence to be expected in e NG (Marchant & Higgins 1993).

NOTES

Other names—Mongolian Plover, Lesser or Mongolian Sand Dotterel, Mongolian Dotterel.

Greater Sand-Plover *Charadrius leschenaultii* VISITOR

Charadrius Leschenaultii Lesson, 1826. *Dict. Sci. Nat.*, éd. Levrault 42: 36.—Pondicherry, India.

A common northern visitor to the NG region from mid-August to early May (mainly late August to late April); a few nonbreeders remain during the austral winter (Bishop 2006). Frequents coastal flats.

Extralimital—breeds in w and c Asia, Mongolia, w China, and Siberia. Winters to e Africa, Madagascar, Arabia, tropical Asia, Bismarcks, Solomons, and Australia (Wiersma 1996).

SUBSPECIES

a. *Charadrius leschenaultii leschenaultii* Lesson

RANGE: Breeds in w China, s Mongolia, s Siberia, and the Altai Mts. Winters to Australasia (Mees 1965, Marchant & Higgins 1993, Wiersma 1996, Bishop 2006).

DIAGNOSIS: Only the nominate form is expected from Region (Marchant & Higgins 1993). The three subspecies are very difficult to distinguish because of the presence of distinct male, female, juvenile, breeding, and nonbreeding plumages.

NOTES

Other names—Large-billed or Large Dotterel, Greater Sand-Plover, Large Sand Dotterel.

Oriental Plover *Charadrius veredus* VISITOR | MONOTYPIC

Charadrius veredus Gould, 1848. *Proc. Zool. Soc. London*: 38.—Port Essington, NT, Australia.

A local and uncommon passage migrant; most records from the dry s coast, mainly the Trans-Fly and Port Moresby environs (Bishop 2006). Frequents open country with scanty grass cover.

Localities—Aru Is[MA], Dolak I, near Kumbe[M], Kurik[K], Bian R[M], Bensbach R[M], Kiunga[J], Tari airstrip[L], Astrolabe Bay[MA], n coast of Huon[CO], and Port Moresby[CO] (Mees 1982b[K], Bishop 2006[M], D. Hobcroft *in litt.*[J], P. Gregory *in litt.*[L]).

Extralimital—breeds on Mongolian and Chinese steppes, wintering to Indonesia and Australia (Mees 1982b, Wiersma 1996).

NOTES

According to Bishop (2006), it appears likely that most individuals arrive in Australia by overflying NG. *Other names*—Asiatic, Eastern, or Oriental Dotterel or Sandplover, Greater Oriental Plover, *Eupoda asiatica, Eupoda veredus, Eupodia veredus.*

Family Jacanidae

Jacanas or Lily-trotters [Family 8 spp/Region 1 sp]

The species of the genus range from the Caribbean, Central America, South America, central and southern Africa, and Madagascar to tropical Asia, the Philippines, Indonesia, New Guinea, and northern and eastern Australia (Jenni 1996). This small family frequents open lagoons and lowland swamps. It is a specialized wader with long toes that are adapted for walking on floating vegetation. Family characters include the elongate nails of all toes, especially the hallux, the polyandrous mating system (in all but *Microparra capensis*), and the boldly marked plumage, exhibiting large blocks of color. Hackett *et mult.* (2008) placed the jacanas in a clade with the painted-snipes (Rostratulidae), and that two-lineage clade as sister to a clade that includes the Plains-wanderer (Pedionomidae) and the seedsnipes (Thinocoridae).

GENUS *IREDIPARRA* MONOTYPIC

Irediparra Mathews, 1911. *Novit. Zool.* 18: 7. Type, by original designation, *Parra gallinacea* Temminck.
    The single species of the genus ranges from southeast Asia to New Guinea and northern and eastern Australia (Jenni 1996). The genus is characterized by the pinkish-fleshy sagittal comb that extends from the beak to the crown and the contrasting black-and-white plumage with golden highlights on the neck and face.

Comb-crested Jacana *Irediparra gallinacea* RESIDENT | MONOTYPIC

Parra gallinacea Temminck, 1828. *Planch. Col. d'Ois.*, livr. 78: pl. 464.—Manado, Sulawesi.
    Synonym: *Parra novae-guinae* Ramsay, 1879. *Proc. Linn. Soc.* (1878) 3: 298.—Vicinity of Port Moresby.
    Synonym: *Hydralector novae hollandiae* Salvadori, 1882. *Orn. Pap. Moluc.* 3: 309.—Australia.
Inhabits lowland wetlands and ponds of the NG lowlands, Aru Is, Misool, ?Salawati[A], Goodenough[CO], and Fergusson I (see Mayr 1941 and Bishop 2006; M. Tarburton's island website[A]). Locally common. One record from Varirata NP at 600 m[CO].
    Extralimital—the Philippines, Borneo, Lesser Sundas, Sulawesi, and n and e Australia (Jenni 1996). Recorded breeding on New Britain (Mayr & Diamond 2001: 373).

NOTES
We follow Mees (1982b) in treating the species as monotypic. *Other names*—Lotusbird, Lily-trotter, Christbird.

Family: Scolopacidae

Curlews, Sandpipers, and Allies [Family 94 spp/Region 33 spp]

The species of the family exhibit a worldwide distribution. Most of the species breed in the far north and migrate to the tropics or temperate Southern Hemisphere during the northern winter. The species inhabit wetlands and open grasslands. These are small to large wading birds with diverse bill morphology, the bill tending to be narrow but either short or long, and sometimes decurved or recurved. This family includes the sandpipers, snipe, woodcock, phalaropes, stints, curlews, godwits, turnstones, and shanks. Cracraft (2013) placed this lineage as sister to a lineage that includes the seedsnipes, painted-snipe, Plains-wanderer, and jacanas.

GENUS *SCOLOPAX*

[Genus 8 spp/Region 1 sp]

Scolopax Linnaeus, 1758. *Syst. Nat.*, ed. 10, 1: 145. Type, by tautonymy, "Scolopax" = *Scolopax rusticola* Linnaeus.

Woodcocks range from eastern North America, northern Eurasia, and Japan, to Sumatra, Java, Sulawesi, Moluccas, and New Guinea (van Gils & Wiersma 1996). The genus is characterized by the heavy body, long straight bill, short neck, large eyes set back on the head, short neck, cryptic barred plumage, and dark barring on the crown. Unlike other shorebirds, woodcocks prefer forested habitats, keeping to dense cover.

New Guinea Woodcock *Scolopax rosenbergii*   ENDEMIC | MONOTYPIC

Scolopax Rosenbergii Schlegel, 1871. *Ned. Tijdschr. Dierk.* (1873) 4: 54.—Arfak Mts, Bird's Head.

Inhabits the floor of montane forest of the Bird's Head, Central Ra, Foja Mts[J], and mts of Huon Penin, 1,500–3,800 m (Mayr & Meyer de Schauensee 1939b, Mayr 1941, Gyldenstolpe 1955a, Bishop 2006, Beehler *et al.* 2012[J]). Should be looked for in the Cyclops, Fakfak, Kumawa, and Wandammen Mts. Most commonly encountered above 2,400 m[CO]. Secretive; easily overlooked.

NOTES

We follow Kennedy *et al.* (2001) in treating this regional isolate as specifically distinct from *S. saturata* of Sumatra and Java. Characters distinguishing the species from *S. saturata* include: longer wing, tail, tarsus; darker and more loosely barred mantle; conspicuous white malar spots; white chin; and dusky brown throat (Kennedy *et al.* 2001). *Other names*—Rufous Woodcock, *Scolopax saturata*.

GENUS *GALLINAGO*   [Genus 17 spp/Region 2 spp]

Gallinago Brisson, 1760. *Orn.* 5: 298. Type, by tautonymy, "Gallinago" Brisson = *Scolopax gallinago* Linnaeus.

Snipes of this genus range through the Americas, Europe, Africa, Madagascar; northern, eastern, and southern Asia; and New Guinea, northern and eastern Australia, and New Zealand (van Gils & Wiersma 1996, Gill *et al.* 2010). The genus is characterized by the heavily marked plumage, dark sagittal and facial stripes, and very long, straight bill. Woodcock (*Scolopax*) species exhibit crown bars rather than sagittal stripes and tend to be barred rather than streaked dorsally.

Latham's Snipe *Gallinago hardwickii*   VISITOR | MONOTYPIC

Scolopax Hardwickii J. E. Gray, 1831. *Zool. Misc.* 1: 16.—Tasmania.

A rare southbound passage migrant, sea level–3,350 m, with a few birds remaining during the northern winter (Beehler *et al.* 1986, Coates 1985, Bishop 2006). Main periods of occurrence are August–October and April–early May (Coates 1985).

Localities—Mount Wilhelmina[MA], Kurik[K], and Nondugl[J]; possible sightings from the Port Moresby area[L,CO]; and Milne Bay[M] (Gyldenstolpe 1955a[J], Mees 1982b[K], Higgins & Davies 1996[M], Bishop 2006[L]).

Extralimital—breeds in the e Primorsky, s Sakhalin, s Kurile Is, and Japan. Winters in e Australia and NZ (Higgins & Davies 1996, van Gils & Wiersma 1996).

NOTES

The species cannot be reliably distinguished from *G. megala* in the field, so Bishop (2006) discounted many of the sight records from e PNG. This is the safest position to take at this point, while acknowledging that the species is a common wintering bird in Australia and that it is unknown whether many of these birds stop over in NG. More effort needs to be made to better document this species' occurrence in the Region. *Other names*—Japanese or Australian Snipe, *Capella hardwickii*.

[Pin-tailed Snipe *Gallinago stenura*

HYPOTHETICAL | MONOTYPIC

Scolopax stenura Bonaparte, 1831. *Ann. Stor. Nat. Bologna* 4: 335.—Bogor, Java.

There have been several reports of field sightings of birds that may have been this species at the Moitaka settling ponds near Port Moresby in October and November (Coates 1985, Hicks 1990).

Extralimital—breeds in north-central and e Russia, from the Ural Mts east to the Sea of Okhotsk. Winters in tropical Asia to e Australia (van Gils & Wiersma 1996). Mist-netted on the n coast of w New Britain (K. D. Bishop *in litt.*).

NOTES

An effort should be made to document the presence of this species in the Region through a diagnostic photo or museum specimen (Bishop 2006).]

Swinhoe's Snipe *Gallinago megala*

VISITOR | MONOTYPIC

Gallinago megala Swinhoe, 1861. *Ibis*: 343.—Between Tanggu and Beijing, n China.

Fairly common on southbound passage and during the northern winter in open moist grasslands, perhaps in greater numbers in w NG (Mayr 1941, Coates 1985, Bishop 2006); on migration it occurs in dry grasslands of the highlands to 3,720 m (Coates 1985). Most often encountered from late August to mid-May (Coates 1985). NG may be an important wintering ground for this species (Bishop 2006).

Extralimital—breeds in sc Siberia, n Mongolia, Amurland, and Ussuriland. Winters in tropical Asia, Indonesia, and n Australia (van Gils & Wiersma 1996).

Other names—Marsh or Chinese Snipe, *Capella megala*.

GENUS *LIMNODROMUS* [Genus 3 spp/Region 2 spp]

Limnodromus Wied, 1833. *Beitr. Naturg. Brasil* 4, Abth. 2: 716. Type, by monotypy, *Scolopax noveboracensis* Gmelin = *Scolopax grisea* Gmelin.

Dowitchers breed in arctic North America and Siberia, wintering to the southern United States, Central America, northern South America, tropical Asia, New Guinea, and northern Australia (van Gils & Wiersma 1996). The genus is characterized by the long and straight bill and the heavily patterned dorsal surface contrasting with the lightly patterned ventral surface, which is enriched with a russet or chestnut background in breeding birds.

Long-billed Dowitcher *Limnodromus scolopaceus*

VAGRANT | MONOTYPIC

Limosa scolopacea Say, 1822. In Long's *Exped. Rocky Mts* 1: 335.—Council Bluffs, Iowa.

Known from a single record of a bird in winter plumage from Aroa Lagoon, nw of Port Moresby, 2–8-Dec-1984 (Anon. 1984, Hicks 1990). Records are known from India, Borneo, Bali, and Victoria, Australia[J] (Bishop 2006, D. Hobcroft *in litt.*, with supporting photos[J]).

Extralimital—breeds in n Siberia, Alaska, and the NW Territories, Canada. Winters mainly in the s United States and Mexico (van Gils & Wiersma 1996); infrequent in the c Pacific (*e.g.*, Hawai`i).

Asian Dowitcher *Limnodromus semipalmatus*

VISITOR | MONOTYPIC

Macrorhamphus semipalmatus Blyth, 1848. *Journ. Asiat. Soc. Bengal* 17(1): 252.—Calcutta, India.

Five records from NG (Bishop 2006). Perhaps more frequent in w NG, as most Australian records are from the nw coast of the continent (Higgins & Davies 1996).

Localities—Wasur NP, Kikori R delta, Angabanga R, and the Moitaka settling ponds (Bishop 2006). The species has been recorded from September to November and in March.

Extralimital—breeds in Siberia, Mongolia, and n Manchuria; winters e India, se Indochina, Indonesia, and e and w Australia (Higgins & Davies 1996).

Other names—Asiatic or Snipe-billed Dowitcher.

Limosa Brisson, 1760. *Orn.* 1: 48; 5: 261. Type, by tautonymy, "Limosa" Brisson = *Scolopax limosa* Linnaeus.

The breeding range for the godwits includes North America, Europe, Siberia, the Russian far east, and Manchuria. Species winter to South America, Africa, tropical Asia, Taiwan, New Guinea, Australia, and New Zealand (van Gils & Wiersma 1996). The genus is characterized by the very long and slightly upturned bill, large size, long legs, and richly patterned breeding plumage.

Black-tailed Godwit *Limosa limosa* VISITOR
Scolopax Limosa Linnaeus, 1758. *Syst. Nat.*, ed. 10, 1: 147.—Sweden.

A common passage migrant, mainly to the s coast; also commonly overwinters in the Trans-Fly; probably present elsewhere in suitable habitat (Bishop 2006).

Localities—Bird's Head[K], Bintuni Bay[J], Dolak I[J], Komolom I[J], Kurik[J], Lampusatu Beach near Merauke[J], Kumbe[L], Bian and Merauke R[J,L], Wasur NP[J], Bensbach R[J], Stephansort (Astrolabe Bay)[CO], Lae[J], Port Moresby[J] (Voous 1963[L], Mees 1982b[K], Bishop 2006[J]). The s coast of NG may constitute an important stopover and wintering location for the species (Bishop 2006).

Extralimital—breeds locally from Iceland to Kamchatka, wintering south to tropical Africa, tropical Asia, and Australia (Higgins & Davies 1996, van Gils & Wiersma 1996).

SUBSPECIES

a. *Limosa limosa melanuroides* Gould
Limosa Melanuroides Gould, 1846. *Proc. Zool. Soc. London*: 84.—Port Essington, NT, Australia.
RANGE: Eastern Siberia, e Mongolia, ne China, and the Russian far east; winters to India, Indochina, Taiwan, Philippines, Indonesia, NG, and Australia (van Gils & Wiersma 1996).
DIAGNOSIS: In comparison to the nominate form, this race is smaller and shorter-billed, and darker rufous ventrally in breeding plumage; nonbreeding plumage of *melanuroides* exhibits a dark mantle and breast (van Gils & Wiersma 1996).

Bar-tailed Godwit *Limosa lapponica* VISITOR
Scolopax lapponica Linnaeus, 1758. *Syst. Nat.*, ed. 10, 1: 147.—Lapland, Sweden.

A rather uncommon passage migrant, mainly on the s coast, but also known on passage from n NG (Bishop 2006). A few overwinter in the Region. Frequents mudflats, beaches, and estuaries; occasionally visits freshwater lagoons and settling ponds, rarely grass airstrips near the coast (Coates 1985). Bishop (2006) suggested that most of this population overflies NG to winter in Australia.

Extralimital—breeds in the high Arctic reaches of Europe, Siberia, and w Alaska, migrating to Europe, w Africa, the Middle East, se Asia, and Australasia (Higgins & Davies 1996, van Gils & Wiersma 1996).

SUBSPECIES

a. *Limosa lapponica baueri* Naumann
Limosa Baueri Naumann, 1836. *Naturgesch. Vög. Deutschl.* 8: 429.—Victoria, Australia.
Synonym: *Limosa lapponica* var. *Novae Zealandiae* Gray, 1846. *Voy. Erebus and Terror*: 13.—New Zealand.
RANGE: Breeds in n and ne Siberia and across Bering Strait into w Alaska. Migrates through e Asia, the Philippines, Greater Sunda Is, and w Oceania. Winters in NZ and Australia. Localities—the Bird's Head[CO], Aru Is[J], Dolak I[J], Bian R[J], Wasur NP[J], Bensbach R[J], Kiunga[Q], Kikori R delta[J], Madang[J], Lae[J], Angabanga R[J], and Rossel I[CO] (Coates 1990: 566[Q], Bishop 2006[J]).
DIAGNOSIS: Rump and uppertail coverts heavily barred brown; relatively long wings and tail (van Gils & Wiersma 1996). Wing: male 199–228, female 216–240 (Higgins & Davies 1996).

Numenius Brisson, 1760. *Orn.* 1: 48; 5: 311. Type, by tautonymy, "Numenius" Brisson = *Scolopax arquata* Linnaeus.

Curlews breed through much of the Arctic and subarctic, plus north-central North America, Europe, and much of Asia; they winter to the tropics and southern temperate zone around the world (van Gils & Wiersma 1996). The genus is characterized by the long and decurved bill, the heavily vermiculated plumage, the small head, and the longish neck.

There is a single record of the Bristle-thighed Curlew (*Numenius tahitiensis*) from Manus Island (Kennerley & Bishop 2001), indicating the species might be looked for along the northern coast (or northern fringing islands) of New Guinea (Bishop 2006).

Little Curlew *Numenius minutus* | VISITOR | MONOTYPIC

Numenius minutus Gould, 1841. *Proc. Zool. Soc. London* (1840): 176.—Maitland, , Australia.

On its southward passage bound for Australia it is an abundant migrant locally on the coast of s NG, especially in the Trans-Fly and around Port Moresby; also occasionally alights on inland airfields in upland valleys (Bishop 2006). Rare on the n coast. On return passage it apparently overflies NG (Bishop 2006). In NG frequents short-grass airfields and playing fields; does not use tidal mudflats. Occurs mostly in the lowlands, occasionally up to 1,100 m and higher (Coates 1985). One was recovered from ice at 4,450 m on the Carstensz Massif (Schodde *et al.* 1975).

Localities—near Sorong[J], Kebar valley[J], Aru Is[MA], Carstensz Massif[J], Dolak I[J], Kurik[J], Wasur NP[J], Bensbach[J], Tabubil[K], Aramia R[J], middle Fly R[J], Baiyer R[J], Astrolabe Bay[MA], and Port Moresby[CO] (Gregory 1995b[K], Bishop 2006[J]). Very large numbers, probably totaling tens of thousands of birds, have been reported from the floodplains of the Bensbach R in October (Finch 1980c) suggesting that the locality lies on a major migration route of the species (Coates 1985, Bishop 2006). Whether s NG is a wintering area remains to be determined. High counts were reported in mid-November.

Extralimital—breeds in c and e Siberia. Migrates through e Asia to the Moluccas, Kai Is, and n and e Australia (Higgins & Davies 1996, van Gils & Wiersma 1996); the main wintering grounds appear to be Australia (Higgins & Davies 1996).

Other names—Little Whimbrel.

Whimbrel *Numenius phaeopus* | VISITOR

Scolopax Phæopus Linnaeus, 1758. *Syst. Nat.*, ed. 10, 1: 146.—Sweden.

Present at coastal sites in the entire NG region throughout year (Bishop 2006). Migrant and wintering birds are common, frequenting mangrove mudflats, exposed reef flats, or coastal beaches. Large numbers roost in mangroves.

Extralimital—breeds in the Arctic, wintering south to coastal areas of the warmer parts of the world (Coates 1985, van Gils & Wiersma 1996).

SUBSPECIES

a. *Numenius phaeopus variegatus* (Scopoli)
Tantalus variegatus Scopoli, 1786. *Delic. Flor. Faun. Insubr.* 2: 92.—Luzon, Philippines.
RANGE: Breeds in e Siberia. Winters in se Asia, NG, Micronesia, w Polynesia, Melanesia, and Australia. Found fairly commonly through all coastal environments (Mayr 1941, Coates 1985). Recorded from the interior uplands at 1,500 m at Kosipi (Bell, in Coates 1985); also at Tabubil (Coates 1990: 566). Schodde *et al.* (1975) reported observation of birds from high valleys of the Carstensz Massif above 3,600 m.
DIAGNOSIS: Underwing heavily barred with dull gray-brown rather than whitish (of nominate form); dark-spotted lower back patch and uppertail coverts barred with white (similar to nominate form); less rich buff overall than *hudsonicus* (van Gils & Wiersma 1996).

NOTES
Other names—Hudsonian Curlew.

Eastern Curlew *Numenius madagascariensis*   VISITOR | MONOTYPIC
Scolopax madagascariensis Linnaeus, 1766. *Syst. Nat.*, ed. 12, 1: 242.—Macassar, Sulawesi.

In NG, mostly found as an uncommon passage migrant on sandy beaches and mudflats along the coast of s NG, with a substantial portion of the world population passing through the Region on its way to its Australian wintering grounds (Bishop 2006). A carcass was recovered at 4,500 m from the glacier on Carstensz Massif (Schodde *et al.* 1975). A bird banded in NSW, Australia, in Jan-1977 was recovered at Marshall Lagoon, just se of Port Moresby, two and a half months later (Coates 1985).

Localities—mouth of the Mimika R[MA], Aru Is[L], Dolak I[J], Komolom I[J], Kumbe R[J], Bian R[J], Lampusatu Beach[J], Merauke[MA], Wasur NP[J], Bensbach R[J], Bula[J], Daru[J], Kikori R delta[J], coast of the Huon Penin[CO], Yule I (Hall Sound)[MA], Port Moresby[MA], and Bootless Bay[J] (Diamond & Bishop 1994[L], Bishop 2006[J]).

Extralimital—breeds in e Siberia, wintering south to Taiwan, Indonesia, NG, and as far as NZ. By far the greatest numbers spend the winter in coastal Australia (Higgins & Davies 1996, van Gils & Wiersma 1996).

Other names—Australian or Far Eastern Curlew.

GENUS *TRINGA*   [Genus 13 spp/Region 7 spp]

Tringa Linnaeus, 1758. *Syst. Nat.*, ed. 10, 1: 148. Type, by tautonymy, "Tringa" = *Tringa ocrophus* Linnaeus.

The species of this genus breed through the entire boreal zone and arctic tundra and winter south to the temperate zone and tropics (van Gils & Wiersma 1996). The genus is characterized by the slim body, small head, longish bill, and finely patterned mantle. The Spotted Redshank (*Tringa erythropus*) is a vagrant to Australia and thus to be expected as a vagrant for the New Guinea region as well. We follow Christidis & Boles (2008), who cited the molecular analysis of Pereira & Baker (2005), who demonstrated that *Heteroscelis* is embedded within *Tringa*.

Common Redshank *Tringa totanus*   VAGRANT
Scolopax Totanus Linnaeus, 1758. *Syst. Nat.*, ed. 10, 1: 145.—Sweden.

Two records from Port Moresby (Bell 1966a, Coates 1972) and two more from w NG, at Sorong (Andrew 1992) and at nearby Jefman I, NW Islands, on 18-Aug-1993 (Redman 2011). The species perhaps is to be expected mainly in w NG, as this vagrant is most often reported from Western Australia (Higgins & Davies 1996).

Extralimital—breeds from Iceland and n Europe to n Manchuria. Winters south to Spain, coastal Africa, Arabia, tropical Asia, and Australia (Higgins & Davies 1996, van Gils & Wiersma 1996).

SUBSPECIES

No specimens are available for study. The form arriving in the Region would be either the nominate form or, more likely based on breeding distribution, *T. t. terrignotae* (van Gils & Wiersma 1996, Simpson & Day 1996). Apparently mainly overflies NG to winter in Australia.

Marsh Sandpiper *Tringa stagnatilis*   VISITOR | MONOTYPIC
Totanus stagnatilis Bechstein, 1803. *Orn. Taschenb. Deutschl.* 2: 292; pl. 29.—Germany.

A fairly common n migrant, with most records from the Trans-Fly and Port Moresby; some birds overwinter. May be expected on southbound passage in suitable habitat anywhere in the Region (Coates 1985, Bishop 2006). Frequents freshwater swamps, shallow freshwater and brackish lagoons, and wet rice-fields; records to 1,400 m[CO].

Localities—Aru Is[K], Biak I[L], Dolak and Komolom I, Kurik, Lampusatu Beach near Merauke, Wasur NP, Bensbach R, Bula Plains, middle Fly R, Kikori R delta, Zenag[CO], and Port Moresby (Diamond & Bishop 1994[K], Bishop 2006, Bishop *in litt.*[L]). A group of 230 was recorded from the Bensbach region on 17-Jul-1982 providing evidence of oversummering in NG during the northern breeding season (Finch *et al.* 1982).

Extralimital—breeds from w Russia and e Ukraine to east-central Siberia, Ussuriland, and ne China. Winters south to Africa, tropical Asia, Philippines, and Australia (van Gils & Wiersma 1996).

Other names—Little Greenshank.

Common Greenshank *Tringa nebularia* VISITOR | MONOTYPIC

Scolopax nebularia Gunnerus, 1767. In Leem, *Beskr. Finm. Lapper.* 251.—Near Trondheim, Norway.

A common passage migrant, with a few nonbreeders stopping over; most records from the s coast of the Mainland. Typically only on salt water, except during August–December passage, when it is found at freshwater sites on occasion (Beehler *et al.* 1986). Some nonbreeding birds remain during the Palearctic nesting season.

Localities—the Bird's Head[K], Bintuni Bay[J], Adi I[J], Aru Is[MA], Paniai Lake[J], Dolak I[L], Kurik[J], Merauke[MA], Lampusatu Beach[J], Bensbach R[J], Tonda[J], Aramia R[J], middle Fly R[J], Daru[MA], Kikori R delta[J], Port Moresby[J], and Amazon Bay[CO] (Gyldenstolpe 1955b[K], Hoogerwerf 1964[L], Bishop 2006[J]).

Extralimital—breeds from n Europe across Asia to Kamchatka. Winters in Spain, Africa, the Middle East, and tropical Asia, east to Australia and NZ (van Gils & Wiersma 1996).

Other names—Greenshank.

Green Sandpiper *Tringa ochropus* VAGRANT | MONOTYPIC

Tringa ocrophus Linnaeus, 1758. *Syst. Nat.*, ed. 10, 1: 149; name subsequently emended to *ochropus*.—Sweden.

One well-described sight record from Manokwari, 11-Jan-1961, by an observer familiar with the species in Europe (Hoogerwerf 1964). Coates (1985) reported an additional possible record from Saidor in Nov-1975. Coates supported recognizing the Manokwari record, noting that the author wrote, "the call of this bird and the dark-coloured inner wing, together with its clear white rump, were characters which were superfluous affirmations of its identification" (Hoogerwerf 1964: 123). The bird was found in a small freshwater pool near an airstrip.

Extralimital—breeds over much of n Eurasia, wintering south to Africa and tropical Asia (van Gils & Wiersma 1996).

Wood Sandpiper *Tringa glareola* VISITOR | MONOTYPIC

Tringa Glareola Linnaeus, 1758. *Syst. Nat.*, ed. 10, 1: 149.—Sweden.

A common passage migrant and winter resident in the Region. Interior records to 1,735 m[CO].

Localities—Waigeo I[MA], Sorong[K], Klamono[K], Manokwari[K], Paniai Lake[J], Biak I[L], Mimika R[MA], Dolak I[J], Kurik[J], Merauke[J], Bensbach R[J], Morehead R[J], Kiunga[CO], middle Fly R[J], Madang[CO], near Bereina[J], Woitape[CO], Port Moresby[CO], and Higaturu[J] (Gyldenstolpe 1955b[K], Bishop 2006[J], Bishop *in litt.*[L]). Habitat mainly freshwater marshes and pools in grassland.

Extralimital—breeds from n Europe across Asia to Kamchatka. Winters throughout Africa and tropical Asia, east to Palau, the Moluccas, and Australia (van Gils & Wiersma 1996).

Grey-tailed Tattler *Tringa brevipes* VISITOR | MONOTYPIC

Totanus brevipes Vieillot, 1816. *Nouv. Dict. Hist. Nat.*, nouv. éd., 6: 410.—Timor, Lesser Sundas.

This northern migrant is the common tattler in NG, both in passage and as a nonbreeding resident, and many nonbreeders remaining during the northern summer (Bishop 2006). Frequents mostly tidal mudflats, sandy beaches, and rocky reefs of the Mainland and all islands, including outer atolls. Uncommon along Trans-Fly coast. Fifty counted at Lea Lea in Jan-1990 (Hicks 1991). Interior records (Wau, Lake Kutubu) should be reassessed, though interior records in passage are possible (B. J. Coates *in litt.*). Stickney (1943) reported specimens from the SE Islands: Doini, Misima, Tagula, and Rossel I.

Extralimital—breeds in c and e Siberia, migrating through Japan and e Asia to Micronesia, se Asia, Solomon Is, coastal Australia, and Norfolk I (Higgins & Davies 1996, van Gils & Wiersma 1996).

NOTES

For generic treatment, we follow Christidis & Boles (2008) and Pereira & Baker (2005). *Other names*— Siberian Tattler, *Heteroscelis brevipes*.

Wandering Tattler *Tringa incana* VISITOR | MONOTYPIC

Scolopax incana Gmelin, 1789. *Syst. Nat.* 1(2): 658.—Moorea I, Society Is.

Scarce in NG, occurring primarily on offshore islands and rocky reefs along the coast, mainly in the e half of the Region. NG and Australia constitute the w limit of its range. NG range also includes Waigeo I, Aru Is, and Manam I (Mayr 1941, see Bishop 2006). To be expected on the smaller SE Islands.

Extralimital—breeds in Alaska and far e Siberia, wintering along the w coast of North America, various islands in the Pacific including the Galapagos, Hawai`i, Solomons, and Vanuatu, as well as the e coast of Australia (Higgins & Davies 1996, van Gils & Wiersma 1996).

NOTES

Earlier treatments considered this and the previous species (*T. brevipes*) as conspecific, thus older reports of this species might actually refer to *brevipes*. *Other names—Heteroscelis incanus*.

GENUS *XENUS* MONOTYPIC

Xenus Kaup, 1829. *Skizz. Entw.-Gesch. Eur. Thierwelt*: 115. Type, by monotypy, *Scolopax cinerea* Güldenstädt.

The single species in the genus breeds from northern Eurasia east to eastern Siberia, and winters to coastal Africa, Arabia, tropical Asia, and Australasia (van Gils & Wiersma 1996). The genus is characterized by the long and recurved bill, small body size, and shortish legs; darting and energetic foraging.

Terek Sandpiper *Xenus cinereus* VISITOR | MONOTYPIC

Scolopax cinerea Güldenstädt, 1775. *Novi Comm. Sci. Imper. Petropol.* 19: 473, pl. 19.—Shore of Caspian Sea near mouth of Terek R.

A regular and common southbound passage migrant through Region; a few overwinter (Coates 1985). Localities—Sorong[J], Bintuni Bay[K], Aru Is[W], Wanggar[MA], Mimika R[MA], Dolak I[J], Merauke[MA], Lampusatu Beach[J], Benbach R[J], Bula[J], Daru[MA], Kikori R delta[J], coastal Huon Penin[CO], Bereina[K], Port Moresby[K], and Fergusson I[K] (Gyldenstolpe 1955b[J], Diamond & Bishop 1994[W], see Bishop 2006[K]). Particularly common on the extensive mudflats of the Trans-Fly, and in smaller numbers on the s coastline of the SE Peninsula (Bishop 2006).

Extralimital—breeds from n Europe to e Siberia. Winters to coastal Africa, Arabia, India, and se Asia to w, n, and coastal Australia (Higgins & Davies 1996, van Gils & Wiersma 1996).

NOTES

Placed in *Tringa* by some authorities. *Tringa terek* is a junior synonym (Monroe 1989, van Gils & Wiersma 1996).

GENUS *ACTITIS* [Genus 2 spp/Region 1 sp]

Actitis Illiger, 1811. *Prodromus Syst. Mamm. Avium*: 262. Type, by subsequent designation (Stejneger 1885. *Bull. U.S. Nat. Mus.* 29: 131), *Tringa hypoleucos* Linnaeus.

The two species of this genus breed throughout the northern temperate and boreal regions of the world, migrating south to Central and South America, Africa, Arabia, tropical Asia, and Australasia (van Gils & Wiersma 1996). This genus is characterized by the small size, medium-length straight bill, rather short legs, pale ventral plumage contrasting with darker mantle, and teetering behavior.

Common Sandpiper *Actitis hypoleucos* VISITOR | MONOTYPIC

Tringa Hypoleucos Linnaeus, 1758. *Syst. Nat.*, ed. 10, 1: 149.—Sweden.

A common northern migrant and overwintering visitor throughout NG in rocky waterside habitats, sea level–3,300 m (Coates 1985). Arrives in mid-July and departs by early May, and a few birds remain during the northern summer (Bishop 2006). Frequents beaches, mudflats, mountain streams, margins of rivers, swamps, and ponds; also visits fields, lawns, and gravel roads (Coates 1985).

Extralimital—breeds from Europe across Asia to Japan. Winters from Africa through tropical Asia to Australia (van Gils & Wiersma 1996).

NOTES
Actitis is treated as masculine (David & Gosselin 2002b). *Other names—Tringa hypoleucos.*

GENUS *ARENARIA* [Genus 2 spp/Region 1 sp]

Arenaria Brisson, 1760. *Orn.* 1: 48; 5: 132. Type, by tautonymy, "Arenaria" Brisson = *Tringa interpres* Linnaeus.

The two turnstone species breed in the Arctic and winter along temperate and tropical coasts (van Gils & Wiersma 1996). The genus is characterized by the distinctive bill (with a slight upturn), pied plumage of breeding adults (white, black, and rufous), compact shape, and short neck.

Ruddy Turnstone *Arenaria interpres*   VISITOR
Tringa Interpres Linnaeus, 1758. *Syst. Nat.*, ed. 10, 1: 148.—Gotland, Sweden.
Widely distributed as a passage migrant and overwintering visitor on coastlines of the NG region from mid-August to mid-April; some birds remain throughout the year. Probably occurs wherever there is suitable habitat. Frequents mostly rocky and stony shores, but also visits beaches and mudflats and at high tide ventures inland to forage on expanses of short grass (Coates 1985, Bishop 2006).

Extralimital—breeds throughout the high Arctic, and winters to the n and s temperate zones and tropics throughout, including islands of the w Pacific, coastal Australia, and NZ (van Gils & Wiersma 1996).

SUBSPECIES

a. *Arenaria interpres interpres* (Linnaeus)
RANGE: Breeds in arctic e Canada, Greenland, n Eurasia, and nw Alaska. Winters from w Europe and Africa to tropical Asia, Australasia, and the sw Pacific (van Gils & Wiersma 1996).
DIAGNOSIS: Slightly larger than *morinella*, and with more dark streaking on crown and less extensive deep chestnut-red on scapulars and upperwing coverts (van Gils & Wiersma 1996).

NOTES
*Other names—*Turnstone.

GENUS *CALIDRIS* [Genus 18 spp/Region 10 spp]

Calidris Merrem, 1804. *Allg. Lit. Zeitung* 2(168): col. 542. Type, by tautonymy, *Tringa calidris* Gmelin = *Tringa canutus* Linnaeus.

The species of this genus breed throughout the arctic regions and winter in tropical and temperate regions (van Gils & Wiersma 1996). This large and diverse genus is characterized by the relatively small size, short neck, small head, short bill, heavily and finely patterned upperparts, and lightly patterned or all-white belly and vent plumage. Gibson & Baker (2012) showed *Tryngites* and *Philomachus* to belong within *Calidris* and recommended they all be assigned to that genus. Because of considerable morphological divergence, we opt to keep these forms as distinct genera at this time. The fanciful breeding plumage of *Philomachus pugnax* and the wonderful courtship displays of *Tryngites subruficollis* merit recognition in monotypic genera.

A field observation was made of a possible Western Sandpiper (*Calidris mauri*) near Merauke in the company of 15 other shorebird species on 11-Nov-1995; the individual was well described but not photographed (S. F. Bailey *in litt.*).

Great Knot *Calidris tenuirostris*   VISITOR | MONOTYPIC
Totanus tenuirostris Horsfield, 1821. *Trans. Linn. Soc. London* 13: 192.—Java.

Uncommon on the SE Peninsula, but huge flocks have been reported from the Trans-Fly of Indonesian NG, where nonbreeders gather on the vast mudflats (Hoogerwerf 1964, Bishop 2006). Frequents tidal mudflats and sandy estuaries (Coates 1985, Bishop 2006). NG is a major stopover during migration for the species, but its significance as a wintering ground is unclear (Bishop 2006).

Localities—Geelvink Bay[CO], the beaches between the Bian and Merauke R[CO], Tabubil[A], near Bereina on the Angabanga R[CO], Port Moresby[MA], Astrolabe Bay[MA], and the Trobriand Is[MA] (Gregory 1994b[A], see Bishop 2006).

Extralimital—breeds in ne Siberia (Verkoyansk Mts, Magadan, Koryak Highlands, and s Chukotskiy Penin). Winters from tropical Asia to the Moluccas and coastal Australia.

Other names—Eastern Knot.

Red Knot *Calidris canutus*   VISITOR
Tringa Canutus Linnaeus, 1758. *Syst. Nat.*, ed. 10, 1: 149.—Sweden.

A southbound and northbound passage migrant in large numbers in the w Trans-Fly, in flocks with *C. tenuirostris* that included as many as 5,000 *C. canutus* (Hoogerwerf 1964). Inexplicably rare everywhere else, with only a smattering of records. Only three records from the SE Peninsula (Bishop 2006), near Bereina (1976, 1977) and Hisiu (1982). A w NG record is known from the Boemi R at the head of Geelvink Bay (Melville 1980). Bishop (2006) noted the species may have been overlooked within large flocks of similar species and furthermore that extensive areas of suitable habitat in NG have never been surveyed. Large numbers have been recorded wintering in Australia.

Extralimital—breeds in the high Arctic of Alaska, Canada, Greenland, and w and c Siberia; winters to South America, Europe, Africa, Indonesia, Australia, and NZ (van Gils & Wiersma 1996).

SUBSPECIES

a. *Calidris canutus ?rogersi* Mathews
Calidris canutus rogersi Mathews, 1913. *Birds Austral.* 3: 270, 273, pl. 163.—Shanghai, China.
RANGE: Breeds on the Chokotskiy Penin and possibly areas farther westward; winters in Australasia (van Gils & Wiersma 1996).
DIAGNOSIS: When compared with the nominate form, *rogersi* is smaller and is paler ventrally, with more white on lower belly (Higgins & Davies 1996, van Gils & Wiersma 1996). Subspecies determination for NG is limited by the few specimens available for study.

NOTES

The fact that *C. c. rogersi* may be the dominant subspecies in Australia suggests its dominant presence in NG (Higgins & Davies 1996). There is also the possibility that the newly described *C. c. piersmai*, which winters in nw Australia, could occur in NG as well (Gill & Donsker 2015). In any case, the subspecies are in need of revision (Higgins & Davies 1996). *Other names*—Knot or Common Knot.

Sanderling *Calidris alba*   VISITOR
Trynga alba Pallas, 1764. In Vroeg, *Cat. Ois., Adumbr.*: 7.—North Sea coast, Netherlands.

Rare in NG. To date known only from the coasts of the w Trans-Fly and Port Moresby, plus a n coast outlier from Lae (Mees 1982b, Coates 1985, Bishop 2006). Scarce but probably widespread along NG's coastal beaches (Coates 1985).

Extralimital—breeds in arctic Canada, Greenland, and Siberia, migrating to warmer coastal regions around the world (van Gils & Wiersma 1996).

NOTES

Subspecies undetermined. Probably the nominate form occurs in Region. *Other names*—*Crocethia alba*.

Red-necked Stint *Calidris ruficollis* VISITOR | MONOTYPIC

Trynga ruficollis Pallas, 1776. *Reise Versch. Prov. Russ. Reichs* 3: 700.—Kulussutai, s Transbaikalia.

A very common passage migrant, with many wintering on the coast and a few nonbreeders remaining during the northern summer (Bishop 2006). Recorded from many Mainland localities, plus Salawati I[MA], Aru Is[MA], Goodenough I[CO], and Misima I[MA]. Most of the world population overwinters in Australia, and NG appears to host a significant proportion as well (Bishop 2006). Frequents mostly tidal mudflats, beaches, salt pans, subcoastal lagoons, and margins of freshwater swamps (Coates 1985).

Extralimital—breeds in north-central and ne Siberia and Alaska, wintering to tropical Asia, Bismarck Arch, Solomon Is, Australia, and NZ (van Gils & Wiersma 1996).

Other names—Rufous-necked Stint, Eastern or Siberian Little Stint, *Erolia ruficollis*.

[Little Stint *Calidris minuta* HYPOTHETICAL | MONOTYPIC

Tringa minuta Leisler, 1812. *Nachtr. Bechstein's Naturgesch. Deutschl.* 1: 74.—Hanau am Main, Germany.

Two sightings from Aroa Lagoon, nw of Port Moresby, in Dec-1979 (Finch 1980a) and Mar-1983 (Finch 1983a), the latter in full breeding dress. Images of the 1983 bird were examined by shorebird experts in 2013, and their assessment was that a positive identification could not be made. As there are a number of records from Australia, the species is to be expected in passage in the Region, but we follow Bishop (2006) in treating this species as hypothetical until better-corroborated records are forthcoming.

Extralimital—breeds above the Arctic Circle, from Norway to north-central Siberia, wintering south to Spain, Turkey, Africa, and India; irregular farther east, although a regular vagrant to Australia (Higgins & Davies 1996, van Gils & Wiersma 1996).

Other names—*Erolia minuta.*]

Long-toed Stint *Calidris subminuta* VISITOR | MONOTYPIC

Tringa subminuta Middendorf, 1853. *Reise Nord.-Ost. Sibir.* 2(2): 222, pl. 19, fig. 6.—Western slopes of the Stanovoye Mts and mouth of the Udá R, Russia.

A regular but rare visitor to NG, where mostly recorded from September to January (Bishop 2006). In Australia, winters mainly along the w coast; thus probably more numerous in w NG (Heron 1978b, 1978c, Higgins & Davies 1996).

Localities—Dolak I[K], Wasur NP[K], Bensbach[CO], and near Port Moresby[CO]. Frequents mostly areas of soft mud on the fringes of freshwater or brackish marshes and lagoons (Coates 1985, Bishop 2006[K]).

Extralimital—breeds in c and e Siberia, migrating to e India, se Asia, and Taiwan, through Indonesia, and wintering from the Philippines to Australia (Higgins & Davies 1996, van Gils & Wiersma 1996).

Other names—Middendorf's Stint, *Erolia subminuta*.

Baird's Sandpiper *Calidris bairdii* VAGRANT | MONOTYPIC

Actodromas (*Actodromas*) *Bairdii* Coues, 1861. *Proc. Acad. Nat. Sci. Phila.* 13: 194.—Fort Resolution, Great Slave Lake, NW Territories, Canada.

Finch (1986b) reported observing a single bird in nonbreeding plumage at Kanosia Lagoon nw of Port Moresby on 24-Nov-1985 in the company of a large flock of *C. acuminata* and a single *Tringa stagnatilis*. A second record, a well-described individual—a lone *Calidris* in a mixed flock of larger shorebirds—was reported from Sorong, w Bird's Head, on 18-Aug-1993 (Redman 2011). Better-documented records are needed for this species. Few records exist for Australasia (Higgins & Davies 1996, Bishop 2006).

Extralimital—breeds in ne Siberia and Alaska, east to nw Greenland; winters to w and s South America; vagrants to Australia (Higgins & Davies 1996, Dickinson & Remsen 2013).

Pectoral Sandpiper *Calidris melanotos*   VISITOR | MONOTYPIC
Tringa melanotos Vieillot, 1819. *Nouv. Dict. Hist. Nat.*, nouv. éd. 34: 462.—Paraguay.

A rare passage migrant through NG from September (rarely late August) until the rains make foraging sites unsuitable in December or January, upon which the birds presumably continue southward (Coates 1985, Bishop 2006). This species is vastly outnumbered by *C. acuminata*. There are only two spring records (Bishop 2006).

Localities—Bensbach R[J], Aviara Beach near Bereina[K], Port Moresby[CO], and Higaturu[CO] (Heron 1978c[K], Bishop 2006[J]). First recorded in the Region by Coates (1973b).

Extralimital—breeds from Siberia east to Hudson Bay; winters mainly in South America, but some travel to Australia and NZ (Higgins & Davies 1996, van Gils & Wiersma 1996), and additional migrant records are from the Caroline Is and Bougainville I (Coates 1985).

Other names—Erolia melanotos.

Sharp-tailed Sandpiper *Calidris acuminata*   VISITOR | MONOTYPIC
Totanus acuminatus Horsfield, 1821. *Trans. Linn. Soc. London* 13: 192.—Java.

A common passage migrant to the NG lowlands, with occasional groups on highland airfields—max 3,720 m[CO]. Northward passage usually much smaller than southbound flight (Coates 1985, Bishop 2006). Recorded from many localities throughout the Mainland from the coast to the interior (Coates 1985, Bishop 2006). Frequents mainly the muddy margins of freshwater marshes; also commonly occurs on areas of short grass such as airfields (Coates 1985). It appears to be mainly a passage migrant to NG, with wintering birds confined mainly to the Trans-Fly (Hoogerwerf 1964).

Extralimital—breeds in north-central and ne Siberia (Lena R delta to the Kolyma R), wintering through Melanesia to New Caledonia and Tonga and south to Australia and NZ (Higgins & Davies 1996, van Gils & Wiersma 1996).

NOTES
Study is needed to determine where and in what numbers this species winters in Region. *Other names—* Siberian or Asian Pectoral Sandpiper, *Erolia acuminata.*

Broad-billed Sandpiper *Calidris falcinellus*   VISITOR
Scolopax Falcinellus Pontoppidan, 1763. *Danske Atlas* 1: 623, pl. 13.—Denmark.

A scarce passage migrant; all records from the s coast of NG. Mainly transient from mid-September to late December and from April to May, though odd records in January, June, and July indicate occasional summering and wintering (Bishop 2006).

Localities—Etna Bay[J], Dolak I[J], Merauke[J], Lampusatu Beach[J], Komolom I[J], Bensbach[J], Bula[J], Lae[CO], Angabanga R[CO], Hisiu[CO], Port Moresby[CO], and Higaturu[CO] (Bishop 2006[J]). Heron (1978b) found it to be occasionally numerous near the mouth of the Angabanga R during southern migration, in flocks of more than 20 individuals. Rare in the Port Moresby environs (Coates 1985).

Extralimital—breeds in isolated populations in n Europe and arctic Siberia, wintering south to Africa and through Asia to Australia and NZ (Higgins & Davies 1996, van Gils & Wiersma 1996).

SUBSPECIES
a. *Calidris falcinellus sibirica* (Dresser)
Limicola falcinellus sibirica Dresser, 1876. *Proc. Zool. Soc. London*: 674.—Siberia and China.
RANGE: Taymyr and the Lena R east to the Kolyma R; winters from ne India to Australia (van Gils & Wiersma 1996).
DIAGNOSIS: Compared to the nominate form this race has brighter rufous edges to feathers of mantle, broader lower supercilium, and narrower upper supercilium (van Gils & Wiersma 1996).

NOTES
We follow Gibson & Baker (2012) in subsuming *Limicola* into *Calidris*. *Other names—Limicola falcinellus.*

Curlew Sandpiper *Calidris ferruginea*　　　　　　　　　VISITOR | MONOTYPIC

Tringa Ferrugineus Pontoppidan, 1763. *Danske Atlas* 1: 624.—Christiansø I, Denmark.

An uncommon southbound passage migrant in small numbers; most records from Port Moresby (Bishop 2006). Most overfly NG, and there is minimal northward stopover (Beehler *et al.* 1986). In NG, occurs mainly during September–November (extreme dates late August to late January) and in much fewer numbers during return northward passage (early April to mid-May); occasionally remains during the austral winter. Frequents the margins of freshwater marshes and coastal mudflats. Peak numbers were 40+ at the Moitaka settling ponds on 25-September and up to 70 at a lagoon at Hisiu on 14-September (Coates 1985). One record from Misima I (Finch 1985).

Extralimital—breeds in the Siberian Arctic, migrating to Africa, Arabia, tropical Asia, Australia, and NZ (van Gils & Wiersma 1996).

NOTES

Use of the specific epithet *ferruginea* instead of *testacea* follows Stresemann (1941). First recorded in the Region by Hoogerwerf (1964). *Other names—Erolia ferruginea, Calidris testacea.*

GENUS *TRYNGITES*　　　　　　　　　　　　　　　　　　MONOTYPIC

Tryngites Cabanis, 1857. *J. Orn.* (1856) 4: 418. Type, by original designation, *Tringa rufescens* Vieillot = *Tringa subruficollis* Vieillot.

The single species of the genus breeds in eastern Siberia, northern Alaska, and arctic Canada, and winters in southern South America (van Gils & Wiersma 1996). The genus is characterized by medium size, small head, abbreviated bill, and grassland habit. It was folded into *Calidris* by Gibson & Baker (2012), but because of distinctive plumage and courtship behavior, we prefer to keep it in a distinct genus.

Buff-breasted Sandpiper *Tryngites subruficollis*　　　　　　VAGRANT | MONOTYPIC

Tringa subruficollis Vieillot, 1819. *Nouv. Dict. Hist. Nat.,* nouv. éd, 34: 465.—Paraguay.

One well-documented sight record of an individual at a settling pond at Higaturu Oil Palm Plantation, Sangara, Oro Prov (near Popondetta), on 26-Sep-1981, in association with a group of *Calidris acuminata* (Mordue 1981).

Extralimital— breeds in e Siberia, n Alaska, and arctic Canada, wintering mostly to South America; vagrant elsewhere. Many reports of singletons from Australia (Higgins & Davies 1996).

Other names—Calidris subruficollis.

GENUS *PHILOMACHUS*　　　　　　　　　　　　　　　　MONOTYPIC

Philomachus Merrem, 1804. *Allg. Lit. Zeitung* 2(168): col. 542. Type, by monotypy, *Tringa pugnax* Linnaeus.

The single species of the genus breeds from northern Europe to easternmost Siberia; it winters south to much of Africa, Arabia, and southern Asia, and is uncommon and irregular elsewhere (van Gils & Wiersma 1996). The genus is characterized by the extreme sexual dimorphism; the flamboyant breeding plumage of the adult male; and the long, brightly colored legs and short, thin bill of the female. It was folded into *Calidris* by Gibson & Baker (2012), but because of the very distinctive male plumage and behavior, we prefer to keep the species in its own genus.

Ruff *Philomachus pugnax*　　　　　　　　　　　　　　VISITOR | MONOTYPIC

Tringa Pugnax Linnaeus, 1758. *Syst. Nat.*, ed. 10, 1: 148.—Sweden.

A rare but regular migrant in e NG; most records to date from the Port Moresby environs (Coates 1985, Bishop 2006). Most often recorded from mid-September to mid-November, occasionally remaining until December and early January; once recorded in early April. Almost all records are from the Moitaka settling ponds near Port Moresby, where from one to four birds occur in a season. Elsewhere the species has been recorded at Bensbach, Hisiu Beach, and near Lea Lea (Bishop 2006), and as a vagrant to Tabubil (P. Gregory *in litt.*).

Extralimital—breeds in the n latitudes of Europe and Asia, wintering in Africa and India, less commonly but regularly to Australia (van Gils & Wiersma 1996). *Other names*—Reeve (female), *Calidris pugnax*.

GENUS *PHALAROPUS* [Genus 3 spp/Region 1 sp]

Phalaropus Brisson, 1760. *Orn.* 1: 50; 6: 12. Type, by tautonymy, "Phalaropus" Brisson = *Tringa fulicaria* Linnaeus.

The species of this genus breed in the arctic regions and western North America, and winter either in southern South America or at sea off western South America, western Africa, Arabia, New Guinea, Indonesia, and the Philippines (van Gils & Wiersma 1996). The genus is characterized by the reverse sexual dimorphism, the swimming habit, the pelagic winter distribution, and the weird twirling feeding method.

Red-necked Phalarope *Phalaropus lobatus* VISITOR | MONOTYPIC

Tringa tobata Linnaeus, 1758. *Syst. Nat.*, ed. 10, 1: 148, 824. [Emended to *lobata* on p. 824.]—Hudson Bay, Canada.

Winters in large numbers off the n coast of NG west to the NW Islands, remaining from October until March (Bishop 2006). Does not venture farther south except as a vagrant. Regional records include: waters of the NW Islands (Dampier Strait south of Waigeo I and near Misool; Beehler field notes), near Adi I off the s shore of the Bird's Neck (Gyldenstolpe 1955b), near the Aru Is, and off the n coast of the NG mainland (north of the Bird's Head, Geelvink Bay, and off the n coast of the w half of PNG; Coates 1985). There are two records from the s coast—one on an inundated ricefield at Kurik, on 24-Jan-1959 (Hoogerwerf 1964); and a dense flock of *ca.* 50 birds on the sea off Round Hill, Central Prov, PNG, in mid-Mar-1968 (Cleland 1968). A dried carcass was collected from the glacial ice at 4,420 m on the Carstensz Massif, Western Ra; and another, freshly dead, was found on scree near the edge of a glacial tarn at 4,500 m on 16-Jan-1972; moreover, individuals and flocks of birds resembling small white ducks swimming at the centers of lakes in the same area were most probably this species (Schodde *et al.* 1975). At sea its distribution is evidently patchy, because often it is not observed during short coastal voyages at times when it may be expected to occur (Coates 1985).

Extralimital—breeds in the arctic and subarctic regions of Eurasia, North America, and Greenland, wintering at sea in three principal areas: off the coast of Peru, southeast of the Arabian Penin, and in the zone between the Philippines, Sulawesi, and n NG (Higgins & Davies 1996).

NOTES

The originally misspelled species epithet (*tobata*) was corrected by Linnaeus, and thus the proper usage today is *lobatus* (David *et al.* 2009). *Other names*—Northern Phalarope.

Family Turnicidae

Buttonquail [Family 16 spp/Region 1 sp]

The species of the family range from Spain and northern and sub-Saharan Africa to tropical Asia, New Guinea, New Britain, the Solomon Islands, and Australia. Buttonquail are distinctive in lacking a crop and a hind toe (Debus 1996). The sex roles are reversed—the female is larger and more brightly colored than the male and is possibly polyandrous. In addition, the female possesses an enlarged trachea and an inflatable bulb in the esophagus for "booming" (Debus 1996). In New Guinea, the single species inhabits the southern lowlands, interior highlands, and Huon Peninsula. It prefers dense grassland and fodder crops (Pratt & Beehler 2014). A small quail-like bird, it is difficult to observe. Hackett *et mult.* (2008) indicated that the family is a sister to a clade that includes the gulls and the Crab-plover (Dromadidae).

Turnix Bonnaterre, 1791. *Tabl. Encyc. Méthod., Orn.* 1: lxxxii, 5. Type, by subsequent designation (G. R. Gray, 1840. *List Gen. Birds*: 63), *Tetrao gibraltaricus* Gmelin = *Tetrao sylvaticus* Desfontaines.

Generic characteristics and geographic range of the genus are as for the family. *T. pyrrhothorax* was recorded from Saibai Island in the northernmost Torres Strait in the late 1800s (Draffan *et al.* 1983). David & Gosselin (2002b) pointed out that the genus name *Turnix* is masculine.

Red-backed Buttonquail *Turnix maculosus*   RESIDENT

Hemipodius maculosus Temminck, 1815. *Pig. et Gall.* 3: 631, 757.—Timor, Lesser Sundas.

Locally distributed in grasslands in c and e NG: Trans-Fly, Eastern Ra, Huon Penin, s watershed of the SE Peninsula, and two SE Islands; sea level–2,500 m. Probably more widely distributed than these records suggest.

Extralimital—the Philippines, Sulawesi, and Lesser Sundas east to New Britain, Dyaul I, Guadalcanal I, and n and e Australia (Mayr & Diamond 2001, Dutson 2011).

SUBSPECIES

a. *Turnix maculosus horsbrughi* Ingram

Turnix horsbrughi Ingram, 1909. *Bull. Brit. Orn. Club* 23: 65.—Yule I, s watershed of the SE Peninsula.
RANGE: Lowlands of s NG. Localities—Dolak I[M], Kurik[K], Merauke distr[CO], Bensbach R[M], Lake Daviumbu[J], Tabubil[Y], Yule I[CO], and near Port Moresby[X] (Rand 1942a[J], Mees 1982b[K], Bell 1984[X], Gregory 1995b[Y], Bishop 2005a[M]).
DIAGNOSIS: Because the female is "cock-feathered," subspecific diagnosis should focus on female plumages. Female *horsbrughi* has a chestnut hindneck, medium-chestnut breast (not as saturated as in *furvus*), and buffy belly (Rand & Gilliard 1967, illustration in Debus 1996). Original description noted that it is smaller and darker than the nominate form; underparts a deep rufous color. A specimen from Okasa, Eastern Ra, was noted to be intermediate between this form and *giluwensis* (Diamond 1972). Bell (1984) reported it as abundant in the savanna grasslands around Waigani near Port Moresby.

b. *Turnix maculosus giluwensis* Sims

Turnix maculosa giluwensis Sims, 1954. *Bull. Brit. Orn. Club* 74: 39.—Northern slopes of Mt Giluwe, Eastern Ra.
RANGE: Mid-montane grasslands of the Eastern Ra. Localities—Tabubil[L], Wahgi valley[CO], Mt Giluwe[J], Awande[DI], Okasa[DI], and Nondugl[K] (Mayr & Gilliard 1954[K], Sims 1956[J], Gregory 1995b[L]).
DIAGNOSIS: Nearest to *furvus*; very dark dorsally; lacks chestnut collar; chin, throat, and belly whitish (Sims 1954, 1956).

c. *Turnix maculosus furvus* Parkes

Turnix maculosa furva Parkes, 1949. *Auk* 66: 84.—Gusika, 16 km n of Finschhafen, Huon Penin.
RANGE: Huon Penin (lowlands only?). Gusika, near Finschhafen (Parkes 1949). Records from Madang lowlands (Coates 1985) probably refer to this form.
DIAGNOSIS: Female lacks chestnut collar and has rich and saturated chestnut underparts (Rand & Gilliard 1967, illustration in Debus 1996). Parkes (1949) noted it is the darkest of the races. Greenway (1973) noted that the holotype is an immature female. The original subspecies epithet *furva* is adjectival and thus should agree with the masculine gender of the genus (S. L. Olson *in litt.*; see Notes below).

d. *Turnix maculosus mayri* Sutter

Turnix maculosa mayri Sutter, 1955. *Verh. Nat. Gesellschaft Basel* 66: 111.—Veina I, near Tagula I, Louisiade Arch, SE Islands.
RANGE: Tagula and adjacent Veina I of the Louisiade Arch.
DIAGNOSIS: Illustration in Debus (1996) shows this form to exhibit a whitish belly and pale chestnut breast and throat.

NOTES

The four races described for the Region are known from small sample sizes and, given the range of plumage variation related to sex and age, these races require additional study, ideally with additional fresh material. *Other names*—Black-backed Bustard Quail, Red-backed Quail, *Turnix maculosa*.

Family Glareolidae

Pratincoles and Coursers
[Family 17 spp/Region 2 spp]

The species of the family range from Spain, Africa, and Madagascar east through Asia to New Guinea and Australia (Maclean 1996). Two species visit the Region as nonbreeders—one from the north and one from the south. Handsome and demurely colored, they are somewhat swallow-like in flight. They run swiftly over their favored habitat of sandy wasteland or short grass. Pratincoles are aerial feeders and have short bills with wide gapes, long pointed wings, deeply forked tails, short legs, and four toes. Absent from the Region, coursers are terrestrial feeders and have longer bills, shorter wings and tails, long legs, and three toes. The Australian Pratincole has features of both groups and is sometimes regarded as a courser. Pratincoles are sister to the Crab-plover (Dromadidae), and the glareolid+dromadid pairing is, in turn, sister to a clade that includes the auks (Alcidae), gulls, and jaegers (Cracraft 2013).

GENUS *STILTIA*
MONOTYPIC

Stiltia G. R. Gray, 1855. *Cat. Gen. Subgen. Birds*: 111. Type, by original designation, *Glareola isabella* Vieillot.

The single species of the genus breeds in northern and central Australia; it winters to New Guinea and eastern and central Indonesia (Maclean 1996). The genus is characterized by the longish legs, absence of a facial pattern, and black flanks. It apparently forms a link between the coursers and the pratincoles (Maclean 1996). Forages for arthropods taken on ground and in the air.

Australian Pratincole *Stiltia isabella*
VISITOR | MONOTYPIC

Glareola isabella Vieillot, 1816. *Analyse*: 69.—Australasia.

A local austral migrant to the Region, appearing in the lowlands and also on highland airfields; most records from south-central and se NG (Coates 1985). Seasonally abundant near Bensbach, where thousands have been seen in the late dry season. Sometimes more than 100 are present at Jackson's Airport (Port Moresby), and up to 80 have been reported from the Lae airfield at Nadzab. Sea level–1,740 m (Coates 1985).

Localities—Kasim on Misool I[K], Waigeo I[MA], Salawati I[MA], Aru Is[MA], Utanata R[MA], Timika airport[L], Digul R[MA], Dolak I[J], Kurik[J], Wasur NP[J], Tonda WMA[J], Bensbach R[J], Bula Plains[J], Kiunga[P], Tabubil[P], Gobi on the Kikori R[Q], Mendi[CO], Tari[CO], Astrolabe Bay[MA], Goroka[CO], Mt Hagen[CO], and Port Moresby[CO] (Mees 1965[K], Bishop 2006[J], S. van Balen/P & J. Ebsworth *in litt.*[L], P. Gregory *in litt.*[P], I. Woxvold *in litt.*[Q]).

Extralimital—main breeding range is inland Australia; these birds migrate in the austral winter to Java, Borneo, and e Indonesia (Maclean 1996). Recorded from New Britain (Mayr & Diamond 2001).

Other names—Australian Courser, Long-legged or Isabelline Pratincole, *Glareola isabella*.

GENUS *GLAREOLA*
[Genus 7 spp/Region 1 sp]

Glareola Brisson, 1760. *Orn.* 1: 48; 5: 141. Type, by tautonymy, "Glareola" Brisson = *Hirundo pratincola* Linnaeus.

The species of the genus range from Europe and Africa east through tropical Asia to NG and Australia (Maclean 1996). The genus is characterized by the shortish legs, long and pointed wings, broad swallow-like bill, and demure plumage.

Oriental Pratincole *Glareola maldivarum*   VISITOR | MONOTYPIC

Glareola (Pratincola) maldivarum J. R. Forster, 1795. *Faunula Indica*, ed. 2: 11.—Open sea at the latitude of the Maldive Is, n Indian Ocean.

Scarce northern migrant in the NG region, occurring mainly between October and April (Bishop 2006). Found in areas of short grass such as airfields (Coates 1985). Most records come from the s watershed. Interior upland records exist for Woitape[CO] and near Tari[CO], with the highest record at 1,670 m[CO]. Also recorded from near Popondetta and Misima I (Coates 1990: 566). A von Rosenberg specimen came from Lakahia I of sw NG (Mayr 1941). One record from June (Coates 1985). Perhaps might be expected to be more common in the west and northwest of the Region.

Extralimital—breeds in east-central Siberia south to tropical Asia, migrating to overwinter throughout Indonesia and southeastward to Australia. A concentration of several million birds was recently discovered overwintering in nw Australia; the implications for NG are that either this species mostly passes over the region or perhaps one or more local concentrations have been overlooked. There are records of vagrants to various Melanesian islands (Dutson 2011).

Other names—Eastern or Collared Pratincole, or Large Indian Pratincole.

Family Laridae

Gulls and Terns   [Family 99 spp/Region 19 spp]

The species of the family range worldwide but are absent from the interior of northern Africa, much of the interior of South America, central Siberia, and central Australia (Burger & Gochfeld 1996, Gochfeld & Burger 1996). We follow Christidis & Boles (2008), Cracraft (2013), and many others in treating the gulls and terns as a single family. Four sub-lineages of this larger clade are recognized—the noddies, skimmers, gulls, and terns. Following Cracraft (2013), we keep the jaegers in a distinct family (Stercorariidae). The gulls are poorly represented in the Region, whereas the terns are abundant and diverse. These two groups follow the same body plan, with modifications. Overall the group is characterized by the short legs, webbed feet, shortish bill, long and pointed wings, and mainly white, gray, and black plumage. The gulls are bulker, thicker-billed, broader-winged, and larger overall. The terns are slighter, slimmer, narrow-winged, and slim billed, with very short legs. The noddies are distinctive in having reverse color patterning (dark body and pale crown). The Laridae is sister to the jaegers and the auks (Cracraft 2013).

Note that a vagrant Black-tailed Gull (*Larus crassirostris*) was reported from the Jaba River delta of Bougainville (Hadden 2004) but remains unknown in the New Guinea region; it might be expected to show up with time.

GENUS *ANOUS*   [Genus 3 spp/Region 2 spp]

Anoüs Stephens, 1826. In Shaw, *Gen. Zool.* 13(1): 139. Type, by subsequent designation (G. R. Gray, 1840. *List Gen. Birds*: 79), *Anoüs niger* Stephens = *Sterna stolida* Linnaeus.

The species of this genus range throughout the tropical seas of the world, but there are few records in the eastern tropical Pacific (Gochfeld & Burger 1996). This very distinctive genus is characterized by dark plumage contrasting with the pale crown, the long, narrow black bill, and the wedge-shaped tail.

Brown Noddy *Anous stolidus*   RESIDENT AND VISITOR

Sterna stolida Linnaeus, 1758. *Syst. Nat.*, ed. 10, 1: 137.—Jamaica.

Common in most NG marine waters, nesting on offshore islands (Coates 1985). Breeding sites in the Region include Bavo I near Port Moresby and Araltamu I of the Siassi Is (Coates 1985). Also a visitor from

Australian waters (Coates & Peckover 2001). Mostly seen offshore and in the vicinity of islets and reefs, occasionally over inshore waters of the Mainland (Coates 1985).

Extralimital—breeds on islands off Central America (Pacific and Caribbean), the tropical Atlantic, throughout the Indian Ocean, and islands of se Asia, Melanesia, Micronesia, and Polynesia (Gochfeld & Burger 1996). Can be found in tropical seas worldwide.

SUBSPECIES

a. *Anous stolidus pileatus* (Scopoli)
Sterna pileata Scopoli, 1786. *Delic. Flor. Faun. Insubr.* 2: 92.—Philippines.
RANGE: Seychelles and Madagascar east to n Australia, Polynesia, Hawai`i, and Easter I (Gochfeld & Burger 1996).
DIAGNOSIS: Forehead and crown bluish gray (Gochfeld & Burger 1996).

NOTES
Other names—Common Noddy.

Black Noddy *Anous minutus* VISITOR (AND RESIDENT?)

Anous minutus Boie, 1844. *Isis von Oken* 37: col. 188.—Raine I, n QLD, Australia.
Seasonally common over most NG offshore waters (Coates 1985); no known record of breeding within the Region, but probably does breed some place in the SE Islands. Nesting confined to tree-covered islands (Beehler *et al.* 1986). Gregory-Smith & Gregory-Smith (1988) reported seeing more than 1,000 birds on Wallai I, 6 km south of Kupiano, off the s coast of the SE Peninsula.

Extralimital—breeds in Indonesia, the Philippines, ne Australia, and Melanesia, Micronesia, and Polynesia; also breeds on islands off Central America, the Caribbean, and the tropical Atlantic to c Africa (Gochfeld & Burger 1996).

SUBSPECIES

a. *Anous minutus minutus* Boie
RANGE: Islands in Melanesia and Polynesia, from the NG region and QLD to the Tuamotu Is (Dickinson & Remsen 2013). Breeds on the Great Barrier Reef and islands in the Bismarck Arch. A sand island in the Conflict Is (between Panasesa and Irai I) had 300+ *A. minutus* and 100+ *A. stolidus* roosting (and nesting?) on it on 29-Sep-2002 (Beehler field notes).
DIAGNOSIS: Minor differences among races in plumage tone and size (Gochfeld & Burger 1996). Wing: male 212–233, female 212–225 (Higgins & Davies 1996).

NOTES
Other names—White-capped Noddy.

GENUS *GYGIS* MONOTYPIC

Gygis Wagler, 1832. *Isis von Oken* 11: col. 1223. Type, by monotypy, *Sterna candida* Gmelin.
The single species of the genus ranges through tropical oceans, breeding on tropical islands (Gochfeld & Burger 1996). The genus is characterized by the small size, all-white plumage, and distinctive upturned black bill.

White Tern *Gygis alba* VISITOR (AND RESIDENT?)

Sterna alba Sparrman, 1786. *Mus. Carls.* 1: no. 11.—Ascension I.
An oceanic visitor to n NG waters, sporadically reported from the coast but regular only on its breeding islands well north of mainland NG. No proven breeding records from Region (Coates 1985).

Extralimital—widespread through tropical warm seas (Gochfeld & Burger 1996).

SUBSPECIES

a. *Gygis alba candida* (Gmelin)
Sterna Candida Gmelin, 1789. *Syst. Nat.* 1(2): 607.—Kiritimati I, Pacific Ocean.

RANGE: Islands of the sw Pacific, from the Caroline Is east to Kiritimati I and south to Tonga and the Society Is (Peters 1934). Nearest breeding records are from island groups north of mainland NG: Wuvulu I, Ninigo Is, Sae Is, Tench I, and Nuguria Is (Coates 1985). Dumbacher *et al.* (2010) reported it near Tube Tube I in the Engineer Is of the SE Islands in late 2009. Hobcroft & Bishop (ms) reported eight birds around Lilius I of the Budibudi Is (SE Islands) in Nov-2009; the birds' behavior suggested possible breeding. Future fieldworkers should check to see whether the species is nesting there. Reported from Ayawi I[MA] in Geelvink Bay.

DIAGNOSIS: Base of bill blue-gray; primaries with dark shafts (Gochfeld & Burger 1996).

NOTES

Gochfeld & Burger (1996) apparently inadvertently interdigitated the distributional ranges of *candida* with the nominate form, which Dickinson & Remsen (2013) indicated correctly is confined to the tropical s Atlantic. *Other names*—Fairy or Atoll Tern, White Noddy, *Anous albus*.

GENUS *CHROICOCEPHALUS* [Genus 10 spp/Region 2 spp]

Chroicocephalus Eyton, 1836. *Cat. Brit. Birds*: 53. Type by subsequent designation (G. R. Gray, 1840. *List Gen. Birds*: 79), *Larus capistratus* Temminck = *Larus ridibundus* Linnaeus.

Small, slim, slender-billed gulls with or without a black head in breeding plumage. We follow Pons *et al.* (2005) and Christidis & Boles (2008) in placing the species *novaehollandiae* and *ridibundus* in *Chroicocephalus* rather than in *Larus*.

Silver Gull *Chroicocephalus novaehollandiae* RARE VISITOR

Larus Novæ-Hollandiae Stephens, 1826. In Shaw, *Gen. Zool.* 13(1): 196.—New South Wales, Australia.

A rare visitor to s NG.

Extralimital—breeds in Australia, New Caledonia, Loyalty Is, and NZ (Burger & Gochfeld 1996).

SUBSPECIES

a. *Chroicocephalus novaehollandiae novaehollandiae* (Stephens)

RANGE: Australia, straying to the NG region rarely. Localities—Lampusatu[M], Daru[MA,Q,K] and Port Moresby[L] (Mayr & Rand 1937[K], Bell 1966b[Q], Finch & Moulten 1979[L], Bishop Nov. 1983[M]). To be expected mainly on the s coast.

DIAGNOSIS: Distinguished from *forsteri* of New Caledonia by smaller size and distinct wing pattern (Johnstone 1982).

NOTES

We follow Higgins & Davies (1996) and Johnstone (1982) rather than Burger & Gochfeld (1996) for subspecific treatment. *Other names*—Red-billed Gull, *Larus novaehollandiae*.

Black-headed Gull *Chroicocephalus ridibundus* VISITOR | MONOTYPIC

Larus ridibundus Linnaeus, 1766. *Syst. Nat.*, ed. 12, 1: 225.—England.

A rare and sporadic northern migrant to Region. Records include: Bintuni Bay on 7-Apr-1989 (Erftemeijer *et al.* 1991); one immature at Biak Harbour on 29–30-Mar-1963 (King 1979); four immature and winter-plumaged birds at Jayapura Harbour on 2-Apr-1963 (King 1979); one subadult at Moitaka settling ponds, Port Moresby, 20–27-Jan-1979 (Finch 1979a); a subadult at the same place 25-Jan–15-Feb-1981 (Finch 1981a); another bird at the Moitaka settling ponds on 25-Mar-1981; and three birds at Lae airstrip on 19-Jan-1985 (Finch 1986c). Coates (1990: 566) also provided records for Misool I, Sorong, and Lae. Also 20 birds at Sorong in Jan–Feb-1986, and birds observed at Waigama, Misool I, in Jan-1986 (Bishop & Diamond 1987). Finally, Gibbs (2009) reported eight at Pelabuan, Biak I (no date).

Extralimital—Breeds widely across Eurasia, from Greenland to Kamchatka; winters south to Africa, Arabia, and tropical Asia (Burger & Gochfeld 1996); one record from Solomon Is (Dutson 2011).

Other names—Common Black-headed Gull, *Larus ridibundus*.

GENUS *GELOCHELIDON*

MONOTYPIC

Gelochelidon C. L. Brehm, 1830. *Isis von Oken* 23: col. 994. Type, by monotypy, *Gelochelidon meridionalis* C. L. Brehm = *Sterna nilotica* Gmelin.

The single species in the genus breeds in coastal North America, Central America, South America, Europe, tropical Asia, eastern China, and Australia (Gochfeld & Burger 1996). The species winters in the tropics and the southern temperate zone. The genus is characterized by the blunt and rather thick black bill, the distinctive foraging behavior (capturing prey on land in flight), the lack of a crest, and the slightly forked tail.

Gull-billed Tern *Gelochelidon nilotica*

VISITOR

Sterna nilotica Gmelin, 1789. *Syst. Nat.* 1(2): 606.—Egypt.

A common visitor to the Region from Australia (and perhaps also from China), mainly between February and September, with occasional birds present in other months (Beehler *et al.* 1986). No evidence for breeding in Region.

Localities—Dolak I[K], Komolom I[K], Lampusatu Beach[K], Merauke[J], Kurik[J], Wasur NP[K], Tonda WMA[K], Bensbach R[CO], near Daru[K], Coutance I[MA], Port Moresby[CO], and Cloudy Bay[MA] (Mees 1982b[J], Bishop 2005[K]). Frequents tidal flats, estuaries, and settling ponds in the lowlands (Coates 1985).

Extralimital—breeds in the Americas, Europe, w Asia, s Asia, e China, and Australia; winters in Central and South America, Africa, the Persian Gulf, tropical Asia, NG, Australia, and NZ (Gochfeld & Burger 1996).

SUBSPECIES

a. *Gelochelidon nilotica ?affinis* (Horsfield)

Sterna affinis Horsfield, 1821. *Trans. Linn. Soc. London* 13(1): 199.—Java.

Synonym: *Gelochelidon nilotica addenda* Mathews, 1912. *Birds Austral.* 2: 331.—China.

RANGE: Breeding in Transbaikalia, Manchuria, and e China (Fuzhou to Hainan), wintering mainly in se Asia (Gochfeld & Burger 1996). Presumably the birds that are found in n and nw NG (*e.g.*, the middle Sepik) are of this form. Ripley (1964a) identified birds he collected from Lake Sentani as *affinis*.

DIAGNOSIS: Bill small, length (from base): 42–47 (n=7). Breeding adult exhibits a mantle that is pale gray (not whitish). We follow Dickinson & Remsen (2013) in using the name *affinis* for the northern population that may winter in nw NG.

b. *Gelochelidon nilotica ?macrotarsa* (Gould)

Sterna macrotarsa Gould, 1837. *Synops. Birds Austral.* 2: pl. 27, f. 2.—Tasmania.

RANGE: Australia. Birds found along the s coast of NG are presumably of this form, as determined by Mees (1992a) and photos in Coates (1985).

DIAGNOSIS: Larger, paler, longer-billed, and longer-legged than *affinis*; bill length 47–58 (n=7). Very pale-mantled, nearly whitish (Mees 1982b, Higgins & Davies 1996).

NOTES

Del Hoyo & Collar (2014) treated *affinis* and *macrotarsa* as distinct species, but Dickinson & Remsen (2013) did not. *Other names—Sterna nilotica.*

GENUS *HYDROPROGNE*

MONOTYPIC

Hydroprogne Kaup, 1829. *Skizz. Entw.-Gesch. Eur. Thierwelt*: 91. Type, by subsequent designation (G. R. Gray, 1846. *List Gen. Birds* 3: [658]), *Sterna caspia* Pallas = *Sterna tschegrava* Lepechin.

The single species of the genus breeds in North America, Eurasia, Africa, Australia, and New Zealand, and winters in Central America, southern Europe, Africa, and Australia (Gochfeld & Burger 1996). The genus is characterized by the robust gull-like body, bulky deep-red bill, rough crest, and very long wings. Flight is powerful and graceful.

Caspian Tern *Hydroprogne caspia* VISITOR | MONOTYPIC

Sterna caspia Pallas, 1770. *Novi Comment. Acad. Sci. Imperial. Petropol.* 14(1): 582, pl. 22, fig. 2.—Caspian Sea.

A scarce visitor to s NG (Pratt & Beehler 2014); less rare in the Trans-Fly but a vagrant to Port Moresby.

Localities—Bian R[L], Lake Murray[CO], Waigani[CO], Port Moresby[CO], and Lake Iaraguma[CO] (Hoogerwerf 1964[L]).

Extralimital—breeds in North America, Europe, Africa, Asia, Australia, and NZ, wintering mainly in the subtropics and tropics but also in interior Australia (Gochfeld & Burger 1996).

NOTES

We follow Gochfeld & Burger (1996) in treating the species as monotypic. The species name *tschegrava* has been officially suppressed (ICZN Opinion 904). *Other names—Sterna caspia, Hydroprogne tschegrava.*

GENUS *THALASSEUS* [Genus 6 spp/Region 2 spp]

Thalasseus Boie, 1822. *Isis von Oken* 1: col. 563. Type, by subsequent designation, (Wagler, 1831. *Isis von Oken* 5: col. 1225), *Sterna cantiaca* Gmelin = *Sterna sandvichensis* Latham.

The species of this genus range from North and Central America, and western South America to Africa, Europe, the Middle East, tropical Asia, Melanesia, Australia, and the tropical western and central Pacific (Gochfeld & Burger 1996). The genus is characterized by the shaggy nuchal crest in breeding individuals, very long wings, long bill, and gray mantle. All species exhibit a white forehead contrasting with a black crown in nonbreeding plumage.

Crested Tern *Thalasseus bergii* RESIDENT AND VISITOR

Sterna Bergii Lichtenstein, 1823. *Verz. Doubl. Zool. Mus. Univ. Berlin*: 80.—Cape of Good Hope, South Africa.

Common along the NG coastline year-round, breeding on offshore islands in the Region (Coates 1985, Pratt & Beehler 2014). Occurs both close inshore and well away from land. One bird banded on Bramble Cay, Torres Strait, on 28 Mar-1980 was recovered at Waima Beach, near Bereina, Central Prov, in 1981 (Coates 1985). Breeds on Bramble Cay (n Torres Strait), but we have no details on breeding colonies in the Region (Coates 1985).

Extralimital—breeds in s Africa, the Red Sea, the Persian Gulf, tropical Asia, and e Australia to tropical islands of the w and c Pacific; winter range extends to adjacent waters (Gochfeld & Burger 1996).

SUBSPECIES

a. *Thalasseus bergii cristatus* (Stephens)
Sterna cristata Stephens, 1826. In Shaw, *Gen. Zool.* 13(1): 146.—China.
RANGE: Malaysia, Philippines, Ryukyu Is, Australia, north to Polynesia. Recorded throughout the Region.
DIAGNOSIS: More white on forehead and darker dorsally than the nominate; smaller and smaller-billed than *velox;* white forehead band of *thalassinus* broader (Gochfeld & Burger 1996).

NOTES

We believe the name "Crested Tern" is descriptive and felicitous, and contrasts well with the following species' name, "Lesser Crested Tern." (We see no need to always deploy both "Lesser" and "Greater" for paired species.) *Other names*—Great, Large, Swift, or Greater Crested Tern, *Sterna bergii.*

Lesser Crested Tern *Thalasseus bengalensis* VISITOR

Sterna bengalensis Lesson, 1831. *Traité Orn.* 8: 621.—Coasts of India.

A common visitor to the coast of s NG, less common elsewhere. Seasonal around Port Moresby (and presumably elsewhere), with larger numbers in October–November and February–March, as if on passage

(Coates 1985, Beehler *et al.* 1986). Mostly seen in the vicinity of islands and reefs, less commonly along Mainland shores.

Extralimital—breeds in n Africa, the Red Sea, Persian Gulf, Maldives, Wallacea, and e Australia (Gochfeld & Burger 1996). Ranges through tropical and subtropical seas from nw Africa, e Africa, and the Bay of Bengal east to n Australia and Bismarck Arch. In spite of indication on range map in Gochfeld & Burger (1996), we do not know of breeding records from the NG region

SUBSPECIES

a. *Thalasseus bengalensis torresii* Gould

Thalasseus Torresii Gould, 1843. *Proc. Zool. Soc. London* (1842): 140.—Port Essington, NT, Australia.

RANGE: Known to breed in n Australia and n Sulawesi (White & Bruce 1986); other breeding locations uncertain (see Gochfeld & Berger 1996, Dickinson & Remsen 2013). In NG occurs mainly from the Aru Is east to the vicinity of Port Moresby (Coates 1985), but has also been recorded off Karkar I and Madang, where fairly common in October (Finch 1981e, 1981g); in Finschhafen in October in flocks of up to 20 birds have been observed (Isles & Menkhorst 1976); the Trobriand Is, with a few in April (Hartley 1984). Present at Port Moresby throughout the year but status unknown, possibly only a visitor from QLD (Coates 1985).

DIAGNOSIS: Probably not distinguishable from the nominate form. Higgins & Davies (1996) noted that the distinction between the nominate form and *torresii* is slight. The latter is larger, with a deeper bill and a paler mantle. Wing 281–321 (Higgins & Davies 1996).

NOTES

Other names—Indian Lesser Crested Tern, *Sterna bengalensis*.

GENUS *STERNULA* [Genus 7 spp/Region 1 sp]

Sternula Boie, 1822. *Isis von Oken* 5: col. 563. Type, by monotypy, *Sterna minuta* Linnaeus = *S. albifrons* Pallas.

We follow Christidis & Boles (2008) and del Hoyo & Collar (2014) in use of the genus *Sternula*. Gochfeld & Burger (1996) subsumed this genus into *Sterna*. The species of this genus breed from the Americas to Europe, Asia, and Australia. Populations winter in the tropics and subtropics. The genus is characterized by small size, white forehead in all plumages, and yellow bill.

Little Tern *Sternula albifrons* VISITOR

Sterna (albifrons) Pallas, 1764. In Vroeg, *Cat. Ois., Adumbr.*: 6.—Maasland, Netherlands.

A common visitor to the Region from September to early May (most depart in March); no NG breeding records to date, though it has bred in New Britain (Coates 1985) and is expected to be a breeder in the NG region (Higgins & Davies 1996: 711–12). In the Region, frequents mostly inshore coastal waters; also occurs locally on lagoons and wide rivers in the lowlands; fairly regular at Moitaka settling ponds near Port Moresby (Coates 1985). The species was fairly common on the lagoons of the Taritatu R in April, flying over the water and feeding in flocks of 10 to 14 individuals (Rand 1942b). In s NG, 100 or more birds were sometimes found perched on exposed reefs (Rand & Gilliard 1967). Diamond & Bishop (1994) reported a first record of the species for the Aru Is, where the birds were in breeding plumage and showed behavioral signs of breeding. One banded as a nestling in Japan on 3-Jun-1975 was subsequently recovered near Orokolo, Gulf Prov, PNG (Coates 1985).

Extralimital—breeds in w Europe and c Africa, e and se Asia, and Micronesia. Also breeds in e Australia (Gochfeld & Burger 1996).

SUBSPECIES

a. *Sternula albifrons sinensis* (Gmelin)

Sterna sinensis Gmelin, 1789. *Syst. Nat.* 1(2): 608.—China.

Synonym: *Sternula placens* Gould, 1871. *Ann. Mag. Nat. Hist.* (4) 8: 192.—Torres Strait.

RANGE: Southeastern Russia, Korea, and Japan; south along the coast of e China and Indochina, and east through the Philippines, Sulawesi, and Lesser Sunda Is to the NG region, Australia, and the Bismarck Arch (Mayr 1941). Records throughout the Region. Sight record from Salawati I (M. Tarburton's island website). DIAGNOSIS: Shafts of three outer primaries white (Gochfeld & Burger 1996). We sink the Australian form *placens* based on the suggestion of Gochfeld & Burger (1996: 657). Wing 178–198 (Higgins & Davies 1996).

NOTES

Some migrants to NG may come from Australian breeding colonies. *Other names*—Least Tern, *Sterna albifrons*.

GENUS *ONYCHOPRION* [Genus 4 spp/Region 3 spp]

Onychoprion Wagler, 1832. *Isis von Oken* 2: col. 277. Type, by monotypy, *Sterna serrata = Sterna fuscata* J. R. Forster.

The breeding range of the species of this genus extends from eastern Siberia and Alaska to the tropical regions of the world. Nonbreeding birds range widely through the tropics and subtropics (Gochfeld & Burger 1996). The characters of the genus include the white forehead, black eyeline, black cap, and gray or blackish mantle. N. David (*in litt.*) noted the genus is masculine.

Grey-backed Tern *Onychoprion lunatus* VAGRANT | MONOTYPIC

Sterna lunata Peale, 1848. *U.S. Explor. Exped.* 8: 277.—Vincennes I, Paumotu Is, Polynesia.

An oceanic species known from Pacific waters north, northeast, and east of the NG region. Possibly a regular visitor to the waters off the n coast. Bailey (1992b) recorded eight individuals on three separate boat trips off Madang (in October), providing a detailed description of his observations. B. J. Coates (*in litt.*) reported seeing this species southwest of Waigeo I on 7-Jun-1990. P. Palmer provided photographs of a bird from Yimas Lake on the Karawari River, Sepik-Ramu, on 26 Jul 2015—an unexpected inland record.

Extralimital—breeds from the Marianas east to Hawai`i and south to the Austral and Tuamotu Is (Gochfeld & Burger 1996). Few records from the Bismarck and Solomon Is (Dutson 2011).

Other names— Spectacled Tern, *Sterna lunata*.

Bridled Tern *Onychoprion anaethetus* RESIDENT

Sterna Anaethetus Scopoli, 1786. *Delic. Flor. Faun. Insubr.* 2: 92.—Panay, Philippines.

Common along most NG coasts, but absent in the open Pacific (Beehler *et al.* 1986). NG records are from both the n and s coasts. Breeding recorded from Port Moresby (Coates 1985); presumably breeds on islands elsewhere in Region (*e.g.*, the SE Islands).

Extralimital—ranges to all tropical coastal seas: Central America, the Caribbean, w Africa, Madagascar and e Africa, Arabia, tropical Asia, Melanesia, and the w, n, and ne coasts of Australia (Gochfeld & Burger 1996).

SUBSPECIES

a. *Onychoprion anaethetus anaethetus* (Scopoli)
RANGE: From the s Ryukyu Is and Taiwan south through the Philippines and Sunda Is to Australia and the NG region. Localities—Misool I, Aru Is, Humboldt Bay, Manam I, Huon Gulf, Port Moresby, and Misima I (Mayr 1941).
DIAGNOSIS: Throat and cheek white, distinct from pale fawn-gray breast and belly; white forehead and eyebrow extends posteriorly beyond eye; color of shoulder darker gray-brown than that of mantle; underwing white; black of loral patch extends out to beak.

NOTES
Other names—Brown-winged Tern, *Sterna anaethetus*.

Sooty Tern *Onychoprion fuscatus*   VISITOR
Sterna fuscata Linnaeus, 1766. *Syst. Nat.*, ed. 12, 1: 228.—Hispaniola.

Found throughout waters of e NG, but no known breeding records for the Region. Pelagic; occurs mostly in deeper offshore waters beyond the outer edge of the reef (Coates 1985, Beehler *et al.* 1986). There is a record of widespread storm-driven birds from late Jun-1967 at Wapenamanda, Mt Hagen, Nomad R, Olsobip, Tabubil, and inland from Bootless Bay (Coates 1985). An adult was caught at Kiunga in Jul-2005 (P. Gregory *in litt.*). I. Woxvold (*in litt.*) captured and weighed more than 20 individuals near Wabo in the lower Purari R in Jul-2012; the birds were emaciated and apparently driven southward across the main body of NG by a large storm north of the Mainland.

Extralimital—breeds on islands throughout warm tropical and subtropical seas of the world; winters in the Pacific mainly south of Hawai`i; in the Atlantic found in a broad band between the Caribbean and w Africa; along the coast of e Africa; between the Malay Penin and w Australia; and from the Solomon Is south to the e coast of Australia (Gochfeld & Burger 1996).

SUBSPECIES

a. *Onychoprion fuscatus serratus* (Wagler)
Sterna serrata Wagler, 1830. *Natursyst. Amphib. Säug. Vogel.*: 88, note 2.—New Caledonia.
RANGE: Coasts of Australia, New Caledonia, NG, Moluccas, Admiralty Is, Bismarck Arch, and Solomon Is (Mayr 1941). Probable nearest breeding site is Bramble Cay in the northernmost Torres Strait (Coates 1985).
DIAGNOSIS: Forehead white, extending into a slight indication of a white fore-eyebrow; black teardrop links eye to base of bill; black cap distinct from dark gray mantle; shoulder of wing nearly black (darker than mantle); primaries and inner rectrices dark gray; throat and breast white; belly and undertail pale gray; bill black; underwing white.

NOTES

Dutson (2011) noted breeding records from s NG. We cannot confirm these, and note that along the NG-Australia border virtually all islands—the only place where this species can breed—are in Australian territory. *Other names—Sterna fuscata.*

GENUS *STERNA*   [Genus 13 spp/Region 3 spp]

Sterna Linnaeus, 1758. *Syst. Nat.*, ed. 10, 1: 137. Type, by tautonymy, "Sterna" = *Sterna hirundo* Linnaeus.

The species of this genus breed worldwide and winter to the tropics and subtropics (Gochfeld & Burger 1996). The genus is characterized by the long, slim bill; lack of a crest; long, narrow, pointed wings; long, deeply forked tail; and very short legs.

Roseate Tern *Sterna dougallii*   RESIDENT
Sterna Dougallii Montagu, 1813. *Orn. Dict. Suppl.*: text and pl. to Tern, Roseate.—Firth of Clyde, Scotland.

Uncommon along the coasts of NG, nesting on offshore islands; never reported from freshwater habitats (Beehler *et al.* 1986). Has bred on Bavo I near Port Moresby (Coates 1985). Records from Misool I[J], Aru Is, Merauke, Huon Gulf, D'Entrecasteaux Arch, and Port Moresby[CO] (Mayr 1941, Mees 1965[J]).

Extralimital—widespread, breeding in local colonies in North America, the Caribbean, Great Britain, islands in the Indian Ocean, e and s Africa, Ryukyu Is, se Asia, Melanesia, and Australia; winters mainly in n South America, w, s, and e Africa, and se Asia, Melanesia, and n Australia (Gochfeld & Burger 1996).

SUBSPECIES

a. *Sterna dougallii bangsi* Mathews
Sterna dougallii bangsi Mathews, 1912. *Birds Austral.* 2: 364.—Fuzhou, China.
RANGE: Breeding colonies on islands in e China, the Ryukyu Is, Philippines, Straits of Malacca, Kai Is, NG region, Solomon Is, New Caledonia (Peters 1934). Also Arabian Sea populations (Gochfeld & Burger 1996). Recorded from the Aru Is (Mayr 1941). It is not clear whether the Bavo I breeding birds are *bangsi* or *gracilis* of Australia.
DIAGNOSIS: Bill red in breeding season and black in the nonbreeding season; mantle very pale gray; cap

entirely black from lores to nape; primaries and tail feathers very pale gray; entire underparts clear white; feet red; tail streamers longer than tips of primaries, in prepared museum specimen; bill long; white of cheek extends above commissure to base of bill posterior to operculum (adult specimen from Solomon Is). The form *bangsi* is most similar to the form *gracilis*. The former differs in the following: bill marginally shorter—33–41 *vs.* 36–45; bill color not diagnostic (Higgins & Davies 1996, B. J. Coates *in litt.*).

b. ***Sterna dougallii ?gracilis*** Gould
Sterna dougallii gracilis Gould, 1845. *Proc. Zool. Soc. London*: 76.—Houtman's Abrolhos, Western Australia.
RANGE: Breeds on islands off the w and n coasts of Australia and on the e coast south to Moreton Bay (Peters 1934). This form possibly visits the sw coast of NG or NW Islands as a nonbreeder.
DIAGNOSIS: Bill slim and long; black when nonbreeding; black crown patch extends far down onto hind-nape; whitish halo adjacent to nape and separating black nape from pale gray mantle; iris hazel; feet coral red (AMNH specimen from Western Australia). Bill marginally longer than that of *bangsi* (see above).

NOTES
Subspecies delimitation needs revisiting (see Higgins & Davies 1996, Burger & Gochfeld 1996). Perhaps best treated as monotypic with individual variation.

Black-naped Tern *Sterna sumatrana*   RESIDENT
Sterna Sumatrana Raffles, 1822. *Trans. Linn. Soc. London* 13: 329.—Sumatra.
Common in offshore waters of NG, nesting on small islands; frequents outer edges of reef and deep offshore waters (Coates 1985). Locally common in the vicinity of breeding islands, otherwise rather scarce. Nesting has been recorded from early April to late December. Prefers small coral islets for its breeding colonies. Breeding localities include islands near Port Moresby and the Amphlett Is, situated in the nw sector of the SE Islands (Coates 1985).
Extralimital—breeds from the islands of the w Indian Ocean to the Maldives, Andamans, and Nicobars, east to the Ryukyus, Micronesia, Melanesia, Polynesia, and n and ne Australia. Nonmigratory (Gochfeld & Burger 1996).

NOTES
We follow the opinions of Baker (1951) and Higgins & Davies (1996) in treating this species as monotypic.

Common Tern *Sterna hirundo*   VISITOR
Sterna Hirundo Linnaeus, 1758. *Syst. Nat.*, ed. 10, 1: 137.—Sweden.
Abundant seasonally along NG's coasts (Coates 1985). The most common tern species found in the Region, it uses coastal and offshore waters of n and e NG; also found at sewage settling ponds (Coates 1985); in s NG it has not been recorded any farther west than Dolak I. Occurs mostly from August or September to late April or early May, but extreme dates of arrival and departure are difficult to determine because of the presence of numbers of nonbreeders throughout the austral winter.
Extralimital—breeds from north-central North America through Eurasia to Kamchatka, wintering south to Central and South America, coastal Africa, tropical Asia, Melanesia, and Australia (Gochfeld & Burger 1996).

SUBSPECIES
a. ***Sterna hirundo longipennis*** Nordmann
Sterna longipennis Nordmann, 1835. In Erman's *Verz. Thier. Pflanz.*: 17.—Mouth of the Kukhtuy R, Okhotsk, e Siberia.
RANGE: Breeds from ne Siberia south to ne China, wintering to se Asia, NG, Melanesia, and Australia (Gochfeld & Burger 1996).
DIAGNOSIS: Black-billed; legs red-brown to blackish brown (Gochfeld & Burger 1996). Nonbreeding birds difficult to diagnose to subspecies (Higgins & Davies 1996).

NOTES
Other names—Eastern Common Tern.

Chlidonias Rafinesque, 1822 (21-February). *Kentucky Gazette* (ns)1(8): 3, col. 5. Type, by monotypy, *Sterna melanops* Rafinesque = *Sterna surinamensis* Gmelin = *Sterna nigra* Linnaeus.

The species of this genus breed from interior North America and Europe east to the Russian far east and northeastern China. One species breeds in Australia. The northern forms winter south to Africa, tropical Asia, Australia, and New Zealand (Gochfeld & Burger 1996). The genus is characterized by small size, very short legs, and gray or blackish breast and mantle in breeding birds.

Whiskered Tern *Chlidonias hybrida* VISITOR

Sterna hybrida Pallas, 1811. *Zoogr. Rosso.-Asiat.* 2: 338.—Southern Volga R, Russia.

A common visitor in February–October from its Australian breeding grounds, with occasional stragglers in other months. Inhabits shallow lagoons and slow-flowing lowland rivers.

Extralimital—breeds in w Europe, s Africa, w Asia, n India and Tibet, and Australia. Winters in Africa, tropical Asia, and Australia (Gochfeld & Burger 1996).

SUBSPECIES

a. *Chlidonias hybrida javanicus* (Horsfield)

Sterna Javanica Horsfield, 1821. *Trans. Linn. Soc. London* 13: 198.—Java.

Synonym: *Hydrochelidon fluviatilis* Gould, 1843. *Proc. Zool. Soc. London* (1842): 140.—Interior of NSW, Australia.

RANGE: Breeds in Australia, migrating to the Moluccas, NG, and se Asia (Dickinson & Remsen 2013). Localities—Salawati I[MA], Sorong[MA], Lake Aimaru[J], the Aru Is[CO], Paniai Lake[K], Danau Bira[L], Lake Sentani[MA], Dolak I[O], Merauke[MA], Bensbach[CO], Lake Murray[M], middle Sepik R[N], Ramu R[MA], Port Moresby[CO] (Junge 1953[K], Gyldenstolpe 1955b[J], Gilliard & LeCroy 1966[N], Bishop 2005a[O], B. J. Coates *in litt.*[M], Beehler field notes 10-Jul-1987[L]).

DIAGNOSIS: Bill dark red, blackish posteriorly to nostril; feet red; belly with charcoal wash, grading to gray on breast and white on chin; narrow white cheek patch below black cap; mantle pale gray; upper surface of tail feathers pale gray; flight feathers gray and white. We follow Mees (1977b) in including *fluviatilis* in *javanicus*.

NOTES

The generic name is now considered to be masculine, but the species epithet *hybrida* is treated as a masculine noun (invariant), not an adjective (David & Gosselin 2002a). *Other names*—Marsh Tern, *Chlidonias hybridus*.

White-winged Tern *Chlidonias leucopterus* VISITOR | MONOTYPIC

Sterna leucoptera Temminck, 1815. *Man. Orn.*: 483.—Lake Geneva, Switzerland.

An uncommon northern visitor to NG (Mayr 1941, Melville 1980, Coates 1985). Mainly a transient, from October to mid-December and again from late March to mid- or late May (extreme dates 5-October and 29-May). A single bird in the company of *C. hybrida* at the Moitaka settling ponds near Port Moresby on 5-Jul-1975 showed the characters of advanced immature plumage; this record appears to be the only record for the northern summer (Coates 1985). Bailey (1992b) reported it as uncommon near shore in the month of October at Madang.

Extralimital—breeds from c and e Europe across to e China and north-central Siberia, wintering south to Africa, the Persian Gulf, tropical Asia, Australia, and NZ (Gochfeld & Burger 1996).

Other names—White-winged Black Tern.

Black Tern *Chlidonias niger* VAGRANT

Sterna nigra Linnaeus, 1758. *Syst. Nat.*, ed. 10, 1: 137.—Near Uppsala, Sweden.

Finch (1986a) reported observing a single individual in breeding plumage at the Moitaka settling ponds near Port Moresby on 18–19-May-1985. The bird was seen by six observers and well described. Subspecies not determined.

Extralimital—Breeds in n North America and Eurasia; winters to Africa and the American tropics (Dickinson & Remsen 2013).

The subspecies found in NG would presumably be the nominate form.

Family Stercorariidae

Skuas and Jaegers [Family 7 spp/Region 4 spp]

The species of the family breed in the far north and far south; in the nonbreeding season they travel the oceans of the world, wintering in the tropical, south temperate, and subantarctic oceans (Furness 1996). Only one species, *Stercorarius pomarinus*, is regularly seen in New Guinea's waters. These are powerful, pelagic, gull-like birds, of either all-dark or pied sooty-and-pale plumage. They are seen only in the open ocean, except in unusual circumstances. All are excellent fliers, specialized in stealing prey from other seabird species. The family is a sister group to the Alcidae (Paton & Baker 2006, Cracraft 2013).

GENUS *STERCORARIUS* [Genus 3 spp/Region 3 spp]

Stercorarius Brisson, 1760. *Orn.* 1: 56; 6: 149. Type, by tautonymy, "Stercorarius" Brisson = *Larus parasiticus* Linnaeus.

The species of this genus breed in the far north and spend the nonbreeding period traveling the open ocean as trans-equatorial migrants, spending the winter in the temperate and tropical waters of the Southern Hemisphere (Furness 1996). The birds of this genus are lithe and graceful but powerful predators. Note the long and pointed wings. Adults exhibit elongate and specialized central tail feathers and a blackish cap contrasting with the pale cheek. Young birds show plumage barring, not streaking. Following del Hoyo & Collar (2014) we recognize *Catharacta* as distinct from *Stercorarius* (*pace* Dickinson & Remsen 2013).

Pomarine Jaeger *Stercorarius pomarinus* VISITOR | MONOTYPIC

Lestris pomarinus Temminck, 1815. *Man. Orn.*: 514.—Arctic regions of Europe.

A fairly common, overwintering, pelagic migrant off the coast of n NG during the northern winter; less common off the s coast. Arrives along n coast in late October and early November (Coates 1970b, Melville 1980, Bailey 1992b).

Extralimital—breeds in the high Arctic of Alaska, Canada, and Russia to easternmost Siberia; winters south to the tropics and subtropics, ranging to s South America, sw Africa, and all the waters of the Australian region (Furness 1996).

Other names—Pomarine Skua.

Arctic Jaeger *Stercorarius parasiticus* VISITOR | MONOTYPIC

Larus parasiticus Linnaeus, 1758. *Syst. Nat.*, ed. 10, 1: 136.—Sweden.

A rare passage migrant from the Arctic. There are fewer than a dozen records of this species from the Region, all from the n coast off the Sepik-Ramu. Arrives in October, apparently earlier than *S. pomarinus* (Bailey 1992b). Greensmith (1975) reported seeing a total of five birds (including four pale morph adults and an immature) between Madang and Karkar I and two birds off the mouth of the Sepik R, in mid-Oct-1974. Finch (1981e, 1981g) observed one at Madang and two near Karkar I in late Oct-1981 (see Coates 1985). Coates (1990: 566) reported a sighting at sea off Port Moresby. Cheshire (2011) combined his sightings of Pomarine and Arctic Jaegers.

Extralimital—breeds in high Arctic of Alaska, Canada, Greenland, Scandanavia, and Russia to easternmost Siberia. Winters in s South America, s Africa, the waters off the s half of Australia, and NZ (Furness 1996).

NOTES

Heavily outnumbered by *S. pomarinus* in the Region. *Other names*—Parasitic Jaeger, Arctic Skua.

Long-tailed Jaeger *Stercorarius longicaudus* VAGRANT
Stercorarius longicaudus Vieillot, 1819. *Nouv. Dict. Hist. Nat.*, nouv. éd. 21: 157.—Northern Europe.

A very rare passage migrant, usually distant from land. Cheshire (2011) recorded the species twice in the Region, once in the Louisiade Is (15 km south of Adele I on 8-Nov-1993), and once in the Bismarck Sea, *ca.* 10 km south of Finschhafen on 24-Sep-1990. The Greensmith record noted in Beehler *et al.* (1986) was, in fact, from the e Solomons, not in the NG region.

Extralimital—breeds in the high Arctic, wintering in the Southern Hemisphere off both coasts of South America and w Africa, and now shown to be a regular migrant off e Australia and NZ but tends to remain farther south than the other species of this genus (Furness 1996).

NOTES

No specimens available, so subspecific identity of birds seen in the Region unknown. By distribution should be the race *pallescens*, which breeds in ne Siberia and North America. *Other names*—Long-tailed Skua.

GENUS *CATHARACTA* [Genus 4 spp/Region 1 sp]

We follow Chu *et al.* (2009) and del Hoyo & Collar (2014) in recognizing this genus as distinct from *Stercorarius* (*pace* Christidis & Boles 2008, Dickinson & Remsen 2013). The adult plumages of this genus can be distinguished from *Stercorarius* by the large size, short tail, bulky body, dark plumage without barring, and absence of elongate specialized central tail feathers. Unlike the jaeger lineage, this genus is found breeding in both the far north and far south. Clearly the two genera are closely allied, but the species of *Catharacta* are so uniform and similar that they merit recognition as a distinct genus. These are large and bulky predators with plumage superficially similar to that of a dark first-year gull but exhibiting a prominent white flash at the base of the primaries.

South Polar Skua *Catharacta maccormicki* VAGRANT | MONOTYPIC
Stercorarius maccormicki H. Saunders, 1893. *Bull. Brit. Orn. Club* 3: 12.—Possession I, Victoria Land, Antarctica.

Sight records from the Coral, Solomon, and Bismarck Seas (Cheshire 2011). The Cheshire records include: "large dark skua," 37 km south of Orangerie Bay on 6-Feb-2000; "positive *maccormicki*," Huon Gulf on 4-Mar-2002; "positive *maccormicki*," 5.5 km north of Bam I on 11-Mar-2002. The latter two were of the pale morph.

Extralimital—breeds on the Antarctic mainland and winters in the Northern Hemisphere, with passage migrants crossing through the Region annually.

Other names—*Stercorarius maccormicki*.

ORDER ACCIPITRIFORMES

The Accipitriformes includes the New World vultures (Cathartidae), Secretarybird (Sagittariidae), Osprey, and the typical diurnal raptors (Cracraft 2013). The work of Hackett *et mult.* (2008) showed that the falcons (Falconiformes) are sister to the parrots and the passerines and distant from the Accipitriformes.

Family Pandionidae

Ospreys MONOTYPIC

We follow the molecular treatment of Wink & Sauer-Gürth (2004) and G. Mayr (2009) in the full familial treatment for the *Pandion* lineage, a treatment also recommended by Debus (1994a) and used by Dickinson & Remsen (2013) and del Hoyo & Collar (2014). G. Mayr (2009) noted pandionid fossils back to the Eocene. Unique family characters include the reversible outer toe and the spiny foot pads (Poole 1994). The Pandionidae is a sister lineage to Accipitridae, the true hawks (Cracraft 2013, Barrowclough *et al.* 2014).

GENUS PANDION MONOTYPIC

Pandion Savigny, 1809. *Descr. Égypte, Hist. Nat.* 1: 69, 95. Type, by monotypy, *Pandion fluviatilis* Savigny = *Falco haliaetus* Linnaeus.

The single species of the genus exhibits a nearly worldwide distribution, though absent from dry continental interiors and desert regions generally (del Hoyo & Collar 2014). The genus is characterized by the eagle-like size, long and narrow but round-tipped wings held in a diagnostic crooked shape, and distinctive hovering while searching for fish.

Osprey *Pandion haliaetus* RESIDENT

Falco Haliaetus Linnaeus, 1758. *Syst. Nat.*, ed. 10: 91.—Sweden.

In small numbers along the Region's coastlines, estuaries, and satellite islands; occasionally found on lowland lakes and rivers (Beehler *et al.* 1986).

Extralimital—nearly worldwide (Stresemann and Amadon 1979, Poole 1994).

SUBSPECIES

a. *Pandion haliaetus cristatus* (Vieillot)

Buteo cristatus Vieillot, 1816. *Nouv. Dict. Hist. Nat.* 4: 481.—New South Wales, Australia.

Synonym: *Pandion haliaetus melvillensis* Mathews, 1912. *Austr. Av. Rec.* 1: 34.—Melville I, NT, Australia.

RANGE: In small numbers along the Region's coastlines, estuaries, and satellite islands; occasionally found on lowland lakes and rivers (Beehler *et al.* 1986). Once observed in the hills ne of Port Moresby at Sirinumu Dam at 540 m (Coates 1985). Also Java and Sulawesi e to Australia, Tasmania, Bismarcks, Solomons, and New Caledonia (Dutson 2011).

DIAGNOSIS: A small form, with a dark breast band and relatively pale crown (Poole 1994).

NOTES

Species-level treatment follows Mayr (1941), Poole (1994), Dickinson & Remsen (2013), and del Hoyo & Collar (2014), rather than Wink *et al.* (2004) and Christidis & Boles (2008), who treated the Australasian form as a distinct species, *P. cristatus. Other names*—Fish Hawk, Eastern Osprey; treated by some authorities as *Pandion cristatus*.

Family Accipitridae

Hawks, Eagles, and Allies
[Family 240 spp/Region 22 spp]

The species of the family exhibit a nearly worldwide distribution. Their body size ranges from smallish to very large. The family comprises the elanine kites, gypaetine vultures, pernine kites, and accipitrine hawks. New Guinea's 24 species can be categorized into several ecological groupings—the goshawks, the kites, the eagles, the buzzards, and the harriers. The goshawks are short-winged, long-tailed bird-eating hawks that often hunt on the wing. The kites are slightly-built open-country birds that consume a wide range of lesser prey and carrion. The buzzards are broad-winged soaring birds of open country. The eagles are a diverse assemblage of forest-dwelling and open-country species that are very large, long-winged, and powerful and often majestic predators. Finally, the harriers are a compact lineage of kite-like open-country species that specialize in preying upon animals inhabiting grasslands and marshes. We adapt a generic sequence primarily from Barrowclough *et al.* (2014), with input from Lerner & Mindell (2005) and Christidis & Boles (2008). Hackett *et mult.* (2008) situated the true hawks as sister to the New World vultures, and that pair as sister to a huge and diverse lineage that includes the owls, rollers, kingfishers, and woodpeckers.

GENUS *ELANUS*
[Genus 4 spp/Region 1 sp]

Elanus Savigny, 1809. *Descr. Égypte, Hist. Nat.* 1: 69, 97. Type, by monotypy, *Elanus caesius* Savigny = *Falco caeruleus* Desfontaines.

The species of this genus range through the Americas, Africa, tropical Asia, New Guinea, and Australia (Thiollay 1994). This very compact genus is characterized by the light frame, unstreaked white-and-gray plumage, black shoulder patch, and distinctive flight with hovering habit. The Black-shouldered Kite (*Elanus axillaris*) has been recorded from Cape York Peninsula, Queensland, and Moa Islands of the Torres Strait and might be expected from southern New Guinea as a vagrant (Draffan *et al.* 1983).

Black-winged Kite *Elanus caeruleus*
RESIDENT

Falco caeruleus Desfontaines, 1789. *Hist. Acad. Roy. Sci., Paris* (1787): 503, pl. 15.—Algiers.

Inhabits the Trans-Fly lowlands west to Dolak I[L], the grassland valleys in the Eastern Ra, the Sepik R, Ramu R, Markham R, and Bulolo R, sea level–3,200 m[J]; no records from the w third of NG (Bell 1968c, see Mees 1982b, Germi *et al.* 2013[L], Sam & Koane 2014[J]).

Extralimital—Spain, Africa, s Asia, se Asia. Mees (1982b) provides a detailed review of the genus in regard to the Australasian forms and their occurrence in NG.

SUBSPECIES

a. *Elanus caeruleus hypoleucus* Gould
Elanus hypoleucus Gould, 1859. *Proc. Zool. Soc. London*: 127.—Near Macassar, Sulawesi.

> Synonym: *Elanus caeruleus wahgiensis* Mayr & Gilliard, 1954. *Bull. Amer. Mus. Nat. Hist.* 103: 332.—Nondugl, Wahgi valley, Eastern Ra.

RANGE: Sumatra east to the Philippines, Sulawesi, and c and e NG (Pratt & Beehler 2014). Localities—Kurik[J], Bensbach R[CO], Daru[J], Koroba[CO], Tari[CO], Pureni[J], Laiagam[CO], Pagwi[J], Wahgi valley[J], Baiyer R[W], Nondugl[J], Banz[J], Kup[J], Togoba[J], Goroka[J], Lufa[J], Astrolabe Bay[L], Wawin[J], Ramu valley[L], Markham valley[CO], Lae[J], Bulolo[J], and Wau[J] (Bell 1968c[J], Mees 1982b[J], S. F. Bailey *in litt.*[L]).

DIAGNOSIS: Upperparts medium gray; undersurface of primaries dark, intermediate, or whitish; underwing patch mainly absent; size large (Mees 1982b).

First recorded in the NG region in 1950 (Wood 1970). We follow Mees (1982b) in subsuming *wahgiensis* into *hypoleucus*. *Other names*—Common Black-shouldered Kite.

GENUS *AVICEDA* [Genus 5 spp/Region 1 sp]

Aviceda Swainson, 1836. *Nat. Hist. Class. Birds* 1: 300. Type, by subsequent designation (Swainson, 1837. *Nat. Hist. Class. Birds* 2: 214), *Aviceda cuculoides* Swainson.

The species of this genus range from tropical Africa and Madagascar to tropical Asia, New Guinea, the Solomon Islands, and northern and northeastern Australia (Thiollay 1994). The genus is characterized by the compact build, the presence of a long and pointed nuchal crest, broad and rounded wings, barring on flanks, and one or more bars on the tail. The African Cuckoo Hawk (*Aviceda cuculoides*) exhibits a plumage remarkably similar to that of *A. subcristata*. We suggest that the westernmost and easternmost vicariants of this small genus exhibit a shared suite of primitive plumage characters.

Pacific Baza *Aviceda subcristata* RESIDENT

Lepidogenys subcristatus Gould, 1838. *Synops. Birds Austral.* 3: pl. [46] and text.—New South Wales, Australia.

Inhabits forest, edge, and scrub through NG and many major satellite islands, sea level–1,700 m[CO].

Extralimital—Lesser Sundas, Moluccas, and Kai Is; n and e Australia; and Bismarck and Solomon Is (Mayr & Diamond 2001, Dutson 2011, Coates & Bishop 1987).

SUBSPECIES

a. *Aviceda subcristata waigeuensis* Mayr
Aviceda subcristata waigeuensis Mayr, 1940. *Amer. Mus. Novit.* 1056: 8.—Waigeo I, NW Islands.
RANGE: Waigeo I.
DIAGNOSIS: As for *stenozona* but larger; wing (female) 314–319 (Rand & Gilliard 1967); male with blackish ventral bars, female with brownish bars.

b. *Aviceda subcristata stenozona* (G. R. Gray)
Baza stenozona G. R. Gray, 1858. *Proc. Zool. Soc. London*: 169.—Aru Is.
RANGE: Salawati[MA], Misool[J], and Batanta I[K]; Aru Is[MA]; also w NG—east in northern watershed to Geelvink Bay, in south to northwest of Port Moresby[MA] (Mees 1965[J], Gyldenstolpe 1955b[K]). Kofiau I birds (Diamond *et al.* 2009) are presumably of this form.
DIAGNOSIS: Wing: male 290–303, female 296–314; barring of underparts brownish black (Rand & Gilliard 1967).

c. *Aviceda subcristata obscura* Junge
Aviceda subcristata obscura Junge, 1956. *Zool. Meded. Leiden* 34: 231.—Biak I, Bay Islands.
RANGE: Biak I.
DIAGNOSIS: Similar to *stenozona* but smaller; wing: male 278–294, female 287–300; upperparts darker and barring on underwing more pronounced (Junge 1956).

d. *Aviceda subcristata megala* (Stresemann)
Baza subcristata megala Stresemann, 1913. *Novit. Zool.* 20: 305 (in key), 307.—Fergusson I, D'Entrecasteaux Arch, SE Islands.
RANGE: The NW Lowlands, Sepik-Ramu, Huon Penin, and n watershed of the SE Peninsula to the Kumusi R; s watershed of SE Peninsula from Port Moresby eastward; also Yapen[MA], Karkar[J], Goodenough[MA], Fergusson[MA], and Normanby I[K] (Diamond & LeCroy 1979[J], LeCroy *et al.* 1984[K]).
DIAGNOSIS: Larger than *stenozona*; wing: male 298–316, female 314–334 (Rand & Gilliard 1967); dark ventral barring broad and complete.

These are thinly defined subspecies that may best be reduced in the future, after additional museum analysis, plus the collection of additional fresh material. *Other names*—Crested Lizard Hawk, Crested Hawk, Crested Baza.

GENUS *HENICOPERNIS* [Genus 2 spp/Region 1 sp]

Henicopernis G. R. Gray, 1859. *Proc. Zool. Soc. London*: 153. Type, by monotypy, *Falco longicauda* Lesson & Garnot.

The species of this genus range from the New Guinea region to New Britain (Mayr & Diamond 2001). The genus is characterized by the long, rounded wings with prominently "fingered" primaries in soaring flight, profusely barred wings and tail, and yellow iris. It apparently belongs to an Australasian lineage of pernine kites that also includes the Square-tailed Kite (*Lophoictinia isura*) and the Black-breasted Buzzard (*Hamirostra melanosternon*; Barrowclough *et al.* 2014). *H. melanosternon* has been recorded from the southwestern islands of the Torres Strait (Draffan *et al.* 1983) and could be expected as a vagrant to southern New Guinea.

Long-tailed Buzzard *Henicopernis longicauda*   ENDEMIC | MONOTYPIC
Falco longicauda Lesson & Garnot, 1828. *Voy. Coquille, Atlas*: pl. 10; 1829. *Zool.* 1: 588.—Manokwari, Bird's Head.

> Synonym: *Henicopernis longicauda fraterculus* Stresemann & Paludan, 1932. *Novit. Zool.* 38: 239.—Serui, Yapen I, Bay Islands.
> Synonym: *Henicopernis longicauda minimus* Junge, 1937. *Nova Guinea* (ns) 1: 150.—Wokan, Aru Is.

Inhabits the forests of the Region, including the Trans-Fly[J], the NW Islands (Waigeo[MA], Batanta[W], Salawati[MA], and Misool[W]), Yapen I[MA], Biak I[MA], Aru Is[MA], and Goodenough[K] and Fergusson[L] I; sea level–3,000 m (Ripley 1964a[W], Germi *et al.* 2013[J], Beehler field notes 27-Mar-1976[K]; Pratt field notes Feb-2003[L]). Also a probable observation from Normanby (Pratt field notes Nov-2003).

Debus (1994a) has synonymized the three described forms, leaving the species monotypic. In support of that, Greenway (1973) noted that both Mayr (1941) and Rand & Gilliard (1967) doubted the validity of the race *fraterculus*, because of minimal size difference and plumage distinction that may relate to foxing of specimens. *Henicopernis* is masculine (David & Gosselin 2002b) but *longicauda* is invariant (a noun in apposition). We do not support use of the group name "honey buzzard" here, because *Henicopernis* is not a member of that lineage and looks no more like a honey buzzard (*Pernis*) than its sister taxa, the Square-tailed and Black-breasted Kites (Barrowclough *et al.* 2014). *Other names*—Long-tailed Honey Buzzard.

GENUS *MACHEIRAMPHUS*   MONOTYPIC

Macheiramphus Bonaparte, 1850. *Rev. et Mag. Zool., Paris* (2)2: 482. Type, by monotypy, *Macheiramphus alcinus* Bonaparte.

The single species of the genus ranges through sub-Saharan Africa, Madagascar, Indonesia, Malaysia, and New Guinea (Thiollay 1994). The genus is characterized by the dark unbarred plumage, long pointed wings, and rough nuchal crest. Deignan (1960) and Brooke & Clancy (1981) showed that the older name of Bonaparte should be retained (*pace* Stresemann & Amadon 1979). The molecular systematic analysis of Barrowclough *et al.* (2014) indicated this genus is sister to the harpy eagle clade.

Bat Hawk *Macheiramphus alcinus*   RESIDENT
Macheiramphus alcinus Bonaparte, 1850. *Rev. et Mag. Zool., Paris* (2)2: 482.—Malacca, se Malay Penin.

Sparsely distributed throughout the Mainland, inhabiting clearings in lowlands and foothill forest; most records from e NG; lowlands to 1,160 m. Rare throughout.

> Extralimital—sub-Saharan Africa, Madagascar, Malaysia, and Indonesia.

a. *Macheiramphus alcinus papuanus* Mayr

Machaerhamphus alcinus papuanus Mayr, 1940. *Amer. Mus. Novit.* 1091: 1.—Kumusi R, n watershed, SE Peninsula.

RANGE: Throughout the Mainland. No insular records. Recent records from w NG are presumed to be of this subspecies. Localities—Kwau village in the Arfak Mts[K], Urisa in Arguni Bay[C], Kobakma village n of Wamena[J], Agats in the Asmat[K,Y], Tiri R in the Mamberamo basin[J], Vanimo[Z], middle Fly R[L], Lake Kutubu[B], Rawlinson Ra in the Huon Penin[A], lower Markham R near Lae[X], Lakekamu basin[W], lower Waria R valley[D], Kumusi R[CO], Veimauri R[CO], Hall Sound[MA], Brown R[CO], Laloki R[CO], Sogeri Plateau[CO], Kumusi R[MA], and Pongani R[CO] (Eastwood 1994[A], Beehler *et al.* 1994[W], Coates & Peckover 2001[Z], Bishop 1987[X], Simpson 1990[L], van Balen *et al.* 2003b[J], van Balen *et al.* 2011[K], Dawson *et al.* 2011[D], Germi *et al.* 2013: Table 2[Y], K. D. Bishop *in litt.*[B], C. Thébaud & B. Mila *in litt.*[C]).

DIAGNOSIS: This subspecies apparently exhibits a reduced crest (Mayr 1940). The NG population is marginally distinct from the adjacent se Asia population (*M. a. alcinus*) in exhibiting considerable white in nuchal crest and on abdomen. As *M. a. papuanus* was apparently based on a single specimen, this is not a surprise. Museum study skins of Asian birds show considerable variation based on age, plumage, and specimen preparation, which encompass the identified differences noted by Mayr. Further study will probably lead to sinking this form.

NOTES

Rarity in w NG is surely a product of incomplete sampling. Undoubtedly the species is thin on the ground but continuously distributed from NG to Malaysia. *Other names*—Bat Falcon, Bat Kite, Bat-eating Buzzard, *Machaerhamphus alcinus*.

GENUS *HARPYOPSIS*   ENDEMIC | MONOTYPIC

Harpyopsis Salvadori, 1875. *Ann. Mus. Civ. Genova* (1876) 7: 682. Type, by monotypy, *Harpyopsis novaeguineae* Salvadori.

The single species of the genus is restricted to the New Guinea mainland (Coates 1985, Pratt & Beehler 2014). The genus is characterized by the forest-interior habit, the owl-like dark facial disks surrounding the large eyes, the short nape ruff, the unpatterned plumage, the obsolete barring of the tail, and the long tarsus. Morphologically, it is very distinct from any other forest raptor in New Guinea. The genus groups with the neotropical *Harpia* and *Morphnus* (Barrowclough *et al.* 2014).

New Guinea Harpy-Eagle *Harpyopsis novaeguineae*   ENDEMIC | MONOTYPIC

Harpyopsis novaeguineae Salvadori, 1876. *Ann. Mus. Civ. Genova* 7: 682.—Andai, Bird's Head.

A scarce but widespread inhabitant of forest interior throughout mainland NG, from the coast to at least 3,200 m (Coates and Peckover 2001). Also apparently in very low numbers within monsoon forest in the Trans-Fly (Mayr & Rand 1937, Bishop 2005a). Has been recorded in forest that had been heavily cutover *ca.* 10 years before the observation (I. Woxvold *in litt.*). Presence is announced by low-pitched and far-carrying crepuscular or nocturnal vocalizations.

NOTES

Currently classified by IUCN as Vulnerable. The species is extirpated from forest that is heavily hunted, but remains in tracts of selectively logged forest that are lightly or never hunted. Can be heard calling from villages where local hunting bans are in place. We propose the species might better be classified as Near Threatened at this time, given the abundance of undisturbed and unhunted forest available for the species, in combination with its very widespread distribution across the Mainland. We conserve the traditional English name universally used in PNG, which properly reflects a close relationship to the harpy eagle clade (Barrowclough *et al.* 2014). *Other names*—New Guinea, Papuan, or Kapul Eagle; Papuan Harpy Eagle.

GENUS *HIERAAETUS* [Genus 5 spp/Region 1 sp]

Hieraaetus Kaup, 1844. *Class. Säugethiere Vögel*: 120. Type, by original designation, *Falco pennatus* Gmelin.

The species of this genus range from Europe and Africa to tropical Asia, New Guinea, and Australia (Debus 1994a). These are small eagles with the look of a buzzard (*Buteo*) when soaring that are characterized by the well-feathered tarsus, pale ventral plumage, and terminal dark tail band. The genus subsumes considerable variation and hence is difficult to characterize generically. Treatment of the single species in New Guinea as a member of *Hieraaetus* acknowledges the morphological distinctions that differentiate the smaller forms from *Aquila* (*fide* R. Schodde *in litt.*), which is apparently sister to *Hieraaetus* (Barrowclough *et al.* 2014). Generic treatment follows Debus (1994a) and the advice of E. C. Dickinson and T. Katzner (*in litt.*) in keeping this species in *Hieraaetus* rather than combining it in *Aquila*.

Pygmy Eagle *Hieraaetus weiskei* MONOTYPIC

Eutolmaetus weiskei Reichenow, 1900. *Orn. Monatsber.* 8: 185.—Astrolabe Ra, 915 m, SE Peninsula.

Sparsely distributed throughout the forests of NG, including the mts of the Bird's Head[W], Central Ra[W,J], and Foja[K], Adelbert[L], and Huon Mts[MA], sea level–1,950 m (Gilliard & LeCroy 1967[L], Mack & Wright 1996[J], Gjershaug *et al.* 2009[W], Beehler *et al.* 2012[K]). Possible sighting on Batanta I (M. Tarburton's island website).

Extralimital—sightings from Halmahera and Seram in the Moluccas; status there is unclear (Coates & Bishop 1997, Gjershaug *et al.* 2009).

NOTES
Our treatment follows the work of Gjershaug *et al.* (2009), who provided compelling morphological and molecular evidence that the Australian and Papuan populations are specifically distinct. *Other names*— earlier authorities treated as Little Eagle, *Hieraaetus morphnoides*.

GENUS *AQUILA* [Genus 11 spp/Region 2 spp]

Aquila Brisson, 1760. *Orn.* 1: 28, 419. Type, by tautonymy, "Aquila" Brisson = *Falco chrysaetos* Linnaeus.

The genus exhibits a nearly cosmopolitan distribution (Debus 1994a) and is characterized by the huge size, very long and broad wings, generally unpatterned dark plumage, powerful talons, feathered tarsus, and prominent yellow cere.

Gurney's Eagle *Aquila gurneyi* RESIDENT | MONOTYPIC

Aquila (?Heteropus) gurneyi G. R. Gray, 1861. *Proc. Zool. Soc. London* (1860): 342, pl. 169.—Bacan I, n Moluccas.

A scarce inhabitant of lowland and hill forest through NG, the NW Islands, Aru Is, Yapen and Biak I, and D'Entrecasteaux Arch, ranging primarily within 100 km of the coast (Mayr 1941, Melville 1980); sea level–1,000 m, max 2,970 m (Coates & Peckover 2001).

Localities—Gag[A], Salawati[MA], Misool[MA,Z], Batanta[B,K], and Waigeo I[MA], Sorong[MA], Aru Is[C], Biak I[D], Yapen I[MA], lower Fly R[CO], Gaima[D], Daru[W], PNG North Coastal Ra[L], between Koroba and Lake Kopiago, upper Hegigio R[J], Tari gap[F], Wahgi valley, Banz[O], Purari basin[J], Adelbert Mts[X], sw of Madang[K], Crater Mt[N], n coast of Huon Penin, Huon Gulf[MA], lower Watut R[J], Lakekamu basin[G], lower Waria R[H], Kokoda[Q], Varirata NP[G], and Goodenough[MA], Fergusson[E], and Normanby I[E] (Mees 1965[Z], 1972[A], Greenway 1966[B], Melville 1980[D], Diamond 1985[L], Mackay 1988[O], Diamond & Bishop 1994[C], Frith & Frith 1992[F], Mack & Wright 1996[N], Dawson *et al.* 2011[H], S. F. Bailey *in litt.* 17-Jul-1994[K], I. Woxvold *in litt.*[J], B. J. Coates *in litt.*[W], Pratt field notes 2003[E], Pratt field notes 1974[X], 2014[Q] Beehler field notes various[G]).

Extralimital—inhabits the Moluccas (White & Bruce 1986, Debus 1994a). Also a record from Boigu I of the n Torres Strait (B. J. Coates *in litt.*).

Other names—*Spizaetus gurneyi*.

Wedge-tailed Eagle *Aquila audax*

Vultur audax Latham, 1801. *Ind. Orn.*, Suppl.: ii.—New South Wales, Australia.
Inhabits the savanna of the Trans-Fly (Bishop 2005a).
Extralimital—Australia.

SUBSPECIES

a. *Aquila audax audax* (Latham)
RANGE: The Trans-Fly of NG and mainland Australia. Localities—Dolak I[J], Kurik[K], Merauke[K], Bian Lakes[J] Wasur NP[J], Bensbach[CO], Wassi Kussa R[J], Gaima[L], mouth of Fly R[J], and Oriomo R[CO] (Rand 1942a[L], Hoogerwerf 1962[K], Bishop 2005a[J]).
DIAGNOSIS: Claws and feet less heavy than those of *fleayi* of Tasmania (Debus 1994a).

NOTES
We follow Schodde *et al.* (2010) in taking 1801 as the date of original publication of the species name, rather than Browning & Monroe (1991), who argued for using 1802 for the Latham publication date. *Other names—Uroaetus audax.*

GENUS *MILVUS*   [Genus 2 spp/Region 1 sp]

Milvus Lacépède, 1799. *Tabl. Méthod. Mamm. Oiseaux*: 4. Type, by tautonymy, *Falco milvus* Linnaeus.
The species of this genus range over the entire Old World but for the driest interior regions of central Asia, Saharan Africa, and the far north of Asia (Thiollay 1994). The genus is characterized by the light body frame, mottled brown plumage, forked tail, floppy flight, and scavenging habit.

Black Kite *Milvus migrans*

Falco migrans Boddaert, 1783. *Tabl. Planch. Enlum.*: 28.—France.
Restricted to c and e NG and the three main D'Entrecasteaux islands; sea level–2,700[J] m (B. J. Coates *in litt.*[J]). Frequents mostly grasslands, roadsides, and other open habitats.
Extralimital—range of the species encompasses nearly the entire Old World but for the drier continental interiors and the n high latitudes; also Australia (Debus 1994a).

SUBSPECIES

a. *Milvus migrans affinis* Gould
Milvus affinis Gould, 1838. *Synops. Birds Austral.* 3: pl. [47, fig. 1 and text].—New South Wales, Australia.
RANGE: The Lesser Sundas, Sulawesi, c and e NG, Goodenough[CO], Fergusson[MA], and Normanby I[K]; also New Britain as vagrant, New Caledonia, and Australia (LeCroy *et al.* 1984[K], Debus 1994a, Coates & Peckover 2001). Apparently absent from the Trans-Fly (Bishop 2005a). Occurs west in the n watershed to Jayapura[CO]; west in the s watershed to Yule I[CO]; and west in the highlands to Tomba Pass of Enga Prov[W], and the lower mts of Ok Tedi[M] (B. J. Coates *in litt.*[W], Bell 1969[M]).
DIAGNOSIS: The pale throat has fine dark streaking, which continues down to the undertail; breast and belly dull red-brown; crown and nape abundantly marked with fine dark streaking; indication of a small dark eyebrow; cheek pale with streaking; back dull dark brown, unstreaked; tail dark with obscure barring and terminal pale tipping; wing coverts form a pale stripe on shoulder.

NOTES
Apparently the species is expanding its range in NG, probably as a result of the growing network of roads; roadside openings and airports serve as favored habitat for the species, which commonly feeds on the carcasses of road-killed Cane Toads. It would be useful to monitor the westward expansion of the range in both the n and s watersheds. *Other names—*Fork-tailed, Black-eared, or Pariah Kite.

GENUS *HALIASTUR* [Genus 2 spp/Region 2 spp]

Haliastur Selby, 1840. *Cat. Gen. Sub-gen. Types Class Aves*: 2 (note), 3. Type, by original designation, *Haliastur pondicerianus* Gmelin = *Falco indus* Boddaert.

The species of this genus range through tropical Asia, eastern China, New Guinea, Melanesia, and Australia (Thiollay 1994). The genus is characterized by the unbarred and rounded tail, and the long, rounded wings with dark tipping contrasting with paler inner remiges.

Whistling Kite *Haliastur sphenurus* RESIDENT | MONOTYPIC
Milvus sphenurus Vieillot, 1818. *Nouv. Dict. Hist. Nat.*, nouv. éd. 20: 564.—New South Wales, Australia.

Inhabits open country and riverine grasslands of the lowlands of NG, to 700 m[CO]; west in the n watershed to Geelvink Bay[J]; and west in the s watershed at least to Timika[K,L]. Also Goodenough I[CO] (Germi et al. 2013[K], S. F. Bailey *in litt.*[J], J. Pap *in litt.*[L]). Apparently absent from the Bird's Head.

Extralimital—Australia and New Caledonia (Dutson 2011).

Other names—Whistling Eagle.

Brahminy Kite *Haliastur indus* RESIDENT
Falco indus Boddaert, 1783. *Tabl. Planch. Enlum.*: 25.—Pondicherry, India.

Forest edge, riparian forest, and nonforest habitats of all NG and its larger satellite islands; sea level–1,000 m, records to 2,700 m in disturbed habitats (Coates & Peckover 2001). The most widespread and commonplace raptor in NG.

Extralimital—tropical Asia to the Bismarcks, Solomons, and coastal n and e Australia (Debus 1994a, Mayr & Diamond 2001).

SUBSPECIES

a. *Haliastur indus girrenera* (Vieillot)
Haliaetus girrenera Vieillot, 1822. *Gal. Ois.* 1: 31, pl. 10.—New South Wales, Australia.
RANGE: The Moluccas, NG and its satellite islands, the Bismarck Arch, and n and e Australia south to n NSW (Mayr 1941, Diamond & LeCroy 1979). Apparently no records from the Trobriand Is (Coates 1985), but should be looked for there.
DIAGNOSIS: Adult exhibits a white head, upper back, throat, and breast; also white tipping to inner tail feathers; mantle dark red-brown; tail and secondaries paler red-brown than mantle; patch on upperwings dark like mantle; primaries dark blackish brown; iris dark brown.

NOTES
Other names—Red-backed Kite, Red-backed or White-headed Sea Eagle, Whistling Eagle.

GENUS *HALIAEETUS* [Genus 8 spp/Region 1 sp]

Haliaeetus Savigny, 1809. *Descr. Égypte, Hist. Nat.* 1: 68, 85. Type, by monotypy, *Haliaeetus nisus* Savigny = *Falco albicilla* Linnaeus.

The species of this genus range through North America, Eurasia, sub-Saharan Africa, Madagascar, New Guinea, the Bismarck Archipelago, and Australia (Thiollay 1994). The genus is characterized by majestic size, very long and broad wings, and white head or tail (or both).

White-bellied Sea-Eagle *Haliaeetus leucogaster* RESIDENT | MONOTYPIC
Falco leucogaster Gmelin, 1788. *Syst. Nat.* 1(2): 257.—New South Wales, Australia.

Inhabits NG's coasts, estuaries, and wetlands associated with large inland waterways as far as the upper Sepik[J] and Taritatu R[K]; also satellite islands—Waigeo I[MA], Aru Is[MA], Biak I[W], Schildpad Is[MA], Karkar I[L], and SE Islands[CO] (Rand 1942b[K], Diamond & LeCroy 1979[L], Bishop ms[W], I. Woxvold *in litt.*[J]); a pair nested at 540 m at Sirinumu Dam (Hunt 1970). Recorded at Kiunga on the Fly R and the Budibudi Is (Hobcroft & Bishop ms).

Extralimital—tropical Asia to the Bismarcks and Australia (Debus 1994a).
Other names—White-breasted Sea-Eagle, White-breasted Fish-eagle.

GENUS *BUTASTUR*   [Genus 4 spp/Region 1 sp]

Butastur Hodgson, 1843. *Journ. Asiat. Soc. Bengal* 12: 311. Type, by original designation, *Circus teesa* Franklin.

The species of this genus breed from equatorial Africa to India, Manchuria and Japan, and winter to tropical Asia and western New Guinea (Debus 1994a). This small genus is characterized by the compact shape, long wings, and uniform, unpatterned, brownish-gray upperparts. The tail has one or more bars, evident on the dorsal surface.

Grey-faced Buzzard *Butastur indicus*   VISITOR | MONOTYPIC
Falco indicus Gmelin, 1788. *Syst. Nat.* 1(2): 264.—Java.

An uncommon and little-known nonbreeding migrant arriving during the northern winter in the NW Islands of Salawati, Waigeo, Saonek, and Efman (Mayr 1941), with one sighting from the Trans-Fly (Wasur NP, Nov-1983, K. D. Bishop *in litt.*). Most likely to be seen at forest edge, open habitats, or migrating (possibly in groups).

Extralimital—breeds in ne Asia (Japan, Korea, ne China, and the Khabarovsk region of e Russia), migrating south in winter to se China, Indochina, Philippines, and Indonesia.

NOTES
Very poorly known in the Region and requires further attention to better determine its status. *Other names*—Frog Hawk (a name also applied to *Accipiter soloensis*).

GENUS *MEGATRIORCHIS*   ENDEMIC | MONOTYPIC

Megatriorchis Salvadori & D'Albertis, 1876. *Ann. Mus. Civ. Genova* 7: 805. Type, by monotypy, *Megatriorchis doriae* Salvadori & D'Albertis.

The single species of the genus is endemic to lowland and hill forest across New Guinea and Batanta I (Debus 1994a). The genus is characterized by the very long and profusely barred tail, short rounded wings, long tarsus, and a distinctive black ocular stripe. Recognition of *Megatriorchis* follows Brown & Amadon (1968) and Debus (1994a). Stresemann & Amadon (1979) placed the species in the genus *Accipiter*. The several characters listed above justify its separation into a separate monotypic genus. This is supported by the molecular systematics of Barrowclough *et al.* (2014).

Doria's Hawk *Megatriorchis doriae*   ENDEMIC | MONOTYPIC
Megatriorchis doriae Salvadori & D'Albertis, 1876. *Ann. Mus. Civ. Genova* 7: 805.—Hall Sound, SE Peninsula.

Known only from NG and Batanta I; lowlands to 1,650 m[CO] (Bishop 1986, Debus 1994a). A rarely encountered denizen of forest interior.

Localities—Batanta I[K], lower Menoo R in the Weyland Mts[F], near Timika[X], Humboldt Bay, Ok Ma[V], above Tabubil[C], Puwani R[G], upper Sepik basin[J], Hunstein Mts[L], Soliabeda[DI], Baiyer valley[A], Keku[B], Stephansort[W], Lolebu[W], Wau[Z], Hall Sound[MA], Astrolabe Ra, Mt Missim[U], and Hall Sound[MA], Varirata NP[C] (PNG Nat. Mus. specimen[A], Stresemann 1923[W], Greenway 1935[Z], Hartert *et al.* 1936[F], Gregory 1995b[V], A. Allison *in litt.*[L], K. D. Bishop *in litt.*[C], Beehler field notes 24-Jul-1984[G], Bishop 1986[K], I. Woxvold *in litt.*[J], S. van Balen *in litt.*[X], Pratt field notes[U], AMNH specimens[B]). The single observation of the species from Karkar I (Finch 1981g: 25) requires confirmation.

NOTES
Original description noted Yule I as the type locality, but this was emended by Mayr (1941) to Hall Sound, which we accept; we doubt the species inhabits Yule I, but note this site was an important offshore base for

collectors. Our retention of the traditional English group name ("Hawk" instead of "Goshawk") reflects the findings of Barrowclough *et al.* (2014) that this species does not belong to the genus *Accipiter*. Other names—Doria's Goshawk, *Accipiter doriae*.

GENUS *CIRCUS* [Genus 14 spp/Region 1 sp]

Circus Lacépède, 1799, *Tabl. Méthod. Mamm. Oiseaux*: 4. Type, by subsequent designation (Lesson, 1828. *Man. Orn.* 1: 105), *Falco aeruginosus* Linnaeus.

The species of this genus range worldwide. The genus is characterized by the owl-like facial disk, slim and elongate body, long wings and tail, tipping flight low over grass, and the white rump patch in most plumages. Our treatment posits there is but a single species found in the Region and that both the breeding population and the Australian migrant birds are conspecific, following Christidis & Boles (2008: 117).

Swamp Harrier *Circus approximans* RESIDENT AND VISITOR

Circus approximans Peale, 1848. *U.S. Explor. Exped.* 8: 64, 308.—Mathuata, Vanua Levu, Fiji.

Inhabits grasslands, marshes, airfields, and wetland edges in c and e NG; sea level–3,800 m[J] (Coates & Peckover 2001[J]). Also a rare migrant from Australia during austral winter, but status uncertain owing to confusion with female and juvenile of the resident race. Few records from w NG and none from the Bird's Head.

Localities—Paniai Lake[W], Timika, Dolak I[O], Kurik[K], Merauke[L], Gaima[M], Lake Habbema[N], Baliem R[N], Tabubil[X], Lake Daviumbu[CO], middle Sepik R[CO], near Mt Giluwe[DI], Kagamuga in Waghi Valley, Nondugl[P], Baiyer valley[CO], Karimui[DI], Okasa[DI], Schrader Ra[DI], lower Markham R[J], Wau[Q], Yule I distr[MA], Astrolabe Bay[MA], Huon Gulf[MA], Aroa R, Wharton Ra[CO], Port Moresby[CO], Popondetta[Z], and China Strait[MA] (Rand 1942a[M], Rand 1942b[N], Junge 1953[W], Gyldenstolpe 1955a[P], Hoogerwerf 1964[L], Mees 1982b[K], Coates 1990: 566[Z], Gregory 1995b[X], Bishop 2005a[O], I. Woxvold *in litt.*[J], Beehler field notes[Q]).

Extralimital—breeds in s Australia, NZ, ?Lord Howe I, Norfolk I, the Kermadec Is, Guadalcanal, Vanuatu, New Caledonia, Loyalty Is, Fiji, Tonga, and Society Is (Stresemann & Amadon 1979, Mayr & Diamond 2001).

SUBSPECIES

a. *Circus approximans spilothorax* Salvadori & D'Albertis
Circus spilothorax Salvadori & D'Albertis, 1876. *Ann. Mus. Civ. Genova* 7: 807.—Yule I, SE Peninsula.
RANGE: All NG east of the Bird's Head, at all elevations.
DIAGNOSIS: Male pied morph is handsomely patterned, with black head, mantle, and wing tips contrasting with bright white underparts and gray tail. Black morph rare, all black with gray tail. Black morph similar to juveniles but may have white rump—difficult to know whether these are early breeding birds in immature plumage. Pale morph female is brown with streaked underparts, pale rump; distinguished from nominate form by tail with clear, dark barring and face often with pale streaking (tail without bands in one specimen). Juvenile and immature are charcoal with brownish cast; can be rusty brown on belly with pale nape patch. Older immature male pied morph is like adult male of nominate race but with darker wing pattern suggestive of adult male pied morph. Note that plumage development and variation are inadequately known for the resident race; plumage sequence likely to pass through several distinct plumages, as in the nominate (Australian) race.

b. *Circus approximans approximans* Peale
Synonym: *Circus approximans gouldi* Bonaparte, 1850. *Consp. Gen. Av.* 1: 34.—New South Wales, Australia.
RANGE: Breeds in e, s, and sw Australia, Tasmania, and NZ. Nonbreeding birds straggle to NG (the Trans-Fly, Eastern Ra, and SE Peninsula); some brown-plumaged birds difficult to assign to race (Christidis & Boles 2008).
DIAGNOSIS: Male similar to female of *spilothorax*, but tail is either all gray or has faint, broken bands; female difficult to separate from the *spilothorax* female—tail often has more obscure bands, sometimes reduced to spots, and face is usually all brown; young birds often rusty brown below, with less pale streaking on nape than resident race. Iris color not helpful to identification.

Some authorities treat the NG form *spilothorax* as a distinct monotypic species (*e.g.*, Coates & Peckover 2001, del Hoyo & Collar 2014); others treat *spilothorax* as a race of the e Asian *C. spilonotus* (*e.g.*, Dickinson & Remsen 2013); yet others (*e.g.*, Brown & Amadon 1968) treat both *spilothorax* and *spilonotus* as races of the widespread Eurasian *C. aeruginosus*. We follow Christidis & Boles (2008) in treating the Australian and NG populations of marsh-harrier as a single species, *C. approximans*. Unconfirmed reports of *C. melanoleucus* from the highlands of NG (*e.g.*, Crome & Swainson 1974) apparently are field misidentifications of the pied morph males of the race *spilothorax*. We follow Baker-Gabb (1979) in not recognizing the subspecies *gouldi*. *Other names*—Pacific, Eastern, Spotted, or Papuan Marsh-Harrier; Australasian or Spotted-backed Harrier; *Circus spilonotus, C. aeruginosus spilothorax*.

GENUS *ACCIPITER* [Genus 46 spp/Region 7 spp]

Accipiter Brisson, 1760. *Orn.* 1: 28, 310. Type, by tautonymy, "Accipiter" Brisson = *Falco nisus* Linnaeus.

The species of this genus range worldwide (Thiollay 1994). The genus is characterized by a distinctive and uniform body plan, with short and broad rounded wings, long narrow tail, rather trim elongate body, long legs, and in most cases a smallish, flat-topped head. These are mainly bird-eating species.

Chinese Sparrowhawk *Accipiter soloensis* VISITOR | MONOTYPIC

Falco Soloënsis Horsfield, 1821. *Trans. Linn. Soc. London* 13: 137.—Solo, Java.

An uncommon wintering visitor to the w half of NG.

Localities—Batanta[J], Waigeo[MA], Gag[MA], and Efman I[MA]; also Koruo Is[M] in the NW Islands, Manokwari[K], Timika[K], Biak I[K], and Dolak I[K] (Beehler field notes 26-Feb-2004[J], Beehler field notes 16-Dec-2011[M], Germi *et al.* 2013[K]). NG encounters are concentrated in open habitats of the coastal lowlands of islands and Mainland of the west.

Extralimital—breeds in Korea and e China; winters south through se Asia to the Moluccas and Lesser Sundas.

NOTES

Should be looked for in sw PNG during the northern winter. *Other names*—Horsfield's Sparrowhawk, Grey Frog Hawk, Frog Hawk, Grey Goshawk.

Variable Goshawk *Accipiter hiogaster* RESIDENT

Falco hiogaster S. Müller, 1841. *Verh. Nat. Gesch. Ned., Land-en Volkenk.* 4: 110, note 3.—Ambon, s Moluccas.

Inhabits forest edge throughout the NG region including most major satellite islands. In NG it ranges up to 1,450 m in disturbed habitats. No records from Mios Num, Karkar, Manam, or Rossel I.

Extralimital—Lesser Sundas (Sumbawa I), Bismarcks, and Solomon Is (Mayr & Diamond 2001).

SUBSPECIES

a. *Accipiter hiogaster griseogularis*

Astur griseogularis G. R. Gray, 1861. *Proc. Zool. Soc. London* (1860): 343.—Bacan and Halmahera I, n Moluccas.

RANGE: Halmahera, Bacan, Tidore, Ternate, and Gebe I (Mees 1972, Stresemann & Amadon 1979).
DIAGNOSIS: Typical morph known only; mantle dark slate; rufous nuchal collar sometimes indistinct; throat gray; ventral plumage vinaceous cinnamon with some narrow whitish abdominal barring; wing: male 217–241, female 251–280 (White & Bruce 1986).

b. *Accipiter hiogaster leucosomus* (Sharpe)

Astur novaehollandiae leucosomus Sharpe, 1874. *Cat. Birds Brit. Mus.* 1: 94 (in key), 119 (ex Schlegel, 1866, *Vog. Ned. Ind.*: 19, 58, pl. 11, fig. 3).—Aiduma I, Triton Bay, Bird's Neck.

RANGE: The entire NG mainland, Misool[J], Waigeo, Salawati, Numfor, Yapen I, Aru Is, Trobriand Is, Woodlark I, East I of Bonvouloir Is (Mayr 1941, Mees 1965[J]). Record from Gag I presumably of this form (Johnstone 2006).

DIAGNOSIS: Typical morph is large; dark-backed; buff-breasted; wing (male) 217 (Rand & Gilliard 1967). This population also includes all-white individuals and all-slate-gray individuals. Slate morph specimen from the Purari R shows obsolete barring on dorsal surface of tail; note that the white morph has been recorded from Heath I.

c. *Accipiter hiogaster misoriensis* (Salvadori)
Urospizias misoriensis Salvadori, 1876. *Ann. Mus. Civ. Genova* 7: 904.—Biak I, Bay Islands.
RANGE: Biak I.
DIAGNOSIS: Similar to *leucosomus* but much smaller; wing (male) 173 (Rand & Gilliard 1967).

d. *Accipiter hiogaster pallidimas* Mayr
Accipiter novaehollandiae pallidimas Mayr, 1940. *Amer. Mus. Novit.* 1056: 10.—Fergusson I, D'Entrecasteaux Arch, SE Islands.
RANGE: Goodenough, Fergusson, and Normanby I[X] (Mayr 1941, Pratt field notes 2003[X]). Greenway (1973) noted that the white morph has not been recorded from these islands. The birds from the Trobriand Is are currently placed in *leucosomus* but may belong here.
DIAGNOSIS: Similar to *leucosomus* but larger and plumage paler; narrowly barred ventrally from throat to vent; breast darker and abdomen much paler; wing (male) 227 (Rand & Gilliard 1967).

e. *Accipiter hiogaster misulae* Mayr
Accipiter novaehollandiae misulae Mayr, 1940. *Amer. Mus. Novit.* 1056: 11.—Misima I, Louisiade Arch, SE Islands.
RANGE: Misima and Tagula I of the Louisiade Arch (Mayr 1941). Presumably the birds from Woodlark I refer to this form.
DIAGNOSIS: Similar to *leucosomus* but larger and more heavily barred ventrally; wing: male 234, female 281 (Rand & Gilliard 1967). Many specimens exhibit plumage that is identical to that of *leucosomus* (Mayr in Greenway 1973).

NOTES
We follow Schodde (1977), Ferguson-Lees *et al.* (2001), and Christidis & Boles (2008) in separating *A. hiogaster* from *A. novaehollandiae* of Australia. They noted the Australian form's larger size and all-gray plumage (in the typical morph), and the abrupt character-break across the Torres Strait. *Other names—* Grey, Varied, White, Vinous-chested, or Rufous-breasted Goshawk, *Accipiter novaehollandiae.*

Brown Goshawk *Accipiter fasciatus*   RESIDENT

Astur Fasciatus Vigors & Horsfield, 1827. *Trans. Linn. Soc. London* 15: 181.—New South Wales, Australia.
    Inhabits savanna, forest edge, and other open habitats of e NG, sea level–1,950 m[CO]. Westward in the n watershed to the w sector of the Sepik-Ramu; in the Central Ra west to Telefomin; and in the s watershed west to the w Trans-Fly. Also an isolated insular population from Rossel I.
    Extralimital—Moluccas (Buru), Lesser Sundas, Australia, New Caledonia, Vanuatu, Kiritimati I, and Rennell and Bellona I (Mayr & Diamond 2001).

SUBSPECIES
a. *Accipiter fasciatus dogwa* Rand
Accipiter fasciatus dogwa Rand, 1941. *Amer. Mus. Novit.* 1102: 1.—Dogwa, Oriomo R, Trans-Fly.
RANGE: The Trans-Fly[MA], from Dolak I[L] and Merauke[L] east to the Oriomo R[MA] (Germi *et al.* 2013[L]).
DIAGNOSIS: Similar to *polycryptus* but paler—whitish below with pale maroon-brown barring; nominate form from Australia has gray throat and dark-edged barring on breast, giving the ventral side a substantially darker appearance. The Australian race *didimus* is very similar, but *dogwa* differs in the dark gray crown and nape—darker than mantle—whereas *didimus* has a medium-gray crown much like mantle or perhaps only slightly darker than mantle; finally, *didimus* exhibits a marginally darker and richer breast pattern than *dogwa.*

b. *Accipiter fasciatus polycryptus* Rothschild and Hartert
Accipiter fasciatus polycryptus Rothschild and Hartert, 1915. *Novit. Zool.* 22: 53.—Sogeri distr, SE
Peninsula.
RANGE: Eastern NG, west in the n watershed to Telefomin[CO], just east of the PNG border.
DIAGNOSIS: Cinnamon barring of breast of adult birds apparently differs from that of *didimus* from
Australia as follows: the NG birds are simply barred white and cinnamon, whereas the Australian birds are
more complexly barred cinnamon-fuscous-white (with a narrow fuscous edge to the posterior edge of the
cinnamon bar).

c. *Accipiter fasciatus rosselianus* Mayr
Accipiter cirrhocephalus rosselianus Mayr, 1940. *Amer. Mus. Novit.* 1056: 12.—Rossel I, Louisiade Arch, SE
Islands.
RANGE: Restricted to Rossel I, Louisiade Arch.
DIAGNOSIS: Originally thought to be a subspecies of *A. cirrocephalus*. Reidentified by Schodde (2015). No
museum specimens of the adult plumage yet available. The two juvenile birds are slightly smaller than
juvenile *A. f. polycryptus*, but are identical in plumage (Schodde 2015). Schodde argued that until adult
birds become available, this form should stand as a valid race of *Accipiter fasciatus*.

NOTES
Other names—Australasian or Australian Goshawk.

Black-mantled Goshawk *Accipiter melanochlamys* ENDEMIC | MONOTYPIC
Urospizias melanochlamys Salvadori, 1876. *Ann. Mus. Civ. Genova* 7: 905.—Hatam, Bird's Head.
> Synonym: *Astur melanochlamys schistacinus* Rothschild and Hartert, 1913, *Novit. Zool.* 20: 482.— Mount Goliath,
> Border Ra.

Inhabits montane forest in NG, with records from the Bird's Head[MA], the Central Ra[MA], and mts of the
Huon Penin[J], 1,200–3,500 m[L]; one record at 600 m[K] (Rand 1942b, Gyldenstolpe 1955b, Ripley 1964a,
Coates and Peckover 2001[K], Freeman *et al.* 2013[J], Sam & Koane 2014[L]).

NOTES
We follow the suggestions of Rand & Gilliard (1967) and Debus (1994a) in treating the species as monotypic (*pace* Gilliard & LeCroy 1970). The plumage of individuals from the Bird's Head is marginally darker
dorsally and ventrally, but we do not think this minor clinal distinction merits recognition. *Other names*—
Black-mantled Sparrowhawk.

Grey-headed Goshawk *Accipiter poliocephalus* NEAR-ENDEMIC | MONOTYPIC
Accipiter poliocephalus G. R. Gray, 1858. *Proc. Zool. Soc. London*: 170.—Aru Is.

Inhabits lowland and hill forest throughout mainland NG, NW Islands, Yapen I, Aru Is, Fergusson, Normanby[J], Misima, and Tagula I[J], sea level–1,500 m (Mayr 1941, Pratt field notes and specimen in
BPBM[J]).

Extralimital—recorded from Saibai I, just off the coast of s NG (n Torres Strait Is of Australia;
Coates & Peckover 2001), but the record was not accepted by the Australian records committee (D.
Hobcroft *in litt.*).

NOTES
A sister species, *A. princeps*, inhabits adjacent New Britain (Mayr & Diamond 2001). Gregory (1995, 1997)
noted an all-dark (blackish-slate) morph in Tabubil as well as an all-gray morph. The species was collected
on Normanby I and Tagula I for the first time by Pratt (specimens at BPBM). An unpublished report from
S. Myers of the species on Biak I from 2004 did not include sufficient detail to be accepted. Moreover, Biak
I is important in the Indonesian bird trade, thus there is opportunity for many non-Biak species to be
observed as escapees, though perhaps not raptors. *Other names*—New Guinea Grey-headed Goshawk.

Collared Sparrowhawk *Accipiter cirrocephalus* RESIDENT

Sparvius cirrocephalus Vieillot, 1817. *Nouv. Dict. Hist. Nat.*, nouv. éd. 10: 329.—New South Wales, Australia.

Inhabits lowland and hill forest, edge, and wooded savanna of NG and outlying islands (see below); sea level–2,500 m[CO], rare above 1,500 m. The named population from Rossel I has been re-identified by Schodde (2015) as belonging to *A. fasciatus*.

Extralimital—Australia (Stresemann & Amadon 1979, Debus 1994a).

SUBSPECIES

a. *Accipiter cirrocephalus papuanus* (Rothschild and Hartert)

Astur cirrhocephalus papuanus Rothschild and Hartert, 1913. *Novit. Zool.* 20: 482.—Snow Mts, Western Ra.

RANGE: All NG, the Aru Is[MA], Salawati[MA], ?Batanta[X], Waigeo[MA], Yapen[MA], and Dolak I[MA] (Beehler field notes[X]).

DIAGNOSIS: Greenway (1973) noted that Condon & Amadon found this form distinct from QLD specimens by the more saturated rufous ventral plumage, the reduced barring, and smaller size. Immature plumage exhibits cinnamon-brown streaking and barring (Rand & Gilliard 1967).

NOTES

A sister species, *A. brachyurus*, inhabits adjacent New Britain and New Ireland (Mayr & Diamond 2001). We follow Debus (1994a) and Christidis & Boles (2008) in reverting to the original spelling, which was inappropriately emended (*fide* R. Schodde *in litt.*). Schodde noted the original spelling has been in general use in Australia over the last century. Sight records from Biak I require confirmation, especially considering the small size of Biak *A. hiogaster* (K. D. Bishop *in litt.*). The species was heard calling on Batanta on 13-Dec-2011 by Beehler (field notes). *Other names*—Australian Sparrowhawk, *Accipiter cirrhocephalus*.

Meyer's Goshawk *Accipiter meyerianus* RESIDENT | MONOTYPIC

Astur Meyerianus Sharpe, 1878. *Journ. Linn. Soc. London, Zool.* 13: 458.—Ansus, Yapen I, Bay Islands.

A very scarce inhabitant of hill and montane forest of n and e NG as well as Yapen[CO], Bagabag, and Karkar I. In NG mainly in foothills to 1,600 m[A], max 2,700 m[A] (Coates & Peckover 2001[A], see Freeman *et al.* 2013).

Localities—Yapen I[MA], Foja Mts[X], Ok Tedi[CO], Hides Ra[M], Carrington R headwaters (upper Strickland basin)[M], Bilbilokabip on the Ok Tedi headwaters[M], Ambua Lodge[Q], Tari gap[L], Schrader Ra[CO], Baiyer R[CO], Karkar I[J], Bagabag I[J], Mt Wilhelm[K], Crater Mt[N], Kratke Mts, Aiyura[CO], YUS uplands[Z], Mt Missim[CO], upper Angabanga R[CO], Veimauri R[CO], and Efogi[CO] (Diamond & LeCroy 1979[J], Diamond 1985[X], Bishop 1987[Q], Frith & Frith 1992[L], Mack & Wright 1996[N], Freeman *et al.* 2013[Z], Sam & Koane 2014[K], I. Woxvold *in litt.*[M]).

Extralimital—Moluccas, Bismarcks, and Solomons (Coates & Bishop 1997, Mayr & Diamond 2001, Dutson 2011).

NOTES

The rarely encountered black morph is difficult to distinguish from the black morph of *Erythrotriorchis buergersi*. *Other names*—Papuan Goshawk.

GENUS *ERYTHROTRIORCHIS* [Genus 2 spp/Region 1 sp]

Erythrotriorchis Sharpe, 1875. *Proc. Zool. Soc. London*: 337. Type, by monotypy, *Falco radiatus* Latham.

Species of the genus inhabit Australia and New Guinea. We follow Debus (1991, 1994a) and Barrowclough *et al.* (2014) in use of this genus for the species *buergersi*. The two species of the genus are robust raptors with pale or cinnamon ventral plumage streaked with blackish, and a patterned mantle washed with rufous.

Chestnut-shouldered Goshawk *Erythrotriorchis buergersi*   ENDEMIC | MONOTYPIC

Astur bürgersi Reichenow, 1914 (February). *Orn. Monatsber.* 22: 29.—Maeanderberg, upper Sepik R, west-central Sepik-Ramu.

Restricted mainly to hill and lower montane forest through e NG, with a single record from w NG (Stresemann 1923); 450–1,580 m.

Localities—Foja and Torricelli Mts[J], Maeanderberg and Lordberg[K], Sau valley[CO], Karimui[DI], Adelbert Mts[T], Junzaing on the Huon Penin[L], Mt Cameron Ra[CO], Neneba[CO], Bubuni[MA], Astrolabe Ra[CO], Isurava[P], and Hydrographer Mts[CO] (Stresemann 1923[K], Mayr 1931[L], Pratt 1983[T], Diamond 1985[J], Pratt field notes 2014[P]).

NOTES
Debus & Edelstam (1994) and Debus *et al.* (1994) have provided additional data on plumages, polymorphism, and nomenclature for this curious species. *Other names*—Buerger's Goshawk or Sparrowhawk, Chestnut-shouldered Hawk, *Accipiter buergersi*.

ORDER STRIGIFORMES

The phylogenomic study of Hackett *et mult.* (2008) indicated that this order is monophyletic, including a clade of two families (Tytonidae and Strigidae) that is sister to the Accipitriformes, Bucerotiformes, Coliiformes, Piciformes, Coraciiformes, among others.

Family Tytonidae

Barn-Owls   [Family 19 spp/Region 4 spp]

Barn-owls are distinct from the typical owls (Strigidae) at the familial level (see Christidis & Boles 2008, Dickinson & Remsen 2013). The species of the family range throughout the tropical and temperate regions of the world, and are absent from the boreal and arctic regions. In New Guinea they occupy virtually all forest and nonforest habitats. Barn-owls are recognized by their heart-shaped facial disks, dark eyes, unstreaked and unbarred plumage, long legs, long rounded wings, short tail, and presence of a serrated comb on the claw of the middle toe (Bruce 1999). These medium-large owls vary in coloration from pale yellow-brown and white to sooty gray. They forage mostly for small mammals in twilight or at night, usually in forest openings or in grassland but also in forest interior. The family is sister to the typical owls.

GENUS *TYTO*   [Genus 16 spp/Region 4 spp]

Tyto Billberg, 1828. *Syn. Faun. Scand.* 1(2): tab A. Type, by monotypy, *Strix alba* Scopoli.

The species of this genus range through South America, central and southern North America, Europe, Africa, Australia, Melanesia, and Polynesia. Characters defining the genus are as for the family.

Sooty Owl *Tyto tenebricosa*   RESIDENT

Strix tenebricosus Gould, 1845. *Proc. Zool. Soc. London*: 80.—Clarence R, NSW, Australia.

Inhabits all NG and Yapen I, sea level–4,000 m. A widespread owl of forest and subalpine shrublands. Extralimital—se Australia.

a. *Tyto tenebricosa arfaki* (Schlegel)
Strix tenebricosa arfaki Schlegel, 1879. *Notes Leyden Mus.* 1: 101.—Arfak Mts, Bird's Head.
RANGE: NG, Yapen I[MA], and ?Waigeo I[J] (Davies 2008[J]). No records from the Trans-Fly.
DIAGNOSIS: Small; wing 254–305 (Rand & Gilliard 1967). Abdomen usually sooty. The plumage pattern of the species is variable, but overall the crown is dark with profuse small white spots, and the ventral plumage is dark with profuse white spotting and obscure barring.

NOTES
By plumage characters, the form *arfaki* from NG is unambiguously allied to the nominate *tenebricosa* from se Australia, exhibiting small white spots on mantle and dark ventral plumage (see Bruce 1999, Norman *et al.* 2012), whereas the population in n QLD (now treated by some authorities as a distinct species, *T. multipunctata*) is distinguished by the large white dorsal spots and the paler ventral plumage. Curiously, Dickinson & Remsen (2013) placed *arfaki* with their species *multipunctata*, but also noted that *arfaki* may be treatable as distinct at the species level. The molecular work of Norman *et al.* (2012) found little genetic divergence among the three Sooty Owl taxa, supporting the case that they all could be classified as a single species, *T. tenebricosa*. Davies's (2008) intriguing report of a bird heard at 450 m elevation on Waigeo I deserves further investigation. *Other names*—Lesser Sooty Owl.

Australian Masked Owl *Tyto novaehollandiae*   RESIDENT
Strix (?) Novæ Hollandiæ Stephens, 1826. In Shaw, *Gen. Zool.* 13(2): 61.—New South Wales, Australia.
Known from a few records from the wooded savanna of the Trans-Fly, sea level–200 m. Considerably larger than *T. delicatula*, which it most closely resembles.
Extralimital—widespread in sw, n, and e Australia and Tasmania; apparently mainly absent from the interior of Australia (Higgins 1999).

SUBSPECIES
a. *Tyto novaehollandiae calabyi* Mason
Tyto novaehollandiae calabyi Mason, 1983. *Bull. Brit. Orn. Club* 103: 126.—Merauke, s NG.
RANGE: The Trans-Fly. Localities—Toerey on the Merauke R[L], Paal Putih[M], Merauke[MA], Bensbach R[J], Tarara[K], Daru[K], and Balimo[N] (Rand 1942a[K], Mees 1982b[L], Hoek Ostende *et al.* 1997[M], BPBM specimen[N], B. J. Coates *in litt.*[J]).
DIAGNOSIS: Closest to *galei* of ne Australia, from which it differs in larger size and darker, dusky, tawny, and less coarsely mottled dorsal plumage (Mason 1983). Wing 303–333 (Rand & Gilliard 1967). Ventral plumage with sparse fine dark spotting. Given the polymorphism and other (age-related?) plumage variation in the species, the described subspecies are suspect.

NOTES
The record of this species from Tari (Hadden 1975a) was subsequently reidentified as *T. longimembris* (Mees 1982b). *Other names*—Masked Owl.

Australian Barn-Owl *Tyto delicatula*   RESIDENT
Strix delicatulus Gould, 1837. *Proc. Zool. Soc. London* (1836): 140.—New South Wales, Australia.
Frequents nonforest habitats of the Sepik-Ramu, Eastern Ra, Huon and SE Peninsulas, and Manam and Karkar I. Sea level–1,680 m[CO].
Extralimital—Lesser Sundas, Australia, New Caledonia, Vanuatu, Fiji, Samoa, Tonga, Niue (Dickinson & Remsen 2013); also Bismarck Arch (Long and Tanga I, possibly also the n coast of New Britain) and Solomon Is (Bishop 1983, Dutson 2011). Vagrant to NZ (Gill *et al.* 2010).

SUBSPECIES
a. *Tyto delicatula meeki* (Rothschild and Hartert)
Strix flammea meeki Rothschild and Hartert, 1907. *Novit. Zool.* 14: 446.—Collingwood Bay, SE Peninsula.
RANGE: Eastern NG. In the n watershed, from the middle Sepik eastward; in the s watershed, from Bereina eastward. Localities—middle Sepik R[CO], Karkar[MA] and Manam I[MA], Wahgi and Baiyer valleys[CO], Okapa[CO],

WauCO, Port MoresbyMA, Collingwood BayMA (see also Gyldenstolpe 1955a, Gilliard & LeCroy 1966).
DIAGNOSIS: Differs from the nominate form from Australia by the much reduced (obsolete) dark barring on the upper surface of the tail feathers.

NOTES
We follow Dickinson & Remsen (2013), whose species-level treatment followed König & Weick (2008) and Wink *et al.* (2009) in recognizing *T. delicatula*. Del Hoyo & Collar (2014) maintained the cosmopolitan *T. alba. Other names*—Barn Owl, Pearly Owl.

Eastern Grass-Owl *Tyto longimembris*   RESIDENT

S[*trix*] *Longimembris* Jerdon, 1839. *Madras J. Lit. Sci.* 10: 86.—Nilgiri Mts, s India.
    Found locally in the Western Ra, Eastern Ra, and mts of the Huon and SE Peninsulas, 1,000–2,500 m. Frequents mid-montane grassland, swampy cane grass, and cultivated areas (Coates 1985). Apparently found at lower elevations on occasion (see below). To be expected from the Border Ra also.
    Extralimital—India, se China, Taiwan, Philippines, Moluccas, Australia, New Caledonia, and Fiji. In Australia it occurs mostly in the lowlands; it has been collected on Thursday I, Torres Strait.

SUBSPECIES
a. *Tyto longimembris papuensis* Hartert
Tyto longimembris papuensis Hartert, 1929. *Novit. Zool.* 35: 103.—Owgarra, 1,830 m, Angabanga R, SE Peninsula.
    Synonym: *Tyto capensis baliem* Ripley, 1964. *Peabody Mus. Nat. Hist. Yale* 19: 37.—Near Wamena, Baliem valley, Western Ra.
RANGE: Largely confined to upland interior grasslands of the Western and Eastern Ra, and mts of the Huon and SE Peninsulas. Localities—Wamena (Baliem valley)L, TariX, Mt GiluweK, Baiyer valleyCO, Jimi valleyCO, Kup in the Wahgi valleyW, NonduglJ, KainantuCO, Huon PeninCO, Owgarra on the upper Angabanga RMA (Mayr & Gilliard 1954W, Gyldenstolpe 1955aJ, Sims 1956K, Ripley 1964aL, Mees 1982b: 87X). Also a record of an active nest in grassland of the upper Ramu R on 27-May-1981 (Stronach 1990); one was observed at 140 m in grassland in the lower Watut R valley in 2010 (I. Woxvold *in litt.*).
DIAGNOSIS: Hartert (1929) noted that this form "differs at a glance by its upperside being more uniform, duller, and paler, with only some very small, tiny white spots near the tips of the feathers." In fact, the dorsal markings are sparsely scattered tiny white slashes, which distinguish this form from QLD birds, on which the markings are larger white triangular-shaped spots (based on examination of AMNH specimens).

NOTES
Species treatment follows Bruce (1999) and Christidis & Boles (2008). Some authorities treat this form as the species *capensis* (Amadon 1959). We follow Dickinson & Remsen (2013) in lumping the race *baliem* into *papuensis*. Stronach (1990) reported observing a pair of this species roosting on the ground in the grasslands of the Bula Plains, Trans-Fly, on 24-Nov-1979, noting he distinguished this species from *T. novaehollandiae* by the distinctive look of the bird's eyes, and by the fact that he found *T. novaehollandiae* not to roost on the ground. This record must remain uncertain until the photographs taken are examined and independently identified. *Other names*—Grass Owl, *Tyto capensis*.

Family Strigidae

Typical Owls [Family 194 spp/Region 6 spp]

The species of the family range worldwide. In New Guinea, the members of the family are found mainly in the lowlands. The range of one species extends up into the mountains to 2,500 m elevation. The most obvious external character is the round shape of the facial disk, compared with the heart-shaped facial disk of barn-owls (Tytonidae). For the New Guinean species, note the pale eyes *vs.* the dark eyes of the barn-owls. Many species exhibit profuse streaking and barring of plumage. The head is large, the legs very short, and the feet zygodactyl (Marks *et al.* 1999).

GENUS *OTUS* [Genus 39 spp/Region 1 sp]

Otus Pennant, 1769. *Indian Zool.*: 3. Type, by monotypy, *Otus bakkamoena* Pennant.

The species of this genus range through Eurasia and Africa and southeastward to Biak Island of western New Guinea (Marks *et al.* 1999). The genus is characterized by small size, the "eared" head plumage, plumage polymorphism, and cryptical plumage patterning of spotting and vermiculation.

Biak Scops-Owl *Otus beccarii* ENDEMIC | MONOTYPIC

Scops beccarii Salvadori, 1876. *Ann. Mus. Civ. Genova* (1875) 7: 906.—Sowek, Biak I, Bay Islands.

Inhabits the forests of Biak I, where it appears to be widespread and fairly common. This is the southeasternmost *Otus* in the Asia-Pacific region, and the only one of its genus in Australasia.

NOTES

Treated by Marshall (1978) as an island race of *Otus magicus*. Our treatment follows Olsen (1999), while acknowledging that *beccarii* is presumably a sister to *magicus*. IUCN lists this species as Endangered; we suggest Near Threatened is its deserved status. The species is easy to locate, and there is abundant forest cover to support the species on Biak and neighboring Supiori. *Other names*—Moluccan Scops-Owl, *Otus magicus*.

GENUS *NINOX* [Genus 33 spp/Region 4 spp]

Ninox Hodgson, 1837. *Madras Journ. Lit. Sci.* 5: 23. Type, by original designation, *Ninox nipalensis* Hodgson = *Strix scutulata lugubris* Tickell.

The species of this genus range from Madagascar and tropical Asia east to Australia and Melanesia (Marks *et al.* 1999). The genus is characterized by the absence of ear tufts, dull brown plumage dorsally, streaking or barring below, and broad, blunt head shape.

Rufous Owl *Ninox rufa* RESIDENT

Athene rufa Gould, 1846. *Proc. Zool. Soc. London*: 18.—Port Essington, NT, Australia.

Sparsely distributed throughout the forests of NG, Waigeo I, and the Aru Is, sea level–1,800 m. Recorded from widely scattered localities in NG. Also ?Batanta I[J] (M. Tarburton's island website[J]).

Extralimital—n Australia.

SUBSPECIES

a. *Ninox rufa humeralis* (Bonaparte)

Athene humeralis Bonaparte, 1850. *Consp. Gen. Av.* 1: 40.—Triton Bay, Bird's Neck.

Synonym: *Noctua aruensis* Schlegel, 1866. *Ned. Tijdschr. Dierk.* 3: 329.—Aru Is.

RANGE: Waigeo I[MA], Gam I[P], Aru Is[MA], and all NG. Localities—Triton Bay[MA], Wandammen Penin[B], Jumbara[C], Kikori[D], Baiyer R[D], Holtekong[J], Uraru in the Purari basin[A], YUS[Z], Junzaing[K], Soputa R nr Popondetta, Goldie R[CO], Varirata NP[P], and Collingwood Bay[B] (PNG Nat. Mus. specimen[A], Mayr 1931[K], Ripley 1964a[J], Freeman *et al.* 2013[Z], Beehler field notes 12-Mar-2005[P], AMNH specimen[B], BPBM specimens[C], K. D. Bishop *in litt.*[D]).

DIAGNOSIS: Wing: male 326–335, female 306–314 mm (Rand & Gilliard 1967). Smaller than *queenslandica* of the e coast of Australia, and darker brown dorsally; also note profuse, fine dull brown barring ventrally and very dark crown. Cheeks blackish, darker than in *rufa* or *marginata* (Mees 1964a, Mason & Schodde 1980). We subsume the form *aruensis* into *humeralis* following Olsen (1999) and Dickinson & Remsen (2013). The form *aruensis* is described based on its small size but is known from a single individual (Mees 1964a).

NOTES
Other names—Rufous Hawk-owl.

Barking Owl *Ninox connivens* RESIDENT

Falco connivens Latham, 1801. *Index Orn.*, Suppl: xii.—Sydney region, NSW, Australia.

Lowlands of the Trans-Fly, e sector of the S Lowlands, Sepik-Ramu, and Huon and SE Peninsulas; also Karkar and Manam I. Frequents savanna, gardens, woodland, and forest in the lowlands (Coates 1985).

Extralimital—n Moluccas and Australia (Higgins 1999).

SUBSPECIES
a. *Ninox connivens assimilis* Salvadori & D'Albertis
Ninox assimilis Salvadori & D'Albertis, 1876. *Ann. Mus. Civ. Genova* (1875) 7: 809.—Mount Epa, Hall Sound, SE Peninsula.

RANGE: Central and e NG. Localities—Merauke, Daru[CO], Balimo[L], Lake Daviumbu[CO], Sepik R[CO], Ramu R, Karkar[K] and Manam I, Hall Sound, Laloki R, Port Moresby[CO], Amazon Bay[J] (Mayr 1941, Bell 1970a[J], Diamond & LeCroy 1979[K], BPBM specimen[L]). Recorded to 1,040 m on Karkar I (Diamond & LeCroy 1979).

DIAGNOSIS: Very small; pale chestnut-brown streaking mixed with some buff and white on abdomen (Mees 1964a, Olsen 1999). Wing 255 (Rand & Gilliard 1967).

NOTES
Other names—Barking Hawk-owl or Barking Boobook.

Southern Boobook *Ninox novaeseelandiae* RESIDENT

Strix novæ Seelandiæ Gmelin, 1788. *Syst. Nat.* 1(1): 296.—Queen Charlotte Sound, South I, NZ.

Savannas and open monsoon woodlands of the Trans-Fly.

Extralimital—Lesser Sunda Is, Australia, Lord Howe I, Norfolk I, and NZ (Higgins 1999, Gill *et al.* 2010).

SUBSPECIES
a. *Ninox novaeseelandiae pusilla* Mayr & Rand
Ninox novaeseelandiae pusilla Mayr & Rand, 1935. *Amer. Mus. Novit.* 814: 3.—Dogwa, Oriomo R, s NG.

RANGE: Known only from the PNG part of the Trans-Fly. Localities—Dogwa[K], Penzara[J], Tarara[J], Bensbach R[K], and Balimo[K] (Rand 1942a[J], Mees 1964a[K], Bishop 2005a).

DIAGNOSIS: Slightly darker, more vinaceous dorsally than the n Australian birds (Mees 1964a, Greenway 1978). Size small (Mees 1964a). Wing 197–205 (Rand & Gilliard 1967).

NOTES
We follow Mees (1964a, 1982a), Higgins (1999), and Christidis & Boles (2008) in treating the NG form as a member of the species *N. novaeseelandiae*. By contrast, Schodde & Mason (1980), Olsen (1999), and Dickinson & Remsen (2013) suggested treating the NG birds as *N. boobook*. *Other names*—Common or Australian Boobook, *N. boobook*.

Papuan Boobook *Ninox theomacha*   ENDEMIC

Spiloglaux theomacha Bonaparte, 1855. *Compt. Rend. Acad. Sci. Paris* 41: 654.—Triton Bay, Bird's Neck.

Mainland NG, Misool I, Waigeo I, the D'Entrecasteaux Arch, Tagula I, and Rossel I; sea level–2,500 m. Frequents forest, edge, and groves of trees in open country. Appears to be less common in the lowlands and apparently absent from the Trans-Fly, where replaced by *N. novaeseelandiae*.

SUBSPECIES

a. *Ninox theomacha theomacha* (Bonaparte)

Synonym: *Noctua Hoedtii* Schlegel, 1871. *Ned. Tijdschr. Dierk.* (1873) 4: 3.—Misool I, NW Islands.

RANGE: Mainland NG, and Misool and Waigeo I[MA]. No records from the Trans-Fly (Coates & Peckover 2001).

DIAGNOSIS: All brown; sooty chocolate-brown dorsally and on head; rich chestnut ventrally (Olsen 1999). Small; wing 176–186 (Rand & Gilliard 1967). Mees (1965: 171) doubted the validity of the form *hoedtii* from Misool and Waigeo of the NW Islands. Our museum research supports this suspicion. Hence *hoedtii* becomes a synonym of the nominate form.

b. *Ninox theomacha goldii* Gurney

Ninox goldii Gurney, 1883. *Ibis*: 171.—Normanby I, D'Entrecasteaux Arch, SE Islands.

RANGE: D'Entrecasteaux Arch[MA].

DIAGNOSIS: Larger than nominate; wing 215–227 (Rand & Gilliard 1967). Belly with much white; brown-and-white barring on flanks (Olsen 1999). Note also white spotting on scapulars and lesser wing coverts; ventral streaking and chestnut spotting; underwing variegated brown and white; primaries obsoletely barred to tip. Although plumage is distinct from Mainland form, the voice is not distinguishable.

c. *Ninox theomacha rosseliana* Tristram

Ninox rosseliana Tristram, 1889. *Ibis*: 557.—Rossel I, Louisiade Arch, SE Islands.

RANGE: Tagula and Rossel I[MA]. Dumbacher *et al.* (2010) recorded the species from Bagaman I in the Calvados chain, just east of Tagula I—it presumably refers to this form.

DIAGNOSIS: Has even more extensive white markings on belly than found in *goldii* (Olsen 1999).

NOTES

Note that a specimen referable to this species in the Natural History Museum, London, labeled "Trobriand," died as a cage bird in the gardens of the Zoological Society of London on 9-Jul-1937. The species thus should be looked for on the Trobriand Is. The relationship of the taxa *goldii* and *rosseliana* warrants investigation. *Other names*—Jungle Hawk-owl, Papuan Boobook Owl, Brown Owl.

GENUS *UROGLAUX*   ENDEMIC | MONOTYPIC

Uroglaux Mayr, 1937. *Amer. Mus. Novit.* 939: 6. Type, by original designation and monotypy, *Athene dimorpha* Salvadori.

The single species of the genus is endemic to the Region. This genus is characterized by the slim, *Accipiter*-like shape, lack of ear tufts, and profuse barring of dorsal surface of tail, back, and wings that contrasts with heavy streaking of the underparts. Also note the long tail, two-thirds the length of the wing, and very heavily-feathered tarsus (Mayr 1937b).

Papuan Hawk-Owl *Uroglaux dimorpha*   ENDEMIC | MONOTYPIC

Athene dimorpha Salvadori, 1874. *Ann. Mus. Civ. Genova* 6: 308.—Sorong, w Bird's Head.

Sparsely but widely distributed through forests of the NG lowlands and hills and Yapen I, sea level–1,500 m. Localities—Sorong[MA], Geelvink Bay[CO], Weyland Mts[CO], Yapen I[MA], near Vanimo[M], Malu near Ambunti, Hunstein Mts[L], Wewak[X], Karawari[N], Gogol R, Gulf Prov, Lae[W], lower Watut R[J], Gabensis[W], Bulolo[W], Port Moresby region[CO], Mt Victoria[CO], Collingwood Bay[CO], Milne Bay[CO], and upper Fly R[K] (Lamothe 1993[W], Shany 1995[M], A. Allison *in litt.*[L], Pratt field notes and specimen at BPBM[K], I. Woxvold *in litt.*[J], T. Ross photo provided to Beehler[X], P. Gregory *in litt.*[N]).

Almost certainly a sister form to the *Ninox* lineage. A white-plumaged nestling was photographed on the grounds of a Wewak hotel (T. Ross *in litt.*). Dickinson & Remsen (2013) and del Hoyo & Collar (2014) gave the English name "Papuan Boobook" to this species, even though that name has long been in use for *Ninox theomacha. Other names*—Papuan Boobook.

ORDER BUCEROTIFORMES

This major lineage is split from the traditional Coraciiformes based on the work of Hackett *et mult.* (2008) and McCormack *et al.* (2012). This new order includes the hornbills, wood-hoopoes (Phoeniculidae), and hoopoes (Upupidae) (Cracraft 2013).

Family Bucerotidae

Hornbills   [Family 53 spp/Region 1 sp]

The species of the family range from southern Africa to tropical Asia east to New Guinea and the Solomon Islands. They are large to very large birds with very large bills, often with a casque atop the bill; conspicuous eyelashes; long, broad wings; long to very long tails; and very short legs (Cracraft 2013). All species are arboreal. This lineage is sister to the ground hornbills (Bucorvidae), and that pair is sister to the wood-hoopoes and hoopoes (Hackett *et mult.* 2008).

GENUS *RHYTICEROS*   [Genus 9 spp/Region 1 sp]

Rhyticeros Reichenbach, 1849. *Avium Syst. Nat.*: pl. L. No species; generic details only. Species added, Bonaparte, 1854. *Ateneo Italiano* 2: 312. Type, by subsequent designation (G. R. Gray, 1855. *Cat. Gen. Subgen. Birds*: 85), *Buceros plicatus* J. R. Forster.

The species of this genus range from Narcodam Island (Andaman Islands) east to New Guinea and the Bismarck and Solomon Islands (Kemp 2001). The genus is characterized by the huge, pale horny bill with a unique ridged casque atop the culmen; black body plumage; and saclike throat pouch.

Blyth's Hornbill *Rhyticeros plicatus*   RESIDENT | MONOTYPIC

Buceros plicatus J. R. Forster, 1781. *Indische Zool.*: 40.—Seram, s Moluccas.
    Synonym: *Buceros ruficollis* Vieillot, 1816. *Nouv. Dict. Hist. Nat.*, nouv. éd. 4: 600.—Waigeo I, NW Islands.
    Synonym: *Rhyticeros plicatus jungei* Mayr, 1937. *Amer. Mus. Novit.* 939: 13.—Madang, e verge Sepik-Ramu.

Inhabits lowland and hill forest throughout NG, the four main NW Islands[MA], Yapen I[MA], and the D'Entrecasteaux Arch[MA] (Normanby I record from LeCroy *et al.* 1984); sea level–500 m, max 1,800 m (Coates & Peckover 2001). Dumbacher *et al.* (2010) provided a record for Skelton I in the SE Islands. The record in Mayr (1941) from Gebe I is probably in error (*fide* Mees 1972).

    Extralimital—The Moluccas and Bismarck Arch (New Britain, New Ireland, New Hanover), and most of the Solomon Is (but not Makira; Mayr & Diamond 2001, Dutson 2011).

NOTES
Treated as monotypic, following Kemp (2001). *Other names*—Papuan Hornbill, Kokomo, *Aceros plicatus*.

ORDER CORACIIFORMES

The traditional order Coraciiformes has been shown to be paraphyletic, with two distinct lineages separated by the Piciformes (Hackett *et mult.* 2008). This reformed order encompasses the kingfishers, motmots (Momotidae), todies (Todidae), ground-roller (Brachypteraciidae), bee-eaters, and rollers (Cracraft 2013). The remaining families in the traditional Coraciiformes (the hornbills, hoopoes, wood-hoopoes, and trogons) are now placed in the Bucerotiformes, and the Piciformes are placed between these two major lineages (McCormack *et al.* 2012), though absent from the Region.

Family Meropidae

Bee-eaters [Family 27 spp/Region 2 spp]

The species of the family range from Africa and Europe east to tropical Asia, New Guinea, the Solomon Islands, and Australia. Characteristics of the family include the specialized aerial foraging habit, the plumage of pastel colors, the long and pointed bill, and the small feet and short legs. The bee-eaters are sister to a clade that includes the rollers, ground-rollers, todies, motmots, and kingfishers (Cracraft 2013).

GENUS *MEROPS* [Genus 24 spp/Region 2 spp]

Merops Linnaeus, 1758. *Syst. Nat.*, ed. 10, 1: 117. Type, by Linnaean tautonymy, *Merops apiaster* Linnaeus.

The species of this genus range from Africa and Europe to New Guinea, Australia, and the Solomon Islands (Fry 2001). The genus is characterized by the brightly colored plumage; distinctive narrow, slightly decurved, and sharply pointed bill; the acrobatic flight; and the black mask.

Blue-tailed Bee-eater *Merops philippinus* RESIDENT | MONOTYPIC

Merops philippinus Linnaeus, 1767. *Syst. Nat.*, ed. 12, 1: *errata* at end of the volume; name for Merops fifth species: 183.—Philippines.

Synonym: *Merops Salvadorii* A. B. Meyer, 1891. *Ibis*: 294.—Kurakakaul, n coast of New Britain.

A localized breeder in the Trans-Fly, Sepik-Ramu, Huon Penin, and SE Peninsula, sea level–150 m. Frequents grasslands, airstrips, and nearby scrub (Coates 1985).

Localities—Dolak I[CO], Princess Marianne Strait[MA], Kurik[L], Benbach R[X], Ifar on Lake Sentani[J], Lake Daviumbu[CO], middle Sepik[K], Ramu R[CO], n coast of Huon[CO], Markham valley[MA], and Popondetta[CO] (Ripley 1964a[J], Gilliard & LeCroy 1966[K], Mees 1982b[L], D. Hobcroft *in litt.*[X]).

Extralimital—tropical Asia east to New Britain, Long, Umboi, and Sakar I (Mayr & Diamond 2001, Dutson 2011).

NOTES

Monotypic (Fry 2001). Some authorities treat this form as being conspecific with *Merops superciliosus*, but we agree with Mees (1982b), who cited the study of Marien (1950) as evidence for maintaining full specific status for *philippinus*, also followed by Dickinson & Remsen (2013). *Other names*—sometimes included in Olive Bee-eater, *Merops superciliosus*.

Rainbow Bee-eater *Merops ornatus*   visitor and local resident | monotypic

Merops ornatus Latham, 1801. *Index Orn.*, Suppl.: xxxv.—New South Wales, Australia.

Throughout NG and satellite islands as an austral migrant (Mayr 1941); sea level–1,500 m, max 3,960 m when migrating over the Bismarck Ra (Beehler field notes). Also breeds around Port Moresby and in the Sepik and Ramu valleys (M. LeCroy *in litt.*). Migrants are present in the Region from early March to early October (Coates & Peckover 2001). Present in small numbers during the remainder of the year in local breeding sites. The main migration route is through Torres Strait. Passage was noted in the Ok Tedi region from 25-March to 25-April, during which time many flocks passed over in a northerly direction toward the Hindenburg Wall (Bell 1969); large flocks were seen heading north at Mt Kaindi in early April (Mackay 1973); and flocks were observed flying east over Murray Pass, 3,000 m, Mt Albert Edward, at a rate of 1,200 birds per hour in early April (Peckover & Filewood 1976).

Extralimital—widespread breeder in Australia; also Moa I in Lesser Sundas. Winters from Lesser Sundas and Sulawesi to the Bismarck Arch (Dutson 2011).

Family Coraciidae

Rollers   [Family 12 spp/Region 1 sp]

The species of the family range from Africa and Spain east through Europe, tropical Asia, and Japan to the Philippines, New Guinea, Solomon Islands, and Australia. Rollers are sister to the ground-rollers of Madagascar. Those two lineages, in turn, are sister to the kingfishers, motmots, and todies (Cracraft 2013).

GENUS *EURYSTOMUS*   [Genus 4 spp/Region 1 sp]

Eurystomus Vieillot, 1816. *Analyse Nouv. Orn. Elem.*: 37. Type, by monotypy, "Rolle des Indes" Buffon = *Coracias orientalis* Linnaeus.

The species of this genus range from western Africa to tropical and northeastern Asia, Borneo, New Guinea, Australia, and the Bismarck and Solomon Islands (Fry 2001). The genus is characterized by the broad, colorful bill, large head, and powerful and graceful flight.

Oriental Dollarbird *Eurystomus orientalis*   resident and visitor

Coracias orientalis Linnaeus, 1766. *Syst. Nat.*, ed. 12, 1: 159.—Java.

Widespread in the Region as a resident and an austral migrant; sea level–1,600 m, one record at 4,500 m (Coates & Peckover 2001). Found in openings, gardens, and forest edge.

Extralimital—From tropical Asia and e China and Korea to the Bismarcks, Solomons, and n and e Australia (Higgins 1999, Mayr & Diamond 2001).

subspecies

a. *Eurystomus orientalis waigiouensis* Elliot

Eurystomus waigiouensis Elliot, 1871. *Ibis*: 203.—Waigeo I, NW Islands.

range: All NG (but scarce along the n coast), Misool, Batanta, Waigeo, Yapen, Karkar, and Bagabag I[CO]; also the D'Entrecasteaux and Louisiade Arch (Mayr 1941). Specimen records suggest that the Mainland resident *waigiouensis* is more numerous in the south and scarce in the north, while *pacificus* is mainly transient in the south and winters in large numbers in the north (Coates 1985). Resident race not yet recorded from Salawati, Biak, Numfor, Mios Num, or Manam I; nor from the Aru or Trobriand Is.

diagnosis: Head blackish brown; ventral and dorsal plumage dark turquoise blue with a green tinge; outer tail feathers with indigo (Fry 2001). Diamond (1972) reported this form can be distinguished from *pacificus* in the field by the darker, richer coloration, the bluer ventral plumage, and the greener dorsal plumage.

b. *Eurystomus orientalis pacificus* (Latham)
Coracias pacifica Latham, 1801. *Index Orn.*, Suppl.: xxvii.—Sydney, NSW, Australia.
RANGE: Breeds in Australia. On migration and during austral winter occurs in Sulawesi, Moluccas, Kai Is, the NG region (including Gebe, Gag, Misool I[Q], Aru Is, Karkar, Bagabag, Fergusson I), the Bismarck Arch (including New Britain and Witu Is), and Solomon Is (Mayr 1941, Mees 1965[Q]). The migratory race is present in the Region from early March to November. It is not possible to distinguish immatures of the two races in the field. Two birds (race not determined, probably *pacificus*) recovered from ice at 4,500 m on the Carstensz Massif had presumably perished while migrating across the high central cordillera (Schodde *et al.* 1975). Migrants arrive around Port Moresby by mid-March and Geelvink Bay by the first week of April. The species becomes noticeably common at Port Moresby from mid-September to early November (Coates 1985).
DIAGNOSIS: Paler than preceding form. Head dull brown; ventral and dorsal plumage dull turquoise with a brown wash; outer tail feathers dull turquoise (Fry 2001).

NOTES
Other names—Dollarbird, Broad-billed Roller.

Family Halcyonidae

Woodland Kingfishers and Kookaburras   [Family 62 spp/Region 23 spp]

The species of the family range from Africa and the Middle East to tropical Asia, the Solomon Islands, Micronesia, and Polynesia. This family includes the kookaburras, paradise-kingfishers, and relatives, which are mainly forest-dwelling and not fish-eating. Species of the family are small to large, with a disproportionally large head, a strong, wedge-shaped bill, short legs, and a prominent tail. Vocalizations are complex and musical, often trilled. We follow Christidis & Boles (2008) in treating this lineage as a full-fledged family and in use of Halcyonidae Vigors, 1825, as the family's name, rather than Dacelonidae Bonaparte, 1841 (Bock 1994). Moyle's (2006) molecular systematic study provided guidance on disposition of the genera as well as boundaries of the familial groupings. The family is sister to Alcedinidae, the river kingfishers (Dickinson & Remsen 2013).

GENUS *TANYSIPTERA*   [Genus 10 spp/Region 9 spp]

Tanysiptera Vigors, 1825. *Trans. Linn. Soc. London* 14: 433, note. Type, by monotypy, *Alcedo dea* Linnaeus, 1766 = *Tanysiptera nais* G. R. Gray, [not *Alcedo dea* Linnaeus, 1758].

The species of this genus range from Australia to New Guinea, New Britain, and Wallacea (Woodall 2001). The genus is characterized by the greatly elongate and narrow central tail feathers, small size, and colorfully patterned body plumage.

Common Paradise-Kingfisher *Tanysiptera galatea*   RESIDENT
Tanysiptera galatea G. R. Gray, 1859. *Proc. Zool. Soc. London*: 154.—Manokwari, Bird's Head.

Frequents forest interior of NG, the NW Islands, and certain fringing islands of the n coast; 0–500 m, max 600 m on the Mainland[co] and 820 m on Karkar I (Diamond & LeCroy 1979). May be absent from the Purari-Kikori region; few records from w sector of the S Lowlands (Etna Bay to Digul R) or n coast of the SE Peninsula. Diamond *et al.* (1977) considered it apparently absent around Vanimo. Beehler & Mack (1999) found it patchy in the Lakekamu basin. Thus it may be locally absent from certain forest tracts and present elsewhere nearby.

Extralimital—Moluccas (Halmahera, Bacan, Buru, and Seram).

a. *Tanysiptera galatea galatea* G. R. Gray

RANGE: Salawati, Batanta, Waigeo, and Gebe I; Bird's Head and Neck; east in the s watershed to Triton Bay; east in the n watershed to the e coast of Geelvink Bay (Siriwo R); absent from Misool I (Mayr 1941).

DIAGNOSIS: Wing 107–111. Dark turquoise cap with brighter edges and sharply defined posterior edge; blackish cheek; blackish-blue mantle; shoulder dark turquoise; base of central tail feathers mainly blue; underwing blue and white.

b. *Tanysiptera galatea minor* Salvadori & D'Albertis

Tanysiptera galatea var. *minor* Salvadori & D'Albertis, 1875. *Ann. Mus. Civ. Genova* 7: 815.—Mount Epa, SE Peninsula.

RANGE: All of s NG east of the Bird's Neck; also the n watershed of the SE Peninsula from the Kumusi R to East Cape.

DIAGNOSIS: Wing 100–108; like nominate but bill and wing length smaller (Rand & Gilliard 1967).

c. *Tanysiptera galatea meyeri* Salvadori

Tanysiptera meyeri Salvadori, 1889. *Orn. Pap. Moluc.*, Agg., pt. 1: 54.—Kafu, east of Aitape, Sepik-Ramu, north-central NG.

RANGE: Northern NG from the Mamberamo east to the Huon Penin. Also Karkar[K], Bagabag[K], and Tarawai I[MA] (Diamond & LeCroy 1979[K]).

DIAGNOSIS: Much more white in the base of the central tail feathers than present in the nominate race (Rand & Gilliard 1967).

d. *Tanysiptera galatea vulcani* Rothschild and Hartert

Tanysiptera hydrocharis vulcani Rothschild and Hartert, 1915. *Novit. Zool.* 22: 42.—Manam I, ne NG.

RANGE: Manam I (Mayr 1941).

DIAGNOSIS: Larger; wing 109–118 (Rand & Gilliard 1967).

NOTES

We here recognize *ellioti*, *riedelii*, *carolinae*, and *rosseliana* as distinct allospecies (see Woodall 2001). *Other names*—Galatea Racket-tail.

Rossel Paradise-Kingfisher *Tanysiptera rosseliana*   ENDEMIC | MONOTYPIC

Tanysiptera rosseliana Tristram, 1889. *Ibis*: 557.   Rossel I, Louisiade Arch, SE Islands.

Rossel I, Louisiade Arch, to 300 m. Inhabits forest interior. Appears widespread and common across the island (Mayr 1941).

NOTES

We raise *rosseliana* to full-species status. It is distinguished from *T. galatea minor* (the geographically nearest sister population) in the following: outer tail feathers all white; basal half of central tail feathers all white; crown rich dark blue, not turquoise; cheek dark blue, not blackish; underwing black, not whitish; mantle deep blue, not blackish; shoulder patch rich dark blue, not spangled turquoise. It is surprising that *Tanysiptera* is present on Rossel but absent from the Trobriands, D'Entrecasteaux, and the rest of the Louisiade Arch. *Other names*—previously treated as a race of *T. galatea*.

Biak Paradise-Kingfisher *Tanysiptera riedelii*   ENDEMIC | MONOTYPIC

Tanysiptera Riedelii J. Verreaux, 1866. *Nouv. Arch. Mus. Paris, Bull.* 2: 21, pl. 3, f. 1.—Biak I, Bay Islands.

Biak I (Mayr 1941, Hoek Ostende *et al.* 1997). Common in all wooded habitats and can be observed at forest edge (Bishop ms).

NOTES

This island sister species of *T. galatea* exhibits a brilliant spangled turquoise hood (encompassing the crown, side of head, and nape); black underwing; spangled turquoise shoulder patch; and largely white basal half of all rectrices. Observed repeatedly in 1997 in n Biak in various wooded habitats, where it was common, unwary, and fairly easy to observe (Beehler & S. van Balen field notes). *Other names*—Riedel's Racket-tail.

Kofiau Paradise-Kingfisher *Tanysiptera ellioti*   ENDEMIC | MONOTYPIC

Tanysiptera ellioti Sharpe, 1870. *Proc. Zool. Soc. London* (1869): 630.—Kofiau I, NW Islands.

Endemic to Kofiau I, where it is common in all wooded habitats (Ripley 1959, Diamond *et al.* 2009, Pratt & Beehler 2014).

NOTES

This is another single-island sister species of *T. galatea*. This form's all-white central tail feathers, without attenuation and without the prominent spatulate tip, are its most distinctive character. Treated by Beehler & Finch (1985) as a subspecies of *galatea*, but recognized as a full species by most other authorities. Found in a remnant patch of coastal forest in Dec-2011 (Beehler field notes). Diamond *et al.* (2009) noted the song to be more plaintive and shorter than that of *galatea*. We suggest its IUCN Red List status should be Near Threatened rather than Vulnerable. *Other names*—Elliot's Paradise-kingfisher, Kafiau Paradise Kingfisher.

Numfor Paradise-Kingfisher *Tanysiptera carolinae*   ENDEMIC | MONOTYPIC

Tanysiptera Carolinae Schlegel, 1871. *Ned. Tijdschr. Dierk.* (1873) 4: 13.—Numfor I, B

Endemic to the island of Numfor (Mayr 1941, Hoek Ostende *et al.* 1997). The species is common in all lowland habitats, including beach vegetation and highly degraded forest. Much of the forest on Numfor has been degraded by logging and subsistence agriculture.

NOTES

A very distinctive island endemic. Entire body plumage—both dorsal and ventral—midnight blue (but for white rump and tail). Feet bright yellow (*pace* Woodall 2001). *Other names*—Caroline Racket-tail.

Little Paradise-Kingfisher *Tanysiptera hydrocharis*   NEAR-ENDEMIC | MONOTYPIC

Tanysiptera hydrocharis G. R. Gray, 1858. *Proc. Zool. Soc. London*: 172.—Aru Is.

Found only in the Aru Is, the Trans-Fly, and near Kiunga, to 300 m. Shares the forest with *T. galatea* and the migrant *T. sylvia*. Apparently occupies drier forests than those occupied by *T. galatea* (Woodall 2001). Song reported to be distinctive.

Localities—Aru Is[MA], Princess Marianne Strait[M], Digul R[CO], Merauke, Wassi Kussa R[MA], Tarara[L], Wuroi[CO], Lake Daviumbu[L], Lake Murray[J], Kiunga[K], lower Fly R[MA] (Rand 1942a[L], Gregory 1997[K], Bishop 2005a[M], B. J. Coates *in litt.*[J]).

Extralimital—a single specimen record from Saibai I in Jan-2010 (B. J. Coates *in litt.*).

NOTES

This is a small version of the widespread *T. galatea*. Distinguished by the small size (wing 85–86), dark blue crown and shoulder, turquoise eyebrow, blackish-blue mantle, white underwing, and diminutive bill. Status on the Aru Is perhaps should be reviewed, as two recent surveys did not record it there (Diamond & Bishop 1994, D. Gibbs *in litt.*). *Other names*—Lesser Paradise Kingfisher, Aru Paradise-kingfisher.

Buff-breasted Paradise-Kingfisher *Tanysiptera sylvia*   RESIDENT AND VISITOR

Tanysiptera sylvia Gould, 1850. In Jardine, *Contr. Orn.*: 105.—Cape York Penin, QLD, Australia.

Resident breeder around Port Moresby, from the upper Angabanga R southeast to the Kemp Welch R; appears to be confined to foothill monsoon forest except for occasional records in second growth fringing nearby lowland rain forest. In the foothill monsoon forest the species coexists with and is outnumbered by *T. galatea*. In addition, Australian breeders migrate to NG to spend the nonbreeding season (Legge *et al.* 2004). Migrants present from late April to early November (Coates & Peckover 2001).

Extralimital—a migratory breeding population inhabits coastal QLD, wintering to NG.

SUBSPECIES

a. *Tanysiptera sylvia sylvia* Gould

RANGE: Breeds in coastal QLD. Winters to NG, in the s watershed from the Setekwa R to the Fly R and Mt Sisa, and thence east to the Lakekamu basin; in the n watershed from Lake Sentani to the middle Sepik.

Localities—Setekwa R[MA], Ifar[MA], near Humboldt Bay[MA], Holtekong[J], Puwani R[Y], Lake Daviumbu[K], Palmer Junction on the Fly R[K], Tabubil[A], Nomad R[M], August R[O], Hunstein Mts[L,N], Moro Camp[Z], Lakekamu basin[X] (Rand 1942a[K], Ripley 1964a[J], Coates 1990: 567[O], Bell 1970b[M], Beehler *et al.* 1994[X], Gregory 1995b[A], Legge *et al.* 2004[N], A. Allison *in litt.*[L], Beehler field notes[Y], K. D. Bishop *in litt.*[Z]).

DIAGNOSIS: The Australian race *sylvia* can be distinguished in the field from endemic NG *salvadoriana* by the following: buff of underparts deeper and richer; area of black on mantle larger; blue on wing and crown deeper and more navy-hued; central tail feathers all white, more attenuated, and pointed.

b. *Tanysiptera sylvia salvadoriana* Ramsay

Tanysiptera salvadoriana Ramsay, 1879. *Proc. Linn. Soc. NSW* 3(3): 259.—Near Port Moresby.

RANGE: The s watershed of the SE Peninsula between the Angabanga and Kemp Welch R. Localities—Deva Deva[CO], Hall Sound, upper Angabanga R[CO], the Aroa R–Mt Cameron area[CO], north of Mt Lawes[CO], foothills of the Astrolabe Ra[CO], Moroka distr, and Kemp Welch R (Mayr 1941)

DIAGNOSIS: Greener dorsally and paler ventrally than nominate. Central tail feathers blue-edged, less attenuated, and not pointed.

NOTES

A question to be addressed is whether there are places in the SE Peninsula where wintering birds from Australia reside next to resident breeding birds. The New Britain form *nigriceps*, which has traditionally been placed within this species, is here considered a distinct species (Dutson 2011).*Other names*—White-tailed or Australian Paradise-Kingfisher.

Red-breasted Paradise-Kingfisher *Tanysiptera nympha* ENDEMIC | MONOTYPIC

Tanysiptera Nympha G. R. Gray, 1840. *Ann. Mag. Nat. Hist.* (1)6: 238.—Bird's Head.

Patchily distributed in hill forest interior of the Bird's Head and Neck, Sepik-Ramu, and nw sector of the SE Peninsula; 500–1,000 m, extremes sea level–1,500 m (Stresemann 1923, Coates & Peckover 2001).

Localities—Bama in the Tamrau Mts[N], Kebar valley[N], Onin Penin[MA], Fakfak Mts[MA], Etna Bay[MA], Rubi[Q], lower Menoo R in the Weyland Mts[J], Wewak coastal hills, Maratambu in the Adelbert Mts[K], near Onga in the Kratke Mts, Huon Penin (Finschhafen[L], Butaueng[L], Braunschweighafen[L], Sattelberg[L], Burrumtal[L], Rawlinson Ra[MA], Ogeramnang[L]), Musom (near Adler R, Huon Gulf), Keku[O], Maratambu[K], Surprise Creek on the upper Watut R[MA], Mt Missim[M], and the lower Waria R[P] (Stresemann 1923[Q], Mayr 1931[L], Hartert *et al.* 1936[J], Gilliard & LeCroy 1967[K], Gilliard & LeCroy 1970[N], Dawson *et al.* 2011[P], Beehler field notes[M], AMNH specimens[O]).

NOTES

Dawson *et al.* (2011) found *T. galatea*, *nympha*, and *danae* in the lower Waria valley. *Other names*—Pink-breasted Paradise Kingfisher.

Brown-headed Paradise-Kingfisher *Tanysiptera danae* ENDEMIC | MONOTYPIC

Tanysiptera Danae Sharpe, 1880. *Ann. Mag. Nat. Hist.* (5) 6: 231.—Milne Bay, SE Peninsula.

Frequents foothill forest of the e sector of the SE Peninsula, from the Waria R to East Cape (in the n watershed) and from the Aroa R to East Cape (in the s watershed); 300–1,000 m (Coates & Peckover 2001). Around Port Moresby, normally occurs only in the hills, above the range of *T. galatea*, but both species were observed near Popondetta, at 150 m elevation, with sightings of *T. danae* outnumbering those of *T. galatea* by five to one (G. Clapp, in Bell 1981).

Localities—the Waria R[MA], Aikora R[J], Baniara[J], Aroa R[MA], Popondetta[CO], Mt Lamington[M], Hydrographer Mts[MA], lower Mambare R[J], Sogeri[CO], Varirata NP[K], Orangerie Bay[L], and Samarai in Milne Bay[MA] (Mayr & Rand 1937[L], many records[K], AMNH specimens[J], BPBM specimen[M]).

Other names—Brown-backed Paradise-kingfisher.

Melidora Lesson, 1830. *Traité d'Orn.* 4: 249. Type, by monotypy, *Melidora euphrosiae* Lesson = *Dacelo macrorrhinus* Lesson.

This endemic and monotypic genus is characterized by the large, broad-based, and strongly hooked beak, turquoise-scalloped crown, and crepuscular behavior.

Hook-billed Kingfisher *Melidora macrorrhina* ENDEMIC

Dacelo macrorrhinus Lesson, 1827. *Bull. Sci. Nat. Férussac* 12: 131.—Manokwari, Bird's Head.

All NG, the NW Islands, and Yapen I; sea level–750 m, max 1,900 m (Freeman & Freeman 2014). Inhabits forest interior, where very difficult to observe. Vocal mainly at dawn and dusk.

SUBSPECIES

a. *Melidora macrorrhina waigiuensis* Hartert
Melidora macrorrhina waigiuensis Hartert, 1930. *Novit. Zool.* 36: 99.—Waigeo I, NW Islands.
RANGE: Waigeo I (Mayr 1941).
DIAGNOSIS: Long-winged: 122–132 (Rand & Gilliard 1967).

b. *Melidora macrorrhina macrorrhina* (Lesson)
RANGE: Misool, Salawati, and Batanta I, the S Lowlands, and the Huon and SE Peninsulas (Mayr 1941).
DIAGNOSIS: Short-winged: 114–123. Pale highlights on black crown of female (Diamond 1972). See photo of male in Coates & Peckover (2001: 108).

c. *Melidora macrorrhina jobiensis* Salvadori
Melidora jobiensis Salvadori, 1880. *Orn. Pap. Moluc.* 1: 502.—Ansus, Yapen I, Bay Islands.
RANGE: Yapen I, NW Lowlands, and the Sepik-Ramu (Mayr 1941).
DIAGNOSIS: Crown of male a deeper blue than for the other forms; crown of female mainly blackish (Rand & Gilliard 1967, Woodall 2001).

NOTES
Other names—Hook-billed Kookaburra.

GENUS *CLYTOCEYX* ENDEMIC | MONOTYPIC

Clytoceyx Sharpe, 1880. *Ann. Mag. Nat. Hist.* (5)6: 231. Type, by original designation and monotypy, *Clytoceyx rex* Sharpe.

This monotypic genus is restricted to the Region, ranging from coastal mangroves to mid-montane forest interior. The genus is characterized by the broad, high, and abbreviated bill, powerful body, and unmarked brown plumage. The bill shape alone is sufficient to warrant the species' placement in a distinct genus. A DNA phylogeny by Moyle (2006) positioned *Clytoceyx* as sister to the kookaburras.

Shovel-billed Kookaburra *Clytoceyx rex* ENDEMIC | MONOTYPIC

Clytoceyx rex Sharpe, 1880. *Ann. Mag. Nat. Hist.* (5)6: 231.—East Cape, NG.

Synonym: *Clytoceyx rex imperator* van Oort, 1909. *Nova Guinea, Zool.* 9: 79.—Lorentz R, at the foot of the Hellwig Mts.
Synonym: *Clytoceyx rex septentrionalis* Paludan, 1935. *Orn. Monatsber.* 43: 54.—Sepik R.

Patchily distributed and uncommon throughout the Mainland, sea level–2,700 m[J] (Stresemann 1923, Mayr 1941, Coates 1985, P. Z. Marki *in litt.*[J]). Frequents forest interior; occurs mainly in the hills and mts, locally in the lowlands in the vicinity of hills.

Localities—Imbai on Anggi Gigi[J], Anggi Lakes[K], Mamberamo R[MA], Bodim[X], Bernhard Camp on the Taritatu R[Q], Alkmaar on the Lorentz R[J,U], Telefomin[T], Black R on the upper Fly R[MA,P], the Tabubil area[Y], Malu on the Sepik R[J], the Kaijende Highlands[O], Ambua[Y], Jimi R[Z], Mt Wilhelm[W], Astrolabe Bay[MA], Purari basin[B], the upper Hegigio R[B], hills near Libano and uplands east of Mt Sisa[B], Baia R[B], the upper Fly R karst[B], YUS[A], Mt Kaindi[A], Surprise Creek on the Watut R[MA], Mt Missim[A], Lakekamu basin[A], Owen Stanley

Ra[Y], Sogeri Plateau[L], and East Cape[MA] (Rand 1942a[P], 1942b[Q], Mayr & Gilliard 1954[Z], Gyldenstolpe 1955b[J], Sims 1956[W], Gilliard & LeCroy 1961[T], Ripley 1964a[X], Hoek Ostende *et al.* 1997[U], Beehler field notes, various[A], Beehler & Sine 2007[O], P. Gregory *in litt.*[Y], B. J. Coates *in litt.*[L], D. Gibbs *in litt.*[K], I. Woxvold *in litt.*[B])

NOTES

We treat this species as monotypic, as there are no substantial plumage differences among the named races, and the size difference between measured museum specimens of the largest form and smallest is less than 10%. The race *imperator* is known from a tiny sample. *Other names*—Shovel-billed Kingfisher, *Dacelo rex*.

GENUS *DACELO* [Genus 4 spp/Region 3 spp]

Dacelo Leach, 1815. *Zool. Misc.* 2: 125. Type, by subsequent designation, *Alcedo gigantea* Latham = *Alcedo novaeguineae* Hermann.

The species of the genus range from Australia to New Guinea. The species mainly inhabit lowland savanna and forest. The genus is characterized by large size, large to very large pointed beak, blue highlighting on wing and rump, and maniacal cackling call.

Blue-winged Kookaburra *Dacelo leachii* RESIDENT

Dacelo leachii Vigors & Horsfield (ex Latham ms), 1827. *Trans. Linn. Soc. London* (1826) 15: 205.—Keppel Bay, Shoalwater Bay, Broad Sound, QLD, Australia.

Woodlands and savanna of the w sector of the S Lowlands, the Trans-Fly, and the Port Moresby region, sea level–600 m. Apparently the species inhabits three noncontiguous sub-ranges in s NG. Occasionally disperses into suburban gardens and plantations.

Extralimital—nw, n, and e Australia.

SUBSPECIES

a. *Dacelo leachii superflua* Mathews
Dacelo leachii superflua Mathews, 1918. *Birds Austral.* 7: 140.—Mimika R, sw NG.
RANGE: The w sector of the S Lowlands, from the Mimika R east toward the PNG border; few records from much of this swath between the Mimika and Merauke. Burrows reported a sight record from Sorong, w Bird's Head (Coates & Peckover 2001). The e terminus of *superflua* is not yet determined, but must be found somewhere near Merauke (Mees 1982b). There may or may not be a distributional gap between *superflua* and *intermedia*.
DIAGNOSIS: Crown streaking exhibits more white than brown. This distinction was confirmed by comparison of museum material at the BMNH.

b. *Dacelo leachii intermedia* Salvadori
Dacelo intermedius Salvadori, 1876. *Ann. Mus. Civ. Genova* 9: 21.—Hall Sound, SE Peninsula.
RANGE: The Trans-Fly and SE Peninsula (Angabanga R to Amazon Bay; Bell 1970a).
DIAGNOSIS: Crown color dominated by dark brown streaking (with a lesser amount of white than *superflua*). See photo in Coates & Peckover (2001: 104). Underparts entirely white without obsolete barring on flanks (Woodall 2001: pl. 11).

Spangled Kookaburra *Dacelo tyro* ENDEMIC

Dacelo tyro G. R. Gray, 1858. *Proc. Zool. Soc. London*: 171.—Aru Is.

Endemic to the Aru Is and the Trans-Fly. Inhabits dry woodland, swamp margins, and wooded savanna, to 300 m. In NG occurs from the Bian R and Habe I (a small island between Merauke and Dolak I) east to the Wassi Kussa R (Coates 1985).

SUBSPECIES

a. *Dacelo tyro tyro* G. R. Gray
RANGE: The Aru Is.
DIAGNOSIS: Adult male has throat white; breast and belly washed with buff; crown and cheek with buff spotting.

b. *Dacelo tyro archboldi* (Rand)

Sauromarptis tyro archboldi Rand, 1938. *Amer. Mus. Novit.* 990: 13.—Tarara, Wassi Kussa R, Trans-Fly.

RANGE: The Trans-Fly, between the Bian R and Wassi Kussa R[CO]. Localities—Kurik[L], Darowin[L], Bensbach R[J], Morehead[MA], Penzara[K], and Tarara[K] (Rand 1942a[K], Mees 1982b[L], B. J. Coates *in litt.*[J]).

DIAGNOSIS: Adult male has all underparts whitish; throat not distinct from breast; crown with whitish spotting. See photo in Coates & Peckover (2001: 104).

NOTES

In places it can be found in the same habitat with *D. leachii* (Coates & Peckover 2001). *Other names*—Aru Giant Kingfisher.

Rufous-bellied Kookaburra *Dacelo gaudichaud* ENDEMIC | MONOTYPIC

Dacelo Gaudichaud Quoy & Gaimard, 1824. *Voy. Uranie, Zool.* 1: 112, pl. 25.—Waigeo I, NW Islands.

All NG, the four main NW Islands[MA], Aru Is[MA], Bay Is (Mios Num and Yapen)[MA], and Sariba and Heath[MA] of the SE Islands; lowlands–750 m, max 1,300 m (Coates & Peckover 2001). The record from Gebe I (Mayr 1941) is in error (Mees 1972). Frequents forest, mangroves, and edge. High-elevation sites include Baiyer R Sanctuary and Mt Bosavi (Coates 1985).

Other names—Rufous-bellied Giant Kingfisher, Gaudichaud's Kingfisher, *Sauromarptis gaudichaud*, *Dacelo gaudichaudi*.

GENUS *TODIRAMPHUS* [Genus 24 spp/Region 7 spp]

Todiramphus Lesson, 1827. *Mém. Soc. Hist. Nat., Paris* 3(3): 420. Type, by subsequent designation (G. R. Gray, 1840. *List. Gen. Birds*: 10), *Todiramphus sacer* Lesson, 1827 = *Alcedo tuta* Gmelin, 1788 [not *Alcedo sacra* Gmelin, 1788].

The species of this genus range from the Horn of Africa to New Zealand and east to the Marquesas (Woodall 2001). The genus is characterized by the black upper mandible (maxilla) and the presence of a dark mask. *Todiramphus* is masculine, and species and subspecies epithets should agree when they are adjectival modifiers (David & Gosselin 2002a). This taxon was formerly included in the genus *Halcyon*. The phylogenetic study of Andersen *et al.* (2015b) demonstrated the rapid speciation of the *T. chloris* complex and recommended specific wholesale taxonomic revision of that lineage, which we have followed here.

Blue-black Kingfisher *Todiramphus nigrocyaneus* ENDEMIC

Halcyon nigrocyanea Wallace, 1862. *Proc. Zool. Soc. London*: 165, pl. 19.—Manokwari, Bird's Head.

Inhabits forest interior through the NG lowlands, Salawati and Batanta I, and Yapen I, with records to 600 m. In the n watershed it ranges east to the lower Markham R (I. Woxvold *in litt.*); in the s watershed, it ranges east to just north of Port Moresby (Mt Cameron and upper Vanapa R)[MA]. Frequents forest interior in the vicinity of streams and swamps (Coates 1985).

SUBSPECIES

a. *Todiramphus nigrocyaneus nigrocyaneus* (Wallace)

RANGE: Salawati and Batanta I; Bird's Head and Neck; w sector of S Lowlands east to Princess Marianne Strait (Mayr 1941).

DIAGNOSIS: White throat patch, blue breast band, and white patch below blue breast band; white patch is small and crescent-shaped in male, large and extending to belly in female (Woodall 2001).

b. *Todiramphus nigrocyaneus quadricolor* (Oustalet)

Cyanalcyon quadricolor Oustalet, 1880 (December). *Le Naturaliste* 2(41): 323.—Eastern coast of Geelvink Bay, between long. 136.5°E and 137°E.

RANGE: Yapen I[MA], the NW Lowlands, and the Sepik-Ramu. Localities—Wanggar R[J], Holtekong[K], Nimbokrang[M], upper Sepik[W], Hunstein Mts[Z], 40 km south of Maprik[N], Oronga in the Adelbert Mts[L], Gogol R[O], Astrolabe Bay[MA], lower Markham R 45 km west of Lae[W] (Hartert *et al.* 1936[J], Ripley 1964a[K], Gilliard &

LeCroy 1967[L], Pearson 1975[N], A. Allison *in litt.*[Z], I. Woxvold *in litt.*[W], various field reports[M], Beehler field notes[O]).

DIAGNOSIS: Male has white throat patch and large rufous patch below blue breast band. See photo in Coates & Peckover (2001: 106).

c. *Todiramphus nigrocyaneus stictolaemus* (Salvadori)

Cyanalcyon stictolaema Salvadori, 1876. *Ann. Mus. Civ. Genova* 9: 20.—Upper Fly R.

RANGE: From the c Trans-Fly east to the Port Moresby environs on the SE Peninsula (Mayr 1941). Localities—Kurik[Q], Morehead[J], Sturt I Camp on the upper Fly R[J], Lakekamu basin[L], Mt Cameron on the upper Vanapa R[MA] (Rand 1942a[J], Mees 1982b[Q], Mack 1998[L]).

DIAGNOSIS: Male throat blue with a little white showing through; breast blue; abdomen blue, with or without a bit of white in center; flanks blackish. Female has throat and abdomen white, broken by a broad blue breast band (similar to female of other two races).

NOTES

Shown by Andersen *et al.* (2015b) to be sister to all other *Todiramphus* species. *Other names*—Black-sided Kingfisher.

Forest Kingfisher *Todiramphus macleayii*   RESIDENT AND VISITOR

Halcyon Macleayii Jardine and Selby, 1830. *Illustr. Orn.* (1)2(7): pl. 101 and text.—Port Essington, NT, Australia.

A resident and also an austral migrant to e and south-central NG. Resident race breeds in the SE Peninsula and along the n coast west to Madang, sea level–700 m. Australian migrants occur locally in e NG (west to Merauke in south and to Madang in north), including the SE Islands as well as the Aru Is, to 1,800 m. Migrants are found in Region from March to October. Frequents savanna, cleared areas, and gardens. In spite of English name, this species does not inhabit forest.

Extralimital—breeds in n and e Australia; migrants from Australia reach the Tanimbar and Kai Is and rarely New Britain, the Solomons, and Norfolk I (Gill *et al.* 2010, Dutson 2011).

SUBSPECIES

a. *Todiramphus macleayii macleayii* (Jardine and Selby)

Synonym: *Halcyon macleayi insularis* von Berlepsch, 1911. *Abh. Senck. Naturf. Ges.* 34: 75.—Trangan, Aru Is.
Synonym: *Cyanalcyon Elisabeth* Heine, 1883. *J. Orn.* 31: 222.—Astrolabe Bay, e verge Sepik-Ramu.

RANGE: Northern Australia; also breeds in e NG, west at least to Hall Sound in the s watershed and to Astrolabe Bay in the n watershed. Some Australian birds of this subspecies may migrate to the Aru Is and the e Lesser Sundas (Mayr 1941).

DIAGNOSIS: Back turquoise blue, with no greenish wash; we follow Mees (1982b) in sinking *elisabeth* and *insularis* into the nominate form.

b. *Todiramphus macleayii incinctus* (Gould)

Halcyon incinctus Gould, 1838. *Synops. Birds Austral.*, app.: 1.—New South Wales, Australia.

RANGE: Breeds in QLD and NSW. Southern breeders migrate to the Kai Is and the S Lowlands, SE Peninsula, and SE Islands—Samarai, Fergusson I, Trobriand Is, Woodlark, Misima, and Tagula I, and New Britain (Mayr 1941). There is a record of this form from Nondugl in the Eastern Ra (Gyldenstolpe 1955a).

DIAGNOSIS: Greenish mantle (Woodall 2001). See photo in Coates & Peckover (2001: 106).

NOTES

Other names—Blue or Macleay's Kingfisher, *Halcyon macleayii*.

Collared Kingfisher *Todiramphus chloris*   RESIDENT

Alcedo Chloris Boddaert, 1783. *Tabl. Planch. Enlum.*: 49.—Buru I, s Moluccas.

Coastal areas of the NW Islands, Bird's Head, and Bird's Neck, at sea level. Frequents mangroves, littoral woodlands, and coastal plantations.

Extralimital—from Red Sea to the Philippines and Indonesia.

a. *Todiramphus chloris chloris* (Boddaert)

RANGE: The Sulawesi region, Moluccas, Lesser Sundas, Misool, Salawati, Gag, Waigeo (Saonek), and w and s coast of the Bird's Head and Neck (Mayr 1941).

DIAGNOSIS: Crown and mantle a bright greenish blue; shoulder, rump, and eyebrow pale blue; loral spot pure white; tail medium blue; bill large and robust. Wing: male 104–121, female 105–118; tail: male 61–74, female 62–73; bill: male 45–56 (Forshaw 1987).

NOTES

Two former races of the Collared Kingfisher have been raised to species status by Andersen *et al.* (2015b); see below. We concur and recognize all three species. They appear to be allopatric: *chloris* in the far west, *sordidus* along coasts facing the Arafura and Coral Seas, and *colonus* restricted to the myriad small islands off the SE Peninsula. The apparent gap in the range between *chloris* and *sordidus* may instead be due to inadequate survey coverage, and both species should be looked for between the Bird's Neck and the Mimika R. These three kingfishers differ in habitat as well: *chloris* occupies a range of coastal habitats, whereas *sordidus* is strictly a mangrove specialist, shunning other types of coastal forest and small islands; *colonus* is an island tramp species, with an ability to colonize small islands—it has not been found on the Mainland, where the larger *sordidus* occurs. In the Region, the western isolate, *T. c. chloris*, is a bright blue-green, much bluer than the other two, olive-green species (*T. sordidus and T. colonus*). *T. c. chloris* is about the same size as *T. sordidus* but much larger than *T. colonus*. *Other names*—Mangrove Kingfisher, *Halcyon chloris*.

Torresian Kingfisher *Todiramphus sordidus*   RESIDENT

Halcyon sordidus Gould, 1842. *Proc. Zool. Soc. London*: 72.—Cape York Penin, QLD, Australia.

   A mangrove specialist of coastal s New Guinea and nw, n, and e Australia.

SUBSPECIES

a. *Todiramphus sordidus sordidus* (Gould)

RANGE: Coast of n Australia, islands of Torres Strait, Aru Is, and the s coast of NG between the Mimika R and East Cape. Few records from Trans-Fly (K. D. Bishop *in litt.*)

DIAGNOSIS: Crown and back dusky olive green; shoulder dull green; wing: male 100–112, female 100–117 (Forshaw 1987). Distinguished from *pilbara* of w Australia by brighter upperparts and broader white collar; *pilbara* is more brownish and slightly paler (Fry *et al.* 1992).

NOTES

Formerly treated as a subspecies of the Collared Kingfisher (*T. chloris*). Elevated to full species status by Andersen *et al.* (2015b). Presumably *sordidus* ranges west of the Mimika R and may meet *chloris* near the Bird's Neck. About the same size as *chloris*, *sordidus* is a darker bird with olive-green dorsal plumage and longer bill and tail. At the other, e end of its range, *T. sordidus* extends right to the tip of the SE Peninsula, but does not venture into the island realm beyond, home of the next species, the similar-looking but much smaller *T. colonus*.

Islet Kingfisher *Todiramphus colonus*   RESIDENT | MONOTYPIC

Halcyon sordidus colonus Hartert, 1896. *Novit. Zool.* 3: 244.—Egum Atoll, Louisiade Arch, SE Islands.

   A dwarf form regionally endemic to the SE Islands, where it is widely distributed on mostly small islands and inhabits all types of coastal forest. Range includes Duchess and Tobweyama I off Normanby I, Bonvouloir Is, Alcester I, Egum Atoll, and Yanaba, north to small islands (*e.g.*, Reu) offshore of Woodlark, and to the Duchateau Is; also found on several larger islands—Misima, Tagula, and Rossel I (Mayr 1941).

NOTES

Formerly treated as a subspecies of *T. chloris*. Elevated to full species status by Andersen *et al.* (2015b). Similar to *sordidus* but much smaller and even darker dorsally, with dusky brownish upperparts; wing: male 89–94, female 88–95 (Forshaw 1987).

Beach Kingfisher *Todiramphus saurophagus*

Halcyon saurophaga Gould, 1843. *Proc. Zool. Soc. London*: 103.—Northern coast of NG west of the mouth of the Mamberamo R (toward Yapen I).

RESIDENT

Patchily distributed in the coastal lowlands of the Bird's Head and n coast of NG; also in the NW Islands, Bay Is, and small islets fringing the SE Peninsula. Frequents trees along the coastline and small islands of the Region.

Extralimital—Moluccas, Bismarcks, and Solomons (Mayr & Diamond 2001).

SUBSPECIES

a. *Todiramphus saurophagus saurophagus* (Gould)
RANGE: The n Moluccas, NW Islands, Bird's Neck, Bay Is, islands off the coast of n NG from Sorong eastward (Seleo, Kairiru, Manam I, islets near Madang), Wasu, Coutance I, patchily along the entire n coast of NG, SE Peninsula (Kupiano), SE Islands (Heath I, D'Entrecasteaux Arch, Egum Atoll, East and Kimuta I[J], islets off Woodlark I), Bismarck Arch, and Solomon Is (Mayr 1941, LeCroy & Peckover 1998[J], Dutson 2011).
DIAGNOSIS: Entire head and underparts white but for narrow and abbreviated dark streak behind eye (Woodall 2001).

NOTES

Other names—Halcyon saurophaga.

Sacred Kingfisher *Todiramphus sanctus*

Halcyon sancta Vigors & Horsfield, 1827. *Trans. Linn. Soc. London* 15: 206.—New South Wales, Australia.

VISITOR

A winter visitor from Australia to all NG and its satellite islands, sea level–2,400 m (Mayr 1941). Frequents openings, woodland edges, and gardens. Northward migration occurs in March and April; migration southward from late September to late October. Occasional individuals remain year-round (Coates 1985).

Extralimital—Breeds in Australia, NZ, Lord Howe I, Norfolk I, New Caledonia, and the Loyalty Is (Coates 1985, Gill *et al.* 2010); winters to the Bismarcks, Solomons, Moluccas, Sulawesi, Lesser Sundas, Java, and Sumatra (Mees 1982b, Higgins 1999).

SUBSPECIES

a. *Todiramphus sanctus sanctus* (Vigors & Horsfield)
RANGE: Breeds in e and s Australia. During migration and in the winter found in n Australia, the Moluccas and Lesser Sundas, the entire NG region, and the Bismarck and Solomon Is (Mayr 1941).
DIAGNOSIS: Crown and back dull blue-green (Woodall 2001).

NOTES

Other names—Halcyon sancta.

GENUS *SYMA*   [Genus 2 spp/Region 2 spp]

Syma Lesson, 1827. *Bull. Sci. Nat. Férussac* 11: 443. Type, by monotypy, *Syma torotoro* Lesson.

Species of the genus range from northeastern Australia to New Guinea. The genus is characterized by the serrated yellow bill. These species were long treated as members of the genus *Halcyon* (*e.g.*, Rand & Gilliard 1967).

Yellow-billed Kingfisher *Syma torotoro*

Syma torotoro Lesson, 1827. *Bull. Sci. Nat. Férussac* 11: 443.—Manokwari, Bird's Head.

RESIDENT

Frequents forest, edge, and mangroves throughout NG, the NW Islands, Aru Is, Yapen I, and D'Entrecasteaux Arch; sea level–700 m, max 1,200 m[CO]. Replaced at higher elevations by the closely allied *S. megarhyncha*.

Extralimital—Cape York Penin, QLD, Australia.

a. *Syma torotoro torotoro* Lesson

Synonym: *Syma torotoro tentelare* Hartert, 1896. *Novit. Zool.* 3: 534.—Aru Is.

Synonym: *Syma torotoro pseutes* Mathews, 1918. *Birds Austral.*, 7: 113.—Mimika R, sw NG.

Synonym: *Syma torotoro brevirostris* Rand, 1938. *Amer. Mus. Novit.* 990: 12.—Tarara, Wassi Kussa R, Trans-Fly, s NG.

Synonym: *Syma torotoro meeki* Rothschild and Hartert, 1901. *Novit. Zool.* 8: 147.—Southeastern NG between Huon Gulf and Brown R.

RANGE: All NG, the four main NW Islands, Aru Is, and Yapen I (Mayr 1941).

DIAGNOSIS: Abdomen whitish; undertail white (Woodall 2001). See photo in Coates & Peckover (2001: 108). We follow Mees (1982b) and Woodall (2001) in subsuming *tentelare*, *brevirostris*, *meeki*, and *pseutes* into the nominate form. Wing (male) 68–78 (AMNH measurements).

b. *Syma torotoro ochracea* Rothschild and Hartert

Syma torotoro ochracea Rothschild and Hartert, 1901. *Novit. Zool.* 8: 148.—Goodenough I, D'Entrecasteaux Arch, SE Islands.

RANGE: D'Entrecasteaux Arch (Mayr 1941).

DIAGNOSIS: Plumage rich and saturated honey brown ventrally, especially on abdomen and undertail; female's black crown patch large (Rand & Gilliard 1967, Woodall 2001). Voice distinctive. Size large and bill large. Wing (male) 79–84.5 (AMNH measurements).

NOTES

Voice of the form *ochracea* is a quavery succession of seven to nine whistled notes in a descending series, not recognizable as that of *S. torotoro* (Pratt & Beehler 2014; Pratt field notes and recordings from 2003; see also D. Gibbs recording from Fergusson I, xeno-canto website, no. 70362). This island form should be studied molecularly to determine its proper status relative to *torotoro* and *megarhyncha*. Hybrids between *torotoro* and the following species may occur rarely (Mees 1982b: 101). *Other names*—Lesser Yellow-billed Kingfisher, *Halcyon torotoro*.

Mountain Kingfisher *Syma megarhyncha*   ENDEMIC

Syma megarhyncha Salvadori, 1896. *Ann. Mus. Civ. Genova* 36: 70.—Moroka, SE Peninsula.

Frequents forest of the Central Ra, mts of the Huon Penin, and possibly the Foja and Adelbert Mts; 1,100–2,200^CO m (records as low as 700 m^CO). Absent from the mts of the Bird's Head. Replaced in the lowlands and hills by the closely allied *S. torotoro*, the transition typically taking place between 600 and 700 m. Record from the Adelbert Mts (Mackay 1991) requires confirmation, as it was not recorded there by previous or subsequent observers.

SUBSPECIES

a. *Syma megarhyncha megarhyncha* Salvadori

Synonym: *Syma megarhyncha wellsi* Mathews, 1918. *Birds Austral.* 7: 113.—Upper Otakwa R, Western Ra.

RANGE: The Central Ra and mts of the SE Peninsula.

DIAGNOSIS: This form is distinguished from *sellamontis* by the presence of dark patches on the distal portion of the bill and the larger size (Rand & Gilliard 1967). See photo in Coates & Peckover (2001: 108). Described form *wellsi* is not distinguishable by measurements or plumage.

b. *Syma megarhyncha sellamontis* Reichenow

Syma sellamontis Reichenow, 1919. *J. Orn.* 67: 334.—Sattelberg, Huon Penin.

RANGE: Mountains of the Huon Penin.

DIAGNOSIS: Ridge of culmen all yellow or nearly all yellow; small size (Woodall 2001).

NOTES

The two subspecies are thinly defined and may warrant lumping. When compared to *S. torotoro*, note that the black eye mark extends from the front to the back of the eye. *Other names*—Mountain Yellow-billed Kingfisher, *Halcyon megarhyncha*.

Family Alcedinidae

River Kingfishers
[Family 24 spp/Region 4 spp]

The species of the family range from Europe to Africa, tropical Asia, and Australia. All are small or very small, with a disproportionally large head, long and weak bill, tiny feet (three-toed in some species), tiny tail, and a prominent pale ear spot. We follow Christidis & Boles (2008) in treating this lineage as a family distinct from the woodland kingfishers (Halcyonidae). We follow the systematic treatment of Moyle (2006) and Moyle *et al.* (2007) for the specific and generic placement within this lineage.

GENUS *ALCEDO*
[Genus 7 spp/Region 1 sp]

Alcedo Linnaeus, 1758. *Syst. Nat.*, ed. 10, 1: 115. Type, by subsequent designation (Swainson, 1820–21 [1821]. *Zool. Illustr.* 1: text to pl. 26), *Alcedo ispida* Linnaeus.

The species of this genus range from Europe, Africa, and Madagascar, east to India, China, southeast Asia, Philippines, New Guinea, and the Bismarck and Solomon Islands (Woodall 2001, Moyle *et al.* 2007). The genus is characterized by a primary diet of fish, the blue dorsal plumage, small size, slim and weak bill, four-toed foot, and tiny abbreviated tail.

Common Kingfisher *Alcedo atthis*
RESIDENT

Gracula Atthis Linnaeus, 1758. *Syst. Nat.*, ed. 10, 1: 109.—Egypt.

Patchily distributed on the n coast of e NG, SE Peninsula, and D'Entrecasteaux and Louisiade Arch. In the n watershed it ranges west to Wewak; in the s watershed it ranges west to the Aroa R (Coates & Peckover 2001). Frequents lower reaches of rivers and creeks, to 100 m.

Extralimital—Europe to Japan and se Asia to the Bismarcks and Solomons (Mayr & Diamond 2001).

SUBSPECIES

a. *Alcedo atthis hispidoides* Lesson

Alcedo hispidoides Lesson, 1837. *Compl. de Buffon* 9: 345.—Buru I, s Moluccas.

RANGE: Sulawesi, Moluccas, ne NG between Wewak and Milne Bay, s watershed of SE Peninsula from the Aroa R eastward, Samarai I, D'Entrecasteaux and Louisiade Arch, and Bismarck Arch (Mayr 1941, Coates 1985, Coates & Peckover 2001). See photo in Coates & Peckover (2001: 103). Mees (1965) could find no confirmation for the records from Misool or Salawati I (*e.g.*, Mayr 1941). Thus there seems to be no evidence of this species from the NW Islands, although *A. atthis* does occur to the west in the Moluccas).

DIAGNOSIS: Throat white; remaining underparts rich honey-rufous; blue (not ocher) behind eye; purple tinge to hindneck and rump (Woodall 2001).

NOTES

It is interesting that this Eurasian species appears to be absent from the w two-thirds of the Mainland and yet the same subspecies inhabits the Moluccas as well as New Britain. This species may be only marginally successful in an environment so rich in kingfishers. *Other names*—River Kingfisher.

GENUS *CEYX*
[Genus ca. 20 spp/Region 3 spp]

Ceyx Lacépède, 1799. *Tabl. Méthod. Mamm. Oiseaux*: 10. Type, by monotypy, *Alcedo tridactyla* Pallas = *Alcedo erithaca* Linnaeus.

The species of this genus range from central Africa and Madagascar to tropical Asia, Sulawesi, Philippines, Australia, New Guinea, and the Bismarck and Solomon Islands (Woodall 2001, Moyle *et al.* 2007, Andersen *et al.* 2013). The genus is characterized by the three-toed foot, tiny size, preference for forest understory, and fish or non-fish diet. Moyle *et al.* (2007) provided molecular evidence for including *azureus, pusillus,* and the *C. lepidus* complex in this genus.

Papuan Dwarf Kingfisher *Ceyx solitarius*   ENDEMIC | MONOTYPIC

Ceyx solitaria Temminck, 1836. *Planch. Col. d'Ois.*: pl. 595, fig. 2.—Triton Bay, Bird's Neck.

Frequents rain forest in the vicinity of streams and pools throughout NG; the NW Islands[MA] but for Salawati and Gebe; the Aru Is[MA]; Biak[MA], Yapen[MA], and Karkar I[K]; and Fergusson and Normanby I[MA] (Diamond & LeCroy 1979[K]); sea level–1,385 m (Freeman & Freeman 2014). Possible record from Goodenough I (reported in M. Tarburton's island website).

NOTES

Andersen *et al.* (2013) provided molecular evidence that the NG population should be treated as a distinct species and that the traditional *lepidus* should be dissected into an array of species-level taxa. This was followed by del Hoyo & Collar (2014). The NG and New Britain forms are phenotypically very distinct from each other (see Coates & Peckover 2001: 102). The NG form exhibits a black bill, white throat, and rich honey-rufous breast. *Other names*—Dwarf Kingfisher, *Ceyx lepidus, Alcedo lepida.*

Azure Kingfisher *Ceyx azureus*   RESIDENT

Alcedo azurea Latham, 1801. *Index Orn.*, Suppl.: xxxii.—New South Wales, Australia.

Frequents wooded margins of rivers, lakes, and tidal mangrove creeks through NG, the NW Islands, Bay Is, Aru Is, Karkar I, and D'Entrecasteaux Arch; sea level–1,000 m, max 1,520[J] m (Coates 1985, Coates & Peckover 2001[J]).

Extralimital—Moluccas to n and e Australia. Replaced in the Bismarck Arch by the closely related *C. websteri.*

SUBSPECIES

a. *Ceyx azureus lessonii* (Cassin)

Alcyone Lessonii Cassin, 1850. *Proc. Acad. Nat. Sci. Phila.* 5: 69.—Manokwari, Bird's Head.

Synonym: *Alcyone azurea wallaceana* Mathews, 1918. *Birds Austral.* 7: 94.—Aru Is.

RANGE: All NG except for the NW Lowlands and Sepik-Ramu; also Misool, Batanta, and Waigeo I; Aru Is; and Fergusson and Normanby I[J] (Mayr 1941, BPBM specimen, collected by Pratt in 2003[J]).

DIAGNOSIS: Underparts a rich ochraceous tan (Rand & Gilliard 1967). See photo in Coates & Peckover (2001: 103).

b. *Ceyx azureus ochrogaster* (Reichenow)

Alcyone ochrogaster Reichenow, 1903. *J. Orn.* 51: 149.—Ramu R, ne NG.

RANGE: Numfor, Biak, and Yapen I; NW Lowlands and Sepik-Ramu; also Kairiru and Karkar I[J] (Mayr 1941, Diamond & LeCroy 1979[J]).

DIAGNOSIS: Underparts paler than in *lessonii*, whiter toward midline (Woodall 2001).

NOTES

Mayr (1941: 86) noted that *wallaceana* was "doubtfully distinct from *lessonii*." We follow Dickinson & Remsen (2013) in subsuming it into *lessonii*. *Other names*—*Alcedo* or *Alcyone azurea.*

Little Kingfisher *Ceyx pusillus*   RESIDENT

Ceyx pusilla Temminck, 1836. *Planch. Col. d'Ois.*, livr. 100: pl. 595, fig. 3.—Triton Bay, Bird's Neck.

Frequents forest pools, rivers, mangroves, and forested swamps through the NG lowlands, NW Islands, Aru Is, and D'Entrecasteaux Arch, sea level–750 m.

Extralimital—Moluccas, Bismarcks, Solomons, and n and e Australia (Higgins 1999, Dutson 2011).

SUBSPECIES

a. *Ceyx pusillus pusillus* Temminck

RANGE: The NW Islands (including Kofiau and Gag); Bird's Head, S Lowlands, and the SE Peninsula; Aru

Is; islands of Torres Strait; and Goodenough, Fergusson, and Normanby I[J] (Mayr 1941, BPBM specimens collected by Pratt in 2003[J]).

DIAGNOSIS: Rump patch a deep blue. No complete or partial blue breast band (Woodall 2001).

b. *Ceyx pusillus laetior* (Rand)
Alcyone pusilla laetior Rand, 1941. *Amer. Mus. Novit.* 1102: 11.—Bernhard Camp, 50 m elevation, Taritatu R, Mamberamo basin, s sector of the NW Lowlands.
RANGE: The NW Lowlands and Sepik-Ramu (Mayr 1941).
DIAGNOSIS: Blue of rump brighter than upper mantle (Woodall 2001).

NOTES
Other names—Alcedo or *Alcyone pusilla.*

ORDER FALCONIFORMES

Hackett *et mult.* (2008) showed the falcons to be distinct from the hawks and eagles at the ordinal level. That study revealed the falcons to be sister to a clade that encompasses both the parrots and the passerines. This novel placement has been recognized by Cracraft (2013) and del Hoyo & Collar (2014).

Family Falconidae

Falcons and Kestrels [Family 64 spp/Region 6 spp]

The world's falcons exhibit a cosmopolitan distribution. In New Guinea, the species range from sea level up to 3,475 m. The family is characterized by the predatory habit, the powerful but stream-lined body, the swift and agile flight, and long narrow tail, which can be fanned.

GENUS *FALCO* [Genus 37 spp/Region 6 spp]

Falco Linnaeus, 1758. *Syst. Nat.*, ed. 10, 1: 88. Type, by subsequent designation, (A.O.U. Committee, 1886), *Falco subbuteo* Linnaeus.

The species of this genus range throughout the world (Debus 1994b). Characters for the genus are generally the same as for the family. Relationships within the genus were addressed by Griffiths (1999).

Spotted Kestrel *Falco moluccensis* RESIDENT
Tinnunculus moluccensis Bonaparte, 1850. *Consp. Gen. Av.* 1: 27.—Ambon, s Moluccas.

In the Region known only from Kofiau and Gag I (Johnstone 2006, Diamond *et al.* 2009) of the NW Islands.

Extralimital—Java, Lesser Sundas, Sulawesi, and Moluccas.

SUBSPECIES
a. *Falco moluccensis ?moluccensis* (Bonaparte)
RANGE: The Moluccas to Kofiau (NW Islands). Presumably the Gag I population is also of this race (see Johnstone 2006).
DIAGNOSIS: Underwing coverts heavily spotted with black; ear coverts dark brown (Clark 1994, Diamond *et al.* 2009). In the adult male (Seram specimen), the nominate form exhibits triangular black spots on the back; pale gray tail with a broad black subterminal band and gray tip; breast and belly heavily marked

with messy black markings; throat plain buff; crown red-brown streaked finely with black; underwing whitish with black spotting. No museum specimens yet exist from Kofiau or Gag.

NOTES

This is a central Indonesian species only recently recorded on the w verge of the Region. Mauro and Wijaya (in Diamond *et al.* 2009) recorded a pair with attendant fledglings—clear evidence of resident status. *Other names*—Moluccan Kestrel.

Nankeen Kestrel *Falco cenchroides* RESIDENT AND VISITOR

Falco Cenchroides Vigors, 1827. *Trans. Linn. Soc. London* 15: 183.—New South Wales, Australia.

A few breed in the high interior of the Western Ra. In addition, Australian breeding birds winter in NG (to 1,680m^{CO}) and the Aru Is.

Extralimital—breeds in Australia, Lord Howe I, and Norfolk I; a visitor to Java, the Moluccas, NZ, and New Caledonia (Debus 1994b, Dutson 2011).

SUBSPECIES

a. *Falco cenchroides baru* Rand

Falco cenchroides baru Rand, 1940. *Amer. Mus. Novit.* 1072: 1.—11 km ne of Mt Wilhelmina, 3,400 m, e sector of the Western Ra.

RANGE: Known from the Lake Habbema–Mt Wilhelmina highlands and the Meren valley of the Carstensz Massif, both of the Western Ra (Schodde *et al.* 1975).

DIAGNOSIS: Adult has a dark gray crown and nape—gray being much darker than that of any Australian birds; side of head and postocular plumage much like crown; mantle deeper red-brown than that of Australian birds; tail gray without rufous tinge; ventral plumage richer and darker (Rand 1940a).

b. *Falco cenchroides cenchroides* Vigors

RANGE: Australia and Christmas I, migrating in nonbreeding season to se Asia, NG, Aru Is, and NZ (Dickinson & Remsen 2013). Localities—Aru Is^{MA}, Merauke^{CO}, Jayapura^{CO}, Bensbach^{L}, Karimui^{DI}, Bomai^{DI}, Baiyer R^{DI}, Tari^{CO}, Olsobip^{CO}, Tabubil^{J}, Markham valley^{CO}, Finschhafen^{CO}, Wau^{K}, Port Moresby^{M}, Popondetta^{CO}, Gurney/Alotau^{CO} (Murray 1988b^{J}, Bishop 2005a^{L}, Beehler field notes^{K}, P. Gregory *in litt.*^{M}).

DIAGNOSIS: Male crown medium gray, offset from buffy pinkish-brown mantle; very sparse streaking. Female crown and mantle buffy pinkish-brown, not very dark; crown sparsely streaked with fine black lines; in both sexes underparts whitish with sparse streaking.

NOTES

Dickinson (2003) did not recognize the race *baru*, stating it was based on migrant birds from Australia, but Dickinson & Remsen (2013) reinstated the race, citing Stresemann & Amadon (1979). Rand (1942b) noted an active nest of *baru* was located on Mount Wilhelmina. *Other names*—Australian Kestrel.

Oriental Hobby *Falco severus* RESIDENT | MONOTYPIC

Falco severus Horsfield, 1821. *Trans Linn. Soc. London* 13: 135.—Java.

Synonym: *Falco severus papuanus* A. B. Meyer and Wiglesworth, 1893. *Abh. Ber. Zool. Mus. Dresden* (1892–93) 4(3): 6.—Huon Gulf, ne NG.

Inhabits forest edge and open wooded habitats throughout NG, Salawati^{MA}, Biak^{J}, and Yapen I^{MA}, to an elevation of 2,000 m at Ambua (Gregory *in litt.*). There is a sight record from Biak I of an adult, which was well observed under excellent conditions, on 5-Jun-1990 (B. J. Coates *in litt.*^{J}). Unconfirmed sight record from Batanta I (M. Tarburton's island website).

Extralimital—Himalayas, s India, and Sri Lanka (where wintering), east to Indochina, Philippines, Java, Sulawesi, Moluccas, Lesser Sundas, New Britain, and the Solomon Is (Mayr & Diamond 2001).

NOTES

We follow Debus (1994b) in treating the species as monotypic (*pace* Stresemann & Amadon 1979, Dickinson & Remsen 2013). The Batanta I record on M. Tarburton's island website requires confirmation. *Other names*—Indian Hobby.

Australian Hobby *Falco longipennis*

Falco longipennis Swainson, 1838. *Anim. Menag.*: 341.—Tasmania.

An Australian breeder that migrates to open country in the NG region, found mainly in the S Lowlands and lowlands of the SE Peninsula (Coates 1985).

Extralimital—breeds in Australia and the Lesser Sundas (Lombok to Timor), wintering to New Caledonia, New Britain, Watom, and Seram (Debus 1994b, Mayr & Diamond 2001, Dutson 2011).

SUBSPECIES

a. *Falco longipennis longipennis* Swainson

RANGE: Breeds in Australia, wintering north and west to e Indonesia, NG, and Melanesia. Localities—Kurik[J], Merauke[CO], the lower Fly R[L], Lake Daviumbu[K], Daru[K], middle Fly R[MA], and Port Moresby[CO] (Rand 1942a[K], Rand & Gilliard 1967[L], Mees 1982b[J]). Also well-described sight records from the Baliem valley (S. F. Bailey *in litt.*) and from Biak I in Jun-1989 (K. D. Bishop *in litt.*).

DIAGNOSIS: Larger and darker than the race *hanieli* from the Lesser Sundas (Debus 1994b). Wing: male 225–250, female 252–288 (Marchant & Higgins 1993).

NOTES

The much paler subspecies *murchisonianus* of n and inland Australia possibly could wander to s NG and should be looked for there in season. *Other names*—Little Falcon.

Brown Falcon *Falco berigora* RESIDENT

Falco Berigora Vigors & Horsfield, 1827. *Trans. Linn. Soc. London* 15: 184.—New South Wales, Australia.

Inhabits open country of NG east of the Bird's Neck (Germi *et al.* 2013); sea level–1,800 m, max 3,000 m (Coates & Peckover 2001). Also Manam and Karkar I[MA].

Extralimital—Australia and Long I (Mayr & Diamond 2001).

SUBSPECIES

a. *Falco berigora novaeguineae* (A. B. Meyer)

Hieracidea novaeguineae A. B. Meyer, 1894. *J. Orn.* 42: 89.—Finschhafen, Huon Penin.

RANGE: All NG east of the Bird's Neck; also Manam[MA], Karkar[J], and Long I[J] (Gyldenstolpe 1955a, Diamond & LeCroy 1979[J], King 1979, Coates 1985, Germi *et al.* 2013: Table 2).

DIAGNOSIS: Overall has reduced spotting when compared to the nominate form; outer web of first six primaries without spotting; conspicuous shaft-streaks dorsally and ventrally (Brown & Amadon 1968). Adults very pale ventrally, with a rufous wash and streaking across breast and thighs (Marchant & Higgins 1993).

NOTES

McDonald (2003) provided evidence that there is considerable age- and sex-related plumage variation that renders racial distinctions problematic. Some authorities have placed this species in the monotypic *Ieracidea* (Peters 1931, Mayr 1941, Rand & Gilliard 1967). *Other names*—Brown Hawk, *Ieracidea berigora*.

[Grey Falcon *Falco hypoleucos* HYPOTHETICAL | MONOTYPIC

Falco hypoleucos Gould, 1841. *Proc. Zool. Soc. London* (1840): 162.—97 km from Swan R, w Australia.

Two sightings: One from the coastal Trans-Fly near Bensbach in Nov-1980 (Stronach 1981a); another in the same period from the Aroa R plains near Port Moresby (Finch 1981b). A rare bird of the Australian desert; by distribution and habitat, not expected to occur in NG.]

Peregrine Falcon *Falco peregrinus* RESIDENT (AND VISITOR)

Falco Peregrinus Tunstall, 1771. *Orn. Brit.*: 1.—England.

A widespread but scarce local resident in NG, the NW Islands, Aru Is, Yapen and Biak I[A], Bam in the Schouten Is, and SE Islands. NG records from sea level to 3,475 m (P. Gregory[A], G. Dutson[A]).

Extralimital—cosmopolitan; includes the Bismarck and Solomon Is (Debus 1994b, Mayr & Diamond 2001, Germi *et al.* 2013, and Dutson 2011).

a. *Falco peregrinus calidus* Latham

Falco calidus Latham, 1790. *Index Orn.*: 41.—India.

RANGE: Breeds through northernmost Eurasia, wintering to the Mediterranean, Africa, Middle East, India, and se Asia to NG (Stresemann & Amadon 1979).

DIAGNOSIS: Similar to *ernesti* but head and mantle dark gray (Rand & Gilliard 1967). Breast with fine dark barring; white ear patch. To date, no NG record of Australian form *macropus*.

b. *Falco peregrinus ernesti* Sharpe

Falco ernesti Sharpe, 1894. *Ibis*: 545.—Mount Dulit, Borneo.

RANGE: The Philippines, Greater Sundas, Bali, NG, Kofiau[J], ?Batanta[L], Salawati[L], Aru Is[MA], Biak[K], Yapen[MA], Normanby[M], Misima[CO], Woodlark[MA], and Tagula I[W]; also the Bismarck and Solomon Is (Sims 1956, LeCroy *et al.* 1984[M], Gilliard & LeCroy 1967, LeCroy & Peckover 1998, Diamond *et al.* 2009[J], Germi *et al.* 2013[K], M. Tarburton's island website[L], Pratt field notes 2004[W]).

DIAGNOSIS: Head, mantle, and most of cheek black; chin and throat pale buff with sparse fine dark streaking; remaining underparts profusely marked with close and narrow black barring, giving the underparts a very dark appearance (Rand & Gilliard 1967, Diamond 1972, Coates & Peckover 2001).

ORDER PSITTACIFORMES

The findings of Hackett *et mult.* (2008) indicate that the parrots are the sister group to the songbirds. The order is characterized by a compact and sturdy body form, short neck, broad and compressed bill, short legs, and strong feet with zygodactyly. This novel ordinal placement was corroborated by Jarvis *et mult.* (2014) and followed by del Hoyo & Collar (2014). We follow Joseph *et al.* (2012) in treating the Psittaciformes as comprising six family-level lineages, three of which inhabit the Region (see also de Kloet & de Kloet 2005).

Family Cacatuidae

Cockatoos and Allies   [Family 21 spp/Region 3 spp]

This is a small parrot family centered on Australia and New Guinea and ranging into Wallacea and the Philippines and east to the Bismarck and Solomon Islands (White *et al.* 2011, Joseph *et al.* 2012). The compact lineage is typified by large size, powerful body, large and powerful bill, presence of a crest, and plumage patterns in mainly white or black and powdery plumage. The family is sister to the Psittacidae, Psittrichasidae, and Psittaculidae (Joseph *et al.* 2011). Arrangement within the family follows White *et al.* (2011).

GENUS *PROBOSCIGER*   MONOTYPIC

Prosciger Kuhl, 1820. *Nova Acta Acad. Caes. Leop. Carol.* 10: 12. Type, by subsequent designation (Salvadori, 1891. *Cat. Birds Brit. Mus.* 20: 102), *Psittacus Goliath* Kuhl = *Psittacus aterrimus* Gmelin.

The single species of the genus ranges from northeastern Australia to New Guinea and the Aru Islands (Rowley 1997). The genus is characterized by the huge and scimitar-like maxilla, the high and ragged crest, and the bare red cheek patch. The molecular work of White *et al.* (2011) placed this genus as sister to *Cacatua*.

Palm Cockatoo *Probosciger aterrimus*   RESIDENT
Psittacus aterrimus Gmelin, 1788. *Syst. Nat.* 1(1): 330.—Aru Is.

Frequents forest, edge, monsoon woodland, and dense savanna throughout the NG lowlands, NW Islands, Aru Is, and Yapen and Sariba I; sea level–750 m, max 1,350 m (Mayr 1941, Coates 1985).

Extralimital—Cape York Penin, QLD, Australia.

SUBSPECIES

a. *Probosciger aterrimus goliath* (Kuhl)
Psittacus Goliath Kuhl, 1820. *Nova Acta Acad. Caes. Leop. Carol.* 10: 92.—Onin Penin, Bird's Neck.
RANGE: Batanta[J], Misool, Salawati, Gam, and Waigeo I; also the Bird's Head and Bird's Neck (Mayr 1941, Gyldenstolpe 1955b[J]).
DIAGNOSIS: Distinct molecularly, phenotypically variable. A rather arbitrary subspecific delineation, based primarily on the molecular treatment of Murphy *et al.* (2007).

b. *Probosciger aterrimus aterrimus* (Gmelin)
Synonym: *Ara Alecto* Temminck, 1835. In Siebold, *Fauna Japonica.*, disc. prelim.: 17.—Waigeo I, NW Islands.
Synonym: *Cacatua intermedia* Schlegel, 1861. *J. Orn.* 9: 380.—Aru Is.
Synonym: *Microglossus aterrimus stenolophus* van Oort, 1911. *Notes Leyden Mus.* 33: 240.—Humboldt Bay and Lake Sentani, n NG.
RANGE: The Aru Is, Yapen I, and all NG east of the Bird's Neck (Mayr 1941, Hoek Ostende *et al.* 1997).
DIAGNOSIS: Variable; individuals of some populations large and some small (Higgins 1999). The study of Murphy *et al.* (2007) did not sample the Yapen population, which presumably belongs in this race by geography.

NOTES

Our consolidation of subspecies follows the recommendations of Forshaw (1989) and Rowley (1997). Using molecular techniques, Murphy *et al.* (2007) indicated that the currently recognized subspecies have minimal validity, but that there are two clades: one for the Bird's Head–Bird's Neck region, and one for the remainder of the Region's range. We have reformulated the subspecific treatment to accord with Murphy *et al.*'s molecular landscape. Another option is to treat the species as variable and monotypic, which future revisers should consider. *Other names*—Great Palm, Great Black, or Goliath Cockatoo.

GENUS *CACATUA*   [Genus 12 spp/Region 2 spp]

Cacatua Vieillot, 1817. *Nouv. Dict. Hist. Nat.*, nouv. éd. 17: 6 (27-Dec-1817). Type, by subsequent designation (Salvadori, 1891. *Cat. Birds Brit. Mus.* 20: 115, 124), *Cacatua cristata* Vieillot = *Psittacus albus* (P. L. S. Müller 1776).

The species of this genus range from the Lesser Sundas, central Philippines, and Sulawesi, east through New Guinea, New Britain, the Solomon Islands, and Australia (Rowley 1997). The genus is characterized by the white plumage, prominent erectile crest, and colored bare eye skin.

Little Corella *Cacatua sanguinea*   RESIDENT
Cacatua sanguinea Gould, 1843. *Proc. Zool. Soc. London* (1842): 138.—Port Essington, NT, Australia.

Recorded in large seasonal flocks from the Merauke, Morehead, and Bensbach regions of the Trans-Fly. In Bensbach frequents forest, trees bordering the river, and nearby grassland. Feeds in ricefields at Kurik (Hoogerwerf 1964). No evidence of migratory movement of the species from Australia. Apparently has decreased in abundance (K. D. Bishop *in litt.*). Not recorded on two recent visits to Bensbach (D. Hobcroft *in litt.*, L. Joseph *in litt.*).

Extralimital—much of the drier regions of Australia (Higgins 1999).

SUBSPECIES

a. *Cacatua sanguinea transfreta* Mees, 1982
Cacatua sanguinea transfreta Mees, 1982. *Zool. Verh. Leiden* 191: 79.—Kurik, s NG.
RANGE: The Trans-Fly of s NG on either side of the PNG border: Dolak I, Kurik, Kumbe, and Bensbach (Mees 1982b, Hoek Ostende *et al.* 1997, K. D. Bishop *in litt.*).

DIAGNOSIS: Differs from *normantoni* of QLD by the buff (not yellow) underwing coloration (Rowley 1997). The fact that the NG form is distinctive supports the notion that these birds are not migrants from Australia.

NOTES
We follow Ford (1985) and Rowley (1997) in treating *sanguinea* as a species distinct from *pastinator* and *tenuirostris*. *Other names*—Bare-eyed Cockatoo or Corella.

Sulphur-crested Cockatoo *Cacatua galerita* RESIDENT
Psittacus galeritus Latham, 1790. *Index Orn.* 1: 109.—Turramurra, New South Wales, Australia.
  Widespread through the NG region, except for some small islands; sea level–1,000 m, max 2,400 m^{CO}. Range includes all NW Islands, Aru Is, Bay Is, and SE Islands.
  Extralimital—widespread across n and e Australia. Introduced to the Moluccas and Palau (Forshaw 1989).

SUBSPECIES
a. *Cacatua galerita triton* Temminck
Cacatua triton Temminck, 1849. *Coup d'Oeil Gen. Poss. Neerl. Ind. Arch.* 3: 405 (note).—Aiduma I, near Triton Bay, Bird's Neck.
RANGE: Inhabits NG and virtually all of the islands in the Region except for the Aru Is. Also introduced to Ambon, Seram Laut Is, Kai Is, and Palau.
DIAGNOSIS: Smaller than *fitzroyi* from nw and north-central Australia, and with broader crest feathers (Rowley 1997) and bluish eye skin.

b. *Cacatua galerita eleonora* (Finsch)
Kakatoe? eleonora Finsch, 1863. *Papageien* 1.—Aru Is.
RANGE: The Aru Is (Hoek Ostende *et al.* 1997).
DIAGNOSIS: Substantially smaller than *triton*. Wing of a single female at the BMNH was 248 *vs.* 310–335 (n=3) for Bird's Head specimens of *triton*. This small form was accepted by Forshaw (1989), following Mees (1972). This race was also recognized by Dickinson & Remsen (2013).

NOTES
Replaced in New Britain by the closely allied *Cacatua ophthalmica* (Mayr & Diamond 2001). *Other names*—Greater Sulphur-crested, White, or Triton Cockatoo.

Family Psittrichasidae

New Guinea Vulturine Parrot and Vasa Parrots [Family 4 spp/Region 1 sp]

This tiny relict family includes the New Guinea Vulturine Parrot (*Psittrichas*) and several species of vasa parrots (*Coracopsis*) of Madagascar and adjacent Indian Ocean islands. Molecular genetic evidence justifies this association, although the two genera are nevertheless somewhat distantly related. The New Guinea Vulturine Parrot has long been recognized for its peculiar morphology; recent molecular studies confirm its distinctive position among the parrots, raising it to familial status (Joseph *et al.* 2012). Dickinson & Remsen (2013) treated the lineage as a subfamily within the Australasian parrots and lories, the Psittaculidae. In contrast, del Hoyo & Collar (2014) treated the lineage as a subfamily situated within the Psittacidae, into which they included all parrot relatives except the cockatoos and Strigopidae. Our positioning of this family within the parrot lineage follows Joseph *et al.* (2012).

Psittrichas Lesson, 1831. *Bull. Sci. Nat. Geol.* 25: 241 (=341). Type, by monotypy, *Psittacus Pecquetii* Lesson = *Banksianus fulgidus* Lesson.

The single species of the genus is confined to the interior and uplands of the New Guinea mainland (Forshaw 1989). The genus is characterized by distinctive cranial osteology (Thompson 1900); the large body with a rather short, squared-off tail; rounded wings; absence of feathering on the forepart of the rather slender head; narrow and elongate bill and skull; and fig-eating specialization (Forshaw 1989, Rowley 1997).

New Guinea Vulturine Parrot *Psittrichas fulgidus*   ENDEMIC | MONOTYPIC
Banksianus fulgidus Lesson, 1830. *Traite Orn.* 1: 181.—Bird's Head.

Patchily distributed through the mts of NG, inhabiting forest canopy in the hills and lower montane forest; ranges up to 1,600 m, extremes 50[K]–2,420[J] m (Mayr 1941, Rand 1942b, Gilliard 1950, Gilliard & LeCroy 1967, Diamond 1972, Coates 1990: 567, Beehler *et al.* 1994[K], Freeman *et al.* 2013[J]). Can be encountered in flat lowland alluvial forest where adjacent to mts. Apparently absent from the Fakfak and Kumawa Mts (J. M. Diamond *in litt.*); also no records from the Cyclops Mts. Uncommon and rather shy; hunted for its red-and-black wing feathers. Specializes on figs.

NOTES
Currently classified by the IUCN Red List as Vulnerable, which we agree is a fair assessment of the threat to it. Although the red-and-black tail and wing feathers are valuable trade goods, the species is a shy canopy-dweller that is difficult to hunt. Our English name highlights the species' geographic range and its most conspicuous feature—the vulturine bare facial skin. *Other names*—Pesquet's, Vulturine, or Bare-headed Parrot.

Family Psittaculidae

Australasian Parrots and Lories   [Family 180 spp/Region 43 spp]

The species of this large and diverse family range from Australasia northeast to Polynesia and northwest through Indonesia to southern Asia and Africa. Distinct lineages within this family that are found in the Region include the tiger-parrots, lories and lorikeets, fig-parrots, hanging parrots, king-parrots and allies, Eclectus Parrot and allies, and pygmy parrots. They are adapted to feeding on nectar, fruits, and seeds, and even lichens and liverworts. They inhabit all wooded environments in the Region, ranging to the timberline. Parrots are compact and often stocky because of the powerful flight muscles that carry them for long distances in search of food; their very short legs are built for scrambling about in trees, and their large heads have strongly hooked beaks for chewing their seed and fruit diet and excavating a nesting chamber. Australasian parrots range from tiny (pygmy parrots) to large (Eclectus Parrot). Our generic sequence follows Joseph *et al.* (2012) overall, but as modified by Schweizer *et al.* (2015) for the lories.

GENUS *PSITTACELLA*   ENDEMIC [Genus 4 spp]

Psittacella Schlegel, 1871. *Ned. Tijdschr. Dierk.* (1873) 4: 35. Type, by subsequent designation (Salvadori, 1880. *Orn. Pap. Moluc.* 1: 145), *Psittacus brehmii* von Rosenberg ms = *Psittacella brehmii* Schlegel.

The four species of *Psittacella* are endemic to the uplands of the mainland of New Guinea (Collar 1997). The genus is characterized by the naked cere, stocky build, red undertail plumage, rounded tail and wings, presence of fine barring or patterning dorsally, very stolid nature, forest interior habit, and diet of seeds of unripe fruit and young leaves (Forshaw 1989). See Joseph et. al (2011a) for molecular systematic review of the genus.

Brehm's Tiger-Parrot *Psittacella brehmii*   ENDEMIC

Psittacella Brehmii Schlegel, 1871. *Ned. Tijdschr. Dierk.* (1873) 4: 35.—Arfak Mts, Bird's Head.

Mountains of the Bird's Head, Central Ra, and Huon Penin; 1,500–2,600 m, extremes 1,150–3,200 m[CO]. Frequents forest, edge, regrowth, grassy glades, and low trees in subalpine shrubland (Gyldenstolpe 1955a). In the Central Ra, replaced at higher elevations by *P. picta*. On the Huon Penin, in the absence of *P. picta*, may range up to the tree line (Diamond 1972).

SUBSPECIES

a. *Psittacella brehmii brehmii* Schlegel
RANGE: Mountains of the Bird's Head (Mayr 1941).
DIAGNOSIS: Male has head dark brown with olive tinge; little yellow in the green of the underparts. Wing: male 118–124, female 119–124 (Rand & Gilliard 1967, Forshaw 1989).

b. *Psittacella brehmii intermixta* Hartert
Psittacella brehmii intermixta Hartert, 1930. *Novit. Zool.* 36: 107.—Mount Goliath, w sector of the Border Ra.
RANGE: The Western Ra and the w and c sectors of the Border Ra, to Mount Goliath (Mayr 1941).
DIAGNOSIS: Male like the nominate, but underparts more yellowish; throat and side of head paler; mantle, back, and uppertail coverts yellowish green barred with black; slightly larger. Wing: male 125–138, female 126–134 (Forshaw 1989).

c. *Psittacella brehmii pallida* A. B. Meyer
Psittacella pallida A. B. Meyer, 1886. *Zeitschr. Ges. Orn.* 3: 3.—Astrolabe Ra, SE Peninsula.
    Synonym: *Psittacella bürgersi* Reichenow, 1918. *J. Orn.* 66: 244.—Schrader Ra, n sector, Eastern Ra.
RANGE: The e sector of the Border Ra (Hindenburg Mts), Eastern Ra, and mts of the SE Peninsula (Forshaw 1989).
DIAGNOSIS: Male crown paler than that of *intermixta*; some birds have a blue wash on abdomen; bill narrower than in nominate *brehmii*. Wing: male 119–130, female 112–130. Female plumage more yellowish than that of nominate; flanks and sides of abdomen yellow barred with black; narrower bill. Immature as for female, but black barring, both above and below, narrower and duller; undertail coverts orange-red tipped with yellowish green; iris yellowish brown (Forshaw 1989) or orange (see photo in Coates & Peckover 2001: 82). We follow suggestion of Gyldenstolpe (1955a) and Gilliard & LeCroy (1968) in placing *buergersi* in synonymy with *pallida* (in agreement with Forshaw 1989, *pace* Dickinson & Remsen 2013).

d. *Psittacella brehmii harterti* (Mayr)
Psittacella brehmii harterti Mayr, 1931. *Mitt. Zool. Mus. Berlin*, 17: 702.—Mongi-Busu, Huon Penin.
RANGE: Mountains of the Huon Penin (Mayr 1941).
DIAGNOSIS: Male as for *pallida* but smaller and with plumage overall less yellowish; head paler and more olive. Wing: male 112–118, female 117–123 (Forshaw 1989).

NOTES
Other names—Brehm's Parrot.

Painted Tiger-Parrot *Psittacella picta*   ENDEMIC

Psittacella picta Rothschild, 1896. *Bull. Brit. Orn. Club* 6: 5.—Mount Victoria, 1,500–2,100 m, Owen Stanley Ra, SE Peninsula.

Frequents subalpine forest and shrublands of the Central Ra only, 2,450 m to timberline, rarely lower (Coates 1985). Absent from outlying ranges. Reported at 1,700 m at Ok Tedi (Coates & Lindgren 1978) and 1,370 m near Efogi, Central Prov, PNG (Donaghey 1970). However, mostly occurs above the elevational range of *Psittacella brehmii*.

SUBSPECIES

a. *Psittacella picta lorentzi* van Oort
Psittacella lorentzi van Oort, 1910. *Notes Leyden Mus.* 32: 212.—Wichmann Mts, 3,000 m, Western Ra.
RANGE: The Western and Border Ra; absent from outlying ranges. Not yet definitively found in the Border Ra of PNG (Gilliard & LeCroy 1961, Gregory & Johnston 1993, Hoek Ostende *et al.* 1997), but the species

has been observed in the Ok Tedi area (Coates & Lindgren 1978). One might expect the form *lorentzi* to range eastward, patchily, to the w verge of the Strickland R gorge (only on the highest summits).

DIAGNOSIS: Both Mayr (1941) and del Hoyo & Collar (2014) treated *lorentzi* as a full species distinct from *P. picta*. Dickinson & Remsen (2013) combined them. Unlike the nominate form and *excelsa*, *lorentzi* has uppertail coverts barred black and yellow, not red. Male otherwise distinguished from *excelsa* by sides of head dusky blue-green trending to black posteriorly; throat blue-green becoming green on breast and yellow-green on abdomen. Female as for *excelsa*, but uppertail coverts barred yellow; side of head and neck blue-green (Rand & Gilliard 1967). See photo of female in Coates & Peckover (2001: 83, lower image). Wing: male 109–122, female 109–116 (Forshaw 1989).

b. *Psittacella picta excelsa* Mayr & Gilliard
Psittacella picta excelsa Mayr & Gilliard, 1951. *Amer. Mus. Novit.* 1524: 6.—Mount Orata, Kubor Mts, Eastern Ra.

RANGE: The Eastern Ra. Presumably only east of the Strickland R gorge.

DIAGNOSIS: Male crown and occiput bright olive brown; cheeks dull brown; throat dull brown; no red-brown on cheek or neck; yellow patches on side of neck only. Female head bright olive brown; throat and cheeks strongly suffused with blue. Wing: male 104–113, female 109–113 (Forshaw 1989).

c. *Psittacella picta picta* Rothschild
RANGE: Mountains of the SE Peninsula, presumably northwest to the Ekuti Divide at the nw terminus of the SE Peninsula.

DIAGNOSIS: Male entire head rich reddish brown, duller on cheek; rump and uppertail coverts chestnut red; yellow collar across hindneck, bordered anteriorly with red-brown; rump with red and black barring; upper breast deep blue in center. Female cheek bluish, distinct from chestnut of crown, nape, and neck collar. See photo in Coates & Peckover (2001: 83, upper image). Wing: male 106–113, female 105–114 (Forshaw 1989).

NOTES
P. p. lorentzi was raised to species level as the Snow Mountain Tiger-parrot by del Hoyo & Collar (2014). It will be important to observe the birds at the point of contact between *P. p. lorentzi* and *P. p. excelsa*—presumably near the Strickland R gorge, which serves as a barrier for a number of montane species. *Other names*—Painted Parrot, Timberline Tiger-Parrot.

Modest Tiger-Parrot *Psittacella modesta*   ENDEMIC
Psitacella modesta Schlegel, 1871. *Ned. Tijdschr. Dierk.* (1873) 4: 36.—Arfak Mts, Bird's Head.

Frequents forest and edge in the mts of the Bird's Head and the Western, Border, and the c sector of the Eastern Ra (east to Mt Hagen), 1,700–2,800 m[CO]. Overlaps range of *Psittacella madaraszi* in Western, Border, and Eastern Ra. In areas of sympatry, *P. modesta* typically occupies the higher elevation (Diamond 1972).

SUBSPECIES
a. *Psittacella modesta modesta* Schlegel
RANGE: Mountains of the Bird's Head.

DIAGNOSIS: Male with feathers of nape and hindneck dull olive yellow with brown margins; no indication of a pale nape collar; throat and breast plain, unbarred pale brownish olive; blue at bend of wing; underwing coverts yellowish green. Female head brown, grading to olive on nape and hindneck; feathers of breast barred orange, edged with dark brown; sides of abdomen barred with yellow and greenish brown (Rand & Gilliard 1967, Collar 1997).

b. *Psittacella modesta collaris* Ogilvie-Grant
Psittacella modesta collaris Ogilvie-Grant, 1914. *Bull. Brit. Orn. Club* 35: 13.—Upper Otakwa R, sw sector of Western Ra.

> Synonym: *Psittacella modesta subcollaris* Rand, 1941. *Amer. Mus. Novit.* 1102: 8.—15 km s of Bernhard Camp, 1,800 m, ne sector of the Western Ra.

RANGE: The Western, Border, and Eastern Ra, from the Otakwa R east to Mt Hagen (Coates 1985).

DIAGNOSIS: Male like *modesta* but with an irregular yellowish collar below brown on hindneck; more rufous-tinged on throat and sides of head. Female like *modesta* but with indistinct yellow markings on hindneck. The form *subcollaris* is not separable from *collaris*.

NOTES
Other names—Modest Parrot, Barred Little Tiger-Parrot.

Madarasz's Tiger-Parrot *Psittacella madaraszi* ENDEMIC
Psittacella madarászi A. B. Meyer, 1886. *Zeitschr. Ges. Orn.* 3: 4, pl. 1, fig. 1.—Astrolabe Ra, SE Peninsula.
Frequents montane forest and edge in the Central Ra, Foja Mts, and mts of Huon Penin, 1,200–2,500 m (records near sea level in hills on s scarp). Overlaps in range with *P. modesta* in the Western, Border, and Eastern Ra. In sympatry, *P. madaraszi* occupies the lower elevation.

SUBSPECIES
a. *Psittacella madaraszi madaraszi* A. B. Meyer
Synonym: *Psittacella madaraszi major* Rothschild, 1936. *Mitt. Zool. Mus. Berlin* 21: 233.—Kunupi, 1,500 m, Weyland Mts, Western Ra.
Synonym: *Psittacella modesta hallstromi* Mayr & Gilliard, 1951. *Amer. Mus. Novit.* 1524: 5.—Yandara, n slope of Mt Wilhelm, Eastern Ra.
RANGE: The Central Ra; presumably the Foja Mts population can also be referred to this form (Beehler *et al.* 2012).
DIAGNOSIS: Male with yellow spotting abundant on hind-crown and nape, extending marginally to cheek and throat. Female with barred nape and hind-crown brightly washed with orange. We combine the forms *major* and *hallstromi* with the nominate because of insufficient size or plumage distinction. These were described based on small samples and insufficient allowance for individual variation.

b. *Psittacella madaraszi huonensis* Mayr & Rand
Psittacella modesta huonensis Mayr & Rand, 1935. *Amer. Mus. Novit.* 814: 3.—Sevia, Huon Penin.
RANGE: The Huon Penin.
DIAGNOSIS: Male with brown of crown more yellowish than found in nominate form. Female with no orange in occiput, nape, or hindneck; less black barring on upperparts (Forshaw 1989).

NOTES
Males of *P. modesta* and *P. madaraszi* are very similar. The latter never exhibits a yellow hind-collar; crown and nape with more profuse pale yellow scaling; green of breast extends higher up to throat; some birds may not be identifiable in the field (Forshaw 1989). *Other names*—Madarasz's Parrot, Plain-breasted Little Parrot.

GENUS *OREOPSITTACUS* ENDEMIC | MONOTYPIC

Oreopsittacus Salvadori, 1877. *Ann. Mus. Civ. Genova* 10: 37. Type, by monotypy, *Trichoglossus arfaki* A. B. Meyer.
The single species of the genus inhabits the major mountain ranges of New Guinea. The genus is characterized by the sharply pointed black maxillary hook to the bill, narrow pointed tail, slim body, unstreaked green body plumage, purple cheek patch, and presence of 14 rectrices (12 in other lorikeets).

Plum-faced Lorikeet *Oreopsittacus arfaki* ENDEMIC
Trichoglossus (*Charmosyna*) *Arfaki* A. B. Meyer, 1874. *Verh. Zool.-Bot. Ges. Wien* 24: 37.—Arfak Mts, 1,070 m, Bird's Head.
Frequents montane and subalpine forest in the mts of the Bird's Head, Central Ra, and Huon Penin; mainly 2,500–3,750 m[J], but locally down to 1,700 m; also a single record in Wau from 1,100 m[K], and collection of the type from 1,070 m in the Arfak Mts (Mayr 1941, Pratt & Beehler 2014[J], Beehler field notes 30-Nov-1978[K]).

SUBSPECIES

a. *Oreopsittacus arfaki arfaki* (A. B. Meyer)

RANGE: The Arfak and Tamrau Mts of the Bird's Head.

DIAGNOSIS: Tail lightly tipped with pale red, and belly and flanks washed with red. Wing: male 72–80, female 72–76 (Forshaw 1989)

b. *Oreopsittacus arfaki major* Ogilvie-Grant

Oreopsittacus arfaki major Ogilvie-Grant, 1914. *Bull. Brit. Orn. Club* 35: 11.—Upper Otakwa R, Nassau Mts, Western Ra.

RANGE: The Western Ra.

DIAGNOSIS: Larger than nominate form; wing: male 84–93, female 79–89. Tail extensively tipped with scarlet (Forshaw 1989). Belly with red wash (orange in female).

c. *Oreopsittacus arfaki grandis* Ogilvie-Grant

Oreopsittacus grandis Ogilvie-Grant, 1895. *Bull. Brit. Orn. Club* 5: 15.—Owen Stanley Ra, SE Peninsula.

RANGE: The Border and Eastern Ra and mts of the Huon and SE Peninsulas, from the Hindenburg Mts eastward (Sims 1956, Gilliard & LeCroy 1961, Coates 1985).

DIAGNOSIS: Abdomen and lower flanks green rather than red (Diamond 1972). See photo in Coates & Peckover (2001: 76).

NOTES

Other names—Whiskered Lorikeet, Plum-faced Mountain Lory.

GENUS *CHARMOSYNA* [Genus 15 spp/Region 9 spp]

Charmosyna Wagler, 1832. *Abh. K. Bayer. Akad. Wiss., Math.-Phys. Kl.* 1: 493. Type, by monotypy, *Ch. papuensis* = *Psittacus papuensis* Gmelin = *Psittacus papou* Scopoli.

The species of this genus range from the Moluccas east through New Guinea to the Bismarcks and Solomons, Vanuatu, New Caledonia, and Fiji (Collar 1997). The genus is characterized by the orange/red bill, pointed tail, and plumage patterned with green, red, yellow, and black/blue. The genus includes a small-size green group (typified by *C. placentis*) and a larger-size red group (typified by *C. josefinae*). The members of the genus exhibit considerable plumage variation. The species are canopy-dwelling nectarivores that are flocking, nomadic, and mainly montane. Schweizer *et al.* (2015) provided a phylogeny of the lorikeets that indicated that *Charmosyna* as traditionally circumscribed is not monophyletic. Additional molecular work that includes additional species from the Region will be required to determine the status of the species included here.

Red-fronted Lorikeet *Charmosyna rubronotata* ENDEMIC

Coriphilus rubronotatus Wallace, 1862. *Proc. Zool. Soc. London*: 165.—Salawati I and w coast of the Bird's Head.

Patchily distributed through lowlands and hills (900 m) of the Bird's Head, NW Lowlands, and Sepik-Ramu (Stresemann 1923); also Salawati and Biak I. Frequents forest, edge, and coconut plantations; occasionally visits flowering trees in open country (Coates 1985).

SUBSPECIES

a. *Charmosyna rubronotata rubronotata* (Wallace)

RANGE: Salawati I, the Bird's Head, NW Lowlands, and Sepik-Ramu (Mayr 1941). Localities—Danau Bira[W], s Mamberamo basin[K], Bodim[J], Puwani R[X], Hunstein Mts[L], Ettapenberg[Z], Malu[Z], Maeanderberg[Z] (Stresemann 1923[Z], Rand 1942b[K], Ripley 1964a[J], A. Allison *in litt.*[L], Beehler field notes 18-Jul-1984[X], 10-Jul-1987[W]).

DIAGNOSIS: Red cap of male small and deep red; ear coverts with a purple hue (Forshaw 1989). See photo in Coates & Peckover (2001: 73).

b. *Charmosyna rubronotata kordoana* (A. B. Meyer)

Trichoglossus (Charmosyna) kordoanus A. B. Meyer, 1874. *Verh. Zool.-bot. Ges. Wien* 24: 38.—Biak I, Bay Islands.

RANGE: Biak I.

DIAGNOSIS: Male distinct from nominate form by the paler and more extensive red patch on crown; ear coverts more blue, less purple. Female not distinguishable from nominate form (Forshaw 1989).

NOTES

This is a little-known species that is often overlooked in the field. The ecological and distributional relationship between this species and *C. placentis* needs further study. It has been reported that where the two live in geographic sympatry, *C. rubronotata* lives at the higher elevation (Beehler *et al.* 1986). *Other names*—Red-spotted Lorikeet, Red-fronted Blue-eared Lory.

Red-flanked Lorikeet *Charmosyna placentis*   RESIDENT

Psittacus placentis Temminck, 1835. *Planch. Col. d'Ois.*, livr. 93 (1834): pl. 553.—Utanata R, sw NG.

Throughout NG, the NW Islands, Aru Is, and Woodlark I; range includes the Trans-Fly (Mees 1982b); mainly at sea level, max 1,600 m (Coates & Peckover 2001). Frequents forest, edge, and areas where trees are in flower.

Extralimital—The Moluccas, Kai Is, Bismarck Arch, and the n Solomon Is (Peters 1937, Mayr 1941, Forshaw 1989, Coates 1985, Mayr & Diamond 2001, Dutson 2011).

SUBSPECIES

a. *Charmosyna placentis intensior* (Kinnear)

Hypocharmosyna placentis intensior Kinnear, 1928. *Bull. Brit. Orn. Club* 48: 84.—Bacan I, n Moluccas.

RANGE: The n Moluccas (Halmahera, Obi, and Ternate)[L] and Gebe I[K] (Mees 1972[K], White & Bruce 1986[L]).

DIAGNOSIS: Relatively small rump spot; ear coverts are blue-violet (Mees 1965).

b. *Charmosyna placentis placentis* (Temminck)

RANGE: The s Moluccas (Ambelau, Seram, Ambon, Seram Laut, Panjang, Tayandu, Kai Is, Aru Is) and s NG (Mayr 1941, Diamond 1972, Hoek Ostende *et al.* 1997).

DIAGNOSIS: Blue rump spot is small; yellow of crown reduced; red chin and throat reduced (Rand & Gilliard 1967). Ear coverts blue.

c. *Charmosyna placentis ornata* Mayr

Charmosyna placentis ornata Mayr, 1940. *Amer. Mus. Novit.* 1091: 1.—Misool I, NW Islands.

RANGE: Kofiau[X], Waigeo, Batanta, Salawati, and Misool I[Q]; Bird's Head; and nw NG east to Sarmi, north-central coast of NW Lowlands (Mayr 1941, Mees 1965[Q], Diamond *et al.* 2009[X]).

DIAGNOSIS: Large rump patch, blue in both sexes; large red throat patch extends down onto breast; green of plumage dark (Mees 1965, Rand & Gilliard 1967, Collar 1997).

d. *Charmosyna placentis subplacens* (P. L. Sclater)

Trichoglossus subplacens P. L. Sclater, 1876. *Proc. Zool. Soc. London*: 519.—Naiabui, Hall Sound, s watershed of SE Peninsula.

RANGE: Northern NG from Sarmi distr (north-central NW Lowlands) east to Milne Bay, and in the s watershed of the SE Peninsula from Hall Sound eastward.

DIAGNOSIS: Rump all green (Mees 1965).

e. *Charmosyna placentis pallidior* (Rothschild and Hartert)

Charmosynopsis placentis pallidior Rothschild and Hartert, 1905. *Novit. Zool.* 12: 253.—Bougainville I, Solomon Is.

RANGE: Woodlark I[MA], Bismarck Arch, and the northernmost Solomon Is (Bougainville and Nuguria Is).

DIAGNOSIS: Similar to *subplacens* but upperparts paler green and ear coverts paler blue (Mees 1965, Rand & Gilliard 1967).

NOTES

The subspecies diagnosis follows Mees (1965), as interpreted by Rand & Gilliard (1967), Forshaw (1989), and Collar (1997). Observation of this species from Karkar I (Finch 1981d, 1981g) is not accepted here. *Other names*—Yellow-fronted Blue-eared Lory, Lowland Lorikeet, Yellow-fronted or Blue-eared Lorikeet.

Red-chinned Lorikeet *Charmosyna rubrigularis*   RESIDENT | MONOTYPIC

Trichoglossus rubrigularis P. L. Sclater, 1881. *Proc. Zool. Soc. London*: 451.—New Britain.
    Synonym: *Hypocharmosyna rubrigularis krakari* Rothschild and Hartert, 1915. *Novit. Zool.* 22: 31.—Karkar I, ne
    NG.

Within the NG Region, known only from Karkar I[K], sea level–1,500 m. Frequents forest. On Karkar I
mostly found from 600 m to the island's summit, but also found uncommonly in the lowlands (Diamond
& LeCroy 1979[K]). Greensmith found the species common on the slopes of the Karkar volcano from at least
1,150 m to the summit at 1,280 m (in Forshaw 1989).
    Extralimital—mts of New Britain, New Ireland (Mayr & Diamond 2001), and possibly New Hanover
(Dutson 2011: 327).

NOTES

Surprisingly, this species is reported as absent from Long I and Umboi I, and yet it is present on Karkar
and New Britain, which bracket Long and Umboi. This checkerboard distribution is puzzling. Diamond &
LeCroy (1979) subsumed *krakari* into the nominate form, making the species monotypic. *Other names*—
Red-chinned Lory.

Pygmy Lorikeet *Charmosyna wilhelminae*   ENDEMIC | MONOTYPIC

Trichoglossus Wilhelminae A. B. Meyer, 1874. *J. Orn.* 22: 55, 56.—Passim and Andai, Bird's Head.
    Forages on flowers in the forest canopy of the mts of the Bird's Head[MA], Central Ra[MA], and Huon
Penin[J]; generally 1,000–2,400 m[J], but in the s watershed it is occasionally found at the base of the mts in
lowland forest near sea level[K] (Freeman *et al.* 2013[J], Beehler field notes[K]). Found mostly in the lower mts
but nomadic, flocks moving widely in search of productive flowering trees (Coates 1985).
    Other names—Wilhelmina's Lorikeet or Pygmy Streaked Lorikeet.

Striated Lorikeet *Charmosyna multistriata*   ENDEMIC | MONOTYPIC

Charmosynopsis multistriata Rothschild, 1911. *Bull. Brit. Orn. Club* 27: 45.—Upper Setekwa R, Western Ra.
    Inhabits the s slopes of the Western, Border, and Eastern Ra and S Lowlands; from the Weyland Mts
southeast to Mt Karimui[J] and Crater Mt[L]; foothills–1,800[K] m (Mack & Wright 1996[L], Collar 1997, Coates
& Peckover 2001, Coates & Peckover 2001[K], Freeman & Freeman 2014[J]). Has been collected as low as 80 m
near Palmer Junction, where several other species normally found in lower montane forest were also present (Rand 1942a). At Ok Tedi it was found at the edge of hill forest at 570 m.
    Localities—Mimika R[MA], Setekwa R[MA], Otakwa R[MA], and upper Fly R (Palmer Junction)[MA], Ok Men-
ga[L], Ok Ma[L], Dablin Creek[L], Tabubil[CO], Mt Bosavi[J], Crater Mt WMA[W], and Mt Karimui[K] (Gregory 1995b[L],
Mack & Wright 1996[W], Pratt field notes 1975[J], B. G. Freeman *in litt.*[K]).

NOTES

Greenway (1978) noted the type was collected on the upper Setekwa R, a tributary of the Otakwa. *Other
names*—Streaked or Yellow Streaked Lory.

Josephine's Lorikeet *Charmosyna josefinae*   ENDEMIC

Trichoglossus Josefinae Finsch, 1873. *Atti Soc. Ital. Sci. Nat. Milano* 15: 427, pl. 7. Arfak Mts, Bird's Head.
    Forages at canopy flowers of forest trees in the mts of the Bird's Head and Western, Border, and East-
ern Ra (Stresemann 1923, Mayr 1941); also Cyclops and Foja Mts, PNG North Coastal Ra, and Mt Bosavi;
700–1800 m, extremes sea level–2,200 m (Coates 1985). Eastern edge of range lies in Bismarck Ra in the n
watershed and Mt Karimui in the s watershed.

SUBSPECIES

a. *Charmosyna josefinae josefinae* (Finsch)
RANGE: Mountains of the Bird's Head, Bird's Neck, and Western and Border Ra (Mayr 1941, Rand 1942b,
Gyldenstolpe 1955b, Collar 1997). Birds from the Ok Tedi region presumably are referable to this form.
DIAGNOSIS: Male crown reddish lilac; lower back red and dusky spot on rump; black patch on abdomen.
Female lower back green (Rand & Gilliard 1967). See photo in Coates & Peckover (2001: 75).

b. *Charmosyna josefinae cyclopum* Hartert
Charmosyna josephinae cyclopum Hartert, 1930. *Novit. Zool.* 36: 104.—Cyclops Mts, ne sector of NW Lowlands, n NG.
RANGE: The Cyclops Mts.
DIAGNOSIS: Black patch on abdomen absent; crown patch black, not blue (Rand & Gilliard 1967).

c. *Charmosyna josefinae sepikiana* Neumann
Charmosyne josephinae sepikiana Neumann, 1922. *Verh. Orn. Ges. Bayern* 15: 235.—Hunstein Top and Lordberg, middle Sepik, Sepik-Ramu.
RANGE: Mountains of the Sepik-Ramu and Eastern Ra. Localities—Hunstein Mts[L], Mt Bosavi[K], Lake Kopiago[CO], Tari[K], Schrader Ra[K], Mt Karimui[J] (Coates & Peckover 2001[K], Freeman & Freeman 2014[J], A. Allison *in litt.*[L]). Birds from the PNG North Coastal Ra (specimens at AMNH & BPBM) presumably are referable to this subspecies.
DIAGNOSIS: Large black belly patch; feathers of crown gray rather than lilac; female with flanks and lower back yellow (Rand & Gilliard 1967).

NOTES
A little-known species whose range is imperfectly delineated. Merits additional study. *Other names—* Josephine's Lory.

Papuan Lorikeet *Charmosyna papou*   ENDEMIC | MONOTYPIC
Psittacus Papou Scopoli, 1786. *Delic. Flor. Faun. Insubr.* 2: 86.—Arfak Mts, Bird's Head.
    Restricted to the Tamrau and Arfak Mts of the Bird's Head.

NOTES
Following Voison & Voison (1997) we treat the Bird's Head form (*C. papou*) as distinct from the three forms from east of the Bird's Head (*P. stellae*). This treatment is based on several prominent plumage characters: (1) *C. papou* lacks a black plumage morph; (2) is sexually monochromatic; (3) exhibits a small yellow patch on each side of breast; (4) exhibits another small yellow patch on each flank; (5) the longer uppertail coverts are green (not red); (6) ventral surface of base of tail feathers is orange-yellow; (7) blue crown patch is narrow and sits above eye; (8) black hind-crown patch is wider than blue patch; (9) posterior to the blue patch are a red nuchal patch, a black band, and a red neck hind-collar; (10) black postocular stripe is absent. This split was also carried out by del Hoyo & Collar (2014). The behavior and vocalizations of this isolated allospecies merit additional study. *Other names—*Fairy Lory.

Stella's Lorikeet *Charmosyna stellae*   ENDEMIC
Charmosyna Stellae A. B. Meyer, 1886. *Zeitschr. Ges. Orn.* 3: 9, pl. 2.—Mountains of the SE Peninsula.
    Inhabits the Central Ra, Adelbert Mts, and mts of the Huon Penin; 1,500 m[CO]–timberline (records down to 1,200 m[CO]). Common in pairs in montane-forest interior; usually seen flying beneath canopy, vocalizing in flight. A passing pair was observed at 1,065 m in upper Fly R karst high-rainfall zone (I. Woxvold *in litt.*).

SUBSPECIES
a. *Charmosyna stellae goliathina* Rothschild and Hartert
Charmosyna stellae goliathina Rothschild and Hartert, 1911. *Novit. Zool.* 18: 160.—Mount Goliath, Border Ra.
RANGE: The Western, Border, and Eastern Ra; also the Adelbert Mts[K] (Pratt 1983[K], Collar 1997).
DIAGNOSIS: Breast all red; longest rectrices are yellow-green, not reddish yellow. See photo in Coates & Peckover (2001: 74).

b. *Charmosyna stellae wahnesi* Rothschild
Charmosyna stellae wahnesi Rothschild, 1906. *Bull. Brit. Orn. Club* 19: 27.—Sattelberg, Huon Penin.
RANGE: Mountains of the Huon Penin[MA].
DIAGNOSIS: Exhibits a broad and complete yellow breast band (Rand & Gilliard 1967).

c. *Charmosyna stellae stellae* A. B. Meyer

Synonym: *Charmosyna atrata* Rothschild, 1898. *Bull. Brit. Orn. Club* 7: 54.—Mount Scratchley, SE Peninsula.

RANGE: Mountains of the SE Peninsula, west to the Angabanga R in the s watershed and to the Herzog Mts in the n watershed[MA].

DIAGNOSIS: Similar to *goliathina*, but longer uppertail coverts are red, not green. The form *atrata* Rothschild was synonymized with this form by Mayr (1941).

NOTES

C. stellae is characterized by the following: (1) presence of both black and red plumage morphs in all subspecies; (2) sexual dichromatism—male with red back patch, and female with yellow back patch (green in the black morph); (3) exhibits no yellow patch on side of breast; (4) exhibits no small yellow patch on flank; (5) basal section of ventral surface of tail feathers is yellowish green; (6) blue patch on hind-crown is large and starts behind eye and extends down toward nape; (7) black bar posterior to blue patch is narrow; (8) posterior to black bar is a large red nape patch (no second black bar present); (9) black postocular stripe is present. One question that needs resolution is the species determination (*papou* vs. *stellae*) of the populations that might exist in the Wandammen, Kumawa, and Fakfak Mts. The English name has long been in use by parrot fanciers. *Other names*—formerly treated as a race of the Papuan Lorikeet, or Fairy Lory, formerly included in *C. papou*.

Fairy Lorikeet *Charmosyna pulchella* ENDEMIC

Charmosyna pulchella G. R. Gray, 1859. *List Birds Brit. Mus., Psittacidae*: 102.—Manokwari, Bird's Head.

Forages at canopy flowers of forest trees in the mts of NG, including virtually all outlying ranges, 500–1,800 m; extremes—foothills at sea level–2,300 m[CO]. Frequents forest and edge; at times wanders to low elevations (Ripley 1964a, Coates 1985). No records from the Wandammen Mts (Diamond & Bishop 2015).

SUBSPECIES

a. *Charmosyna pulchella pulchella* G. R. Gray

Synonym: *Charmosinopsis bella* De Vis, 1900. *Ann. Queensland Mus.* 5: 12, pl. 8.—Wharton Ra?, SE Peninsula.

RANGE: Mountains of the Bird's Head, the Central Ra, the PNG North Coastal Ra, Adelbert Mts, and mts of Huon Penin (all ranges but for the Foja and Cyclops Mts). Records for mt ranges subsequent to Mayr (1941) include: Fakfak Mts (Gibbs 1994), North Coastal Ra (Diamond specimens at AMNH), and Adelbert Mts (Pratt 1983).

DIAGNOSIS: Breast red with small yellow streaks; rump patch dull blue (Collar 1997). We subsume *bella* into the nominate form (Forshaw 1989). See photos in Coates & Peckover (2001: 75)—both birds featured are the nominate form, although the left-hand bird exhibits the dark postocular patch typical of *rothschildi*.

b. *Charmosyna pulchella rothschildi* (Hartert)

Charmosynopsis pulchella rothschildi Hartert, 1930. *Novit. Zool.* 36: 105.—Cyclops Mts, ne sector of NW Lowlands, n NG.

RANGE: The Cyclops[MA] and Foja Mts[J], and mts of the upper Mamberamo near Doormanpaad[MA] (Beehler & Prawiradilaga 2010[J]).

DIAGNOSIS: Dark crown patch joined to rear edge of supercilium, large green breast patch punctuated by fine yellow streaks, dull purplish belly, and no blue rump; female more extensively green below (Collar 1997).

NOTES

There is considerable plumage variation subsumed in these two subspecies, and this merits additional scrutiny. *Other names*—Little Red Lorikeet, Little Red Lory.

Neopsittacus Salvadori, 1875. *Ann. Mus. Civ. Genova* 7: 761. Type, by monotypy, *Nanodes musschenbroekii* Schlegel.

The two species of the genus are endemic to the mountains of the New Guinea region. This small genus is characterized by the mainly green plumage, contrasting reddish-orange breast and belly, and yellow/orange-streaked cheek.

Yellow-billed Lorikeet *Neopsittacus musschenbroekii*   ENDEMIC

Nanodes Musschenbroekii Schlegel (ex von Rosenberg ms), 1871. *Ned. Tijdschr. Dierk.* (1873) 4: 34.— Hatam, Arfak Mts, Bird's Head.

> Synonym: *Neopsittacus musschenbroekii major* Neumann, 1924. *Orn. Monatsber.* 32: 38.—Schrader Ra, n sector, Eastern Ra.
> Synonym: *Neopsittacus musschenbroekii medius* Stresemann, 1936. *Mitt. Zool. Mus. Berlin* 21: 231.—Mount Sumuri, Weyland Mts, Western Ra.

Frequents canopy flowers of forest and edge in mts of the Bird's Head, Central Ra, Foja Mts[J], and the Huon Penin, 1,100–3,650 m[K] (Beehler *et al.* 2012[J], Beehler field notes 5-Aug-1987[K]). Occasionally found feeding together with the smaller but very similar *N. pullicauda*, which inhabits a higher overall elevational zone. Mist-netted at 3,650 m at Lake Omha near Mt Scratchley (Beehler field notes).

NOTES

We treat the species as monotypic, which subsumes subspecies based on some minor variations in size and coloration of facial streaking. Individual populations are quite variable, especially in the patterning of the red-orange on the breast and belly. See also Mees (1964) for a slightly different treatment. The two species of the genus can be distinguished in the field with care, but often overflying birds can be assigned only to genus. *Other names*—Yellow-billed Mountain Lory, Musschenbroek's Lorikeet.

Orange-billed Lorikeet *Neopsittacus pullicauda*   ENDEMIC | MONOTYPIC

Neopsittacus pullicauda Hartert, 1896. *Novit. Zool.* 3: 17.—Mount Victoria, SE Peninsula.

> Synonym: *Neopsittacus muschenbrocki alpinus* Ogilvie-Grant, 1914. *Bull. Brit. Orn. Club* 35: l2.—Upper Otakwa R, 2,450 m, Western Ra.
> Synonym: *Neopsittacus pullicauda socialis* Mayr, 1931. *Mitt. Zool. Mus. Berlin* 17: 700.—Saruwaged Ra, Huon Penin.

Frequents canopy flowers of forest and edge in higher mts of the Central Ra and Huon Penin; mainly 2,100–3,800 m, records as low as 800 m (Coates & Peckover 2001). Absent from the Bird's Head. Frequents mossy forest. In the lower parts of its elevational range often found in same locality and sometimes in same feeding tree as *N. musschenbroekii*.

NOTES

We treat the species as variable and clinal, with no recognized subspecies, which are only thinly differentiated (see Fig. 2b). Birds are smaller and paler in the west and larger and more richly colored in the southeast. *Other names*—Orange-billed Mountain Lory, Emerald Lorikeet.

GENUS *LORIUS*   [Genus 6 spp/Region 2 spp]

Lorius Vigors, 1825. *Zool. Journ.* 2: 400. Type, by original designation, *Psittacus domicella* Linnaeus, 1758. *Domicella* Wagler, 1832, was formerly used for this genus, since *Larius* Boddaert, 1783, applicable to *Eclectus*, had priority; however, Boddaert's name was officially suppressed in 1970 (ICZN Opinion 938), leaving *Lorius* Vigors, 1825, as the correct name for the genus.

The species of this genus range from the Moluccas east to New Guinea, New Britain, New Ireland, and the eastern Solomon Islands (Collar 1997). The genus is characterized by the large head, compact body, rounded tail, and plumage boldly patterned in red, green, black, blue, and yellow.

Black-capped Lory *Lorius lory*

Psittacus Lory Linnaeus, 1758. *Syst. Nat.*, ed. 10, 1: 100.—Bird's Head.

Typically found in pairs foraging at flowering trees through NG, the NW Islands, and the Bay Is; sea level–1,200 m, records up to 1,800 m[J] (Coates & Peckover 2001[J]). Frequents forest, edge, and clearings in forest. Absent from the SE Islands.

SUBSPECIES

a. *Lorius lory lory* (Linnaeus)

Synonym: *Lorius lory major* Rothschild and Hartert, 1901. *Novit. Zool.* 8: 66.—Waigeo I, NW Islands.

RANGE: The four main NW Islands and the Bird's Head (Mayr 1941, Mees 1965).

DIAGNOSIS: Underwing coverts mostly red (Rand & Gilliard 1967). Blue-black of underparts connects broadly with nape band of the same color (Coates & Peckover 2001). See photo in Coates & Peckover (2001: 71, upper-right image). Synonymy of *major* with *lory* follows Mees (1965) and Forshaw (1989).

b. *Lorius lory erythrothorax* Salvadori

Lorius erythrothorax Salvadori, 1877. *Ann. Mus. Civ. Genova*, 10: 32.—Mount Epa, Hall Sound, SE Peninsula.

Synonym: *Lorius erythrothorax rubiensis* A. B. Meyer, 1893. *Abh. Ber. Zool. Mus. Dresden* (1892–93) 4(3): 10.—Rubi, Bird's Neck.

RANGE: The Bird's Neck, S Lowlands, and SE Peninsula (Rand 1942a, Forshaw 1989).

DIAGNOSIS: Blue-black of abdomen extends barely onto breast; blue band on hindneck narrow; red band on upper mantle broad (Rand & Gilliard 1967). See photo in Coates & Peckover (2001: 71, upper-left image).

c. *Lorius lory cyanauchen* (S. Müller)

Psittacus cyanauchen S. Müller, 1841. *Verh. Nat. Gesch. Ned., Land-en Volkenk.* 4: 107.—Biak I, Bay Islands.

RANGE: Biak I (Mayr 1941).

DIAGNOSIS: No red band on nape; hindneck mauve-blue (Rand & Gilliard 1967, Forshaw 1989).

d. *Lorius lory jobiensis* (A. B. Meyer)

Domicella lori var. *jobiensis* A. B. Meyer, 1874. *Sitzungsber. Akad. Wiss. Wien* 70, Abth. 1: 229, 231.—Yapen I, Bay Islands.

RANGE: Yapen and Mios Num I (Mayr 1941).

DIAGNOSIS: Larger than *salvadorii*; wing 170–183 (Rand & Gilliard 1967). A rosy tinge to red of breast (Forshaw 1989).

e. *Lorius lory viridicrissalis* de Beaufort

Lorius cyanauchen viridicrissalis de Beaufort, 1909. *Nova Guinea* 5, livr. 3: 403.—Lake Sentani, Humboldt Bay, n NG.

RANGE: Northern NG from the Mamberamo R to the PNG border (Forshaw 1989).

DIAGNOSIS: Similar to *salvadorii*, but band on hindneck more blackish blue.

f. *Lorius lory salvadorii* A. B. Meyer

Lorius salvadorii A. B. Meyer, 1891. *Abh. Ber. Zool. Mus. Dresden* (1890–91) 3(4): 6.—Astrolabe Bay, e verge Sepik-Ramu.

RANGE: Northern NG from the PNG border to Astrolabe Bay, encompassing the Sepik-Ramu.

DIAGNOSIS: Underwing coverts mostly blue-black; blue-black of abdomen extends only onto posterior edge of breast; wing 152–175 mm (Rand & Gilliard 1967). Band on hindneck dark blue (Forshaw 1989).

g. *Lorius lory somu* Diamond

Lorius lory somu Diamond, 1967. *Amer. Mus. Novit.* 2284: 4.—Soliabeda, Eastern Ra.

RANGE: The s side of the Border and Eastern Ra, from Ok Tedi (Coates & Lindgren 1978) to Karimui[DI], Bomai[DI], and Soliabeda[DI] in the Purari drainage. Possibly a foothills and uplands form. The subspecies *erythrothorax* apparently extends east and west of this form, but in the S Lowlands and not the uplands—a rather unusual subspecific distribution, but not unique to this example.

DIAGNOSIS: Similar to *erythrothorax* in extent of blue-violet on belly and exhibiting red underwing, but differs in absence of dark band across hindneck—the neck is all red instead (Diamond 1967).

Our racial treatment follows Forshaw (1989). More consolidation of subspecies may be in order. *Other names*—Western Black-capped Lory, *Domicella lory.*

Purple-bellied Lory *Lorius hypoinochrous*   RESIDENT

Lorius hypoinochrous G. R. Gray, 1859. *List Birds Brit. Mus., Psittacidae*: 49.—Tagula I, Louisiade Arch, SE Islands.

Inhabits the SE Peninsula, west to Lae in the n watershed, and Cape Rodney in the s watershed; also the D'Entrecasteaux, Trobriand, and Louisiade Is. Frequents forest, edge, and coconut plantations. On the Mainland confined to lowlands and foothills; on Goodenough I to 1,600 m (Mayr & van Deusen 1956).

Extralimital—Bismarck Arch (Dutson 2011).

SUBSPECIES

a. *Lorius hypoinochrous devittatus* Hartert
Lorius hypoenochrous devittatus Hartert, 1898. *Novit. Zool.* 5: 530.—Fergusson I, D'Entrecasteaux Arch, SE Islands.
RANGE: The SE Peninsula, east of the Angabanga R (in the s watershed) and Huon Gulf (in the n watershed); D'Entrecasteaux Arch, Trobriand Is, Woodlark group (Woodlark, Dugumenu, Gawa), Long I, and the Bismarck Arch (Mayr 1941).
DIAGNOSIS: Lacks black tips found on greater coverts of underwing of nominate form (Collar 1997). See photos in Coates & Peckover (2001: 71).

b. *Lorius hypoinochrous hypoinochrous* G. R. Gray
RANGE: Misima and Tagula I, Louisiade Arch (Mayr 1941).
DIAGNOSIS: Black tipping on greater coverts of underwing (Collar 1997); breast paler red than upper abdomen (Rand & Gilliard 1967).

c. *Lorius hypoinochrous rosselianus* Rothschild and Hartert
Lorius hypoenochrous rosselianus Rothschild and Hartert, 1918. *Novit. Zool.* 25: 312.—Mt Rossel, Rossel I, Louisiade Arch, SE Islands.
RANGE: Rossel I (Mayr 1941).
DIAGNOSIS: Pinker red ventrally, streaked violet (Collar 1997).

NOTES
Other names—Eastern Black-capped Lory, *Domicella hypoinochroa.*

GENUS *PSITTEUTELES*   [Genus 2 spp/Region 1 sp]

Psitteuteles Bonaparte, 1854. *Rev. et Mag. Zool., Paris*, (2)6: 157. Type, by subsequent designation (G. R. Gray, 1855. *Cat. Gen. Subgen. Birds*: 88), *Psittacus versicolor* Vigors = *Trichoglossus versicolor* Lear.

The species of this genus range from the Central Ranges and mountains of the Huon Peninsula of New Guinea to northern Australia. This genus of small lorikeets is characterized by the red cap and mainly green plumage. Schweizer *et al.* (2015) demonstrated the non-monophyly of the genus as typically construed. The type species for the genus, *P. versicolor*, is basal to a large clade that includes *P. goldiei* as an additional basal lineage (Schweizer *et al.* 2015: fig. 2). Given the basal affinities of both of these species, and their distant relationship to *P. iris*, we suggest keeping *goldiei* and *versicolor* in *Psitteuteles* until more is learned of their proper positions in the lories. By contrast, it is clear that *P. iris* better belongs with the genus *Trichoglossus.*

Goldie's Lorikeet *Psitteuteles goldiei*   ENDEMIC | MONOTYPIC

Trichoglossus Goldiei Sharpe, 1882. *Journ. Linn. Soc. London, Zool.* 16: 317, 426.—Astrolabe Ra, SE Peninsula.

Inhabits the Central Ra and mts of the Huon Penin[J], 1,500–2,750 m[CO], occasionally in lowlands at the base of the mts (Mayr 1941, Freeman *et al.* 2013[J]). Inhabits the canopy of forest and planted eucalypts in upland interior towns (*e.g.*, Porgera, Mt Hagen, Goroka). Flocking and nomadic. Often found in flowering trees with other small lorikeets.

NOTES

Coates (1985) and Freeman *et al.* (2013) reported this species from the Huon Penin, based on sight records and audio recordings. *Other names*—Red-capped Streaked Lory or Lorikeet.

GENUS *TRICHOGLOSSUS* [Genus ca. 11 spp/Region 1 sp]

Trichoglossus Stephens, 1826. In Shaw, *Gen. Zool.* 14(1): 129. Type, by subsequent designation (Lesson, 1828. *Man. Orn.* 2: 147), *Psittacus haematodus* Linnaeus.

The species of this genus range from Sulawesi, the southern Moluccas, Mindanao, and Pohnpei south to New Guinea, the Solomon Islands, Vanuatu, New Caledonia, and northern and eastern Australia (Collar 1997). This small group of diversely plumaged species is characterized by the rather long pointed tail, orange bill, slim form, red iris, and large head.

Rainbow Lorikeet *Trichoglossus haematodus* RESIDENT

Psittacus haematod[*us*] Linnaeus, 1771. *Mantissa Plant.*: 524.—Ambon, s Moluccas.

Throughout the NG region, including many satellite islands; sea level–1,800 m, rarely to 2,440 m[CO]. Largely absent from the SE Islands. Frequents a wide variety of wooded habitats, from forest to suburban gardens.

Extralimital—e Indonesia (s Moluccas and Lesser Sundas), Bismarck and Solomon Is, Vanuatu and New Caledonia, and n and e Australia (Collar 1997).

SUBSPECIES

a. *Trichoglossus haematodus haematodus* (Linnaeus)

   Synonym: *Trichoglossus haematodus intermedius* Rothschild and Hartert, 1901. *Novit. Zool.* 8: 70.—Near Madang, e verge Sepik-Ramu.

   Synonym: *Trichoglossus haematodus berauensis* Cain, 1955. *Ibis* 97: 433.—Manokwari, Bird's Head.

RANGE: Eastern Indonesia: Buru, Seram, Ambon, Seram Laut, Watubela, and Tayandu; the four main NW Islands; Numfor and Yapen I; and w NG, in the n watershed east from Bird's Head to Astrolabe Bay (Madang); in the s watershed east to the upper Purari R (Mayr 1941, White & Bruce 1986, Forshaw 1989).

DIAGNOSIS: Yellow ear crescent present; green of nape matches that of mantle; belly patch is blue and breast is blue-barred. Upper back with some orange-red spotting. Mees (1965) has synonymized *berauensis* with this form. We follow the suggestion of Forshaw (1989) in sinking *intermedius*.

b. *Trichoglossus haematodus massena* Bonaparte

Trichoglossus haematodus massena Bonaparte, 1854. *Rev. et Mag. Zool.* (2)6: 157.—Vanuatu.

   Synonym: *Trichoglossus aberrans* Reichenow, 1918. *J. Orn.* 66: 439.—Bismarck Arch.

   Synonym: *Trichoglossus haematodus micropteryx* Stresemann, 1922. *J. Orn.* 70: 407.—Sattelberg, Huon Penin.

RANGE: The SE Peninsula, west to the Huon Penin in the n watershed and to Lake Kutubu in the s watershed; also the Schouten Is, and Kairiru, Manam[K], Karkar[K], Bagabag[K], and Budibudi Is (Diamond & LeCroy 1979[K]). Distribution in the SE Islands—only in the isolated Budibudi Is but nowhere else—is remarkable. Extralimital range includes the Bismarck Arch (except range of *flavicans*), the Solomon Is, and Vanuatu (Dutson 2011).

DIAGNOSIS: The ear crescent is greenish rather than yellow; fine and sparse blue barring on breast; green belly patch. Upper back without (or with obscure) red spotting. Plumage overall paler than nominate form; nuchal collar more greenish; narrower edging to breast feathering; bill small. See photo in Coates & Peckover (2001: 72, middle photo).

c. *Trichoglossus haematodus rosenbergii* Schlegel

Trichoglossus Rosenbergii Schlegel, 1871. *Ned. Tijdschr. Dierk.* (1873) 4: 9.—Biak I, Bay Islands.

RANGE: Biak I.

DIAGNOSIS: Entire nape and nuchal area forms a broad bright yellow patch on back of head and nape. See photo in Coates & Peckover (2001: 72, bottom image).

d. *Trichoglossus haematodus nigrogularis* G. R. Gray

Trichoglossus nigrogularis G. R. Gray, 1858. *Proc. Zool. Soc. London*: 183.—Aru Is.

    Synonym: *Trichoglossus caeruleiceps* D'Albertis & Salvadori, 1879. *Ann. Mus. Civ. Genova* 14: 41.—Katau R, Trans-Fly, s NG.

    Synonym: *Trichoglossus haematodus brooki* Ogilvie-Grant, 1907. *Bull. Brit. Orn. Club* 19: 102.—Type an aviary bird said to have come from Swangi I, off the s coast of Trangan I, Aru Is.

RANGE: The e Kai Is; Aru Is; and s NG between Princess Marianne Strait and the Fly R.

DIAGNOSIS: Entire crown and sides of head pale to medium blue; breast pale, with narrow barring; abdomen blackish green (Forshaw 1989). We follow the suggestion of Forshaw (1989) in sinking *caeruleiceps*. This form is close to *moluccanus* and perhaps should by lumped with it. Mees (1982b) provided a differing view, based on size (which we think is a less important character in this species, which exhibits hyper-variable plumage across geographies).

NOTES

Each local population shows some distinctive features, making designation of subspecies difficult (Cain 1955, Diamond 1972). We lump *brooki* into *nigrogularis* to await collection of wild birds from a known locality. We place *micropteryx* within *massena* following Dickinson (2003). Although Diamond & LeCroy (1979: 480) listed this species as occurring on Manam I, it is unclear which race was involved. In the SE Islands, this lory seems to be known at present only from the Budibudi Is, where it was recorded by D. Hobcroft and K. D. Bishop (ms); we wonder whether this population, and that of *Charmosyna placentis* on Woodlark, might have arrived by human agency, as there are no source populations nearby. The Misima record (Mayr 1941) is something of a mystery, because *T. haematodus* is absent there. LeCroy & Peckover (1998) provided evidence that the specimens were possibly collected on Kimuta I instead; however, this lory was not seen recently on Kimuta either (Pratt field notes 2004).

    This is one of those widespread and hyper-variable lineages for which species boundaries are problematic. A full molecular analysis is needed. Dickinson & Remsen (2013) treated this assemblage as five species (but only one from Region), adding *T. forsteni*, *T. weberi*, *T. capistratus*, and *T. rubritorquis*. Further splits were made by del Hoyo & Collar (2014): *T. rosenbergii* of Biak I and *T. moluccanus* from Australia. The lack of conformance between these two world compendia underscores the lack of clarity on this complex lineage. *Other names*—Coconut or Rainbow Lory, Green-naped, Red-collared, or Red-breasted Lorikeet, *Trichoglossus rosenbergii*.

GENUS *EOS*   [Genus 6 spp/Region 2 spp]

Eos Wagler, 1832. *Abh. K. Bayer. Akad. Wiss., Math.-Phys. Kl.* 1: 494. Type, by subsequent designation (G. R. Gray, 1840. *List. Gen Birds*: 52), *E. indica* (Gmelin) = *Psittacus histrio* P. L. S. Müller.

    The species of this genus range from small islands north of Sulawesi and the Moluccas east to the Northwestern Islands and Bay Islands (Collar 1997). They are small and compact, slim, orange-billed, red-and-black lories variously marked with violet-blue. They occur mostly on small islands.

Violet-necked Lory *Eos squamata*   RESIDENT

Psittacus Squamatus Boddaert, 1783. *Tabl. Planch. Enlum.*: 42 (ex Daubenton, *Pl. Enlum.*: 684).—Gebe I, NW Islands.

    A tramp species of the NW Islands. Found in all wooded island habitats, foraging on canopy flowers. Extralimital—the n Moluccas.

SUBSPECIES

a. *Eos squamata squamata* Boddaert

    Synonym: *Eos squamata attenua* Ripley, 1957. *Postilla* Yale Peabody Mus. 31: 2.—Kamoa I, Schildpad Is, north of Misool, NW Islands.

RANGE: Gebe[MA], Gag[M], Kofiau[K], Waigeo[MA], Ayu[J], and Batanta I[MA]; Schildpad Is[J]; also small islands near Misool I[L] (Ripley 1964a[J], Mees 1965[L], Johnstone 2006[M], Diamond *et al.* 2009[K]).

DIAGNOSIS: Exhibits variable amounts of violet wash on the throat and nape. Crown entirely red, bluish collar, and reduced dark coloration on the belly (Mees 1965). The form *attenua* was subsumed into the nominate form by Mees (1965).

NOTES

No records from Salawati I. One might question whether this small-island tramp species actually does occur on the larger islands of Waigeo and Batanta (as historically reported) or instead was collected on adjacent islets. It is abundant on Kofiau (Diamond *et al.* 2009), but seems to have been collected only on fringing islets off Misool (Mees 1965). *Other names*—Moluccan Red Lory, Moluccas Red Lory.

Black-winged Lory *Eos cyanogenia* ENDEMIC | MONOTYPIC

Eos cyanogenia Bonaparte, 1850. *Compt. Rend. Acad. Sci. Paris* 30: 135.—Numfor I, Bay Islands.

Frequents forest and edge on Biak, Numfor, Manim, and Mios Num I[L] of Geelvink Bay (Peters 1937[L], Mayr 1941). On Biak it is fairly common in small flocks in forest patches and plantations. Probably found at all elevations.

Other names—Biak Red Lory.

GENUS *PSEUDEOS* [Genus 2 spp/Region 1 spp]

Pseudeos Peters, 1935. *Proc. Biol. Soc. Wash.* 48: 68. Type, by original designation and monotypy, *Eos fuscata* Blyth.

The two species of the genus inhabit New Guinea, plus Salawati and Yapen Islands and the Solomons (Schweizer *et al.* 2015). The genus is characterized by the medium size, orange bill, and lack of blue or violet in plumage.

Dusky Lory *Pseudeos fuscata* ENDEMIC | MONOTYPIC

Eos fuscata Blyth, 1858. *Journ. Asiat. Soc. Bengal* 27: 279.—Manokwari, Bird's Head.

Synonym: *Eos incondita* A. B. Meyer, 1886. *Zeitschr. Ges. Orn.* 3: 6, pl. 1, fig. 2.—Manokwari, w Bird's Head.

Frequents forest, edge, clearings, and plantations through NG, Salawati[MA], and Yapen[MA] I, sea level–2,700 m[J] (Sam & Koane 2014[J]). Most common in the foothills. Often flies long distances between roosting and feeding places. Has been seen flying over the main cordillera of NG (Bell 1979b), and in the evening, flocks have been observed flying to the mts from the lowlands (Rand 1942b). Can be encountered in very loud, chattering flocks in feeding trees, the sound carrying more than a km.

NOTES

Includes three plumage morphs—yellow, orange, and red (Pratt & Beehler 2014). Photos of the red and yellow morphs appear in Coates & Peckover (2001: 70). *Other names*—Dusk-Orange or White-rumped Lory.

GENUS *CHALCOPSITTA* [Genus 4 spp/Region 3 spp]

Chalcopsitta Bonaparte, 1850 [February]. *Compt. Rend. Acad. Sci. Paris* 30: 134; *Consp. Gen. Av.* 1: 3 (after 15-Apr-1850). Type, by subsequent designation (G. R. Gray 1855. *Cat. Gen. Subgen. Birds*: 86), *Psittacus ater* Scopoli.

The species of this genus inhabit New Guinea and the Solomon Islands. The genus is characterized by the distinctive shallow wingbeat, dry, high-pitched, shrieking flight call, and dark ocular patch of bare skin.

Black Lory *Chalcopsitta atra*   ENDEMIC

Psittacus ater Scopoli, 1786. *Delic. Flor Faun. Insubr.* 2: 87.—New Guinea.

The Bird's Head, Bird's Neck, and NW Islands (Batanta, Misool, Salawati). Inhabits lowland forest and coastal and riparian openings to 100 m elevation (Forshaw 1989, Coates & Peckover 2001).

SUBSPECIES

a. *Chalcopsitta atra bernsteini* von Rosenberg
Chalcopsitta Bernsteini von Rosenberg, 1861. *J. Orn.* 9: 46.—Misool I, NW Islands.
RANGE: Misool I (Mees 1965).
DIAGNOSIS: Forehead typically tinged red; thighs dark red (Ripley 1964a, Rand & Gilliard 1967). Rump bluer; presumed to be a stable population of hybrid origin [*atra* × *insignis*] (Mees 1965). Whether this merits subspecies rank or should be considered simply "intergradient" is unclear. Misool I is on a shallow shelf that was contiguous with the Onin Penin in the Quaternary (at the last low sea-level stand).

b. *Chalcopsitta atra atra* (Scopoli)
RANGE: Batanta[MA], Salawati[MA], and adjacent Arar I[J]; also w part of Bird's Head (Ripley 1964a[J]). Lowland and savanna forest at sea level. Localities on the Bird's Head—Sorong, Erkwero, Jeflio, Mega, Loewelala (Gyldenstolpe 1955b).
DIAGNOSIS: Face, forehead, and thighs black (Gyldenstolpe 1955b, Rand & Gilliard 1967). See photo in Coates & Peckover (2001: 69).

c. *Chalcopsitta atra insignis* Oustalet
Chalcopsitta insignis Oustalet, 1878. *Bull. Assoc. Sci. France* (1)21: 247.—Amberpon I, off e Bird's Head.
>   Synonym: *Chalcopsitta spectabilis* van Oort, 1908. *Notes Leyden Mus.* 30: 127.—Mamberiok Penin, head of Geel-
>   vink Bay.
RANGE: The w coast of Geelvink Bay (Amberpon I[MA]), the Onin Penin[MA] (Fakfak and Kaukas), and Kambala[J] on the Bomberai Penin (Gyldenstolpe 1955b[J]).
DIAGNOSIS: Red on forehead, cheek, thighs, underwing coverts, and concealed bases of neck and breast feathers (Rand & Gilliard 1967). Easternmost birds may intergrade with *C. scintillata* at the head of Geelvink Bay—hybrid described as *C. atra 'spectabilis'*.

NOTES
Other names—Rajah Lory.

Brown Lory *Chalcopsitta duivenbodei*   ENDEMIC | MONOTYPIC

Chalcopsittacus Duivenbodei Dubois, 1884. *Bull. Mus. Roy. d'Hist. Nat. Belgique* 3: 113. pl. 5.—Tana Mera, n NG.
>   Synonym: *Chalcopsittacus duyvenbodei syringanuchalis* Neumann, 1915. *Orn. Monatsber.* 23: 179.—Near Madang,
>   e verge Sepik-Ramu.
The NW Lowlands and Sepik-Ramu, from the e shore of Geelvink Bay east to Astrolabe Bay; lowlands–150 m[CO] (Rand 1942b).

NOTES
The form *intermedius* was synonymized by Mayr (1941). It will be interesting to determine whether the sw edge of the range of this lory reaches the nw edge of the range of *scintillata* in s Geelvink Bay, and whether there is any hybridization at that point of contact. Forshaw (1989) noted the dubious value of the named subspecies *syringanuchalis*. *Other names*—Duyvenbode's Lory.

Yellow-streaked Lory *Chalcopsitta scintillata*   ENDEMIC

Psittacus sintillatus Temminck, 1835. *Planch. Col. d'Ois.*, livr. 96: pl. 569; emended to *scintillatus* by Temminck, 1839. *Table Method.* (1838): 61.—Triton Bay, Bird's Neck.

Inhabits s NG from the Bird's Neck (Triton Bay and head of Geelvink Bay) east to the Kemp Welch R, including the Trans-Fly; also the Aru Is; lowlands–800 m[CO]. Frequents forest, edge, and wooded savanna (Coates & Peckover 2001).

a. *Chalcopsitta scintillata scintillata* (Temminck)

RANGE: The e half of the Bird's Neck, the S Lowlands, and the Trans-Fly. Ranges from the head of Geelvink Bay (Waropen, Rubi, Jaur, and Mesan) and s NG from Triton Bay[K] to the Fly R (Mees 1982b, Hoek Ostende *et al.* 1997[K]). This form grades toward *chloroptera* east of the Lorentz R.

DIAGNOSIS: Underwing coverts red, with minimal green (Collar 1997).

b. *Chalcopsitta scintillata rubrifrons* Gray

Chalcopsitta rubrifrons G. R. Gray, 1858. *Proc. Zool. Soc. London*: 182, pl. 135.—Aru Is.

RANGE: The Aru Is.

DIAGNOSIS: Exhibits broader, more orange-colored streaking on breast; amount of red on forehead variable (Rand & Gilliard 1967).

c. *Chalcopsitta scintillata chloroptera* (Salvadori)

Chalcopsittacus chloropterus Salvadori, 1876. *Ann. Mus. Civ. Genova* 9: 15.—Coast of se NG opposite Yule I.

RANGE: The c and e sector of the S Lowlands and s watershed of the SE Peninsula; ranges from the the the Kemp Welch R westward to the upper Lorentz R (Mayr 1941, Forshaw 1989). This form and the nominate apparently intergrade between the Fly and Lorentz R (Forshaw 1989).

DIAGNOSIS: Exhibits mainly green or green and red underwing coverts (Rand & Gilliard 1967).

NOTES

These three subspecies are thinly differentiated and merit additional examination to distinguish individual variation from true geographic differentiation. The s and e parts of Geelvink Bay and the Bird's Neck region are worth visiting to determine the extent of the ranges of the three Mainland species of *Chalcopsitta*—is this an area of ongoing interspecific hybridization or introgression? Spelling of the species name follows Mees (1982b) and the supporting recommendation of N. David (*in litt.* 28-May-2012), who noted the original spelling *sintillatus* was an unambiguous misspelling, which can be corrected under the ICZN 1999 Art. 32.5.1. *Other names*—Greater Streaked Lory, Yellowish-streaked Lory.

GENUS *PSITTACULIROSTRIS*   ENDEMIC [Genus 3 spp]

Psittaculirostris J. E. and G. R. Gray, 1859. *Cat. Mamm. Birds Brit. Mus.*: 42. Type, by monotypy, *Psittacula desmarestii* Lesson, 1830 = *Psittacus desmarestii* Dumont, 1826.

The species of this genus inhabit the Northwestern Islands and all of lowland New Guinea. Closely allied to *Cyclopsitta*, the three species are confined to New Guinea lowland and hill forest (Forshaw 1989). The genus is characterized by the chunky body, very large head, feathered cere, very short spiky tail, generally green plumage, brightly colored head patterning (yellow, orange, blue, turquoise, black), elongate feathering of the ear coverts, prominently notched bill, and specialization on fig seeds. Presumably the three species of the genus form a superspecies (Forshaw 1989).

Large Fig-Parrot *Psittaculirostris desmarestii*   ENDEMIC

Psittacus Desmarestii Dumont, 1826. *Dict. Sci. Nat.*, éd. Levrault 39: 89.—Manokwari, Bird's Head.

Inhabits the NW Islands (less Waigeo), the Bird's Head, S Lowlands, Eastern Ra, and SE Peninsula. Absent from the NW Lowlands, Sepik-Ramu, Huon and n watershed of the SE Peninsula to the Kumusi R, where replaced by congeners (*P. salvadorii* and *P. edwardsii*). Inhabits mainly lowland forest; in places penetrates into interior mid-montane basins (to 1,650 m[CO]). Frequents forest and edge.

SUBSPECIES

a. *Psittaculirostris desmarestii blythii* (Wallace)

Cyclopsitta blythii Wallace, 1864. *Proc. Zool. Soc. London*: 284.—Misool I, NW Islands.

RANGE: Misool I (Mees 1965, Hoek Ostende *et al.* 1997).

DIAGNOSIS: Cheek and crown rich orange; nape green; breast band dull orange-brown with thin blue upper edge. Blue spot below eye absent in adults but present in young birds (Mees 1965, Forshaw 1989).

b. *Psittaculirostris desmarestii occidentalis* (Salvadori)

Cyclopsittacus occidentalis Salvadori, 1876. *Ann. Mus. Civ. Genova* (1875) 7: 910.—Salawati I, NW Islands.

> Synonym: *Cyclopsitta desmarestii intermedia* van Oort, 1909. *Notes Leyden Mus.* 30: 229.—Sekru, Onin Penin, Bird's Neck.

RANGE: Salawati and Batanta I, w coast of the Bird's Head, and Onin and Bomberai Penin (Mayr 1941, Gyldenstolpe 1955b, Hoek Ostende *et al.* 1997).

DIAGNOSIS: Cheeks, ear coverts, and throat deep yellow instead of green; no blue on occiput; blue spot below eye present but less extensive and paler, more greenish in color. See photo in Coates & Peckover (2001: 80). We follow the suggestion of Forshaw (1989) in combining *intermedia* with *occidentalis*. By contrast, Collar (1997) recognized *intermedia* as distinct. In a third viewpoint, Dickinson & Remsen (2013) synonymized *intermedia* with the nominate form, which follows Mayr (1941). This nonconformity of views is perhaps to be expected, given the remarkable amount of color variation exhibited by regional isolates of this polytypic lowland species.

c. *Psittaculirostris desmarestii desmarestii* (Dumont)

RANGE: The e part of the Bird's Head east to Triton Bay and the e coast of Geelvink Bay (the Waropen). Range includes the Wandammen Penin (Mayr 1941, Gyldenstolpe 1955b).

DIAGNOSIS: Blue subocular spot; cheek washed with green; crown deep orange (fading posteriorly); throat collar blue and red; nape green (Collar 1997).

d. *Psittaculirostris desmarestii godmani* (Ogilvie-Grant)

Cyclopsittacus godmani Ogilvie-Grant, 1911 (March). *Bull. Brit. Orn. Club*: 27: 67.—Upper Mimika R, sw NG.

RANGE: The S Lowlands from the Mimika R to the Fly R, where it intergrades with *cervicalis* (Rand 1942a). Localities—Mimika R[MA], Baliem valley (1,500 m)[A], upper Fly R/Black R[CO], and Mt Bosavi[CO] (Coates & Peckover 2001[A]).

DIAGNOSIS: Male with no blue on occiput or below eye; crown orange-red; face and cheek bright yellow; nuchal patch pale orange; pale blue breast band and flanks. Female with green nuchal patch (Collar 1997). The form *meeki* Rothschild and Hartert was synonymized by Mayr (1941).

e. *Psittaculirostris desmarestii cervicalis* (Salvadori & D'Albertis)

Cyclopsittacus cervicalis Salvadori & D'Albertis, 1876. *Ann. Mus. Civ. Genova* (1875) 7: 811.—Mount Epa, Hall Sound, SE Peninsula.

RANGE: The e sector of the S Lowlands; also the SE Peninsula, west to the Kumusi R in the n watershed and west to the Fly R in the s watershed (where it intergrades with *godmani*). Also the se sector of the Eastern Ra. Localities—Kuli in the Wahgi valley (1,500 m)[A], Bomai[DI], Karimui[DI] (1,100 m), Purari basin[J], lower Eloa R and Lakekamu basin[K], the Mainland near Yule I[CO], the upper Brown R[CO], Goldie R[CO], Rigo[CO], Hydrographer Mts[CO], and Kumusi R[MA] (Beehler 1992[K], Coates & Peckover 2001[A], I. Woxvold *in litt.*[J]).

DIAGNOSIS: Crown and cheek orange; occiput and hindneck blue; darker blue band across breast; lower breast variably tinged with orange-buff (Forshaw 1989, Collar 1997).

NOTES

Split into three species by del Hoyo & Collar (2014), who added Yellow-naped Fig-parrot (*P. godmani*) and Red-faced Fig-parrot (*P. cervicalis*). We are resistent to do this without molecular data and given there are five well-marked populations. The plumage variation is so substantial that we are unable to agree on where to establish the species breaks. In addition, abutting populations (which del Hoyo & Collar split as species) show considerable intergradation. *Other names*—Golden-headed, Red-faced, or Yellow-naped Fig-parrot, Desmarest's Figparrot.

Edwards's Fig-Parrot *Psittaculirostris edwardsii*   ENDEMIC | MONOTYPIC

Cyclopsittacus Edwardsii Oustalet, 1885. *Ann. Sci. Nat., Zool.* (6)19(3): 1.—Kafu, east of Aitape, nw Sepik-Ramu.

Inhabits the c sector of n NG, from Jayapura east to the Huon Gulf and lower Markham R. Frequents the canopy of forest, edge, and partly cleared areas, sea level–800 m. Records from the Jimi R uplands (Gyldenstolpe 1955a).

Localities—Humboldt Bay[MA], Tami R[J], Vanimo[CO], Maprik[K], Utu[CO], Astrolabe Bay[MA], Jimi R[L,M], Finisterre Ra[N], Simbang on the Huon Gulf[MA], Finschhafen[CO], and lower Markham R[Z] (Mayr & Gilliard 1954[M], Gyldenstolpe 1955a[L], Ripley 1964a[J], Pearson 1975[K], Forshaw 1989[N], Beehler field notes[Z]). Comprehensive multiyear surveys in the YUS area on the n slopes of the Saruwaged Ra (Huon) did not encounter this fig-parrot (Beehler field notes).

Salvadori's Fig-Parrot *Psittaculirostris salvadorii*   ENDEMIC | MONOTYPIC

Cyclopsittacus Salvadorii Oustalet, 1880. *Bull. Assoc. Sci. France* (2)1: 172.—The Waropen, ne coast of Geelvink Bay, e sector of the NW Lowlands.

Endemic to the NW Lowlands, sea level–400 m. Records from the e shore of Geelvink Bay[M], Biri village in the w Mamberamo basin[L], Danau Bira of the n Van Rees Mts[L], Bernhard Camp (Taritatu R)[Q], Kwerba in the sw foothills of the Foja Mts[K], and Nimbokrang[J,Z] (Rand 1942b[Q], Diamond 1985[L], Forshaw 1989[M], Beehler *et al.* 2012[K], J. Mittermeier photo[J], L. Petersson photo[Z]). Eastern verge of its range unknown, but apparently somewhere just west of the Cyclops Mts. *P. edwardsii* occurs immediately to the east.

NOTES

Of interest is the behavior of the species of *Psittaculirostris* at contact zones. This would include the n coast west of the Cyclops Mts, where *salvadorii* and *edwardsii* may meet, and perhaps in the Bird's Neck, where *salvadorii* and *desmarestii* may meet. Dickinson & Remsen (2013) noted that the e edge of the range of *P. salvadorii* extends to the Sepik R. We know of no records for *P. salvadorii* in the Sepik basin. *Other names*—Whiskered Fig-Parrot.

GENUS *CYCLOPSITTA*   [Genus 2 spp/Region 2 spp]

Cyclopsitta Reichenbach, 1850. *Avium Syst. Nat.*: 82. Type, by subsequent designation (Pucheran, 1853. *Mammif. Ois.* In Hombron & Jacquinot, *Voyag. Pôle Sud*), *Psittacula diophthalma* Hombron and Jacquinot.

The species of this genus inhabit New Guinea and the humid forests of eastern Australia (Collar 1997). The genus is characterized by a specialization in seed-eating (mainly figs), the small stubby body, short tail, and primarily green plumage with colorful patterning on the head. *Opopsitta* is a junior synonym to *Cyclopsitta* (Collar 1997). We thus follow Storr (1973) and Schodde (1978) in reinstating *Cyclopsitta*. *Cyclopsittacus* was an unjustified emendation of *Cyclopsitta* by Sundevall, 1872 (Schodde 1978). These incorrect generic names are not treated as new generic epithets by the ICZN Code (Art. 51.3.1), and thus the names of the naming authorities using the incorrect generic name are not put in parentheses once the genus name has been corrected [*e.g.*, see below, *Cyclopsitta gulielmitertii fuscifrons* Salvadori; not *Cyclopsitta gulielmitertii fuscifrons* (Salvadori)].

Orange-breasted Fig-Parrot *Cyclopsitta gulielmitertii*   ENDEMIC

Psittacula gulielmi III Schlegel, 1866. *Ned. Tijdschr. Dierk.* 3: 252.—West coast of the Bird's Head and Salawati I.

Found in lowland forest on the Mainland, as well as Salawati and the Aru Is (Mayr 1941); sea level–300 m, max 1,100 m[CO]. May be patchily distributed in part of the n watershed of PNG. Frequents forest, edge, and openings. In many localities in the s watershed this is the only fig-parrot. For instance, Mack & Wright (1996) reported the species common in flocks at 1,100 m at Crater Mt WMA, whereas *C. diophthalma* was only occasionally encountered. In such places *C. gulielmitertii* is often common, sometimes abundant. However, on the n side, from the Huon Penin westward, where other fig-parrot species are present and often common, it seems to be local or rare (Coates 1985).

SUBSPECIES

a. *Cyclopsitta gulielmitertii gulielmitertii* (Schlegel)

RANGE: Salawati and the w coast of the Bird's Head (Mayr 1941).

DIAGNOSIS: A large form; male with forepart of crown deep blue; no yellow on inner edges of base of wing quills; bright yellow cheek with a black "comma"; breast abruptly orange, offset from yellow throat; wing

91–98; female with orange ear patch; black crown and cheek spot interrupted by yellow patch at base of bill that traces below eye to connect with ear patch; throat blue and green (Rand & Gilliard 1967, Forshaw 1989, Collar 1997). See photo in Coates & Peckover 2001: 81, top right image).

b. *Cyclopsitta gulielmitertii melanogenia* (von Rosenberg)
Psittacula melanogenia von Rosenberg, 1866. *Ned. Tijdschr. Dierk.* 3: 330.—Aru Is.
RANGE: The Aru Is (Hoek Ostende *et al.* 1997).
DIAGNOSIS: Similar to *fuscifrons*, but female exhibits paler, more greenish breast; wing 74–80 (Rand & Gilliard 1967, Forshaw 1989). Mees (1980) corrected authorship to von Rosenberg.

c. *Cyclopsitta gulielmitertii fuscifrons* Salvadori
Cyclopsittacus fuscifrons Salvadori 1876. *Ann. Mus. Civ. Genova* 9: 14.—Upper Fly R.
RANGE: The S Lowlands from the Mimika R to the Fly R, and including the Trans-Fly. Localities—Kurik, Wasur NP, Morehead, Gaima, Tarara, Penzara, and Black R Camp (Rand 1942a, Bishop 2005a).
DIAGNOSIS: Similar to *suavissima*, but forehead and forecrown dark brown; male cheek whitish with large black spot, sharply offset by orange breast; yellow present on inner webs of wing quills; wing 72–88 (Rand & Gilliard 1967, Forshaw 1989, Collar 1997).

d. *Cyclopsitta gulielmitertii nigrifrons* Reichenow
Cyclopsittacus nigrifrons Reichenow, 1891. *J. Orn.* 39: 217.—Sepik R.
RANGE: The NW Lowlands and Sepik-Ramu, from west of the Mamberamo R to the Sepik R (Mayr 1941).
DIAGNOSIS: Male with forehead and forecrown black; throat, lores, and side of head pale yellow except for a black mark on cheek; breast to upper abdomen orange; female with cheeks yellow, bordered behind by a conspicuous black band and below by a greenish-blue band; ear coverts orange; breast greenish; immature similar to adult female but has orange on side of throat; wing 91 (Coates 1985, Forshaw 1989, Collar 1997).

e. *Cyclopsitta gulielmitertii amabilis* Reichenow
Cyclopsittacus amabilis Reichenow, 1891. *J. Orn.* 39: 432.—Saparako, Huon Gulf, ne NG.
  Synonym: *Opopsitta nigrifrons ramuensis* Neumann, 1915. *Orn. Monatsber.* 23: 180.—Ramu valley, at the foot of the Bismarck Ra, Sepik-Ramu, ne NG.
RANGE: The n coast of e NG from the Ramu R to Milne Bay (Mayr 1941).
DIAGNOSIS: Male with forehead and forecrown bluish black; side of head, throat, breast, and upper abdomen pale yellow; black cheek spot present; female with forecrown blackish, extending as a black mask across eye and to ear; throat pale yellow; breast and upper abdomen orange but gradually grading into the paler cheek patch; smaller, wing 79–87 (Rand & Gilliard 1967, Forshaw 1989). Form *macilwraithi* Rothschild was synonymized by Mayr (1941). See photo of *amabilis* in Coates & Peckover (2001: 81, top-left image).

f. *Cyclopsitta gulielmitertii suavissima* P. L. Sclater
Cyclopsitta suavissima P. L. Sclater, 1876. *Proc. Zool. Soc. London*: 520, pl. 54.—Naiabui, Hall Sound, s watershed of SE Peninsula.
RANGE: The e sector of the S Lowlands and the s watershed of the SE Peninsula, from the Purari R to the Kemp Welch R (Diamond 1972).
DIAGNOSIS: On male, note very large black cheek spot set against whitish cheek; black spot on cheek nearly reaches base of bill; breast orange; forehead and forecrown dark blue; ear coverts yellowish white; yellow edging present in inner webs of base of wing quills. Female has black face with white subocular stripe; orange ear patch and orange breast (Rand & Gilliard 1967, Coates 1985, Collar 1997). Wing 73–81 (Forshaw 1989). See photo in Coates & Peckover (2001: 81, bottom image).

NOTES
Dickinson & Remsen (2013) broke this polytypic species into three component species, and del Hoyo & Collar (2014) recognized four species, following the work of Schnitker (2007). The result is the following: *C. gulielmitertii*: Salawati I and Bird's Head and Neck; *C. nigrifrons*: NW Lowlands, Sepik-Ramu, n watershed of SE Peninsula (an additional split being eastern *C. amabilis*); *C. melanogenia*: S Lowlands, Aru Is, s watershed of SE Peninsula. We agree that a breakup of this variable species may be in order, but further study is required, particularly with respect to forms *gulielmitertii* and *nigrifrons*. We have subsumed

ramuensis into *amabilis* because of its intermediate plumage and undefined range, perhaps the product of contact between *amabilis* and *nigrifrons* (Rand & Gilliard 1967). A museum specimen of the nominate form in Amsterdam is labeled "Misool" but lacks other pertinent data, and thus should not be considered an accepted record for this island (Mees 1980). Comprehensive surveys in the YUS area on the n slopes of the Saruwaged Ra (Huon) did not encounter this species. *Other names*—William's or Black-cheeked Fig Parrot; Blue-fronted, Black-fronted, Creamy-breasted, and Dusky-cheeked Fig-parrots; *Opopsitta gulielmiterti, Psittaculirostris gulielmiterti, Cyclopsitta nigrifrons, C. amabilis,* and *C. suavissima.*

Double-eyed Fig-Parrot *Cyclopsitta diophthalma*   RESIDENT
Psittacula diophthalma Hombron and Jacquinot, 1841. *Ann. Sci. Nat., Zool.* (2)16: 318.—Triton Bay, Bird's Neck.

Frequents forest, edge, and openings through NG, the NW Islands, Aru Is, D'Entrecasteaux Arch, and Tagula I (Coates 1985). Patchily distributed on the Mainland, and apparently absent from the lowlands of the s watershed in many localities (Coates 1985). Sea level–1,600 m[J]; max 2,590 m in Wapenamanda gorge[K] (Gyldenstolpe 1955a[J], Beehler field notes 18-Jul-2002[K]). Mainly only in mts in s watershed (Coates 1985). This species and *C. gulielmitertii* seem to replace each other locally. Also several additional records from the middle and lower Purari basin (I. Woxvold *in litt.*).

Extralimital—patchily along the e coast of Australia.

SUBSPECIES
a. *Cyclopsitta diophthalma diophthalma* (Hombron and Jacquinot)
   Synonym: *Cyclopsittacus coccineifrons* Sharpe, 1882. *J. Linn. Soc. London, Zool.* 16: 318.—Astrolabe Ra, SE Peninsula.
   Synonym: *Cyclopsittacus festetichi* Madarász, 1902. *Termés. Füzetek* 25: 350.—Astrolabe Bay, e verge Sepik-Ramu.
RANGE: Kofiau, Misool, Waigeo, and Salawati I, and all NG except the c and w sectors of the S Lowlands and the Trans-Fly (Mayr 1941).
DIAGNOSIS: The nominate form is distinctive in the following characters: male has forehead and crown red, grading to yellow posteriorly; a pale blue spot in front of eye; green band behind eye; cheek red, the red margined with blue posteriorly. Female is like male, but cheek buffy (rather than red) with blue margin, and thin red stripe from base of bill to behind eye forming a subocular line (Collar 1997). We follow Mayr (1941) and Forshaw (1989) in subsuming *coccineifrons* and *festetichi* into the nominate form. For another view, see Diamond (1972: 152). An image of this form appears in Coates & Peckover (2001: 82).

b. *Cyclopsitta diophthalma aruensis* (Schlegel)
Psittacula diophthalma aruensis Schlegel, 1874. *Mus. Pays-Bas* 3 (*Psittaci*): 33.—Aru Is.
RANGE: The Aru Is and s NG from the Mimika R to the Fly R (Mayr 1941, Hoek Ostende *et al.* 1997).
DIAGNOSIS: Male with a red cap and cheek and no blue eyespot. Female's face blue, lacking any red (Collar 1997).

c. *Cyclopsitta diophthalma virago* Hartert
Cyclopsittacus virago Hartert, 1895. *Novit. Zool.* 2: 61.—Fergusson I, D'Entrecasteaux Arch, SE Islands.
RANGE: Goodenough, Fergusson, and Normanby I[J] (Mayr 1941, Pratt field notes 2003[J]).
DIAGNOSIS: Male much like *aruensis*. Female head green with blue forehead with a small red spot within it (Collar 1997).

d. *Cyclopsitta diophthalma inseparabilis* Hartert
Cyclopsittacus inseparabilis Hartert, 1898. *Bull. Brit. Orn. Club* 8: 9.—Tagula I, Louisiade Arch, SE Islands.
RANGE: Tagula I.
DIAGNOSIS: Male and female like female of *virago*, but side of head green, without any blue wash (Rand & Gilliard 1967).

NOTES
As with the preceding species, this polytypic species should perhaps be dissected into several component species in the Region. Other very distinctive populations occur in Australia. Molecular analysis may help to discern the most substantial genetic breaks. Dickinson & Remsen (2013) and del Hoyo & Collar (2014)

both recognized but a single species in this lineage. *Other names*—Two-eyed, Dwarf, Blue-faced, or Red-faced Fig Parrot, *Opopsitta* or *Psittaculirostris diophthalma*.

GENUS *LORICULUS* [Genus 13 spp/Region 1 sp]

Loriculus Blyth, 1849. *Journ. Asiat. Soc. Bengal* 19: 236. Type, by subsequent designation (G. R. Gray, 1855. *Cat. Gen. Subgen. Birds*: 88), *Psittacus galgulus* Linnaeus.

The species of this genus range from Sri Lanka and India east to Luzon, Sumatra, Borneo, Sulawesi, Flores, Halmahera, New Guinea, New Britain, and New Ireland (Collar 1997). The genus is characterized by tiny size, short rounded tail, extremely fine and pointed bill, and naked cere; note also the habits of roosting head down and of the female carrying nesting materials by storing them amid the rump feathers (Forshaw 1989).

Orange-fronted Hanging Parrot *Loriculus aurantiifrons* ENDEMIC

Loriculus aurantiifrons Schlegel, 1871. *Ned. Tijdschr. Dierk.* (1873) 4: 9.—Misool I, NW Islands.

Throughout the lowlands of NG, Misool, Waigeo, Karkar, Goodenough, and Fergusson I (Stresemann 1923, Mayr 1941); sea level–300 m, max 1,600 m[CO]. Frequents rain forest, edge, and partly cleared areas. Widely distributed on the Mainland but apparently no records from the Trans-Fly; may be absent from some suitable localities.

SUBSPECIES

a. *Loriculus aurantiifrons aurantiifrons* Schlegel
RANGE: Misool I (Mees 1965, Hoek Ostende *et al.* 1997).
DIAGNOSIS: Male forehead yellow or orange-yellow, extending onto forecrown; this crown patch larger than in other subspecies. Wing (male) 65–69 (Forshaw 1989).

b. *Loriculus aurantiifrons meeki* Hartert
Loriculus aurantiifrons meeki Hartert, 1895. *Novit. Zool.* 2: 62.—Fergusson I, D'Entrecasteaux Arch, SE Islands.
Synonym: *Loriculus aurantiifrons batavorum* Stresemann, 1913. *J. Orn.* 61: 602.—Upper Otakwa R, Western Ra.
RANGE: Mainland NG, Waigeo[J], Karkar[K], Goodenough[M], and Fergusson I. Localities—Otakwa R[J], Jaya-pura[Y], Kiunga[N], Ok Tedi R[Z], Black R[X], Angoram[O], the upper Jimi R[L], and near Alotau[L] (Mayr 1941, Rand 1942a[X], 1942b[Y], Gyldenstolpe 1955a[L], 1955b[J], Mees 1965, Bell 1970e[M], Diamond 1972, Diamond & LeCroy 1979[K], Gregory 1995b[N], P. Gregory *in litt.*[Z], Beehler field notes[O]).
DIAGNOSIS: Male forehead spot smaller and yellower than in nominate, not orange; wing 69–76. Female has yellowish-brown bases to feathers of forehead and forecrown; wing 69–75 (Forshaw 1989). We found no substantial characters distinguishing *batavorum* from *meeki*.

NOTES

We follow Dutson (2011) in treating the Bismarck form *tener* as specifically distinct from *aurantiifrons* (*pace* Mayr & Diamond 2001)—they are presumably sister species. *L. aurantiifrons* probably should be treated as monotypic. Variation in size and color of crown spot is minimal. Current known ranges leave gaps in distribution. In the north the gap extends from the Sepik east to Madang; in the south the gap is between the Setekwa R and Lake Kutubu. Field work is needed to determine whether these gaps are real or products of inadequate sampling. *Other names*—Bat Lorikeet, Papuan Hanging-Parrot.

GENUS *ALISTERUS* [Genus 3 spp/Region 2 spp]

Alisterus Mathews, 1911. *Novit. Zool.* 18: 13. Replacement name for *Aprosmictus* Gould, 1865 (not Gould, 1843). Type, by original designation, *Psittacus cyanopygius* Vieillot = *Psittacus scapularis* Lichtenstein.

The species of this genus range from southeastern Australia to New Guinea and west to Halmahera, Ambon, Seram, the Sula Islands, Peleng, and Buru Island (Forshaw 1989). The genus is characterized by the predominantly red-and-green plumage, very long tail, small bill and head, and distinctive graceful flight (Forshaw 1989).

Moluccan King-Parrot *Alisterus amboinensis*   RESIDENT

Psittacus amboinensis Linnaeus, 1766. *Syst. Nat.*, ed. 12, 1: 141.—Ambon, s Moluccas.

Restricted in the Region to the NW Islands, Bird's Head, Bird's Neck, and Weyland Mts; sea level–1,200 m, occasionally higher (Mayr 1941).

Extralimital—west to Halmahera, Seram, Buru, and the Sula Is (Collar 1997).

SUBSPECIES

a. *Alisterus amboinensis dorsalis* (Quoy & Gaimard)

Psittacus (Platycercus) dorsalis Quoy & Gaimard, 1830. *Voy. Astrolabe, Zool.* 1: 234, pl. 21, fig. 3.—Manokwari, Bird's Head.

RANGE: Waigeo[MA], Gam[MA], Salawati[MA], and Batanta I[MA], and the Bird's Head east to the s Bird's Neck (Etna Bay) and the Weyland Mts (Hartert *et al.* 1936), where it meets the range of *A. chloropterus*.

DIAGNOSIS: Lacks pink on undersides of lateral tail feathers; head and breast dark red (Forshaw 1989).

NOTES

Questions to be addressed by future researchers: Do the two *Alisterus* species hybridize in the Weyland Mts? If not, do they sort ecologically? *Other names*—Amboina Kingparrot.

Papuan King-Parrot *Alisterus chloropterus*   ENDEMIC

Aprosmictus chloropterus Ramsay, 1879. *Proc. Linn. Soc. NSW* 3: 251.—Goldie R, SE Peninsula.

Frequents the middle and lower stories of the forest interior of NG east of the Bird's Neck, from the foothills to 1,800 m, max 2,800 m[K] (Beehler field notes[K]); absent from the Bird's Head and Neck, where replaced by the closely related *A. amboinensis*.

SUBSPECIES

a. *Alisterus chloropterus callopterus* (D'Albertis & Salvadori)

Aprosmictus callopterus D'Albertis & Salvadori, 1879. *Ann. Mus. Civ. Genova* 14: 29.—Upper Fly R.
    Synonym—*Aprosmictus wilhelminae* Ogilvie-Grant, 1911. *Bull. Brit. Orn. Club* 27: 83.—Kapare R, 520 m, Bird's Neck.

RANGE: The Western, Border, and Eastern Ra (Diamond 1972, Forshaw 1989).

DIAGNOSIS: Male similar to the nominate form but with a narrow blue band across the upper mantle, not extending up onto hindneck and nape. Female similar to nominate form, but throat and breast have some dirty-red infusion.

b. *Alisterus chloropterus moszkowskii* (Reichenow)

Aprosmictus moszkowskii Reichenow, 1911. *Orn. Monatsber.* 19: 82.—Tana, mouth of the Mamberamo R, NW Lowlands.

RANGE: The NW Lowlands and Sepik-Ramu east to Aitape, also including the Foja Mts[J], and Bewani and Torricelli Mts[K] (Mayr 1941, Beehler *et al.* 2012[J], J. M. Diamond specimens at AMNH[K]). Range of this subspecies must abut that of *callopterus* all along the n scarp of the Central Ra.

DIAGNOSIS: Male similar to *callopterus*. Female red-headed like male and very different from female of nominate form; blue band across upper mantle absent or only slightly indicated; mantle and back dull green; sides of breast green. Immature similar to female, but yellowish-green wing band narrower and duller; breast strongly marked with green (Forshaw 1989).

c. *Alisterus chloropterus chloropterus* (Ramsay)

RANGE: Mountains of the Huon and SE Peninsulas; nw terminus of range presumably on the e verge of the Kratke Mts.

DIAGNOSIS: Male: blue of nape extends up to hind-crown and down onto upper mantle. Female: head and throat entirely green (though throat washed brownish); wing lacks bright yellow-green chevron found in females of other races.

Observed in the Adelbert Mts, subspecies unknown (Pratt 1983). *Other names*—Green-winged King Parrot.

GENUS *APROSMICTUS* [Genus 2 spp/Region 1 sp]

Aprosmictus Gould, 1842. *Birds Austral.* 5(8): text to pl. 17, 18. Type, by subsequent designation (Gray, 1846. *List Gen. Birds* 2: 408), *Psittacus erythropterus* Gmelin.

The species of this genus range patchily through the Lesser Sundas (Wetar, Timor, Roti), the Trans-Fly of New Guinea, and northern and eastern Australia (Collar 1997). The genus, similar to *Alisterus*, is characterized by the shorter and more squared-off tail, bright green body plumage, and bright red feathering on the wing (Forshaw 1989).

Red-winged Parrot *Aprosmictus erythropterus* RESIDENT

Psittacus erythropterus Gmelin, 1788. *Syst. Nat.* 1(1): 343.—Endeavor R, n QLD, Australia.

Found in open savanna of the s Trans-Fly, sea level–100 m (Rand 1942a). Sometimes in large flocks, suggesting nomadic movements.

Extralimital—n and e Australia.

SUBSPECIES

a. *Aprosmictus erythropterus coccineopterus* (Gould)

Ptistes coccineopterus Gould, 1865. *Handbook Birds Austral.* 2: 37.—Port Essington, NT, Australia.

> Synonym: *Aprosmictus erythropterus papua* Mayr & Rand, 1936. *Mitt. Zool. Mus. Berlin* 21: 241.—Wuroi, Oriomo R, Trans-Fly.

RANGE: The Trans-Fly, from Dolak I to the Oriomo R. Also n Australia from King Sound to the Cape York Penin. Also Lake Daviumbu (Bishop 2005a).

DIAGNOSIS: Our review of specimens at the AMNH confirms the suggestion of Forshaw (1989) that the subspecies *papua* is a synonym of *coccineopterus*, and we find *coccineopterus* only very thinly distinct from the nominate form.

NOTES

Mees (1982b) suggested treating the species as monotypic, based on an analysis by Frith & Hitchcock (1974), who lumped *coccineopterus* with the nominate form. We suggest future research to revisit the status of the subspecies in Australia.

GENUS *ECLECTUS* MONOTYPIC

Eclectus Wagler, 1832. *Abh. K. Bayer. Akad. Wiss., Math.-Phys. Kl.* 1: 495. Type, by subsequent designation (Gray, 1840. *List Gen. Birds*: 52), *Psittacus grandis* Gmelin = *Psittacus roratus* (P. L. S. Müller).

The single species of the genus ranges from Sumba and the Moluccas east to New Guinea, the Bismarck and Solomon Islands, and the Cape York Peninsula of Queensland (Forshaw 1989). The genus, apparently close to *Geoffroyus* in terms of cranial osteology (Thompson 1900), is characterized by the stocky body, large head and bill, squared-off tail, long round-tipped wings, and notched maxilla. The extreme plumage dichromatism between the sexes, greater than that of any other parrot genus, is notable (Forshaw 1989). The generic name *Larius* has been suppressed (ICZN Opinion 938).

Eclectus Parrot *Eclectus roratus* RESIDENT

Psittacus roratus P. L. S. Müller, 1776. *Natursyst.*, Suppl.: 77.—Ambon, s Moluccas.

Frequents virtually all wooded habitats throughout the Region, except for some e islands; sea level–1,000 m, max 1,900 m^{CO}. Absent from Misima, Rossel, and Woodlark I (Mayr 1941).

Extralimital—Moluccas, Lesser Sundas, ne Australia, and the Bismarck and Solomon Is (Mayr & Diamond 2001, Dutson 2011).

SUBSPECIES

a. *Eclectus roratus polychloros* (Scopoli)

Psittacus polychloros Scopoli, 1786. *Delic. Flor. Fauna Insubr.*: 87.—New Guinea.

> Synonym: *Psittacus pectoralis* P. L. S. Müller, 1776. *Natursyst.*, Suppl.: 78.—Onin Penin, Bird's Neck. [Now considered a synonym of the nominate form, *E. r. roratus* (see Collar 1997: 395).]
> Synonym: *Eclectus polychlorus* var. *aruensis* G. R. Gray, 1858. *Proc. Zool. Soc. London*: 182.—Aru Is.
> Synonym: *Larius roratus biaki* Hartert, 1932 (February). *Nova Guinea* 15: 448.—Biak I, Bay Islands.
> Synonym: *Eclectus roratus maforensis* Rothschild, 1932 (December). *Novit. Zool.* 38: 203.—Numfor I, Bay Islands.

RANGE: Mainland NG[MA]; all the NW Islands; Aru Is and Bay Is; D'Entrecasteaux and Trobriand Is; and Tagula I (Mayr 1941, Mees 1972, Johnstone 2006, Diamond *et al.* 2009). Also Southeast Is of Indonesia (where introduced) and Kai Is (Coates & Bishop 1997).

DIAGNOSIS: The widespread form *polychloros* is distinguished by its large size; wing (male) 240–279 (Forshaw 1989); male with yellowish tail tips; female breast and abdomen blue with reduced purple and undertail coverts red (Collar 1997).

NOTES

We have examined specimens of *aruensis* and find them identical to *polychloros*. We follow the suggestion of Forshaw (1989: 220) in synonymizing *biaki* with *polychloros*. In addition, the form *maforensis*, considered "doubtfully distinct" by Rand & Gilliard (1967), is subsumed into *polychloros*. Other names—Red-sided Eclectus Parrot, Red-sided Parrot, Kalanga, *Larius roratus*.

GENUS *GEOFFROYUS*　　　　　　　　　　　　　　　[Genus 3 spp/Region 2 spp]

Geoffroyus Bonaparte, 1850. *Consp. Gen. Av.* 1: 6. Type, by tautonymy, *Psittacus geoffroyi* Bechstein.

　　The species of this genus range from the Lesser Sundas and Moluccas to New Guinea and the Bismarck and Solomon Islands (Collar 1997). Forshaw (1989) noted that the genus is characterized by the medium size, stocky build, pronounced sexual dimorphism, squared-off tail, pointed wings, naked cere, and lack of a maxillary notch. Note that the skull is similar to that of *Eclectus* (Thompson 1900).

Red-cheeked Parrot *Geoffroyus geoffroyi*　　　　　　　　　　　RESIDENT

Psittacus Geoffroyi Bechstein, 1811. In Latham, *Allgem. Uebers. Vog.* 4: 103, pl. 21.—Timor, Lesser Sundas.

　　Frequents lowland forest, gardens, and savanna in NG, the Aru Is, NW Islands, Bay Is, and SE Islands, to 800 m, rarely to 1,240 m[M] (Freeman & Freeman 2014[M]). Absent from the Trobriand Is and Woodlark I.

　　Extralimital—Lesser Sundas, Moluccas, including the Tanimbar and Kai Is, south to ne QLD (Collar 1997).

SUBSPECIES

a. *Geoffroyus geoffroyi pucherani* Souancé

Geoffroyus Pucherani Souancé, 1856. *Rev. et Mag. Zool.* (2)8: 218.—Triton Bay, Bird's Neck.

RANGE: All the NW Islands, and the Bird's Head, east to Etna Bay in the s watershed and to the head of Geelvink Bay in the n watershed (Mayr 1941, Johnstone 2006).

DIAGNOSIS: Similar to *minor*, but reddish-brown marking on wing coverts absent or slight; less bronze-brown suffusion on mantle; rump and lower back very dark brownish-red; underwing coverts dark blue (Forshaw 1989).

b. *Geoffroyus geoffroyi jobiensis* (A. B. Meyer)

Pionias Pucherani var. *jobiensis* A. B. Meyer, 1874. *Sitzungsber. Akad. Wiss. Wien* 70, Abt. 1: 225.—Yapen I, Bay Islands.

RANGE: Yapen and Mios Num I (Mayr 1941).

DIAGNOSIS: Similar to *minor* but with paler blue underwing coverts and much reduced reddish-brown markings on upperwing coverts; lower back brighter red, less brownish than in *minor*; males have red of forecrown extending back onto hind-crown (Forshaw 1989). Mayr (1941) mentioned that the population from Mios Num I may be diagnosable. We have not seen the museum material for that island population.

c. *Geoffroyus geoffroyi aruensis* (G. R. Gray)

Psittacus aruensis G. R. Gray, 1858. *Proc. Zool. Soc. London*: 183.—Aru Is.

 Synonym: *Geoffroyus orientalis* A. B. Meyer, 1891. *Abh. Ber. Zool. Mus. Dresden* (1890–91) 3(4): 4.—Bussum and Jakama, near Finschhafen, Huon Penin.

RANGE: The Aru Is; s NG (Mimika R east to Milne Bay) and e NG and Huon Penin southeast to Milne Bay; also Goodenough, Fergusson, and Normanby I[L] (Mayr 1941, LeCroy *et al.* 1984[L]).

DIAGNOSIS: Similar to *floresianus* (e Lesser Sundas) but overall plumage paler, especially on underparts, which are more yellowish; underwing coverts variable, but generally darker than those of *floresianus* (Forshaw 1989). Rump plumage green, not red (Diamond 1972). See photos in Coates & Peckover (2001: 85).

d. *Geoffroyus geoffroyi mysorensis* (A. B. Meyer)

Pionias Pucherani var. *mysorensis* A. B. Meyer, 1874. *Sitzungsber. Akad. Wiss. Wien* 70, Abt. 1: 225.—Biak I, Bay Islands.

RANGE: Biak and Numfor I (Mayr 1941).

DIAGNOSIS: Male: similar in plumage to *minor* but with violet-blue of crown extending down over hindneck and red of face down onto upper throat; no bronze-brown tinge on mantle; more extensive reddishbrown markings on upperwing coverts; rump and lower back darker red and less brownish. Female: brown of crown extending down over hindneck; other differences as in male (Forshaw 1989). The Numfor population may be marginally distinctive, probably not worthy of diagnosis (noted in Mayr 1941). Dickinson & Remsen (2013) noted that the proper subspecies spelling is *mysorensis*.

e. *Geoffroyus geoffroyi minor* Neumann

Geoffroyus personatus minor Neumann, 1922. *Verh. Orn. Ges. Bayern* 15: 235.—Upper Ramu R, e sector of the Sepik-Ramu.

RANGE: Northern NG from the Mamberamo R to Astrolabe Bay, the upper Ramu R, and along n coast of Huon Penin to Wasu (Collar 1997).

DIAGNOSIS: Male: similar to *aruensis*, but lower back and rump brownish red; red of face darker and less rose-hued; mantle lightly washed with bronze-brown. Female: head slightly darker brown compared to female *aruensis*; other differences as for male (Forshaw 1989).

f. *Geoffroyus geoffroyi sudestiensis* De Vis

Geoffroyus sudestiensis De Vis, 1890. *Ann. Rept. Brit. New Guinea*, 1888–89: 58.—Tagula I, Louisiade Arch, SE Islands.

RANGE: Misima and Tagula I, Louisiade Arch (Mayr 1941).

DIAGNOSIS: Adult like *aruensis*, but overall plumage slightly more yellowish; no reddish-brown markings on wing coverts; crown and nape of female are dark green (Forshaw 1989).

g. *Geoffroyus geoffroyi cyanicarpus* Hartert

Geoffroyus aruensis cyanicarpus Hartert, 1899. *Novit. Zool.* 6: 81.—Rossel I, Louisiade Arch, SE Islands.

RANGE: Rossel I (Mayr 1941).

DIAGNOSIS: Male: similar to *sudestiensis*, but overall plumage darker green; cheeks and ear coverts strongly washed with mauve-blue; blue edge to wing from bend to outermost primary. Female: similar to *sudestiensis*, but plumage darker green; crown and nape brown slightly tinged with green; blue edge to wing from bend to outermost primary (Forshaw 1989).

NOTES

A careful review of the populations of the species across its broad geographic range, preferably with molecular input, would be worthwhile.

Blue-collared Parrot *Geoffroyus simplex*   ENDEMIC

Pionias simplex A. B. Meyer, 1874. *Verh. Zool.-Bot. Ges. Wien* 24: 39.—Arfak Mts, 1,070 m, Bird's Head.

 Ranges through the mts of NG; 800–1,800 m, extremes sea level–2,300 m—not infrequently descends to the foothills and nearby lowlands. Frequents forest and edge. No records from the Cyclops Mts or mts of Wandammen and the Huon Penin. Highly nomadic, moving widely in search of preferred mast. Favors the nuts of the abundant mid-montane oak *Castanopsis acuminatissima*.

a. *Geoffroyus simplex simplex* (A. B. Meyer)
RANGE: Mountains of the Bird's Head (Mayr 1941).
DIAGNOSIS: Male's collar pale blue (Rand & Gilliard 1967). Underwing coverts pale blue (Forshaw 1989).

b. *Geoffroyus simplex buergersi* Neumann
Geoffroyus simplex bürgersi Neumann, 1922. *Verh. Orn. Ges. Bayern* 15: 235.—Maeanderberg, upper Sepik R, west-central Sepik-Ramu.
RANGE: Throughout the Central Ra^MA; also the Adelbert Mts (Pratt 1983). Presumably the populations in the Foja Mts, PNG North Coastal Ra, and Adlebert Mts are referable to this race.
DIAGNOSIS: Male's collar gray, glossed with violet, and more extensive on hindneck; underwing coverts deep blue (Rand & Gilliard 1967). See photo in Coates & Peckover (2001: 85).

NOTES
There are unconfirmed records from the mts of the Huon Penin. *Other names*—Lilac-collared or Simple Parrot.

GENUS *TANYGNATHUS* [Genus 4 spp/Region 1 sp]

Tanygnathus Wagler, 1832. *Abh. K. Bayer. Acad. Wiss., Math.-Phys. Kl.* 1: 501. Type, by subsequent designation (Gray, 1840. *List. Gen. Birds*: 52), *Tanygnathus macrorhynchus* (Linnaeus) Wagler = *Psittacus megalorynchos* Boddaert.

The species of this genus range from Luzon and Palawan in the Philippines south to Sulawesi and Sumba, and east to Buru, Halmahera, and the Northwestern Islands (Forshaw 1989). The genus is characterized by the very large bill, large head, tapered body leading to a rather narrow, nearly pointed tail, no prominent notch in the maxilla, naked cere, and slight sexual dimorphism (Forshaw 1989). Illustrations in Collar (1997) do not capture the proper body form of species of this genus.

Great-billed Parrot *Tanygnathus megalorynchos*   RESIDENT
Psittacus megalorynchos Boddaert, 1783. *Tabl. Planch. Enlum.*: 45.—Northwestern Islands [here restricted].

Primarily a small-island specialist found w of the Bird's Head. Occurs in lowland and coastal forest and woodland on all the main islands, plus many small fringing islets, near sea level; travels between small islands to forage and roost (Collar 1997, Coates & Bishop 1997). Perhaps also visits w coast of the Bird's Head to forage, but no recent such records known.

Extralimital—Moluccas, Talaud and Sangihe Is, Flores, n and c Moluccas, Timor, Semao I, and Tanimbar Is (Collar 1997, Coates & Bishop 1997).

SUBSPECIES
a. *Tanygnathus megalorynchos megalorynchos* (Boddaert)
RANGE: Talaud and Sangihe Is; Obi; n Moluccas; all of the NW Islands (Gyldenstolpe 1955b, Johnstone 2006, Diamond *et al.* 2009); also islets off the w coast of the Bird's Head (Mayr 1941). Localities—Waigama and Adoa (Misool I; Mees 1965). Putative Bird's Head sightings (Beehler *et al.* 1986) presumably in error, based on a confusion between Sorong I and Sorong town, or perhaps following Gyldenstolpe's (1955b: 198) listing of the species as recorded from the Bird's Head by other authorities.
DIAGNOSIS: Head bright green, without any bluish cast; underparts green with a yellowish suffusion; underwing and sides of breast yellowish; and primaries with a bluish cast (Forshaw 1989). Shoulder blackish with yellow edging.

NOTES
Range of this species in the Region poorly understood and in need of additional clarification. Presumably this is a tramp species of small islands, mobile and irregular in its distribution. Boddaert's original spelling stands as correct (Beehler & Finch 1985, E. C. Dickinson *in litt.*). *Other names*—Large-billed, Moluccan, or Island Parrot.

Micropsitta Lesson, 1831. *Traité d'Orn.* 8: 646. Type, by monotypy, *Psittacus (Psittacula) pygmaeus* Quoy & Gaimard [not *Psittacus pygmaeus* Gmelin] = *Micropsitta chloroxantha* Oberholser.

The species of this genus are restricted to New Guinea and the Bismarck and Solomon Islands. This very distinctive genus is characterized by the tiny size, creeper-like bark-foraging behavior, abbreviated tail with stiffened feather-shaft tips that allow woodpecker-like trunk creeping, long toes, and peculiar flattened bill. Diet is apparently of vegetable material scraped from the surface of bark—presumably lichens and/or liverworts (Forshaw 1989).

Yellow-capped Pygmy Parrot *Micropsitta keiensis* RESIDENT
Nasiterna keiensis Salvadori, 1876. *Ann. Mus. Civ. Genova* 7: 984.—Kai Is.

Inhabits lowland forest of the NW Islands, the Bird's Head and Neck, Aru Is, and w sectors of the S Lowlands and the Trans-Fly. Records to 750 m (P. Gregory *in litt.*).
Extralimital—Kai Is.

SUBSPECIES
a. *Micropsitta keiensis chloroxantha* Oberholser
Micropsitta chloroxantha Oberholser, 1917. *Proc. Biol. Soc. Wash.* 30: 126.—Manokwari, Bird's Head.
Replacement name for *Psittacus pygmeus* Quoy & Gaimard, 1830. *Voy. Astrolabe* 1: 232; preoccupied by *Psittacus pygmaeus* Gmelin, 1788.
    Synonym: *Micropsitta keiensis sociabilis* Greenway, 1966. *Amer. Mus. Novit.* 2258: 8.—Mount Besar, Batanta I, NW Islands.
RANGE: All the NW Islands (Mayr 1941, Greenway 1966, Johnstone 2006); also the Bird's Head and Bird's Neck.
DIAGNOSIS: Male: center of breast and abdomen variably marked with orange-red; remainder of underparts yellowish green. Female: entire underparts yellowish green; crown greenish yellow. Forshaw (1989) and Mees (1972) postulated that *sociabilis* is not distinct from this form.

b. *Micropsitta keiensis keiensis* (Salvadori)
Nasiterna keiensis Salvadori, 1876. *Ann. Mus. Civ. Genova* (1875) 7: 984.—Kai Is.
    Synonym: *Nasiterna pygmaea viridipectus* Rothschild, 1911. *Bull. Brit. Orn. Club* 27: 45.—Upper Setekwa R, w sector of S Lowlands.
RANGE: The Kai and Aru Is[MA], and w sector of the S Lowlands east to Mt Bosavi[L] and the Purari basin[K]; range includes the Trans-Fly (Coates 1990: 567[L], I. Woxvold *in litt.*[K]).
DIAGNOSIS: No red in plumage; green of underparts darker than that of *chloroxantha* (Rand & Gilliard 1967).

NOTES
Presumed to be a sister species of *geelvinkiana* and *pusio*, as the geographic ranges of the three are generally complementary.

Geelvink Pygmy Parrot *Micropsitta geelvinkiana* ENDEMIC
Nasiterna pygmaea Geelvinkiana Schlegel, 1871. *Ned. Tijdschr. Dierk.* (1873) 4: 7.—Numfor I, Bay Islands.

Numfor and Biak Is, Geelvink Bay, sea level–150 m, perhaps higher (Collar 1997). Inhabits primary and secondary forest and woodland regrowth adjacent to local gardens.

SUBSPECIES
a. *Micropsitta geelvinkiana geelvinkiana* (Schlegel)
RANGE: Numfor I, Geelvink Bay (Hoek Ostende *et al.* 1997).
DIAGNOSIS: Male: crown purple-blue, extending to face and neck; substantial yellow-orange midline slash down breast to abdomen. Female: dark blue crown patch small, and ventral slash absent.

b. *Micropsitta geelvinkiana misoriensis* (Salvadori)
Nasiterna misoriensis Salvadori, 1876. *Ann. Mus. Civ. Genova* (1875) 7: 909.—Biak I, Bay Islands.

RANGE: Biak I.

DIAGNOSIS: Male: crown and face dark brown; nape with yellow line (Forshaw 1989). Female: feathers of crown brown edged with blue; no yellow in nape (Collar 1997).

Buff-faced Pygmy Parrot *Micropsitta pusio* RESIDENT

Nasiterna pusio P. L. Sclater, 1866. *Proc. Zool. Soc. London* (1865): 620, pl. 35.—Duke of York Is, New Ireland.

Frequents forest and edge in n NG and the SE Peninsula; west in north to Geelvink Bay; in s watershed ranges west to the Kikori R and Lake Kutubu[K] (Schodde & Hitchcock 1968[K], Mack & Wright 1996). Also islands off the n coast of PNG and selected SE Islands. Sea level–500 m, max 900 m.

Extralimital—Bismarck Arch (Mayr & Diamond 2001, Dutson 2011).

SUBSPECIES

a. *Micropsitta pusio beccarii* (Salvadori)
Nasiterna beccarii Salvadori, 1876. *Ann. Mus. Civ. Genova* 8: 396.—Wairor, e coast of the Bird's Head.
RANGE: Northern NG from the w coast of Geelvink Bay (Momi, at e foot of Arfak Mts[MA]) to the Kumusi R; Manam[MA], Karkar[MA], Bagabag[K], and Umboi I[L] (Diamond & LeCroy 1979[K], Dutson 2011[L]). Captured in Lobo, Triton Bay, and Sewicki Lake, Arguni Bay; and observed at Kaimana in Oct–Nov-2014 (C. Thébaud & B. Mila *in litt.*); these are sw range extensions for the species and within the range of *M. keiensis*.
DIAGNOSIS: In comparison to nominate form, overall plumage darker, with less yellow on underparts; forehead and sides of head darker brown; wing (male) 59–66 (Forshaw 1989).

b. *Micropsitta pusio pusio* (P. L. Sclater)
RANGE: The e sector of the S Lowlands (Kikori and Lake Kutubu); the SE Peninsula from the Lakekamu basin to Milne Bay; also New Britain, Duke of York Is, and Witu Is (Mayr 1941, Dutson 2011).
DIAGNOSIS: Plumage paler than that of *beccarii*. Wing (male) 62–69 (Forshaw 1989). See photo in Coates & Peckover (2001: 78, all three images).

c. *Micropsitta pusio harterti* Mayr
Micropsitta pusio harterti Mayr, 1940. *Amer. Mus. Novit.* 1091: 2.—Fergusson I, D'Entrecasteaux Arch, SE Islands.
RANGE: Fergusson I.
DIAGNOSIS: Wing (male) 59–63. When compared to the nominate form, note less yellow on underparts; throat washed with blue; head markings duller (Forshaw 1989).

d. *Micropsitta pusio stresemanni* Hartert
Micropsitta pusio stresemanni Hartert, 1926. *Novit. Zool.* 33: 130.—Mount Riu, Tagula I, Louisiade Arch, SE Islands.
RANGE: Misima and Tagula I.
DIAGNOSIS: Wing (male) 62–69. Similar to *harterti* but with more yellow on underparts and larger size (Forshaw 1989).

NOTES
Other names—Buffy-faced Pygmy Parrot.

Red-breasted Pygmy Parrot *Micropsitta bruijnii* RESIDENT

Nasiterna bruijnii Salvadori, 1875. *Ann. Mus. Civ. Genova* 7: 715, 907, pl. 21.—Arfak Mts, Bird's Head.

A diminutive and often overlooked inhabitant of the mts of NG, 1,600–2,300 m[CO]; occasionally down to 300 m[K], and one record at sea level[CO] (I. Woxvold *in litt.*[K]). Frequents forest, edge, and shade trees in coffee plantations. No records from the Cyclops Mts. Recently recorded in the Kumawa Mts (Diamond & Bishop 2015)

Extralimital—Buru and Seram in the s Moluccas, New Britain, New Ireland, Bougainville, Kolombangara, and Guadalcanal (Mayr & Diamond 2001, Dutson 2011).

a. *Micropsitta bruijnii bruijnii* (Salvadori)

RANGE: Mountains of the Bird's Head[MA]; the Fakfak and Wandammen Mts[MA] of the Bird's Neck; the Western, Border, and Eastern Ra[MA]; Foja Mts[J], PNG North Coastal Ra[K], and Adelbert Mts[W]; and mts of the Huon and SE Peninsulas[MA] (Pratt 1983[W], Beehler *et al.* 2012[J], Diamond specimens at AMNH[K]).

DIAGNOSIS: Male crown reddish buff; nape, hindneck, and collar deep blue; lower side of head and throat orange-buff; breast, belly, and undertail red-orange; flanks green (Rand & Gilliard 1967, Collar 1997).

NOTES

So far there have been no records of this species sharing forest habitat with *Micropsitta pusio* in the foothills. There is apparently a distinct population at Tabubil (south-central Border Ra) that remains undescribed; the forecrown is paler, with a yellow wash (Murray 1988a; see photo in Coates & Peckover 2001: 79). Gregory (1997, 2007a) noted observing a flock of 150 *M. bruijnii* near Tabubil. There is also the account of a flock of 650 seen in Nov-1992 flying near Ok Tedi reported in Collar (1997). Gregory (2007a: 107) stated: "That was a remarkable year for them in this area and flocks of hundreds were seen on several occasions." Certainly this mass flocking phenomenon of this bizarre little bird merits study, especially since so little is understood regarding its diet and ecology. *Other names*—Rose-breasted or Mountain Pygmy-Parrot.

ORDER PASSERIFORMES

Hackett *et mult.* (2008) showed that the songbirds are monophyletic and sister to a lineage that includes the parrots and allies. The songbirds+parrots clade is sister to the falcons. Barker *et al.* (2004) identified three major clades within the Passeriformes—the rock wrens (Acanthisittidae), the suboscines, and the oscines. Details of the family relationships within the Passeriformes are provided by Barker *et al.* (2004), Jønsson *et al.* (2011b), Alström *et al.* (2013), Aggerbeck *et al.* (2014), and Cracraft (2014).

Family Pittidae

Pittas [Family 29 spp/Region 3 spp]

The species of the family range from tropical Africa east through tropical Asia to New Guinea, the Bismarck and Solomon Islands, and Australia. Species inhabit lowland and hill forest interior in the Region. Pittas are mainly terrestrial but usually vocalize from an elevated perch. They are compact, with rounded wings, very short tail, long legs, and colorful plumage. The 10th primary is fully developed, but the 10th secondary is vestigial (Schodde & Mason 1999). Seen rarely but heard frequently, pittas are the only suboscines in the Region (Barker *et al.* 2004). Treatment for the family follows Irestedt *et al.* (2006).

GENUS *ERYTHROPITTA* [Genus 6 spp/Region 1 sp]

Erythropitta Bonaparte, 1854. *Ateneo Italiano* 2(11): 317. Type, by subsequent designation, *Pitta Macklotii* Temminck (Gray 1855: 144).

The species of the genus range from Borneo and the Philippines to New Guinea and northeastern Australia. They are characterized by the unpatterned face and throat; breast band distinct from belly patch; belly all red; and no prominent turquoise shoulder patch (Schodde & Mason 1999). Juvenile plumage is brownish. We recognize *E. erythrogaster* as a single species, a more conservative treatment than that of Irestedt *et al.* (2013), who divided the lineage into multiple species (see below).

Red-bellied Pitta *Erythropitta erythrogaster* RESIDENT (AND VISITOR?)

Pitta erythrogaster Temminck, 1823. *Planch. Col. d'Ois.*, livr. 36: pl. 212.—Manila, Luzon, Philippines.

Frequents the floor of lowland rain-forest interior throughout NG including the Trans-Fly, the NW Islands, Aru Is, Yapen I, D'Entrecasteaux Arch, and Rossel I, sea level–1,250 m^{CO} (one record to 1,680 m^{CO}).

Extralimital—the Philippines and Sulawesi to ne Australia. Cape York (QLD) birds are presumed to migrate across the Torres Strait to winter in NG, although evidence is scant (Higgins *et al.* 2001). Also ranges into the Bismarck Arch (Coates 1990, Mayr & Diamond 2001, Dutson 2011).

SUBSPECIES

a. *Erythropitta erythrogaster bernsteini* (Junge)
Pitta erythrogaster bernsteini Junge, 1958. *Ardea* 46: 88.—Gebe I, NW Islands.
RANGE: Gebe I.
DIAGNOSIS: A blue-backed race marginally larger than *cyanonota* of Ternate (n Moluccas), slightly paler dorsally, and with blue of breast silvery (Erritzøe 2003). Note rufous-brown head; blue back, mantle and wings; and narrow breast band.

b. *Erythropitta erythrogaster macklotii* (Temminck)
Pitta Macklotii Temminck, 1834. *Planch. Col. d'Ois.*, livr. 92: pl. 547.—Triton Bay, Bird's Neck.

Synonym: *Pitta kuehni* Rothschild, 1899. *Bull. Brit. Orn. Club*, 10: 3.—Kasiui I, se of Seram I, s Moluccas.

Synonym: *Pitta macklotii oblita* Rothschild & Hartert, 1912. *Novit. Zool.* 19: 197.—Avera, upper Aroa R, s watershed of the SE Peninsula.

Synonym: *Pitta mackloti aruensis* Rothschild & Hartert, 1901. *Novit. Zool.* 8: 63.—Wokan, Aru Is.

RANGE: The four main NW Islands, Yapen I, Aru Is, Bird's Head and Neck, S Lowlands, Trans-Fly[J], and the SE Peninsula east in the n watershed to the Kumusi R, and in the s watershed east to Cloudy Bay (Mayr 1941, Bishop 2005a[J]).

DIAGNOSIS: Crown fuscous brown, in some instances with dark markings; nape and hind-crown rich red-brown, sometimes trending to golden brown (duller than that of *habenichti*); back dull green, rump dull blue, blackish band anterior to red of breast. See photos in Coates (1990: 14) and Coates & Peckover (2001: 115). We follow Dickinson & Christidis (2003) in sinking *oblita* into *macklotii*. We follow Erritzøe (2003) in placing *kuehni* in synonymy with *macklotii*. We fold *aruensis* into *macklotii* because of minimal size differences and unstable plumage differences (Rand & Gilliard 1967, Erritzøe 2003).

c. *Erythropitta erythrogaster habenichti* (Finsch)

Pitta Habenichti Finsch, 1912. *Orn. Monatsber.* 20: 102.—Near Bogia, e sector of the Sepk-Ramu.

RANGE: The NW Lowlands and the Sepik-Ramu.

DIAGNOSIS: Nape patch dull orange-red (Irestedt *et al.* 2013). The Yapen population may better be assigned to this race.

d. *Erythropitta erythrogaster loriae* (Salvadori)

Pitta Loriae Salvadori, 1890. *Ann. Mus. Civ. Genova* 29: 579.—Suau I, off coast of Cloudy Mts, Milne Bay, SE Peninsula.

RANGE: The tip of the SE Peninsula, west along the n coast as far as the Kumusi R, and along the s coast to Cloudy Bay.

DIAGNOSIS: Crown brown but heavily streaked with black; nape dark chestnut; back greenish; face dark brown; throat blackish.

e. *Erythropitta erythrogaster finschii* (Ramsay)

Pitta (Erythropitta) finschii Ramsay, 1884 *Proc. Linn. Soc. NSW* 9: 864.—Fergusson I, D'Entrecasteaux Arch, SE Islands.

RANGE: Fergusson and Goodenough I, D'Entrecasteaux Arch (Mayr 1941). Sight record from Normanby I presumably referable to this form (eBird report from Sibonai on 14-Jun-2014 by B. Barkley & E. Enbody).

DIAGNOSIS: Similar to *loriae*, but back and wings bluish and head dark brown (Rand & Gilliard 1967). Note also dull brown throat and black lower throat band anterior to blue of breast.

f. *Erythropitta erythrogaster meeki* (Rothschild)

Pitta meeki Rothschild, 1898. *Bull. Brit. Orn. Club* 8: 6.—Rossel I, Louisiade Arch, SE Islands.

RANGE: Rossel I (Mayr 1941)

DIAGNOSIS: Similar to *macklotii*, but head paler, hindneck pale brown, throat dusky gray—not black; narrow black collar to upper breast; crown without black; mantle greenish.

g. *Erythropitta erythrogaster digglesi* Krefft

Pitta digglesi Krefft, 1869. *Ibis*: 350.—Cape York, Australia, or New Guinea.

RANGE: Breeds in far ne QLD, and apparently some of these birds migrate north to the Trans-Fly for the nonbreeding season (Schodde & Mason 1999, Dickinson & Christidis 2014).

DIAGNOSIS: Paler and brighter colored than resident *macklotii* from the S Lowlands (Schodde & Mason 1999).

NOTES

Irestedt *et al.* (2013) provided a detailed overview of the phylogeography of this species. Their study indicated this species colonized the Region from the west and that many insular forms may constitute distinct species, the details of which must be worked out in future studies. Within the NG region, they recognized the Moluccan Pitta, *E. rufiventris* (for the population *bernsteini*); Habenicht's Pitta, *E. habenichti*; Papuan Pitta, *E. macklotii* (including *loriae*); D'Entrecasteaux Pitta, *E. finschii*; and Louisiade Pitta, *E. meeki*. *Other names*—Blue-breasted Pitta, *Pitta erythrogaster*.

Pitta Vieillot, 1816. *Analyse Nouv. Orn. Elem.*: 42. Type, by subsequent designation (G. R. Gray, 1855. *Cat. Gen. Subgen. Birds*: 43), *Corvus triostegus* Sparrman = *Corvus brachyurus* Linnaeus.

The species of this genus range from tropical Africa to Australia (Erritzøe 2003). The genus is characterized by the terrestrial habit, patterned plumage exhibiting artfully situated blocks of color, very abbreviated tail, compact globular body, and loud musical vocalization. Treatment follows Irestedt *et al.* (2006).

Hooded Pitta *Pitta sordida*   RESIDENT

Turdus sordidus P. L. S. Müller, 1776. *Natursyst.*, Suppl.: 143; based on "Merle des Philippines" of Buffon (Daubenton), 1765–1775, *Planches Enlum.* 1: pl. 89.—Philippines.

Inhabits the floor of forest interior throughout NG and many satellite islands; sea level–500 m, max 1,200 m[CO]. Widespread in NG except for parts of the Trans-Fly where perhaps replaced by *P. versicolor* (Coates 1990); also inhabits the NW Islands, Aru Is, Adi I, Bay Is, and Karkar I, but absent from the SE Islands.

Extralimital—Crown, Long, and Tolokiwa I, extending westward to the Philippines and tropical Asia, including the Himalayas (Coates 1990, Mayr & Diamond 2001). In hill and lower montane forest on Karkar (Diamond & LeCroy 1979). Appears to be nomadic during dry periods (Coates 1990).

SUBSPECIES

a. *Pitta sordida novaeguineae* P. L. S. Müller & Schlegel
Pitta novae-guineae P. L. S. Müller & Schlegel, 1845. In Temminck, *Verh. Nat. Gesch. Nederl. Overz. Bezit., Zool.* (1842), Pitta: 19.—Manokwari, Bird's Head. Replacement name for *Pitta atricapilla* Quoy & Gaimard, 1830. *Voy. Astrolabe, Zool.* 1: 258, pl. 8, fig. 3.
    Synonym: *Pitta atricapilla hebetior* Hartert, 1930. *Novit. Zool.* 36: 92.—Karkar I.
RANGE: Through all the NW Islands; all NG; also Karkar and Long I (Mayr 1941). Within the Trans-Fly recorded from the Merauke R and Bian Lakes and from Tarara east to the Fly R (Bishop 2005b).
DIAGNOSIS: Glittering green breast and black hood; greenish-blue border to red belly patch; black spot anterior to red of belly. The race *hebetior* of Karkar I is not separable (Diamond & LeCroy 1979). See photos in Coates (1990: 18–19) and Coates & Peckover (2001: 114). Wing (male) 99–114 (n=8).

b. *Pitta sordida goodfellowi* C. M. N. White
Pitta novaeguineae goodfellowi C. M. N. White, 1937. *Bull. Brit. Orn. Club* 57: 136.—Silbattabatta, Aru Is.
RANGE: The Aru Is (Mayr 1941).
DIAGNOSIS: Similar to *novaeguineae* but smaller and darker (Rand & Gilliard 1967). Wing (male) 96–99 (n=4).

c. *Pitta sordida mefoorana* Schlegel
Pitta Novae Guineae Mefoorana Schlegel, 1874. *Mus. Pays-Bas* 3, *Rev. Coll.*, Monogr. 37 (Pitta, Revue): 8.—Numfor I, Bay Is.
RANGE: Numfor I.
DIAGNOSIS: Broad blue border to red belly patch; black spot atop belly patch; bright green border to black hood at top of breast (Erritzøe 2003).

d. *Pitta sordida rosenbergii* Schlegel
Pitta Rosenbergii Schlegel, 1871. *Ned. Tijdschr. Dierk.* (1873) 4: 16.—Biak I, Bay Is.
RANGE: Biak I.
DIAGNOSIS: No black spot atop large red belly patch; blue border to belly patch reduced anteriorly (Erritzøe 2003).

NOTES
Other names—Black-headed Pitta.

Noisy Pitta *Pitta versicolor*
Pitta versicolor Swainson, 1825. *Zool. Journ.* 1: 468.—New South Wales, Australia.

Recorded in all seasons from the Trans-Fly; also a single sight record from near Port Moresby. Occurs in NG as a winter visitor from Australia and probably also as a breeding resident locally in the Trans-Fly (Coates 1990). Frequents dense monsoon woodland, mangrove forest and gallery forest; lowlands only (Coates 1990).

Extralimital—breeds in e Australia north to some of the more substantial islands of the Torres Strait, where it nests from January to April (Higgins *et al.* 2001). Reported to occur on passage only on smaller islands from September to December, and never to be seen in the early dry season (Draffan *et al.* 1983).

SUBSPECIES

a. *Pitta versicolor simillima* Gould
Pitta simillima Gould, 1868. *Proc. Zool. Soc. London*: 76.—Cape York, n QLD, Australia.
RANGE: Northern QLD, islands of the Torres Strait, and the Trans-Fly. Localities—Kurik in early June[Z], Katau R[CO], Daru I[CO], near Bensbach R in July–October[J], also in November, late December, and throughout April[CO] (Rand 1942a, Finch 1980c, 1982b; Mees 1982b[Z], Hicks 1985a; Bishop 2005a, W. S. Peckover, pers. comm., K. D. Bishop[J]). Also Bell (1968b) reported observing this species in monsoon gallery forest at Eriama Creek near Port Moresby on 16–17-Aug-1966[CO]. Heard calling at Bensbach on a recent visit in late August in response to heavy rains (D. Hobcroft *in litt.*). Also calling there Sep-2009 (P. Gregory *in litt.*)
DIAGNOSIS: Very similar to the nominate form but small; wing 113–129 (Higgins *et al.* 2001); bright green dorsally and bright yellow-brown ventrally (Rand & Gilliard 1967, Schodde & Mason 1999, Erritzøe 2003). White bar on fourth to sixth primaries reduced and usually does not extend to third primary (Schodde & Mason 1999). See photos in Coates & Peckover (2001: 114).

NOTES
Other names—Blue-winged or Buff-breasted Pitta.

Family Ptilonorhynchidae

Bowerbirds   [Family 21 spp/Region 13 spp]

Members of the family range throughout the Region as well as through much of Australia. Absent from the Southeastern Islands. The New Guinean species are mainly found in the hills and mountains, although two species (*Sericulus ardens* and *Ailuroedus buccoides*) are primarily lowland dwelling. Two species (*Chlamydera cerviniventris* and *C. lauterbachi*) prefer nonforest scrub habitats, but the remainder inhabit forest interior and range upward into montane forest. All of the sexually dichromatic species are polygamous and bower building. Two of the five monochromatic species are monogamous and pair bonding, whereas the other three are polygamous and bower building. All are highly frugivorous and some are fig specialists. The family supports four distinct lineages: catbirds, the Toothbilled Bowerbird (*Scenopoeetes dentirostris*), maypole bowerbirds, and avenue bowerbirds. The males of the polygamous species are noted mimics of the songs of other bird species as well as a wide array of other sounds in the environment. The molecular studies of Barker *et al.* (2004) indicated this family is a sister group to the Climacteridae, a placement corroborated by Jønsson *et al.* (2011b). Gilliard (1969) pointed out that the species with the most beautifully plumaged males exhibit the smallest and least-developed bowers and that the most elaborate bowers are constructed by the plainest species.

Ailuroedus Cabanis, 1851. *Mus. Hein.* 1: 213, note. Type, by monotypy, *Ptilonorhynchus Smithii* Vigors & Horsfield = *Lanius crassirostris* Paykull.

The species of this genus range from eastern Australia to the Region (Frith & Frith 2004). The genus is characterized by the powerful ivory-colored bill with the hooked maxilla, large head, green dorsal plumage, ventral spotting, powerful grasping claws, pair-bonded and monogamous nesting behavior, and fig-eating habit.

White-eared Catbird *Ailuroedus buccoides*   ENDEMIC
Kitta buccoides Temminck, 1836. *Planch. Col. d'Ois.*, livr. 97 (1835): pl. 575.—Triton Bay, Bird's Neck.

Inhabits the interior of lowland and hill forest throughout NG, the NW Islands, and Yapen I; sea level–800 m, max 1,300 m (Freeman & Freeman 2014). No records from the Trans-Fly, where replaced by *A. melanotis*. Replaced at higher elevations by *A. melanotis* but overlaps locally with that species. No records from Misool I.

SUBSPECIES
a. *Ailuroedus buccoides buccoides* (Temminck)
Synonym: *Ailuroedus buccoides oorti* Rothschild & Hartert, 1913. *Novit. Zool.* 20: 526.—Waigeo I, NW Islands.
Synonym: *Ailuroedus buccoides cinnamomeus* Mees, 1964. *Zool. Meded. Leiden* 40: 126.—Lorentz R, sw NG.
RANGE: Waigeo, Batanta, and Salawati I[MA]; also the Bird's Head and Neck, the sw extremity of the NW Lowlands, and the S Lowlands east to Lake Kutubu (Frith & Frith 2004).
DIAGNOSIS: Ventral plumage pale ocher with large black spotting; crown dark brown, typically tinged with greenish (Mees 1964c: 127). Some variation in darkness of crown and ocher wash on breast. We include *oorti* as a synonym following Mees (1964c). In addition, we consider the form *cinnamomeus* best combined with the nominate form, which tends toward paler ventrally in the west, and darker and richer ventrally in the east. Mees (1964c) made the case for *cinnamomeus* by focusing unnecessarily on the holotype of the nominate form, which apparently is unusually pale. We warn against focusing on holotypes rather than topotypical series, which encompass natural variation.

b. *Ailuroedus buccoides geislerorum* A. B. Meyer
Aeluroedus geislerorum A. B. Meyer, 1891. *Abh. Ber. Zool. Mus. Dresden* (1890–91) 3(4): 12.—Near Madang, e verge of the Sepik-Ramu.
RANGE: Yapen I and the NW Lowlands, Sepik-Ramu, Huon Penin, and the n watershed of the SE Peninsula (Mayr 1941, Rand 1942b, Mayr 1931, Gilliard & LeCroy 1967).
DIAGNOSIS: Crown pale tan-brown; pure white ear covert feathering extends forward onto lower lores; breast with reduced ocher wash. See photos in Coates (1990: 382) and Coates & Peckover (2001: 206).

c. *Ailuroedus buccoides stonii* Sharpe
Aeluraedus stonii Sharpe, 1876. *Nature* 14: 339.—Laloki R, SE Peninsula.
RANGE: The s watershed of the Eastern Ra and SE Peninsula east to Amazon Bay. Not recorded from the Trans-Fly (Bishop 2005a).
DIAGNOSIS: Crown dark blackish brown; underparts washed heavily with ocher, marked with small black spots, mainly across breast. See photos in Coates (1990: 383) and Coates & Peckover (2001: 206).

NOTES
Other names—White-throated or Least Catbird.

Black-eared Catbird *Ailuroedus melanotis*   ENDEMIC
Ptilonorhynchus melanotis G. R. Gray, 1858. *Proc. Zool. Soc. London*: 181.—Aru Is.

Mainland NG, Misool I, and the Aru Is; Mainland populations range typically between 900 and 1,800 m, to a max of 2,250 m[CO]; but one population inhabits the Trans-Fly lowlands, and is also found in lowland forest on the Aru Is and Misool I. Known to occur throughout the Central Ra and all peripheral ranges except for the Cyclops Mts.

Extralimital—ne Australia.

SUBSPECIES

a. *Ailuroedus melanotis arfakianus* A. B. Meyer

Ailuroedus arfakianus A. B. Meyer, 1874. *Sitzungsber. Akad. Wiss. Wien* 69: 82.—Arfak Mts, Bird's Head

> Synonym: *Ailuroedus crassirostris misoliensis* Mayr & Meyer de Schauensee, 1939. *Proc. Acad. Nat. Sci. Phila.* 91: 152.—Tip, Misool I, NW Islands.

RANGE: Misool I[MA] and the Bird's Head (Mayr 1941, Mees 1965). The population from the Kumawa Mts apparently refers to this form (see Diamond 1985: 81). Populations in the Fakfak and Wandammen Mts (Diamond 1985) are currently uncollected and undiagnosed but may refer to this form.

DIAGNOSIS: Distinct from the nominate form by the following: the dark-plumaged throat with black spotting on a background that is dirty-white to buff; chest not marked blackish but darker green marked with narrow, pale buff feather centers. The pale postocular patch is larger and white, and the large spots on the black crown are paler than in nominate form (Frith & Frith 2004). Mees (1965: 195–96) made clear his doubt on the validity of the race *misoliensis*. We follow Mees's suggestion and subsume *misoliensis* into *arfakianus* because of only minor size difference and minimal plumage distinctions.

b. *Ailuroedus melanotis melanotis* (G. R. Gray)

RANGE: Lowland forest of the Aru Is[MA] and Trans-Fly—from Kurik[J] east to the Oriomo R (Rand 1942a, Mees 1982b[J], Bishop 2005a). Inhabits lowlands, unlike other populations.

DIAGNOSIS: Dark plumage on head and dark edges of upper breast feathers are blackish. Large buff spots on nape and breast are rather medium buff, not too dark; greenish wash on belly. Adult male exhibits the very large bill (Frith & Frith 2004, Ogilvie-Grant 1915). Mees (1982b) reviewed this form in detail, comparing birds from the Aru Is with those from the adjacent Mainland. He found a measurable difference in size (Mainland birds smaller) but he considered the size difference not substantial enough to warrant subspecific distinction.

c. *Ailuroedus melanotis facialis* Mayr

Ailuroedus crassirostris facialis Mayr, 1936. *Amer. Mus. Novit.* 869: 4.—Otakwa R, Western Ra.

RANGE: The s slopes of the Western Ra, east to where it presumably meets with the form *melanocephalus*, which occurs at least as far west as the s slopes of the Eastern Ra. Currently (Frith & Frith 2004) there is a large gap in knowledge of subspecific assignment on the s slopes of the Central Ra (between the Baliem R and Karimui). We presume the gap will be filled by *facialis* from the west and *melanocephalus* from the east.

DIAGNOSIS: Similar to nominate subspecies, but throat darker, less white, more buff. Spotting on crown and upper back much darker and more numerous; more green in the smudgy breast band (Frith & Frith 2004).

d. *Ailuroedus melanotis jobiensis* Rothschild

Aeluroedus jobiensis Rothschild, 1895. *Bull. Brit. Orn. Club* 4: 26.—?The Waropen, w sector of the NW Lowlands (Mayr 1962).

> Synonym: *Ailuroedus melanotis guttaticollis* Stresemann, 1922. *Orn. Monatsber.* 30: 35.—Hunstein Mts, Sepik-Ramu.

RANGE: The n slopes of the Western, Border, and Eastern Ra; and the Foja, Cyclops, Bewani, Torricelli, Prince Alexander, and Adelbert Mts (Diamond 1985); also mts of the Sepik and Jimi R (Rand 1942b, Coates 1990).

DIAGNOSIS: Similar to the nominate form, but pale spotting on blackish crown tending toward buff, and chin, throat, and upper breast blackish with fine buff spotting; remaining underparts darker than those of *melanotis* (Frith & Frith 2004). See photos in Coates (1990: 384) and Coates & Peckover (2001: 206).

e. *Ailuroedus melanotis astigmaticus* Mayr

Ailuroedus melanotis astigmaticus Mayr, 1931. *Mitt. Zool. Mus. Berlin* 17: 647.—Ogeramnang, Huon Penin.

RANGE: Mountains of the Huon Penin.

DIAGNOSIS: Crown blackish with a few narrow white lines rather than spots; dark blackish collar with abundant pale buff spots.

f. *Ailuroedus melanotis melanocephalus* Ramsay

Aeluroedus melanocephalus Ramsay, 1883. *Proc. Linn. Soc. NSW* 8: 25.—Astrolabe Ra, SE Peninsula.

RANGE: Mountains of the SE Peninsula, west in the s watershed to Mt Karimui[DI], and west in the n watershed to the Herzog Mts and head of Huon Gulf.

DIAGNOSIS: Similar to the nominate form, but underparts generally darker, blacker on chest and throat, and more ocher below. Crown blacker, and buff spots (some are streaks) slightly smaller and sparser; greenish wash on belly; lacks pale spots on wing coverts. Differs from *arfakianus* by buff crown spots not white ones (Frith & Frith 2004). See photo in Coates & Peckover (2001: 206).

NOTES

We follow Frith & Frith (2004), Higgins *et al.* (2006), and Christidis & Boles (2008) in separating the Australian *A. crassirostris* as a full species distinct from *melanotis* (*pace* Dickinson & Christidis 2014). There is a great deal of minor and clinal plumage variation in the NG forms, which has led to the description of many races. We have combined the races that are the most thinly defined and least compelling in terms of geography. *Other names*—Spotted Catbird; included as races of Green Catbird, *A. crassirostris*, by some authorities.

GENUS *AMBLYORNIS*   ENDEMIC [Genus 5 spp]

Amblyornis Elliot, 1872. *Ibis*: 113. Type, by monotypy, *Ptilorynchus inornatus* Schlegel.

The species of this genus can be found in virtually all of New Guinea's mountain ranges, but are absent from the Region's offshore islands. The genus is characterized by the compact form, blunt but unhooked bill, olive-brown plumage, and the circular bower contruction. We do not follow the molecular phylogeny of Kusmierski *et al.* (1997), which interdigitates *Prionodura* and *Archboldia* among species of *Amblyornis*. Such treatment is contradicted by morphology, bower construction, behavior, distribution, and vocalization data. David & Gosselin (2002b) noted that *Amblyornis* is to be treated as feminine.

Streaked Bowerbird *Amblyornis subalaris*   ENDEMIC | MONOTYPIC

Amblyornis subalaris Sharpe, 1884. *Journ. Linn. Soc. London, Zool.* 17: 408.—Astrolabe Ra, SE Peninsula.

Inhabits lower montane forest of the e sector of the SE Peninsula; 670–1,200 m, max 1,500 m[CO]. In the s watershed it ranges from the Angabanga R (northwest of Port Moresby) east to Milne Bay; in the n watershed it ranges from Mt Suckling east to the mts of Milne Bay Prov. Inhabits the interior of lower montane forest dominated by oaks (*Lithocarpus* and *Castanopsis*).

NOTES

The w terminus of the range of this species in the n and s watersheds has not been defined in detail and should be delineated by future fieldworkers. Within its limited range, this species occurs sympatrically with *A. macgregoriae*, which occupies forest at a higher elevation. The bower of *A. macgregoriae* is built on the crest of a ridge in forest interior. By contrast, that of *A. subalaris* is placed on the slope below the ridge crest, with the main bower decoration wall and openings facing downhill. *Other names*—Striped Bowerbird, Eastern Gardenerbird.

MacGregor's Bowerbird *Amblyornis macgregoriae*   ENDEMIC

Amblyornis macgregoriae De Vis, 1890. *Ann. Rept. Brit. New Guinea*, 1888–89: 61.—Musgrave Ra, 2,150–2,750 m, SE Peninsula.

Throughout the Central Ra, Adelbert Mts, and Mt Bosavi; 1,600–3,300 m, min 1,050 m[CO]. Inhabits the interior of montane forest; heard far more often than seen. Forages in canopy and displays on ground at its ridge-crest bower.

SUBSPECIES

a. *Amblyornis macgregoriae mayri* Hartert

Amblyornis inornatus mayri Hartert, 1930. *Novit. Zool.* 36: 30.—Weyland Mts, Western Ra.

RANGE: The Western Ra, Border Ra east to the Hindenburg Mts of extreme w PNG (Schodde & McKean 1973b), and presumably farther east to the Strickland R gorge.

DIAGNOSIS: The larger subspecies, with the longer tail and crest. Male tail 87–98; male crest 62–74 (Frith & Frith 2004).

b. *Amblyornis macgregoriae macgregoriae* De Vis

> Synonym: *Amblyornis inornatus aedificans* Mayr 1931, *Mitt. Zool. Mus. Berlin* 17: 648.—Dawong, Herzog Mts, SE Peninsula.
>
> Synonym: *Amblyornis macgregoriae kombok* Schodde and McKean, 1973. *Emu* 73: 53.—Minj-Nona Divide, 2,140 m, Kubor Mts, Eastern Ra.
>
> Synonym: *Amblyornis macgregoriae nubicola* Schodde and McKean, *Emu* 73: 55.—Mount Wadimana, Milne Bay Prov, SE Peninsula.
>
> Synonym: *Amblyornis macgregoriae amati* Pratt, 1983. *Emu* (1982) 82: 121.—Mount Mengam, Adelbert Mts, ne NG.
>
> Synonym: *Amblyornis macgregoriae lecroyae* Frith and Frith, 1997. *Bull. Brit. Orn. Club* 117: 201.—Mount Bosavi, 1,400 m, south-central NG.

RANGE: The Eastern Ra and mts of the SE Peninsula, presumably from the Strickland R gorge east to Milne Bay. Also the Adelbert Mts.

DIAGNOSIS: Marginally smaller than *mayri*; male tail 78–93; male crest 49–69 (Gilliard & LeCroy 1961, Frith & Frith 2004). See photos in Coates (1990: 392–96).

NOTES

Gillard & LeCroy (1961: 73) made a study of the available museum material and concluded that it "seems best to use tail length and crest length alone to differentiate races." Our study of museum material supports the thinking of Gilliard & LeCroy that plumage color is not useful in determining subspecies, only measurements. For other views, see Schodde & McKean (1973b) and Frith & Frith (2004). Frith & Frith recognized seven races, based on size and coloration of body plumage. We recognize two subspecies and treat a third form (*A. germana*) as a distinct species. The population *germana* of the Huon Penin is here treated as a full species based on the distinctive bower construction (Coates & Peckover 2001, Beehler field notes).

Benz (2011)found that this species displayed only shallow genetic structure across its geographic range. *Other names*—Macgregor's Gardenerbird, Crested Bowerbird; sometimes written as MacGregor's Bowerbird.

Huon Bowerbird *Amblyornis germana* ENDEMIC | MONOTYPIC

Amblyornis subalaris germanus Rothschild, 1910. *Bull. Brit. Orn. Club* 27: 13.—Rawlinson Ra, Huon Penin.
> Mountains of the Huon Penin, 1,660–2,940 m (Beehler field notes from YUS elevational transect).

NOTES

The distinctive bower of this regional isolate supports its elevation to species status (Coates & Peckover 2001: 209, Pratt & Beehler 2014: plate 67a). Whereas *A. macgregoriae* places its bower on the crest of the ridge, the bower of *A. germana* is placed on the slope below the ridge crest. In addition, this bower faces downslope, showing a decorated bower "face" that is absent from the circular bower of *A. macgregoriae*. The platform of the bower of *A. macgregoriae* is constructed of moss and is defined by a moss perimeter. By contrast, the bower of *A. germana* exhibits a perimeter of sticks that surrounds a mushroom-shaped head of sticks atop the central moss wall, the wall being the site of much decoration with small colored items. With its decorated wall, the bower of *A. germana* is more similar to that of *A. subalaris*. Plumage and morphometric distinctions of this species include small size, and short wing and crest (Frith & Frith 2004). Unpublished molecular analysis indicates this isolate is highly divergent from *A. macgregoriae* (Benz 2011); this is paralleled in differentiation of display behavior and courtship vocalization (B. W. Benz *in litt.*). *Other names*—earlier treatments included this species within *A. macgregoriae*.

Golden-fronted Bowerbird *Amblyornis flavifrons* ENDEMIC | MONOTYPIC

Amblyornis flavifrons Rothschild, 1895. *Novit. Zool.* 2: 480; (Fig. *Novit. Zool.* 3, pl. 1, figs. 3 and 4.).—Foja Mts, NW Lowlands.

Inhabits the interior of montane forest of the Foja Mts, east of the Mamberamo R and north of the Taritatu R, 1,000–2,100 m, (Diamond 1982, Beehler & Prawiradilaga 2010, Beehler *et al.* 2012).

Absent from the Van Rees Mts, which are to the west of the Mamberamo R (J. M. Diamond *in litt.*). Male crest is very distinctive, with the golden-orange extending to the forehead. In addition, the bower is distinctive because of its blue decorations and the lack of an elevated perimeter wall that typifies the circular terrestrial base of the bower of *A. macgregoriae*. The adult male displays with a small blue berry held in the tip of its beak. *Other names*—Yellow-fronted Gardenerbird or Bowerbird, Golden-maned Gardener.

Vogelkop Bowerbird *Amblyornis inornata* ENDEMIC | MONOTYPIC
Ptilorhynchus inornatus Schlegel, 1871. *Ned. Tijdschr. Dierk.* (1873) 4: 51.—Hatam, Arfak Mts, e Bird's Head.
   The Bird's Head and Neck: Arfak, Tamrau, and Wandammen Mts, 1,000–2,000 m. Populations observed from the Fakfak and Kumawa Mts may refer to this species (see below).

NOTES
Plain-plumaged and sexually monochromatic *Amblyornis* populations are known from the Kumawa and Fakfak Mts (s Bird's Neck region; Diamond 1985, Gibbs 1994); they are morphologically referable to this species, but as their maypole bower is more like the bower of *A. macgregoriae* than that of *A. inornata*, their specific identity merits further study (Diamond 1985, Gibbs 1994). Gibbs makes the case that these should be treated as a distinct species. Molecular work by Uy & Borgia (2000) show these populations to be very closely related. Diamond & Bishop (2015) show bower design is variable locally and involves learning. Two questions to resolve are (1) whether the bower contruction is uniform within each of these two outlying ranges, and then (2) whether they are uniform across the two ranges. Such bower information will be useful in informing future decision on species limits. *Other names*—Gardener or Plain Bowerbird, Brown Gardenerbird, *Amblyornis inornatus*.

GENUS *ARCHBOLDIA* ENDEMIC | MONOTYPIC

Archboldia Rand, 1940. *Amer. Mus. Novit.* 1072: 9. Type, by original designation, *Archboldia papuensis* Rand.
   The single species of the genus is endemic to the highlands of New Guinea. The genus is characterized by the distinctive black plumage, the peculiar flat crown evident in both sexes, the erect golden forehead crest of the adult male, and the fern-mat bower construction, distinct from the bowers of *Amblyornis*. Molecular studies by Kusmierski *et al.* (1997) and Benz (2011) indicated that this species is very close to the species in the genus *Amblyornis* (both phylogenies place *papuensis* within *Amblyornis*, though whether based on the same sequence data we are not aware). That said, abundant behavior, plumage, morphological, and bower-related distinctions warrant retention of this genus (see Frith & Frith 2004). This is an instance in which molecular results do not agree with behavior, plumage, and morphology. In this particularly conflicting instance, we side with the latter.

Archbold's Bowerbird *Archboldia papuensis* ENDEMIC
Archboldia papuensis Rand, 1940. *Amer. Mus. Novit.* 1072: 9.—Bele R, 2,200 m, 18 km n of Lake Habbema, Western Ra.
   Patchily distributed in higher mts of the Western and Eastern Ra; 2,600–2,900 m, extremes 1,750–3,660 m[CO]; no records to date from the Border Ra. Inhabits beech and podocarp forest and frost-disturbed high-montane forest with *Pandanus* and scrambling bamboo (*Nastus productus*) on shallow slopes and plateaus.

SUBSPECIES

a. *Archboldia papuensis papuensis* Rand
RANGE: Patchily distributed through high uplands of the Western Ra, from Paniai Lake east to the Baliem R, 2,870–3,660 m (Frith & Frith 2004). Localities—Bobairo[K] at Paniai Lake, Ilaga[J], Bele R (Ibele)[Q], 9 km northeast of Lake Habbema[Q], 18 km southwest of Bernhard Camp[Q] (Rand 1942b[Q], Junge 1952[K], Ripley 1964a[J]). DIAGNOSIS: Tail shorter than that of *sanfordi*. Tail of adult female nominate 122–130 (n=5) *vs.* 132–149 (n=9) for *sanfordi*. All plumages apparently paler, sooty gray, less black than those of *sanfordi*.

b. *Archboldia papuensis sanfordi* Mayr & Gilliard

Archboldia papuensis sanfordi Mayr & Gilliard, 1950. *Amer. Mus. Novit.* 1473: 1.—Southwestern slopes of Mt Hagen, 6 km w of Tomba, Eastern Ra.

RANGE: Patchily distributed in the w sector of the Eastern Ra. Localities—Tari Gap[L], Karius Ra[L], s and sw slopes of Mt Giluwe[J], and near Tomba on the sw slopes of Mt Hagen[K]; 1,800–2,800 m (Mayr & Gilliard 1954[K], Sims 1956[J], Frith & Frith 1988b[L]).

DIAGNOSIS: Larger, longer-tailed, and shorter-billed than the nominate form (Frith & Frith 2004). Apparently slightly blacker, less gray than nominate (all plumages). Sibley & Monroe (1990) treated *sanfordi* as a distinct species. See photos in Coates (1990: 385–91) and Coates & Peckover (2001: 207).

NOTES

Should also be looked for in the Star Mts westward to Mt Mandala of the w sector of the Border Ra. There is an unconfirmed local informant's report of this species from the Kubor Mts (R. W. Campbell in Coates 1990), so it should be looked for there as well. It is currently known from few localities; presumably future fieldworkers will locate the species in many additional sites on the Central Ra in years to come. Easily overlooked in its difficult habitat. *Other names*—Sanford's or Tomba Bowerbird, *Archboldia sanfordi*.

GENUS *SERICULUS*   [Genus 4 spp/Region 3 spp]

Sericulus Swainson, 1825. *Zool. Journ.* 1(4): 476. Type, by monotypy, *Meliphaga chrysocephalus* Lewin.

The species of this genus range from southeastern Australia to lowland and hill forest of western and central New Guinea. The genus is characterized by very distinctive male and female plumages. The male has specialized golden or orange head and nape feathering; its remaining plumage is dominated by black and gold. The female is plain brownish above and paler below, with dark scalloping ventrally. A systematic analysis of the genus, based on molecular results, was provided by Zwiers *et al.* (2008).

Masked Bowerbird *Sericulus aureus*   ENDEMIC | MONOTYPIC

Coracias aurea Linnaeus, 1758. *Syst. Nat.*, ed. 10, 1: 108.—Bird's Head.

Inhabits the canopy of hill forest in w and n NG: mts of the Bird's Head, Bird's Neck (Fakfak Mts and mts of the Wandammen Penin)[MA], n scarp of the Western, Border, and Eastern Ra (Weyland Mts east to the Jimi R[L]), Foja Mts[K], and the Torricelli and Prince Alexander Mts[J] of the PNG North Coastal Ra, 850–1,400 m (Diamond 1985[J], Frith & Frith 2004[L], Beehler *et al.* 2012[K]).

Localities—Tamrau Mts[Z], Bivak October in the Arfak Mts[J], 4 km southwest of Bernhard Camp[K], 6 km southwest of Bernhard Camp[K], Mt Turu[L], Mt Nibo[L], Ruti on the Jimi R[CO] (Rand 1942b[K], Gyldenstolpe 1955b[J], Diamond 1969[L], Beehler field notes[Z]). Probably also to be expected from the Fakfak Mts, from which there is a 19th-century trade skin record (Diamond 1985). No records from the n scarp of the Border Ra but to be expected there.

NOTES

We follow Lenz (1999) and Frith & Frith (2004) in treating *S. aureus* as distinct from *S. ardens*, and we agree with Frith & Frith (2004) that the three NG species of *Sericulus* represent sister species. All three construct similar bowers, although *S. ardens* locates its bower in lowland forest, whereas the other two place theirs in lower montane forest. The male plumages of the three species exhibit stepped increase in melanism: *ardens < aureus < bakeri*. *Other names*—Black-faced Golden Bowerbird, *Xanthomelus aureus*.

Flame Bowerbird *Sericulus ardens*   ENDEMIC | MONOTYPIC

Xanthomelus ardens D'Albertis & Salvadori, 1879. *Ann. Mus. Civ. Genova* 14: 113.—Upper Fly R.

Southern NG and the Trans-Fly, from the Wataikwa R east to the upper Fly R and Mt Bosavi, south to the Trans-Fly (Frith & Frith 2004). Inhabits interior canopy of lowland and foothill forest, sea level–760 m[CO].

Localities—Wataikwa R[MA], Lorentz R[J], Endrick R[J], between the Kumbe and Merauke R[Z], Bian Lakes[Z], Kiunga area[J], Elevala R[Z], Black R[K], Tarara[Z], Nomad R[J], and Mt Bosavi[J] (Rand 1942a[K], Bell 1970d, Bishop 2005a, b[Z], Frith & Frith 2004[J]).

The male of *S. ardens* exhibits the following color differences from its sister form *S. aureus*: (1) bill pale horn-colored rather than black and gray; (2) face and throat orange and yellow rather than black; (3) primary coverts yellow rather than black; (4) tibial feathering dull brown rather than black; (5) mantle plumage red-orange rather than yellow-orange. The female of *S. ardens* exhibits (1) greater reddish suffusion about the head, and (2) reduced scalloping in the throat and upper breast (Frith & Frith 2004). The two species presumably meet at the Mimika and Wataikwa R, where males with intermediate plumage occur. *Other names*—Flamed or Golden Bowerbird, Golden Regent Bowerbird.

Fire-maned Bowerbird *Sericulus bakeri* ENDEMIC | MONOTYPIC

Xanthomelus bakeri Chapin, 1929. *Amer. Mus. Novit.* 367: 1.—Adelbert Mts, *ca.* 915 m, above Maratambu village, e sector of the Sepik-Ramu.

Inhabits the interior of lower montane forest in the Adelbert Mts, nnw of Madang, in a narrow elevational band between 900 and 1,450 m.

Localities—Keki Lodge[J], Memenga forest[K], Ilebaguma[CO], near Wanuma[K], and Kowat[B] (Gilliard & LeCroy 1967[K], Beehler field notes[J], Pratt *in litt.*[B]).

NOTES
Probably occupies one of the smallest geographic ranges of any bird species of mainland NG. Common at Keki Lodge, but rare between Wanuma and Mt Mengam; its Red List status remains unclear. Probably not yet threatened, even though a considerable portion of the foothills of the Adelberts have been selectively logged. Evidently can adapt to human disturbance, so long as upland forest habitat remains. *Other names*—Adelbert or Beck's Bowerbird, *Xanthomelus bakeri*.

GENUS *CHLAMYDERA* [Genus 5 spp/Region 2 spp]

Chlamydera Gould, 1837. *Birds Austral.* 1: text to pl. 3, note. Type, by monotypy, *Calodera maculata* Gould.

The species of this genus range through much of Australia and patchily through mainland New Guinea. The genus is characterized by sexual monochromatism (but for a small lilac crest in males of three species); generally dull plumage, darker dorsally and with obscure streaking on the throat and neck; blackish-gray bill; polygamous mating system; and the avenue-type bower construction.

Yellow-breasted Bowerbird *Chlamydera lauterbachi* ENDEMIC

Chlamydodera lauterbachi Reichenow, 1897. *Orn. Monatsber.* 5: 24.—Yagei, upper Ramu, e sector of the Sepik-Ramu.

Patchily distributed through grasslands and open nonforest habitats in lowlands and uplands from the Western Ra east to the n slopes of the Huon Penin, sea level–1,770 m[CO]. In the west most records come from the s watershed. In the east all records come from the n side of the Central Ra. There are two records from the n slopes of the mts of the Huon Penin (Gilliard & LeCroy 1968), where it meets *C. cerviniventris*. Absent from the Bird's Head. One record from the n base of the Bewani Mts. The range is patchy and seemingly relictual. Westernmost record is from the mts at the w verge of the Mamberamo basin.

Sympatric with *C. cerviniventris* at Aiome, Marienberg, Ramu valley, southeast of Hupai, n Finisterre Ra, and Yawan in the Saruwaged Ra (Gilliard & LeCroy 1968, Bailey 1992a, Frith & Frith 2004, 2008, R. Jensen *in litt.*).

SUBSPECIES

a. *Chlamydera lauterbachi uniformis* Rothschild
Chlamydera lauterbachi uniformis Rothschild, 1931. *Novit. Zool.* 36: 250.—Siriwo R, 72 km from mouth, w Mamberamo basin, NW Lowlands.
RANGE: Very patchily distributed in w and c NG. Localities—Siriwo R[MA], Kamura R[MA], Digul R[MA], Puwani R at the n scarp of the Bewani Mts[Z], middle Sepik[K], Marienberg[CO], Kanganaman[K], Wahgi valley[CO], upper Ramu R[M], Sepik-Wahgi Divide, Minj[J], Awande[DI], Asaro valley[CO], Okasa[DI], Miarosa[DI] (Mayr & Camras

1938[M], Sims 1956[J], Gilliard & LeCroy 1966[K], Beehler field notes Jul-1984[Z]). Both subspecies may meet in the Baiyer valley (see below).

DIAGNOSIS: Crown is olive green, lacking the distinctive coppery-rose bloom of the nominate form. See photos in Coates (1990: 405–6).

b. *Chlamydera lauterbachi lauterbachi* (Reichenow)
RANGE: The Ramu valley east to the n scarp of the Huon Penin. Also perhaps the e Sepik and Baiyer R. Localities—Aiome[L], Bogadjim[L], upper Ramu R[K], n foothills of the Finisterre Ra[L], Yawan in YUS[W] (Mayr & Camras 1938[K], Gilliard & LeCroy 1968[L], Beehler field notes[W]). Image of a male at a bower at Baiyer R Sanctuary in Coates (1990: 404, photo 406) showed an individual with the bright copper cap diagnostic of this form. But on the following pages (406–7, photo 410) one finds a male at a bower in the Baiyer valley with the olive crown of the nominate race. This is not surprising, as the Baiyer drains into the e Sepik and may be a zone of overlap, as it is for *Paradisaea minor* and *P. raggiana*. A definitive image of a red-crowned bird displaying in the bower of a *C. cerviniventris* near Aiome is featured in Gilliard (1969: color fig. VIII, facing p. 71); this bird was collected and is at AMNH.
DIAGNOSIS: Crown exhibits a coppery-rose bloom. See photo in Coates (1990: 404).

NOTES
Other names—Lauterbach's or Yellow-bellied Bowerbird.

Fawn-breasted Bowerbird *Chlamydera cerviniventris*　　　　　RESIDENT | MONOTYPIC
Chlamydera cerviniventris Gould, 1850. In Jardine, *Contr. Orn.*: 106. (also *Proc. Zool. Soc. London*, 1851: 201).—Cape York, n QLD, Australia.

　　　Largely confined to coastal areas of e NG, west in the n watershed to Jayapura, and west in the s watershed to Dolak I of the Trans-Fly (Bishop 2005a). Also known from the Bird's Head (Hoogerwerf 1964) and a few interior populations in the mts of the Huon and SE Peninsulas. Sea level–1,000 m, max 1,800 m (Coates & Peckover 2001, Bishop 2005a). Inhabits patches of scrub and trees in savanna, grassland, and forest edge.

　　　Localities—Kebar valley[Y], Ransiki[Y], Cyclops Mts[X], Ifar on Lake Sentani[J], Humboldt Bay[MA], Dolak I[M], Kurik[Q], Wasur NP[M], Wanggo R[Z], Erambu[Z], Penzara[K], Tarara[K], Lake Daviumbu[K], Gaima on e bank of Fly R[K], Oriomo R[MA], Jimi valley[CO], Madang[MA], Aiyura[CO], Dawong[L], Wau[CO], Garaina[CO], Hall Sound, Sogeri Plateau[CO], Mt Lamington[CO], Collingwood Bay[CO], Mt Orian[CO], Agaun[CO], Alotau[CO], and Samarai (Mayr 1931[L], Rand 1942a[K], 1942b[X], Mees 1964b[Z], 1982b[Q], Ripley 1964a[J], Hoogerwerf 1971[Y], Bishop 2005a[M]). An old De Vis record from Tagula I, Louisiade Arch (Mayr 1941), is almost certainly an error.

　　　Extralimital—Cape York Penin, QLD.

NOTES
It will be interesting to see how the two species of *Chlamydera* interact in areas of overlap. Does one find females regularly visiting bowers of both species? Is there hybridization? *Other names*—Buff-breasted Bowerbird.

Family Climacteridae

Australian Treecreepers　　　　　　　　　　　　　　　　[Family 7 spp/Region 1 sp]

This family is confined to Australia and New Guinea. In shape and habits, the species of Climacteridae resemble the creepers of Eurasia and North America, except that they do not use the tail as a prop when creeping. Ants are the primary food; flight is direct, with rapid wingbeats, and with a swoop toward the end of the flight; toes are long, powerfully clawed, and hemi-syndactyl (Schodde & Mason 1999). Molecular evidence suggests that this family's nearest living relatives are the bowerbirds (Barker *et al.* 2004, Jønsson *et al.* 2011b).

Cormobates Mathews, 1922. *Austral. Avian Rec.* 5: 6. Type, by original designation, *Certhia leucophaea* Latham.

Species of the genus range from eastern Australia to New Guinea. The birds feed and breed in dispersed pairs; egg background color is white. Note presence of the tomial notch on the maxilla and the distinctive sex-specific cheek patch (Schodde & Mason 1999). *Cormobates* is treated as feminine (David & Gosselin 2002b).

Papuan Treecreeper *Cormobates placens*   ENDEMIC

Climacteris placens P. L. Sclater, 1874. *Proc. Zool. Soc. London* (1873): 693.—Hatam, Arfak Mts, Bird's Head.

Patchily distributed in montane forest of the mts of the Bird's Head, the Western Ra, the Border Ra, the w sector of the Eastern Ra, and the mts of the SE Peninsula; 1,250–2,600 m, max 3,000 m[CO]. Absent from the e sector of the Eastern Ra and the outlying ranges (Diamond 1972) but for a record on the ne slopes of Mt Wilhelm (Sam & Koane 2014).

SUBSPECIES

a. *Cormobates placens placens* (P. L. Sclater)

Synonym: *Climacteris placens steini* Mayr, 1936. *Amer. Mus. Novit.* 869: 5.—Mount Sumuri, Weyland Mts, Western Ra.

Synonym: *Climacteris placens inexpectata* Rand, 1940. *Amer. Mus. Novit.* 1072: 11.—9 km ne of Lake Habbema, 2,800 m, Western Ra.

RANGE: Mountains of the Bird's Head, and the Western, Border, and Eastern Ra east to Tari Gap[J] (Gilliard & LeCroy 1961, Ripley 1964a, Frith & Frith 1992[J]). Largely absent east of Tari Gap. Sight records from ne slopes of Mt Wilhelm may refer to this subspecies (Sam & Koane 2014).

DIAGNOSIS: Male has throat whitish merging to buffy gray of upper breast; belly with ochraceous wash; flanks dark brownish gray with buff streaking; lower abdomen spotted buff; undertail is dark-barred (Noske 2007). Neither *steini* nor *inexpectata* is substantially distinct from the nominate form, and hence both are included here.

b. *Cormobates placens meridionalis* (Hartert)

Climacteris placens meridionalis Hartert, 1907. *Bull. Brit. Orn. Club* 21: 27.—Owgarra, Angabanga R, SE Peninsula.

RANGE: Mountains of the SE Peninsula as far west as Wau.

DIAGNOSIS: Male ventral plumage distinctive: entirely dull and dark buff from throat to vent, with indistinct dark edgings to flank feathers; throat rather dark and with speckling.

NOTES

Keast (1957) considered this species to be conspecific with *leucophaea* of Australia. We follow Christidis & Boles (2008) in considering the NG form as distinct from the Australian species. The level of molecular divergence between the visually quite different white-throated western form and the buff-throated form from the SE Peninsula should be checked. *Other names*—White-throated Treecreeper, Papuan Barkcreeper, *Climacteris placens* or *leucophaea*.

Family Maluridae

Fairywrens
[Family 32 spp/Region 6 spp]

Species of the family range widely through Australia and extend to the Aru Is and New Guinea. Fairywrens are small, perky, insect eaters that generally inhabit thickets. The New Guinean species have colonized the lowlands to 2,900 m and occupy grassland, forest edge, and forest interior; they are most numerous in shrubbery and brushy ecotones. All except *Sipodotus wallacii* cock their tails and give high scolding notes when disturbed. Most of the Region's species occur in small, sedentary groups. The passerine phylogeny of Barker *et al.* (2004) placed this lineage as sister to the pardalotes (Pardalotidae) of Australia and the honeyeaters (Meliphagidae); this overall clade is sister to the large Australasian corvine assemblage. Norman *et al.* (2007) and Jønsson *et al.* (2011b) provided additional molecular evidence of the sister relationship between Maluridae and Meliphagidae. We follow Driskell *et al.* (2011) for the generic sequence. Note also that Lee *et al.* (2012) provided a phylogeny not inconsistent with that of Driskell *et al.* (2011), but with several New Guinean forms absent. These two studies revealed three basal clades of fairywrens, all broadbilled: *Sipodotus*, *Clytomyias*, and *Chenorhamphus*. That these older lineages are all endemic to the forests of New Guinea is biogeographically noteworthy.

GENUS *SIPODOTUS*   ENDEMIC | MONOTYPIC

Sipodotus Mathews, 1928. *Bull. Brit. Orn. Club* 48: 83. Type, by original designation, *Todopsis wallacii* G. R. Gray.

This monotypic genus is endemic to the Region. The genus is characterized by the elongate bill, uncocked tail, all-white underparts, and white wing bars. It is sister to *Clytomyias* and *Chenorhamphus* (Driskell *et al.* 2011).

Wallace's Fairywren *Sipodotus wallacii*   ENDEMIC | MONOTYPIC

Todopsis wallacii G. R. Gray 1862. *Proc. Zool. Soc. London* (1861): 429, 434, pl. 43, fig. 2.—Misool I, NW Islands.
> Synonym: *Todopsis coronatus* Gould, 1878. *Birds Austral.* 8.—Aru Is. Name no longer preoccupied by *Malurus coronatus* Gould, 1858.

Patchily distributed in lowland forest and dense second growth across the NG mainland; also the Aru Is[MA], Misool I[MA], ?Salawati[J], and Yapen I[MA] (M. Tarburton's island website[J]); mainly in hills from 100 to 800 m, but also found in adjacent lowlands and locally up to *ca.* 1,200 m[CO]. Perhaps largely absent from the e sector of the Sepik-Ramu (though one record from the Prince Alexander Mts; M. LeCroy *in litt.*), from much of the expanse between the Kikori and the Lakekamu R, and from the n sector of the SE Peninsula to the Huon Penin (Coates 1990).

Localities—Tamrau Mts[W], Arfak Mts[MA], 4 km southwest of Bernhard Camp[Q], Digul R[CO], Mimika R[CO], Foja Mts[Z], Jayapura[Q], Merauke R[Y], Bian Lakes[Y], Wuroi[Y], Lake Daviumbu[K], upper Fly R[J], Palmer Junction on Fly R[K], Nomad R[U], Kiunga area[V], Elevala R[Y], Tabubil[C], Aure R[J], Hunstein Mts[L], Crater Mt[X], Purari basin[J], Brown R[CO], Varirata NP[B], Sibium Mts[A], Kumusi R[D] (Rand 1942a[K], 1942b[Q], Bell 1970b[U], Gilliard & LeCroy 1970[W], Clapp 1987[A], Gregory 1995b[V], Mack & Wright 1996[X], Bishop 2005b[Y], Beehler *et al.* 2012[Z], P. Gregory *in litt.*[C], A. Allison *in litt.*[L], I. Woxvold *in litt.*[J], D. Hobcroft *in litt.*[B], AMNH specimen[D]).

NOTES
We follow Mayr (1941) and Rand & Gilliard (1967) in not recognizing the Aru I form. Material examined in Leiden confirms this decision. Mayr (1941) cited De Vis regarding a possible record from Goodenough I, probably in error. Range of this species in c PNG needs to be better delineated. *Other names*—Wallace's Wren Warbler, Blue-capped Wren, *Todopsis wallacii*.

GENUS *CLYTOMYIAS*

ENDEMIC | MONOTYPIC

Clytomyias Sharpe, 1879. *Notes Leyden Mus.* 1: 31. Type, by original designation, *Clytomyias insignis* Sharpe.

This monotypic genus is restricted to the mountains of New Guinea. The genus is characterized by the lack of blue, white, or black in plumage, the cocked tail, red-brown hood, flattened and broadened bill, and prominent rictal bristles. It is sister to *Chenorhamphus* (Driskell *et al.* 2011).

Orange-crowned Fairywren *Clytomyias insignis*

ENDEMIC

Clytomyias insignis Sharpe, 1879. *Notes Leyden Mus.* 1(10): 31.—Tjobonda, Arfak Mts, Bird's Head.

Mountains of the Bird's Head, Central Ra, and Huon Penin; 1,700–2,700 m, extremes 1,200–3,000 m[CO]. Inhabits patches of dense regrowth and thickets of scrambling bamboo in montane forest.

SUBSPECIES

a. *Clytomyias insignis insignis* Sharpe
RANGE: Mountains of the Bird's Head.
DIAGNOSIS: Throat whitish; rest of underparts pale cream (Rand & Gilliard 1967). Mantle gray-washed.

b. *Clytomyias insignis oorti* Rothschild & Hartert
Clytomyias insignis oorti Rothschild & Hartert, 1907. *Novit. Zool.* 14: 460.—Head of Aroa R, s watershed of the SE Peninsula.
RANGE: The Central Ra and mts of the Huon Penin. Presumably from the Weyland Mts east to Milne Bay. Includes a sight record by Diamond & Bishop from Mt Sisa (K. D. Bishop *in litt.*).
DIAGNOSIS: Similar to nominate form, but underparts darker, and white in throat reduced (Rand & Gilliard 1967). Mantle buff-washed. See photo in Coates (1990: 97) and Coates & Peckover (2001: 127).

NOTES
Other names—Rufous Wren Warbler.

GENUS *CHENORHAMPHUS*

ENDEMIC [Genus 2 spp]

Chenorhamphus Oustalet, 1878. *Bull. Hebd. Assoc. Sci. France* 21: 248. Type, by monotypy, *Chenorhamphus cyanopectus* Oustalet = *Todopsis grayi* Wallace.

The species of this genus are characterized by the broad and flattened bill, weak sexual dichromatism, and prominent rictal bristles. Recognition of this genus as distinct from *Malurus* follows Driskell *et al.* (2011).

Broad-billed Fairywren *Chenorhamphus grayi*

ENDEMIC | MONOTYPIC

Todopsis grayi Wallace, 1862. *Proc. Zool. Soc. London*: 166.—Sorong, w Bird's Head.
> Synonym: *Chenorhamphus piliatus* Reichenow, 1920. *J. Orn.* 68: 399.—Maeanderberg, upper Sepik R, west-central Sepik-Ramu.

Salawati[MA], the Bird's Head and Neck[MA], the NW Lowlands, and the Sepik-Ramu[MA]; sea level–1,100 m, with records to 1,600 m[J] (Stresemann 1923, Coates & Peckover 2001[J]). Frequents small natural openings in rain forest created by streams, fallen trees and landslides; avoids openings in second growth and forest disturbed by human activity.

Localities—Sorong[MA], Sainkedoek and Kalawos in the Tamrau Mts[J], Koor R[J], Andai[L], Mt Moari[L], Wa Samson[MA], Siwi[MA], Rumberpon I[CO], near Triton Bay[O], Menoo R in the Weyland Mts[M], Kasonoweja and Pionierbivak in the Mamberamo basin[J], Van Rees Mts[J], Danau Bira[J], Foja Mts[J], Bodim[K], Cyclops Mts[J], Humboldt Bay[MA], Thurnwald Ra[CO], Puwani R near Vanimo[Q], Maeanderberg[MA], Hunstein Mts[N], near Pagwi[CO], Bewani[J], Torricelli[J] Mts, Prince Alexander Mts[CO] (Hartert *et al.* 1936[M], Ripley 1964[K], Diamond 1981b[J], LeCroy & Diamond 1995[L], A. Allison *in litt.*[N], Beehler field notes[Q], C. Thébaud & B. Mila *in litt.*[O]). Unconfirmed report from Batanta I (M. Tarburton's island website).

Other names—Broad-billed Wren, Broad-billed Wren Warbler.

Campbell's Fairywren *Chenorhamphus campbelli*   ENDEMIC | MONOTYPIC

Malurus campbelli Schodde, 1982. *The Fairywrens*: 32, pl. 3.—Mount Bosavi, south-central NG.

The c sector of the S Lowlands from Kiunga[J] to Mt Bosavi, 100–800 m (Schodde & Weatherly 1982, 1983, Gregory 1995b[J]). Observed at Lake Murray in 2014 (B. J. Coates *in litt.*). Possible sighting from Nomad R by H. L. Bell (Coates 1990). With additional field work, we expect the range of this species to be extended eastward and westward. Apparently very cryptic and difficult to locate in the field (D. Hobcroft *in litt.*).

NOTES

Treated as a subspecies of *C. grayi* by Beehler & Finch (1985) and LeCroy & Diamond (1995). We now follow Schodde (1984) and Driskell *et al.* (2011) in recognizing this isolated vicariant form as a full species, based on molecular distance. Note that because of timing of publications, the lectotype of *C. campbelli* is the male individual in a photo by Campbell, designated as the type by Schodde & Weatherly (1983) subsequent to the publication of the name by Schodde & Weatherly (1982). See Schodde (1984) and LeCroy & Diamond (1995: 192). The species has been looked for without success in the Purari basin (I. Woxvold *in litt.*) and in the Kikori basin (J. M. Diamond and K. D. Bishop *in litt.*).

GENUS *MALURUS*   [Genus 11 spp/Region 2 spp]

Malurus Vieillot, 1816. *Analyse Nouv. Orn. Elem.*: 44. Type, by monotypy, *Motacilla cyanea* Latham.

The species of this genus range widely though New Guinea and Australia (Rowley & Russell 2007). The genus is characterized by reduced rictal bristles, strong sexual dichromatism, presence of black or bright blue in body plumage of male, and cocked tail (Driskell *et al.* 2011).

Emperor Fairywren *Malurus cyanocephalus*   ENDEMIC

Todus cyanocephalus Quoy & Gaimard, 1830. *Voy. Astrolabe, Zool.* 1: 227; 1833 *Atlas*: pl. 5, fig. 4.—Manokwari, Bird's Head.

Inhabits dense undergrowth associated with forest edge and regenerating thickets through NG, Salawati I, Aru Is, and Biak and Yapen I; sea level–750 m[CO], max 1,000 m. Absent from the ne sector of the Mainland from the Ramu R to Milne Bay.

SUBSPECIES

a. *Malurus cyanocephalus cyanocephalus* (Quoy & Gaimard)

> Synonym: *Malurus cyanocephalus dohertyi* Rothschild & Hartert, 1903. *Novit. Zool.* 10: 477.—Takar, NW Lowlands.

RANGE: Salawati I[MA], the Bird's Head and Neck, the NW Lowlands, and the Sepik-Ramu. Presumably the population on Yapen I (King 1979) is of this race.
DIAGNOSIS: Blue mantle feathers of male with a turquoise hue; crown feathers slightly darker. Includes *dohertyi* (Mayr 1941, Ripley 1964a).

b. *Malurus cyanocephalus bonapartii* (G. R. Gray)

Todopsis bonapartii G. R. Gray, 1859. *Proc. Zool. Soc. London*: 156.—Aru Is.
RANGE: The Aru Is[MA], S Lowlands, Trans-Fly, and s watershed of the SE Peninsula to Amazon Bay (Bishop 2005b).
DIAGNOSIS: Blue of male darker, more cobalt and less turquoise, both on crown and on mantle, than in nominate. See photos of male and female in Coates (1990: 91) and Coates & Peckover (2001: 127).

c. *Malurus cyanocephalus mysorensis* (A. B. Meyer)

Todopsis mysorensis A. B. Meyer, 1874. *Sitzungsber. Akad. Wiss. Wien* 69: 74.—Biak I, Bay Is.
RANGE: Biak I.
DIAGNOSIS: Crown of male rich cobalt blue; mantle feathers a deep cobalt blue, marginally darker than mantle of *bonapartii*. Female crown a darker blue than in nominate female (Rand & Gilliard 1967).

It is odd that the two darker-mantled forms can be found in the north (Biak I) and the south (Aru Is and s Mainland), separated by the paler-mantled nominate form. If these two dark-mantled forms were adjacent geographically, we would not hesitate to combine them. *Other names*—Blue Wren Warbler, Imperial Wren, *Todopsis cyanocephala*.

White-shouldered Fairywren *Malurus alboscapulatus*   ENDEMIC

Malurus alboscapulatus A. B. Meyer, 1874. *Sitzungsber. Akad. Wiss. Wien* 69: 496.—Arfak Mts, *ca.* 1,070 m, Bird's Head.

Throughout the NG lowlands and mid-montane grasslands; sea level–2,150 m, max 2,750 m[CO]. Widespread, except in west of range, where apparently patchy. Inhabits a low mix of shrubbery and grass in clearings, old gardens, savanna, and marshes; often present in the vicinity of villages (Coates 1990).

SUBSPECIES

a. *Malurus alboscapulatus alboscapulatus* A. B. Meyer
RANGE: The Bird's Head.
DIAGNOSIS: Female black on mantle but distinguished from male by white underparts (with incomplete black collar); amount of black on flanks varies.

b. *Malurus alboscapulatus aida* Hartert
Malurus alboscapulatus aida Hartert, 1930. *Novit. Zool.* 36: 78.—Ifar, Lake Sentani, ne sector of the NW Lowlands.
    Synonym: *Malurus alboscapulatus balim* Rand, 1940. *Amer. Mus. Novit.* 1072: 5.—Baliem R, 1,600 m, Western Ra.
    Synonym: *Malurus alboscapulatus randi* Junge, 1952. *Zool. Meded. Leiden* 31: 248.—Paniai Lake, Western Ra.
RANGE: The Kumawa Mts of the Bird's Neck (Diamond & Bishop 2015), the NW Lowlands, plus the n slopes of the Western and Border Ra east to Telefomin.
DIAGNOSIS: Female black like male but for white eyebrow and wings with a brownish cast (Rowley & Russell 2007).

c. *Malurus alboscapulatus lorentzi* van Oort
Malurus lorentzi van Oort, 1909. *Nova Guinea, Zool.* 9: 91.—Lorentz R, sw NG.
    Synonym: *Malurus alboscapulatus dogwa* Mayr & Rand, 1935. *Amer. Mus. Novit.* 814: 11. Wuroi, Oriomo R,
        Trans-Fly, s NG.
RANGE: Southern NG and the s watershed of the Western and Border Ra from the Mimika R east to the Trans-Fly. Range includes Dolak I, Bian Lakes, and Lake Daviumbu (Bishop 2005a).
DIAGNOSIS: Female with gray-brown crown; narrow white eyebrow and broken eye-ring; no white on scapulars; mantle mid-brown; cream-white underparts; cinnamon flanks (Rowley & Russell 2007). See photo of female *lorentzi* in Coates (1990: 93–94) and Coates & Peckover (2001: 126, lower image).

d. *Malurus alboscapulatus kutubu* Schodde and Hitchcock
Malurus alboscapulatus kutubu Schodde and Hitchcock, 1968. *CSIRO Wildl. Res. Tech. Pap.* 13: 42.— Lake Kutubu, s watershed of Eastern Ra.
RANGE: Uplands of the s watershed of the Eastern Ra, between Lake Kutubu and Mt Giluwe.
DIAGNOSIS: Female sooty black with brown wings (Schodde & Hitchcock 1968).

e. *Malurus alboscapulatus naimii* D'Albertis
Malurus naimii D'Albertis, 1876. *Ann. Mus. Civ. Genova* (1875) 7: 827.—Mon, Yule I, SE Peninsula.
    Synonym: *Musciparus tappenbecki* Reichenow, 1897. *Orn. Monatsber.* 5: 25.—Yagei, upper Ramu, e sector of
        the Sepik-Ramu.
    Synonym: *Malurus alboscapulatus mafulu* Mayr & Rand, 1935. *Amer. Mus. Novit.* 814: 10.—Mafulu, mts of SE Penin.
RANGE: The Sepik-Ramu, e sector of the Eastern Ra, the S Lowlands east of the Kikori R, and the nw sector of the s watershed of the SE Peninsula. Includes the Adelbert Mts and presumably the PNG North Coastal Ra, and the w sector of the Huon Penin. Occurs west of the range of *moretoni*.
DIAGNOSIS: Female similar to the female nominate but has more black on flanks (Rowley & Russell 2007). Specimens of females from Astrolabe Bay variably marked with white and black ventrally, but throat all white.

f. *Malurus alboscapulatus moretoni* De Vis
Malurus moretoni De Vis, 1892. *Ann. Rept. Brit. New Guinea*, 1890–91: 97.—Bartle Bay, SE Peninsula.
RANGE: The SE Peninsula, west in the n watershed to the Huon Penin; west in the s watershed to Port Moresby (Mayr 1941).
DIAGNOSIS: Adults as for *aida* but small, and immatures with more white in underparts; wing 46; tail 36–47 (Rand & Gilliard 1967). See photos of adult female in Coates (1990: 95) and Coates & Peckover (2001: 126, upper image).

NOTES
Building on the work of Schodde & Weatherly (1982), we have synonymized a series of very similar subspecies based on museum analysis. Adult male plumage is identical throughout. The gradation of immature to adult female plumage confounds analysis of the variable females across populations. It is likely that we are still recognizing too many races. Fergusson I sight record requires confirmation; lack of subsequent sightings there suggests this record is erroneous (see Mackay & Mackay 1974b). *Other names*—Black and White Wren Warbler, White-shouldered Wren.

Family Meliphagidae

Honeyeaters   [Family 178 spp/Region 65 spp]

The species of the family range from the Lesser Sundas eastward to Micronesia, Polynesia, Melanesia, Australia, and New Zealand. This species-rich family dominates the avifauna in much of Australasia. The largest family of birds in the Region, it is diverse in size, color, and habits. It includes both the crow-like Giant Wattled Honeyeater and the tiny Elfin Myzomela—among the largest and smallest songbirds in the Region. Different genera approximate sunbirds, warblers, flycatchers, and jays. Some possess bare wattles and facial skin of various bright colors. Bill shape and size vary considerably. Most produce unmelodious vocalizations. The honeyeaters are generalized songbirds that possess a brush-tipped tongue and attendant musculature for feeding on nectar and pollen (Schodde & Mason 1999). They are often found to be pugnacious and dominant foragers at nectar-producing trees. According to Driskell & Christidis (2004), Barker *et al.* (2004), Gardner *et al.* (2010), and Jønsson *et al.* (2011b) the meliphagid lineage is sister to the Pardalotidae of Australia, and secondarily to the Maluridae of Australia and New Guinea. This overall clade is a sister to the core Corvoidea, which includes the crows, birds of paradise, and monarchs. We present a generic sequence of the Meliphagidae based on Joseph *et al.* (2014b: Fig. 1), further informed by the results of Nyári & Joseph (2011: Fig. 1), Driskell & Christidis (2004: fig 4), Gardner *et al.* (2010: Fig. 1), and Andersen *et al.* (2014a).

GENUS *MYZOMELA*   [Genus 30 spp/Region 9 spp]

Myzomela Vigors & Horsfield, 1827. *Trans. Linn. Soc. London* 15: 318. Type, by original designation, *Meliphaga cardinalis* Vigors & Horsfield = *Certhia sanguinolenta* Latham.

The species of this genus range from Sulawesi east to New Guinea, Australia, Melanesia, and Polynesia (Higgins *et al.* 2008). This species-rich but morphologically compact genus is characterized by the tiny size, fine and decurved bill, dark iris, and plumage patterned with red, black, olive, and brown. Schodde & Mason (1999) also noted the conspicuous presence of red in the plumage, the abbreviated and squared-off tail, egg background color white with reddish-brown spotting, and shallow medial furcation of tongue.

Red-collared Myzomela *Myzomela rosenbergii* ENDEMIC

Myzomela Rosenbergii Schlegel, 1871. *Ned. Tijdschr. Dierk.* (1873) 4: 38.—Arfak Mts, Bird's Head.

Widespread in the mts of NG and Goodenough I; 1,200–3,700 m, extremes 600–3,950 m[CO]. Inhabits montane forest, edge, and alpine shrubbery. No records from the Wandammen Mts.

SUBSPECIES

a. *Myzomela rosenbergii rosenbergii* Schlegel

Synonym: *Myzomela rosenbergii wahgiensis* Gyldenstolpe, 1955. *Arkiv f. Zool.* (2)8(1): 155.—Weiga, s slope of Sepik-Wahgi Divide, Wahgi valley, Eastern Ra.

RANGE: Montane forests of NG, including the mts of the Bird's Head[MA], the Kumawa[A] and Fakfak Mts[K], Central Ra[MA], Foja[J] and Cyclops Mts[MA], PNG North Coastal Ra[L], Adelbert Mts[B], and mts of Huon[MA] (Pratt 1983[B], Gibbs 1994[K], Beehler *et al.* 2012[J], J. M. Diamond *in litt.*[A], specimens at AMNH & BPBM[L]).

DIAGNOSIS: Smaller and shorter-billed than *longirostris*; female dorsal plumage dark sepia brown with pale cinnamon striping; rump and throat red; lores and chin blackish; breast dull tawny brown (Rand & Gilliard 1967). Bill (male) 16–19 (Gyldenstolpe 1955b). The form *wahgiensis* is included here, following Gilliard & LeCroy (1961) and Diamond (1972).

b. *Myzomela rosenbergii longirostris* Mayr & Rand

Myzomela rosenbergii longirostris Mayr & Rand, 1935. *Amer. Mus. Novit.* 814: 12.—Goodenough I, D'Entrecasteaux Arch, SE Islands.

RANGE: Mountains of Goodenough I.

DIAGNOSIS: Larger and longer-billed than the nominate form. Male's red feathers deeper red; female very different, with head and upperparts olive brown with red on nape, hindneck, mantle, and rump (LeCroy & Peckover 1999). Bill (male) 24–25 (LeCroy & Peckover 1999). These multiple and striking differences suggest species-level differentiation, but this decision awaits ongoing molecular study (B. W. Benz *in litt.*).

NOTES

Other names—Red-collared Honeyeater, Black-and-Red Myzomela.

White-chinned Myzomela *Myzomela albigula* ENDEMIC

Myzomela albigula Hartert, 1898. *Bull. Brit. Orn. Club* 8: 20.—Rossel I, Louisiade Arch, SE Islands.

An island endemic confined to the SE Islands, sea level–300 m. Inhabits mainly small islands in the Louisiades and islets west and northwest of the Louisiades. The only large island inhabited is Rossel.

SUBSPECIES

a. *Myzomela albigula pallidior* Hartert

Myzomela pallidior Hartert, 1898. *Bull. Brit. Orn. Club* 8: 21.—Kimuta I, Louisiade Arch, SE Islands (see LeCroy & Peckover 1998).

RANGE: Islets in the SE Islands: Alcester I[M], Bonvouloir Is[MA], Conflict Is[K], Deboyne Is[MA], and Kimuta I[J], Good[L], Haszard[L], Panapompom[L], Rara[L], and Itamarina I[L] (LeCroy & Peckover 1998[J], Dumbacher *et al.* 2010[L], Beehler field notes 28-Sep-2002[K], Pratt *et al.* 2006[M]). Apparently does not inhabit Misima I (LeCroy & Peckover 1998).

DIAGNOSIS: Male overall paler than the nominate form; white of throat extends up onto lower cheek; breast and belly with a whitish midline streak; ventral plumage obscurely dark-streaked.

b. *Myzomela albigula albigula* Hartert

RANGE: Rossel I.

DIAGNOSIS: Male crown dark drab gray-brown with obscure mottling; mantle slightly paler, with reduced mottling; cheek drab gray-brown; chin and throat off-white, with obscure pink stripe up midline; breast and belly drab gray-brown with paler mottling; border between breast band and throat is sharp and well demarcated, with the darkest portion of the breast band abutting the white of the throat.

NOTES

Other names—White-chinned Honeyeater, White-throated Myzomela.

Dusky Myzomela *Myzomela obscura* RESIDENT

Myzomela obscura Gould, 1843. *Proc. Zool. Soc. London* (1842): 136.—Port Essington, NT, Australia.

Confined to the s coast of the Bird's Head and Neck, S Lowlands, SE Peninsula, Aru Is, and Adi and Biak I. Frequents mangroves, woodland, scrub, edge, and village environs (*e.g.*, Daru, Port Moresby, and Alotau; Coates 1990). On the SE Peninsula inhabits the s watershed and the e sector of the n watershed. Found mostly in the lowlands, max 600 m[CO]. An old von Rosenberg record for Misool I was dismissed by Mees (1965) as from the Aru Is.

Extralimital—n and e Australia and Moluccas; also islands in Torres Strait.

SUBSPECIES

a. *Myzomela obscura fumata* (Bonaparte)

Ptilotis fumata Bonaparte, 1850. *Consp. Gen. Av.* 1: 392.—Utanata R, sw NG.

Synonym: *Myzomela obscura aruensis* Kinnear, 1924. *Bull. Brit. Orn. Club* 44: 69.—Aru Is.

RANGE: The Aru Is, the w Bird's Head[J], and s NG east to Milne Bay (Mayr 1941, Gyldenstolpe 1955b[J]); also along the n coast of the SE Peninsula from Milne Bay northwest to Cape Nelson and Popondetta. Should be looked for on Salawati I, following a report from there (Beckers *et al.* in M. Tarburton's island website). The Misool I record (Mayr 1941) apparently is an error (Mees 1965).

DIAGNOSIS: Overall plumage a cold gray-brown dorsally and ventrally; darker than nominate, and blackish throat patch reduced (Higgins *et al.* 2008). The form *aruensis* is identical to this form. See photograph of *fumata* in Coates (1990: 241).

b. *Myzomela obscura rubrobrunnea* A. B. Meyer

Myzomela rubrobrunnea A. B. Meyer, 1874. *Sitzungsber. Akad. Wiss. Wien* 70: 203.—Biak I, Bay Is.

RANGE: Biak I

DIAGNOSIS: Darker than nominate, with a brownish-red wash to plumage and a reddish throat stripe (Higgins *et al.* 2008). This taxon was treated as a full species by Mayr & Meyer de Schauensee (1939a), who stated that it was better to "list this form as a good species, until the relationship ... is better understood." Mayr (1941) subsequently indicated the uncertainty of treating this taxon as a subspecies of *obscura*, by putting the species name *obscura* in parentheses. We await molecular results before elevating this form to species.

NOTES

Other names—Dusky or Red-brown Honeyeater, Moluccan Myzomela.

Ruby-throated Myzomela *Myzomela eques* ENDEMIC

Cinnyris eques Lesson & Garnot, 1827. *Bull. Sci. Nat. Férussac* 11: 386.—Waigeo I, NW Islands.

Mainland NG and the NW Islands; sea level–500 m, max 1,310 m (Stresemann 1923, Freeman & Freeman 2014). Widely distributed on the Mainland except, apparently, for much of the Trans-Fly. Inhabits rain forest and edge; most often found at flowering trees.

SUBSPECIES

a. *Myzomela eques eques* (Lesson & Garnot)

RANGE: Misool, Salawati, and Waigeo I; also the Bird's Head and Neck (Mayr 1941).

DIAGNOSIS: Male dorsally dark dun brown, darkest on crown; throat and breast dark dun brown, the former marked by a bright red midline stripe that extends up onto chin; belly slightly paler than breast, dark of breast forming an obscure band.

b. *Myzomela eques nymani* Rothschild & Hartert

Myzomela eques nymani Rothschild & Hartert, 1903. *Novit. Zool.* 10: 223.—Simbang, Huon Penin.

RANGE: The S Lowlands, Trans-Fly, and all of the SE Peninsula. Within the Trans-Fly recorded from Wasur NP and along the Maro R, and in PNG from Sturt I Camp on the Fly R (Bishop 2005a). Within the S Lowlands of PNG the only records are from Palmer Junction and in the area around Kopi Camp on the Kikori R (Bishop 2005b).

DIAGNOSIS: Male very similar to nominate form but for broader red throat patch. Mayr (1941) spelled this race "*nymanni*," which is in error. The race is named for E.O.A. Nyman, who collected on the Huon

Penin, and the subspecies was correctly spelled *nymani* in the original description by Rothschild & Hartert (LeCroy 2011).

c. *Myzomela eques primitiva* Stresemann & Paludan
Myzomela eques primitiva Stresemann & Paludan, 1932. *Novit. Zool.* 38: 143.—Gratlager, middle Sepik, Sepik-Ramu.
RANGE: The n sector of the Bird's Neck, the NW Lowlands, and the Sepik-Ramu.
DIAGNOSIS: Darker plumaged than nominate form and with a smaller red throat patch.

d. *Myzomela eques karimuiensis* Diamond
Myzomela eques karimuiensis Diamond, 1967. *Amer. Mus. Novit.* 2284: 8.—Karimui, 1,110 m, Eastern Ra.
RANGE: Known only from Karimui, in the se sector of the Eastern Ra. Probably will be found throughout the uplands of the Purari drainage.
DIAGNOSIS: Darkest plumage of all races, with throat patch deeper red (Diamond 1967). An examination of the holotype indicates this is a thinly delineated form. By distribution, one would expect the s watershed of the Eastern Ra to support a continous population of *nymani*, which ranges across the s side of the island. The biogeographic placement of *karimuiensis* is curious. Perhaps this form is a product of gene flow between the darker *primitiva* from the n watershed and *nymani* from the s watershed. Alternatively, it may be that this form is a dark ecomorph—a product of the high-rainfall environment in which it lives in the upper Purari region.

NOTES
The form *cineracea*, from New Britain and Umboi I, we treat as specifically distinct from *eques*, following Mayr & Diamond (2001). We have chosen to alter the traditional English name from Red-throated Myzomela (*cf.* Beehler *et al.* 1986) to Ruby-throated Myzomela in order to highlight this species' most striking plumage feature. *Other names*—Red-throated Honeyeater or Myzomela.

Red Myzomela *Myzomela cruentata* RESIDENT
Myzomela cruentata A. B. Meyer, 1874. *Sitzungsber. Akad. Wiss. Wien* 70: 202.—Arfak Mts, Bird's Head.
    The Central Ra and virtually all uplands of NG and Yapen I, 750–1,450 m[CO]. Outlying ranges include the Arfak[J,K], Fakfak[L], Foja[M], Cyclops[J], PNG North Coastal Ra[O], Adelbert Mts[P,Q], and mts of the Huon Penin[CO] (Hartert 1930[J], Gyldenstolpe 1955a[K], Gilliard & LeCroy 1967[P], Pratt 1983[Q], Gibbs 1994[L], Beehler *et al.* 2012[M], specimens at AMNH & BPBM[O]). In Nov 2014 seen twice at canopy flowers at 50 m near Karawawi at the base of the s Kumawa Mts (C. Thébaud & B. Mila *in litt.*). No records from the Wandammen Mts. Inhabits forest, edge, and, less commonly, eucalypt savanna (Coates 1990). Seen mainly at canopy flowers in forest.
    Extralimital—Bismarck Arch (Mayr & Diamond 2001).

SUBSPECIES
a. *Myzomela cruentata cruentata* A. B. Meyer
RANGE: Yapen I and all NG (Mayr 1941).
DIAGNOSIS: Male plumage bright, glossy crimson, brightest on rump and uppertail coverts (Higgins *et al.* 2001).

NOTES
Other names—Red-tinted Myzomela, Red Honeyeater.

Papuan Black Myzomela *Myzomela nigrita* ENDEMIC
Myzomela nigrita G. R. Gray, 1858. *Proc. Zool. Soc. London*: 173.—Aru Is.
    Uplands throughout NG; lowlands of the Trans-Fly; Waigeo I, Aru Is, Yapen and Mios Num I; also the D'Entrecasteaux Arch, Louisiade Arch, and Woodlark I; 600–1,000 m, extremes sea level–1,250 m[CO]. On NG mainland a species of forest interior canopy. No records from the Fakfak, Kumawa, Wandammen, or Cyclops Mts.

SUBSPECIES

a. *Myzomela nigrita steini* Stresemann & Paludan

Myzomela nigrita steini Stresemann & Paludan, 1932. *Orn. Monatsber.* 40: 14.—Waigeo I, NW Islands.

RANGE: Waigeo I. Reported from Batanta by S. Hogberg in Mar-2006 (M. Tarburton's island website).

DIAGNOSIS: Both sexes hen-plumaged. Molecularly not divergent (B. W. Benz *in litt.*).

b. *Myzomela nigrita nigrita* G. R. Gray

Synonym: *Myzomela meyeri* Salvadori, 1881. *Orn. Pap. Moluc.* 2: 292.—Rubi, Bird's Neck. Replacement name for *Myzomela erythrocephala*, Meyer, 1874, *Sitzungsber. Akad. Wiss. Wien* 70: 204; preoccupied by *Myzomela erythrocephala* Gould, 1840.

Synonym: *Myzomela nigrita louisiadensis* Hartert, 1898. *Novit. Zool.* 5: 527.—Tagula I, Louisiade Arch, SE Islands.

RANGE: All NG, the Aru Is[MA], Yapen I[MA], and the Louisiade Arch[MA]. Within the s Trans-Fly recorded from along the Merauke R and Bian Lakes, and in PNG from Bensbach, Morehead, Tarara, and Wuroi (K. D. Bishop *in litt.*).

DIAGNOSIS: Male plumage shiny black; female dull buffy brown with rose wash on face. The form *meyeri* is not sufficiently distinct to merit recognition; slight cline in size. The form *louisiadensis* was described from a small series exhibiting size intermediate between *pluto* and *nigrita*. It does not merit recognition as it is only 8% larger as measured by compared wing length.

c. *Myzomela nigrita pluto* Forbes

Myzomela pluto W. A. Forbes, 1879. *Proc. Zool. Soc. London*: 266.—Mios Num I, Bay Is.

RANGE: Mios Num I.

DIAGNOSIS: Larger than all other populations. Wing (male) 63.5–68.6 (Higgins *et al.* 2008).

d. *Myzomela nigrita forbesi* Ramsay

Myzomela forbesi Ramsay, 1880. *Proc. Linn. Soc. NSW* (1879) 4: 469.—Fergusson I, D'Entrecasteaux Arch, SE Islands.

RANGE: Fergusson, Goodenough, Normanby[J,K], and Dobu I (Mayr 1941, LeCroy *et al.* 1984[J], Pratt specimen at BPBM[K]).

DIAGNOSIS: Male with broad red patch atop crown. Female with some red on crown (Higgins *et al.* 2008).

NOTES

We have carried out some consolidation of races. Our choice of English name reflects the need to distinguish this species geographically from *M. pammelaena* of the Bismarck Arch. *Other names*—Black Myzomela, Carbon or Blackened Honeyeater.

Red-headed Myzomela *Myzomela erythrocephala*   RESIDENT

Myzomela erythrocephala Gould, 1840. *Proc. Zool. Soc. London* (1839): 144.—King Sound, nw Australia.

Inhabits mangroves in the Aru Is and coastal Mainland from the w sector of S Lowlands, Trans-Fly, and s watershed of SE Peninsula eastward to Domara, east of Cape Rodney (Tolhurst 1988); near sea level[CO].

Extralimital—n tier of Australia. The form *dammermani* of the Lesser Sundas is now treated as a distinct species (Dickinson & Christidis 2014).

SUBSPECIES

a. *Myzomela erythrocephala infuscata* Forbes

Myzomela infuscata Forbes, 1879. *Proc. Zool. Soc. London*: 263.—Aru Is.

RANGE: The Aru Is and s NG between the Mimika R and the mouth of the Fly R; also Daru (Mayr 1941). The Triton Bay record is apparently in error (see Mees 1982b: 151). Within the s Trans-Fly recorded from Dolak I (Nash unpubl.), Wasur NP (K. D. Bishop field notes), Bensbach (P. Gregory *in litt.*), and Daru I (K. D. Bishop *in litt.*).

DIAGNOSIS: Mantle of male blackish brown; blackish-brown breast band 7 mm wide and contrasting with the dark gray-brown toward the lower part of the breast and flanks. Size large; wing (male) 61–64 (Schodde & Mason 1999).

b. *Myzomela erythrocephala ?erythrocephala* Gould

RANGE: Northern Territory (Australia), n QLD, islands of Torres Strait, and apparently the SE Peninsula, Hall Sound east to Domara, near Cape Rodney (Mayr 1941).

DIAGNOSIS: Mantle and narrow breast band sooty gray-brown; mid-breast and flanks paler buff gray. Size small; wing (male) 57–62 (Schodde & Mason 1999). Schodde & Mason (1999) withhold judgment on whether the birds from the SE Peninsula merit being treated as the nominate form, suggesting they may be assignable to *infuscata*.

NOTES

Other names—Mangrove Red-headed Myzomela or Honeyeater.

Elfin Myzomela *Myzomela adolphinae* ENDEMIC | MONOTYPIC

Myzomela adolphinae Salvadori, 1876. *Ann. Mus. Civ. Genova* (1875) 7: 946.—Arfak Mts, Bird's Head.

Patchily distributed through the mts of the Mainland: the Bird's Head, the e sector of the Border Ra, the Eastern Ra, the mts of the SE Peninsula, the PNG North Coastal Ra[J], and the mts of Huon Penin (Mayr 1941, Diamond specimens at AMNH[J]); 1,100–1,950 m, min 500 m[CO]. Apparently absent from the Western Ra and the w and c sectors of the Border Ra. Inhabits montane forest and edge, and tall trees in native gardens. Often seen in the vicinity of upland village environs.

Localities—Tamrau Mts[L], Arfak Mts[MA], Kyaka Territory[DI], Baiyer valley[DI], Nondugl[DI], Kubor Mts[J], YUS[L], Junzaing[K], Ogeramnang[K], and Wau[M], Varirata NP[M] (Mayr 1931[K], Mayr & Gilliard 1954[J], Beehler field notes[L], P. Gregory *in litt.*[M]). A single sighting from the Adelbert Mts (Mackay 1991) requires confirmation, as it was not recorded there by Pratt (1983).

NOTES

The absence of this species from the entire w half of the main body of NG is unusual; a similar distribution is found in *Zosterops novaeguineae*. Our English name highlights the tiny and sprite-like nature of this smallest of the genus. *Other names*—Mountain Red-headed Honeyeater or Myzomela, Red-headed Mountain Myzomela, Mountain Myzomela.

Sclater's Myzomela *Myzomela sclateri* RESIDENT | MONOTYPIC

Myzomela sclateri Forbes, 1879. *Proc. Zool. Soc. London*: 265, pl. 25, fig. 2.—Palakuru/Credner I, off n coast of New Britain.

A small-island specialist recorded in the Region only from Karkar I[MA], n of Madang (e Sepik-Ramu), where it can be found in all wooded habitats. Absent from the NG mainland, but to be looked for on any small island off the n coast of NG. On Karkar, mainly 1,000–1,830 m, but visits flowering trees at sea level. Apparently does not occur on Bagabag I (Diamond & LeCroy 1979).

Extralimital—small islands off the coasts of n New Britain: Crown, Long, Umboi, Tolokiwa, Bali, Witu, islands in Kimbe Bay, Talele, Watom, Nanuk, and Palakuru I (Coates 1990, Dutson 2011).

NOTES

This is a classic tramp species that inhabits species-poor oceanic islands (Diamond 1974). *Other names*—Scarlet-bibbed Myzomela or Honeyeater, Scarlet-throated Honeyeater.

GENUS *XANTHOTIS* [Genus 4 spp/Region 2 spp]

Xanthotis Reichenbach, 1852. *Handb. Spec. Orn., Abth.* 2, Meropinae, continuatio IX: 139. Type, by monotypy, *Xanthotis flaviventris* Reichenbach = *Myzantha flaviventer* Lesson.

The species of this genus range from New Guinea to northeastern Australia (Higgins *et al.* 2008, Andersen *et al.* 2014a). The genus is characterized by the complex facial pattern of bare skin and the subocular, postocular, and auricular streaks; the bill is medium length and pointed, with a slight decurve. *Xanthotis* is treated as masculine (David & Gosselin 2002b).

Tawny-breasted Honeyeater *Xanthotis flaviventer* RESIDENT

Myzantha flaviventer Lesson, 1828. *Man. Orn.* 2: 67.—Manokwari, Bird's Head. Replacement name for *Philedon chrysotis* Lesson & Garnot, preoccupied.

Inhabits forest and edge of lowlands and hills through NG, the NW Islands, the Aru Is, Yapen I, and the D'Entrecasteaux and Trobriand Is, sea level–1,600 m (Mayr 1941, Freeman & Freeman 2014).

Extralimital—n Torres Strait Is (Boigu and Saibai), and Cape York Penin, QLD.

SUBSPECIES

a. *Xanthotis flaviventer fusciventris* Salvadori

Xanthotis fusciventris Salvadori, 1876. *Ann. Mus. Civ. Genova* (1875) 7: 947.—Batanta I, NW Islands.

RANGE: Waigeo and Batanta I (Mayr 1941).

DIAGNOSIS: Similar to the nominate form but with more extensive bright yellow and ocher wash on breast, and dull and cold gray-brown belly and undertail; dorsal plumage washed heavily with bright green (see Greenway 1966: 21).

b. *Xanthotis flaviventer flaviventer* (Lesson)

> Synonym: *Xanthotis chrysotis austera* Ripley, 1957. *Postilla* Yale Peabody Mus. 31: 4.—Tamulol, Misool I, NW Islands.
> Synonym: *Xanthotis rubiensis* A. B. Meyer, 1884. *Zeitschr. Ges. Orn.* 1: 289.—Rubi, head of Geelvink Bay, Bird's Neck.

RANGE: Misool and Salawati I, Bird's Head, and Bird's Neck (Mayr 1941).

DIAGNOSIS: Chin gray; throat and breast brown with a strong yellow wash; belly rich brown without yellow wash; crown and back dark brown with a greenish wash; underwing bright rufous; face with bare skin below and behind eye; also black feathers below eye, yellow ear stripe, and white postocular streak. Mees (1965) synonymized the form *austera* with this form.

c. *Xanthotis flaviventer saturatior* (Rothschild & Hartert)

Ptilotis chrysotis saturatior Rothschild & Hartert, 1903. *Novit. Zool.* 10: 445.—Wanumbai, Kobroor I, Aru Is.

> Synonym: *Xanthotis chrysotis giulianettii* Mayr, 1931. *Mitt. Zool. Mus. Berlin*, 17: 663.—Avera, Aroa R, s watershed of the SE Peninsula.
> Synonym: *Meliphaga flaviventer tararae* Salomonsen, 1966a. *Breviora* Mus. Comp. Zool. Harvard 254: 6.—Tarara, Wassi Kussa R, Trans-Fly.

RANGE: The Aru Is[MA], the Trans-Fly, the S Lowlands, and s watershed of the SE Penin to the Aroa R. Also ranges into islands of the Torres Strait (Saibai, Boigu, and Daru I). Range of this race includes Lake Kutubu, Wahgi valley, Aseki, and presumably around Karimui (although see Diamond 1972: 374 for another interpretation). Widespread in Trans-Fly (Bishop 2005a).

DIAGNOSIS: Yellow and green wash entirely lacking; throat pale gray; breast buff-tan, grading to richer tan on flanks; dorsal plumage olive brown; face pattern as for nominate. Nape distinctly spotted off-white; abdomen faintly spotted off-white (Schodde & Mason 1999). We follow Rand (1938b) and Schodde & Mason (1999) in subsuming the form *tararae* into *saturatior*. See photos of *saturatior* in Coates (1990: 276) and Coates & Peckover (2001: 177).

d. *Xanthotis flaviventer meyeri* Salvadori

Xanthotis meyeri Salvadori, 1876. *Ann. Mus. Civ. Genova* (1875) 7: 947.—Yapen I, Bay Is. Not *Ptilotis pyrrotis* Lesson, 1840 (indeterminable).

> Synonym: *Xanthotis flaviventer philemon* Stresemann, 1921. *Anz. Orn. Ges. Bayern* 1: 35.—Malu, Sepik R, c sector of the Sepik-Ramu.

RANGE: Yapen I[MA], the NW Lowlands, and the Sepik-Ramu. Presumably includes the n foothills of the Western, Border, and Eastern Ra.

DIAGNOSIS: Duller and plainer than other races; entire dorsal surface dark drab brown; crown and sides of head blackish brown (Higgins *et al.* 2008). Crown dun brown; mantle slightly richer; no yellow or green in plumage; no ventral mottling; bare facial patch reduced; yellow postocular streak reduced and flame orange in some birds. Intergrades with *madaraszi* eastward into Morobe Prov (R. Schodde, as reported in Coates 1990).

e. *Xanthotis flaviventer madaraszi* (Rothschild & Hartert)

Ptilotis chrysotis madaraszi Rothschild & Hartert, 1903. *Novit. Zool.* 10: 446.—Simbang, Huon Penin.

RANGE: The Huon Penin, Markham valley, and the upper Watut R (south to Wau).

DIAGNOSIS: Similar to *meyeri* but paler dorsally, and with olive on mantle (Higgins *et al.* 2008). Facial stripe small, and face dull.

f. *Xanthotis flaviventer visi* (Hartert)

Ptilotis visi Hartert, 1896. *Novit. Zool.* 3: 15.—Mailu distr, 40 km w of Orangerie Bay.

    Synonym: *Xanthotis chrysotis kumusii* Mayr 1931 *Mitt. Zool. Mus. Berlin* 17: 663.—Kumusi R, SE Peninsula.

RANGE: The SE Peninsula, northwest to Morobe in the n watershed and west in the s watershed to the Aroa R, where it meets the easternmost edge of the range of *saturatior*.

DIAGNOSIS: Similar to *saturatior* but with much larger yellow facial stripe, more rufescent mantle; slight yellow-green wash on breast; belly rich brown; crown and nape with yellow wash that contrasts the rufous mantle. We include the race *kumusii* in *visi* based on Schodde & Mason (1999) and Higgins *et al.* (2008).

g. *Xanthotis flaviventer spilogaster* (Ogilvie-Grant)

Ptilotis spilogaster Ogilvie-Grant, 1896. *Ibis*: 251.—Fergusson I, D'Entrecasteaux Arch, SE Islands.

RANGE: The D'Entrecasteaux and Trobriand Is (Mayr 1941).

DIAGNOSIS: Similar to *filiger* of QLD but with obscure pale yellowish-white spots on breast and belly; white-feathered face patch covers bare skin; white feathers encompass rear half of ocular and all of post-ocular region; reduced yellow streak below the white patch; crown dun brown.

NOTES

This variable species shows considerable regional clinal variation in body plumage and facial pattern that is difficult to capture in distinct subspecies. We have reduced the named subspecies to a workable number (following, in part, Schodde & Mason 1999: 228). Further study may reduce this number even more. The race *mayeri* Rothschild, treated by LeCroy (2011), had been synonymized by Mayr (1941) into the race *rubiensis*, which here is synonymized into the nominate form. This checklist does not treat forms that were already in synonymy by the time of Mayr (1941). With regard to the discussion of LeCroy (2011), there is no need for an additional subspecies covering the Weyland Mts, given the availability of the races *meyeri* (n watershed of w NG) and *saturatior* (s watershed of w NG). Placement of this and the following species in *Xanthotis* follows Mayr (1941) and Higgins *et al.* (2008). *Other names*—Brown or Tawny-breasted Xanthotis, *Meliphaga flaviventer*, *Xanthotis chrysotis*.

Spotted Honeyeater *Xanthotis polygrammus*   ENDEMIC

Ptilotis polygramma G. R. Gray, 1862. *Proc. Zool. Soc. London* (1861): 429.—Waigeo I, NW Islands.

    Inhabits rainforest and foothill monsoon forest through NG and the NW Islands, 500-1,400 m[CO]; also found in lowlands around Port Moresby and the Trans-Fly, where it ventures into dense savanna (Coates 1990). No records from the Wandammen, Fakfak, or Kumawa Mts.

SUBSPECIES

a. *Xanthotis polygrammus polygrammus* (G. R. Gray)

RANGE: Waigeo I.

DIAGNOSIS: Much reduced yellow auricular feather stripe. Blackish crown; whitish speckling on nape, and pale spotting on mantle and wing coverts.

b. *Xanthotis polygrammus poikilosternos* A. B. Meyer

Xanthotis poikilosternos A. B. Meyer, 1874. *Sitzungsber. Akad. Wiss. Wien* 70: 112 Andai, Bird's Head.

    Synonym: *Xanthotis polygramma kuehni* Hartert, 1930. *Novit. Zool.* 36: 49.—Misool I, NW Islands.

RANGE: Misool I, Salawati I, the Bird's Head, and the s scarp of the Central Ra east at least to the Border Ra of PNG (Mayr 1941, Gilliard & LeCroy 1961). Unconfirmed sight record from Batanta I (M. Tarburton's island website); the species should be looked for there—either this race or the nominate form.

DIAGNOSIS: Crown dark olive with blackish streaking; bright yellow tuft at bend of wing; auricular patch very rich yellow; throat off-white mottled with pale gray. Mantle without the prominent pale spotting. We follow Mees (1965) in subsuming *kuehni* into *poikilosternos*.

c. *Xanthotis polygrammus septentrionalis* Mayr

Xanthotis polygramma septentrionalis Mayr, 1931. *Mitt. Zool. Mus. Berlin* 17: 665.—Maeanderberg, upper Sepik R, west-central Sepik-Ramu.

RANGE: The NW Lowlands and Sepik-Ramu; includes the Cyclops and Adelbert Mts as well as the mts of the Sepik region. The Foja Mts birds have not been assigned to race but may belong here (Beehler *et al.* 2012).

DIAGNOSIS: Auricular stripe essentially absent. Very similar to *lophotis*.

d. *Xanthotis polygrammus lophotis* Mayr

Xanthotis polygramma lophotis Mayr, 1931. *Mitt. Zool. Mus. Berlin* 17: 664.—Junzaing, Huon Penin.
> Synonym: *Xanthotis polygramma candidior* Mayr & Rand, 1935. *Amer. Mus. Novit.* 814: 15.—Wuroi, Oriomo R, Trans-Fly.

RANGE: Mountains of the Huon and SE Peninsulas, the Eastern Ra, and the Trans-Fly. Range includes Karimui[DI], Wasur NP, Bensbach, Lake Daviumbu, and the Maro R of the Trans-Fly (Bishop 2005a).

DIAGNOSIS: Feathering of crown with abundant blackish streaks, making crown very dark (marginally darker than *poikilosternos*). A thinly delineated subspecies. See photos in Coates (1990: 278) and Coates & Peckover (2001: 176).

NOTES
Other names—Spotted Xanthotis, Gray's or Many-spotted Honeyeater, *Meliphaga polygramma*.

GENUS *PHILEMON* [Genus 15 spp/Region 5 spp]

Philemon Vieillot, 1816. *Analyse Nouv. Orn. Elem.*: 47. Type, by subsequent designation (G. R. Gray, 1840. *List Gen. Birds*: 15), *Merops moluccensis* Gmelin.

The species of this genus range from Timor and Halmahera east to New Guinea, the Bismarck Archipelago, Australia, and New Caledonia (Higgins *et al.* 2001). The genus is characterized by the medium to large size, very drab and dull gray-brown plumage (in many instances with dull silvery highlights or patches), strong and pointed black bill, blackish facial skin, and elongate body form. Young birds show yellow flecking on sides of breast.

Meyer's Friarbird *Philemon meyeri* ENDEMIC | MONOTYPIC

Philemon meyeri Salvadori, 1878. *Ann. Mus. Civ. Genova* 12: 339.—Rubi, head of Geelvink Bay, Bird's Neck.

Inhabits rain forest, swamp forest, edge, and tall second growth through NG west to the Bird's Neck in the Fakfak and Kumawa Mts[J] (Stresemann 1923, Diamond 1985[J]); sea level–500 m, max 1,200 m. It seems odd that there are no records of this widespread lowland species from the Bird's Head. Reports by birders from Batanta (Matheve & van den Schoor 2009) and Salawati (Beckers *et al.* in M. Tarburton's island website) should be confirmed. There are Trans-Fly records from Wasur NP, Bian Lakes, the Bensbach area, and Morehead; widespread in the S Lowlands (Bishop 2005a).

NOTES
For notes on the issues surrounding the naming of this species, see LeCroy (2011: 120–21).
Other names—Dwarf Friarbird.

Brass's Friarbird *Philemon brassi* ENDEMIC | MONOTYPIC

Philemon brassi Rand, 1940. *Amer. Mus. Novit.* 1072: 13.—Bernhard Camp, 50 m elevation, Taritatu R, s sector of the NW Lowlands.

Known from only the Mamberamo basin of the NW Lowlands, where it has been recorded from various sectors of this vast inland depression: Bernhard Camp on the Taritatu R, Tirawiwa R and Logari R of the Wapoga watershed, the lower Mamberamo R, and possibly on the Rouffaer R (Rand 1942b, Mack & Alonso 2000: 127, Higgins *et al.* 2008). Known from riparian tall cane grass and riparian second growth.

Discovered by the Third Archbold Expedition in 1939, this remains one of NG's little-known species and merits additional study.

Little Friarbird *Philemon citreogularis* RESIDENT

Tropidorhynchus citreogularis Gould, 1837. *Synops. Birds Austral.* 1: pl. 13, fig. 1; also 1837, *Proc. Zool. Soc. London* (1836): 143.—Interior of NSW, Australia.

Resident in the Trans-Fly, inhabiting monsoon forest and savanna woodland (Mees 1982b, Bishop 2005a).

Extralimital—n, e, and se Australia.

SUBSPECIES

a. *Philemon citreogularis papuensis* Mayr & Rand

Philemon citreogularis papuensis Mayr & Rand, 1935. *Amer. Mus. Novit.* 814: 15.—Dogwa, Oriomo R, s NG.

RANGE: The Trans-Fly. Localities—Kurik[J], Kumbe R[J], Wasur NP, Bensbach, Dogwa on the Oriomo R[J], and Tarara[K] (Rand 1942a[K], Mees 1982b[J]).

DIAGNOSIS: Smaller and darker than the nominate form, with bolder white feather tips on breast; iris gray, not brown (Higgins *et al.* 2008). Bare skin around eye pale blue-gray (Mees 1982b). The subspecies name is correctly spelled *papuensis* (Dickinson & Christidis 2014), not *papuanus* (*pace* Salomonsen 1967: 406). Wing 118–124 (Rand & Gilliard 1967).

NOTES

Other names—Yellow-throated Friarbird, Little Leatherhead.

Helmeted Friarbird *Philemon buceroides* RESIDENT

Philedon buceroides Swainson, 1837. *Anim. Menag.*: 325.—Timor, Lesser Sundas.

A commonplace and vocal resident in the lowlands and hills of NG, the NW Islands, Aru Is, Yapen I, Adi and Daru I, D'Entrecasteaux Arch, Trobriand Is, and Tagula I; sea level–1,000 m, max 2,170 m (Coates & Peckover 2001). At higher elevations confined to town and garden habitats of interior upland valleys. In the lowlands can be found in riparian forest.

Extralimital—Lesser Sundas and n and ne Australia.

SUBSPECIES

a. *Philemon buceroides novaeguineae* (S. Müller)

Tropidorhynchus Novae Guineae S. Müller, 1843. *Verh. Nat. Gesch. Ned., Land-en Volkenk.* 1: 153.—Utanata R, Bird's Neck.

> Synonym: *Philemon novaeguineae brevipennis* Rothschild & Hartert, 1913. *Novit. Zool.* 20: 513.—Southern slopes of the Western Ra. LeCroy (2011) notes the type series of this race was probably collected from the Setekwa R, a tributary of the Otakwa R, w sector of the S Lowlands.
> Synonym: *Philemon novaeguineae fretensis* Salomonsen, 1966. *Breviora* Mus. Comp. Zool. Harvard 254: 9.—Hall Sound, s watershed, SE Peninsula.
> Synonym: *Philemon novaeguineae trivialis* Salomonsen, 1966. *Breviora* Mus. Comp. Zool. Harvard 254: 9.—Collingwood Bay, n watershed of SE Peninsula.

RANGE: Kofiau, Misool, Salawati, Batanta, and Waigeo I; the Bird's Head and all s NG; also the SE Peninsula: all of the s watershed and the n watershed west to the Kumusi R (Mayr 1941).

DIAGNOSIS: Head plumage dark, and knob on bill small (Rand & Gilliard 1967). Mees (1982b) synonymized *brevipennnis*, showing it had essentially the same type locality as *novaeguineae*. Diamond (1972) synonymized *fretensis*. We follow LeCroy's (2011) suggestion that *trivialis* is based on trivial distinctions and merits folding into *novaeguineae*. See photos of *novaeguineae* in Coates (1990: 253) and Coates & Peckover (2001: 168).

b. *Philemon buceroides aruensis* (A. B. Meyer)

Tropidorhynchus aruensis A. B. Meyer, 1884. *Zeitschr. Ges. Orn.* 1: 216.—Aru Is.

RANGE: The Aru Is.

DIAGNOSIS: Very large knob atop bill; otherwise thinly distinct from the nominate form (Mees 1982b). Rand & Gilliard (1967) mistakenly stated that this form exhibits a small knob.

c. *Philemon buceroides jobiensis* (A. B. Meyer)

Tropidorhynchus jobiensis A. B. Meyer, 1874. *Sitzungsber. Akad. Wiss. Wien* 70: 113.—Ansus, Yapen I, Bay Is.

RANGE: Yapen I and n NG from the Mamberamo R east to Salamaua (s verge of Huon Gulf). The races *jobiensis* and *novaeguineae* probably meet and intergrade between Salamaua and the Kumusi R.

DIAGNOSIS: Prominent pale hindneck ruff, and much-reduced knob on bill.

d. *Philemon buceroides subtuberosus* Hartert

Philemon novaeguineae subtuberosus Hartert, 1896. *Novit. Zool.* 3: 238.—Fergusson I, D'Entrecasteaux Arch, SE Islands.

RANGE: The D'Entrecasteaux Arch and Trobriand Is (Mayr 1941)

DIAGNOSIS: Very similar to nominate form, but knob atop bill reduced slightly.

e. *Philemon buceroides tagulanus* Rothschild & Hartert

Philemon novaeguineae tagulanus Rothschild & Hartert, 1918. *Novit. Zool.* 25: 319.—Tagula I, Louisiade Arch, SE Islands.

RANGE: Tagula I.

DIAGNOSIS: Substantially the smallest race; bill small; knob on bill small.

NOTES

Although many subspecies have been described across the range of the species, the racial variation in the NG region is rather thin and clinal, and only the forms from Tagula I and the Aru Is are particularly distinctive. Higgins *et al.* (2008) noted that the populations from the Lesser Sundas have pale plumage, whereas those from NG have darker plumage. Schodde & Mason (1999) and Christidis & Boles (2008) lumped the species *novaeguineae* into *buceroides*, but with reservations. We follow these authorities, but recommend detailed molecular analysis to resolve whether to split or combine the two lineages. This species is one of a suite of nonforest birds that have colonized upland interior valleys cleared for cultivation (Beehler 1978c). Sight records reported by K. D. Bishop (*in litt.*) on Biak I in 1989 and 1990 require corroboration. *Other names*—Leatherhead, New Guinea Friar Bird, *Philemon novaeguineae*.

Noisy Friarbird *Philemon corniculatus* RESIDENT

Merops corniculatus Latham, 1790. *Index Orn.* 1: 276.—Endeavor R, QLD, Australia. [See Schodde & Mason (1999: 293) for designation of a neotype of *corniculatus* and fixation of the type locality.]

A resident of the savanna of the Trans-Fly. Inhabits low, open savanna.

Extralimital—e Australia.

SUBSPECIES

a. *Philemon corniculatus corniculatus* (Latham)

Synonym: *Philemon corniculatus ellioti* Mathews, 1912. *Novit. Zool.* (1911) 18: 423.—Mount Elliot, QLD, Australia.

RANGE: Trans-Fly savannas and ne QLD (Higgins *et al.* 2001). Localities—Tarara, Morehead R[CO], Dogwa and Wuroi on the Oriomo R (Bishop 2005a). Neither K. D. Bishop nor D. Hobcroft found this species on visits to Bensbach, but P. Gregory saw one *ca.* 15 km south of Bensbach Lodge in 2009.

DIAGNOSIS: Higgins *et al.* (2008) subsumed the race *ellioti* into the nominate form, noting that it is an intergrade between this form (of ne QLD) and *monachus*, of e and se Australia. The nominate form is browner and paler than *monachus*. Schodde & Mason (1999) noted the mantle is mid-fawn and the sides are dull fawn spotted white.

NOTES

Other names—Bald Friarbird.

GENUS *LICHMERA*

Lichmera Cabanis, 1851. *Mus. Hein.* 1: 118. Type, by subsequent designation (Bonaparte, 1854. *Compt. Rend. Acad. Sci. Paris* 38: 263), *Glyciphila ocularis* Gould.

The species of this genus range from the Lesser Sundas and the Moluccas to New Guinea, Australia, Vanuatu, the Loyalty Islands, and New Caledonia (Higgins *et al.* 2008). The genus is characterized by the slim, decurved, medium-length, and pointed bill, the dark brown iris, and the rather slight body. It is very similar to the genus *Meliphaga* but for the slimmer, more decurved bill, smaller size, and very drab plumage overall.

Olive Honeyeater *Lichmera argentauris*   RESIDENT | MONOTYPIC
Ptilotis argentauris Finsch, 1871. *Abh. Nat. Verh. Bremen* (1870) 2: 364.—Waigeo I, NW Islands.
    Synonym: *Stigmatops chloris* Salvadori, 1878. *Ann. Mus. Civ. Genova* 12: 337.—Gebe I, NW Islands.
Inhabits wooded habitats and gardens in the NW Islands: Gebe[MA], Waigeo[MA], Misool[MA], Wai[J], and Haitlal I[K]; also the Schildpad Is[MA] (Mees 1965[K], Beehler field notes 13-Mar-2005[J]).

    Extralimital—the Moluccas: Halmahera, Damar, and Lusaolate I off N Seram (Coates & Bishop 1997, Higgins *et al.* 2008).

NOTES
The species is treated as monotypic by Higgins *et al.* (2008). It is surprising that this species is absent from Gag and Kofiau I. *Other names*—Plain Olive, Silver-spangled, or Silver Honeyeater.

Brown Honeyeater *Lichmera indistincta*   RESIDENT
Meliphaga indistincta Vigors & Horsfield, 1827. *Trans. Linn. Soc. London* 15: 315.—King George Sound, Western Australia.
    Locally common in savanna and woodland of the Trans-Fly and Aru Is, near sea level (Bishop 2005a). Extralimital—w, n, and e Australia (Higgins *et al.* 2001).

SUBSPECIES

a. *Lichmera indistincta nupta* (Stresemann)
Stigmatops indistincta nupta Stresemann, 1912. *Novit. Zool.* 19: 344.—Manien I, Aru Is.
RANGE: The Aru Is.
DIAGNOSIS: Similar to *ocularis* but darker dorsally; (largely) lacks the yellow postocular tuft, and has a larger area of bare skin around eye (Higgins *et al.* 2008). Also note yellowish wash to chin.

b. *Lichmera indistincta ocularis* (Gould)
Glyciphila (?) *ocularis* Gould, 1838. *Synops. Birds Austral.* 4, app.: 6.—New South Wales, Australia.
RANGE: Northern and e Australia; s NG from Merauke to Lake Daviumbu in the middle Fly R (Bishop 2005a); also Saibai I (Torres Strait).
DIAGNOSIS: Compared to the nominate form of nw and c Australia, *ocularis* is longer-billed, darker-capped, darker-breasted, and with bold margins to remiges and rectrices (Higgins *et al.* 2008). Compared to *nupta*, *ocularis* has a paler mantle, more prominent postocular tuft, and a smaller area of bare ocular skin.

NOTES
Mees (1982b) considered the Trans-Fly population to be allied with the Australian nominate form, following Frith & Hitchcock (1974), who considered all the Australian birds to be the nominate form. Schodde & Mason (1999) subsequently dissected the Australian birds into three subspecies. *Other names*—Australian Brown Honeyeater, Warbling Honeyeater.

Silver-eared Honeyeater *Lichmera alboauricularis*   ENDEMIC
Stigmatops albo-aricularis Ramsay, 1878. *Proc. Linn. Soc. NSW* (1878–79) 3(1): 75.—Heath I, Milne Bay, SE Islands.
    Very patchily distributed, inhabiting scrubby habitats near the coast or along larger rivers of the NW Lowlands, the Sepik-Ramu, and the e half of the SE Peninsula; also on a few small coastal islands (Coates

& Peckover 2001). Near sea level. Preferred habitat seems to be tall cane grass near water, but the species is also found in village and plantation environs (Higgins *et al.* 2008).

SUBSPECIES

a. *Lichmera alboauricularis olivacea* Mayr
Lichmera alboauricularis olivacea Mayr, 1938. *Zool. Ser., Field Mus. Nat. Hist.* 20: 468.—Ifar, Lake Sentani, n NG.
RANGE: Patchily distributed in n NG—the Mamberamo R; the n coast of w NG east of the mouth of the Mamberamo R to Sentani; the middle and lower Sepik R[J]; and the lower Ramu R[CO] (Gilliard & LeCroy 1966[J]).
DIAGNOSIS: Similar to nominate form, but dorsal plumage darker.

b. *Lichmera alboauricularis alboauricularis* (Ramsay)
RANGE: Coastal lowlands of the SE Peninsula, southeast from Popondetta on the n coast, and from Bereina eastward on the s coast. SE Islands: Mailu, Heath[MA], and Doini I[MA] (Higgins *et al.* 2008). Common and noisy in Alotau, but very local in Port Moresby (P. Gregory *in litt.*).
DIAGNOSIS: Crown olive with obscure dark markings; back brown-olive with mottling; bare eye skin dotted with small white featherlets and a small triangular ear patch of pale yellow; throat and breast profusely patterned with brownish triangular spots with a buff-olive pale background; sides of breast dull brownish; overall looks dark and mottled. See photos in Coates (1990: 297) and Coates & Peckover (2001: 182).

NOTES
Other names—Eared or White-spangled Honeyeater.

GENUS *ENTOMYZON* MONOTYPIC

Entomyzon Swainson, 1825. *Zool. Journ.* 1: 480. Type, by original designation, "blue-faced Grakle" Latham = *Gracula cyanotis* Latham, 1801.

The single species of the genus ranges from eastern and northern Australia to the Trans-Fly of New Guinea (Higgins *et al.* 2008). The genus is characterized by the large size, distinct blue eye skin, and white, green, and black plumage pattern.

Blue-faced Honeyeater *Entomyzon cyanotis* RESIDENT
Gracula cyanotis Latham, 1801. *Index Orn.*, Suppl.: 29.—Sydney, NSW, Australia.

Inhabits open woodland, riverine woodland and forest, and tall scrub in the s Trans-Fly. Extralimital—n and e Australia.

SUBSPECIES

a. *Entomyzon cyanotis griseigularis* van Oort
Entomiza cyanotis griseigularis van Oort, 1909. *Nova Guinea, Zool.* 9(1): 97.—Merauke, s NG. Replacement name for *Entomyzon cyanotis harterti* Robinson and Laverock, an intergrade with the nominate race (Schodde & Mason 1999: 275).
RANGE: The s Trans-Fly and Cape York Penin, Australia. Localities—Kumbe[M], Merauke distr[J], Wasur NP[K], Bensbach area[K], and Tarara[L] and Penzara[L] on the Wassi Kussa R (Bangs & Peters 1926[J], Rand 1942a[L], Mees 1982[M], K. D. Bishop *in litt.*[K]).
DIAGNOSIS: Much smaller than the nominate form, with a slightly larger pale patch on the underwing (Higgins *et al.* 2008). Wing (male) 143–150, compared with 146–160 for race *albipennis* of nw and north-central Australia and 150–162 for the nominate form (Schodde & Mason 1999).

NOTES
Other names—White-quilled Honeyeater.

GENUS *MELITHREPTUS*

Melithreptus Vieillot, 1816. *Analyse Nouv. Orn. Elem.*: 46. Type, by subsequent designation (Gadow, 1884. *Cat. Birds Brit. Mus.* 9: 204), "Heorotaire fuscalbin" Vieillot = *Certhia lunata* Vieillot.

The species of this genus range throughout Australia north to southern and eastern New Guinea (Higgins *et al.* 2008). The genus, an apparent sister form to *Entomyzon* (Joseph *et al.* 2014b), is characterized by the black cap, white postocular crescent, colored eye skin, bright green mantle, and white or whitish ventral plumage.

White-throated Honeyeater *Melithreptus albogularis*   RESIDENT

Melithreptus albogularis Gould, 1848. *Birds Austral.* 4(30): pl. 74; also 1848. *Proc. Zool. Soc. London* (1847) 15: 220.—Port Essington, NT, Australia.

A common inhabitant of savannas and open country of the Trans-Fly, Markham R valley, and parts of the SE Peninsula, sea level–800 m.

Extralimital—n and e Australia.

SUBSPECIES

a. *Melithreptus albogularis albogularis* Gould

RANGE: The Trans-Fly from Kurik[J] east to Wasur NP[K], and the Bensbach area east to Wuroi on the Oriomo R[CO]; Kaiapit and the middle Markham valley[CO]; Baroka near Bereina[CO], east to Kupiano[CO]; inland to the upper Aroa R[CO], the Sogeri Plateau[CO], and Port Moresby[CO]; the upper Musa valley[L] and Pongani-Oro Bay[M] (Clapp 1979[L], 1980b[M] Mees 1982b[J], Bishop 2005a[K]). Also nw and n Australia.

DIAGNOSIS: Mantle is bright yellowish citrine; size small. Wing (male) 69–73. By comparison, wing (male) for race *inopinatus* of e Australia is 74–78 (Schodde & Mason 1999).

NOTES

Dickinson & Christidis (2014: 161, note 18) split out a distinct subspecies, *vinitinctus*, based on the DNA differences of birds from the Cape York Penin (Toon *et al.* 2010), that may encompass the birds from the SE Peninsula. This requires further study. *Other names*—White-chinned Honeyeater.

GENUS *GLYCICHAERA*   MONOTYPIC

Glycichaera Salvadori, 1878. *Ann. Mus. Civ. Genova* 12: 335. Type, by original designation, *Glycichaera fallax* Salvadori.

The one species in the genus is confined to New Guinea and northeastern Queensland. The genus is characterized by the warbler-like size, bill shape, and body form, and the uniquely drab olive and green plumage. We note that the molecular phylogenies of neither Driskell & Christidis (2004) nor Joseph *et al.* (2014) supported Schodde & Mason's (1999) supposition that the genus should be merged into *Timeliopsis*.

Green-backed Honeyeater *Glycichaera fallax*   RESIDENT

Glycichaera fallax Salvadori, 1878. *Ann. Mus. Civ. Genova* 12: 335.—Naiabui, Hall Sound, SE Peninsula.

Inhabits forest, edge, and regrowth throughout NG, the NW Islands, Aru Is, and Yapen I; sea level–850 m, max 1,300 m (Pratt & Beehler 2014).

Extralimital—Cape York Penin, QLD.

SUBSPECIES

a. *Glycichaera fallax fallax* Salvadori

Synonym: *Glycichaera poliocephala* Salvadori, 1878. *Ann. Mus. Civ. Genova* 12: 336.—Andai, Bird's Head.
Synonym: *Sericornis sylvia* Reichenow, 1899. *Jour. f. Orn.* 47: 118.—Madang, e verge of the Sepik-Ramu.

RANGE: All mainland NG, Misool I, the Aru Is, and Yapen I (Mayr 1941).

DIAGNOSIS: Head and neck grayish olive brown, slightly paler on lores, cheek, and ear coverts; pale yellowish chin and throat with olive mottling; mantle dark olive; breast dull yellowish (Higgins *et al.* 2008). The form *sylvia*, recognized by Rand & Gilliard (1967), was subsumed into the nominate race by Mayr (1941), Salomonsen (1967), and Gilliard & LeCroy (1968). We merge *poliocephala* into the nominate form, as its putative color differences are minimal and variable. See photos of the nominate form in Coates (1990: 298) and Coates & Peckover (2001: 183).

b. *Glycichaera fallax pallida* Stresemann & Paludan
Glycichaera fallax pallida Stresemann & Paludan, 1932. *Orn. Monatsber.* 40: 15.—Waigeo I, NW Islands.
RANGE: Waigeo[MA], Batanta[L], and ?Salawati I[J] (Greenway 1966[L], Beehler *et al.* 1986[J]).
DIAGNOSIS: Similar to nominate form but grayer, with contrasting back color, and with paler ventral plumage (Higgins *et al.* 2008). Salawati record requires corroboration.

NOTES
Other names—Green-backed Straightbill, Grey-headed or White-eyed Honeyeater, *Timeliopsis fallax*.

GENUS *PTILOPRORA*   ENDEMIC [Genus 6 spp]

Ptiloprora De Vis, 1894. *Ann. Rept. British New Guinea 1893–94*, 94: 103. Type, by subsequent designation (Salomonsen, 1967. In Paynter, *Check-list Birds World* 12: 412, footnote), *Ptilotis guisei* De Vis.

The six species of the genus are endemic to the island of New Guinea. The genus is characterized by the slim, decurved and pointed black bill; compact size; dark-streaked crown, auriculars, and mantle; and red, green, or pale gray (not dark brown) iris.

Leaden Honeyeater *Ptiloprora plumbea*   ENDEMIC | MONOTYPIC
Ptilotis plumbea Salvadori, 1894. *Ann. Mus. Civ. Genova* (2)14: 151.—Moroka, s watershed of SE Peninsula.
  Synonym: *Meliornis schistacea* De Vis, 1897. *Ibis* (7) 3: 381.—Southeastern New Guinea. [Suppressed by ICZN
    Opinion 684.]
  Synonym: *Ptiloprora plumbea granti* Mayr, 1931. *Bull. Brit. Orn. Club* 51: 59.—Upper Otakwa R, Western Ra.
Very uncommon and locally distributed through the Central Ra of NG, 1,000–2,100 m. More records from the w, nw, and se parts of the Central Ra, and few records from the c segment.
  Localities—Otakwa R[MA], Unchemchi in the Hindenburg Mts[J], Ok Tedi[Y], Lepa Ridge[P], n slope Mt Wilhelm[B], Aseki[CO], Dawong[K], Mt Missim[X], Bellavista[CO], Woitape[CO], Efogi[CO], Moroka[MA], Mt Maguli[L], and Suria[Z] (Greenway 1935[X], Mayr 1931[K], Gilliard 1950[L], Gilliard & LeCroy 1961[J], Beehler 1988[P], Gregory 1995b[Y], Beehler field notes[Z], P. Z. Marki *in litt*[B]). Notably scarce and local.

NOTES
The two described subspecies apparently represent endpoints of an east-west cline. Variation is negligible and described subspecies are not recognized here. Schodde (1978) noted that the suppressed form *Meliornis schistacea* is now identifiable as this form (but is no longer available post-suppression). *Other names*— Leaden Streaked Honeyeater.

Yellow-streaked Honeyeater *Ptiloprora meekiana*   ENDEMIC
Ptilotis meekiana Rothschild & Hartert, 1907. *Novit. Zool.* 14: 482.—Head of the Aroa R, 1,400 m, s watershed of the SE Peninsula.
  Patchily distributed in montane forest of the n slope of the Western Ra, and mts of the Huon and SE Peninsulas; 1,500–2,400 m, extremes 1,300–2,800 m[CO]. Also a few records from the Eastern Ra.

SUBSPECIES
a. *Ptiloprora meekiana occidentalis* Rand
Ptiloprora meekiana occidentalis Rand, 1940. *Amer. Mus. Novit.* 1072: 13.—Bele R, 2,200 m, 18 km n of Lake Habbema, Western Ra.
RANGE: The n slope of the Western Ra and patchily in the Eastern Ra. Localities—Ilaga[J], 9 km northeast of Lake Habbema[K], Bele R Camp[K], Ambua[L], Rondon Ridge[M], Mt Karimui[N], Lepa Ridge[P], ne slopes of Mt Wilhelm[O], and Mt Michael[DI] (Rand 1942b[K], Ripley 1964a[J], Beehler field notes 21-Aug-1982[P], P. Gregory *in*

litt.[M], Sam & Koane 2014[O], B. G. Freeman *in litt.*[N], K. D. Bishop *in litt.*[L]).

DIAGNOSIS: Marginally larger than the nominate form, and ventral plumage brighter, more yellowish and greenish. Wing: male 82–92, female 77–82 (Rand 1940a).

b. *Ptiloprora meekiana meekiana* (Rothschild & Hartert)

RANGE: Mountains of the Huon and SE Peninsulas. Localities—Ogeramnang[J], Aseki[CO], Mt Missim[K], upper Mambare R[CO], Mt Tafa[CO], Efogi[CO] (Mayr 1931[J], Greenway 1935[K]).

DIAGNOSIS: Grayer and duller, marginally smaller, and perhaps longer-billed than *occidentalis* (Higgins *et al.* 2008). Wing: male 82–88, female 71–76 (Rand 1940a).

NOTES

These two races may be part of an east-west cline. *Other names*—Olive-streaked, Meek's Streaked, or Yellowish-streaked Honeyeater.

Rufous-sided Honeyeater *Ptiloprora erythropleura*   ENDEMIC

Ptilotis erythropleura Salvadori, 1876. *Ann. Mus. Civ. Genova* (1875) 7: 949.—Arfak Mts, Bird's Head

Inhabits montane forest and edge on the Bird's Head and Neck (Fakfak, Kumawa Mts) and the Western Ra, 1,300–2,500 m (Beehler *et al.* 1986). Endemic to Indonesian NG.

SUBSPECIES

a. *Ptiloprora erythropleura erythropleura* (Salvadori)

RANGE: The Tamrau and Arfak Mts of the Bird's Head (Gyldenstolpe 1955b, Gilliard & LeCroy 1970).

DIAGNOSIS: Iris red or reddish-brown. Slightly paler ventrally than *dammermani*.

b. *Ptiloprora erythropleura dammermani* Stresemann & Paludan

Ptiloprora erythropleura dammermani Stresemann & Paludan, 1934. *Orn. Monatsber.* 42: 44.—Mount Sumuri, 2,200 m, Weyland Mts, Western Ra.

 Synonym: *Ptiloprora guisei incerta* Junge, 1952. *Zool. Meded. Leiden* 31: 249.—Bobairo, Paniai Lake distr, Western Ra.

RANGE: The Bird's Neck and the Western Ra. Localities—Fakfak Mts[Y, Z], ?Kumawa Mts[Z], Kunupi[K], Mt Sumuri[K], Enarotali[U], Ilaga[J], 9 km northeast of Lake Habbema[Q], Bele R Camp[Q] (Hartert *et al.* 1936[K], Rand 1942b[Q], Ripley 1964a[J], Melville 1980[U], Diamond & Bishop 2015[Z], Gibbs 1994[Y]).

DIAGNOSIS: Iris greenish or green-yellow rather than red; plumage essentially identical to nominate but perhaps marginally richer and darker ventrally. Diamond (1985) noted the birds from the Kumawa Mts had a green iris, which ties that population to this form. But Diamond also noted a yellow wash ventrally, unlike this form. The yellow wash in *P. guisei* indicates an immature plumage. Gibbs (1994) observed a species of *Ptiloprora* in the Fakfak Mts, and postulated it might represent a novel race of *perstriata*, as the size was too big for *erythropleura*. Gibbs (1994) preliminarily placed the form in *erythropleura*. Diamond & Bishop (2015) indicate both Fakfak and Kumawa Mts populations are most similar to *dammermani* but perhaps diagnosible as distinct.

NOTES

Salomonsen (1967) noted that the putative race *incerta*, based on a single specimen, is probably a hybrid between *P. erythropleura dammermani* and *P. perstriata*. *Other names*—Rufous-sided Streaked Honeyeater, Red-sided Honeyeater.

Grey-streaked Honeyeater *Ptiloprora perstriata*   ENDEMIC | MONOTYPIC

Ptilotis perstriata De Vis, 1898. *Ann. Rept. Brit. New Guinea*, 1896–97: 86.—Wharton Ra, SE Peninsula.

 Synonym: *Ptilotis erythropleura lorentzi* van Oort, 1909 *Nova Guinea Zool.* 9: 95.—Hellwig Mts, e sector of the Western Ra.

 Synonym: *Ptiloprora guisei praedicta* Hartert, 1930. *Novit. Zool.* 36: 49.—Mount Wondiwoi, Wandammen Penin, Bird's Neck.

Inhabits forest, edge, and subalpine shrubland through the NG Central Ra and mts of the Wandammen Penin. East of the Strickland R gorge this species is found only above 2,500 m[CO]. West of the Strickland, where *P. guisei* is absent, *perstriata* can be found as low as 1,650 m[CO]. Absent from the Huon Penin. Also apparently absent from the e sector of the Eastern Ra and the nw sector of the SE Peninsula (it is found on Mt Albert Edward and presumably southeast of there).

NOTES

The elevational relationship between this species and its sister *P. guisei* would bear closer examination, especially in light of ongoing impacts of climate change on species' elevational ranges. Discussed by LeCroy (2011), *Ptilotis precipua nigritergum* Rothschild & Hartert was synonymized by Mayr (1941) into the race *lorentzi* (itself now synonymized into the nominate form). Male wing size in Table 9 of Diamond (1969) showed that the range of the nominate form (87–106) entirely encompasses the range of *praedicta* (102–105). Plumage distinctions are nonexistent. Choice of English name is based on the single distinguishing character of this species, which is the gray streaking on the back; this species is not black-backed. *Other names*—Black-backed or Many-streaked Honeyeater, Black-backed Streaked Honeyeater.

Mayr's Streaked Honeyeater *Ptiloprora mayri* ENDEMIC

Ptiloprora guisei mayri Hartert, 1930. *Novit. Zool.* 36: 49.—Cyclops Mts, ne sector of NW Lowlands.
　　Inhabits montane forest of outlying n ranges of the NW Lowlands and Sepik-Ramu only: Foja, Cyclops, and Bewani Mts, 1,200–2,170 m.

SUBSPECIES

a. *Ptiloprora mayri acrophila* Diamond
Ptiloprora mayri acrophila Diamond, 1969. *Amer. Mus. Novit.* 2362: 46.—Mount Menawa, Bewani Mts, n NG.
RANGE: The Foja[J] and Bewani[K] Mts (Diamond 1969[K], Beehler & Prawiradilaga 2010[J]).
DIAGNOSIS: Pale edgings of body feathering have an olive wash (Higgins *et al.* 2008). Note also obscure olive wash to cheek and whitish-gray chin and throat.

b. *Ptiloprora mayri mayri* Hartert
RANGE: The Cyclops Mts.
DIAGNOSIS: Pale edges of feathering generally have a gray wash rather than olive wash (Higgins *et al.* 2008).

NOTES

Perhaps the two subspecies of *P. mayri* are no more than large subspecies of *P. guisei* (Rand & Gilliard 1967). This merits study using molecular tools. Our English name highlights that this bird is a member of the genus *Ptiloprora*. *Other names*—Mayr's Honeyeater.

Rufous-backed Honeyeater *Ptiloprora guisei* ENDEMIC | MONOTYPIC

Ptilotis (?) *guisei* De Vis, 1894. *Ann. Rept. Brit. New Guinea*, 1893–94: 103.—Mount Maneao, SE Peninsula.
　　Synonym: *Ptiloprora guisei umbrosa* Mayr, 1931. *Mitt. Zool. Mus. Berlin*, 17: 666.—Schrader Ra, 2,000 m, n sector of the Eastern Ra.
An abundant inhabitant of montane forest and edge through the Eastern Ra, Mt Bosavi, Adelbert Mts, and mts of the Huon and SE Peninsulas. In the presence of its sister form *P. perstriata*, this species ranges from 1,750 to 2,700 m (P. Z. Marki *in litt.*); in the absence of *P. perstriata* (as on the Huon Penin), *P. guisei* ranges up to the tree line.

NOTES

We treat the form as monotypic. An examination of topotypical material at the AMNH supports this view. *Ptilotis praecipua* Hartert, discussed in LeCroy (2011), was synonymized by Mayr (1941). *P. guisei* apparently does not range westward across the Strickland R gorge, which serves as a geographic barrier for various upland songbirds (*e.g.*, *Astrapia mayeri*, *Melionyx princeps*). *Other names*—Red-backed, Green-eyed, or Guise's Honeyeater, Brown-backed Streaked Honeyeater.

Pycnopygius Salvadori, 1880. *Ann. Mus. Civ. Genova* 16: 78. Type, by monotypy, *Pycnonotus (?) stictocephalus* Salvadori.

The species of this genus are endemic to the Region. They inhabit forest canopy and edge, from the lowlands to 2,000 m (Pratt & Beehler 2014). Species of the genus exhibit a streaked crown and obscure mottling or streaking on the breast; plumage either predominantly gray or dull brown; and bill rather short and straight.

Plain Honeyeater *Pycnopygius ixoides*   ENDEMIC

Ptilotis ? *ixoides* Salvadori, 1878. *Ann. Mus. Civ. Genova* 12: 338.—Sorong, w Bird's Head.

Inhabits forest canopy throughout mainland NG; sea level–1,200 m, max 1,400 m[CO]. No records from the Trans-Fly, but widespread through the S Lowlands of PNG (Bishop 2005b).

SUBSPECIES

a. *Pycnopygius ixoides ixoides* (Salvadori)

Synonym: *Pycnopygius ixoides cinereifrons* Salomonsen, 1966a. *Breviora* Mus. Comp. Zool. Harvard 254: 8.—
   3 km below junction of Black and Palmer R, 100 m, upper Fly R.

RANGE: From the Bird's Head east to the head of Geelvink Bay (Menoo R), and the s slopes of the Central Ra east probably to Port Moresby.

DIAGNOSIS: Crown patterned with gray and blackish-brown scalloping; mantle dull brown with some olive edges; throat plain buff grading to a slightly richer buff breast with some obscure scalloping; belly and vent more saturated with dun brown. The race *cinereifrons* is a synonym of the nominate form (LeCroy 2011). See photo of *ixoides* in Coates (1990: 273).

b. *Pycnopygius ixoides proximus* (Madarász)

Ptilotis proxima Madarász, 1900. *Orn. Monatsber.* 8: 3.—Erima, Astrolabe Bay, ne NG.

Synonym: *Ptilotis simplex* Reichenow, 1915. *J. Orn.* 63: 126.—Maeanderberg, upper Sepik R, west-central
   Sepik-Ramu.

RANGE: The NW Lowlands and Sepik-Ramu.

DIAGNOSIS: As for nominate form, but overall grayer-washed without the brown or yellow tint; dull brown with a gray tone dorsally. The form *simplex* was recognized by Salomonsen (1967) but had already been placed in synonymy by Mayr (1941).

c. *Pycnopygius ixoides unicus* Mayr

Pycnopygius ixoides unicus Mayr, 1931. *Mitt. Zool. Mus. Berlin* 17: 666.—Sattelberg, 600 m, Huon Penin.

RANGE: The Huon Penin and the Herzog Mts and Kuper Ra of the nw sector of the SE Peninsula. (Mayr 1931, Greenway 1935)

DIAGNOSIS: More obvious olive tone to gray of crown and neck; margins of remiges and rectrices more buff; belly plumage tending toward *finschi*.

d. *Pycnopygius ixoides finschi* (Rothschild & Hartert)

Ptilotis finschi Rothschild & Hartert, 1903. *Novit. Zool.* 10: 448.—Mountains of the SE Peninsula.

RANGE: The SE Peninsula, on the n coast east of the Kumusi R and on the s coast east of Port Moresby (few records). For notes on collecting locality of the holotype for this form, see LeCroy (2011).

DIAGNOSIS: Distinctive; ventral plumage bright buff rufous; dorsal plumage slightly more rufous brown. The population with the warmest brown wash, especially on the belly (almost rusty).

NOTES

Other names—Brown, Nondescript, or Olive-brown Honeyeater.

Streak-headed Honeyeater *Pycnopygius stictocephalus*   ENDEMIC | MONOTYPIC

Pycnonotus (?) *stictocephalus* Salvadori, 1876. *Ann. Mus. Civ. Genova* 9: 34.—Naiabui, Hall Sound, SE Peninsula.

Frequents riparian forest edge, tall trees in village gardens, and woodlands through NG, Salawati I[MA], and the Aru Is[MA]; sea level–500 m, locally to 970 m[CO]. Widespread in the S Lowlands and the Trans-Fly (Stresemann 1923, Mees 1982b, Bishop 2005a).

Other names—Streak-capped Honeyeater, Spangle-crowned Honeyeater

Marbled Honeyeater *Pycnopygius cinereus*   ENDEMIC

Ptilotis cinerea P. L. Sclater, 1874. *Proc. Zool. Soc. London* (1873): 693.—Hatam, Arfak Mts, Bird's Head.

Inhabits montane forest canopy of the Bird's Head, Central Ra, Adelbert Mts, and Huon Penin; 1,000–2,000 m, min 500 m[CO]. Frequent in fragmented upland woodlots near Wapenamanda in Enga Prov (D. Hobcroft *in litt.*).

SUBSPECIES

a. *Pycnopygius cinereus cinereus* (P. L. Sclater)

Synonym: *Pycnopygius cinereus dorsalis* Stresemann & Paludan, 1934. *Orn. Monatsber.* 42: 44.—Kunupi, 1,300 m, Weyland Mts, Western Ra.

RANGE: Mountains of the Bird's Head and the Western Ra.

DIAGNOSIS: Mottling of ventral plumage obsolete; no pale or whitish spotting on throat or breast. The form *dorsalis* is only marginally darker dorsally than the nominate form and thus not recognized here.

b. *Pycnopygius cinereus marmoratus* (Sharpe)

Ptilotis marmorata Sharpe, 1882. *Journ. Linn. Soc. London, Zool.* 16: 319, 438.—Moroka distr, s watershed of SE Peninsula.

RANGE: The Border Ra east to the mts of the SE Peninsula; also the Adelbert Mts[J] and mts of the Huon Penin[MA] (Pratt 1983[J]).

DIAGNOSIS: Ventral plumage heavily mottled with abundant pale scalloping; whitish midline stripe on belly. See photo in Coates (1990: 275).

NOTES

Other names—Gray Honeyeater, Gray-fronted, Grayish-brown Honeyeater.

GENUS *CONOPOPHILA*   [Genus 3 spp/Region 1 sp]

Conopophila Reichenbach, 1852. *Handb. Spec. Orn., Abth.* 2, Meropinae, continuatio IX: 119. Type, by subsequent designation (G. R. Gray, 1855. *Cat. Gen. Subgen. Birds*: 24), *Entomophila ? albogularis* Gould.

The species of the genus range from Australia to southern and eastern New Guinea (Higgins *et al.* 2008). The genus is characterized by the very abbreviated blackish bill, white belly, plain plumage entirely lacking bars or streaks, and yellow in the wing.

Rufous-banded Honeyeater *Conopophila albogularis*   RESIDENT | MONOTYPIC

Entomophila ? albogularis Gould, 1843. *Proc. Zool. Soc. London* (1842): 137.—Port Essington, NT, Australia.

Synonym: *Conopophila albogularis mimikae* Mathews, 1924. *Birds Austral.* 11: 390.—Mimika R, w sector of the S Lowlands.

The Aru Is[MA], w Bird's Head, se Bird's Neck, S Lowlands, s watershed of SE Peninsula, Markham valley, and Sepik basin (Bishop 2005a, 2005b, Higgins *et al.* 2008). Frequents trees, shrubs, reedbeds, mangroves, gardens, and scrub (Coates 1990). Mainly in lowlands; max 600 m.

Localities: Sorong[J], Bugi[M], Daru[M], Gaima[M], Lake Daviumbu[M], Kiunga[K], Tabubil[K], Turama R[L], e Sepik[CO], middle Markham R[CO], Sogeri Plateau[CO], Port Moresby distr[CO], upper Musa R[CO], and Alotau[CO] (Rand 1942a[M], Gyldenstolpe 1955b[J], Burrows 1993[L], I. Woxvold *et al.* 2015[K]). Seems to be expanding its range into forested regions by invading towns and similar second-growth habitat, such as at Kiunga and Tabubil (I. Woxvold *in litt.*). First appeared in Kiunga in 2009 (P. Gregory *in litt.*)

Extralimital—n and ne Australia. Also present on islands of the Torres Strait.

The species is treated as monotypic following Schodde & Mason (1999). *Other names*—Rufous-breasted or White-throated Honeyeater.

GENUS RAMSAYORNIS   [GENUS 2 SPP/REGION 1 SP]

Ramsayornis Mathews, 1912. *Austral. Avian Rec.* 1: 115. Type, by original designation, *Gliciphila subfasciata* Ramsay = *Glyciphila modesta* G. R. Gray.

The species of this genus range from northern Australia to southern, eastern, and northwestern New Guinea (Higgins *et al.* 2008). The genus is characterized by the small size, pale ventral plumage with incomplete barring, and unmarked white throat and undertail. These birds build a very deep and pendulous, semi-hooded nest (Schodde & Mason 1999). *Ramsayornis* is masculine.

Brown-backed Honeyeater *Ramsayornis modestus*   RESIDENT | MONOTYPIC
Glyciphila modesta G. R. Gray, 1858. *Proc. Zool. Soc. London*: 174.—Aru Is.

Very patchily distributed around NG's coastline: the NW Islands, n sector of the NW Lowlands, north-central and ne Sepik-Ramu, Aru Is, the Trans-Fly, e sector of the S Lowlands, the SE Peninsula (only east of Popondetta in n watershed), and the D'Entrecasteaux Arch. Mainly lowlands; max 600 m[CO]. Inhabits scrub and woodland along watercourses, mangroves, and town gardens.

Localities—Waigeo[MA], Batanta[K], Salawati I[MA], Sorong[K], Aru Is[K], Mamberamo[CO], Kurik[J], Bian Lakes[J], Wasur NP[J], Bensbach[J], Kairiru I[CO], Wewak[CO], the middle and lower Fly R[CO], Lake Daviumbu[J], near Madang[L], Port Moresby[CO], Varirata NP[CO], Mambare R[CO], Alotau[CO], D'Entrecasteaux Is[MA] (Gyldenstolpe 1955b[K], Oliver & Hopkins 1989[L], Bishop 2005a[J]). Reported to migrate from n QLD in winter and may disperse to e NG. First reported from Normanby I by Pratt (field notes 2003).

Extralimital—ne Australia.

Other names—Modest Honeyeater.

GENUS *MELIPOTES*   ENDEMIC [Genus 4 spp]

Melipotes P. L. Sclater, 1874. *Proc. Zool. Soc. London* (1873): 695, pl. 56. Type, by monotypy, *Melipotes gymnops* P. L. Sclater.

The four species of this genus are endemic to the mountains of mainland New Guinea. Representatives of the genus can be found on all main and outlying ranges of New Guinea (except the Adelbert Mountains). The characters defining the genus include the large and prominent eye patch of bare yellow or orange skin; the ability to flush the eye patch a deep red; the lack of prominent territorial vocalization (unique in the family); the abbreviated black bill; and the nearly exclusively frugivorous habit. The genus is presumably closely related to *Macgregoria*. In spite of the superficial similarity of the orbital patch exhibited by the two groups, the structure differs. Whereas for *Melipotes* the patch is bare eye skin, with some ventral wattling in two species, for *Macgregoria* the eye patch is a loose, fleshy wattle that is appended to the face only at the eye-ring.

Western Smoky Honeyeater *Melipotes gymnops*   ENDEMIC | MONOTYPIC
Melipotes gymnops P. L. Sclater, 1874. *Proc. Zool. Soc. London* (1873): 695, pl. 56.—Hatam, Arfak Mts, Bird's Head.

Mountains of the Bird's Head and Neck, 1,200–2,700 m. Includes the Tamrau, Arfak, and Wandammen Mts (Diamond 1985). Inhabits montane forest interior and edge. Range may also include the Fakfak Mts (see Diamond & Bishop 2015).

Gibbs (1994) noted that the local representative of this genus he observed in the Fakfak Mts seems closest to this species, whereas Diamond (1985) noted the birds he observed in the Fakfak Mts appeared closest to *fumigatus*. Gibbs's observations further complicate matters, because he described a bird with a pendent facial wattle that appears similar to that of *M. carolae* from the Foja Mts. For the latest thinking on this, see discussion of Diamond & Bishop (2015). Our English name distinguishes the species from the following, more widespread, form. *Other names*—Western or Arfak Melipotes, Arfak or Spot-bellied Honeyeater.

Common Smoky Honeyeater *Melipotes fumigatus*   ENDEMIC
Melipotes fumigatus A. B. Meyer, 1886. *Zeitschr. Ges. Orn.* 3: 22, pl. 4, fig. 1.—Mount Maguli, s watershed, SE Peninsula.

The Central Ra, Kumawa and Cyclops Mts, PNG North Coastal Ra, and Mt Bosavi (Stresemann 1923, Mayr 1941, Diamond 1985); 1,000–4,200 m. Replaced by close relatives in the mts of Bird's Head, mts of Wandammen Penin, Foja Mts, and mts of Huon Penin. Inhabits montane forest, edge, disturbed areas, and gardens. Ubiquitous in montane forest of the Central Ra. Usually seen singly, occasionally in pairs or small groups. Isolated population from the Fakfak Mts may refer to this species, *M. gymnops*, or *M. carolae*.

SUBSPECIES
a. *Melipotes fumigatus kumawa* Diamond
Melipotes fumigatus kumawa Diamond, 1985. *Emu* 85: 82.—Southern slopes of the Kumawa Mts, 1,430 m, Bird's Neck.
RANGE: The Kumawa and Fakfak Mts of the s Bird's Neck (Diamond & Bishop 2015).
DIAGNOSIS: Lower breast and belly boldly marked with white crescents; tips of upperwing coverts distinctly white; pale underwing patch large and whitish; chin patch dark (Diamond 1985).

b. *Melipotes fumigatus goliathi* Rothschild & Hartert
Melipotes gymnops goliathi Rothschild & Hartert, 1911. *Bull. Brit. Orn. Club* 29: 34.—Mount Goliath, Border Ra.
RANGE: The Western, Border, and Eastern Ra; also the Cyclops Mts and PNG North Coastal Ra[J] (Mayr 1941, Diamond 1985[J]).
DIAGNOSIS: Darker overall than nominate form, and more heavily spotted on breast and belly, with diffuse pale feather edgings. Pale gray chin patch distinctly offset from obscure dark breast band. See photo in Coates (1990: 265).

c. *Melipotes fumigatus fumigatus* A. B. Meyer
RANGE: Mountains of the SE Peninsula (Mayr 1941).
DIAGNOSIS: Plumage gray and pale; pale chin patch evident (Higgins *et al.* 2008).

NOTES
Diamond (1985) observed a population of *Melipotes* in the Fakfak Mts that he tentatively referred to this species. Gibbs's (1994) observation of this population led him to believe it was closest to *gymnops*. This Fakfak population thus remains to be determined and diagnosed, once a series is collected. Given the Bird's Neck's biogeographic affiliation with the Bird's Head, we would normally expect this form to be more closely allied to *gymnops*. However, given the population in the adjacent Kumawa Mts has been allied with *fumigatus* (Diamond 1985: 82), we must await the results of a future field trip to the Fakfak Mts. *Other names*—Smoky or Eastern Smoky Honeyeater, Common Melipotes.

Wattled Smoky Honeyeater *Melipotes carolae*   ENDEMIC | MONOTYPIC
Melipotes carolae Beehler, Prawiradilaga, de Fretes, and Kemp, 2007. *Auk* 124: 1002.—Bog Camp, Foja Mts, 1,650 m, NW Lowlands.

Inhabits upland forests of the Foja Mts, 1,150–2,150 m (Beehler *et al.* 2007). Apparently absent from the Van Rees Mts.

Gibbs (1994) observed a population of *Melipotes* in the Fakfak Mts with pendent facial wattles, which are diagnostic for *M. carolae*. The Fakfak form needs to be collected and described (see Diamond & Bishop 2015).

Spangled Honeyeater *Melipotes ater* ENDEMIC | MONOTYPIC
Melipotes ater Rothschild & Hartert, 1911. *Bull. Brit. Orn. Club* 29: 13.—Rawlinson Ra, Huon Penin.
Mountains of the Huon Penin, 1,200–3,300 m[CO]. Inhabits canopy of forest interior, feeding mainly on fruit. Has been seen traveling through the forest canopy in parties of more than 20 (Beehler field notes).
Other names—Huon Melipotes, Huon Honeyeater.

GENUS *MACGREGORIA* ENDEMIC | MONOTYPIC

Macgregoria De Vis, 1897. *Ibis*: 251. Type, by monotypy, *Macgregoria pulchra* De Vis.
Endemic to the subalpine zone of the Central Ranges. Occurs in small, fragmented populations in western, central, and eastern New Guinea. The genus is characterized by the very large size, velvety-black body plumage, loose semicircular orbital wattle attached only at the eye-ring, bright ocher patch in the flight feathers, and the musical sound produced by the wing when gliding. Classified as a honeyeater by Cracraft & Feinstein (2000). It was long treated as a primitive bird of paradise (see Frith & Beehler 1998).

Giant Wattled Honeyeater *Macgregoria pulchra* ENDEMIC
Macgregoria pulchra De Vis, 1897. *Ibis*: 251, pl. 7.—Mount Scratchley, SE Peninsula.
Patchily distributed in small, relict populations in the highest reaches of the Western and Border Ra and mts of the SE Peninsula; 3,200–3,500 m, extremes 2,700–4,000 m[CO]. Largely restricted to patches of subalpine woodland, although in certain seasons it can be found sparingly at slightly lower elevations in upper montane forest. Distribution appears to be related to the availability of subalpine woodland dominated by the conifer *Dacrycarpus compactus* (Podocarpaceae). Absent from many seemingly suitable patches of habitat in the Victor Emanuel, Kaijende, Hagen, Giluwe, Kubor, Bismarck, Kratke, and Ekuti ranges.

SUBSPECIES
a. *Macgregoria pulchra carolinae* Junge
Macgregoria pulchra carolinae Junge, 1939. *Nova Guinea* (ns)3: 82.—Oranje Mts, 3,800 m, Western Ra.
RANGE: Discontinuously in the Western Ra and Border Ra. Localities—Kemabu Plateau[J], Lake Habbema[K,Q], near Mt Wilhelmina[K], Mt Mandala[R], Star Mts of PNG[L] (Rand 1942b[K], Ripley 1964a[J], Barker & Croft 1977[L], Beehler 1983[Q], K. D. Bishop *in litt.*[R]; see Frith & Beehler 1998).
DIAGNOSIS: Very similar to but shorter-tailed than the nominate form. Tail of adult male (n=10) 128–141 (Frith & Beehler 1998).

b. *Macgregoria pulchra pulchra* De Vis
RANGE: Mountains of the SE Peninsula, 2,800–3,700 m. Localities—Mt Albert Edward[MA], Murray Pass[MA], Neowa basin[J], Mt Scratchley[MA], Lake Omha[K], Mt Batchelor[MA], and Mt Victoria[MA] (Safford & Smart 1996[J], Beehler 1991[K]).
DIAGNOSIS: Long-tailed. Tail of adult male (n=18) 156–168 (Frith & Beehler 1998).

NOTES
This is a specialist frugivore that could be impacted by ongoing climate change. Diet is mainly or entirely of fruit. The species and its habitat merit additional detailed study (see Beehler 1991: 221–43). The western form (*carolinae*) is marginally distinct from the nominate, and perhaps the species could be considered monotypic, with the tail-length differences clinal in nature. The weight measurements (340–357 g) provided in Frith & Beehler (1998) for the subspecies *carolinae* are almost certainly in error (more likely 240–257 g). Our English name features the large size of this aberrant honeyeater, and its striking eye wattle. *Other names*—Macgregor's Bird of Paradise or Honeyeater.

Melilestes Salvadori, 1876. *Ann. Mus. Civ. Genova* (1875) 7: 950. Type, by original designation, *Ptilotis megarhynchus* G. R. Gray.

The single species of the genus is endemic to the Region. The genus is characterized by the combination of the very long and decurved bill, serrated tomia, orange iris, and plain olive-brown plumage. The genus exhibits the longest bill of the family.

Long-billed Honeyeater *Melilestes megarhynchus*   ENDEMIC

Ptilotis megarhynchus G. R. Gray, 1858. *Proc. Zool. Soc. London*: 174.—Aru Is.

Mainland NG, the four main NW Islands, Aru Is, and Yapen I; sea level–1,500 m, max 2,120 m (Stresemann 1923, Coates & Peckover 2001). Inhabits rain forest and edge.

SUBSPECIES

a. *Melilestes megarhynchus megarhynchus* (G. R. Gray)

Synonym: *Melilestes megarhynchus brunneus* Salomonsen, 1966a. *Breviora* Mus. Comp. Zool. Harvard 254: 1.— Siwi, Arfak Mts, Bird's Head.

RANGE: The Aru Is[MA], Misool[MA] and Salawati I[MA], Bird's Head and Neck, S Lowlands, and Huon and SE Peninsulas. Within the Trans-Fly this species has been recorded from the Bian Lakes[J], Wuroi on the Oriomo R, and Gaima (K. D. Bishop *in litt.*[J], Bishop 2005a).

DIAGNOSIS: Ventral plumage pale gray-brown with yellow-olive tinge, merging to dull olive brown on the flanks (Higgins *et al.* 2008). Diamond (1972) noted the ventral plumage is more olive and less gray than that of *vagans*. We subsume the race *brunneus* into the nominate form following Gilliard & LeCroy (1970) and Higgins *et al.* (2008).

b. *Melilestes megarhynchus vagans* (Bernstein)

Arachnothera vagans Bernstein, 1864. *J. Orn.* 12: 405.—Waigeo I, NW Islands.

Synonym: *Melilestes megarhynchus stresemanni* Hartert, 1930. *Novit. Zool.* 36: 45.—Jayapura, Humboldt Bay, n NG.

RANGE: Batanta[J], Waigeo[MA], and Yapen I[MA]; also the NW Lowlands and Sepik-Ramu (Greenway 1966[J]).

DIAGNOSIS: Crown gray (not olive), with blackish streaking; chin pale gray. We combine *stresemanni* into *vagans* because of insufficient character distinction, as suggested by Greenway (1966). See photos of *vagans* in Coates (1990: 271) and Coates & Peckover (2001: 174).

GENUS *TIMELIOPSIS*   ENDEMIC [Genus 2 spp]

Timeliopsis Salvadori, 1876. *Ann. Mus. Civ. Genova* (1875) 7: 963. Type, by original designation, *Timeliopsis trachycoma* Salvadori = *Euthyrhynchus griseigula* Schlegel.

The two species of the genus are endemic to the Region. The genus inhabits forest interior of the lowlands and uplands of New Guinea, and is characterized by the wholly plain, unstreaked olive or brown plumage, the straight and sharply pointed beak, and the orange iris.

Olive Straightbill *Timeliopsis fulvigula*   ENDEMIC

Euthyrhynchus fulvigula Schlegel, 1871. *Ned. Tijdschr. Dierk.* 4 (1873): 40.—Arfak Mts, Bird's Head.

Mountains of the Bird's Head and Neck, Central Ra, Foja Mts, PNG North Coastal Ra, and Huon Penin; 1,400–2,200 m, extremes 750–2,700 m[CO]. A widespread but uncommon inhabitant of the interior of montane forest.

SUBSPECIES

a. *Timeliopsis fulvigula fulvigula* (Schlegel)

RANGE: Mountains of the Bird's Head. The population from the Fakfak Mts (Gibbs 1994) has not been diagnosed to subspecies but may belong here.

DIAGNOSIS: Buffy ventrally, with reduced olive wash.

b. *Timeliopsis fulvigula meyeri* (Salvadori)

Euthyrhynchus meyeri Salvadori, 1896. *Ann. Mus. Civ. Genova* (2)16: 97.—Moroka, s watershed of SE Peninsula.

      Synonym: *Timeliopsis fulvigula montana* Mayr, 1931. *Mitt. Zool. Mus. Berlin* 17: 659.—Mount Goliath, Border Ra.

      Synonym: *Timeliopsis fulvigula fuscicapilla* Mayr, 1931. *Mitt. Zool. Mus. Berlin* 17: 658.—Junzaing, Huon Penin.

RANGE: The Central Ra and mts of the Huon Penin. Populations from the Foja Mts[J] and PNG North Coastal Ra[K] have not been diagnosed to subspecies but may belong here (Beehler *et al.* 2012[J], AMNH & BPBM specimens[K]).

DIAGNOSIS: Flanks exhibit a strong olive-green wash, which extends up onto breast. The races *montana* and *fuscicapilla* were subsumed into *meyeri* by Diamond (1972). See photo of *meyeri* in Coates & Peckover (2001: 183).

NOTES

Other names—Olive or Mountain Straight-billed Honeyeater, Mountain Straightbill.

Tawny Straightbill *Timeliopsis griseigula*   ENDEMIC

Euthyrhynchus griseigula Schlegel, 1871. *Ned. Tijdschr. Dierk.* (1873) 4: 39.—Sorong, w Bird's Head.

    Rare and patchily distributed in lowland rain-forest interior through the NG lowlands, sea level–800 m[CO], in the Bird's Head and Neck, NW Lowlands, Sepik-Ramu, S Lowlands, and SE Peninsula. Apparently no records along the n coast from the PNG border east to Madang; on the s side, no records from the e end of the Bird's Neck east to Lohiki, Gulf Prov, but for sightings in the upper Fly and upper Strickland (Coates 1990, Higgins *et al.* 2008, Diamond & Bishop 2015).

SUBSPECIES

a. *Timeliopsis griseigula griseigula* (Schlegel)

RANGE: The Bird's Head, NW Lowlands, and Sepik-Ramu. Localities—Sorong[MA], Sainkedoek[J], Bomberai Penin[CO], middle Menoo R[K], Bodim[L], Humboldt Bay[CO], Tami R[K], and lower Gogol R[M] (Hartert *et al.* 1936[K], Mayr & Meyer de Schauensee 1939b[J], Ripley 1964a[L], Bailey 1992a[M]).

DIAGNOSIS: Brownish olive dorsally, with a rufous tinge to uppertail coverts, remiges, and tail; ventral plumage buff with a yellowish-olive wash, slightly more saturated than in *fulviventris* (Higgins *et al.* 2008).

b. *Timeliopsis griseigula fulviventris* (Ramsay)

Plectorhyncha ? fulviventris Ramsay, 1882. *Proc. Linn. Soc. NSW* 6: 718.—Mountains of the SE Peninsula.

      Synonym: *Acrocephalus cervinus* De Vis, 1897. *Ibis*: 386.—Boirave, se NG. [*cf.* Iredale, 1956, *Birds New Guinea* 2: 152 and pl. 24, fig. 10.]

RANGE: The SE Peninsula and the e sector of the S Lowlands; west in the s watershed to Kiunga; west in the n watershed to the Watut R near the head of Huon Gulf (these two endpoint populations have not been formally diagnosed at this point). Localities—Kiunga[J], Lohiki[CO], lower Watut R[CO], Lakekamu basin[K], Brown R[CO], Kumusi R[MA], and Orangerie Bay[CO] (Beehler & S. Pruett-Jones: specimen deposited at Wau Ecology Institute[K], K. D. Bishop *in litt.*[J]).

DIAGNOSIS: More buffy rufous than nominate form; rufous olive to rufous brown dorsally, and with, in some cases, stronger rufous tinge on rump and tail; ventral plumage buff brown to pale rufous (Higgins *et al.* 2008). See photo in Coates & Peckover (2001: 183).

NOTES

The race *fulviventris* is thinly defined. The species is rare and elusive and thus probably has been overlooked in many localities. We suspect its distribution is more widespread than indicated by our range descriptions. Sighting from Batanta I of a pair on 19-Jul-1994 (S. F. Bailey *in litt.*). *Other names*—Tawny or Grey-throated Straight-billed Honeyeater, Lowland Straightbill.

Meliphaga Lewin, 1808. *Birds of New Holland*: 4, pl. 5. Type, by subsequent designation (G. R. Gray, 1840. *List Gen. Birds*: 15), *Meliphaga chrysotis* Lewin = *Ptilotis lewinii* Swainson.

The species of this genus range from the Lesser Sundas through New Guinea to Australia (Coates & Bishop 1997, Higgins *et al.* 2008). The genus is characterized by the lack of any bare ocular skin and the presence of a distinct white or yellow auricular patch, a narrow whitish or yellowish rictal streak, distinctive gape skin, and plain olive-green dorsal plumage. Situation of the genus within the family follows Driskell *et al.* (2004). Species sequence within the genus follows Norman *et al.* (2007). For the unpracticed, most of the species (with notable exceptions) are essentially impossible to identify with confidence in the field and also are very difficult in the hand or museum. The myriad subspecies described exemplify the subjective ability of practitioners to "see" differences in virtually every sample collected. Rand (1936a, 1941) created many new subspecies for the species of *Meliphaga*. We have synonymized many forms, but there is probably more work to be done toward full consolidation. Much of the described variation is a product of foxing, preparation of the study skin, and local individual variation related to age and sex. That said, the uniformity of plumages overlays and masks geographic phylogenetic structure within species, making the task of identifying and sorting out subspecies that much more difficult (Norman *et al.* 2007). Joseph *et al.* (2014) generated a phylogeny that divides the genus into two clades—the smaller clade including the species *aruensis*, *lewinii*, and *notata*, and the larger clade including the remaining species. Given the similarity of plumages and behaviors across the two groupings, we take the conservative approach and keep them all as *Meliphaga*, which is in accord with Dickinson & Christidis (2014).

Puff-backed Meliphaga *Meliphaga aruensis* ENDEMIC | MONOTYPIC

Ptilotis aruensis Sharpe, 1884. *Rept. Zool. Coll. Voy. Alert*: 19.—Aru Is.
  Synonym: *Ptilotis aruensis sharpei* Rothschild & Hartert, 1903. *Novit. Zool.* 10: 442.—Manokwari, Bird's Head.
  Mainland NG; Misool[J], Salawati[K,L,M], Batanta[MA], and Waigeo I[MA]; the Aru Is[MA], Yapen I[CO] (and nearby Keboi I)[CO], the D'Entrecasteaux Arch[MA], and Kiriwina I in the Trobriand Is[MA] (Mees 1965[J], Beckers *et al.* in M. Tarburton's island website[K], Matheve & van den Schoor 2009[L], specimen at BMNH[M]). Widespread on the Mainland (including the Trans-Fly; Bishop 2005a), inhabiting forest and edge; sea level–1,200, max 1,580 m[CO]. Salawati I material was collected by Frost in 1934 and is in the collection at the BMNH (specimen no. 1934.10.21-1889-89). First recorded on Normanby by Pratt in 2003 (specimens at BPBM).

NOTES
Species characteristics: greenish olive dorsally with a slight brown tint; rump tuft thick and prominent, with black and white fringing (of birds in the hand); lores olive; rictal streak pale yellow; ear patch medium yellow, large, and squarish; some blackish behind eye; bill rather blunt; size large; unmarked underparts; underwing creamy yellow-olive. Norman *et al.* (2007) suggested that there is a "southeastern form" that merits species status. We find this problematic since the two described subspecies (lumped here) both range into the SE Peninsula, segregating by watershed (one north, one south). So at this point, we follow Higgins *et al.* (2008) in not erecting a new species at this time given our inadequate data and confusing nonconformance of this molecular form with the minimal plumage variation. Wing (male) 85–94. We do not recognize the race *sharpei* because it is subsumed in the range of variation within the nominate form. The species shows considerable variation from population to population, which makes delineation of geographic subspecies difficult. *Other names*—Large-tufted Honeyeater.

Scrub Meliphaga *Meliphaga albonotata* ENDEMIC | MONOTYPIC

Ptilotis albonotata Salvadori, 1876. *Ann. Mus. Civ. Genova* 9: 33.—Naiabui, Hall Sound, SE Peninsula.
  Synonym: *Meliphaga montana auga* Rand, 1936. *Amer. Mus. Novit.* 872: 6.—Mafulu, SE Peninsula.
  Synonym: *Meliphaga montana setekwa* Rand, 1936. *Amer. Mus. Novit.* 872: 6.—Upper Setekwa R, Western Ra.
  Synonym: *Meliphaga montana gretae* Gyldenstolpe and Gilliard, 1955. *Arkiv f. Zool.* (2)8: 166.—Nondugl, Wahgi valley, Eastern Ra.
Throughout the hills and lower mts of mainland NG, in foothills to 1,950 m[CO]. Most common in scrub and garden habitats in the uplands. Rare in the lowlands. Apparently absent from forest interior. No

records from Yapen I, the NW Lowlands, or the Sepik-Ramu. Lowland records from Nomad, 15 km north of Kiunga, and the lower Kikori R (Bishop 2005b).

NOTES
Species characteristics: white auricular and pale yellow rictal streak; dorsal surface olive with a strong greenish tinge; bright yellow-green edging to flight feathers; underwing color creamy yellow; medium-size; medium-length dark bill; underparts rather dark but without mottling. *Other names*—Scrub or Southern White-eared Meliphaga, White-marked or Diamond Honeyeater.

Mimic Meliphaga *Meliphaga analoga*   ENDEMIC | MONOTYPIC
Ptilotis analoga Reichenbach, 1852. *Handb. Spec. Orn., Abth.* 2, Meropinae, continuatio IX: 103, tab. 467.—Triton Bay, Bird's Neck.
> Synonym: *Ptilotis longirostris* Ogilvie-Grant, 1911. *Bull. Brit. Orn. Club*, 29: 27.—Wamma I, Aru Is.
> Synonym: *Meliphaga analoga flavida* Stresemann & Paludan, 1932. *Novit. Zool.* 38: 147.—Yapen I, Bay Is.
> Synonym: *M. analoga connectens* Salomonsen, 1966a. *Breviora* Mus. Comp. Zool. Harvard 254: 5.—Madang, e verge of the Sepik-Ramu.
> Synonym: *Meliphaga mimikae rara* Salomonsen, 1966a. *Breviora* Mus. Comp. Zool. Harvard 254: 4.—Bernhard Camp, 50 m elevation, Taritatu R, Mamberamo basin, s sector of the NW Lowlands.
> Synonym: *Meliphaga analoga papuae* Salomonsen, 1966a. *Breviora* Mus. Comp. Zool. Harvard 254: 4.—Wuroi, Oriomo R, Trans-Fly, s NG.

Mainland NG, the four main NW Islands, Aru Is, Bay Is (Mios Num, Yapen), Kairiru and Muschu I[CO] (Mayr 1941); sea level–1,250 m, max 1,450 m (Coates & Peckover 2001). Edge of forest and second growth. No records from the n watershed of the SE Peninsula. In Trans-Fly observed at Kurik, Bensbach, and Lake Daviumbu (K. D. Bishop *in litt.*).

NOTES
The apparent minor regional variation is mainly based on specimen preparation, age, and perhaps sex. Species characteristics: entire dorsal surface greenish olive; face slightly darker; rictal streak and auricular patch pale yellow; auricular medium-size and rather broad; lores olive green; breast band obscure and darkish olive gray; no mottling of underparts; underwing mottled pale yellow; rump tuft medium-size and with pale tipping but no blackish; iris pale brown. The form "*Meliphaga mimikae rara*," described from a single specimen that Rand had already determined to belong to the species *M. analoga*, is placed in synonymy here (see LeCroy 2011: 71), along with four other forms that do not have merit. The Salomonsen subspecies were synonymized by Diamond (1972). *Other names*—Mimic or Allied Honeyeater, Yellow-spotted Meliphaga.

Tagula Meliphaga *Meliphaga vicina*   ENDEMIC | MONOTYPIC
Ptilotis analoga vicina Rothschild & Hartert, 1912. *Novit. Zool.* 19: 203.—Tagula I, Louisiade Arch, SE Islands.
> Endemic to Tagula I of the Louisiade Arch, sea level–800 m (Pratt & Beehler 2014). Abundant in forest and edge of the lowlands and hills.

NOTES
A large member of the *Meliphaga analoga* complex, with a longish bill. This is the only member of the group in the Louisiade Arch. Bill long and slim, rictal streak pale yellowish; ear spot rich yellow and small; underparts unstreaked, with an olive-yellow wash but no mottling; underwing creamy yellow, with an ochraceous tint; iris dark brown to gray; bill black or brownish black. Male: wing 88–96; tail 72–79; exposed culmen 18–20; tarsus 23. Female: wing 83–86; tail 68–71; exposed culmen 15.5–16.0 (Coates 1990). Molecular analysis should determine whether this form should be subsumed into *M. analoga. Other names*—Tagula or Louisiades Honeyeater, Sudest Meliphaga.

Mountain Meliphaga *Meliphaga orientalis*   ENDEMIC

Ptilotis flavirictus orientalis A. B. Meyer, 1894. *J. Orn.* 42: 92.—Southeastern Peninsula.

Forest interior of the uplands of NG and Waigeo I; 550–1,750 m, common above 1,300 m (Coates & Peckover 2001, Diamond & Bishop 2015). No records from the Cyclops Mts.

SUBSPECIES

a. *Meliphaga orientalis facialis* Rand
Meliphaga orientalis facialis Rand, 1936. *Amer. Mus. Novit.* 872: 16.—Siwi, Arfak Mts, Bird's Head.
RANGE: Waigeo I, the Bird's Head and Neck, and the s scarp of the Central Ra east to the se sector of the Eastern Ra (Okapa). Birds from the Fakfak and Kumawa Mts (Diamond 1985) are presumably of this race.
DIAGNOSIS: Facial mask less pronounced than for nominate form (Higgins *et al.* 2008). Ventral mottling reduced. See photo in Coates (1990: 288).

b. *Meliphaga orientalis citreola* Rand
Meliphaga analoga citreola Rand, 1941. *Amer. Mus. Novit.* 1102: 14.—6 km sw of Bernhard Camp, 1,200 m, ne sector of the Western Ra.
   Synonym: *Meliphaga orientalis becki* Rand, 1936. *Amer. Mus. Novit.* 872: 17.—Zakaheme, 1,220 m, Huon Penin.
RANGE: The n scarp of the Western and Border Ra; northern coastal ranges: Foja, Bewani, Torricelli, Prince Alexander, and Adelbert Mts, and mts of Huon; also the Herzog Mts and Kuper Ra of the nw sector of the SE Peninsula. Birds of the Wandammen Penin (Diamond 1985) may belong here. Diamond (1985) also noted a possible observation of this species in the montane forest of Yapen I.
DIAGNOSIS: Notable yellow wash to plumage; reduced ventral mottling (Higgins *et al.* 2008); wing (male) 71–81 (Rand & Gilliard 1967). Small size; bill long, narrow, and blackish; similar to nominate form but for olive lores and postocular; reduced ventral mottling; small auricular spot. Pratt's (1983) examination of the populations supports the lumping of *becki*. See photo of *citreola* in Coates (1990: 288). Diamond's discussion (1969: 43–45) suggested the lumping of *becki* and *facialis*. We find *facialis* to exhibit a distinctive facial coloration (paler than nominate).

c. *Meliphaga orientalis orientalis* (A. B. Meyer)
RANGE: Mountains of the SE Peninsula, east of the preceding two forms. Specimens (Austral. Nat. Wildl. Coll.) from Aseki (s watershed) appear to belong here (Coates 1990).
DIAGNOSIS: Wing (male) 73–79 (Rand & Gilliard 1967). Overall size small; bill small, all dark, and slim; crown olive green with some obscure darker speckling; lores and postocular blackish; rictal streak pale yellow; auricular patch medium yellow, small, and teardrop shaped; mantle and back bright olive green; underwing bright yellow; entire underparts darkish with fine obscure mottling throughout; yellowish wash along midline.

NOTES
The races of this species are thinly defined and perhaps merit sinking. *Other names*—Hill-forest or Mountain Honeyeater, Small Spot-breasted Meliphaga.

White-eared Meliphaga *Meliphaga montana*   ENDEMIC

Ptilotis montana Salvadori, 1880. *Ann. Mus. Civ. Genova* 16: 77.—Arfak Mts, Bird's Head.

Foothill forest on the Bird's Head and Bird's Neck (Diamond & Bishop 2015) and the n watershed of the main body of NG as far east as the Sibium Mts (Clapp 1987) and Goodenough Bay (Austral. Nat. Wildl. Coll. specimen, collector R. Schodde); also Batanta and Yapen I and the Fakfak, Kumawa, Van Rees, and Foja Mts (Diamond 1985); 500–1,500 m[CO]. There is a single Austral. Nat. Wildl. Coll. specimen (collected by W. B. Hitchcock) from the s coast of the SE Peninsula at Cape Rodney (Coates 1990). Apparently the range of this species in the n watershed turns the corner at Milne Bay and extends west onto the s scarp to where it meets *M. mimikae* southeast of Port Moresby.

SUBSPECIES

a. *Meliphaga montana montana* (Salvadori)
   Synonym: *Meliphaga montana germanorum* Hartert, 1930. *Novit. Zool.* 36: 47.—Cyclops Mts, ne sector of NW Lowlands, n NG.

Synonym: *Meliphaga montana sepik* Rand, 1936. *Amer. Mus. Novit.* 872: 7.—Hunstein Mts, middle Sepik, north-central NG.

Synonym: *Meliphaga montana huonensis* Rand, 1936. *Amer. Mus. Novit.* 872: 8.—Junzaing, Huon Penin.

Synonym: *Meliphaga montana aicora* Rand, 1936. *Amer. Mus. Novit.* 872: 9.—Aicora R, Oro Prov, SE Peninsula.

Synonym: *Meliphaga montana margaretae* Greenway, 1966. *Amer. Mus. Novit.* 2258: 22.—Mount Besar, 850 m, Batanta I, NW Islands.

RANGE: Uplands of the Bird's Head and Neck, Batanta I[J], and the n watershed of NG; the PNG North Coastal Ra; and the s watershed of the far e sector of the SE Peninsula to Cape Rodney (Greenway 1966[J]). Presumably the populations in the Fakfak, Kumawa, and Foja Mts pertain to this expanded form (Diamond 1985).

DIAGNOSIS: Indistinct yellowish streaking on belly; indistinct greenish collar; dark-crowned.

b. *Meliphaga montana steini* Stresemann & Paludan

Meliphaga montana steini Stresemann & Paludan, 1932. *Novit. Zool.* 38: 222.—Yapen I, 850 m, Bay Is.

RANGE: Hill forest on Yapen I.

DIAGNOSIS: Browner dorsally than nominate, but very thinly defined and perhaps not meriting subspecific recognition.

NOTES

Meliphaga montana auga, *gretae*, and *setekwa* are junior synonyms of *M. albonotata* and treated in that account. The forms *margaretae*, *germanorum*, *sepik*, *aicora*, and *huonensis* are here subsumed into nominate *montana* for lack of objective diagnostic characters. The form *steini* needs further museum review. It will be interesting to look at how *M. montana* and *M. mimikae* segregate in sympatry in the far SE Peninsula. *Other names*—Forest White-eared Meliphaga, White-eared Mountain Meliphaga, Forest Honeyeater.

Mottled Meliphaga *Meliphaga mimikae*   ENDEMIC | MONOTYPIC

Ptilotis mimikae Ogilvie-Grant, 1911. *Bull. Brit. Orn. Club* 29: 27.—Mimika R, w sector of the S Lowlands.

Synonym: *Meliphaga mimikae granti* Rand, 1936. *Amer. Mus. Novit.* 872: 4.—Mafulu, 1,250 m, s watershed of the SE Peninsula.

Synonym: *Meliphaga mimikae bastille* Diamond, 1967. *Amer. Mus. Novit.* 2284: 12.—Karimui, 1,110 m, se sector of Eastern Ra.

Hill forest interior of the s watershed of NG; 500–1,150 m, max 1,400 m[CO] (Diamond 1972). There is a single record from the n scarp of the SE Peninsula in the Hydrographer Mts (Coates 1990). Replaced over most of the n watershed and the Bird's Head by *M. montana*.

NOTES

Species plumage characters: dorsal surface olive green with a distinct yellowish wash; rictal streak and auricular patch pale yellow, but rictal slightly paler; auricular patch medium-size; forehead and lores dark olive green; underparts heavily mottled on breast and belly; underwing dull ocher yellow (diagnostic); bill medium-size and shape (type series). Wing (male) 86–95. The two named clinal populations are merged with the nominate form because of inadequately distinctive characters. Shown to be the sister form of *M. montana* of the n watershed (Norman *et al.* 2007). In the SE Peninsula, both forms cross the main range and apparently meet. The form "*Meliphaga mimikae rara*," based on a single specimen from the n watershed of NG, is in fact an atypical specimen of *M. analoga* (see LeCroy 2011: 71) and has been placed in synonymy under that species. The race *bastille* is the same size as the nominate, and plumage distinctions are based on pecularities of specimen preparation. *Other names*—Spot-breasted Meliphaga; Spot-breasted, Large Spot-breasted, or Mottle-breasted Honeyeater.

Yellow-gaped Meliphaga *Meliphaga flavirictus*   ENDEMIC

Ptilotis flavirictus Salvadori, 1880. *Ann. Mus. Civ. Genova* 16: 76.—Fly R (240 km upstream).

Mainland NG, sea level–1,400 m[CO], but apparently sparsely distributed in uplands. Rarely seen because of its canopy-dwelling habits. In the SE Peninsula, no records southeast of Wau in the n watershed and no records east of Port Moresby in the s watershed. Also may be absent from the n coast east of the mouth of the Sepik (but these range gaps may be a product of inadequate sampling). Best located when foraging at flowers in the forest canopy.

a. *Meliphaga flavirictus crockettorum* Mayr & Meyer de Schauensee

Meliphaga flavirictus crockettorum Mayr & Meyer de Schauensee, 1939. *Proc. Acad. Nat. Sci. Phila.* 91: 142.—Bamoskabu, s flank of Tamrau Mts, Bird's Head.

RANGE: The Bird's Head and Neck, NW Lowlands, and Sepik-Ramu; also a record from the Herzog Mts (Wau), and from the westernmost sector of the S Lowlands (Otakwa R).

Localities—Tamrau Mts[J], Otakwa R[MA], Bernhard Camp on the Taritatu R[K], middle Sepik R[L], and Wau[MA] (Mayr & Meyer de Schauensee 1939b[J], Rand 1942b[K], Gilliard & LeCroy 1966[L]).

DIAGNOSIS: Bill slim; wing (male) 74.5 (Mayr & Meyer de Schauensee 1939b). Rich olive-green dorsal plumage; forehead and crown as for mantle; rictal streak broad and richer yellow than ear patch, which is pale yellow and rounded; lores dark green; rump tuft thick and green-tipped; bill blackish.

b. *Meliphaga flavirictus flavirictus* (Salvadori)

RANGE: The S Lowlands, from the Trans-Fly east to the s watershed of the SE Peninsula. Localities—Sturt I[J] and Oroville Camp[J] (Fly R), Gaima[J], Elevala R[K], Mengino[DI], Karimui[DI], Lakekamu basin[L], and monsoon forest near Port Moresby[CO] (Rand 1942a[J], P. Gregory *in litt.*[K], Beehler field notes[L]).

DIAGNOSIS: Dorsally pale gray-olive; side of head grayer; auricular patch very pale yellow; rictal streak prominent and same color as ear patch. Wing (male) 77–82 (Rand 1936a). Bill brownish horn-colored in many specimens. See photos in Coates (1990: 295) and Coates & Peckover (2001: 182).

NOTES

A small and slim-billed canopy dweller that is seldom encountered. This species exhibits the most prominent yellow gape flange and rictal streak of any meliphaga. Small, pale yellow ear patch (paler than rictal streak). Yellowish-tinged legs also diagnostic (gray in other meliphagas). Best found by voice (Pratt & Beehler 2014). *Other names*—Yellow-gaped Honeyeater.

Elegant Meliphaga *Meliphaga cinereifrons*   ENDEMIC | MONOTYPIC

Meliphaga gracilis cinereifrons Rand, 1936. *Amer. Mus. Novit.* 872: 20.—Rona, s watershed of the SE Peninsula.

    Synonym: *Meliphaga gracilis stevensi* Rand, 1936. *Amer. Mus. Novit.* 872: 20.—Confluence of the Bulolo and Watut R, Morobe Prov, n sector of SE Peninsula.

Restricted to the SE Peninsula: in the s watershed from Malalaua to Milne Bay; in the n watershed from the Bulolo valley southeast to Milne Bay. Mainly a hill-forest species, typically to 600 m, max 1,200 m[CO].

NOTES

Wing (male) 75–82 (Coates 1990). We follow Higgins *et al.* (2008) in treating the form as monotypic. See several photos in Coates (1990: 291–94). The species can be distinguished with difficulty from other small forms, such as *M. gracilis*, with which this form has traditionally been clustered. Separated by Coates (1990) and subsequently by the molecular studies of Norman *et al.* (2007). The species is distinguished by the following: dark sides of face and often blackish hind edge to the large yellow ear spot; ear spot large and round, often with a projection reaching upward behind the eye; gape skin orange; thin yellowish-white rictal streak; underwing coverts with a dilute ochraceous wash; variable pale eye-ring; iris ranges from dark brown to gray. The form *stevensi* of the n watershed of the SE Peninsula is molecularly anomalous, with some samples allying with *cinereifrons* and others with *analoga*. That said, topotypical *stevensi* collected by Herbert Stevens in the upper Watut drainage are not morphologically distinct from nominate *cinereifrons*, and the form is not recognized here (*pace* Rand 1936). We look forward to further field and molecular analysis that might unravel this enigma. *Other names*—Elegant Honeyeater; formerly treated as a race of Graceful Meliphaga, *M. gracilis*.

Graceful Meliphaga *Meliphaga gracilis*   RESIDENT

Ptilotis gracilis Gould, 1866. *Proc. Zool. Soc. London*: 217.—Cape York Penin, QLD, Australia.

    The S Lowlands and the Trans-Fly, from the Bird's Neck east to the Purari R delta; also the Aru Is (Coates 1990). Inhabits mangroves, swamp forest, edge, scrub, woodland, and gardens; sea level–300 m, max 800 m (Pratt & Beehler 2014).

    Extralimital—Cape York Penin and ne QLD.

a. *Meliphaga gracilis gracilis* (Gould)

RANGE: Southern NG from Triton Bay east to the Purari R; Aru Is; also Cape York Penin, QLD. Records from Dolak I, Kurik, Wasur NP, Bian Lakes, Bensbach, and Lake Daviumbu (Bishop 2005b).

DIAGNOSIS: A very small form with a slender bill; wing (male) 71–77 (Rand & Gilliard 1967). Greenish-olive upperparts; unmottled ventral plumage, with yellowish wash on center of belly; underwing pale yellow; auricular spot small and rounded; rictal streak reduced and pale yellow; chin pale washed with gray-olive tint. Differs from *imitatrix* of n QLD by the paler olive mantle and paler breast with a pale creamy-yellow cast (Schodde & Mason 1999).

NOTES

The named populations *cinereifrons* and *stevensi*, originally allied with the species *gracilis*, have been shown to belong to a distinct species, *M. cinereifrons*, from molecular studies by Norman *et al.* (2007). *Other names*—Slender-billed Meliphaga, Graceful Honeyeater, Lesser Yellow-spotted Honeyeater.

GENUS *GAVICALIS* [Genus 3 spp/Region 1 sp]

Gavicalis Schodde & Mason, 1999. *Dir. Austral. Birds, Passerines*: 229. Type, by original designation, *Melithreptus virescens* Vieillot.

The species of this genus range from Australia to coastal and insular New Guinea. They are characterized by medium size, dorsal plumage darker than ventral plumage, ventral streaking, medium-length bill, prominent narrow black mask, yellow subocular plume, white auricular plume, and no bare ocular skin (Schodde & Mason 1999). Deployment of this genus for these species in the place of *Lichenostomus* follows Nyári & Joseph (2011).

Varied Honeyeater *Gavicalis versicolor* RESIDENT

Ptilotis versicolor Gould, 1843. *Proc. Zool. Soc. London* (1842): 136.—Cape York Penin, QLD, Australia.

Coastal NG, the NW Islands, Yapen I, selected SE Islands, and many small islets on the coastal shelf; near sea level. Inhabits coastal vegetation, particularly coconut groves, village shrubbery, gardens, and mangroves. Abundant in the n coastal towns of Jayapura, Vanimo, Wewak, and Madang. Ranges from treetop to shrubbery levels (Coates 1990).

Extralimital—coastal n QLD.

SUBSPECIES

a. *Gavicalis versicolor sonoroides* (G. R. Gray)

Ptilotis sonoroides G. R. Gray, 1862. *Proc. Zool. Soc. London* (1861): 428.—Waigeo I, NW Islands.

  Synonym: *Meliphaga virescens intermedia* Mayr & Rand, 1935. *Amer. Mus. Novit.* 814: 15.—Samarai I, Milne Bay.

  Synonym: *Meliphaga versicolor vulgaris* Salomonsen, 1966a. *Breviora* Mus. Comp. Zool. Harvard 254: 5.—Finschhafen, Huon Penin.

RANGE: The Schildpad Is, the four main NW Islands, and Yapen I; the entire n coast of NG from the Bird's Head (Sorong) to Milne Bay; also the Killerton Is and Doini, Samarai, Fergusson[J], and Normanby I[K] (Mayr 1941, LeCroy *et al.* 1984[J], Pratt field notes[K]).

DIAGNOSIS: Dorsal coloration more gray-brown than that of nominate form; ventral plumage background dull off-white to grayish white, with only a faint yellow wash on breast. We follow suggestion of Higgins *et al.* (2008) in synonymizing the races *vulgaris* and *intermedia* into *sonoroides*. See photo in Coates (1990: 278).

b. *Gavicalis versicolor versicolor* (Gould)

RANGE: Coastal s NG between Timika and Milne Bay, islands of Torres Strait, and coastal ne QLD to Townsville (Coates 1990, Higgins *et al.* 2008). Localities—south of Timika in mangroves[L], Dolak I, Wasur NP, Gaima, Aramia wetlands, and Daru (Bishop 2005a, S. van Balen *in litt.*[L]).

DIAGNOSIS: Substantial yellow wash on throat and breast; mantle washed greenish (Higgins *et al.* 2008).

Traditionally considered conspecific with *virescens* (Mayr 1941, Rand & Gilliard 1967). Salomonsen (1967) treated as specifically distinct. Ford (1978) and Schodde & Mason (1999) discussed geographic variation and species limits among regional populations. *Other names*—Yellow-streaked Honeyeater, *Meliphaga* or *Lichenstomus versicolor*.

GENUS *PTILOTULA* [Genus 6 spp/Region 1 sp]

Ptilotula Mathews, 1912. *Novit. Zool.* 18 (1911): 414. Type, by monotypy, *Ptilotis flavescens* Gould.

The species of this genus inhabit Australia and extend marginally into the Southeastern Peninsula of New Guinea. The genus is characterized by the small size, smallish black bill, yellow- or olive-washed plumage, pale ventral plumage, no bare ocular skin, and twin auricular plumes (either yellow and black or white and black). Deployment of this genus for this species in the place of *Lichenostomus* follows Nyári & Joseph (2011).

Yellow-tinted Honeyeater *Ptilotula flavescens* RESIDENT

Ptilotis flavescens Gould, 1840. *Proc. Zool. Soc. London* (1839): 144.—Derby, far nw sector of Western Australia.

Restricted to the rain-shadow savanna near Port Moresby, sea level–450 m. NG range extends from the vicinity of the Aroa R southeast to Port Moresby and inland to the Sogeri Plateau (Sirinumu Dam; Coates 1990). Inhabits eucalypt savanna; also frequents town gardens but only where eucalypts are present. Absent from the Trans-Fly.

Extralimital—inhabits a broad band across n Australia, but generally absent from the Cape York Penin.

a. *Ptilotula flavescens flavescens (Gould)*

Synonym: *Ptilotis germana* Ramsay, 1878. *Proc. Linn. Soc. NSW* 3: 2.—Laloki R, Port Moresby distr, SE Peninsula.

RANGE: Savannah country of the s watershed of the SE Peninsula and much of n Australia (Higgins *et al.* 2008).

DIAGNOSIS: When compared to *melvillensis* (Melville I, NT, Australia), plumage of belly shows reduced yellow wash; crown and nape more gray (Higgins *et al.* 2008). Forehead pale dusty yellow; breast and flanks lightly streaked pale gray; face and throat concolorous (Schodde & Mason 1999). We follow Higgins *et al.* (2008) in lumping *germana* into the nominate form. See photos of *flavescens* in Coates (1990: 281–82).

Treatment follows Mayr (1941), *pace* Salomonsen (1967), who considered the NG form to be conspecific with *P. fusca*. *Other names*—Pale-yellow or Yellowish Honeyeater, *Lichenostomus flavescens*.

GENUS *CALIGAVIS* [Genus 3 spp/Region 2 spp]

Caligavis Iredale, 1956. *Birds New Guinea* 2: 150. Type, by original designation, *Ptilotis obscura* De Vis.

The species of this genus inhabit forest interior in New Guinea and eastern Australia. The genus is defined by the prominent periocular pattern of bare skin and black, yellow, and white facial plumes; body plumage is olive green (Schodde & Mason 1999). Deployment of this genus for these species after the disarticulation of *Lichenostomus* follows Nyári & Joseph (2011).

Obscure Honeyeater *Caligavis obscura* ENDEMIC | MONOTYPIC

Ptilotis obscura De Vis, 1897. *Ibis*: 383.—Mount Scratchley, SE Peninsula.

Synonym: *Meliphaga obscura viridifrons* Salomonsen, 1966. *Breviora* Mus. Comp. Zool. Harvard 254: 7.—Bamoskaboe, 700 m, Karoon, Tamrau Mts, Bird's Head.

Patchily distributed in hill forest on the Mainland: the Bird's Head, NW Lowlands, Central Ra, Foja Mts[j], and PNG North Coastal Ra[K]; mainly 200–1,100 m, extremes 75[J]–1,400 m[CO] (Beehler *et al.* 2012[J], I. Woxvold *in litt.*[J], specimens at AMNH[K]). Range includes the Weyland Mts[L], Van Rees Mts, upper Fly, Elevala R, Kikori R (Hartert *et al.* 1936[L], K. D. Bishop *in litt.*). No records from the Fakfak, Kumawa, Wandammen, or Cyclops Mts; also apparently absent from ne NG from the mouth of the Sepik southeast to Morobe station.

NOTES

The minor differences defining *viridifrons* do not merit recognition. Salomonsen described this race from material collected by S. Dillon Ripley in 1938, which had been meticulously reviewed already by Mayr & Meyer de Schauensee (1939b). *Other names*—Lemon-cheeked Honeyeater, *Lichenostomus* or *Oreornis obscurus, Meliphaga obscura.*

Black-throated Honeyeater *Caligavis subfrenata*   ENDEMIC | MONOTYPIC

Ptilotis subfrenata Salvadori, 1876. *Ann. Mus. Civ. Genova* (1875) 7: 948.—Hatam, Arfak Mts, Bird's Head.
     Synonym: *Ptilotis salvadorii* Hartert, 1896. *Novit. Zool.* 3: 531.—Mount Victoria, Owen Stanley Ra, SE Peninsula.
     Synonym: *Xanthotis melanolaema* Reichenow, 1915 (early). *J. Orn.* 63: 127.—Schrader Ra, n sector of the Eastern Ra.
     Synonym: *Ptilotis salvadorii utakwensis* Ogilvie-Grant, 1915 (late). *Ibis,* Jubilee Suppl. 2: 71.—Upper Otakwa R, sw
        sector of Western Ra.
Montane forests of the Bird's Head, Central Ra, and Huon Penin; 2,000–3,500 m, extremes 1,070–3,680 m[CO]. Mainly found in forest interior in the upper stories of the forest.

NOTES

Treated here as monotypic. Rand & Gilliard (1967) stated the various described subspecies are "very similar." The differences are of marginal significance, most likely clinal (paler, more olive in west, darker and green in the east). It is interesting to note the lack of subspecific distinction of the Bird's Head form from those of the Central Ra (K. D. Bishop *in litt.*). *Other names*—Black-fronted or Sub-bridled Honeyeater, *Meliphaga subfrenata, Lichenostomus* or *Oreornis subfrenatus.*

GENUS *OREORNIS*   ENDEMIC | MONOTYPIC

Oreornis van Oort, 1910. *Notes Leyden Mus.* 32: 214. Type, by monotypy, *Oreornis chrysogenys* van Oort.
    The single species of the genus is endemic to a few subalpine regions in the highest reaches of the Western Ranges. The genus is characterized by the large size, predominantly olive-green plumage, the long and broad orangish-golden-feathered malar stripe, and the high-elevation subalpine habitat. This distinctive species has not yet been the subject of molecular studies, and its relationships and placement within honeyeaters is unresolved. By plumage, the genus is apparently close to *Caligavis.*

Orange-cheeked Honeyeater *Oreornis chrysogenys*   ENDEMIC | MONOTYPIC

Oreornis chrysogenys van Oort, 1910. *Notes Leyden Mus.* 32: 215.—Oranje Mts, 4,150 m, Western Ra.
    High subalpine forest and alpine shrublands in the Western Ra; 3,250–4,150 m, min 2,450 m. Known only from the highest ranges west of the Baliem R gorge.
    Localities—Water-val Bivak of the upper Ilaga valley[J], 7 km east of Mt Wilhelmina[K], and Lake Habbema[K] (Rand 1942b[K], Ripley 1964a[J]).

NOTES

Should be looked for in the Mt Mandala highlands east of the Baliem valley. Ripley (1964a) reported obtaining specimens at 2,440 m. This seems very low, and was probably a result of a bad guess of where specimens were collected by a local hunter. *Other names*—*Lichenostomus* or *Meliphaga chrysogenys.*

Melidectes P. L. Sclater, 1874. *Proc. Zool. Soc. London* (1873): 694. Type, by monotypy, *Melidectes torquatus* P. L. Sclater.

The species of this genus inhabit montane forest of nearly all of New Guinea's major mountain ranges but do not occur off the Mainland. We follow Coates (1990) and Anderson *et al.* (2014a) in considering the Bismarck Honeyeater (*Vosea whitemanensis*) and the bearded honeyeaters (*Melionyx*) as generically distinct (*pace* Higgins *et al.* 2008, Dickinson & Christidis 2014). The genus is characterized by the complex facial patterning of bare skin and colored feathering, the presence of gular wattling, the rufous undertail coverts, and the green edging to remiges and rectrices. Unlike the species of *Melionyx*, the species of *Melidectes* are infamous vocalizers—among the great vocal gymnasts of the New Guinea montane forest. Unlike *Vosea*, all species of this genus exhibit gular wattling and predominantly gray plumage with pale scalloping of the mantle. *Vosea whitemanensis*, a New Britain endemic often placed in *Melidectes* (*e.g.*, Dickinson & Christidis 2014), clusters with *Myzomela*, not *Melidectes* (Andersen *et al.* 2014a).

Ornate Melidectes *Melidectes torquatus* ENDEMIC

Melidectes torquatus P. L. Sclater, 1874. *Proc. Zool. Soc. London* (1873): 694, pl. 55.—Hatam, Arfak Mts, Bird's Head.

Mountains of the Bird's Head, the Central Ra, Adelbert Mts, and mts of the Huon Penin; 1,100–1,800 m, extremes 900–2,200 m (Coates & Peckover 2001). Inhabits forest and edge. Conspicuous in tall shade trees in garden habitats in upland interior montane valleys, including town environs such as Mt Hagen.

SUBSPECIES

a. *Melidectes torquatus torquatus* P. L. Sclater

Synonym: *Melidectes torquatus nuchalis* Mayr, 1936. *Amer. Mus. Novit.* 869: 7.—Weyland Mts, 1,500 m, Western Ra.

Synonym: *Melidectes torquatus mixtus* Rand, 1941. *Amer. Mus. Novit.* 1102: 14.—Baliem R, 1,600 m, Western Ra.

RANGE: Mountains of the Bird's Head, the Western Ra, and the Border Ra, east at least to the Victor Emanuel Mts and perhaps as far as the Strickland R gorge.

DIAGNOSIS: No throat wattles; wing 115 (Rand & Gilliard 1967).

b. *Melidectes torquatus polyphonus* Mayr

Melidectes torquatus polyphonus Mayr, 1931. *Mitt. Zool. Mus. Berlin*, 17: 660.—Dawong, Herzog Mts.

RANGE: The Central Ra, presumably from the Strickland R gorge east to the nw sector of the SE Peninsula (near Wau); also the Adelbert Mts (Pratt 1983).

DIAGNOSIS: Larger than the nominate form; slightly darker dorsally; superciliary stripe rufous; mantle black-brown; hindneck rufous brown; darker ventrally; gular wattles absent (Rand & Gilliard 1967, Higgins *et al.* 2008). Only thinly distinct from the nominate form. See photos in Coates (1990: 264) and Coates & Peckover (2001: 171). Wing 120 (Rand & Gilliard 1967).

c. *Melidectes torquatus cahni* Mertens

Melidectes torquatus cahni Mertens, 1923. *Senckenbergiana* 5: 229.—Kulungtufu, Huon Penin.

RANGE: Mountains of the Huon Penin.

DIAGNOSIS: Small and dark dorsally; ventral plumage pale rufous; gular wattle small but obvious (Higgins *et al.* 2008). Thinly distinct from the nominate form. Wing 117 (Rand & Gilliard 1967).

d. *Melidectes torquatus emilii* A. B. Meyer

Melidectes Emilii A. B. Meyer, 1886. *Zeitschr. Ges. Orn.* 3: 22.—Mount Maguli, SE Peninsula.

RANGE: Mountains of the SE Peninsula.

DIAGNOSIS: Reduced white throat patch; prominent orange gular wattle joined to gape wattle; ventral plumage strongly rufous (Higgins *et al.* 2008). LeCroy (2011) noted that the type locality may be incorrect. See her discussion of the mysteries of the source of the type material.

The forms *polyphonus* and *cahni* merit additional examination and perhaps should be synonymized with the nominate form. We take that step with *nuchalis* and *mixtus*, which are not reliably distinguishable from nominate *torquatus*. *Other names*—Ornate Honeyeater, Ornamental Melidectes, Cinnamon-breasted Wattlebird.

Cinnamon-browed Melidectes *Melidectes ochromelas*   ENDEMIC | MONOTYPIC

Melirrhophetes ochromelas A. B. Meyer, 1874. *Sitzungsber. Akad. Wiss. Wien* 70: 111.—Hatam, Arfak Mts, Bird's Head.

> Synonym: *Melirrhophetes batesi* Sharpe, 1886. *Nature* 34: 340.—Sogeri distr, s watershed of SE Peninsula.
> Synonym: *Melidectes ochromelas lucifer* Mayr, 1931. *Mitt. Zool. Mus. Berlin* 17: 661.—Ogeramnang, Huon Penin.

Patchily distributed through the mts of the far west, north, northeast, and southeast, and largely absent in the middle sections of NG. Recorded from the Bird's Head[MA], Wandammen Penin[MA], Foja Mts[J], Weyland Mts[MA], c Western Ra[MA], and mts of Huon[MA] and SE Peninsulas[MA]; 1,600–2,420[K] m (Hartert *et al.* 1936, Beehler *et al.* 2012[J], Freeman *et al.* 2013[K]). Inhabits lower montane to mid-montane forest and edge. Replaced at higher elevations by *M. belfordi* in the SE Peninsula, *M. foersteri* on the Huon Penin, and *M. leucostephes* on the Bird's Head. Generally scarce or absent, though may be locally common.

A comparison of specimens collected in the same year (1929) of topotypical material from the three races confirms that the species should be treated as monotypic. Much of the variation is due to variable degrees of foxing and specimen preparation. A record from Ubaigubi, south-central sector of the Eastern Ra (*PNG Bird Soc. Newsl.* 214: 4) requires corroboration. *Other names*—Cinnamon-browed Honeyeater, Mid-Mountain Melidectes.

Vogelkop Melidectes *Melidectes leucostephes*   ENDEMIC | MONOTYPIC

Melirrhophetes leucostephes A. B. Meyer, 1874. *Sitzungsber. Akad. Wiss. Wien* 70: 110.—Hatam, Arfak Mts, Bird's Head.

> Mountains of the Bird's Head and Neck, including the Tamrau, Arfak, Fakfak[K] and Kumawa[J] Mts; montane forest and edge, 1,000–1,800 m (Diamond 1985[J], Gibbs 1994[K]). Very shy and difficult to observe in the Arfak Mts.
> *Other names*—Vogelkop Honeyeater, Arfak Melidectes, White-fronted Melidectes or Honeyeater.

Huon Melidectes *Melidectes foersteri*   ENDEMIC | MONOTYPIC

Melirrhophetes foersteri Rothschild & Hartert, 1911. *Bull. Brit. Orn. Club* 29: 12.—Rawlinson Ra, e sector of the Huon Penin.

> Frequents upland forest through the mts of the Huon Penin, 1,600[CO]–3,700[L] m (Beehler field notes 1975[L]). Inhabits canopy vegetation of forest interior. Rarely mist-netted. Recent surveys on the YUS transect did not detect this species below 2,300 m (Beehler field notes).

As with *M. leucostephes* of the Bird's Head, this species is very shy and difficult to observe in the field, quite unlike the vocal, curious, confiding, and omnipresent *M. belfordi* of the Central Ra. *Other names*—Huon Honeyeater, Huon Wattled Honeyeater, Foerster's Melidectes.

Belford's Melidectes *Melidectes belfordi*   ENDEMIC

Melirrhophetes belfordi De Vis, 1890. *Ann. Rept. British New Guinea*, 1888–89: 60.—Mount Knutsford, SE Peninsula.

> A very conspicuous inhabitant of montane forest of the Central Ra; 1,600–3,350 m[co], extremes 1,400–3,800 m[co]. The very similar *M. rufocrissalis* and *M. ochromelas* usually occupy lower elevations when they coexist with *M. belfordi*. This is the only *Melidectes* occurring higher than 2,400 m in the Central Ra. Populations in some areas (*e.g., M. b. griseirostris* of the Mt Goliath area of the Western Ra) are of hybrid

origin (*M. belfordi* × *M. rufocrissalis*) but exhibit a stable phenotype (Stresemann 1923). The species is absent from all outlying ranges, including Mt Bosavi.

a. *Melidectes belfordi joiceyi* (Rothschild)

Melirrhophetes belfordi joiceyi Rothschild, 1921. *Novit. Zool.* 28: 285.—Kunupi, 1,830 m, Weyland Mts, Western Ra.

> Synonym: *Melidectes belfordi kinneari* Mayr, 1936. *Bull. Brit. Orn. Club* 57: 42.—Otakwa R, sw slopes of the Western Ra.

RANGE: The w and c sector of the Western Ra (Hartert *et al.* 1936).
DIAGNOSIS: Small, with distinct greenish-olive edging to feathers of mantle (Rand & Gilliard 1967). The form *kinneari* not distinct.

b. *Melidectes belfordi belfordi* (De Vis)

> Synonym: *Melidectes leucostephes brassi* Mayr & Rand, 1936. *Mitt. Zool. Mus. Berlin* 21: 247.—Mount Tafa, 2,000 m, SE Peninsula.

RANGE: The e sector of the Western Ra; also the Border and Eastern Ra, and mts of the SE Peninsula.
DIAGNOSIS: Larger than preceding form and back dull gray (Rand & Gilliard 1967). Edging to feathers of mantle gray, not olive green as in *joiceyi*. The low-elevation form *brassi* is included here; plumage identical, size marginally smaller—described as a nongeographic "elevational subspecies." See photos of the nominate form in Coates (1990: 260) and Coates & Peckover (2001: 170).

c. *Melidectes belfordi griseirostris* (Rothschild & Hartert)

Melirrhophetes belfordi griseirostris Rothschild & Hartert, 1911. *Bull. Brit. Orn. Club* 29: 34.—Mount Goliath, Border Ra.

RANGE: Mount Goliath, in the s watershed of the w sector of the Border Ra.
DIAGNOSIS: Bill pale gray; also note presence of throat and gape wattles. We agree with LeCroy (2011) that this is not "intergradient" but instead a stable population of hybrid origin and thus merits recognition.

d. *Melidectes belfordi schraderensis* Gilliard & LeCroy

Melidectes belfordi schraderensis Gilliard & LeCroy, 1968. *Amer. Mus. Novit.* 2343: 33.—Mount Kominjim, 2,530 m, Schrader Ra, n sector of the Eastern Ra.
RANGE: Schrader Ra.
DIAGNOSIS: Orbital skin lemon yellow rather than blue or blue-green (Higgins *et al.* 2008). Bill black; eyebrow and ear patch white; small gape wattle present. Probably of hybrid origin.

Pure *M. belfordi* exhibits a black forehead, a black bill that is shorter than that of *M. rufocrissalis*, a deep blue face patch, and a white eyebrow; and lacks an orange gular wattle. Hybridization with *M. rufocrissalis* probably occurs extensively at mid-montane levels where the natural forest habitat has been extensively cleared for gardening by humans. In the Wahgi region a zone of hybridization occurs between *ca.* 1,900 and 2,300 m, with birds above 2,300 m closest to Belford's, while those below 2,100 m are very close to pure *M. rufocrissalis* (Mayr & Gilliard 1952, Gilliard 1959). We have moved the form *stresemanni*, a subspecies of hybrid origin, to *M. rufocrissalis*, as it exhibits many more characters of *rufocrissalis* than *belfordi*. *Other names*—Belford's Honeyeater, Black-bill.

Yellow-browed Melidectes *Melidectes rufocrissalis*   ENDEMIC

Melirrhophetes rufocrissalis Reichenow, 1915. *J. Orn.* 63: 126.—Schraderberg, Sepik Mts, n sector of the Eastern Ra.

The Border Ra, Eastern Ra, and mts of northernmost SE Peninsula, 1,200–2,300 m, from Mt Goliath east to the s slopes of the Wau valley. Evidence of hybridization with *M. belfordi* throughout its range wherever habitats have been disturbed. Center of distribution seems to be the Border and Eastern Ra (Rand & Gilliard 1967). *M. rufocrissalis* inhabits lower and mid-montane forest and edge, and trees in gardens. Occurs from about 1,100 m up to 2,400 m. Replaced at higher elevations by *M. belfordi* but hybridizes with that species where the two meet, especially in disturbed areas; for comments on this phenomenon see *M.*

belfordi account (see also Mayr & Gilliard 1952, Gilliard 1959, Gilliard & LeCroy 1968, Diamond 1972, and Beehler & Sine 2007).

SUBSPECIES
a. *Melidectes rufocrissalis rufocrissalis* (Reichenow)
RANGE: The Border and Eastern Ra, from the Star Mts east to the Bismarck and Kubor Mts (Higgins *et al.* 2008). The Mt Bosavi population may belong here.
DIAGNOSIS: Forehead white. See photo in Coates & Peckover (2001: 170).

b. *Melidectes rufocrissalis gilliardi* Diamond
Melidectes rufocrissalis gilliardi Diamond, 1967. *Amer. Mus. Novit.* 2284: 9.—Mount Karimui, south-central sector of the Eastern Ra.
    Synonym: *Melidectes rufocrissalis thomasi* Diamond, 1969. *Amer. Mus. Novit.* 2362: 55.
RANGE: The S watershed of the Eastern Ra, from Aseki west to Okapa and Karimui[DI].
DIAGNOSIS: Forehead black. Of hybrid origin. See photo in Coates (1990: 261). Note that in Coates's photo the white in the lores is probably pollen encrusted from the bird feeding on flowers. The subspecies name *gilliardi* is no longer a junior homonym of *Melidectes fuscus gilliardi* Gyldenstolpe, because of our shift, in this publication, of the species *fuscus* into the genus *Melionyx*. The name *gilliardi*, having been created subsequent to 1961, once again becomes available following the Code (ICZN 1999, Art. 59.4).

c. *Melidectes rufocrissalis stresemanni* Mayr
Melidectes leucostephes stresemanni Mayr, 1931. *Mitt. Zool. Mus. Berlin* 17: 713.—Dawong, Herzog Mts.
RANGE: The nw sector of the SE Peninsula—Herzog Mts[J] and south into the Central Ra around Bulldog Road[K], Mt Kaindi[K], and Ekuti Divide[K] (Mayr 1931[J], Beehler field notes and specimens at Harvard[K] and BPBM[K]).
DIAGNOSIS: Very similar to the preceding form. This form, *stresemanni*, has traditionally been placed with *M. belfordi*, but we note that many of the plumage characters of this subspecies agree most closely with the species *rufocrissalis*, and hence we move this form to *rufocrissalis*. Forehead dark; orbital skin pale green; bill gray; eyebrow and ear patch yellow (not white). Of hybrid origin, but apparently stable. We recognize subspecies populations of hybrid origin when there is evidence of plumage stability in available collections (*pace* Higgins *et al.* 2008). See photo in Coates (1990: 261, photo 239).

NOTES
Pure *M. rufocrissalis* exhibits a white forehead, pale blue-gray bill that is longer than that of *M. belfordi*, a pale green face patch, a yellow eyebrow, and a distinct orange gular patch. Apparently endemic to c NG, with pure populations of *M. belfordi* occurring to both the west and southeast of the entire range of *rufocrissalis*. The species distributions of this obvious species pair do not align with Diamond's (1972) dropout model for montane speciation. *Other names*—Reichenow's Melidectes, Yellow-browed Honeyeater, Yellow-browed Wattlebird.

GENUS *MELIONYX*   ENDEMIC [Genus 3 spp]

Melionyx Iredale, 1956. *Birds New Guinea* 2: 141. Type, by original designation, *Acanthochoera fusca* De Vis.
    The three species of this genus are endemic to the higher mountains of the Central Ranges of New Guinea. The genus, traditionally embedded within *Melidectes*, is characterized by the overall sooty-gray plumage, lack of green feather edges to the remiges and rectrices, the very slim, decurved and pointed bill, the small colored ocular skin patch, and the absence of gular wattles. Sadly, none of the various molecular systematic studies of the Meliphagidae included a sample of *Melionyx*. We place the genus after *Melidectes* by tradition, but its form and behavior suggest it belongs elsewhere in the family. This could be another case where aggressive flower feeders converge in plumage color and pattern. We treat the generic name as masculine.

Sooty Honeyeater *Melionyx fuscus*   ENDEMIC | MONOTYPIC

Acanthochoera fusca De Vis, 1897. *Ibis*: 383.—Mount Scratchley, SE Peninsula.
> Synonym: *Melidectes fuscus occidentalis* Junge, 1939. *Nova Guinea* (ns)3: 59.—Wichmann Mts, Western Ra.
> Synonym: *Melidectes fuscus gilliardi* Salomonsen, 1966a. *Breviora* Mus. Comp. Zool. Harvard 254: 10.—Mount Wilhelm, 3,350 m, Eastern Ra.

Inhabits mossy upper montane forest, edge, and alpine shrubbery of the Central Ra; mainly above 2,600 m to the tree line, min 1,700 m[CO]. Common in e NG, rare in w NG.

NOTES

The populations distributed across the Central Ra form a size cline, from small in the west to large in the SE Peninsula. There are no detectable plumage or soft-part distinctions. The largest birds are 4% larger than the smallest birds (Salomonsen 1966a). We therefore sink the two size subspecies and treat the species as monotypic. Diamond (1972) had already sunk *gilliardi*, the middle population in the cline. The eye-skin coloration is not diagnostic for these thinly defined subspecies. Skin color is variable. The main bare ocular patch is usually bluish white, but some are identified as ivory-colored. The postocular spot is typically brick red, but ranges to orange or yellow. The westernmost specimen at the AMNH (Bele R) has a blue-white ocular patch with an orange postocular spot, and the easternmost specimen (Mt Albert Edward) exhibits the same color pattern. *Other names*—Sooty Melidectes, *Melidectes fuscus*.

Long-bearded Honeyeater *Melionyx princeps*   ENDEMIC | MONOTYPIC

Melidectes princeps Mayr & Gilliard, 1951. *Amer. Mus. Novit.* 1524: 13.—Mount Wilhelm, 3,600 m, Eastern Ra.

Restricted to the high subalpine zone of the Eastern Ra above 2,750 m[CO]. Inhabits mossy subalpine forest, shrublands, and copses near the timberline.

Localities—Omyaka and Kai-ingri Camps in the Kaijende Highlands[L], Mt Giluwe[CO], Mt Hagen[J], Minj R[J], the Kubor Mts (Mt Orata, Mt Kinkain, and Minj-Nona Divide)[K], Mt Wilhelm in the Bismarck Ra[K], and Mt Michael[DI] (Mayr & Gilliard 1954[K], Sims 1956[J], Diamond 1972, Beehler & Sine 2007[L]).

NOTES

Presumably ranges west to the Strickland R gorge. Should also be looked for to the north of Porgera in the high central divide around Yakopi Nalenk summit (north of the Lagaip R). There is probably a considerable geographic gap separating this species and the range of its sister form, *M. nouhuysi*, known from the Star Mts of the Border Ra. There are no high ranges that continuously link the habitats of these two species in the intervening zone. We suggest this species be reclassified as Near Threatened or Least Concern rather than Vulnerable by the IUCN Red List. *Other names*—Long-bearded Melidectes, *Melidectes princeps*.

Short-bearded Honeyeater *Melionyx nouhuysi*   ENDEMIC | MONOTYPIC

Melirrhophetes nouhuysi van Oort, 1910. *Notes Leyden Mus.* 32: 215.—Oranje Mts, Western Ra.

Subalpine forest and shrublands of the Western and Border Ra, 3,050–4,500 m. Occurs from Carstensz Massif east to the Star Mts of PNG.

Localities—the high plateau west of Ilaga[J], Lake Habbema[K], northeast of Mt Wilhelmina[K], and Mt Capella[L] (Rand 1942b[K], Ripley 1964a[J], Gregory & Johnston 1993[L]).

NOTES

Higgins *et al.* (2008) reported that the eastern population exhibits a bill longer than that of the western birds. Eastern terminus of range unknown but probably does not extend to the Strickland R, because of absence of high-elevation habitats. Apparently absent from the Victor Emanuel Mts (Gilliard & LeCroy 1961). *Other names*—Short-bearded Melidectes, *Melidectes nouhuysi*.

Family Acanthizidae

Australasian Warblers and Allies
[Family 59 spp/Region 20 spp]

Species of the family range from the Malay Peninsula and tropical Asia east to New Guinea, Australia, Melanesia, and New Zealand (Gregory 2007b). The greatest diversity is found in New Guinea and Australia. These birds are the ecological equivalents of the Old World warblers of Eurasia and Africa but are unrelated and instead are closest to fairywrens (Maluridae) and honeyeaters (Meliphagidae). All Australasian warblers are insectivorous and build dome-shaped nests. Most are dull-plumaged, though a few species have bright-plumaged males; the tail has 12 rectrices; wings are rounded and have 10 primaries and a vestigial 10th secondary (Schodde & Mason 1999). Norman *et al.* (2009b) cluster *Acanthiza*, *Gerygone*, and *Sericornis* as a clade, and placed them as sister to a cluster that includes the fairywrens and the honeyeaters. See also Gardner *et al.* (2010) on the molecular treatment of the group. Cracraft (2013) placed the Acanthizidae as sister to the pardalotes (Pardalotidae); in turn, the acanthizid+pardalotid clade is sister to the honeyeaters.

GENUS *PACHYCARE*
ENDEMIC | MONOTYPIC

Pachycare Gould, 1876. *Birds New Guinea* 3: pl. 15. Type, by monotypy, *Pachycephala flavogrisea* Meyer.

The single species of the genus is endemic to the uplands of the New Guinea mainland (Boles 2007). Novel familial placement follows Norman *et al.* (2009). The genus is characterized by the distinctive size (large for a warbler); bright yellow face, forehead, and underparts; and white spotting on inner secondaries. The placement is provisional, and additional study is needed considering the unique features of the genus. There are some things about it that are warbler-like: the roofed nest on the ground, foraging mode, and white subterminal spots in the tail. Other aspects are whistler-like: plumage color and loud ringing song. This bird was long placed within the Pachycephalidae (*e.g.*, Mayr 1941). Schodde & Christidis (2014) created a new subfamily, the Pachycareinae, to house this genus. Both *Pachycare* and *Oreoscopus* of Australia have been placed in monogeneric subfamilies, which, together, are considered sister to the remaining genera of the Acanthizidae.

Goldenface *Pachycare flavogriseum*
ENDEMIC

Pachycephala flavogrisea A. B. Meyer, 1874. *Sitzungsber. Akad. Wiss. Wien* 69: 495.—Arfak Mts, Bird's Head.

Most mts of NG; 800–1,500 m, extremes 400–1,920 m (Coates 1990, Freeman & Freeman 2014, Diamond & Bishop 2015). Inhabits the middle and upper stories of hill and lower montane forest interior. As yet, no records from the Kumawa, Cyclops, or Adelbert Mts, and only an audio-detection from the Foja Mts.

SUBSPECIES

a. *Pachycare flavogriseum flavogriseum* (A. B. Meyer)
RANGE: Mountains of the Bird's Head and Wandammen Penin (Mayr 1941).
DIAGNOSIS: Underparts deeper and richer than Spectrum Yellow (Smythe 1975, Beehler & Prawiradilaga 2010). The four subspecies can be ranked from deep orange to yellow ventrally as follows: *lecroyae* > *subaurantium* > *flavogriseum* > *subpallidum*.

b. *Pachycare flavogriseum subaurantium* Rothschild & Hartert
Pachycare flavogrisea subaurantia Rothschild & Hartert, 1911. *Orn. Monatsber.* 19: 157.—Upper Otakwa R, Western Ra.
> Synonym: *Pachycare flavogrisea randi* Gilliard, 1961. *Amer. Mus. Novit.* 2031: 2.—6 km sw of Bernhard Camp, 1,200 m, ne sector of the Western Ra.
RANGE: The Western, Border, and Eastern Ra (Hartert *et al.* 1936, Gilliard & LeCroy 1961).
DIAGNOSIS: Slightly richer yellow ventrally than the nominate form (Smythe 1975, Beehler & Prawiradilaga

2010). See photo in Coates & Peckover (2001: 159). The form *randi* is considered a synonym based on review of material from the AMNH. It is intermediate between this form and *subpallidum*.

c. *Pachycare flavogriseum lecroyae* Beehler & Prawiradilaga
Pachycare flavogriseum lecroyae Beehler & Prawiradilaga, 2010. *Bull. Brit. Orn. Club* 130: 282.—Mount Menawa, Bewani Mts, n NG.
RANGE: The Bewani and Torricelli Mts of the PNG North Coastal Ra.
DIAGNOSIS: Underparts deep orange (Spectrum Orange in Smythe 1975, see Beehler & Prawiradilaga 2010). Richest and most orange ventrally of all the subspecies; see figure 2c, comparing this form with *subaurantium*.

d. *Pachycare flavogriseum subpallidum* Hartert
Pachycare flavogrisea subpallida Hartert, 1930. *Novit. Zool.* 36: 53.—Bihagi, head of the Mambare R, SE Peninsula.
RANGE: Mountains of the Huon and SE Peninsulas, westward in the s watershed probably to Okasa near the e terminus of the Eastern Ra[DI], and in the n watershed grading into *subaurantium* in the Sepik Mts. The "Bihagi country," where the type was obtained, is in the Kokoda valley, near the site of the current village of Kanga (residents of which are Biagge speakers). This is where the Mambare's headwaters lie, at the n foot of Mt Victoria.
DIAGNOSIS: Underparts bright yellow but with no orange tinge (= Spectrum Yellow; Smythe 1975, Beehler & Prawiradilaga 2010). Color not as saturated ventrally as the nominate form.

NOTES
The name *Pachycare* is neuter (David & Gosselin 2002b). Rand & Gilliard (1967) stated that the ventral plumage color fades with time. This is contradicted by the AMNH material collected by Diamond in the PNG North Coastal Ra in 1966, which remains as deeply orange-tinted as when originally collected. The race *randi* was described by Gilliard from material collected on the Third Archbold Expedition. This material had been carefully reviewed by Rand and Mayr a quarter century before and was not described as distinct at that time. It is thus not surprising this is now considered a synonym. Heard in the Foja Mts, the species remains uncollected there and without visual description (Beehler & Prawiradilaga 2010). It should also be looked for in the Adelbert Mts. *Other names*—Dwarf Whistler, *Pachycare flavogrisea*.

GENUS *CRATEROSCELIS* ENDEMIC [Genus 3 spp]

Crateroscelis Sharpe, 1883. *Cat. Birds Brit. Mus.* 7: 507 (key), 590. Type, by subsequent designation, *Myiothera murina* Temminck ms = *Brachypteryx murinus* P. L. Sclater.

This endemic genus is characterized by the terrestrial habit, melodious song, long legs, compact body form, and abbreviated tail. Preliminary molecular results indicated the genus may be paraphyletic, with *nigrorufa* embedded within the *Sericornis* lineage (Benz 2011). This move is not supported by plumage, external morphology, or vocalization.

Rusty Mouse-Warbler *Crateroscelis murina* ENDEMIC

Brachypteryx murinus P. L. Sclater, 1858. *Journ. Linn. Soc. London, Zool.* 2: 158.—Triton Bay, Bird's Neck.

Lowland and lower montane forest across NG, the NW Islands, Yapen I, and the Aru Is, sea level–1,700 m (at higher elevations where *C. nigrorufa* is absent; Pratt & Beehler 2014).

SUBSPECIES
a. *Crateroscelis murina capitalis* Stresemann & Paludan
Crateroscelis murinus capitalis Stresemann & Paludan, 1932. *Orn. Monatsber.* 40: 15.—Waigeo I, NW Islands.
    Synonym: *Crateroscelis murina fumosa* Ripley, 1957. *Postilla* Yale Peabody Mus. 31: 3.—Tamulol, Misool I, NW Islands.
RANGE: Misool[MA], Waigeo[MA] and Batanta I[J] (Greenway 1966[J]).
DIAGNOSIS: Dark brown crown; ocher-washed underparts; whitish throat; deep buff flanks. We disagree with Greenway (1966), who suggested *capitalis* is not valid. Our comparison of museum material supports

retention of this form. Misool birds show no distinct characters warranting subspecies recognition. Mees (1965) synonymized *fumosa* with the nominate form, but we found it closer to *capitalis*.

b. *Crateroscelis murina murina* (P. L. Sclater)
RANGE: Salawati and Yapen I, and all of NG (Mayr 1941) except for the middle Fly R region (the range of *pallida*).
DIAGNOSIS: Crown blackish brown, darker than mantle; throat whitish; smudgy breast band of rich fawn; flanks as for breast band; wing 61 (Rand & Gilliard 1967). Crown plumage is distinct between the subspecies *murina* and *capitalis*. See photos of the nominate subspecies in Coates (1990: 99) and Coates & Peckover (2001: 129).

c. *Crateroscelis murina monacha* (G. R. Gray)
Alcippe monacha G. R. Gray, 1858. *Proc. Zool. Soc. London*: 175.—Aru Is.
RANGE: The Aru Is.
DIAGNOSIS: White below, with a fawn wash on lower throat and buff brown on flanks and undertail; crown darker than mantle.

d. *Crateroscelis murina pallida* Rand
Crateroscelis murina pallida Rand, 1938. *Amer. Mus. Novit.* 991: 2.—Eastern bank of the middle Fly R, opposite Sturt I, S Lowlands.
RANGE: Known from the middle and lower Fly R (Lake Daviumbu and Sturt I) and the Bian Lakes of south-central NG (Bishop 2005a).
DIAGNOSIS: Crown olive brown; underparts largely white without the fawn wash (Rand & Gilliard 1967).

NOTES
Other names—Lowland or Chanting Mouse-warbler, or Lowland Mouse-babbler.

Bicoloured Mouse-Warbler *Crateroscelis nigrorufa* ENDEMIC | MONOTYPIC
Sericornis nigro-rufa Salvadori, 1894. *Ann. Mus. Civ. Genova* (2)14: 151.—Moroka, s watershed of SE Peninsula.

> Synonym: *Crateroscelis nigrorufa blissi* Stresemann & Paludan, 1934. *Orn. Monatsber.* 42: 46.—Kunupi, Weyland Mts, Western Ra.

Inhabits mid-montane forest of the Central Ra and mts of the Huon Penin, 1,220–2,500 m[co] (Mayr 1931, Rand 1942b, Gyldenstolpe 1955a). Apparently patchily distributed. Possibly overlooked from seemingly suitable localities because of the very narrow elevational range in any particular mountainside (Diamond 1972). On particular mountainsides, this species occurs at elevations higher than *C. murina* and lower than *C. robusta*.

NOTES
Preliminary molecular evidence suggests this species should be transferred to *Sericornis* (Benz 2011), but plumage and morphology do not support such a move—plumage and morphology agree very well with the other species of *Crateroscelis*, which forms a compact genus. The species differs from the other two *Crateroscelis* in that it forages in the understory as well as on the forest floor and it lives in small groups rather than as dispersed pairs; however, its song resembles that of the other mouse-warblers. The supposedly well-defined subspecies in fact exhibit a great deal of overlapping variability, such that samples from east and west include individuals that exhibit the whole range of variation. We compared adult males from the Western Ra to males from the SE Peninsula; the ventral plumage was identical, and dorsally the Western Ra birds were only marginally darker. A Western Ra juvenile had a lot of chocolate-brown coloration on its flanks, and the original subspecies diagnosis was probably based on young birds, which seem to be darker. The nw ridge of Mt Karimui was surveyed by Diamond in the 1960s and then again by Freeman & Freeman in 2013; the first survey did not find this species, whereas the second survey did (Diamond 1972, Freeman & Freeman 2014). Beehler *et al.* (2012) noted an unconfirmed sighting from the Foja Mts. Also unconfirmed sightings have recently been made in the Arfak Mts by F. Lambert and M. van Beirs on separate field trips (F. Lambert *in litt.*). The species should be looked for there by future field workers. *Other names*—Mid-mountain Mouse-babbler or Mouse-warbler, Black-backed Mouse-Warbler.

Mountain Mouse-Warbler *Crateroscelis robusta*   ENDEMIC

Gerygone robusta De Vis, 1898. *Ann. Rept. Brit. New Guinea*, 1896-1897: 84.—Wharton Ra, SE Peninsula.

Inhabits forest interior of the mts of the Bird's Head, the Fakfak Mts[L], Kumawa Mts[K], Central Ra, Wandammen Mts[MA], Foja Mts[W], Cyclops Mts, PNG North Coastal Ra, and mts of the Huon and SE Peninsulas; 1,250–3,680 m[CO] (Diamond 1985[K], Gibbs 1994[L], Beehler *et al.* 2012[W]; see Gregory 2007b). Mainly above 1,700 m (Coates & Peckover 2001). Not recorded from the Adelbert Mts. Populations in the Fakfak and Kumawa Mts have not been diagnosed to subspecies.

SUBSPECIES

There appear to be three plumage groupings of subspecies (see Beehler & Prawiradilaga 2010). Whether these three groups should be treated as full species is unclear, mainly because of the strangely interdigitated distributions of the three, which are atypical of the geographic ranges of youthful sister species (see Beehler & Prawiradilaga 2010).

I. White-throated Forms: (1) plumage sexually dichromatic; (2) white-throated; (3) male with a prominent dark breast band.

a. *Crateroscelis robusta diamondi* Beehler & Prawiradilaga
Crateroscelis robusta diamondi Beehler & Prawiradilaga, 2010. *Bull. Brit. Orn. Club* 130: 280.—Bog Camp, Foja Mts, 1,652 m, Mamberamo basin, nw NG.
RANGE: Uplands of the Foja Mts. Might also be looked for in the nearby Van Rees Mts.
DIAGNOSIS: Sexually dimorphic. Male with throat snowy white; upperparts uniform dark olive brown; narrow white belly patch along midline; female similar but plumage pattern washed out (Beehler & Prawiradilaga 2010). A white-throated form inhabits the Kumawa Mts (Diamond & Bishop 2015).

b. *Crateroscelis robusta robusta* (De Vis)
> Synonym: *Crateroscelis robusta pratti* Engilis and Cole, 1997. *Occ. Pap. Bishop Mus.* 52: 7.—2.3 km n, 0.4 km
> w of Agaun, Dumae Creek, 1,525 m, 9.908°S, 149.383°E, SE Peninsula.
RANGE: The Eastern Ra and mts of the Huon and SE Peninsulas. Ranges from Tari Gap east to Milne Bay (Gilliard & LeCroy 1968). Western terminus of range presumably the Strickland R gorge, west of which one finds buff-throated form *sanfordi*.
DIAGNOSIS: Male dull gray-brown dorsally; ventrally with dark gray-brown breast band, white throat, and whitish patch on midline of belly. Female much like *deficiens*, with breast band reduced to dull brown-gray wash on breast and flanks; no white midline patch. See photos of male and apparent female in Coates (1990: 102–3) and Coates & Peckover (2001: 128). Race *pratti* is thinly defined and was not recognized by Gregory (2007b) or Benz (2011).

II. Buff-throated Forms: (1) plumage of male and female similar; (2) entire underparts uniformly dull rufous buff; (3) dark breast band absent; (4) distinctive white throat absent; (5) iris color: male brown, orange-brown, gold rimmed with brownish red, or dull yellow ringed with rose; female rusty brown or pale gray-brown.

c. *Crateroscelis robusta sanfordi* Hartert
Crateroscelis sanfordi Hartert, 1930. *Novit. Zool.* 36: 81.—Mount Wondiwoi, Wandammen Penin, Bird's Neck.
RANGE: The Wandammen Mts, Western Ra, and Border Ra, east to the Strickland R gorge.
DIAGNOSIS: Dorsal plumage rich dark brown; underparts paler than dorsal plumage; palest on throat and grading from dull fawn-buff to rich buff brown on flanks; no indication of a breast band.

d. *Crateroscelis robusta bastille* Diamond
Crateroscelis robusta bastille Diamond, 1969. *Amer. Mus. Novit.* 2362: 18.—Mount Nibo, Torricelli Mts, 1,450 m, n NG.
RANGE: The Bewani and Torricelli Mts of the PNG North Coastal Ra (Diamond 1969).
DIAGNOSIS: Dorsal plumage dull dark olive; ventral surface paler than that of the nominate form (Diamond 1969).

III. Gray-breasted Forms: (1) plumage of male and female similar; (2) underparts pale below, with a whitish throat and olive flanks; (3) no dark breast band; (4) iris color: male red, female brown.

e. *Crateroscelis robusta peninsularis* Hartert

Crateroscelis robusta peninsularis Hartert, 1930. *Novit. Zool.* 36: 82.—Lehuma, Arfak Mts, Bird's Head.
> Synonym: *Crateroscelis robusta ripleyi* Mayr & Meyer de Schauensee, 1939. *Proc. Acad. Nat. Sci. Phila.* 91: 121.—
> Bon Kourangen, Tamrau Mts, Bird's Head.

RANGE: Mountains of the Bird's Head (Gilliard & LeCroy 1970). Possibly the populations from Kumawa and Fakfak Mts belong here.

DIAGNOSIS: Underparts with more white (Rand & Gilliard 1967). A thinly defined race. Gilliard & LeCroy (1970) and Gregory (2007b) recognize the Tamrau population as *ripleyi*. If accepted, such would be the only named form endemic to that mountain range, which is biogeographically improbable.

f. *Crateroscelis robusta deficiens* Hartert

Crateroscelis robusta deficiens Hartert, 1930. *Novit. Zool.* 36: 81.—Cyclops Mts, ne sector of the NW Lowlands, n NG.

RANGE: The Cyclops Mts.

DIAGNOSIS: Upperparts deep brown; uppertail coverts less rufous; underparts with considerable wash of dirty gray-olive (Rand & Gilliard 1967). Very close to the female plumage of the nominate form. Distinguished from *peninsularis* by the slightly warmer brown coloration of plumage throughout. Not well distinguished.

NOTES

This species, with its two levels of infraspecific variation, is more complex than most, and will benefit from detailed molecular systematic investigation currently underway. In his doctoral thesis, Benz (2011) presented a molecular phylogeography of *C. robusta*, finding that phenotypic similarities of throat and breast color between the e Central Ra subspecies *robusta* and the multiple subspecies from the Bird's Head and Cyclops Mountains were "not due to close phylogenetic affinity (*i.e.*, long-distance dispersal or vicariance), but rather retention of ancestral character states among terminal populations" (Benz 2011: 94). His analyses also "recovered two primary phylogroups" that were deeply divergent: a western one, *sanfordi*, and an eastern one, *robusta*. He advocated elevating the western phylogroup to species status, "Sanford's Mouse-Warbler." This research awaits further sampling and analyses before being published, but we draw attention to the results here as an indication of the future systematics of this complex species group. *Other names*—High-mountain Mouse-warbler, Mountain Mouse-babbler.

GENUS *SERICORNIS*   [Genus 12 spp/Region 7 spp]

Sericornis Gould, 1838. *Synops. Birds Austral.* 4: pl. 58. Type, by original designation, *Acanthiza frontalis* Vigors & Horsfield.

The species of this genus range from Australia to New Guinea (Gregory 2007b). These species, which are uniform in body form, are characterized by the slight build, medium-length bill, unpatterned olive or brown mantle, pale throat (with or without markings), and medium-length tail (with or without dark subterminal bar). They are most similar in form to birds of the genus *Gerygone*, but differ in their understory habits, somewhat heavier build, lack of distinct plumage pattern, and song of short repetitive musical phrases. No comprehensive molecular study of the New Guinea *Sericornis* has yet been conducted.

Pale-billed Scrubwren *Sericornis spilodera*   ENDEMIC

Entomophila? spilodera G. R. Gray, 1859. *Proc. Zool. Soc. London*: 155.—Manokwari, Bird's Head.

Inhabits forest throughout NG, Batanta and Waigeo I, the Aru Is, and Yapen I; 250–1,200 m[CO], extremes sea level–1,650 m[CO]. Widely distributed on the Mainland in the foothills; also found in the lowlands of the Trans-Fly[CO].

SUBSPECIES

a. *Sericornis spilodera ferrugineus* Stresemann & Paludan
Sericornis spilodera ferruginea Stresemann & Paludan, 1932. *Orn. Monatsber.* 40: 16.—Waigeo I,
NW Islands.

> Synonym: *Sericornis spilodera batantae* Mayr, 1986. *Check-list Birds World* 11: 423.—Mount Besar, 850 m, Batanta
> I, NW Islands. Replacement name for *Sericornis spilodera intermedia* Greenway, 1966.

RANGE: Waigeo[MA] and Batanta I (Greenway 1966).
DIAGNOSIS: Forehead rufous; ear coverts grayish rufous; very pale yellowish-white underparts, with spotting on throat grayish and almost obsolete (Rand & Gilliard 1967). The population from Batanta is slightly
grayer atop the crown than *ferrugineus*, but not warranting named status. Ventral plumage is identical.

b. *Sericornis spilodera spilodera* (G. R. Gray)
RANGE: The Bird's Head, Yapen I, the Foja Mts[J], and the n slopes of the Western, Border, and Eastern Ra,
east to Astrolabe Bay (Mayr 1941, Beehler *et al.* 2012[J]).
DIAGNOSIS: Crown blackish; ear coverts dull fuscous; back and wing coverts dull gray-green; tail dark
rufous brown; throat white, with rather diffuse gray spotting; abdomen washed with yellow; flanks olive
with some obscure streaking (Rand & Gilliard 1967). See photo in Coates & Peckover (2001: 129).

c. *Sericornis spilodera aruensis* Ogilvie-Grant
Sericornis aruensis Ogilvie-Grant, 1911. *Bull. Brit. Orn. Club* 29: 29.—Wokan, Aru Is.
RANGE: The Aru Is.
DIAGNOSIS: Distinct from all other subspecies. Overall pale; crown and cheek pale cinnamon; throat pure
white without dark spotting; mantle pale olive, wings concolorous; flanks washed pale yellow.

d. *Sericornis spilodera granti* (Hartert)
Aethomyias spilodera granti Hartert, 1930. *Novit. Zool.* 36: 85.—Otakwa R, sw NG.
RANGE: The s slopes of the Western and Border Ra (Mayr 1941). Birds captured near Triton Bay on the s
coast of the Bird's Neck and at 1,100 m in the s Kumawa Mts may be of this subspecies (C. Thébaud & B.
Mila *in litt.*).
DIAGNOSIS: Similar to *guttatus* but crown rich dark brown, forehead with a rusty wash, slightly darker than
mantle; cheek with a brownish wash; throat whitish, heavily marked with blackish spotting; breast and
belly smudged yellowish and olive; flanks olive.

e. *Sericornis spilodera guttatus* (Sharpe)
Aethomyias guttata Sharpe, 1882. *Journ. Linn. Soc. London, Zool.* 16: 432.—Sogeri distr., SE Peninsula.

> Synonym: *Sericornis spilodera wuroi* Mayr, 1937. *Amer. Mus. Novit.* 904: 15.—Wuroi, Oriomo R, Trans-Fly, s NG.

RANGE: The Trans-Fly, presumably the e sector of the S Lowlands, the e sector of the Eastern Ra, and the
mts of the Huon and SE Peninsulas. In the n watershed presumably west toward Astrolabe Bay, and in the
s watershed west toward the PNG border. Exact boundary of w edge of range in north and south unknown
(Mayr 1941, Diamond 1972).
DIAGNOSIS: Similar to *spilodera*, but crown and mantle greenish; forehead and cheek tinged brown (Rand
& Gilliard 1967). The form *wuroi* is not distinct from this form. See photo in Coates (1990: 104). *Other
names*—Pale-billed Sericornis.

Sericornis beccarii and *Sericornis nouhuysi* and hybrid populations

We believe that the Perplexing Scrubwren (*Sericornis virgatus*) is not a valid species, but instead is the
result of massive hybridization between *S. nouhuysi* and *S. beccarii*. The result is a hybrid swarm; this
situation is much like what has happened with *Melidectes belfordi* and *M. rufocrissalis*. We believe that *S.
nouhuysi* and *S. beccarii* have introgressed and generated stable populations of hybrid origin in various foothill regions of n and w NG (see Coates 1990: 104–6). Many of the hybrid populations have been
named as subspecies of "*Sericornis virgatus*." Being of hybrid origin, these populations were properly (if
unwittingly) referred to a species named "Perplexing Scrubwren" (see Mayr 1941).

In our treatment here, we place all the various hybrid forms formerly in *S. virgatus* into one or the
other of the parent forms—*nouhuysi* or *beccarii*. Our placement is determined by which parental characters dominate that hybrid form.

True *S. beccarii* is a lowland and foothill species and *nouhuysi* is a lower-montane and montane species. Look for hybrid populations to be most common at elevations where the two parent species would meet (in the lower montane zone above 900 m). Because of the complexity of the crossing and presumed backcrossing, the range of intermediate forms is bewildering and difficult to diagnose. Genes from *beccarii* seem to dominate in the west and in the lowlands, whereas the *nouhuysi* genes dominate eastward and in the higher mountains. That said, we don't believe there are any hybrid forms east of the Sepik-Ramu (in the north) or the Purari R (in the south), for that seems to be the e geographic range limit of one of the parental species, *S. beccarii*. We have no molecular proof of our hypothesis. For the various different views on the status of *S. virgatus*, see Mayr (1937a), Rand (1941), Rand & Gilliard (1967), Diamond (1969, 1985), Gilliard & LeCroy (1970), Coates (1990), Gregory (2007b), and Dickinson & Christidis (2014). No revising authority has ever agreed with the previous published treatment (although ours is very close to Coates's assessment). Certainly this is one of the great species-delineation challenges of nonmolecular ornithology in NG, along with the *Meliphaga analoga* complex. We assume it could be clarified by molecular analysis of a selection of samples from across the ranges of *S. nouhuysi* and *S. beccarii*.

Tropical Scrubwren *Sericornis beccarii*   RESIDENT

Sericornis beccarii Salvadori, 1874. *Ann. Mus. Civ. Genova* 6: 79.—Wokan, Aru Is.

Inhabits lowland rain forest of the Aru Is and the w sector of the S Lowlands; also foothills of the w two-thirds of NG, including various outlying ranges. "Pure" *S. beccarii* can be found in the Aru Is, the lowlands of south-central NG, and n QLD. Hybrid populations are found in uplands in c, n, and w NG.

Extralimital—ne QLD, Australia.

SUBSPECIES

a. *Sericornis beccarii beccarii* Salvadori

RANGE: The Aru Is.

DIAGNOSIS: Crown dark olive brown, not substantially different from mantle coloration; rump a richer reddish brown; forehead and lores blackish, with a small white loral spot before each eye; whitish upper and lower eyelid; a blackish mark at the bend of the wing, marked with white spotting on the tertials; throat whitish; obscure dusky breast band; center of breast and belly with yellowish wash; flanks olive-washed; tail as for rump color. This is a nonhybrid form upon which the species is based.

b. *Sericornis beccarii cyclopum* Hartert

Sericornis magnirostris cyclopum Hartert, 1930. *Novit. Zool.* 36: 83.—Cyclops Mts, ne sector of NW Lowlands, n NG.

Synonym: *Sericornis beccarii weylandi* Mayr, 1937. *Amer. Mus. Novit.* 904: 11.—Kunupi, Weyland Mts, Western Ra.
Synonym: *Sericornis beccarii wondiwoi* Mayr, 1937. *Amer. Mus. Novit.* 904: 11.—Mount Wondiwoi, Wandammen Penin, Bird's Neck.

RANGE: The Weyland Mts, Wandammen Mts, Foja Mts, and Cyclops Mts (Hartert *et al.* 1936). Montane forest, 600–1,600 m.

DIAGNOSIS: A hybrid form with some minor regional variation. Similar to *randi* but white supraloral spot and white eyespots small and obscure; blackish line surrounding white supraloral spot narrow; throat whitish, obscurely streaked with gray; breast with slight yellowish wash but mottled gray; ear coverts tinged cinnamon; dark bend of wing with typical white wing bars; crown and mantle olive brown. We have merged *weylandi* and *wondiwoi* into this form because the distinctions are minimal, and assuming all three are of hybrid origin, such minor variation is to be expected.

c. *Sericornis beccarii randi* Mayr

Sericornis beccarii randi Mayr, 1937. *Amer. Mus. Novit.* 904: 10.—Wuroi, Oriomo R, Trans-Fly, s NG.

RANGE: A pure (nonhybrid) form of the Trans-Fly. Localities—Maro R, Wasur NP, Bensbach, Tarara, and Wuroi (Bishop 2005a).

DIAGNOSIS: Nearly identical to the nominate form, with broken white eye-ring and white spot above bill, but upperparts greener; upper throat white, midthroat spotted with black; blackish lores; breast and belly yellow, breast with gray streaks; undertail yellowish; bend of wing blackish, marked with incomplete white wing bars more prominent than those of the nominate form. A nonhybrid form typical of the species.

Truly "typical" *beccarii* in the Region can be found in the Aru Is and the Fly R region. The characters for this species are the dark spot at the bend of the wing with the incomplete white wing barring, the white eyebrow and lid, and the whitish throat and underparts pale-washed with yellow. Dickinson & Christidis (2014) placed these forms into *S. magnirostra*, following the very inclusive treatment of Schodde & Mason (1999). An undescribed population attributed to this species has been observed on Waigeo (I. Mauro field observation). *Other names*—Beccari's Scrubwren or Sericornis.

Large Scrubwren *Sericornis nouhuysi* ENDEMIC

Sericornis arfakiana nouhuysi van Oort, 1909. *Nova Guinea Zool.* 9, (1): 90.—Hellwig Mts, e sector of the Western Ra.

Mountains of the Bird's Head, Central Ra, Yapen I, Fakfak and Kumawa Mts, PNG North Coastal Ra, and mts of the Huon Penin; *ca.* 600–2,500 m, max 1,200–3,750 m. It appears that only hybrid-swarm populations occur in the Kumawa, Fakfak, and PNG North Coastal Ra. Expect to see typical *nouhuysi* above 1,700 m in most ranges of NG from the Bird's Head to the Foja Mts and all the Central Ra. We expect to find the purest forms of *nouhuysi* on the Huon and SE Peninsulas, east of the range of *beccarii*, with which *nouhuysi* seems to have hybridized widely in the west. See also Diamond & Bishop (2015: 319).

SUBSPECIES

a. *Sericornis nouhuysi cantans* Mayr
Sericornis magnirostris cantans Mayr, 1930. *Orn. Monatsber.* 38: 177.—Arfak Mts, Bird's Head. Replacement name for *Sericornis arfakiana* Salvadori, 1876, *Ann. Mus. Civ. Genova* (1875) 7: 962; preoccupied by *Gerygone* (?) [= *Sericornis*] *arfakiana* Salvadori, ib.: 960.
Synonym: *Sericornis beccarii imitator* Mayr, 1937. *Amer. Mus. Novit.* 904: 12.—Siwi, Arfak Mts, Bird's Head.
RANGE: Mountains of the Bird's Head, above 1,400 m, and presumably also including the populations from the Fakfak and Kumawa Mts (Diamond 1985).
DIAGNOSIS: This group of populations of hybrid origin is variable. Buffy-rusty wash of face extends onto cheek; dark bend of wing with obsolete buff-spotted wing bar; dark brown crown; buff-olive obscure breast band; belly with pale yellow-olive wash. Placed by Mayr (1941) in the species *beccarii*; by Rand & Gilliard (1967) in *virgatus*; and by Gregory (2007b) in *nouhuysi*. The form *imitator* is combined here with *cantans*; *imitator* is a hybrid population from a slightly lower elevation than *cantans* but the two forms meet and hybridize in the mts of the Bird's Head and these constitute nongeographic subspecies.

b. *Sericornis nouhuysi jobiensis* Stresemann & Paludan
Sericornis magnirostris jobiensis Stresemann & Paludan, 1932. *Novit. Zool.* 38: 230.—Yapen I, Bay Is.
RANGE: Yapen I.
DIAGNOSIS: Cinnamon on forehead and face; dark and less olive dorsally; pale below, with dark mottling across breast; indication of two vestigial wing bars at bend of wing. Of hybrid origin with greater gene contribution from *nouhuysi*.

c. *Sericornis nouhuysi nouhuysi* van Oort
RANGE: The Western Ra (Weyland, Nassau, and Oranje Mts).
DIAGNOSIS: A large form; rich rusty face and chin; lores deep rust; overall plumage rich and dark; throat strongly buff; dorsal surface rich brown; iris dark red or brown. A nonhybrid form of pure *nouhuysi* genes. Inhabits elevations above zone where *beccarii* genes intrude.

d. *Sericornis nouhuysi idenburgi* Rand
Sericornis beccarii idenburgi Rand, 1941. *Amer. Mus. Novit.* 1102: 11.—6 km sw of Bernhard Camp, 1,200 m, Tariratu R, ne sector of the Western Ra.
RANGE: Known only from the slopes above the Taritatu R, between 850 and 1,200 m elevation.
DIAGNOSIS: Upperparts dark brown; the dark underparts heavily washed with brownish olive and obscurely streaked on throat and breast; forecrown and face with rusty wash; the pale eyebrow and white lower lid much reduced or absent; dark patch at bend of wing, marked by narrow white or grayish tips forming obsolete wing bars. Presumably of hybrid origin, with predominant *nouhuysi* phenotype.

e. *Sericornis nouhuysi virgatus* (Reichenow)

Crateroscelis virgata Reichenow, 1915. *J. Orn.* 63: 128.—Maeanderberg, 600 m, upper Sepik R, west-central Sepik-Ramu.

> Synonym: *Sericornis virgatus boreonesiotictus* Diamond, 1969. *Amer. Mus. Novit.* 2362: 21.—Mount Somoro, 1,360 m, Torricelli Mts, Sepik-Ramu.

RANGE: The n slopes of the Eastern Ra where they drain into the Sepik basin, and the PNG North Coastal Ra (Gregory 2007b). Also Mt Bosavi, Mt Sisa, and presumably the s slopes of the Eastern Ra (Coates 1990). DIAGNOSIS: This is a hybrid form sharing many plumage characters of the parent species *nouhuysi*. Note pale cinnamon face; dull greenish-olive dorsal surface; ventral plumage very pale off-white with gray-buff wash; indication of a dark breast band; throat with obsolete pale streaking; indication of obsolete spotting in form of vestigial wing bars; tail brown. See photo in Coates & Peckover (2001: 130). The form *boreonesiotictus* does not differ substantially from *virgatus*.

f. *Sericornis nouhuysi pontifex* Stresemann

Sericornis arfakiana pontifex Stresemann, 1921. *Anz. Orn. Ges. Bayern* 1: 34.—Lordberg, Sepik Mts, Eastern Ra.

> Synonym: *Sericornis magnirostris stresemanni* Mayr, 1930. *Orn. Monatsber.* 38: 177.—Schrader Ra, n sector of the Eastern Ra. Replacement name for *Sericornis arfakiana rufescens* Stresemann, 1921; preoccupied by *Gerygone? [= Sericornis] rufescens* Salvadori, 1876.

RANGE: The Border and Eastern Ra mainly at higher elevations, above the hybrid zone. This form is pure *nouhuysi*. DIAGNOSIS: Pale tips of alula and coverts usually lacking; bill very pale (Gregory 2007b). Yellowish olive ventrally, breast and throat mottled with rufous, flanks darker; tail dark and without rufous; breast with obsolete streaking evident. See photo of nesting bird from Tari Gap in Coates (1990: 107).

g. *Sericornis nouhuysi adelberti* Pratt

Sericornis nouhuysi adelberti Pratt, 1983. *Emu* (1982) 82: 120.—Mount Mengem, 1,500 m, Adelbert Mts, n NG.

RANGE: Adelbert Mts. DIAGNOSIS: Dullest of the races of this species; face buffy rufous; throat very pale; breast and belly pale olive (Pratt 1983). Apparently pure *nouhuysi*.

h. *Sericornis nouhuysi oorti* Rothschild & Hartert

Sericornis arfakiana oorti Rothschild & Hartert, 1913. *Novit. Zool.* 20: 503.—Bihagi, head of Mambare R, SE Peninsula.

> Synonym: *Sericornis nouhuysi monticola* Mayr & Rand, 1936. *Mitt. Zool. Mus. Berlin* 21: 246.—Mount Albert Edward, 3,680 m, SE Peninsula.

RANGE: Mountains of the Huon and SE Peninsulas (Mayr 1931). DIAGNOSIS: A population with pure *nouhuysi* genes. Rich brownish on head, tail, and wings; more olive on back; buffy rufous on chin and sides of head; underparts yellowish olive; flanks more olive; undertail rufous. We fold *monticola* into *oorti* because the former is no more than the larger alpine populations of *oorti* (described as a high-elevation subspecies). This work does not recognize nongeographic ("elevational") subspecies for reasons stated in the Introduction. There is a size cline with elevation in the mts of the SE Peninsula that may also include some shift in plumage coloration. This is of interest biologically, and should be studied in the field and molecular lab, but a subspecies epithet is not needed in such an instance. See photo in Coates (1990: 106) and in Coates & Peckover (2001: 130).

NOTES

The core characters of the species include: (1) no pale/whitish markings on face or bend of wing; (2) rusty in face; (3) large bill; (4) olive brown dorsally and pale olive ventrally with obscure mottling; (5) large size. Some of the former "*S. virgatus*" have been placed here in *nouhuysi*; these are populations of hybrid origin (*nouhuysi × beccarii*). An undescribed hybrid swarm exists on Mt Bosavi (specimens at BPBM). *Other names*—Mountain, Large Mountain, or Noisy Scrubwren, Large Mountain Sericornis.

Vogelkop Scrubwren *Sericornis rufescens*

ENDEMIC | MONOTYPIC

Gerygone? rufescens Salvadori, 1876. *Ann. Mus. Civ. Genova* (1875) 7: 961.—Hatam, Arfak Mts, Bird's Head
   Recorded from the Bird's Head (Arfak and Tamrau Mts) and Bird's Neck (Kumawa Mts, Fakfak Mts; Diamond 1985, Diamond & Bishop 2015), 1,000–2,000 m (Coates & Peckover 2001).

NOTES
Small; dull rusty wash on face; belly with pale yellowish wash; flanks with olive wash; crown and mantle dull brown; tail rusty brown, with dark subterminal band on some outer rectrices. *Other names*—Arfak Scrubwren, Arfak or Rufous Sericornis.

Buff-faced Scrubwren *Sericornis perspicillatus*

ENDEMIC | MONOTYPIC

Sericornis perspicillata Salvadori, 1896. *Ann. Mus. Civ. Genova* (2)16: 99.—Moroka, s watershed of SE Peninsula.
   Synonym: *Sericornis nigroviridis* Miller, 1964. *Auk* 81: 2.—Edie Creek, 2,130 m, west of Wau, SE Peninsula.
Inhabits forest and edge of the Central Ra, Foja Mts[J], PNG North Coastal Ra[K], Mt Bosavi[CO], Adelbert Mts[L], and mts of the Huon Penin[MA]; 1,700–2,600 m, extremes 850–2,800 m[CO] (Diamond 1985[K], Pratt 1983[L], Beehler *et al.* 2012[J]). Replaced on the Bird's Head by the closely allied *S. rufescens*. Not recorded from the Bird's Neck or the Cyclops Mts.

NOTES
S. perspicillatus was considered by Mayr (1941) possibly to be a subspecies of *S. rufescens*. The type of *S. nigroviridis* Miller, 1964, was determined to be a melanistic individual of *perspicillatus* (Beehler 1978a). *Other names*—Buff-faced Sericornis, Black and Green Sericornis.

Papuan Scrubwren *Sericornis papuensis*

ENDEMIC

Acanthiza papuensis De Vis, 1894. *Ann. Rept. Brit. New Guinea*, 1893–94: 102.—Mount Maneao, SE Peninsula.
   Inhabits montane forest through the Central Ra, Foja Mts, and Huon Penin (Diamond 1985), 2,000–3,500 m[CO]. Netted once at 850 m (Beehler 1980b).

SUBSPECIES
a. *Sericornis papuensis buergersi* Stresemann
Sericornis bürgersi Stresemann, 1921. *Anz. Orn. Ges. Bayern* 1: 34.—Schraderberg, Sepik Mts, n sector of the Eastern Ra.
RANGE: The n slopes of the Central Ra from the far west (Weyland Mts) east to the Schrader Ra; also Foja Mts[J] (Gyldenstolpe 1955a, Diamond 1985, Beehler *et al.* 2012[J]).
DIAGNOSIS: More brownish-washed dorsally, and more brownish ocher below (Gregory 2007b).

b. *Sericornis papuensis meeki* Rothschild & Hartert
Sericornis meeki Rothschild & Hartert, 1913. *Novit. Zool.* 20: 503.—Mount Goliath, Border Ra.
RANGE: The s slopes of the Border Ra and e sector of Western Ra (Mt Goliath, Hellwig Mts). Presumably this subspecies ranges through the s watershed of the Western, Border, and Eastern Ra.
DIAGNOSIS: Greenish olive dorsally; distinct rufous wash on lores, cheek, and sides of throat; abdomen dark olive; blackish subterminal band on outer rectrices.

c. *Sericornis papuensis papuensis* (De Vis)
RANGE: The Eastern Ra and mts of the Huon[J] and SE Peninsulas; from Mt Karimui[DI] eastward (Freeman *et al.* 2013[J]).
DIAGNOSIS: Generally olive brown, dark dorsally, pale ventrally (Gregory 2007b). Note slight rusty wash in lores. Iris dark brown or dark gray-brown. See photo in Coates (1990: 109) and Coates & Peckover (2001: 130).

NOTES
There is considerable nongeographic plumage variation in this species, perhaps mainly related to age. That, combined with the very poor preparation of the tiny specimens, makes diagnosis of proper

geographic variation difficult. We have left these subspecies untouched, with the hope that some future museum worker will properly tackle this group. *Other names*—Olive Scrubwren, Papuan Sericornis.

Grey-green Scrubwren *Sericornis arfakianus*   ENDEMIC | MONOTYPIC

Gerygone? arfakiana Salvadori, 1876. *Ann. Mus. Civ. Genova* (1875) 7: 960.—Arfak Mts, Bird's Head
   Synonym: *Sericornis olivacea* Salvadori, 1896. *Ann. Mus. Civ. Genova* (2)16: 100.—Moroka, s watershed of
    SE Peninsula.

Inhabits forest and edge of the mts of NG; 1,200–1,400 m, extremes 670–1,780 m[CO,U] (Freeman & Freeman 2014[U]). Records for peripheral ranges since Mayr (1941) include the Fakfak[J] and Foja Mts[K], PNG North Coastal Ra[L], and Adelbert Mts[M] (Beehler *et al.* 2012[K], Gibbs 1994[J], Gilliard & LeCroy 1967[M], specimens at AMNH & BPBM[L]). No records from the Kumawa Mts.

NOTES
Some individuals of the e populations are more olive-washed ventrally, but there is considerable variation in this trait. *Other names*—Grey-green Sericornis, Dusky or Olive Scrubwren.

GENUS *ACANTHIZA*   [Genus 14 spp/Region 2 spp]

Acanthiza Vigors & Horsfield, 1827. *Trans. Linn. Soc. London* 15: 224. Type, by original designation, *Motacilla pusilla* Shaw.

   The species of this genus range from New Guinea to Australia (Gregory 2007b). The genus is characterized by tiny size, abbreviated pale-tipped dark tail, distinctive speckled face patch, and tiny pointed bill. Most species are communal in habit; nests are domed and without a pendent "tail" (Schodde & Mason 1999). Transfer of the species *cinerea* from *Gerygone* to *Acanthiza* follows the molecular phylogeny of Nyári & Joseph (2012) and our species sequence for the genus is based on the phylogeny presented in Nyári & Joseph (2012: fig. 3).

New Guinea Thornbill *Acanthiza murina*   ENDEMIC | MONOTYPIC

Gerygone murina De Vis, 1897. *Ibis*: 377.—Mount Scratchley, SE Peninsula.

   Inhabits upper montane and subalpine forests of the Central Ra; 2,500 m to timberline, min 1,930 m[CO]. Found in forest and edge. Most common at higher elevations.

   Other names—Papuan Thornbill, De Vis Tree Warbler.

Grey Thornbill *Acanthiza cinerea*   ENDEMIC | MONOTYPIC

Gerygone? cinerea Salvadori, 1876. *Ann. Mus. Civ. Genova* (1875) 7: 958.—Hatam, Arfak Mts, Bird's Head.

   Inhabits the canopy of montane forest of the Bird's Head, Wandammen Mts[K], Fakfak Mts[L], Central Ra[MA], PNG North Coastal Ra[M], and possibly the Foja Mts[J] (Diamond 1985[K], Gibbs 1994[I], Beehler *et al.* 2012[J], specimens at AMNH[M]). Ranges from 1,000 m (Wandammen Penin) to 2,800 m[CO]; mainly found above 2,000 m[CO].

NOTES
The very brief tail with the subterminal band, the abbreviated and narrow bill, and the grizzling of the forehead support the repositioning of this species into *Acanthiza*. Should be looked for in the high-elevation forests of the Huon Penin. *Other names*—Mountain or Grey Gerygone Warbler.

GENUS *GERYGONE*   [Genus 16 spp/Region 7 spp]

Gerygone Gould, 1841. In *Grey's Journ. Two Exped. Discov. Northwest. Austral.* 2: 417, note. Replacement name for *Psilopus* Gould, 1838; preoccupied by *Psilopus* Meigen. Type, by subsequent designation (G. R. Gray, 1840. *List Gen. Birds*: 22), *Psilopus albogularis* Gould = *Psilopus olivaceus* Gould.

   The species of this genus range from the Malay Peninsula and southernmost Indochina east to

New Guinea, Australia, New Zealand, and Norfolk Island (Gregory 2007b). The genus is characterized by canopy-dwelling habit, small size, medium-length tail, repeated musical series, "warbler" body form, warbler-like bill, and a pendent "tailed" nest (Schodde & Mason 1999). Systematics of the genus follows Nyári & Joseph (2012).

Yellow-bellied Gerygone *Gerygone chrysogaster* ENDEMIC
Gerygone chrysogaster G. R. Gray, 1858. *Proc. Zool. Soc. London*: 174.—Aru Is.
Inhabits the canopy of lowland and foothill forest throughout NG, the NW Islands, Aru Is, Mios Num I, and Yapen I, sea level–650 m (highest 1,220 m at Baiyer R Sanctuary; I. Woxvold *in litt.*).

SUBSPECIES

a. *Gerygone chrysogaster neglecta* Wallace
Gerygone neglecta Wallace, 1865. *Proc. Zool. Soc. London*: 475.—Waigeo I, NW Islands.
RANGE: Waigeo and Batanta I (Mayr 1941).
DIAGNOSIS: Entire ventral surface washed very pale yellow, strongest on flanks; greenish olive dorsally; maxilla apparently horn-colored. Mees (1965) recommended placing the Batanta population with this form.

b. *Gerygone chrysogaster notata* Salvadori
Gerygone notata Salvadori, 1878. *Ann. Mus. Civ. Genova* 12: 344.—Wa Samson R, w Bird's Head.
  Synonym: *Gerygone neglecta dohertyi* Rothschild & Hartert, 1903. *Novit. Zool.* 10: 473.—Kapaur, Onin Penin, Bird's Neck.
  Synonym: *Gerygone chrysogaster leucothorax* Mayr, 1940. *Amer. Mus. Novit.* 1091: 2.—Wanggar, head of Geelvink Bay, Bird's Neck.
RANGE: The Bird's Head and Neck, and Salawati I[K] (Mees 1965[K]). The Misool I record (Mayr 1941) is likely in error, and the specimen is probably from Aru Is instead (Mees 1965).
DIAGNOSIS: Crown dull brown; mantle richer brown; rump and tail reddish brown; throat and breast off-white; cheek gray; belly whitish; cheek pale buff; bill horn-colored (Gregory 2007b). We lumped *leucothorax* and *dohertyi* with this form, as we find no substantial plumage distinctions to identify them.

c. *Gerygone chrysogaster chrysogaster* G. R. Gray
RANGE: Yapen I[MA], Mios Num I[J], the Aru Is[MA], and all NG east of the Bird's Neck. The Mios Num specimen was collected by von Rosenberg on 9-May-1869 (specimen in Mus. Leiden[J]).
DIAGNOSIS: Olive brown dorsally; wing and tail browner; cheek gray-buff; a narrow pale supraloral stripe present; chin, throat, and breast grayish white (Gregory 2007b). N. Collar (*in litt.*) noted prominent pale wing bars. See photo in Coates (1990: 112–13) and Coates & Peckover (2001: 133).

NOTES
Other names—Yellow-bellied Flyeater or Fairy-warbler.

White-throated Gerygone *Gerygone olivacea* RESIDENT
Psilopus olivaceus Gould, 1838. *Synops. Birds Austral.* 4: pl. 61.—New South Wales, Australia.
Confined to savanna around Port Moresby[CO]; sea level to 650 m atop Sogeri Plateau[CO]. Inhabits *Eucalyptus* and *Melaleuca* woodland.
    Extralimital—n and e Australia.

SUBSPECIES

a. *Gerygone olivacea cinerascens* Sharpe
Gerygone cinerascens Sharpe, 1878. *Journ. Linn. Soc. London, Zool.* 13: 494.—Port Moresby, SE Peninsula.
RANGE: The s coast of the SE Peninsula[MA]. Relict population between the Tapala-Malalaua area and Rigo, and probably east to Cape Rodney (Coates 1990). Also Cape York Penin, QLD.
DIAGNOSIS: Very much like the nominate form; slightly smaller; paler and grayer dorsally; tail spots slightly narrower (Gregory 2007b). Schodde & Mason (1999) noted the following plumage traits: mantle pale mid-gray, with a hint of olive in some individuals; white tips on tail medium-size; pale band in base of tail dull white and ill-defined or obscure.

NOTES
Other names—White-throated Warbler, White-throated Gerygone Warbler.

Green-backed Gerygone *Gerygone chloronota* RESIDENT

Gerygone chloronotus Gould, 1843. *Proc. Zool. Soc. London* (1842): 133.—Port Essington, NT, Australia.

A common and vocal inhabitant of hill forest interior throughout NG, the NW Islands, and Aru Is, sea level–1,500 m.

Extralimital—n Australia.

SUBSPECIES

a. *Gerygone chloronota chloronota* Gould

Synonym: *Pseudogerygone cinereiceps* Sharpe, 1886. *Nature* 34: 340.—Sogeri distr, Astrolabe Ra, SE Peninsula.

Synonym: *Gerygone aruensis* Büttikofer, 1893. *Notes Leyden Mus.* 15: 259.—Aru Is.

RANGE: All NG, Salawati[K], Batanta[L], and Waigeo I[MA], and the Aru Is[MA] (Mees 1965[K], Greenway 1966[L]). Occurs in the lowlands of the Trans-Fly[CO]. Coates (1990) postulated that the species is absent from the n watershed of the SE Peninsula as well as east of Port Moresby. That said, Pratt heard this species at Isurava, inland from Kokoda (field notes Feb-2014). Also breeds in n Australia (Schodde & Mason 1999). Widespread in the Trans-Fly (Bishop 2005a).

DIAGNOSIS: NG forms are not distinct from the nominate form from the NT, Australia. The form *darwini* of n Western Australia has a stronger yellow wash both dorsally and ventrally (Schodde & Mason 1999).

NOTES

David & Gosselin (2002a) overruled Beehler & Finch (1985), who had asserted that *chloronotus* was a noun in apposition. David & Gosselin noted that the species epithet is based on an adjectival Greek form, and thus should modify the feminine *Gerygone*. *Other names*—Grey-headed Gerygone, Green-backed Warbler, *Gerygone chloronotus*.

Fairy Gerygone *Gerygone palpebrosa* RESIDENT

Gerygone palpebrosa Wallace, 1865. *Proc. Zool. Soc. London*: 475.—Aru Is.

Inhabits forest and edge in foothills, lower mts, and occasionally in the lowlands of NG, NW Islands, Aru Is, and Yapen I, sea level–1,460 m[CO].

Extralimital—e QLD.

SUBSPECIES

a. *Gerygone palpebrosa palpebrosa* Wallace

Synonym: *Gerygone inconspicua* Ramsay, 1878. *Proc. Linn. Soc. NSW* (1879) 3: 116.—Laloki R, SE Peninsula.

RANGE: Misool, ?Salawati, and Waigeo I, the Aru Is, Bird's Head and Neck, S Lowlands, and SE Peninsula (Mayr 1941, Mees 1965). The original Salawati record was cited by Beehler *et al.* (1986) but without documentation.

DIAGNOSIS: Male with olive-green crown matching mantle; lores and throat blackish; white loral spot and white subocular stripe. We merge *inconspicua* into this form because of minimal distinctions between the two. See photo of the nominate form in Coates (1990: 115) and Coates & Peckover (2001: 133).

b. *Gerygone palpebrosa wahnesi* (A. B. Meyer)

Pseudogerygone wahnesi A. B. Meyer, 1899. *Orn. Monatsber.* 7: 144.—Bongu, Astrolabe Bay.

RANGE: Yapen I, the NW Lowlands, the Sepik-Ramu, and the n watershed of the SE Peninsula to the Kumusi R, where it intergrades with the nominate form (Mayr 1941, Leiden specimens from 15-Apr-1903 and 28-Mar-1903).

DIAGNOSIS: Male with black crown, face, and throat; white loral spot and white subocular stripe.

c. *Gerygone palpebrosa tarara* Rand

Gerygone palpebrosa tarara Rand, 1941. *Amer. Mus. Novit.* 1102: 11.—Tarara, Wassi Kussa R, s NG.

RANGE: The Trans-Fly, from Dolak I east to Lake Daviumbu and the mouth of the Fly R.

DIAGNOSIS: Male with green crown and blackish face washed with green-brown (Gregory 2007b).

Other names—Black-headed Gerygone Warbler, Black-throated Warbler, Fairy Warbler.

Large-billed Gerygone *Gerygone magnirostris* RESIDENT

Gerygone magnirostris Gould, 1843. *Proc. Zool. Soc. London* (1842): 133.—Greenhill I, Port Essington, NT, Australia.

All lowlands of NG, the NW Islands, Aru Is, Bay Is, Manam I, Karkar I, Daru I, and the D'Entrecasteaux and Louisiade Arch; mainly lowlands but in some locations to 1,200 m^{CO} (resident at Baiyer R). Inhabits mangrove swamps, forest edge, and scrub; most commonly found near water (Coates 1990).

Extralimital—n and ne Australia; also islands in Torres Strait, including Dam, Boigu, and Saibai (Gregory 2007b).

SUBSPECIES

a. *Gerygone magnirostris conspicillata* (G. R. Gray)

Microeca conspicillata G. R. Gray, 1859. *Proc. Zool. Soc. London*: 156.—Manokwari, Bird's Head.

> Synonym: *Ethelornis magnirostris cobana* Mathews, 1926. *Bull. Brit. Orn. Club* 47: 40.—Mimika R, w sector of the S Lowlands. Replacement name for *Zosterops* [= *Gerygone*] *fusca* Bernstein, 1864; preoccupied by *Psilopus* [= *Gerygone*] *fuscus* Gould, 1838.

> Synonym: *Gerygone magnirostris occasa*, Ripley, 1957. *Postilla* Yale Peabody Mus. 31: 3.—Kofiau I, NW Islands.

RANGE: Kofiau^J, Waigeo^{MA}, Batanta^{MA}, and Salawati I^{MA}; also the Bird's Head, the n sector of the Bird's Neck (Wandammen Penin)^{MA}, and the w extremity of the S Lowlands (Ripley 1957^J).

DIAGNOSIS: Off-white throat and belly; buff sides of breast; dun brown dorsally; white eye-ring; whitish undertail. Diamond *et al.* (2009) noted the questionable status of *occasa* based on a visit to Kofiau in 1986. The forms *cobana* and *occasa* are within the range of variation of *conspicillata* and thus are combined here.

b. *Gerygone magnirostris brunneipectus* (Sharpe)

Pseudogerygone brunneipectus Sharpe, 1879. *Cat. Birds Brit. Mus.* 4: 221.—Aru Is.

> Synonym: *Pseudogerygone conspicillata mimikae* Ogilvie-Grant, 1915. *Ibis*, Jubilee Suppl. 2: 168.—Mouth of the Mimika R, w sector of the S Lowlands.

RANGE: The s sector of the Bird's Neck, the Aru Is^{MA}, Daru I^{MA}, and all s NG^{MA}; also islands in Torres Strait (Gregory 2007b).

DIAGNOSIS: Buff on sides of breast; pale buff undertail; white throat; broken pale eye-ring; dun brown dorsally.

c. *Gerygone magnirostris hypoxantha* Salvadori

Gerygone hypoxantha Salvadori, 1878. *Ann. Mus. Civ. Genova* 12: 345.—Biak I, Bay Is.

RANGE: Biak I.

DIAGNOSIS: Underparts entirely pale yellow (chin to undertail); loral streak white; upperparts uniformly gray-brown; broken eye-ring white (Gregory 2007b). Suggested by some authorities to merit species status. According to S. van Balen (*in litt.* 2012), the song of this form is quite similar to that of the Australian race *cairnsensis*. K. D. Bishop (*in litt.*) has studied the birds of Biak comprehensively and is in accord with van Balen in considering *hypoxantha* best treated as a subspecies of *G. magnirostris* (Mayr & Meyer de Schauensee 1939a, Mayr 1941, Rand & Gilliard 1967, Mayr 1986, Dickinson & Christidis 2014; *pace* Gregory 2007b).

d. *Gerygone magnirostris affinis* A. B. Meyer

Gerygone affinis A. B. Meyer, 1874. *Sitzungsber. Akad. Wiss. Wien* 70: 116.—Yapen I, Bay Is. [Citation is on p. 116, not p. 117; M. LeCroy *in litt.*]

RANGE: Yapen, Manam, and Karkar I^J, the NW Lowlands, the Sepik-Ramu, and the n watershed of the SE Peninsula (Mayr 1941, Diamond & LeCroy 1979^J).

DIAGNOSIS: Olive dorsally; crown distinctly gray; broken white eye-ring; cheek buff-olive; throat off-white; belly with pale yellow wash; outer rectrices with white and subterminal dark band.

e. *Gerygone magnirostris rosseliana* Hartert

Gerygone magnirostris rosseliana Hartert, 1899. *Novit. Zool.* 6: 79.—Rossel I, Louisiade Arch, SE Islands.

    Synonym: *Gerygone rosseliana onerosa* Hartert, 1899. *Novit. Zool.* 6: 209.—Misima I, Louisiade Arch, SE Islands.

    Synonym: *Gerygone magnirostris proxima* Rothschild & Hartert, 1918. *Novit. Zool.* 25: 319.—Fergusson I, D'Entrecasteaux Arch, SE Islands.

    Synonym: *Gerygone magnirostris tagulana* Rothschild & Hartert, 1918. *Novit. Zool.* 25: 318.—Tagula I, Louisiade Arch, SE Islands.

RANGE: Goodenough, Fergusson, and Normanby I[J]; also Misima, Rossel, and Tagula I (Mayr 1941, Pratt field notes[J]).

DIAGNOSIS: Dull brown dorsally; broken white eye-ring; obscure buff breast band, most prominent on sides of breast; pale yellow wash on abdomen; throat off-white; iris brown. We have included *proxima*, *onerosa*, and *tagulana* here because of lack of distinctive plumage characters.

NOTES

We have chosen to sink a number of subspecies that were recognized by Mayr (1986) and Gregory (2007b), given the lack of objectively defining characters possessed by this species. It is remarkable that such a drab and featureless species could generate such a profusion of subspecies descriptions. *Other names*—Large-billed Fairy-warbler, Brown-breasted Flyeater.

Mangrove Gerygone *Gerygone levigaster*   RESIDENT

Gerygone levigaster Gould, 1843. *Proc. Zool. Soc. London* (1842): 133.—Port Essington, NT, Australia.

    Mainly the s coast of NG, from the s shore of the Bird's Neck east to Port Moresby and Rigo in the SE Peninsula; also the Bintuni Bay mangroves. Inhabits low mangroves (*Avicennia*) and perhaps also fringes of creeks in floodplain forest (Coates 1990). In the Purari R basin some evidence of movement upriver during the se tradewinds season, when found in riparian forest at open-canopy tributary edges (I. Woxvold *in litt.*).

    Extralimital—n and e Australia.

SUBSPECIES

a. *Gerygone levigaster pallida* Finsch

Gerygone pallida Finsch, 1898. *Notes Leyden Mus.* 20: 134.—Triton Bay, Bird's Neck.

RANGE: Head of Bintuni Bay and coastal s NG where there is suitable mangrove habitat. Localities—Bintuni Bay[J], Triton Bay, mouth of the Mimika R, Komolom I[L], Daru, mouth of the Oriomo R, Lese Oalai[CO], near Port Moresby[CO], and Rigo[CO] (Mayr 1941, Erftemeijer *et al.* 1991[J], Bishop 2005a[L]).

DIAGNOSIS: Browner above than the nominate form of Australia; lacks white at bases and near tips of outer webs of outer tail feathers (Higgins & Peter 2002). See photo in Coates (1990: 117).

NOTES

Formerly considered a population of the widespread Australian species *G. fusca* (Mayr 1941); our treatment follows Schodde (1975), Ford (1981a), Schodde & Mason (1999), and Gregory (2007b). *Other names*—Mangrove Warbler, Buff-breasted Flyeater.

Brown-breasted Gerygone *Gerygone ruficollis*   ENDEMIC | MONOTYPIC

Gerygone ruficollis Salvadori, 1876. *Ann. Mus. Civ. Genova* (1875) 7: 959.—Hatam, Arfak Mts, Bird's Head.

    Synonym: *Gerygone insperata* De Vis, 1892. *Ann. Rept. Brit. New Guinea*, 1890–91: 94.—Mount Suckling, SE Peninsula.

Mountains of the Bird's Head, Central Ra, Fakfak[J] and Foja Mts[K], and Huon Penin; 1,450–2,450 m, extremes 900–3,300 m[CO] (Mayr 1941, Gibbs 1994[J], Beehler *et al.* 2012[K]). Inhabits canopy of forest, edge, casuarina groves, shade trees in villages and towns, and alpine shrublands (Coates 1990).

NOTES

Considerable individual variation leads us to treat this species as monotypic. The AMNH collection holds adult specimens from both ends of NG that are identical in plumage (*e.g.*, AMNH no. 606887 from the Arfak Mts and AMNH no. 420297 from Murray Pass). Diamond possibly heard this species at 1,300 m on Karkar I (Diamond & LeCroy 1979). *Other names*—Treefern Gerygone Warbler.

Family Pomatostomidae

Australasian Babblers   [Family 5 spp/Region 2 spp]

The species of the family inhabit Australia and New Guinea (Matthew 2007). The family is characterized by the superficial similarity to the *Pomatorhinus* babblers of tropical Asia: the body slim, the bill longish and decurved, and the tail long, fan-shaped, and 12-feathered. Osteologically, the vestigial and convoluted internasal septum is distinctive (Schodde & Mason 1999). Both New Guinean species are gregarious, living in family groups that often join mixed-species foraging parties, especially the black-and-brown flocks of the lowland forests of New Guinea (Diamond 1987). Both species build large, complex, covered nests that are also used as group roosts. Breeding is communal (Schodde & Mason 1999). The phylogeny of Barker *et al.* (2004) and Aggerbeck *et al.* (2014) showed this lineage to be distinct and isolated, sister to the passerid and corvid assemblages, along with the Orthonychidae (Jønsson *et al.* 2011b).

GENUS *GARRITORNIS*   ENDEMIC | MONOTYPIC

Garritornis Iredale, 1956. *Birds New Guinea* 2: 79. Type, by original designation, *Pomatorhinus isidorei* Lesson.

The single species of this endemic genus is characterized by its very sociable behavior, uniform rufous-brown plumage, the decurved yellowish bill, the longish fanned tail, rain-forest habitat, and the diagnostic pendent nest that is tended cooperatively and used as a communal roosting place (Matthew 2007). We retain this genus distinct from *Pomatostomus*, owing in part to the genetic results of Edwards & Wilson (1990), who found that *isidorei* was the most divergent of all the Australasian babblers.

Papuan Babbler *Garritornis isidorei*   ENDEMIC | MONOTYPIC

Pomatorhinus Isidorei Lesson, 1827. *Dict. Sci. Nat.*, éd. Levrault 50: 37.—Manokwari, e Bird's Head.
> Synonym: *Pomatorhinus isidori calidus* Rothschild 1931 *Novit. Zool.* 36: 266.—Siriwo R, w Mamberamo basin, nw NG.

Inhabits rain forest and gallery forest of the lowlands of NG, Misool[MA], and ?Waigeo I (Ripley 1959); sea level–300 m[CO], max 500 m[J] (Beehler *et al.* 1986[J]). Perhaps absent along the Mainland's n watershed from the Huon Penin to Morobe and apparently absent from the s Trans-Fly, though present in the S Lowlands (Bishop 2005b).

NOTES

We treat the species as monotypic; designation of the putative subspecies was based on slight variation in throat coloration that does not merit recognition (see Mees 1980: 3, also comment in Matthew 2007: 334). It is strange that there is only one specimen from Waigeo I, which was taken by a field collector working apart from his scientific employer (Ripley 1959). A number of previous expeditions to the island failed to secure this vocal and conspicuous flocking species. It is thus necessary to reconfirm its presence on Waigeo.
Other names—New Guinea, Isidore's, or Rufous Babbler; Isidore's Rufous Babbler, *Pomatostomus isidori*.

GENUS *POMATOSTOMUS*   [Genus 4 spp/Region 1 sp]

Pomatostomus Cabanis, 1851. *Mus. Hein.* 1: 83. Type, by subsequent designation (G. R. Gray, 1855. *Cat. Gen. Subgen Birds*: 45), *Pomatorhinus temporalis* Vigors & Horsfield.

The species of this genus range from Australia to New Guinea (Matthew 2007). The genus is characterized by the decurved bill; the neatly patterned facial plumage much resembling that of a *Pomatorhinus* babbler; the very social habit; the savanna-woodland habitat preference; and the distinctive domed stick nest that is tended cooperatively (Schodde & Mason 1999).

Grey-crowned Babbler *Pomatostomus temporalis*   RESIDENT

[*Pomatorhinus*] *Temporalis* Vigors & Horsfield, 1827. *Trans. Linn. Soc. London* (1826) 15: 330.—Shoalwater Bay, QLD, Australia.

Inhabits savanna and monsoon woodland of the Trans-Fly, from Dolak I east to the Oriomo R (Rand 1942a, Bishop 2005a).

Extralimital—w, n, e, and se Australia (Dickinson & Christidis 2014).

SUBSPECIES

a. *Pomatostomus temporalis temporalis* (Vigors & Horsfield)

  Synonym: *Pomatorhinus temporalis strepitans* Mayr & Rand, 1935. *Amer. Mus. Novit.* 814: 6.—Dogwa, Oriomo R, Trans-Fly, s NG.

RANGE: The Trans-Fly, between the Digul R and Oriomo R (Mees 1982b). Also e Australia.

DIAGNOSIS: Upper breast white with rufescence confined to belly and lower breast (Matthew 2007). Forehead dark, separating whitish eyebrows; breast creamy white (Schodde & Mason 1999). See photo in Coates (1990: 77). We follow Mees (1982b) and Matthew (2007) in subsuming *strepitans* into the nominate form. Our comparison of QLD material with that from s NG indicated no significant plumage distinction.

NOTES

Other names—Australian or Northern Babbler.

Family Orthonychidae

Logrunners   [Family 3 spp/Region 1 sp]

The species of the family range from the uplands of New Guinea to the uplands of eastern Australia. This tiny and peculiar lineage is characterized by its terrestrial habit, rounded whirring wings, its ground-scraping foraging method involving lateral movement of the foot and leg, abbreviated bill, woodpecker-like pointed rectrices, and distinctive leg musculature (Schodde & Mason 1999). Nests are domed. Barker *et al.* (2004) showed that this small lineage, along with the Pomatostomidae, is sister to the Passerida+"core Corvoidea", a huge assemblage of songbirds including many families of Australasian songbirds plus many families of Eurasian songbirds. Jønsson *et al.* (2011b) confirmed this sister position.

GENUS *ORTHONYX*   [Genus 3 spp/Region 1 sp]

Orthonyx Temminck, 1820. *Man. Orn.* éd. 2, 1: 81. No species; generic details only. Type, by subsequent monotypy (Ranzani, 1822. *Elem. Zool.* 3: 19), *Orthonyx temminckii* Ranzani.

The species of this genus inhabit the uplands of eastern Australia and New Guinea (Joseph *et al.* 2001, Boles 2007). The genus is characterized by small to medium size, shortish bill, 10 feathers in tail, spine-tipped rectrices, and terrestrial habit.

Papuan Logrunner *Orthonyx novaeguineae*   ENDEMIC

Orthonyx Novae Guineae A. B. Meyer, 1874. *Sitzungsber. Akad. Wiss. Wien* 69: 83.—Arfak Mts, Bird's Head

Patchily distributed, with records from the mts of the Bird's Head, Western Ra, Eastern Ra, and mts of the SE Peninsula, 1,200–2,900 m[CO]. Rarely encountered. Forages on the floor of montane forest interior.

SUBSPECIES

a. *Orthonyx novaeguineae novaeguineae* A. B. Meyer

RANGE: The Bird's Head (Tamrau and Arfak Mts)[MA]. Localities—above Waibeem village in the Tamrau Mts[J], Hatam and Bivak October in the Arfak Mts[K] (Gyldenstolpe 1955b[K], Beehler field notes[J]).

DIAGNOSIS: Narrower black margins to feathers of upperparts; sides of breast paler gray than in *victorianus*. Joseph *et al.* (2001) noted a deep molecular break between this form and *victorianus*.

b. *Orthonyx novaeguineae victorianus* van Oort
Orthonyx temminckii victoriana van Oort, 1909. *Notes Leyden Mus.* 30: 234.—Mount Victoria, SE Peninsula.
> Synonym: *Orthonyx temminckii dorsalis* Rand, 1940. *Amer. Mus. Novit.* 1074: 2.—Bele R, 2,200 m, 18 km n of Lake Habbema, Western Ra.

RANGE: From the Western Ra patchily east to the mts of the SE Peninsula. Localities—Ilaga[M], 9 km northeast of Lake Habbema[L], Bele R Camp[L], 18 km southwest of Bernhard Camp[L], Tari Gap[N], Tomba Pass[O], Mt Missim[J], Mt Albert Edward[CO], Mt Tafa[K], Mt Scratchley[MA], Mt Knutsford[MA], Mt Victoria[MA], and near Tetebedi[CO] (Greenway 1935[J], Mayr & Rand 1937[K], Rand 1942b[L], Ripley 1964a[M], Frith & Frith 1992[N], K. D. Bishop *in litt.*[O]).

DIAGNOSIS: Similar to *novaeguineae* but exhibiting less white ventrally, and rump darker (Rand & Gilliard 1967). We follow Joseph *et al.* (2001) in subsuming *dorsalis* into *victorianus*. See photos in Coates (1990: 59–60).

NOTES
We follow Joseph *et al.* (2001) and Boles (2007) in treating the NG populations of *Orthonyx* as a species distinct from the Australian populations. *Orthonyx* is masculine (David & Gosselin 2002b). Some authors have split off *victorianus* based on genetic distance. *Other names*—earlier authorities have lumped the NG forms with the Australian Logrunner, *Orthonyx temminckii*.

Family Cnemophilidae

Satinbirds   ENDEMIC [Family 3 spp]

The species of this compact and enigmatic lineage can be found only in the Central Ranges of New Guinea. Formerly considered a distinct subfamily within the birds of paradise (Paradisaeidae), it was shown by Cracraft & Feinstein (2000) to be only distantly related, in spite of the superficial similarity of the species and their polygynous court-display behavior. Barker *et al.* (2004) provided a phylogeny that indicated this group is sister to Callaeidae (the New Zealand wattlebirds), and these two are sister to the Melanocharitidae (berrypeckers and longbills). This is confirmed by Jønsson *et al.* (2011b), though see Irestedt & Ohlson (2008) and Aggerbeck *et al.* (2014) for conflicting placements. These three sexually dimorphic species are shy inhabitants of montane forest, where they subsist on a diet largely or entirely of fruit. The most comprehensive treatment of the group is found in Frith & Beehler (1998).

GENUS *CNEMOPHILUS*   ENDEMIC [Genus 2 spp]

Cnemophilus De Vis, 1890. *Ann. Rept. Brit. New Guinea*, 1888–89: 62. Type, by monotypy, *C. macgregorii* De Vis.

The species of this genus are endemic to the Central Ranges of New Guinea. The genus is characterized by the compact shape; short and rounded wings; short, rather broad bill; and strong sexual dichromatism, including the brightly colored mouth skin and presence of silky or iridescent feathers in the adult male. In overall appearance the females resemble female *Amblyornis* bowerbirds.

Loria's Satinbird *Cnemophilus loriae*   ENDEMIC | MONOTYPIC
Loria loriae Salvadori, 1894. *Ann. Mus. Civ. Genova* (2)14: 151.—Moroka, s watershed of SE Peninsula.
> Synonym: *Loria loriae amethystina* Stresemann 1934, *Orn. Monatsber.* 42: 144.—Schraderberg, Sepik Mts, n sector of the Eastern Ra.

Synonym: *Loria loriae inexpectata* Junge, 1939. *Nova Guinea* (ns)3: 77.—Bijenkorf, Oranje Mts, Western Ra. Inhabits montane forest and edge the length of the Central Ra; 1,800–2,400 m, extremes 1,200–2,800 m[CO] (Stresemann 1923, Hartert *et al.* 1936, Mayr 1931, Rand 1942b, Gyldenstolpe 1955a, Diamond 1972, Coates 1990, Frith & Beehler 1998).

NOTES

We treat the species as monotypic. The variation within the populations (*e.g.*, color of the gloss of the inner secondaries) is such that recognizing these forms is not merited. *Other names*—Loria's Bird of Paradise, *Loria loriae.*

Crested Satinbird *Cnemophilus macgregorii*   ENDEMIC

Cnemophilus macgregorii De Vis, 1890. *Ann. Rept. Brit. New Guinea*, 1888-1889: 62.—Mount Knutsford, Owen Stanley Ra, SE Peninsula.

    Endemic to upper montane forest of the Central Ra, from the Western Ra (north of Lake Habbema) east to the c sector of the SE Peninsula (northeast of Port Moresby); 2,600–3,500 m, extremes 2,300–3,650 m[CO]. Inhabits upper montane forest and subalpine shrubbery. Sight records from near Lake Habbema area by Davies (2008) in Jul-2007 and by Matheve *et al.* (2009) on 24-Jul-2009.

SUBSPECIES

a. *Cnemophilus macgregorii sanguineus* Iredale

Cnemophilus macgregorii sanguineus Iredale, 1948. *Austral. Zool.* 11: 162.—Kumdi, Mt Hagen distr, Eastern Ra.

      Synonym: *Cnemophilus macgregorii kuboriensis* Mayr & Gilliard, 1954. *Bull. Amer. Mus. Nat. Hist.* 103: 361.—Mount Orata, Kubor Mts, Eastern Ra.

RANGE: From the Kaijende Highlands (south of Porgera) south and east to the Bismarck Ra. Localities—Kaijende Highlands[W], Tari Gap[M], Mt Giluwe[CO], Kumul Lodge[X], Tomba[X], Mt Hagen[K], Kumdi[J], Mt Orata in the Kubor Mts[K], Weiga in the Wahgi valley[J], Minj R, the Bismarck Ra[CO], and Mt Karimui[DI] (Mayr & Gilliard 1954[K], Gyldenstolpe 1955a[J], Sims 1956[L], Frith & Beehler 1998[M], Beehler & Sine 2007[W], Beehler field notes[X]). DIAGNOSIS: Dorsal plumage deep reddish orange. We follow Diamond (1972), Frith & Beehler (1998), and others in sinking *kuboriensis*. See photos of *sanguineus* in Coates (1990: 424–25) and Coates & Peckover (2001: 217). See comparative photo of three described forms in figure 2d. This does not include the undiagnosed population in the Western Ra, which is probably distinct.

b. *Cnemophilus macgregorii macgregorii* De Vis

RANGE: Mountains of the SE Peninsula, from the Ekuti Divide[CO] (south of Wau) to Mt Knutsford[CO] (northeast of Port Moresby). Recorded from the Bulldog Road[J], Mt Albert Edward[J], Murray Pass[K], Mt Tafa[K], Mt Musgrave[MA], and Mt Knutsford[MA] (Mayr & Rand 1937[K], Bell 1971b[L], specimens at BPBM[J]). The se terminus of the range not known. DIAGNOSIS: Male is yellow-orange dorsally, distinctly paler and yellower than *sanguineus*. See photos in Coates (1990: 424).

NOTES

There is a need to diagnose the population from the Western Ra. There is at least one individual collected from w NG, originally preserved as an anatomical specimen (from Jabogema, north of Lake Habbema; B. Poulson *in litt.*). Neither the e nor w boundary of the geographic range of this species has been adequately delineated. Thus it should be looked for southeast of Mt Knutsford and west of the Baliem in w NG (should be expected in the Carstensz highlands), as well as in the Star Mts of the Indonesian-PNG border. *Other names*—Crested, Multi-crested, or Sickle-crested Bird of Paradise.

Loboparadisea Rothschild, 1896. *Bull. Brit. Orn. Club* 6: 15. Type, by monotypy, *Loboparadisea sericea* Rothschild.

    The single species of the genus is endemic to the Central Ranges of New Guinea. The genus, evidently the sister to *Cnemophilus*, is characterized by the male's bulbous and bifurcate maxillary wattle, and the mottled ventral plumage of the female and juvenile plumages.

Yellow-breasted Satinbird *Loboparadisea sericea*   ENDEMIC

Loboparadisea sericea Rothschild, 1896. *Bull. Brit. Orn. Club* 6: 16.—Probably from the Weyland Mts, Western Ra [obtained as a trade skin on Kurudu I, east of Yapen I].

    The Western, Border, and Eastern Ra, plus the northernmost mts of the SE Peninsula, 625–2,000 m[CO]. Patchily distributed, uncommon or absent from seemingly appropriate habitats. Confined to the canopy of the interior of lower montane and mid-montane forest.

SUBSPECIES

a. *Loboparadisea sericea sericea* Rothschild

RANGE: The Western Ra, Border Ra, and Eastern Ra. Localities—Kunupi in the Weyland Mts[K], Gebroeders Mts[P], Otakwa R[MA], WakWak (foot of Western Ra)[P], Mt Goliath[MA], Ok Tedi area[CO], Deikimdikin in the Victor Emanuel Mts[W], Telefomin[CO], Baia R headwaters at the foot of the Karius Ra[L], Kubor Mts[Z], Lepa Ridge[V], Kopa in the Jimi valley[J], Mt Karimui[DI], Mt Sisa[L], Mt Bosavi[A], Soliabeda[DI], Crater Mt[U], Guyebi on the n slope of the Bismarck Ra[P], nw slopes of Mt Wilhelm[X], and Buntibasa distr in the Kratke Mts[P] (Hartert *et al.* 1936[K], Gyldenstolpe 1955b[J], Gilliard & LeCroy 1961[W], Mack & Wright 1996[U], Frith & Beehler 1998[Z], Sam & Koane 2014[X], J. M. Diamond & K. D. Bishop *in litt.*[L], Beehler field notes[V], Pratt field notes and BPBM specimen[A], AMNH specimens[P]).

DIAGNOSIS: Dorsal plumage duller than that of *aurora*, the crown darker and browner; the male's bill wattle ranges from yellowish (in west) to pale green (in east). No significant difference in size (Frith & Beehler 1998). See photos in Coates (1990: 428). See comparative photo of the two described forms in figure 2e.

b. *Loboparadisea sericea aurora* Mayr

Loboparadisea sericea aurora Mayr, 1930. *Orn. Monatsber.* 38: 147.—Dawong, Herzog Mts, SE Peninsula.

RANGE: The range of this form generally lies east of the Eastern Ra, in the nw sector of the SE Peninsula, east of the Tauri-Watut Gap, but see notes below on recent record from Mt Wilhelm. Localities—Aseki-Oiwa area[K], Dawong in Herzog Mts[MA], and Wau[J] (Greenway 1935[J], Austral. Nat. Wildl. Coll.[K]).

DIAGNOSIS: These two subspecies are thinly defined and may reflect clinal variation from west to east. *L. s. aurora* exhibits a pale blue bill wattle; the mantle is marginally paler than that of *sericea* and has a yellowish wash, and the crown exhibits an olive wash (not brown-washed; Frith & Beehler 1998).

NOTES

The racial distinctions need additional study. Wattle color may be definitive, apparently ranging from yellow to green to blue (west, central, east). P. Z. Marki (*in litt.*) netted two adult males on the n slopes of Mt Wilhelm in October 2015. These birds exhibited a blue wattle and olive wash to the crown, indicative of the eastern form *aurora*. One option is to treat the species as clinal and monotypic, with some minor differences between the western and eastern birds. Perhaps could be expected to range further eastward down the SE Peninsula. *Other names*—Yellow-breasted, Wattled, Wattle-billed, or Shield-billed Bird of Paradise.

Family Melanocharitidae

Berrypeckers and Longbills

ENDEMIC [Family 11 spp]

This small endemic family comprises a lineage of fruit eaters that was formerly situated in the Dicaeidae and a lineage of nectarivores that was traditionally placed in the Meliphagidae. These are small forest dwellers that mainly inhabit the lower and middle stories of the forest, though occasionally are found in the canopy of fruiting and flowering trees. All species possess an olive plumage. The family is diverse morphologically and difficult to characterize. Barker *et al.* (2004) provided the molecular evidence for this novel treatment. Jønsson *et al.* (2011b) had the Melanocharitidae as an isolated lineage with closest affinities to the Cnemophilidae. Schodde & Christidis (2014) created the subfamily Toxorhamphinae to encompass the two species of *Toxorhamphus*, and the Oedistomatinae for the two species of *Oedistoma*. Their paper noted that these two lineages are deeply divergent from other Melanocharitidae as well as from each other.

GENUS *MELANOCHARIS*

ENDEMIC [Genus 5 spp]

Melanocharis P. L. Sclater, 1858. *Proc. Linn. Soc. London, Zool.* 2: 157. Type, by monotypy, *Dicaeum niger* Lesson.

The five species of the genus are endemic to the forests of the Region. The genus is characterized by sexual dichromatism and reverse sexual size dimorphism (adult males are smaller than adult females); male dorsal plumage blue-black (or olive green); abbreviated fruit-eating bill; and distinctive cup nest bound with spider silk and decorated with lichens.

Obscure Berrypecker *Melanocharis arfakiana*

ENDEMIC | MONOTYPIC

Dicaeum arfakianum Finsch, 1900. *Notes Leyden Mus.* 22: 70.—Arfak Mts, Bird's Head.

This little-known species is very rare but widespread in the foothills of the Mainland. The only two historical museum specimens were taken from Moris in the Arfak Mts (Bird's Head) and at Matsika (950 m) along the upper Angabanga R., nw of Port Moresby (Mayr 1941, Schodde 1978). Two more specimens referable to this species were taken 10 km north of Tabubil and at Mt Robinson in Jul-1994, in the Border Ra of PNG (Gregory 1997, B. Whitney *in litt.*). Locations of sight records include the Tabubil area (Mt Robinson, Ok Menga, Dablin Creek, 680–950 m)[J,L,M], the Adelbert Mts (Keki Lodge)[K], Boana at 950 m[J] (nw of Lae), and the lower mts along the Kokoda Track (670–1,100 m) including Efogi, Ower's Corner, and Varirata NP[CO] (B. Whitney *in litt.*[J], Gregory 1997[L], 2008b[M], Beehler field notes 12-Jul-2012[K]).

NOTES

The species is best identified by the horn-colored bill, the bright orange gape, the yellow pectoral tuft, the shortish unmarked tail, and the olive-green dorsal plumage. The squeaky and jumbled song is similar to that of *M. nigra*. The apparent rarity and patchy distribution suggests that it may be either (1) a disappearing species, compressed into a narrow elevational band between two closely allied, ecologically similar congeners, or (2) a canopy-frequenting species that may have been overlooked in most places (Schodde 1978). It was observed foraging on banana figs in a large canopy tree at Keki Lodge in the Adelbert Mts (Beehler field notes). It may be a strict canopy dweller, which would explain it never being netted and rarely being seen.

Data from the two new specimens collected by B. Whitney are provided here. Specimen 1: 19-Jul-1994. 10 km by road north of Tabubil, Western Prov, PNG, 770 m. Male, testis 1 × 1.5mm, little fat, no molt, stomach with insect parts. Bill: mandible entirely orange; maxilla dark slate with a faint orange base; iris pale brown; eye-ring (fleshy orbital skin) dull yellowish; legs olive green, soles slightly yellower; shot in subcanopy of tall forest using tape playback. Pectoral tuft with yellow wash; belly with slight yellow wash; mantle greenish olive; crown grayish olive; throat and breast grayish olive. Tail 33.5, wing (unflattened) 59, bill from base 11.5, tarsus 18. Specimen 2: 20-Jul-1994. Male. Mt Robinson, *ca.* 15 km north of Tabubil, Western Prov, PNG. No specimen label data. Plumage description as for specimen 1. Tail 34, wing 58, bill from base 11.5, tarsus 16.5. *Other names*—Arfak Berrypecker.

Black Berrypecker *Melanocharis nigra* ENDEMIC

Dicaeum niger Lesson, 1830. *Voy. Coquille, Zool.* (1826) 1: 673.—Manokwari, Bird's Head.

All NG, the main NW Islands but for Batanta I, Yapen I, Mios Num I, and the Aru Is; sea level–1,200 m, max 1,600 m (Coates & Peckover 2001). Widespread on the Mainland except for the Trans-Fly (*fide* Bishop 2005a). Inhabits forest and edge.

SUBSPECIES

a. *Melanocharis nigra pallida* Stresemann & Paludan

Melanocharis nigra pallida Stresemann & Paludan, 1932. *Novit. Zool.* 38: 151.—Waigeo I, NW Islands.

RANGE: Waigeo I.

DIAGNOSIS: Male's ventral plumage entirely medium gray and upperparts all black (Rand & Gilliard 1967).

b. *Melanocharis nigra nigra* (Lesson)

RANGE: Misool and Salawati I, the Bird's Head, and the Bird's Neck (Mayr 1941).

DIAGNOSIS: As for *pallida*, but male's ventral plumage darker gray (Rand & Gilliard 1967). Edges of remiges in the male are black, not olive as compared with next subspecies (Diamond 1972).

c. *Melanocharis nigra chloroptera* Salvadori

Melanocharis chloroptera Salvadori, 1876. *Ann. Mus. Civ. Genova* (1875) 7: 987.—Aru Is.

RANGE: The Aru Is[MA] and s NG[MA] from the Mimika R to the e highlands of PNG and Purari R. Localities—Sturt I Camp[J], Lake Daviumbu[J], Palmer Junction Camp[J], Black R Camp[J], Karimui[DI], Bomai[DI], Soliabeda[DI], and Mt Karimui[DI] (Rand 1942a[J]).

DIAGNOSIS: Similar to the nominate form, but male's wing coverts and remiges edged green, contrasting with the black of the dorsal plumage (Rand & Gilliard 1967). Ventral plumage of male is gray, not black (Diamond 1972).

d. *Melanocharis nigra unicolor* Salvadori

Melanocharis unicolor Salvadori, 1878. *Ann. Mus. Civ. Genova* 12: 333.—Yapen I, Bay Is.

RANGE: Yapen and Mios Num I[MA]; also the NW Lowlands, Sepik-Ramu, and SE Peninsula; west in the s watershed of the SE Peninsula to Putei (northeast of Kerema).

DIAGNOSIS: Male entirely black dorsally and ventrally, except for white underwing and pectoral tufts (Rand & Gilliard 1967). See photos in Coates (1990: 316–17).

NOTES

Diamond (1972) noted that it would be useful to see what happens when the distinctive gray-bellied race (*nigra*) meets the black-bellied race (*unicolor*) in the n and s watersheds. Do these forms mix and interbreed or do they act as good species? These contact zones should be looked for in the Waropen region of the e shores of Geelvink Bay (in the north), and in the lowlands east of the Purari R (in the south).

Mid-mountain Berrypecker *Melanocharis longicauda* ENDEMIC

Melanocharis longicauda Salvadori, 1876. *Ann. Mus. Civ. Genova* (1875) 7: 942.—Arfak Mts, Bird's Head.

Inhabits forest interior of the mts of NG, 700–1,900 m[CO]. There is one extraordinary record of an immature male in the lowlands near Magori village, Amazon Bay, Central Prov (Coates 1990). No records from the Cyclops Mts. Appears to be absent from the Adelbert Mts (Gilliard & LeCroy 1967, Pratt 1983).

SUBSPECIES

a. *Melanocharis longicauda longicauda* Salvadori

Synonym: *Melanocharis longicauda chloris* Stresemann & Paludan, 1934. *Orn. Monatsber.* 42: 45.—Kunupi, 1,200 m, Weyland Mts, Western Ra.

Synonym: *Melanocharis longicauda umbrosa* Rand, 1941. *Amer. Mus. Novit.* 1102: 15.—6 km sw of Bernhard Camp, 1,200 m, ne sector of the Western Ra.

RANGE: Mountains of the Bird's Head, Bird's Neck (Wandammen Mts), and Western and Border Ra. We presume this form ranges east to the Strickland R gorge.

DIAGNOSIS: Male's ventral plumage pale olive gray washed with yellow, especially on the flanks (Rand & Gilliard 1967). Putative difference for the race *chloris* was a product of specimen preparation (see photos of *chloris* and *longicauda*, fig. 2f). This form is clinal, with ventral plumage slightly darker in east. We thus sink *umbrosa*, as we are treating the minor plumage variation as part of a west-east cline.

b. *Melanocharis longicauda captata* Mayr

Melanocharis longicauda captata Mayr, 1931. *Mitt. Zool. Mus. Berlin* 17: 668.—Junzaing, Huon Penin.

RANGE: The Eastern Ra (Wahgi valley to Okapa) and mts of the Huon Penin.

DIAGNOSIS: Similar to *orientalis*, but white spot on inner web of outer tail feathers is larger, forming a bar (Rand & Gilliard 1967). See photo in Coates & Peckover (2001: 191).

c. *Melanocharis longicauda orientalis* Mayr

Melanocharis longicauda orientalis Mayr, 1931. *Mitt. Zool. Mus. Berlin* 17: 669.—Avera, upper Aroa R, s watershed of the SE Peninsula.

RANGE: Mountains of the SE Peninsula.

DIAGNOSIS: Similar to male of nominate but with a smaller white patch on inner web of outer tail feathers Diamond (1972).

d. *Melanocharis longicauda* undescribed form(s)

RANGE: The Fakfak and Kumawa Mts of the Bird's Neck.

DIAGNOSIS: The population in the Fakfak Mts is apparently a distinct race or new species, with a satin-white breast and with a yellow wash on throat and breast (Gibbs 1994). C. Thébaud & B. Mila (*in litt.*) and Diamond & Bishop (2015) also reported a similar form from the Kumawa Mts.

NOTES

Populations in the Foja Mts[J] and PNG North Coastal Ra[K] have yet to be assigned to subspecies (Beehler *et al.* 2012[J]; specimens at Bogor, AMNH, and BPBM[K]). *Other names*—Lemon-breasted, Yellow-bellied, or Long-tailed Berrypecker.

Fan-tailed Berrypecker *Melanocharis versteri*   ENDEMIC

Pristorhamphus versteri Finsch, 1876. *Proc. Zool. Soc. London* (1875): 642.—Arfak Mts, Bird's Head.

Inhabits forest interior of all NG ranges exceeding 1,800 m; elevational range extends from 1,750 m to timberline, min 1,250 m[CO]. Visits tree-fern shrublands in alpine grassland. No records from the Fakfak, Kumawa, or Adelbert Mts, ranges that do not exceed 1,800 m.

SUBSPECIES

a. *Melanocharis versteri versteri* (Finsch)

RANGE: Mountains of the Bird's Head.

DIAGNOSIS: Male ventral plumage medium gray; throat quite dark; white patch in tail 32–38 mm long (Rand & Gilliard 1967).

b. *Melanocharis versteri meeki* (Rothschild & Hartert)

Pristorhamphus versteri meeki Rothschild & Hartert, 1911. *Bull. Brit. Orn. Club* 29: 36.—Mount Goliath, 1,525 m, Border Ra.

RANGE: The Western and Border Ra and Foja Mts.

DIAGNOSIS: Ventral plumage of male darker than all other forms. The race *albescens* Rothschild & Hartert was synonymized with this form by Mayr (1941). It is discussed by LeCroy (2010). Diagnosis of Foja Mts male based on color image provided by C. Milensky and color images of museum specimens on deposit in the Mus. Bogoriense in Cibinong (D. Prawiradilaga *in litt.*).

c. *Melanocharis versteri maculiceps* (De Vis)

Sarganura maculiceps De Vis, 1898. *Ann. Rept. Brit. New Guinea*, 1896–97: 87.—Wharton Ra, SE Peninsula.
  Synonym: *Pristorhamphus versteri virago* Stresemann, 1923. *Arch. Naturgesch.* 89, Abt. A, Heft 7: 68.—Schraderberg, Sepik Mts, n sector of the Eastern Ra.

RANGE: Mountains of north-central NG (Cyclops, Bewani[A], and Torricelli Mts, and Schrader Ra), Mt Hagen, the Okapa area, and mts of the Huon and SE Peninsulas (Mayr 1941, Coates 1990, specimens at AMNH & BPBM[A]).

DIAGNOSIS: Male plumage exhibits palest ventral coloration and largest tail patches of any race. See photo in Coates & Peckover (2001: 190). We combine *virago* into *maculiceps*, as it is an intermediate form linking the pale *maculiceps* and the dark *meeki*.

NOTES

The species' presence in the Wandammen Mts is based on a sight record (J. M. Diamond *in. litt.*); the subspecies is undetermined.

Streaked Berrypecker *Melanocharis striativentris*   ENDEMIC

Melanocharis striativentris Salvadori, 1894. *Ann. Mus. Civ. Genova* (2)14: 150.—Moroka, s watershed of SE Peninsula.

    Inhabits mid-montane forests of the Central Ra and Huon Penin; 1,150–2,300 m, extremes 550–2,600 m[CO].

SUBSPECIES

a. *Melanocharis striativentris axillaris* (Mayr)

Neneba striativentris axillaris Mayr, 1931. *Mitt. Zool. Mus. Berlin* 17: 670.—Upper Otakwa R, Western Ra.
RANGE: The Western Ra (Hartert *et al.* 1936).
DIAGNOSIS: Axillaries and underwing coverts white; underparts heavily streaked; no white in tail feathers (Rand & Gilliard 1967). Belly with considerable yellow-white wash. LeCroy (2010: 105) presented information on location of where the holotype was collected.

b. *Melanocharis striativentris striativentris* Salvadori

    Synonym: *Neneba prasina* De Vis, 1897. *Ibis*: 384.—Neneba, n slope of Mt Scratchley, SE Peninsula.
    Synonym: *Neneba striativentris chrysocome* Mayr, 1931. *Mitt. Zool. Mus. Berlin* 17: 670.—Junzaing, Huon Penin.
    Synonym: *Melanocharis striativentris albicauda* Mayr & Gilliard, 1952. *Amer. Mus. Novit.* 1577: 6.—Bihagi, head of Mambare R, n watershed of Owen Stanley Ra, SE Peninsula.
RANGE: The Border Ra, Eastern Ra, and mts of the Huon and SE Peninsulas. Localities—Dablin Creek[L], Rondon Ridge[L], Nondugl[J], Weiga[J], Awande[DI], Miarosa[DI], Okasa[DI], Mt Karimui[DI], Junzaing[K], Ogeramnang[K], Garaina[M], Mt Missim[M], Agaun[M] (Mayr 1931[K], Gyldenstolpe 1955a[J], P. Gregory *in litt.*[L], BPBM specimens[M]).
DIAGNOSIS: Similar to *axillaris*, but axillaries and underwing coverts gray-brown, ventral plumage less heavily streaked, and belly pure yellow. See photos of *striativentris* in Coates (1990: 321) and Coates & Peckover (2001: 192). Schodde (1978) noted that *prasina* and *albicauda* are synonyms of the nominate form—their distinction of white in the tail is based on age. Note that the type localities for the two are essentially the same place, but just given different names. The form *albicauda* does indeed have a small patch of white at the base of the outer tail feathers, but it is so unsubstantial that we agree with Schodde's sinking of the form. The nature of the plumage of this species and the individual variation makes it difficult to properly assign subspecies. We thus include the marginally distinct *chrysocome* here with the nominate form.

NOTES

Other names—Striated or Green Berrypecker, *Neneba striativentris*.

GENUS *RHAMPHOCHARIS*   ENDEMIC [Genus 2 spp]

Rhamphocharis Salvadori, 1876. *Ann. Mus. Civ. Genova* (1875) 7: 943. Type, by monotypy, *Rhamphocharis crassirostris* Salvadori.

    The two species of the genus are endemic to the uplands of New Guinea. The genus is characterized by the long bill (longer than the bill of any *Melanocharis*), the ventral spotting of the female plumage (dark spots in one species, pale spots in the other), and the plain, hen-plumaged male. The western form, *crassirostris*, shows clear relationships with the genus *Melanocharis*. Undoubtedly, *Melanocharis* and *Rham-*

phocharis are sister lineages, and some authorities have lumped them into a single genus (see Sibley & Monroe 1990). For many years *Rhamphocharis* was treated as monospecific. Reexamination of the plumages, size, and bill shape of the western populations led us to split that smaller, short-billed form from the larger, longer-billed eastern form with the very distinctive and peculiar dark-brown female plumage with profuse white spotting.

Thick-billed Berrypecker *Rhamphocharis crassirostris*   ENDEMIC

Rhamphocharis crassirostris Salvadori, 1876. *Ann. Mus. Civ. Genova* (1875) 7: 943.—Hatam, Arfak Mts, Bird's Head

Inhabits montane forest of the Bird's Head, Western Ra, and Border Ra, 1,150–2,300 m. The population in the Foja Mts is most likely this species. Range break between this species and the following is presumed to be the Strickland R gorge. This species, split from *piperata* for the first time, is little known, with few museum specimens available for study. As with its eastern counterpart, it probably forages for small, soft fruits of canopy trees and also gleans small invertebrates from leaves and epiphytes in the middle stages of the forest. But distinctive bill shape may allow a different foraging behavior from its close relative.

SUBSPECIES

a. *Rhamphocharis crassirostris crassirostris* Salvadori
RANGE: Mountains of the Bird's Head east to Oranje Mts (Mt Goliath, Treubbivak). Foja Mts population may refer to this form (Beehler *et al.* 2012).
DIAGNOSIS: Male and female are smaller, shorter-tailed, and shorter-billed than *R. piperata*. Female exhibits a whitish throat and white on belly, as well as profuse dark brown splotches and streaks on a pale ventral background; bill apparently blackish in both sexes.

b. *Rhamphocharis crassirostris interposita* Mees
Rhamphocharis crassirostris interposita Mees, 1964. *Zool. Verh. Leiden* 66: 20.—Star Mts, Border Ra.
RANGE: Star Mts (Mt Antares). May also include an unsexed subadult bird, wing 70.5 and weight 18 g, collected in the Hindenburg Ra (Gilliard & LeCroy 1961). Coates and Lindgren (see Gregory 1995b) may have observed the distinctive female of this species at Mt Binnie (near Tabubil), reporting dark-streaked flanks and white pectoral tufts. In the same general area, Woxvold *et al.* (2015) observed a bird that also may have been the female of this form ("pale below with dark streaking on the sides of the breast.").
DIAGNOSIS: As for the nominate, but bill shorter and white spots on undersurface of tail larger (Mees 1964b). As the female plumage is more indicative of the species, it is a shame that both birds collected of *interposita* exhibit the plain male plumage. The distinctive female plumage has been observed (see above) but not collected.

NOTES
We here split the traditional *R. crassirostris* into two species: the western form *R. crassirostris* and the eastern form *R. piperata*. The western species, *crassirostris*, is distinctive in its short blackish bill (15.2–16.1, n=6), more like that of a typical *Melanocharis*, its small size, and its short tail (45–46, n=2). The adult female plumage exhibits pale ventral background plumage mottled with dark olive smudging; it shows an indication of a pale throat and pale abdomen, both with some blurry dark smudging. The adult male plumages for *R. crassirostris* and *R. piperata* are very similar. See comparison photos in figs. 3a–b.

Spotted Berrypecker *Rhamphocharis piperata*   ENDEMIC | MONOTYPIC

Ptilotis piperata De Vis, 1898. *Ann. Rept. Brit. New Guinea*, 1896–97: 86.—Suku, Vanapa valley, SE Peninsula.
Synonym: *Rhamphocharis crassirostris viridescens* Mayr, 1931. *Mitt. Zool. Mus. Berlin*, 17: 715.—Dawong, Herzog Mts, SE Peninsula.

Inhabits montane forest of the Eastern Ra and the Huon and SE Peninsulas; 1,150–2,300 m, extremes 850–2,700 m[CO]. Presumably ranges west to the Strickland R gorge. Forages for small, soft fruits of canopy trees and also gleans small invertebrates from leaves and epiphytes in the middle stages of the forest. Rarely netted.

Localities—Tari Gap[U], Baia R headwaters[K], foot of Karius Ra[K], Ambua[J], Murmur Pass[J], Kubor Mts[Z], Weiga[L], Nondugl[L], Miarosa[DI], Mt Michael[DI], Mengino[DI], Mt Karimui[DI], Dawong[X], Huon Penin[CO], YUS transect[Q], Mt Missim[U], upper Eloa R[W], and Bulldog Road[U] (Mayr 1931[X], Mayr & Gilliard 1954[Z], Gyldenstolpe 1955a[L], Beehler 1980b[W], Freeman *et al.* 2013[Q], K. D. Bishop *in litt.*[J], I. Woxvold *in litt.*[K], various field sightings[U]). The species appears irruptive (D. Hobcroft *in litt.*).

NOTES

The adult female is very distinctive. Note the peculiar, longish, drooping bill (17.5–22, n=8) with horn-colored base of the mandible in both sexes, the female's dark throat with abundant small pale spots, the entirely dark breast and abdomen with profuse pale spotting, and the larger size. Tail length 46–55 (n=9). See comparison photos of museum specimens of *R. crassirostris* and *R. piperata* in figure 3a, and photos of a live individual of *R. piperata* in Coates (1990: 323). The very plain and unmarked males of the two species differ most diagnostically by the bill shape and length. English name features the unique pale spotting. *Other names*—traditionally included within *R. crassirostris*.

GENUS *OEDISTOMA*   ENDEMIC [Genus 2 spp]

Oedistoma Salvadori, 1876. *Ann. Mus. Civ. Genova* (1875) 7: 952. Type, by original designation, *Oedistoma pygmaeum* Salvadori.

The two species of the genus are endemic to the Region. The genus is characterized by the diminutive size; slim, decurved, and sharply pointed bill; abbreviated tail; yellow eye-ring; and olive and gray plumage.

Spectacled Longbill *Oedistoma iliolophus*   ENDEMIC

Melilestes iliolophus Salvadori, 1876. *Ann. Mus. Civ. Genova* (1875) 7: 951.—Mios Num and Yapen I, Bay Is.

All NG, Waigeo I, Mios Num I, Yapen I, and the D'Entrecasteaux Arch, 100–1,750 m[CO]. Widespread on the Mainland, but few records from the Trans-Fly: Bian Lakes, Sturt I, Lake Daviumbu (K. D. Bishop *in litt.*, Bishop 2005a). Frequents forest interior, occasionally edge (Coates 1990).

SUBSPECIES

a. *Oedistoma iliolophus cinerascens* (Stresemann & Paludan)
Toxorhamphus iliolophus cinerascens Stresemann & Paludan, 1932. *Novit. Zool.* 38: 144.—Waigeo I, NW Islands.
RANGE: Waigeo I.
DIAGNOSIS: Crown gray rather than dark olive green (Rand & Gilliard 1967).

b. *Oedistoma iliolophus iliolophus* (Salvadori)
　　Synonym: *Melilestes affinis* Salvadori, 1876. *Ann. Mus. Civ. Genova* (1875) 7: 952.—Arfak Mts, Bird's Head.
　　Synonym: *Toxorhamphus iliolophus flavus* Mayr & Rand, 1935. *Amer. Mus. Novit.* 814: 13.—Wuroi, Oriomo R,
　　　Trans-Fly.
RANGE: Yapen I, Mios Num I, and all NG (Mayr 1941).
DIAGNOSIS: Dorsally greenish olive, slightly darker upon crown; chin dull olive gray; throat dull olive gray; smudgy obsolete breast band slightly darker than throat; belly pale yellowish olive; flanks with slightly more yellowish tint. See photo in Coates & Peckover (2001: 188). Wing (male) 66, 67(3), 68(2). We subsume the races *affinis* and *flavus* into the nominate form, based on a review of relevant topotypical material.

c. *Oedistoma iliolophus fergussonis* (Hartert)
Melilestes fergussonis Hartert, 1896. *Novit. Zool.* 3: 237.—Fergusson I, D'Entrecasteaux Arch, SE Islands.
RANGE: The three main D'Entrecasteaux Is (Mayr 1941).
DIAGNOSIS: The largest race. Flanks with a prominent bright yellow wash; mantle with a grayish tint to olive coloration. Wing (male) 68. The epithet *fergussonis* is invariant (Dickinson & Christidis 2014).

NOTES

The specific epithet *iliolophus* is here considered a masculine noun in apposition, the ending of which does not change to agree in gender with its genus, which is neuter (Beehler & Finch 1985, Gregory 2008b, *pace* Salomonsen 1967, see also LeCroy 2011: 47). Subspecies epithets must agree with the neuter genus. Our English name highlights the yellow eye-ring. *Other names*—Plumed or Grey-bellied Longbill, Dwarf Honeyeater, *Toxorhamphus iliolophus*.

Pygmy Longbill *Oedistoma pygmaeum* ENDEMIC
Oedistoma pygmaeum Salvadori, 1876. *Ann. Mus. Civ. Genova* (1875) 7: 952.—Profi, Arfak Mts, Bird's Head.

Mainland NG, Waigeo and Misool I, and the D'Entrecasteaux Arch; sea level–700 m, max 1,300 m. Frequents forest and edge; also dense monsoon woodland in the Trans-Fly (Coates 1990).

SUBSPECIES

a. *Oedistoma pygmaeum pygmaeum* Salvadori
> Synonym: *Oedistoma pygmaeum flavipectus* Salomonsen, 1966a. *Breviora* Mus. Comp. Zool. Harvard 254: 2.— Wuroi, Oriomo R, Trans-Fly, s NG.
> Synonym: *Oedistoma pygmaeum olivascens* Salomonsen, 1966a. *Breviora* Mus. Comp. Zool. Harvard 254: 2.— Upper Watut R, nw sector of the SE Peninsula.
> Synonym: *Oedistoma pygmaeum waigeuense* Salomonsen, 1966a. *Breviora* Mus. Comp. Zool. Harvard 254: 2.— Waigeo I, NW Islands.

RANGE: Misool and Waigeo I and all NG (Mayr 1941). Widespread in the Trans-Fly of PNG (Bishop 2005a).
DIAGNOSIS: Uniformly greenish olive dorsally, slightly paler on rump; throat dirty off-white; breast pale olive with yellow tint; belly with pale yellow wash; birds from sw NG with brighter flanks. We follow, in part, Diamond (1972) in synonymizing the three Salomonsen forms listed above.

b. *Oedistoma pygmaeum meeki* (Hartert)
Anthreptes meeki Hartert 1896, *Novit. Zool.* 3: 239.—Fergusson I, D'Entrecasteaux Arch, SE Islands.
RANGE: Goodenough[MA], Fergusson[MA], and Normanby I[J] (Pratt field notes 2003[J]).
DIAGNOSIS: Whitish ventrally; crown gray (Rand & Gilliard 1967).

NOTES

We sink *flavipectus* and *olivascens* following Diamond (1972). We subsume *waigeuense* based on our own museum analysis. Salomonsen described as novel material that had been pored over by a range of museum ornithologists in prior decades. Each is presumably based on a single atypical specimen, upon which a new name was constructed, and with no acknowledgment that populations exhibit substantial nongeographic variation related to age, sex, and specimen preparation. *Other names*—Pygmy Honeyeater.

GENUS *TOXORHAMPHUS* ENDEMIC [Genus 2 spp]

Toxorhamphus Stresemann, 1914. *Novit. Zool.* 21: 394. Type, by original designation, *Cinnyris Novae Guineae* Lesson.

The two species in the genus are endemic to the Region. The genus is characterized by the small size, the very abbreviated tail, the very long, decurved, and pointed bill, and the yellowish and olive-washed ventral plumage.

Yellow-bellied Longbill *Toxorhamphus novaeguineae* ENDEMIC
Cinnyris Novae-Guineae Lesson, 1827. *Dict. Sci. Nat.*, éd. Levrault 1: 22.—Manokwari, Bird's Head.

The NG lowlands, from the Bird's Head eastward in the s watershed to the Lakekamu basin (w sector of SE Peninsula) and eastward in the n watershed to the n scarp of the Huon Penin; also the NW Islands, Aru Is, and Yapen I, sea level–500 m generally and to 1,200 m in the n watershed[CO] (Stresemann 1923, Mayr 1941). Absent from the Trans-Fly. Inhabits forest and edge. In the Central Ra, replaced at higher elevations by *T. poliopterus*.

a. *Toxorhamphus novaeguineae novaeguineae* (Lesson)

RANGE: The four main NW Islands and Yapen I; also w NG, east in the n watershed as far as the n scarp of the Huon Penin[J], and east in the s watershed as far as Etna Bay (Mayr 1941, Beehler field notes of bird mist-netted on the YUS transect[J]).

DIAGNOSIS: Plumage olivaceous ventrally (Rand & Gilliard 1967). See photos in Coates & Peckover (2001: 189).

b. *Toxorhamphus novaeguineae flaviventris* (Rothschild & Hartert)

Melilestes novae-guineae flaviventris Rothschild & Hartert, 1911. *Bull. Brit. Orn. Club* 28: 44.—Kobror, Aru Is.

RANGE: The Aru Is[MA] and s NG between the Otakwa R and the Lakekamu R[J,K] (Beehler *et al.* 1994[J]; specimens at USNM[K]). Birds netted at 1,100 m in the Kumawa Mts in Nov-2014 may be of this subspecies (C. Thébaud & B. Mila *in litt.*).

DIAGNOSIS: Similar to the nominate form but more yellowish, less olivaceous ventrally (Rand & Gilliard 1967).

NOTES

Other names—Green-crowned, Canary, or New Guinea Longbill, Canary Bowbill or False-sunbird.

Slaty-headed Longbill *Toxorhamphus poliopterus*   ENDEMIC | MONOTYPIC

Melilestes poliopterus Sharpe, 1882. *Journ. Linn. Soc. London, Zool.* 16: 318.—Astrolabe Ra, SE Peninsula.

> Synonym: *Toxorhamphus poliopterus septentrionalis* Mayr & Rand, 1935. *Amer. Mus. Novit.* 814: 14.—Hompua, 1,070 m, Huon Penin.

> Synonym: *Toxorhamphus poliopterus maximus* Rand, 1941. *Amer. Mus. Novit.* 1102: 13.—15 km sw of Bernhard Camp, 1,500 m, ne sector of the Western Ra.

The Central Ra, Adelbert Mts, Mt Bosavi, and mts of Huon Penin; 500–2,000 m, extremes 300–2,450 m (Gilliard & LeCroy 1967, Pratt 1983, Coates & Peckover 2001). Widespread on the Central Ra west to the Weyland Mts (absent from the Bird's Head[MA]). Frequents mainly forest and edge, sometimes second growth and gardens.

NOTES

We follow Diamond (1972) in synonymizing *septentrionalis* into the nominate form. The two remaining subspecies represent a minor east-west cline in size and saturation of plumage colors. Since the size difference is less than 10% of the mean wing length (68.5 *vs.* 73.5 = 7%), and the color difference is minimal, we subsume *maximus* into the nominate form. *Other names*—Grey-winged Longbill or Bowbill, Slaty-chinned Longbill, Grey-winged False-sunbird.

Family Paramythiidae

Painted Berrypeckers   ENDEMIC [Family 3 spp]

This family of three species is endemic to the uplands of mainland New Guinea. It comprises three brightly colored fruit eaters of the canopy of montane forest and subalpine shrublands (Coates 2008). The species consume fruits and other plant matter, and apparently arthropods at least occasionally (B. J. Coates *in litt.*). Barker *et al.* (2004) placed the group sister to the orioles (Oriolidae). Jønsson *et al.* (2011b) situated this lineage as sister to the whipbirds (Psophodidae). Aggerbeck *et al.* (2014) wedged this lineage between the orioles and whipbirds. The two genera of this lineage were considered to be aberrant bulbuls by Harrison & Parker (1966).

GENUS *OREOCHARIS*

Oreocharis Salvadori, 1876. *Ann. Mus. Civ. Genova* (1875) 7: 939. Type, by monotypy, *Oreocharis stictoptera* Salvadori = *Parus arfaki* A. B. Meyer.

This monotypic genus is confined to the uplands of New Guinea (Coates 2008). The genus is characterized by the smallish and compact body form, the tit-like male plumage, the frugivorous habit, and the strong sexual dichromatism. Typically found in vocal, monospecific flocks.

Tit Berrypecker *Oreocharis arfaki*   ENDEMIC | MONOTYPIC

Parus (?) *Arfaki* A. B. Meyer, 1875 (1-April). *Sitzungsber. Abh. Nat. Ges. Isis Dresden*: 75.—Arfak Mts, Bird's Head.

> Synonym: *Oreocharis arfaki bloodi* Gyldenstolpe, 1955. *Arkiv f. Zool.* (2)8(1): 174.—Weiga, s slope of Sepik-Wahgi Divide, Wahgi valley, Eastern Ra.

Inhabits canopy of montane forest and edge of most mts of NG: mts of the Bird's Head, Wandammen Mts, Central Ra, Foja Mts[L], PNG North Coastal Ra, and mts of the Huon Penin; 2,200–2,700 m[K], extremes 400–3,650 m (Gyldenstolpe 1955b, Diamond 1985, Coates & Peckover 2001[K], Beehler *et al.* 2012[L]). A flock was observed at a fruit tree at 400 m at Kanga, Kokoda valley (Beehler field notes 13-Aug-1986). Mistnetted at 3,650 m at Lake Omha, Mt Scratchley (Beehler field notes 12-Aug-1987).

NOTES

We follow Gilliard & LeCroy (1970) in not recognizing the race *bloodi*. *Other names*—Painted, Arfak, or New Guinea Berrypecker.

GENUS *PARAMYTHIA*   ENDEMIC [Genus 2 spp]

Paramythia De Vis, 1892. *Ann. Rept. Brit. New Guinea*, 1890–91: 95. Type, by monotypy, *Paramythia montium* De Vis.

The two species of the genus are restricted to the Central Ranges and mountains of the Huon Peninsula. The genus is characterized by the frugivorous habit, the distinctively patterned plumage, and the lack of prominent vocalizations. The genus epithet is feminine (David & Gosselin 2002a). *Montium* is a noun in genitive case, and invariant. The adjectival names (*olivacea* and *brevicauda*) must agree with the gender of the genus. We recognize two species in the genus.

Western Crested Berrypecker *Paramythia olivacea*   ENDEMIC | MONOTYPIC

Paramythia montium olivaceum van Oort, 1910. *Notes Leyden Mus.* 32: 213.—Oranje and Hellwig Mts, 3,500 m, e sector of the Western Ra.

> Synonym: *Paramythia montium alpinum* Salomonsen, 1961. *Amer. Mus. Novit.* 2067: 5.—6 km e of Mt Wilhelmina summit, 3,600 m, Oranje Mts, Western Ra.
> Synonym: *Paramythia montium occidentis* Mees, 1964. *Zool. Verh. Leiden* 66: 22.—Hellwig Mts, 2,400 m, e sector of the Western Ra.

High-elevation forests of the Western Ra.

Localities—Mt Sumuri[J], Ilaga[K], 7 km east of Mt Wilhelmina[L], Lake Habbema[L], Bele R Camp[L], Treubbivak of the Hellwig Mts[M] (Hartert *et al.* 1936[J], Rand 1942b[L], Ripley 1964a[K], Mees 1964[M]).

NOTES

P. olivacea is distinguished from *P. montium* by the following characters: (1) black bar on nape; (2) abbreviated all-white eyebrow; (3) flanks all blue, lacking yellow patch; 4) back plumage yellowish olive rather than green; (5) greenish-yellow undertail feathering; and (6) crest longer and recurved rather than shorter and decurved. We note that not all of these characters show on the plate of these two forms in Coates (2008: 348, pl. 24). We treat this newly recognized species as monotypic and do not support recognition of elevational races, such as *alpina*, which lack a geographic range component (see Rand 1936b, LeCroy 2010: 117). The form *occidentis* was described by Mees (1964b) as an elevational subspecies, which we are reluctant to recognize, as it is based on an elevational cline in size and is nongeographic. Presumably, the break between the geographic ranges of these two high-elevation species is the Baliem R gorge. The type locality mentioned both the Oranje and Hellwig Mts. The latter are essentially the se sector of the former, although the true delineation of the Oranje Mts is vague and often includes the highlands east of the Baliem R gorge. *Other names*—Crested Berrypecker; traditionally treated as a race of *P. montium*.

Eastern Crested Berrypecker *Paramythia montium*   ENDEMIC | MONOTYPIC
Paramythia montium De Vis, 1892. *Ann. Rept. Brit. New Guinea*, 1890–91: 95.—Mount Suckling, SE Peninsula.

> Synonym: *Paramythia montium brevicauda* Mayr & Gilliard, 1954. *Bull. Amer. Mus. Nat. Hist.* 103: 372.—Saruwaged Ra, Huon Penin.

High mts of the e sector of the Border Ra, the Eastern Ra, and mts of the Huon and SE Peninsulas; 2,450 m to timberline[CO]. Inhabits upper and high montane forest and alpine thickets. Generally common, becoming abundant at higher elevations (Coates 1990). The w boundary of the range of this species is at least west to the mts near the PNG border (Gilliard & LeCroy 1961), but may range as far as the mts just east of the Baliem R gorge. This needs to be checked.

NOTES
We measured tails of the two described forms (*montium* and *brevicauda*) included in this eastern sister species: nominate tail 83–98; *brevicauda* tail 82–93.5. Given the substantial overlap in tail length, we do not recognize *brevicauda*. *Other names*—traditionally treated as Crested Berrypecker.

Family Psophodidae

Whipbirds and Allies   [Family 5 spp/Region 1 sp]

A small and uniform family of birds of the understory or ground that inhabit Australia and New Guinea. Plumage characters that link typical forms include the black throat, the white malar stripe, and the crest and long tapered tail. Jønsson *et al.* (2011b) placed this small group as sister to the Paramythiidae, and that pair as a clade sister to the Eulacestomatidae. By contrast, Aggerbeck *et al.* (2014) placed the psophodid+paramythiid clade sister to the orioles.

GENUS *ANDROPHOBUS*   ENDEMIC | MONOTYPIC
Androphobus Hartert and Paludan, 1934. *Orn. Monatsber.* 42: 46. Type, by original designation and monotypy, *Androphilus viridis* Rothschild & Hartert.

    This monotypic genus is endemic to the uplands of New Guinea. The sole member of the genus is characterized by its small size, lack of a crest, olive-green plumage, and white malar stripe contrasting with the blackish throat (Hartert *et al.* 1936). The plumage pattern is evidence of the close relationship to the Australian *Psophodes*. Jønsson *et al.* (2011b) showed the two to be sister lineages.

Papuan Whipbird *Androphobus viridis*   ENDEMIC | MONOTYPIC
Androphilus viridis Rothschild & Hartert, 1911. *Bull. Brit. Orn. Club* 29: 33.—Mount Goliath, Border Ra.

    A shy, very inconspicuous and rarely encountered inhabitant of thick undergrowth in montane forest interior of the Western, Border, and Eastern Ra, from the Weyland Mts east to Tari Gap; 1,400–2,800 m (see Boles 2007).

    Localities—Weyland Mts[J], Treubbivak[K], Ilaga[L], 9 km northeast of Lake Habbema[M], Bele R Camp[M], Mt Goliath[MA], and below Tari Gap[N] (Hartert *et al.* 1936[J], Rand 1942b[M], Ripley 1964a[L], G. Versteeg specimen[K], Hicks 1992[N]).

    Other names—Green-backed Babbler.

Family Cinclosomatidae

Jewel-babblers and Quail-thrushes

[Family 11 spp/Region 5 spp]

These are colorful and vocal ground dwellers that in the Region are strictly forest dwelling. Plumage is soft, and all exhibit a black mask. Norman *et al.* (2009b) and Jønsson *et al.* (2011b) showed that the Eupetidae of Dickinson (2003) and the Psophodidae of Christidis & Boles (2008) both included two disparate lineages. The work of Norman *et al.* and Jønsson *et al.* showed that *Ptilorrhoa* and *Cinclosoma* are sister forms constituting the Cinclosomatidae, distinct from *Androphobus* and *Psophodes*, which are situated in the Psophodidae, an arrangement followed by Dickinson & Christidis (2014). The Cinclosomatidae is variously placed between the Cracticidae and Pachycephalidae (Norman *et al.* 2009b) or as sister to a clade that includes the Paramythiidae and the Psophodidae (Jønsson *et al.* 2011b) or between Campephagidae and Pachycephalidae (Dickinson & Christidis 2014). Aggerbeck *et al.* (2014) situated the Cinclosomatidae as sister to *Falcunculus* (Falcunculidae) and that pair sister to the Australian bellbirds (Oreoicidae).

GENUS *PTILORRHOA*

ENDEMIC [Genus 4 spp]

Ptilorrhoa Peters, 1940. *Auk* 57: 94. Type, by original designation, *Eupetes caerulescens* Temminck.

The four species of the genus are endemic to the Region. They inhabit the floor of the forest interior from the lowlands to the highlands (Boles 2007). The genus is characterized by its strictly terrestrial habit, loud vocalizations, prominent white throat or cheek patch, and richly patterned plumage of rufous, blue, white, and black.

Spotted Jewel-babbler *Ptilorrhoa leucosticta*

ENDEMIC

Eupetes leucostictus P. L. Sclater, 1874. *Proc. Zool. Soc. London* (1873): 690: pl. 52.—Hatam, Arfak Mts, Bird's Head.

Mountains of the Bird's Head, the Central Ra, and those outlying ranges that exceed 1,500 m elevation; found from 1,750 to 2,400 m, extremes 1,200–2,700 m[CO]. No records from the Fakfak, Kumawa, or Adelbert Mts. Inhabits the floor of montane forest. Replaced at lower elevations by *P. castanonota*. Unlike all other members of the genus, sexually monochromatic.

SUBSPECIES

a. *Ptilorrhoa leucosticta leucosticta* (P. L. Sclater)
RANGE: Mountains of the Bird's Head.
DIAGNOSIS: Mantle and rump olive; crown and nape red-brown; breast ashy blue; black triangle on upper breast spotted white (Rand & Gilliard 1967, Boles 2007).

b. *Ptilorrhoa leucosticta mayri* (Hartert)
Eupetes leucostictus mayri Hartert, 1930. *Novit. Zool.* 36: 87.—Mount Wondiwoi, Wandammen Penin, Bird's Neck.
RANGE: Wandammen Mts.
DIAGNOSIS: Similar to nominate form, but back and rump red-brown like crown and nape; crescent of black at top of breast (Boles 2007).

c. *Ptilorrhoa leucosticta centralis* (Mayr)
Eupetes leucostictus centralis Mayr, 1936. *Amer. Mus. Novit.* 869: 1.—Weyland Mts, 1,800 m, Western Ra.
RANGE: The w half of the Central Ra, from the Weyland Mts east to the PNG border.
DIAGNOSIS: Similar to *mayri* but breast greenish olive; upperparts red-brown; black throat spot with tiny white speckling (Rand & Gilliard 1967). Dorsal red-brown quite variable. Reviewing specimens from Araboebivak, Paniai Lake, in the w sector of the Western Ra, we found the mantle of one male very bright red-brown but another dull red-brown.

d. *Ptilorrhoa leucosticta sibilans* (Mayr)

Eupetes leucostictus sibilans Mayr, 1931. *Mitt. Zool. Mus. Berlin* 17: 691.—Cyclops Mts, ne sector of NW Lowlands, n NG.

RANGE: Cyclops Mts.

DIAGNOSIS: Similar to *loriae* but breast band olive green; mantle olive; crown bright chestnut; crescent of black atop breast (Rand & Gilliard 1967, Boles 2007).

e. *Ptilorrhoa leucosticta menawa* Diamond

Ptilorrhoa leucosticta menawa Diamond, 1969. *Amer. Mus. Novit.* 2362: 19.—Mount Menawa, Bewani Mts, 1,375 m, PNG North Coastal Ra, Sepik-Ramu.

RANGE: Bewani Mts.

DIAGNOSIS: When compared to *loriae*, exhibits a more extensive olive wash on underparts that is bright; also note bright olive mantle (Diamond 1969).

f. *Ptilorrhoa leucosticta loriae* (Salvadori)

Eupetes loriae Salvadori, 1896. *Ann. Mus. Civ. Genova* (2)16: 102.—Moroka, s watershed of SE Peninsula.

RANGE: Mount Hagen and the Schrader Ra east to the mts of the SE Peninsula.

DIAGNOSIS: Similar to nominate (green-mantled), but black triangle on upper breast without white spots; breast and flanks with a dull greenish wash; abdomen blue (Rand & Gilliard 1967). See photo in Coates & Peckover (2001: 122).

g. *Ptilorrhoa leucosticta amabilis* (Mayr)

Eupetes leucostictus amabilis Mayr, 1931. *Mitt. Zool. Mus. Berlin* 17: 691.—Junzaing, Huon Penin.

RANGE: Mountains of the Huon Penin (Mayr 1931).

DIAGNOSIS: Like nominate (green-backed), but underparts mostly blue (Rand & Gilliard 1967).

NOTES

The Foja Mts population has not been diagnosed to subspecies (Beehler *et al.* 2012). *Other names*—High Mountain Eupetes, *Eupetes leucostictus.*

Blue Jewel-babbler *Ptilorrhoa caerulescens*   ENDEMIC

Eupetes caerulescens Temminck, 1836. *Planch. Col. d'Ois.*, livr. 97 (1835): pl. 574.—Triton Bay, Bird's Neck.

Lowlands and hills of NG, Misool, and Salawati I, sea level–850[J] m (I. Woxvold *in litt.*[J]). Replaced by the closely allied *P. geislerorum* between the n slopes of the e sector of the Huon Penin and the n watershed of the SE Peninsula. One record from the Trans-Fly, from Wuroi (Bishop 2005a). Note that this species and *P. geislerorum* are both present inland from Astrolabe Bay in the Adelbert Mts. Inhabits forest and regrowth; also found locally in gallery forest. Replaced at higher elevations by *P. castanonota*, with little or no overlap (Coates 1990). Freeman *et al.* (2013) found it common on the n slopes of the c sector of the mts of the Huon Penin (the YUS transect).

SUBSPECIES

a. *Ptilorrhoa caerulescens caerulescens* (Temminck)

RANGE: Misool[J,L], Salawati[L], the Bird's Head, and much of the Bird's Neck; east in the n watershed to the Wanggar distr[K], and in the s watershed to Etna Bay (Mayr 1941, Hartert *et al.* 1936[K], Ripley 1964a[J], Mees 1965[L]). In the Wanggar distr at the head of Geelvink Bay apparently occurring together with *nigricrissus* without any evidence of hybridization (Mayr 1941).

DIAGNOSIS: Female similar to male, but white of throat reaches eye (Rand & Gilliard 1967).

b. *Ptilorrhoa caerulescens nigricrissus* (Salvadori)

Eupetes nigricrissus Salvadori, 1876. *Ann. Mus. Civ. Genova* 9: 36.—Naiabui, Hall Sound, SE Peninsula.

RANGE: Southern NG from Etna Bay of the Bird's Neck east to Milne Bay.

DIAGNOSIS: Similar to *neumanni*, but female exhibits a slim white supercilium; undertail coverts partly marked with black (Rand & Gilliard 1967). See photos in Coates (1990: 63–65) and Coates & Peckover (2001: 123).

c. Ptilorrhoa caerulescens neumanni (Mayr & Meyer de Schauensee)
Eupetes caerulescens neumanni Mayr & Meyer de Schauensee, 1939. *Proc. Acad. Nat. Sci. Phila.* 91: 122.—
Cyclops Mts, ne sector of NW Lowlands, n NG.
RANGE: The NW Lowlands, Sepik-Ramu, and n slopes of the w and c sectors of the Huon Penin (Freeman
et al. 2013). A specimen collected by R. Beck from Keku s of Madang, and now in the AMNH collection is
a blue-plumaged female that can be assigned here, not to *geislerorum* (M. LeCroy *in litt.*). The range of this
form and of *P. geislerorum* on the Huon Penin requires further refinement.
DIAGNOSIS: As for the nominate form, but black of lores joins to black of ear coverts below eye (Rand &
Gilliard 1967). Female with a white supercilium (Boles 2007).

NOTES
David & Gosselin (2002a) noted that the trinomial *nigricrissus* is an invariant noun and thus should not be
altered to modify the feminine generic epithet. *Other names*—Lowland Jewel-babbler or Eupetes.

Dimorphic Jewel-babbler *Ptilorrhoa geislerorum*   ENDEMIC | MONOTYPIC
Eupetes geislerorum A. B. Meyer, 1892. *J. Orn.* 40: 259.—Butaueng, Huon Penin.
    Apparently occurs in two noncontiguous populations: (1) in the Adelbert Mts of the ne sector of
the Sepik-Ramu; (2) on the n slope of the mts of the e sector of the Huon Penin, east to Collingwood
Bay along the n watershed of the SE Peninsula. This species seems to be an inhabitant of hill forest to
1,220m[CO]. Where this species and the preceding meet near Madang, they apparently sort out by elevation,
this form upslope from the preceding (Coates 1990).
    Localities—Adelbert Mts[K], Sattelberg[CO], Finschhafen[J], Malolo[J], upper Watut R[CO], near Garaina[CO],
Collingwood Bay[MA] (Mayr 1931[J], Gilliard & LeCroy 1967[K]). Absent from YUS in the c sector of the n slopes
of the Huon Penin (Freeman *et al.* 2103).

NOTES
See photo of female in Coates & Peckover (2001: 122). We follow Coates (1990) and Boles (2007) in treating *geislerorum* and *caerulescens* as distinct species. The relationship between the two needs further study
in the field. Male of *geislerorum* exhibits a dull olive crown distinct from powder blue of neck and mantle;
indication of a pale blue eyebrow; black eye stripe contiguous with black necklace. Female *geislerorum* is
entirely lacking in blue in plumage; and has a pale eyebrow; olive-brown crown and mantle; black mask
contiguous with black of necklace; breast and belly rich chestnut; black spotting on undertail. On the n
scarp of the Huon Penin on the YUS transect, Freeman *et al.* (2013) found only *P. caerulescens neumanni*
and no evidence of *P. geislerorum*, which is known from northwest and southeast of this area. Greenway
(1935: 55) noted that H. Stevens collected specimens of both *geislerorum* and *P. castanonota* at 700 m in
the upper Watut (at the same elevation though apparently not on the same date or same camp). *Other
names*—Brown-headed Jewel-babbler.

Chestnut-backed Jewel-babbler *Ptilorrhoa castanonota*   ENDEMIC
Eupetes castanonotus Salvadori, 1876. *Ann. Mus. Civ. Genova* (1875) 7: 966.—Mount Morait, Tamrau Mts,
Bird's Head.
    Mountains of NG and Batanta and Yapen I. Distribution includes mts of the Bird's Head; the Fakfak,
Kumawa, and Wandammen Mts; the Foja Mts, PNG North Coastal Ra, Adelbert Mts, and mts of the
Huon Penin (Diamond 1985). No records from the Cyclops Mts. Elevational range 900–1,450 m, extremes
300–1,580 m[CO]. Inhabits hill and lower montane forest (Diamond 1985). In most of NG replaced at lower
elevations by *P. caerulescens* and higher elevations by the *P. leucosticta* (Coates 1990).

SUBSPECIES
a. Ptilorrhoa castanonota gilliardi (Greenway)
Eupetes castanonotus gilliardi Greenway, 1966. *Amer. Mus. Novit.* 2258: 14.—Forested summit, Mt Besar,
850 m, Batanta I, NW Islands.
RANGE: Batanta I.

DIAGNOSIS: Male resembles the nominate form; female darker, crown almost black, and back more olive; underparts duller; feathers of lower belly tipped white (Boles 2007). Most distinctive feature is the very dark mantle and crown of the female.

b. *Ptilorrhoa castanonota castanonota* (Salvadori)
RANGE: The Bird's Head (Gilliard & LeCroy 1970). Bird's Neck birds may belong here (Diamond & Bishop 2015).
DIAGNOSIS: Entire back and rump of the female is chestnut-colored; supercilium blue (Rand & Gilliard 1967). Male exhibits a broad blue supercilium; crown reddish brown; mantle chestnut (Boles 2007). Underparts of female a bright and pale blue (toward sky blue).

c. *Ptilorrhoa castanonota saturata* (Rothschild & Hartert)
Eupetes castanonotus saturatus Rothschild & Hartert, 1911. *Orn. Monatsber.* 19: 157.—Setekwa R, Western Ra.
RANGE: The s slope of the Western and Border Ra, presumably east at least to the Strickland R gorge; this form may range eastward as far as Mt Karimui, where it meets the race *pulchra* (Diamond 1972).
DIAGNOSIS: Back dark chestnut; underparts deeper blue, with a purple cast (Rand & Gilliard 1967, Boles 2007).

d. *Ptilorrhoa castanonota uropygialis* (Rand)
Eupetes castanonotus uropygialis Rand, 1940. *Amer. Mus. Novit.* 1074: 2.—6 km sw of Bernhard Camp, 1,200 m, ne sector of the Western Ra.
RANGE: The n slope of the Western and Border Ra and PNG North Coastal Ra.
DIAGNOSIS: Female exhibits chestnut mantle, bordered posteriorly by dark olive brown; lower back and rump blue, not chestnut; supercilium blue (Rand 1940b, Boles 2007).

e. *Ptilorrhoa castanonota buergersi* (Mayr)
Eupetes castanonotus bürgersi Mayr, 1931. *Mitt. Zool. Mus. Berlin* 17: 691.—Lordberg, Sepik Mts, Eastern Ra.
RANGE: The n scarp of the Eastern Ra (Lordberg and Etappenberg), Adelbert Mts, and probably w terminus of the Finisterre Ra (Gilliard & LeCroy 1967, Coates 1990).
DIAGNOSIS: Most like *saturata*, but dorsal surface brighter and ventral plumage paler (Rand & Gilliard 1967).

f. *Ptilorrhoa castanonota par* (Meise)
Eupetes castanonotus par Meise, 1930. *Orn. Monatsber.* 38: 17.—Sattelberg, Huon Penin.
RANGE: Mountains of the Huon Penin (Mayr 1931).
DIAGNOSIS: Most similar to *saturata*, but anterior edge of mantle brighter chestnut, and female with paler blue to whitish eye stripe (Rand & Gilliard 1967).

g. *Ptilorrhoa castanonota pulchra* (Sharpe)
Eupetes pulcher Sharpe, 1882. *Journ. Linn. Soc. London, Zool.* 16: 319.—Taburi distr, Astrolabe Ra, SE Peninsula.
RANGE: Mountains in the SE Peninsula and the e sector of the Eastern Ra; west to the Herzog Mts in the n watershed, and west to Mt Karimui in the s watershed (Diamond 1972).
DIAGNOSIS: Eye stripe of female cream-colored, and undertail coverts blackish (Rand & Gilliard 1967, Boles 2007). See photo of adult male in Coates & Peckover (2001: 122).

NOTES
The various subspecies require further review, with the aid of additional fresh material from the field. *Other names*—Mid-mountain Eupetes, *Eupetes castanonotus*.

Cinclosoma Vigors & Horsfield, 1827. *Trans. Linn. Soc. London* (1826) 15: 219. Type, by monotypy, *Turdus punctatus* Latham = *Turdus punctatus* Shaw.

The species of this genus inhabit Australia and New Guinea (Boles 2007). The genus is characterized by the thrush-like appearance, plump body, longish rounded tail, terrestrial habit, loud whistled vocalization, and complexly patterned plumage of chestnut, dull brown, black, and white. Males of all species exhibit a black throat and white facial striping (Schodde & Mason 1999). Quail-thrushes are typically birds of sclerophyll forest and arid habitats in Australia, but the sole New Guinea species inhabits rain forest.

Painted Quail-thrush *Cinclosoma ajax* ENDEMIC
Eupetes ajax Temminck, 1836. *Planch. Col. d'Ois.*, livr. 97 (1835): pl. 573.—Triton Bay, Bird's Neck.

Very patchily distributed. Known from five areas of the south and far west: (1) the w and se shore of Geelvink Bay and foothills of the Weyland Mts; (2) the s Bird's Neck[MA]; (3) the upper Fly and Strickland R (I. Woxvold *in litt.*); (4) the middle Fly R and Trans-Fly[MA]; and (5) the s watershed of the SE Peninsula from Milne Bay west along the s foothills to Port Moresby and Hall Sound[MA]; sea level–800 m[CO]. Inhabits the floor of forest interior. Perhaps mainly a foothill and hill forest species, though present in the lowlands both in the Trans-Fly and in the vicinity of hills (Coates 1990).

SUBSPECIES

a. *Cinclosoma ajax ajax* (Temminck)
RANGE: The sw sector of the Bird's Head and Neck, and the sw sector of the NW Lowlands in foothills of the Weyland Mts. Western and se coast of Geelvink Bay; also the s coast of the Bird's Neck. Localities—Momi[MA], Nabire[K], Lengguru R at Triton Bay[L], and lower Menoo R in Weyland Mts[J], at 300 m (Hartert *et al.* 1936[J], Stein collection at AMNH[K], C. Thébaud & B. Mila *in litt.*[L]).
DIAGNOSIS: Male olive brown dorsally; rich rufous brown on flanks (Boles 2007). Wing (male) 110 (Rand & Gilliard 1967).

b. *Cinclosoma ajax alare* Mayr & Rand
Cinclosoma ajax alaris Mayr & Rand, 1935. *Amer. Mus. Novit.* 814: 6.—Wuroi, Oriomo R, s NG.
    Synonym: *Cinclosoma ajax muscalis* Rand, 1940. *Amer. Mus. Novit.* 1074: 2.—8 km below Palmer Junction, 80 m elevation, upper Fly R.
RANGE: Known only from the lower and upper Fly and Strickland R. Localities—Wuroi[MA], Tarara[J], Morehead R[CO], Palmer R[J], Oriomo R[MA], Nomad R[K], Mt Bosavi[LU], gently rolling lowland forests to the west and to the north of Kiunga[N], also in the upper reaches of the Elevala R (Rand 1942a[J], Bell 1970b[K], K. D. Bishop *in litt.*[N]).
DIAGNOSIS: Male flanks pale tan; wing (male) 100–108. Northern birds, described as *muscalis*, are slightly darker dorsally; that form is combined here because otherwise the two forms are identical. One must assume the variation from pale to dark relates to the cline in rainfall from dry in the south to wet in the north in the drainage.

c. *Cinclosoma ajax goldiei* (Ramsay)
Eupetes goldiei Ramsay, 1879. *Proc. Linn. Soc. NSW* 3: 303.—Port Moresby, SE Peninsula.
RANGE: The s watershed of the SE Peninsula, from Hall Sound to Milne Bay (Mayr 1941).
DIAGNOSIS: Male very pale dorsally; flanks bright; female with reduced spotting in wing; small; wing (male) 97–106. See photos of male and female in Coates & Peckover (2001: 124).

NOTES
The three recognized forms are thinly defined. *Other names*—Ajax or False Quail-thrush.

Family Machaerirhynchidae

Boatbills

[Family 2 spp/Region 2 spp]

The species of this newly erected family range throughout New Guinea and extend into northeastern Queensland (Schodde & Mason 1999, Aggerbeck *et al.* 2014). Inhabiting lowland and highland forest of the Mainland as well as the Northwestern and Aru Islands, the two species are vocal and active foragers in the middle and upper stories of forest interior. Note the flattened and broadened bill, distinctive yellow pigmentation, and the small and neatly-constructed cup nest. Anatomically, the aperturate palate and narrowed ectethmoids with vestigial lachrymals further distinguish the group (Schodde & Mason 1999). Historically, the species have been allied with the monarchs. Jønsson *et al.* (2010b) showed this lineage as being sister to two clades: one including the woodswallows (Artamidae) and butcherbirds (Cracticidae), the other including the berryhunters (*Rhagologus*, Rhagologidae). Norman *et al.* (2009b) located this lineage as sister to the Cracticidae and Artamidae, which is supported by the phylogeny of Kearns *et al.* (2012) and Jønsson *et al.* (2011b).

GENUS *MACHAERIRHYNCHUS*

[Genus 2 spp/Region 2 spp]

Machaerirhynchus Gould, 1851. *Birds Austral.*, Suppl. pt. 1: plate and text; also 1851, *Proc. Zool. Soc. London* (1850): 277. Type, by monotypy, *Machaerirhynchus flaviventer* Gould.

The species of this genus range through the lowlands and highlands of New Guinea, and one species extends its range into northeastern Queensland. The genus is characterized by the flattened and broadened fly-catching bill, tiny size, narrow tail, and black, yellow, and white plumage.

Yellow-breasted Boatbill *Machaerirhynchus flaviventer*   RESIDENT

Machaerirhynchus flaviventer Gould, 1851. *Birds Austral.*, Suppl. 1: plate and text.—Cape York, QLD, Australia.

Inhabits forests of NG, the NW Islands, and Aru Is; sea level–800 m, max 1,300 m[CO]. Found mainly in the lowlands and hills. No records from Batanta I.

Extralimital—ne Australia.

SUBSPECIES

a. *Machaerirhynchus flaviventer albigula* Mayr & Meyer de Schauensee

Machaerirhynchus flaviventer albigula Mayr & Meyer de Schauensee, 1939. *Proc. Acad. Nat. Sci. Phila.* 91: 128.—Siwi, Arfak Mts, Bird's Head.

RANGE: Misool I, the Bird's Head and Neck, NW Lowlands, and Sepik-Ramu; east to Astrolabe Bay (Mayr 1941).

DIAGNOSIS: Male with extensive white chin, rich yellow throat and breast extending up to form a yellow cheek and ear patch; small white streak above black lores; black crown and mantle; rump black with green infusion; much white barring on wing. Wing (male) 59–62 (Rand & Gilliard 1967).

b. *Machaerirhynchus flaviventer albifrons* G. R. Gray

Machaerirhynchus albifrons G. R. Gray, 1862. *Proc. Zool. Soc. London* (1861): 429, pl. 43, fig. 1.—Waigeo I, NW Islands.

RANGE: Waigeo and Salawati I (Mayr 1941).

DIAGNOSIS: White of throat confined to chin; forehead and eyebrow stripe yellow in male (Rand & Gilliard 1967).

c. *Machaerirhynchus flaviventer xanthogenys* G. R. Gray
Machaerirhynchus xanthogenys G. R. Gray, 1858. *Proc. Zool. Soc. London*: 176.—Aru Is.
RANGE: The Aru Is[MA], S Lowlands[J], Trans-Fly[K], and s watershed of the SE Peninsula, east to Milne Bay[MA] (Bishop 2005a[K], 2005b[J]).
DIAGNOSIS: Similar to *novus*, but forehead and eyebrow of male yellow. Ear and cheek yellow. Bill small, 16–17 mm (Rand & Gilliard 1967). Wing (male) 61–63 (Mayr & Rand 1937). See photo of a male of this form in Coates & Peckover (2001: 148).

d. *Machaerirhynchus flaviventer novus* Rothschild & Hartert
Machaerirhynchus flaviventer novus Rothschild & Hartert, 1912. *Novit. Zool.* 19: 200.—Kumusi R, SE Peninsula.
RANGE: The n coast of the SE Peninsula from Holnicote Bay northwest to the Huon Penin (Heldsbach Coast) and the Watut valley (Mayr 1941).
DIAGNOSIS: Eyebrow of male white; white of throat extending nearly to black of shoulder (Rand & Gilliard 1967).

NOTES
Other names—Yellow-breasted Flatbill Flycatcher, Boat-billed Flycatcher, Wherrybill.

Black-breasted Boatbill *Machaerirhynchus nigripectus* ENDEMIC

Macherirhynchus nigripectus Schlegel, 1871. *Ned. Tijdschr. Dierk.* (1873) 4: 43.—Arfak Mts, Bird's Head.
    Inhabits forest and regrowth in the mts of the Bird's Head and Neck (Fakfak Mts), Central Ra, Foja Mts, Adelbert Mts, and mts of the Huon Penin; 1,300–2,600 m, extremes 850–2,750 m[CO]. No records from the Kumawa Mts.

SUBSPECIES
a. *Machaerirhynchus nigripectus nigripectus* Schlegel
RANGE: Mountains of the Bird's Head (Mayr 1941).
DIAGNOSIS: Small. Wing (male) 60; tail (male) 54 (Rand & Gilliard 1967); female mantle sooty gray.

b. *Machaerirhynchus nigripectus saturatus* Rothschild & Hartert
Machaerirhynchus nigripectus saturatus Rothschild & Hartert, 1913. *Novit. Zool.* 20: 498.—Mount Goliath, Border Ra.
RANGE: The Western, Border, and Eastern Ra (Mayr 1941, Diamond 1972).
DIAGNOSIS: Wing (male) 66; tail (male) 65; female mantle deeper black (Rand & Gilliard 1967). See photo of male of this form in Coates & Peckover (2001: 148).

c. *Machaerirhynchus nigripectus harterti* van Oort
Machaerirhynchus nigripectus harterti van Oort, 1909. *Notes Leyden Mus.* 30: 235.—Owen Stanley Ra, SE Peninsula.
RANGE: Mountains of the Huon and SE Peninsulas (Mayr 1941). The population in the Adelbert Mts is also presumably of this form (Pratt 1983).
DIAGNOSIS: Size equivalent to *saturatus*; female mantle plumage duller black (Rand & Gilliard 1967).

NOTES
Three thinly defined subspecies. The two most similar forms inhabit the far west and far east, and a darker-backed female (*saturatus*) inhabits the middle zone. The Fakfak[J], Foja, and Adelbert Mts[K] populations remain undiagnosed (Gibbs 1994[J], Beehler *et al.* 2012[K]). *Other names*—Black-breasted Flatbill Flycatcher.

Family Cracticidae

Butcherbirds and Allies [Family 12 spp/Region 7 spp]

We follow Russell and Rowley (2009) and Kearns *et al.* (2012) in keeping the Cracticidae as a family distinct from the Artamidae. Species of the family are restricted to Australia, New Guinea, and fringing islands. The family is characterized by the predatory habit, the strong, hooked bill, and black and white featuring prominently in the plumage. The New Guinean assemblage includes five butcherbirds and two peltopses. The phylogeny of Barker *et al.* (2004) indicated this family is sister to the woodswallows (Artamidae). Fuchs *et al.* (2006) provided a phylogeny that indicated that the Cracticidae+Artamidae clade is sister to the tropical Asian ioras (Aegithinidae). By contrast, the phylogeny of Kearns *et al.* (2012) indicated that the butcherbirds are sister to the African bushshrikes (Malaconotidae), and that the Artamidae are sister to that pair of lineages. Jønsson *et al.* (2011b) clustered *Machaerirhynchus, Artamus, Peltops,* and *Cracticus,* and showed this cracticine clade to be sister to a large clade the includes the African bushshrikes and the vangas (Vangidae). Aggerbeck *et al.* (2014) showed that same cracticine cluster to be sister to the cuckooshrikes (Campephagidae). Schodde & Mason (1999) and Christidis & Boles (2008) united the butcherbirds and woodswallows in the Artamidae. Schodde & Christidis (2014) elevated the lineage encompassing *Peltops* to subfamily status, the Peltopsinae.

GENUS *PELTOPS*   ENDEMIC [Genus 2 spp]

Peltops Wagler, 1829. *Isis von Oken* 22(6): col. 656. Type, by original designation and monotypy, *Eurylaimus blainvillii* Garnot.

The two species of the genus are endemic to the Region. They inhabit forest openings from the lowlands to mid-montane uplands (Stresemann 1923). This compact genus is characterized by the black, white, and red plumage, upright posture, and canopy fly-catching habit. The genus *Peltops*, long considered an aberrant flycatcher, is now shown to be a member of the Cracticidae (Sibley & Ahlquist 1984, Barker *et al.* 2004, Kearns *et al.* 2012).

Lowland Peltops *Peltops blainvillii*   ENDEMIC | MONOTYPIC

Eurylaimus Blainvillii Lesson & Garnot, 1827. *Voy. Coquille, Zool., Atlas* 1(3): pl. 19, fig. 2.—Manokwari, Bird's Head.

Inhabits canopy clearings and edges of open spaces through lowland forest on NG, Misool, Salawati, and Waigeo I (Mayr 1941); sea level–600 m, locally to 1,170 m (near Baiyer R Sanctuary; I. Woxvold *in litt.*). Replaced at higher elevations by *P. montanus.* Trans-Fly observations from the Bian Lakes and Maro R (Bishop 2005b). A single bird seen once on the Killerton Is of Oro Prov, 500 m off the Mainland coast, was presumably a vagrant (Clapp 1980a). No records from Batanta I.

NOTES

This species is distinguished from *P. montanus* by the the larger bill, smaller white patch on mantle, the narrower and more teardrop-shaped postocular patch, and the slow and strange nonbird-like clicking vocalizaton (Pratt & Beehler 2014). Kearns *et al.* (2012: fig. 1) indicated a confused molecular story for the two *Peltops* species, with putative *blainvillii* morphotypes appearing in both species' molecular lineages. This merits further study. It is perhaps indicative of some hybridization between the species. *Other names*—Clicking or Lowland Shieldbill, Lowland Peltops-flycatcher.

Mountain Peltops *Peltops montanus*

Peltops blainvillii montanus Stresemann, 1921. *Anz. Orn. Ges. Bayern* 1: 35.—Hunstein Mts, middle Sepik, north-central NG.

Endemic to the mts of NG (Mayr 1941); mainly 750–1,680 m, extremes: 600–2,800 m (Coates 1990). Records from peripheral ranges include the Bird's Head[MA], Fakfak[J], Kumawa[M], Wandammen[M], and Foja Mts[J], PNG North Coastal Ra[J,N], Adelbert Mts[O], and Huon Penin[MA] (Gilliard & LeCroy 1967[O], Gibbs 1994[J], Beehler *et al.* 2012[L], J. M. Diamond *in litt.*[M], specimens at AMNH & BPBM[N]). No records from the Cyclops Mts. Frequents canopy openings and edges of clearings in montane forest. Replaced at lower elevations by *Peltops blainvillii*, apparently without overlap, but the findings of Kearns *et al.* (2012) indicate there may be introgression between the two.

NOTES
Other names—Tinkling or Singing Shieldbill, Peltops, or Peltops-flycatcher.

GENUS *CRACTICUS*

[Genus 7 spp/Region 5 spp]

Cracticus Vieillot, 1816. *Analyse Nouv. Orn. Elem.*: 37. Type, by monotypy, "Cassican-Calybé" Buffon = *Rhamphastos cassicus* Boddaert.

The species of this genus range from Australia north to the New Guinea region. The genus is characterized by predominantly black or pied plumage; the prominent long, conical, bluish-white hooked bill; and the powerful, rich, and complex vocalizations. We follow Russell & Rowley (2009) and Kearns *et al.* (2012) in including the Australian Magpie ("*Gymnorhina*") in this genus. Dickinson & Christidis (2014) recognized *Gymnorhina*, and also the genus *Melloria* for *Cracticus quoyi*, based on the phylogeny—but not the taxonomic recommendations—of Kearns *et al.* (2012). However, we believe this fine degree of splitting is unnecessary and awkward, necessitating the placement of *quoyi*, a butcherbird in every respect, in a genus separate from *Cracticus*. Regarding the classifying of the Australian Magpie as a terrestrial butcherbird, the observation of Kearns *et al.* (2012: 948) is worth repeating: "There are many similar examples where striking phenotypic differences among closely related species have been hypothesized to result from strong selective pressures on ecological traits that drive phenotypic adaptation to novel environments or ecological resources" (followed by a list of examples).

Black Butcherbird *Cracticus quoyi*

Barita Quoyi Lesson & Garnot, 1827. *Bull. Sci. Nat. Férussac* 10: 289.—Manokwari, Bird's Head.

Mainland NG, the NW Islands, Aru Is, Yapen I, Dolak I, and Daru I; sea level–750 m, max 2,200 m[CO]. Inhabits forest interior, mangroves, and occasionally forest edge.

Extralimital—n and ne Australia, including Boigu and Saibai I.

SUBSPECIES

a. *Cracticus quoyi quoyi* (Lesson & Garnot)
RANGE: Misool, Salawati, Waigeo, and Yapen I, and all NG except for the Trans-Fly (Mayr 1941).
DIAGNOSIS: Smaller than *alecto*; wing (male) 169–192 (Higgins *et al.* 2006). Bill narrower than that of *alecto* (Schodde & Mason 1999). See photos in Coates (1990: 379).

b. *Cracticus quoyi alecto* Schodde & Mason
Cracticus quoyi alecto Schodde & Mason, 1999. *Dir. Austral. Birds, Passerines*: 534.—Merauke, s NG.
RANGE: The Trans-Fly, Aru Is, Daru I, and islands of n Torres Strait (Boigu and Saibai). Field observations from Dolak I and Wasur NP (Bishop 2005a).
DIAGNOSIS: Larger than nominate form; wing (male) 175–197 (Higgins *et al.* 2006); immature plumage dull black (Schodde & Mason 1999).

NOTES
The molecular work of Kearns *et al.* (2012) provided evidence that the nominate NG population is distinct enough from the Australian populations (for which the name *spaldingi* 1878 has priority) as measured

by their molecular signature that the two may constitute cryptic species (both are all black). Whether the populations that constitute the taxon *alecto* in s NG are allied with the nominate form in NG or the Australian populations was not studied by Kearns *et al.* (2012). Until this issue can be resolved, we take the traditional course of keeping all NG and Australian populations in the same species.

Black-backed Butcherbird *Cracticus mentalis*   RESIDENT
Cracticus mentalis Salvadori and D'Albertis, 1876. *Ann. Mus. Civ. Genova* (1875) 7: 824.—Nicura, Hall Sound, SE Peninsula.

Trans-Fly and Port Moresby savannas, to 600 m^{CO}. Inhabits open woodland, gardens, and scrub. Extralimital—ne QLD.

SUBSPECIES

a. *Cracticus mentalis mentalis* Salvadori and D'Albertis
RANGE: The dry zones of s NG: the Trans-Fly from Dolak I to Lake Daviumbu and Balimo (Bishop 2005a), and the SE Peninsula from Hall Sound to Kupiano and inland to the Sogeri Plateau (Coates 1990).
DIAGNOSIS: Larger than the Australian *kempi*, with a putatively broader black patch on the back (Russell & Rowley 2009). Examination of topotypical material at AMNH did not confirm the stability of this dorsal plumage difference. The dorsal coloration may be age-related. Wing (male) 148–156; bill (male) 37–40 (Higgins *et al.* 2006).

NOTES
Other names—White-throated Butcherbird.

Hooded Butcherbird *Cracticus cassicus*   ENDEMIC
Rhamphastos cassicus Boddaert, 1783. *Tabl. Planch. Enlum.*: 38.—Bird's Head.

All NG and most satellite islands (except the Louisiades), sea level–1,450 m^{CO}; includes the Aru Is, NW Islands (including Gebe but not Kofiau or Gag), Bay Is, islands off the n coast (Kairiru, Muschu), and various SE Islands. Frequents edge, partly cleared areas, tall riparian trees in forest, and gardens (Coates 1990); on Goodenough I ranges into open savanna (Bell 1970e).

SUBSPECIES

a. *Cracticus cassicus cassicus* (Boddaert)
RANGE: All mainland NG; Misool, Gebe, Salawati, Batanta, and Waigeo I; Aru Is; Biak, Numfor, Kurudu, and Yapen I; and Kairiru, Muschu and Basilaki I (Mayr 1941). In the Trans-Fly from the Maro R east to the Aramia R (Bishop 2005a).
DIAGNOSIS: Shorter-winged and smaller than *hercules*; wing (male) 170 (Rand & Gilliard 1967).

b. *Cracticus cassicus hercules* Mayr
Cracticus cassicus hercules Mayr, 1940. *Amer. Mus. Novit.* 1091: 3.—Kaileuna I, Trobriand Is.
RANGE: Kiriwina, Kitava, and Kaileuna I; also Goodenough, Fergusson, and Normanby I^{J,K} (Mayr 1941, Pratt field notes^{J}, LeCroy *et al.* 1984^{K}).
DIAGNOSIS: Longer-winged and much larger than nominate; wing (male) 185 (Rand & Gilliard 1967). Some young birds (and females?) tend to exhibit more black on the back.

NOTES
Type description for the species is found on p. 38, not 83 (Mees 1965, *pace* Mayr 1941). *Other names*—Black-headed or Black-backed Butcherbird.

Tagula Butcherbird *Cracticus louisiadensis*   ENDEMIC | MONOTYPIC
Cracticus louisiadensis Tristram, 1889. *Ibis*: 555.—Tagula I, Louisiade Arch, SE Islands.

Tagula and Sabara I^{J} of the Louisiade Arch (Mayr 1941, Pratt field notes^{J}). Common in forest and edge, sea level–500 m (Pratt & Beehler 2014). May also inhabit fringing islands associated with Tagula.

Molecular studies showed this species to be a sister form to the widespread *Cracticus cassicus* (Kearns *et al.* 2012). *Other names*—Louisiade, Sudest, or White-rumped Butcherbird.

Australian Magpie *Cracticus tibicen* RESIDENT

C[*oracias*] *tibicen* Latham, 1801. *Index Orn.*, Suppl.: xxvii.—New South Wales, Australia.

Forages on the ground in small groups in the Trans-Fly savannas, near sea level. Inhabits mixed savanna woodland and open savanna.

Extralimital—widespread across Australia. Introduced to NZ, Fiji, and the Solomon Is (Guadalcanal).

SUBSPECIES

a. *Cracticus tibicen papuanus* (Bangs & Peters)

Gymnorhina tibicen papuana Bangs & Peters, 1926. *Bull. Mus. Comp. Zool.* Harvard 67: 431.—Princess Marianne Strait, s NG.

RANGE: The Trans-Fly, from Princess Marianne Strait east to Dogwa on the Oriomo R[MA] (Mees 1982b, Bishop 2005a). Tree savanna near the coast.

DIAGNOSIS: The bill is quite stout anteriorly; male plumage apparently variable; some males white from nape to rump, whereas others have a black lower back (L. Joseph *in litt.*). Original description (Bangs & Peters 1926) noted the narrow white nuchal patch; feathers of tibia white with black at the base only; wing (male) 230–255; tail (male) 121–130 (Higgins *et al.* 2006).

NOTES

We follow Christidis & Boles (2008) and Russell & Rowley (2009) in subsuming this species into the genus *Cracticus*. Kearns *et al.* (2012) provided molecular data showing that this species is sister to *Cracticus quoyi*. Dickinson & Christidis (2014) maintained *Gymnorhina* as a distinct genus based mainly on morphology and behavior. *Other names*—*Gymnorhina tibicen*.

Family Artamidae

Woodswallows [Family 11 spp/Region 3 spp]

We treat the Artamidae as a freestanding family, following Rowley & Russell (2009) and Kearns *et al.* (2012). The species of the family range from tropical Asia to New Guinea, Australia, Vanuatu, New Caledonia, and Fiji. Woodswallows are medium-size, compact, aerial predators that exhibit a short and nearly conical bill, triangular wings, and an abbreviated, squared-off tail. They are very sociable and have a complex, chattering song. In New Guinea, species range throughout the lowlands to as high as 2,800 m. Plumage is generally sooty gray, usually patterned with white. Note pale blue gray bill in all species. Woodswallows prefer open sites   mountain ridges, forest clearings, towns, and savannas—where there are exposed perches from which they sally out after flying insects. Schodde & Mason (1999) and Christidis & Boles (2008) united the Artamidae and Cracticidae into a single family. The phylogeny of Kearns *et al.* (2012) placed the woodswallows as sister to the clade that includes the bushshrikes (Malaconotidae), the helmetshrikes (Prionopidae), and the peltopses and the butcherbirds (Cracticidae). See discussion in the family account for the Cracticidae.

Artamus Vieillot, 1816 (April). *Analyse Nouv. Orn. Elem.*: 41. Type, by monotypy, "Langraien" (Buffon) = *Lanius leucorhynchus* Linnaeus.

The species of this genus range from the Andaman Islands and Malay Peninsula to Australia, New Caledonia, and Fiji (Rowley & Russell 2009). Generic characteristics are as for the family.

White-breasted Woodswallow *Artamus leucorynchus* RESIDENT

Lanius leucoryn[chus] Linnaeus, 1771. *Mantissa Plant.*: 524.—Manila, Luzon, Philippines. [We follow Rowley & Russell (2009) who argue that prevailing usage justifies acceptance of the *leucorynchus* rather than *leucoryn* (*pace* Dickinson & Christidis 2014).]

Inhabits open areas with scattered trees throughout NG, the NW Islands, Aru Is, and D'Entrecasteaux Arch, sea level–800 m (Pratt & Beehler 2014). K. D. Bishop (*in litt.*) reported observing the species on Biak I in 1989 and 1990.

Extralimital—India east to New Caledonia, Vanuatu, Fiji, and Australia. Absent from the Bismarck Arch and Solomon Is (Dutson 2011).

SUBSPECIES

a. *Artamus leucorynchus leucopygialis* Gould
Artamus leucopygialis Gould, 1842. *Birds Australia* 4, 2: pl. 33.—Namoi R, NSW.
RANGE: The Moluccas and Kai Is; all NG; Misool[MA], Salawati[MA], Batanta[MA], Waigeo[MA], Gag[G], and Gebe I[J]; Aru Is[MA]; Goodenough[MA], Fergusson[MA], and Normanby I[L]; Amphlett Is[MA]; Tubatuba and Bentley of the Engineer Is[CO]; Samarai I[CO]; and islands off the s coast of the SE Peninsula: Loupom and Mailu off Amazon Bay and Fisherman's and Haidana near Port Moresby[CO] (Mees 1972[J], LeCroy *et al.* 1984[L], Johnstone 2006[G]). Also n and e Australia, including islands of the Torres Strait (Yam, Yorke, Saibai, and Boigu I[CO]. Widespread in the Trans-Fly (Bishop 2005a).
DIAGNOSIS: Crown, throat, and chin intermediate in color between those of the darker nominate race and *musschenbroekii* from the Tanimbar Is, and the paler *humei* of Cocos, Andaman Is (Rowley & Russell 2009). Population from the middle Sepik exhibits a pale gray (not whitish) breast (AMNH specimens collected by Gilliard in 1954).

NOTES

Rowley & Russell (2009) indicated the original Linnaean spelling of *leucorynchus* is correct (*pace* Dickinson & Christidis 2014). *Other names*—White-breasted Swallow-shrike, White-rumped Woodswallow, *Artamus leucoryn.*

Great Woodswallow *Artamus maximus* ENDEMIC | MONOTYPIC

Artamus maximus A. B. Meyer, 1874. *Sitzungsber. Akad. Wiss. Wien* 69: 203.—Hatam, Arfak Mts, Bird's Head.

> Synonym: *Artamus maximus wahgiensis* Gyldenstolpe, 1955. *Ark. f. Zool.* (2)8(1): 121.—Dagie, Wahgi valley, Eastern Ra.

Mountains of NG[J], 600–2,800 m[K] (Diamond 1985[J], Coates & Peckover 2001[K]). Inhabits clearings with dead trees or high dead limbs for open perches. No records from the Cyclops or Kumawa Mts (Diamond 1985). Coates & Peckover (2001) reported a single record from the lowlands.

NOTES

We follow Sims (1956) in not recognizing the race *wahgiensis*. A sister form, *A. insignis*, inhabits the Bismarck Arch (Mayr & Diamond 2001). *Other names*—Giant, Greater, Papuan, New Guinea, or Black-breasted Wood-swallow.

Black-faced Woodswallow *Artamus cinereus* PRESUMED RESIDENT

Artamus cinereus Vieillot, 1817. *Nouv. Dict. Hist. Nat.*, nouv. éd. 17: 297.—Southern coast of Western Australia opposite Archipelago of the Recherche.

Apparently a scarce resident in the Trans-Fly, from Princess Marianne Strait east to the Wassi Kussa R (Bangs & Peters 1926), though to date there is no proof of nesting in NG. No observations of migratory movements across the Torres Strait. A report of the species from Wau (Symons 1992) is anomalous. Photos of the bird are not identifiable to species.

Extralimital—widespread across Australia.

SUBSPECIES

a. *Artamus cinereus normani* (Mathews)
Austrartamus melanops normani Mathews, 1923. *Birds Austral.* 10(4): 255.—Normanton, QLD, Australia.
RANGE: The Trans-Fly of NG and ne Australia (Rowley & Russell 2009). Localities—Princess Marianne Strait on 6-Nov-1923[J]; Kurik in Jul-1960 and Mar-1961[K]; 80 km inland along the Kumbe R on 12-Oct-1960[CO]; Bula plains, 23-Oct-1969[L], early Oct-1979, and 25-Nov-1980[N]; and 70 km east of Morehead on 1-Nov-1980[CO] (Bangs & Peters 1926[J], Hoogerwerf 1964[K], Lindgren 1971[L], Stronach 1981b[N], Mees 1982b).
DIAGNOSIS: Compared to the nominate form, the facial mask is reduced (not reaching throat); less extensive black rump patch; vent and undertail coverts white (not black) and sharply demarcated from belly; and broader white tips to rectrices (Rowley & Russell 2009).

NOTES

Use of the name *normani* Mathews for this population, rather than *hypoleucus* Sharpe or *albiventris* Gould, is explained in Schodde & Mason (1999: 567). For another view see Mees (1982b). *Other names*—Gray, Gray-breasted, or White-bellied Wood-swallow.

Family Rhagologidae

Berryhunters ENDEMIC | MONOTYPIC

The single species of the family is confined to the upland forests of New Guinea. For decades, *Rhagologus* was considered to be an aberrant whistler (Hartert *et al.* 1936, Mayr 1941, Rand & Gilliard 1967, Beehler & Finch 1985). Jønsson *et al.* (2011b) demonstrated the isolated affinities of *Rhagologus*, placing it sister to the tropical Asian ioras (Aegithinidae) and far from the whistlers (Pachycephalidae). More recently, Aggerbeck *et al.* (2014) placed *Rhagologus* sister to the clade that includes *Peltops*, *Cracticus*, and *Artamus*. Schodde & Christidis (2014) formally erected a family for this single species and provided an initial diagnosis. Although the sexually dimorphic and mottled plumages of *R. leucostigma* are unlike those of any other whistler, in body form and aspect this species does, indeed, have a whistler-like appearance. That said, its primarily fruit eating habit is distinctive, and its remarkably beautiful slurred song is also unlike any whistler vocalization. The English name "berryhunter" is derived from a translation of the generic epithet.

GENUS *RHAGOLOGUS* ENDEMIC | MONOTYPIC

Rhagologus Stresemann & Paludan, 1934. *Orn. Monatsber.* 42: 45. Type, by monotypy, *Pachycephala leucostigma* Salvadori.

This monotypic genus is endemic to the uplands of mainland New Guinea. Form and habits are as for the family. Female plumage is brighter than that of male. Also unique is heavy mottling on breast and belly of female (and male of one subspecies).

Mottled Berryhunter *Rhagologus leucostigma*   ENDEMIC

Pachycephala leucostigma Salvadori, 1876. *Ann. Mus. Civ. Genova* (1875) 7: 933.—Arfak Mts, Bird's Head.

A secretive but vocal inhabitant of upland forest interior of the Bird's Head, Central Ra, and Huon Penin; 1,500–2,550 m, extremes 820–2,900 m[CO]. Inhabits lower montane forest and occasionally regrowth. Very difficult to observe in the interior of forest subcanopy, but once its lovely, slurred musical series is learned it can be quickly identified. Most easily observed when foraging for small fruit in forest subcanopy.

SUBSPECIES

a. *Rhagologus leucostigma leucostigma* (Salvadori)

> Synonym: *Rhagologus leucostigma novus* Rand, 1940. *Amer. Mus. Novit.* 1072: 7.—Kunupi, 1,500 m, Weyland Mts, Western Ra.

RANGE: Mountains of the Bird's Head and the w third of the Central Ra, meeting the range of the distinct population *obscurus* in the w Border Ra.
DIAGNOSIS: Male is rufous faced, not gray faced as in *obscurus*; in addition, male is generally olive and brown rather than the gray of *obscurus*, with fine barring of flanks. Female plumages of the two races are much more similar, but the nominate female has a darker rufous face patch that is of smaller extent, and underparts are buff tinged (Rand & Gilliard 1967, Boles 2007). We subsume *novus* into the nominate form, as that minor plumage variation is easily encompassed within this form, which is very distinct from the eastern form, *obscurus*.

b. *Rhagologus leucostigma obscurus* Rand

Rhagologus leucostigma obscurus Rand, 1940. *Amer. Mus. Novit.* 1072: 7.—Mafulu, 1,250 m, SE Peninsula.
RANGE: The Border Ra, Eastern Ra, and mts of the SE Peninsula. The race extends west in the n watershed to the w sector of the Border Ra; in the s watershed west to the c sector of the Border Ra (Mt Goliath). Localities—15 km southwest of Bernhard Camp on the Taritatu R[J], Mt Goliath[MA], below Kumul Lodge[K], Rondon Ridge[K], Ambua[K], Mt Hagen[L], Awande[DI], Okasa[DI], Mt Michael[DI], Mengino[DI], Karimui[DI], YUS[P], Junzaing[M], upper Eloa R[N], and Mt Missim[P] (Rand 1942b[J], Mayr 1931[M], Mayr & Gilliard 1954[L], P. Gregory *in litt.*[K], Beehler 1980b[N], Beehler field notes[P]).
DIAGNOSIS: Male plumage is lead gray and entirely lacks rufous or brown in plumage but for the dull rufous undertail. Compared to the nominate form, this female has a larger rufous face patch and exhibits more prominent scalloping on breast and belly, with white of underparts lacking buffy tinge. See photos of male and female in Coates & Peckover (2001: 159).

NOTES

Other names—Mottled or Red-vented Whistler, *Pachycephala leucostigma*.

Family Campephagidae

Cuckooshrikes and Trillers   [Family 93 spp/Region 16 spp]

The species of the family range from Africa east to tropical Asia and thence to New Guinea, Australia, and the tropical Pacific (Taylor 2005). Cuckooshrikes are small to medium-large arboreal songbirds. They flick their wings after alighting; the function of this peculiar behavior is unknown. Several species carry out male-female duets, and most are conspicuous vocalists. Characteristics of this very distinctive family include the strong-looking black bill, well-rounded head, longish tapered tail, slim body form, dense and soft plumage, and small and weak feet (Schodde & Mason 1999). Anatomic features include a single well-developed and trabeculate humeral fossa; fully aperturate nasal cavity; and a well-developed interpalatine process, among other features (Higgins *et al.* 2006). The group falls into the core Corvoidea and is sister to a clade that includes a broad range of Australasian songbirds (Barker *et al.* 2004). Biogeography and molecular systematics of the family have been reviewed by Fuchs *et al.* (2007) and Jønsson *et al.* (2010c). We follow the phylogeny and taxonomic treatment presented therein.

Coracina Vieillot, 1816 (April). *Analyse Nouv. Orn. Elem.*: 37. Type, by subsequent designation (Cabanis, 1851. *Mus. Hein.* 1: 62), "Choucari" Buffon = *Corvus papuensis* Gmelin.

The species of this genus range from Australasia and Melanesia west through tropical Asia to Africa and Madagascar. The genus is diverse in size, morphology, and plumage, and it is difficult to find characters that consistently distinguish the species of *Coracina* from those of *Edolisoma*. Furthermore, the phylogeny of Jønsson *et al.* (2010c) shows that the species they combined to form the genus *Coracina* are quite divergent when compared with one another (unlike the more uniform *Edolisoma*), and we suspect that future systematic study may further partition them into additional genera. We follow Fuchs *et al.* (2007) and Jønsson *et al.* (2010c) for treatment of the genus.

Stout-billed Cuckooshrike *Coracina caeruleogrisea* RESIDENT | MONOTYPIC

Campephaga caeruleogrisea G. R. Gray, 1858. *Proc. Zool. Soc. London*: 179.—Aru Is.
   Synonym: *Campephaga strenua* Schlegel, 1871. *Ned. Tijdschr. Dierk.* (1873) 4: 44.—Arfak Mts, Bird's Head.
   Synonym: *Coracina caeruleogrisea adamsoni* Mayr & Rand, 1936. *Mitt. Zool. Mus. Berlin* 21: 245.—Mafulu, 1,250 m, SE Peninsula.

Throughout hill forest and lower mts of NG, Yapen I[MA], and the Aru Is[MA]; sea level–1,600 m, max 2,450 m[CO]. Reaches the lowlands at Sturt I Camp on the Fly R (Bishop 2005a). Inhabits forest midstory and edge (Pratt & Beehler 2014).

Extralimital—recently recorded with photographic evidence from Halmahera I, Moluccas (Bashari & van Balen 2014).

NOTES
Three thinly defined subspecies have been combined. Size difference (wing length) less than 10%. Color differences not detected in a large AMNH specimen series of topotypical material for the three described forms. *Other names*—Stout-billed Greybird, Blue-grey Cuckooshrike.

Hooded Cuckooshrike *Coracina longicauda* ENDEMIC | MONOTYPIC

Graucalus longicauda De Vis, 1890. *Ann. Rept. Brit. New Guinea*, 1888–89: 59.—Musgrave Ra, 2,150–2,750 m, SE Peninsula.
   Synonym: *Coracina longicauda grisea* Junge, 1939. *Nova Guinea* (ns) 3: 5.—Hellwig Mts, 2,600 m, e sector of the Western Ra.

The Central Ra and mts of the Huon Penin; 1,800–3,600 m, extremes 1,300–3,700 m[CO]. Inhabits upper montane forest. Found in noisy family parties foraging in the canopy of forest and edge.

NOTES
Treated here as monotypic. Described subspecies relate to a cline in size that should not be enshrined in formal names. The difference in wing length of named subspecies is 12 mm (equivalent to 6%). *Other names*—Long-tailed Cuckooshrike, Black-hooded Greybird.

Barred Cuckooshrike *Coracina lineata* RESIDENT AND VAGRANT

Ceblepyris lineatus Swainson, 1825. *Zool. Journ.* 1: 466.—Moreton Bay, QLD, Australia.

Very patchily distributed in the mts of NG: Bird's Head, Bird's Neck, Foja Mts, w Trans-Fly, e Border Ra, PNG North Coastal Ra, Adelbert Mts, Huon Penin, SE Peninsula; also Waigeo I, Aru Is, Numfor I; 600–1,450 m. Inhabits hill forest, edge, and partly cleared areas. A fig specialist. Hover-gleans ripe fruit from the canopy of fruiting figs. Curiously absent from much of both the Western Ra and Eastern Ra.

Extralimital—e Australia, Bismarcks (New Britain, Lolobau, New Ireland), and n Solomons (Mayr & Diamond 2001, Dutson 2011).

SUBSPECIES

a. *Coracina lineata axillaris* (Salvadori)
Graucalus axillaris Salvadori, 1876. *Ann. Mus Civ. Genova* (1875) 7: 925.—Mansema, Arfak Mts, Bird's Head.

RANGE: Waigeo I[MA], Gam I[M], the Aru Is[K], and mts of the Bird's Head[MA] and Bird's Neck[N], Foja Mts, PNG North Coastal Ra[L], Adelbert Mts[J], Huon Penin[MA], and SE Peninsula[MA] (Gilliard & LeCroy 1967[J], Diamond & Bishop 1994[K], AMNH specimens[L], Beehler field notes 12-Mar-2005[M], Diamond & Bishop 2015[N]). Hill forest, 400–1,300 m, occasionally lower. Sightings only from the Aru Is (Diamond & Bishop 1994); Foja Mts record from a single observation (Beehler *et al.* 2012). Recently observed well at Kaimana on the Bird's Neck in Oct-2014 (C. Thébaud & B. Mila *in litt.*).

DIAGNOSIS: Male exhibits plain gray (unbarred) underparts; female heavily barred ventrally (Rand & Gilliard 1967, Taylor 2005). Diamond & Bishop (1994) noted the Aru Is population is closest to this form.

b. *Coracina lineata maforensis* (A. B. Meyer)
Campephaga maforensis A. B. Meyer, 1874. *Sitzungsber. Akad. Wiss. Wien* 69: 386.—Numfor I, Bay Is.
RANGE: Numfor I (Mayr 1941).
DIAGNOSIS: Male has a black belly, distinctly barred with very thin white lines; a powder-blue forehead (see illustration in Pratt & Beehler 2014); and is the most distinctive form of this species. Note that the female also exhibits the pale forehead. The illustration of this form in Taylor (2004: pl. 2) fails to show the male's blackish underparts with fine white scoring.

c. *Coracina lineata lineata* (Swainson)
RANGE: Breeds through e Australia. Once recorded from Merauke as a vagrant (van Oort 1909). Also possible Australian migrants observed at Ok Tedi and the upper Fly (Coates 1990).
DIAGNOSIS: Sexes alike, with breast, belly, and undertail barred finely with black and white (Taylor 2004). Lores black in male *vs.* dark gray in female; and rectrices mid-gray grading abruptly to black at tip (Schodde & Mason 1999). Note this form gives a vocalization quite distinct from the e NG population—a jay-like *yaanhh!* rather than the sad and musical *weeyur*. The two may be distinct at the species level (but see Jønsson *et al.* 2010c).

NOTES
The species status of the very distinctive population from Numfor I bears further study. *Other names*—Yellow-eyed Cuckoo-shrike, Yellow-eyed Greybird.

Boyer's Cuckooshrike *Coracina boyeri*   ENDEMIC
Campephaga Boyeri G. R. Gray, 1846. *List Gen. Birds* 1: [283].—Triton Bay, Bird's Neck.
    Inhabits forest, edge, and mangroves throughout NG, Misool, Salawati, Yapen, and Daru I; sea level–1,100 m, max 1,450 m.

SUBSPECIES
a. *Coracina boyeri boyeri* (G. R. Gray)
RANGE: Salawati[J], Misool[MA], and Yapen I[MA]; also the Bird's Head and n NG east to the SE Peninsula (Collingwood Bay), where it intergrades with *subalaris* (Taylor 2005, Pratt & Beehler 2014, J. M. Diamond & K. D. Bishop *in litt.* reported seen 1983 and 1986[J]).
DIAGNOSIS: Underwing rich rufous; female exhibits whitish lores and silvery-blue crown (Rand & Gilliard 1967). See photo of adult male in Coates & Peckover (2001: 119).

b. *Coracina boyeri subalaris* (Sharpe)
Graculus subalaris Sharpe, 1878. *Mitt. Zool. Mus. Dresden* 1(3): 364.—Fly R.
RANGE: The S Lowlands, Trans-Fly, and s watershed of the SE Peninsula; also Daru. Trans-Fly sightings include the Maro R, Wasur NP, Bian Lakes, Morehead, and the Fly R (Bishop 2005a).
DIAGNOSIS: Underwing pale cinnamon; female exhibits gray (not whitish) lores and dull blue-gray (not silvery-blue) crown (Rand & Gilliard 1967). Male apparently indistinguishable from nominate form.

NOTES
Other names—Rufous-underwing Greybird or Cuckooshrike.

Black-faced Cuckooshrike *Coracina novaehollandiae*   VISITOR AND RESIDENT

Turdus novae Hollandiae Gmelin, 1789. *Syst. Nat.* 1(2): 814.—Adventure Bay, Tasmania.

Widespread austral migrant to NG and the SE Islands. Especially common in s NG and in interior upland valleys, sea level–1,830 m[CO]. A small population breeds in the Port Moresby region (Coates 1990). Migrants found from April to October (Coates 1990). Frequents mainly open savanna and cultivated lands. Migration between Australia and NG may occur on a broad front across the Coral Sea (Draffan *et al.* 1983). In NG has been recorded from many localities on the Mainland, especially in s areas; also e satellite islands. Expected as a migrant in the Aru Is given its migration to the Moluccas (Coates & Bishop 1997) but not recorded there yet (Diamond & Bishop 1994).

Extralimital—breeds in Australia; also a visitor to the Lesser Sundas, Bismarcks, and Solomons (Schodde & Mason 1999). Straggler to NZ (Gill *et al.* 2010).

SUBSPECIES

a. *Coracina novaehollandiae novaehollandiae* (Gmelin)
RANGE: Breeding confined to Tasmania and islands in Bass Strait. Nonbreeding vagrant to the Region (Mayr 1941).
DIAGNOSIS: Gray of breast much the same tone as gray of mantle; *subpallida* exhibits a whitish-gray breast; *melanops* has a slightly longer and broader bill (Schodde & Mason 1999, Taylor 2005). Wing (male) 189–202 (Higgins *et al.* 2006).

b. *Coracina novaehollandiae melanops* (Latham)
Corvus melanops Latham, 1801. *Index Orn.*, Suppl. 2: xxiv.—Sydney, NSW, Australia.
   Synonym: *Coracina novaehollandiae didimus* Mathews, 1912. *Austral. Avian Rec.*, 1: 42.—Melville I, NT, Australia.
RANGE: Breeds across much of Australia (all but range of *subpallida* of w Australia); small breeding population in Port Moresby region (Coates 1990: 36). As an Australian migrant to NG[MA], Misool I[Q], Bay Is[MA], Karkar I[K], SE Islands[CO], Bismarck Arch[CO], and Nissan I[CO] (Mees 1965[Q], Diamond & LeCroy 1979[K]). Also as an Australian migrant to Wallacea (Coates & Bishop 1997). Bishop (2005a) noted only one sighting in the Trans-Fly (Sturt I), but it is often common along the Fly R north of Kiunga at passage periods (P. Gregory *in litt.*).
DIAGNOSIS: Similar to but longer-billed than the nominate form; bill 21–25 *vs.* 19–21 (Schodde & Mason 1999, Taylor 2005). Wing (male) 183–214 (Higgins *et al.* 2006).

NOTES
Other names—Australian Greybird.

White-bellied Cuckooshrike *Coracina papuensis*   RESIDENT

Corvus papuensis Gmelin, 1788. *Syst. Nat.* 1(1): 371.—Manokwari, Bird's Head.

Widespread through NG, the NW Islands (but not Waigeo I); the Aru Is; Adi, Biak, Yapen, and Tagula I; sea level–1,650 m[CO]. Inhabits forest edge, open woodland, gardens, mangroves, and suburban areas (Coates 1990).

Extralimital—n and e Australia, Moluccas, and Bismarck and Solomon Is. Also Torres Strait Is as a visitor (Draffan *et al.* 1983, Mayr & Diamond 2001).

SUBSPECIES

a. *Coracina papuensis papuensis* (Gmelin)
   Synonym: *Campephaga melanolora* G. R. Gray, 1860. *Proc. Zool. Soc. London*: 353.—Bacan I and Ternate, n Moluccas.
   Synonym: *Coracina papuensis intermedia* Rothschild, 1931. *Novit. Zool.* 36: 267.—Upper Setekwa R, sw NG.
RANGE: Kofiau, Gag[J], Gebe, Misool, Salawati, Batanta, Biak[K], and Yapen I (Mayr 1941, Johnstone 2006[J], van Balen & Bishop ms[K]); on Mainland from the Bird's Head east along the n coast to the head of Huon Gulf, where it intergrades with *angustifrons*, and along the s coast as far as the Lorentz R of sw NG. Also n Moluccas. Observed on Biak I by K. D. Bishop (*in litt.*) in Oct-1989 and Jun-1990.
DIAGNOSIS: Gray breast and lower throat match cheek and mantle; chin paler (Taylor 2005). Race *melan-olora* not sufficiently distinct from nominate form (Taylor 2005). The form *intermedia* exhibits plumage

intermediate between the nominate form and *angustifrons*. We don't believe the level of distinction merits formal recognition.

b. *Coracina papuensis hypoleuca* (Gould)
Graucalus hypoleucus Gould, 1848. *Proc. Zool. Soc. London*: 38.—Port Essington, NT, Australia.
RANGE: Northern Australia and the Aru Is (Mayr 1941).
DIAGNOSIS: Entire underparts whitish but for a faint gray band smudged across upper breast; wing (male) 151–153; bill (male) 21.5–25.5.

c. *Coracina papuensis angustifrons* (Sharpe)
Graucalus angustifrons Sharpe, 1876. *Journ. Linn. Soc. London, Zool.* (1878) 13: 81.—Port Moresby, SE Peninsula.

> Synonym: *Coracina papuensis oriomo* Mayr & Rand, 1936. *Mitt. Zool. Mus. Berlin* 21: 244.—Wuroi, Oriomo R, Trans-Fly, s NG.

RANGE: The Trans-Fly, the c and e sectors of the S Lowlands, and the SE Peninsula, west in the north to Huon Gulf, in the south at least to Wasur NP (Bishop 2005a). Also perhaps c NG—Baiyer R and Lake Kutubu (*fide* Diamond 1972, Taylor 2005).
DIAGNOSIS: Much like *hypoleuca* but with very pale gray continuing down onto abdomen. Note also thicker, blacker lores in male (R. Schodde *in litt.*). See photos in Coates (1990: 36–37) and Coates & Peckover (2001: 118). The form *meekiana* Rothschild & Hartert was synonymized into this form by Mayr (1941).

d. *Coracina papuensis louisiadensis* (Hartert)
Graucalus hypoleucus louisiadensis Hartert, 1898. *Novit. Zool.* 5: 524.—Tagula I, Louisiade Arch, SE Islands.
RANGE: Tagula I.
DIAGNOSIS: Much like *hypoleuca* and *angustifrons*, but wing shorter and bill longer (Rand & Gilliard 1967). Wing (male) 142–149; bill (male) 24.0–29.0.

NOTES
Other names—Papuan Greybird; Little, Papuan, or White-breasted Cuckoo-shrike.

GENUS *CAMPOCHAERA*   ENDEMIC | MONOTYPIC

Campochaera "*Salvadori*" in Sharpe, 1878. *Mitt. K. Zool. Mus. Dresden* 3: 363. Type, by monotypy, *Campephaga sloetii* Schlegel.

This monotypic genus is endemic to the mainland of New Guinea and is characterized by the slight body, small size, and plumage patterned with black, white, and yellow-orange (Rand & Gilliard 1967).

Golden Cuckooshrike *Campochaera sloetii*   ENDEMIC
Campephaga sloetii Schlegel, 1866. *Nederl. Tijdschr. Dierk.* 3: 253.—Seleh, sw cape of the Bird's Head.

The Bird's Head and Neck, NW Lowlands, S Lowlands, and w sector of the s watershed of the SE Peninsula eastward to Port Moresby; sea level–800 m, max 1,240 m (Freeman & Freeman 2014). Inhabits lowland and hill forest (Coates 1990). Apparently no records from the s bank of the Sepik R in foothills of Border and Eastern Ra (Stresemann 1923), but known from the foothills of the Western Ra, on the s verge of the Mamberamo basin and on the Ramu R. Absent from the n and e sectors of the SE Peninsula, the Huon Penin, and from all but the nw and se sectors of the Sepik-Ramu.

SUBSPECIES
a. *Campochaera sloetii sloetii* (Schlegel)
RANGE: The Bird's Head and the n watershed patchily east to the Ramu R (Sam & Koane 2014). Localities—Seleh on the w coast of the Bird's Head[J], Sorong[J], Klamono[J], north of Nabire[P], Danau Bira[W], Bernhard Camp[Q], Poee on Lake Sentani[K], Wewak[CO], Ramu R[L] (Rand 1942b[Q], Gyldenstolpe 1955b[J], Ripley 1964a[K], King 1979[P], Sam & Koane 2014[L], Beehler field notes 10-Jul-1987[W]). The Bürgers expedition to the Sepik did not obtain this form (Stresemann 1923).
DIAGNOSIS: Crown gray (Pratt & Beehler 2014).

b. *Campochaera sloetii flaviceps* Salvadori

Campochaera flavicipitis Salvadori, 1879. *Ann. Mus. Civ. Genova* (1880) 15: 38, note.—Fly R. [Name emended by Salvadori, 1881.]

RANGE: Southern NG from the Mimika R east to Port Moresby. Localities—Mimika R[MA], Karimui[DI], Bomai[DI], Soliabeda[DI], 30 km above D'Albertis Junction[J] and Black R Camp[K] (Fly R), Dablin Creek[L], Ok Ma[L], Nomad R[N], Ningerum area[J], Kiunga area[J], Elevala R[J], Ketu R[J], lower Kikori R[J], Lakekamu basin[M], Mt Epa[MA], Hall Sound[MA], Moroka[CO], Sogeri Plateau[MA], and Goldie R[MA] (Rand 1942a[K], Bell 1970b[N], Beehler *et al.* 1994[M], P. Gregory *in litt.*[L], Bishop 2005b[J]).

DIAGNOSIS: Crown of both sexes olive (Pratt & Beehler 2014).

NOTES
Should be looked for in the Ramu valley east to Madang. *Other names*—Orange Cuckooshrike.

GENUS *LALAGE*   [Genus 18 spp/Region 4 spp]

Lalage Boie, 1826. *Isis von Oken* 19(10): col. 973. Type, by monotypy, *Turdus orientalis* Gmelin = *Turdus niger* Forster.

   The species of this genus range from islands in the western Indian Ocean and tropical Asia east to Australia (Dickinson & Christidis 2014). Species of the genus are small and slim, sexually dichromatic, and primarily dark above and pale below, often with fine barring.

White-winged Triller *Lalage tricolor*   RESIDENT (AND VISITOR?) | MONOTYPIC

Ceblepyris tricolor Swainson, 1825. *Zool. Journ.* 1: 467.—Sydney, NSW.

   Patchily distributed from the middle Sepik R[CO], Trans-Fly[CO], Markham valley[CO], Port Moresby[CO], and savannas in the e sector of the n watershed of the SE Peninsula[CO]; to 850 m[CO]. Proven breeders known only from Port Moresby. Others are probably migrants from Australia (Coates 1990).

   Extralimital—Throughout Australia. Australian birds are migratory in the s parts of their range, moving north in winter, apparently reaching NG from time to time, though not as a regular visitor.

NOTES
We follow Jønsson *et al.* (2010c), who treated *tricolor* and *sueurii* of c and e Indonesia as specifically distinct. This is in accord with White & Bruce (1986) and Schodde & Mason (1999). *L. tricolor*, as now defined, is monotypic. *Other names*—formerly considered conspecific with *L. sueurii*, now the White-shouldered Triller.

Black-browed Triller *Lalage atrovirens*   ENDEMIC | MONOTYPIC

Campephaga (*Lalage*) *atrovirens* G. R. Gray, 1862. *Proc. Zool. Soc. London* (1861): 430.—Misool I, NW Islands.

   Inhabits forest, edge, and gardens of the Bird's Head and Neck, NW Lowlands, Sepik-Ramu, east to the n slope of the Huon Penin; also the NW Islands: Kofiau[J], Misool[MA], Salawati[MA], and Waigeo[MA] (Diamond *et al.* 2009[J]). Sea level–1,400 m[CO]. No records from Batanta I. Occurs in the s watershed at Lake Kopiago of the w sector of the Eastern Ra (George & Coates 1970). Diamond (1985) found it in the Fakfak and Kumawa Mts of the Bird's Neck.

NOTES
We have separated the form *leucoptera*, from Biak I, as a distinct island allospecies (see below). *L. atrovirens* should be looked for on Yapen I. *Other names*—White-rumped Triller.

Biak Triller *Lalage leucoptera*   ENDEMIC | MONOTYPIC

Campephaga leucoptera Schlegel, 1871. *Ned. Tijdschr. Dierk.* (1873) 4: 45.—Biak I, Bay Is.
    Widespread in forest and edge on Biak I (Mayr 1941).

NOTES

Here treated as specifically distinct from *L. atrovirens.* Defining characters include (1) male's black hood extends down over entire cheek and auricular; (2) male with single extensive white wing patch; (3) male with black rump; and (4) female with unbarred ventral plumage (Taylor 2005). S. van Balen (*in litt.* 2012) noted that the vocalization is distinct from that of *L. atrovirens.* Song is a series of loud, medium-pitched, clear two-note whistles at a rate of *ca.* one phrase per second (*fide* K. D. Bishop): *whir-whee! whir-whee! whir-whee!* As the bird becomes more excited, the song achieves a bubbling quality. *Other names*—treated by many authors as conspecific with *L. atrovirens.*

Varied Triller *Lalage leucomela*   RESIDENT

Campephaga Leucomela Vigors & Horsfield, 1827. *Trans. Linn. Soc. London* 15: 215.—Southeastern QLD, Australia.
    Inhabits forest, edge, and shade trees in openings of the S Lowlands and the SE Peninsula; west in the n watershed to the Markham valley; west in the s watershed to the Mimika R; also the Aru and Trobriand Is, and D'Entrecasteaux and Louisiade Arch (Mayr 1941). Occurs mostly in the lowlands and hills, locally up to 1,800 m^{CO}. Replaced in w and n NG by *L. atrovirens.*
    Extralimital—Kai Is, Bismarck Arch, and n and ne Australia (Mayr & Diamond 2001).

SUBSPECIES

a. *Lalage leucomela polygrammica* (G. R. Gray)
Campephaga polygrammica G. R. Gray, 1858. *Proc. Zool. Soc. London*: 179.—Aru Is.
RANGE: The Aru Is, S Lowlands, ?Trans-Fly, and SE Peninsula, northwest to the Huon Gulf area and Lae.
DIAGNOSIS: Underparts gray, with rufescent wash restricted to undertail (Rand & Gilliard 1967). Male has a vestigial white eyebrow, black rump with white mottling, and strongly barred underparts; female sooty gray above, barred on rump, more richly colored ventrally, with heavy barring (Taylor 2005). Birds in the Trans-Fly may show plumage characters related to intergradation with the Australian population *yorki*; *i.e.*, in some museum specimens the ventral barring may be largely obsolete (see Schodde & Mason 1999: 588).

b. *Lalage leucomela obscurior* Rothschild & Hartert
Lalage karu obscurior Rothschild & Hartert, 1917. *Bull. Brit. Orn. Club* 37: 16.—Fergusson I, D'Entrecasteaux Arch, SE Islands.
    Synonym: *Lalage karu trobriandi* Mayr, 1936. *Amer. Mus. Novit.* 869: 1.—Kiriwina I, Trobriand Is, SE Islands.
RANGE: D'Entrecasteaux Arch and Trobriand Is (Mayr 1941).
DIAGNOSIS: Rufescent wash extends up onto breast, especially in female (Rand & Gilliard 1967, Taylor 2005). See photo in Coates & Peckover (2001: 121). We do not recognize *trobriandi*, which exhibits a 4% difference in wing length and very minor plumage differences.

c. *Lalage leucomela pallescens* Rothschild & Hartert
Lalage karu pallescens Rothschild & Hartert, 1917. *Bull. Brit. Orn. Club* 37: 17.—Tagula I, Louisiade Arch, SE Islands.
RANGE: Misima and Tagula I (LeCroy & Peckover 1998).
DIAGNOSIS: Similar to *obscurior*, but underparts paler (Rand & Gilliard 1967).

NOTES

Other names—Pied or White-browed Triller.

Edolisoma Jacquinot and Pucheran, 1853. *Voy. Pôle Sud, Zool.* 3: 69. Type, by original designation, *Edolisoma marescoti* = *Lanius melas* Lesson.

    The species of this genus range from the Philippines and eastern Indonesia east to Micronesia and south to New Guinea, Melanesia, and Australia (Jønsson *et al.* 2010c, Dickinson & Christidis 2014). This generic name, long relegated to synonymy, has been resurrected to encompass a distinct generic lineage identified by the molecular systematic analysis of Jønsson *et al.* (2010c), who demonstrated that the well-known genus *Coracina* was, in fact, not monophyletic. All of the species treated below were formerly encompassed by *Coracina*. This resurrected genus (*e.g.*, Mayr 1941) is initially distinguished by its molecular affinities (Fuchs *et al.* 2007, Jønsson *et al.* 2010c). By superficial morphology, the genus is distinguished by the smallish and rather trim body form, strong sexual dimorphism, and distinctive duetting vocalizations. We call the members of the genus *Edolisoma* "cicadabirds," to distinguish them from the cuckoo-shrikes. *Edolisoma* is neuter.

Black-bellied Cicadabird *Edolisoma montanum*   ENDEMIC

Campephaga montona A. B. Meyer, 1874. *Sitzungsber. Akad. Wiss. Wien* 69: 386.—Arfak Mts, Bird's Head.
    Widespread through the mts of NG. Inhabits canopy of forest and edge; 1,000–2,450 m, extremes 770–2,800 m[CO]. No records from the Wandammen Mts.

SUBSPECIES

a. *Edolisoma montanum montanum* (A. B. Meyer)
    Synonym: *Edoliisoma montana minus* Rothschild & Hartert, 1907. *Novit. Zool.* 14: 464.—Bihagi, head of the
        Mambare R, SE Peninsula.
RANGE: The Bird's Head, Central Ra, Cyclops Mts, Adelbert Mts[J], and mts of the Huon and SE Peninsulas (Mayr 1941, Pratt 1983[J]).
DIAGNOSIS: Black of chin and throat reduced; size small (Diamond 1969). Wing (male) 124–142. There is an east-west size cline in the Central Ra that renders *minus* obsolete.

b. *Edolisoma montanum bicinia* (Diamond)
Coracina montana bicinia Diamond, 1969. *Amer. Mus. Novit.* 2362: 16.—Mount Nibo, Torricelli Mts, n NG.
RANGE: The Cyclops, Torricelli, and Bewani Mts (Diamond 1969). The population from the Foja Mts has not been diagnosed but probably belongs here.
DIAGNOSIS: Black of chin and throat of female extensive; large; wing (male) 140–152 (Diamond 1969).

NOTES
Sightings from the Fakfak[J], Kumawa[K], and Foja Mts[L]; these populations undiagnosed (Gibbs 1994[J], Beehler *et al.* 2012[L], J. M. Diamond *in litt.*[K]). *Other names*—Black-bellied Greybird, *Coracina montana*.

Papuan Cicadabird *Edolisoma incertum*   ENDEMIC | MONOTYPIC

Campephaga incerta A. B. Meyer, 1874. *Sitzungsber. Akad. Wiss. Wien* 69: 387.—Ansus, Yapen I, Bay Is.
    Throughout hills and lower mts of NG[J]; also ?Misool[K], Batanta[J], Waigeo[MA], Mios Num[MA], and Yapen I[MA], from the foothills to 1,800 m (Diamond 1985[J], M. Tarburton's island website[K]). The species' range includes the Arfak[MA], Wandammen[J], Fakfak[J], Kumawa[J], Weyland[MA], Foja[J], Van Rees[J], Cyclops Mts[MA], PNG North Coastal Ra[J]; Sepik[MA], Adelbert[G], Saruwaged[MA], and Herzog Mts[MA]; and mts of the SE Peninsula[MA] (Gilliard & LeCroy 1967[G], Diamond 1985[J]). A species of upland forest and edge. Mees (1965) and Taylor (2005) noted that old records for Batanta and Misool require corroboration—which Diamond (1985) provided for Batanta, but not for Misool. The Misool specimen record from Ripley (1959) was subsequently reidentified as *E. schisticeps*. Recent unconfirmed sighting from Misool (M. Tarburton's island website).

NOTES
Treated as distinct from *morio* of Sulawesi I by White & Bruce (1986) and Taylor (2005). Male exhibits black lores, cheek, ear, and throat; gray-blue of breast and belly saturated; background color of male with a strong blue tone, distinct from that of female, which is more gray. Female gray-throated and lacking black

on face. Primaries of both sexes blackish in closed wing, with an indication of gray edging to some interior primaries of the female. The female is much like the males of *E. schisticeps* and *E. tenuirostre*. The latter is larger, with a large pale gray patch on the secondary coverts. *E. schisticeps* male exhibits a broader black tip to central rectrices and much gray in the secondaries, forming a patch that combines with the secondary coverts. Female of nominate *morio* from Sulawesi barred brown and black ventrally; male of *morio* much like *incertum*. The form *sharpei* Rothschild & Hartert was synonymized by Mayr (1941). Mack & Wright (1996) reported that this species is absent from Crater Mt, replaced by *E. schisticeps*. *Other names*— Sharpe's Cicadabird, Black-shouldered Cuckooshrike or Cicadabird, Müller's Greybird, *Coracina morio*.

Common Cicadabird *Edolisoma tenuirostre*   RESIDENT AND VISITOR

Gracaulus tenuirostris Jardine, 1831. *Edinburgh Journ. Nat. Geog. Sci.* (ns)3: 211.—New South Wales, Australia.

An Australian migrant to most of the Region. Mainland breeding population (*aruense*) confined to the Trans-Fly and sw NG, apparently. There are insular breeding populations on the Aru Is and Waigeo, Biak, Numfor, Misima, Tagula, and Rossel I (and perhaps also on Fergusson and Woodlark I); sea level–1,460ᴶ m (Beehler field notes YUS transectᴶ). Inhabits edge, gardens, and open country with trees. On most if not all islands, due to lack of competition from congeners, the species commonly also occurs within forest (Coates 1990).

Extralimital—Sulawesi, Timor, Moluccas, n and e Australia, Bismarck Arch, Solomon Is, and Micronesia (Mayr & Diamond 2001, Dutson 2011). A passage migrant through the Torres Strait Is (Draffan *et al.* 1983).

SUBSPECIES

a. *Edolisoma tenuirostre nehrkorni* Salvadori
Edoliisoma nehrkorni Salvadori, 1890. *Orn. Pap. Moluc.*, Agg.: 91.—Waigeo I, NW Islands.
RANGE: Waigeo I. Known only from the type specimen.
DIAGNOSIS: Male has black ear coverts, chin, and throat (Taylor 2005).

b. *Edolisoma tenuirostre numforanum* (Peters and Mayr)
Coracina tenuirostris numforana Peters and Mayr, 1960. *Check-list Birds World* 9: 187.—Numfor I, Bay Is. Replacement name for *Edoliosoma neglectum* Salvadori,1879. *Ann. Mus. Civ. Genova* (1880) 15: 36 (not *Volcivora neglecta* Hume, 1877).
RANGE: Numfor I.
DIAGNOSIS: Female with ventral barring reduced to a few dark scallops across breast, mainly on sides. Immature male with unmarked pale cinnamon-buff underparts. Male with inner edges of secondaries without white edging.

c. *Edolisoma tenuirostre meyerii* Salvadori
Edoliosoma meyerii Salvadori, 1878. *Ann. Mus. Civ. Genova* 12: 327.—Biak I, Bay Is.
RANGE: Biak I.
DIAGNOSIS: Male is very dark blue-gray dorsally; black from chin to breast (as for *E. incertum* male); female underparts deep buff to rust-colored, with a small amount of dark spotting; crown and nape blue-gray grading to brown on the back; cheek dark gray with pale flecking.

d. *Edolisoma tenuirostre aruense* Sharpe
Edoliisoma aruense Sharpe, 1878. *Mitt. Zool. Mus. Dresden* 1(3): 369.—Lutor, Aru Is.
RANGE: The Aru Is, the Trans-Fly, and the w and c sectors of the S Lowlands, from the Mimika R east to Daru (Mees 1982b, Bishop 2005a).
DIAGNOSIS: Plumage not distinct from nominate form, but size very small (Taylor 2005). Wing (male) 114–124. Inner edges of some secondaries with prominent white edging. See photo in Coates & Peckover (2001: 120).

e. *Edolisoma tenuirostre tenuirostre* (Jardine)
 Synonym: *Edoliisoma müllerii* Salvadori 1876. *Ann. Mus. Civ. Genova* 7: 927.—Utanata R, sw NG. Replacement name for *Ceblepyris plumbea* S. Müller, 1843, *Verh. Nat. Gesch. Ned., Land-en Volkenk.* 1: 189; preoccupied by *Ceblepyris plumbea* Wagler, 1827.

RANGE: Breeds in e Australia; winters throughout NG region. Insular records from Kofiau, Misool, and Salawati I, Karkar I[J], D'Entrecasteaux Arch, and Woodlark I (Mayr 1941, Diamond & LeCroy 1979[J]). Perhaps shares habitat with local insular breeding populations—in other words, there may be both breeding populations as well as wintering visitors of *E. t. tenuirostre* in the SE Islands). See Notes section below for further distributional discussion.

DIAGNOSIS: Throat dark blue-gray; black facial mask of male limited to lores and eye stripe, not extending onto throat; female dull buff gray above, and finely barred below with buff on off-white background. Wing (male) 130–136 (n=5).

f. *Edolisoma tenuirostre tagulanum* Hartert
Edoliosoma amboinense tagulanum Hartert, 1898. *Novit. Zool.* 5: 524.—Tagula I, Louisiade Arch, SE Islands.
RANGE: Tagula and Misima Is (Mayr 1941).
DIAGNOSIS: Like *aruensis* but large-billed (Rand & Gilliard 1967). Female with dense fine barring on breast. Wing (male) 123–129 (n=5).

g. *Edolisoma tenuirostre rostratum* Hartert
Edoliosoma rostratum Hartert 1898, *Bull. Brit. Orn. Club* 8: 20.—Rossel I, Louisiade Arch, SE Islands.
RANGE: Rossel I.
DIAGNOSIS: Bill large and heavily hooked (Rand & Gilliard 1967). Black mask of male extends to throat and extensively onto ear (Taylor 2005). Female with sides of neck and breast variably barred with black, this barring nearly obsolete in some specimens (Rand & Gilliard 1967).

NOTES
The relative distributions of breeding and wintering birds is problematic and requires additional study, ideally using molecular techniques. Possibly there are both migrants and local residents of this species sharing islands (*e.g.*, Fergusson and Woodlark I; see Rothschild & Hartert 1896). As highlighted by Schodde & Mason (1999), the complexity of variation combined with the uncertain distribution of local resident populations makes clarification of variation in this species problematic, indeed. *Other names*—Long-billed Greybird, Cicadabird, *Coracina tenuirostris*.

Grey-headed Cicadabird *Edolisoma schisticeps* ENDEMIC
Campephaga schisticeps G. R. Gray, 1846. *List Gen. Birds* 1: 283, based on *Voy. Pôle Sud*: pl. 10, fig. 1.— Triton Bay, Bird's Neck.
    Throughout w and c NG; also the NW Islands, Yapen I, and D'Entrecasteaux Arch; 150–1,450 m[CO], down to sea level in hilly country. Absent from the n watershed of NG from the Sepik R east to Milne Bay and absent from the s watershed between Port Moresby and Milne Bay.

SUBSPECIES
a. *Edolisoma schisticeps schisticeps* (G. R. Gray)
RANGE: The Bird's Head and Neck (Fakfak and Kumawa Mts)[J], Salawati[J] and Misool I[MA] (Diamond 1985[J]).
DIAGNOSIS: Female with crown gray, ear coverts and throat rufous, chin pale rusty, breast and belly rich chestnut. Male with blue-gray throat, black lores, and blackish-gray ear patch.

b. *Edolisoma schisticeps poliopsa* Sharpe
Edoliisoma poliopsa Sharpe, 1882. *Journ. Linn. Soc. London, Zool.* 16: 318.—Moroka distr, s watershed of SE Peninsula.
RANGE: Southern NG, from the Kapare R east at least as far as Port Moresby (Mayr 1941, Diamond 1972). Not in the Trans-Fly (Mees 1982b: 124, Bishop 2005b).
DIAGNOSIS: Female with slate-gray (not rusty) crown, nape, and ear coverts; chin flecked with gray.

c. *Edolisoma schisticeps reichenowi* Neumann
Edoliisoma schisticeps reichenowi Neumann, 1917. *Orn. Monatsber.* 25: 153.—Etappenberg, Sepik R, north-central NG.

Synonym: *Edoliisoma schisticeps moszkowskii* Neumann, 1917. *Orn. Monatsber.* 25: 154.—Pamwi, lower Mamberamo R, NW Lowlands.

RANGE: The NW Lowlands and Sepik-Ramu. Localities—Pamwi in the lower Mamberamo[MA], Bodim[M], Bernhard Camp[L], Jayapura[L], Baliem valley[M], Telefomin[K], Ettapenberg[MA], upper Sepik[J] (Rand 1942b[L], Gilliard & LeCroy 1961[K], Ripley 1964a[M], I. Woxvold *in litt.*[J]).

DIAGNOSIS: Female with crown and back rufous, lores and chin whitish; male with chin blackish (Rand & Gilliard 1967, Taylor 2005).

d. *Edolisoma schisticeps vittatum* **Rothschild & Hartert**
Edolisoma schisticeps vittatum Rothschild & Hartert, 1914. *Novit. Zool.* 21: 5.—Goodenough I, D'Entrecasteaux Arch, SE Islands.
RANGE: The three main D'Entrecasteux Is[MA,J] (Normanby record from Pratt field notes[J]).
DIAGNOSIS: Female's rich red-brown underparts barred finely with black; crown, chin, and ear coverts gray (Rand & Gilliard 1967).

NOTES
The record of *E. schisticeps poliopsa* from Sibil in the Border Ra (Mees 1964b: 16) is a misidentification of *E. incertum* (Mees 1982b: 124). *Other names*—Gray's Cuckooshrike or Greybird.

Black Cicadabird *Edolisoma melas*   ENDEMIC
Lanius melas Lesson, 1828. *Man. Orn.* 1: 128.—Manokwari, Bird's Head.
    Throughout NG, the NW Islands, Aru Is, Yapen and Daru I; sea level–750 m, max 1,250 m[CO]. Inhabits forest and edge.

SUBSPECIES
a. *Edolisoma melas waigeuense* **Stresemann & Paludan**
Edolisoma melan waigeuense Stresemann & Paludan, 1932. *Orn. Monatsber.* 40: 17.—Waigeo I, NW Islands.
RANGE: Waigeo I.
DIAGNOSIS: Female underparts deep tawny, with some obsolete dark barring on flanks and indication of obsolete barring on breast; black lores; dorsal plumage darker than ventral plumage.

b. *Edolisoma melas batantae* **Gyldenstolpe and Mayr**
Edolisoma melas batantae Gyldenstolpe and Mayr, 1955. *Arkiv f. Zool.* (2)8(2): 389.—Wailibit, Batanta I, NW Islands.
RANGE: Batanta I.
DIAGNOSIS: Female as for the females of *tommasonis* of Yapen I and *waigeuense* of Waigeo I, but lacking the obsolete barring of these two; wing and tail more blackish than any other form (Gyldenstolpe 1955b: 389).

c. *Edolisoma melas melas* **(Lesson)**
Synonym: *Edolisoma melan goodsoni* Mathews, 1928. *Novit. Zool.* 34: 373.—Trangan I, Aru Is.
Synonym: *Edoliisoma melas meeki* Rothschild & Hartert, 1903. *Novit. Zool.* 10: 207.—Milne Bay, SE Peninsula.
RANGE: Salawati I, the Aru Is, and all mainland NG, including the Trans-Fly[J] (Mayr 1941, Bishop 2005b[J]).
DIAGNOSIS: Male with blue-green gloss; female rich reddish brown without barring; paler ventrally, darker dorsally. The race *goodsoni* is a synonym of the nominate form. Diamond (1972: 198) noted that the defining characters of *meeki* and the nominate form exhibit a checkerboard pattern along the s watershed of NG, demonstrating that these two forms exhibit a pattern of variation that cannot be properly captured by the geographic subspecies concept. We thus combine them.

d. *Edolisoma melas tommasonis* **Rothschild & Hartert**
Edoliisoma melas tommasonis Rothschild & Hartert, 1903. *Novit. Zool.* 10: 206.—Ansus, Yapen I, Bay Is.
RANGE: Yapen I.
DIAGNOSIS: Female plumage bright red-brown; crown as for mantle; breast color quite saturated (Rand & Gilliard 1967). Ventral plumage with obsolete barring. Marginally brighter and richer than the nominate form. A thinly defined race.

David & Gosselin (2002a) showed the name change to *melaena* made by Peters (1960) was in error. The species epithet *melas* is invariant. C. A. Green (*in litt.*) in 1989 observed what he identified as a male-female pair of this species on Biak I and provided a detailed and credible description. Should be looked for there by future fieldworkers. *Other names*—Black Greybird, New Guinea Cuckoo-shrike, *Coracina melaena*.

Family Neosittidae

Sittellas [Family 3 spp/Region 2 spp]

The three species of the family are restricted to Australia and New Guinea. Sittellas are small nuthatch-like birds that live in tight flocks and forage for insects over branches and limbs of forest trees, continually uttering high twittering notes. In New Guinea they inhabit montane forest. A foraging flock moves rapidly through tree crowns, searching all sides of branches; this is unique among New Guinea's birds (Coates 1990). One species is highly polymorphic (with differing patterns of black, gray, brown, white); the other is blackish with unique pink feathering on face and tail tip. The bill is dagger-shaped and slightly upturned; legs are very short; wing has 10 primaries and 10 secondaries, the 10th secondary vestigial (Schodde & Mason 1999). Use of the family name Neosittidae follows Bock (1994: 210–11). Jønsson *et al.* (2011b) placed this lineage sister to the core Corvoidea, without a close relative. Aggerbeck *et al.* (2014) placed this lineage as sister to *Eulacestoma*.

GENUS *DAPHOENOSITTA* [Genus 3 spp/Region 2 spp]

Daphoenositta De Vis, 1897. *Ibis*: 380. Type, by monotypy, *Daphoenositta miranda* De Vis.

The species of this genus inhabit upland New Guinea and Australia. They are nuthatch-like, short-legged, with plumage varied; the feet and eye-ring are waxy yellow (Schodde & Mason 1999).

Papuan Sittella *Daphoenositta papuensis* ENDEMIC
Sitta papuensis Schlegel, 1871. *Ned. Tijdschr. Dierk.* 4(1873): 47.—Arfak Mts, Bird's Head.

Inhabits montane forest canopy of the Bird's Head, Central Ra, and Huon Penin[J]; 1,400–2,200 m, extremes 1,075–2,650 m[CO] (Freeman *et al.* 2013[J]).

SUBSPECIES

a. *Daphoenositta papuensis papuensis* (Schlegel)
RANGE: Arfak Mts. Sight record from the Fakfak Mts (Rheindt 2012) may refer to this form.
DIAGNOSIS: Male exhibits black atop head and on neck, and a darked-streaked pale throat; female is white-hooded (Rand & Gilliard 1967, Noske 2007).

b. *Daphoenositta papuensis alba* (Rand)
Neositta papuensis alba Rand, 1940. *Amer. Mus. Novit.* 1072: 10.—15 km sw of Bernhard Camp, 1,800 m, ne sector of the Western Ra.
    Synonym: *Neositta papuensis intermedia* Junge, 1952. *Zool. Meded. Leiden* 31: 249.—Bobairo, Paniai Lake, Western Ra.
RANGE: The n slopes of the Western and Border Ra. Perhaps should be expected east to the Bismarck Ra.
DIAGNOSIS: Male head and neck all white (Rand & Gilliard 1967).

c. *Daphoenositta papuensis toxopeusi* (Rand)
Neositta papuensis toxopeusi Rand, 1940. *Amer. Mus. Novit.* 1072: 11.—Bele R, 2,200 m, 18 km n of Lake Habbema, Western Ra.
    Synonym: *Neositta papuensis wahgiensis* Gyldenstolpe, 1955. *Arkiv f. Zool.* (2)8(1): 153.—Above Nondugl, Wahgi valley, Eastern Ra.

RANGE: The s watershed of the Western, Border, and Eastern Ra.

DIAGNOSIS: Male with head and neck streaked brown, black, and white all over. A male from Mt Hagen (AMNH specimen no. 705735) not distinct from topotypical specimen from Bele R (no. 342336).

d. *Daphoenositta papuensis albifrons* (Ramsay)

Sittella albifrons Ramsay 1883. *Proc. Linn. Soc. NSW* (1884) 8: 24.—Astrolabe Ra, s watershed of the SE Peninsula.

RANGE: Mountains of the SE Peninsula. The population from the Huon Penin (Freeman *et al.* 2013) may ally with this form.

DIAGNOSIS: Similar to *toxopeusi*, but male with a more dark-grizzled crown.

NOTES

We follow Mayr (1941), Greenway (1967), and Dickinson & Christidis (2014) in treating the NG populations as a species distinct from those of Australia. R. Schodde (*in litt.*) noted the lack of a wing flash in the forms from NG as well as the montane habitat preference. Treating the NG populations as distinct brings us back to the traditional treatment (*e.g.*, Mayr 1941, Rand & Gilliard 1967; *pace* Schodde & Mason 1999 and Christidis & Boles 2008). This also is in alignment with the recognition of the NG populations of *Cormobates* (Climacteridae) as distinct. We consolidate subspecies because of the many plumages, evidence of polymorphism, and inadequate museum collections. The extent of geographic variation may be clouded by sex- and age-related plumage variation plus plumage polymorphism. The sight record for the Adelbert Mts (Mackay 1991) requires confirmation, as it was not recorded there by Pratt (1983). The NG populations of the species require much additional work. The complexity of this lineage in NG is probably considerable, given the situation outlined for Australia by Schodde & Mason (1999). Published molecular analyses are not available for the NG region for comparing NG *vs.* Australian birds. *Other names*—Varied Sittella, *Neositta papuensis*, *Daphoenositta* or *Neositta chrysoptera*.

Black Sittella *Daphoenositta miranda*   ENDEMIC | MONOTYPIC

Daphoenositta miranda De Vis, 1897. *Ibis*: 380.—Mount Scratchley, SE Peninsula.

> Synonym: *Daphaenositta miranda frontalis* van Oort, 1910. *Notes Leyden Mus.* 32: 214.—Hellwig Mts, 2,600 m, e sector of the Western Ra.

> Synonym: *Daphoenositta miranda kuboriensis* Mayr & Gilliard, 1952. *Amer. Mus. Novit.* 1577: 5.—Mount O-Mar, Kubor Mts, Eastern Ra.

Inhabits the canopy branches of upper montane forest in the Central Ra; 2,000–3,700 m, mainly above 2,450 m[CO].

NOTES

We treat the species as monotypic; the described populations were based on extent of reddish pink on chin, which ranges from extensive in the west to less extensive in the east and intermediate in c NG. No indications of breaks in this cline. The clinal variation appears to relate to a single plumage character. *Other names*—Pink-faced Nuthatch.

Family Oreoicidae

Australian Bellbirds

The three species of the family are restricted to Australia and New Guinea. The family was formally erected by Schodde & Christidis (2014), based on the molecular systematic work of various research laboratories (Jønsson & Fjeldså 2006, Driskell *et al.* 2007, Jønsson *et al.* 2007, 2010b, 2011b, and Norman *et al.* 2009b). These ground-foraging songbirds are otherwise diverse in morphology and plumage. Schodde & Christidis (2014) noted all exhibit a crest, but that character does not legitimately extend to *Aleadryas*. All three are capable vocalists, although of the three, *Aleadryas* is the least accomplished. Schodde & Christidis (2014) noted that *Aleadryas* and *Oreoica* exhibit terete orbital processes directed ventrally over the temporal fossae. *Ornorectes* should be checked to see whether it exhibits this presumably derived osteological character. For a detailed diagnosis of the lineage, see Schodde & Christidis (2014: 508). Jønsson *et al.* (2011b) showed this lineage to be basal and distantly related to the core Corvoidea. Aggerbeck *et al.* (2014) indicated relationships to other obscure lineages such as *Falcunculus* (Falcunculidae) and *Cinclosoma* (Cinclosomatidae).

GENUS *ALEADRYAS* ENDEMIC | MONOTYPIC

Aleadryas Iredale, 1956. *Birds New Guinea* 2: 106. Type by monotypy, *Pachycephala rufinucha* P. L. Sclater.

This monotypic genus is restricted to the uplands of New Guinea (Jønsson & Fjeldså 2006; Driskell *et al.* 2007, Jønsson *et al.* 2007, 2010b, 2011b; Dumbacher *et al.* 2008; Norman *et al.* 2009b; Schodde & Christidis 2014). The genus is characterized by creeper-like foraging behavior; head plumage of rufous, gray, and yellow; and the pale iris. Norman *et al.* (2009) found this genus to be sister to *Oreoica* (they did not sample *Ornorectes*), whereas Jønsson *et al.* (2011b) showed it as sister to *Ornorectes cristatus*.

Rufous-naped Bellbird *Aleadryas rufinucha* ENDEMIC

Pachycephala rufinucha P. L. Sclater, 1874. *Proc. Zool. Soc. London* (1873): 692.—Hatam, Arfak Mts, Bird's Head.

Inhabits forest and regrowth through the mts of NG; 1,750–2,600 m, extremes 1,200–3,600 m[CO]. No records from the Cyclops or Kumawa Mts.

SUBSPECIES

a. *Aleadryas rufinucha rufinucha* (P. L. Sclater)
RANGE: Mountains of the Bird's Head (Mayr 1941).
DIAGNOSIS: Forehead all slate gray, with pale gray lores; red-brown hind-crown patch narrow; ventral flanks olive green.

b. *Aleadryas rufinucha niveifrons* (Hartert)
Pachycephala rufinucha niveifrons Hartert, 1930. *Novit. Zool.* 36: 57.—Mount Wondiwoi, Wandammen Penin, Bird's Neck.
RANGE: The Bird's Neck (Wandammen Mts), Western, Border, and Eastern Ra, east to Okapa. Apparently populations in the PNG North Coastal Ra and Adelbert Mts are of this form (see Pratt 1983).
DIAGNOSIS: Distinctive white forehead spot; broader red-brown patch on hind-crown than nominate; upperparts olive green. See photos of this form in Coates (1990: 218–19); note in image on p. 218 that one member of the pair has the large white forehead spot, whereas the other, sitting upon the nest, has a much -reduced forehead patch, mainly gray, demonstrating local plumage variation (or sexual dichromatism) that could confound subspecies diagnosis.

c. *Aleadryas rufinucha gamblei* (Rothschild)

Pachycephala gamblei Rothschild, 1897. *Bull. Brit. Orn. Club* 7: 22.—Mount Cameron, Owen Stanley Ra, SE Peninsula.

> Synonym: *Pachycephala rufinucha lochmia* Mayr, 1931. *Mitt. Zool. Mus. Berlin*, 17: 674.—Ogeramnang, Saruwaged Ra, Huon Penin.
> Synonym: *Pachycephala rufinucha prasinonota* Mayr, 1931. *Mitt. Zool. Mus. Berlin*, 17: 674.—Dawong, Herzog Mts, SE Peninsula.

RANGE: Mountains of the Huon and SE Peninsulas.

DIAGNOSIS: Similar to *niveifrons*, but chestnut patch on hind-crown does not extend onto nape. See photo in Coates & Peckover (2001: 164). The hind-crown patch is narrowest in the nominate form, medium in *gamblei*, and most extensive in *niveifrons*. Three thinly delineated subspecies.

NOTES

We follow Schodde & Christidis (2014) in placing this aberrant form into the Oreoicidae. We follow Dickinson (2003) in subsuming *prasinonota* into *gamblei*. Other combinations are a product of our museum analysis. The undiagnosed population in the Fakfak Mts is known only from sightings (Gibbs 1994). The Foja Mts population also remains undiagnosed. *Other names*—Rufous-naped Whistler, *Pachycephala rufinucha*.

GENUS *ORNORECTES*   ENDEMIC | MONOTYPIC

Ornorectes Iredale, 1956. *Birds New Guinea* 2: 93. Type, by original designation, *Rectes cristata* Salvadori.

Formerly situated within *Pitohui* (Rand & Gilliard 1967), this monotypic and endemic genus is characterized by the prominent erectile crest and the long and unusual tolling-bell song. For resurrection of the genus, see Dumbacher (2014). The phylogenies of both Dumbacher *et al.* (2008) and Jønsson *et al.* (2011b) placed this genus as sister to *Aleadryas* and the *Ornorectes+Aleadryas* cluster as sister to *Oreoica*.

Piping Bellbird *Ornorectes cristatus*   ENDEMIC | MONOTYPIC

Rectes cristata Salvadori, 1876. *Ann. Mus. Civ. Genova* 7: 930.—Mount Morait, n coast of the Bird's Head.

> Synonym: *Pitohui cristatus arthuri* Hartert, 1930. *Novit. Zool.* 36: 61.—Cyclops Mts, ne sector of the NW Lowlands.
> Synonym: *Pitohui cristatus kodonophonos* Mayr, 1931. *Mitt. Zool. Mus. Berlin* 17: 676.—Aroa R, s watershed of the SE Peninsula.

Sparsely distributed through the hill forests of NG; 400–1,300 m[CO], near sea level in the foothills and at the mouth of the Fly R. Apparently absent from the n watershed from the mouth of the Sepik R southeastward to Milne Bay, although heard at Isurava, inland from Kokoda (Pratt field notes 2014). Also apparently absent from the SE Peninsula east of Port Moresby.

Localities—Mount Morait on the Bird's Head[MA], Fakfak Mts[U], Kumawa Mts[Z], lower Menoo R in Weyland Mts[W], Danau Bira[A], Bernhard Camp on Taritatu R[Q], Bian Lakes[K], Cyclops Mts[MA], Oriomo R[MA], Nena–upper May R divide[L], Ok Ma[Y], upper Fly R karst[L], Mount Bosavi foothills[L], lower Kikori R[J], Lake Kutubu[CO], Sepik Mts[CO], Karimui[DI], Aroa R[MA], and Varirata NP[CO] (Hartert *et al.* 1936[W], Rand 1942b[Q], Gibbs 1994[U], Beehler field notes 12-Jul-1987[A], Bishop 2005a[J], P. Gregory *in litt.*[Y], K. D. Bishop *in litt.*[K], I. Woxvold *in litt.*[L], C. Thébaud & B. Mila *in litt.*[Z]). See photos of this species in Coates (1990: 233).

NOTES

We treat the species as monotypic. The age- and sex-related plumage variation swamps any proper geographic variation. Should be looked for in the Adelbert Mts. *Other names*—Crested Pitohui, *Pitohui cristatus*.

Family Eulacestomatidae

Ploughbills

ENDEMIC | MONOTYPIC

The single species of the family is endemic to the montane forests of the Central Ranges of New Guinea. Family status was formally established by Schodde & Christidis (2014), based on the molecular systematic work of Jønsson *et al.* (2007, 2011b) and Aggerbeck *et al.* (2014), among others. The phylogeny of Norman *et al.* (2009b) situated the genus *Eulacestoma* as sister to a cluster that included *Colluricincla*, *Pseudorectes*, *Melanorectes*, and *Pachycephala* (all of Pachycephalidae). By contrast, Jønsson *et al.* (2011b) showed this genus to be sister to the core Corvoidea, along with *Daphoenositta* (Neosittidae), with affinities to the Cinclosomatidae, Paramythiidae, and Psophodidae. The single species is sexually dichromatic and the male exhibits a prominent pink gape wattle absent in the female and the immature bird. The bill is laterally compressed, very strong, and hooked. The species favors bamboo thickets in wet montane forests, where it apparently employs its powerful bill to extract arthropods from crevices and stems. For details of taxonomic diagnosis, see Schodde & Christidis (2014: 509).

GENUS *EULACESTOMA*

ENDEMIC | MONOTYPIC

Eulacestoma De Vis, 1894. *Ann. Rept. Brit. New Guinea*, 1893–94: 102. Type, by monotypy, *Eulacestoma nigropectus* De Vis.

Range and characters are as for the family. The single species in the genus is endemic to New Guinea's Central Ranges.

Wattled Ploughbill *Eulacestoma nigropectus*

ENDEMIC | MONOTYPIC

Eulacestoma nigropectus De Vis, 1894. *Ann. Rept. Brit. New Guinea*, 1893–94: 102.—Mount Maneao, SE Peninsula.

> Synonym: *Eulacestoma nigropectus clara* Stresemann & Paludan, 1934. *Orn. Monatsber.* 42: 44.—Kunupi, 2,000 m, Weyland Mts, Western Ra.

Inhabits forest and adjoining ecologically disturbed areas with dense regrowth, especially scrambling bamboo, ranging throughout the Central Ra from 2,100 to 2,500 m, extremes 1,250–2,850 m (Coates & Peckover 2001).

Localities—Kunupi in the Weyland Mts[L], Bele R Camp[Q], Mt Goliath[MA], Kaijende Highlands[Z], Hides Ra[K]; Maruba R[K], Tagari headwaters[K], Ambua[M], Tari Gap[J], Mt Giluwe[CO], Tomba[J], Kumul Lodge area[M], Mt Hagen[W], Kubor Mts[W], Mt Karimui[DI], Mt Kaindi[Z], Bulldog Road[Z], and Mt Maneao[MA] (Hartert *et al.* 1936[L], Mayr & Gilliard 1954[W], Rand 1942b[Q], I. Woxvold *in litt.*[K], K. D. Bishop *in litt.*[J], BPBM specimens[Z], P. Gregory *in litt.*[M]).

NOTES

We follow Mayr & Gilliard (1954) and Diamond (1972) in treating the form as monotypic. *Other names*—Wattled Shrike-tit, Ploughshare Tit.

Family Pachycephalidae

Whistlers and Allies [Family 56 spp/Region 20 spp]

The members of the family range from tropical Asia to Micronesia, Polynesia, and Australia. This family includes the true whistlers (*Pachycephala* and *Melanorectes*) and shrikethrushes (*Colluricincla*, *Pseudorectes*). Recent molecular phylogenies have justified the removal of four genera long placed in the Pachycephalidae (Dumbacher *et al.* 2008, Jønsson *et al.* 2010a, 2010b, 2011b; Dumbacher 2014). Formerly included were: *Eulacestoma*, now in its own family, Eulacestomatidae; *Aleadryas* and *Ornorectes*, now in Oreoicidae; and *Pitohui*, a genus recently shown to be a member of the oriole family, Oriolidae (Jønsson *et al.* (2010a). In general, whistlers and shrikethrushes are compact and robust, with a powerful, hooked, bill. They are noted songsters and stolid foragers.

GENUS *COLLURICINCLA* [Genus 5 spp/Region 3 spp]

Colluricincla Vigors & Horsfield, 1827. *Trans. Linn. Soc. London* 15: 213. Type, by monotypy, *Colluricincla cinerea* = *Turdus harmonicus* Latham.

The species of this genus range from New Guinea to Australia (Dickinson & Christidis (2014). The genus is characterized by the plain (or dull), sexually monomorphic plumage, obscure ventral streaking, shrike-like bill, medium-long tail, and distinctive musical song. Deiner *et al.* (2011) provided a phylogeography of *Colluricincla megarhyncha*.

Little Shrikethrush *Colluricincla megarhyncha* RESIDENT

Muscicapa megarhyncha Quoy & Gaimard, 1830. *Voy. Astrolabe, Zool.* 1: 172, pl. 3, fig. 1.—Manokwari, Bird's Head.

Inhabits the interior of forest and regrowth throughout the Mainland and most larger satellite islands; sea level–1,400 m, max 2,300 m[CO]. NG range includes the NW Islands, Aru Is, Bay Is, and SE Islands.

Extralimital—islands in s Torres Strait and n and e Australia.

SUBSPECIES

Despite its plain coloration, the Little Shrikethrush exhibits considerable geographic variation that traditionally has been enshrined in more than 20 subspecies in the Region (Mayr 1941). As such, the species has been a test case for researchers investigating patterns of phylogeographic differentiation in lowland birds for the NG region (Stresemann 1923, Mayr 1944, 1967; Ford 1979; Deiner *et al.* 2011), and it deserves special attention here.

The recent molecular study by Deiner *et al.* (2011) heavily sampled populations across the NG region (except for Biak). Their results greatly clarified the phylogeography of *Colluricincla megarhyncha* by demonstrating deep divergence among populations and defining regions of endemism. Unfortunately for the taxonomy of *C. megarhyncha*, some of the well-supported Mainland clades identified by genetic research of Deiner *et al.* (2011) are phenotypically indistinguishable, leaving the systematist at a loss as to how and where to define species boundaries (see Notes, below).

Also, the phylogeography of the Little Shrikethrush is made even more complicated by the fact that its range extends into Australia, where its populations are also varied, and may constitute more than one species (Ford 1979, Schodde & Mason 1999, Nyári & Joseph 2012). Unfortunately, Deiner *et al.* (2011) did not examine Australian populations beyond Cape York Penin, and thus we are left with an incomplete picture of the evolution of *C. megarhyncha* across its range.

Below we follow, with some modification, the organizing principles provided by Rand & Gilliard (1967) that most likely originated with the work of Mayr (1944, 1967). We caution the reader that this is a placeholder treatment based only on the limited range in character states of plumage and bill coloration. Furthermore, this color variation evidently provides little or no clue to underlying phylogenetic relationships between some populations (Deiner *et al.* 2011). The mtDNA phylogeny of Deiner *et al.* (2011: fig. 4) should likewise be viewed as only a first step toward a phylogeographic understanding of this complex lineage—one that does not provide a practical successor to our legacy treatment of subspecies and subspecies groups (Rand & Gilliard 1967). See the discussion of the problem in the Notes section below.

I. Rusty-breasted Forms: (1) upperparts brownish or grayish olive; (2) ventral plumage rusty, with throat patch paler, with or without streaking or mottling; (3) bill horn-colored (Rand & Gilliard 1967).

a. *Colluricincla megarhyncha megarhyncha* (Quoy & Gaimard)
Synonym: *Myiolestes aruensis* G. R. Gray, 1858. *Proc. Zool. Soc. London*: 180.—Aru Is.
Synonym: *Pinarolestes megarhynchus misoliensis* Meise, 1929. *Abh. Ber. Mus. Tierk. Dresden* (1927–29) 17(4): 18.—Misool I, NW Islands.
Synonym: *Pinarolestes megarhynchus goodsoni* Hartert, 1930. *Novit. Zool.* 36: 59.—Merauke, s NG.
Synonym: *Myiolestes megarhynchus ferrugineus* Hartert and Paludan, 1936. *Mitt. Zool. Mus. Berlin* 21: 206.— Lower Menoo R, head of Geelvink Bay, Bird's Neck.
Synonym: *Myiolestes megarhynchus wuroi* Mayr & Rand, 1936. *Mitt. Zool. Mus. Berlin* 21: 247.—Wuroi, Oriomo R, s NG.
Synonym: *Myiolestes megarhynchus palmeri* Rand, 1938. *Amer. Mus. Novit.* 991: 10.—Palmer R, 3.2 km below its junction with the Black R, upper Fly R.
RANGE: Misool and Salawati I[MA]; the Aru Is[MA]; the Bird's Head and Neck, and the S Lowlands east to the Fly R, including the Trans-Fly.
DIAGNOSIS: Generally rusty brown, darker and duller dorsally; crown, nape, and back rich medium brown; tail slightly more rusty; wings as for tail. Underparts generally warm fawn; slightly paler on chin; some populations (Salawati I birds) show obscure streaking on upper breast. The forms *misoliensis*, *ferruginea*, and *aruensis* are identical to the nominate race, based on our examination of topotypical materials at the BMNH (see Mees 1965 on *misoliensis*). Mees (1982b) synonymized *goodsoni*, *wuroi*, and *palmeri*. The Trans-Fly birds are marginally paler and more smoothly tinted below than the surrounding populations. Also the Trans-Fly birds are more washed out and paler than those from Cape York. The form *normani* from the Cape York Penin region is a well marked subspecies but lacks a molecular signature distinct from the Trans-Fly population (Deiner *et al.* 2011, fig. 4).

b. *Colluricincla megarhyncha affinis* (G. R. Gray)
Myiolestes affinis G. R. Gray, 1862. *Proc. Zool. Soc. London* (1861): 430.—Waigeo I, NW Islands.
RANGE: Waigeo I.
DIAGNOSIS: A dull olive-gray race, like *obscura*, but with obscure, fine barring on the lower underparts, unique to this subspecies; in life probably more greenish. Bill horn-colored in specimens; plumage generally dull, olive grayish brown; crown dark gray-brown; mantle and back olive brown; edges to flight feathers rusty; throat pale gray with obscure speckling; smudgy gray-brown breast band, extending onto flanks; belly with a dull whitish patch extending to the undertail. Duller and less rusty ventrally than *batantae*.

c. *Colluricincla megarhyncha batantae* (Meise)
Pinarolestes megarhynchus batantae Meise, 1929. *Abh. Ber. Mus. Tierk. Dresden* (1927–29) 17(4): 18.— Batanta I, NW Islands.
RANGE: Batanta I, ranging to the summit of Mt Besar (Greenway 1966).
DIAGNOSIS: Upperparts dark olive brown; throat buff; underparts rufous with a gray suffusion; throat and breast well streaked and mottled with gray; bill brown or dark brown. There is considerable individual variation in the amount of streaking on the throat and breast. Finally, the head plumage of the females is paler and more greenish (see also Mayr & Meyer de Schauensee 1939b and Mees 1965). Greenway (1966) provided two important learning points for museum researchers. He compared fresh material of this subspecies against birds collected 100 years before, and noted that those old birds were "no longer recognizeable" and had faded to resemble the nominate subspecies from the Bird's Head. They had become much paler and more rufous—a classic case of museum foxing. Moreover, Greenway (1966: 18) noted that this form has two color morphs, highlighting the issue of individual variation that needs to be accounted for when diagnosing races of this confusing species.

d. *Colluricincla megarhyncha tappenbecki* Reichenow

Colluricincla tappenbecki Reichenow, 1899. *J. Orn.* 47: 118.—Madang, e verge of the Sepik-Ramu.

RANGE: The e sector of the Sepik-Ramu and the c and e sectors of the Eastern Ra. Localities—Lake Kutubu, middle Sepik, Ramu R, Mt Giluwe, Astrolabe Bay, and Wahgi valley (Diamond 1972).

DIAGNOSIS: Bill horn-colored; crown dark brown; mantle slightly paler and richer; rump a rich rusty color; tail like rump; underparts entirely fawn brown, darkest in obscure breast band, grading paler toward belly; throat pale with abundant dark streaks; background color of throat paler than breast; in some birds the obscure dark streaking reaches the breast.

e. *Colluricincla megarhyncha madaraszi* (Rothschild & Hartert)

Pinarolestes megarhyncha madaraszi Rothschild & Hartert, 1903. *Novit. Zool.* 10: 100.—Sattelberg, Huon Penin.

> Synonym: *Myiolestes megarhynchus neos* Mayr, 1931. *Mitt. Zool. Mus. Berlin* 17: 716.—Malalo, near Salamaua, SE Peninsula.

RANGE: The e sector of both the S Lowlands and Eastern Ra, east to the nw sector of the SE Peninsula. Also the Huon Penin.

DIAGNOSIS: Throat gray; bill dark gray (Boles 2007). The form *neos* is not distinct.

f. *Colluricincla megarhyncha despecta* (Rothschild & Hartert)

Pinarolestes megarhyncha despectus Rothschild & Hartert, 1903. *Novit. Zool.* 10: 100.—Milne Bay, SE Peninsula.

RANGE: The s coast of the SE Peninsula from Milne Bay west at least to the Purari R.

DIAGNOSIS: Bill pale bluish slate, not blackish; lower breast lacks the fine dark streaks; otherwise nearly identical to *maeandrina*. See photos in Coates (1990: 221–23) and Coates & Peckover (2001: 165).

g. *Colluricincla megarhyncha superflua* (Rothschild & Hartert)

Pinarolestes megarhynchus superfluus Rothschild & Hartert, 1912. *Novit. Zool.* 19: 205.—Kumusi R, SE Peninsula.

RANGE: The n coast of the SE Peninsula from Collingwood Bay northwest presumably to the Herzog Mts.

DIAGNOSIS: Similar to *despecta* but for prominent broad dark diffuse streaking on breast band; throat also heavily mottled with dark buff brown.

II. Dark-billed Forms: (1) dorsal plumage grayish olive; (2) ventral plumage pale gray to gray-brown; (3) ochraceous ventral wash found in group I lacking except in *maeandrina*; (4) bill black in male (Rand & Gilliard 1967).

h. *Colluricincla megarhyncha melanorhyncha* (A. B. Meyer)

Myiolestes melanorhynchus A. B. Meyer, 1874. *Sitzungsber. Akad. Wiss. Wien* 69: 494.—Biak I, Bay Is.

RANGE: Biak I.

DIAGNOSIS: Dorsal plumage dark olive brown; wings rufous; underparts paler, more olive and less rufous; streaking on underparts reduced or absent; bill blackish or very dark brown. A distinctive subspecies because of the rufous wings (hopefully not just a juvenile trait). Because its song is reputed to be whistler-like, genetic screening may yield surprises.

i. *Colluricincla megarhyncha obscura* (A. B. Meyer)

Rectes obscura A. B. Meyer, 1874. *Sitzungsber. Akad. Wiss. Wien* 69: 390.—Ansus, Yapen I, Bay Is.

RANGE: Yapen I.

DIAGNOSIS: Upperparts brownish gray; underparts gray with slight brownish tinge; throat streaked dark gray; breast mottled dark brown (Boles 2007). Lacks rufous wash to flanks that is found in *hybridus*.

j. *Colluricincla megarhyncha hybridus* (Meise)

Pinarolestes megarhynchus hybridus Meise, 1929. *Abh. Ber. Mus. Tierk. Dresden* (1927–29) 17(4): 17.—Tana Merah, n NG.

> Synonym: *Myiolestes megarhynchus idenburgi* Rand, 1940. *Amer. Mus. Novit.* 1072: 9.—4 km sw of Bernhard Camp, 850 m, ne sector of the Western Ra.

RANGE: The NW Lowlands and n slopes of the Western Ra.

DIAGNOSIS: Head grayish olive; underparts pale, slightly buff, breast more heavily streaked (Boles 2007).

LeCroy (2010) corrected the type locality error made by Rand. For trinomial spelling and usage, see David & Gosselin (2002a: 36).

k. *Colluricincla megarhyncha maeandrina* (Stresemann)
Pinarolestes megarhynchus maeandrinus Stresemann, 1921. *Anz. Orn. Ges. Bayern* 1: 36.—Maeanderberg, upper Sepik R, west-central Sepik-Ramu.
RANGE: The PNG North Coastal Ra, n slope of e sector of Western Ra, Border Ra, and w sector of Eastern Ra. Localities—Baliem valley, Telefomin, Maeanderberg, and Yellow R.
DIAGNOSIS: Male exhibits a blackish bill (Rand & Gilliard 1967). Obscure streaking of throat abundant and making throat relatively dark; streaks trend onto breast; some obscure narrow dark streaking onto lower breast, which is dull pale buff (without the rusty coloration of *tappenbecki*). See photo in Coates & Peckover (2001: 165).

III. Olive-and-Gray Forms: (1) mantle olive to olive green; (2) crown darker and grayer; (3) underparts whitish to grayish, with breast and flanks gray-brown or gray, and abdomen or undertail tinged yellow; streaking moderately distinct on throat and sometimes elsewhere; (4) bill black in both sexes; birds of this group resemble a larger edition of *Pachycephala simplex* (Rand & Gilliard 1967).

l. *Colluricincla megarhyncha fortis* (Gadow)
Pachycephala fortis Gadow, 1883. *Cat. Birds Brit. Mus.* 8: 369.—D'Entrecasteaux Arch, SE Islands.
   Synonym: *Pachycephala fortis trobriandi* Hartert, 1896. *Novit. Zool.* 3: 236.—Kiriwina, Trobriand Is.
RANGE: The Trobriand Is and D'Entrecasteaux Arch (Mayr 1941).
DIAGNOSIS: The rusty-brown wash found in members of group I absent entirely, replaced by gray and olive green; crown dark gray; mantle and back dark olive green; upper surface of tail slightly browner; throat patch large and mottled white and gray; obscure breast band gray-buff; belly yellowish buff with more pale yellow toward midline; bill blackish; size large. The Trobriand population is not distinct.

m. *Colluricincla megarhyncha discolor* De Vis
Colluricincla discolor De Vis, 1890. *Ann. Rept. Brit. New Guinea*, 1888–89: 60.—Tagula I, Louisiade Arch, SE Islands.
RANGE: Tagula I.
DIAGNOSIS: Similar to *fortis*, but crown brownish gray, mantle olive with a more brownish wash than in *fortis*, and abdomen with a duller wash, with less yellow.

NOTES
Our treatment follows Mayr (1944, 1967), Rand & Gilliard (1967), Ford (1979), and Mees (1982b). We also note the phylogeographic analysis of Ford (1979), Deiner *et al.* (2011), and Nyári & Joseph (2013). Deiner *et al.* (2011) provided molecular support for eight molecularly divergent lineages within this assemblage in the Region. The Deiner lineages are as follows, from highly divergent to less divergent: (1) Sepik-Ramu+Huon, (2) Trans Fly+Cape York, (3) Batanta+Salawati+Bird's Head+Western Ra, (4) Waigeo I, (5) NW Lowlands I Yapen I, (6) Tagula I, (7) D'Entrecasteaux Is, and (8) S Lowlands I SE Peninsula. There is good correlation between these molecular entities and the phenotypic groupings of the insular populations (nos. 4, 6, 7) and for certain Mainland populations (2, 5), and these can all be readily identified visually. However, the entities covering most of Mainland NG—nos. 1, 3, and 8—are not really separable by plumage and bill coloration. *C. megarhyncha* seems to be another cryptically plumaged species that is colored with such a limited palette that true phylogeographic patterns are obscured or confounded by minor or nonexistent plumage variation.

   There is, indeed, considerable plumage variation within this lineage, but much of it is clinal, graded, and subtle, and adequate subspecies consolidation proves very difficult. We have synonymized to a minor degree, but mainly leave the major threads of this Gordian knot to be unraveled by more courageous future museum workers, aided by comprehensive and finely detailed molecular systematic results. We agree with Deiner *et al.* that several good species and a number of proper subspecies may lurk within the existing *Colluricincla megarhyncha* complex, but the solution of the riddle is beyond the capability of this analysis.

   With regard to the many named subspecies, the problem is exacerbated by: (1) museum foxing, (2) obscure sexual dichromatism, (3) inadequate topotypical series, (4) nongeographic variation, and (5) the over-description of thinly diagnosed forms. Moreover, this species, along with the *Meliphaga analoga* complex and several *Sericornis* forms, appears to provide a "trap" for museum workers seeking to define subspecies. These variable cryptic species exhibit no nonclinal phenotypic characters, and the minor and diffuse plumage variation for some perverse reason leads to the creation of an excess of subspecies names, each with virtually no reliable and objective distinction from its sister forms. In all of these instances, multiple species hide behind a veil of minimal plumage distinction, and yet, practitioners tend to parse out thinly defined subspecies, and fail to properly detect and diagnose the cryptic species present.

We believe that *C. megarhyncha* merits dissection into component species once clearly delineated, but we mourn the present lack of alignment between external features and the molecules. Perhaps the systematics of *C. megarhyncha* will be clarified by further genetic study, fresh examination of plumage coloration of live birds of known age and sex, and other lines of research such as behavioral and ecological inquiry at localities where the phylogenetic entities meet. Finally, this complex must be studied whole, examining the populations in Australia as well as those in the Region. Nyári & Joseph (2013) demonstrated that there is phylogenetic structuring of the populations in Australia as substantial as that Deiner *et al.* found for the NG region. *Other names*—Rufous Shrike-thrush, Brown Shrike-flycatcher.

Sooty Shrikethrush *Colluricincla tenebrosa* ENDEMIC | MONOTYPIC

Pachycephala tenebrosa Rothschild, 1911. *Bull. Brit. Orn. Club* 29: 20.—Mount Goliath, Border Ra.
> Synonym: *Pachycephala tenebrosa atra* Rothschild, 1931. *Novit. Zool.* 36: 260.—Gebroeders Mts, 1,830 m, w sector of the Western Ra.

A rarely encountered inhabitant of understory shrubbery within montane forest in the Western, the Border, and the w and north-central sectors of the Eastern Ra, 1,450–2,150 m (Pratt & Beehler 2014).

Localities—Kunupi in the Weyland Mts[K], upper Siriwo R in the Gebroeders Mts[MA], Menoo R in the Weyland Mts[K], 15–18 km southwest of Bernhard Camp[Q], Mt Goliath[MA], Ok Tedi (Mt Robinson[B], Bilbilokabip[A]), Unchemchi in the Hindenburg Mts[L], uplands between Mt Sisa and Hegigio R[J], and the w Schrader Ra[MA] (Hartert *et al.* 1936[K], Rand 1942b[Q], Gilliard & LeCroy 1961[L], Gregory 1995[B], I. Woxvold *in litt.*[J], specimen in Mus. Victoria[A]).

NOTES

Similar to a very dark *C. megarhyncha*. Voice much like that of *C. megarhyncha* (I. Woxvold *in litt.*). Our placement in *Colluricincla* follows our own museum research, LeCroy (2010), and Dickinson & Christidis (2014). The nomenclatural confusion between this form and the Morningbird of Palau (currently *Pachycephala tenebrosa* Hartlaub) has been considerable. See LeCroy (2010) and Dumbacher (2014) for details on the confusion and its subsequent clarification. The two subspecies of this NG endemic were described based on atypical (extreme) museum specimens of a species that shows considerable individual variation, perhaps associated with age (increasing melanism). Two specimens of adult males in the AMNH collection (no. 302275 from the Weyland Mts and no. 659478 from Mt Goliath) are identical. Moreover, the ranges of the named subspecies made no biogeographic sense (both occurring west and east, and north and south). We include a comparison photo of topotypical examples of the two described forms (fig. 3c). *Other names*—Sooty Whistler, *Colluricincla umbrina* (Rothschild), *Pachycephala tenebrosa* Rothschild (not *Rhectes* [*Pachycephala*] *tenebrosus* Hartlaub = the Morningbird).

Grey Shrikethrush *Colluricincla harmonica* RESIDENT

Turdus harmonicus Latham, 1801. *Index Orn.*, Suppl.: 41.—Sydney, NSW, Australia.
> Inhabits forest edge, lightly wooded habitats in towns, and scrub of the coastal Trans-Fly, the e sector of the S Lowlands, the Sepik-Ramu, the Eastern Ra, and the SE Peninsula. There is a recent sighting from Kiunga (D. Hobcroft *in litt.*). Patchily distributed in the open interior valleys of the Eastern Ra and the Sepik-Ramu; sea level–700 m, max 1,700 m[CO]. Apparently this species' range is expanding in NG. Now regularly observed at Ambua (K. D. Bishop *in litt.*).
> Extralimital—Australia.

a. *Colluricincla harmonica superciliosa* Masters

Colluricincla superciliosa Masters, 1876. *Proc. Linn. Soc. NSW* 1: 50.—Cape Grenville, n QLD, Australia.
    Synonym: *Colluricincla brunnea tachycrypta* Rothschild & Hartert, 1915. *Novit. Zool.* 22: 60.—Milne Bay, se NG.

RANGE: The SE Peninsula west along the s coast to Dolak I, and along the n coast to the Ramu R. Also interior upland valleys of the Wahgi region. Localities—Dolak I[J], Kurik[Q], Wasur NP[J], Bian Lakes[J], Bensbach[J], Tabubil[K], Lake Daviumbu[J], Mabaduan[L], Tarara[L], Wewak[CO], Ambua, Goroka[CO], Mt Hagen town[CO], Kainantu[CO], Aiyura[CO], Markham valley[CO], Bulolo valley[CO], Bereina[CO], Sogeri Plateau[CO], Port Moresby[CO], and Alotau[M] (Rand 1942a[L], Mees 1982b[Q], Gregory 1997[K], Bishop 2005a[J], Pratt field notes[M]). Also n QLD.

DIAGNOSIS: Similar to *C. h. brunnea* of Australia, but slightly grayer, with a diffuse white eyebrow extending behind the eye, and with some faint dark streaks on chin and throat (Boles 2007). We follow MacDonald (1973) and Ford & Parker (1974) in sinking *tachycrypta*. Note also the useful discussion of Mees (1982b) regarding this form and its Australian relatives. See photos in Coates (1990: 225).

NOTES
Other names—Whistling Shrike-thrush, Grey Shrike-flycatcher.

GENUS PSEUDORECTES   ENDEMIC [Genus 2 spp]

Pseudorectes Sharpe, 1877. *Cat. Birds Brit. Mus.* 3: 287. Type, by subsequent designation (Mathews, 1930. *Syst. Avium Austral.*: 641), *Rectes ferrugineus* Bonaparte.

    The species of this genus inhabit lowland forests of the Region. The work of Dumbacher *et al.* (2008), Dumbacher (2014), and Jønsson *et al.* (2007, 2010a, 2010b, 2011b), has led to the dissection of *Pitohui*. We follow Dumbacher (2014) and Dickinson & Christidis (2014) in recognizing the genus *Pseudorectes*. This genus of two endemic species is characterized by sturdy build, rufous plumage, breast paler than mantle, and a strong bill. No obvious unique plumage or morphological characters define the lineage. Shown to cluster with *Colluricincla* (Norman *et al.* 2009b), it is alternatively included in that genus (Jønsson *et al.* 2010b). We depart from the historic names Rusty and White-bellied Pitohuis, used in Pratt & Beehler (2014), and adopt the English name "shrikethrush" for these two species, following the current trend (Dickinson & Christidis 2014).

Rusty Shrikethrush *Pseudorectes ferrugineus*   ENDEMIC

Rhectes ferrugineus Bonaparte, 1850. *Compt. Rend. Acad. Sci. Paris* 31: 563.—Triton Bay, Bird's Neck.
    Throughout NG, the NW Islands, Aru Is, and Yapen I; sea level–800 m, max 1,100 m[CO]. Inhabits forest, edge, and mature regrowth. This is a common flocking species that often forms the core of NG's remarkable black-and-brown mixed-species bird flocks (Diamond 1987).

SUBSPECIES

a. *Pseudorectes ferrugineus ferrugineus* (Bonaparte)
RANGE: Misool and Salawati I[MA]; the Bird's Head and Neck; and the S Lowlands and Trans-Fly east to the Purari R.
DIAGNOSIS: Crown and mantle brown with a rufous wash; rump and tail rufous; underparts pale rusty buff, slightly paler on throat; bill black; male and female alike (Boles 2007). Wing 140 (Rand & Gilliard 1967).

b. *Pseudorectes ferrugineus leucorhynchus* (G. R. Gray)
Rectes leucorhynchus G. R. Gray, 1862. *Proc. Zool. Soc. London* (1861): 430.—Waigeo I, NW Islands.
    Synonym: *Pitohui ferrugineus fuscus* Greenway, 1966. *Amer. Mus. Novit.* 2258: 19.—Batanta I, NW Islands.
RANGE: Waigeo and Batanta I[MA].
DIAGNOSIS: Plumage darker than for nominate form; bill yellowish white; wing long, 140–154 (Rand & Gilliard 1967).

c. *Pseudorectes ferrugineus brevipennis* (Hartert)

Rhectes ferrugineus brevipennis Hartert, 1896. *Novit. Zool.* 3: 534.—Aru Is.

RANGE: The Aru Is.

DIAGNOSIS: Darker than nominate form but paler than *holerythrus* (Boles 2007). Wing short (128).

d. *Pseudorectes ferrugineus holerythrus* (Salvadori)

Rectes holerythra Salvadori, 1878. *Ann. Mus. Civ. Genova* 12: 474.—Yapen I, Bay Is.

RANGE: Yapen I[MA], the NW Lowlands, and the w and c sectors of the Sepik-Ramu, east to the Sepik R. At w verge of range, at the head of Geelvink Bay and at the base of the Weyland Mts, this race intergrades with *ferrugineus*; also grades consistently paler along the n coast of NG moving eastward, until a pure *clarus* population is reached east of the Sepik R (see Rand 1942b: 492).

DIAGNOSIS: Plumage dark (Diamond 1972). Mayr (1941) subsumed the form *heurni* into this subspecies. Rand (1942b) referred to his material from Bernhard Camp and Jayapura as *heurni*, and thus those populations belong here.

e. *Pseudorectes ferrugineus clarus* (A. B. Meyer)

Rhectes ferrugineus clarus A. B. Meyer, 1894. *J. Orn.* 42: 91.—Finschhafen, Huon Penin.

RANGE: The e sector of the Sepik-Ramu, the lowlands of the Huon Penin, the e sector of the S Lowlands, and the SE Peninsula. West in the n watershed to the Sepik R, and west in the s watershed to the Purari R (Mees 1982b).

DIAGNOSIS: Distinguished from the nominate form by richer, more reddish-brown dorsal coloration; but underparts as for nominate. See photos in Coates (1990: 230–31) and Coates & Peckover (2001: 167).

NOTES

Thinly defined subspecies, probably merit additional consolidation. *Other names*—Rusty Pitohui, *Pitohui ferrugineus*

White-bellied Shrikethrush *Pseudorectes incertus*　　　　　　　ENDEMIC | MONOTYPIC

Pitohui incertus van Oort, 1909. *Nova Guinea Zool.* 9: 94.—Lorentz R, sw NG.

Known only from high-rainfall lowland forest enclaves in the c sector of the S Lowlands: (1) between Kiunga and Palmer Junction (including Oroville Camp and Elevala R), upper Fly R (K. D. Bishop *in litt.*); (2) Omtiek Bivak I, 90 km upstream from the mouth of the Lorentz R, se Indonesian NG; sea level–100 m (Diamond & Raga 1978). Inhabits seasonally inundated forest with a broken canopy on flat interior alluvial plain. The species appears to be common in the known localities, and this habitat is thought not to be under threat.

Other names—White-bellied or Mottle-breasted Pitohui, *Pitohui incertus*.

GENUS *MELANORECTES*　　　　　　　　　　　　　　　ENDEMIC | MONOTYPIC

Melanorectes Sharpe, 1877. *Cat. Birds Brit. Mus.* 3: 289. Type, by monotypy, *Rhectes nigrescens* Schlegel.

The genus comprises a single species endemic to New Guinea. The genus is characterized by strong sexual dichromatism, a heavy, shrike-like hooked bill with a prominent tooth, and all-black male plumage. The phylogenies of Jønsson *et al.* (2007, 2010b) and Dumbacher *et al.* (2008) place this species sister to *Pachycephala* and have necessitated the resurrection of the name *Melanorectes* (Dumbacher 2014) and the change of the common name to "shrikethrush." The species was formerly placed in the genus *Pitohui*.

Black Shrikethrush *Melanorectes nigrescens*　　　　　　　　　　　ENDEMIC

Rectes nigrescens Schlegel, 1871. *Ned. Tijdschr. Dierk.* (1873) 4: 46.—Arfak Mts, Bird's Head.

Inhabits forest interior in the mts of NG; 1,600–2,200 m, extremes 1,000–2,600 m[CO]. No records from the Cyclops Mts, PNG North Coastal Ra, or Adelbert Mts.

a. *Melanorectes nigrescens nigrescens* (Schlegel)

RANGE: Mountains of the Bird's Head and Bird's Neck (Diamond & Bishop 2015).

DIAGNOSIS: Male slaty black; female with crown tinged gray and ventral plumage fulvous but gray-tinged (Rand & Gilliard 1967).

b. *Melanorectes nigrescens wandamensis* (Hartert)

Pitohui nigrescens wandamensis Hartert, 1930. *Novit. Zool.* 36: 59.—Mount Wondiwoi, Wandammen Penin, Bird's Neck.

RANGE: Uplands of the Wandammen Penin.

DIAGNOSIS: Similar to the nominate form, but female plumage brighter and a deeper rufous brown, with crown as for mantle (Hartert 1930, Rand & Gilliard 1967). Also male abdomen black rather than dark gray of nominate form (Boles 2007).

c. *Melanorectes nigrescens meeki* (Rothschild & Hartert)

Pitohui meeki Rothschild & Hartert, 1913. *Novit. Zool.* 20: 507.—Mount Goliath, Border Ra.

Synonym: *Pitohui nigrescens bürgersi* Stresemann, 1922. *J. Orn.* 70: 406.—Schraderberg, Sepik Mts, n sector of the Eastern Ra.

RANGE: The Western, Border, and Eastern Ra (Mayr 1941, Diamond 1972).

DIAGNOSIS: Female is more richly plumaged than other races, with a reddish wash. The description of this form was based on a single adult female specimen (LeCroy 2010). See photos in Coates (1990: 235) and Coates & Peckover (2001: 165).

d. *Melanorectes nigrescens harterti* Reichenow

Melanorhectes harterti Reichenow, 1911. *Orn. Monatsber.* 19: 184.—Interior of Huon Penin.

RANGE: Mountains of the Huon Penin (Mayr 1931).

DIAGNOSIS: Female as for nominate but with crown lacking the gray wash.

e. *Melanorectes nigrescens schistaceus* (Reichenow)

Rhectes nigrescens schistaceus Reichenow, 1900. *Orn. Monatsber.* 8: 187.—Aroa R, s watershed of the SE Peninsula.

RANGE: Mountains of the SE Peninsula (Mayr 1941).

DIAGNOSIS: Male grayer than other races; female more gray-tinged ventrally (Boles 2007).

NOTES

Thinly drawn subspecies based on level of saturation of brown and black in female and male plumage, respectively. Sight records only from Fakfak[G] and Foja Mts[B] (Gibbs 1994[G], Beehler *et al.* 2012[B]). *Other names*—Black Pitohui, Black Whistler, *Pitohui nigrescens*.

GENUS *PACHYCEPHALA* [Genus 40 spp/Region 14 spp]

Pachycephala Vigors, 1825. *Trans. Linn. Soc. London* 14: 444. Type, by original designation, *Muscicapa pectoralis* Latham.

The species of this genus range from eastern India to the Philippines, New Guinea, Australia, Melanesia, and Polynesia (Boles 2007). The genus is characterized by the compact, large-headed form, stubby bill, stolid upright arboreal foraging behavior, and pleasant many-phrased whistled song; sexual dichromatism is predominant in the genus (Schodde & Mason 1999). Sequence of the species follows Jønsson *et al.* (2010b), as further informed by Andersen *et al.* (2014b) and Joseph *et al.* (2014a) for the *P. pectoralis* complex.

Regent Whistler *Pachycephala schlegelii* ENDEMIC

Pachycephala Schlegelii Schlegel (ex von Rosenberg ms), 1871. *Ned. Tijdschr. Dierk.* (1873) 4: 43.—Arfak Mts, Bird's Head.

Forests of virtually all mts of NG; 1,850–3,650 m, min 1,300 m^CO^. Mainly found in the lower and middle stories of the interior of montane forest, where a common and vocal species. No records from the Adelbert Mts.

SUBSPECIES

a. *Pachycephala schlegelii schlegelii* Schlegel

RANGE: Mountains of the Bird's Head and Wandammen Penin. Possibly the populations observed in the Fakfak and Kumawa Mts (Gibbs 1994, Diamond & Bishop 2015) belong here.

DIAGNOSIS: The pale throat patch of the female (with white and gray speckling) is larger and whiter than that of *obscurior*.

b. *Pachycephala schlegelii obscurior* Hartert

Pachycephala schlegeli obscurior Hartert, 1896. *Novit. Zool.* 3: 15.—Eafa distr, Owen Stanley Ra, SE Peninsula.

> Synonym: *Pachycephala schlegelii viridipectus* Hartert and Paludan, 1936. *Mitt. Zool. Mus. Berlin*, 21: 203.—Kunupi, Weyland Mts, Western Ra.
> Synonym: *Pachycephala schlegelii cyclopum* Hartert, 1930. *Novit. Zool.* 36: 54.—Cyclops Mts, ne sector of NW Lowlands, n NG.

RANGE: Throughout the Central Ra; also the Cyclops Mts and mts of the Huon Penin; presumably the populations from the Foja[J] and Bewani Mts[K] are of this form (Beehler *et al.* 2012[J], specimens at AMNH & BPBM[K]).

DIAGNOSIS: The pale throat patch of the female (with white and gray speckling) is grayer and bounded posteriorly by a broader breast band than in the nominate form. See photo in Coates & Peckover (2001: 161).

NOTES

Sympatric with *P. soror* between 1,450 and 2,000 m, and with *P. balim* in middle elevations of the Western Ra (at *ca.* 2,000 m). *Other names*—Schlegel's Whistler.

Sclater's Whistler *Pachycephala soror* ENDEMIC

Pachycephala soror P. L. Sclater, 1874. *Proc. Zool. Soc. London* (1873): 692.—Hatam, Arfak Mts, Bird's Head.

Mountains of the Bird's Head and Neck, Central Ra, Mt Bosavi, Adelbert Mts, Huon Penin, and Goodenough I (Diamond 1985); 1,100–1,900 m, extremes 150–2,450 m^CO^. Inhabits lower montane forest interior. No records from the Wandammen, Foja, or Cyclops Mts, nor from the PNG North Coastal Ra.

SUBSPECIES

a. *Pachycephala soror soror* P. L. Sclater

RANGE: Mountains of the Bird's Head (Mayr 1941).

DIAGNOSIS: Male's tail feathers black, edged green (Rand & Gilliard 1967).

b. *Pachycephala soror octogenarii* Diamond

Pachycephala soror octogenarii Diamond, 1985. *Emu* 85: 78.—Southern watershed of Kumawa Mts, 1,160 m, Bird's Neck.

RANGE: Kumawa Mts. Diamond (1985) noted that the Fakfak Mts population is "considered an unnamed race of *P. soror* closest to *octogenarii*." This was confirmed by Gibbs (1994), who revisited the Fakfak Mts in 1992. Diamond & Bishop (2015) suggested *octongenarii* may be worthy of species status.

DIAGNOSIS: Male head dark gray rather than jet black; black collar very narrow; edges of flight feathers olive. Fakfak population very similar, but posterior border of black collar is ocher.

c. *Pachycephala soror klossi* Ogilvie-Grant

P[*achycephala*] *soror klossi* Ogilvie-Grant, 1915. *Ibis*, Jubilee Suppl. 2: 88.—Upper Otakwa R, Western Ra.

RANGE: The Western, Border, and Eastern Ra[DI]; also the Adelbert Mts[J] and mts of the Huon Penin[K] (Mayr 1931[K], Gilliard & LeCroy 1967[J]).

DIAGNOSIS: Male tail feathers black entirely (Rand & Gilliard 1967, Diamond 1972). See photos in Coates & Peckover (2001: 160).

d. *Pachycephala soror bartoni* Ogilvie-Grant

P[achycephala] soror bartoni Ogilvie-Grant, 1915. *Ibis*, Jubilee Suppl. 2: 89.—Uplands of SE Peninsula.
Synonym: *Pachycephala soror remota* Mayr & Van Deusen, 1956. *Amer. Mus. Novit.* 1792: 4.—Top Camp, e slopes of Goodenough I, D'Entrecasteaux Arch.

RANGE: Mountains of the SE Peninsula[MA] and Goodenough I.

DIAGNOSIS: Male tail feathers olive green with black shaft streaks; wing 85–92 (Rand & Gilliard 1967, Diamond 1972). LeCroy (2010) argued for the validity of *remota*, based on a substantial size difference. We note *remota* was based on a collection of one adult male and one juvenile male only. Beehler collected additional material from Goodenough I in 1976, including four adult males with wings of 86.5, 91.5(2), and 95.0. Adding the single specimen from Mayr & van Deusen (1956), wing 96, this gives a mean wing length of 92.1 for *remota*. Measurements for *bartoni* were 85–91, which we give a mean of 88. The percentage difference in the means is 4.6%, under our minimum cutoff for a valid size subspecies.

NOTES
Other names—Hill Golden Whistler.

Louisiade Whistler *Pachycephala collaris*   ENDEMIC

Pachycephala collaris Ramsay, 1878. *Proc. Linn. Soc. NSW* 3: 74.—Teste I, Dumoulin Is, southeast of Basilaki I, SE Islands.

Known only from the SE Islands. Formerly treated as a subspecies of *P. pectoralis*. The elevation of *collaris* to species rank follows the work of Andersen *et al.* (2014b), whose molecular phylogeny of the golden whistler complex demonstrated that the *collaris* clade is the most divergent of the core *pectoralis* lineage. For an additional molecular analysis of the golden whistlers that supports this split, see Jønsson *et al.* (2014).

SUBSPECIES

a. *Pachycephala collaris collaris* Ramsay

Synonym: *Pachycephala pectoralis misimae* Rothschild & Hartert, 1918. *Novit. Zool.* 25: 311.—Kimuta I [LeCroy & Peckover 1998], Louisiade Arch, SE Islands.

RANGE: Teste I in the Dumoulin Is, Panasesa I in the Conflict Is[K], Calavados Is, East and Hastings I in the Bonvouloir Is, Alcester and Egum I near Woodlark I, Deboyne Is, Kimuta I, and Renard I (Mayr 1941, Beehler field notes 29-Sep-2002[K]). Misima record (Mayr 1941) is probably in error; specimen apparently from Kimuta I (LeCroy & Peckover 1998).

DIAGNOSIS: We follow Galbraith (1956) and Mayr (1967) in subsuming *misimae* into this form. Tail green; bill long; female with upperparts bright brownish olive; breast band bright rusty; abdomen rich yellow; wing quills edged brown; wing 95; culmen 24 (Rand & Gilliard 1967).

b. *Pachycephala collaris rosseliana* Hartert

Pachycephala rosseliana Hartert, 1898. *Bull. Brit. Orn. Club* 8: 8.—Rossel I, Louisiade Arch, SE Islands.

RANGE: Rossel I.

DIAGNOSIS: Similar to *collaris*, but female with breast and abdomen bright but paler yellow and breast band narrower and more olive. Bill of both forms black. Type description noted the male is somewhat intermediate between *melanura* and *collaris*; tail blackish with an olive base; feathers of thighs black at base.

NOTES

Andersen *et al.* (2014b) sequenced 10 genes to generate a new phylogeny for the *Pachycephala pectoralis* complex. Their findings inform our treatment here, which is to recognize three species of golden whistler for the Region: *collaris* of the SE Islands, *balim* from a restricted range in the Western Ra, and *melanura* from coastal NG. The Andersen work placed *pectoralis* in Australia and *citreogaster* in the Bismarck Arch. See also the treatment of Jønsson *et al.* (2008), who provided a molecular phylogeny of the golden whistler complex. Schodde & Mason (1999: 443–44) outlined the historical treatment of this complex. Jønsson *et al.* (2014) discussed taxon cycles of avian lineages in Australasia and included the golden whistlers as a prime example. That paper demonstrated the molecular distinctiveness of *P. collaris* and *P. balim*.

Diagnosis of this insular species from its sister forms is difficult. The males share yellow nape and breast, and the drab females exhibit a whitish throat, tan breast band, and yellow breast. *P. collaris* differs by its longer bill and green tail. Andersen *et al.* (2014b) sampled both races of *collaris* and found substantial divergence between them, comparable to species-level differences found among other species pairs in the golden whistler complex. *Other names*—formerly treated as a subspecies of *P. pectoralis* or *P. citreogaster*.

Baliem Whistler *Pachycephala balim* ENDEMIC | MONOTYPIC
Pachycephala pectoralis balim Rand, 1940. *Amer. Mus. Novit.* 1072: 8.—Baliem R, 1,600 m, Western Ra.

The species is restricted to montane forest on the slopes of the e sector of the Western Ra south of the Baliem valley, 1,600–2,400 m (Rand 1942b) and in casuarinas and shade trees in village environs on the floor of the Baliem valley (Archbold *et al.* 1942, Ripley 1964a). Full extent of the distribution of this strangely restricted form is unknown. Rand (1942b) reported that this species is "fairly common in the forest at 2,400 m, where *P. schlegelii* was rare … and these species seem to replace each other, with slight overlap."

NOTES

Andersen *et al.* (2014b) elevated this population to species status, which was foreshadowed by Sibley & Monroe (1990). Both sexes of *P. balim* have green edges to the wing feathers (these are silvery gray in *P. melanura*) and a short bill. Female has crown brownish-washed; mantle dull olive green; edges of secondaries greenish; male has a rather broad black breast band. Rand (1940a) stated that this upland NG form was closest to *P. melanura spinicaudus*; however, that form is overall paler, exhibits gray edges to flight feathers, and is black-tailed. This apparently relictual montane interior population, nestled between populations of *P. schlegelii* (higher) and *P. soror* (lower), is nothing if not a biogeographic anomaly. It should be studied in detail in the field, and should be looked for east and west of its currently known range.

Other names—formerly a subspecies of the Common Golden Whistler, *P. pectoralis*, which has been split into several species.

Mangrove Golden Whistler *Pachycephala melanura* RESIDENT
Pachycephala melanura Gould, 1843. *Proc. Zool. Soc. London* (1842): 134.—Derby, Western Australia.

Patchily distributed along the coastlines of the Mainland in mangroves and other inundated forest: Bintuni Bay, south of Timika[J], the coastal Trans-Fly to the s coast of the SE Peninsula, from Madang (Gogol lowlands and Tab I[K]) to the Huon Penin, and Killerton Is[L] off the n coast of the SE Peninsula (Clapp 1980a[L], Bailey 1992a[K], van Balen & Rombang 1998[J]). Also in forest and scrub of the SE Is.

Extralimital—n coast of Australia, Torres Strait Is, coastal Bismarck Arch and fringing islets. If the form *whitneyi* is assignable to this species, then the species' range also includes some small islands se of Bougainville I, in the Solomon Is (Whitney, Momalufu, and Akiri I; Boles 2007: 427, Dutson 2011).

SUBSPECIES

a. *Pachycephala melanura spinicaudus* (Pucheran)
Pteruthius spinicaudus Pucheran, 1853. *Voy. Pôle Sud, Zool.* 3: 58.—Warrior I, Torres Strait.
RANGE: Cape York, islands of the Torres Strait, the Bird's Neck (Bintuni Bay), and the S Lowlands east to the Purari R delta—including Dolak I, Merauke, and Daru (Bishop 2005a). Mangroves, other coastal formations and second growth, particularly on small islands off the coast.
DIAGNOSIS: Female has crown grayish, back dull gray-olive, wing grayish (not greenish), belly dull yellow-buff. Male exhibits a broad yellow nuchal collar, bright mid-citrine mantle, and very broad black breast band (Schodde & Mason 1999). See photo in Coates & Peckover (2001: 162). For a discussion of variation in the population of the Torres Strait region, see Schodde & Mason (1999: 446). The subspecies name *spinicaudus* is treated as a noun in apposition and thus is invariant (David & Gosselin 2002a).

b. *Pachycephala melanura dahli* Reichenow

Pachycephala melanura dahli Reichenow, 1897. *Orn. Monatsber.* 5: 178.—Credner Is, adjacent to the Duke of York Is, Bismarck Arch.

> Synonym: *Pachycephala pectoralis fergussonis* Mayr, 1936. *Amer. Mus. Novit.* 869: 2.—Fergusson I, D'Entrecasteaux Arch, SE Islands.

RANGE: Madang islets; also coast and coastal islets of the SE Peninsula, west to the Purari R delta in the s watershed and west in the n watershed to the Fly Is[K] of the s Huon Gulf; also Fergusson[MA], Duchess and Tobweyama I[M], and the Killerton Is[CO] (Pratt specimens at BPBM[M], Beehler field notes 25-Oct-1978[K]). It was not found by Pratt (field notes 2003) on the larger islands of Fergusson and Normanby, where it had been reported previously, raising the possiblity that the specimen records originated from the small fringing islets off the main larger islands. Also found along the coastlines of New Britain and New Ireland, many small islands in the Bismarck Arch, Nissan I, and Bougainville I (Dutson 2011).

DIAGNOSIS: Female with a medium-bright yellow belly; obscure narrow buff breast band, and whitish-gray throat with brownish speckling; crown grayish; edges of secondaries grayish. Male with edges of secondaries gray; tail black; short bill. The race *fergussonis* was lumped into *dahli* by Galbraith (1956), followed by Mayr (1967: 25).

NOTES

Unconfirmed observation from Salawati I (M. Tarburton's island website). Andersen *et al.* (2014b) recognized *P. melanura* as a species-level taxon that inhabits coastal Australia and NG (mainly in mangroves). Differs from *P. collaris* in silvery-gray (not green) margins to secondaries. Male's tail always has black dorsal surface (not green). Female's gray-tan breast band and tail partly black; underparts vary from bright yellow like male's to quite pale in birds living along the s coast (Pratt & Beehler 2014). *P. citreogaster* (*e.g.*, of Dickinson 2003) is restricted to the Bismarck Arch by Andersen *et al.* (2014b). *Other names*—Black-tailed Whistler.

Golden-backed Whistler *Pachycephala aurea*   ENDEMIC | MONOTYPIC

Pachycephala aurea Reichenow, 1899. *Orn. Monatsber.* 7: 131.—Ramu R, ne NG.

Very patchily distributed but widespread on the Mainland. No records from the Bird's Head and Neck, the NW Lowlands, or the e sector of the SE Peninsula (Stresemann 1923).

Localities—the Menoo R in the Weyland Mts[L], Mimika and Setekwa R[MA], Ok Tedi (5 km downstream from Tabubil)[J], Telefomin[CO], Nomad R[M], middle Sepik R[CO], upper Ramu R[MA], Lake Kopiago[CO], Nembi valley (sw of Mendi)[CO], Purari R[K], and a few localities in the mts and lowlands of the SE Peninsula: Lakekamu basin[W], Garaina[CO], and Naoro[CO] (Hartert *et al.* 1936[L], Bell 1970b[M], Beehler *et al.* 1994[W], K. D. Bishop *in litt.*[J], I. Woxvold *in litt.*[K]). Highest record is 1,300 m at Lake Kopiago (Campbell 1979). Inhabits second growth, shrubbery, or a mixture of bushes and reedbeds, on small river islands, river flats, and the fringes of lakes; also *Albizia* shade trees in a tea plantation on the floor of an upland valley (at Garaina). Found mainly in the hills, but also found in some lowland localities (Coates 1990, Beehler *et al.* 1994). *Other names*—Golden-yellow Whistler, Yellow-backed Whistler.

Vogelkop Whistler *Pachycephala meyeri*   ENDEMIC | MONOTYPIC

Pachycephala meyeri Salvadori, 1890. *Orn. Pap. Moluc., Agg.*: 104.—Arfak Mts, Bird's Head. Replacement name for *Pachycephala affinis* Meyer, 1884, *Zeitsch. Ges. Orn.* 1: 200; preoccupied by *Pachycephala affinis* Meyer, 1874.

The Arfak[J] and Tamrau Mts[K] of the Bird's Head, 970–1,520 m (Mayr & Meyer de Schauensee 1939b[K], Gyldenstolpe 1955b[J]). Sight record from the Foja Mts (Diamond 1985, Beehler *et al.* 2012) needs corroboration.

NOTES

A little-known species badly needing additional field and molecular study. *Other names*—Grey-crowned or Meyer's Whistler.

Lorentz's Whistler *Pachycephala lorentzi* ENDEMIC | MONOTYPIC

Pachycephala schlegelii lorentzi Mayr, 1931. *Mitt. Zool. Mus. Berlin* 17: 673.—Hellwig Mts, 2,600 m, e sector of the Western Ra. Replacement name for *Poecilodryas caniceps pectoralis* van Oort, 1910, *Notes Leyden Mus.* 32: 213; preoccupied by *Muscicapa* [= *Pachycephala*] *pectoralis* Latham, 1801.

The Western[J] and Border Ra[K], east to the Hindenburg[L] and Victor Emanuel Mts[L], 1,600–3,800 m (Rand 1942b[J], Gilliard & LeCroy 1961[L], Ripley 1964a, Gregory & Johnston 1993[K]).

Brown-backed Whistler *Pachycephala modesta* ENDEMIC

Poecilodryas modesta De Vis, 1894. *Ann. Rept. Brit. New Guinea*, 1893–94: 101.—Mount Maneao, 1,525 m, SE Peninsula.

Inhabits montane forest and subalpine shrublands of the Border Ra, Eastern Ra, and Huon and SE Peninsulas; 1,750–3,600 m, min 1,130 m[CO]. No records from west of the PNG border. Sighting from the Adelbert Mts (Mackay 1991) merits corroboration, as it was not recorded there by Pratt (1983).

SUBSPECIES

a. *Pachycephala modesta telefolminensis*

Pachycephala modesta telefolminensis Gilliard & LeCroy, 1961. *Bull. Amer. Mus. Nat. Hist.* 123: 61.—Mount Ifal, 2,225 m, Victor Emanuel Mts, Border Ra.

RANGE: The e sector of the Border Ra, including the Victor Emanuel and Hindenburg Mts.
DIAGNOSIS: Crown dark slate gray; mantle dark dun brown with a gray wash; throat off-white; obsolete breast band narrow and dirty gray.

b. *Pachycephala modesta modesta* (De Vis)

> Synonym: *Pachycephala hypoleuca* Reichenow, 1915. *J. Orn.* 63: 125.—Schraderberg, Sepik Mts, n sector of the
> Eastern Ra.

RANGE: The Eastern Ra and mts of the Huon and SE Peninsulas (Mayr 1941).
DIAGNOSIS: Crown slate gray; mantle dun brown; throat off-white; breast band obscure and buff gray.

NOTES

The form *hypoleuca* and the nominate *modesta* include material that broadly overlaps, and hence the two are combined here. The form *Pachycephala moroka* Rothschild & Hartert was synonymized with the nominate by Mayr (1941); see LeCroy (2010) for additional information on the specimen on which *moroka* was based. The two recognized races are but thinly defined. Sister to *P. lorentzi* (Anderson *et al.* 2013). *Other names*—Modest Whistler, Grey Mountain Whistler.

Island Whistler *Pachycephala phaionota* RESIDENT | MONOTYPIC

Myiolestes phaionotus Bonaparte, 1850. *Consp. Gen. Av.* 1: 358.—Banda Is, s Moluccas.

A small-island tramp species of the Moluccas and the westernmost sector of the Region. Apparently confined to wooded coastal vegetation on tiny fringing islets of the NW Islands, the Aru Is, and islets off Biak I in Geelvink Bay (Mayr 1941).

Localities—Kofiau I[J], Schildpad Is[MA], Wai I[K], islets off Sorong[MA], Pulau Babi of the Aru Is[L], and Rani I in Geelvink Bay; also islets fringing Misool[MA], Salawati[MA], and Waigeo[MA] (Ripley 1964a[J], Diamond & Bishop 1994[L], Beehler field notes 29-Feb-2004[K]). The specimen from "Waigama" on Misool (Mees 1965) was probably taken on an adjacent fringing small island.

Extralimital—Mayu, Ternate, Tidore, Damar, Mare, Moti, Banda, Seram Laut, Tayandu, and Kai Is (White & Bruce 1986).

NOTES

We follow David & Gosselin (2002a) in treatment of the species epithet as adjectival, hence *phaionota* to agree with the feminine *Pachycephala*. We follow White & Bruce (1986) in treating the species as monotypic. The form *stresemanni* Janey from Majau I in the Moluccas was not recognized by Mees (1965). *Other names*—Island Robin-whistler, *Pachycephala phaionotum* or *phaionotus*.

Rusty Whistler *Pachycephala hyperythra*

Pachycephala hyperythra Salvadori, 1876. *Ann. Mus. Civ. Genova* (1875) 7: 932.—Arfak Mts, Bird's Head.

Inhabits hill forest patchily throughout NG, 600–1,300 m (extremes 200–1,400 m). Range includes mts of the Bird's Head and Neck, the Foja Mts, PNG North Coastal Ra, mts of the Huon Penin, and Mt Bosavi (Stresemann 1923). Seems to be absent east of Port Moresby and from the n watershed of the SE Peninsula (except for a record from the Bulolo valley; BPBM specimens).

SUBSPECIES

a. *Pachycephala hyperythra hyperythra* Salvadori

RANGE: The Bird's Head, Bird's Neck, w sector of the n scarp of the Western Ra, and Foja Mts[B] (Beehler *et al.* 2012[B]). Localities—Onin, Bomberai, and Wandammen Penin[MA, K], and Kunupi and lower Menoo R in the Weyland Mts[J] (Hartert *et al.* 1936[J], Diamond & Bishop 2015[K]).

DIAGNOSIS: Face and cheek brown, and throat patch small and indistinct; secondaries edged rufous; back olive brown; dull breast band distinct from paler abdomen.

b. *Pachycephala hyperythra sepikiana* Stresemann

Pachycephala hyperythra sepikiana Stresemann, 1921. *Anz. Orn. Ges. Bayern* 1: 36.—Maeanderberg, upper Sepik R, west-central Sepik-Ramu.

RANGE: The n scarp of the Western, Border, and Eastern Ra. Localities—various points near Bernhard Camp on the Taritatu R[J], Depapre, and the upper Jimi R[K] (Rand 1942b[J], Beehler field notes 19-Aug-1982[K]).

DIAGNOSIS: When compared to the nominate form, this race is dark-faced and gray-cheeked; back more olive; white throat patch larger and distinct; and breast and abdomen more ochraceous. See photo in Coates (1990: 215).

c. *Pachycephala hyperythra reichenowi* Rothschild & Hartert

Pachycephala hyperythra reichenowi Rothschild & Hartert, 1911. *Orn. Monatsber.* 19: 178.—Sattelberg, Huon Penin.

RANGE: Mountains of the Huon Penin (Mayr 1931).

DIAGNOSIS: Similar to *sepikiana*, but back darker and browner and breast duller ochraceous (Rand & Gilliard 1967).

d. *Pachycephala hyperythra salvadorii* Rothschild

Pachycephala salvadorii Rothschild, 1897. *Bull. Brit. Orn. Club* 7: 22.—Moroka distr, s watershed of SE Peninsula. Replacement name for *Pachycephala sharpei* Salvadori, 1896, *Ann. Mus. Civ. Genova* (2)16: 88; preoccupied by *Pachycephala sharpei* A. B. Meyer, 1885.

RANGE: Mountains of the SE Peninsula and the s sector of the Eastern Ra. Localities—Lake Kutubu[CO], Mt Bosavi[J], Karimui[DI], Bomai[DI], Soliabeda[DI], Mt Karimui[DI], upper Watut R[J], lower Wau valley[K], and Moroka distr[MA] (Pratt field observation[K], specimen at BPBM[J]).

DIAGNOSIS: Throat patch gray; overall plumage dull; cheek and crown gray; lacking in bright highlights (Diamond 1972). See photo in Coates & Peckover (2001: 164).

NOTES

See LeCroy (2010: 16–17) for additional information on the type specimen for the race *salvadorii*. The four recognized subspecies are quite variable and as currently designated are not very satisfactory, but include sufficient plumage characters to remain recognized here (see Pratt & Beehler 2014 for illustrations of three races; note how different the nominate race [5W] looks). Be aware that there is considerable within-population variation. The form found in the foothills of the upper Fly has not been assigned to subspecies; it exhibits a fairly large white throat patch and is quite rich rusty above (P. Gregory *in litt.*, *pace* Rand 1942a). The species is badly represented in museum collections and merits further field study. *Other names*—Rufous-breasted, Rusty-breasted, or Brownish Whistler.

Grey Whistler *Pachycephala simplex*   RESIDENT

Pachycephala simplex Gould, 1843. *Proc. Zool. Soc. London* (1842): 135.—Port Essington, NT, Australia.

Throughout NG, the NW Islands, Aru Is, Yapen and Mios Num I, D'Entrecasteaux Arch, and Tagula I. Inhabits forest, second growth, edge. Sea level–1,400 m, max 1,550 m[CO].

Extralimital range extends west to the Kai Is and to n and ne Australia.

SUBSPECIES

a. *Pachycephala simplex griseiceps* G. R. Gray

Pachycephala griseiceps G. R. Gray, 1858. *Proc. Zool. Soc. London*: 178.—Aru Is.

> Synonym: *Pachycephala griseiceps perneglecta* Hartert, 1930. *Novit. Zool.* 36: 56.—Near the Otakwa R, w sector of the S Lowlands.
> Synonym: *Pachycephala griseiceps waigeuensis* Stresemann & Paludan, 1932. *Novit. Zool.* 38: 153.—Waigeo I, NW Islands.
> Synonym: *Pachycephala griseiceps gagiensis* Mayr, 1940. *Amer. Mus. Novit.* 1091: 3.—Gag I, NW Islands.

RANGE: The NW Is but for Kofiau[MA]; the Bird's Head and Neck[MA]; Aru Is[MA]; S Lowlands, Trans-Fly, and s watershed of SE Peninsula east to Port Moresby (Greenway 1966). Hybridizes with *dubia* in the Port Moresby area (Rand & Gilliard 1967).

DIAGNOSIS: Crown grayish; mantle olive brown; throat dull whitish with obsolete gray streaks; abdomen with a yellow wash. We combine *perneglecta*, *waigeuensis*, and *gagiensis* into this form because of lack of workable plumage characters to distinguish these four described forms (see Mees 1965). It seems much of the described variation (which is minimal) is based on differing specimen preparation. See photo in Coates & Peckover (2001: 163).

b. *Pachycephala simplex jobiensis* A. B. Meyer

Pachycephala griseiceps var. *jobiensis* A. B. Meyer, 1874. *Sitzungsber. Akad. Wiss. Wien* 69: 394.—Yapen I, Bay Is.

> Synonym: *Pachycephala miosnomensis* Salvadori, 1879. *Ann. Mus. Civ. Genova* 15: 46.—Mios Num I, Bay Is.

RANGE: Yapen and Mios Num I[MA], NW Lowlands, and Sepik-Ramu, hybridizing with *dubia* in the hinterland of Astrolabe Bay (Rand & Gilliard 1967).

DIAGNOSIS: Similar to *griseiceps*, but mantle browner and belly more richly washed with yellow.

c. *Pachycephala simplex dubia* Ramsay

Pachycephala dubia Ramsay, 1879. *Proc. Linn. Soc. NSW* 4: 99.—Laloki R, s watershed of the SE Peninsula. Replacement name for *Eopsaltria* (?) *brunnea* Ramsay, 1877, *Proc. Linn. Soc. NSW* 1: 391.

RANGE: The Huon and SE Peninsulas; in the s watershed from Port Moresby eastward; in the n watershed from the Huon Penin eastward; also the three main islands of the D'Entrecasteaux Arch (Mayr 1941).

DIAGNOSIS: This form, as with the Australian nominate form, entirely lacks yellow wash in plumage; back olive brown; throat whitish with some dark streaking; breast band distinct and pale brown, with some obscure streaks; abdomen white, washed with pale buff. See photos in Coates (1990: 214).

d. *Pachycephala simplex sudestensis* (De Vis)

Eopsaltria simplex sudestensis De Vis, 1892. *Ann. Rept. Brit. New Guinea* 1892: 96.—Tagula I, Louisiade Arch, SE Islands.

RANGE: Tagula I.

DIAGNOSIS: Plumage as for *griseiceps*, but larger and longer-billed (Boles 2007). Wing 88; bill 20 (Rand & Gilliard 1967). This is a thinly defined form that may merit combination with *griseiceps*. The form *alberti* Hartert was synonymized by Mayr (1941) into this form (see also LeCroy 2010).

NOTES

We follow Schodde & Mason (1999) and Christidis & Boles (2008) in considering the *simplex* and *griseiceps* groups as members of a single polytypic species. Subsequent workers may find these comprise two distinct lineages worthy of splitting. Note also that the race *sudestensis* is separated from the other yellow-washed races by the brown-washed race *dubia* on the SE Peninsula. *Other names*—Brown or Grey-headed Whistler or Thickhead, *Pachycephala griseiceps*.

Black-headed Whistler *Pachycephala monacha* ENDEMIC

Pachycephala ? monacha G. R. Gray, 1858. *Proc. Zool. Soc. London*: 179.—Aru Is.

The Aru Is and patchily distributed across the uplands of NG and the Huon Penin; absent from the Bird's Head and Bird's Neck; 550–1,980 m[CO], min 175 m (Beehler field notes). Most common in interior upland valleys. Hybridizes with *P. leucogastra* on the Sogeri Plateau and at Rigo, Bereina, Mafulu, and Baroka on the s slopes of the SE Peninsula. Inhabits dense regrowth along rivers and creeks, and shade trees in open agricultural mid-montane valley habitats. Also common in casuarina and *Albizia* trees lining roads in PNG highland valleys such as Minamb, Tari, and Wahgi (K. D. Bishop *in litt.*) and at Dablin Creek, near Tabubil (Gregory *in litt.*).

SUBSPECIES

a. *Pachycephala monacha monacha* G. R. Gray
RANGE: The Aru Is.
DIAGNOSIS: Mantle of male brownish black, not deep black.

b. *Pachycephala monacha lugubris* Salvadori
Pachycephala monacha lugubris Salvadori, 1881. *Ann. Mus. Civ. Genova* (1878) 12(21): 332.—Utanata R, w sector, Western Ra.
> Synonym: *Pachycephala dorsalis* Ogilvie-Grant, 1911. *Bull. Brit. Orn. Club* 29: 26.—Mimika R, w sector of the S Lowlands.

RANGE: The s watershed of the Western, Border, and Eastern Ra; the Sepik-Ramu; and uplands of the Huon and SE Peninsulas. Localities—mouth of the Mimika R[MA], Utanata R[MA], Modijo in Charles Louis Mts[L], Wamena in Baliem valley[J,Q], Puwani R at n base of Bewani Mts[U], Telefomin[K], Dablin Creek[N], Torricelli Mts[CO], Arip valley in drainage of upper Fly R[CO], Tari[CO], Wahgi valley[CO], Baiyer valley[CO], Tabebuga[CO], Nondugl[M], Dagie[M], upper Jimi valley[CO], Ramu and Simbai valleys[CO], Asaro valley[CO], Okasa[DI], Lufa[DI], Karimui[DI], Bomai[DI], Obura[CO], Aiyura valley[CO], Huon Penin[CO], Snake R[MA], Bulolo valley[CO], Wau[U], Mafulu[MA], Sogeri Plateau[CO], and Agaun[CO] (Rand 1942b[Q], Gyldenstolpe 1955a[M], Gilliard & LeCroy 1961[K], Ripley 1964a[J], King 1979[L], Beehler field notes[U], P. Gregory *in litt.*[N]).
DIAGNOSIS: Dorsal plumage blacker than nominate (Boles 2007). Use of the name *lugubris* follows Mees (1994) and Boles (2007).

NOTES

Following Mayr (1941), Rand & Gilliard (1967), Schodde & Mason (1999), and Boles (2007), we treat this taxon as a full species, distinct from *rufiventris* of Australia and *leucogastra*. Hybrids between *P. monacha* and *P. leucogastra* have been recorded from the Sogeri Plateau, Rigo, Bereina, Mafulu, and Boroka (Coates 1990). Dickinson & Christidis (2014: 188 note 6) argue for use of the subspecies epithet *dorsalis* instead of the older *lugubris*. M. LeCroy (*in litt.*) noted: "It is true that Salvadori (1881: 233) first placed this manuscript name in synonymy, although he there considered it a nomen nudum." Stresemann (1923) noted how *lugubris* and *monacha* males differ and thus validates *lugubris* Salvadori, 1881 (as Dekker & Quaiser 2006 have stated). *P. dorsalis* was not named by Ogilvie-Grant until 1911, long after Salvadori listed it. Salvadori noted that Mueller had a female, and that is the specimen correctly considered the holotype of *lugubris* by Dekker & Quaiser (2006: 8), but the name was validated using males of *monacha* and *lugubris*. *Other names*—Black-backed Whistler.

White-bellied Whistler *Pachycephala leucogastra* ENDEMIC

Pachycephala leucogastra Salvadori and D'Albertis, 1876. *Ann. Mus. Civ. Genova* (1875) 7: 822.—Mount Epa, SE Peninsula.

Very patchily distributed, scarce, and restricted to e NG: c sector of the Sepik-Ramu, e sector of the S Lowlands, s watershed of the SE Peninsula, and Rossel I; sea level–1,200 m (Coates & Peckover 2001). Inhabits mainly wet *Eucalyptus* savanna in the hills; also mangroves, tall trees in savanna and in rubber plantations (Coates 1990). Behavior is similar to that of *P. monacha*; presumably a sister to that species.

a. *Pachycephala leucogastra leucogastra* Salvadori & D'Albertis

RANGE: Known from three Mainland areas: the n sector of the Sepik basin, the mouth of the Purari R, and the s coast of the SE Peninsula. Localities—Forok near Wewak[J], Lumi[J], lower Purari R[CO], Hall Sound[MA], Bereina[CO], Hisiu[CO], Mafulu on the Aroa R[CO], Vanapa R[CO], Sogeri Plateau[CO], Varirata NP[K], Mt Epa[MA], Port Moresby[MA], Efogi[CO], and Rigo[CO] (Bell 1984[J], K. D. Bishop *in litt.*[K]).

DIAGNOSIS: Male's mantle and nape medium gray, distinct from black crown and face; wings and tail dark brown (Boles 2007); also pectoral area whitish, not grayish. Female underparts white. Male from the mouth of the Purari has upperparts darker gray than typical nominate *leucogastra* and tail black—perhaps evidence of hybridization with *P. monacha* (Coates 1990).

b. *Pachycephala leucogastra meeki* Hartert

Pachycephala meeki Hartert, 1898. *Bull. Brit. Orn. Club* 8: 15.—Rossel I, Louisiade Arch, SE Islands.

RANGE: Rossel I.

DIAGNOSIS: Male darker dorsally than nominate and with gray at sides of breast. Three plumages well represented in collection: the black-collared male; the gray-collared female with underparts tinged with buff; and the "imm"?, which is streaked below, gray-buff dorsally, and with no collar.

NOTES

The Rossel Is race is quite distinct and is geographically far removed from the Mainland populations; its systematic status should be examined further. *Pachycephala leucogastra* forms a superspecies with *P. rufiventris*, *P. griseonota*, *P. monacha*, and *P. arctitorquis* (Boles 2007). Past treatments (*e.g.*, Beehler & Finch 1985) have considered this species to be conspecific with *P. rufiventris*. Specimens from Lumi & Wewak in the foothills of the c PNG North Coastal Ra have not yet been diagnosed to subspecies. The distribution appears relictual. See Notes section of the preceding species regarding interspecific hybridization. *Other names—Pachycephala leucogaster.*

Family Oriolidae

Orioles, Pitohuis, and Figbirds   [Family 35 spp/Region 8 spp]

The species of the family range from Europe and Africa east to China, Korea, Philippines, Indonesia, New Guinea, Australia, and historically New Zealand (Walther & Jones 2008, Zuccon & Ericson 2012). Orioles are medium-large arboreal fruit eaters with melodious songs. They inhabit forest and open wooded country from lowlands to middle elevations. Family characters: bill slender but strong; tarsus short; body slim; tail narrow and long (Higgins *et al.* 2006). Familial systematics follows Jønsson *et al.* (2010a). The phylogeny of Jønsson *et al.* (2011b) and Zuccon & Ericson (2012) supports the position of *Pitohui* as sister to the orioles, and as a result the true pitohuis are now placed in a new subfamily, the Pitohuinae, within the Oriolidae (Schodde & Christidis 2014, Dickinson & Christidis 2014: 638–40). In addition, this family now includes the extinct frugivore *Turnagra* from New Zealand (Zuccon & Ericson 2012).

GENUS *PITOHUI*   ENDEMIC [Genus 4 spp]

Pitohui Lesson, 1831. Traité *Orn.*: 375. Type, by subsequent designation (Sharpe, 1877. *Cat. Birds Brit. Mus.* 3: 283), *Lanius kirhocephalus* Lesson & Garnot. Date of original publication of name corrected in Dickinson *et al.* (2011: 119).

The four species of the genus are endemic to the Region. The traditional genus has recently been dissected, based on molecular analysis (Dumbacher *et al.* 2008, Dumbacher 2014), and the revised *Pitohui* contains only the forms that were formerly in the species traditionally known as *kirhocephalus* and *dichrous*. Treatment of the various new genera split from the traditional *Pitohui* follows Dumbacher (2014). As currently construed, *Pitohui* is characterized as a medium-billed, uncrested, vocal and flocking form that inhabits lowlands and lower mountains. The subspecies treatment is based on the framework created by Mayr (1941), but we follow Dumbacher *et al.* (2008) in splitting the traditional *P. kirhocephalus* and grouping the subspecies into three distinct species lineages: *kirhocephalus, cerviniventris,* and *uropygialis.* This split is a departure from the noncommittal and unchanged single-species arrangement of Dumbacher (2014). However, we feel the break up of this highly subspecies-rich species is long overdue, and we base our split on the recommendations made by Dumbacher *et al.* (2008: 779–80): "The confusing variation within this species has been long recognized (Beehler *et al.* 1986) and results from three highly differentiated genetic groups within *P. kirhocephalus* that distinguish the north-coast, south-coast, and West Papuan Island clades (Dumbacher *et al.* 2008). Other analyses directed at biogeography continue to support further splitting *P. kirhocephalus* (J. M. Dumbacher unpublished data) and the data presented here provide strong support for each of these major groupings." Even more splitting may be possible with the future screening of populations, for there are approximately nine phenotypic groupings of variable pitohui subspecies. Certain populations of *Pitohui* are chemically-defended (Dumbacher *et al.* 1992).

Northern Variable Pitohui *Pitohui kirhocephalus*   ENDEMIC

Lanius kirhocephalus Lesson & Garnot, 1827. *Voy. Coquille, Atlas*: pl. 11; 1829. *Zool.* 1: 633.—Manokwari, Bird's Head.

Inhabits forest and especially edge and regrowth of the e coast of the Bird's Head and Neck, NW Lowlands, and Sepik-Ramu; also Yapen I and small coastal islands in Geelvink Bay (Stresemann 1923); lowlands–1,100 m, max 1,500 m[CO]. Typically most common in lowland scrub, but in foothills can be found sharing habitat with *P. dichrous.*

SUBSPECIES

a. *Pitohui kirhocephalus decipiens* (Salvadori)
Rectes decipiens Salvadori, 1878. *Ann. Mus. Civ. Genova* 12: 473.—Kapaur, Onin Penin, Bird's Neck.
RANGE: Onin Penin (Kapaur and Sekru) of the Bird's Neck.
DIAGNOSIS: Sexually dichromatic; male with a blackish head; female with head not so dark; wings and tail blackish.

b. *Pitohui kirhocephalus kirhocephalus* (Lesson & Garnot)
    Synonym: *Pitohui kirhocephalus salvadorii* Meise, 1929. *Abh. Ber. Mus. Tierk. Dresden* (1927–29) 17(4): 20.—Warbusi, w coast of Geelvink Bay. [Based on a population of intergrades between *kirhocephalus* and *dohertyi.*]
RANGE: The ne coast of the Bird's Head south to Momi (Geelvink Bay).
DIAGNOSIS: Gray-hooded; mantle and back rich reddish brown; wings and tail fuscous; underparts rich fawn-buff; bill pale or brownish (not black).

c. *Pitohui kirhocephalus dohertyi* Rothschild & Hartert
Pitohui dohertyi Rothschild & Hartert, 1903. *Novit. Zool.* 10: 95.—Roon I, Geelvink Bay, off e coast of Bird's Head.
RANGE: Fringing islands and area around the Wandammen Penin, sw Geelvink Bay.
DIAGNOSIS: Head black (male) or chocolate brown (female); wings and tail black; body plumage rich dark brown; bill black. Rump brown, not black. Male is very similar to *P. dichrous,* but the brown body plumage is darker and more maroon-washed.

d. *Pitohui kirhocephalus rubiensis* (A. B. Meyer)
Rhectes rubiensis A. B. Meyer, 1884. *Sitzungsber. Abh. Nat. Ges. Isis Dresden,* Abh. 1: 32.—Rubi, head of Geelvink Bay, Bird's Neck.
    Synonym: *Pitohui kirhocephalus stramineipectus* van Oort, 1907. *Notes Leyden Mus.* 29: 73.—Lobo, Triton Bay, Bird's Neck.
    Synonym: *Pitohui kirhocephalus carolinae* Junge, 1952. *Zool. Meded. Leiden* 31: 248.—Etna Bay, Bird's Neck.
    Synonym: *Pitohui kirrhocepalus adiensis* Mees, 1964. *Zool. Meded. Leiden* 40: 126.—Adi I, Bird's Neck.
RANGE: The c and w sectors of the Bird's Neck, head of Geelvink Bay (to Etna Bay and Triton Bay); also Adi I.

DIAGNOSIS: Similar to *decipiens*, but male with head and tail paler, and female somewhat paler than male (Rand & Gilliard 1967).

e. *Pitohui kirhocephalus brunneivertex* Rothschild

Pitohui jobiensis brunneivertex Rothschild, 1931. *Novit. Zool.* 36: 262.—Mount Derimapa, 1,525 m, Gebroeders Mts, w sector of the Western Ra.

RANGE: The se coast of Geelvink Bay (Siriwo to Wanggar R; Hartert *et al.* 1936).

DIAGNOSIS: Head dull brown, slightly offset from rich red-brown upperparts and bright rufous underparts; iris red-brown; bill horn-colored. LeCroy (2010: 57) noted the holotype is from Mt Derimapa, not the Siriwo R as noted by Mayr (1941).

f. *Pitohui kirhocephalus jobiensis* (A. B. Meyer)

Rectes jobiensis A. B. Meyer, 1874. *Sitzungsber. Akad. Wiss. Wien* 69: 205.—Ansus, Yapen I, Bay Is.

RANGE: Yapen and Kurudu I, Bay Is.

DIAGNOSIS: Bill pale horn; plumage entirely rich red-brown, breast and belly slightly brighter; iris dark red. Sexually monochromatic.

g. *Pitohui kirhocephalus meyeri* Rothschild & Hartert

Pitohui meyeri Rothschild & Hartert, 1903. *Novit. Zool.* 10: 96.—Takar, n NG.

Synonym: *Pitohui kirhocephalus proteus* Hartert, 1932. *Nova Guinea, Zool.* 15: 469.—Ifar, Sentani Lake, n NG.

RANGE: The NW Lowlands.

DIAGNOSIS: Head and throat pale olive brown; upperparts rufous brown; underparts dull buffy; sexes similar. This is a much duller race than *senex* or *brunneicaudus*, which replace it to the east in the Sepik-Ramu. The Archbold specimens from Bernhard Camp are variable—either darker and richer, or paler—perhaps from a hybrid swarm? We follow Rand (1942b) and Ripley (1964a) in sinking the race *proteus* into *meyeri*.

h. *Pitohui kirhocephalus senex* Stresemann

Pitohui kirhocephalus senex Stresemann, 1922. *Orn. Monatsber.* 30: 8.—Maeanderberg, upper Sepik R, west-central Sepik-Ramu.

RANGE: The w sector of the Sepik-Ramu, in the upper Sepik valley, intergrading with *brunneicaudus* along the middle Sepik R (Gilliard & LeCroy 1966).

DIAGNOSIS: Male and female alike. Similar to *brunneicaudus*, but underparts slightly paler; hood buff gray; back and rump rich red-brown; underparts pale fawn; bill horn.

i. *Pitohui kirhocephalus brunneicaudus* (A. B. Meyer)

Rhectes brunneicaudus A. B. Meyer, 1891. *Abh. Ber. Zool. Mus. Dresden* (1890–91) 3(4): 10.—Near Madang, e verge of the Sepik-Ramu.

RANGE: The e sector of the Sepik-Ramu, between the lower Sepik, Astrolabe Bay, and the upper Ramu; also the n coast of the Huon Penin, east to Wasu (Gilliard & LeCroy 1966, 1967).

DIAGNOSIS: Head and throat buff gray; back deep reddish chestnut; wings and tail grayish brown; underparts ochraceous rufous or rufous buff; iris deep maroon to dark brown; bill brownish gray; sexes similar. Underparts richer fawn-buff than in *senex*. See photo in Coates (1990: 227) and Coates & Peckover (2001: 166).

NOTES

The original 20 races found in Rand & Gilliard (1967) are separated here into three distinct species. *Other names*—traditionally treated as part of the Variable Pitohui, a more-encompassing *P. kirhocephalus.*

Raja Ampat Pitohui *Pitohui cerviniventris*   ENDEMIC

Rectes cerviniventris G. R. Gray, 1862. *Proc. Zool. Soc. London* (1861): 430.—Waigeo I, NW Islands.

Inhabits the n NW Islands: Batanta, Sagewin, Waigeo, and Gam I.

SUBSPECIES

a. *Pitohui cerviniventris pallidus* van Oort

Pitohui cerviniventris pallidus van Oort, 1907. *Notes Leyden Mus.* 29: 71.—Batanta I, NW Islands.

RANGE: Batanta and Sagewin I[MA].

DIAGNOSIS: Similar to the nominate race but paler above and below; also larger; wing 100 (Rand & Gilliard 1967). Head, throat, upper breast, back, rump all pale gray buff; wings and tail slightly browner and darker; breast and belly pale fawn; bill black; iris dark brown. Sexually monochromatic.

b. *Pitohui cerviniventris cerviniventris* (G. R. Gray)

RANGE: Waigeo and Gam I[MA].

DIAGNOSIS: Upperparts pale gray-brown, darker and more brownish on wings and tail; paler and grayer on top and sides of head; throat grayish brown; breast, abdomen, and undertail coverts ferruginous; wing 95 (Rand & Gilliard 1967). Sexually monochromatic. These two forms are very close.

NOTES

Dumbacher *et al.* (2008) demonstrated that *cerviniventris* and *pallidus* are sister forms, whose sister clade includes *meridionalis* and *brunneiceps*. It is interesting to note that the two pitohui species in the NW Islands separate themselves across the Sagewin Strait—this species to the north, and *P. uropygialis* to the south. The strait is a biogeographic barrier for *Cicinnurus magnificus* and *respublica* and other species pairs. Although very narrow, it must be deep and old. *Other names*—formerly treated as a component of the Variable Pitohui, *P. kirhocephalus*.

Southern Variable Pitohui *Pitohui uropygialis*   ENDEMIC

Rectes uropygialis G. R. Gray, 1862. *Proc. Zool. Soc. London* (1861): 430.—Misool I, NW Islands.

Inhabits forest and edge of the s NW Islands (Salawati and Misool) and Aru Is; also w Bird's Head, S Lowlands, and s watershed of the SE Peninsula. Presumably also on the n coast of the e sector of the SE Peninsula.

SUBSPECIES

a. *Pitohui uropygialis uropygialis* (G. R. Gray)

Synonym: *Rectes tibialis* Sharpe, 1877. *Cat. Birds Brit. Mus.* 3: 285.—Sorong distr, w Bird's Head.

RANGE: Misool and Salawati I[MA] and the w half of the Bird's Head (Sorong, Loewelala).

DIAGNOSIS: Sexually dichromatic; male with black head, breast, wings, rump, and tail; otherwise rich and dark reddish brown; this plumage closely approximates the plumage of *P. dichrous*, but *uropygialis* is larger and darker and richer brown. Female apparently with hood chocolate brown and not extending down onto breast as much as that of male. The form *tibialis* belongs here as a synonym (Mees 1965).

b. *Pitohui uropygialis aruensis* (Sharpe)

Rectes aruensis Sharpe, 1877. *Cat. Birds Brit. Mus.* 3: 285.—Aru Is.

RANGE: The Aru Is.

DIAGNOSIS: Male similar to *nigripectus* but abdomen more richly colored chestnut; breast chestnut mixed with black; female head and throat dusky dark brown, not black. Wing 117 (Rand & Gilliard 1967).

c. *Pitohui uropygialis nigripectus* van Oort

Pitohui aruensis nigripectus van Oort, 1909. *Nova Guinea, Zool.* 9: 93.—Lorentz R, sw NG.

RANGE: The w and west-central sectors of the S Lowlands, between the Mimika R and the upper Eilanden R.

DIAGNOSIS: Male back chestnut; abdomen ferruginous; rest of plumage black; female with gray-brown head; breast and abdomen ferruginous; bill black; wing 125 (Rand & Gilliard 1967).

d. *Pitohui uropygialis brunneiceps* (D'Albertis & Salvadori)

Rectes brunneiceps D'Albertis & Salvadori, 1879. *Ann. Mus. Civ. Genova* 14: 70.—Middle Fly R, c S Lowlands.

RANGE: The c and e sectors of the S Lowlands; s flanks of the Eastern Ra; also the Trans-Fly; from Bian Lakes[J] east to Sturt I[K], Palmer Junction[K], Soliabeda[DI], Karimui[DI], Bomai[DI], Purari R[MA], and Vailala R[MA] (Rand 1942a[K], Bishop 2005a[J]).

DIAGNOSIS: Head and throat gray-brown to dark brown; back rusty rufous; wings and tail brown; underparts rich ochraceous buff; iris red to red-brown; bill black; legs and feet gray to dark brown; sexes similar. Immatures resemble the adults.

e. *Pitohui uropygialis meridionalis* (Sharpe)

Rectes meridionalis Sharpe, 1888. *Ibis*: 437.—Astrolabe Ra, SE Peninsula.

RANGE: The SE Peninsula, in the s watershed from the Lakekamu basin southeast to Milne Bay; in the n watershed from Milne Bay northwest to Awaiama Bay and perhaps considerably farther; should be looked for in Collingwood Bay and the Mambare and Musa deltas, although Rand & Gilliard (1967) reported the species absent from these lowland forests. Dawson *et al.* (2011) did not report it from the lower Waria R.

DIAGNOSIS: Male with head, throat, upper breast (and sometimes lower breast), wings, lower rump, upper-tail coverts, and tail black; back rusty chestnut; underparts rufous buff; bill black. Female like *brunneiceps*. The male of this race is very similar to *P. dichrous* but is larger (wing 122–130, compared with 100–109 for *P. dichrous*), the bill is heavier, and the uppertail coverts and lower rump are black, not rufous.

NOTES

Other names—formerly treated as subspecies of the Variable Pitohui, *P. kirhocephalus*.

Hooded Pitohui *Pitohui dichrous*   ENDEMIC | MONOTYPIC

Rhectes dichrous Bonaparte, 1850. *Compt. Rend. Acad. Sci. Paris* 31: 563.—Triton Bay, Bird's Neck.

   Synonym: *Pitohui dichrous monticola* Rothschild, 1904. *Bull. Brit. Orn. Club* 14: 79.—Upper Aroa R, SE Peninsula.

Inhabits forest and edge throughout the uplands of mainland NG and Yapen I[MA]; locally at sea level but usually in hills and higher: 600–1,700 m, max 2,000 m[CO]. Recorded near sea level at: Jayapura[CO], Bian Lakes[J], the middle Fly R[CO], Sturt I[J], Madang[CO], Huon Penin[CO], Lae[CO], Hall Sound[CO], and Hisiu[CO] (Bishop 2005a[J]). On the YUS transect on the n scarp of the Huon Penin, *P. dichrous* and *P. kirhocephalus* shared foothill forest and scrub between 140 and 950 m (Beehler field notes).

NOTES

We follow Rand & Gilliard (1967: 437) in treating this species as monotypic.

GENUS *SPHECOTHERES*   [Genus 3 spp/Region 1 sp]

Sphecotheres Vieillot, 1816. *Analyse Nouv. Orn. Elem.*: 42. Type, by monotypy, *Sphecotheres viridis* Vieillot.

   The species of this genus range from the Lesser Sundas to Australia. The genus is characterized by the distinctive bare skin surrounding the eye, sexual dichromatism, fig-eating habit, and colonial nesting.

Australasian Figbird *Sphecotheres vieilloti*   RESIDENT

Sphecotera Vieilloti Vigors & Horsfield, 1827. *Trans. Linn. Soc. London* 15: 215.—Keppel Bay, near Rock-hampton, QLD, Australia.

   Frequents savanna, town gardens, and mangroves on the SE Peninsula; lowlands mainly, but found up to *ca.* 500 m on the Sogeri Plateau. Also a single sighting from the south-central S Lowlands (Coates 1990).

   Extralimital—n and e Australia, Kai Is, and islands in s Torres Strait.

SUBSPECIES

a. *Sphecotheres vieilloti salvadorii* Sharpe

Sphecotheres salvadorii Sharpe, 1877. *Cat. Birds Brit. Mus.* 3: 224, pl. 12.—Port Moresby, SE Peninsula.

RANGE: The SE Peninsula, from Bereina southeast to Kupiano in the s watershed; Alotau and Collingwood Bay; also a single sight record at the Bamu R in the lower Fly (Bell 1967a). Localities—Bamu R, Bereina, Port Moresby, Sogeri Plateau, Kupiano, and Rabaraba (Coates 1990). Unconfirmed sightings from Wau (Macadam NP) and the Trobriand Is (*PNG Bird Soc. Newsl.* 201: 23 and 209: 8); the first is almost certainly in error, and the second perhaps more plausible but still requiring corroboration.

DIAGNOSIS: Similar to the nominate form but smaller. Male's lower breast, upper belly, and flanks mostly yellow with slight orange tinge; all rectrices except central pair with extensive white tipping (Walther & Jones 2008). See photos in Coates (1990: 368).

Species-level treatment follows Schodde & Mason (1999) and Christidis & Boles (2008).
Other names—Figbird, Green Figbird, *Sphecotheres viridis.*

GENUS *ORIOLUS* [Genus 28 spp/Region 3 spp]

Oriolus Linnaeus, 1766. *Syst. Nat.*, ed. 12, 1: 160. Type, by tautonymy, *Oriolus galbula* Linnaeus = *Coracias oriolus* Linnaeus.

The species of this genus range from Australia and New Guinea west to Europe and Africa (Walther & Jones 2008). The genus is characterized by the frugivorous habit, accomplished vocal ability, sturdy build, and strong bill.

Olive-backed Oriole *Oriolus sagittatus* RESIDENT (AND VISITOR?)

C[*oracias*] *Sagittata* Latham, 1801. *Index Orn.*, Suppl.: xxvi.—Sydney, NSW, Australia.

Breeds in wooded savannas of the Trans-Fly, from Merauke to the Oriomo R, near sea level (Bishop 2005a). Perhaps also a seasonal visitor from Australia. Records include specimens and sight records in May, July, and August, and specimens in December and February, and a breeding record for February. Ten of the 14 specimens collected at Dogwa, Oriomo R, by members of Archbold Expedition in February, were immatures—an unusually high proportion (Mayr & Rand 1937). The species is an uncommon visitor to the s Torres Strait at the end of the dry season (Draffan *et al.* 1983) and has been recorded from Murray I in July and early August (Ingram 1976), suggesting that there may be some movement between Australia and NG. That said, the population sampled in the Trans-Fly has been described as a local subspecies, not the nominate form from e Australia, which is migratory (Schodde & Mason 1999). One at Varirata NP in Aug-2001 (P. Gregory *in litt.*)

Extralimital—n and e Australia; one record from New Britain (Dutson 2011).

SUBSPECIES

a. *Oriolus sagittatus magnirostris* van Oort
Oriolus sagittata magnirostris van Oort, 1910. *Notes Leyden Mus.* 32: 82.—Merauke, s NG.
RANGE: The Trans-Fly. Localities—Maro R[J], Merauke[MA], Erambu[J], Bensbach[J], Tarara[K], Penzara[K], Dogwa on the Oriomo R[J], and Daru[J] (Rand 1942a[K], Bishop 2005a[J]).
DIAGNOSIS: Small; pale; uniform, heavy wedge-shaped streaks dorsally (most notable on crown); tail spots small; wing: male 140–145, female 135–142 (Walther & Jones 2008). Schodde & Mason (1999) noted heavy and uniformly cuneate-shaped markings on the mantle in both sexes; remiges of adult males washed citrine; upper surfaces of rectrices in both sexes washed citrine.

NOTES
Other names—White-bellied, Australian, Green-backed, or Green Oriole

Brown Oriole *Oriolus szalayi* ENDEMIC | MONOTYPIC

Mimeta szalayi Madarász, 1900. *Termes. Füzet.* 24: 76, 80.—Finschhafen, Huon Penin.

A common inhabitant of forest and edge through NG and the four main NW Islands[MA]; sea level–1,200 m, max 1,850 m[CO]. Trans-Fly records include Princess Marianne Strait, Wasur NP, Bian Lakes, Bensbach, and Wuroi (Bishop 2005a). A record from the Trobriand Is (*PNG Bird Soc. Newsl.* 209: 8) was probably a misidentification of a juvenile *Philemon buceroides* (Coates 1990).

NOTES
In the Trans-Fly apparently does not coexist with its two sister species (*O. flavocinctus* and *O. sagittatus*). *Other names*—Striated or New Guinea Oriole.

Green Oriole *Oriolus flavocinctus*   RESIDENT | MONOTYPIC

[*Mimetes*] *flavo-cinctus* P. P. King, 1826. *Surv. Intertrop. Coasts Austral.* 2: 418.—Northern Territory, Australia.

> Synonym: *M*[*imeta*] *mülleri* Bonaparte, 1850. *Consp. Gen. Av.* 1: 346.—Dourga R, Princess Marianne Strait, Trans-Fly.

Breeds in the Aru Is[MA] and the s Trans-Fly from Dolak I east to the Mai Kussa R, near sea level. Inhabits monsoon forest and riparian scrub forest (Coates 1990).

Localities—Dolak I[L], Roma I[MA], Dourga R[MA], Kurik[K], Merauke[L], Wasur NP[L], Bensbach R[CO], Wassi Kussa R[MA], and Bugi—3 km east of the Mai Kussa R[J] (Rand 1942a[J], Mees 1982b[K], K. D. Bishop 2005a[L]).

Extralimital—n and ne Australia and Lesser Sunda Is.

NOTES

We follow Keast (1956) and Mees (1964b, 1982b) in treating the form as monotypic (*pace* Schodde & Mason 1999 and Walther & Jones 2008). A rare gray morph is known from sight records and from a specimen taken from Bensbach, now in the PNG Nat. Mus. (B. J. Coates *in litt.*). Map in Higgins *et al.* (2006) showing range to include the SE Peninsula is in error. *Other names*—Australian Yellow, Yellow, or Yellow-bellied Oriole.

Family Rhipiduridae

Fantails   [Family 46 spp/Region 14 spp]

The species of this family range from tropical Asia to Melanesia, Polynesia, and New Zealand. Fantails are a distinct and compact family dominated by the genus, *Rhipidura*. Barker *et al.* (2004) first discovered the sister relationship between the *Rhipidura* fantails and *Chaetorhynchus*, later confirmed by Nyári *et al.* (2009) and Norman *et al.* (2009b). The clade formed by those two genera are sister to a large lineage that includes, among others, the crows, monarchs, and birds of paradise. We place *Chaetorhynchus* at the base of the Rhipiduridae. Given the large size of the genus *Rhipidura*, we are happy to maintain this distinct and speciose lineage as a full-fledged family, a practice that is followed by others (*e.g.*, Dickinson 2003, Nyári *et al.* 2009). Fantails are distinct in the following characters: tiny size, slim body, long fanned tail, small head, broad abbreviated bill, and short rounded wing. Our treatment here is based on the molecular phylogeny of Nyári *et al.* (2009). Most recently, Schodde & Christidis (2014) created the Lamproliinae as a subfamily of the Rhipiduridae to encompass the genera *Chaetorhynchus* and *Lamprolia*, formalizing their phylogenetic relationship despite obvious differences in appearance and behavior.

GENUS *CHAETORHYNCHUS*   ENDEMIC | MONOTYPIC

Chaetorhynchus A. B. Meyer, 1874. *Sitzungsber. Akad. Wiss. Wien* 69: 493. Type, by monotypy, *Chaetorhynchus papuensis* A. B. Meyer.

The single species of the genus is endemic to the uplands of New Guinea. The genus is characterized by the glossy all-black plumage with some metallic highlighting; the heavy short rictal bristles; the strong abbreviated and hooked bill; the white patch at the bend of the wing, and the complex and beautiful song. This monotypic genus is traditionally classified as a member of the Dicruridae. We follow Jønsson *et al.* (2011b) in situating *Chaetorhynchus* as sister to *Lamprolia*, with the pair as sister to *Rhipidura*. Coates (1990) described various aspects of posture and behavior of *Chaetorhynchus* that link it to the fantails, and Schodde & Christidis (2014) listed multiple anatomical characters that are rhipidurid rather than dicrurid. We have adopted the name Drongo Fantail to reflect the bird's true affinity.

Drongo Fantail *Chaetorhynchus papuensis*   ENDEMIC | MONOTYPIC

Chaetorhynchus papuensis A. B. Meyer, 1874. *Sitzungsber. Akad. Wiss. Wien* 69: 493.—Arfak Mts, Bird's Head.

A noisy and active inhabitant of midstory forest interior of mts throughout NG; 600–1,460 m, extremes 200–1,600 m[CO] (Coates & Peckover 2001). Range includes the Tamrau[MA], Arfak[MA], Fakfak[J], and Kumawa Mts[K], Central Ra[MA], Wandammen[MA] and Foja Mts[L], PNG North Coastal Ra[M], Adelbert Mts[N], and Huon Penin[MA]; also mts of SE Peninsula[MA] (Gibbs 1994[J], Gilliard & LeCroy 1967[N], Beehler *et al.* 2012[L], specimens at AMNH & BPBM[M], J. M. Diamond *in litt.*[K]). No records from the Cyclops Mts.

Other names—Mountain or Papuan Mountain Drongo, Pygmy Drongo, Papuan Silktail.

GENUS *RHIPIDURA*   [Genus 44 spp/Region 13 spp]

Rhipidura Vigors & Horsfield, 1827. *Trans. Linn. Soc. London* 15: 246. Type, by subsequent designation (G. R. Gray, 1840. *List Gen. Birds*: 32), *Muscicapa flabellifera* Gmelin = *Muscicapa fuliginosa* Sparrman.

The species of this genus range from Australia, New Zealand, New Guinea, and Melanesia, north and west to Micronesia and tropical Asia (Taylor 2006). The genus is characterized by the very slight body, long cocked and fanned tail, small head, abbreviated bill, prominent rictal bristles, and the short, rounded wing. Our treatment here is informed by the molecular phylogeny of the genus by Nyári *et al.* (2009) but with some modification to clade sequence based on morphological characters (Mayr 1986, Boles 2006).

The Grey Fantail (*Rhipidura albiscapa*), of Australian woodlands, should be looked for in southern New Guinea as a rare migrant; it is easily confused with the similar *R. phasiana*, which is a mangrove specialist. Look for *R. albiscapa* in non-mangrove habitats.

Willie Wagtail *Rhipidura leucophrys*   RESIDENT

Turdus leucophrys Latham, 1801. *Index Orn.*, Suppl.: xlv.—Sydney, NSW, Australia.

Throughout NG and most nearby islands; sea level–1,300 m, max 2,800 m. Inhabits savanna, coasts, riverbanks, clearings, gardens, and town environs (Coates 1990).

Extralimital—Buru, Seram, all Australia; also the Bismarcks and most Solomon Is (Mayr & Diamond 2001, Higgins *et al.* 2006).

SUBSPECIES

a. *Rhipidura leucophrys melaleuca* (Quoy & Gaimard)

Muscipeta melaleuca Quoy & Gaimard, 1830. *Voy. Astrolabe, Zool.* 1: 180.—Carteret Harbor, New Ireland, Bismarck Arch.

RANGE: Mainland NG, all NW Islands[MA], the Aru Is, Kairiru, Karkar[J], and Bagabag I[J]; Goodenough, Fergusson, and Normanby I[CO]; also the Bismarck Arch, Solomon Is, and Moluccas (Mayr 1941, Diamond & LeCroy 1979[J]). Absent from the Louisiade Arch.

DIAGNOSIS: Underparts plain white; wing (male) 99–105 (Higgins *et al.* 2006). Boles (2006) also noted this race has a relatively large bill. Schodde & Mason (1999) noted the following characters: size large; bill long and broad; rictal bristles stout, reaching the tip of the bill.

NOTES

The Australian race *picata* does not range northward into s NG (*fide* Schodde & Mason 1999, Boles 2006, *pace* Mayr 1941). *Other names*—Black-and-white Fantail or Flycatcher, White-browed Fantail.

Black Thicket-Fantail *Rhipidura maculipectus*   ENDEMIC | MONOTYPIC

Rhipidura maculipectus G. R. Gray, 1858. *Proc. Zool. Soc. London*: 176.—Aru Is.

The NW Islands (Salawati and Batanta), Aru Is, Bird's Head, Bird's Neck, S Lowlands, and w half of the s watershed of the SE Peninsula to the Port Moresby area, possibly to Orangerie Bay; also Daru I (Coates 1990). In the Geelvink Bay region east to Warbusi. Range includes the Trans-Fly (Rand 1942a). Lowlands only. Inhabits swamp forest, mangrove forest, dense undergrowth, and other thick vegetation on damp ground (Coates 1990). Mees (1965) questioned the old record from Misool I. Future fieldworkers need to look for the species on Misool.

White-bellied Thicket-Fantail *Rhipidura leucothorax* ENDEMIC

Rhipidura leucothorax Salvadori, 1874. *Ann. Mus. Civ. Genova* 6: 311.—Andai, w coast of the Bird's Head.

A vocal inhabitant of dense thickets and tangles at forest edge of the NG mainland; also Kairiru and Muschu I; sea level–750 m, extreme 1,350 m[CO]. Few records from the Trans-Fly.

SUBSPECIES

a. *Rhipidura leucothorax leucothorax* Salvadori
> Synonym: *Rhipidura leucothorax clamosa* Diamond, 1967. *Amer. Mus. Novit.* 2284: 7.—Soliabeda, 600 m, s watershed of Eastern Ra.

RANGE: Western, c, and e NG; east to Astrolabe Bay in the n watershed, and east to Port Moresby in the s watershed; also Kairiru and Muschu I[CO]. Trans-Fly records from Lake Daviumbu and Aramia R (Bishop 2005a).

DIAGNOSIS: Dorsal plumage ranges from dull blackish brown (patchy in some specimens) to very dull slate brown. The form *clamosa* from the Karimui region includes specimens that compare well with topotypical specimens of the nominate form. See photos in Coates (1990: 125).

b. *Rhipidura leucothorax episcopalis* Ramsay

Rhipidura episcopalis Ramsay, 1878. *Proc. Linn. Soc. NSW* 2: 371.—Southern coast of the SE Peninsula.

RANGE: The SE Peninsula west along the s watershed to Kapa Kapa, and along the n watershed to the middle Sepik[CO], where it intergrades with *leucothorax*.

DIAGNOSIS: Back dull brownish with a slight rufescent tinge.

NOTES

The minor subspecific variation may be clinal, and perhaps the species should be treated as monotypic. *Other names*—Black-throated or White-breasted Thicket Fantail.

Sooty Thicket-Fantail *Rhipidura threnothorax* ENDEMIC

Rhipidura threnothorax S. Müller, 1843. *Verh. Nat. Gesch. Ned., Land-en Volkenk.* 1: 185 (note).—Triton Bay, Bird's Neck.

Widespread in lowland and foothill forest on NG, the NW Islands, Aru Is, and Yapen I; sea level–800 m, max 1,240 m[J] (Freeman & Freeman 2014[J]). Inhabits the understory of shaded forest interior.

SUBSPECIES

a. *Rhipidura threnothorax threnothorax* S. Müller

RANGE: The Aru Is[MA], Misool[J], Waigeo[MA], and Salawati I[MA]; also throughout Mainland. Lowland forest up to 900 m, occasionally to 1,100 m (Ripley 1964a[J], Mees 1965). Widespread in Trans-Fly (Bishop 2005a).

DIAGNOSIS: Crown brown. See photos in Coates (1990: 121).

b. *Rhipidura threnothorax fumosa* Schlegel

Rhipidura fumosa Schlegel, 1871. *Ned. Tijdschr. Dierk.* (1873) 4: 42.—Yapen I, Bay Is.

RANGE: Yapen I.

DIAGNOSIS: Crown blackish; mantle deeper brown than Mainland forms.

Rufous-backed Fantail *Rhipidura rufidorsa* ENDEMIC

Rhipidura rufidorsa A. B. Meyer, 1874. *Sitzungsber. Akad. Wiss. Wien* 70: 200.—Passim and Rubi, e coast of the Bird's Head, and Yapen I, Bay Is.

Inhabits the middle stories of forest interior through NG, Misool, and Yapen I; sea level–650 m, max 850 m[CO]. Widespread on the Mainland except for perhaps much of the Trans-Fly.

SUBSPECIES

a. *Rhipidura rufidorsa rufidorsa* Meyer

RANGE: Misool[MA] and Yapen I[MA], the Bird's Head[MA], and most of NG, east in the n watershed apparently to Astrolabe Bay, and in the s watershed to the Fly R (perhaps farther); turnover zone between this form and *kubuna* lies somewhere between the Fly R and the Port Moresby region. Absent from the Trans-Fly.

DIAGNOSIS: Rufous of mantle and rump moderately bright (Rand & Gilliard 1967).

b. *Rhipidura rufidorsa kubuna* Rand
Rhipidura rufidorsa kubuna Rand, 1938. *Amer. Mus. Novit.* 991: 9.—Kubuna, SE Peninsula.
RANGE: The s watershed of the SE Peninsula, ranging westward at least to the Purari R (Diamond 1972).
DIAGNOSIS: Rufous of mantle dull and pale, tinged grayish; rump pale (Rand 1938b). See photos in Coates (1990: 127) and Coates & Peckover (2001: 137).

c. *Rhipidura rufidorsa kumusi* Mathews
Rhipidura rufidorsa kumusi Mathews, 1928. *Novit. Zool.* 34: 373.—Kumusi R, n watershed of the SE Peninsula.
RANGE: The n coast of the SE Peninsula between the Kumusi R and Collingwood Bay[CO]; presumably ranges northwestward in the n watershed to Astrolabe Bay to meet the nominate form.
DIAGNOSIS: Rufous of back most vivid of the three races (Rand & Gilliard 1967).

NOTES
Three thinly defined forms. They would be subsumed as clinal, except the brightest and dullest forms both appear in the SE Peninsula (bright in n watershed *vs.* dull in s watershed). Apparently there are three distinct vocal dialects; these are found in the Sepik, the upper Fly, and the Port Moresby region (K. D. Bishop *in litt.*, P. Gregory *in litt.*). *Other names*—Grey-breasted Rufous Fantail, Chestnut-backed Fantail.

Dimorphic Fantail *Rhipidura brachyrhyncha*   ENDEMIC | MONOTYPIC
Rhipidura brachyrhyncha Schlegel, 1871. *Ned. Tijdschr. Dierk.* (1873) 4: 42.—Arfak Mts, Bird's Head.
   Synonym: *Rhipidura devisi* North, 1898. *Proc. Linn. Soc. NSW* (1897) 22: 444.—Mount Scratchley, SE Peninsula.
      Replacement name for *Rhipidura albicauda* De Vis, 1897, *Ibis*: 375; preoccupied by *Rhipidura albicauda* North, 1895.
Inhabits mt forest of the Bird's Head[MA], Central Ra[MA], Foja Mts[J], and Huon Penin[MA] (Diamond 1985[J]); 2,000–3,680 m, extremes 1,160 m–timberline[CO]. The report from the Adelbert Mts (Mackay 1991) is probably in error. This is a high-elevation species not expected in the Adelberts. It was not recorded there by Pratt (1983).

NOTES
This dichromatic species is rather variable, and individual variation exceeds the described variation between the subspecies. We have found individuals in museum collections from the two disparate topotypical localities that are identical. *Other names*—Dimorphic Rufous Fantail.

Rufous Fantail *Rhipidura rufifrons*   RESIDENT AND VISITOR
Muscicapa rufifrons Latham, 1801. *Index Orn.*, Suppl.: 1[=50].—Sydney, NSW, Australia.
   A winter visitor from Australia to the Trans-Fly, the middle Fly, and the SE Peninsula; also a resident population in the SE Islands. On mainland NG, the Australian migrants in the Trans-Fly occur in forest, including monsoon forest (Coates 1990). Resident island populations inhabit forested habitats on the Killerton, D'Entrecasteaux, Bonvouloir, and Louisiade groups of the SE Islands. Found only near sea level on the Mainland but ranges up to 1,600 m on Goodenough I (Beehler field notes).
   Extralimital—n Moluccas, Micronesia, Solomon Is, Temotu Is, and e Australia from Cape York southward (Boles 2006, Dickinson & Christidis 2014).

SUBSPECIES
a. *Rhipidura rufifrons rufifrons* (Latham)
RANGE: Winter visitor from se Australia to s NG. Localities—Maro R[K], Wasur NP[K], Bensbach[K], Gaima on the lower Fly R[J], Lake Daviumbu[K], near Kiunga[L], Port Moresby, Orangerie Bay (Mayr 1941, Rand 1942a[J], Bishop 2005a[K], P. Gregory *in litt.*[L]). A passage migrant through the Torres Strait Is (Draffan *et al.* 1983).
DIAGNOSIS: Forehead and line above eye rufous, lower eyelid white; rufous of rump extends substantially onto tail feathers (Rand & Gilliard 1967). Face dull brown; crissum cinnamon, this wash extending up onto belly; dirty-gray tipping to rectrices (Schodde & Mason 1999).

b. *Rhipidura rufifrons intermedia* North

Rhipidura rufifrons intermedia North, 1902. *Victorian Naturalist* 19: 101.—Bellenden Ker Ra, n QLD, Australia.

RANGE: Eastern QLD north to Cairns; birds of this form apparently winter north to the Trans-Fly (Schodde & Mason 1999).

DIAGNOSIS: Tips of rectrices white, not dirty gray; belly whitish (Schodde & Mason 1999).

c. *Rhipidura rufifrons louisiadensis* Hartert

Rhipidura louisiadensis Hartert, 1899. *Novit. Zool.* 6: 78.—Rossel I, Louisiade Arch, SE Islands.

RANGE: Goodenough[CO], Fergusson[CO], Duchess, and Tobweyama I[J]; Irai and Panasesa I of the Conflict Is[K]; Hastings and East I[MA]; Kimuta[L]; Deboyne, Renard, and Rossel I[CO] (LeCroy & Peckover 1998[L], Pratt field notes[J], Beehler field notes 29-Sep-2002[K]). Birds from the Killerton Is have not been diagnosed but probably belong here. Record for Misima is probably in error for adjacent Kimuta I (LeCroy & Peckover 1998). This bird is typically found on small SE Islands, with the exception of Goodenough and Rossel I. On Goodenough I, the species inhabits both mangroves (Bell 1970f) and mid-montane oak forest (Mayr & Van Deusen 1956). Found islandwide on Rossel (Pratt field notes 2004). It was not encountered on Fergusson or Normanby by Pratt (field notes 2003) but was found on offshore islets.

DIAGNOSIS: Tips of rectrices white, not gray (Rand & Gilliard 1967). Chin and throat white; breast band strongly scaled posteriorly; ear coverts very dark brown; rectrices rufous only at base (Boles 2006). See photo in Coates & Peckover (2001: 139).

Arafura Fantail *Rhipidura dryas* RESIDENT

Rhipidura dryas Gould, 1843. *Proc. Zool. Soc. London* (1842): 132.—Port Essington, NT, Australia.

The NW Islands and Aru Is; also a Mainland record from mangroves of the Mimika R of sw NG; possibly also in mangroves northwest of Port Moresby; lowlands only (Boles 2006). Distribution on mainland of s NG requires additional field study.

Extralimital—Lesser Sundas, Banda Is, Kai Is, s Moluccas, and n Australia (Boles 2006).

SUBSPECIES

a. *Rhipidura dryas squamata* S. Müller

Rhipidura squamata S. Müller, 1843. *Verh. Nat. Gesch. Ned., Land-en Volkenk.* 1: 184 (note).—Banda Is, s Moluccas.

Synonym: *Rhipidura squamata henrici* Hartert, 1918. *Bull. Brit. Orn. Club* 38: 59.—Kasiui I, southeast of Seram.

RANGE: Kofiau[J], Misool[MA], Wai[K], Salawati[MA], Haitlal[L], and Waigeo[MA] I; Schildpad Is[MA]; also Babi I in the Aru Is[MA] (Mees 1965[L], Diamond *et al.* 2009[J], Beehler field notes 13-Mar-2005[K]). Extralimital—Banda, Seram Laut, Tayandu, and Kai Is (White & Bruce 1986). Sighting from Batanta I (S. Hogburg observation, reported on M. Tarburton's island website).

DIAGNOSIS: Crown, nape, and shoulder dun gray; back and rump bright rufous; black bandit mask; broad black breast bar, spotted posteriorly; white chin and cheek; upper surface of wings and tail dun gray (based on K. D. Bishop photo). We follow White & Bruce (1986) in lumping the thinly defined *henrici* (*pace* Boles 2006).

b. *Rhipidura dryas streptophora* Ogilvie-Grant

Rhipidura streptophora Ogilvie-Grant, 1911. *Bull. Brit. Orn. Club* 29: 25.—Mimika R, w sector of the S Lowlands.

RANGE: Known only from the Mimika R of sw NG. Probably this form is the "rufous fantail" reported from mangroves from the Gulf of Papua east to Hisiu, nw of Port Moresby (*fide* P. Gregory). The species should be looked for in the mangroves of south-central NG.

DIAGNOSIS: Forehead, crown, and back washed with bright rufous; flanks dark; undertail coverts richly colored (Boles 2006).

NOTES

For this species-level treatment, we follow Schodde & Mason (1999). This species is distinguished from the preceding by the long, more strongly graduated tail; the narrower rufous patches at base of tail; and more extensive white tipping to outer rectrices (Boles 2006). Which of these two cryptic sibling species inhabits the coastal mangroves of the e sector of the S Lowlands and the s coast of the SE Peninsula needs clarification. *Other names*—formerly included within *R. rufifrons*.

Black Fantail *Rhipidura atra*   ENDEMIC

Rhipidura atra Salvadori, 1876. *Ann. Mus. Civ. Genova* (1875) 7: 922.—Mount Moari, Arfak Mts, Bird's Head.

Inhabits the mts of NG; also reported from Waigeo I (Rand & Gilliard 1967). Ranges from 1,000 to 2,150 m, extremes 700–3,200 m (Coates 1990, Coates & Peckover 2001). Frequents clearings and treefalls in forest interior and dense regrowth at forest edge.

SUBSPECIES

a. *Rhipidura atra atra* Salvadori
RANGE: Waigeo I[J], mts of the Bird's Head, Wandammen Mts, throughout the Central Ra, and mts of Huon Penin (Mayr 1941, Rand & Gilliard 1967[J]). Undiagnosed populations sighted in the Fakfak[K] and Kumawa[L] Mts may belong here (Gibbs 1994[K], J. M. Diamond *in litt.*[L]). Record from Waigeo I requires corroboration.
DIAGNOSIS: Female plumage duller and paler than that of the following form. See photo in Coates (1990: 131) and Coates & Peckover (2001: 138).

b. *Rhipidura atra vulpes* Mayr
Rhipidura atra vulpes Mayr, 1931. *Mitt. Zool. Mus. Berlin* 17: 684.—Cyclops Mts, ne sector of NW Lowlands.
RANGE: The Cyclops Mts[MA] and PNG North Coastal Ra[J] (specimens at AMNH & BPBM[J]). Presumably birds from the Foja[K] and Adelbert Mts[L] refer to this form (Pratt 1983[L], Beehler *et al.* 2012[K]).
DIAGNOSIS: Female plumage richer and deeper rufous; darker dorsally.

NOTES

The race *vulpes* is diagnosable but thinly defined by plumage. In contrast to the conclusions drawn on plumage characters, molecular genetic research by Benz (2011: 132) found *R. atra* is "composed of 3 geographically structured phylogroups including a highly divergent Vogelkop lineage that is sister to a Papuan Peninsula lineage in the east and a broadly distributed lineage that extends from the Watut/Tauri R valley west to the Wandammen Mountains, including outlying sky-island populations along the north coast." Here is another case of a uniformly plumaged bird demonstrating hidden but strong phylogeographic structuring.

Friendly Fantail *Rhipidura albolimbata*   ENDEMIC | MONOTYPIC

Rhipidura albo-limbata Salvadori, 1874. *Ann. Mus. Civ. Genova* 6: 312.—Hatam, Arfak Mts, Bird's Head.
   Synonym: *Rhipidura albo-limbata lorentzi* van Oort, 1909. *Nova Guinea, Zool.* 9: 85.—Hellwig Mts, e sector of the Western Ra.
   Synonym: *Rhipidura auricularis* De Vis, 1890. *Ann. Rept. Brit. New Guinea*, 1888–89: 59.—Musgrave Ra, SE Peninsula.

Inhabits forests of virtually all mts of NG, 1,750 m–timberline (records down to 1,130 m; Coates & Peckover 2001). Widespread through the Central Ra; also mts of the Bird's Head[MA] and Bird's Neck (Fakfak[J] and Kumawa Mts[K]), Foja Mts[L], Cyclops Mts[MA], PNG North Coastal Ra[M], and mts of the Huon Penin[MA] (Gibbs 1994[J], Beehler *et al.* 2012[L], specimens at AMNH & BPBM[M], J. M. Diamond *in litt.*[K]). No records from the Wandammen Mts. The sole Adelbert Mts sighting from Mackay (1991) warrants corroboration, as the species was not recorded there by Pratt (1983).

NOTES

We treat this species as monotypic. The described subspecies are based on nongeographic differences in elevation. Higher-elevation birds are darker and larger. We believe it is better to treat this as environmentally driven variation in phenotype.

Chestnut-bellied Fantail *Rhipidura hyperythra*   ENDEMIC

Rhipidura hyperythra G. R. Gray, 1858. *Proc. Zool. Soc. London*: 176.—Aru Is.

All NG, the Aru Is, and Yapen I; foothills–1,500, max 1,750 m^{CO}; patchily distributed in the lowlands, usually adjacent to main ranges. Additional lowland populations are isolated in the Oriomo, Morehead, and Tarara R of the Trans-Fly. No records from the Fakfak, Kumawa, or Wandammen Mts.

SUBSPECIES

a. *Rhipidura hyperythra hyperythra* G. R. Gray

> Synonym: *Rhipidura Mülleri* A. B. Meyer, 1874. *Sitzungsber. Akad. Wiss. Wien* 69: 502.—Lobo, Triton Bay, Bird's Neck. Replacement name for *Rhipidura rufiventris* S. Müller, 1843, *Verh. Nat. Gesch. Ned., Land-en Volkenk.* 1: 194 (note); preoccupied by *Platyrhynchus* [= *Rhipidura*] *rufiventris* Vieillot 1818.

RANGE: The Aru Is and w and c NG, from the Bird's Head east in the n watershed to Astrolabe Bay (where intergrading with *castaneothorax*), and in the s watershed presumably as far as the Purari R. Also Yapen I (Mayr 1941, Diamond 1972).

DIAGNOSIS: White spots in outer tail feathers 6–14 mm (Diamond 1972). The form *muelleri* is merged with the nominate form because the range of extent of white on the chin is variable, encompassing both forms.

b. *Rhipidura hyperythra castaneothorax* Ramsay

Rhipidura castaneothorax Ramsay, 1879. *Proc. Linn. Soc. NSW* 3: 270.—Goldie R, SE Peninsula.

RANGE: The SE Peninsula, northwest in the n watershed to the Huon Penin and northwest in the s watershed to the Purari R of the e sector of the S Lowlands.

DIAGNOSIS: Similar to the nominate form, but white tips to outer rectrices larger, 15–20 mm (Rand & Gilliard 1967, Diamond 1972). See photos in Coates (1990: 134) and Coates & Peckover (2001: 137).

Mangrove Fantail *Rhipidura phasiana*   RESIDENT | MONOTYPIC

Rhipidura phasiana De Vis, 1885. *Proc. Roy. Soc. Queensland* 1: 158.—Kimberley, mouth of Norman R, nw QLD, Australia.

Inhabits mangroves of the Trans-Fly from Dolak I to the PNG border; the s watershed of the SE Peninsula from Bereina (northwest of Port Moresby) east to Galley Reach; also apparently the Aru Is (Diamond & Bishop 1994).

Extralimital—coastal n and nw Australia.

NOTES

We follow Boles (2006) in treating the species as monotypic. Most, or all, historical records of "Grey Fantail" in NG are referable to *R. phasiana*, the Mangrove Fantail. What Mayr (1986) treated as the single species *R. fuliginosa* has subsequently been dissected into three species: *R. fuliginosa* of NZ, *R. albiscapa* of Australian woodlands, and *R. phasiana* of mangrove habitats of n Australia and s NG (see Schodde & Mason 1999, Gill *et al.* 2010). The Australian *R. albiscapa* may occur in the Fly R region as a migrant. There is an unconfirmed sight record of what may have been either a resident *R. phasiana* or a migrant *R. albiscapa* from Dolak I and Bensbach in the Trans-Fly (see Coates 1990: 133, K. D. Bishop *in litt.*). This need for species-level clarification also applies to the sight record of a "Grey Fantail" by Diamond & Bishop (1994) from the Aru Is. *Other names*—Pheasant Fantail.

Northern Fantail *Rhipidura rufiventris*   RESIDENT

Platyrhynchos rufiventris Vieillot, 1818. *Nouv. Dict. Hist. Nat.*, nouv. éd. 27: 21.—Timor, Lesser Sundas.

All NG and most satellite islands, sea level–1,600 m (Pratt & Beehler 2014). Inhabits forest and edge, and can be found foraging in garden regrowth (Coates 1990). Apparently absent from Numfor I; also absent from Karkar and the Trobriand Is, Woodlark, and Rossel I (Mayr 1941).

Extralimital—Lesser Sundas, Moluccas, n tier of Australia, and the Bismarck Arch (Coates 1990, Mayr & Diamond 2001, Dutson 2011).

a. *Rhipidura rufiventris vidua* Salvadori & Turati

Rhipidura vidua Salvadori and Turati, 1874. *Ann. Mus. Civ. Genova* 6: 313.—Kofiau I, NW Islands.

RANGE: Kofiau I.

DIAGNOSIS: Abdomen pure white and breast band with conspicuous white markings (see Ripley 1959). Diamond *et al.* (2009) observed the subspecies in the wild, and noted the clean patterning and the darkness of the gray of the plumage.

b. *Rhipidura rufiventris gularis* S. Müller

Rhipidura gularis S. Müller, 1843. *Verh. Nat. Gesch. Ned., Land-en Volkenk.* 1: 185 (note).—Utanata R, sw NG.

RANGE: All NG, the NW Islands (Batanta, Gag, Gebe, Misool, Salawati, Waigeo), Bay Is (Mios Num, Yapen); Manam, Fergusson, and Goodenough I, and recently collected by T. K. Pratt from Normanby; also apparently found on Boigu I off the s bulge of NG in Australian territory (Mayr 1941, Schodde & Mason 1999). Mayr (1986) noted that the Gebe and Gag I birds refer to this form. Johnstone (2006) confirmed Mayr's statement regarding the Gag I birds. Widespread in the Trans-Fly (Bishop 2005a).

DIAGNOSIS: Abdomen with a buff wash; breast band with few or no pale markings (Rand & Gilliard 1967). Compared to *R. r. isura* of n Australia, plumage of *gularis* is generally darker (Higgins *et al.* 2006). See photos in Coates (1990: 137).

c. *Rhipidura rufiventris kordensis* A. B. Meyer

Rhipidura kordensis A. B. Meyer, 1874. *Sitzungsber. Akad. Wiss. Wien* 70: 201.—Biak I, Bay Is.

RANGE: Biak I.

DIAGNOSIS: Very distinctive. Dorsal plumage and breast band blackish, not gray; abdomen pure white (Rand & Gilliard 1967).

d. *Rhipidura rufiventris nigromentalis* Hartert

Rhipidura setosa nigromentalis Hartert, 1898. *Novit. Zool.* 5: 525.—Tagula I, Louisiade Arch, SE Islands.

RANGE: Tagula and Misima I (Mayr 1941).

DIAGNOSIS: Similar to *gularis* but white tips to outer tail feathers 30–40 mm; chin black.

NOTES

Other names—White-throated or Banded Fantail; Northern Flycatcher.

Family Dicruridae

Drongos [Family 22 spp/Region 1 sp]

The species of this family range from sub-Saharan Africa and tropical Asia, east to eastern China, New Guinea, the Bismarck and Solomon Islands, and northern and eastern Australia (Rocamora & Yeatman-Bertholet 2009). Members of the family are typically all-black songbirds that perch upright in openings and sally out to snatch flying insects. Current treatment places all the members of the family into a single genus, *Dicrurus*. *Chaetorhynchus papuensis*, traditionally considered a primitive drongo, is now treated as a relative of the fantails, family Rhipiduridae (Barker *et al.* 2004). The phylogeny of Norman *et al.* (2009) placed *Dicrurus* as sister to the Corcoracidae. By contrast, the phylogeny of Jønsson *et al.* (2011b) placed the drongos as sister to the clade that includes fantails, monarchs, birds of paradise, and crows. Aggerbeck *et al.* (2014) provided yet another permutation that placed the drongos as sister to a broad corvine assemblage that excluded the fantails.

Dicrurus Vieillot, 1816 (14-April). *Analyse Nouv. Orn. Elem.*: 41. Type, by subsequent designation (G. R. Gray, 1841. *List. Gen. Birds*, ed. 2: 47), *Corvus balicassius* Linnaeus.

The geographic distribution of the species of this genus is as for the family. The genus is characterized by the all-black or all-gray plumage, the upright posture, presence of rictal bristles, the sallying habit, and the prominent black bill. Anatomical features include an extended maxillary plate in the roof of the palate, thickened nasal bars, a large temporal fossa, and a long and double zygomatic process (Higgins *et al.* 2006: 33). Pasquet *et al.* (2007) provide a preliminary phylogeny of the genus, indicating *bracteatus* and *megarhynchus* (from New Ireland) are sister forms.

Spangled Drongo *Dicrurus bracteatus*   RESIDENT
Dicrurus bracteatus Gould, 1843. *Proc. Zool. Soc. London*: 132.—Moreton Bay, QLD, Australia.

Inhabits forest, edge, and wooded openings of all NG and most satellite islands, sea level–1,450 m^{CO} (at highest elevations only in disturbed habitats). The Australian population is abundant as a dry-season visitor to the Trans-Fly. Its insular range includes the Aru Is, NW Islands, Bay Is, Trobriand Is, D'Entrecasteaux Arch, and Tagula I.

Extralimital—islands of the Torres Strait and the Bismarck Arch (Dutson 2011), west to the Moluccas, south to n and e Australia, and east to isolated populations in the Solomon Is (Guadalcanal and Makira). A major passage migrant in the Torres Strait Is (Draffan *et al.* 1983). We consider the Philippine populations to belong to a distinct species (Dickinson & Christidis 2014).

SUBSPECIES

a. *Dicrurus bracteatus atrocaeruleus* G. R. Gray
Dicrurus bracteatus atrocaeruleus G. R. Gray, 1861. *Proc. Zool. Soc. London* (1860): 354.—Bacan and Halmahera I, n Moluccas.
RANGE: Bacan, Halmahera, and Kofiau I (Diamond *et al.* 2009).
DIAGNOSIS: This is a large form, with a long wing (168–181) and long tail (150–165) that is moderately forked; bill long and heavy; frontal feathers well developed; throat and breast with large spangles; color generally bluish black with purple iridescence (Rand & Gilliard 1967, Rocamora and Yeatman-Bertholet 2009).

b. *Dicrurus bracteatus carbonarius* (Bonaparte)
Edolius carbonarius Bonaparte, 1850. *Consp. Gen. Av.* 1: 352.—Triton Bay, Bird's Neck.
  Synonym: *Dicrurus bracteatus ultramontanus* Stresemann, 1923. *Arch. Naturgesch.* 89, Abt. A, 7: 47.—Aru Is.
    Replacement name for *Dicrurus assimilis* G. R. Gray, 1858, *Proc. Zool. Soc. London*: 178 (not *Corvus* [= *Dicrurus*] *adsimilis* Bechstein, 1794).
  Synonym: *Chibia propinqua* Tristram, 1889. *Ibis*: 556.—D'Entrecasteaux Arch.
  Synonym: *Chibia carbonaria dejecta* Hartert, 1898. *Novit. Zool.* 5: 522.—Tagula I, Louisiade Arch, SE Islands.
RANGE: All NG; the four main NW Islands plus Gebe I; Yapen, Numfor, and Biak I; Aru Is; Trobriand Is; Goodenough, Fergusson, and Normanby I; Tagula I; Torres Strait Is (Mayr 1941). No records from Gag or Mios Num I, and absent from Misima, Rossel, and Woodlark I.
DIAGNOSIS: Gloss of plumage strongly blue-purple (Rand & Gilliard 1967, Diamond 1972). Short-winged and short-tailed: wing (male) 144–159, tail (male) 125–149 (Higgins *et al.* 2006). Schodde & Mason (1999) noted that the hackles at the side of the neck are reduced; tail moderately forked; tips of longest rectrices greatly upcurled. See photos of this form in Coates (1990: 143–45) and Coates & Peckover (2001: 141, upper image).

c. *Dicrurus bracteatus bracteatus* Gould
RANGE: Breeds from c QLD south to NSW, Australia (Schodde & Mason 1999). As a migrant found in the Trans-Fly between Dolak I and Lake Daviumbu (Bishop 2005a). A single bird in breeding condition obtained at Bugi, s NG, on 3-Jan-1937 (Rand 1938b). Once observed near Port Moresby; possibly breeds in the c Trans-Fly (Coates 1990). Migrants also reported from Kiunga airstrip and Lake Murray (B. J. Coates *in litt.*).

DIAGNOSIS: Gloss of plumage strongly green, and on average larger than *carbonarius* (Rand & Gilliard 1967). Schodde & Mason (1999) noted frontal bristles reduced; side-neck hackles moderately well developed and acute; and tail rather shallowly forked. Wing (male) 153–173 (Higgins *et al.* 2006). See photo of this form in Coates & Peckover (2001: 141, lower image).

NOTES
We follow Christidis & Boles (2008) in treating the NG forms as the species *bracteatus* rather than *hottentotus* (*pace* Mayr & Diamond 2001). We follow Rocamora & Yeatman-Bertholet (2009) in synonymizing the subspecific forms *ultramontanus*, *propinquuus*, and *dejectus*. *Other names*—King Crow, Fishtail, *Dicrurus hottentottus*, *D. carbonarius*.

Family Ifritidae

Ifrits   ENDEMIC | MONOTYPIC

This obscure lineage of a single species has been erected as a family by Schodde & Christidis (2014). The phylogeny of Norman *et al.* (2009b) situated the genus as sister to the monarchs. By contrast, the phylogeny of Jønsson *et al.* (2011b) placed *Ifrita* as sister to a cluster that includes the monarchs, crows, and shrikes. Aggerbeck *et al.* (2014) placed *Ifrita* as a sister to a clade that included the birds of paradise and melampittas. It is a compact, creeper-like species with buff plumage and a black-and-blue cap. Birds travel in pairs within foraging flocks, and are vocal and sociable. The species is remarkable in employing chemical defense (toxic feathers) in the manner of a pitohui (Dumbacher *et al.* 2000).

GENUS *IFRITA*   ENDEMIC | MONOTYPIC

Ifrita Rothschild, 1898. *Bull. Brit. Orn. Club* 7: 53. Type, by monotypy, *Ifrita coronata* Rothschild = *Todopsis kowaldi* De Vis.

The single species of the genus is endemic to the high mountains of New Guinea. The genus is characterized by the creeping habit, blue-and-black crown feathering, and presence of poisonous alkaloids in its skin and feathers.

Blue-capped Ifrit *Ifrita kowaldi*   ENDEMIC | MONOTYPIC
Todopsis kowaldi De Vis, 1890. *Ann. Rept. Brit. New Guinea*, 1888–89: 59.—Owen Stanley Ra, SE Peninsula.
   Synonym: *Ifrita kowaldi brunnea* Rand, 1940. *Amer. Mus. Novit.* 1074: 2.—Kunupi, Weyland Mts, Western Ra.
A vocal, sociable, and confiding inhabitant of montane forest interior of the Central Ra and Huon Penin; 2,000–2,900 m^{CO}, extremes 1,460–3,680 m (Mayr & Rand 1937). Absent from the Bird's Head.

NOTES
The form *brunnea* does not hold up against an examination of a large number of specimens, which exhibit considerable nongeographic variation in color of dorsal and ventral plumage, and at best any real variation is apparently clinal. The two forms were based on the "slightly duller" *vs.* "slightly richer" plumage principle, which relies on differences that may change over time due to fading and foxing. It is best to be suspicious of subspecies entirely based on such a measure. An *ifrit* is a supernatural being in Arabic lore, in a class of infernal jinn noted for their strength and cunning (Wikipedia). Given the Ifrit is one of the very few birds on earth with toxic feathers deployed for chemical defense (Dumbacher *et al.* 2000), this is an appropriate name. *Other names*—Blue-capped Ifrita, Blue-capped Babbler, Ifrit.

Family Paradisaeidae

Birds of Paradise [Family 41 spp/Region 37 spp]

The species of the family are found mainly in the Region, but extend west to the northern Moluccas (the Halmahera/Bacan region) and south to eastern Australia. In New Guinea, members of the family inhabit forests at all elevations, but the greatest concentration of species can be found in the mid-montane zone (1,500–2,500 m). This morphologically diverse family is most famous for its remarkable nuptial plumages and elaborate courtship displays. Most species are somewhat crow-like or starling-like, with a powerful bill and feet, and strongly undulating flight. Characters defining the family include 12 rectrices and 10 primaries; strong bill for prying open capsular fruit and excavating bark and epiphytes for insects; powerful feet used for leaping and clambering about arboreally; the frugivorous habit; and foraging mainly in forest canopy (Schodde & Mason 1999). Most females are black-and-brown plumaged, typically with ventral barring. Males are varied and typically iridescent, with the following colors in order of predominance: black, brown, yellow, orange, green, red, and white. Rather than pleasant songs, the males have loud, repeated advertisement calls of cawings, screeches, or powerful whistles. Nearly all polygamous species are sexually dimorphic—males are elaborately plumed, but females are dull and cryptic. Males of these forms faithfully attend display courts or perches, where they attract and mate with females. In these species, females alone build the nest and raise the offspring. The monogamous species are sexually monomorphic and lack specialized nuptial plumage; in these, both sexes attend the nest.

As currently circumscribed, the Paradisaeidae no longer encompasses the satinbirds (now the Cnemophilidae) or the monotypic genus *Macgregoria* (moved to the honeyeaters, Meliphagidae). This family exemplifies an explosive adaptive radiation, driven in part by strong sexual selection. This is an important group of seed dispersers, and there is evidence of bird-plant mutualisms involving members of the family (Beehler 1983, Beehler & Dumbacher 1996). Many species join mixed flocks of brown and black forest birds foraging for arthropod prey. Spelling of the family name Paradisaeidae is conserved (ICZN Opinion 2294). Various publications providing molecular systematic analyses have informed the most recent classification of the family (*e.g.*, Nunn & Cracraft 1992, Irestedt *et al.* 2009). Our treatment combines the results of Frith & Beehler (1998) with the more recent molecular results, especially Irestedt *et al.* (2009). Placement of the family within the core corvoids follows Jønsson *et al.* (2011b).

Following Gilliard (1969), Cooper & Forshaw (1977), and Frith & Beehler (1998), we continue to use "bird of paradise" rather than the novel and less felicitous "bird-of-paradise," which is now advocated by Dickinson & Christidis (2014) and Gill & Donsker (2015). Our belief is that the hyphenation is neither necessary nor appropriate, given the history of usage over several centuries.

GENUS *PHONYGAMMUS* MONOTYPIC

Phonygammus Lesson & Garnot, 1826. *Bull. Sci. Nat. Férussac* 8: 110. Type, by monotypy, *Barita Keraudrenii* Lesson & Garnot.

The single recognized species of the genus ranges from the Cape York Peninsula, Queensland, to the New Guinea region. The genus is characterized by the distinctive shaggy crest and throat hackles and the diagnostic coiled trachea that rests atop the breast and which can be seen just under the skin when the body feathering is parted. For the past few decades, this genus has been subsumed in *Manucodia* by numerous authors (*e.g.*, Diamond 1972, Frith & Beehler 1998), but the trend recently, influenced by molecular evidence, has been to regard *Phonygammus* as generically distinct from its sister lineage *Manucodia* (Cracraft & Feinstein 2000, Irestedt *et al.* 2009). Given the substantial differences in male vocalization and elevational distribution between populations within *P. keraudrenii*, there may be more than one species of *Phonygammus*. This issue needs to be further explored molecularly and in the field. If *P. keraudrenii* is split, then reversion to the original group name "trumpetbird" for the genus would be easier and less wordy than "trumpet manucode." *Phonygammus* is masculine.

Trumpet Manucode *Phonygammus keraudrenii* RESIDENT
Barita Keraudrenii Lesson & Garnot, 1826. *Bull. Sci. Nat. Férussac* 8: 110.—Manokwari, Bird's Head.

Patchily distributed across mainland NG, the Aru Is, and the D'Entrecasteaux Arch. Lowland-dwelling populations are found mainly (or solely) in the s watershed. Montane-dwelling forms inhabit the Central Ra, most n ranges (Arfak, Fakfak, Van Rees, Foja, Bewani, Torricelli, Adelbert Mts), and the mts of Goodenough, Fergusson, and Normanby I (Stresemann 1923, Frith & Beehler 1998). The various populations are distributed from the lowlands to 2,000 m[CO]. No records from the Kumawa, Wandammen, or Cyclops Mts, nor from the n slopes of the Western and Border Ra. Also appears to be absent from the well-surveyed Huon Penin. Scarce or absent from many localities, especially the n watershed of w NG. A shy but vocal inhabitant of forest interior, seen only rarely—mainly when foraging at fig trees at forest edge or in gardens.

Extralimital—Cape York Penin of Australia and islands on the n side of Torres Strait (Boigu, Saibai).

SUBSPECIES

I. Lowland Forms

a. *Phonygammus keraudrenii keraudrenii* (Lesson & Garnot)
RANGE: The Bird's Head, Bird's Neck (Onin Penin), and westernmost NG. Sea level–1,000 m (Mayr 1941).
DIAGNOSIS: A small form, wing (male) 144–166 (n=24) (Frith & Beehler 1998). Generally a steely-green and blue form with wings and tail purple (Rand & Gilliard 1967).

b. *Phonygammus keraudrenii jamesii* Sharpe
Phonygama jamesii Sharpe, 1877. *Cat. Birds Brit. Mus.* 3: 181.—Aleya, Hall Sound, s watershed of the SE Peninsula.
> Synonym. *Manucodia keraudrenii aruensis* Cracraft, 1992. *Cladistics* 8: 10.—Wanoem Bay, Kobror I, Aru Is.
> Synonym: *Phonygammus keraudrenii diamondi* Cracraft, 1992. *Cladistics* 8: 12.—Awande, near Okapa, Eastern Ra.
RANGE: The Aru Is[MA], the S Lowlands, the Trans-Fly, and the w sector of the SE Peninsula. Inhabits lowland forest, up to 600 m. Also Saibai and Boigu I of northernmost Torres Strait (Schodde & Mason 1999).

Localities—Mimika R[MA], upper Digul R[J], Maro R[J], Wasur NP[J], Bian Lakes[J], Bensbach[J], Morehead[J], Wuroi[J], Sturt I[L], Lake Daviumbu[L], Lake Murray[K], Palmer Junction[L], Black R Camp[L], Kiunga[Q], Ok Menga[Q], Nomad R[N], Karimui[DI], Awande[DI], Okasa[DI], Lakekamu basin[M], and Hall Sound[MA] (Rand 1942a[L], Bell 197b[N], Portelli 2013[K], Beehler *et al.* 1994[M], Gregory *in litt.*[Q], Bishop 2005a[J]). We presume this lowland form ranges further eastward in the lowlands of the s watershed between Port Moresby and Milne Bay, but this needs to be confirmed in the field.
DIAGNOSIS: Compared to the nominate form *jamesii* is larger, wing (male) 155–171 (n=30; Frith & Beehler 1998), and with head tufts longer (Rand & Gilliard 1967). Schodde & Mason (1999) noted the following: "gloss to body plumage rich deep bluish green in both sexes, grading to violet-blue over wings and tail." Lowland dwelling; note distinctive *ooo-oh* male song. This and the nominate form perhaps constitute a vocally distinct lowland-dwelling species.

The race *gouldii* from Cape York exhibits a strong "oil green" gloss to plumage (Schodde & Mason 1999). The race *aruensis* is marginally distinct in dorsal iridescence (purplish rather than greenish)—not meriting subspecies status. Museum material from the Aru Is had been available since the 1850s and studied by a wide range of taxonomists (including Mayr), who found no substantial distinction from topotypical *jamesii* or the nominate form. The museum material that Cracraft named as *diamondi* was studied in detail by Diamond (1972) and assigned to *jamesii*.

II. Upland Forms

c. *Phonygammus keraudrenii neumanni* Reichenow

Phonygammus neumanni Reichenow, 1918. *J. Orn.* 66: 438.—Lordberg, 1,500 m, Sepik Mts, n watershed of the Eastern Ra.

> Synonym: *Manucodia keraudrenii adelberti* Gilliard & LeCroy, 1967. *Bull. Amer. Mus. Nat. Hist.* 138: 72.—
> Nawawu, Adelbert Mts, ne NG.

RANGE: Uplands of the n slopes of the Central Ra of c and e NG as well as the Bewani and Adelbert Mts of the n coast of PNG.

DIAGNOSIS: Shortest occipital tufts of any race (Frith & Beehler 1998). The form *adelberti* is not substantially distinct from *neumanni*.

d. *Phonygammus keraudrenii purpureoviolaceus* A. B. Meyer

Phonygama purpureo-violacea A. B. Meyer, 1885. *Zeitschr. Ges. Orn.* 2: 375, pl. 15.—Astrolabe Ra, SE Peninsula.

> Synonym: *Phonygammus keraudrenii mayri* Greenway, 1942. *Proc. New Engl. Zool. Club* 19: 51.—Wau, 1,159 m,
> Morobe Prov, SE Peninsula.

RANGE: Mountains of the SE Peninsula and e sector of the Eastern Ra, 700–1,800 m. Localities—Kassam Pass[DI], Aseki[J], Mt Missim[K], Wau[K], Owgarra[L], Avera[L], and Mt Maguli[M] (Gilliard 1950[M], R. Schodde *in litt.*[J], specimens at BPBM[K], specimens at AMNH[L]).

DIAGNOSIS: Longest occipital tufts of any Mainland race (Frith & Beehler 1998). Mantle purplish violet, not steel green (Rand & Gilliard 1967). Specimens from the c and e sectors of the Eastern Ra were variously referred to three different races (Diamond 1972). At that time, Diamond did not recognize that there were lowland and upland forms and was not privy to the distinctive vocal dialects. See a photo in Coates (1990: 437) and Coates & Peckover (2001: 218). Plumage distinctions between the populations are thin and difficult to characterize.

III. Island Form

e. *Phonygammus keraudrenii hunsteini* Sharpe

Phonygama hunsteini Sharpe, 1882. *Journ. Linn. Soc. London, Zool.*: 442.—Normanby I, D'Entrecasteaux Arch, SE Islands.

RANGE: Rare in upland forests of Fergusson, Goodenough and Normanby I (Mayr 1941).

DIAGNOSIS: Distinct from mainland forms. Largest of any population; wing (male) 179–194 (Frith & Beehler 1998). Iridescence duller than other races. Mantle dark blue-purple not steel green; head tufts longer and greenish, less bluish (Rand & Gilliard 1967). The central tail feathers are somewhat twisted, reminiscent of those of *Manucodia comrii*.

NOTES

Overall, the descriptions of the many races of this species were based on size, head plume length, and iridescent coloration of contour plumage. Recognizing five races may be excessive. We have clustered the subspecies into highland, lowland, and island groupings, which may constitute three cryptic biological species, identifiable by male advertisement song. We await molecular analysis to confirm this. The adult male advertisement vocalization of the NG upland populations is a peculiar long, tremulous, throaty note that diminishes while remaining on pitch. The lowland NG and Australian populations give a high-pitched descending musical disyllable without tremolo (*ooo-oh* or *uuu-uhh*). Females of all forms give a scratchy cough. In support of the distinctness of the lowland and upland species, Beehler noted that the upland form exists in the upper Biaru R at 1,000 m, a day's walk from the alluvial lowlands of the Lakekamu basin, where the lowland form is found (both in the same watershed, with no distributional barriers separating the two). *Other names*—Trumpetbird, *Manucodia keraudrenii*.

Manucodia Boddaert, 1783. *Tabl. Planch. Enlum.*: 39, no. 634. Type, by monotypy, *Manucodia chalybea* Boddaert = *Paradisea chalybata* J. R. Forster.

The species of this genus are endemic to the Region. All species are sexually monochromatic; males are slightly larger than females. They are forest dwelling and shy, but vocal. All species possess elongate and looped trachea for sound production (Frith & Beehler 1998, Pratt & Beehler 2014). The form and shape of this looped trachea among the populations bears additional study and may aid in species identification (*e.g.*, *jobiensis* vs. *chalybatus*). The genus is characterized by glossy purple-and-blue-highlighted plumage, red iris, strong legs and powerful feet, looped trachea, and a fruit diet that features mainly figs. Subsequent to the assertions of Filewood & Peckover (1978) and Beehler & Finch (1985), the ICZN ruled that the genus name is masculine (see Dickinson & Christidis 2014: 251: note 11).

Curl-crested Manucode *Manucodia comrii*   ENDEMIC | MONOTYPIC

Manucodia comrii P. L. Sclater, 1876. *Proc. Zool. Soc. London*: 459, pl. 42.—Fergusson I, D'Entrecasteaux Arch, SE Islands.

> Synonym: *Manucodia comrii trobriandi* Mayr, 1936. *Amer. Mus. Novit.* 869: 3.—Kaileuna, Trobriand Is, SE Islands.
Restricted to the major D'Entrecasteaux and Trobriand Is of the SE Islands, sea level–2,200 m[CO]. Includes the main islands plus Wagifa and Dobu I (Frith & Beehler 1998). Inhabits forest at all elevations, littoral woodland, and gardens. Fairly common to abundant. Usually seen singly or in pairs, sometimes in small congregations in feeding trees. Feeds on fruit. Active, vocal, and conspicuous; unwary and inquisitive; males often perch in the open while singing, posturing oddly. The xylophone-like duets are among the most beautiful bird songs in the NG forest.

NOTES

Described subspecies *trobriandi* is based solely on size (Mayr 1936, Frith & Beehler 1998). Comparison of mean wing measurements between the two (244 *vs.* 226; data from Frith & Beehler 1998) gives a difference of 7.9%, which is below our minimum size threshold of 10%. Note also that the ranges of the wing-length measurements of the two named races overlap by 6 mm. This is in accord with Cracraft (1992), who did not recognize *trobriandi*. Given the very low coralline aspect of the Trobriands, we suspect that they may have been connected to the D'Entrecasteaux Arch during low sea level as recently as 18,000 years ago (Chivas *et al.* 2001, Naish *et al.* 2009). The looped trachea of the adult male is the longest of any species of manucode, crossing the sternal keel and terminating and looping back down by the right side of the body at the level of the cloaca (Beddard 1891). *Other names*—Curl-breasted Manucode.

Crinkle-collared Manucode *Manucodia chalybatus*   ENDEMIC | MONOTYPIC

Paradisea Chalybata J. R. Forster, 1781. *Zool. Ind.*, *Faunula Indica*: 40.—Arfak Mts, Bird's Head.

Ranges through rain forest and other wooded habitats in the hills and lower mts of NG and Misool I; 600–1,500 m, extremes sea level–1,750 m[CO]. This is the common manucode of human-altered habitats in upland valleys, such as at Wau and Baiyer R. Fairly common in forested hills and lower mts above 500 m but usually scarce or absent in the lowlands. We seriously doubt the population is as patchily distributed as depicted by the range map in Frith & Frith (2009)—and this applies to most of their range maps for the manucodes. Probably present but overlooked from most hill forest sites. Localities: Waigama on Misool I[I], Arfak[H], Wandammen[H], Fakfak Mts[K], Kumawa Mts[J], near Bernhard Camp[L], Bokondini[M], Cyclops Mts[H], Sturt I Camp[N], Palmer Junction Camp[N], Black R Camp[N], Lake Kutubu[O], Baiyer R[P], Karimui[DI], Soliabeda[DI], Mt Karimui[DI], Maratambu[R], Memenga forest[R], YUS[P], Sattelberg[Q], Junzaing[Q], Ogeramnang[Q], Finschhafen[Q], Wau[P], Varirata NP[P] (Hartert 1930[H], Mayr 1931[Q], Rand 1942a[N], 1942b[L], Ripley 1964a[M], Mees 1965[J], Gilliard and LeCroy 1967[R], Schodde & Hitchcock 1968[O], Gibbs 1994[K], Beehler field notes[P], Diamond *in litt.*[J]).

NOTES

Nearly identical to *Manucodia jobiensis*, though it is ecologically distinct, occupying hill forest rather than lowland forest. Too little is known of these two taxa to say much more. Even specimens in the museum are challenging to distinguish in the hand. We can speculate that we are witnessing the evolution of lowland

and upland sister forms with *M. chalybatus* and *jobiensis*, as we see with populations of *Phonygammus*. The culmen of *chalybatus* is distinctive—the proximal terminus of the culmen (where it meets the feathers of the forecrown) is flattened and rounded and somewhat bulbous proximally. By contrast, the structure for *jobiensis* is narrow and inserts into the forecrown feathers in a narrowing point (see fig. 3d). Less reliable is the purple sheen to the crown of *chalybatus*. Finally, the feathering of *chalybatus* on throat, upper breast and cheek possesses a greenish and velvety crinkling at a scale finer than the blue-tinged patterning on the breast and throat of the male *jobiensis*. *Other names*—Green-breasted, Green, or Crinkle-breasted Manucode, *Manucodia chalybata*.

Jobi Manucode *Manucodia jobiensis* ENDEMIC | MONOTYPIC

Manucodia jobiensis Salvadori, 1876. *Ann. Mus. Civ. Genova* (1875) 7: 969.—Wonapi, Yapen I, Bay Is.
    Synonym: *Manucodia rubiensis* A. B. Meyer, 1885. *Zeitschr. Ges. Orn.* 2: 374.—Rubi, Geelvink Bay, Bird's Neck.
Yapen I, the NW Lowlands, the Sepik-Ramu, and the w verge of the S Lowlands: the NG range extends in the n watershed from the e shore of Geelvink Bay eastward to the mouth of the Ramu R, and in the s watershed from the e sector of the Bird's Neck to the Setekwa R (Stresemann 1923). Sea level–760 m. Apparently absent from the w Bird's Neck (Bomberai and Onin Penin) and the Bird's Head. Inhabits forest and edge in the lowlands and hills. Replaced at higher elevations by the closely related *M. chalybatus*.
    Localities—Rubi[MA], Siriwo[MA], Waropen[MA], Wanggar R[J], Mimika R[CO], Setekwa R[MA], Bernhard Camp[K], Tami R[L], Puwani R[M], Hunstein valley[N], upper Ramu R[CO], and Astrolabe Bay[O] (Stresemann 1923[O], Hartert et al. 1936[J], Rand 1942b[L], Ripley 1964a[K], A. Allison *in litt.*[N], Beehler field notes 20-Jul-1984[M]).

NOTES

This is the least understood of the manucodes. This species merits attention in the field, in order to better delineate field characters in plumage, behavior, and voice. We follow Cracraft (1992) and Frith & Frith (2009) in not recognizing the form *rubiensis*. The best way of identifying the species in the hand is the narrow and keeled proximal terminus of the culmen, where it meets the skull; this structure in *chalybatus* is widened and rounded, the culmen flattened and meeting the feathering of the forecrown in a U rather than the V that typifies *jobiensis* (see fig. 3d). One might expect the looping of the trachea to be distinct from species to species and hence a possible character for distinguishing this from *M. chalybatus*. Should be checked on live birds in the field. *Other names*—Allied Manucode.

Glossy Manucode *Manucodia ater* ENDEMIC

Phonygama ater Lesson, 1830. *Voy. Coquille, Zool.* 1: 638.—Manokwari, Bird's Head.
    Widespread in lowland forest and wooded savanna of the Region (including the Trans-Fly), ranging through the NW Islands, the Aru Is, the Mainland, islands off the coast of the SE Peninsula (Samarai, Sariba, Mailu, Yule), and Tagula I of the Louisiades (Mayr 1941); from sea level to the foothills (Gilliard 1956), max 900 m[CO]. This is the common manucode of human-altered habitats in the lowlands (I. Woxvold *in litt.*).

SUBSPECIES

a. *Manucodia ater ater* (Lesson)
    Synonym: *Manucodia ater subalter* Rothschild & Hartert,1929. *Bull. Brit. Orn. Club* 49: 110.—Dobo, Aru Is.
RANGE: All NG; widespread in the Trans-Fly[K]; also Gebe[G]; Misool, Salawati, Batanta, Gam, and Waigeo I; Aru Is; and some e fringing islands: Yule, Samarai, Sariba, and Mailu[CO] (Mayr 1941, Mees 1972[G], Bishop 2005b[K]).
DIAGNOSIS: Smaller than *alter*, and smaller-billed. We follow Cracraft (1992) in sinking *subalter* into the nominate form. Island populations and those at the se tip of New Guinea tend to be more purplish (Gilliard 1956). Male's tracheal loop short, extending less than halfway down the sternum. Song throughout is a high-pitched, prolonged whistle sounding like an electronic tone.

b. *Manucodia ater alter* Rothschild & Hartert
Manucodia ater alter Rothschild & Hartert, 1903. *Novit. Zool.* 10: 84.—Tagula I, Louisiade Arch, SE Islands.
RANGE: Tagula I.

DIAGNOSIS: Similar to the nominate form but larger, with more massive bill, and gloss on ventral plumage more purple. Male's tracheal loop long, extending to distal edge of sternum. Song a deep, swelling hum. Mean measurements for adult males of the two races, based on data in Frith & Beehler (1998): *ater* bill (from base) 39 *vs. alter* bill 42; *ater* tail 150 *vs. alter* tail 165. A long-tailed form.

NOTES
We recommend field and molecular study of the Tagula population to determine whether it has differentiated sufficiently from Mainland populations to be raised to species level. *Other names*—Glossy-mantled or Black Manucode, *Manucodia atra*.

GENUS *PTERIDOPHORA*   ENDEMIC | MONOTYPIC

Pteridophora A. B. Meyer, 1894. *Bull. Brit. Orn. Club* 4: 11. Type, by monotypy, *Pteridophora alberti* A. B. Meyer.

The single species of the genus is endemic to the Central Ranges of New Guinea. The genus is characterized by the male's twin long and enameled occipital plumes. Additionally, the gray plumage of the female is unique within the family. Diamond (1972) subsumed this genus into *Lophorina*, a treatment not subsequently followed (*e.g.*, Frith & Beehler 1998, Dickinson & Christidis 2014). Irestedt *et al.* (2009) placed this genus as the sister to *Parotia*.

King of Saxony Bird of Paradise *Pteridophora alberti*   ENDEMIC | MONOTYPIC
Pteridophora alberti A. B. Meyer, 1894. *Bull. Brit. Orn. Club* 4: 11.—Weyland Mts, Western Ra.
   Synonym: *Pteridophora alberti buergersi* Rothschild, 1931. *Novit. Zool.* 36: 253.—Schrader Ra, n sector of the Eastern Ra.
   Synonym: *Pteridophora alberti hallstromi* Mayr & Gilliard, 1951. *Amer. Mus. Novit.* 1524: 12.—Southwestern slope of Mt Hagen, Eastern Ra.
Inhabits montane forest and edge of the Western, Border, and Eastern Ra, east as far as the Kratke Mts[CO], 1,500–2,850 m[CO] (Hartert *et al.* 1936).

NOTES
We treat the species as monotypic, following Diamond (1972) and Cracraft (1992). *Other names*—Enamelled Bird of Paradise, Kissaba.

GENUS *PAROTIA*   ENDEMIC [Genus 6 spp]

Parotia Vieillot, 1816. *Analyse Nouv. Orn. Elem.*: 35. Type, by monotypy, "Sifilet" Buffon = *Paradisea sefilata* J. R. Forster.

The species of this genus are endemic to the uplands of New Guinea. The genus is characterized by the presence of six spatulate-tipped, erectile head wires in the adult male. Diamond (1972) included this genus into his expanded *Lophorina*. This treatment has not been widely followed. A local informant has reported a population of long-tailed *Parotia* from the Cyclops Mts, and it should be looked for there (Beehler & Prawiradilaga 2010).

Western Parotia *Parotia sefilata*   ENDEMIC | MONOTYPIC
Paradisea sefilata J. R. Forster, 1781. *Spec. Fauna Ind., Ind. Zool*: 40.—Arfak Mts, Bird's Head.
   Mountains of the Bird's Head and Wandammen Penin, 1,200–1,800 m (Hartert 1930, Gyldenstolpe 1955b). Confined to montane forest interior.

NOTES
This species should be looked for in the Kumawa Mts (Bomberai Penin) and Fakfak Mts (Onin Penin) of the sw Bird's Neck region. *Other names*—Arfak Parotia, Arfak Six-wired Bird of Paradise.

Wahnes's Parotia *Parotia wahnesi*   ENDEMIC | MONOTYPIC

Parotia wahnesi Rothschild, 1906. In Foerster & Rothschild, *Two New Birds of Paradise*: 2.—Rawlinson Ra, Huon Penin.

Inhabits montane forest of the Huon Penin and Adelbert Mts, 1,100–1,700 m^{CO}. In the Adelbert Mts it has been found from 1,300 m up to the highest summit, at 1,600 m, where two females were collected and a subadult male observed (Pratt 1983).

NOTES

The Adelbert Mts population of *P. wahnesi*, apparently rare and local when first discovered in 1974, may no longer be present there. J. M. Diamond and K. D. Bishop (*in litt.*) apparently heard the bird vocalizing at about 1,525 m in the Adelberts in 2004. This is apparently the last record for this isolated population. Several more recent searches for the species in the Adelberts have failed to find it. If, indeed, it is no longer present, its disappearance may be the result of a combination of factors: forest loss, hunting of birds at their display courts, and possible impacts from ongoing climate change. *Other names*—Wahnes' Six-wired Bird of Paradise, Huon Parotia.

Lawes's Parotia *Parotia lawesii*   ENDEMIC | MONOTYPIC

Parotia lawesii Ramsay, 1885. *Proc. Linn. Soc. NSW* 10: 247.—Mount Maguli, SE Peninsula.

 Synonym: *Parotia lawesi fuscior* Greenway, 1934. *Proc. New Engl. Zool. Club* 14: 2.—Mount Missim, 1,646 m, Morobe Prov, SE Peninsula.
 Synonym: *Parotia lawesii exhibita* Iredale, 1948. *Austral. Zool.* 11: 162.—Hoiyevia, Mt Hagen distr, Eastern Ra.

Inhabits montane forest interior of the e verge of the Border Ra (Oksapmin), the Eastern Ra, and the nw half of the SE Peninsula; 1,200–1,900 m, extremes 500–2,300 m^{CO} (Frith & Beehler 1998). Replaced by *P. helenae* in the e third of the SE Peninsula.

NOTES

This species and *P. helenae* come into contact in both the n and s watersheds of the SE Peninsula. It would be interesting to study the nature of interspecific interactions in these contact zones. The races *exhibita* of the PNG central highlands and *fuscior* of the Herzog Mts are synonyms (Diamond 1972, Schodde & McKean 1973a, Schodde & Mason 1974, and Cracraft 1992). Perhaps the most interesting (and unstudied) fact related to this species is the ability of the adult male to change its iris color from mainly cobalt blue to mainly yellow. It apparently has two parts to its iris, the inner blue and the outer yellow, and it can make one or the other predominant. Both authors have observed handheld males accomplish this feat. The distributional record from Oksapmin, on the w side of the Strickland R gorge is notable (BPBM specimen no. BBM-NG 99832). Perhaps the lower elevational range of this genus means such a break in the mountains is not as effective as it is for higher-elevation species (*e.g.*, *Melionyx princeps*). *Other names*—Lawes's Six-wired Bird of Paradise.

Eastern Parotia *Parotia helenae*   ENDEMIC | MONOTYPIC

Parotia helenae De Vis, 1897. *Ibis*: 390.—Neneba, upper Mambare R, 1,220 m, north of Mt Scratchley, SE Peninsula.

The e sector of the SE Peninsula, west in the n watershed to the Waria R, and in the south probably only west to the Keveri Hills, just west of Mt Suckling; 1,100–1,500 m, min 500 m.

Localities—the upper Waria R/Bubu R, Hydrographer Mts, Isurava on the Kokoda Track^{K}, upper Mambare R (Neneba, Biagge), Agaun, Mt Moiba (north of Mt Simpson), and possibly also the s watershed from Mt Dayman west to Mt Suckling (Schodde & McKean 1973a, Coates 1990, Pratt field notes^{K}). DIAGNOSIS: Supra-narial tuft of the adult male is bronzy brown, not silvery white; the assertion that the dorsal plumage of female is less russet (Frith & Beehler 1998) is not borne out by the museum material reviewed at AMNH. In the female, the best diagnostic characters are the sharply keeled culmen and the more extensively exposed (unfeathered) distal portion of the culmen: in *helenae* 18.3–19.5 (n=3) exposed *vs.* 11.5–14.8 (n=6) in *lawesii* (see fig. 3e). When looked at from the side, the shape of the maxilla leading up the the ridge of the culmen appears slightly concave dorsally for *helenae* and convex for *lawesii*.

Schodde & McKean (1973a) treated the southeasternmost *Parotia* population, *helenae*, as a full species, a treatment we follow here. *Other names*—Helen's Parotia.

Carola's Parotia *Parotia carolae* ENDEMIC

Parotia carolae A. B. Meyer, 1894. *Bull. Brit. Orn. Club* 4: 6.—Weyland Mts, Western Ra.

Inhabits the interior of montane forest of the Western, Border, and Eastern Ra, 1,100–2,000 m[CO]. Range overlaps with *P. lawesii* in the Eastern Ra; in this area of overlap *P. carolae* seems to be more common in the n watershed. I. Woxvold (*in litt.*) reported a record from 500 m in the upper Fly; also visits Ok Menga at 450 m (P. Gregory *in litt.*).

SUBSPECIES

a. *Parotia carolae carolae* A. B. Meyer
RANGE: The westernmost sector of the Western Ra, including the Weyland Mts east to the Paniai Lake region (Hartert *et al.* 1936).
DIAGNOSIS: Chin of adult male dusky olive brown smudged blackish, with paler tips to elongate whiskers surrounding the malar area; eye-ring coppery; throat buff with a dull golden sheen (Rand & Gilliard 1967, Frith & Beehler 1998).

b. *Parotia carolae meeki* Rothschild
Parotia carolae meeki Rothschild, 1910. *Bull. Brit. Orn. Club* 27: 35.—Otakwa R, 686 m, Western Ra.
RANGE: The Western and Border Ra, from Paniai Lake east to the PNG Border.
DIAGNOSIS: Chin and sides of throat blackish in adult male (Frith & Beehler 1998).

c. *Parotia carolae chalcothorax* Stresemann
Parotia carolae chalcothorax Stresemann, 1934. *Orn. Monatsber.* 42: 145.—Doormanpaad, highlands south of the Mamberamo R, ne sector of the Western Ra.
RANGE: Apparently confined to the Doorman Mts on the n verge of the Western Ra, overlooking the Mamberamo basin.
DIAGNOSIS: Upperparts of the adult male exhibit a bright coppery sheen; underparts very coppery as well; long feathers of the lores brownish (Frith & Beehler 1998).

d. *Parotia carolae chrysenia* Stresemann
Parotia carolae chrysenia Stresemann, 1934. *Orn. Monatsber.* 42: 147.—Lordberg, 1,500 m, Sepik Mts, Eastern Ra.
    Synonym: *Parotia carolae clelandiorum* Gilliard, 1961. *Amer. Mus. Novit.* 2031: 5.—Telefomin, 1,525 m, Victor
      Emanuel Mts, Border Ra. [Subspecies name corrected by LeCroy 2015.]
RANGE: The Border and Eastern Ra, from the Victor Emanuel and Hindenburg Mts to the Schrader Ra[DI], Bismarck Ra[CO], and Ubaigubi[CO] (Stresemann 1923, Frith & Beehler 1998).
DIAGNOSIS: Upperparts of adult male jet black (Frith & Beehler 1998). See photos in Coates (1990: 473) and Coates & Peckover (2001: 230).

NOTES

Should be looked for in the Kratke Mts, east of Crater Mt. *P. c. clelandiorum* is merged with *chrysenia* because the minor plumage differences described belie the rather chaotic individual variation, given the inadequate number of specimens (two) in adult male plumage available to study. Also, this form may be a product of comparisons of fresh material with older, foxed material. Mainly based on a 6% (10 mm) difference in length of the male wing. Irestedt *et al.* (2009) situated this species as sister to the main cluster of core *Parotia* species. *Other names*—Queen Carola's Six-wired Bird of Paradise, Queen Carola's Parotia.

Bronze Parotia *Parotia berlepschi* ENDEMIC | MONOTYPIC

Parotia berlepschi Kleinschmidt, 1897. *Orn. Monatsber.* 5: 46.—Foja Mts, Mamberamo basin, nw NG.

Confined to the uplands of the Foja Mts, east of the upper Mamberamo, 1,200–1,600 m. Inhabits interior of montane forest (Beehler & Prawiradilaga 2010, Beehler *et al.* 2012).

Traditionally treated as a race of *carolae* (Mayr 1941, Diamond 1985). New information on voice and iris color of male and female, among other characters, support splitting the form from *carolae* (Frith & Frith 2009, Beehler *et al.* 2012, Scholes *et al.* ms). *Other names*—Foja Parotia.

GENUS *SELEUCIDIS*   ENDEMIC | MONOTYPIC

Seleucidis Lesson, 1835. *Hist. Nat. Ois. Parad.*, Synopsis: 28, pl. 35. Type, by monotypy, *Seleucidis acanthilis* Lesson = *Paradisea melanoleuca* Daudin.

This monotypic genus is confined to the lowlands of New Guinea and Salawati Island. The genus is characterized by the long, powerful, bayonet-like bill, the large velvety breast shield, bright yellow bloomers, and the 12 recurved tail wires exhibited by the adult male. *Seleucidis* is masculine (David & Gosselin 2002b). Dickinson & Christidis (2014: 252, note 11) acted as first revisers and chose *Seleucidis* over *Seleucides*, as these were apparent dual original spellings. Irestedt *et al.* (2009) placed this monotypic genus sister to a clade that includes *Ptiloris*, *Lophorina*, *Semioptera*, and *Drepanornis*.

Twelve-wired Bird of Paradise *Seleucidis melanoleucus*   ENDEMIC | MONOTYPIC
Paradisea melanoleuca Daudin, 1800. *Traité Orn.* 2: 278.—Salawati I or Bird's Head.
Synonym: *Seleucides ignotus auripennis* Schlüter, 1911. *Falco* 7: 2.—Wewak, w Sepik-Ramu.
Inhabits flat alluvial lowland forest on NG and Salawati I. Range includes the Bird's Head and Neck, NW Lowlands, Sepik-Ramu, S Lowlands, and the w two-thirds of the s watershed of the SE Peninsula (Stresemann 1923). In the Trans-Fly known only from the lower Digul R (Bangs & Peters 1926). Absent from hilly areas, including the n coast and e sector of the SE Peninsula, the w sector of the Bird's Neck, and the Huon Penin. Inhabits swamp forest, especially with sago palm and pandanus, and sometimes nearby alluvial rain forest. Occurs only in the lowlands, mostly near sea level, but up to 180 m in places (Frith & Beehler 1998).

The species is best treated as monotypic as there are minimal plumage or size differences between the named races. The absence of this species from the entire n watershed of the SE Peninsula plus the e sector of the s watershed of the SE Peninsula is interesting, in light of the apparent absence from one or both watersheds of this peninsular region of a suite of other lowland species: *Chalcopsitta scintillata*, *Ducula mullerii*, *Malurus cyanocephalus*, and *Poecilodryas hypoleuca* (J. M. Diamond *in litt.*). *Other names*—*Seleucidis melanoleuca*.

GENUS PTILORIS   [Genus 4 spp/Region 2 spp]

Ptiloris Swainson, 1825. *Zool. Journ.* 1(4): 479. Type, by monotypy, *Ptiloris paradiseus* Swainson.

The species of this genus range from New Guinea (throughout) south to eastern Australia. This genus is characterized by compact body, elongate neck, powerful legs and claws, broad wings, short tail, iridescent breast shield, powerful decurved chisel-like bill, and loud rustling produced by wings of the male in flight. We are not willing to fold *Lophorina* into *Ptiloris*, because of the conflicting list of morphological characters (see Frith & Beehler 1998: 56–58, *pace* Irestedt *et al.* 2009). Certainly we agree that the two genera are sisters (Frith & Beehler 1998: 64), but *Lophorina* is distinctive in exhibiting the large black head plume, the winged breast shield, the abbreviated bill, the medium-long tail, and the peek-a-boo display by the male.

Magnificent Riflebird *Ptiloris magnificus*   RESIDENT

Falcinellus magnificus Vieillot, 1819. *Nouv. Dict. Hist. Nat.*, nouv. éd. 28: 167.—Manokwari, Bird's Head.

Inhabits forests of lowlands and hills of w and c NG; sea level–700 m, max 1,200 m (Stresemann 1923, Pratt & Beehler 2014). Trans-Fly records only from the Bian Lakes and Wuroi (Bishop 2005a). Not found on any satellite islands. Fairly common, but shy and wary and heard far more than seen; the adult male is particularly difficult to observe.

Extralimital—Cape York Penin, QLD, Australia.

SUBSPECIES

a. *Ptiloris magnificus magnificus* (Vieillot)

RANGE: From the Bird's Head eastward in the n watershed to the mouth of the Sepik R (Beehler & Swaby 1991), and eastward in the s watershed to the Purari R (Frith & Beehler 1998)

DIAGNOSIS: Distinguished from *alberti* of Australia by the following: long wing and tarsus; bill less curved; adult male with prominent maroon sheen ventrally; and longer central rectrices (Frith & Frith 2009).

NOTES

Voice is best means of separating the two NG species of riflebirds. The male *P. magnificus* gives twin powerful rising whistles—*whoiieet? whoit?*—sometimes extended to four notes. Cues in plumage and morphology can help distinguish *magnificus* from *intercedens*; differences exhibited by *magnificus* include the following: the longer bill, on average; base of culmen with a 15 mm, feather-free, rounded and flattened ridge where the proximal extent of the culmen inserts into the skull; flank plumes longer and extending beyond the tail. See comparative image of the culmen shape of the two species in figure 3f. In the Purari R basin, I. Woxvold (*in litt.*) has recorded mainly the call of *P. magnificus*; however, a "hybrid" call was recorded on several occasions, with the call structure and pitch like those of *magnificus*, but one or more of the notes having rolled *r* sounds, giving a growling quality reminiscent of the vocalization of *P. intercedens*. *Other names*—Albert Riflebird, *Craspedophora* or *Lophorina magnifica*.

Growling Riflebird *Ptiloris intercedens*   ENDEMIC | MONOTYPIC

Ptiloris intercedens Sharpe, 1882. *Journ. Linn. Soc. London, Zool.*, 16: 444.—Milne Bay, SE Peninsula.

Inhabits forest interior of the lowlands and hills of the e third of NG; west in the s watershed to the Purari R, and west in the n watershed to the Adelbert Mts; sea level–1,400 m, max 1,740 m (Coates & Peckover 2001).

NOTES

Our treatment follows Beehler & Swaby (1991), who noted the very distinctive advertisement vocalizations of the two *Ptilorus* forms in NG and suggested that these represent two distinct biological species. The male of *P. intercedens* delivers a harsh growling *crrraiy-crrrrow!* with the first note rising, the second descending. In addition, *P. intercedens* (all plumages) exhibits a narrow-ridged and feathered culmen, with only a few mm of the ridge exposed and the forehead feathering meeting at the culmen far from the base (Frith & Beehler 1998; see bills compared in fig. 3f). Photos of this species can be found in Coates & Peckover (2001: 227). Future fieldworkers should compare the mating displays of the two in the field to determine whether there are substantial differences. In addition, it would be good for fieldworkers to study the interactions of territorial males of these two at points where their ranges meet in NG: in the Purari in the south and in the lower Ramu in the north. B. J. Coates (*in litt.*) noted that in 1967 he heard a male give a distinctive vocalization east of Yalumet on the n scarp of the Saruwaged Ra: a two-note series of whistles, the first upslurred, the second, downslurred. Beehler, working a few km west on the YUS transect, has recorded only the typical vocalizations of *intercedens*. What Coates may have heard is the atypical vocalization of a young male. *Other names*—traditional treatments lump this form into *P. magnificus*.

Lophorina Vieillot, 1816. *Analyse Nouv. Orn. Elem.*: 35. Type, by monotypy, "Le Superbe" Buffon = *Paradisea superba* J. R. Forster.

The single species of the genus is endemic to the upland forests of New Guinea. The genus is characterized by the huge velvety head plume and the winged breast shield of the male. Based on female plumage characters, Diamond (1972) lumped an array of related taxa into this genus (species that we currently place in the following genera: *Parotia*, *Ptiloris*, *Cicinnurus*, *Astrapia*, and *Pteridophora*). The purpose of his lumping was to highlight the close relationships among these species. While we agree that these lineages are closely related, we resist this path, given that each lineage has an abundance of characters that define it, making the delineation of unambiguous genera straightforward. We do not follow Irestedt *et al.* (2009), who embedded this genus within *Ptiloris*. See discussion of this issue in the account for the genus *Ptiloris*.

Superb Bird of Paradise *Lophorina superba*   ENDEMIC

Paradisea Superba J. R. Forster, 1781. *Ind. Zool., Faunula Indica*: 40 (based on Daubenton, *Planch. Enlum.*: 632).—Arfak Mts, Bird's Head.

Mountains of the Bird's Head, the Central Ra, mts of Wandammen and Huon Penin, and Adelbert Mts; 1,650–1,900 m, extremes 1,000–2,300 m[CO]. Inhabits lower montane forest, edge, ecologically disturbed areas, and even *Casuarina* and *Castanopsis* woodlots in interior highland valleys (*e.g.*, Tari valley). No records from the Cyclops Mts, Foja Mts, or PNG North Coastal Ra.

SUBSPECIES

a. *Lophorina superba superba* (Pennant)

RANGE: Mountains of the Bird's Head.

DIAGNOSIS: The adult male lacks the black spots on the feathers of the breast shield present in all other races. Additional diagnosis for all races refers solely to the adult female plumage. Plumage of head and side of throat entirely blackish in adult female; underparts tinged faintly with buff and barred with blackish brown (Frith & Beehler 1998). Back and wing plumage dull brown, without rufescent wash to wing edgings. Vocalization of the male is so distinct from that of eastern birds (from PNG) as to be unrecognizable—a shrill *yia, yia ... yia-yia-yia*.

b. *Lophorina superba niedda* Mayr

Lophorina superba niedda Mayr, 1930. *Orn. Monatsber.* 38: 179.—Mount Wondiwoi, Wandammen Penin, Bird's Neck.

RANGE: Mountains of the Wandammen Penin.

DIAGNOSIS: Underparts of adult female darker and more ochraceous than in nominate form (Gilliard 1969).

c. *Lophorina superba feminina* Ogilvie-Grant

Lophorina superba feminina Ogilvie-Grant, 1915. *Ibis*, Jubilee Suppl. 2: 27.—Otakwa R, Nassau Mts, Western Ra.

RANGE: The Western and Border Ra, from the Weyland Mts east to the Victor Emanuel and Hindenburg Mts (Hartert *et al.* 1936, Ripley 1964a). Includes the mts south of the Mamberamo basin (Doormanpaad).

DIAGNOSIS: Adult female exhibits a broad white supercilium that extends narrowly behind the eye and around the nape to the other eye; crown buff-speckled. Wing edgings rufescent, contrasting with the dull olive-brown mantle.

d. *Lophorina superba latipennis* Rothschild

Lophorina minor latipennis Rothschild, 1907. *Bull. Brit. Orn. Club* 19: 92.—Rawlinson Ra, Huon Penin.

> Synonym: *Lophorina superba connectens* Mayr, 1930. *Orn. Monatsber.* 38: 180.—Dawong, Herzog Mts, SE Peninsula.
> Synonym: *Lophorina superba addenda* Iredale, 1948. *Austral. Zool.* 11: 162.—Mount Hagen distr, Eastern Ra.

RANGE: The Eastern Ra, from the mts east of the Strickland R gorge southeast to the mts of the w sector of the SE Peninsula, where it presumably meets *L. s. minor*. Also the Adelbert Mts (Pratt 1983) and mts of the Huon Penin.

DIAGNOSIS: Head of adult female dark brown, chin whitish; broad white supercilium; white streaking on forehead, crown, and nape; mantle olive brown (Frith & Beehler 1998). See photo in Coates (1990: 465) and Coates & Peckover (2001: 230). We synonymize *addenda* and *connectens*, following Diamond (1972) and Cracraft (1992).

e. *Lophorina superba minor* Ramsay

Lophorina superba minor Ramsay, 1885. *Proc. Linn. Soc. NSW* 10: 242.—Astrolabe Ra, SE Peninsula.

RANGE: Mountains of the SE Peninsula. The nw terminus of this subspecies' range is uncertain, but evidently southeast of Wau.

DIAGNOSIS: Female blacker-plumaged than *latipennis*, with head and throat blackish brown and upperparts rich dark chestnut. Supercilium largely absent, appearing as a small postocular pale streak. Little or no pale marking on the nape (Frith & Beehler 1998). See photo of female in Coates (1990: 465).

f. *Lophorina superba sphinx* Neumann

Lophorina superba sphinx Neumann, 1932. *Orn. Monatsber.* 40: 121.—Type locality unknown.

RANGE: Unknown. Mayr (1962) hypothesized this form may inhabit mts of the far southeast of SE Peninsula, but Schodde (1978) demonstrated that it is not found there.

DIAGNOSIS: Based on a single immature male, collected from an unknown locality. "Female" plumage more reddish brown on upperparts than that of *minor;* eye stripe even less extensive; and forehead and neck without white flecking (Frith & Beehler 1998). The unique specimen is considerably larger than *minor* (Frith & Frith 1997).

NOTES

The molecular work of J.A. Shearer *et al.* (unpubl.) shows a major break between *L. s. minor* of the SE Peninsula *vs.* the other lineages to the west. In addition, there is a lesser molecular break separating the populations of the Bird's Head and Wandammen Penin from those of the main body of NG. A third molecular break separates *feminina* to the west of the Strickland R gorge and *latipennis* east of the gorge (J. Cracraft *in litt.*). "*Lophorina superba pseudoparotia*" Stresemann (1934) is, in fact, a hybrid between *L. superba* and *Parotia carolae* (Frith & Frith 1996). *Other names*—Lesser Superb Bird of Paradise.

GENUS *DREPANORNIS* ENDEMIC [Genus 2 spp]

Drepanornis P. L. Sclater, 1873 (31-July). *Nature* 8: 192; *idem*, 1873, *Proc. Zool. Soc. London*: 560 (Aug-1873). Replacement name for *Drepanephorus* P. L. Sclater, 1873, *nec* Egerton, 1872 (Pisces). Type, by original designation, *Drepanephorus albertisi* P. L. Sclater.

The two species of the genus are endemic to New Guinea. The range is restricted to selected lowlands and uplands of New Guinea (Frith & Beehler 1998). The genus is characterized by the exceedingly long, narrow, and decurved bill, the bare facial patch, the buffy rounded tail, and the dull brown-and-gray plumage that lacks black or metallic plumage prevalent in species of *Epimachus*. The phylogeny of Irestedt *et al.* (2009) situated *Drepanornis* as sister to a clade that includes *Ptiloris*, *Lophorina*, and *Semioptera*.

Black-billed Sicklebill *Drepanornis albertisi* ENDEMIC

Drepanephorus albertisi P. L. Sclater, 1873. *Nature* 8: 151 and *Proc. Zool. Soc. London*: 558, pl. 47.—Hatam, Arfak Mts, Bird's Head.

Patchily distributed through mts of the Bird's Head and Neck, Western Ra, Foja Mts, Eastern Ra, Huon Penin, and SE Peninsula; 1,100–1,900 m, extremes 600–2,250 m^{CO}. Absent from many localities (or overlooked). Inhabits the canopy of forest interior in the lower and mid-montane zones, occasionally selectively logged forest, and rarely forest edge; uncommon and locally absent from seemingly suitable habitat. Territorial and sedentary. A single local informant's report from the Adelbert Mts (Pratt 1983) warrants corroboration. In spite of its loud and melodious voice, this is perhaps the most difficult bird of paradise to locate.

a. *Drepanornis albertisi albertisi* (P. L. Sclater)

RANGE: The Bird's Head and Neck. Localities—Tamrau Mts[J], Hatam in the Arfak Mts[MA], Wandammen Mts[MA] (Beehler field notes[J]). Populations from Fakfak[K], Kumawa[L], and Foja Mts[M] presumably refer to this form but remain undiagnosed (Diamond 1985[L], Gibbs 1994[K], Beehler *et al.* 2012[M]).

DIAGNOSIS: Tail rich rufous brown.

b. *Drepanornis albertisi cervinicauda* P. L. Sclater

Drepanornis albertisi cervinicauda P. L. Sclater, 1884. *Proc. Zool. Soc. London* (1883): 578.—Vicinity of Port Moresby, SE Peninsula.

> Synonym: *Drepanornis albertisi inversa* Rothschild, 1936. *Mitt. Zool. Mus. Berlin* 21: 188.—Kunupi, Weyland Mts, Western Ra.

RANGE: Ranges patchily through the Central Ra. Localities—Weyland Mts[J], Moro in the Kikori basin[O], Lordberg[K], Nipa[CO], Ambua[L], Karimui[DI], Mt Karimui[DI], Dawong[M], Mt Missim[L], Suria[L], near Uberi[Q], near Port Moresby[MA], and Sibium Mts[N] (Stresemann 1923[K], Mayr 1931[M], Hartert *et al.* 1936[J], Gilliard 1950[Q], Clapp 1987[N], Beehler 1987[L], K. D. Bishop *in litt.*[O]).

DIAGNOSIS: Tail pale cinnamon.

c. *Drepanornis albertisi geisleri* A. B. Meyer

Drepanornis geisleri A. B. Meyer, 1893. *Abh. Ber. Zool. Mus. Dresden* (1892–93) 4(3): 15.—Sattelberg, Huon Penin.

RANGE: Mountains of the Huon Penin.

DIAGNOSIS: Most similar to the dark-tailed birds of the Bird's Head, and distinct from the adjacent populations from the Central Ra, which are pale-tailed. We split this form out solely because we do not believe the geographically disparate Bird's Head and Huon birds are sister forms.

NOTES

We treat the species as polytypic only because the tail coloration of the Central Ra population is so obviously distinct from that of the Bird's Head and Huon. *Other names*—Buff-tailed Sicklebill, *Epimachus albertisi*.

Pale-billed Sicklebill *Drepanornis bruijnii*　　　　　ENDEMIC | MONOTYPIC

Drepanornis Bruijnii Oustalet, 1880. *Ann. Sci. Nat.* (6)9(5): 1; and *Bull. Assoc. Sci. France* (1880): 172.—Northern Waropen, w coast of Geelvink Bay, NW Lowlands.

The NW Lowlands and w sector of the Sepik-Ramu. Restricted to the lowlands and low hills of nw and north-central NG, including the Waropen distr, the Mamberamo basin, the hills and lowlands of the n border region, and the w verge of the Sepik basin; lowlands to 175 m. First recorded in PNG by Whitney (1987), who located several singing males near Vanimo, West Sepik Prov, in 1983.

Localities—Wanti, Waropen distr[K]; Biri, w Mamberamo basin[K]; Pionierbivak, lower Mamberamo[K]; Sarmi, n Mamberamo[K]; Noau[Z]; Kasiwa[Z]; Kwerba[W]; Papasena[W]; Wensudu on the n coast[K]; Witriwai R, n coast[K]; Takar[K]; Walckenaers Bay[K]; Bodim[J]; Holtekong[J]; Humboldt Bay[C], Tami R[MA]; Vanimo[L]; near Krissa[X]; Puwani, n base of the Bewani Mts[X]; Utai, s base of the Bewani Mts[B] (Ripley 1964a[J], Diamond 1981a[K], Beehler & Beehler 1986[X], Whitney 1987[L], Beehler *et al.* 2012[W], van Balen *et al.* 2009[Z], specimen at BPBM[B], AMNH specimens[C]). Singing males have been found in selectively logged forest and low karst hills as well as lowland alluvial forest (Beehler & Beehler 1986).

Other names—White-billed or Lowland Sicklebill, *Epimachus bruijnii*.

GENUS *EPIMACHUS*　　　　　　　　　　　　　　　ENDEMIC [Genus 2 spp]

Epimachus Cuvier, 1816. *Règne Anim.* 1 (1817): 407. Type, by monotypy, *Upupa magna* Gmelin = *Promerops fastuosus* Hermann.

The two species of the genus are endemic to the mountainous uplands of New Guinea (Frith & Beehler 1998). The genus is characterized by the very long lanceolate tail, the strongly decurved bill, and the iridescent green and purple plumage patches on the mantle and crown of the mainly blackish plumage of the adult male. It was combined with *Drepanornis* by Diamond (1972) but separated by Frith & Beehler (1998). For a list of distinguishing generic characters, refer to Frith & Beehler (1998: 56–58). The phylogeny of Irestedt *et al.* (2009) situated this genus sister to a clade that includes *Paradigalla* and *Astrapia*, implying that the long, curved bill has been independently acquired in *Epimachus*.

Brown Sicklebill *Epimachus meyeri* ENDEMIC | MONOTYPIC

Epimachus meyeri Finsch and A. B. Meyer, 1885. *Zeitschr. Ges. Orn.* 2: 380.—Mount Maguli, Owen Stanley Ra, SE Peninsula.

> Synonym: *Falcinellus meyeri albicans* van Oort, 1915. *Zool. Meded. Leiden* 1: 228.—Treubbivak, 2,366 m, Treub Mts, e sector of the Western Ra.
>
> Synonym: *Epimachus meyeri megarhynchus* Mayr & Gilliard, 1951. *Amer. Mus. Novit.* 1524: 10.—Gebroeders Mts, 1,830–2,150 m, w sector of the Western Ra.
>
> Synonym: *Epimachus meyeri bloodi* Mayr & Gilliard, 1951. *Amer. Mus. Novit.* 1524: 10.—Mount Hagen, Eastern Ra.

Inhabits the canopy of montane forest and edge throughout the Central Ra; 1,900–2,900 m, extremes 1,525–3,200 m[CO]. Absent from the Bird's Head and Huon Penin. Where sympatric with *E. fastosus* it does not occur below *ca.* 2,100 m, the lower elevations being occupied by *E. fastosus*.

NOTES

The populations vary clinally, with the flank plumages of the adult male paler and whiter, and body size becoming smaller, as one trends westward (see fig. 3g, showing described races *albicans*, *bloodi*, and nominate). This rather diffuse and clinal trend does not merit subspecies designation. *Other names*—Meyer's Sicklebill, Brown Sickle-billed Bird of Paradise, Gray Sabre-tailed Bird of Paradise.

Black Sicklebill *Epimachus fastosus* ENDEMIC | MONOTYPIC

Promerops fastuosus Hermann, 1783. *Tab. Aff. Anim.*: 194 (based on *Planch. Enlum.*: pls. 638 39). —NG, restricted to the Arfak Mts by Hartert, 1930, *Novit. Zool.* 36: 33.

> Synonym: *Epimachus fastuosus stresemanni* Hartert, 1930. *Novit. Zool.* 36: 34.—Schraderberg, Sepik Mts, n sector of the Eastern Ra.
>
> Synonym: *Falcinellus striatus atratus* Rothschild & Hartert, 1911. *Novit. Zool.* 18: 160.—Mount Goliath, Border Ra.
>
> Synonym: *Epimachus fastosus ultimus* Diamond, 1969. *Amer. Mus. Novit.* 2362: 31.— Summit of Mt Menawa, Bewani Mts, Sepik-Ramu.

Inhabits interior of mid-montane forest of the Bird's Head[MA], Wandammen Penin[MA], Western, Border, and Eastern Ra (Weyland Mts[MA] east to the Bismarck Ra[J,CO] and Mt Karimui[DI]), Foja Mts[K], PNG North Coastal Ra[CO], and Mt Bosavi[CO], 1,300–2,550 m[CO] (Stresemann 1923[J], Beehler *et al.* 2012[K]). In places where *Epimachus meyeri* is found with *E. fastosus*, the latter species inhabits mainly a lower elevation (the upper limit of the *fastosus* range meeting the lower limit of *meyeri*). Distribution in the Central Ra appears to be continuous, not patchy.

NOTES

The described forms *stresemanni*, *atratus*, and *ultimus* are best subsumed into the nominate form; geographic variation is slight and there is ample within-population variation. For instance, bill length, an important character defining *ultimus*, is probably the product of clinal variation and small sample size. Focusing on characters upon which the argument for the subspecies was based (Diamond 1969: 31–36), we compare the bill length of female *ultimus* (70–73, mean 71; n=10) with that of the nominate population (66–77, mean 72; n=53). The entire range of bill length for *ultimus* is encompassed by the range of the nominate form. Likewise, the tail (mean 320) of female *ultimus* is only 8% longer than the mean of the race with the second-longest female tail, *atratus*. That does not meet our 10% rule. Spelling of the species epithet was emended by the original author to *fastosus* on a later page (p. 202) of the same publication, in which the author made his original intention clear (David *et al.* 2009). *Other names*—Greater Sicklebill, Black Sickle-billed Bird of Paradise, *Epimachus fastuosus*.

Paradigalla Lesson, 1835. *Hist. Nat. Ois. Parad.*: 242. Type, by monotypy, *P. carunculata* Lesson.

The two species of the genus are restricted to the uplands of the Bird's Head, Bird's Neck (?), Western Ranges, Border Ranges, and Eastern Ranges. The genus is characterized by the bright yellow forehead wattles and attendant colored gape wattles, the narrow bill, and the sexually monochromatic black plumage. Irestedt *et al.* (2009) situated *Paradigalla* as sister to *Astrapia*. Dickinson & Christidis (2014) subsumed *Paradigalla* into *Astrapia*.

Long-tailed Paradigalla *Paradigalla carunculata*   ENDEMIC | MONOTYPIC
Paradigalla carunculata Lesson, 1835. *Hist. Nat. Ois. Parad.*: 242.—Arfak Mts, Bird's Head

Inhabits forest of the Bird's Head (Arfak Mts) and possibly the Bird's Neck (?Fakfak Mts of the Onin Penin; Gibbs 1994).

NOTES

Treated by Mayr (1941) as conspecific with *P. brevicauda*. Characters that define *carunculata* are the two-tone (blue-red) gape wattle and the long, wedge-shaped tail. Should be looked for in the Tamrau Mts of the Bird's Head and Wondiwoi Mts of the Bird's Neck in spite of not being recorded by initial field workers. A *Paradigalla* population was observed in the uplands of the Fakfak Mts (Gibbs 1994). It is distinguished from the Arfak-dwelling *P. carunculata* by the pale yellow-white facial wattle, more swollen and paler blue malar wattle, the apparent lack of the red malar wattle, and the short square-cut tail extending 3–4 cm beyond the wingtip. Not observed by Diamond & Bishop (2015) in the Fakfak or Kumawa Mts. Specimens are required for full diagnosis. *Other names*—Wattled Bird of Paradise.

Short-tailed Paradigalla *Paradigalla brevicauda*   ENDEMIC | MONOTYPIC
Paradigalla brevicauda Rothschild & Hartert, 1911. *Novit. Zool.* 18: 159.—Mount Goliath, Border Ra.
> Synonym: *Paradigalla intermedia* Ogilvie-Grant, 1913. *Bull. Brit. Orn. Club* 31: 105.—Otakwa R, 1,680 m, Nassau Mts, Western Ra.

Inhabits montane forest and edge of the Western, Border, and Eastern Ra, from the Weyland Mts to the Bismarck Ra (Stresemann 1923, Hartert *et al.* 1936, Ripley 1964a); 1,570–2,380 m, extremes 1,400–2,580 m[CO]. Absent from the Bird's Head and the SE Peninsula of PNG. Should be looked for in the Kratke Mts of the easternmost sector of the Eastern Ra.

NOTES

Distinguished from *P. carunculata* by the distinctive wattles and abbreviated tail. The gular wattle is all blue rather than blue and pinkish red. Description of the subspecies *intermedia* was based on immature specimens (Frith & Beehler 1998). *Other names*—Short-tailed Wattlebird, Blue-and-yellow Wattled Bird of Paradise.

GENUS *ASTRAPIA*   ENDEMIC [Genus 5 spp]

Astrapia Vieillot, 1816. *Analyse Nouv. Orn. Elem.*: 36. Type, by monotypy, *Paradisea nigra* Gmelin.

The species of this genus are endemic to the uplands of New Guinea—the mountains of the Bird's Head, the Central Ranges, and the mountains of the Huon Peninsula. This distinctive and morphologically uniform genus is characterized by the small, slim bill; the greatly elongate central rectrices; the male's absence of flank plumes; the presence of specialized iridescent green feathering on throat, breast, crown, or nape; and the rustling wings of the adult male. On two separate occasions (Brass 1956: 144, Anon. 1976: 2) a long-tailed black bird, postulated to be a species of *Astrapia*, has been reported sighted in the mountain forests of Goodenough Island. Expeditions by various fieldworkers have failed to confirm its presence on the island (see Beehler 1986: 81–105). Species sequence follows Irestedt *et al.* (2009).

Arfak Astrapia *Astrapia nigra*   ENDEMIC | MONOTYPIC
Paradisea nigra Gmelin, 1788. *Syst. Nat.* 1(1): 401.—Arfak Mts, e Bird's Head.

The Arfak and Tamrau Mts of the Bird's Head, 1,700–2,450 m. Inhabits the interior of montane forest. Uncommon and difficult to observe. Probably ranges to the highest summits of the Arfak Mts (*ca.* 2,950 m) and adjacent Tamrau Mts.

NOTES

Field observations of this species from the uplands of the Tamrau Mts were made by J. MacKinnon (above Kebar, Nov-1980) and Beehler and A. Safford (above Waibeem, 2,095 m, 7-Sep-2006). We retain the English name Arfak Astrapia for historical reasons. *Other names*—Black Astrapia, Arfak Bird of Paradise.

Splendid Astrapia *Astrapia splendidissima*   ENDEMIC
Astrapia splendidissima Rothschild, 1895. *Novit. Zool.* 2: 59, pl. 5.—Probably the Weyland Mts (*fide* Mayr 1941: 171).

Inhabits montane and subalpine forest and edge of the Western Ra, Border Ra, and northwestern-most fringe of the Eastern Ra, 1,750–3,450 m; ranges up to tree line on the higher summits. Known e limit of its range is the headwaters of the Korosameri R, north of the Lagaip R, in the w verge of the Eastern Ra (B. W. Benz *in litt.*).

SUBSPECIES

a. *Astrapia splendidissima splendidissima* Rothschild
RANGE: The w sector of Western Ra, including the Charles Louis Mts, Weyland Mts, and uplands of Paniai Lake[J] (Mayr 1941, Junge 1952[J]).
DIAGNOSIS: See following race.

b. *Astrapia splendidissima helios* Mayr
Astrapia splendidissima helios Mayr, 1936. *Amer. Mus. Novit.* 869: 3.—Mount Goliath, w sector of Border Ra.
   Synonym: *Astrapia splendidissima elliottsmithorum* [corrected] Gilliard, 1961. *Amer. Mus. Novit.* 2031: 3.—Mount Ifal, 2,200 m, Victor Emanuel Mts. [LeCroy (2014) demonstrated the proper construction of the subspecies epithet should be plural genitive.]
RANGE: The Western and Border Ra, from the c Western Ra east to the Victor Emanuel Mts and w Sepik Mts (Ripley 1964a, Gilliard & LeCroy 1961, B. W. Benz *in litt.*).
DIAGNOSIS: The race *helios* differs from the nominate form in the following ways. (1) crown, neck, and dorsal collar of male more bluish and less golden green; (2) spatulate tips of central rectrices of male broader; (3) female slightly darker dorsally; (4) plumages of both sexes of this form exhibit extensive white bases to underside of outer primaries (Frith & Beehler 1998: 255). These two forms are but thinly differentiated and, in fact, the species could easily be treated as monotypic. We follow Cracraft (1992) in subsuming *elliottsmithorum* into this form. Gilliard (1961: 3) noted that in order to observe the plumage distinction exhibited by *elliottsmithorum* one must observe the specimen "with the eye nearly parallel to the plumage"—perhaps an indication that one would be looking for a marginal plumage distinction, rather than an entrenched genetic difference.

NOTES

It will be interesting to determine whether this species and *A. mayeri* meet and hybridize in the Central Ra north of the Strickland R gorge. One might look for such a contact zone north of the confluence of the Lagaip and Strickland R in forests above 2,200 m (Beehler & Sine 2007). *Other names*—Splendid Bird of Paradise.

Ribbon-tailed Astrapia *Astrapia mayeri*   ENDEMIC | MONOTYPIC
Astrapia mayeri Stonor, 1939 (February). *Bull. Brit. Orn. Club* 59: 57.—Mount Hagen, Eastern Ra.
   Synonym: *Taeniaparadisea macnicolli* Kinghorn, 1939 (December). *Austral. Zool.* 9: 295.—West and northwest of Mt Hagen, Eastern Ra.
   Synonym: *Astrapia recondita* Kuroda, 1943. *Bull. Biogeogr. Soc. Japan* 13: 33.—Type locality unknown; probably from the region around Mt Hagen [not Morobe Prov as guessed by the author].

Hybrid: *Astrarchia barnesi* Iredale, 1948. *Austral. Zool.* 2: 160.—Mount Hagen [hybrid *A. mayeri* × *A. stephaniae*]. Restricted to a small w segment of the Eastern Ra, from Mt Hagen[CO] and Mt Giluwe[CO] northwest to the Kaijende Highlands[J], Tari Gap[CO], and probably at least to the e verge of the Strickland R gorge, 2,120[J]–3,400 m[CO] (Beehler & Sine 2007[J]). Its entire range encompasses an area no more than 200 × 100 km. The w extremity of the range is not yet defined. Hybrids are common between *A. mayeri* and *A. stephaniae* on the sw slopes of Mt Hagen and at Tari Gap, at the lower elevations of *A. mayeri's* range (below 2,450 m). This species is replaced east of Mt Hagen by *A. stephaniae*. In the absence of this syntopic congener, *A. mayeri* ranges to its lowest recorded elevations (2,120 m) at Lake Tawa on the sw verge of the Kaijende Highlands (Beehler & Sine 2007).

NOTES

Should be looked for along the high n scarp of the Central Ra north of Porgera; here this species may come into contact with *Astrapia splendidissima*. *Other names*—Ribbontailed Bird of Paradise.

Stephanie's Astrapia *Astrapia stephaniae* ENDEMIC

Astrarchia Stephaniae Finsch and A. B. Meyer, 1885. *Zeitschr. Ges. Orn.* 2: 378.—Mount Maguli, SE Peninsula.

Confined to the c and e sectors of Eastern Ra and the mts of the SE Peninsula; 1,500–2,800 m, extremes 1,700–3,500 m[CO]. Inhabits montane forest, edge, and disturbed forest. Overlaps in range and hybridizes with *A. mayeri* westward from Mt Hagen and Mt Giluwe to the westernmost edge of *stephaniae's* distribution. Where sympatric with the Ribbon-tailed Astrapia it is replaced by that species at higher elevations, generally above 2,450 m.

SUBSPECIES

a. *Astrapia stephaniae feminina* Neumann

Astrapia feminina Neumann, 1922. *Verh. Orn. Ges. Bayern* 15: 236.—Schraderberg, Sepik Mts, n sector of the Eastern Ra.

RANGE: The Schrader Ra, Sepik-Wahgi Divide, and Bismarck Ra of PNG (Stresemann 1923, Mayr 1941, Mayr & Gilliard 1954).

DIAGNOSIS: Adult female is similar to the nominate form but exhibits less contrast between the color of the crown/nape and the color of the back. Also, the nape is more bluish black in adult males. Finally, the wing-to-tail ratio is lower than for the nominate (Frith & Beehler 1998).

b. *Astrapia stephaniae stephaniae* (Finsch & A. B. Meyer)

Synonym: *Astrapia stephaniae ducalis* Mayr, 1931. *Mitt. Zool. Mus. Berlin* 17: 711.—Dawong, Herzog Mts, SE Peninsula.

RANGE: The s scarp of the Eastern Ra and all uplands of the SE Peninsula. Ranges from Nipa[CO] and Tari Gap[CO] southeast to the Kuper Ra[J] and both the n and s watersheds of the Ekuti Divide and Owen Stanley Ra as far east as Mt Dayman[K] (Greenway 1935[J], see map in Frith & Beehler 1998[K]).

DIAGNOSIS: Female of the nominate form exhibits ventral plumage that is more rufous, and mantle coloration more olive brown. (Rand & Gilliard 1967). See photo in Coates (1990: 454).

NOTES

A. s. feminina likely merits submersion into the nominate form, making this species monotypic. We follow Gilliard & LeCroy (1968) in subsuming *ducalis* into the nominate form. *Other names*—Princess Stephanie's Bird of Paradise or Astrapia.

Huon Astrapia *Astrapia rothschildi* ENDEMIC | MONOTYPIC

Astrapia rothschildi Foerster, 1906. In Foerster & Rothschild, *Two New Birds of Paradise*: 2.—Rawlinson Ra, Huon Penin.

Inhabits montane forests of the Huon Penin, 1,460–3,500 m[CO] (Mayr 1931, Draffan 1977b). More common at higher elevations. Quiet and unwary. Frequents mainly the canopy but comes lower to feed on fruit of *Schefflera* vines.

Other names—Rothschild's Astrapia, Huon Bird of Paradise.

Cicinnurus Vieillot, 1816. *Analyse Nouv. Orn. Elem.*: 35. Type, by monotypy, "Manucode" Buffon = *Paradisea regia* Linnaeus.

We follow Beehler & Finch (1985), Frith & Beehler (1998), and Dickinson & Christidis (2014) in folding *Diphyllodes* into *Cicinnurus*. The genus is endemic to the Region, ranging through the Northwestern Islands, Aru Islands, and Yapen Island, and throughout the lowlands, hills, and lower mountains of New Guinea. Characters defining the genus include the curled central tail wires of the adult male, small bill, and small body size. Most traditional treatments in the 20th century (*e.g.*, Gilliard 1969, Cooper & Forshaw 1977) used *Cicinnurus* for a single species (*regius*) and *Diphyllodes* for the species *magnificus* and *respublica*. Diamond (1972) recommended folding these two genera into *Lophorina*. The phylogeny of Irestedt *et al.* (2009) combined the traditional *Cicinnurus* and *Diphyllodes* into a clade that is sister to the plumed *Paradisaea* birds of paradise.

King Bird of Paradise *Cicinnurus regius*   ENDEMIC

Paradisea regia Linnaeus, 1758. *Syst. Nat.*, ed. 10, 1: 110.—Aru Is.

Inhabits forest interior of lowlands and hills of NG, the Aru Is, and Misool, Salawati, and Yapen I; sea level–300 m, max 1,150 m^{CO}. Absent from much of the Trans-Fly, where the habitat is largely unsuitable. Overlaps broadly in elevational range with *Cicinnurus magnificus*. No confirmed records from Batanta or Waigeo I.

SUBSPECIES

a. *Cicinnurus regius regius* (Linnaeus)

Synonym: *Paradisea Rex* Scopoli, 1786. *Del Faun. et Flor. Insubr.*, pt. 2: 88.—Sorong distr, w sector of the Bird's Head. [Based on Sonnerat, 1776, *Voy. Nouv. Guinée*: 156, pl. 95.]

RANGE: The Aru Is, Misool and Salawati I, and all NG except for the NW Lowlands and Sepik-Ramu (Stresemann 1923, Frith & Beehler 1998). The Trans-Fly records range from the Maro R in the west to the Oriomo R in the east (Bishop 2005a).

DIAGNOSIS: Forehead crest of adult male yellow-orange rather than red and the orange extending well up onto crown; supraocular mark a round spot rather than a slash. See photo of male in Coates (1990: 487–89) and Coates & Peckover (2001: 233). We follow Mees (1964b) in synonymizing *rex* here.

b. *Cicinnurus regius coccineifrons* Rothschild

Cicinnurus regius coccineifrons Rothschild, 1896. *Novit. Zool.* 3: 10.—Yapen I, Bay Is.

Synonym: *Cicinnurus regius similis* Stresemann, 1922. *J. Orn.* 70: 405.—Near Madang, e verge of the Sepik-Ramu.

Synonym: *Cicinnurus regius cryptorhynchus* Stresemann, 1922. *J. Orn.* 70: 405.—Taua, lower Mamberamo R, NW Lowlands.

Synonym: *Cicinnurus regius gymnorhynchus* Stresemann, 1922. *J. Orn.* 70: 405.—Heldsbach Coast, near Finschhafen, Huon Penin.

RANGE: The n coast of NG from the Waropen distr of e Geelvink Bay east to the n scarp of the Huon Penin (Frith & Beehler 1998).

DIAGNOSIS: Adult male has a reddish (not yellowish-orange) forehead crest and the supraocular dark mark a narrow slash rather than a spot.

NOTES

Subspecific treatment follows Mees (1964b, 1982b) and Frith & Beehler (1998). We recognize these two forms, from the n and s watersheds of NG, with some reservation. There is substantial nongeographic variation in male dorsal plumage. Females exhibit no geographic variation. M. LeCroy (*in litt.*) suggested that the shape of the supraocular spot may, in part, be a product of skin preparation, so care must be taken in using this particular character. *Other names*—Little King Bird of Paradise.

Magnificent Bird of Paradise *Cicinnurus magnificus*   ENDEMIC

Paradisea Magnifica J. R. Forster, 1781. *Spec. Faun. Ind., Ind. Zool.*: 40 (based on Daubenton, *Pl. Enlum.*: pl. 631).—Arfak Mts, Bird's Head.

Throughout forested uplands of NG, Salawati, ?Misool, and Yapen I; from rugged foothills near sea level to 1,450 m, max 1,780 m^{CO}. Rare or absent in flat lowland alluvial forest. Replaced on Batanta and Waigeo I by *C. respublica*. Inhabits forest interior and is rarely seen except when feeding at fruiting trees or at display grounds. Mees (1965) presented the case for excluding Misool I from the range of this species, in spite of early reports to the contrary (*e.g.*, Mayr 1941). J. M. Diamond (*in litt.*) reported that a local informant described the species as occurring on Misool. Should be looked for there.

SUBSPECIES

a. *Cicinnurus magnificus magnificus* (Pennant)

RANGE: Salawati I, Bird's Head, and Bird's Neck (Wandammen and Onin Penin). Population sighted in the Kumawa Mts undiagnosed but likely belongs here by geography (J. M. Diamond *in litt.*, C. Thébaud & B. Mila *in litt.*). No confirmed record from Misool I (Mees 1965) but likely to be found there.

DIAGNOSIS: Crown of male brown (Rand & Gilliard 1967). Scapulars and inner secondaries clay-colored (Gilliard 1969). Edges of flight feathers yellowish (Frith & Beehler 1998).

b. *Cicinnurus magnificus chrysopterus* (Elliot)

Diphyllodes speciosus var. *chrysopterus* Elliot, 1873. *Monogr. Birds Parad.*: pl. 13 and text.—Yapen I, Bay Is.

Synonym: *Diphyllodes magnificus intermedius* Hartert, 1930. *Novit. Zool.* 36: 36.—Upper Setekwa R, Western Ra.

RANGE: Yapen I, the NW Lowlands (including Foja and Cyclops Mts), the Sepik-Ramu (including the PNG North Coastal Ra), and the w sector of the S Lowlands, east to the Fly R (Stresemann 1923, Frith & Beehler 1998). Observed at Danau Bira at 350 m in hilly lowlands (Beehler field notes 10-Jul-1987).

DIAGNOSIS: Very similar to nominate form but for its darker crown and more orange edging to the flight feathers (Frith & Beehler 1998). We sink *intermedius* into this form following Cracraft (1992).

c. *Cicinnurus magnificus hunsteini* (A. B. Meyer)

Diphyllodes Hunsteini A. B. Meyer, 1885. *Zeitschr. Ges. Orn.* 2: 389, pl. 21.—Mount Maguli, SE Peninsula.

Synonym: *Diphyllodes magnificus extra* Iredale, 1950. *Birds of Paradise and Bower Birds*: 111.—Mount Hagen distr, Eastern Ra.

RANGE: The e sector of the S Lowlands, Huon Penin, and SE Peninsula; in the s watershed from the Fly R eastward. Also includes interior populations from the Bismarck Ra, Wahgi valley region, and Mt Hagen (Coates 1990).

DIAGNOSIS: Crown of male with an orange wash (Diamond 1972). Note also edge of flight feathers a bright, rich orange (Frith & Beehler 1998). See photos in Coates (1990: 496–507).

NOTES

Thinly defined subspecies. We suggest an additional review of the races. *Other names—Diphyllodes magnificus.*

Wilson's Bird of Paradise *Cicinnurus respublica*   ENDEMIC | MONOTYPIC

Loph[orina] respublica Bonaparte, 1850 (February). *Compt. Rend. Acad. Sci. Paris* 30: 131, 291.Waigeo I, NW Islands.

Waigeo and Batanta I. Inhabits interior upland forest, 300–1,000 m (Frith & Beehler 1998). *Other names—*Waigeu Bird of Paradise, *Diphyllodes respublica.*

GENUS *PARADISORNIS*   ENDEMIC | MONOTYPIC

Paradisornis A. B. Meyer, 1885. *Zeitschr. Ges. Orn.* 2: 385. Type, by monotypy, *Paradisornis rudolphi* Finsch.

The single species in the genus is endemic to the mid-montane zone of the eastern and east-central sectors of the Central Ranges of New Guinea. The genus is characterized by the distinctively patterned black and cobalt-blue plumage; the blue pectoral plumes of the male; and the broken white eye-ring in all plumages. *Paradisornis* is closely allied to *Paradisaea*, but this genus is distinct in more than a dozen plumage, behavioral, and vocal characters, detailed in Frith & Beehler (1998: 61–62). The phylogeny of Irestedt *et al.* (2009) demonstrated that *Paradisornis rudolphi* is sister to the *Paradisaea* clade.

Blue Bird of Paradise *Paradisornis rudolphi*   ENDEMIC

Paradisornis Rudolphi Finsch and A. B. Meyer, 1885. *Zeitschr. Ges. Orn.* 2: 385, pl. 20.—Mount Maguli, SE Peninsula.

Inhabits montane forest of the Eastern Ra and SE Peninsula, 1,100–2,000 m[CO]. Ranges from the Karius Ra, Tari, and Kompiam southeast to the SE Peninsula, including the c Owen Stanley Ra, east to Mt Obree. Apparently patchily distributed and absent from some areas that appear suitable for the species. Inhabits forest and woodland regrowth. Common in the Tari valley, where it can be found in village woodlots (when not hunted for its plumes). I. Woxvold (*in litt.*) reported the species at Laite (*ca.* 25 km southwest of Tari), on the e slopes of the Karius Ra. Presumably not to be expected west of the Strickland R gorge. The se terminus of the range of this species in the Owen Stanley Ra is not yet delineated.

SUBSPECIES

a. *Paradisornis rudolphi margaritae* (Mayr & Gilliard)

Paradisaea rudolphi margaritae Mayr & Gilliard, 1951. *Amer. Mus. Novit.* 1524: 11.—Kimil R, 30 km wnw of Nondugl, Wahgi valley, Eastern Ra.

RANGE: From the e sector of the Eastern Ra (Karimui) westward toward the Strickland R gorge; w terminus of range unknown; but at least to the e slopes of the Karius Ra. Localities—Laite on the e slopes of the Karius Ra[N], Tari[CO], Kompiam[J], Mt Giluwe[J], Mt Hagen[K], Sepik-Wahgi Divide[J], Lepa Ridge[M], Kimil R in the Wahgi valley[K], Kuli in the Kubor Mts[I], the Bismarck Ra[CO], and Karimui[DI] (Mayr & Gilliard 1954[K], Gyldenstolpe 1955a[L], Frith & Beehler 1998[J], Beehler field notes[M], I Woxvold *in litt.*[N]).

DIAGNOSIS: Female exhibits blackish barring across breast and onto flanks as well as abdomen (Frith & Beehler 1998). See photo in Coates & Peckover (2001: 239). The two subspecies meet near Mt Karimui. It is uncertain which form inhabits Crater Mt (Mack & Wright 1996).

b. *Paradisornis rudolphi rudolphi* Finsch

Synonym: *Paradisea rudolphi ampla* Greenway, 1934. *Proc. New Engl. Zool. Club* 14: 1.—Mount Missim, 1,737 m, Morobe distr, SE Peninsula

RANGE: From the e sector of the Eastern Ra (Awande[DI] and Okapa[DI]) east to the Herzog Mts and Kuper Ra and thence southeastward down the Owen Stanley Ra to the upper Kumusi and Brown R, Kokoda, and Mt Obree (Frith & Beehler 1998). Patchily distributed. Absent from places that appear ideal for the species (Pratt field notes from the Kokoda Track).

DIAGNOSIS: Female exhibits faint or obsolete barring on breast and flanks but more prominent and regular black barring on center of lower abdomen.

NOTES

The e and w boundaries of the species' range are poorly delineated and need further study. The two subspecies are thinly differentiated. *Paradisaea bloodi* Iredale, 1948 (*Austral. Zool.* 11: 161.—Minyip, Mt Hagen, Eastern Ra) is a hybrid, *Paradisornis rudolphi* × *Paradisaea raggiana. Other names*—Prince Rudolph's Bird of Paradise, *Paradisaea rudolphi*.

GENUS *PARADISAEA*   ENDEMIC [Genus 6 spp]

Paradisaea Linnaeus, 1758. *Syst. Nat.*, ed. 10, 1: 110. Type, by subsequent designation (G. R. Gray, 1840, *Cat. Gen. Subgen. Birds*: 39), *Paradisaea apoda* Linnaeus.

The species of this genus are endemic to the Region, ranging throughout the lowland forests and hills of mainland New Guinea as well as to the Northwestern Islands, Yapen Island, and the D'Entrecasteaux Archipelago. The genus is characterized by the iridescent green throat patch of the male, the abundant erectile pectoral plumes (red, orange, yellow, or white), and the brown wing feathers (male and female).

In spite of its clear affinities to *Paradisaea*, we split *Paradisornis* from this compact group because of its many unique features. This is supported by the phylogeny of Irestedt *et al.* (2009). The spelling *Paradisaea* (rather than *Paradisea*) has been conserved (ICZN Opinion 2294).

Emperor Bird of Paradise *Paradisaea guilielmi*   ENDEMIC | MONOTYPIC

Paradisea Guilielmi Cabanis, 1888. *J. Orn.* 36: 119.—Sattelberg, Huon Penin.

Mountains of the Huon Penin, 450–1,500 m^{CO}. Inhabits upland forest, old gardens, and shade coffee plantations. Replaced at lower elevations by *P. raggiana* in the e part of range, and by *P. minor* in the w sector (hybridization with both of these species is documented).

Other names—Emperor of Germany Bird of Paradise, White-plumed Bird of Paradise.

Red Bird of Paradise *Paradisaea rubra*   ENDEMIC | MONOTYPIC

Paradisea rubra Daudin, 1800. *Traité Orn.* 2: 271.—Waigeo I, NW Islands.

Restricted to Waigeo, Batanta, Gam, and Saonek I of the NW Islands, sea level–600 m (Frith & Beehler 1998). Inhabits forest and forest edge in the lowlands and hills.

NOTES

In the NW Islands, both this species and Wilson's Bird of Paradise inhabit only the islands north of Sagewin Strait. By contrast, the Lesser and Magnificent Birds of Paradise inhabit only islands south of Sagewin Strait, the narrow channel that separates Batanta I and Salawati I. It thus appears that these two species pairs, *Paradisaea rubra/P. minor* and *Cicinnurus respublica/C. magnificus*, speciated across this small but deep water barrier. *Other names*—*Uranornis rubra*.

Goldie's Bird of Paradise *Paradisaea decora*   ENDEMIC | MONOTYPIC

Paradisea decora Salvin & Godman, 1883. *Ibis*: 131.—Fergusson I, D'Entrecasteaux Arch, SE Islands.

Restricted to the upland forests of Fergusson and Normanby I (D'Entrecasteaux Arch); in hills to approximately 600 m, absent above that elevation. Inhabits hill forest and edge. Follows the hills down to near the coast in a few rugged locations. It is surprising that this island endemic is absent from adjacent Goodenough I.

Lesser Bird of Paradise *Paradisaea minor*   ENDEMIC

Paradisea minor Shaw, 1809. *Gen. Zool.* 7(2): 486.—Manokwari, Bird's Head.

Inhabits forest and edge on Misool I, Yapen I, and w and n NG: the Bird's Head and Neck, the Weyland Mts, and the n watershed of NG east to the upper Ramu and the n flank of the Huon Penin at least to Singarokai. Not documented from Salawati I but should be looked for there, as one was reported heard (Dutson & Tindige 2006: 26). Biogeographically, its presence is to be expected on Salawati I. Sea level–1,000 m, max 1,550 m^{CO}.

SUBSPECIES

a. *Paradisaea minor minor* Shaw

> Synonym: *Paradisaea minor pulchra* Mayr & Meyer de Schauensee, 1939. *Proc. Acad. Nat. Sci. Phila.* 91: 151.—Tip, Misool I, NW Islands.
> Synonym: *Paradisea Finschi* A. B. Meyer, 1885. *Zeitschr. Ges. Orn.* 2: 383.—Karau, between Aitape and the mouth of the Sepik, n NG.

RANGE: Misool I, the Bird's Head and Neck, NW Lowlands, Sepik-Ramu, and n scarp of Huon east at least to Singarokai (Stresemann 1923).

DIAGNOSIS: Flank plumes pale golden-yellow (Rand & Gilliard 1967); wing (male) 180–196; flank plumes shorter than those of *jobiensis* (Frith & Beehler 1998). See photos in Coates (1990: 519–22). The plumes of the population *finschi* are quite variable; some are brighter than the nominate form, but others are identical and not separable. We follow Mees (1965) in subsuming *pulchra* into this form.

b. *Paradisaea minor jobiensis* Rothschild

Paradisea minor jobiensis Rothschild, 1897. *Bull. Brit. Orn. Club* 6: 46.—Yapen I, Bay Is.

RANGE: Yapen I.

DIAGNOSIS: Flank plumage more abundant and luxuriant than that of nominate and can be up to 560 mm long (Rand & Gilliard 1967). Also slightly larger than nominate form; wing (male) 183–210.

NOTES

Hybridization with *P. raggiana* occurs not uncommonly in the upper Ramu valley and Finisterre Ra and occasionally in the Baiyer valley. The incidence of hybridization with *P. apoda* is unknown (it presumably contacts that species somewhere between Etna Bay of the s Bird's Neck and the Mimika R). The name *albescens* is based on an artifact and thus has no standing under the ICZN Code (M. LeCroy *in litt.*). *Other names*—Little Emerald Bird of Paradise.

Greater Bird of Paradise *Paradisaea apoda*   ENDEMIC | MONOTYPIC

Paradisaea apoda Linnaeus, 1758. *Syst. Nat.*, ed. 10, 1: 110.—Aru Is.

   Synonym: *Paradisea apoda* Linnaeus, var. *novae guineae* D'Albertis & Salvadori, 1879. *Ann. Mus. Civ. Genova* 14: 96.—Middle Fly R, c sector of the S Lowlands.

Inhabits forest of the Aru Is and the w half of the S Lowlands, from the Mimika R east to the Fly R, sea level–920 m (Frith & Beehler 1998). Trans-Fly records include the lower Digul R (Bangs & Peters 1926), Kurik, Wasur NP, Bian Lakes, Lake Daviumbu, and Wuroi (Bishop 2005a). In south-central NG, *P. apoda* overlaps with *P. raggiana* (hybrids known from Morehead); orange-plumed hybrids also seen at a *P. apoda* lek often visited by birders north of Kiunga. At the w verge of *P. apoda*'s range it may meet populations of *P. minor*, but details of overlap and occurrence of hybrids are unknown. Occurs in lowlands and hills, mainly in the interior. An attempt to reconfirm this species' occurrence at Nomad R on the Strickland (*cf.* Mackay 1966) was unsuccessful (Frith & Beehler 1998).

NOTES

We follow Cracraft (1992) in treating the species as monotypic. *Other names*—Great Bird of Paradise.

Raggiana Bird of Paradise *Paradisaea raggiana*   ENDEMIC

Paradisea raggiana P. L. Sclater, 1873. *Proc. Zool. Soc. London*: 559.—Orangerie Bay, SE Peninsula.

   Inhabits all forested habitats in the e half of s NG and the e third of n NG, sea level–1,800 m[CO]; rare in flat alluvial lowlands, and most common between 500 and 1,000 m. Found in the e sector of the S Lowlands and the Huon and SE Peninsulas. The species ranges west in the n watershed to the upper Ramu R (Bailey 1992a) and the n coast of the Huon to the vicinity of Wasu. It extends west in the s watershed to the Wahgi R (interior), and the Fly R, where it hybridizes with *P. apoda*. Trans-Fly records include Tarara, Wuroi, and Aramia R (Bishop 2005a). Reaches the PNG border near Ningerum (Frith & Beehler 1998). An isolated sw population inhabits relict rain forest in the Trans-Fly, at Morehead. In the upper Fly R it over-laps locally in range with *P. apoda*, where hybridization occurs. Hybrid birds found in leks at Yakyu village, just west of the PNG border, 85 km east-southeast of Merauke (Dewi Pramitha *in litt.*)

   The northern race, *P. r. augustaevictoriae*, with its yellow back and yellowish-orange flank plumes, suggests, at first glance, extensive gene flow with the Lesser Bird of Paradise, *P. minor*; yet in *augustae-victoriae* the breast cushion is more fully developed than in *P. r. salvadorii* or *raggiana*—a feature poorly developed in *P. minor*. Since this plumage ornament features strongly in display, it may serve as an isolating mechanism between *P. r. augustaevictoriae* and *P. minor* where they come into contact.

SUBSPECIES

a. *Paradisaea raggiana salvadorii* Mayr & Rand

Paradisaea apoda salvadorii Mayr & Rand, 1935. *Amer. Mus. Novit.* 814: 11.—Vanumai, Central Prov, s watershed of the SE Peninsula.

RANGE: Southern NG, from the Oriomo and Fly R east to Cloudy Bay (east of Kupiano). Also extends into the upper Purari R to the interior Wahgi valley, from where it sometimes wanders, via deforested habitat, across "hybrid gap" into the Baiyer valley, in the n watershed (where *P. minor* dominates).

DIAGNOSIS: Mantle brown (not yellow) in both sexes. Ventral surface of flank plumes scarlet (Rand & Gilliard 1967). See photos in Coates (1990: 508–13) and Coates & Peckover (2001: 236, 237).

b. *Paradisaea raggiana augustaevictoriae* Cabanis
Paradisea Augustae Victoriae Cabanis, 1888. *J. Orn.* 36: 119.—Finschhafen, Huon Penin.
> Synonym: *Paradisea granti* North, 1906. *Victorian Naturalist* 22: 156.—German New Guinea? [presumably from near Morobe station, n coast of the SE Peninsula].

RANGE: The coast of the Huon Gulf and the Markham R valley northwest up to the upper Ramu, where it hybridizes with *P. minor*. Ranges southeast to the Wafa, Bulolo, and Wau valleys, thence into the Waria valley and coastally to the lower Mambare R (Frith & Beehler 1998). Presumably also meets and hybridizes with *P. minor* on the n coast of the Huon on the n flank of the Saruwaged Ra, west of Wasu.
DIAGNOSIS: Flank plumes of adult male are apricot orange; mantle and back to rump washed heavily with yellow; yellow throat collar narrow (Rand & Gilliard 1967). See photo in Coates (1990: 515) and Coates & Peckover (2001: 237).

c. *Paradisaea raggiana intermedia* De Vis
Paradisea intermedia De Vis, 1894. *Ann. Rept. Brit. New Guinea*, 1893–94: 105.—Kumusi R, SE Peninsula.
RANGE: The c sector of the n watershed of the SE Peninsula, from Collingwood Bay (where the plumes are a little more reddish) to Holnicote Bay, the Kumusi R, and the lower Mambare R (Mayr 1941).
DIAGNOSIS: Similar to *salvadorii*, but both sexes exhibit yellow on mantle and back, with yellow streaking down to uppertail coverts. Ventral surface of flank plumes scarlet (Rand & Gilliard 1967).

d. *Paradisaea raggiana raggiana* P. L. Sclater
RANGE: The s watershed of the e sector of the SE Peninsula, from Cloudy Bay to Milne Bay; thence on the n coast from Milne Bay northwest to Cape Vogel (Frith & Beehler 1998).
DIAGNOSIS: Mantle of adult male washed with yellow, fading as it grades into the sepia of the lower back; ventral surface of flank plumes deep scarlet (Rand & Gilliard 1967).

NOTES
The described form *granti* is an intergradient of *augustaevictoriae* and *intermedia*, and apparently found in the boundary between the ranges of these two on the n coast of the SE Peninsula. P. Gregory and K. D. Bishop (*in litt.*) have observed a lek that included this species and *P. apoda*, 17 km north of Kiunga. *Other names*—Count Raggi's or Red-plumed Bird of Paradise.

Family Melampittidae

Melampittas   ENDEMIC [Family 2 spp/Region 2 spp]

Schodde & Christidis (2014) elevated this lineage to familial status based on molecular distance and morphology. The family comprises two all-black terrestrially dwelling songbirds of the interior of upland or montane forest. We place each species in its own genus following Schodde & Christidis (2014). Between the two, there is a substantial disparity in size and robustness of bill and legs (*Melampitta lugubris* being very gracile and *Megalampitta gigantea* very robust). The phylogeny of Barker *et al.* (2004) showed that *Melampitta* and *Megalampitta* are sister forms. Familial placement of this cluster is problematic (Sibley & Ahlquist 1987, Nunn & Cracraft 1992, Barker *et al.* 2004, Norman *et al.* 2009b, Jønsson *et al.* 2011b). We place them as sister to the birds of paradise (Paradisaeidae), following Jønsson *et al.* (2011b). See Schodde & Christidis (2014) for a detailed diagnosis of the family.

Melampitta Schlegel, 1871. *Ned. Tijdsch. Dierk.* (1873) 4: 47. Type, by monotypy, *Melampitta lugubris* Schlegel.

The single species of the genus is endemic to the Region. It inhabits the floor of montane forest interior. The genus is characterized by the long legs, all-black adult plumage, and terrestrial habit. The affinities of this species have long been uncertain. Placement with the birds of paradise was suggested by Sibley & Ahlquist (1987). Placement near the Cnemophilidae followed comments by Frith & Frith (1990). The phylogeny of Barker *et al.* (2004) placed the genus as sister to the clade that includes *Grallina* and *Monarcha* (Monarchidae). Jønsson *et al.* (2011b) placed *Melampitta* as the sister to the birds of paradise within a larger clade that also encompasses *Struthidea* and *Corcorax* (both of Corcoracidae), a placement supported by Aggerbeck *et al.* (2104). Schodde & Christidis (2014) placed the Greater Melampitta in its own genus, *Megalampitta*, primarily based on morphology.

Lesser Melampitta *Melampitta lugubris*   ENDEMIC | MONOTYPIC

Melampitta lugubris Schlegel, 1871. *Ned. Tijdschr. Dierk.* (1873) 4: 47.—Arfak Mts, Bird's Head.
>   Synonym: *Mellopitta lugubris rostrata* Ogilvie-Grant, 1913. *Bull. Brit. Orn. Club* 31: 104.—Otakwa R, Western Ra.
>   Synonym: *Melampitta lugubris longicauda* Mayr & Gilliard 1952. *Amer. Mus. Novit.* 1577: 1.—Mount Tafa, 2,400
>   m, near the Wharton Ra, SE Peninsula.

An elusive but vocal ground dweller that inhabits the mts of the Bird's Head[MA], Central Ra[MA], and Huon Penin[MA]; mainly 2,000–2,800 m, extremes 1,150–3,500 m[CO].

NOTES

Treated here as monotypic. The size characters of the putative subspecies *rostrata* and *longicauda* do not merit recognition (see Boles 2007). Probably clinal. *Other names*—Lesser Blackwit, Lesser False Pitta.

Megalampitta Schodde & Christidis, 2014. *Zootaxa* 3786(5): 513. Type, by original designation, *Mellopitta gigantea* Rothschild, 1899.

The single species of the genus is a large, ground-dwelling, all-black songbird with a distinctive vocalization and sinkhole-dwelling roosting habits (Diamond 1983). We follow Schodde & Christidis (2014) in placing this species in its own monotypic genus, based on the substantial differences in body form, plumage, and habit of this species when compared with *Melampitta lugubris*. The latter is slight of form, with slim legs, small bill, short tail, plush forehead plumage, and sexual dimorphism in iris color. *M. gigantea* has powerful legs, a substantial pitohui-like body form, a long ragged tail, and a powerful, pitohui-like bill. The two are similar in exhibiting a brown-and-black immature plumage. This is perhaps a shared derived feature that indicates sister status. Museum study skins often feature infestation of feather mites on the face of *M. gigantea*. Note also the worn and degraded tail feathers. The genus name is feminine (Schodde & Christidis 2014).

Greater Melampitta *Megalampitta gigantea*   ENDEMIC | MONOTYPIC

Mellopitta gigantea Rothschild, 1899. *Orn. Monatsber.* 7: 137.—Mount Moari, Arfak Mts, Bird's Head.

A little-known ground dweller of NG's upland forests. Patchily distributed; known from scattered lower montane localities: the Bird's Head, Bird's Neck, s scarp of Western and Border Ra, Torricelli Mts in the PNG North Coastal Ra, the Kikori basin, and the e sector of the SE Peninsula, 500–1,400 m (Diamond 1983, 1985). The Torricelli Mts locality is documented by a specimen at the BPBM. Almost certainly more widely distributed than the above range description indicates; should be looked for in karst terrain. Frequents the forest floor and understory in the vicinity of limestone sinkholes, where the species apparently roosts and nests (Diamond 1983), though also has been found in areas free of karst and sinkholes (Gregory 1997).

Localities—Mt Moari, Arfak Mts[MA]; Siwi, e Bird's Head[J]; Fakfak Mts[K]; Kumawa Mts[K]; Lobo, Triton Bay[J]; Mbuta Lake and Avona, Etna Bay[L]; Otakwa R[MA]; Setekwa R[J]; near Tabubil, Mt Robinson[M,O]; Ok Ma[N],

Ok Menga[N], and P'nyang Camps[P], upper Fly R; Gobe and Moro, Kikori basin[P]; Mt Somoro, Torricelli Mts[J]; and Boneno, near Mt Mura, 48 km northwest of Mt Simpson[J] (Diamond 1983[J], 1985[K], Gregory 1996[M], Gregory 1997[O], P. Gregory *in litt.*[N], I. Woxvold *in litt.*[P], C. Thébaud & B. Mila *in litt.*[L]).

Other names—*Melampitta gigantea*.

Family Monarchidae

Monarchs [Family 96 spp/Region 21 spp]

The species of the family range from Africa, tropical Asia, and Japan, south and east to Australia, New Caledonia, Micronesia, and Polynesia. The New Guinean species are confined mainly to forest habitats in lowland and lower montane forest; none are recorded higher than 2,300 m. Barker *et al.* (2004) showed that *Monarcha* and *Grallina* form a clade, which in turn is sister to *Melampitta* (Melampittidae). These three are shown to be sister to *Struthidea* and *Corcorax* (both of the Corcoracidae). Jønsson *et al.* (2011b) and Aggerbeck *et al.* (2014) placed the monarchs as sister to the shrikes and crows. Usage of the younger name Monarchidae, as opposed to the older Myiagridae, follows Bock (1994), based on prevailing usage. Characters distinguishing the monarchs include noisy behavior; contrasting plumage patterns—often black and white; light body form; broadened and flattened bill; tendency to join mixed feeding flocks; and the deep cup nest bound externally with spider web (Schodde & Mason 1999). Our treatment follows Andersen *et al.* (2015a), who have expanded on the work of Filardi & Smith (2005) and Filardi & Moyle (2005). Additional molecular systematic work that touches on the monarchs includes that by Fabre *et al.* (2014), which takes a primary focus on the genus *Myiagra*. A history of familial systematic treatment of the monarchs and relatives is provided by Higgins *et al.* (2006).

GENUS *GRALLINA* [Genus 2 spp/Region 2 spp]

Grallina Vieillot, 1816. *Analyse Nouv. Orn. Elem.*: 42. Type, by monotypy, *Grallina melanoleuca* Vieillot = *Corvus cyanoleucus* Latham.

The species of this genus range from Australia to New Guinea. The species are characterized by striking black-and-white plumage, large size, slim bill, terrestrial habits, mud-cup nest, and loud call (Orenstein 1975, Schodde & Mason 1999). Both species forage on the ground for insects and give conspicuous displays while calling. Both parents, and perhaps other individuals, share nesting duties. The genus was recently shown to belong to the Monarchidae, and thus the family Grallinidae (*e.g.*, Rand & Gilliard 1967, Beehler & Finch 1985) is no longer recognized. Andersen *et al.* (2015a) provided a phylogeny that situates *Grallina* as sister to the *Arses+Myiagra* clade.

Torrentlark *Grallina bruijnii* ENDEMIC | MONOTYPIC

Grallina bruijnii Salvadori, 1876. *Ann. Mus. Civ. Genova* (1875) 7: 929.—Arfak Mts, Bird's Head.

Forages along small, rocky, swift-flowing forest streams in the mts of NG; 400–2,300 m, max 2,800 m[CO]. Range includes the Tamrau and Arfak Mts[MA], Central Ra[MA], Foja Mts[J], PNG North Coastal Ra[K], Adelbert Mts[L], and mts of the Huon Penin[MA] (Pratt 1983[L], Beehler *et al.* 2012[J], AMNH specimens[K]). No records from the Fakfak, Kumawa, Wandammen, or Cyclops Mts. Occasionally seen in forest well away from water, presumably moving from one stream to another.

NOTES

The species epithet has been spelled variously (*cf.* Mathews 1930 *vs.* Mayr 1941). Original published citation is *Grallina bruijnii* Salvadori (*pace* Mayr 1941, 1962). *Other names*—*Pomareopsis bruijni*.

Magpielark *Grallina cyanoleuca*

RESIDENT

C[*orvus*] *cyanoleucus* Latham, 1801. *Index Orn.*, Suppl.: xxv.—Sydney, NSW, Australia.

A local resident in the s Trans-Fly; records from the Maro R, Wasur NP, and Bensbach (Bishop 2005a). Frequents open savanna, grassy floodplain, and the margins of waterways (Coates 1990). First recorded from NG by Lindgren (1971).

Extralimital—widespread in Australia; also Timor; has been recorded from the Tayandu I, s Moluccas. In Australia locally nomadic, occurring in the s Torres Strait as an irregular winter visitor. Not recorded to be a passage migrant through the Torres Strait Is (Draffan *et al.* 1983) but has been recorded from Boigu I in the northernmost Torres Strait Is. Possibly an occasional visitor from Australia.

SUBSPECIES

a. *Grallina cyanoleuca ?neglecta* Mathews
Grallina cyanoleuca neglecta Mathews, 1912. *Novit. Zool.* 18(3): 372.—Parry Creek, Wyndham, Western Australia.
RANGE: Northern tier of Australia. The Trans-Fly population needs to be confirmed as of this subspecies, which is likely, given the birds of the Torres Strait Is are of this form (Schodde & Mason 1999).
DIAGNOSIS: Schodde & Mason (1999) noted the following: a small form with a proportionally short tail and long wing; wing (male) 166–176; tail (male) 108–112.

NOTES

M. LeCroy (*in litt.*) noted that there is a specimen in the Carnegie Museum in Pittsburgh labeled as taken at Narinuma village on the Laloki R, near Port Moresby, by Shelly Denton on 20-Sep-1883. This is the only evidence of its possible occurrence in the savanna country of the SE Peninsula. The species epithet *cyanoleuca* has been conserved by the ICZN (Opinion 2240). *Other names*—Australian Magpie-lark.

GENUS *ARSES*   [Genus 4 spp/Region 2 spp]

Arses Lesson, 1831, *Traité Orn.* 5: 387. Type, by subsequent designation (G. R. Gray, 1840. *List Gen. Birds*: 31), *Muscicapa telescopthalmus* Lesson & Garnot.

The species of this genus range from New Guinea to northeastern Australia (Clement *et al.* 2006). The genus is characterized by the blue eye wattle, pied plumage, black crown contrasting the nape plumage, and sexual dichromatism. Note also the distinctive foraging habit of gleaning their food by spiraling upward around trunks and larger limbs in scansorial fashion (Schodde & Mason 1999). *Arses* is sister to *Myiagra*, and the *Arses*+*Myiagra* clade is sister to the main radiation of monarchs (Filardi & Moyle 2005, Fabre *et al.* 2014, Andersen *et al.* 2015a).

Frilled Monarch *Arses telescopthalmus*

ENDEMIC

Muscicapa telescopthalmus Lesson & Garnot, 1827. *Voy. Coquille, Atlas* 1(5): pl. 18, fig. 1; 1829. *Zool.* 1(13): 593.—Manokwari, Bird's Head.

Inhabits forest through NG, the NW Islands, and Aru Is; sea level–1,200 m, max 1,500 m^{CO}. Absent from the NW Lowlands and Sepik-Ramu, where it is replaced by *A. insularis*.

SUBSPECIES

a. *Arses telescopthalmus telescopthalmus* (Lesson & Garnot)
RANGE: Misool and Salawati I^{MA} and the Bird's Head and Neck.
DIAGNOSIS: Male with a black crown, face, and throat; narrow white nuchal collar; black mantle; white rump and back; white breast and belly. Female with blackish cap and dark gray face; chestnut nape and breast band; bright chestnut mantle; whitish chin and white belly. Wing (male) 80–86 (Rand & Gilliard 1967). Extent of black on chin of male varies (Schodde & Mason 1999).

b. *Arses telescopthalmus batantae* Sharpe
Arses batantae Sharpe, 1879. *Notes Leyden Mus.* (1878) 1: 21.—Batanta I, NW Islands.
RANGE: Waigeo and Batanta I (Mayr 1941, see Greenway 1966).
DIAGNOSIS: Female very pale-breasted; buff of breast gradually grades to white on abdomen.

c. *Arses telescopthalmus aruensis* Sharpe
Arses aruensis Sharpe, 1879. *Notes Leyden Mus.* (1878) 1: 22.—Aru Is.
RANGE: The Aru Is.
DIAGNOSIS: Mantle plumage of female quite dark but with a rich reddish-brown tinge, distinct from bright red-brown of nape.

d. *Arses telescopthalmus harterti* van Oort
Arses telescophthalmus harterti van Oort, 1909. *Nova Guinea, Zool.* 9: 86.—Lorentz R, sw NG.
RANGE: The w sector of the S Lowlands and Trans-Fly[J] east to the Purari R and Okasa[DI] (Bishop 2005a[J]). Also Boigu I of Torres Strait (Rand 1942a, Schodde & Mason 1999).
DIAGNOSIS: Male with reduced white patch on back; rump shows a mix of white and gray. Female with black cap and ear patch; ocher collar, throat, chin, and breast; and mantle and back olive brown without the rich red-brown tone.

e. *Arses telescopthalmus henkei* Meyer
Arses henkei Meyer, 1886. *Zeitschr. Ges. Orn.* 3: 16, pl. 3, figs 1 and 2.—Astrolabe Ra, s watershed of the SE Peninsula.
    Synonym: *Arses lauterbachi* Reichenow, 1897. *Orn. Monatsber.* 5: 161.—Finschhafen, Huon Penin.
RANGE: The SE Peninsula and mts of the Huon Penin. This form intergrades with *harterti* in the e sector of the S Lowlands, approaching the Purari R (Schodde & Mason 1999). In the n watershed, *henkei* meets *A. insularis* on the n slopes of the Huon Penin.
DIAGNOSIS: Female with a pale cinnamon breast grading into very pale tan abdomen; chin whitish. See photos of male and female of this form in Coates (1990: 164) and Coates & Peckover (2001: 145). We combine *lauterbachi* into the older form, *henkei*, whereas Schodde & Mason (1999: 502) considered *henkei* an intergradient form, to be folded into *harterti*.

NOTES
Following Coates (1990) and Schodde & Mason (1999), we treat *insularis* of n NG from Yapen I to Madang as a full species, based on the distinctive male plumage. Further evidence in support of this arrangement is given by Freeman *et al.* (2013), who found both *telescopthalmus* and *insularis* in forest on the YUS transect on the n scarp of the Saruwaged Ra, Huon Penin. An apparent hybrid adult male *insularis* × *telescopthalmus* was netted and photographed in 2012 on the YUS transect of the n scarp of the Huon Penin (Beehler field notes). More study of this population and its relation to others is needed. A closely allied form (*lorealis*) on Cape York Penin is sometimes considered a race of *telescopthalmus*; we follow Clement *et al.* (2006) and Christidis & Boles (2008) in treating the two as distinct. However, recent molecular research by Fabre *et al.* (2014) and Andersen *et al.* (2015a) found that both *lorealis* and *kaupi*, another n QLD taxon, were imbedded within *telescopthalmus* and that this combined clade was sister to *insularis*. To break the taxon *telescopthalmus* into four species in order to maintain species status for *lorealis* and *kaupi* is not practical at this time, as to do so would yield two species-level taxa in New Guinea (one in the west and northwest, the other in the east and south) that are not readily diagnosable and whose geographic ranges cannot be clearly delimited. More research is needed to resolve this complex issue. We follow Schodde & Mason (1999) in spelling of the species epithet and original authorship. *Other names*—Frilled or Frill-necked Flycatcher, New Guinea Frilled Monarch, White-lored Flycatcher, *A. telescophthalmus*.

Ochre-collared Monarch *Arses insularis*   ENDEMIC | MONOTYPIC
Monarcha insularis A. B. Meyer, 1874. *Sitzungsber. Akad. Wiss. Wien* 69(1): 395.—Ansus, Yapen I, Bay Is.
    Inhabits forest on Yapen I, the NW Lowlands, and the Sepik-Ramu, east at least to Singarokai and Yalumet[CO] of the n slope of the Huon Penin, where it coexists with *telescopthalmus* (Freeman *et al.* 2013). I. Woxvold (*in litt.*) also reported the two species coexisting in the lower Markham valley. Ranges to 1,070 m[CO].
    Localities—Danau Bira[L], Bernhard Camp[M], Kasiwa[X], Bodim[W], upper Sepik[J], Baiyer R[CO]; Oronga, Maratambu, and Memenga forest in the Adelbert Mts[K]; the Karawari R[Q], the lower Markham R[J], Astrolabe Bay[MA], YUS transect[L], and Wasu[CO] (Rand 1942b[M], Ripley 1964a[W], Gilliard & LeCroy 1967[K], van Balen *et al.* 2009[X], P. Gregory *in litt.*[Q], Beehler field notes 11-Jul-1987[L], I. Woxvold *in litt.*[J]).

See photo in Coates & Peckover (2001: 145). *Other names*—formerly treated as a race of *A. telescopthalmus*.

GENUS *MYIAGRA* [Genus 15 spp/Region 6 spp]

Myiagra Vigors & Horsfield, 1827. *Trans. Linn. Soc. London* 15: 250. Type, by subsequent designation (G. R. Gray, 1840. *List Gen. Birds*: 32), *Myiagra rubeculoides* Vigors & Horsfield = *Todus rubecula* Latham.

The species of this genus range from the Moluccas and Lesser Sundas to Micronesia, New Guinea, Australia, the Bismarcks and Solomons, Fiji, and Oceania (Clement *et al.* 2006). The genus is characterized by the flat-topped crown, dark mantle of the male, distinctive sexual dichromatism, and habit of rapidly vibrating the tail. Fabre *et al.* (2014) provided a detailed molecular phylogeny of the genus. Placement within the family (along with *Arses*) follows Andersen *et al.* (2015a). Both papers treat *Arses* and *Myiagra* as sisters.

Shining Flycatcher *Myiagra alecto* RESIDENT
Drymophila alecto Temminck, 1827. *Planch. Col. d'Ois.*, livr. 72: pl. 430, fig. 1 and text.—Ternate, n Moluccas.

Widespread in the NG region and its satellite islands; sea level–500 m, max 1,220 m[CO] (Diamond & LeCroy 1979). Mainly found in scrubby habitat near water. Inhabits mangroves, shrubbery, and edge (Coates 1990).

Extralimital—the Moluccas; Tanimbar; n and e Australia, islands in the Torres Strait; and the Bismarck Arch (Coates 1990, Mayr & Diamond 2001, Dutson 2011).

SUBSPECIES

a. *Myiagra alecto chalybeocephala* (Lesson & Garnot)
Muscicapa chalybeocephalus Lesson & Garnot, 1828. *Voy. Coquille, Atlas*: pl. 15, fig. 1.—Port Praslin, New Ireland.
> Synonym: *Piezorhynchus alecto novae-guineensis* Mathews, 1928. *Bull. Brit. Orn. Club* 48: 92.—Mimika R, w sector of the S Lowlands.

RANGE: The four main NW Islands[MA], plus Kofiau[X], Gag[G], and Gebe[K]; Biak[MA], Numfor[MA], Kurudu[MA], and Yapen I[MA]; all NG except for range of *wardelli*; islands off the n coast: Wallis[CO], Tarawai[CO], Kairiru[CO], Muschu[CO], Schouten[CO], Karkar[J], Bagabag[J], and Madang islets[CO]; also the Bismarck Arch (Mees 1972[K], Diamond & LeCroy 1979[J], Johnstone 2006[G], Diamond *et al.* 2009[X]).
DIAGNOSIS: Wing longer than for nominate (Clement *et al.* 2006). Female with glossy blue-black hood and rich and bright rufous-brown mantle. See photos of male and female of this form in Coates (1990: 173) and Coates & Peckover (2001: 149).

b. *Myiagra alecto rufolateralis* (G. R. Gray)
Piezorhynchus rufolateralis G. R. Gray, 1858. *Proc. Zool. Soc. London*: 176.—Aru Is.
RANGE: The Aru Is.
DIAGNOSIS: Female dorsal surface very dark, with black of crown bleeding onto mantle. Schodde & Mason (1999) noted this race exhibits a bill that is short and broad, characters also exhibited by *chalybeocephala*.

c. *Myiagra alecto wardelli* (Mathews)
Piezorhynchus nitidus wardelli Mathews, 1911. *Bull. Brit. Orn. Club* 27: 99. Cooktown, QLD.
RANGE: The Trans-Fly[J], including Daru I[MA]; also ne Australia (Schodde & Mason 1999, Bishop 2005a[J]).
DIAGNOSIS: Female upperparts much like those of *chalybeocephala* but for dark gray-brown collar posterior to shiny blue-black crown and nape.

d. *Myiagra alecto manumudari* (Rothschild & Hartert)
Monarcha chalybeocephalus manumudari Rothschild & Hartert, 1915. *Novit. Zool.* 22: 43.—Manam I, Sepik-Ramu region.
RANGE: Manam I.
DIAGNOSIS: Female exhibits a very pale rufous dorsal surface and tail.

e. *Myiagra alecto lucida* G. R. Gray

Myiagra lucida G. R. Gray, 1858. *Proc. Zool. Soc. London*: 176.—Tagula I, Louisiade Arch, SE Islands.
RANGE: The D'Entrecasteaux Arch; Amphlett and Trobriand Is; and Woodlark, Joanet, Misima, and Tagula I (Mayr 1941).
DIAGNOSIS: Rather large; large-billed; female with very pale upperparts (Clement *et al.* 2006).

NOTES

Inclusion of this and the following species in *Myiagra* follows Schodde & McKean (1976), Clement *et al.* (2006), and Fabre *et al.* (2014), but for another view see Mees (1982b), who made the case for placing *alecto* in its own genus, *Piezorhynchus*. Schodde & Mason (1999) provided alternative detailed hypotheses for subspeciation in the species *alecto*. Fabre *et al.* (2014) and Andersen *et al.* (2015a) found deep divisions within the subspecies clades. *Other names*—Common Shining Flycatcher, Shining Monarch Flycatcher, Glossy Flycatcher, Shining Myiagra.

Restless Flycatcher *Myiagra inquieta*   RESIDENT

Turdus inquietus Latham, 1801. *Index Orn.*, Suppl.: xl.—Sydney, NSW, Australia.
    The Trans-Fly. Frequents scrub and partly submerged trees and sedge beds bordering rivers in grassy floodplain and open savanna (Coates 1990).
    Extralimital—n, e, and sw Australia and Saibai I (Higgins *et al.* 2006).

SUBSPECIES

a. *Myiagra inquieta nana* (Gould)

Seïsura nana Gould, 1870. *Ann. Mag. Nat. Hist.* (4)6: 224.—Northern Australia.
RANGE: The s Trans-Fly (Dolak I) inland to Lake Ambuve[L], Lake Pangua[J], Obo[J], and Lake Daviumbu[K] (Gregory *et al.* 1996c[L], Gregory 1997[J], Bishop 2005a[K]). Also e and n Australia and Torres Strait Is.
DIAGNOSIS: The population *nana* is substantially smaller than *inquieta*. Statements that the mantle plumage of the former is very glossy blue-black while that of the latter is duller and less blackish are not supported by a review of the museum specimens. Wing (male) 86–96 (Higgins *et al.* 2006).

NOTES

Authorities vary on whether to elevate *nana* to species level, as advocated by Schodde & Mason (1999) and followed by Clement *et al.* (2006) and Dickinson & Christidis (2014). In light of the recent molecular research of Andersen *et al.* (2015a), who showed that the form *nana* is nested within the *inquieta* lineage, we prefer to choose a conservative course at this time and instead follow Higgins *et al.* (2006) and Christidis & Boles (2008) in not recognizing this split. *Other names*—Paperbark Flycatcher, *Seisura inquieta*, *Myiagra nana*.

Broad-billed Flycatcher *Myiagra ruficollis*   RESIDENT

Platyrhynchos ruficollis Vieillot, 1818. *Nouv. Dict. Hist. Nat.*, nouv. éd. 27: 13.—Timor, Lesser Sundas.
    The S Lowlands and the s watershed of the SE Peninsula; also Daru I and the Aru Is. Inhabits mangrove forest of coastal localities, but also can be found locally in monsoon forest and scrub near water (Coates 1990).
    Extralimital—n Australia, Timor, and islands of Torres Strait (including Boigu and Deliverance I), Sulawesi, and some Lesser Sunda Is (including Sumba, Timor, and Tanimbar; Coates 1990, Higgins *et al.* 2006).

SUBSPECIES

a. *Myiagra ruficollis mimikae* Ogilvie-Grant

Myiagra mimikae Ogilvie-Grant, 1911. *Bull. Brit. Orn. Club* 29: 26.—Mouth of the Mimika R, w sector of the S Lowlands.
RANGE: Northern QLD, islands of the Torres Strait, Aru Is, Dolak I[J], Daru[J], and s NG from the Mimika R[MA] to the Laloki R[MA] (Bishop 2005a[J]). Johnson & Richards (1994) caught this species at the edge of a flooded oxbow above D'Albertis Junction on the Fly R.
DIAGNOSIS: Similar to the nominate form but larger, paler, and less richly toned (Clement *et al.* 2006).

Schodde & Mason (1999) reported "rich glossy leaden" mantle of male and "pallid leaden mantle" of female; wing: male 71–75, female 66–70.

NOTES
Other names—Broad-billed Monarch or Myiagra, Broad-billed Myiagra Flycatcher.

Biak Black Flycatcher *Myiagra atra* ENDEMIC | MONOTYPIC
Myiagra atra A. B. Meyer, 1874. *Sitzungsber. Akad. Wiss. Wien* 69(1): 498.—Numfor and Biak I, Bay Is.
Inhabits the middle and upper stories of forest, forest edge, and secondary forest right down to the coast on Biak and Numfor of the Bay Is (Mayr 1941, K. D. Bishop *in litt.*). A specimen was taken from adjacent Rani I (Clement *et al.* 2006); also seen on Owi I in 1993 (P. Gregory *in litt.*).
Other names—Black Myiagra Flycatcher, Biak Myiagra or Flycatcher.

Leaden Flycatcher *Myiagra rubecula* RESIDENT AND VISITOR
Todus rubecula Latham, 1801. *Index Orn.*, Suppl.: xxxii.—Sydney, NSW, Australia.
Resident in the s Bird's Neck, S Lowlands, SE Peninsula, and SE Islands, to 500 m. In addition, Australian migrants are known from the Trans-Fly but also possibly show up elsewhere in the Region.
Extralimital—breeding range includes n and e Australia. Two old records from the Bismarcks (Dutson 2011) are presumably vagrants from Australia.

SUBSPECIES

a. *Myiagra rubecula rubecula* (Latham)
RANGE: Breeds in se Australia (Schodde & Mason 1999). On migration and in the nonbreeding season found in n QLD and s NG. So far, recorded with certainty only from the Trans-Fly, including Merauke[J], Gaima[K], Sturt I[K], the lower and middle Fly R[MA], Lake Daviumbu[K], and Daru[MA] (Mees 1982b[J], Bishop 2005a[K]). Note that the published sight records from Kumil R, Aiyura, near Bulolo, and Lae are most likely *M. cyanoleuca* (Coates 1990).
DIAGNOSIS: Upperparts and lores dark; throat iridescent (Clement *et al.* 2006). Wing: male 77–82, female 75–81 (Higgins *et al.* 2006). See photo of male and female in Coates (1990: 169).

b. *Myiagra rubecula papuana* Rothschild & Hartert
Myiagra rubecula papuana Rothschild & Hartert, 1918. *Novit. Zool.* 25: 317.—Kumusi R, SE Peninsula.
RANGE: The S Lowlands, Trans-Fly[J], and SE Peninsula[MA], west in the s watershed from Milne Bay to the Bird's Neck (Triton Bay), and in the n watershed from Milne Bay[CO] to the Kumusi R[CO] (Rand 1942a, Mees 1982b[J]). Also apparently this form inhabits Boigu I in the n Torres Strait (Schodde & Mason 1999).
DIAGNOSIS: Small and short-billed; male with pale dorsal plumage, dull lores, and dull and less iridescent throat than conspecifics. In addition, Schodde & Mason (1999) noted the following: lores and face of male leaden gray; female with uniformly mid-cinnamon-russet throat and breast. Wing: male 73–79, female 73–78 (Higgins *et al.* 2006).

c. *Myiagra rubecula sciurorum* Rothschild & Hartert
Myiagra rubecula sciurorum Rothschild & Hartert, 1918. *Novit. Zool.* 25: 318.—Rossel I, Louisiade Arch, SE Islands.
RANGE: The SE Islands—Fergusson, Dobu, and Normanby I[J]; Conflict Is; and Misima, Tagula, and Rossel I (Mayr 1941, LeCroy *et al.* 1984[J]).
DIAGNOSIS: Similar to nominate form but smaller, wing (male) 75; crown and throat darker and glossier (Rand & Gilliard 1967).

NOTES
Races difficult to separate in practice. *Other names*—Leaden Monarch or Myiagra, Leaden Myiagra Flycatcher.

Satin Flycatcher *Myiagra cyanoleuca*   VISITOR | MONOTYPIC
Platyrhynchos cyanoleucus Vieillot, 1818. *Nouv. Dict. Hist. Nat.*, nouv. éd. 27: 11.—Sydney, NSW, Australia.

A nonbreeding migrant from Australia, wintering in the e two-thirds of NG. Noted in interior high-land valleys: Baiyer R[CO], Wau[P], Sattelberg[L] (to 1,200 m); in some lowland sites: Lorentz R[MA], Jayapura[CO], Lake Daviumbu[J], Iniok village in the upper Sepik[K], Ramu R[CO], Port Moresby[MA], Karkar and Manam I[MA]; and in the SE Islands: Trobriand[CO], Woodlark[MA], Goodenough[MA], Fergusson[MA], Misima[CO], and Tagula[MA] (Mayr 1931[L], K. D. Bishop *in litt.*[J], I. Woxvold *in litt.*[K], Beehler field notes 13-Aug-1982[P]). Few records from w NG. Present in the Region mainly from early March to October, though there are records from Manam I in December and January (Rothschild & Hartert 1915b).

Extralimital—breeds in e Australia; winter records from the Bismarck Arch[CO] and NZ (Gill *et al.* 2010). The species is a passage migrant through the Torres Strait Is (Draffan *et al.* 1983).

NOTES
Other names—Satin Myiagra or Monarch, Satin Myiagra Flycatcher.

GENUS *SYMPOSIACHRUS*   [Genus 20 spp/Region 7 spp]

Symposiachrus Bonaparte, 1854. *Compt. Rend. Acad. Sci. Paris* 38: 650. Type, by original designation, *Drymophila trivirgata* Temminck

The species of this newly resurrected genus range from the Lesser Sundas and the Moluccas to New Guinea, Australia, and the Bismarck and Solomon Islands (Andersen *et al.* 2015a). The genus is character-ized by a light and compact body form, medium-size bill, medium-length tail, and strongly patterned male plumage, typically with black, gray, white, and buff. Molecular trees pertaining to this genus include those of Filardi & Smith (2005), Filardi & Moyle (2005), and particularly Andersen *et al.* (2015a).

Fantailed Monarch *Symposiachrus axillaris*   ENDEMIC
Monarcha axillaris Salvadori, 1875. *Ann. Mus. Civ. Genova* 7: 921.—Profi, Arfak Mts, 1,035 m, Bird's Head.

Inhabits lower montane forest through the mts of NG and Goodenough I; 1,400–2,200 m, extremes 700–2,650 m (Coates & Peckover 2001). No records from the Cyclops Mts.

SUBSPECIES

a. *Symposiachrus axillaris axillaris* (Salvadori)
RANGE: The Arfak[J], Fakfak[K], Kumawa[L], Wandammen[MA], and Foja Mts[M]; Western Ra[J]; also PNG North Coastal Ra[N] (Gyldenstolpe 1955b[J], Diamond 1985[M], Gibbs 1994[K], specimens at AMNH & BPBM[N], C. Thébaud & B. Mila *in litt.*[L]). Specimens from Bernhard Camp (head of the Mamberamo) have extensive white axillars. Captured at 1,100 m in the s Kumawa Mts (C. Thébaud & B. Mila *in litt.*).
DIAGNOSIS: White pectoral tufts large.

b. *Symposiachrus axillaris fallax* (Ramsay)
Rhipidura fallax Ramsay, 1885. *Proc. Zool. Soc. London* (1884): 580.—Astrolabe Ra, SE Peninsula.
RANGE: The Border Ra east to the SE Peninsula[MA]; also the Adelbert Mts[J], mts of the Huon Penin[MA], and Goodenough I[K] (Rand & Gilliard 1967[K], Gilliard & LeCroy 1967[J]).
DIAGNOSIS: White pectoral tufts small. See photo in Coates & Peckover (2001: 142).

NOTES
Two very similar subspecies, but the pectoral tufts do stand out prominently in the nominate form, making museum diagnosis straightforward. *Other names*—Black Monarch, Black Monarch-flycatcher, *Monarcha axillaris*.

Rufous Monarch *Symposiachrus rubiensis*

ENDEMIC | MONOTYPIC

Tchitrea rubiensis A. B. Meyer, 1874. *Sitzungsber. Akad. Wiss. Wien* 69: 494.—Rubi, head of Geelvink Bay, Bird's Neck.

Inhabits lowland forest of w and c NG, from the e Bird's Head (e flank of the Arfak Mts) to the Sepik-Ramu in the n watershed; in the s watershed from Triton Bay east to the Strickland basin (Nomad R); sea level–175 m. Few records from the s scarp.

Localities—Andai[M], Momi[MA], Windesi[MA], near Teluk Sebakor on the Bomberai Penin[J], Rubi[MA], Windesi[MA], Lobo on Triton Bay[M], Setekwa R area[MA], near Nabire on the Boemi R[K], Mamberamo R[MA], Humboldt Bay[MA], Nomad R[J], Utai[N], Puwani R[L], Sepik R[MA], Ramu R[MA], Adelbert Mts[O], Gogol R[L], Sempi near Madang[N], and Keku[N] (Stresemann 1923[M], Melville 1980[K], Diamond 1985[J], Beehler field notes 18-Jul-1984[L], specimen at AMNH[N], specimen at BPBM[O]). Absent from the w and s Bird's Head. Generally uncommon and patchily distributed.

NOTES

We follow Andersen *et al.* (2015a) in placing this species into *Symposiachrus*, as sister to *S. axillaris*. A published record (Ripley 1964a: 61) from 1,370 m in the Baliem valley was a misidentification of female *Arses telescopthalmus* (Yale specimen examined). *Other names*—Rufous Monarch-flycatcher, *Monarcha rubiensis*.

Hooded Monarch *Symposiachrus manadensis*

ENDEMIC | MONOTYPIC

Muscicapa manadensis Quoy & Gaimard, 1830. *Voy. Astrolabe, Zool.* 1: 174, pl. 3, fig. 3.—Manokwari, Bird's Head.

Patchily distributed through hill forest and adjacent lowlands of the NG mainland; mainly hilly lowlands to 375 m, max 1,200 m[CO]. Absent from all NG islands. Also apparently absent from the lower Fly R and Trans-Fly but present at Kiunga and Nomad R[CO]. Possibly mainly a hill-forest species ranging only marginally out from the hilly zone into alluvial lowland forest (Coates 1990). Occurs in lowland riparian forest on the Karawari R (P. Gregory *in litt.*).

Other names—Black and White Monarch Flycatcher.

Spot-winged Monarch *Symposiachrus guttula*

ENDEMIC | MONOTYPIC

Muscicapa guttula Lesson, 1828. *Man. Orn.*: 591, pl. 16, fig. 2.—Manokwari, Bird's Head.

All NG, the main NW Islands[K] (plus Sagewin[CO], Gag[J], and Gebe[L]); the Aru Is; Mios Num, Yapen, Kairiru[CO], Vokeo[CO], Misima, and Tagula I; and the D'Entrecasteaux Arch (Mayr 1941, Mees 1972[L], Diamond 1985[K], Johnstone 2006[J]). Sea level–800 m, max 1,200 m. Widespread on the Mainland, inhabiting forest interior. Kofiau I record withdrawn (see below).

NOTES

Symposiachrus is apparently masculine; however, the specific epithet *guttula* is a noun in apposition and invariant. The Beccari specimen from Kofiau I has been restudied and was found to be an immature *S. julianae* (Diamond *et al.* 2009). *Other names*—Spot-wing Monarch Flycatcher, *Monarcha guttula*, *M. guttulus*.

Kofiau Monarch *Symposiachrus julianae*

ENDEMIC | MONOTYPIC

Monarcha julianae Ripley, 1959. *Postilla* Yale Peabody Mus. 38: 9.—Kofiau I, NW Islands.

Endemic to Kofiau I (NW Islands), where widespread in forest (Diamond *et al.* 2009). Observed in recent visits by Diamond & Bishop (in 1986), Mauro & Wijaya (in 2002 and 2007), Beehler (in 2011), and Bishop (in 2012). Kofiau has been selectively logged and currently has no protected area. Although the species is apparently tolerant of selective logging, the current status of forest habitats on Kofiau is poorly known. Whether the species is in decline is unknown.

NOTES

We suggest the species should be reclassified from Vulnerable to Near Threatened by the IUCN Red List. We situate this species following *S. guttula* as we believe the plumage characters of *julianae* indicate it is sister to the widespread *S. guttula*. *Other names*—Black-backed Monarch, Kofiau Monarch Flycatcher, Kafiau Monarch, *Monarcha julianae*.

Spectacled Monarch *Symposiachrus trivirgatus* RESIDENT AND VISITOR

Drymophila trivirgata Temminck, 1826. *Planch. Col. d'Ois.*, livr. 70: pl. 418, fig. 1.—Timor, Lesser Sundas.
   Synonym: *Monarcha bernsteinii* Salvadori, 1878. *Ann. Mus. Civ. Genova* 12: 322.—"Salawati I" [in error; = Ambon]. [A junior synonym of *S. t. nigrimentum* from Ambon. See Notes section below.]
An Australian migrant to the Trans-Fly and a resident in the SE Islands.

Extralimital—breeds in e Australia and the islands of the Torres Strait, Timor, Lesser Sundas, and Moluccas. Migratory in s parts of its Australian range. In the n Torres Strait recorded from Yam, Yorke, Stephens, Nepean, Darnley, Dauar, and Murray I (Ingram 1976, Draffan *et al.* 1983).

SUBSPECIES

a. *Symposiachrus trivirgatus gouldii* (G. R. Gray)

Monarcha trivirgatus gouldii G. R. Gray, 1861. *Proc. Zool. Soc. London* (1860): 352.—New South Wales, Australia.
RANGE: Breeds in e Australia, south of the n QLD range of *melanorrhous* (Schodde & Mason 1999, Clement *et al.* 2006). A migrant to the Torres Strait Is, with a few traveling to the Trans-Fly (Draffan *et al.* 1983). Localities—Merauke[J], Wasur NP[J], Bensbach[J], Lake Daviumbu[K], and Tarara[K] (Rand 1942a[K], Bishop 2005a[J]). A scarce but regular visitor, mainly from May to October, with some nonbreeders remaining during the rest of the year (Coates 1990). Immatures collected at Lake Daviumbu, Tarara, and Wando (Coates 1990). Also a sighting from Varirata NP, presumably of this race (P. Gregory *in litt.*).
DIAGNOSIS: Similar to *melanorrhous* from ne QLD but exhibits deeper gray upperparts (Clement *et al.* 2006). Mantle mid-leaden gray; no black on wing coverts; rump dark leaden; breast mid-rust, washing onto the white belly (Schodde & Mason 1999). No pale upper fringe to black mask, as is present in *melanopterus*.

b. *Symposiachrus trivirgatus melanopterus* (G. R. Gray)

Monarcha melanoptera G. R. Gray, 1858. *Proc. Zool. Soc. London*: 178.—Round I, near Tagula I, Louisiade Arch, SE Islands.
RANGE: Normanby I[K]; also Panasesa I (Conflict Is)[J], Alcester, East, and Hastings I; Misima, Kimuta, Rossel, and Tagula I (Mayr 1941, Beehler field notes 29-Sep-2002[J], Pratt specimen coll.[K]). In addition, Dumbacher *et al.* (2010) recorded it on Good, Nare, Haszard, Panapompom, and Rara I in the SE Islands.
DIAGNOSIS: Quite distinctive. Wing coverts black, not gray; pale gray crown and nape; pale upper fringe to black mask; rufous underparts paler.

NOTES

The form *bernsteinii*, described from a single female specimen putatively collected on Salawati I, is apparently based on a specimen taken from Ambon I (G. F. Mees determination noted in Dekker 2003) and thus becomes a junior synonym of the Moluccan form *nigrimentum*. Neither Diamond nor Bishop recorded this species from Salawati or elsewhere in the NW Islands in recent field trips (J. M. Diamond *in litt.*). This solves a lingering mystery of this form and its weird distribution. There are no valid described forms of any avian subspecies unique to Salawati I. It was thus logical to be suspicious of the provenance of *bernsteinii*. Salawati's close geographic situation immediately west of the Bird's Head, and the shallow-water channel separating it, and its relatedly low-elevation insular aspect, make it an unlikely place to support a distinct population of *S. trivirgatus*. Apparently, *bernsteinii* is not based on a specimen from the Region, and thus should be removed from the list of named forms from NG. This simplifies the breeding geography of this species in the NG region. It also resolves Diamond's (1985) issue of sympatry with the similar *S. guttula* on Salawati (*M. trivirgatus* does not occur). *Other names*—Spectacled Monarch Flycatcher; Spectacled, Black-fronted, or White-bellied Flycatcher; *Monarcha trivirgatus*.

Biak Monarch *Symposiachrus brehmii*   ENDEMIC | MONOTYPIC
Monarcha Brehmii Schlegel, 1871. *Ned. Tijdschr. Dierk.* (1873) 4: 14.—Biak I, Bay Is.
   Biak I.

NOTES
A species that merits additional field study. Its current status is uncertain. Has been observed sparingly by most visiting observers. IUCN treats the species as Endangered. We suggest Vulnerable is a more realistic classification for the species, given the decent condition of large tracts of forest on Biak and adjacent Supiori today. *Other names*—Biak Monarch Flycatcher, *Monarcha brehmii*.

GENUS *CARTERORNIS*   [Genus 3 spp/Region 1 sp]

Carterornis Mathews, 1912. *Austral. Av. Rec.* 1: 111. Type, by original designation, *Monarcha leucotis* Gould.
   Andersen *et al.* (2015a) placed *chrysomela* and the Australian *leucotis* as a clade that is sister to their narrowly circumscribed *Monarcha* and advocated reinstatement of the genus name *Carterornis* for this clade. The species of this genus range from the Moluccas, New Guinea, and northeastern Australia to the Bismarck Archipelago (Dickinson & Christidis 2014). Characters defining the genus include the white-and-black facial pattern, the large white or yellow shoulder patch, and the habit of foraging atop canopy foliage. Otherwise the genus is similar to *Symposiachrus*. See Andersen *et al.* (2015a) for details of the placement of this genus in a richly sampled phylogeny.

Golden Monarch *Carterornis chrysomela*   RESIDENT
Muscicapa chrysomela Lesson & Garnot, 1827. *Voy. Coquille, Atlas*: pl. 18, fig. 2.—New Ireland. [Proper authorship of the species clarified by Dickinson *et al.* 2011.]
   A common, vocal, and active inhabitant of forest canopy throughout NG, the NW Islands, Aru Is, Biak I, and D'Entrecasteaux Arch; sea level–750 m, max 1,250 m[CO].
   Extralimital—various Bismarck islands but (strangely) not New Britain (Mayr & Diamond 2001, Dutson 2011).

SUBSPECIES

a. *Carterornis chrysomela melanonotus* (P. L. Sclater)
Monarcha melanonotus P. L. Sclater, 1877. *Proc. Zool. Soc. London*: 100.—Arfak Mts, Bird's Head.
RANGE: Misool, Salawati, Batanta^J, and Waigeo I, and the Bird's Head and Neck (Mayr 1941, Greenway 1966^J).
DIAGNOSIS: Female dark olive dorsally, with a short white stripe below eye; throat and breast rich brownish olive; lower breast and belly olive with bright yellowish wash. Male with black above bill extending to eye, chin, and throat; marked with a white spot before eye; crown orange-yellow grading to deep yellow on nape; mantle glossy blue-black; rump yellow; deep yellow upperwing coverts form a large patch. Male's black back patch large (Rand & Gilliard 1967).

b. *Carterornis chrysomela aurantiacus* (A. B. Meyer)
Monarcha melanonotus aurantiacus A. B. Meyer, 1891. *Abh. Ber. Zool. Mus. Dresden* (1890–91) 3(4): 9.— Near Madang, e verge of the Sepik-Ramu.
RANGE: The NW Lowlands and the Sepik-Ramu (Mayr 1941).
DIAGNOSIS: Similar to *melanonotus*, but male with deep orange on forehead and cheek, and deep yellow-orange on upper breast.

c. *Carterornis chrysomela aruensis* (Salvadori)
Monarcha aruensis Salvadori, 1874. *Ann. Mus. Civ. Genova* 6: 309.—Aru Is.
RANGE: The S Lowlands and the Trans-Fly east to the Purari basin (Karimui); also the Aru Is (Mayr 1941, Diamond 1972).
DIAGNOSIS: Female entirely olive green dorsally including rump; belly olive yellow; breast smudged olive with yellowish wash. Male exhibits an extensive black back patch and a reduced yellow shoulder slash.

d. *Carterornis chrysomela kordensis* (A. B. Meyer)

Monarcha kordensis A. B. Meyer, 1874. *Sitzungsber. Akad. Wiss. Wien* 69(1): 202.—Biak I, Bay Is.

RANGE: Biak I.

DIAGNOSIS: Female plumage very distinctive: bright orange-yellow head, face, and underparts; olive mantle and back; yellow rump; yellow patch on upperwing (upper coverts). Male with deep orange hood and breast. Male's black back patch reduced (Rand & Gilliard 1967).

e. *Carterornis chrysomela nitidus* (De Vis)

Poecilodryas nitida De Vis, 1897. *Ibis*: 376.—Boirave, Orangerie Bay, SE Peninsula.

Synonym: *Monarcha chrysomela praerepta* C. M. N. White, 1935. *Bull. Brit. Orn. Club* 56: 38.—Fergusson I, D'Entrecasteaux Arch, SE Islands.

RANGE: The Huon Penin, e sector of the S Lowlands, SE Peninsula, and three main islands of the D'Entrecasteaux Arch (Mayr 1941).

DIAGNOSIS: Mayr (1986) combined *praereptus* with this form. This is a small subspecies; male plumage more yellow and less orange (Rand & Gilliard 1967). See photos of male and female of this form in Coates (1990: 161–62) and Coates & Peckover (2001: 142).

NOTES

The Biak form is so distinctive it merits consideration for species status. *Carterornis* is masculine, hence use of the masculine *aurantiacus* and *nitidus*. Coates (1990) noted the intriguing distribution of this species, which parallels those of *Rallina tricolor* and *Gymnocrex plumbeiventris*. All three inhabit NG and New Ireland but are absent from intervening New Britain. *Other names*—Black-and-gold Monarch, Black and Yellow Monarch Flycatcher, *Monarcha chrysomela*.

GENUS MONARCHA   [Genus 6 spp/Region 3 spp]

Monarcha Vigors & Horsfield, 1827. *Trans. Linn. Soc. London* 15: 254. Type, by monotypy, *Muscipeta carinata* Swainson = *Muscicapa melanopsis* Vieillot, 1818.

The species of the genus range from Sulawesi, the Moluccas, New Guinea, and Australia to the Bismarck, Solomon, and Micronesian Islands. Andersen *et al.* (2015a) more narrowly defined this genus to include only the following three species for the Region. The genus is characterized by the simply patterned adult plumage, typically with gray, cinnamon, and black; also note the lack of sexual dichromatism. We follow Mees (1982b) and David & Gosselin (2002a) in treatment of *Monarcha* as masculine. For systematic placement of the genus see Fabre *et al.* (2014) and Andersen *et al.* (2015a).

Islet Monarch *Monarcha cinerascens*   RESIDENT

Drymophila cinerascens Temminck, 1827. *Planch. Col. d'Ois.*, livr. 72: pl. 430, fig. 2.—Timor, Lesser Sundas.

A tramp species that inhabits coastal habitats on islets off the NG Mainland and in the NW Islands, Bay Is, Aru Is, and the SE Islands; also Karkar and Manam I; rarely found on the NG mainland along the n coast between the Bird's Head and Lae. Occurs up to 1,460 m on Karkar and to the highest elevations on Bagabag I (Stresemann 1923, Diamond & LeCroy 1979).

Extralimital—mainly on islets in the Bismarcks, n Solomons, Lesser Sundas, Sulawesi, Moluccas, and Timor (Mayr & Diamond 2001, Dutson 2011)

SUBSPECIES

a. *Monarcha cinerascens inornatus* (Garnot)

Muscicapa inornata Lesson, 1828. *Man. Orn.*: 191, Pl. 16, Fig. 1.—Manokwari, Bird's Head.

RANGE: Misool and Waigeo I; n coast of Bird's Head from Sorong to Manokwari; and Aru Is (Mayr 1941). The population on Kofiau I[D] likely belongs here (Diamond *et al.* 2009[D]). Birds on Adi I off the s side of the Bird's Neck apparently belong with this race (Gyldenstolpe 1955b).

DIAGNOSIS: Crown and mantle darker than face and throat; breast and belly medium chestnut. Mees (1965) treated the Misool I population as allied to the nominate form, not *inornatus*; this is in disagreement with our treatment, which aligns with Coates *et al.* (2006).

b. *Monarcha cinerascens steini* Stresemann & Paludan

Monarcha cinerascens steini Stresemann & Paludan, 1932. *Novit. Zool.* 38: 196.—Numfor I, NW Islands.

RANGE: Numfor I.

DIAGNOSIS: Gray of face and throat/upper breast very whitish pearly gray; mantle slightly darker; chestnut of breast and belly of medium saturation.

c. *Monarcha cinerascens geelvinkianus* A. B. Meyer

Monarcha geelvinkianus A. B. Meyer, 1884. *Sitzungsber. Abh. Nat. Ges. Isis Dresden,* Abh. 1: 23.—Yapen I, Bay Is.

RANGE: Yapen, Biak, and Mios Korwar I (Mayr 1941). Presumably inhabits mainly islets of Geelvink Bay. Its presence on Yapen perhaps needs checking, as that is a species-rich land-bridge island.

DIAGNOSIS: Chestnut of breast and belly of a dark and rich saturation.

d. *Monarcha cinerascens fuscescens* A. B. Meyer

Monarcha fuscescens A. B. Meyer, 1884. *Sitzungsber. Abh. Nat. Ges. Isis Dresden*, Abh. 1: 23.—Yamna I, off n coast of the NW Lowlands.

RANGE: Islands off the coast of the NW Lowlands between the Mamberamo R and Humboldt Bay (Mayr 1941).

DIAGNOSIS: A very large form; breast and belly of pale rufous tan rather than chestnut.

e. *Monarcha cinerascens impediens* Hartert

Monarcha cinerascens impediens Hartert, 1926. *Novit. Zool.* 33: 40.—Feni I, New Ireland.

> Synonym: *Monarcha cinerascens nigrirostris* O. Neumann, 1929. *J. Orn.* 77: 197.—Coast of Huon Gulf, Huon Penin.

RANGE: Tarawai I east to Karkar[J], Manam[J], and Bagabag I[J]; Tami Is; n scarp of the Huon Penin; also the Bismarck Arch (except for range of *perpallidus*) and Solomon Is (Mayr 1941, Diamond & LeCroy 1979[J], Mayr & Diamond 2001, Dutson 2011). Probably also includes birds on small islands off the n coast of the SE Peninsula (Lababia I and presumably the Fly Is).

DIAGNOSIS: Somewhat variable in bill and plumage coloration, hence the combination of *nigrirostris* into *impediens* is the best treatment (see Diamond & LeCroy 1979). This form is characterized by generally darker plumage (Clement *et al.* 2006). Very dull and drab; brown of breast and belly dull; mantle gray with a brownish wash.

f. *Monarcha cinerascens rosselianus* Rothschild & Hartert

Monarcha cinerascens rosselianus Rothschild & Hartert, 1916. *Novit. Zool.* 23: 297.—Rossel I, Louisiade Arch, SE Islands.

RANGE: Fergusson and Goodenough I, Amphlett and Trobriand Is, Woodlark I, Bonvouloir Is, Renard, Kimuta[J] (not Misima), Tagula, and Rossel I (Mayr 1941, LeCroy & Peckover 1998[J]).

DIAGNOSIS: Throat and upper breast more blue-gray than those of *impediens*; brown of belly richer, more saturated; mantle with a blue-gray wash; slight pale wash around eye forming an obscure eyebrow. On Rossel, Pratt collected birds with black feathering at the base of the bill (specimens at BPBM).

NOTES

This species exhibits considerable individual variation. Mees (1965) suggested sinking *inornatus* into the nominate form. *M. cinerascens* and *M. melanopsis* are closely related; note there are specimens from the SE Islands that appear intermediate between the two. Our English name highlights this tramp species' preference for small islands. *Other names*—Island Monarch, Island Grey-headed Monarch Flycatcher.

Black-faced Monarch *Monarcha melanopsis*   VISITOR | MONOTYPIC

Muscicapa melanopsis Vieillot, 1818. *Nouv. Dict. Hist. Nat.*, nouv. éd. 21: 450.—Sydney, NSW, Australia.

Migrant from breeding grounds in e Australia to c, e, and s NG, as well as the Aru and Trobriand Is, D'Entrecasteaux Arch[L,MA], and Tagula I (Mayr 1941, LeCroy *et al.* 1984[L]). Occurs in forest and scrub to 800 m. On mainland NG, recorded west in the s watershed to Kurik[J] in the Trans-Fly[K] (Mees 1982b[J], Bishop 2005a[K]), and west in the n watershed to Huon Gulf[MA] (Finschhafen). Also records from Star[CO] and Hindenburg Mts[N] and Kiunga[M] (Higgins *et al.* 2006[N], D. Hobcroft *in litt.*[M]). Found in the Region mainly from early March to mid-October, with some immatures remaining through the austral summer (Coates 1990). Diamond & Bishop (1994) provided a first record for the Aru Is.

Extralimital—breeds in coastal e Australia from ne QLD south to Victoria; winters in Australia north to n QLD. During migration occurs on islands in the Torres Strait (including Boigu, Yam, Yorke, Nepean, Stephens, and Bramble Cay), though the lack of a heavy passage there suggests that migration to and from Australia probably takes place on a broader front.

Other names—Grey-winged Monarch Flycatcher, Black-faced or Pearly-winged Flycatcher.

Black-winged Monarch *Monarcha frater*   RESIDENT AND VISITOR

Monarcha frater P. L. Sclater, 1874. *Proc. Zool. Soc. London* (1873): 691.—Hatam, Arfak Mts, Bird's Head.

Inhabits forest and edge through the hills and mts of NG; 400–1,600 m but locally near sea level at Wasu on the Huon Penin[CO]. Absent from the NW and SE Islands. Presumably some Australian birds winter in NG (Higgins *et al.* 2006).

Extralimital—breeds on Cape York Penin, QLD; those birds apparently winter in NG. Recorded as a passage migrant in the Torres Strait (Draffen *et al.* 1983).

SUBSPECIES

a. *Monarcha frater frater* P. L. Sclater

RANGE: Mountains of the Bird's Head (Mayr 1941).

DIAGNOSIS: Black forecrown, chin, and upper throat; periocular and postocular feathering gray. Pale spot before eye distinct from black of lores. Black on face reduced when compared to other NG forms (Higgins *et al.* 2006).

b. *Monarcha frater kunupi* Hartert & Paludan

Monarcha frater kunupi Hartert and Paludan, 1934. *Orn. Monatsber.* 42: 45.—Kunupi, Weyland Mts, Western Ra.

RANGE: The w sector of the Western Ra. Birds from the Foja Mts may belong here (Beehler *et al.* 2012).

DIAGNOSIS: Black of face not entirely encompassing eye; inner secondaries all black; longer uppertail coverts black (Rand & Gilliard 1967).

c. *Monarcha frater periophthalmicus* Sharpe

Monarcha periophthalmicus Sharpe, 1882. *Journ. Linn. Soc. London, Zool.* 16: 318 (and 430).—Moroka distr, s watershed of SE Peninsula.

RANGE: The e sector of the Western Ra, the Border Ra, Eastern Ra, Adelbert Mts[J], and Huon and SE Peninsulas (Gilliard & LeCroy 1967[J]). Presumably the birds from the PNG North Coastal Ra[K] refer to this form (specimens at AMNH & BPBM[K]).

DIAGNOSIS: Forecrown, lores, chin, and periocular and postocular feathering black; black of forecrown grading into gray of crown; inner secondaries edged gray; uppertail coverts with little or no black (Rand & Gilliard 1967). A photo can be found in Coates (1990: 152).

d. *Monarcha frater ?canescens* Salvadori

Monarcha canescens Salvadori, 1876. *Ann. Mus. Civ. Genova* (1875) 7: 991.—Near Somerset, Cape York, QLD, Australia.

RANGE: Cape York and the humid forests of ne QLD (Schodde & Mason 1999). Apparently winters in s NG (Draffan *et al.* 1983); this needs confirmation.

DIAGNOSIS: Black mask reduced; black throat extensive; belly deep chestnut and dorsum pale pearly gray (Schodde & Mason 1999).

NOTES

Also observed in the Fakfak Mts, subspecies undetermined (Gibbs 1994). It would be valuable to know where the Australian birds winter in NG. *Other names*—Black-winged Monarch Flycatcher.

Family Corvidae

Crows and Allies [Family 126 spp/Region 3 spp]

The species of the family range worldwide, absent only from the Southern Cone of South America, much of the high Arctic, and the central Sahara (dos Anjos 2009). The greatest diversity of forms inhabits tropical Asia. The family includes jays, magpies, choughs, treepies, and others, but only true crows (genus *Corvus*) are found in New Guinea. These are large; black, brown, or gray; and have heavy bills and powerful legs. The species in the Region inhabit forest, savanna, and coastal areas, and range up to about 1,500 m elevation. According to the molecular phylogenies of Barker *et al.* (2004) and Jønsson *et al.* (2011b), the crows are a sister group to the shrikes (Laniidae).

GENUS *CORVUS* [Genus 47 spp/Region 3 spp]

Corvus Linnaeus, 1758. *Syst. Nat.*, ed. 10, 1: 105. Type, by tautonymy, *Corvus corax* Linnaeus.

The species of the genus range worldwide (dos Anjos 2009). The genus is characterized by the large size, robust bill, uniform black or black-and-gray plumage, and raucous vocalizations. We follow dos Anjos (2009: 622) in uniting *Gymnocorvus* with *Corvus*, as there are a number of plumage characters that link *C. tristis* with *C. fuscicapillus*. Jønsson *et al.* (2012) provided the molecular evidence for placing *Gymnocorvus* in *Corvus*. These authors also showed *tristis* and *fuscicapillus* to be sister species (but with weak statistical support) and linked them with the Australian white-eyed species of *Corvus* (see also Haring *et al.* 2012).

Brown-headed Crow *Corvus fuscicapillus*   ENDEMIC | MONOTYPIC
Corvus fuscicapillus G. R. Gray, 1859. *Proc. Zool. Soc. London*: 157.—Aru Is.
   Synonym: *Corvus megarhynchus* Bernstein, 1864. *J. Orn.* 12: 107.   Waigeo I, NW Islands.
This species' peculiar relictual distribution includes forest on the Aru Is[MA], Waigeo[J] and Gam I[MA] of the NW Islands; and three sites in the NW Lowlands: Taua, on the lower Mamberamo R[MA]; Kasiwa, west of the Mamberamo R[K]; and Nimbokrang, west of Sentani[W] (Ripley 1964a[J], van Balen *et al.* 2009[K], S. F. Bailey *in litt.*[W]). Sea level–500 m. It mainly inhabits forest and also mangroves, but rarely open habitats and never open coastal habitats. Reported to be common in the forest interior of Waigeo I (Davies 2008).

NOTES
The described island subspecies, *megarhynchus*, based on inadequate comparative material, is not worth maintaining (*fide* Debus 2009: 622). C. A. Green (*in litt.*) observed a single bird with an arched upper mandible on Yapen I that he tentatively identified as this species. This is distributionally consonant with the current range of the species just to the east in the lower Mamberamo R. Should be looked for by future fieldworkers visiting Yapen I. *Other names*—Brown-capped Crow.

Grey Crow *Corvus tristis*   ENDEMIC | MONOTYPIC
Corvus tristis Lesson & Garnot, 1827. *Bull. Sci. Nat. Férussac* 10: 291.—Manokwari, Bird's Head.

This odd, vocal species inhabits NG, Salawati, Batanta, Adi[J], and Yapen I; also the D'Entrecasteaux Arch including Normanby I[K]; sea level–1,500 m[CO] (Mayr 1941, Gyldenstolpe 1955b[J], Pratt field notes 2003[K]). Also on small coastal islands (Ron, Kairiru). Trans-Fly records include Dolak I, Bian Lakes, Sturt I, and Oriomo R (Bishop 2005a). Typically encountered in family groups.

NOTES
This unusual species is most closely allied to the ecologically similar *Corvus fuscicapillus*. These two species share the following characters: (1) blue iris (2) atypical brown plumage, and (3) forest-dwelling habit. *Other names*—Bare-eyed Crow, *Gymnocorvus tristis*.

Torresian Crow *Corvus orru*   RESIDENT

Corvus orru Bonaparte, 1850. *Consp. Gen. Av.* 1: 385.—Aiduma I, near Triton Bay.

Frequents verges of lowland coastal forest, regrowth, and plantations on the Mainland and most satellite islands, sea level–650 m. Its presence up to 1,200 m along the Kokoda Track, SE Peninsula, in association with villages and gardening, seems to be a recent development. In the Region not found in rain-forest interior. Absent from the Aru Is, Karkar, and Rossel I; also no records for Manam or Mios Num.

Extralimital—n two-thirds of Australia, Lesser Sundas, and Moluccas; also islands of Torres Strait.

SUBSPECIES

a. *Corvus orru orru* Bonaparte

   Synonym: *Corvus salvadorii* Finsch, 1884. Mitt. *Orn. Vereins Wien*: 109.—Port Moresby, SE Peninsula.

RANGE: Obi I in the n Moluccas; patchily through NG; the four main NW Islands as well as the Schildpad Is, Gag, Gebe[J], and Kofiau; Adi, Yapen, Numfor, Biak, Tarawai, Kairiru, Muschu, D'Entrecasteaux, and Trobriand Is, Tubatuba, Bentley in the Engineer Is, Woodlark, Misima, and Tagula I[CO]; also small islands off the s coast of the SE Peninsula: Loupom, Mailu, Fisherman's, Haidana (Mayr 1941, Mees 1972[J], Melville 1980).
DIAGNOSIS: Slightly shorter-winged and shorter-tailed than *ceciliae* of Australia (Debus 2009). Nominate form (male) wing 320–340, tail 170–185 (Schodde & Mason 1999); *ceciliae* (male) wing 342–380, tail 179–214 (Higgins *et al.* 2006).

NOTES

We follow Vaurie (1962) in considering *salvadorii* a synonym of the nominate form. Mees (1982b) corrected the date of publication and type locality. *Other names*—Australian or Papuan Crow.

Family Laniidae

Shrikes   [Family 34 spp/Region 2 spp]

The species in this family range throughout the world but for South America, southern Central America, and Australia (Yosef 2008). Shrikes are predatory songbirds with strong, hooked bills. The molecular research of Barker *et al.* (2004) showed that this family is sister to the Corvidae. This is corroborated by Jønsson *et al.* (2011b).

GENUS *LANIUS*   [Genus 30 spp/Region 2 spp]

Lanius Linnaeus, 1758. *Syst. Nat.*, ed. 10, 1: 93. Type, by subsequent designation (Swainson, 1824. *Zool. Journ.* [1825] 1: 294), *Lanius excubitor* Linnaeus.

The species of this genus range from North America to Eurasia and southeastward to eastern Indonesia and New Guinea (Yosef 2008). The genus is characterized by the very distinctive body form (large head, strong hooked bill, narrow rounded tail), presence of a cap or mask, and predaceous habit.

Brown Shrike *Lanius cristatus*   VAGRANT

Lanius cristatus Linnaeus, 1758. *Syst. Nat.*, ed. 10, 1: 93.—Bengal, e India and Bangladesh.

There are two records from the w sector of Indonesian NG of this vagrant from se Asia.

Extralimital—breeds in c and e Siberia south to se China; winter range extends from s Asia east to the Philippines and Indonesia (Yosef 2008).

SUBSPECIES

a. *Lanius cristatus lucionensis* Linnaeus

Lanius lucionensis Linnaeus, 1766. *Syst. Nat.*, ed. 12, 1: 135.—Luzon, Philippines.

RANGE: A record from "Manokwari" on 6-Feb-1963 (Rand & Gilliard 1967: 612). M. LeCroy (*in litt.*) notes: "This is probably the specimen collected at Oransbari by Larry P. Richards and now in the BPBM, noted in a letter from E. T. Gilliard to A. L. Rand on 7 Feb. 1964." The record was published (Thompson 1964). Also a record of a bird photographed at Nabire on 28-Dec-2014 (J. Pap *in litt.*, confirmed by the authors). DIAGNOSIS: Male forecrown pale gray; back gray-brown; throat white; breast and belly fawn-buff. Female paler than male; mask blackish brown; supercilium less distinct (Yosef 2008). Young birds exhibit fine scalloping ventrally and dorsal plumage dull brown from crown to lower back; mask dark brown.

NOTES
Other names—Red-tailed Shrike.

Long-tailed Shrike *Lanius schach* RESIDENT
Lanius schach Linnaeus, 1758. *Syst. Nat.*, ed. 10, 1: 94.—Guangzhou, se China.

Inhabits gardens and upland grassland edges of c and e NG, ranging west to near the PNG border, 400–2,650 m^{CO}.

Extralimital—c Asia and India to e China, thence south to the Philippines and Indonesia (Yosef 2008).

SUBSPECIES
a. *Lanius schach stresemanni* Mertens
Lanius schach stresemanni Mertens, 1923. *Senckenbergiana* 5: 228.—Kulungtufu, 1,000 m, Saruwaged Ra, Huon Penin.
RANGE: The Border Ra, Eastern Ra, mts of the Huon Penin, and mts of the SE Peninsula from Telefomin in the west to Mafulu, northwest of Port Moresby, in the east. Seems to be absent west of Telefomin and southeast of Port Moresby. No records from Indonesian NG.
DIAGNOSIS: Very similar to *longicaudatus* from Thailand but with a gray nape and upper mantle (Coates 1990, Yosef 2008). See photos in Coates (1990: 50) and Coates & Peckover (2001: 121).

NOTES
Should be looked for east and west of current range boundaries described above. We presume the species is expanding its range. It is probably a relatively young arrival in the Region, though not terribly recent—as evidenced by the presence of an endemic subspecies. *Other names*—Schach, Red-backed, Black-headed, or Black-capped Shrike.

Family Petroicidae

Australasian Robins [Family 49 spp/Region 26 spp]

The species of this family range from Australia and New Guinea to the Solomons, Fiji, Samoa, and New Zealand. This is a diverse group of insect eaters that includes the terrestrial ground-robins, others that live high in the forest canopy, a mangrove specialist, and one that exclusively frequents streamsides. Usage of the family-level name Petroicidae rather than Eopsaltriidae (Beehler & Finch 1985, Mayr 1986) is based on original designation of the family name, not age of the erection of the oldest generic name in the family (Bock 1994). The family is characterized by the upright (sometime hunched) posture, pale flash in the open wing, and small size. Schodde & Mason (1999) noted that these robins are the Region's "perch and pounce" insect eaters. They have the distinctive habit of perching on vertical trunks in a way no other group does. We follow Jønsson *et al.* (2011b) in placement of this family in the "transitory Oscines." Two papers provide important molecular phylogenetic information on genera within the family: Loynes *et al.* (2009) and Christidis *et al.* (2011). Our treatment follows mainly the Christidis work, which recognized one new genus and resurrected several old Mathews genera long out of use. Christidis *et al.* did this mainly because the more inclusive genera were shown to be paraphyletic. Barker *et al.* (2004) found the Petroicidae to be an early divergence within the Passerida, which includes the warblers and thrushes, among others. Cracraft (2014) split the family into the Eopsaltriidae (*Drymodes, Eopsaltria, Microeca,* and relatives) and the Petroicidae (*Petroica, Amalocichla, Pachycephalopsis,* and allies).

Amalocichla De Vis, 1892. *Ann. Rept. British New Guinea*, 1890–91: 95. Type, by monotypy, *Amalocichla sclateriana* De Vis.

The two species in the genus are endemic to the Region, inhabiting the interior of montane and sub-alpine forest. The members of the genus are characterized by the terrestrial habit, the long legs, abbreviated tail, obscure thrush-like plumage, melodious whistled song, whitish throat patch, and whitish patch before eye. The genus has variously been situated in the thrushes (Mayr 1941, Rand & Gilliard 1967), Australasian robins (Beehler & Finch 1985), and Australasian warblers (Olson 1987). For placement of the genus within the Petroicidae, we refer to Norman *et al.* (2009) and Christidis *et al.* (2011). Note that Christidis *et al.* (2011) found very deep branching between the two species of *Amalocichla*, commensurate with genus-level branching elsewhere within the family; however, they did not propose raising *incerta* to its own genus.

Greater Ground-Robin *Amalocichla sclateriana*   ENDEMIC

Amalocichla sclateriana De Vis, 1892. *Ann. Rept. Brit. New Guinea*, 1890–91: 95.—Owen Stanley Ra, SE Peninsula.

Recorded from the Western Ra and high mts of the Huon and SE Peninsulas, 2,700–3,900 m (Coates & Peckover 2001). No records yet confirm this species from the Border or Eastern Ra. This huge gap may be a product of inadequate survey, as evidenced by the recent mist-netting of this species on the Huon Penin (Freeman *et al.* 2013). There are possible audial encounters from Tari Gap and the Kaijende Highlands (Beehler 1993b). Inhabits the floor of upper montane and subalpine forest interior. Locally fairly common, otherwise scarce or rare. Shy, retiring, seldom seen, and little known. Best found by its distinctive mournful vocalization.

SUBSPECIES

a. *Amalocichla sclateriana occidentalis* Rand
Amalocichla sclateriana occidentalis Rand, 1940. *Amer. Mus. Novit.* 1074: 1.—9 km ne of Lake Habbema, 2,800 m, Western Ra.
RANGE: The Western Ra.
DIAGNOSIS: Brown of crown, back, wing coverts, and rump rich chestnut; throat pale but mottled with brown and dark brown; indication of a dark-smudged breast band; belly off-white. Bill length difference cited by Rand (1940b) and Rand & Gilliard (1967) is trivial: 32 for *occidentalis vs.* 29–31 for the nominate.

b. *Amalocichla sclateriana sclateriana* De Vis
RANGE: The highest mts of the Owen Stanley Ra. Localities—Mt Kumbak/Bulldog Road[J], Mt Albert Edward, Murray Pass[K], and Lake Omha near Mt Scratchley[L] (Mayr & Rand 1937[K]; Beehler field notes 31-Jul-1986[L]; specimens at BPBM, nos. 184634 & 184635[J]).
DIAGNOSIS: Upperparts paler and more olive, and underparts grayer, more olive, and less mottled, than for *occidentalis* (Boles 2007).

NOTES
A single individual of this species mist-netted in the Saruwaged Ra of the Huon Penin in 2011 has not been assigned to subspecies (Freeman *et al.* 2013) and should be investigated further. So little is understood about the distribution and plumage variation of this cryptically plumaged robin that we hesitate to draw any conclusions about its systematics and phylogeography. Should be searched for, using playback, in the high subalpine forests of the Border and Eastern Ra. *Other names*—Greater New Guinea Thrush, Cloudforest Velvet-thrush, Sclater's False-thrush.

Lesser Ground-Robin *Amalocichla incerta*   ENDEMIC

Eupetes incertus Salvadori, 1876. *Ann. Mus. Civ. Genova* (1875) 7: 967.—Arfak Mts, Bird's Head.

Mountains of NG; 1,750–2,500 m, extremes 900–2,750 m (Coates & Peckover 2001). Widely but locally distributed through the Central Ra; also the Bird's Head, Kumawa and Wandammen Mts, Foja

Mts, PNG North Coastal Ra, and mts of the Huon Penin. Inhabits the floor of montane forest interior. Common but easily overlooked unless the song is known. Much better represented in museums than the preceding species.

SUBSPECIES

a. *Amalocichla incerta incerta* (Salvadori)
RANGE: The Arfak Mts of the Bird's Head (Mayr 1941). Should be looked for in the Tamrau Mts, though not recorded by Ripley, Gilliard, or Beehler on their field trips there.
DIAGNOSIS: Breast and sides of belly buffy fawn; throat whitish. Marginally distinct from following subspecies by the warmer tone ventrally and dorsally, but each population exhibits considerable variation in color. Kumawa Mts birds may be distinct (Diamond & Bishop 2015).

b. *Amalocichla incerta brevicauda* (De Vis)
Drymoedus brevicauda De Vis, 1894. *Ann. Rept. Brit. New Guinea*, 1893–94: 103.—Mount Maneao, SE Peninsula.
> Synonym: *Amalocichla incerta olivacentior* Hartert, 1930. *Novit. Zool.* 36: 85.—Mount Wondiwoi, Wandammen Penin, Bird's Neck.

RANGE: Mountains of the Bird's Neck (Wandammen Penin), the Central Ra, and mts of the Huon Penin; also the Foja Mts[J] and PNG North Coastal Ra[K] (Beehler *et al.* 2012[J], specimens at AMNH & BPBM[K]).
DIAGNOSIS: Breast and flanks dull olive brown, entirely lacking the rich buffy fawn found in the nominate form. Also note: dorsal plumage deeper brown with reduced rufous; white throat feathers tipped black; breast and flanks more gray-brown (Boles 2007). A review of the breadth of variation in adult plumage for topotypical *brevicauda* and *olivacentior* demonstrates no reliable means of distinguishing these two forms, hence their combination. See excellent photo of the form from the Huon Penin (Coates 1990: 188), which best matches the description for the western (nominate) subspecies.

NOTES
We suggest that future revisers consider treating the species as monotypic—variation probably entirely clinal. *Other names*—Lesser New Guinea Thrush, Rusty Velvet-thrush, Dubious False-thrush.

GENUS *PACHYCEPHALOPSIS*　　　　　　　　ENDEMIC [Genus 2 spp]

Pachycephalopsis Salvadori, 1879. *Ann. Mus. Civ. Genova* 15: 48, note 3. Type, by monotypy, *Pachycephala hattamensis* A. B. Meyer.

The two species of the genus are endemic to the uplands of New Guinea. The genus is characterized by the large head, compact form, whitish iris, and white throat patch. It is a sister lineage to the genus *Amalocichla* (Christidis *et al.* 2011).

Green-backed Robin *Pachycephalopsis hattamensis*　　　　　　ENDEMIC
Pachycephala hattamensis A. B. Meyer, 1874. *Sitzungsber. Akad. Wiss. Wien* 69: 391.—Hatam, Arfak Mts, 1,020 m, Bird's Head.

Mainly upland forests of w NG: mts of the Bird's Head, Bird's Neck, and Yapen I, and the Western Ra, Foja Mts, and Mt Sisa; 760–1,650 m, locally to 2,000 m[CO]. It is sympatric with the closely allied *P. poliosoma* throughout most of its distribution in the Central Ra. Inhabits lower stories of the closed interior of hill and lower montane forest.

SUBSPECIES

a. *Pachycephalopsis hattamensis hattamensis* (A. B. Meyer)
> Synonym: *Pachycephalopsis hattamensis axillaris* Mayr, 1931. *Bull. Brit. Orn. Club* 51: 59.—Upper Otakwa R, Western Ra.

RANGE: Mountains of the Bird's Head and the Western Ra (Gyldenstolpe 1955b, Melville 1980). Population from the Foja Mts presumably is referable to this form (Beehler *et al.* 2012). Eastern terminus of the range of this form unknown.
DIAGNOSIS: No diffuse gray collar separating white chin from olive breast plumage.

b. *Pachycephalopsis hattamensis ernesti* Hartert

Pachycephalopsis hattamensis ernesti Hartert, 1930. *Novit. Zool.* 36: 69.—Wondiwoi Mts, Wandammen Penin, Bird's Neck.

RANGE: The Wandammen Mts, w coast of Geelvink Bay.

DIAGNOSIS: Presence of a diffuse gray collar separating the white chin from the olive breast plumage (Rand & Gilliard 1967).

c. *Pachycephalopsis hattamensis insularis* Diamond

Pachycephalopsis hattamensis insularis Diamond, 1985. *Emu* 85: 77.—Near Ambidiru village, Yapen I, Bay Is.

RANGE: Yapen I uplands (Diamond 1985).

DIAGNOSIS: Reduced white patch on chin; uppertail coverts olive, not brown; tail darker than that of *ernesti*; back more olive and less yellow-washed; perhaps smaller (Diamond 1985, Boles 2007).

d. *Pachycephalopsis hattamensis lecroyae* Boles

Pachycephalopsis hattamensis lecroyae Boles, 1989. *Bull. Brit. Orn. Club* 109: 120.—Magidobo, 6.18°S, 142.66°E, 1,450 m, Mt Sisa, s slopes of Eastern Ra.

RANGE: The sw sector of the Eastern Ra. Known from a single site: Magidobo, on Mt Sisa (Boles 2007).

DIAGNOSIS: Darker overall than the nominate form; throat and chin white; upper breast lacking the gray band; undertail coverts more orange; axillaries more olive, with only a trace of orange (Boles 1989, 2007).

NOTES

The species should be looked for in the Border Ra. Currently there is a substantial distributional gap on the Central Ra on either side of the PNG border. *Other names*—Green Thicket-flycatcher.

White-eyed Robin *Pachycephalopsis poliosoma*   ENDEMIC

Pachycephalopsis poliosoma Sharpe, 1882. *Journ. Linn. Soc. London, Zool.* 16: 318.—Astrolabe Ra, SE Peninsula.

Central Ra and mts of the Sepik-Ramu and Huon Penin; 700–1,700 m, extremes 400–2,200 m^{CO}. Replaced on the Bird's Head by the closely allied *Pachycephalopsis hattamensis*, but sympatric with that species in the Western, Border and Eastern Ra. *P. poliosoma* (usually) replaces *P. hattamensis* above 975 m where their ranges overlap (Diamond 1985).

SUBSPECIES

a. *Pachycephalopsis poliosoma approximans* (Ogilvie-Grant)

Pachycephala poliosoma approximans Ogilvie-Grant, 1911. *Bull. Brit. Orn. Club* 29: 26.—Iwaka R, Western Ra.

> Synonym: *Pachycephala poliosoma albigularis* Rothschild, 1931. *Novit. Zool.* 36: 260.—Gebroeders Mts, w sector of the Western Ra.
> Synonym: *Pachycephalopsis poliosoma balim* Rand, 1940. *Amer. Mus. Novit.* 1074: 4.—Baliem R, 1,600 m, Western Ra.

RANGE: The s watershed of the Western, Border, and Eastern Ra, east to Okapa (Boles 2007) but not as far east as the Kratke Mts, which support the form *hypopolia*.

DIAGNOSIS: White throat patch medium-size (21 mm diameter); ventral plumage medium gray; mantle medium slate gray with slight bluish cast. See photo in Coates & Peckover (2001: 157). The forms *balim* and *albigularis* are not substantially distinct from this form.

b. *Pachycephalopsis poliosoma hunsteini* (Neumann)

Pachycephala poliosoma hunsteini Neumann, 1922. *Verh. Orn. Ges. Bayern* 15: 237.—Hunstein Mts, middle Sepik, north-central NG.

> Synonym: *Pachycephalopsis poliosoma idenburgi* Rand, 1940. *Amer. Mus. Novit.* 1074: 5.—6 km sw of Bernhard Camp, 1,200 m, ne sector of the Western Ra.

RANGE: The n slopes of the Western, Border, and Eastern Ra; also PNG North Coastal Ra^J (Mayr 1941, AMNH & BPBM specimens^J).

DIAGNOSIS: White throat patch much reduced to pale gray chin patch (8–10 mm); ventral plumage dark gray to dark medium gray; mantle dark slate gray with no bluish cast.

c. *Pachycephalopsis poliosoma hypopolia* Salvadori
Pachycephalopsis hypopolia Salvadori, 1899. *Boll. Mus. Zool. Torino* 15 (=14), no. 360: 2.—Sattelberg, Huon Penin.
RANGE: The Adelbert Mts[J] and mts of the Huon Penin[K] (Mayr 1931[K], Gilliard & LeCroy 1967[J]). Also the Kratke Mts of the e verge of the Eastern Ra (Shaw Mayer specimen from Buntibasa distr).
DIAGNOSIS: White throat patch much reduced to pale gray chin patch (8 mm); ventral plumage darkish medium gray; mantle dark slate gray with no bluish cast.

d. *Pachycephalopsis poliosoma poliosoma* Sharpe
RANGE: The Herzog Mts and mts of the SE Peninsula (Mayr 1941).
DIAGNOSIS: White throat patch medium-size (18 mm); ventral plumage medium gray; mantle medium slate gray with slight bluish cast. See photo in Coates (1990: 201).

NOTES
This species comprises four rather thinly drawn subspecies. *Other names*—Eastern White-eyed Robin, White-throated Thicket-flycatcher.

GENUS *PETROICA*   [Genus 12 spp/Region 2 spp]

Petroica Swainson, 1829. *Zool. Illustr.* (2)1(8): pl. 36 and text. Type, by monotypy, *Muscicapa multicolor* Gmelin.
　　The species of this genus range from Australia and New Guinea to the Solomon Islands, Fiji, Samoa, and New Zealand (Boles 2007). The genus is characterized by plumage patterned in black, gray, red, pink, and white; and diminutive blackish bill. Schodde & Mason (1999) noted the following characters: small; sexually dimorphic; forehead spot present; whitish stripes in wing and tail; fledglings brown with pale shaft streaking and dusky speckling. The phylogeny generated by Christidis *et al.* (2011) placed this genus as sister to *Eugerygone*, and the pair as sister to *Pachycephalopsis* and *Amalocichla*.

Subalpine Robin *Petroica bivittata*   ENDEMIC
Petroeca bivittata De Vis, 1897. *Ibis*: 376.—Mount Scratchley, 3,720 m, SE Peninsula.
　　Patchily distributed in the Central Ra and mts of the Huon Penin[J], 2,700–3,900 m[CO] (Freeman *et al.* 2013[J]). Inhabits high montane forest and subalpine shrubland.

SUBSPECIES
a. *Petroica bivittata caudata* Rand
Petroica bivittata caudata Rand, 1940. *Amer. Mus. Novit.* 1072: 5.—9 km ne of Lake Habbema, 2,850 m, Western Ra.
RANGE: The Western Ra and probably the Border Ra. Localities—Carstensz Massif[J], Kemabu Plateau[K], Lake Habbema[L], Meren valley[J], Yellow valley[J], 9 km northeast of Lake Habbema[L] (Rand 1942b[L], Ripley 1964a[K], Schodde *et al.* 1975[J]). Birds reported from the Border Ra (Star Mts[M]) probably belong here (Gregory & Johnston 1993[M]). The break between these two subalpine races could be the Strickland R gorge.
DIAGNOSIS: White in tail much reduced.

b. *Petroica bivittata bivittata* De Vis
RANGE: High mts of the Eastern Ra and the SE Peninsula. Localities—Kaijende Highlands[K,W], Lamende Ra[J], Mt Giluwe[J], Mt Hagen[CO], Mt Wilhelm[X], Mt Albert Edward[L], English Peaks[CO], Mt Scratchley[MA], Mt Knutsford[CO], Mt Victoria[CO], and Mt Thumb[CO] (Mayr & Rand 1937[L], Sims 1956[J], Beehler & Sine 2007[K], Beehler *et al.* 2011[W], P. Gregory *in litt.*[X]). Birds observed in the mts of the Huon Penin (Freeman *et al.* 2013) might belong with this race.
DIAGNOSIS: Male has outer pair of tail feathers with broad white tips and a white outer web (Rand & Gilliard 1967). See photo in Coates & Peckover (2001: 152).

The populations in the Border Ra and the Huon Penin remain undiagnosed. Since *Petroica archboldi* is clearly the "alpine" robin, we believe *P. bivittata* is best called "Subalpine Robin." *Other names*—Alpine, New Guinea, or Cloud-forest Robin, Mountain Robin-flycatcher.

Snow Mountain Robin *Petroica archboldi*   ENDEMIC | MONOTYPIC
Petroica archboldi Rand, 1940. *Amer. Mus. Novit.* 1072: 5.—Mount Wilhelmina, 4,100 m, Western Ra.

Restricted to the highest peaks of the Western Ra. Recorded from Mt Wilhelmina[J] and the Carstensz Massif[K], and presumably Idenburg Top[K]; 3,850–4,550[K] m (Rand 1942b[J], Schodde *et al.* 1975[K]). Forages on rocky slopes and cliffs on talus slopes; also in high valleys among rocky tundra and alpine heath (Rand 1940a, 1942b, Schodde *et al.* 1975).

NOTES

This species has the highest elevational range of any bird in the Region. It would be valuable to resurvey the elevational distribution of the species on Mt Wilhelmina, which was first surveyed in 1938. How much has the range altered over those seven decades? For a species that lives only above 3,850 m, even a 500 m shift upward in elevational range would have a substantial impact on population size, because of the resultant dissection and diminution of available alpine habitat. Observed near the Freeport Mine above Tembagapura in 2000 (just below 4,200 m) and in 2011 (at 4,200 m) by S. van Balen (van Balen & Noske 2014). In the most recent observation, the species was seen perching on mining equipment at the mine itself. To be looked for on the summits of Mts Goliath and Mandala. Even though this species is an inhabitant of the Snow Mts (now Western Ra), we think when applied adjectivally, the name should be "Snow Mountain Robin" in the same manner that Mountain Chickadee is based on this bird's affinity for mountains. *Other names*—Alpine or Snow Robin, Rock Robin-flycatcher, Archbold's New Guinea Robin, Snow Mountains Robin.

GENUS *EUGERYGONE*   ENDEMIC | MONOTYPIC

Eugerygone Finsch, 1901. *Notes Leyden Mus.* 22: 200. Type, by original designation and monotypy, *Pseudogerygone rubra* Sharpe.

This monotypic genus is endemic to the uplands of mainland New Guinea. The genus is characterized by small size, sexual dimorphism, white flash in wing in both sexes, and distinctive wing posturing while foraging. Treated by Mayr (1941) as a warbler. Keast (1977) noted its possible close relationship with the genus *Petroica*. The molecular phylogeny of Christidis *et al.* (2011) placed *Eugerygone* as sister to *Petroica*.

Garnet Robin *Eugerygone rubra*   ENDEMIC | MONOTYPIC
Pseudogerygone rubra Sharpe, 1879. *Notes Leyden Mus.* (1878) 1: 29.—Tjobonda, Arfak Mts, Bird's Head.

Synonym: *Eugerygone rubra saturatior* Mayr, 1931. *Mitt. Zool. Mus. Berlin* 17: 678.—Junzaing, Huon Penin.

Inhabits forest and edge through the mts of NG; 1,700–2,500 m, extremes 1,400–3,680 m[CO]. Widely distributed but uncommon and local in the Arfak Mts[MA], the Central Ra[MA], Kumawa Mts[J], Foja Mts[J], and mts of the Huon Penin[MA] (Diamond 1985[J]).

NOTES

Our review of the extensive material in the AMNH indicates there is no substantive geographic variation in this species. We thus treat it as monotypic. Report of this species in the Fakfak Mts by Coates (1990) was apparently in error. Not recorded there by Diamond (1985) or Gibbs (1994). *Other names*—Redbacked Warbler.

GENUS *DEVIOECA*

Devioeca Mathews, 1925, *Bull. Brit. Orn. Club* 45: 93. Type, by original designation, *Microeca papuana* A. B. Meyer.

The single species of the genus is endemic to the upland forests of New Guinea. A small canopy dweller of cloud forest, it is stout-bodied and stumpy-tailed, yellow ventrally and yellow-green dorsally (Schodde & Mason 1999). The resurrection of this monotypic genus was the result of molecular analysis by Christidis *et al.* (2011).

Papuan Flyrobin *Devioeca papuana*
ENDEMIC | MONOTYPIC

Microeca papuana A. B. Meyer, 1875. *Sitzungsber. Abh. Nat. Ges. Isis Dresden*: 75.—Arfak Mts, Bird's Head.

New Guinea mts; 1,750–2,600 m, extremes 1,100–3,500 m (Coates & Peckover 2001). Range includes the mts of the Bird's Head[MA], Wandammen Mts[MA], Fakfak Mts[G], Central Ra[MA], Foja Mts[J], PNG North Coastal Ra (Mt Menawa, Bewani Mts)[S], Mt Bosavi[CO], and mts of the Huon Penin[MA] (Gibbs 1994[G], Beehler *et al.* 2012[J]; specimens at AMNH[S]). Inhabits montane forest, edge, and small clearings.

Other names—Canary Flyrobin, Canary Robin, Montane or Yellow Flycatcher, Canary Flycatcher, Yellow Microeca Flycatcher, *Microeca papuana*.

GENUS *KEMPIELLA*
[Genus 2 spp/Region 2 spp]

Kempiella Mathews, 1913, *Austral. Avian Rec.* 1: 109. Type, by original designation and monotypy, *Kempiella kempi* Mathews.

The species of this genus range from the New Guinea region to Cape York Peninsula, Australia. They are small-bodied and long-tailed and have yellowish feet, reduced rictal bristles, wings rounded, and 10th primary disproportionally long (Schodde & Mason 1999). Christidis *et al.* (2011) disarticulated the traditional *Microeca*, in which *Kempiella* was once placed, because they found that *Monachella* was embedded in that cluster.

Yellow-legged Flyrobin *Kempiella griseoceps*
RESIDENT

Microeca griseoceps De Vis, 1894. *Ann. Rept. Brit. New Guinea*, 1893–94: 101.—Mount Maneao, SE Peninsula.

> Synonym: *Microeca griseiceps occidentalis* Rothschild & Hartert, 1903. *Novit. Zool.* 10: 471.—Warmendi, Arfak Mts, Bird's Head.
> Synonym: *Kempiella kempi* Mathews, 1913. *Austral. Avian Rec.* 2: 12.—Cape York, n QLD, Australia.
> Synonym: *Microeca poliocephala* Reichenow, 1915. *J. Orn.* 63: 124.—Lordberg, 1,500 m, Sepik Mts, Eastern Ra.

Inhabits upper stories of forest interior of NG mts (600–2,300 m[CO]) and lowlands of the Trans-Fly at Tarara and Wuroi (Bishop 2005a). Diamond & Bishop (1994) provided a first record for the Aru Is. No records, to date, for the mts of the Huon Penin or the Kumawa, Wandammen, or Foja Mts.

Localities—Warmendi on the Bird's Head[MA], Fakfak Mts[G], Weyland Mts[MA], 6 km southwest of Bernhard Camp[L], Cyclops Mts[MA], Doormanpaad[MA], Tarara[K], PNG North Coastal Ra (Mt Somoro, Torricelli Mts)[S], Telefomin[J], Ok Tedi[CO], Oriomo R[MA], Lordberg[MA], Schrader Ra[CO], Tabibuga[CO], and Okasa[DI], Mt Maneao[MA] (Rand 1942a[K], 1942b[L]; Gilliard & LeCroy 1961[J]; Gibbs 1994[G]; specimens at AMNH[S]). The sole Adelbert Mts record from Mackay (1991) warrants confirmation.

Extralimital—Cape York Penin, QLD.

NOTES

We reviewed topotypical material at the AMNH and found sufficient nongeographic variation to warrant treating this form as monotypic. See also Diamond (1972) and Boles (2007), as well as Schodde & Mason (1999) for other viewpoints. *Other names*—Yellow-legged Flycatcher or Robin, Yellow or Little Yellow Flycatcher, Grey-headed Microeca, *Microeca griseoceps*.

Olive Flyrobin *Kempiella flavovirescens*   ENDEMIC | MONOTYPIC
Microeca? flavo-virescens G. R. Gray, 1858. *Proc. Zool. Soc. London*: 178.—Aru Is.
    Synonym: *Zosterops cuicui* De Vis, 1897. *Ibis*: 384.—Boirave, Orangerie Bay, SE Peninsula.
Mainland NG, NW Islands (Batanta, Misool, Waigeo), Aru Is, and Yapen I (Mayr 1941); sea level–1,000 m[K],
max 1,500 m[CO] (Coates & Peckover 2001[K]). Widespread in the Trans-Fly, including Morehead, Tarara, and
Wuroi (Bishop 2005a). Inhabits rain forest and monsoon forest, also teak plantations, occurring mainly in
the lowlands and lower hills (Coates 1990).

NOTES

A large specimen series reviewed at the AMNH confirmed that the form *cuicui* is synonymous with the
nominate form. *Other names*—Orange-chinned Flyrobin, Olive-yellow or Olive Flycatcher or Robin, Olive
Microeca Flycatcher, *Microeca flavovirescens*.

GENUS *MONACHELLA*   MONOTYPIC

Monachella Salvadori, 1874. *Ann. Mus. Civ. Genova* 6: 82. Type, by monotypy, *Monachella saxicolina* Salvadori = *Muscicapa Mülleriana* Schlegel.
    This monotypic genus endemic to New Guinea and New Britain inhabits all upland and montane
regions of New Guinea. The genus is characterized by the strictly riparian habit and the black cap, wings,
and tail contrasting with the pale gray back. The species is a noisy riparian specialist (Orenstein 1975). The
phylogeny of Christidis *et al.* (2011) found *Monachella* nested between *Microeca flavigaster* and *Devioeca
papuana*.

Torrent Flycatcher *Monachella muelleriana*   RESIDENT
Muscicapa Mülleriana Schlegel, 1871. *Nederl. Tidjschr. Dierk.* (1873) 4: 40.—Triton Bay, Bird's Neck.
    New Guinea, from the foothills to 2,130 m (Coates & Peckover 2001). Widely distributed on the
Mainland. No records from the Bird's Neck (Wandammen, Fakfak, and Kumawa Mts) or Adelbert Mts,
though likely to be present in those areas. Frequents substantial fast-flowing open streams with boulders.
Occurs mainly in the hills and lower mts; also present in the lowlands in the vicinity of hills (Coates 1990).
    Extralimital—New Britain, where apparently very local and known only from four rivers in the east
of the island (Mayr & Diamond 2001, Dutson 2011).

SUBSPECIES

a. *Monachella muelleriana muelleriana* (Schlegel)
RANGE: All NG foothills.
DIAGNOSIS: Paler than *coultasi* of New Britain; back pale gray; rump whitish; breast white (Boles 2007).
See photo in Coates & Peckover (2001: 150). The De Vis record from Goodenough I (Mayr 1941) is not
accepted here (Coates 1990).

NOTES

Other names—Torrent Robin or Flyrobin, River or Grey-and-white Flycatcher.

GENUS *MICROECA*   [Genus 3 spp/Region 2 spp]

Microeca Gould, 1841. *Proc. Zool. Soc. London* (1840): 172. Type, by monotypy, *Microeca assimilis* Gould.
    The three species of this genus range from Australia to the Tanimbar Is and inhabit New Guinea
lowland forests, openings, and the savannas of the south. These three species are all that are left of the formerly more expansive genus, as a result of the molecular phylogeny of Christidis *et al.* (2011). This newly
circumcribed *Microeca* is characterized by the open-country habit, washed-out plumage, and dark legs.

Jacky Winter *Microeca fascinans* RESIDENT
Loxia fascinans Latham, 1801. *Index Orn.*, Suppl.: xlvi.—Sydney, NSW, Australia.

Patchily distributed in the s watershed of the SE Peninsula, generally centered on the Port Moresby region. Distribution in NG, which is similar to that of *Gerygone olivacea* and *Ptilotula flavescens*, suggests that this species is relictual (see Bell 1982e). Inhabits eucalypt savanna, lowlands only. Has been in decline and may no longer be extant in NG. A 2013 field survey of the savanna north of Port Moresby did not record the species (L. Joseph *in litt.*).

Extralimital—widely distributed in Australia.

SUBSPECIES

a. *Microeca fascinans zimmeri* Mayr & Rand
Microeca leucophaea zimmeri Mayr & Rand, 1935. *Amer. Mus. Novit.* 814: 7.—Port Moresby, SE Peninsula.
RANGE: The rain-shadow section of the s watershed of the SE Peninsula: Bereina southeast to Port Moresby (Coates 1990).
DIAGNOSIS: Plumage has a yellowish wash (Boles 2007).

NOTES
Other names—Jackywinter, Australian Brown Flycatcher, *Microeca leucophaea*.

Lemon-bellied Flyrobin *Microeca flavigaster* RESIDENT
Microeca flavigaster Gould, 1843. *Proc. Zool. Soc. London* (1842): 132.—Port Essington, NT, Australia.

Widespread in open habitats of the s watershed of NG and the Aru Is[W] (Diamond & Bishop 1994[W]); more locally distributed through the n watershed; sea level–1,460 m (Coates & Peckover 2001). Distribution in NG is patchy. Inhabits eucalypt savanna, open woodland, forest edge, rubber plantations, and mangroves (Coates 1990).

Extralimital—n and ne Australia.

SUBSPECIES

a. *Microeca flavigaster laeta* Salvadori
Microeca laeta Salvadori, 1878. *Ann. Mus. Civ. Genova* 12: 323.—Wandammen Penin, Bird's Neck.
RANGE: The n watershed of the Bird's Neck (Wandammen Mts), NW Lowlands, Border Ra (Telefomin[J]), and Sepik-Ramu[CO], east to Madang[MA] (Gilliard & LeCroy 1961[J], see Boles 2007).
DIAGNOSIS: Similar to *tarara* but with a more yellow wash and bill more blackish (Rand & Gilliard 1967). Mantle "brownish olive with a citrine cast; superciliary stripe washed yellowish" (Schodde & Mason 1999).

b. *Microeca flavigaster tarara* Rand
Microeca flavigaster tarara Rand, 1940. *Amer. Mus. Novit.* 1074: 3.—Tarara, Wassi Kussa R, Trans-Fly.
RANGE: The S Lowlands, Trans-Fly[CO], and westernmost sector of the SE Peninsula[CO]. Localities—Mimika R[MA], Dolak I[J], Kurik[L], Wasur NP[J], Bensbach[J], Tarara[J], Wuroi[J], Wassi Kussa R[MA], upper Fly R[K], Purari R delta[CO], and Malalaua[CO] (Mees 1982b[L], Gregory 1997[K], Bishop 2005a[J]). Record from Markham R may be of this race or *flavissima*. Also seen in vicinity of Tabubil, but these birds were quite yellow below and perhaps belong with *laeta*.
DIAGNOSIS: Similar to *flavissima* but for darker olive of dorsal plumage and paler yellow of ventral plumage (Rand & Gilliard 1967). Presumably the Aru Is population can be allied to this form. Mees (1982b) suggested revisiting the validity of *tarara* with fresh and adequate material. Clearly a thinly defined subspecies.

c. *Microeca flavigaster flavissima* Schodde & Mason
Microeca flavigaster flavissima Schodde & Mason, 1999. *Dir. Austral. Birds, Passerines*: 376.—Moitaka, Port Moresby, SE Peninsula. [Because *Microeca flavigaster terraereginae* Mathews, 1912, *Novit. Zool.* 18 (1911): 303.—Cairns, n QLD, was based on a holotype that is intergradient between populations of Cape York Penin and central-eastern QLD, that subspecific name is invalid. Schodde & Mason (1999) renamed the population as above.]
RANGE: Northern QLD and the SE Peninsula, west to the Markham R valley (in north) and Kemp Welch

R (in south). Localities—Wau[P], Yule I[MA], Kemp Welch R[MA], Varirata NP[CO], Moitaka[MA], Port Moresby[MA], Oro Bay[CO], upper Musa valley[CO], Pongani[CO], Raba Raba[P], and Alotau[CO] (Beehler field notes[P]).
DIAGNOSIS: Pale base of mandible conspicuous (Rand & Gilliard 1967). Superciliary stripe washed yellowish; mantle brownish olive with a citrine cast; throat white washed yellow; face mid-gray washed yellow (Schodde & Mason 1999).

NOTES
Apparently a sister form to *M. hemixantha* of the Tanimbar Is (Schodde & Mason 1999). *Other names—* Yellow-bellied Robin, Lemon-breasted Microeca Flycatcher, Yellow-bellied Flycatcher.

GENUS *DRYMODES* [Genus 3 spp/Region 1 sp]

Drymodes Gould, 1841. *Proc. Zool. Soc. London*: 170. Type, by monotypy, *Drymodes brunneopygia* Gould.
    The species of this genus range from southwestern, southern, and far northeastern Australia to New Guinea (Boles 2007). The genus is characterized by the strictly terrestrial habit, the cocking up of the long tail, the long pale legs, twin wing bars, and diagonal dark eye slash. It has been shown to be sister to the cluster that includes *Tregellasia*, *Peneothello*, and *Poecilodryas*, among others (Christidis *et al.* 2011). The genus epithet, *Drymodes*, is feminine (David & Gosselin 2002b).

Papuan Scrub-Robin *Drymodes beccarii* ENDEMIC
Drymoedus beccarii Salvadori, 1876. *Ann. Mus. Civ. Genova* (1875) 7: 965.—Profi, Arfak Mts, Bird's Head.
    Inhabits the floor of forest interior of foothills and lower mts of NG; found in the lowlands near the mouth of the Fly R; also the Aru Is. Sea level–1,450 m[CO]. No records from the Fakfak Mts, Kumawa Mts, mts of the Huon Penin, or the n watershed of the c sector of the SE Peninsula (Coates 1990).

SUBSPECIES
a. *Drymodes beccarii beccarii* Salvadori
    Synonym: *Drymaoedus brevirostris* De Vis, 1897. *Ibis*: 386.—Boirave, Orangerie Bay, SE Peninsula.
RANGE: The Aru Is[MA] and all NG except for the NW Lowlands and PNG North Coastal Ra. Populations from the Arfak[MA] and Wandammen[MA] Mts belong here. Within the Indonesian Trans-Fly recorded from along the Merauke R (Silvius *et al.* 1989); within the PNG Trans-Fly recorded from Morehead, Wuroi on the Oriomo R, Sturt Camp, and north to Lake Daviumbu (Bishop 2005a).
DIAGNOSIS: Underparts variably whitish or washed with pale buff. Mantle rusty brown with a warm tone.

b. *Drymodes beccarii nigriceps* Rand
Drymodes superciliaris nigriceps Rand, 1940. *Amer. Mus. Novit.* 1074: 1.—4 km sw of Bernhard Camp, 850 m, ne sector of the Western Ra.
RANGE: Known only from the upper Mamberamo (Bernhard Camp)[J], Cyclops Mts[MA], PNG North Coastal Ra[K], and Adelbert Mts[L] (Rand 1942b[J], Pratt 1983[L], specimens at AMNH & BPBM[K]). Birds from the Foja Mts remain undiagnosed.
DIAGNOSIS: Dingy and dusky below; mantle dull brown without rusty tone.

NOTES
A review of variation in the collection at AMNH showed that plumage is quite variable, with individual populations often harboring both pale and dark forms. Our solution of two subspecies is rather unsatisfactory geographically, with one widespread form broken by a marginally distinct population in the middle of NG. Another option would be to treat the species as monotypic, exhibiting considerable chaotic plumage variation. Christidis *et al.* (2011: 10) demonstrated that the NG lineage is distinct from *D. superciliaris* of Australia, and they recommended elevation of *beccarii* to species level. However, broader molecular sampling within NG is needed to better clarify the phylogeography and systematics of the NG populations with respect to the Australian ones. *Other names—*Northern Scrub-Robin, *Drymodes superciliaris.*

Heteromyias Sharpe, 1879, *Cat. Birds Brit. Mus.* 4: 239. Type, by monotypy, *Poecilodryas? Cinereifrons* Ramsay.

The species of this genus range from northeastern Queensland to the uplands of New Guinea. The species dwell in the understory of lowland and hill forest interior. Characters include the prominent white flash in the wing, the long pale eyebrow, and the all-whitish underparts. The legs are long and pale. Our circumscription of this genus follows Christidis *et al.* (2011), except that we add *brachyurus* (lacking in the study of Christidis *et al.*), which shares several generic characters, including the long pinkish legs, and differs from *Poecilodryas* in this respect. The genus epithet is masculine (David & Gosselin 2002b).

Ashy Robin *Heteromyias albispecularis*   ENDEMIC

Pachycephala albispecularis Salvadori, 1876. *Ann. Mus. Civ. Genova* (1875) 7: 931.—Arfak Mts, Bird's Head.

The Arfak and Tamrau Mts of the Bird's Head[MA] and the Kumawa Mts of the Bird's Neck[K], as well as the Foja Mts[J]; 850–2,400 m[CO] (Stresemann 1923, Diamond 1985[K], Beehler *et al.* 2012[J]). Should be looked for in the Fakfak and Wandammen Mts.

SUBSPECIES

a. *Heteromyias albispecularis albispecularis* (Salvadori)
RANGE: Mountains of the Bird's Head and the Kumawa Mts.
DIAGNOSIS: No black on gray head; eyebrow pale gray; white triangular subocular spot present; breast pale gray; bend of wing without black marking. Apparently the birds from the Kumawa Mts differ slightly by the more ocher wash on the belly (Diamond 1985: 77). Diamond (1985) stated that the Kumawa birds exhibit a white chin, but C. Thébaud & B. Mila (*in litt.*) noted that three adult-looking birds captured in Nov-2014 in the s Kumawa Mts at 1,100 m exhibited the gray chin typical of the nominate form. Kumawa Mts birds may be distinct (Diamond & Bishop 2015).

b. *Heteromyias albispecularis* undescribed form
RANGE: The Foja Mts
DIAGNOSIS: Crown, nape, mantle, wings plain slaty gray; large triangular white subocular mark with the longest vertex pointed at the cheek and the base aligned with the side of the cheek; lores dark; and no indication of a pale eyebrow (photos of live bird; specimens in Mus. Bogoriense in Cibinong). This isolated form may merit full species status, in line with treatment of other Foja Mts endemics—*Amblyornis flavifrons*, *Melipotes carolae*, and *Parotia berlepschi*.

NOTES

We split *H. armiti* from the traditional *H. albispecularis* of Mayr (1941). *H. albispecularis* can be distinguished as follows: crown, hindneck, chin, and ear coverts gray rather than black; pale line from above eye to side of nape pale gray rather than white; in all cases the black plumage exhibited by *armiti* is replaced by gray in *albispecularis*. Plumage of *albispecularis* much closer to *H. cinereifrons* of n QLD. See comparative photo of the three forms (fig. 3h). Voice distinct from that of *armiti*: 8–12 identical notes delivered in rapid fashion (many per second), without any of the cadence or hesitation characteristic of *H. armiti*. Other names—Ground Thicket-flycatcher; Black-cheeked Robin, *Poecilodryas albispecularis*.

Black-capped Robin *Heteromyias armiti*   ENDEMIC

Poecilodryas armiti De Vis, 1894. *Ann. Rept. Brit. New Guinea*, 1893–94: 101.—Mount Maneao, SE Peninsula.

A vocal but shy and reclusive inhabitant of montane forest interior of the Central Ra to 2,700 m[J] as well as the Adelbert Mts and mts of the Huon Penin (P. Z. Marki *in litt.*[J]). Forages in low vegetation and on the ground.

SUBSPECIES

a. *Heteromyias armiti rothschildi* Hartert
Heteromyias albispecularis rothschildi Hartert, 1930. *Novit. Zool.* 36: 70.—Mount Goliath, Border Ra.
      Synonym: *Heteromyias albispecularis centralis* Rand, 1940. *Amer. Mus. Novit.* 1074: 4.—18 km sw of Bernhard
         Camp, 2,150 m, ne sector of the Western Ra.

RANGE: The Western, Border, and Eastern Ra (Mayr 1941).

DIAGNOSIS: Crown, hindneck, lores, ear coverts, and small spot on chin (variable in size) black; conspicuous white line from over eye to side of nape; back olive brown; extensive ochraceous wash on flanks; wing (male) 98 (Rand & Gilliard 1967). See photo in Coates (1990: 190). Future revisers may wish to consider subsuming this form into the very similar *armiti*.

b. *Heteromyias armiti armiti* (De Vis)

> Synonym: *Heteromyias albispecularis atricapilla* Mayr, 1931. *Mitt. Zool. Mus. Berlin* 17: 681.—Ogeramnang, Saruwaged Ra, Huon Penin.

RANGE: The Adelbert Mts[J] and mts of the Huon and SE Peninsulas (Mayr 1931, Pratt 1983[J], Diamond 1985).

DIAGNOSIS: Very similar to *rothschildi*, but forehead gray, crown brownish black, back greenish olive, rump reddish brown, and wing (male) 102 (Rand & Gilliard 1967).

NOTES

We have chosen to split the black-capped *armiti*+*rothschildi* clade from the gray-capped *albispecularis* of the Bird's Head and Bird's Neck. The two differ in plumage (see fig. 3h) and voice. *H. armiti* has a distinctive black-and-white facial pattern lacking in the washed-out *albispecularis*. In addition, *H. armiti*'s voice is a rapid and cadenced series of high-pitched notes with a weird herky-jerky quality. Each note is identical, but the pace of delivery has an uneven quality. Very distinctive (various recordings by Beehler from YUS transect). *Other names*—Ground Thicket-flycatcher; Black-cheeked Robin, *Poecilodryas albispecularis*.

Black-chinned Robin *Heteromyias brachyurus*   ENDEMIC

Leucophantes brachyurus P. L. Sclater, 1874. *Proc. Zool. Soc. London* (1873): 691, pl. 53.—Andai, e foothills, Arfak Mts, Bird's Head.

> Frequents interior of lowland and foothill forests of the Bird's Head and Neck, Yapen I, NW Lowlands, and Sepik-Ramu (Stresemann 1923).

SUBSPECIES

a. *Heteromyias brachyurus brachyurus* (P. L. Sclater)

RANGE: The Bird's Head and Neck (Wandammen Mts), and the sw sector of the NW Lowlands. Localities—Andai[J], Mt Moari[K], and Majubi[J] (Gyldenstolpe 1955b[J], AMNH specimen[K]).

DIAGNOSIS: Crown black; mantle deep slaty gray (Rand & Gilliard 1967).

b. *Heteromyias brachyurus albotaeniata* (A. B. Meyer)

Amaurodryas albotaeniata A. B. Meyer, 1874. *Sitzungsber. Akad. Wiss. Wien* 69: 498.—Yapen I, Bay Is.

> Synonym: *Poecilodryas brachyura dumasi* Ogilvie-Grant, 1915. *Ibis*, Jubilee Suppl. 2: 163.—Near Humboldt Bay, n NG.

RANGE: Yapen I[MA], the NW Lowlands, and the Sepik-Ramu, east to the mouth of the Sepik R and upper Ramu R[CO]; sea level–950 m (I. Woxvold *in litt.*). Localities—lower Menoo R of the Weyland Mts[K], Mamberamo R[J], Bodim[M], near Humboldt Bay[MA], upper May R[L], Sepik R[MA], and Bundi[CO] (Hartert *et al.* 1936[K], Rand 1942b[J], Ripley 1964a[M], I. Woxvold *in litt.*[L]).

DIAGNOSIS: Dorsal plumage black, like crown (Rand & Gilliard 1967). See photo in Coates (1990: 190).

NOTES

N. David (*in litt.*) confirms the species epithet is an adjective and should agree in gender with the masculine genus. The form *dumasi* agrees well with specimens of *albotaeniata* and hence is combined with that subspecies. *Other names*—White-breasted Flycatcher, New Guinea Robin or Flyrobin, Short-tailed Flycatcher-robin, *Poecilodryas brachyura*.

Plesiodryas, Mathews, 1920. *Birds Austral.* 8: 185. Type by original designation. Replacement name for *Megalestes* Salvadori, 1875; preoccupied by *Megalestes* Selys-Longchamps.

This monotypic genus is endemic to the uplands of the Mainland (Christidis *et al.* 2011). The single species is an aberrant canopy dweller with a high-pitched tone-like song. Ventral plumage exhibits contrasting black throat and whitish abdomen. The generic epithet is feminine (N. David *in litt.*). We follow Christidis *et al.* (2011) in recognizing this genus.

Black-throated Robin *Plesiodryas albonotata*   ENDEMIC

Megalestes albonotatus Salvadori, 1875. *Ann. Mus. Civ. Genova* 7: 770.—Arfak Mts, Bird's Head.

Inhabits the upper stories of montane forest and edge through the mts of the Bird's Head, Central Ra, and Huon Penin; 1,600–2,300 m, extremes 1,150–2,750 m^{CO}.

SUBSPECIES

a. *Plesiodryas albonotata albonotata* (Salvadori)
RANGE: Mountains of the Bird's Head (Mayr 1941).
DIAGNOSIS: Ventral plumage darker slate gray than that of *griseiventris* and with a compact white belly patch

b. *Plesiodryas albonotata griseiventris* (Rothschild & Hartert)
Poecilodryas (Megalestes) albonotata griseiventris Rothschild & Hartert, 1913. *Novit. Zool.* 20: 496.—Mount Goliath, Border Ra.
RANGE: The Western, Border, and Eastern Ra, east to the Kratke Mts (Rand 1942b, Gyldenstolpe 1955a, Diamond 1972, Coates 1990).
DIAGNOSIS: Lower breast and belly slate gray, with white confined to vent (Rand & Gilliard 1967).

c. *Plesiodryas albonotata correcta* (Hartert)
Poecilodryas albonotatus correctus Hartert, 1930. *Novit. Zool.* 36: 68.—Mount Cameron, Owen Stanley Ra, SE Peninsula.
RANGE: Mountains of the Huon and SE Peninsulas, west in main ranges to Aseki (Coates 1990).
DIAGNOSIS: Similar to the nominate form, but black of throat less extensive; also white belly patch elongate and bracketed by gray flank feathering. See photo in Coates (1990: 192) and Coates & Peckover (2001: 155).

NOTES
It is curious that the two white-bellied forms are found at the two ends of NG, separated by a gray-bellied form. *Other names*—Black-throated Flycatcher or Thicket-flycatcher, Black-bibbed Robin.

GENUS *POECILODRYAS*   [Genus 3 spp/Region 1 sp]

Poecilodryas Gould, 1865. *Handb. Birds Austral.* 1: 287. Type, by subsequent designation (Sharpe, 1879. *Cat. Birds Brit. Mus.* 4: 240, 242), *Petroica? cerviniventris* Gould.

The species of this genus inhabit northern Australia and New Guinea. The three species can be characterized by small size, pied plumage, and loud musical vocalization. The genus is evidently a close relative of *Heteromyias*, to which the species *brachyurus* has been transferred. Our treatment follows Christidis *et al.* (2011).

Black-sided Robin *Poecilodryas hypoleuca*   ENDEMIC

Petroica hypoleuca G. R. Gray, 1859. *Proc. Zool. Soc. London*: 155.—Manokwari, Bird's Head.

Lowlands and hills of the Bird's Head and Neck, NW Lowlands, Sepik-Ramu, S Lowlands, Huon Penin, and w sector of the SE Peninsula. Sea level–300 m, max 1,200 m^{CO}. Absent from the Trans-Fly.

a. *Poecilodryas hypoleuca hypoleuca* (G. R. Gray)

RANGE: Misool and Salawati I[MA]; the Bird's Head; the s watershed of the Bird's Neck; the S Lowlands; Maro R and Sturt I of the Trans-Fly (Bishop 2005a); and the s watershed of the SE Peninsula, east to Port Moresby (Mayr 1941).

DIAGNOSIS: Dorsal plumage black without a brownish wash (Rand & Gilliard 1967). See photos in Coates (1990: 191) and Coates & Peckover (2001: 154).

b. *Poecilodryas hypoleuca steini* Stresemann & Paludan

Poecilodryas hypoleuca steini Stresemann & Paludan, 1932. *Novit. Zool.* 38: 157.—Waigeo I, NW Islands.

RANGE: Waigeo I. One sighting from Batanta I (Dutson & Tindige 2006); should be looked for there, and subspecies ascertained.

DIAGNOSIS: Upperparts washed with brown (Rand & Gilliard 1967).

c. *Poecilodryas hypoleuca hermani* Madarász

Poecilodryas hermani Madarász, 1894. *Bull. Brit. Orn. Club* 3: 47.—Bongu, Finisterre Ra, Huon Penin.

RANGE: The NW Lowlands[J], Sepik-Ramu[K], Huon Penin[MA], and nw sector of the SE Peninsula east to the lower Waria R[L] (Rand 1942b[J], Gilliard & LeCroy 1967[K], Dawson *et al.* 2011[L]).

DIAGNOSIS: Similar to the nominate, but with a broader white area above the black lores, a larger white wing patch, and dorsal plumage a rich deep black (Rand & Gilliard 1967). This form is only thinly distinct from the nominate form.

NOTES

Not collected or observed from much of the se sector of the SE Peninsula. *Other names*—Black and White Flycatcher, Black-sided Flyrobin, Pied Robin.

GENUS *PENEOTHELLO*   [Genus 5 spp/Region 5 spp]

Peneothello Mathews, 1920. *Birds Austral.* 8: 185. Type, by original designation, *Poecilodryas? sigillata* De Vis.

The five species of the genus all inhabit New Guinea, with one (*pulverulenta*) extending its range into northern Australia (Loynes *et al.* 2009, Christidis *et al.* 2011). The species of the genus are characterized by the blackish bill, legs, and wings, and plumage of black, gray, blue-gray, and white. The name *Peneothello* is feminine (David & Gosselin 2002b).

White-rumped Robin *Peneothello bimaculata*   ENDEMIC

Myiolestes ? bimaculatus Salvadori, 1874. *Ann. Mus. Civ. Genova* 6: 84.—Putat, Arfak Mts, Bird's Head.

Patchily distributed through lower montane forest of the Bird's Head, the Bird's Neck, the main body of NG, Yapen I, and the SE Peninsula (Stresemann 1923); 700–1,100 m, extremes 300–1,700 m[CO]. Width of elevational range at any one location is very narrow. Apparently absent from the hills and lower mts of the n watershed of the Western and Border Ra and the w sector of the Eastern Ra (between the Weyland Mts and the Bismarck Ra).

SUBSPECIES

a. *Peneothello bimaculata bimaculata* (Salvadori)

RANGE: The Bird's Head, the Van Rees and Foja Mts[Q], Yapen I[Q], and the s slopes of the Central Ra throughout. Localities—mts of the Bird's Head[MA], sw foothills of the Western Ra[DI], Weyland Mts[MA], Dablin Creek above Tabubil[J], hills above the Ok Ma R[J], upper Fly R karst[K], Mt Bosavi[CO], Karimui[DI], Bomai[DI], upper Eloa R[L], Mt Cameron in the Astrolabe Ra[MA], and Veimauri[M] (Beehler 1980b[L], Diamond 1985[Q], Gregory *in litt.*[M], K. D. Bishop *in litt.*[J], I. Woxvold *in litt.*[K]). Netted at 972 m in Oct-2014 uphill from Triton Bay in the Bird's Neck (C. Thébaud & B. Mila *in litt.*).

DIAGNOSIS: White lower belly and undertail. See photo in Coates (1990: 198).

b. *Peneothello bimaculata vicaria* (De Vis)

Paecilodryas vicaria De Vis, 1892. *Ann. Rept. Brit. New Guinea*, 1890–91: 94.—Mount Suckling, SE Peninsula.

RANGE: The Adelbert Mts, mts of the Huon Penin, and the n watershed of the mts of the SE Peninsula. Localities—Oronga[J] in the Adelbert Mts, Sattelberg[K], the lower Mambare R[MA], Mt Suckling[MA], and Mt Dayman[CO] (Mayr 1931[K], Gilliard & LeCroy 1967[J]).

DIAGNOSIS: Abdomen and undertail black (Boles 2007).

NOTES
Other names—White-rumped Thicket-flycatcher.

White-winged Robin *Peneothello sigillata*   ENDEMIC

Poecilodryas ? sigillata De Vis, 1890. *Ann. Rept. Brit. New Guinea*, 1888–89: 59.—Mount Victoria, SE Peninsula.

High elevations of the Central Ra and mts of the Huon Penin; 2,400–3,700 m, extremes 2,150–3,900 m[CO]. Inhabits understory of montane forest and adjacent subalpine shrubbery. Replaced at lower elevations by *P. cyanus*.

SUBSPECIES
a. *Peneothello sigillata quadrimaculata* (van Oort)
Poecilodryas quadrimaculatus van Oort, 1910. *Notes Leyden Mus.* 32: 213.—Hellwig Mts, 2,600 m, e sector of the Western Ra.
RANGE: The Western Ra (Rand 1942b).
DIAGNOSIS: Conspicuous white patch on each side of breast. See photo in Coates & Peckover (2001: 156).

b. *Peneothello sigillata sigillata* (De Vis)
Synonym: *Peneothello sigillatus hagenensis* Mayr & Gilliard, 1952. *Amer. Mus. Novit.* 1577: 4.—Summit Camp, Mt Hagen, Eastern Ra.
RANGE: The Border Ra, Eastern Ra, and mts of the SE Peninsula.
DIAGNOSIS: No white patch on each side of breast; white inner secondaries tipped black; wing (male) 94–100 (Rand & Gilliard 1967). We combine *hagenensis* with this form, as the black tipping to white inner secondaries is present on specimens from Tomba. See photos in Coates (1990: 197) and Coates & Peckover (2001: 156).

c. *Peneothello sigillata saruwagedi* (Mayr)
Poecilodryas sigillata saruwagedi Mayr, 1931. *Mitt. Zool. Mus. Berlin* 17: 680.—Mongi-Busu, Saruwaged Ra, Huon Penin.
RANGE: Mountains of the Huon Penin.
DIAGNOSIS: Similar to the nominate form, but with more black on tips of secondaries; also slightly smaller; wing (male) 92–96 (Rand & Gilliard 1967).

NOTES
Other names—White-winged Flyrobin, Thicket-robin, or Thicket-flycatcher; Black Flycatcher-robin.

Mangrove Robin *Peneothello pulverulenta*   RESIDENT

Myiolestes pulverulentus Bonaparte, 1850. *Consp. Gen. Av.* 1: 358.—Utanata R, sw NG.
Found in various coastal localities of the NG mainland; the Aru Is[MA]; and several interior records from the Fly and Sepik R[CO]. A lowland dweller inhabiting coastal mangroves and interior swamp forest and reedbeds.

Extralimital—coastal nw, n, and ne Australia.

SUBSPECIES
a. *Peneothello pulverulenta pulverulenta* (Bonaparte)
RANGE: Coastal mangroves of the s Bird's Head, Bird's Neck, NW Lowlands, nw sector of the Sepik-Ramu, S Lowlands, s coast of the SE Peninsula, and one patch on the n coast of the SE Peninsula. Localities—Noisaroe at head of Geelvink Bay[MA], Mimika R[MA], Utanata R[MA], coast of Geelvink Bay[MA], Kurik[J],

Maro R[J], Wasur NP[J], Lake Daviumbu[M], Daru[J], Humboldt Bay[MA], middle Sepik[K], Purari R delta[CO], Hall Sound[M], Hisiu[CO], Killerton Is[CO], East Cape[MA], and Bona Bona I[L] (Rand 1942a[M], Gilliard & LeCroy 1966[K], Bishop 2005a[J], D. Hobcroft *in litt.*[L]).

DIAGNOSIS: Distinguished from *leucura* of the Aru Is and ne Australia by broader bill and paler breast (Schodde & Mason 1999). Very thinly defined. We would not recognize the subspecies but for the work of Nyári & Joseph (2013), who demonstrated three distinct molecular lineages that match up geographically with the described subspecies.

b. *Peneothello pulverulenta leucura* (Gould)

Eopsaltria leucura Gould, 1869 (1-August). *Birds Austral.*, Suppl. pt. 5: plate and text.—Cape York district, QLD, Australia.

RANGE: The Aru Is and coastal ne Australia (Schodde & Mason 1999).

DIAGNOSIS: See diagnosis for nominate form.

NOTES

We follow Christidis *et al.* (2011) in generic placement of this outlier species. See also Schodde & Mason (1999). See photo in Coates (1990: 196). *Other names—Peneonanthe* or *Eopsaltria pulverulenta*.

Smoky Robin *Peneothello cryptoleuca*   ENDEMIC

Poecilodryas cryptoleucus Hartert, 1930. *Novit. Zool.* 36: 67.—Lehuma, Arfak Mts, Bird's Head.

Mountains of the Bird's Head and Neck (Kumawa Mts), the w sector of the Western Ra (Weyland Mts), and the Foja Mts, 1,400–2,200 m. Inhabits understory of the interior of montane forest.

SUBSPECIES

a. *Peneothello cryptoleuca cryptoleuca* (Hartert)

RANGE: Mountains of the Bird's Head and the Foja Mts[K] (Beehler & Prawiradilaga 2010[K]).

DIAGNOSIS: The palish patch on the abdomen more obscure and mantle plumage slightly paler than in *albidior*. Diamond (1985) suggested the Foja population was closer to *albidior*, but our specimens (now in Mus. Bogoriense in Cibinong) show otherwise.

b. *Peneothello cryptoleuca maxima* Diamond

Peneothello cryptoleucus maximus Diamond, 1985. *Emu* 85: 77.—Southern slopes of the Kumawa Mts, 1,400 m, Bird's Neck.

RANGE: The Kumawa Mts.

DIAGNOSIS: Very distinctive. Larger than other subspecies, and underparts with white extending from throat to undertail.

c. *Peneothello cryptoleuca albidior* (Rothschild)

Poecilodryas cryptoleucus albidior Rothschild, 1931. *Novit. Zool.* 36: 263.—Gebroeders Mts, w sector of Western Ra.

RANGE: The Western Ra. Localities—Mt Sumuri in the Weyland Mts[K], Gebroeders Mts[MA], and Ilaga[J] (Hartert *et al.* 1936[K], Ripley 1964a[J]).

DIAGNOSIS: As for the nominate form, but abdomen patch paler, mantle darker. This form, thinly defined, may best be combined with the nominate form.

NOTES

Other names—Grey Thicket-flycatcher, Grey Robin.

Blue-grey Robin *Peneothello cyanus*   ENDEMIC

Myiolestes ? *cyanus* Salvadori, 1874. *Ann. Mus. Civ. Genova* 6: 84.—Hatam, Arfak Mts, Bird's Head.

Inhabits understory of montane forest interior in the mts of NG; 1,550–2,400 m, extremes 900–2,750 m[CO]. Widespread in the Central Ra; also the Arfak, Fakfak (?), Foja, Cyclops, Torricelli, and Adelbert Mts, mts of the Huon Penin, and Mt Bosavi.

a. *Peneothello cyanus cyanus* (Salvadori)

RANGE: Mountains of the Bird's Head.

DIAGNOSIS: Plumage pale, with no black.

b. *Peneothello cyanus subcyanea* (De Vis)

Poecilodryas subcyanea De Vis, 1897. *Ibis*: 377.—Mountains of the SE Peninsula.

> Synonym: *Poecilodryas cyana atricapilla* Hartert and Paludan, 1934. *Orn. Monatsber.* 42: 45.—Kunupi, Weyland Mts, Western Ra.

RANGE: The Wandammen Mts[MA]; Central Ra east to Milne Bay[MA]; Cyclops[MA], Torricelli[J,K], and Adelbert Mts[L]; and mts of the Huon Penin[MA] (Mayr 1941, Pratt 1983[L], specimens at AMNH[J] & BPBM[K]). Populations of the Fakfak[M] and FojaMts[N] undiagnosed but likely belong here too (Gibbs 1994[M], Beehler *et al.* 2012[N]), but see Diamond & Bishop (2015: 323).

DIAGNOSIS: Plumage dark and crown blackish.

NOTES

The species epithet *cyanus* is invariant (David & Gosselin 2002b). The PhD thesis of Benz (2011) offered a rare look at the phylogeography of certain species of montane birds in NG based on intensive sampling across populations, irrespective of subspecies boundaries. While our inspection of geographic variation in plumage coloration of *P. cyanus* revealed only two thinly defined races (this study), the analyses of mitochondrial sequence data by Benz (2011: 9) "recovered three primary clades within *P. cyanus*, the distributions of which correspond to currently recognized subspecies and are consistent with prominent biogeographic boundaries including the Vogelkop Peninsula [Bird's Head] and Strickland River valley." Benz's recognized subspecies are *cyanus*, *atricapilla* (which we don't recognize), and *subcyanea*. This is an example where plumage color is highly conserved in a plain-looking species. *Other names*—Slaty Robin, Slaty Thicket-flycatcher.

GENUS *GENNAEODRYAS*   ENDEMIC | MONOTYPIC

Gennaeodryas Mathews, 1920. *Birds Austral.* 8: 186. Type, by original designation, *Eopsaltria placens* Ramsay.

This monotypic genus is endemic to the uplands of New Guinea. The genus is characterized by the bright orange legs and the distinctive yellow, green, and gray plumage. Placement of the species *placens* into a monotypic genus follows Christidis *et al.* (2011). Their Fig. 4 placed this lineage basal to *Tregellasia* and *Eopsaltria*.

Banded Yellow Robin *Gennaeodryas placens*   ENDEMIC | MONOTYPIC

Eopsaltria placens Ramsay, 1879. *Proc. Linn. Soc. NSW* 3: 272.—Goldie R, SE Peninsula.

Inhabits interior of hill forest on Batanta I and several widely separated areas on the Mainland: the Bird's Neck, Weyland Mts, Adelbert Mts, Eastern Ra, and the s watershed of the SE Peninsula; mainly 500–800 m, extremes 100–1,450 m[CO].

Localities—Mt Besar of Batanta I[Q], Fakfak Mts[K,Z], Kumawa Mts[K], mts of the Wandammen Penin[K], lower Menoo R in the Weyland Mts[J], Kikori R (Moro, Kantobo, Gobe[W], Darai Plateau)[W], Lake Kutubu[K], Mt Bosavi[K], Crater Mt[A], Karimui[DI], Bomai[DI], Soliabeda[DI], Keki Lodge[L] in the Adelbert Mts, Keku[M], Astrolabe Bay[MA], Veimauri[CO], Angabanga R[CO], Goldie R[MA], Kubuna[U], Aroa R[K], and Kotoi distr[N] (Hartert *et al.* 1936[J], Mayr & Rand 1937[U], Greenway 1966[O], Diamond 1985[K], Gibbs 1994[Z], Mack & Wright 1996[A], Beehler field notes 25-May-1999[L], I. Woxvold *in litt.*[W], Beck specimen at AMNH[M], A. S. Anthony specimen at AMNH[N]). Probably overlooked in some areas where its elevational range is narrow.

Other names—Olive-yellow Robin or Flycatcher, Yellow Thicket-flycatcher, *Poecilodryas placens*.

Tregellasia Mathews, 1912. *Austral. Avian Rec.* 1: 110. Type, by original designation, *Eopsaltria capito* Gould.

The species of this genus range from eastern Australia to the New Guinea region. The genus is characterized by small and compact body, white lores, and yellow abdomen contrasting with olive-green mantle. Schodde & Mason (1999) also noted the flattened bill and stout rictal bristles, and juveniles with plain, russet-washed plumage. This lineage is sister to *Eopsaltria* of Australia (Christidis *et al.* 2011).

White-faced Robin *Tregellasia leucops*   RESIDENT

Leucophantes leucops Salvadori, 1875. *Ann. Mus. Civ. Genova* 7: 921.—Profi and Mori, Arfak Mts, Bird's Head.

Inhabits forest interior of the mts of NG, the lowlands of the Trans-Fly, and Yapen I[J] (Mayr 1941, Diamond 1985[J]); 600–1,650 m, extremes sea level–2,200 m (Coates & Peckover 2001). Widespread but easily overlooked.

Extralimital—Cape York Penin, QLD.

SUBSPECIES

a. *Tregellasia leucops leucops* (Salvadori)
RANGE: Mountains of the Bird's Head and Bird's Neck (Mayr 1941, Diamond & Bishop 2015).
DIAGNOSIS: Forehead black; also black encircles eye and a black line extends from eye to gape; large white spot between eye and nostril; chin and upper throat indistinctly washed with white; crown grayish; bill mainly black (Rand & Gilliard 1967).

b. *Tregellasia leucops mayri* (Hartert)
Poecilodryas leucops mayri Hartert, 1930. *Novit. Zool.* 36: 67.—Mount Wondiwoi, Wandammen Penin, Bird's Neck.
RANGE: The n sector of the Bird's Neck (Wandammen Mts) and the Weyland Mts of the w sector of the Western Ra (Hartert *et al.* 1936).
DIAGNOSIS: Very similar to the nominate form, but no black line from eye to gape (Hartert 1930). White spot in front of eye continuous with whitish chin.

c. *Tregellasia leucops nigroorbitalis* (Rothschild & Hartert)
Poecilodryas leucops nigro-orbitalis Rothschild & Hartert, 1913. *Novit. Zool.* 20: 497.—Upper Otakwa R, Western Ra.
RANGE: The Fakfak Mts, Kumawa Mts, and s slope of the Western Ra. Six birds captured in the s Kumawa Mts were aligned with this race (C. Thébaud & B. Mila *in litt.*), not the nominate form (*pace* Diamond 1985).
DIAGNOSIS: Similar to *mayri* but with more white on cheek, and crown blacker. The birds from the s Kumawas had a large white spot in front of the eye continuous with lores and chin (C. Thébaud & B. Mila *in litt.*).

d. *Tregellasia leucops heurni* (Hartert)
"*Poecilodryas*" *leucops heurni* Hartert, 1932. *Nova Guinea, Zool.* 15: 467.—Doormanpaad Bivak, Mamberamo Mts, Western Ra.
RANGE: The c and e sectors of the n watershed of the Western Ra; by distribution, the populations of Yapen I and the Foja Mts are probably of this form (see Diamond 1985).
DIAGNOSIS: Forehead all white and continuous with white throat patch; white encircles eye (Rand & Gilliard 1967).

e. *Tregellasia leucops nigriceps* (Neumann)
Poecilodryas leucops nigriceps Neumann, 1922. *Verh. Orn. Ges. Bayern* 15: 237.—Hunstein Mts, middle Sepik, north-central NG.
RANGE: The s and n slopes of the Border Ra: Victor Emanuel Mts and mts of the Sepik (Mayr 1941).
DIAGNOSIS: Similar to *heurni* but white of throat much reduced. Thinly defined.

f. *Tregellasia leucops melanogenys* (A. B. Meyer)

Poecilodryas melanogenys A. B. Meyer, 1893. *Abh. Ber. Zool. Mus. Dresden* (1892–93) 4(3): 12.—Sattelberg, Huon Penin.

RANGE: Hills of n NG: Cyclops Mts, PNG North Coastal Ra[D], Adelbert Mts[G], and mts of the Huon Penin[M] (Mayr 1931[M], Gilliard & LeCroy 1967[G], Diamond 1985[D]).

DIAGNOSIS: Chin yellow; nape washed with olive (Boles 2007). Ear coverts black; white encircles eye. See photo in Coates & Peckover (2001: 155).

g. *Tregellasia leucops auricularis* (Mayr & Rand)

Microeca leucops auricularis Mayr & Rand, 1935. *Amer. Mus. Novit.* 814: 7.—Wuroi, Oriomo R, Trans-Fly, s NG.

RANGE: The Trans-Fly.

DIAGNOSIS: Similar to *wahgiensis*, but throat pure white; ear coverts, face, and forehead white (Rand & Gilliard 1967).

h. *Tregellasia leucops wahgiensis* Mayr & Gilliard

Tregellasia leucops wahgiensis Mayr & Gilliard, 1952. *Amer. Mus. Novit.* 1577: 2.—Base of Mt Orata (behind Kup), Kubor Mts, Eastern Ra.

RANGE: The Eastern Ra from the Wahgi valley region east to the w sector of the SE Peninsula; east to the Mambare R in the n watershed and east to the Aroa R in the s watershed (Gyldenstolpe 1955a, Diamond 1972).

DIAGNOSIS: Similar to *nigriceps* and *melanogenys* but white about eye more extensive; crown olive green (blackish in Wau birds); bill more extensively yellow (Rand & Gilliard 1967). Diamond (1972) noted white chin, yellow throat, white forehead, and pale straw-yellow bill with black terminus to maxilla. See photo in Coates (1990: 195).

i. *Tregellasia leucops albifacies* (Sharpe)

Poecilodryas albifacies Sharpe, 1882. *Journ. Linn. Soc. London, Zool.* 16: 318.—Sogeri, SE Peninsula.

RANGE: Mountains of the e sector of the SE Peninsula, on the n coast as far west as the Mambare R and on the s coast as far west as Port Moresby and Varirata NP[J] (K. D. Bishop *in litt.*[J]).

DIAGNOSIS: Like *wahgiensis*, but forehead and maxilla black (Rand & Gilliard 1967). See photo in Coates (1990: 195) and Coates & Peckover (2001: 155).

NOTES

This is a frustrating set of minor subspecies, each with one or two minor unitary (nongradient) differences, so they remain standing. The overall variation from west to east is not substantial. *Other names*—White-faced Flycatcher, Little or White-throated Robin, White-faced Yellow Robin.

Family Alaudidae

Larks   [Family 93 spp/Region 1 sp]

The species of the family range worldwide but for Greenland and much of South America. Most species inhabit Africa and Eurasia. The single New Guinean species is a small, cryptic, ground-dwelling bird of open grasslands. It walks or runs over the ground, searching for insects and seeds. The flight is strong. Many species perform song flights in the breeding season. The larks are sister to a clade that includes the white-eyes (Zosteropidae), leaf-warblers (Phylloscopidae), bulbuls (Pycnonotidae), swallows (Hirundinidae), and *Garrulax* laughing-thrushes (Barker *et al.* 2004). Placement of the family after Petroicidae follows Barker *et al.* (2004). Included in the Sylvioidea, a cluster of insectivore families, by Alström *et al.* (2013).

Mirafra Horsfield, 1821. *Trans. Linn. Soc. London* 13: 159. Type, by monotypy, *Mirafra javanica* Horsfield.

The species of this genus range through Madagascar, Africa, tropical Asia, New Guinea, and Australia (Juana *et al.* 2004). The genus is characterized by a sturdy conical bill, brown-streaked dorsal plumage, whitish underparts with streaking on the breast, and pink legs.

Horsfield's Bushlark *Mirafra javanica* RESIDENT

Mirafra Javanica Horsfield, 1821. *Trans. Linn. Soc. London* 13: 159.—Java.

Locally distributed through NG. Inhabits regions where there are extensive areas of short tussock-forming grass. Found mostly in the lowlands, but inhabits mid-montane valleys up to 1,680 m (Coates 1990).

Extralimital—se Asia to Australia. We follow Dickinson & Christidis (2014) in treating the African populations as a distinct species, *M. cantillans*.

SUBSPECIES

a. *Mirafra javanica aliena* Greenway
Mirafra javanica aliena Greenway, 1935. *Proc. New Engl. Zool. Club* 14: 50.—Junction of the Bulolo and Watut R, 690 m, n watershed of the SE Peninsula.

> Synonym: *Mirafra javanica sepikiana* Mayr, 1938. *Zool. Ser. Field Mus. Nat. Hist.* 20: 466.—Marienberg, Sepik R, north-central NG.
> Synonym: *Mirafra javanica timoriensis* Mayr, 1944. *Bull. Amer. Mus. Nat. Hist.* 83: 154.—Dili, East Timor, Lesser Sundas.

RANGE: The Lesser Sunda Is (Timor, Sawu) and also patchily distributed through NG. Localities—Dolak I[J], Kurik[J], Kaisa on the Kumbe R[K], Merauke[MA], Wasur NP[J], Bensbach R[J], Bula plains[CO], lower Sepik (Marienberg)[MA], Baiyer valley[CO], Wahgi valley[CO], Kelanoa-Sialum[CO], Markham valley[CO], Mumeng[CO], Bulowat on the Watut R[MA], Bulolo[CO], Wau[CO], Aroa R[CO], Rigo[CO], Port Moresby[CO], Fisherman's I[CO] (Mees 1982b[K], Bishop 2005a[J]).

DIAGNOSIS: Dorsally heavily marked with blackish streaking; crown, secondaries and wing coverts blackish with buff feather edges; throat plain buff white; breast band rich buff with dark markings; belly buff and unmarked. The forms *sepikiana* and *timoriensis* not distinguishable from NG form *aliena* (Mees 1982b). The NG form is distinct from *horsfieldii*, the population in e QLD (Higgins *et al.* 2006).

NOTES

The species is possibly also present at Simbai (1,740 m), Bismarck Ra (Majnep & Bulmer 1977). Rand & Gilliard (1967) noted the population from the Trans-Fly may deserve subspecific diagnosis. They mention seeing a single diagnosable specimen from Merauke. Mees (1982b) showed this not to be so. *Other names*—Singing or Australasian Bushlark, Horsfield's Lark.

Family Hirundinidae

Swallows and Martins [Family 84 spp/Region 6 spp]

The species of the family range worldwide but for northern Canada, northern Siberia, and Greenland. The family is well defined and distinctive from any other bird family. Swallows are aerial birds only superficially similar to swifts or treeswifts but with a more tapered body and very different flight, with the wings bending more at the wrist. While swallows spend most of the day hawking for insects, they are also often seen perching on wires and bare branches (unlike swifts) and on the ground collecting nesting material. In the Region, they are found primarily in open country of the lowlands and mid-montane valleys. Sequence placement is based on Barker *et al.* (2004) and Alström *et al.* (2013). Of the families inhabiting the region, the swallows are an isolated group with affinities to a wide array of songbirds, including several babbler lineages, white-eyes (Zosteropidae), leaf-warblers (Phylloscopidae), and bulbuls (Pycnonotidae), among others (Alström *et al.* 2013).

GENUS *RIPARIA*

Riparia T. Forster, 1817. *Synop. Cat. Brit. Birds*: 17. Type, by monotypy and tautonymy, *Riparia europaea* Forster = *Hirundo riparia* Linnaeus.

The species of this genus range through North America, Europe, Africa, and temperate Asia, in many instances wintering south into the tropics (Turner 2004). The genus is characterized by the slightly forked tail, the plain gray-brown upperparts, and lack of recurved barbs on the outer primaries.

Sand Martin *Riparia riparia*   VAGRANT

Hirundo riparia Linnaeus, 1758. *Syst. Nat.*, ed.10, 1: 192.—Sweden.

Two sight records from NG. Presumably a straggling migrant from Asia, to be expected during the northern fall or winter. NG records: (1) two birds, "almost certainly" this species, seen flying over a creek mouth north of Sialum, ne coast of the Huon Penin, on 5-Feb-1977 (Bell 1979a); (2) a party of two or three birds seen hawking over the surface of a reed-fringed lagoon in the e Wahgi valley on 3-Jan-1979 (Thompson 1979).

Extralimital—nearly worldwide in distribution, although absent from Australia. Breeds in the Northern Hemisphere; winters south to South America, Africa, and tropical Asia, rarely to the Philippines and Borneo. Frequents open country, particularly along rivers and marshes. One unconfirmed sight record from New Britain (Dutson 2011).

SUBSPECIES

a. *Riparia riparia ?ijimae* (Lönnberg)

Clivicola riparia ijimae Lönnberg, 1908. *Journ. Coll. Sci. Imp. Univ. Tokyo* 23(14): 38.—Tretiya Padj, Sakhalin, Russia.

RANGE: Breeds in ne Asia and winters in the islands of se Asia, apparently very rarely to NG. Should be looked for in the NW Islands and the Bird's Head in the proper season.

DIAGNOSIS: Darker brown than nominate form; gray-brown breast band separating white throat from white underparts; back uniformly dark gray-brown (Turner 2004). Marginally distinct from other east Asian forms.

NOTES

Other names—Collared Sand Martin, Bank Swallow.

GENUS *HIRUNDO*

Hirundo Linnaeus, 1758. *Syst. Nat.*, ed. 10, 1: 191. Type, by subsequent designation (Swainson, 1837. *Nat. Hist. and Classif. Birds* 2: 340), *Hirundo rustica* Linnaeus.

The species of this genus range from North America and Africa to Eurasia and Australia (Turner 2004, Dor *et al.* 2010). The genus, which is poorly delineated from sister genera, can best be characterized by the dark rump, dorsal coloration mainly or entirely dark blue with a metallic sheen, unstreaked ventral plumage, and deeply forked tail.

Barn Swallow *Hirundo rustica*   VISITOR

Hirundo rustica Linnaeus, 1758. *Syst. Nat.*, ed. 10, 1: 191.—Sweden.

Regular Asian migrant to NG, fairly common during the northern autumn and winter in lowlands of the west, the Sepik, and around Port Moresby, less regularly elsewhere. Recorded to 1,740 m[co]. Frequents wetland areas.

Extralimital—nearly worldwide in distribution, breeding in the temperate Northern Hemisphere, migrating south to the tropics.

SUBSPECIES

a. *Hirundo rustica gutturalis* Scopoli

Hirundo gutturalis Scopoli, 1786. *Delic. Flor. Faun. Insub.* 2: 96.—Panay I, Philippines.

RANGE: Breeds in e Siberia, n China, Korea, and Japan. During migration and during the northern winter found over the entire Indo-Australian region from India east to NG and n Australia (Mayr 1941).
DIAGNOSIS: Rich chestnut forehead and throat; black lower throat collar offset by whitish-buff breast and belly. Chestnut of throat bleeds onto black collar. Schodde & Mason (1999) noted the following: dark breast band narrowed to a thin line; size small; short and shallowly forked tail; wing (male) 112–120.

NOTES
Other names—the Swallow, Common Swallow.

Pacific Swallow *Hirundo tahitica*   RESIDENT
Hirundo tahitica Gmelin, 1789. *Syst. Nat.* 1(2): 1016.—Tahiti.
    All NG and many satellite islands; sea level–2,200 m; max 2,700 m (Coates & Peckover 2001). Frequents open areas, wetlands, and settled areas (Coates 1990).
    Extralimital—tropical Asia, Ryukyu Is, through c Pacific exclusive of range of *H. neoxena*. Range includes the Bismarck Is, Bougainville and Buka, and islands east to Tahiti and south to Vanuatu and New Caledonia (Coates 1990, Mayr & Diamond 2001, Dutson 2011).

SUBSPECIES
a. *Hirundo tahitica frontalis* Quoy & Gaimard
Hirundo frontalis Quoy & Gaimard, 1830. *Voy. Astrolabe, Zool.*, 1: 204; *Atlas*: pl. 12, fig. 1.—Manokwari, Bird's Head.
RANGE: Northern and w NG, the NW Islands, Aru Is, Bay Is, and islands off the n coast: Tarawai, Walis, Kairiru, Muschu, Karkar, and Bagabag (Coates 1990).
DIAGNOSIS: Blue-black dorsally, and more gray ventrally than *albescens* (Turner 2004). Forehead very dark chestnut, darker than chestnut of throat, which grades into dirty gray on upper breast; pale middle of lower breast and belly.

b. *Hirundo tahitica albescens* Schodde & Mason
Hirundo tahitica albescens Schodde & Mason, 1999. *Dir. Austral. Birds, Passerines*: 667, 668.—Margarida, Amazon Bay, SE Peninsula.
RANGE: Southern and e NG. West in the s watershed to Wasur NP in the Trans-Fly (K. D. Bishop *in litt.*); in the n watershed the w boundary of the range is unknown (Schodde & Mason 1999). Range includes the Amphlett Is[CO], and Goodenough[MA], Fergusson[MA], Normanby[L], Samarai, and Bentley I[CO]; vagrant to Misima I[P] (LeCroy *et al.* 1984[L], Pratt field notes 2004[P]).
DIAGNOSIS: Paler ventrally than *frontalis*, with large pale tail spots (Turner 2004). Downy bases of mantle feathers extensively whitish; lower ventral plumage off-white, washed pale fawn toward vent (Schodde & Mason 1999). See photo in Coates & Peckover (2001: 116).

NOTES
Forms a superspecies with *Hirundo neoxena* of Australia. *Other names*—House or Hill Swallow.

[Welcome Swallow *Hirundo neoxena*   HYPOTHETICAL
Hirundo neoxena Gould, 1843. *Birds Austral.* 9: pl. 13.—Tasmania.
    A possible vagrant to NG from Australia. One seen with other swallow species at Aroa lagoon, Central Prov, on 8-Nov-1980 (Finch 1980d); and two at Hisiu, Central Prov, 28-Oct–17-Nov-1984, with one still present on 8-December (Finch 1984). Both of these records are perhaps later than would be expected for the occurrence of this species (Coates 1990). An additional record comes from the Moitaka settling ponds on 30-Mar-1988 (Hicks & Gregory-Smith 1990). Free-flying individuals can be difficult to separate from *H. tahitica*.
    Extralimital—breeds in Australia, NZ, Lord Howe, Norfolk, Chatham, and Kermadec Is (Turner 2004). Also occurs on islands in the s Torres Strait, north at least to Thursday I. Partially migratory during winter in s part of range, and migrates to New Caledonia (Dutson 2011).]

Cecropis Boie, 1826, *Isis von Oken* 19(10): col. 971. Type, by subsequent designation (Salvadori, 1881, *Orn. Pap. Moluc.* 2: 1), *Hirundo capensis* Gmelin = *Hirundo cucullata* Boddaert.

The species of this genus range from Africa and Europe east to Japan and tropical Asia, with migrants wandering to Australia and New Guinea. The genus is characterized by the deeply forked tail, rufous rump, and patterned blue, rufous, and white plumage.

Red-rumped Swallow *Cecropis daurica* VISITOR

Hirundo daurica Laxmann, 1769. *Kongl. Vet. Akad. nya Handl.* 30: 209.—Mount Schlangen, Altai, Russia. [Proper authority for this species epithet follows Dickinson 2003.]

An uncommon migrant to NG, with records scattered across the Mainland. Coates (1990) posited that the species has undergone a notable change of status in the NG region in recent years, from being an extreme rarity (it was unrecorded before 1974) to a regular visitor during the northern winter.

Extralimital—breeds in parts of Africa, s Europe, s, c, and e Asia, and Japan. Northern populations winter south to the tropics. Formerly a vagrant to ne Australia but now found more or less annually (P. Gregory *in litt.*). Also records from Bismarcks and n Solomons (Dutson 2011).

SUBSPECIES

a. *Cecropis daurica ?japonica* (Temminck and Schlegel)
Hirundo alpestris japonica Temminck and Schlegel, 1845. In Siebold, *Fauna Japonica*, *Aves*: 33, pl. 11.—Japan. [For date of publication we follow Dickinson (2003), *pace* Peters (1960).]
RANGE: Breeds in e Russia, ne China, Korea, and Japan; winters to n Australia (Turner 2004). Localities—Manokwari[Z], Ningerum, Kiunga, Tabubil[W], Daru, Liddle R (foothills of Central Ra, *ca.* 215 m, in the upper Strickland R)[J], Angoram, Tari[L], Nadzab, Port Moresby, Aroa lagoon, Kanosia lagoon, Waigani swamp, Marshall lagoon, Alotau, Normanby I, and Fergusson I, all during the northern winter (Coates 1990, Gregory 2007a[Z], S. F. Bailey *in litt.*[L], I. Woxvold *in litt.*[J], Gregory *in litt.*[W]).
DIAGNOSIS: Presumed to be *japonica* by Schodde & Mason (1999). This form is distinguished by the faint streaks on the rusty rump; heavily streaked underparts, predominating on the throat and lower throat; and the broadly interrupted hind-collar (Turner 2004). Differs from nominate form by larger size of dark steaks on flanks.

NOTES

Cecropis striolata, the sister species to *C. daurica*, is not known from the NG region (Schodde & Mason 1999). It breeds in Indochina, the Philippines, and Indonesia but is sedentary. The birds observed in the far west by Gregory *et al.* (1996b) are assumed to be *daurica*, not *striolata*. *Other names—Hirundo daurica*.

Petrochelidon Cabanis, 1851. *Mus. Hein.* 1: 47. Type, by subsequent designation (Gray, 1855. *Cat. Gen. Subgen. Birds*: 13), *Hirundo melanogaster* Swainson = *Hirundo pyrrhonota* Vieillot.

The species of this genus range from the Americas to Africa, tropical Asia, and Australia. The genus is characterized by the unforked or minimally forked tail, the predominance of navy blue dorsally, the presence of some rufous in plumage, and the pale rump patch in most species. Mees (1982b) made the case for keeping *nigricans* and *ariel* in this genus, and molecular analysis supports this (Sheldon *et al.* 2005).

Tree Martin *Petrochelidon nigricans* VISITOR

Hirundo nigricans Vieillot, 1817. *Nouv. Dict. Hist. Nat.*, nouv. éd. 14: 523.—Hobart, Tasmania.

A common Australian migrant, wintering especially to n NG. NG birds found mainly from late February to October, with a small number of birds remaining in the Port Moresby area during the remainder of the year (Coates 1990). Frequents mainly grasslands and open areas, including wetlands. Mainly in the lowlands, occasionally to 1,830 m. Widespread but for the Milne Bay region (Mayr 1941, Coates 1990, Bishop 2005a).

Extralimital—Breeds in e Indonesia and much of Australia. Winters in n Australia, NG, the Bismarcks, and the Solomons (Turner 2004, Dutson 2011).

a. *Petrochelidon nigricans nigricans* (Vieillot)
RANGE: Breeds in Tasmania. Winters north to the Aru and Kai Is, NG, Bismarck Arch, and Solomon Is (Mayr 1941). Specimens examined at AMNH from the Cyclops Mts, Daru, and Kumusi R are attributable to this form.
DIAGNOSIS: Forehead rufous; abdomen off-white with some indication of very fine dark streaking; rump dirty white with obsolete dark streaking. Wing 105–111 (Schodde & Mason 1999). Determining subspecies of wintering birds is very difficult, as plumage distinctions are minor.

b. *Petrochelidon nigricans ?neglecta* Mathews
Petrochelidon neglecta neglecta Mathews, 1912. *Novit. Zool.* 18: 301.—Fitzroy R, nw Australia.
RANGE: Breeds through mainland Australia, wintering north into NG apparently (Schodde & Mason 1999). Specimens examined at AMNH from Lake Daviumbu, Cyclops Mts, and Bioto Creek are attributable to this subspecies.
DIAGNOSIS: Forehead creamy rufous; underparts with ocher wash on abdomen; smaller than nominate form, wing 100–107 (Schodde & Mason 1999). A marginal race, perhaps meriting combination with nominate form.

NOTES
Coates (1990: 29) posited the species may be a breeder in the NG region. This requires confirmation. *Other names*—Australian Tree Martin, *Hirundo nigricans*.

Fairy Martin *Petrochelidon ariel*   VISITOR | MONOTYPIC
Collocalia Ariel Gould, 1842. *Birds Austral.* 9: pl. 15.—New South Wales, Australia.
    A rare visitor from Australia to the Fly R and Port Moresby.
    Localities—Bensbach R, Bula plains, Balamuk[K], Tabubil, Nomad R[K], Tari, Kanosia, and Aroa R; records from May, August, September, October, and November (Coates 1990, Bell 1968a[K]). Sometimes in flocks (Murray *et al.* 1987).
    Extralimital—breeds widely in Australia. Winters north to the Torres Strait Is (Schodde & Mason 1999). *Other names*—*Hirundo ariel*.

Family Pycnonotidae

Bulbuls   [Family 130 spp/Region 1 introduced sp]

The species of the family range from Africa and the Middle East to tropical Asia, Japan, and Indonesia. Certain species are popular cage birds and have been introduced widely. A single species has been recorded from the Region, certainly as an escapee from a human-transported population related to the Indonesian bird trade. Placement of the family follows Barker *et al.* (2004).

GENUS *PYCNONOTUS*   [Genus 34 spp/Region 1 introduced sp]

Pycnonotus "Kuhl" Boie, 1826. *Isis von Oken* 19(10): col. 973. Type, by monotypy, *Turdus capensis* Linnaeus.
    The species of this genus range from Africa to the Middle East and tropical Asia to Sundaland (Fishpool & Tobias 2005). This is a diverse assemblage of tree-dwelling songbirds; many species are crested, and all exhibit a medium-length bill and medium-long tail. Plumage generally includes black, brown, green, white, and red/yellow; many species exhibit a facial pattern. Bulbuls are noted songsters.

Sooty-headed Bulbul *Pycnonotus aurigaster*

Turdus aurigaster Vieillot, 1818. *Nouv. Dict. Hist. Nat.*, nouv. éd. 20: 258.—Java.

Apparently a small population has become established in Biak town on Biak I. First reported for Biak town on 28-Jan-1995 by Holmes (1997) and then by S. van Balen (*in litt.* 2012), who suggested they were introduced from Java or s Sulawesi, where the species is feral. Recently recorded from Sentani[j] and Timika[k] (G. Dutson *in litt.*[j], J. Pap *in litt.*[k]). Known from sightings only.

Extralimital—native range is mainland se Asia and Java. Naturalized in s Sulawesi (Fishpool & Tobias 2005).

SUBSPECIES

a. *Pycnonotus aurigaster ?aurigaster* (Vieillot)

RANGE: Native to Java and Bali. Introduced to Singapore, Sumatra, s Sulawesi, Biak town, Sentani, and Timika.

DIAGNOSIS: The nominate form can be identified by the golden-yellow undertail feathers, whitish tail tips, black crest, and dark gray-brown wings (Fishpool & Tobias 2005). Identified in NG from free-flying birds rather than specimens.

Family Phylloscopidae

Leaf-Warblers   [Family 77 spp/Region 3 spp]

This family group, originally recognized by Jerdon in 1863 (Bock 1994; see also Dickinson & Christidis 2014: 497), has arisen once again from the recent disarticulation of the old and all-encompassing "Sylviinae" of Peters's *Check-list*. It is a product of the refinements brought about by molecular systematics (*e.g.*, Alström *et al.* 2006, 2013). The family includes the genera *Phylloscopus* and *Seicercus*. Members of this family are tiny tree-dwelling warblers with delicate features; plumage is generally dominated by olive, gray, green, buff, chestnut, and yellow. Placement of the family follows Barker *et al.* (2004) and Alström *et al.* (2013). Among lineages inhabiting the Region, the white-eyes (Zosteropidae) are the nearest relative to the leaf-warblers.

GENUS *SEICERCUS*   [Genus 48 spp/Region 3 spp]

Seicercus Swainson, 1837. *Nat. Hist. Class. Birds* 2: 84, 259, fig. 229a. Type, by monotypy, *Cryptolopha auricapilla* Swainson = *Sylvia burkii* Burton.

The species of this genus range from Africa and western Europe east to Siberia and Alaska, and also from tropical Asia south to New Guinea and the Solomon Islands (Dickinson & Christidis 2014). Some northern forms winter to the south. This newly enlarged genus (encompassing some species of *Phylloscopus*) is characterized by small size, compact body form, and thin and abbreviated bill; also note the plain or yellow-washed plumage, darker dorsally, paler ventrally; usually with a dark eyeline and a pale supercilium, and in some sublineages a prominent pale eye-ring and black crown striping. Olsson *et al.* (2005) and Alstrom *et al.* (2013) provided results for an initial phylogeny of a selection of species in the genus, including *Seicercus poliocephalus*.

The Arctic Warbler, a Palearctic migrant long known as *Phylloscopus borealis*, has been dissected into several species, all of which are now placed in the genus *Seicercus* (Saitoh *et al.* 2008, 2010). One of the split-off forms, known as the Kamchatka Leaf-Warbler (*Seicercus examinandus*) winters in Wallacea (Coates 1990). This species has been recorded as a vagrant to the Kaniet (=Anchorite) Islands just north of the New Guinea region (Mayr 1955) and possibly to e Indonesia (Dickinson & Christidis 2014); it should be looked for in the Region as a rare vagrant during the northern winter season.

Island Leaf-Warbler *Seicercus poliocephalus*

Gerygone? poliocephala Salvadori, 1876. *Ann. Mus. Civ. Genova* (1875) 7: 960.—Arfak Mts, Bird's Head.

Inhabits canopy vegetation of forest and edge through the mts of NG, Yapen, Karkar, Fergusson, and Goodenough I; 1,400–2,400 m, min 640 m[CO].

Extralimital—Bismarck and Solomon Is (Coates 1990, Mayr & Diamond 2001, Dutson 2011), and west to Halmahera, Seram, and Buru (Bairlein 2006, Dickinson & Christidis 2014).

SUBSPECIES

a. *Seicercus poliocephalus poliocephalus* (Salvadori)

RANGE: Mountains of the Bird's Head and Wandammen Penin (Mayr 1941). Fakfak and Kumawa Mts populations may belong here (Diamond 1985). Birds from the s Kumawas appear close to this form but exhibit a yellowish rather than off-white eyebrow (C. Thébaud & B. Mila *in litt.*).
DIAGNOSIS: Crown blackish olive, without the pale median stripe (Rand & Gilliard 1967). Off-white eyebrow; dusky postocular; pale lower cheek and throat; bright yellow wash on breast; mantle brighter greenish than crown.

b. *Seicercus poliocephalus albigularis* (Hartert and Paludan)

Phylloscopus trivirgatus albigularis Hartert and Paludan, 1936. *Mitt. Zool. Mus. Berlin* 21: 218.—Mount Derimapa, Gebroeders Mts, w sector of the Western Ra. Replacement name for *Phylloscopus trivirgatus albigula* Rothschild, 1931, *Novit. Zool.* 36: 262; preoccupied by *Phylloscopus indicus albigula* Hesse, 1912.
> Synonym: *Phylloscopus trivirgatus paniaiae* Junge, 1952. *Zool. Meded. Leiden* 31: 248.—Araboebivak, Paniai Lake, Western Ra.

RANGE: The w sector of the Western Ra (Weyland Mts and Paniai Lake). The Yapen I population may belong here (Diamond 1985).
DIAGNOSIS: Dark crown with pale median streak; chin, throat, and side of head whitish (Rand & Gilliard 1967).

c. *Seicercus poliocephalus cyclopum* (Hartert)

Phylloscopus trivirgatus cyclopum Hartert, 1930. *Novit. Zool.* 36: 65.—Cyclops Mts, ne sector of NW Lowlands, n NG.
RANGE: The Cyclops Mts.
DIAGNOSIS: Like *giulianettii* but upperparts brighter and more yellowish green; median line atop crown yellow-washed (Rand & Gilliard 1967).

d. *Seicercus poliocephalus giulianettii* (Salvadori)

Gerygone giulianettii Salvadori, 1896. *Ann. Mus. Civ. Genova* 36 [=(2)16]: 81.—Moroka, s watershed of SE Peninsula.
RANGE: The Central Ra from the Baliem R[A] east to the mts of the SE Peninsula[MA]; also the PNG North Coastal Ra[K], Adelbert Mts[P], mts of Huon Penin[MA], and uplands of Karkar I[J]; 1,200–2,400 m (Rand 1942b[A], Diamond & LeCroy 1979[J], Pratt 1983[P], Diamond 1985[K]). Foja Mt population undiagnosed but possibly belongs here (Diamond 1985).
DIAGNOSIS: Similar to *albigularis* but dorsal plumage more greenish, less brownish olive (Rand & Gilliard 1967). See photo in Coates & Peckover (2001: 134).

e. *Seicercus poliocephalus hamlini* (Mayr & Rand)

Phylloscopus trivirgatus hamlini Mayr & Rand, 1935. *Amer. Mus. Novit.* 814: 8.—Goodenough I, D'Entrecasteaux Arch.
RANGE: Upland forests of Goodenough[MA] and Fergusson I[J] (Pratt field observation[J]). A single undiagnosed museum specimen (BPBM no. 184541) collected by Pratt presumably refers to this race.
DIAGNOSIS: Like *giulianettii*, but sides of crown much blacker and ventral plumage much more saturated with yellow.

NOTES

The distinctive insular forms from Biak and Numfor I, traditionally placed in *S. poliocephalus*, are here treated as two distinct species (see following species accounts). Dickinson & Christidis (2014) included these island populations in an NG-wide species that they called *S. maforensis*. They used *maforensis* as the species name because it is older than the name *poliocephalus* and has priority. Since we separate out the

Biak I species as distinct, the Dickinson & Christidis treatment does not apply here. *Other names*—Mountain or New Guinea Leaf-Warbler, *Phylloscopus poliocephalus*, *P. trivirgatus*.

Numfor Leaf-Warbler *Seicercus maforensis* ENDEMIC | MONOTYPIC

Gerygone maforensis A. B. Meyer, 1874. *Sitzungsber. Akad. Wiss. Wien* 70: 119.—Numfor I, Bay Is.
Numfor I.

NOTES

We give this very distinctive insular form full species status. Characters distinguishing the species from *S. poliocephalus* include: (1) entirely olive-gray crown and face; (2) crown stripe lacking; (3) sides of head gray with indistinct and abbreviated eyebrow stripe; (4) mantle grayish olive, only marginally brighter than crown; (5) obscure yellowish tips of greater coverts forming an obscure or obsolete wing bar; (6) primaries and secondaries edged with bright yellow-green; (7) throat and belly dirty white; (8) breast smudged gray; (9) olive flanks; (10) upper surface of tail dark brown; (11) overall plumage very dull and plain; (12) lower mandible horn-colored in museum specimens from Leiden and AMNH (Rand & Gilliard 1967, Bairlein 2006). It is a priority to deploy molecular techniques to examine the status of the forms on Numfor and Biak against those from the Mainland. *Other names*—until recently, treated as a subspecies of *Phylloscopus poliocephalus*; also formerly treated as *Phylloscopus trivirgatus*.

Biak Leaf-Warbler *Seicercus misoriensis* ENDEMIC | MONOTYPIC

Phylloscopus trivirgatus misoriensis Meise, 1931. *Novit. Zool.* 36: 318, note 1.—Biak I, Bay Islands. Replacement name for *Seicercus? trochiloides* Salvadori, 1876; preoccupied by *Acanthiza trochiloides* Sundervall, 1838.
Biak I.

NOTES

We give this very distinctive insular form full species status. Characters distinguishing this species from *S. poliocephalus* include the following: (1) upperparts dark olive green, duller on head; (2) median and greater wing coverts with yellowish tips creating two indistinct wing bars; (3) cheeks and throat yellow; (4) rest of underparts olive green, grayer on sides, and yellower on center of body; (5) photographs of living birds show a yellowish bill and legs (Rand & Gilliard 1967, S. van Balen *in litt.*). In addition, S. van Balen (*in litt.*) has noted that Biak birds are very different from any relatives, not only morphologically but behaviorally. The Biak photos seem to show well-developed rictal bristles; song much higher pitched than that of populations of *S. poliocephalus*. *Other names*—traditionally treated as a subspecies of *Phylloscopus poliocephalus/trivirgatus*.

Family Zosteropidae

White-eyes [Family 123 spp/Region 9 spp]

The white-eyes range from Africa and tropical Asia to Australia and many Pacific islands. White-eyes are tiny flocking birds that feed mainly on flowers and fruit in the forest canopy. The lineage is quite homogeneous and compact. Barker *et al.* (2004) show that the lineage is sister to the true timaliid babblers, and that that the clade including white-eyes and babblers is sister to the *Sylvia* warblers. This placement is confirmed by the phylogeny of Alström *et al.* (2006). Gelang *et al.* (2009) showed that *Zosterops* falls entirely within the babbler clade. Moyle *et al.* (2009) confirmed the lineage is a tight monophyletic cluster that exemplifies explosive insular speciation. The genus *Zosterops* is shown to be sister to species of *Stachyris* and *Yuhina* (Gelang *et al.* 2009). We follow Dickinson & Christidis (2014) in recognizing an expanded Zosteropidae (encompassing *Yuhina*) as a freestanding family. Placement of the family follows Barker *et al.* (2004) and Alström *et al.* (2013). Within the Region, the closest family to the Zosteropidae is the Phylloscopidae (Alström *et al.* 2013).

Zosterops Vigors & Horsfield, 1827. *Trans. Linn. Soc. London* 15: 234. Type, by subsequent designation (Lesson, 1828. *Man. Orn.* 1: 286), *Motacilla maderaspatana* Linnaeus.

The species of this genus range from Africa, Madagascar, the Indian Ocean islands, and tropical Asia east to the Russian far east and to Micronesia, New Guinea, Australia, Samoa, Lifou, Norfolk Island, and New Zealand (van Balen 2008). The genus is characterized by small size, compact body form, small pointed bill, white eye-ring, and primarily olive, yellow, and white plumage. We follow the species sequence of van Balen (2008). Phylogenetic relationships of some of the Region's species are treated in Moyle *et al.* (2009), but much more work is desired, as the widespread regional species display curious patterns of distribution, suggesting unappreciated phylogeographic and systematic complexity. The genus epithet is treated as masculine (ICZN 1999, Art. 30.1.4.3).

Lemon-bellied White-eye *Zosterops chloris* RESIDENT

Z[*osterops*] *chloris* Bonaparte, 1850. *Consp. Gen. Av.* 1: 398.—Banda Is, s Moluccas.

A small-island tramp species known from islets fringing the main NW Islands and the Aru Is (Mayr 1941).

Extralimital—small islands of c and e Indonesia and Sulawesi (van Balen 2008).

SUBSPECIES

a. *Zosterops chloris chloris* Bonaparte

RANGE: Babi, Karang, and Enu I of the Aru Is; Schildpad Is near Misool; Wai I (north of Batanta; Beehler field notes 29-Feb-2004, 13-Mar-2005); Soa I near Little Kai; small islands from Kai to Gissar and Banda Is (Mayr 1941). A tramp species restricted to small islands in e Indonesia. Mees (1965) noted that the von Rosenberg specimens supposedly originating from Misool are of suspect provenience.

DIAGNOSIS: Crown and upperparts yellowish olive; throat bright yellow; rest of underparts various shades of yellow; flanks yellow-olive; cheek yellow-olive; white eye-ring (van Balen 2008).

Other names—Moluccan, Yellow, or Mangrove White-eye, some treatments included *Z. citrinellus.*

[Pale White-eye *Zosterops citrinellus* HYPOTHETICAL

Z[*osterops*] *citrinella* Bonaparte, 1850. *Consp. Av.* 1: 398.—Timor, Lesser Sundas.

No records from the NG region to date, but inhabits nearby islands of the Torres Strait and may be expected to stray to Daru I or the coastal Trans-Fly.

Extralimital—Torres Strait Is and small islands off Cape York, QLD; also the Lesser Sundas (van Balen 2008).

SUBSPECIES

a. *Zosterops citrinellus albiventris* Reichenbach

Zosterops albiventris Reichenbach, 1852. *Handb. Spec. Orn., Abth.* 2, Meropinae, continuatio IX: 92 (based on Hombron & Jacquinot, *Voy. Pôle Sud*: pl. 19, fig. 3).—Warrior I, Torres Strait.

RANGE: The Torres Strait Is (Warrior, Deliverance, Cairncross); small islands on the coast of Cape York, QLD (Mayr 1941); and islands of the Lesser Sundas (Bangs & Peters 1926, Higgins *et al.* 2006, van Balen 2008).

DIAGNOSIS: "Hardly distinguishable from the nominate form, but does exhibit a heavier and larger bill" (van Balen 2008). Plumage soft, and bill long, large, and heavy (Higgins *et al.* 2006).

NOTES

Zosterops citrinellus inhabits the small islands in the Torres Strait south of the Trans-Fly and thus could be expected as a vagrant to coastal s NG (Draffan *et al.* 1983). We follow LeCroy (2011: 16) in treating the species name as an adjective (*pace* David & Gosselin 2002a), which must agree in gender with the genus, which is masculine. *Other names*—Ashy-bellied White-eye, *Zosterops citrinella.*]

Black-fronted White-eye *Zosterops atrifrons*

Zosterops atrifrons Wallace, 1864. *Proc. Zool. Soc. London* (1863): 493.—Manado, n Sulawesi.

Inhabits hill forest of the Bird's Head, Bird's Neck, S Lowlands, Huon Penin[M], and SE Peninsula, 500–1,700 m (Freeman & Freeman 2014[M]). Range of this species is exclusive of the range of the following (sister) species.

Extralimital—Moluccas to Sulawesi (though, surprisingly, absent from the NW Islands; Coates & Bishop 1997).

SUBSPECIES

a. *Zosterops atrifrons chrysolaemus* Salvadori

Zosterops chrysolaema Salvadori, 1876. *Ann. Mus. Civ. Genova* (1875) 7: 954.—Arfak Mts, Bird's Head.

> Synonym: *Zosterops minor tenuifrons* Greenway, 1934. *Proc. New Engl. Zool. Club* 14: 3.—Wau, 1,128 m,
> Morobe distr, SE Peninsula.

RANGE: Mountains of the Bird's Head east through the s watershed of the Western, Border, and Eastern Ra, east to range of *delicatulus*. Also mts of the n watershed of the SE Peninsula. Presumably the undiagnosed populations observed in the Wandammen, Fakfak, and Kumawa Mts belong here (Diamond 1985).

DIAGNOSIS: Forehead blackish (Tamrau Mts, plus the Bird's Neck and eastward) or dark green (Arfak Mts); lores with blackish wash; eye-ring broad; throat patch deep yellow and truncated; alula dark brown.

b. *Zosterops atrifrons gregarius* Mayr

Zosterops minor gregaria Mayr, 1933. *Orn. Monatsber.* 41: 53.—Sattelberg, Huon Penin.

RANGE: The Adelbert Mts and mts of the Huon Penin (Mayr 1931, Gilliard & LeCroy 1967).

DIAGNOSIS: Forehead greenish; lores darkish; eye-ring rather narrow but bright and distinct; throat medium yellow, closer to nominate than *delicatulus* to the southeast; back bright yellow-green like that of nominate. A specimen collected by Beck at Hamboa (Finisterre Ra) at 915 m on 12-Feb-1929 has characters that tend more toward *chrysolaemus*, giving an indication of possible introgression. Also, the song of the Huon birds is the same as that of *chrysolaemus* from Wau, indicating the close relationship.

c. *Zosterops atrifrons delicatulus* Sharpe

Zosterops delicatula Sharpe, 1882, *Journ. Linn. Soc. London, Zool.* 16: 318 and 440.—Astrolabe Ra, SE Peninsula.

> Synonym: *Zosterops delicatula pallidogularis* Stresemann, 1931. *Mitt. Zool. Mus. Berlin* 17: 222.—Fergusson I,
> D'Entrecasteaux Arch, SE Islands.

RANGE: The s watershed of the mts of the SE Peninsula[L]; and Goodenough[MA], Fergusson[MA], and Normanby I[P] (Gilliard 1950[L], Pratt field notes 2003[P]).

DIAGNOSIS: Forehead, lores, forecrown, and forecheek blackish; eye-ring broad and prominent; mantle dark greenish, distinct from bright yellow-green rump; throat pale yellow without any rich orange wash as found in *chrysolaemus*. See photo in Coates & Peckover (2001: 195).

NOTES

We have placed the populations with the dark foreheads and lores and broad eye-rings into *Z. atrifrons*. Fieldworkers need to look at character distributions displayed by populations between the e Huon uplands and the Snake R mts/Mt Shungol of the n Herzog Mts for evidence of hybridization with *Z. minor*. We have no doubt that *atrifrons* and *minor* are sister forms and closely related, but we think the character differences are of a magnitude to merit species-level distinction. *Z. a. delicatulus* from the SE Peninsula is virtually identical to *Z. a. sulaensis* from the Sula Is of Wallacea. It would be worthwhile to conduct molecular studies of populations of *minor* and *atrifrons* near their points of geographic contact in the hills east of the Bird's Neck region, and in ne NG between the e terminus of the PNG North Coastal Ra east to the mts of the Huon Penin and the Herzog Mts.

Zosterops sharpei Finsch (*Z. minor sharpei* in Mayr 1941: 217) described in error as from the Aru Is, is a junior synonym of *Z. a. atrifrons* and actually based on specimens from the n peninsula of Sulawesi (see Mees 1961: 64). *Other names*—earlier authorities typically combined the forms assembled here with those of *Z. minor*.

Green-fronted White-eye *Zosterops minor* ENDEMIC

Zosterops albiventer minor A. B. Meyer, 1874. *Sitzungsber. Akad. Wiss. Wien* 70: 115.—Yapen I, Bay Is.

A hill-forest species that ranges from Yapen I through the n watershed of NG from the head of Geelvink Bay east presumably to the e sector of the PNG North Coastal Ra; 750–1,400 m.

SUBSPECIES

a. *Zosterops minor rothschildi* Stesemann & Paludan
Zosterops minor rothschildi Stesemann and Paludan, 1934. *Orn. Monatsber.* 42: 44.—Mount Derimapa, Gebroeders Mts, w sector of the Western Ra.
RANGE: The w sector of the Western Ra.
DIAGNOSIS: Throat rich deep yellow as in *minor*; eye-ring medium width; lores, all upperparts bright yellow-green; iris reddish brown; The broader eye-ring may be evidence of some introgression with *Z. atrifrons* of the Bird's Neck. The unique type was taken by Fred Shaw Meyer on 29-Jun-1930, from Mt Derimapa, Gebroeders Mts. Bill black, feet gray; female adult ova not enlarged (holotype, AMNH). For additional discussion of this type, see LeCroy (2011).

b. *Zosterops minor minor* A. B. Meyer
RANGE: Yapen I and uplands of the NW Lowlands, Sepik-Ramu, and n scarp of the c sector of the Central Ra; also the Foja Mts[J], Cyclops Mts[K], and PNG North Coastal Ra[L] (Mayr 1941, Hartert 1930[K], Beehler *et al.* 2012[J], specimens at AMNH[L]).
DIAGNOSIS: Narrow and indistinct white eye-ring; rich egg-yolk-yellow throat patch offset from white upper breast; very bright yellow-green upperparts; rump quite bright; lack of any indication of darkening or blackening of the forehead. See photos in Coates (1990: 331) and Coates & Peckover (2001: 195).

NOTES

Species characters: forehead green; eye-ring narrow or indistinct; throat bright yellow. In general, we follow the treatment of Mees (1961) in recognizing *Z. atrifrons* as a widespread form ranging from Sulawesi to the SE Peninsula, but we split off as a distinct local species the populations *minor* and *rothschildi* as a northwestern vicariant that we recognize as the species *Z. minor*. Mees treated both distinct groups as *atrifrons*, and van Balen (2008) treated the two distinct NG groupings as *minor*, distinct from the Indonesian *atrifrons*. The geographic range of *minor* is defined by recognized NG distributional barriers and thus is biogeographically plausible. *Other names*—traditionally subsumed within Black-fronted White-eye, *Z. atrifrons*.

Tagula White-eye *Zosterops meeki* ENDEMIC | MONOTYPIC

Zosterops meeki Hartert, 1898. *Novit. Zool.* 5: 528.—Tagula I, Louisiade Arch, SE Islands.

Mainly in upland forests of Tagula I in the Louisiade Arch, sea level–700 m. Canopy dwelling and difficult to observe. Song and habitat preference similar to the presumed sister species *Z. atrifrons*.
Other names—White-throated White-eye.

Biak White-eye *Zosterops mysorensis* ENDEMIC | MONOTYPIC

Zosterops mysorensis A. B. Meyer, 1874. *Sitzungsber. Akad. Wiss. Wien* 70: 116.—Korido, Biak I, Bay Is.

An uncommon and unobtrusive inhabitant of forest on Biak I, sea level–675 m (van Balen 2008).

NOTES

This species and *Z. atriceps* of Halmahera, Moluccas, both exhibit a white throat and both inhabit small offshore islands. These are putative sister forms (van Balen 2008). *Other names*—Soepiori White-eye.

Capped White-eye *Zosterops fuscicapilla* ENDEMIC | MONOTYPIC

Zosterops fuscicapilla Salvadori, 1876. *Ann. Mus. Civ. Genova* (1875) 7: 955.—Arfak Mts, Bird's Head.

Frequents forest and edge of the Bird's Head[J], Bird's Neck (Fakfak[J], Kumawa[K], and Wandammen Mts[J]), Western and Border Ra[MA], and westernmost sector of the Eastern Ra east to Tari[CO]; also Foja Mts[J], Cyclops Mts[MA], and PNG North Coastal Ra[J]; 1,200–1,850 m, extremes 750–2,200 m[CO] (Diamond 1985[J], Diamond & Bishop 2015[K]).

Zosterops fuscicapilla is here treated as monotypic, as we have elevated *Z. crookshanki* to species status (see following account). Based on David & Gosselin's (2002a: 38) interpretation of species epithets ending in *capilla*, this species name should remain invariant. Female wing chord: 55.5, 56.5, 59. *Other names*—Western Mountain or Yellow-bellied Mountain White-eye, *Zosterops fuscicapillus*.

Oya Tabu White-eye *Zosterops crookshanki*   ENDEMIC | MONOTYPIC

Zosterops fuscicapilla crookshanki Mayr & Rand, 1935. *Amer. Mus. Novit.* 814: 16.—Goodenough I, D'Entrecasteaux Arch, SE Islands.
RANGE: Montane forests of Goodenough and Fergusson I.

NOTES

We treat this upland island endemic as a full species. Reasons for distinguishing it from *Z. fuscicapilla* include the following: (1) isolated island range far east of nearest population of true *fuscicapilla*; true *fuscicapilla* occurs on no islands anywhere in its range; (2) broader white eye-ring; (3) pearl-gray iris *vs.* brown iris in *fuscicapilla*; (4) greenish crown and hind-crown on Goodenough, but blackish on Fergusson; (5) alula green, not black. Female wing (AMNH specimen no. 222113) 58.5; tail 41.5; tarsus 17; bill from base: 14 (wing chord not distinct from that of *fuscicapilla*). Known from only two specimens in the type series collected on Goodenough I and three specimens collected on Fergusson I by T. K. Pratt. There appears to be a difference between the two island populations in the color of the crown, but we consider the species monotypic for the present. Oya Tabu is the local name for Mt Kilkerran, where the white-eye was discovered on Fergusson I. It translates approximately as "Forbidden (or Sacred) Mountain," perhaps because the islands' peaks are shunned by the people of the D'Entrecasteaux Arch because they are the home of traditional spirits (Brass 1956).

New Guinea White-eye *Zosterops novaeguineae*   ENDEMIC

Zosterops novaeguineae Salvadori, 1878. *Ann. Mus. Civ. Genova* 12: 341.—Arfak Mts, Bird's Head.
     Range discontinuous: Mts of the Bird's Head (above 1,000 m?) and Bird's Neck (Kumawa Mts but not the Fakfaks[J], 600–1,200 m); lowlands of the Aru Is, Trans-Fly, and lower Sepik (at Awar); e sector of the Eastern Ra; and mts of the Huon and SE Peninsulas; 1,200–2,400 m, extremes 750–2,600 m[K] (Diamond 1985[J], Coates & Peckover 2001[K]). Absent from all n coastal ranges and from the w two-thirds of the Central Ra. Inhabits upland forest and edge and garden habitats. Montane populations of the SE Peninsula are replaced at lower elevations in the hills by *Z. atrifrons*.

SUBSPECIES

a. *Zosterops novaeguineae novaeguineae* Salvadori
RANGE: Mountains of the Bird's Head and Bird's Neck (Kumawa Mts).
DIAGNOSIS: Eye-ring narrow but complete and white; throat lemon yellow and with an indistinct posterior border of gray-green; forehead yellow-green; lores greenish; crown to rump bright yellow-green; bend of wing with pale yellow spot; alula dark brown.

b. *Zosterops novaeguineae wuroi* Mayr & Rand
Zosterops novaeguineae wuroi Mayr & Rand, 1935. *Amer. Mus. Novit.* 814: 16.—Wuroi, Oriomo R, Trans-Fly, s NG.
     Synonym: *Zosterops novaeguineae aruensis* Mees, 1953. *Zool. Meded. Leiden* 32: 26.—Wokan and Kobroor, Aru Is.
RANGE: The Aru Is and the Trans-Fly.
DIAGNOSIS: Small; greener above than other races; lores and forehead greenish like crown; eye-ring medium broad; throat lemon yellow, not extensive, and grading into gray-white of breast. We have combined *aruensis* into this form because of minimal plumage differences.

c. *Zosterops novaeguineae wahgiensis* Mayr & Gilliard
Zosterops novaeguineae wahgiensis Mayr & Gilliard, 1951. *Amer. Mus. Novit.* 1524: 14.—Nondugl, Wahgi valley, Eastern Ra.

Synonym: *Zosterops novaeguineae shaw-mayeri* Mayr & Gilliard 1951: 14.—Yandara, se of Mt Wilhelm summit, Eastern Ra.

RANGE: The Eastern Ra (van Balen 2008).

DIAGNOSIS: Throat medium yellow, extending well down onto upper breast; eye-ring medium width; no blackish in lores or forehead; pale spot at bend of wing; alula dark brown. Mees (1961) subsumed the race *shawmayeri* into *wahgiensis*. This lumped race was described from a single specimen (as was *magnirostris*, by Mees). See photo in Coates (1990: 331) and Coates & Peckover (2001: 195).

d. *Zosterops novaeguineae magnirostris* Mees

Zosterops novaeguineae magnirostris Mees, 1955. *Zool. Meded. Leiden* 34: 153.—Awar, ne Sepik-Ramu.

RANGE: Known only from Awar on the n coast of NG opposite Manam I, ne coast of the Sepik-Ramu (Mees 1955); also possibly in the Adelbert Mts (Pratt 1983).

DIAGNOSIS: Size small; large bill; pale yellow throat with greenish tinge; yellow median streak down mid-line of belly (van Balen 2008). It is difficult to know what to make of this race known from but a single specimen. New races, in general, should not be erected from single specimens of birds out of range, especially when the race has been judiciously passed over for description by earlier researchers. Nearest other populations come from the Huon Penin (to the southeast) and the Bismarck Ra (south-southeast).

e. *Zosterops novaeguineae oreophilus* Mayr

Zosterops novaeguineae oreophila Mayr, 1931. *Mitt. Zool. Mus. Berlin* 17: 671.—Junzaing, Huon Penin.

RANGE: Mountains of the Huon Penin (Mayr 1931).

DIAGNOSIS: Dull yellow-green on forehead and above lores; eye-ring two-tone, white, with upper half grayish; yellow of throat pale and not very extensive, grading posteriorly into gray-white of upper breast; undertail pale yellow; pale spot at bend of wing; alula dark brown.

f. *Zosterops novaeguineae crissalis* Sharpe

Zosterops crissalis Sharpe, 1884. *Cat. Birds Brit. Mus.* 9: 165.—Astrolabe Ra, s watershed of the SE Peninsula.

RANGE: Mountains of the SE Peninsula.

DIAGNOSIS: Upperparts strongly washed with yellow, with forehead distinctly yellow near base of bill; eye-ring fairly broad and prominent; pale spot at bend of wing; alula brown; rump bright yellow-green. See photo in Coates (1990: 331).

NOTES

Any green-crowned Mainland populations above 1,600 m can be safely identified as this species by the lemon-yellow throat patch, the posterior edge of which grades into the gray-white of the upper breast (but sharply divided in birds from the Bird's Head and in lowlands). Also distinguished by song. *Other names*—Mountain or Papuan White-eye, New Guinea Mountain White-eye.

Louisiade White-eye *Zosterops griseotinctus*   RESIDENT

Zosterops griseotincta G. R. Gray, 1858. *Proc. Zool. Soc. London*: 175.—Duchateau Is, Louisiade Arch, SE Islands.

Scattered across the SE Islands region, mainly on islets, but including one large island, Rossel; range encompasses small islands off Normanby I, the Deboyne Is, Conflict Is, Bonvouloir Is, and Louisiades; sea level–300 m.

Extralimital range includes the Bismarck Arch: Crown, Long, Tolokiwa, Nauna, and Nissan I (Mayr & Diamond 2001, Dutson 2011).

SUBSPECIES

a. *Zosterops griseotinctus longirostris* Ramsay

Zosterops longirostris Ramsay, 1879. *Proc. Linn. Soc. NSW* 3: 288.—Heath I, off the e tip of the SE Peninsula.

RANGE: Islets in the e and n sectors of the SE Islands, including Heath I, East and Hastings I of the Bonvouloir Is, and Alcester I, near Woodlark (Mayr 1941).

DIAGNOSIS: Bill long; feet dark; culmen 19 (Rand & Gilliard 1967).

b. *Zosterops griseotinctus griseotinctus* G. R. Gray

Synonym: *Zosterops aignani* Hartert, 1899. *Novit. Zool.* 6: 210.—Kimuta I, Louisiade Arch.

RANGE: Kimuta I and the Deboyne Is, Duchateau I, and Conflict Is[K] of the SE Islands (Mayr 1941, LeCroy & Peckover 1998, Beehler field notes 29-Sep-2002[K]). Population on Panasesa I of the Conflict Is remarkably dense—the dawn chorus of this species is overpowering when heard from a boat offshore (Beehler field notes 29-Sep-2002).

DIAGNOSIS: Bill shorter than that of *longirostris*; culmen 17 (Rand & Gilliard 1967). Race *aignani* combined with nominate following Mees (1961) and van Balen (2008). For additional background on the type specimens and actual location of collection on Kimuta I, an islet off Misima, see LeCroy (2011).

c. *Zosterops griseotinctus pallidipes* De Vis

Zosterops pallidipes De Vis, 1890. *Ann. Rept. Brit. New Guinea*, 1888–89: 60.—Rossel I, Louisiade Arch, SE Islands.

RANGE: Rossel I.

DIAGNOSIS: Feet pale, flesh-colored (Rand & Gilliard 1967); darker lores and more yellowish overall than nominate (van Balen 2008).

NOTES

The species ranges westward to Duchess and Tobweyama I (off Normanby I); these undiagnosed populations (Pratt specimens at BPBM) should be looked for elsewhere in the D'Entrecasteaux and Trobriand Is. See Linck *et al.* (2015) for a molecular phylogeography of the populations of this species.

Other names—Island White-eye, Louisiades White-eye, *Zosterops griseotincta*.

Family Acrocephalidae

Reed-Warblers [Family 59 spp/Region 2 spp]

The species of the family range from Britain and Europe, eastern and southern Africa, Madagascar, and the Mascarene Islands, east to Siberia, tropical Asia, and the Pacific. The family includes seven genera: *Acrocephalus, Arundinax, Hippolais, Chloropeta, Iduna, Calamonastides*, and *Nesillas* (Fregin *et al.* 2009, Alström *et al.* 2006, 2011, Dickinson & Christidis 2014). Many species breed in the north temperate zone and winter southward (Fregin *et al.* 2009). Generally medium-size songbirds, typically unstreaked, all are drably colored in buffs and browns, darker dorsally. Many are quite vocal. Among local families, this lineage is closest to the Locustellidae, the grassbirds (Alström *et al.* 2013).

GENUS *ACROCEPHALUS* [Genus 42 spp/Region 2 spp]

Acrocephalus J. A. & J. F. Naumann, 1811. *Naturgesch. Land-Wasser-Vögel Nördl. Deutschl., Nachtrag*: 199. Type, by subsequent designation (G. R. Gray, 1840. *List Gen. Birds*: 21), *Turdus arundinaceus* Linnaeus.

The species of this genus range from western Europe east to Siberia and south to Africa, Australia, and the tropical Pacific (Bairlein 2006, Cibois *et al.* 2011). The genus is characterized by very plain plumage, pale eyebrow, pale underparts, and loud and harsh song. Cibois *et al.* (2001) have noted that the molecular evidence indicates there have been several independent colonizations of the Pacific by this lineage. Fregin *et al.* (2009) provided additional molecular systematic detail for the genus as a whole.

Oriental Reed-Warbler *Acrocephalus orientalis* VISITOR | MONOTYPIC

Salicaria turdina orientalis Temminck & Schlegel, 1847. In Siebold, *Fauna Japonica, Aves*: 50, pl. 20b.—Japan.

A rare northern migrant to the Region, ranging from near sea level up to 550 m. Should be looked for between September and May (Coates 1990). The records include one specimen from the Kebar valley, Bird's Head, early 1962 (LeCroy 1969, Hoogerwerf 1971); also specimens probably from Biak I and the Aru Is (LeCroy 1969); a sight record from Aroa lagoon, Central Prov, 1-Dec-1979 (Finch 1980b); an unsubstantiated sighting attributed to *A. arundinaceus*, Bensbach R, 17-Nov-1985 (Finch 1988a); a singing bird at Lake Ambuve, in the middle Fly R (Gregory *et al.* 1996c); and a bird captured and photographed at Waigani swamp, Central Prov, 16-Oct-1986 (in *Muruk* 3: 53–54, this bird was claimed, erroneously, to have been *Locustella fasciolata*; see Coates 1990). Frequents grasses and thickets in moist areas and near water. (Coates 1990).

Extralimital—breeds from se Siberia to Japan and se China, wintering to Bangladesh, Indochina, Philippines, and Indonesia east to the Lesser Sundas (Bairlein 2006).

NOTES

Acrocephalus orientalis should be looked for in the westernmost of the NW Islands. Larger than the following species (which breeds in NG), *A. orientalis* is identified by the following: wing 75–95; bill more stout; pink gape; cold brown mantle; longer and more conspicuous pale superciliary stripe; paler underparts, with an indication of dark streaking on lower throat and breast; lower mandible pinkish or yellowish pink (Coates 1990). Also note the distinctive song of croaking, guttural, scratchy notes, very different from song of *A. australis* (D. Hobcroft *in litt.*). *Other names*—Eastern Great Reed-warbler.

Australian Reed-Warbler *Acrocephalus australis*   RESIDENT (AND VISITOR?)

Calamoherpe australis Gould, 1838. In Lewin, *Nat. Hist. Birds NSW*, index to synonyms to pl. 18.—Paramatta, NSW, Australia.

Common though patchily distributed throughout the NG lowlands (mainly in s NG) and a number of mid-montane valleys; sea level–2,300 m. Inhabits stands of reeds fringing wetlands; also tall dense grass, bamboo, thick scrub, and tall cane grass near water or in damp situations

Extralimital—west to Buru and Sumba, Timor, Australia, Long I, Umboi I, New Britain, New Ireland, and Solomon Is (Coates 1990, Mayr & Diamond 2001, Higgins *et al.* 2006, Dutson 2011, Dickinson & Christidis 2014).

SUBSPECIES

a. *Acrocephalus australis sumbae* Hartert

Acrocephalus stentoreus sumbae Hartert, 1924. *Treubia* 6: 21.—Nangamesi Bay, near Waingapu, Sumba, Lesser Sundas.

Synonym: *Acrocephalus meyeri* Stresemann (ex Neumann ms), 1924. *Orn. Monatsber.* 32: 168.—Toriu R, Gazelle Penin, New Britain.

RANGE: Widely distributed but very local in NG. Localities—Anggi Lakes[MA], Paniai Lake[CO], Baliem valley[CO], Ilaga[L], Mamberamo R[CO], Lorentz R[MA], Dolak I[J], Merauke[K], Lake Daviumbu[MA], Bensbach R[CO], ?Chambri Lakes[CO], Lake Kopiago[CO], near Pureni[CO], Kandep Lakes[CO], Wahgi and Baiyer valleys[CO], Ukarumpa[CO], Herzog Mts[CO], Popondetta[CO], Port Moresby[CO], Agaun[CO], and Orangerie Bay[MA] (Ripley 1964a[L], Mees 1982b[K], Silvius & Taufik 1989[J]). Few records from the n lowlands. Extralimital—west to Buru and Sumba (Watson 1986); northeast to the Bismarck Arch and east to the Solomon Is (Dutson 2011).

DIAGNOSIS: Includes *meyeri* of the Bismarck Arch and Solomon Is (see Watson 1986). Wing 61–73; tail 57–69; bill from base 19.8–22.4; tarsus 24–26; upperparts and sides of head buff brown; crown duller; rump brighter; pale eyebrow; throat and breast buffy white, palest on throat and center of belly; flanks buffy; lower mandible yellowish orange; inside of mouth yellow or salmon; iris brown; legs gray (Coates 1990). See photo in Coates & Peckover (2001: 134).

b. *Acrocephalus australis ?australis* (Gould)

RANGE: Possibly an occasional migrant from Australia to s NG, as evidenced by presence in the Torres Strait Is (Schodde & Mason 1999).

DIAGNOSIS: Mantle mid-brown; underparts tawny to cream; wing pointed; wing 67–73 (Schodde & Mason 1999).

NOTES

Taxonomic status of the species of breeding reed-warblers in NG has been problematic (Mayr 1941, Watson 1986, Beehler & Finch 1985, Coates 1990). Currently, *A. australis* is considered, on molecular evidence, as distinct from *A. stentoreus* of Eurasia, Africa, and Wallacea (Bairlein 2006). The taxon to which the NG breeding population is assigned is *sumbae*. Samples of *sumbae* from the Solomon Is are genetically nested within *australis*, not *stentoreus* (Cibois *et al.* 2011), but DNA samples of *sumbae* from NG have not yet been analyzed. Bairlein (2006) also noted that the race *sumbae* has been combined into *celebensis* by some authors. *Other names*—various authories at times treated this form as *A. arundinaceus* or *A. stentoreus*.

Family Locustellidae

Grassbirds and Allies [Family 57 spp/Region 5 spp]

This small, novel family, recognized because of recent molecular systematics, includes the genera *Bradypterus*, *Megalurus*, *Schoenicola*, and *Locustella* and ranges across tropical Africa, Eurasia, and Australasia (Alström *et al.* 2011). Bock (1984) showed that the name Locustellidae had priority over the name Megaluridae. Members are cryptically plumaged, medium-small, thicket-dwelling birds, difficult to distinguish as a group from the preceding family. Among regional families, the grassbirds are most closely allied to Acrocephalidae, the reed-warblers (Alström *et al.* 2013).

GENUS *LOCUSTELLA* [Genus 24 spp/Region 1 sp]

Locustella Kaup, 1829. *Skizz. Entw.-Gesch. Europ. Thierwelt*: 115. Type, by tautonymy, *Sylvia locustella* Latham = *Motacilla naevia* Boddaert.

The species of this genus range from western Europe to Siberia and eastern Asia, wintering south to Africa and tropical Asia (Bairlein 2006, Alström *et al.* 2011). The genus is characterized by the very dull buff-gray plumage, paler on the belly; and the lack of a blackish eyebrow but usually an indication of a whitish eyebrow. Morphologically, it is difficult to distinguish from some species of *Acrocephalus*, which typically exhibit a more strongly patterned face and crown.

Gray's Grasshopper-Warbler *Locustella fasciolata* VISITOR

Acrocephalus fasciolatus G. R. Gray, 1861. *Proc. Zool. Soc. London* (1860): 349.—Bacan I, n Moluccas.

A regular winter visitor to the NG region, with most records from the w half of the island, including the Bird's Head and Neck, NW Islands, and Bay Is. Few records from e NG. Winters in rank forest-edge regrowth, grasslands, and abandoned gardens, sea level–1,800 m. Present in the Region from early October to April (Coates 1990).

Localities—Gebe[Z,M], Misool[M], Salawati[M], Waigeo[M], Kumawa Mts[A], Biak[Q], Yamna[CO], Manokwari[MA], and Yapen I[Q]; also Kebar valley[Y], Bird's Head[CO], Onin Penin[MA], Kunupi in Weyland Mts[L], Paniai Lake[W], Nabire[Y], Yamna I[MA], Depapare, Kiunga[U], upper Fly R[K], Bewani foothills[CO], Lake Kutubu[J], and Waigani swamp[CO] (Hartert *et al.* 1936[L], Junge 1953[W], Mees 1965[M], 1972[Z], King 1979[Q], Finch 1985[U], Diamond 1986[Y], Gregory 1997[K], Diamond & Bishop 2015[A], K. D. Bishop *in litt.*[J]).

Extralimital—breeds from Siberia east to n Japan and Manchuria; migrates through Japan and China, wintering in the Philippines, Sulawesi, and Moluccas. One specimen record from New Britain, 24-Dec-1988 (W. E. Boles *in litt.*). Also a record from New Ireland (Diamond 1986).

NOTES

We follow Dickinson & Christidis (2014) in considering the birds wintering in NG to be the monotypic *L. fasciolata*, not *L. amnicola*. Coates (1990) noted that a record from Waigani swamp on 16-Oct-1986 (*Muruk* 3:53-54) was apparently an *Acrocephalus* misidentified as a *Locustella*. *Other names*—Gray's Warbler.

Megalurus Horsfield, 1821. *Trans. Linn. Soc. London* 13: 158. Type, by monotypy, *Megalurus palustris* Horsfield.

The species of this genus range from eastern Russia, Japan, and China south to New Guinea, Australia, and New Zealand (Bairlein 2006). The genus is characterized by the whitish ventral plumage, brownish dorsal surface with prominent darker streaking, and the long tail with pointed rectrices. All possess a pale eyebrow. Following the guidance of Alström *et al.* (2011) and P. Alström (*in litt.*), we maintain a single genus for the four grassbirds of NG (*pace* Dickinson & Christidis 2014). This is the conservative treatment. We await additional molecular analysis before taking the step to dissect this well-known genus.

Tawny Grassbird *Megalurus timoriensis*   RESIDENT

Megalurus timoriensis Wallace, 1864. *Proc. Zool. Soc. London* (1863): 489.—Timor, Lesser Sundas.

A widespread Australasian species that within the Region is known only from the lowland grasslands of the middle Fly region of NG.

Extralimital—the Philippines, Timor, Sulawesi, and the Moluccas, south to n and e Australia. Inhabits tall, dense grass in open country.

SUBSPECIES

a. *Megalurus timoriensis muscalis* (Rand)

Megalurus timoriensis muscalis Rand, 1938. *Amer. Mus. Novit.* 991: 4.—Lake Daviumbu, middle Fly R, s NG.

RANGE: Known only from grasslands of the middle Fly R and the Trans-Fly of Indonesian NG. Localities—Dolak I[J], Kurik[J], Wasur NP[J], Bian Lakes[J], Paal Putih[K], Bensbach[J], Lake Daviumbu[J] (Bishop 2005a[J], Hoogerwerf specimen in the Mus. Leiden[K]).

DIAGNOSIS: Small; wing (male) 64; crown with pronounced streaking (Rand & Gilliard 1967). See photo in Coates & Peckover (2001: 135).

NOTES

We follow Schodde & Mason (1999), Bairlein (2006), and Dickinson & Christidis (2014) in separating the NG species (*M. macrurus*) from the widespread *M. timoriensis*. Species characters are outlined by Schodde & Mason (1999: 702): species *timoriensis* has underparts with warm tawny cast, flanks unstreaked, immatures without yellowish-washed underparts, tail proportionally short (tail-to-wing ratio 1.15 to 1.38); species *macrurus* has underparts washed cold pale gray-brown, flanks variably dusky-streaked, immatures washed ventrally with yellow, tail proportionally long (tail-to-wing ratio 1.40 to 1.52). *Other names—Cincloramphus timoriensis.*

Papuan Grassbird *Megalurus macrurus*   RESIDENT

Sphenoeacus macrurus Salvadori, 1876. *Ann. Mus. Civ. Genova* 9: 35.—Naiabui, Hall Sound, SE Peninsula.

All NG but for much of the c sector of the S Lowlands and the Trans-Fly. Inhabits lowlands, middle elevations, and alpine zone (4,000 m[CO]). Apparently co-occurs with *M. timoriensis* in the middle Fly R region of PNG without intergradation (Schodde & Mason 1999). Observed at 1,950 m in grasslands on the n slope of Mt Oiamadawa'a on Goodenough I on 21-Mar-1976 (Beehler field notes). Perhaps absent from much of NG's lowland grassland zones (Sepik, Markham, Fly).

Extralimital—Presumably the birds in the Bismarck Arch refer to this species (Bairlein 2006, Dutson 2011).

SUBSPECIES

a. *Megalurus macrurus stresemanni* (Hartert)

Megalurus timoriensis stresemanni Hartert, 1930. *Novit. Zool.* 36: 79.—Kofo, Anggi Lake, Bird's Head.

RANGE: Mid-montane grasslands of the Arfak Mts of the Bird's Head. Known only from fern groves at the Anggi Lakes (1,900 m).

DIAGNOSIS: Similar to *macrurus* but with a heavy ocher wash on flanks (Rand & Gilliard 1967).

b. *Megalurus macrurus macrurus* (Salvadori)

Synonym: *Megalurus timoriensis mayri* Hartert, 1930. *Novit. Zool.* 36: 79.—Ifar, Lake Sentani, n NG.

Synonym: *Megalurus macrurus wahgiensis* Mayr & Gilliard, 1951. *Amer. Mus. Novit.* 1524: 9.—Tomba, s slope of Mt Hagen, Eastern Ra.

RANGE: Inhabits lowland and mid-montane grasslands throughout the main body of NG east of the Bird's Neck and not including the Huon Penin.

DIAGNOSIS: Flanks unstreaked; crown reddish brown and unstreaked or with obsolete streaking. Rather variable within and between regions.

c. *Megalurus macrurus alpinus* (Mayr & Rand)

Megalurus timoriensis alpinus Mayr & Rand, 1935. *Amer. Mus. Novit.* 814: 8.—Mount Albert Edward, 3,680 m, SE Peninsula.

Synonym: *Megalurus timoriensis montanus* Mayr & Gilliard, 1951. *Amer. Mus. Novit.* 1524: 9.—Summit grasslands, Mt Hagen, Eastern Ra.

RANGE: All alpine grasslands of the Central Ra.

DIAGNOSIS: Largest form; heavily streaked on flanks and mantle; plumage dark (Bairlein 2006). We accept this widespread alpine form with reservations.

d. *Megalurus macrurus harterti* (Mayr)

Megalurus timoriensis harterti Mayr, 1931. *Mitt. Zool. Mus. Berlin*, 17: 686.—Ogeramnang, Saruwaged Ra, Huon Penin.

RANGE: Mid-montane and alpine grasslands of Huon Penin (Mayr 1931).

DIAGNOSIS: Like *macrurus* but upperparts and flanks much darker (Rand & Gilliard 1967). A doubtful race that probably should be combined with *macrurus*.

NOTES

Noted by Sims (1956) and Mees (1982b) to be badly over-split in the Region. We have reduced from seven to four races based on a museum review. This is one of the widespread NG species that shows clinal variation with geography and elevation; in addition, the plain buff and brown plumage with substantial individual variation has led to over-splitting. *Other names*—*Cincloramphus macrurus*; formerly treated as races of *Megalurus timoriensis*.

Fly River Grassbird *Megalurus albolimbatus*   ENDEMIC | MONOTYPIC

Poodytes albo-limbatus D'Albertis & Salvadori, 1879. *Ann. Mus. Civ. Genova* 14: 87.—Fly R, near Lake Daviumbu.

Known from two areas in the Fly region: the middle Fly at Lakes Ambuve[J], Daviumbu[K], Owa[M], Pangua[J] (Rand 1942a[K], Gregory *et al.* 1996c[J], Gregory 1997[M]); and at Bensbach, in the far south near the PNG border (Gregory 1997). In addition, there are sight records from Wasur NP, where N. Stronach reported the species breeding in swamps associated with Rawa Biru and Ukra swamp (Bishop 2005a). At Lake Daviumbu, the species inhabits mixed areas with clumps of reeds, masses of floating ricegrass, and open stands of lotus lilies growing along fringing channels and waterways with water 2–3 m deep (Rand 1938b). At the Bensbach R it frequents thick stands of sedge growing in water at the edge of small inlets and wider stretches of the lower river (Finch 1980e). Appears to prefer wetter habitats than *M. timoriensis*.

At Bensbach it seems to definitely favor beds of *Cyperus*, an aquatic sedge. Beds of this plant seem historically diminished in the Bensbach area, and the bird is very local (D. Hobcroft *in litt.*). Its habitat appears to be threatened by grazing by the exotic invasive Rusa Deer.

NOTES

Should be looked for on the Indonesian side of the border in suitable habitat. *Other names*—Fly River Grass Warbler, D'Albertis's Grassbird, *Poodytes albolimbatus*.

Little Grassbird *Megalurus gramineus*   RESIDENT

Sphenoeacus gramineus Gould, 1845. *Proc. Zool. Soc. London*: 19.—Tasmania.

The single known breeding population in the Region is recorded from Paniai Lake in the Western Ra, at 1,600 m. Eleven birds were collected from Paniai Lake and Majepa/Tigi Lake (Junge 1953). In addition, there are undocumented reports from Dec-1994 and Apr-1995 in the middle Fly R of the S Lowlands (Gregory *et al.* 1996c).

Extralimital—widespread in sw, e, and se Australia (Higgins *et al.* 2006).

SUBSPECIES

a. *Megalurus gramineus papuensis* Junge
Megalurus gramineus papuensis Junge, 1952. *Zool. Meded. Leiden* 31: 248.—Paniai Lake, Western Ra.
RANGE: Paniai Lake, w sector of Western Ra.
DIAGNOSIS: When compared to birds from sw Australia, the NG birds are tinged with more rusty brown dorsally, especially on the forehead and rump; forehead lacking in blackish streaks (Junge 1953). Bairlein's (2006) note that underparts are unstreaked (apparently copied from Rand & Gilliard 1967) is evidently an error, and based upon confusion over use of the term "front" for forehead as opposed to breast (see Junge 1952: 248, 1953: 47).

NOTES

Other names—Little Grass Warbler, Striated Marshbird, *Poodytes gramineus.*

Family Cisticolidae

Cisticolas [Family 139 spp/Region 2 spp]

This is another family extracted from the Sylviinae of Peters's *Check-list* based on molecular systematic studies (Barker *et al.* 2004). This family ranges from Africa, Madagascar, and southern Europe east to China and Japan, tropical Asia, New Guinea, the Bismarck Archipelago, and Australia (Ryan 2006). The family includes the genera *Cisticola, Prinia, Heliolais, Orthotomus, Bathmocercus,* and *Apalis* (Alström *et al.* 2006). Members are small, vocal warblers of grassland and shrubbery and, in certain instances, forest. Characters include: wings rounded, tail mainly long and narrow, weak legs, and a finely pointed bill (Ryan 2006). Placement of the family follows Barker *et al.* (2004). This lineage is most closely related to the grassbirds (Locustellidae), reed-warblers (Acrocephalidae), and Malagasy warblers (Bernieridae; Alström *et al.* 2013).

GENUS *CISTICOLA* [Genus 45 spp/Region 2 spp]

Cisticola Kaup, 1829. *Skizz. Entw.-Gesch. Europ. Thierwelt*: 119. Type, by tautonymy, *Sylvia cisticola* Temminck.

The species of this genus range from Africa, Madagascar, and western Europe, east to China, Indochina, New Guinea, and Australia (Ryan 2006). The genus is characterized by the small size, grassland habitat, narrow round-tipped tail, and pale and unstreaked ventral plumage contrasting with the brownish (typically streaked) dorsal plumage. *Cisticola* is masculine (David & Gosselin 2002b: 262).

Zitting Cisticola *Cisticola juncidis* RESIDENT

Sylvia Juncidis Rafinesque, 1810. *Caratteri Ale. Nuovi Gen. Spec. Anim. Sicilia* (1809) (I): 6 (spec. 10).—Roccella, Italy.

Recorded in the Region only from the coastal plain of the Trans-Fly. Added to the NG list from sight records and specimens taken in the Bensbach region of the Western Prov (Finch 1980f). Inhabits open grassland on the subcoastal floodplain (Coates 1990).

Localities—Dolak I, Wasur NP, Bensbach, and Morehead[P] (Bishop 2005a, PNG Nat. Mus. specimens[P]).
Extralimital—Australia through Indonesia to Europe and Africa.

SUBSPECIES

a. *Cisticola juncidis ?laveryi* Schodde & Mason

Cisticola juncidis laveryi Schodde & Mason, 1979. *Emu* 79: 52.—Bobowala, *ca.* 40 km s of Ayr, QLD, Australia.

RANGE: Northeastern Australia; the population from the NG Trans-Fly may refer to this race.

DIAGNOSIS: Similar to the Australian *normani* but more heavily streaked dorsally in nonbreeding plumage; breeding females have small tail mirrors (Ryan 2006). Type description notes: in nuptial plumage male crown dull brown, the feathers distinctly edged off-white; mantle broadly striped dusky black; rump cinnamon (Schodde & Mason 1979). Coates (1990) noted that two specimens in the PNG Nat. Mus. (collected by N. Stronach), apparently adult males in nuptial plumage, are similar to this form. When compared to *normani*, the NG birds exhibit a paler crown, are indistinctly mottled darker brown (not solid darkish brown), and have the hindneck slightly paler; wing 49, tail 35–36 (Coates 1990).

NOTES

Taxonomic and distributional status of this species in s NG needs additional study. N. Stronach (*in litt.*) noted that the Trans-Fly birds had song flights and song that are very different from those of African and European birds, which are much weaker in both. *Other names*—Fan-tailed Warbler, Streaked Fantail Warbler, Common or Fan-tailed Cisticola.

Golden-headed Cisticola *Cisticola exilis* RESIDENT

Malurus exilis Vigors & Horsfield, 1827, ex Latham ms. *Trans. Linn. Soc. London* 15: 223.—Sydney, NSW, Australia.

All NG, in lowlands and mid-montane valleys; also Bagabag and Manam I and the D'Entrecasteaux Arch. Sea level–1,400 m, max 2,660 m at Tari Gap. Within the Region, widespread in e and s NG, including the Trans-Fly; apparently scarce in w NG—but present on the Bird's Head and in the Baliem valley and doubtless elsewhere (Coates 1990). Inhabits savanna and open grassland, usually with grass height of more than 1 m (Coates 1990).

Extralimital—India, s China, and se Asia to Australia and the Bismarck Arch (Dutson 2011).

SUBSPECIES

a. *Cisticola exilis diminutus* Mathews

Cisticola exilis diminuta Mathews, 1922. *Birds Austral.* 9: 373.—Paterson Creek, Cape York, QLD, Australia.

RANGE: Northern QLD and islands of the Torres Strait; mainland NG including the Trans Fly[K]; Kairiru, Muschu, Bagabag[J], Manam, Fergusson, Goodenough, and Normanby I[L]; also Madang islets but not, apparently, Karkar (Mayr 1941, Diamond & LeCroy 1979[J], Mees 1982b[K], LeCroy *et al.* 1984[L]).

DIAGNOSIS: Richer rufous dorsally than nominate; also smaller, and with a plain rump year-round (Ryan 2006). Schodde & Mason (1999) provided the following characters: crown of nuptial male rusty rufous; mantle rich rusty rufous in all plumages; back moderately streaked; flanks washed rufous buff; size small: wing (male) 45–48. See photo of a breeding male in Coates (1990: 87) and nonbreeding individual in Coates & Peckover (2001: 135).

NOTES

Other names—Bright-capped or Golden Cisticola.

Family Sturnidae

Starlings and Mynas   [Family 114 spp/Region 8 spp]

The species of the family range from Europe and Africa east to tropical Asia and the tropical Pacific. Two species have been widely introduced around the world—Common Starling (*Sturnus vulgaris*) and Common Myna (*Acridotheres tristis*). Members of the family are slim to robust birds with a moderately heavy bill and strong legs. New Guinea's native species inhabit forest, woodland, edge, and gardens, from lowlands to middle elevations. All resident species are primarily fruit eating, although insects are also consumed. We follow the sequence of Feare & Craig (1998). Jønsson *et al.* (2011b) showed sturnids to be sister to the *Sylvia* warblers and the swallows, which accords moderately well with the finding of Barker *et al.* (2004). Details of the generic phylogeny are provided by Zuccon *et al.* (2006).

GENUS *APLONIS*   [Genus 24 spp/Region 5 spp]

Aplonis Gould, 1836. *Proc. Zool. Soc. London*: 73. Type, by subsequent designation (G. R. Gray, 1840. *List Gen. Birds*: 40), *Aplonis fusca* Gould.

The species of this genus range from Bangladesh east to the Philippines, New Guinea, and Pacific islands as far as Samoa, Tonga, and Rarotonga (Craig & Feare 2009). The genus is characterized by the uniform dark or dull plumage, often glossy black; the medium-size dark bill, with culmen strongly curved; and the slim and compact form; the tail varies from abbreviated to elongate. The generic spelling *Aplonis* is conserved (ICZN Opinion 2285; *fide* R. Schodde *in litt.*). *Aplonis* is feminine (David & Gosselin 2002b).

Metallic Starling *Aplonis metallica*   RESIDENT (AND VISITOR?)

Lamprotornis metallicus Temminck, 1824. *Planch. Col. d'Ois.*, livr. 45: pl. 266.—Ambon, s Moluccas.

A flocking and colonial-nesting canopy fruit-eater that ranges through NG and virtually all satellite islands; sea level–1,000 m[CO], infrequently higher. Frequents forest, edge, gardens, and trees in clearings (Coates 1990). Upland records on the Mainland include an adult female found dead at Lake Habbema (3,225 m)[J], Ilaga R (2,290 m)[K], and Mt Giluwe (2,130 m)[L] (Rand 1942b[J], Ripley 1964a[K], Sims 1956[L]).

Extralimital—Moluccas, islands of Torres Strait, ne and e QLD, Bismarck Arch, and Solomon Is (Mayr & Diamond 2001, Higgins *et al.* 2006, Dutson 2011).

SUBSPECIES

a. *Aplonis metallica metallica* (Temminck)
RANGE: The Moluccas, all NG (including the Trans-Fly[CO]), the four main NW Islands[MA] (plus Gag[G] and Kofiau[D]), Aru Is[MA], Biak I[Z], Yapen and Mios Num I[MA], Sarmi I[J], also Tarawai[CO]; all Schouten Is; Kumamba[MA], Karkar[MA], Bagabag[K], and Sariba I[CO]; D'Entrecasteaux Arch[MA]; Amphlett, Engineer, and Duchateau Is[CO]; and Woodlark[MA], Misima[MA], Tagula[MA], and Rossel I[MA] (Ripley 1964a[J], Diamond & LeCroy 1979[K], Johnstone 2006[G], Diamond *et al.* 2009[D], van Balen & Bishop ms[Z]). Also ne Australia, Bismarck Arch, and Solomon Is. Birds from ne Australia apparently migrate to NG (Coates 1990, Higgins *et al.* 2006). No records for Gebe or Manam I.
DIAGNOSIS: Strong iridescence of plumage: crown glossed purple and nape glossed satin green (Craig & Feare 2009); large size: wing (male) 106–114 (Schodde & Mason 1999).

b. *Aplonis metallica inornata* (Salvadori)
Calornis inornata Salvadori, 1880. *Ann. Mus. Civ. Genova* 16: 194.—Biak I, Bay Is.
RANGE: Biak and Numfor I[MA]. Observed on Owi I on 13-Jan-1994 (P. Gregory *in litt*).
DIAGNOSIS: Smaller than nominate form, and iridescence much duller (Rand & Gilliard 1967); lacks green on nape; head tinged purple (Craig & Feare 2009).

NOTES
Other names—Shining or Colonial Starling.

Yellow-eyed Starling *Aplonis mystacea*   ENDEMIC | MONOTYPIC

Calornis mystacea Ogilvie-Grant, 1911. *Bull. Brit. Orn. Club* 29: 28.—Parimau, Mimika R, w sector of the S Lowlands.

Restricted to the lowland forests of the S Lowlands and the w sector of the SE Peninsula. Ranges from the e sector of the Bird's Neck (head of Geelvink Bay) east to the Lakekamu basin (Hartert *et al.* 1936, Diamond & Raga 1976, Beehler & Bino 1995). A little-known species, which may have been overlooked in some areas because of its similarity to *A. metallica*, with which it sometimes associates.

Localities—Wanggar R[W], Mimika R[MA], Agats[Z], Kiunga area[L], Ok Ma[Y], Elevala R[J], Palmer Junction Camp[P], Black R Camp[P], Oroville Camp[P], upper Strickland R[K], Turama R[J], lower Kikori R[J], Purari basin[K], Lakekamu basin[X] (Hartert *et al.* 1936[W], Rand 1942a[P], Beehler & Bino 1995[X], Gregory 1995b[Y], 1997[L], Burrows 1993[J], van Balen *et al.* 2011[Z], I. Woxvold *in litt.*[K]).

NOTES

Can be distinguished from the very similar *A. metallica* by the yellow iris, the blackish (not greenish) throat, the all-black head without the greenish-purplish highlighting on the neck and nape, the shorter less-pointed tail, and the prominent naral tuft sometimes visible even in flight (P. Gregory *in litt.*). Also has a contact call distinct from that of Metallic Starling (D. Hobcroft *in litt.*).

Singing Starling *Aplonis cantoroides*   RESIDENT | MONOTYPIC

Calornis cantoroides G. R. Gray, 1862. *Proc. Zool. Soc. London* (1861): 431.—Misool I, NW Islands.

All NG, NW Islands, Aru Is, and some SE Islands. Mainly near sea level, but occasionally to 1,740 m in open upland interior valleys (extreme 1,980 m; Coates & Peckover 2001). Frequents forest edge, clearings, gardens, and cultivated areas (Coates 1990). Within s Trans-Fly recorded from Wasur NP east to Daru I and the Fly R delta (Bishop 2005a).

Localities—Batanta, Misool, Salawati, and Waigeo I[MA]; Adi I[L]; Paniai Lake[M]; upper Wanggar R[W]; Baliem valley[CO]; Bernhard Camp on the Taritatu R[Q]; Wasur NP[J]; Fly R delta[J]; Daru[J]; Mendi[CO]; Baiyer valley[CO]; Nondugl[K]; Wahgi valley[X]; Kundiawa[CO]; Goroka[CO]; islands off n coast: Yamna[MA], Tarawai[MA], Karkar[Z], Madang islets[CO]; Sogeri Plateau[CO]; e satellite islands of Tubatuba in the Engineer Is and Renard I in the Conflict Is[CO]; islands off the se coast: Loupom and Mailu I[CO]; and Normanby, Woodlark, Misima, and Tagula I[MA] (Hartert *et al.* 1936[W], Rand 1942b[Q], Mayr & Gilliard 1954[X], Gyldenstolpe 1955a[K], 1955b[L], Diamond & LeCroy 1979[Z], Melville 1980[M], Bishop 2005a[J]).

Extralimital—widespread through the Bismarcks and Solomons (Mayr & Diamond 2001) and recorded from islands of the n Torres Strait (Higgins *et al.* 2006).

Long-tailed Starling *Aplonis magna*   ENDEMIC

Lamprotornis magnus Schlegel, 1871. *Ned. Tijdschr. Dierk.* (1873) 4: 18.—Soëk, Biak I, Bay Is.

Widespread on Biak and Numfor I, sea level–650 m. Observed on Owi I on 11-Mar-1991 by D. Gibbs (2009, and *in litt.*).

SUBSPECIES

a. *Aplonis magna brevicauda* (van Oort)
Macruropsar magnus brevicauda van Oort, 1908. *Notes Leyden Mus.* 30: 70.—Numfor I, Bay Is.
RANGE: Numfor I.
DIAGNOSIS: Much shorter-tailed than nominate form. Tail (male) 129–155 (n=8; birds measured at AMNH). Craig & Feare (2009) stated that this form is less glossy than the nominate, but that is not borne out by examination of adult male museum specimens.

b. *Aplonis magna magna* (Schlegel)
RANGE: Biak I.
DIAGNOSIS: Long-tailed. Tail (male) 180–252 (n=6; AMNH). Tail of this form is 52% longer than that of *brevicauda*.

Moluccan Starling *Aplonis mysolensis*   RESIDENT | MONOTYPIC
Calornis mysolensis G. R. Gray, 1862. *Proc. Zool. Soc. London* (1861): 431.—Misool I, NW Islands.

The NW Islands: Misool[MA], Gebe[MA], Kamoa[J] (Schildpad Is), Pecan I (off Waigama, Misool I)[Q], Kofiau[D], Waigeo[J], Salawati[MA], Batanta[MA], Ayu Is[J], Wai I[K], and other small islets among these larger ones[MA] (Ripley 1964a[J], Mees 1965[Q], Diamond *et al.* 2009[D], Beehler field notes 29-Feb-2004[K]). Presumably more prevalent on the small islets, the favored habitat of this insular tramp species.

Extralimital—Moluccas: Banggai and Sula Is, Morotai, Halmahera, Ternate, Bacan, and Obi (Craig & Feare 2009).

NOTES
Treated as monotypic following Craig & Feare (2009). *Other names*—Island Starling.

GENUS *MINO*   [Genus 3 spp/Region 2 spp]

Mino Lesson, 1827. *Bull. Sci. Nat. Férussac* 10: 159. Type, by monotypy, *Mino dumontii* Lesson.

The species of this genus range throughout the lowlands and lower hills of New Guinea and the Bismarck and Solomon Islands (Craig & Feare 2009). The genus is characterized by the orange-yellow bill with the maxilla without a terminal hook, the presence of a white or yellow rump and undertail, the abbreviated tail, and the bulky shape.

Yellow-faced Myna *Mino dumontii*   ENDEMIC | MONOTYPIC
Mino Dumontii Lesson, 1827. *Bull. Sci. Nat. Férussac* 10: 159.—Manokwari, Bird's Head.
> Synonym: *Mino dumonti violaceus* Berlepsch, 1911. *Abh. Senckenb. Naturf. Ges.* 34: 62.—Near Madang, e verge
> of the Sepik-Ramu.

Mainland NG, Salawati, Batanta, Waigeo, Yapen, Daru, and Aru Is (Mayr 1941); sea level–750 m, max 1,800 m[CO]. Widespread throughout the s Trans-Fly (Bishop 2005a). The absence of this species on e satellite islands is peculiar (Coates 1990). Inhabits forest, edge, and tall trees in clearings (Coates 1990). Melville (1980) and K. D. Bishop (*in litt.*) observed the species on Biak I near the airport in 1976 and 1992. More recently, free-flying birds were observed by S. van Balen in Biak town, presumably escaped cage birds (Biak I is an important center for the bird trade in e Indonesia because of its status as a hub for airline operations).

NOTES
We follow Craig & Feare (2009) in treating this species as monotypic. A sister species (*M. kreffti*) is found in the Bismarck and Solomon Is (*pace* Mayr & Diamond 2001). *Other names*—Orange-faced Grackle or Myna.

Golden Myna *Mino anais*   ENDEMIC
Sericulus Anaïs Lesson, 1839. *Rev. Zool.*: 44.—New Guinea.

Mainland NG and Salawati I, sea level–570 m[CO]. No records from the Trans-Fly or the n coast of the SE Peninsula.

SUBSPECIES

a. *Mino anais anais* (Lesson)
RANGE: Salawati[MA] and the w sector of the Bird's Head[MA]. Exact extent of the geographic distribution of this distinctive form on the Mainland needs additional study.
DIAGNOSIS: Crown and nape entirely black. Yellow plumage of hindneck and breast pale (Craig & Feare 2009). Bare skin around eye yellowish.

b. *Mino anais orientalis* (Schlegel)
Gracula anais orientalis Schlegel, 1871. *Ned. Tijdschr. Dierk.* (1873) 4: 52.—Bondey, w coast, Geelvink Bay, opposite Roon I, e Bird's Head.
RANGE: The e sector of the Bird's Head and Neck (Kambala[J], Kapaur and Sekru[MA], Onin Penin[J]), Yapen

I[MA], NW Lowlands[MA], Sepik-Ramu[K], and the n scarp of the Huon Penin[L] (Stresemann 1923[K], Gyldenstolpe 1955b[J], Beehler field notes[L]).

DIAGNOSIS: Yellow crown and black nape (Craig & Feare 2009).

c. *Mino anais robertsoni* D'Albertis
Mino robertsoni D'Albertis, 1877. *Ann. Mus. Civ. Genova* 10: 12.—Fly R.
RANGE: The S Lowlands (Kapare and Mimika R) east to the s watershed of the SE Peninsula (Mayr 1941, Coates 1990).
DIAGNOSIS: Crown and nape yellow-orange (Craig & Feare 2009). See photo in Coates & Peckover (2001: 203).

NOTES
The record from Yapen I reported by Mayr (1941) has been discounted in detail by J. M. Diamond (*in litt.*). It would be worth studying the relative distribution of the ranges of the nominate form and *orientalis* on the Bird's Head. If the two forms are found together without intergradation they may need to be treated as distinct species. *Other names*—Golden-breasted Myna.

GENUS *ACRIDOTHERES* [Genus 10 spp/Region 1 hypothetical sp]

Acridotheres Vieillot, 1816, *Analyse Nouv. Orn. Elem.*: 42. Type, by subsequent designation (Gray, 1840. *List Gen. Birds*: 40), *Paradisea tristis* Linnaeus.

The species of this genus range from Afghanistan and Kazakhstan east to China and south to Java. The genus comprises robust starlings patterned in black, gray, brown, and white. Note the pale white patch on the folded wing and the dark tail with white tipping; species often have a crested forehead (Amadon 1943). The species *A. tristis* has been introduced widely.

[**Common Myna** *Acridotheres tristis*   HYPOTHETICAL

Paradisea tristis Linnaeus, 1766. *Syst. Nat.*, ed. 12, 1: 167.—Probably Pondicherry, se India.

Sight records of multiple birds from Alotau (Milne Bay Prov) between 1997 and 2005 (D. Mitchell *in litt.*). No recent sightings, and apparently this very localized introduced population has disappeared.

Extralimital—Native to s Asia. Introduced populations inhabit the Solomon Is and n QLD, so should be expected to invade and settle in the Region before too long. Successful breeding recorded in the Torres Strait Is in the 1980s by Draffan *et al.* (1983). A common urban human commensal in Australian towns.

SUBSPECIES
a. *Acridotheres tristis ?tristis* (Linnaeus)
RANGE: Native range—Afghanistan, Pakistan, and Turkmenistan east to India, Burma, and se Asia. Introduced to sites in North America, e South Africa, the Middle East, Madagascar, Hong Kong, Brunei, Sumatra, Taiwan, Japan, Australia, Solomon Is, Vanuatu, New Caledonia, Fiji, and other Pacific islands (Craig & Feare 2009). Today a common town bird in n QLD. The population on Bougainville is declining and may now be gone (Coates 1990, P. Gregory *in litt.*).
DIAGNOSIS: Back, breast, and flanks dull medium gray-brown (not blackish brown as in the Sri Lankan form *melanosternus*).

NOTES
Other names—Indian Myna.]

GENUS *STURNUS* [Genus 2 spp/Region 1 hypothetical sp]

Sturnus Linnaeus, 1758. *Syst. Nat.*, ed. 10, 1: 167. Type, by tautonymy, "Sturnus" = *Sturnus vulgaris* Linnaeus.

The genus has a native range of Eurasia, wintering to North Africa (Craig & Feare 2009). *S. vulgaris* has been introduced to North America, including Mexico, the Caribbean, South Africa, Australia, and New

Zealand. The genus is characterized by the compact body, iridescent blackish plumage, abbreviated and squared tail, yellow pointed bill, a bare area of skin around the eye, rictal bristles absent or mostly so, and eggs unspotted; it is primarily a terrestrial forager (Amadon 1943).

[**Common Starling** *Sturnus vulgaris*   HYPOTHETICAL
Sturnus vulgaris Linnaeus, 1758. *Syst. Nat.*, ed. 10, 1: 167.—Sweden.

There are three NG records: an adult in breeding plumage at Moitaka settling ponds near Port Moresby on 18–19-Oct-1970 (Coates 1970c); three adults at the same place a few weeks later (Layton 1971); a single bird at Moitaka on 14-Aug-1985, reported by Finch (1988b). Presumably vagrants from Australia, perhaps stowaways on a freighter. Failed to establish in Region.

Extralimital—occurs naturally in temperate Europe and w Asia. Introduced to Australia, NZ, South Africa, and North America and has spread to the Kermadec and Norfolk Is, Fiji, Tonga, Vanuatu, and subantarctic Macquarie I. In Australia found mainly in the southeast, where abundant, occasionally north to Cape York Penin. Disperses widely. Rare in far n QLD (P. Gregory *in litt.*). Possible records from Vanuatu and Loyalty Is (Dutson 2011).

SUBSPECIES

a. *Sturnus vulgaris ?vulgaris* Linnaeus
RANGE: Native to w Europe. Introduced widely; few records from the Region (see above).
DIAGNOSIS: Brown dorsally, darker and with violet iridescence on forehead and crown; wing dark brown; tail brown (Craig & Feare 2009). The individuals encountered in the Region presumably refer to the nominate form, which was introduced to Australia (Schodde & Mason 1999).

NOTES
Other names—European, Eurasian, or Northern Starling; Starling.]

Family Turdidae

Thrushes   [Family 159 spp/Region 2 spp]

The species of the family range through the New and Old World. The true thrushes lie within the Passerida and are sister to the true Muscicapidae; the Muscicapidae+Turdidae clade has as its sister the Cinclidae (Barker *et al.* 2004). Members of this family are medium-size and largely ground dwelling; plumage is variable but often cryptically patterned or brown; many species are prominent vocalists. Placement of the family follows Barker *et al.* (2004).

GENUS *ZOOTHERA*   [Genus 15 spp/Region 1 sp]

Zoothera Vigors, 1832. *Proc. Zool. Soc. London*: 172. Type, by monotypy, *Zoothera monticola* Vigors.

The species of this genus range from western Russia and the Himalayas east to Siberia and south to Indonesia, New Guinea, Australia, and Melanesia (Collar 2005). The genus is characterized by the cryptic terrestrial habit, sturdy build, and medium-length bill.

Russet-tailed Thrush *Zoothera heinei*   RESIDENT
Oreocincla Heinei Cabanis, 1851. *Mus. Hein.* 1: 6.—Southern QLD, Australia. [For date of original reference see Quaisser & Nicolai 2006, *pace* Schodde & Mason 1999: 642.]

Hill forests of NG, 500–1,700 m^{CO}. Rarely in adjacent lowlands. Rather few records. Inhabits the forest floor of primary hill forest interior. Very shy and rarely seen. Most often encountered through mist-netting.

Extralimital—e QLD, ne NSW, Mussau I, and Choiseul I (Mayr & Diamond 2001, Higgins *et al.* 2006, Dickinson & Christidis 2014).

a. *Zoothera heinei papuensis* (Seebohm)

Geocichla papuensis Seebohm, 1881. *Cat. Birds Brit. Mus.* 5: 158, pl. 9.—Southeastern Peninsula.

RANGE: Presumably ranges through hill forest of all the mts of NG; records include the Kumawa Mts[A], the upper Otakwa R[MA], the Foja Mts[F], Bewani Mts[U], mts of the upper Strickland R[V], and the Agogo Ra of the PNG s highlands[V], Mt Karimui[DI], Okasa[DI], Sena R[DI], Baiyer R[J], Adelbert Mts[P], Sattelberg[G], YUS[Z], Mt Missim[X], upper Eloa R[W], Lakekamu basin[Q], Mafulu[H], Varirata NP[X] (Mayr 1931[G], Mayr & Rand 1937[H], Beehler 1980b[W], Pratt 1983[P], Beehler *et al.* 1994[Q], Beehler *et al.* 2012[F], Freeman *et al.* 2013[Z], Beehler field notes[X], K. D. Bishop *in litt.*[J], J. M. Diamond, *in litt.*[A], I. Woxvold *in litt.*[V], BPBM specimen no. 109994[U]).

DIAGNOSIS: Crown dark olive brown scaled with blackish feather edgings; brown of back becomes more red-brown on rump and with reduced black feather edging from anterior to posterior; throat and chin all white; black scaling of buff feathers forms an indistinct collar on the upper breast; outer tail feathers off-white, distinct from brown inner tail feathers. Smaller and darker than nominate form of e Australia (Collar 2005). See photos in Coates (1990: 56) and Coates & Peckover (2001: 117).

NOTES

Formerly treated as conspecific with *Z. dauma*. We follow Schodde & Mason (1999) and Collar (2005) in splitting this form from *dauma*. Relationship of the NG form to the two Australian ones should be further investigated by molecular techniques. *Other names*—White's, Ground, or Scaly Thrush; *Zoothera dauma*.

GENUS *TURDUS* [Genus 80 spp/Region 1 sp]

Turdus Linnaeus, 1758. *Syst. Nat.*, ed. 10, 1: 168. Type, by subsequent designation (G. R. Gray, 1840. *List. Gen. Birds*: 27), *Turdus viscivorus* Linnaeus.

The species of this genus range from the Americas to Africa, Eurasia, New Guinea, and Polynesia (Collar 2005). The genus is characterized by a distinct and invariant body form: medium size, medium-length tail, medium-length bill; as well as a distinctive musical song and terrestrial habit in most cases (though many feed on fruit in trees). Plumage is variable, but mainly rufous, brown, gray, white, and black, and is plain or spotted below.

Island Thrush *Turdus poliocephalus* RESIDENT

Turdus poliocephalus Latham, 1801. *Index Orn.*, Suppl.: xliv.—Norfolk I.

Highest mts of the Central Ra, Huon Penin, and Karkar and Goodenough I; mainly 2,300–4,500 m[CO]. On the Mainland and Goodenough I the species inhabits montane forest, shrublands, alpine grassland, and rocky outcrops (Coates 1990).

Extralimital—Christmas I, Sumatra, and Taiwan to Samoa, Norfolk I, and Lord Howe I; also Bismarck Arch, and Bougainville I (Coates 1990, Mayr & Diamond 2001). New Britain record from Bishop & Jones (2001).

SUBSPECIES

a. *Turdus poliocephalus versteegi* Junge

Turdus poliocephalus versteegi Junge, 1939. *Nova Guinea* (ns)3: 8.—Kajan Mts, Oranje Mts, Western Ra.

RANGE: The Western Ra. Presumably also ranges eastward through the Border Ra to the Strickland R gorge.

DIAGNOSIS: Breast and belly very dark chocolate brown; paler hood.

b. *Turdus poliocephalus papuensis* (De Vis)

Merula papuensis De Vis, 1890. *Ann. Rept. Brit. New Guinea*, 1888–89: 60.—Mount Victoria, SE Peninsula.

> Synonym: *Turdus poliocephalus keysseri* Mayr, 1931. *Mitt. Zool. Mus. Berlin* 17: 692.—Mongi-Busu, Saruwaged Ra, Huon Penin.
>
> Synonym: *Turdus poliocephalus carbonarius* Mayr & Gilliard, 1951. *Amer. Mus. Novit.* 1524: 7.—Mount Wilhelm, Eastern Ra.
>
> Synonym: *Turdus poliocephalus erebus* Mayr & Gilliard, 1952. *Amer. Mus. Novit.* 1577: 7. Replacement name

for *Turdus poliocephalus carbonarius* Mayr & Gilliard, *nec Turdus carbonarius* Lichtenstein = *Platycichla flavipes* Vieillot.

RANGE: The Eastern Ra and mts of the Huon and SE Peninsulas. Presumably extends west to the Strickland R gorge (Mayr 1941, Diamond 1972). Also Karkar I (Diamond & LeCroy 1979).

DIAGNOSIS: Smaller and slightly paler on mantle and breast than *versteegi*. See photos in Coates (1990: 55) and Coates & Peckover (2001: 117). The forms *keysseri* and *erebus* (= *carbonarius*) are not reliably distinct from *papuensis*. Considerable variation in plumage related to age (and sex?) perhaps explains the creation of these obsolescent subspecies.

c. *Turdus poliocephalus canescens* (De Vis)

Merula canescens De Vis, 1894. *Ann. Rept. Brit. New Guinea*, 1893–94: 105.—Goodenough I, D'Entrecasteaux Arch, SE Islands.

RANGE: Highlands of Goodenough I above 1,500 m.

DIAGNOSIS: Head pale tan-gray all over, forming a hood. Rest of body blackish (Collar 2005). A single specimen was collected in 1981 by H. Sakulas at 1,600 m on Goodenough I (deposited in collection at Wau Ecology Inst.; Beehler field notes).

NOTES

The nonmolecular systematic review of this widespread species by Peterson (2007) suggested treating the NG forms as three distinct species. We await molecular analysis to better clarify the issue. *Other names—* New Guinea Blackbird.

Family Muscicapidae

Old World Flycatchers   [Family 303 spp/Region 4 spp]

The species of the family range from northern Eurasia south to Africa and east to the Philippines and Indonesia. Barker *et al.* (2004) placed this lineage as a sister to the true thrushes (Turdidae). The current Muscicapidae includes the bushchats and rock-thrushes, formerly placed in the true thrushes (Sangster *et al.* 2010). The family as now construed includes considerable diversity in form and habit.

GENUS *CALLIOPE*   [Genus 4 spp/Region 1 sp]

Calliope Gould, 1836. *Birds Europe* 2: pl. 118 and text. Type, by monotypy and tautonomy, *Calliope lathami* Gould = *Motacilla calliope* Pallas.

The four species of this newly resurrected genus (Sangster *et al.* 2010) breed from western China and the Himalayas east to eastern Russia. Birds winter southward, as far as mainland southeast Asia and the Philippines. The genus is characterized by the small and compact body form, small tail, complex song, minor or major sexual dimorphism, and either reddish or black throat of the male. The genus is difficult to circumscribe, given presence of monomorphic and dimorphic forms and the abundance of small robin-like species in the Old World.

Siberian Rubythroat *Calliope calliope*   VAGRANT | MONOTYPIC

Motacilla Calliope Pallas, 1776. *Reise Versch. Prov. Russ. Reichs.* 3: 697.—Between Yenesei and the Lena R, e Russia.

Recorded once from the nw sector of the SE Peninsula: a bird in female plumage observed near a creek on Mt Kaindi at 1,750 m, above Wau on the Mt Kaindi Road, on 4-Apr-1983 (Finch 1983d).

Extralimital—breeds in n Asia, wintering south to Indochina and the Philippines.

NOTES

Moved from *Luscinia* to *Calliope* by Sangster *et al.* (2010), based on a molecular phylogeny. *Other names—* Rubythroat, *Erithacus* or *Luscinia calliope.*

GENUS *SAXICOLA* [Genus 13 spp/Region 1 sp]

Saxicola Bechstein, 1802. *Orn. Taschenb.*: 216. Type, by subsequent designation (Swainson, 1827. *Zool. Journ.* 3, 10: 172), *Motacilla rubicola* Linnaeus. Date of original publication follows Dickinson *et al.* 2011.

The species of this genus range from western Europe and Africa east to Siberia and southeast to New Guinea and the Bismarck Archipelago (Collar 2005). The genus is characterized by the small size and compact form, plumage dimorphism, bold plumage patterns in the males, and preference for open and shrubby habitats; birds are typically seen perched on a shrub (Collar 2005).

Pied Bushchat *Saxicola caprata* RESIDENT

Motacilla Caprata Linnaeus, 1766. *Syst. Nat.*, ed. 12, 1: 335.—Luzon, Philippines.

Patchily distributed in lowlands of NG but widespread in the uplands east of the Bird's Neck; sea level–2,850 m[CO]. Expanding its range in open agricultural habitats. Inhabits grasslands, gardens, and edges of airfields (Coates 1990).

Extralimital—central Asia, India, and se Asia to Long I, Watom I, New Britain, and New Ireland (Mayr & Diamond 2001, Dutson 2011).

SUBSPECIES

a. *Saxicola caprata aethiops* (P. L. Sclater)

Poecilodryas aethiops P. L. Sclater, 1880. *Proc. Zool. Soc. London*: 66, pl. 7, fig. 1.—Kabakadai, New Britain.
    Synonym: *Saxicola caprata belensis* Rand, 1940. *Amer. Mus. Novit.* 1072: 4.—Baliem R, 1,600 m, Western Ra.
    Synonym: *Saxicola caprata wahgiensis* Mayr & Gilliard, 1951. *Amer. Mus. Novit.* 1524: 8.—Mafulu, SE Peninsula.
RANGE: The e sector of the Western Ra; the Border and Eastern Ra; the ne sector of the NW Lowlands and the Sepik-Ramu; the Huon and SE Peninsulas; also New Britain (Mayr 1931, Bell 1970a, Coates 1990). J. Pap (*in litt.* 2015) reported several sight records from the Timika area of the w sector of the S Lowlands. Westernmost upland record is from Ilaga (Ripley 1964a). Absent from many grassland areas in w NG.
DIAGNOSIS: Males across Melanesia are uniform (with some variation in size apparently based on elevation). By contrast, females show considerable (age-related?) within-population plumage variation, making a clear delineation of subspecies difficult. We thus recognize but a single subspecies for the NG region. The modal plumage for this variable form is not terribly different from that of the nearest Moluccan form, *S. c. cognatus*, from the Babar Is. Topotypical material of *aethiops* differs from *cognatus* as follows (female plumage): darker dorsally and more heavily streaked ventrally.

NOTES

Saxicola is masculine, but the species epithet *caprata* is invariant (David & Gosselin 2002b). The population(s) in the Region shows some variation (male size and female plumage), but a confusion of names and type localities has rendered a taxonomic mess, and for the time it is best to treat the NG populations as a single race, until additional fresh material supports multiple subspecies. The statement by Collar (2005) that the male plumage of the form *belensis* lacks a white wing spot is not correct. Female plumage patterns are confusing and may be related to age. Presumably this species is a rather new immigrant to the Region. *Other names*—Pied Chat, Pied Bush Chat.

GENUS *MONTICOLA* [Genus 13 spp/Region 1 sp]

Monticola Boie, 1822. *Isis*: col. 552. Type, by subsequent designation (G. R. Gray, 1847. *Gen. Birds* 1 (1849): 220), *Turdus saxatilis* Linnaeus.

The species of this genus range from western Europe, Madagascar, and Africa east through Asia to Indonesia and the Philippines (Collar 2005). This very uniform genus is distinguished by sexual dichromatism, with the male exhibiting blue and rufous brown in the plumage, and the female much duller, scaly below, and darker above.

Blue Rock-thrush *Monticola solitarius* VAGRANT
[*Turdus*] *Solitarius* Linnaeus, 1758. *Syst. Nat.*, ed. 10, 1: 170.—Italy.

Three sight records from the Region. This is a solitary species that perches conspicuously on large rocks, buildings, cliffs, or dead trees in open country, cleared areas, and towns (Coates 1990).

Extralimital—breeds across Eurasia and North Africa from Spain and Morocco to China, Korea, and Japan. Partly migratory, wintering south to n Africa and tropical Asia east to Taiwan, the Philippines, and the Moluccas; rarely to Palau (Coates 1990). A single record from e Australia (Higgins *et al.* 2006).

SUBSPECIES

a. *Monticola solitarius ?philippensis* P. L. S. Müller
Turdus Philippensis P. L. S. Müller, 1776. *Natursyst.*, Suppl.: 145.—Philippines.
RANGE: Breeds from ne Russia to Korea, China, Japan, and Taiwan; winters south to Hainan, Indochina, Philippines, Indonesia, and Palau (Ripley 1964b). Sight record of an adult male at Paga Hill, Port Moresby, from 7–19-Jan-1986 (Hicks & Finch 1987). Also a single bird was observed by Gregory *et al.* (1996a) at the airstrip at Manokwari on 11-Jan-1994; finally, an individual was observed at the university campus at Manokwari on 21–22-Oct-2003 (van Balen & Noske 2006).
DIAGNOSIS: Male with rufous breast and undertail; female darker than other races; throat and face dull powdery blue-gray, with mottling (Collar 2005).

GENUS *MUSCICAPA* [Genus 23 spp/Region 1 sp]

Muscicapa Brisson, 1760. *Orn.* 1, 32: 2: 357, pl. 5, fig. 3. Type, by tautonymy, "Muscicapa" Brisson = *Motacilla striata* Pallas.

The species of this genus range from western Europe and Africa east to Siberia, China, and Indochina. They winter to Africa and tropical Asia (Taylor 2006). The genus is characterized by the small and compact "flycatcher" body form, the dull plumage of grays, buffs, and white, the upright posture, and the rather abbreviated bill.

Grey-streaked Flycatcher *Muscicapa griseisticta* VISITOR | MONOTYPIC
Hemichelidon griseisticta Swinhoe, 1861. *Ibis*: 330.—Near Tanggu, n China.

Recorded during the northern winter in the NW Islands, the Bird's Head, and in the n watershed of w NG, as far east as the Border Ra of PNG (Victor Emanuel Mts), sea level–1,400 m. Recorded in NG between 6-October and 27-April (Coates 1990).

Localities—Gebe[X], Gag[A], Salawati[W], Misool[MA], Pelee[Z], Waigeo[MA], and Batanta I[W]; Andai[MA], Manokwari[MA], Sainkedoak[MA], mts of the Wandammen Penin[W]; Biak[L], Adi[K], Mapia[MA], and Yapen I[MA]; Foja Mts[B], Danau Bira[X], Telefomin[X], Ok Menga[J], Tabubil[J], and foothills of the Bewani Mts[CO] (Gyldenstolpe 1955b[K], Diamond 1981c[X], 1985[W], Gregory 1997[J], Johnstone 2006[A], Beehler *et al.* 2012[B], van Balen & Bishop ms[L], Beehler field notes 19-Dec-2011[Z]).

Extralimital—breeds in e Siberia and ne China; winters through Moluccas, Sulawesi, Philippines, and e Indonesia.

Other names—Gray-spotted or Spot-breasted Flycatcher.

Family Dicaeidae

Flowerpeckers

[Family 45 spp/Region 3 spp]

We follow Cracraft (2014) in treating the flowerpeckers as distinct from the sunbirds (Nectariniidae) at the family level (*pace* Jønsson *et al.* 2011b). The species are characterized by very small size, colorful (or very dull) plumage, abbreviated bill and tail, and specialized adaptations for feeding on mistletoe fruits. The family includes species that range from tropical Asia east to the Solomon Islands and Australia. The flowerpeckers are important dispersers, especially of mistletoes. Placement of the family follows Barker *et al.* (2004).

GENUS *DICAEUM*   [Genus 39 spp/Region 3 spp]

Dicaeum Cuvier, 1816. *Règne Animal.* 1 (1817): 410. Type, by subsequent designation (G. R. Gray, 1840. *List Gen. Birds*: 13), *Certhia erythronotum* Gmelin = *Certhia erythronotus* Latham = *Certhia cruentata* Linnaeus.

The species of this genus range from tropical Asia to New Guinea, Australia, and the Solomon Islands (Cheke & Mann 2008). The genus is characterized by the abbreviated but sharply pointed bill; tiny and compact body; abbreviated tail; and specialized association with mistletoes. Barker *et al.* (2004) placed this genus sister to the Nectariniidae.

Olive-crowned Flowerpecker *Dicaeum pectorale*   ENDEMIC
Dicaeum pectorale S. Müller, 1843. *Verh. Nat. Gesch., Land-en Volkenk.* 1: 162.—Triton Bay, Bird's Neck.

The Bird's Head and Neck and the NW Islands. Mainland distribution extends east along the s coast at least to Triton Bay, and on the n coast to the head of Geelvink Bay (lower Menoo R). Forest, edge of forest, and regrowth up to 1,500 m (record to 2,350 m; Cheke & Mann 2008).

SUBSPECIES

a. *Dicaeum pectorale ignotum* Mees
Dicaeum pectorale ignotum Mees, 1964. *Zool. Meded. Leiden* 40(15): 128.—Gebe I, NW Islands.
RANGE: Gebe I. May also include the population on Gag I (Johnstone 2006).
DIAGNOSIS: When compared to the nominate form, more olive above and paler and less yellowish on center of belly and undertail; wing and tail slightly longer (Cheke & Mann 2008).

b. *Dicaeum pectorale pectorale* S. Müller
RANGE: The four main NW Islands, the Bird's Head, and the Bird's Neck; the Kofiau I population remains undiagnosed but probably aligns with this form (Mayr 1941, Diamond *et al.* 2009).
DIAGNOSIS: See diagnosis for *ignotum.*

NOTES

We treat *D. pectorale* as distinct from *D. geelvinkianum*, following Mayr (1941), Gregory (2008b), and our own character analysis, which compared the Bird's Neck population *D. p. pectorale* with the adjacent population of *D. geelvinkianum* from the e verge of the Bird's Neck (*D. g. obscurifrons*). Male plumage of the species *pectorale* differs in the following characters: crown lacking a red patch and instead entirely olive; rump lacking red and instead olive. Female plumage distinct in also lacking any red on crown or rump and in underparts having a predominantly whitish rather than yellowish wash. A biogeographic reason for treating this form as specifically distinct is the sharp parapatric break in plumage characters at a mainland vicariance boundary well known for a wide range of species pairs. *Other names*—Pectoral, Shining, or Papuan Flowerpecker.

Red-capped Flowerpecker *Dicaeum geelvinkianum*

Dicaeum geelvinkianum A. B. Meyer, 1874. *Sitzungsber. Akad. Wiss. Wien* 70: 120.—Ansus, Yapen I, Bay Is.

All NG east of the Bird's Neck, and most satellite islands (excepting the NW Islands, Aru Is, Mios Num, Trobriand Is, and Woodlark I; Mayr 1941, Stresemann 1923); sea level–1,500 m, max 2,350 m^{CO}. Inhabits forest, edge, garden habitats, and some urban areas (Coates 1990). This species is replaced on the Aru Is by a non-sister species, *D. hirundinaceum*, and from the Bird's Neck westward to the NW Islands by sister species *D. pectorale*.

Extralimital—Australian islands offshore Mainland NG: Boigu and Saibai I. (Higgins *et al.* 2006).

SUBSPECIES

a. *Dicaeum geelvinkianum maforense* Salvadori

Dicaeum maforense Salvadori, 1876. *Ann. Mus. Civ. Genova* (1875) 7: 944.—Numfor I, Bay Is.

RANGE: Numfor I.

DIAGNOSIS: Dorsal plumage dull olive with no blue gloss; crown of male with dark and obscure red patch; rump patch similar; male breast patch large, pale red, and diffuse; wing (male) 52 (Rand & Gilliard 1967).

b. *Dicaeum geelvinkianum misoriense* Salvadori

Dicaeum misoriense Salvadori, 1876. *Ann. Mus. Civ. Genova* (1875) 7: 945.—Korido, Biak I, Bay Is.

RANGE: Biak I.

DIAGNOSIS: Differs from the similar *maforense* in the reduced and patchy red throat patch of male; red rump patch of female of this form may be larger and brighter than that of female *maforense*.

c. *Dicaeum geelvinkianum geelvinkianum* A. B. Meyer

RANGE: Yapen and Kurudu I, Bay Is (Mayr 1941).

DIAGNOSIS: Male's breast patch compact and bright red; red crown patch a bit irregular but bright; rump patch as for crown patch; whitish chin patch distinct, offset from dull olive of cheeks and breast; belly dull pale whitish yellow; mantle with slight indication of blue-black sheen. Female as for male, but white chin and red breast patch absent.

d. *Dicaeum geelvinkianum obscurifrons* Junge

Dicaeum geelvinkianum obscurifrons, 1952. *Zool. Meded. Leiden* 31(22): 249.—Paniai Lake, Western Ra.

RANGE: The e verge of the Bird's Neck (ne of Etna Bay) east to Paniai Lake of the w end of the Western Ra (Rand & Gilliard 1967, Cheke & Mann 2008).

DIAGNOSIS: Most like *maforense*, but larger size, and red breast patch brighter red; wing (male) 58 (Rand & Gilliard 1967). Duller and browner than *diversum* from farther east in the Western and Border Ra (Cheke & Mann 2008).

e. *Dicaeum geelvinkianum diversum* Rothschild & Hartert

Dicaeum geelvinkianum diversum Rothschild & Hartert, 1903. *Novit. Zool.* 10: 215.—Lower Mamberamo R, nw NG.

> Synonym: *Dicaeum geelvinkianum setekwa* Rand, 1941. *Amer. Mus. Novit.* 1102: 14.—Setekwa R, 600 m, sw sector of the Western Ra.
> Synonym: *Dicaeum geelvinkianum centrale* Rand, 1941. *Amer. Mus. Novit.* 1102: 15.—Baliem R, 1600 m, east-central sector of the Western Ra.

RANGE: The Western Ra and w and c sectors of the Border Ra.

DIAGNOSIS: Male is like nominate, but mantle with more blue-black gloss, and crown less patchy and more intense red. The form *setekwa* is nearly identical, but some females show a tiny remnant reddish patch on breast; *centrale* is slightly larger but does not merit recognition, as it is clinal. The described form *simillimum* Hartert was synonymized with this form by Mayr (1941).

f. *Dicaeum geelvinkianum rubrigulare* D'Albertis & Salvadori

Dicaeum rubrigulare D'Albertis & Salvadori, 1879. *Ann. Mus. Civ. Genova* 14: 74.—middle Fly R, S Lowlands.

RANGE: Known from the middle and lower Fly R, from Palmer Junction to the mouth, at Gaima (Rand & Gilliard 1967).

DIAGNOSIS: Similar to *rubrocoronatum*, but red breast patch larger and extending onto throat; chin white (Rand & Gilliard 1967).

g. *Dicaeum geelvinkianum albopunctatum* D'Albertis & Salvadori

Dicaeum albo-punctatum D'Albertis & Salvadori, 1879. *Ann. Mus. Civ. Genova* 14: 75.—Katau R, s NG.

RANGE: Daru I and the Trans-Fly, from the Oriomo R west at least to Kurik and Merauke; range of this form probably extends west to the Digul R (Mees 1982b).

DIAGNOSIS: Similar to *rubrigulare*, but red breast patch still larger and covers chin. There seems to be a steep cline along the flow of the Fly R, with the southernmost birds having a very large red chin and breast patch, the northernmost birds more typical of the widespread *diversum* and the middle Fly birds intermediate.

h. *Dicaeum geelvinkianum rubrocoronatum* Sharpe

Dicaeum rubrocoronatum Sharpe, 1876. *Nature*: 339.—Port Moresby, SE Peninsula.

RANGE: Karkar[J], Bagabag[J], and Manam I, the Sepik-Ramu, e sector of the S Lowlands, e sector of the Eastern Ra, and Huon and SE Peninsulas (Mayr 1941, Diamond & LeCroy 1979[J]).

DIAGNOSIS: Similar to *diversum* but mantle of male blacker and more purple-glossed; wing (male) 53 (Rand & Gilliard 1967). Mantle a deep glossy violet; red throat patch of the male rather small (Diamond 1972). See photos in Coates (1990: 309) and Coates & Peckover (2001: 187).

i. *Dicaeum geelvinkianum violaceum* Mayr

Dicaeum geelvinkianum violaceum Mayr, 1936. *Amer. Mus. Novit.* 869: 6.—Goodenough I, D'Entrecasteaux Arch.

RANGE: D'Entrecasteaux Arch (including Dobu; Mayr 1941).

DIAGNOSIS: Similar to *rubrocoronatum*, but dorsal plumage duller and less glossy; wing 53 (Rand & Gilliard 1967). Cheke & Mann (2008) noted also the following: paler and duller dorsally, with a purple-violet gloss; red in plumage darker; ventral plumage more gray-washed; abdomen grayish olive.

j. *Dicaeum geelvinkianum nitidum* Tristram

Dicaeum nitidum Tristram, 1889. *Ibis*: 555.—Tagula I, Louisiade Arch, SE Islands.

RANGE: Tagula and Misima I (Mayr 1941).

DIAGNOSIS: Compared with *rubrocornatum*, red of breast spot extends further down breast; abdomen and flanks more yellowish (Cheke & Mann 2008). Wing (male) 56 (Rand & Gilliard 1967). Some authorities (*e.g.*, Rand & Gilliard 1967, Cheke & Mann 2008) have treated this and *rosseli* as a distinct species (*D. nitidum*).

k. *Dicaeum geelvinkianum rosseli* Rothschild & Hartert

Dicaeum geelvinkianum rosseli Rothschild & Hartert, 1914. *Bull. Brit. Orn. Club* 35: 32.—Rossel I, Louisiade Arch, SE Islands.

RANGE: Rossel I, Louisiade Arch.

DIAGNOSIS: Similar to preceding but larger and paler; wing (male) 59 (Rand & Gilliard 1967). Cheke & Mann (2008) noted the following: shoulder glossed blue-green; dorsally more olive green and less blackish than *nitidum*.

NOTES

The plumage characters are to a degree clinal; birds are plainer in the west and are more colorful—with larger and brighter red patches and glossy black backs—in the c and e Mainland. We follow the suggestion of Rand & Gilliard (1967: 578) and Coates (1990) in treating *nitidum* as a subspecies of the widespread and variable *D. geelvinkianum* (*pace* Salomonsen 1967), because the degree of character differentiation seems to be no greater than for some of the mainland subspecies. A molecular investigation would better inform the phylogeography of this complex species. *Other names*—includes Louisiade Flowerpecker, *Dicaeum nitidum*.

Mistletoebird *Dicaeum hirundinaceum*

Motacilla hirundinacea Shaw, 1792. *Nat. Misc.* 4: pl. 114.—Australia.
    In the Region known only from the Aru Is.
    Extralimital—throughout Australia, and Kai and Tanimbar Is (Higgins *et al.* 2006).

SUBSPECIES

a. *Dicaeum hirundinaceum ignicolle* G. R. Gray

Dicaeum ignicolle G. R. Gray, 1858. *Proc. Zool. Soc. London*: 173.—Aru Is.

RANGE: The Aru Is. The sight record from near Morehead in the Trans-Fly (Bellchambers *et al.* 1994) is now thought be an error (see *Muruk* 7: 96).

DIAGNOSIS: Male differs from nominate form of Australia in having dull yellowish underparts with olive-green flanks (Cheke & Mann 2008).

NOTES

Iredale (1956) reported a specimen from the "head of the Gulf of Papua." This specimen, collected by T. F. Beven and registered with the Australian Museum in Sydney, cannot be found (Coates 1990). This surprising Mainland record requires corroboration. This taxon needs a review; the pink-bellied populations in the s Moluccas are very distinctive (D. Hobcroft *in litt.*). *Other names*—Australian or Mistletoe Flowerpecker.

Family Nectariniidae

Sunbirds   [Family 137 spp/Region 3 spp]

We follow Cracraft (2014) in treating the Nectariniidae as a full family distinct from Dicaeidae, the flowerpeckers (*pace* Ericson & Johansson 2003, Barker *et al.* 2004). This traditional treatment of the sunbirds includes species that range from Africa through tropical Asia, east to the Solomon Islands and Australia. Central Africa and India are the regions of greatest species richness. The diet is nectar and arthropods (primarily spiders). Members of the family are characterized by small size, colorful plumage, long and decurved bill, and specialized adaptations to feeding on flowers. Situation of the family follows Barker *et al.* (2004).

GENUS *LEPTOCOMA*   [Genus 5 spp/Region 1 sp]

Leptocoma Cabanis, 1851. *Mus. Hein.* 1: 105. Type, by original designation, *Nectarinia hasseltii* Temminck = *Certhia brasiliana* Gmelin.
    The species of this genus range from tropical Asia southeast to the Bismarck Archipelago (Cheke & Mann 2008). The genus is characterized by the short tail, the dark mantle, the metallic green crown, and the iridescent and metallic blackish throat. Generic treatment follows Cheke & Mann (2008).

Black Sunbird *Leptocoma aspasia*   RESIDENT

Cinnyris aspasia Lesson & Garnot, 1828. *Voy. Coquille, Atlas* (1830) 7: pl. 30, fig. 4; *Zool.* 15: 676.—Manokwari, Bird's Head.
    Synonym: *Cinnyris sericeus* Lesson, 1827. *Dict. Sci. Nat.*, éd. Levrault 1: 21.—Near Manokwari, Bird's Head.
      Replacement name for *C*[*erthia*] *sericea* Bechstein, 1811 (see Hachisuka 1952: 46; LeCroy 2010: 137).
Throughout NG and its satellite islands, mainly at low elevations, but in disturbed habitats to 1,200 m. Frequents mainly rain forest and edge. Absent from Rossel I.
    Extralimital—Sulawesi to the Moluccas and the Bismarck Arch (Mayr & Diamond 2001).

SUBSPECIES

a. *Leptocoma aspasia mariae* Ripley

Nectarinia sericea mariae Ripley, 1959. *Postilla* Yale Peabody Mus. 38: 13.—Kofiau I, NW Islands.

RANGE: Kofiau I (Ripley 1964a).

DIAGNOSIS: Similar to *cochrani*, but male exhibits throat glossed pansy violet; greener cap than *auriceps*; wing coverts and rump yellowish blue-green; female brighter yellow ventrally than *cochrani* (Cheke & Mann 2008). A thinly defined form perhaps meriting subsuming into the form *cochrani*.

b. *Leptocoma aspasia auriceps* (G. R. Gray)

Nectarinia auriceps G. R. Gray, 1861. *Proc. Zool. Soc. London* (1860): 348.—Bacan and Ternate I.

RANGE: Gebe I (Mees 1972).

DIAGNOSIS: Male plumage with throat bluish; cap golden green; rump and wing coverts steel blue. Female with yellow underparts paler than those of *mariae* (Ripley 1959).

c. *Leptocoma aspasia cochrani* (Stresemann & Paludan)

Cinnyris sericeus cochrani Stresemann & Paludan, 1932 (January). *Orn. Monatsber.* 40: 15.—Waigeo I, NW Islands.

RANGE: Waigeo and Misool I[MA]; birds on Salawati and Batanta I[L] probably belong here (Greenway 1966[L]).

DIAGNOSIS: Slightly smaller than the nominate form; male with a bright purple-blue throat; differs from *caeruleogula* of Umboi I and New Britain in having wing coverts and rump bright green or blue-green with yellow suffusion (Cheke & Mann 2008). Mayr (1941) and Rand (1967) left unresolved the subspecific affinities of the insular populations from Salawati and Batanta I, and this lapse was not resolved by Cheke & Mann (2008).

d. *Leptocoma aspasia aspasia* (Lesson & Garnot)

    Synonym: *Cinnyris salvadorii* Shelly, 1877. *Monogr. Sunbirds*: 105 (pl. 35, fig. 2).—Yapen I, Bay Is.

    Synonym: *Nectarinia sericea bergmanii* Gyldenstolpe, 1955. *Arkiv f. Zool.* 8: 353.—Adi I, off Bomberai Penin, Bird's Neck.

    Synonym: *Cinnyris sericea vicina* Mayr, 1936. *Amer. Mus. Novit.* 869: 5.—Simbang, Huon Penin.

RANGE: Adi, Yapen, and Kurudu I, and all NG (Mees 1965, Greenway 1966); also included in range are Karkar[J], Bagabag[J], and Manam I (though birds slightly larger); and Doini I[CO] (Diamond & LeCroy 1979[J]).

DIAGNOSIS: Male head glossy bluish green; greenish-blue gloss on rump, uppertail coverts, and tail; throat patch purple; female wings and mantle olive gray or olive green (Cheke & Mann 2008). Rand (1967) subsumed *bergmanii* and *vicina* into the nominate form.

e. *Leptocoma aspasia aspasioides* (G. R. Gray)

Nectarinia aspasia aspasioides G. R. Gray, 1861. *Proc. Zool. Soc. London* (1860): 348.—Ambon, s Moluccas.

    Synonym. *Chalcostetha chlorocephala* Salvadori, 1874. *Ann. Mus. Civ. Genova* 6: 78.—Wokan, Aru Is.

RANGE: The Aru Is and Moluccas—Seram, Ambon, Nusa Lau, Watubela.

DIAGNOSIS: Male long-billed; differs from nominate in having greenish crown, a more blue-greenish rump and wing coverts, and dark blue throat; female has olive-gray underparts (Cheke & Mann 2008).

f. *Leptocoma aspasia maforensis* (A. B. Meyer)

Chalcostetha aspasia var. *maforensis* A. B. Meyer, 1874. *Sitzungsber. Akad. Wiss. Wien* 70: 123.—Numfor I, Bay Is.

RANGE: Numfor I.

DIAGNOSIS: Male with a green back and golden crown (Cheke & Mann 2008).

g. *Leptocoma aspasia mysorensis* (A. B. Meyer)

Chalcostetha aspasia var. *mysorensis* A. B. Meyer, 1874. *Sitzungsber. Akad. Wiss. Wien* 70: 124.—Biak I, Bay Is.

RANGE: Biak I and Schouten Is.

DIAGNOSIS: Longer-billed than *cornelia*; male typically with blackish, unglossed ventral plumage; some specimens have some violet gloss in throat; more blue-green dorsally than nominate (Cheke & Mann 2008).

h. *Leptocoma aspasia nigriscapularis* (Salvadori)

Hermotimia nigriscapularis Salvadori, 1876. *Ann. Mus. Civ. Genova* (1875) 7: 937.—Mios Num I, Bay Is.

RANGE: Mios Num and Rani I (13 km south of Biak).

DIAGNOSIS: Male distinctive in having second and third rows of scapulars black; primaries velvety black (Cheke & Mann 2008).

i. *Leptocoma aspasia cornelia* (Salvadori)
Hermotimia cornelia Salvadori, 1878. *Atti R. Ac. Sci. Torino* 13: 319.—Tarawai I, n NG.
RANGE: Tarawai I.
DIAGNOSIS: Larger than nominate; male with crown, back, and rump glossed blue-green; throat glossed purple-violet (Cheke & Mann 2008).

j. *Leptocoma aspasia veronica* Mees
Leptocoma sericea veronica Mees, 1965. *Ardea* 53: 46.—Liki I, nw NG.
RANGE: Liki I in the Kumamba Is (north of Sarmi, off the n coast of the NW Lowlands).
DIAGNOSIS: A large form; male exhibits slightly more bluish-tinged green crown; wing coverts, underparts, and rump steel blue (Cheke & Mann 2008).

k. *Leptocoma aspasia christianae* (Tristram)
Cinnyris christianae Tristram, 1889. *Ibis*: 555.—Misima I.
RANGE: The D'Entrecasteaux Arch; Trobriand, Amphlett, and Marshall Bennett Is; also Woodlark, Tagula, and Misima I (Mayr 1941).
DIAGNOSIS: Large and long-billed; male crown dark green; throat glossed blue; female more yellow below than nominate (Cheke & Mann 2008). See photos in Coates (1990: 304-305) and Coates & Peckover (2001: 186).

NOTES
For generic treatment we follow Rand (1967) and Cheke & Mann (2008). For reasons against resurrecting the name *sericea* following the shift in generic treatment to *Leptocoma*, see detailed argument by LeCroy (2010: 137), which we follow here. The numerous subspecies are based mainly on color and iridescence of the male's glossy black plumage and on size and bill length. Whether this variability reflects significant genetic differentiation is unknown. *Other names—Nectarinia sericea, N. aspasia, Leptocoma sericea.*

GENUS *CINNYRIS* [Genus 51 spp/Region 2 spp]

Cinnyris Cuvier, 1816. *Régne Animal.* 1 (1817): 411. Type, by subsequent designation (G. R. Gray, 1855, *Cat. Gen. Subgen. Birds*: 19), *Certhia splendida* Shaw = *Certhia coccinigaster* Latham.
    The species of this genus range from western and southern Africa east through tropical Asia to New Guinea, Australia, and the Bismarck and Solomon Islands. The genus contains a diverse assemblage of African and Asian sunbirds, with large and small forms, long-tailed and short-tailed forms; but all are sexually dichromatic and exhibit some form of breast banding or other distinction between the throat and breast plumage.

Olive-backed Sunbird *Cinnyris jugularis* RESIDENT
Certhia jugularis Linnaeus, 1766. *Syst. Nat.*, ed. 12, 1: 185.—the Philippines.
    Throughout NG and all satellite islands except Gebe and Woodlark I, the Louisiade and Trobriand Is, and much of the range of *C. idenburgi* (NW Lowlands and Sepik-Ramu). Mainly a lowland species but under disturbed conditions ranges up to 1,580 m (Coates & Peckover 2001). Frequents second-growth forest and edge, scrub, coconut groves, and gardens (Coates 1990).
    Extralimital—the Andaman and Nicobar Is, insular se Asia to ne Australia, and the Bismarck and Solomon Is (Mayr & Diamond 2001). Inhabits many islands in Torres Strait.

SUBSPECIES
a. *Cinnyris jugularis frenatus* (S. Müller)
Nectarinia frenata S. Müller, 1843. *Verh. Nat. Gesch. Ned., Land-en Volkenk.* 1: 173.—Triton Bay, Bird's Neck.
    Synonym: *Cyrtostomus frenatus valia* Mathews, 1929. *Bull. Brit. Orn. Club* 50: 11.—Goodenough I,
        D'Entrecasteaux Arch. Replacement name for *Cyrtostomus frenatus olivaceus* Mathews (not *Cinnyris*
        olivaceus Smith, 1839).
RANGE: The Sula Is, n Moluccas, all NW Islands but for Gebe[K,L,M], the Aru Is[CO], Bay Is[CO], and D'Entrecasteaux Arch (Goodenough[MA], Fergusson[MA], and Normanby[O]), and Mainland but for range of

C. idenburgi. The distribution of this subspecies includes Manam[CO], Karkar[J], Bagabag[J], the Madang islets[CO], Port Moresby islets (Motupore, Loloata, Daugo, Idihi)[CO], Dobu I, and Wamea I of the Amphlett Is[CO] (Greenway 1966[L], Diamond & LeCroy 1979[J], LeCroy *et al.* 1984[O], Johnstone 2006[M], Diamond *et al.* 2009[K]). No records from the Trobriands, Louisiade Arch, or Woodlark I.

DIAGNOSIS: Male exhibits more extensive metallic gloss on breast shield and a more obvious malar stripe compared to *ornatus* from much of Sundaland (Cheke & Mann 2008). Schodde & Mason (1999) noted the following: mantle "bright citrine-olive" and belly "rich mid-yellow" in both sexes; male with full mid-yellow malar and postocular stripe; size large: wing (male) 54–58. See photos in Coates & Peckover (2001: 186).

NOTES
Cinnyris is masculine (David & Gosselin 2002b). We follow the suggestion of Cheke & Mann (2008) in splitting off *C. idenburgi* as distinct from *jugularis* as well as from the other black-bellied forms (*C. teysmanni* and the *C. clementiae* group). See details in the following account. There are museum specimens of typical male *C. jugularis* from Jayapura, Wewak, Stephansort, Oronga in the Adelbert Mts (Gilliard & LeCroy 1967), and Keku (R. Beck specimen at AMNH). This is clear evidence that *C. jugularis* and *C. idenburgi* share considerable territory and are at least partially sympatric. *Other names*—Yellow-bellied, Black-throated, or Yellow-breasted Sunbird, *Nectarinia jugularis*, *Cyrtostomus frenatus*.

Rand's Sunbird *Cinnyris idenburgi*   ENDEMIC | MONOTYPIC

Cinnyris jugularis idenburgi Rand, 1940. *Amer. Mus. Novit.* 1072: 12.—Bernhard Camp, 50 m elevation, Taritatu R, Mamberamo basin, s sector of the NW Lowlands.

Inhabits the interior drainage of the Mamberamo basin; apparently also in the Sepik basin as well as that of the middle Ramu[CO]. Species range is poorly delineated, based on few collections or field encounters: Bernhard Camp on the Taritatu R, Karawari airstrip in the lower Sepik, base of the Schrader Ra, and Aiome in the middle Ramu (Rand 1942b, Coates 1990).

NOTES
We follow the suggestion of Cheke & Mann (2008) in splitting off *C. idenburgi* as distinct from *jugularis* as well as from the Moluccan black-bellied forms (*teysmanni* and *clementiae*). The range of *idenburgi* is entirely enclosed within that of the yellow-bellied *C. jugularis frenatus* and also is geographically isolated from the Moluccan black-bellied forms. Bill and tail are smaller than in *jugularis* (Rand 1940a). The adult male exhibits: darker plumage than yellow-bellied *jugularis*; no yellow eyebrow; blackish face; tail black with only an indication of small terminal gray marks in the ventral surface of the outer rectrices; entire underparts except pectoral tufts glossy bluish black; pectoral tufts bright orange; crown dark greenish olive; darker back tinged with golden brown (Rand 1940a). The female differs in having darker, brownish-washed upperparts; smaller, grayish terminal marks on the outer tail feathers; and a whitish throat (Rand 1940a). There are specimens of a male and female sunbird from Aiome in the middle Ramu in the collection of the AMNH. The male is identical to topotypical *idenburgi*. The female, an apparent adult, tends toward typical female *C. j. frenatus*, with bright ventral plumage including the throat and chin, a pale yellow eyebrow, and whitish tail spotting (Gilliard & LeCroy 1968). It is unknown whether they were a mated pair; perhaps not, and instead they give evidence that the two species coexist. Pratt and J. A. Anderton observed a typical male *idenburgi* at Karawari airstrip in the lower Sepik in 2011. The full distribution of *idenburgi* and its interaction with *jugularis* merits study. *Other names*—formerly treated as a subspecies of *C. jugularis*.

Family Passeridae

Old World Sparrows [Family 38 spp/Region 2 introduced spp]

The species of this family range from western Europe and Africa east to Siberia, China, Japan, and tropical Asia. The family is represented in the New Guinea region by two species in the genus *Passer*. Birds of the family are common, adaptable, and primarily terrestrial; most species build bulky and untidy nests in trees, shrubs, or buildings. Barker *et al.* (2004) placed this lineage as sister to a large assemblage that includes the wagtails and Fringillidae, among others.

GENUS *PASSER* [Genus 23 spp/Region 2 introduced spp]

Passer Brisson, 1760. *Orn.* 1: 36. Type, by subsequent designation, "Passer" Brisson = *Fringilla domestica* Linnaeus.

Two species have been introduced (or self-introduced) to the Region since 1975. The species of this genus range through Europe, Africa, and Asia, and have been introduced to the Americas and Australasia (Summers-Smith 2009). The genus is characterized by the compact body, large head, small conical bill, sexual dichromatism, pale ocular stripe of the female, and diagnostic black chin patch of the male (absent in some aberrant species). Barker *et al.* (2004) placed the genus in a clade that includes *Prunella* (Prunellidae), *Ploceus* (Ploceidae), *Motacilla* (Motacillidae), and *Fringilla* (Fringillidae).

House Sparrow *Passer domesticus* INTRODUCED

Fringilla domestica Linnaeus, 1758. *Syst. Nat.*, ed. 10, 1: 183.—Sweden.

There are scattered records for NG of this widespread nonnative species.

Extralimital—Eurasia and n Africa; has spread or been introduced nearly worldwide, including e Australia, Vanuatu, and New Caledonia (Dutson 2011).

SUBSPECIES

a. *Passer domesticus ?domesticus* (Linnaeus)

RANGE: The nominate form's native haunts include w and n Europe east to Siberia. Localities—Sorong[L], Biak I[Q], Eastern Ra (Tabubil[J], Mt Hagen[J]), Wasu on n coast of Huon[K], Nadzab[K], and Port Moresby[L] (B. J. Coates *in litt.*[J], Holmes 1997[L], Beehler field notes[K], van Balen & Bishop ms[Q]). Probably more widely distributed than these records indicate. Introduced (or self-introduced) from Australia to PNG, and perhaps more actively introduced in Indonesian NG. Currently well established in Port Moresby (K. D. Bishop *in litt.*). Status elsewhere poorly documented. First record was of four birds at Kila Kila, Port Moresby, on 30-Dec-1976 (Ashford 1978). Also a female on Paga Hill, Port Moresby, 10–12-Jun-1986 (Hicks 1986). Six found at Mendi in 2004 (P. Gregory in litt.). Observed at Tabubil, Mt Hagen, and Kiunga (K. D. Bishop *in litt.*). Coates (1990) supposed these birds arrived as stowaways on ships sailing from Australia to PNG. Ten birds were observed by Hicks (1989a) on Yule I on 24-Dec-1989. First recorded as a breeding resident in the Torres Strait Is in 1978 (Draffan *et al.* 1983).

DIAGNOSIS: According to Schodde & Mason (1999), the nominate form inhabits e Australia. Presumably this is the form that has colonized PNG. In male, note the pale postocular spot and the gray (not white) cheek, and pale-streaked mantle (Summers-Smith 2009). Schodde & Mason (1999) noted the following: male with rich ash-gray cap and rump, rich rufous-chestnut nuchal band, pale ash-gray cheek and ear covert; female with deep sandy-gray mantle with black streaking.

NOTES

One can assume that this species may come to settle in most or all communities in the Region within the next decade. *Other names*—English Sparrow.

Eurasian Tree Sparrow *Passer montanus*   INTRODUCED

Fringilla montana Linnaeus, 1758. *Syst. Nat.*, ed. 10, 1: 183.—Northern Italy.

   Introduced (or self-introduced) to NG.

   Extralimital—Europe and Asia; introduced to se Australia, New Britain, and Guadalcanal (Dutson 2011). Also locally in North America.

SUBSPECIES

a. *Passer montanus ?malaccensis* A. J. C. Dubois

Passer montanus malaccensis A. J. C. Dubois, 1887. *Fauna. Ill. Vert. Belg. Ois.* 1: 572.—Malacca, Malay Penin.

RANGE: The native range of this subspecies extends from Nepal east to Burma, Thailand, Indochina, and Bali. Presumably birds from Indonesia made their way (carried on ships) east to NG. Current range includes the NW Islands, Bird's Head, Bird's Neck, Bay Is, Western Ra, NW Lowlands, Border Ra, Sepik-Ramu, S Lowlands, Trans-Fly, and SE Peninsula. Localities—Salawati[L], Batanta[L], Sorong[W,L], Manokwari[W], Lobo[X], Kaimana[X], Biak[M,Q], Numfor[W], Tembagapura[U], Wamena[W,A], Amamapare near Timika[Z], Nimbokrang[A], Jayapura[J], Sentani[W], upper Sepik[K], Kerema[K], Tabubil[A,V], Lake Murray[J], Lae[K] and Port Moresby[B] (Holmes 1989[M], 1997[J], S. F. Bailey *in litt.*[Q], Hornbuckle & Merrill 2004[L], van Balen *et al.* 2011[Z], van Balen & Noske 2014[U], Woxvold *et al.* 2015[V], G. Dutson *in litt.*[W], B. J. Coates *in litt.*[B], I. Woxvold *in litt.*[K], C. Thébaud & B. Mila *in litt.*[X], P. Gregory *in litt.*[A]).

DIAGNOSIS: Similar to but smaller than *P. m. saturatus* of e Asia, often with heavier streaking dorsally (Summers-Smith 2009).

NOTES

For a discussion of the subspecific provenience of the Australia birds, see Higgins *et al.* (2006). *Other names*—Tree Sparrow or European Tree Sparrow.

Family Estrildidae

Mannikins, Parrotfinches, and Allies   [Family 132 spp/Region 17 spp]

The species of the family range from sub-Saharan Africa east through tropical Asia, Australasia, and the Pacific Islands. These are small, chubby, seed-eating birds, with a heavy, pointed bill, and a thin, tapering tail (Higgins *et al.* 2006). Many species are patterned and colorful. Most inhabit grasslands and travel in flocks. Bentz (1979) collected detailed anatomical data on the Ploceidae and Estrildidae to inform their relationships. The sequence of genera follows that of Christidis & Boles (2008), who discuss initial molecular work surrounding this lineage. Groth (1998) found the Ploceidae, Estrildidae, and Prunellidae to form a clade.

GENUS *OREOSTRUTHUS*   ENDEMIC | MONOTYPIC

Oreostruthus De Vis, 1898. *Ibis*: 175. Type, by monotypy, *Oreospiza fuliginosa* De Vis. Replacement name for *Oreospiza* De Vis, 1897, *Ibis*: 388; preoccupied by *Oreospiza* Ridgway, 1896, and *Oreospiza* Keitel, 1857.

   The single species of the genus is endemic to the Central Ranges of New Guinea. The genus is characterized by the plump body, the blunt conical red bill, and the predominantly brown plumage with red rump and red wash on flanks.

Mountain Firetail *Oreostruthus fuliginosus*   ENDEMIC | MONOTYPIC

Oreospiza fuliginosa De Vis, 1897. *Ibis*: 389.—Mount Scratchley, SE Peninsula.

   Synonym: *Oreostruthus fuliginosus pallidus* Rand, 1940. *Amer. Mus. Novit.* 1072: 14.—Lake Habbema, 3,225 m, Western Ra.

Synonym: *Oreostruthus fuliginosus hagenensis* Mayr & Gilliard, 1954. *Bull. Amer. Mus. Nat. Hist.* 103: 372.—Southern slope of Mt Hagen, 3,050 m, Eastern Ra.

High mts of all the Central Ra; 2,700–3,700 m, min 2,200 m^{CO}. Range extends from the Tembagapura Road^J (Western Ra) east to Myola^{CO} and Mt Knutsford^{MA} of the SE Peninsula (Beehler field notes^J). Frequents mainly grassy glades in forest and shrubs bordering alpine grassland, but in the Hindenburg Mts, in the absence of such habitat it is found in pure forest (Rand 1942b, Coates 1990).

NOTES

We do not believe the trivial clinal variation in plumage intensity merits subspecies recognition. Plumage ranges from rich and dark in the southeast to pale and washed-out in the northwest (Payne 2010). *Other names*—Crimson-sided or Red-sided Mountain Finch.

GENUS *NEOCHMIA* [Genus 4 spp/Region 1 sp]

Neochmia G. R. Gray, 1849. *Gen. Birds* 2: 369. Type, by original designation, *Fringilla phaeton* Hombron & Jacquinot.

The species of this genus range from Australia to southern New Guinea (Payne 2010). The genus is characterized by the long, pointed tail, the broad conical bill, and red plumage. As defined by Payne (2010), this is a poorly circumscribed genus with much plumage variation among the member species.

Crimson Finch *Neochmia phaeton* RESIDENT

Fringilla phaeton Hombron & Jacquinot, 1841. *Ann. Sci. Nat. Paris* (2)16: 314.—Raffles Bay, NT, Australia.

Inhabits grasslands, wet reedbeds, ricefields, and gardens in the Trans-Fly and southernmost c sector of the S Lowlands; at sea level. Now present in the Sentani area of the ne sector of the NW Lowlands. These are probably escaped cage birds.

Extralimital—n and ne Australia.

SUBSPECIES

a. *Neochmia phaeton evangelinae* D'Albertis & Salvadori

Neochmia evangelinae D'Albertis & Salvadori, 1879. *Ann. Mus. Civ. Genova* 14: 89.—Lower Fly R, S Lowlands.

RANGE: The Trans-Fly, the south-central S Lowlands, and Cape York Penin, QLD (Schodde & Mason 1999). Localities—Dolak I^{CO}, Digul R^K, Kurik^K, between the Maro and Bian R^K, Merauke^{CO}, Wasur NP^K, Gaima^X, Dogwa^M, Lake Murray^{CO}, Lake Daviumbu^K, Rumgenai^L, Oriomo R^{CO}, Balimo^{CO}, Aramia R^K (Mayr & Rand 1937^M, Rand 1942a^X, Gregory 1997^L, Bishop 2005a^K). Recently recorded by H. van der Kerkhof at the Doyo Baru airstrip, *ca.* 7 km west of Sentani, and by P. Gregory at Nimbokrang (in 2015), on the n coast (probably offspring of escaped cage birds). Also photographed at Sentani in 2005 (posted at the Internet Bird Collection). These birds match this race.

DIAGNOSIS: Schodde & Mason (1999) noted: belly off-white; crissum dark gray. See photo in Coates (1990: 335).

NOTES

Other names—Blood, Red-tailed, or Red-faced Finch, Australian Firefinch.

GENUS *ERYTHRURA* [Genus 10 spp/Region 2 spp]

Erythrura Swainson, 1837. *Nat. Hist. Class. Birds* 2: 280 (originally misspelled *Erythura*, but universally emended). Type, by monotypy (Temminck, 1821. *Planch. Col. d'Ois.*, livr. 96), *Erythura viridis* Swainson, 1837 = *Fringilla specura* Temminck = *Loxia prasina* Sparrman.

The species of this genus range from southeast Asia east to New Guinea, Australia, Micronesia, Fiji, Vanuatu, and the Loyalty Islands (Payne 2010). The genus is characterized by the conical beak, pointed tail, and brightly patterned plumage with blue and green (but in many cases also with red and buff). Most plumages exhibit a facial pattern or mask.

Blue-faced Parrotfinch *Erythrura trichroa* RESIDENT
Fringilla trichroa Kittlitz, 1833. *Mem. Pres. Acad. Imp. Sci. St. Pétersbourg* 2: 8, tab. 10.—Kosrae, Caroline Is.

Hills and mts of NG, islands off the n coast, and SE Islands, 750–3,700 m (P. Z. Marki *in litt.*) Occasionally near sea level in hills. Frequents forest edge, glades, and gardens. A generalist forest finch that congregates at seeding bamboo (Diamond 1972, Mack & Wright 1996).

Extralimital—Sulawesi to Micronesia, Bismarcks, Solomons, Vanuatu, Loyalty Is, and ne Australia; widespread in the Bismarcks (Coates 1990, Mayr & Diamond 2001, Higgins *et al.* 2006, Dutson 2011).

SUBSPECIES

a. *Erythrura trichroa sigillifer* (De Vis)
Lobospingus sigillifer De Vis, 1897. *Ibis*: 389.—Mountains of se NG; apparently Mt Scratchley, SE Peninsula.
RANGE: All mts of NG; Manam[MA], Karkar[MA], Goodenough[MA], Fergusson[J], and Tagula I[MA]; New Britain, New Ireland, Crown, Long, Tolokiwa, and Umboi I; also Feni Is and Cape York Penin, QLD (Stresemann 1923, Dutson 2011, specimen at BPBM[J]).
DIAGNOSIS: Slightly less yellow-toned above than the nominate form of the Caroline Is; also in some cases with a bluish wash ventrally (Payne 2010).

NOTES
Pratt collected the species on 14-Feb-2003 at 1,375 m on Mt Kilkerran, Fergusson I (BPBM no. 184542), a new island record. *Other names*—Blue-headed Parrotfinch, Blue-faced Finch.

Papuan Parrotfinch *Erythrura papuana* ENDEMIC | MONOTYPIC
Erythrura trichroa papuana Rothschild & Hartert, in Hartert, 1900. *Novit. Zool.* 7: 7.—Arfak Mts, Bird's Head.

Mountains of NG; 1,200–2,600 m, min 500 m. Widely but patchily distributed through the mts of the Bird's Head, Central Ra, and Adelbert Mts. Apparently absent from the mts of the Huon Penin (Freeman *et al.* 2013).

Localities—Tamrau Mts[MA], Arfak Mts[MA], Paniai Lake[CO], Thurnwald Ra[CO], upper Sepik R[CO], Ambua[J], Awande[DI], Okasa[DI], Adelbert Mts[K], Aseki[CO], Mt Missim[X], Bulldog Road[X], Mt Tafa[Z], and Agaun[CO] (Mayr & Rand 1937[Z], Pratt 1983[K], K. D. Bishop *in litt.*[J], Beehler field notes[X]). Inhabits forest canopy where found foraging at figs and *Castanopsis* oaks (Payne 2010). No records from the e sector of the Western Ra or from the w two-thirds of the Eastern Ra (Payne 2010). Present in good numbers at Rondon Ridge, above Mt Hagen town, during a *Castanopsis* masting event (D. Hobcroft *in litt.*).

NOTES
No racial variation has been described for this species, though Pratt (1983) noted that a single male specimen from the Adelbert Mts has a large patch of turquoise in the center of its breast, unlike males from elsewhere. Further material is required to determine whether this is an aberration or a diagnosable subspecies. LeCroy (2013: 112) confirmed the species name was given by Rothschild & Hartert within a paper authored by Hartert, and hence the authorities here differ from that given in Mayr (1941) and elsewhere. Given the rather elusive nature of the species, one might expect to find it in a number of ancillary ranges where it is currently not recorded, and it should be searched for by future fieldworkers visiting the mts of the Bird's Neck, the Foja Mts, the PNG North Coastal Ra, and mts of the Huon Penin. *Other names*—Large-billed Parrotfinch, Papuan Parrot-Mannikin.

GENUS *LONCHURA* [Genus 29 spp/Region 13 spp]
Lonchura Sykes, 1832. *Proc. Zool. Soc. London*: 94. Type, by subsequent designation (Sharpe, 1890. *Cat. Birds Brit. Mus.* 13: 326), *Fringilla nisoria* Sykes = *Loxia punctulata* Linnaeus.

The species of this genus range from tropical India to the northern Solomon Islands (Payne 2010). The genus is characterized by the small and compact body; the short, pointed tail; the large, conical pale blue-gray bill; and the plumage patterned in black, brown, buff, ocher, and white. Species are mainly grassland seedeaters. Our sequence is adapted from Payne (2010) and Dickinson & Christidis (2014).

[Scaly-breasted Mannikin *Lonchura punctulata*   HYPOTHETICAL

Loxia punctulata Linnaeus, 1758. *Syst. Nat.*, ed. 10, 1: 173.—Calcutta, India.

Coates (1990) noted that this species was reported from the Port Moresby distr in 1966 and 1968 and again as recently as 1982–83 (*PNG Bird Soc. Newsl.* 11: 1, 31: 2, and 201: 37). The first record, from Brown R, was apparently a misidentification of *L. tristissima*. The other records, from the Port Moresby suburbs, may have been of escaped cage birds. No recent reports. We await a credible record before adding to the list for the Region.

Extralimital—tropical Asia to e Indonesia. Introduced successfully to e Australia (Higgin *et al.* 2006). *Other names*—Spice or Nutmeg Mannikin, Spice Finch, Scaly-breasted Munia.]

Black-faced Mannikin *Lonchura molucca*   RESIDENT | MONOTYPIC

Loxia molucca Linnaeus, 1766. *Syst. Nat.*, ed. 12, 1: 302.—Ambon, s Moluccas.

Kofiau and Gag I (NW Islands). Presumably can be found in any grassland area on Kofiau. Discovered by Johnstone (2006) on Gag I in 1997 and by Diamond *et al.* (2009) on Kofiau I in 2002.

Extralimital—Halmahera, the Lesser Sunda Is, and Sulawesi (Payne 2010).

NOTES

We follow Dickinson & Christidis (2014) in treating the species as monotypic. *Other names*—Moluccan Munia or Mannikin, Black-faced Munia.

Streak-headed Mannikin *Lonchura tristissima*   NEAR-ENDEMIC

Munia tristissima Wallace, 1865. *Proc. Zool. Soc. London*: 479.—Bird's Head.

Patchily distributed throughout the NG lowlands and Karkar I; mainly sea level–1,000 m; max 1,700 m on Karkar I (Diamond & LeCroy 1979). Inhabits the narrow grassy fringes of streams or trails through forest, grassy clearings mixed with second-growth bushes, and old gardens. Does not occur in open grassland (Coates 1990). Replaced by sister form *L. leucosticta* in an area of s NG from the Lorentz R east to the Kikori R region and perhaps farther eastward. Also absent from the easternmost sector of the SE Peninsula. Evidence of hybridization between these two sister species has been found in the Lorentz R area (Mees 1958) and near Port Moresby (Coates 1990, Restall 1997). Note that interspecific hybridization is commonplace in this genus (Payne 2010).

Extralimital—Umboi I (Dutson 2011).

SUBSPECIES

a. *Lonchura tristissima tristissima* (Wallace)
RANGE: The Bird's Head and Neck.
DIAGNOSIS: Paler; dark brown on mantle; wing coverts without dark tips; upper rump with narrow black bar; pale streaking on crown and face; upper rump with blackish bar (Payne 2010).

b. *Lonchura tristissima hypomelaena* Stresemann & Paludan
Lonchura tristissima hypomelaena Stresemann & Paludan, in Hartert *et al.*, 1934. *Orn. Monatsber.* 42: 43.—Kunupi, 1,200 m, Western Ra.
RANGE: The w sector of the Western Ra: Weyland and Charles Louis Mts[J] (Rand & Gilliard 1967, King 1979[J], LeCroy 1999).
DIAGNOSIS: Underparts, face, and tail darker than those of the nominate form; wing coverts with dark tips; upper rump with broad black bar; underparts blackish (Payne 2010).

c. *Lonchura tristissima calaminoros* (Reichenow)
Munia calaminoros Reichenow, 1916. *Orn. Monatsber.* 24: 169.—Augustahafen, Sepik-Ramu, ne NG.
RANGE: The w sector of the S Lowlands, the NW Lowlands, Sepik-Ramu[J], n watershed of the SE Peninsula, and s watershed of the SE Peninsula east of Port Moresby (Gilliard & LeCroy 1966[J]). The range of this form may need adjustment once more material is made available. Relative ranges of this form and *hypomelaena* need additional work (see LeCroy 1999: 217–18).

DIAGNOSIS: Lacks black bar on upper rump; note whitish wing bar on greater upperwing coverts (Payne 2010). See photo in Coates & Peckover (2001: 196).

d. *Lonchura tristissima ?bigilalei* Restall

Lonchura tristissima bigilalae Restall, 1995. *Bull. Brit. Orn. Club* 115: 148.—Type locality unknown but in general vicinity of Port Moresby. [Name amended by Snow (1997) to *bigilalei*; see also LeCroy 1999.]

RANGE: The s watershed of the SE Peninsula; exact extent unknown, but mapped by Restall (1997) to range from the Lakekamu basin southeast to Rigo.

DIAGNOSIS: Lacks the black bar on upper rump; longest uppertail coverts brown, not black; ventral plumage paler brown; flanks spotted white and buff. Apparently of hybrid origin (Payne 2010). Restall (1997) followed the nomenclatural amendment of Snow (1997). Furthermore, LeCroy (1999) provided additional clarifying data on nomenclature and ranges. Dickinson (2003) and Dickinson & Christidis (2014) did not accept the name emendations proposed by Snow (1997), based on a differing interpretation of the ICZN Code, but M. LeCroy (*in litt.*) noted that Art. 31.1.2 allows for this sort of emendation, correcting an obvious *lapsus*. Thus status is uncertain (see Dickinson & Christidis 2014).

NOTES

We follow Mayr (1941, 1968) in keeping this species and *L. leucosticta* distinct at the species level. We do this because the juvenile plumages of the two are distinct, and the two forms behave as good species over virtually their entire ranges. Details of hybridization (in the Lorentz R and the sw sector of the SE Peninsula) is limited. No evidence of widespread introgression (but see Coates 1990). The species has apparently been recorded at Kiunga (R. Hicks field notes in LeCroy 1999), which, as LeCroy noted, is near the type locality of *leucosticta*. It is possible these two species meet in the upper Fly. This should be investigated. More details are needed before we take the major step of combining these species, which have stood distinct for more than a century. *Other names*—Streak-headed Munia.

White-spotted Mannikin *Lonchura leucosticta*   ENDEMIC | MONOTYPIC

Munia leucosticta D'Albertis & Salvadori, 1879. *Ann. Mus. Civ. Genova* 14: 88.—Middle Fly R.
> Synonym: *Lonchura leucosticta moresbyae* Restall, 1995. *Bull. Brit. Orn. Club* 115: 149.—Type locality unknown; presumably in the Port Moresby environs.

The S Lowlands and the Trans-Fly, from the Lorentz R east to the Kikori R region (upper Kikori; K. Aplin specimen) and perhaps eastward. Frequent at Bensbach and Kiunga (D. Hobcroft *in litt.*, Gregory 1997). Inhabits grassland and wetlands near forest edge (Coates 1990, Payne 2010). We follow Mayr (1941), Rand (1942a), Rand & Gilliard (1967), and Restall (1997) in keeping this species distinct from the preceding (*pace* Coates 1990, LeCroy 1999, Payne 2010).

NOTES

Male exhibits profuse streaking and spotting of white on crown, face, and back; white eyebrow; buff-brown belly and breast; and extensive yellowish rump abutting brown of back (no black bar). Clearly sister to *L. tristissima*, but exhibiting substantial distinctions in plumage and habit. Coates (1990) and Payne (2010) subsumed this form into *L. tristissima*. We prefer to keep the two as separate species (Restall 1997), in spite of the presence of what appears to be a population of hybrid origin between the upper Purari R and the Port Moresby region (here allied with *tristissima*). Apparent hybrids also reported from the w verge of the range of *leucosticta* (Lorentz R). Evidence of hybridization, per se, does not mandate combining the two taxa in question, as evidenced by the example of the hybridization between *Melidectes belfordi* and *rufocrissalis* as well as *Paradisaea raggiana* and *apoda*. We note that *L. leucosticta* exhibits foraging habits distinct from those of *tristissima*. Rand & Gilliard (1967) wrote, regarding *leucosticta*: "feeds in half-submerged grass mats on the edges of Fly River lagoons." See photo of individuals from the c sector of the S Lowlands in Coates & Peckover (2001: 196).

Following, in part, LeCroy (1999), we choose not to recognize the form *moresbyae* for several reasons: (1) it was described from a single specimen of unknown age and sex; (2) the specimen was collected from an unknown locality, (3) by an unknown collector; (4) it is marginally distinct from the nominate form; and (5) because it is supposedly of hybrid origin, but based on a single specimen, we cannot know whether it is a stable phenotype worthy of a subspecies name. We would be willing to resurrect this form once a sexed and aged series that exhibits the proposed distinctions is collected from a known locality. If and when resurrected, we believe the name properly should be *moresbyensis*, as it was named after its location—the city of Port Moresby—not the man, Captain John Moresby (see LeCroy 1999). *Other names—* some authorities treat as conspecific with *L. tristissima*.

Grand Mannikin *Lonchura grandis*   ENDEMIC

Munia grandis Sharpe, 1882. *Journ. Linn. Soc. London, Zool.* 16: 319.—Taburi, Astrolabe Ra, SE Peninsula.

Locally distributed in the NW Lowlands, Sepik-Ramu, e sector of the Eastern Ra, and the SE Peninsula. From lowlands to as high as 1,280 m (Stresemann 1923, Diamond 1972, Coates & Peckover 2001). Absent from the entire s watershed of NG west of Okapa (Purari drainage). In the n watershed ranges from the Mamberamo basin east to Milne Bay.

SUBSPECIES

a. *Lonchura grandis destructa* (Hartert)

Munia grandis destructa Hartert, 1930. *Novit. Zool.* 36: 42.—Ifar, Lake Sentani, n NG.

> Synonym: *Lonchura grandis heurni* Hartert, 1932. *Nova Guinea, Zool.* 15: 477.—Batavia Bivak, Mamberamo R, nw NG.

RANGE: The NW Lowlands (Coates 1990).

DIAGNOSIS: Similar to the nominate form but mantle browner (Rand & Gilliard 1967). Rump reddish orange; uppertail coverts orange; tail yellow; wing (male) 56–62 (Payne 2010). The form *heurni* does not merit separation from *destructa*.

b. *Lonchura grandis grandis* (Sharpe)

> Synonym: *Munia grandis ernesti* Stresemann, 1921. *Anz. Orn. Ges. Bayern* 1: 33.—Seerosensee, lower Sepik valley, n NG.

RANGE: The Sepik-Ramu[J], Eastern Ra[DI], and SE Peninsula[CO,DI] (Gilliard & LeCroy 1966[J]).

DIAGNOSIS: Rump reddish gold to yellowish gold, grading to chrome orange on uppertail coverts; mantle rufous brown; wing (male) 52–55 (Payne 2010). The form *ernesti* is not distinct from this form. See photos in Coates (1990: 338) and Coates & Peckover (2001: 198).

NOTES

Other names—Great-billed Mannikin, Grand Munia.

Grey-banded Mannikin *Lonchura vana*   ENDEMIC | MONOTYPIC

Munia vana Hartert, 1930. *Novit. Zool.* 36: 42.—Kofo, Anggi Giji Lake, 2,000 m, Arfak Mts, Bird's Head.

Known from a single collection site and a few sight records from highland grasslands of the e Arfak Mts. Most observations from the grasslands of Anggi Giji Lake at 1,900 m. Apparently also a record from the Kebar valley of the Tamrau Mts (I. Mauro, online report).

NOTES

True extent of range unknown and needs further study. Presumably the species inhabits wet grasslands in both Anggi Giji and Anggi Gita—sister lakes—though Gibbs (2009) reported that it is unknown from Anggi Gita. Note that Kofo is Mayr's collecting locality on Anggi Giji (not on Anggi Gita, *pace* LeCroy 2013). *Other names*—Arfak Mannikin, Grey-banded Munia.

Grey-headed Mannikin *Lonchura caniceps*   ENDEMIC

Munia caniceps Salvadori, 1876. *Ann. Mus. Civ. Genova* 9: 38.—Naiabui, Hall Sound, SE Peninsula.

The SE Peninsula: in the n watershed found between Lae and Dyke Ackland Bay; in the s watershed between Malalaua and Kupiano (Coates 1990, Payne 2010). The Lae record is from King (1979). Inhabits grassland and savanna, particularly moist places (Coates 1990). Includes lowland and upland populations that are subspecifically distinct.

SUBSPECIES

a. *Lonchura caniceps caniceps* (Salvadori)

RANGE: Lowlands along the s coast of the SE Peninsula, from Malalaua east at least to Kupiano (Gilliard 1950, Coates 1990).

DIAGNOSIS: The three subspecies are very distinct. This form exhibits a prominent black breast and belly; dorsally similar to *kumusii* but darker-mantled, and pale gray of head sharply offset from dark brown of mantle. See photos in Coates (1990: 339–40) and Coates & Peckover (2001: 197).

b. *Lonchura caniceps scratchleyana* (Sharpe)

Munia scratchleyana Sharpe, 1898. *Bull. Brit. Orn. Club* 7: 60.—Mount Albert Edward, SE Peninsula.

RANGE: Mid-montane grasslands of the s watershed of the SE Peninsula, 1,000–2,080 m. Known from the Auga, Vanapa, Angabanga and Aroa valleys and at Myola (Hicks 1987, Coates 1990).

DIAGNOSIS: Unlike the two lowland subspecies: throat, breast, and abdomen buffy pale rust; lower belly black; crown and nape patterned with dark and paler flecks of brown and buff; mantle buffier than that of *kumusii*.

c. *Lonchura caniceps kumusii* (Hartert)

Munia caniceps kumusii Hartert, 1911. *Bull. Brit. Orn. Club* 27: 47.—Kumusi R, n coast of the SE Peninsula.

RANGE: Lowlands of the n watershed of the SE Peninsula from Lae[J] to Safia[K], upper Musa R (King 1979[J], Coates 1990[K]).

DIAGNOSIS: Most similar to the nominate form. Differs in the dull gray-brown breast and belly, with black undertail; mantle paler than that of nominate, and the distinction between pale of crown and nape versus the dark of mantle is less prominent.

NOTES

We follow Payne (2010) in subspecific treatment. Population in the uplands of the n watershed of the SE Peninsula (Garaina, Chirima valley) undetermined but probably is assignable to *scratchleyana* (Payne 2010). The population from Myola, in the uplands of the s watershed, presumably belongs here as well (Payne 2010). Peckover & Filewood (1976) noted this species has hybridized with *L. castaneothorax* in the Port Moresby region. *Other names*—Grey-headed Munia.

Grey-crowned Mannikin *Lonchura nevermanni*   ENDEMIC | MONOTYPIC

Lonchura nevermanni Stresemann, 1934. *Orn. Monatsber.* 42: 101.—Merauke, s NG.

Restricted to the Trans-Fly and southernmost S Lowlands, near sea level (Rand 1942a, Bell 1967, Mees 1982b, Coates 1990, Johnson & Richards 1994, K. D. Bishop *in litt.*). Within the Trans-Fly seemingly nomadic and recorded only irregularly and only in small flocks in tall grass, reed-fringed lagoons, floating mats of grass, and ricefields (Coates 1990).

Localities—Mapa[J], Kurik[J], Merauke[CO], and Bian Lakes[J] of Indonesian NG; PNG records from Bensbach[J], Morehead[J], Aramia R[J], Lake Daviumbu[CO], and Balimo[CO] (Bishop 2005a[J]). A hybrid with *L. stygia* is known from Kurik[K] (Mees 1982b[K]).

Other names—Grey-crowned Munia, White-crowned Mannikin.

Hooded Mannikin *Lonchura spectabilis*   RESIDENT

Donacicola spectabilis P. L. Sclater, 1879. *Proc. Zool. Soc. London*: 449, pl. 37, fig. 2.—New Britain.

Inhabits grasslands locally from the ne sector of the NW Lowlands, the Sepik-Ramu, the Eastern Ra, the Huon Penin, and the n sector of the SE Peninsula; sea level–2,450 m, max 3,000 m (Coates & Peckover 2001). The species occurs from Jayapura to Guari.

Extralimital range includes Long, Tolokiwa, Umboi, New Britain, and Watom I (Coates 1990).

SUBSPECIES

a. *Lonchura spectabilis mayri* (Hartert)

Munia spectabilis mayri Hartert, 1930. *Novit. Zool.* 36: 42.—Ifar, Lake Sentani, ne sector of the NW Lowlands, n NG.

Synonym: *Lonchura spectabilis sepikensis*, Jonkers and Roersma, 1990. *Dutch Birding* 12(1): 22–25.—A few km s of Urimo Cattle Station, Sepik plains, *ca.* 60 km ssw of Wewak, East Sepik Prov, n NG.

RANGE: Lake Sentani east to the e sector of the Sepik-Ramu (Rand 1942b).

DIAGNOSIS: Mantle paler brown than that of *wahgiensis*; rump more golden; underparts whitish. The form *sepikensis*, based on a photo of an unsexed and unaged free-flying individual, cannot be determined. We ally that named form here for geographic reasons. Plumage variation in this species is problematic in c PNG, as there may be age-related plumage changes that confound establishment of stable characters for proper subspecies.

b. *Lonchura spectabilis wahgiensis* Mayr & Gilliard
Lonchura spectabilis wahgiensis Mayr & Gilliard, 1952. *Amer. Mus. Novit.* 1577: 7.—Kegsugl, s slope of Mt Wilhelm, Eastern Ra.

Synonym: *Lonchura spectabilis gajduseki* Diamond, 1967. *Amer. Mus. Novit.* 2284: 14.—Karimui, 1,110 m, Eastern Ra.

RANGE: The Eastern Ra, from Lake Kopiago east to mts of the Huon and SE Peninsulas (Mayr 1931, Mayr & Gilliard 1954, Diamond 1972, Coates 1990). Presumably found in both the n and s watersheds.

DIAGNOSIS: Mantle dark red-brown; uppertail coverts yellow; underparts whitish to pale buff (buff in some specimens); bill dark gray (Payne 2010). See photos in Coates (1990: 343) and Coates & Peckover (2001: 197). Following Payne (2010) and Dickinson & Christidis (2014), we fold *gajduseki* into *wahgiensis*.

NOTES
Restall (1997) pointed out that this species exhibits both geographic and nongeographic plumage variation. Restall (1997: pl. 13) illustrated some of this nongeographic variation, which includes barring of the flanks in some individuals, as well as multiple shades of ventral color from white to deep chestnut. The described form *sepikensis* exemplifies this apparent polymorphism. Larger series of specimens from certain localities and allied molecular analysis of variation in these variable populations may help us better understand what is happening with this curious species. *Other names*—New Britain Mannikin, New Guinea Munia, Mayr's Mannikin.

Chestnut-breasted Mannikin *Lonchura castaneothorax*   RESIDENT
Amadina castaneothorax Gould, 1837. *Synops. Birds Austral.* 2: pl. 21, fig. 2.—New South Wales, Australia.
    Locally common in the w sector of the NW Lowlands, the w sector of the Western Ra, the ne sector of the NW Lowlands, the n sector of the Sepik-Ramu, and the Huon and SE Peninsulas; also Manam I and the D'Entrecasteaux Arch; sea level–1,800 m^{CO}.
    Extralimital—n and e Australia (with records from the far s and far sw), including islands of the Torres Strait (Coates 1990, Higgins *et al.* 2006). Has been introduced to New Caledonia, Vanuatu, and Polynesia (Coates 1990, Dutson 2011).

SUBSPECIES
a. *Lonchura castaneothorax uropygialis* Stresemann & Paludan
Lonchura castaneothorax uropygialis Stresemann & Paludan, in Hartert *et al.* 1934. *Orn. Monatsber.* 42: 43.—Lower Menoo R, 300 m, head of Geelvink Bay, Bird's Neck.
RANGE: The westernmost sector of the NW Lowlands.
DIAGNOSIS: Face all black; rump and uppertail rufous (Payne 2010). Note also extensive fine barring on flanks (Higgins *et al.* 2006). Schodde & Mason (1999) noted that the form *uropygialis* is only thinly differentiated from the nominate form of Australia.

b. *Lonchura castaneothorax boschmai* Junge
Lonchura castaneithorax boschmai Junge, 1952. *Zool. Meded. Leiden* 31(22): 249.—Araboebivak, Paniai Lake, Western Ra.
RANGE: The Western Ra. Localities—Paniai Lake and Araboe R (Payne 2010).
DIAGNOSIS: Nearest to *ramsayi*. Streaks on head darker and broader; rump and uppertail coverts straw yellow; breast darker, and this color extends farther down to lower breast; flanks chestnut; white of abdomen restricted (Junge 1952).

c. _Lonchura castaneothorax sharpii_ (Madarász)
Donacicola sharpii Madarász, 1894. _Bull. Brit. Orn. Club_ 3: 47.—Bongu, nw sector of the Huon Penin.
RANGE: The n watershed of c and e NG from Humboldt Bay east to Astrolabe Bay, the Markham valley, Lae, Mumeng, Bulolo, and Wau[J]; also Manam I (Coates 1990, Pratt field notes[J]).
DIAGNOSIS: Forehead, crown, and nape pale gray or whitish; rump rufous chestnut (Payne 2010).

d. _Lonchura castaneothorax ramsayi_ Delacour
Lonchura castaneothorax ramsayi Delacour, 1943. _Zoologica_ 28: 84.—Port Moresby, SE Peninsula. Replacement name for _Donacola nigriceps_ Ramsay, 1877; preoccupied by _Spermestes (Lonchura) nigriceps_ Cassin, 1852.
RANGE: The SE Peninsula from the Kumusi R (in the north) and Bereina (in the south) southeastward to Milne Bay; also Goodenough and Normanby I[CO] (Gilliard 1950).
DIAGNOSIS: Head and throat black; feathers on crown and hindneck brown with pale spotting; rump rufous, becoming golden yellow on uppertail coverts (Payne 2010). See photo in Coates (1990: 345).

NOTES
Coates (1990) & K. Stryjewski (_in litt._) reported hybridization between _L. grandis_ and this species. _Other names_—Chestnut-breasted Munia, Chestnut Finch.

Black Mannikin _Lonchura stygia_   ENDEMIC | MONOTYPIC
Lonchura stygia Stresemann, 1934. _Orn. Monatsber._ 42: 102.—Mandum, upper Bian R, Trans-Fly.
    Restricted to the s Trans-Fly, where generally rare and infrequently encountered: upper Bian R[K], Wasur NP[K], Bensbach R[K], Lake Murray[J], Lake Owa[L], and Lake Daviumbu[X] (Rand 1942a[X], Gregory 1997[L], Portelli 2013[J], K. D. Bishop _in litt._[K]), near sea level. Inhabits reeds and floating mats of ricegrass on lagoons and swamps, and tall grass in nearby savanna; also visits standing rice crops at Kurik (Mees 1982b). Hybridization with _L. nevermanni_ sometimes occurs in the wild, with a published record from Kurik (Mees 1982b). Two individuals were observed at Kiunga airstrip 17-May-2005 (P. Gregory _in litt._). A flock of 20 birds was observed feeding on seeding grasses on 8-Sep-2014 at Kiunga airstrip (D. Hobcroft _in litt._).

NOTES
Mees (1982b) discussed the possible sister relationship between this species and _nevermanni_, and concluded that the fact of hybridization is trumped by the distinct juvenile plumages of the two forms. _Other names_—Black Munia.

Black-breasted Mannikin _Lonchura teerinki_   ENDEMIC
Lonchura teerinki Rand, 1940. _Amer. Mus. Novit._ 1072: 14.—Bele R, 18 km n of Lake Habbema, 2,200 m, Western Ra.
    Mid-montane grasslands of the c section of the Western Ra, in both the n and s watersheds, 1,200–2,300 m.

SUBSPECIES
a. _Lonchura teerinki teerinki_ Rand
RANGE: The Baliem valley of the s watershed of the Western Ra. Localities—Bele R Camp[Q], Baliem R Camp[Q], 9 km northeast of Lake Habbema[Q], Wamena[J], Kurulu[J], Holuwon[U] (Rand 1942b[Q], Ripley 1964a[J], Beehler _et al._ 1986[U]).
DIAGNOSIS: Mantle and crown dull dark brown. See photo in Coates & Peckover (2001: 198).

b. _Lonchura teerinki mariae_ Ripley
Lonchura teerinki mariae Ripley, 1964. _Bull. Peabody Mus. Nat. Hist. Yale_ 19: 74.—Bokondini, 1,280 m, Western Ra.
RANGE: The n watershed of the Western Ra. Known only from the grasslands of the Bokondini area, 50 km northwest of Wamena and on the n side of a large ridge from the main Grand Valley of the Baliem (Ripley 1964a). Waters of the Bokondini valley flow north into the Mamberamo basin.
DIAGNOSIS: Crown blackish and mantle deeper and richer brown than that of nominate; black of lower

flanks, thighs, and undertail more truly black (Ripley 1964a). J. Lai and K. Zyskowski compared the type series at Yale in 2012 with specimens from the type series of the nominate form from AMNH and confirmed to M. LeCroy that the subspecific distinctions noted by Ripley remained diagnostic (LeCroy 2013).

NOTES
Other names—Grand Valley Mannikin, Black-breasted Munia.

Eastern Alpine Mannikin *Lonchura monticola*   ENDEMIC | MONOTYPIC

Munia monticola De Vis, 1897. *Ibis*: 387.—Mount Scratchley, SE Peninsula.
    Synonym: *Lonchura monticola myolae* Restall, 1995. *Bull. Brit. Orn. Club* 115: 145.—Mounts Scratchley and
      Knutsford, Owen Stanley Ra, SE Peninsula.
High mts of the SE Peninsula, 2,080[L]–3,900 m[CO] (Hicks 1987[L]). Known from the Wharton and Owen Stanley Ra, including the Myola grasslands (south-southeast of Mt Victoria). Inhabits grasslands at high elevations, especially in the vicinity of shrubbery and rock outcrops.
    Localities—Mt Albert Edward[J], Murray Pass[J], English Peaks[K], Lake Myola[CO], Mt Knutsford[L] (Mayr & Rand 1937[J], LeCroy 2013[L], Beehler field notes[K]).

NOTES
Following LeCroy (1999, 2013), we do not recognize the form *myolae*, which is based either on nongeographic variation or on possible micro-variation across a minor geographic discontinuity. Coates (1990) stated that there was evidence of hybridization between *L. monticola* and *L. spectabilis* near Guari. *Other names*—Alpine Mannikin, Eastern Alpine Munia.

Western Alpine Mannikin *Lonchura montana*   ENDEMIC | MONOTYPIC

Lonchura montana Junge, 1939. *Nova Guinea* (ns)3: 67.—Oranje Mts, 4,150 m, Western Ra.
    The Western and Border Ra, east to the Star Mts of PNG (Mt Capella); 3,200–3,800 m, extremes 2,130–4,150 m[CO] (Ripley 1964a, Gregory & Johnston 1993, Coates & Peckover 2001). The easternmost record (Mt Capella) is confirmed by a specimen in the Australian Museum in Sydney, collected by Tim Flannery, which appears to be not racially distinct from the Western Ra population (Coates 1990, Gregory & Johnston 1993). Inhabits alpine grassland and edges of subalpine shrubbery; also visits cultivated upper mid-montane fields (Coates 1990).
    Localities—Ilaga valley[J],7 km east of Mt Wilhelmina[Q], Lake Habbema[Q], Mt Capella[L] (Rand 1942b[Q], Ripley 1964b[J], Gregory & Johnston 1993[L]).
    Other names—Snow Mountain Mannikin, Western Alpine Munia.

Family Motacillidae

Wagtails and Pipits   [Family 67 spp/Region 4 spp]

The species of the family range worldwide. In New Guinea, wagtails and pipits live in open habitats at various elevations, foraging in grassy or barren areas or streamsides. The New Guinean pipits are year-round residents; the wagtails winter from Siberia. Slim, ground-feeding birds, they are characterized by the slender, pointed bill, slender body, long tail, long legs and, in some species, an elongate hind claw (Rand & Gilliard 1967). Barker *et al.* (2004) placed this lineage in the Passerida, sister to the Fringillidae and allies.

GENUS *MOTACILLA* [Genus 12 spp/Region 2 spp]

Motacilla Linnaeus, 1758. *Syst. Nat.*, ed. 10, 1: 184. Type, by tautonymy, "Motacilla" = *Motacilla alba* Linnaeus.

The species of this genus exhibit breeding ranges that extend from Madagascar and southern Africa to Eurasia, tropical Asia, and Japan. Species winter to Africa, tropical Asia, and Australia (Tyler 2004). The genus is characterized by the slim body form, long narrow tail, white outer rectrices, terrestrial habit, and fine, straight bill.

Eastern Yellow Wagtail *Motacilla tschutschensis* VISITOR

Motacilla flava tschutschensis Gmelin, 1789. *Syst. Nat.* 1(2): 962.—Coast of Chukotski Penin, e Russia.

An uncommon but regular northern migrant to NG. Most common in the lowlands of w NG, less so in the Western Ra and in the lowlands of e NG. A wagtail of swamp edge and wet grasslands (Coates 1990). Records from all elevations[CO]. Dates of occurrence: 7-October to 24-April.

Localities—Kofiau[M], Waigeo I[CO], Jeflio[L] (Bird's Head), Wasior[CO], Paniai Lake[CO], Carstensz Massif[CO], Mimika R[L], Timika[CO], Kurik[J], Merauke[J], Wasur NP[J], Bensbach R[J], Lake Daviumbu[J], Lake Wangbin[CO], Ok Menga[CO], Lakes Pangua and Owa[K], Aiome[CO], near Bereina[CO], Hisiu[CO], Aroa R[CO], Laloki R[CO], Waigani swamp[CO], and Sangara[CO] (Gyldenstolpe 1955b[L], Gregrory 1997[K], Bishop 2005a[J], Diamond *et al.* 2009[M]). Diamond & Bishop (1994) provided a first record for the Aru Is.

Extralimital—breeds in the Russian far east and w Alaska; winters to se Asia, straggling to Australia; vagrant to Bismarcks and Bougainville (Dutson 2011).

SUBSPECIES

One subspecies has been identified, and two additional may be expected, based on Australian records (see Schodde & Mason 1999, Tyler 2004).

a. *Motacilla tschutschensis tschutschensis* Gmelin

Synonym: *Motacilla flava simillima* Hartert, 1905. *Vog. Pal. Fauna*, 1: 289.—Sulu Is (on migration), presumably breeding on Kamchatka.

RANGE: Breeds in the Chukotski Penin, e Anadyrland, Kamchatka, n Kuril Is, Commander Is, and w Alaska. Wintering to se Asia and the Philippines to Wallacea and n Australia (Tyler 1999). In the Region, the only specimen record is from the Mimika R, at the foot of the w sector of the Western Ra (Mayr 1941). Also recorded from Australia (Schodde & Mason 1999).

DIAGNOSIS: Nonbreeding male plumage exhibits gray crown with olive wash; olive mantle; off-white supercilium; mottled ear coverts; buff throat trending to yellowish on breast; and broken breast band of buff brown. Female dull gray crown, olive brown mantle, cream-colored eyebrow, whitish throat grading to pale yellow on breast, breast band buff brown; first-year birds grayer dorsally and whiter ventrally, with a dusky necklace indicated (Schodde & Mason 1999). Alström & Mild (2004) synonymized *simillima* into this form.

b. *Motacilla tschutschensis ?taivana* (Swinhoe)

Budytes taivanus Swinhoe, 1863. *Proc. Zool. Soc. London*: 274, 334.—Taiwan.

RANGE: Breeds in e Siberia. Might be expected in NG, as it has been identified as a migrant to n Australia by Schodde & Mason (1999). Noted to winter to the East Indies (Mayr 1960).

DIAGNOSIS: Nonbreeding male: crown and mantle dull citrine, eyebrow yellow, cheek and ear citrine, underparts pale yellow, deepening on belly, apparently chin and throat pale yellow; female: dull citrine on crown and mantle, yellowish eyebrow, pale yellow underparts, grading to deep yellow on belly and under-tail, brownish wash forms breast band; first-year birds: grayer dorsally and paler and creamier ventrally, with a more distinct dusky necklace (Schodde & Mason 1999).

c. *Motacilla tschutschensis ?macronyx* (Stresemann)

Budytes flava macronyx Stresemann, 1920. *Avif. Macedon.*: 76.—Vladivostok, e Russia.

RANGE: Breeds in Ussuriland and Amurland west to Transbaikalia and south to ne Mongolia (Mayr 1960). Might be expected in NG, as has been identified as a migrant to n Australia by Schodde & Mason (1999). Mayr (1960) noted this form winters to Indochina, Malaya, and Sumatra.

DIAGNOSIS: Nonbreeding male: dark gray crown with brownish-green wash, mantle greenish-brown, small cream-colored eyebrow, dark gray cheek and ear, white chin and upper throat, pale yellow breast with remnant gray breast band; female: gray crown with brownish-cream wash, mantle greenish-brown, whitish eyebrow, white throat, pale yellow breast, narrow grayish breast band (Schodde & Mason 1999).

NOTES

We follow Pavlova *et al.* (2003) and Banks *et al.* (2004) in splitting this lineage from *M. flava*. Classification of the yellow wagtails (*M. flava*, in the broad sense) is complex and unresolved (Pavlova *et al.* 2003, Dickinson & Christidis 2014). *Other names*—earlier treatments considered NG visitors to be Yellow or Green-headed Wagtail, *Motacilla flava*.

Grey Wagtail *Motacilla cinerea*   VISITOR

Motacilla Cinerea Tunstall, 1771. *Orn. Brit.*: 2.—Wycliffe, Yorkshire, England.

A common northern migrant to NG, wintering along mt streams and in upland bogs of the n watershed; uncommon in the s watershed; also in the NW Islands (Misool, Batanta, Waigeo), and Biak[W], Yapen, and Karkar I[CO] (Mayr 1941, S. F. Bailey *in litt.*[W]); 600–2,500 m, occasionally in lowlands (Coates 1990). Season of occurrence mainly from late September to late April, though has been recorded in the Wahgi valley as early as late August (Gyldenstolpe 1955a).

Extralimital—breeds in w Europe, North Africa, and Asia east to Japan; winters south to e Africa and tropical Asia, east to Wuvulu and New Ireland (Dutson 2011).

SUBSPECIES

a. *Motacilla cinerea cinerea* Tunstall

    Synonym: *Parus caspicus* Gmelin, 1774. *Reise durch Russland* 3: 104, pl. 20, fig. 2.—Enseli, n Iran.
    Synonym: *Motacilla melanope* Pallas, 1776. *Reise Versch. Prov. Russ. Reichs.* 3: 696.—East of Lake Baikal, e Russia.
    Synonym: *Pallenura robusta* Brehm, 1857. *J. Orn.* 5: 32.—Japan.

RANGE: Breeds in Eurasia from Norway east to Siberia, the Himalayas, and China; winters south to Africa, tropical Asia, and NG (Diamond 1972).

DIAGNOSIS: In nonbreeding birds, the pale yellow of breast and belly contrasts with white throat, which may exhibit some dirty-gray streaking on sides; eyebrow complete to lores (Tyler 2004). Based on discussion of Tyler (2004: 785), we combine the races *robusta* and *melanope* with the nominate form.

NOTES

For reasons for use of *M. cinerea* vs. *M. caspica*, see Schodde & Bock (2008). *Other names*—*Motacilla caspica*.

GENUS *ANTHUS*   [Genus 43 spp/Region 2 spp]

Anthus Bechstein, 1805. *Gemein. Nat. Deutschl.* 2: 247, 302, and 465. Type, by subsequent designation (Selby, 1825. *Illust. Brit. Orn.*: xxix), *Alauda pratensis* Linnaeus.

The species of this genus range from the Americas to Africa, Asia, New Guinea, Australia, and New Zealand (Tyler 2004). The genus is characterized by the terrestrial habit, cryptic plumage of browns and buffs, slim straight bill, and the elongate hallux claw.

Australian Pipit *Anthus australis*   RESIDENT

Anthus australis Vieillot, 1818. *Nouv. Dict. Hist. Nat.* (ns) 26: 501.—Sydney, NSW, Australia.

Grasslands of e NG. Distribution within the Region restricted to grasslands of the Eastern Ra, mts of westernmost SE Peninsula, and lowland grasslands of the Markham and Ramu R (Stronach 1990, Bailey 1992a). Inhabits mainly areas of short grass such as airfields, playing fields, and pastures, where often seen on roads; mainly 1,200–2,200 m, max 3,000 m, but at lower elevations in the Ramu and Markham (Coates 1990).

Extralimital—Australia (Schodde & Mason 1999).

a. *Anthus australis exiguus* Greenway

Anthus australis exiguus Greenway, 1935. *Proc. New Engl. Zool. Club* 14: 53.—Wau, Morobe distr, SE Peninsula.

RANGE: Confined mainly to upland interior grasslands of the Eastern Ra and the SE Peninsula, southeast to Gauri. Localities—Timika[M], Tari[CO], Laiagam[CO], Wabag[CO], Mendi[CO], Kup[J], Nondugl[J], Kegsugl[J], Wahgi valley[J], Baiyer valley[K], Jimi valley[CO], Goroka[CO], Kainantu[CO], Okapa[DI], Tarabo[DI], Dumpu[CO], Gusap[L], the Ramu and Markham R lowlands[X], Mumeng[CO], upper Watut R[CO], Wau[N], Kunimaipa valley[CO], and the Aibala valley near Guari[CO] (Greenway 1935[N], Mayr & Gilliard 1954[J], Gyldenstolpe 1955a[K], Stronach 1990[L], Bailey 1992a[X], S. van Balen *in litt.*[M]). Mees (1965) noted a Wallace specimen mentioned by Gray from Misool I. This was dismissed by Mayr (1941)—perhaps a mislabeled trade skin (but see Mees 1965: 177).

DIAGNOSIS: When compared to *rogersi* of QLD, this form is darker dorsally, and more heavily streaked on breast (Tyler 2004).

NOTES

Earlier authors treated this population as a form of *Anthus richardsi*. We follow Schodde & Mason (1999) and Dickinson & Christidis (2014) in treating the form as *Anthus australis* (*pace* Christidis & Boles 2008, who treated the NG, NZ, and Australian populations as the species *novaeseelandiae*). *Other names*—Richard's Pipit, *Anthus richardsi*, *Anthus novaeseelandiae*.

Alpine Pipit *Anthus gutturalis* ENDEMIC

Anthus gutturalis De Vis, 1894. *Ann. Rept. Brit. New Guinea*, 1893–94: 103.—Mount Maneao, SE Peninsula.

Inhabits grassy and rocky openings of the highest mts of the Central Ra and the Huon Penin, 2,550–4,600 m[CO]. Present on most summits with well-developed alpine grasslands as well as grassy or swampy alpine and subalpine basins below summit regions. In some instances inhabits grassy clearings in nearby upper montane forest.

SUBSPECIES

a. *Anthus gutturalis wollastoni* Ogilvie-Grant

Anthus wollastoni Ogilvie-Grant, 1913. *Bull. Brit. Orn. Club* 31: 105.—Otakwa R, 3,600 m, Western Ra.

   Synonym: *Anthus gutturalis rhododendri* Mayr, 1931. *Mitt. Zool. Mus. Berlin* 17: 692.—Mongi-Busu, 2,600 m, Saruwaged Ra, Huon Penin.

RANGE: The Western Ra, Border Ra, Eastern Ra, and mts of the Huon Penin.

DIAGNOSIS: Dark spotting on side of throat not forming a single dark stripe. Underparts mottled and dark. Differences between *wollastoni* and the form *rhododendri* marginal at best; probably merit combination.

b. *Anthus gutturalis gutturalis* De Vis

RANGE: Mountains of the SE Peninsula

DIAGNOSIS: Spotting on side of throat forms a solid dark stripe or irregular dark patch. Underparts paler and with more reduced pattern than in *wollastoni*. See photo in Coates & Peckover (2001: 116).

NOTES

Other names—New Guinea Pipit.

PART III

Carola's Parotia, illustration by John Anderton

Bibliography

Note to readers: This bibliography includes some publications that are not cited in the text but which have informed the preparation of this work.

Aggerbeck, M., J. Fjeldså, L. Christidis, P.-H. Fabre, & K.A. Jønsson. 2014. Resolving deep lineage divergences in core corvoid passerine birds supports a proto-Papuan island origin. *Molec. Phylog. Evol.* 70: 272–85.

Ali, S., & S.D. Ripley. 1983. *Handbook of the Birds of Indian and Pakistan.* Compact Edition. Oxford Univ. Press, New Delhi.

Alström, P., P.G.P. Ericson, U. Olsson, & P. Sundberg. 2006. Phylogeny and classification of the avian superfamily Sylvoidea. *Molec. Phylog. Evol.* 38: 381–97.

Alström, P., S. Fregin, J.A. Norman, P.G.P. Ericson, L. Christidis, & U. Olsson. 2011. Multilocus analysis of a taxonomically densely sampled dataset reveal extensive non-monophyly in the avian family Locustellidae. *Molec. Phylog. Evol.* 58: 513–26.

Alström, P., & K. Mild. 2003. *Pipits and Wagtails of Europe, Asia, and North America.* A. and C. Black, London.

Alström, P., U. Olsson, & Lei Fu-min. 2013. A review of the recent advances in the systematics of the avian superfamily Sylvioidea. *Chinese Birds* 4(2): 99–131.

Amadon, D. 1942. Birds collected during the Whitney South Sea Expedition. 49. Notes on some non-passerine genera. 1. *Amer. Mus. Novit.* 1175: 1–11.

Amadon, D. 1943. The genera of starlings and their relationships. *Amer. Mus. Novit.* 1247: 1–9.

Amadon, D. 1959. Remarks on the subspecies of the Grass Owl, *Tyto capensis. J. Bombay Nat. Hist. Soc.* 56: 344–46.

Amadon, D. 1962. Sturnidae; Cracticidae. In E. Mayr & J.C. Greenway, eds., *Check-list of Birds of the World,* vol. 15: 75–120; 166–71. Mus. Comp. Zool., Harvard Univ., Cambridge, MA.

Amadon, D. 1978. Remarks on the taxonomy of some Australasian raptors. *Emu* 78: 115–18.

Andersen, M.J., P.A. Hosner, C.E. Filardi, & R.G. Moyle. 2015a. Phylogeny of the monarch flycatchers reveals extensive paraphyly and novel relationships within a major Australo-Pacific radiation. *Molec. Phylog. Evol.* 83: 118–36.

Andersen, M.J., A. Naikatini, R.G. Moyle. 2014a. A molecular phylogeny of Pacific honeyeaters (Aves: Meliphagidae) reveals extensive paraphyly and an isolated Polynesia radiation. *Molec. Phylog. Evol.* 71: 308–15.

Andersen, M.J., A.S. Nyári, I. Mason, L. Joseph, J.P. Dumbacher, C.E. Filardi, & R.G. Moyle. 2014b. Molecular systematics of the world's most polytypic bird: *Pachycephala pectoralis/melanura* (Aves: Pachycephalidae) species complex. *Zool. Journ. Linn. Soc.* 170: 566–88.

Andersen, M.J., C.H. Oliveros, C.E. Filardi, & R.G. Moyle. 2013. Phylogeography of the Variable Dwarf-Kingfisher *Ceyx lepidus* (Aves: Alcedinidae) inferred from mitochondrial and nuclear DNA sequences. *Auk* 130: 118–31.

Andersen, M.J., H. Shult, A. Cibois, J.-C. Thibault, C. Filardi, & R. Moyle. 2015b. Rapid diversification and secondary sympatry in Australo-Pacific kingfishers (Aves: Alcedinidae: *Todiramphus*). *Royal Society Open Science,* 2015 2 140375. doi:10.1098/rsos.140375.

Andrew, P. 1992. *The Birds of Indonesia: A Checklist (Peters' Sequence).* Indonesian Orn. Soc., Jakarta.

dos Anjos, L 2009. Corvidae (crows). In J. del Hoyo, A. Elliott, & D. Christie, eds., *Handbook of the Birds of the World,* vol. 14: 494–641. Lynx Edicions, Barcelona.

Anon. 1976. [Sighting of unidentified long-tailed black bird from Goodenough Island]. *PNG Bird Soc. Newsl.* 118: 2.

Anon. 1984. [Long-billed Dowitcher *Limnodromus scolopaceus* sighting.] *PNG Bird Soc. Newsl.* 212: 6.

Anon. 2012. *Paradisaea* Linnaeus, 1758, and Paradisaeidae Swainson, 1825 (Aves): names conserved. *Bull. Zool. Nomencl.* 69: 77–78.

Ap-Thomas, M, & J. Ap-Thomas. 1976. Birds of Motupore Island, Bootless Inlet. *PNG Bird Soc. Newsl.* 121: 6.

Archbold, R., A.L. Rand, & L.J. Brass. 1942. Summary of the 1938–1939 New Guinea expedition. Results of the Archbold Expeditions no. 41. *Bull. Amer. Mus. Nat. Hist.* 89: 197–288.

Archibald, G.W., & C.D. Meine. 1996. Gruidae (cranes). In J. del Hoyo, A. Elliott, & J. Sargatal, eds., *Handbook of the Birds of the World*, vol. 3: 60–89. Lynx Edicions, Barcelona.

Argeloo, M., & R. Dekker. 1996. Bulwer's Petrel in Indonesia. *Kukila* 8: 13–14.

Ashford, R.W. 1978. First record of House Sparrow for Papua New Guinea. *Emu* 78: 36.

Auber, L. 1934. Der Rassenkreis *Chalcopsittacus duivenbodei* Dubois. *Anz. Orn. Ges. Bayern* 2: 313–15.

Austin, J.J., V. Bretagnolle, & E. Pasquet. 2004. A global molecular phylogeny of the small *Puffinus* shearwaters and implications for systematics of the Little-Audubon's Shearwater complex. *Auk* 121: 847–64.

Bailey, S.F. 1992a. Bird observations in lowland Madang Province. *Muruk* 5: 111–38.

Bailey, S.F. 1992b. Seabirds in Madang Province, Papua New Guinea, September–November 1989. *Emu* 93: 223–32.

Bairlein, F. 2006. Sylviidae (Old World warblers). In J. del Hoyo, A. Elliott, & D. Christie, eds., *Handbook of the Birds of the World*, vol. 11: 492–709. Lynx Edicions, Barcelona.

Baker, A. 1975. Systematics and evolution of Australian oystercatchers. *Emu* 74: 277.

Baker, A.J., Y. Yatsenko & E.S. Tavares. 2012. Eight independent nuclear genes support monophyly of the plovers: the role of mutational variance in gene trees. *Molec. Phylog. Evol.* 65: 631–41.

Baker, R.H. 1948. Report on collections of birds made by the United States Naval Medical Research Unit No. 2 in the Pacific War area. *Smithsonian Misc. Coll.* 107(15): 1–74.

Baker, R.H. 1951. The avifauna of Micronesia, its origin, evolution and distribution. University of Kansas Publ., *Mus. Zool.* 3(1): 1–359.

Baker-Gabb, D. 1979. Remarks on the taxonomy of the Australasian Harrier (*Circus approximans*). *Notornis* 28: 325–29.

van Balen, S. 2008. Zosteropidae (white-eyes). In J. del Hoyo, A. Elliott, & D. Christie, eds., *Handbook of the Birds of the World*, vol. 13: 402–85. Lynx Edicions, Barcelona.

van Balen, S., & R. Noske. 2006. Around the archipelago [Blue Rock-thrush *Monticola solitarius*]. *Kukila* 13: 83–88.

van Balen, S., & R. Noske. 2014. Around the archipelago [Papua sightings of Snow Mountain Robin *Petroica archboldi*]. *Kukila* 17: 162–63.

van Balen, S., R. Noske, & A. Supriatna. 2009. Around the archipelago [*Cyclopsitta gulielmiterti*]. *Kukila* 14: 73–85.

van Balen, S., R. Noske, & A. Supriatna. 2011. Around the archipelago [*Macheiramphus alcinus*]. *Kukila* 15: 126–43.

van Balen, S., & W.M. Rombang. 1998. *The Birds of the PT Freeport Indonesia Contract of Work Mining and Project Area, Irian Jaya, Indonesia.* Biodiversity Study Series Vol. 7. PT Freeport Indonesia, Jakarta, and PT Hatfindo Prima, Bogor, Indonesia.

van Balen, S., S. Suryadi, & D. Kalo. 2003a. Birds of the Dabra area, Mamberamo Basin, Papua, Indonesia. In S.J. Richards & S. Suryadi, eds., *A biodiversity assessment of Yongsu-Cyclops Mountains and the southern Mamberamo basin, Papua, Indonesia. RAP Bull. Biol. Assess.* 25: 84–88, 172–78. Conservation International, Washington, DC.

van Balen, S., S. Suryadi, & D. Kalo. 2003b. A second sighting of Bat Hawk in Irian Jaya (Papua). *Kukila* 12: 70–71.

Bangs, O., & J.L. Peters. 1926. A collection of birds from southwestern New Guinea (Merauke coast and inland). *Bull. Mus. Comp. Zool. Harvard* 67: 421–34.

Banks, R.C., C. Cicero, J.L. Dunn, A.W. Kratter, P.C. Rasmussen, J.V. Remsen, J.D. Rising, & D.F. Stotz. 2004. Forty-fifth supplement to the American Ornithologists' Union Checklist of North American Birds. *Auk* 121: 985–95.

Baptista, L., P.W. Trail, & H.M. Horblit. 1997. Columbidae (pigeons). In J. del Hoyo, A. Elliott, & J. Sargatal, eds., *Handbook of the Birds of the World*, vol. 4: 60–245. Lynx Edicions, Barcelona.

Barker, F.K., G.F. Barrowclough, & J.G. Groth. 2001. A phylogenetic hypothesis for passerine birds: taxonomic and biogeographic implications of an analysis of nuclear DNA sequence data. *Proc. B Roy. Soc. Lond.* 269: 295–308.

Barker, F.K., A. Cibois, P. Schikler, J. Feinstein, & J. Cracraft. 2004. Phylogeny and diversification of the largest avian radiation. *Proc. Nat. Acad. Sci. USA.* 101: 11040–45.

Barker, W.R., & J.R. Croft. 1977. The distribution of Macgregor's Bird of Paradise. *Emu* 77: 219–22.

Barrowclough, G.F., J.G. Groth, J.E. Lai, & S.M. Tsang. 2014. The phylogenetic relationships of the endemic genera of Australo-Papuan hawks. *J. Raptor Res.* 48: 36–43.

Barrowclough, G.F., J.G. Groth, & L. Mertz. 2006. The RAG-1 exon in the avian order Caprimulgiformes: phylogeny, heterozygosity, and base composition. *Molec. Phylog. Evol.* 41: 238–48.

Bashari, H., & S. van Balen. 2014. First record of Stout-billed Cuckooshrike *Coracina caruleogrisea* in Wallacea, a remarkable range extension from New Guinea. *Bull. Brit. Orn. Club* 134: 302–4.

Beaufort, L.F. de. 1909. Birds from Dutch New Guinea. *Nova Guinea, Zool.* 5: 419.

Beddard, F.E. 1891. Ornithological notes [tracheal morphology of the Curl-crested Manucode *Manucodia comrii*]. *Ibis* (6)3: 512–14.

Beehler, B.M. 1978a. The status of *Sericornis nigroviridis*. *Condor* 80: 115–16.

Beehler, B.M. 1978b. *Upland Birds of Northeastern New Guinea*. Wau Ecology Institute Handbook No. 5, Wau, Papua New Guinea.

Beehler, B.M. 1978c. Historical changes in the avifauna of the Wau valley, Papua New Guinea. *Emu* 78: 61–64.

Beehler, B.M. 1978d. Notes on the mountain birds of New Ireland. *Emu* 78: 65–70.

Beehler, B.M. 1980a. Black Swan *Cygnus atratus*: a new species for the New Guinea region. *PNG Bird Soc. Newl* 173/174: 3.

Beehler, B.M. 1980b. Notes on some birds of the upper Eloa River, Papua New Guinea. *Emu* 80: 87–88.

Beehler, B.M. 1982. Birds in the lower Eloa River valley. *PNG Bird Soc. Newsl.* 187/188: 4–6.

Beehler, B.M. 1983. Frugivory and polygamy in birds of paradise. *Auk* 100: 1–12.

Beehler, B.M. 1987. Ecology and behavior of the Buff-tailed Sicklebill (Paradisaeidae: *Epimachus albertisi*). *Auk* 104: 48-55.

Beehler, B.M. 1988. Miscellaneous observations of birds in Irian Jaya and Papua New Guinea. *Muruk* 3: 51–52.

Beehler, B.M. 1991. *A Naturalist in New Guinea* [includes field notes on life history of the Giant Wattled Honeyeater *Macgregoria pulchra*]. Univ. of Texas, Austin.

Beehler, B.M., ed. 1993a. A biodiversity analysis for Papua New Guinea. Part 2: 1–434. *Papua New Guinea Conservation Needs Assessment*. Biodiversity Support Program, Washington, DC.

Beehler, B.M. 1993b. Does the Greater Ground-Robin *Amalocichla sclateriana* inhabit Tari Gap? *Muruk* 6: 19.

Beehler, B.M., & L.E. Alonso. 2001. Southern New Ireland, Papua New Guinea: a biodiversity assessment. *RAP Bull. Biol. Assess.* 21: 1–101. Conservation International, Arlington, VA.

Beehler, B.M., & C.H. Beehler. 1986. Observations on the ecology and behavior of the Pale-billed Sicklebill. *Wilson Bull.* 98: 505–15.

Beehler, B.M., & R. Bino. 1995. Yellow-eyed Starling *Aplonis mystacea* in Central Province, Papua New Guinea. *Emu* 68–70.

Beehler, B.M., J.M. Diamond, N. Kemp, E. Scholes III, C. Milensky, & T.G. Laman. 2012. Avifauna of the Foja Mountains of western New Guinea. *Bull. Brit. Orn. Club* 132: 84–101.

Beehler, B.M., & J.P. Dumbacher. 1990. Interesting observations of birds at Varirata National Park, June–July 1989. *Muruk* 4: 111–12.

Beehler, B.M., & J.P. Dumbacher. 1996. More examples of fruiting trees visited predominantly by birds of paradise. *Emu* 96: 81–88.

Beehler, B.M., & B.W. Finch. 1981. Observations [Papuan Nightjar *Eurostopodus papuensis*]. *PNG Bird Soc. Newsl.* 181/182: 40.

Beehler, B.M., & B.W. Finch. 1985. Species check-list of birds of New Guinea. *Austral. Orn. Monogr.* 1: 1–124.

Beehler, B.M., & A.J. Kirkman (eds.). 2013. *Lessons Learned from the Field: Achieving Conservation Success in Papua New Guinea*. Conservation International, Arlington, VA.

Beehler, B.M., & A.L. Mack. 1999. Constraints to characterising spatial heterogeneity in a lowland forest avifauna in New Guinea. In N.J. Adams and R.H. Slotow, eds., *Proc. 22nd Internat. Orn. Congr.*: 2569–79. Durban, South Africa.

Beehler, B.M., & J.L. Mandeville. 2016. History of ornithological exploration of New Guinea. In W.E. Davis, H.L. Recher, W.E. Boles, & J.A. Jackson, eds., *Contributions to the History of Australasian Ornithology* 3. Memoir no. 20. Nuttall Orn. Club, Cambridge, MA.

Beehler, B.M., T.K. Pratt, & D.W. Zimmerman. 1986. *Birds of New Guinea*. Princeton Univ. Press, Princeton, NJ.

Beehler, B.M., & D.M. Prawiradilaga. 2010. New taxa and new records of birds from the north coastal ranges of New Guinea. *Bull. Brit. Orn. Club* 130: 277–85.

Beehler, B.M., D.M. Prawiradilaga, Y. de Fretes, & N. Kemp. 2007. A new species of smoky honeyeater (Meliphagidae: *Melipotes*) from western New Guinea. *Auk* 124: 1000–1009.

Beehler, B.M., J.B. Sengo, C. Filardi, & K. Merg. 1994. Documenting the lowland rainforest avifauna in Papua New Guinea—effects of patchy distributions, survey effort, and methodology. *Emu* 95: 149–61.

Beehler, B.M., & R. Sine. 2007. Birds of the Kaijende Highlands, Enga Province, Papua New Guinea. In S. Richards, ed., A rapid biodiversity assessment of the Kaijende Highlands, Enga Province, Papua New Guinea. *RAP Bull. Biol. Assess.* 45: 47–51. Conservation International, Arlington, VA.

Beehler, B.M., R. Sine, & L. Legra. 2011. Birds of the Kaijende Highlands—the high elevation avifauna revisited. In S.J. Richards & B.G. Gamui, eds., Rapid biological assessments of the Nakanai Mountains and the Upper Strickland Basin: surveying the biodiversity of Papua New Guinea's sublime karst environments. *RAP Bull. Biol. Assess.* 60: 203–10. Conservation International, Arlington, VA.

Beehler, B.M., & R.J. Swaby. 1991. Phylogeny and biogeography of the *Ptiloris* riflebirds (Aves: Paradisaeidae). *Condor* 93: 738–45.

Bell, H.L. 1966a. Some wader observations from New Guinea. *Emu* 66: 32.

Bell, H.L. 1966b. Observations [first record of the Silver Gull *Chroicocephalus novaehollandiae* from New Guinea]. *PNG Bird Soc. Newsl.* 5: 1.

Bell, H.L. 1967a. Bird life of the Balimo Sub-district, Papua. *Emu* 67: 57–79.

Bell, H.L. 1967b. Some distribution notes on New Guinea highland birds. *Emu* 67: 211–14.

Bell, H.L. 1968a. The Fairy Martin, a new bird for New Guinea. *Emu* 68: 5.

Bell, H.L. 1968b. The Noisy Pitta in New Guinea. *Emu* 68: 92–94.

Bell, H.L. 1968c. Some distribution notes on New Guinea highland birds. *Emu* 68: 211–14.

Bell, H.L. 1969. Field notes on the birds of the Ok Tedi River drainage, New Guinea. *Emu* 69: 193–211.

Bell, H.L. 1970a. Field notes on the birds of Amazon Bay, Papua. *Emu* 70: 23–26.

Bell, H.L. 1970b. Field notes on the birds of the Nomad River Sub-district, Papua. *Emu* 70: 97–104.

Bell, H.L. 1970c. Observations. *PNG Bird Soc. Newsl.* 55: 1.

Bell, H.L. 1970d. The Flamed Bowerbird *Sericulus aureus*. *Emu* 70: 64–68.

Bell, H.L. 1970e. Additions to the avifauna of Goodenough Island, Papua. *Emu* 70: 179–82.

Bell, H.L. 1971a. Observations [Telefomin area]. *PNG Bird Soc. Newsl.* 66: 3.

Bell, H.L. 1971b. Field notes on birds of Mount Albert Edward, Papua. *Emu* 71: 13–19.

Bell, H.L. 1979a. Possible occurrence of Sand Martin at Finschhafen. *PNG Bird Soc. Newsl.* 155: 12.

Bell, H.L. 1979b. The effect on rainforest birds of planting teak *Tectona grandis* in Papua New Guinea. *Austral. Wildlife Res.* 6: 305–18.

Bell, H.L. 1981. Information on New Guinea kingfishers Alcedinidae. *Ibis* 123: 51–61.

Bell, H.L. 1982a. A bird community of lowland rainforest in New Guinea. 1. Composition and density of the avifauna. *Emu* 82: 24–41.

Bell, H.L. 1982b. A bird community of lowland rainforest in New Guinea. 2. Seasonality. *Emu* 82: 65–74.

Bell, H.L. 1982c. A bird community of lowland rainforest in New Guinea. 3. Vertical distribution of the avifauna. *Emu* 82: 143–62.

Bell, H.L. 1982d. A bird community of lowland rainforest in New Guinea. 4. Birds of secondary vegetation. *Emu* 82: 217–24.

Bell, H.L. 1982e. Abundance and seasonality of the savanna avifauna at Port Moresby, Papua New Guinea. *Ibis* 124: 252–74.

Bell, H.L. 1983. A bird community of lowland rainforest in New Guinea. 5. Mixed-species feeding flocks. *Emu* 82 (1982): 256–75.

Bell, H.L. 1984a. New or confirmatory information on some species of New Guinean birds. *Austral. Bird Watcher* 10: 209–28.

Bell, H.L. 1984b. A bird community of lowland rainforest in New Guinea. 6. Foraging ecology and community structure of the avifauna. *Emu* 84: 142–58.

Bellchambers, K., E. Adams, & S. Edwards. 1994. Observations of some birds of coastal and lowland Western Province, Papua New Guinea. *Muruk* 6(3): 28–39.

van Bemmel, A.C.V. 1947. Two small collections of New Guinea birds. *Treubia* 19: 1–45.

Bentz, G.D. 1979. The appendicular myology and phylogenetic relationships of the Ploceidae and Estrildidae (Aves: Passeriformes). *Bull. Carnegie Mus. Nat. Hist.* 15: 1–25.

Benz, B.W. 2011. Deciphering the evolutionary history of the montane New Guinea avifauna: comparative phylogeography and insights from paleodistributional modeling in a dynamic landscape. PhD diss., Univ. of Kansas, Lawrence, KS.

Berggy, J. 1978. Bird observations in Madang Province. *PNG Bird Soc. Newsl.* 148: 9–20.

Besnard, G., J.A.M. Bertrand, B. Delahaie, Y.X.C. Bourgeois, E. Lhuillier, & C. Thébaud. 2015. Valuing museum specimens: high-throughput DNA sequencing using historical collections of New Guinea crowned pigeons (*Goura*). *Biol. J. Linn. Soc.* doi:10.1111/bij.12494.

Birks, S.M., & S.V. Edwards. 2002. A phylogeny of megapodes (Aves: Megapodiidae) based on nuclear and mitochondrial DNA sequences. *Molec. Phylog. Evol.* 23: 408–21.

Bishop, K.D. 1977. Further bird observations from Baiyer River. *PNG Bird Soc. Newsl.* 138: 7–14.

Bishop, K.D. 1978. Birds of the lower Jimmi valley, WHP. *PNG Bird Soc. Newsl.* 145: 17–21.

Bishop, K.D. 1983. Some notes on non-passerine birds of west New Britain. *Emu* 83: 235–41.

Bishop, K.D. 1984. A preliminary report on the reserves of south-east Irian Jaya (Pulau Kimaam [Dolak], Wasur, Rawa Biru, Kumbe-Merauke and Danau Bian). WWF/IUCN Project no. 1528.

Bishop, K.D. 1986. Doria's Hawk *Megatriorchis doriae* on Batanta Island, Irian Jaya. *Kukila* 2: 85.

Bishop, K.D. 1987. Interesting bird observations in Papua New Guinea. *Muruk* 2: 52–57.

Bishop, K.D. 1995. Yellow-billed Spoonbill (*Platalea flavipes*): a new species for Papua New Guinea. *Muruk* 7: 48.

Bishop, K.D. 2005a. A review of the avifauna of the TransFly Eco-region: the status, distribution, habitats and conservation of the region's birds. TransFly Ecoregion Action Program, WWF Project no. 9S0739.02.

Bishop, K.D. 2005b. A review of the avifauna of New Guinea's southern lowland forests and freshwater swamp forests: the status, distribution, habitats, and conservation of the region's birds. High Conservation Value Forests, WWF Project no. PG0033.0.

Bishop, K.D. 2006. Shorebirds in New Guinea: their status, conservation, and distribution. *Stilt* 50: 103–34.

Bishop, K.D. MS. An annotated checklist of brids of Biak-Supiori and satellite islands. Unpublished ms, 19 pp.

Bishop, K.D., & J.M. Diamond. 1987. The Black-headed Gull *Larus ridibundus* in Irian Jaya. *Kukila* 3: 45–46.

Bishop, K.D., & D.N. Jones. 2001. Montane birds of the Nakanai Mountains, west New Britain. *Emu* 101: 205–20.

Björklund, M. 1994. Phylogenetic relationships among Charadriiformes: reanalysis of previous data. *Auk* 111: 825–32.

Black, A. 1986. The taxonomic affinity of the New Guinean Magpie *Gymnorhina tibicen papuana*. *Emu* 86: 65–70.

Blakers, M., S.J.J.F. Davies, P.N. Reilly. 1984. *The Atlas of Australian Birds.* Melbourne Univ. Press, Carlton, Victoria.

Bock, W.J. 1963. Relationships between the birds of paradise and bower birds. *Condor* 65: 91–125.

Bock, W.J. 1994. History and nomenclature of avian family-group names. *Bull. Amer. Mus. Nat. Hist.* 222: 1–281.

Bogert, C. 1937. The distribution and migration of the Long-tailed Cuckoo (*Urodynamis taitensis* Sparrman). *Amer. Mus. Novit.* 933: 1–12.

Boles, W.E. 1979. The relationships of the Australo-Papuan flycatchers. *Emu* 79: 107–10.

Boles, W.E. 1981. The subfamily name of the monarch flycatchers. *Emu* 81: 50.

Boles, W.E. 1989. A new subspecies of the Green-backed Robin *Pachycephalopsis hattamensis*, comprising the first record from Papua New Guinea. *Bull. Brit. Orn. Club* 109: 119–21.

Boles, W.E. 2006. Rhipiduridae (fantails). In J. del Hoyo, A. Elliott, & D. Christie, eds., *Handbook of the Birds of the World*, vol. 11: 200–243. Lynx Edicions, Barcelona.

Boles, W.E. 2007. Orthonychidae (logrunners); Eupetidae (jewel-babblers); Pachycephalidae (whistlers); Petroicidae (Australasian robins). In J. del Hoyo, A. Elliott, & D. Christie, eds., *Handbook of the Birds of the World*, vol. 12: 338–489. Lynx Edicions, Barcelona.

Bonilla, A.J., E.L. Braun, & R.T. Kimball. 2010. Comparative molecular evolution and phylogenetic utility of 30-UTRs and introns in Galliformes. *Molec. Phylog. Evol.* 56: 536–42.

Bostock, N. 2000. *Recurvirostra novaehollandiae* in Papua. *Kukila* 11: 152.

Bourne, W.R.P. 1998. Observations of seabirds. *Sea Swallow* 47: 23–36.

Bourne, W.R.P., & T.J. Dixon. 1971. Observations of seabirds 1967–1969. *Sea Swallow* 22: 29–60.

Brass, L.J. 1956. Results of the Archbold Expeditions no. 75. Summary of the Fourth Archbold Expedition to New Guinea (1953). *Bull. Amer. Mus. Nat. Hist.* 111: 77-152.

Braun, M.J., & C.J. Huddleston. 2009. A molecular phylogenetic survey of caprimulgiform nightbirds illustrates the utility of non-coding sequences. *Molec. Phylog. Evol.* 53: 948–60.

Brooke, R.K., & P.A. Clancy. 1981. The authorship of the generic and specific name of the Bat Hawk. *Bull. Brit. Orn. Club* 101: 371–72.

Brooks, T.M., & K.M. Helgen. 2010. A standard for species. *Nature* 467: 540–41.

Brown, D.M., & C.A. Toft. 1999. Molecular systematics and biogeography of the cockatoos (Psittaciformes: Cacatuidae). *Auk* 116: 141–57.

Brown, L., & D. Amadon. 1968. *Eagles, Hawks, and Falcons of the World.* 2 vols. Country Life Books, Feltham, Middlesex, UK.

Browning, M.R., & B.L. Monroe. 1991. Classifications and corrections of the dates of issue of some publications containing descriptions of North American birds. *Arch. Nat. Hist.* 183: 381–405.

Bruce, M.D. 1999. Tytonidae (barn owls). In J. del Hoyo, A. Elliott, & J. Sargatal, eds., *Handbook of the Birds of the World*, vol. 5: 34–75. Lynx Edicions, Barcelona.

Bulmer, R. 1967. Why is the cassowary not a bird? A problem of zoological taxonomy among the Karam of the New Guinea highlands. *Man* 2: 5–25.

Burger, J., & M. Gochfeld. 1996. Laridae (gulls). In J. del Hoyo, A. Elliott, & J. Sargatal, eds., *Handbook of the Birds of the World*, vol. 3: 572–623. Lynx Edicions, Barcelona.

Burrows, I. 1993. Some notes on birds seen in the Turama River area, Gulf Province. *Muruk* 6(1): 28–32.

Cadée, G.C. 1985. Some data on seabird abundance in Indonesian waters, July/August 1984. *Ardea* 73: 183–88.

Cadée, G.C. 1989. Seabirds in the Banda Sea in February/March 1985. *Marine Research in Indonesia* 27: 19–34.

Cain, A.J. 1955. A revision of *Trichoglossus haematodus* and of the Australian platycercine parrots. *Ibis* 97: 432–79.

Campbell, R. 1975. Observations [Woitape, Central District]. *PNG Bird Soc. Newsl.* 113: 6–7.

Campbell, R. 1979. Birds of the Tari District, Southern Highlands Province. *PNG Bird Soc. Newsl.* 159: 9–14.

Campbell, R. 1981. Observations of birds from the western side of the Southern Highlands. *PNG Bird Soc. Newsl.* 179/180: 37–40.

Campbell, R., & R. Mackay. 1982. The more interesting birds of Southern Highlands Province. *PNG Bird Soc. Newsl.* 197/198: 4–8.

Carboneras, C. 1992a. Procellariidae (shearwaters). In J. del Hoyo, A. Elliott, & J. Sargatal, eds., *Handbook of the Birds of the World*, vol. 1: 216–57. Lynx Edicions, Barcelona.

Carboneras, C. 1992b. Hydrobatidae (storm-petrels). In J. del Hoyo, A. Elliott, & J. Sargatal, eds., *Handbook of the Birds of the World*, vol. 1: 258–71. Lynx Edicions, Barcelona.

Carboneras, C. 1992c. Sulidae (gannets and boobies). In J. del Hoyo, A. Elliott, & J. Sargatal, eds., *Handbook of the Birds of the World*, vol. 1: 312–25. Lynx Edicions, Barcelona.

Carboneras, C. 1992d. Anatidae (ducks, geese, and swans). In J. del Hoyo, A. Elliott, & J. Sargatal, eds., *Handbook of the Birds of the World*, vol. 1: 536–628. Lynx Edicions, Barcelona.

Chantler, P. 1999. Apodidae (swifts). In J. del Hoyo, A. Elliott, & J. Sargatal, eds., *Handbook of the Birds of the World*, vol. 5: 388–457. Lynx Edicions, Barcelona.

Cheke, R.A., & C.F. Mann. 2008. Nectariniidae (sunbirds); Dicaeidae (flowerpeckers). In J. del Hoyo, A. Elliott, & D. Christie, eds., *Handbook of the Birds of the World*, vol. 13: 196–321; 350–89. Lynx Edicions, Barcelona.

Cheshire, N. 2010. Procellariiformes observed around Papua New Guinea including the Bismarck Archipelago from 1985 to 2007. *South Australian Ornithologist* 36: 9–24.

Cheshire, N. 2011. Observations of Long-tailed Skua *Stercorarius longicaudus*, South Polar Skua *Stercorarius maccormicki,* and other skus in Papua New Guinea waters. *Muruk* 10: 57–60.

Chivas, A.R., A. Garcia, & S. van der Kaars. 2001. Sea-level and environmental changes since the last interglacial in the Gulf of Carpenteria, Australia: a overview. *Quatern. Internat.* 83: 19–46.

Christidis, L., & W.E. Boles. 1994. The taxonomy and species of birds of Australia and its territories. *RAOU Monograph* 2: 1–112.

Christidis, L., & W.E. Boles. 2008. *Systematics and Taxonomy of Australian Birds.* CSIRO Publishing, Collingwood, Victoria.

Christidis, L., M. Irestadt, D. Rowe, W.E. Boles, & J.A. Norman. 2011. Mitochondrial and nuclear DNA phylogenies reveal a complex evolutionary history in the Australasian robins (Passeriformes: Petroicidae). *Molec. Phylog. Evol.* 61: 726–38.

Christidis, L., R. Schodde, D.D. Shaw, & S.F. Maynes. 1990. Relationships among the Australo-Papuan parrots, lorikeets, and cockatoos (Aves: Psittaciformes): protein evidence. *Condor* 93: 302–17.

Chu, P.C., S.K. Eisenschenk, & Z. Shao-tong. 2009. Skeletal morphology and the phylogeny of skuas (Aves: Charadriiformes, Stercorariidae). *Zool. Journ. Linn. Soc.* 157: 612–21.

Cibois, A., J.S. Beadell, G.R. Graves, E. Pasquet, B. Slikas, S.A. Sonsthagen, J.-C. Thibault, & R.C. Fleischer. 2011. Charting the course of reed-warblers across the Pacific islands. *J. Biogeogr.* 38: 1963–75.

Cibois, A., J.-C. Thibault, C. Bonillo, C.E. Filardi, D. Watling, & E. Pasquet. 2014. Phylogeny and biogeography of the fruit doves (Aves: Columbidae). *Molec. Phylog. Evol.* 70: 442–53.

Clapp, G.E. 1979. Preliminary report on the flycatcher *Microeca flavigaster* and the honeyeater *Melithreptus albogularis* found in eucalyptus trees in the Northern Province of Papua New Guinea. *PNG Bird Soc. Newsl.* 159: 6–8.

Clapp, G.E. 1980a. Some thoughts on the land and non-marine birds of Killerton Islands in the Northern Province of Papua New Guinea. *PNG Bird Soc. Newsl.* 167/168: 6–23.

Clapp, G.E. 1980b. Speculation on the occurrence of some birds in the Musa valley and Oro Bay/Pongani *Eucalyptus* and associated savanna areas of the Northern Province of Papua New Guinea: recent invaders or relict populations from the last glaciation. *PNG Bird Soc. Newsl.* 173/174: 20–35.

Clapp, G.E. 1981. Further selected unusual bird observations in Popondetta town, 1979–1980. *PNG Bird Soc. Newsl.* 185/186: 13–27.

Clapp, G.E. 1986. Birds of Mt Scratchley summit and environs: 3520 metres asl in south-eastern New Guinea. *Muruk* 1(3): 4–14.

Clapp, G.E. 1987. Birds of the lower Sibium Mountains, Papua New Guinea. *Muruk* 2: 45–52.

Clark, W.S. 1994. Oriental Falconidae (Oriental falcons). In J. del Hoyo, A. Elliott, & J. Sargatal, eds., *Handbook of the Birds of the World*, vol. 2: 216–77. Lynx Edicions, Barcelona.

Clarke, M.M. 1982. Notes on a visit to Bensbach, 6th to 13th November 1982. *PNG Bird Soc. Newsl.* 195/196: 18–20.

Cleere, N. 1998. Nightjars: A guide to the nightjars, nighthawks, and their relatives. Yale University Press, New Haven, CT, USA.

Cleere, N. 1999. Caprimulgidae (nightjars). In J. del Hoyo, A. Elliott, & J. Sargatal, eds., *Handbook of the Birds of the World*, vol. 5: 302–87. Lynx Edicions, Barcelona.

Cleere, N. 2010. *Nightjars, Potoos, Frogmouths, Oilbird and Owlet-nightjars of the World.* WildGuides Ltd, Old Basing, Hampshire.

Cleere, N., A.W. Kratter, D.W. Steadman, M.J. Braun, C.J. Huddleston, C.E. Filardi, & G.C.L. Dutson. 2007. A new genus of frogmouth (Podargidae) from the Solomon Islands—results from a taxonomic review of *Podargus ocellatus inexpectatus* Hartert 1901. *Ibis* 149(2): 271–86.

Cleland, E. 1968. Observations [phalaropes *Phalaropus*]. *PNG Bird Soc. Newsl.* 29: 4.

Cleland, E. 1969. Observations [Brown River and Mogubu]. *PNG Bird Soc. Newsl.* 49: 2.

Clement, P., P.A. Gregory, C.W. Moeliker. 2006. Monarchidae (monarchs [species accounts]). In J. del Hoyo, A. Elliott, & D. Christie, eds., *Handbook of the Birds of the World*, vol. 11: 280–329. Lynx Edicions, Barcelona.

Coates, B.J. 1970a. Observations [first PNG record of Flesh-footed Shearwater *Ardenna carneipes*]. *PNG Bird Soc. Newsl.* 51: 1.

Coates, B.J. 1970b. Observations [Pomarine Jaeger *Stercorarius pomarinus* in PNG]. *PNG Bird Soc. Newsl.* 54: 3.

Coates, B.J. 1970c. The Common Starling—a bird new to New Guinea. *PNG Bird Soc. Newsl.* 59: 3.

Coates, B.J. 1972. A Redshank near Port Moresby—a new bird for New Guinea. *PNG Bird Soc. Newsl.* 81: 3.

Coates, B.J. 1973a. Birds observed on Mount Albert Edward, Papua. *PNG Bird Soc. Newsl.* 84: 3–7.

Coates, B.J. 1973b. The Pectoral Sandpiper—a new bird for Papua New Guinea. *PNG Bird Soc. Newsl.* 92: 3–4.

Coates, B.J. 1974. Observations [Moitaka settling ponds]. *PNG Bird Soc. Newsl.* 96: 2.

Coates, B.J. 1985. *The Birds of Papua New Guinea, Including the Bismarck Archipelago and Bougainville.* Vol. 1, Non-passerines. Dove Publications, Alderley, Queensland.

Coates, B.J. 1990. *The Birds of Papua New Guinea, Including the Bismarck Archipelago and Bougainville.* Vol. 2, Passerines. Dove Publications, Alderley, Queensland.

Coates, B.J. 1995. Maned Duck [Australian Wood-Duck] *Chenonetta jubata* near Port Moresby: the first record for the New Guinea region. *Muruk* 7: 73–74.

Coates, B.J. 2008. Paramythiidae (painted berrypeckers). In J. del Hoyo, A. Elliott, & D. Christie, eds., *Handbook of the Birds of the World*, vol. 13: 340–49 Lynx Edicions, Barcelona.

Coates, B.J., & K.D. Bishop. 1997. *A Field Guide to the Birds of Wallacea: Sulawesi, Moluccas and Lesser Sundas*. Dove Publications, Alderley, Queensland.

Coates, B.J., G.C.L. Dutson, & C.E. Filardi. 2006. Monarchidae (monarchs [family introduction]). In J. del Hoyo, A. Elliott, & D. Christie, eds., *Handbook of the Birds of the World*, vol. 11: 244–79. Lynx Edicions, Barcelona.

Coates, B.J., & E. Lindgren. 1978. *Ok Tedi Birds*. Report of a preliminary survey of the avifauna of the Ok Tedi area, Western Province, Papua New Guinea. Ok Tedi Environmental Task Force, Ok Tedi Development Co., and Office of Environment and Conservation, Papua New Guinea, Boroko, Papua New Guinea.

Coates, B.J., & W.S. Peckover. 2001. *Birds of New Guinea and the Bismarck Archipelago*. Dove Publications, Alderley, Queensland.

Coates, B.J., & G.W. Swainson. 1978. Notes on the birds of Wuvulu Island. *PNG Bird Soc. Newsl.* 145: 8–10.

Coates, B.J., G. Yates, & L.W. Filewood. 1973. Observations [Hooded Pitta *Pitta sordida*]. *PNG Bird Soc. Newsl.* 84: 3.

Collar, N. 1996. Otididae (bustards). In J. del Hoyo, A. Elliott, & J. Sargatal, eds., *Handbook of the Birds of the World*, vol. 3: 240–75. Lynx Edicions, Barcelona.

Collar, N. 1997. Psittacidae (parrots). In J. del Hoyo, A. Elliott, & J. Sargatal, eds., *Handbook of the Birds of the World*, vol. 4: 280–479. Lynx Edicions, Barcelona.

Collar, N. 2005. Turdidae (thrushes). In J. del Hoyo, A. Elliott, & D. Christie, eds., *Handbook of the Birds of the World*, vol. 10: 514–807. Lynx Edicions, Barcelona.

Collins, C.T., & R.K. Brooke. 1976. A review of the swifts of the genus *Hirundapus* (Aves: Apodidae). *Contrib. Sci. L.A. County Mus.* 282: 1–21.

Colman, P.H. 1967. List of birds from the Lower Watut valley, 10 km north-west of Bulolo, at about 780 meters, collected between 25th July and 26th August 1967. *PNG Bird Soc. Newsl.* 24: 3–4.

Condon, H.T. 1975. *Checklist of the Birds of Australia*, Part 1. Royal Austral. Orn. Union, Melbourne, Victoria.

Cooke, R. 1976. Observations [White-tailed Tropicbird *Phaethon lepturus*]. *PNG Bird Soc. Newsl.* 126: 2.

Cooper, W.R. 1988. High altitude scrubfowl. *Muruk* 3: 4.

Cooper, W.T., & J.M. Forshaw. 1977. *The Birds of Paradise and Bowerbirds*. Collins, Sydney.

Cracraft, J. 1992. The species of the birds-of-paradise (Paradisaeidae): applying the phylogenetic species concept to a complex pattern of diversification. *Cladistics* 8: 1–43.

Cracraft, J. 2013. Avian higher-level relationships and classification: nonpasseriforms. In E.C. Dickinson & JV. Remsen, eds., *The Howard and Moore Complete Checklist of the Birds of the World*, 4th ed., vol. 1, Non-passerines: xxi–xlvii. Aves Press, Eastbourne, UK.

Cracraft, J. 2014. Avian higher-level relationships and classification: passeriforms. In E.C. Dickinson & L. Christidis, eds., *The Howard and Moore Complete Checklist of the Birds of the World*, 4th ed., vol. 2, Passerines: xxii–xlv. Aves Press, Eastbourne, UK.

Cracraft, J., & J. Feinstein. 2000. What is not a bird of paradise? Molecular and morphological evidence places *Macgregoria* in the Meliphagidae and the Cnemophilinae near the base of the corvoid tree. *Proc. B Roy. Soc. Lond.* 267: 233–41

Craig, A.J.F.K., & C.J. Feare. 2009. Sturnidae (starlings). In J. del Hoyo, A. Elliott, & D. Christie, eds., *Handbook of the Birds of the World*, vol. 14: 654–759. Lynx Edicions, Barcelona

Cramp, S., chief ed. 1977. *Handbook of the Birds of Europe, the Middle East and North Africa. The Birds of the Western Palearctic*. Vol. 1, Ostriches to Ducks. Oxford Univ. Press, Oxford, UK.

Cramp, S., chief ed. 1983. *Handbook of the Birds of Europe, the Middle East and North Africa. The Birds of the Western Palearctic*. Vol. 3, Waders to Gulls. Oxford Univ. Press, Oxford, UK.

Crome, F.H., & G.W. Swainson. 1974. Sight record of the Pied Harrier in northern New Guinea. *Emu* 74: 103.

David, N., E.C. Dickinson, & S.M.S. Gregory. 2009. Justified corrections to avian names under Article 32.5.1.1 of the International Code of Zoological Nomenclature. *Zootaxa* 2217: 56–66.

David, N., & M. Gosselin. 2002a. Gender agreement of avian species names. *Bull. Brit. Orn. Club* 122: 14–49.

David, N., & M. Gosselin. 2002b. The grammatical gender of avian genera. *Bull. Brit. Orn. Club* 112: 257–82.

Davies, C. 2008. [Waigeo Island, March–April 2007; Mt Trikora, Indonesian New Guinea, 2–17 July 2007.] *Muruk* 9(1): 33–47.

Davies, S.J.J.F. 2002. *Ratites and Tinamous*. Oxford Bird Families of the World. Oxford Univ. Press, Oxford, UK.

Dawson, J., C. Turner, O. Pileng, A. Farmer, C. McGary, C. Walsh, A. Tamblyn, & C. Yosi. 2011. Bird communities of the lower Waria valley, Morobe Province, Papua New Guinea: a comparison between habitat types. *Trop. Conserv. Sci.* 4: 317–48.

Debus, S.J.S. 1991. An annotated list of New South Wales. *Austral. Birds* 24: 72–89.

Debus, S.J.S. 1994a. Accipitridae (hawks and eagles [selected species accounts]). In J. del Hoyo, A. Elliott, & J. Sargatal, eds., *Handbook of the Birds of the World*, vol. 2: 52–205. Lynx Edicions, Barcelona.

Debus, S.J.S. 1994b. Australasian Falconiformes. In J. del Hoyo, A. Elliott, & J. Sargatal, eds., *Handbook of the Birds of the World*, vol. 2: 216–77. Lynx Edicions, Barcelona.

Debus, S.J.S. 1996. Turnicidae (buttonquail). In J. del Hoyo, A. Elliott, & J. Sargatal, eds., *Handbook of the Birds of the World*, vol. 3: 44–59. Lynx Edicions, Barcelona.

Debus, S.J.S. 2009. Australasian Corvidae (crows). In J. del Hoyo, A. Elliott, & D. Christie, eds., *Handbook of the Birds of the World*, vol. 14: 621–22. Lynx Edicions, Barcelona.

Debus, S.J.S., & C. Edelstam. 1994. Etymology, type locality, and morphology of the Chestnut-shouldered (Bürgers') Goshawk *Erythrotriorchis buergersi*. *Austral. Bird Watcher* 15: 380–82.

Debus, S.J.S., C. Edelstam, & D.A. Mead. 1994. The black morph of the Chestnut-shouldered (Bürgers') Goshawk *Erythrotriorchis buergersi*. *Austral. Bird Watcher* 15: 212–17.

Deignan, H.G. 1960. The oldest name for the Bat-eating Pern. *Bull. Brit. Orn. Club* 80: 121.

Deignan, H.G. 1964. Orthonychinae; Timaliinae. In E. Mayr & R.A. Paynter Jr., eds., *Check-list of Birds of the World*, vol. 10: 228–39; 240–455. Mus. Comp. Zool., Harvard Univ., Cambridge, MA.

Deiner, K., A.R. Lemmon, A.L. Mack, R.C. Fleischer, & J.P. Dumbacher. 2011. A passerine bird's evolution corroborates the geologic history of the island of New Guinea. *PLoS One* 6(5): e19479. doi:10.1371/journal.pone.0019479.

Dekker, R.W.R.J. 2003. Type specimens of birds in the National Museum of Natural History, Leiden. Part 2, Passerines: Eurylaimidae–Eopsaltriidae (Peters Sequence). *NNM Tech. Bull.*, 6: 1–142.

Dekker, R.W.R.J., & C. Quaisser. 2006. Type specimens of birds in the National Museum of Natural History, Leiden. Part 3, Passerines: Pachycephalidae–Corvidae (Peters Sequence). *NNM Tech. Bull.*, Leiden 9: 1–77.

De Pietri, V.L. 2013. Interrelationships of the Threskiornithidae and the phylogenetic position of the Miocene ibis 'Plegadis' paganus from the Saint-Gérand-le-Puy area in central France. *Ibis* 155: 544–60.

Diamond, J.M. 1966. Zoological classification system of a primitive people. *Science* 151: 1102–4.

Diamond, J.M. 1967. New subspecies and records of birds from the Karimui Basin, New Guinea. *Amer. Mus. Novit.* 2284: 1–17.

Diamond, J.M. 1969. Preliminary results of an ornithological exploration of the North Coastal Range, New Guinea. *Amer. Mus. Novit.* 2362: 1–67.

Diamond, J.M. 1972. *Avifauna of the Eastern Highlands of New Guinea*. Publ. no 12. Nuttall Orn. Club, Cambridge, MA.

Diamond, J.M. 1974. Colonization of exploded volcanic islands by birds: the supertramp strategy. *Science* 184: 803–6.

Diamond, J.M. 1981a. *Epimachus bruijnii*, the Lowland Sickle-billed Bird-of-Paradise. *Emu* 81: 82–86.

Diamond, J.M. 1981b. Distribution, habits, and nest of *Chenorhamphus grayi*, a malurid endemic to New Guinea. *Emu* 81: 97–100.

Diamond, J.M. 1981c. *Muscicapa griseisticta* wintering in New Guinea. *Emu* 81: 170.

Diamond, J.M. 1982. Rediscovery of the Yellow-fronted Bowerbird. *Science* 216: 431–34.

Diamond, J.M. 1983. *Melampitta gigantea*: possible relation between feather structure and underground roosting habits. *Condor* 85: 89–91.

Diamond, J.M. 1985. New distributional records and taxa from the outlying mountain ranges of Irian Jaya. *Emu* 85: 65–91.

Diamond, J.M. 1986. First record of the Large Grasshopper Warbler *Locustella fasciolata* from islands east of New Guinea. *Emu* 86: 249.

Diamond, J.M. 1987. Flocks of brown and black New Guinean bird: a bicolored mixed-species foraging association. *Emu* 87: 201–11.

Diamond, J.M., & K.D. Bishop. 1994. New records and observations from the Aru Islands, New Guinea region. *Emu* 94: 41–45.

Diamond, J.M., and K.D. Bishop. 2015. Avifauna of the Kumawa and Fakfak Mountains, Indonesian New Guinea. *Bull. Brit. Orn. Club* 135: 292–331.

Diamond, J.M., & M. LeCroy. 1979. Birds of Karkar and Bagabag Islands, New Guinea. *Bull. Amer. Mus. Nat. Hist.* 164: 469–531.

Diamond, J.M., I. Mauro, K.D. Bishop, & L. Wijaya. 2009. The avifauna of Kofiau Island, Indonesia. *Bull. Brit. Orn. Club* 129: 165–81.

Diamond, J.M., & M.N. Raga. 1976. Some birds rarely seen in Papua New Guinea. *PNG Bird Soc. Newsl.* 127: 12.

Diamond, J.M., & M.N. Raga. 1978. The Mottled-breasted Pitohui *Pitohui incertus*. *Emu* 78: 49–53.

Diamond, J.M., M.N. Raga, & J. Wiakabu. 1977. Report on a bird survey in the proposed Vanimo Timber Area. Publ. 77/10: 467–531. Wildlife Branch, Papua New Guinea, Department of Natural Resources.

Dickinson, E.C., ed. 2003. *The Howard and Moore Complete Checklist of the Birds of the World*. 3rd ed. Christopher Helm, London.

Dickinson, E.C., & L. Christidis, eds. 2014. *The Howard and Moore Complete Checklist of the Birds of the World*. 4th ed. Vol. 2, Passerines. Aves Press, Eastbourne, UK.

Dickinson, E.C., L.K. Overstreet, R.J. Dowsett, & M.D. Bruce. 2011. *Priority! The Dating of Scientific Names in Ornithology*. Aves Press, Northampton, UK.

Dickinson, E.C., & F.F.J.M. Pieters. 2011. Some bibliographic findings on "Muséum d'Histoire naturelle des PaysBas. Revue méthodique et critique des collections déposées dans cet établissement" (1862–1881), edited by Hermann Schlegel. *Zool. Biblio*. 1(3): 116–35.

Dickinson, E.C., & J.V. Remsen, eds. 2013. *The Howard and Moore Complete Checklist of the Birds of the World*. 4th ed. Vol. 1, Non-passerines. Aves Press, Eastbourne, UK.

Donaghey, R. 1970. Additions to Efogi list. *PNG Bird Soc. Newsl*. 57: 1–2.

Donaghey, R., & C. Lawrence. 1969. Observations [Mt Victoria]. *PNG Bird Soc. Newsl*. 39: 3–4.

Dor, R., R.J. Safran, F.H. Sheldon, D.W. Winkler, & I.J. Lovette. 2010. Phylogeny of the genus *Hirundo* and the Barn Swallow subspecies complex. *Molec. Phylog. Evol*. 56: 409–18.

Dorst, J., & J.-L. Mougin. 1979. Pelecaniformes. In E. Mayr & G.W. Cottrell, eds., *Check-list of Birds of the World*, vol. 1, 2nd ed.: 155–93. Mus. Comp. Zool., Harvard Univ., Cambridge, MA.

Downes, M.C. 1968. Observations [Australian Bustard *Adeotis australis*]. *PNG Bird Soc. Newsl*. 37: 2.

Draffan, R.D.W. 1976. Owlet nightjar *Aegotheles cristatus*. *PNG Bird Soc. Newsl*. 122: 9–10.

Draffan, R.D.W. 1977a. Some birds of the Finschhafen District, January to May 1977. *PNG Bird Soc. Newsl*. 133: 7–8.

Draffan, R.D.W. 1977b. List of some birds seen at Ogeramnang Airstrip, 1600 metres, Pindiu Sub District, 18–19 May 1977. *PNG Bird Soc. Newsl*. 133: 8.

Draffan, R.D.W., S.T. Garnett, & G.J. Malone. 1983. Birds of the Torres Strait: an annotated list and biogeographic analysis. *Emu* 83: 207–34.

Driskell, A.C., & L. Christidis. 2004. Phylogeny and evolution of the Australo-Papuan honeyeaters (Passeriformes, Meliphagidae). *Molec. Phylog. Evol*. 31: 943–60.

Driskell, A.C., J.A. Norman, S. Pruett-Jones, M. Mangall, S. Sonsthagen, & L. Christidis. 2011. A multigene phylogeny examining evolutionary and ecological relationships in the Australo-Papuan wrens of the subfamily Malurinae (Aves). *Molec. Phylog. Evol*. 60: 480–85.

Dumbacher, J.P. 2014. A taxonomic revision of the genus *Pitohui* Lesson, 1831 (Oriolidae), with historical notes on names. *Bull. Brit. Orn. Club* 134: 19–22.

Dumbacher, John P., B.M. Beehler, Thomas F. Spande, H. Martin Garraffo, John W. Daly 1992. Homobatrachotoxin in the Genus Pitohui: Chemical Defense in Birds? *Science* 258: 799–801.

Dumbacher, J.P., K. Deiner, L. Thompson, & R.C. Flesicher. 2008. Phylogeny of the avian genus *Pitohui* and the evolution of toxicity in birds. *Molec. Phylog. Evol*. 49: 774–81.

Dumbacher, J.P., B.A. Iova, D. Mindell, P. Gibert, & T. Bozic. 2010. Surveys of birds and bird diseases in multiple small islands of the Louisiade Archipelago, Milne Bay Province: a pilot project. Unpublished.

Dumbacher, J.P., T.K. Pratt, & R.C. Fleischer. 2003. Phylogeny of the owlet-nightjars (Aves: Aegothelidae) based on mitochondrial DNA sequence. *Molec. Phylog. Evol*. 29: 540–49.

Dumbacher, J.P., T.F. Spande, & J.W. Daly. 2000. Batrachotoxin alkaloids from passerine birds: a second toxic bird genus (*Ifrita kowaldi*) from New Guinea. *Proc. Nat. Acad. Sci. USA* 97: 12970–75.

Dutson, G. 2011. *Birds of Melanesia: Bismarcks, Solomons, Vanuatu, and New Caledonia*. Princeton Univ. Press, Princeton, NJ.

Dutson, G., & K. Tindige. 2006. [West Papua bird tour, April 2006.] *Muruk* 9(1): 1–33.

Dwyer, P.D. 1981. Two species of megapode laying in the same mound. *Emu* 81: 173–74.

Dyson, W. 1973. Observations [Sogeri Plateau]. *PNG Bird Soc. Newsl*. 88: 2.

Eastwood, C. 1994. First PNG nesting record for Bat Hawk *Macheiramphus alcinus*. *Muruk* 6(3):12–14.

Edwards, S.V., & A.C. Wilson. 1990. Phylogenetically informative length polymorphism and sequence variability in mtDNA of Australian babblers (*Pomatostomus*). *Genetics* 126: 695–711.

Elliott, A. 1992a. Pelecanidae (pelicans). In J. del Hoyo, A. Elliott, & J. Sargatal, eds., *Handbook of the Birds of the World*, vol. 1: 290–311. Lynx Edicions, Barcelona.

Elliott, A. 1992b. Ciconiidae (storks). In J. del Hoyo, A. Elliott, & J. Sargatal, eds., *Handbook of the Birds of the World*, vol. 1: 436–65. Lynx Edicions, Barcelona.

Elliott, A. 1994. Megapodiidae (megapodes). In J. del Hoyo, A. Elliott, & J. Sargatal, eds., *Handbook of the Birds of the World*, vol. 2: 278–309. Lynx Edicions, Barcelona.

Engilis, A., & R.E. Cole. 1997. Avifaunal observations from the Bishop Museum Expedition to Mount Dayman, Milne Bay Province, Papua New Guinea. *Bishop Mus. Occas. Pap.* 52: 1–19.

Erftemeijer, P., & G. Allen. 1989. Bird observations at Danau Kurumoi, Irian Jaya. *Kukila* 4: 153–54.

Erftemeijer, P., G. Allen, Zuwendra, & S. Kosamah. 1991. Birds of the Bintuni Bay region, Irian Jaya. *Kukila* 5: 85–98.

Ericson, P.G.P., L. Christidis, M. Irestadt, & J.A. Norman. 2002. Systematic affinities of the lyrebirds (Passeriformes: *Menura*), with a novel classification of the major groups of passerine birds. *Molec. Phylog. Evol.* 25: 53–62.

Ericson, P.G.P., I. Envall, M. Irestedt, & J.A. Norman. 2003a. Inter-familial relationships of the shorebirds (Aves: Charadriiformes) based on nuclear DNA sequence data. *BMC Evol. Biol.* 3: 16.

Ericson, P.G.P., M. Irestedt, & U.S. Johansson. 2003b. Evolution, biogeography, and patterns of diversification in passerine birds. *J. Avian Biol.* 34: 3–15.

Ericson, P.G.P., & U.S. Johansson. 2003. Phylogeny of Passerida (Aves: Passeriformes) based on nuclear and mitochondrial sequence data. *Molec. Phylog. Evol.* 29: 126–38.

Erritzøe, J. 2003. Pittidae (pittas). In J. del Hoyo, A. Elliott, & D. Christie, eds., *Handbook of the Birds of the World*, vol. 8: 106–61. Lynx Edicions, Barcelona.

Erritzøe, J., C.F. Mann, F. Brammer, & R.A. Fuller. 2012. *Cuckoos of the World*. A. & C. Black, London.

Fabre, P.-H., M. Moltensen, J. Fjeldså, M. Irestedt, J.-P. Lessard, & K.A. Jønsson. 2014. Multiple waves of colonization by monarch flycatchers (*Myiagra*, Monarchidae) across the Indo-Pacific and their implications for coexistence and speciation. *J. Biogeogr.* 41: 274–86.

Falla, R.A. 1940. The genus *Pachyptila*. *Emu* 40: 218–36.

Feare, C. J., & A.J.F.K. Craig. 1998. *Starlings and Mynas*. A. & C. Black, London.

Fenton, M.B. 1975. Acuity of echolocation in *Collocalia hirundinacea*. *Biotropica* 7: 1–7.

Ferguson-Lees, J., D.A. Christie, K. Franklin, D. Mead, & P. Burton. 2001. *Raptors of the World*. Christopher Helm, London.

Filardi, C.E., & R.G. Moyle. 2005. Single origin of a pan-Pacific bird group and upstream colonization of Australasia. *Nature* 438: 216–19.

Filardi, C.E., & C.E. Smith. 2005. Molecular phylogenetics of monarch flycatchers (genus *Monarcha*) with emphasis on Solomon Island endemics. *Molec. Phylog. Evol.* 37: 776–88.

Filewood, L.W.C. 1974. A new bird [Red-rumped Swallow *Cecropis daurica*] for PNG. *PNG Bird Soc. Newsl.* 104: 3–4.

Filewood, L.W.C., & W.S. Peckover. 1978. Scientific names used in *Birds of New Guinea and Tropical Australia* and the *Handbook of New Guinea Birds*. *Wildlife in Papua New Guinea* 78/12.

Finch, B.W. 1979a. First Papua New Guinea record of Black-headed Gull. *PNG Bird Soc. Newsl.* 151: 1.

Finch, B.W. 1979b. Second Australasian record of Red-rumped Swallow *Hirundo daurica*. *PNG Bird Soc. Newsl.* 151: 2–5.

Finch, B.W. 1979c. Observations [first record of Pintail *Anas acuta* for New Guinea]. *PNG Bird Soc. Newsl.* 151: 8.

Finch, B.W. 1980a. First record of Little Stint *Calidris minuta* for Papua New Guinea. *PNG Bird Soc. Newsl.* 163/164: 10–15.

Finch, B.W. 1980b. First PNG record of the Great Reed Warbler *Acrocephalus arundinaceus orientalis* at Aroa Lagoon, Central Province. *PNG Bird Soc. Newsl.* 163/164: 16–20.

Finch, B.W. 1980c. Bensbach. *PNG Bird Soc. Newsl.* 171/172: 17–33.

Finch, B.W. 1980d. Welcome Swallow *Hirundo neoxena* at Aroa lagoon, Central Province, a new species for the New Guinea region. *PNG Bird Soc. Newsl.* 173/174: 4–5.

Finch, B.W. 1980e. Rediscovery of the Fly River Grassbird *Megalurus albolimbatus* in the Bensbach River system, Western Province. *PNG Bird Soc. Newsl.* 173/174: 6–12.

Finch, B.W. 1980f. Discovery of the Zitting Cisticola *Cisticola juncidis* in the Bensbach floodplain [Bula-Balamuk] Western Province. *PNG Bird Soc. Newsl.* 173/174: 13–17.

Finch, B.W. 1981a. Second Black-headed Gull *Larus ridibundus* at Moitaka Sewage Farm. *PNG Bird Soc. Newsl.* 175/176: 7.

Finch, B.W. 1981b. Sight record of a possible Grey Falcon *Falco hypoleucus* on the Aroa Plains [Central Province]. *PNG Bird Soc. Newsl.* 175/176: 12–15.

Finch, B.W. 1981c. The survivors—unusual species as relics in Port Moresby's patches of gallery forest along the Laloki River plus a comprehensive list of the species found in the Laloki valley near Port Moresby. *PNG Bird Soc. Newsl.* 175/176: 24–46.

Finch, B.W. 1981d. Some observations in the Madang Province. *PNG Bird Soc. Newsl.* 183/184: 7–27.

Finch, B.W. 1981e. Boat trip from Madang to Karkar. *PNG Bird Soc. Newsl.* 183/184: 21.

Finch, B.W. 1981f. Asiatic Dowitcher *Limnodromus semipalmatus* at Moitaka Settling Ponds: second record for the New Guinea region. *PNG Bird Soc. Newsl.* 183/184: 36–39.

Finch, B.W. 1981g. Boat trip from Madang to Karkar. Appendix B and Appendix C. *PNG Bird Soc. Newsl.* 183/184: 22–29 [Includes reports of Doria's Hawk *Megatriorchis doriae* and Red-flanked Lorikeet *Charmosyna placentis* from Karkar I.]

Finch, B.W. 1982a. Sight record of the Obscure Berrypecker *Melanocharis arfakiana* near Owers Corner, Central Province. *PNG Bird Soc. Newsl.* 187/188: 11–12.

Finch, B.W. 1982b. Christmas at Bensbach. *PNG Bird Soc. Newsl.* 189/190: 19–40.

Finch, B.W. 1982c. White-eyed Duck *Aythya australis* nesting at Kanosia Lagoon, Central Province. *PNG Bird Soc. Newsl.* 193/194: 9.

Finch, B.W. 1982d. Black-winged Stilt (*Himantopus himantopus*) nesting at Kanosia Lagoon, Central Province—a new breeding species for the island of New Guinea. *PNG Bird Soc. Newsl.* 193/194: 10.

Finch, B.W. 1982e. Red-kneed Dotterel (*Erythrogonys cinctus*) at Kanosia Lagoon, first record for Central Province. *PNG Bird Soc. Newsl.* 193/194: 11.

Finch, B.W. 1982f. Fairy Martin *Hirundo ariel* at Kanosia Lagoon, first record for Central Province. *PNG Bird Soc. Newsl.* 193/194: 12.

Finch, B.W. 1982g. Gray's Grasshopper-Warbler *Locustella fasciolata*. *PNG Bird Soc. Newsl.* 197/198: 40.

Finch, B.W. 1983a. Little Stint *Calidris minuta* at Aroa Lagoon, Central Province. Second record for New Guinea region. *PNG Bird Soc. Newsl.* 201/202: 3–4.

Finch, B.W. 1983b. The northern race of Little Ringed Plover *Charadrius dubius curonicus* in the Port Moresby region. *PNG Bird Soc. Newsl.* 201/202: 9–10.

Finch, B.W. 1983c. Party of seven Striated Swallows *Hirundo (daurica) striolata* at Moitaka/Waigani Swamp. *PNG Bird Soc. Newsl.* 201/202: 14.

Finch, B.W. 1983d. Siberian Rubythroat *Erithacus calliope*. A new species for the Australasian region. *PNG Bird Soc. Newsl.* 201/202: 31.

Finch, B.W. 1983e. The shearwaters of Cape Suckling and the possibility of Hutton's Shearwater soon being added to the New Guinea list. *PNG Bird Soc. Newsl.* 203–206: 5–7.

Finch, B.W. 1984. Welcome Swallows *Hirundo neoxena* at Hisiu Lagoon—second record for the New Guinea region. *PNG Bird Soc. Newsl.* 212: 4–5.

Finch, B.W. 1985. Noteworthy observations in Papua New Guinea and Solomons. *PNG Bird Soc. Newsl.* 215: 6–12.

Finch, B.W. 1986a. Black Tern *Chlidonias niger*, at Moitaka Settling Ponds, Central Province—first record for the New Guinea region. *Muruk* 1(1): 26–28.

Finch, B.W. 1986b. Baird's Sandpiper *Calidris bairdii* at Kanosia Lagoon—first record for the New Guinea region. *Muruk* 1(3): 17–19.

Finch, B.W. 1986c. Black-headed Gulls *Larus ridibundus* at Lae airstrip, Morobe Province. Third record for mainland Papua New Guinea. *Muruk* 1(3): 20.

Finch, B.W. 1988a. Great Reed Warbler *Acrocephalus arundinaceus* at Bensbach—second record for Papua New Guinea. *Muruk* 3: 3.

Finch, B.W. 1988b. Eurasian Starling *Sturnus vulgaris* at Moitaka. *Muruk* 3: 4.

Finch, B.W., & D. Gillison. 1988. Rediscovery of a resident population of Eurasian Coot *Fulica atra* in Papua New Guinea. *Muruk* 3: 6. [Second author's surname incorrectly spelled "Gillieson" in original publ.]

Finch, B.W., L. Howell, & A.H. Howell. 1982. Observations [Marsh Sandpiper *Tringa stagnatilis*]. *PNG Bird Soc. Newsl.* 193/194: 39.

Finch, B.W., & P.G. Kaestner. 1990. Probable American Golden Plover *Pluvialis dominica* at Moitaka Settling Ponds. *Muruk* 4: 106–8.

Finch, B.W., & G. Moulten. 1979. Observations [Silver Gull *Chroicocephalus novaehollandiae*]. *PNG Bird Soc. Newsl.* 151: 10.

Fishpool, L., & J. Tobias. 2005. Pycnonotidae (bulbuls). In J. del Hoyo, A. Elliott, & D. Christie, eds., *Handbook of the Birds of the World*: vol. 10: 124–251. Lynx Edicions, Barcelona.

Folch, J. 1992. Casuariidae (cassowaries). In J. del Hoyo, A. Elliott, & J. Sargatal, eds., *Handbook of the Birds of the World*, vol. 1: 90–97. Lynx Edicions, Barcelona.

Ford, J. 1975. Systematics and hybridization of figbirds *Sphecotheres*. *Emu* 75: 163–71.

Ford, J. 1978. Intergradation between the Varied and Mangrove Honeyeaters. *Emu* 78: 71–74.

Ford, J. 1979. Subspeciation, hybridization, and relationships in the Little Shrike-thrush *Colluricincla megarhyncha* of Australia and New Guinea. *Emu* 79: 195–210.

Ford, J. 1981a. Morphological and behavioural evolution in the *Gerygone fusca* complex. *Emu* 81: 57–81.

Ford, J. 1981b. Evolution, distribution, and stage of speciation in the *Rhipidura fuliginosa* complex in Australia. *Emu* 81: 128–44.

Ford, J. 1981c. Hybridization and migration in Australian populations of the Little and Rufous-breasted Bronze-Cuckoos. *Emu* 81: 209–22.

Ford, J. 1982. Hybrid phenotypes in male Figbirds *Sphecotheres viridis* in Queensland. *Emu* 82: 126–30.

Ford, J. 1983. Speciation in the ground-thrush complex *Zoothera dauma* in Australia. *Emu* 83: 141–51.

Ford, J. 1985. Species limits and phylogenetic relationships in corellas of the *Cacatua pastinator* complex. *Emu* 85: 163–80.

Ford, J., & S.A. Parker. 1974. Distribution and taxonomy of some birds from south-western Queensland. *Emu* 74: 177–94.

Forshaw, J.M. 1987. *Kingfishers and Related Birds.* Vol. 2, Alcedinidae, *Halcyon* to *Tanysiptera*. Lansdowne, Sydney.

Forshaw, J.M. 1989. *Parrots of the World.* 3rd ed. Lansdowne Editions, Willoughby, NSW, Australia.

Freeman, B.G., A. Class, J. Mandeville, S. Tomassi, & B.M. Beehler. 2013. Ornithological survey of the mountains of the Huon Peninsula, Papua New Guinea. *Bull. Brit. Orn. Club* 133: 2–16.

Freeman, B.G., & A.C. Freeman. 2014. Rapid upslope shifts in New Guinean birds illustrate strong distributional responses of tropical montane species to global warming. *Proc. Nat. Acad. Sci. USA* 111: 4490–94.

Fregin, S., M. Haase, U. Olsson, & P. Alström. 2009. Multilocus phylogeny of the family Acrocephalidae (Aves: Passeriformes)—the traditional taxonomy overthrown. *Molec. Phylog. Evol.* 52: 866–78.

Friedmann, H. 1968. The evolutionary history of the avian genus *Chrysococcyx*. *Bull. U.S. Nat. Mus.* 165: 1–137.

Frith, C.B. 1979. Ornithological literature of the Papuan Subregion 1915 to 1976, an annotated bibliography. *Bull. Amer. Mus. Nat. Hist.* 164: 379–465.

Frith, C.B., & B.M. Beehler. 1998. *The Birds of Paradise.* Oxford Univ. Press, Oxford, UK.

Frith, C.B., & D.W. Frith. 1987. The Logrunner *Orthonyx temminickii* (Orthonychidae), at Tari Gap, Southern Highlands Province, Papua New Guinea. *Muruk* 2: 61–62.

Frith, C.B., & D.W Frith. 1988a. The Chestnut Forest-rail *Rallina rubra* (Rallidae), at Tari Gap, Southern Highlands Province, Papua New Guinea, and its vocalisations. *Muruk*: 3: 48–50.

Frith, C.B., & D.W. Frith. 1988b. Discovery of nests and the egg of Archbold's Bowerbird. *Austral. Bird Watcher* 14: 262–76.

Frith, C.B., & D.W. Frith. 1990. Nesting biology and relationships of the Lesser Melampitta *Melampitta lugubris*. *Emu* 90: 65–73.

Frith, C.B., & D.W. Frith. 1992. Annotated list of birds in western Tari Gap, Southern Highlands, Papua New Guinea, with some nidification notes. *Austral. Bird Watcher* 14: 262–76.

Frith, C.B., & D.W. Frith. 1996. The unique type specimen of the bird of paradise *Lophorina superba pseudoparotia* Stresemann 1934 (Paradisaeidae): a hybrid *Lophorina superba* × *Parotia carolae*. *J. Orn.* 137: 515–21.

Frith, C.B., & D.W. Frith. 1997. Biometrics of birds of paradise (Aves: Paradisaeidae) with observation on interspecific and intraspecific variation and sexual dimorphism. *Mem. Queensland Mus.* 42: 159 212.

Frith, C.B., & D.W. Frith. 2004. *The Bowerbirds.* Oxford Univ. Press, Oxford, UK.

Frith, C.B., & D.W. Frith. 2008. *Bowerbirds: Nature, Art, History.* Frith & Frith Books, Malanda, Queensland.

Frith, C.B., & D.W. Frith. 2009. Paradisaeidae (birds of paradise). In J. del Hoyo, A. Elliott, & D. Christie (eds.). 2009. *Handbook of the Birds of the World*, vol. 14: 404–92. Lynx Edicions, Barcelona.

Frith, C.B., & D.W. Frith. 2010. *Birds of Paradise: Nature, Art, History.* Frith & Frith Books, Malanda, Queensland.

Frith, H.J. 1982a. *Waterfowl in Australia.* Rev. ed. Rigby, Adelaide, South Australia.

Frith, H.J. 1982b. *Pigeons and Doves of Australia.* Rigby, Adelaide, South Australia.

Frith, H.J., F.H.J. Crome, & B.K. Brown. 2006. Aspects of the biology of the Japanese snipe *Gallinago hardwickii*. *Austral. J. Ecol.* 2: 341–68.

Frith, H.J., F.H.J. Crome, & T.O. Wolfe. 1976. Food of fruit pigeons in New Guinea. *Emu* 76: 49–58.

Frith, H.J., & W.B. Hitchcock. 1974. Fauna survey of the Port Essington District, Cobourg Peninsula, Northern Territory of Australia, Birds. *CSIRO Div. Wildl. Res. Tech. Paper* 28: 109–78.

Fry, C.H. 1969. The evolution and systematics of the bee-eaters (Meropidae). *Ibis* 111: 557–92.

Fry, C.H. 1980. The evolutionary biology of kingfishers (Alcedinidae). *Living Bird* 18: 113–60.

Fry, C.H. 2001. Meropidae (bee-eaters); Coraciidae (rollers). In J. del Hoyo, A. Elliott, & J. Sargatal, eds., *Handbook of the Birds of the World*, vol. 6: 286–341; 342–77. Lynx Edicions, Barcelona.

Fry, C.H., K. Fry, & A. Harris. 1992. *Kingfishers, Bee-eaters, and Rollers*. Princeton Univ. Press, Princeton, NJ.

Fuchs, J., C. Cruaud, A. Couloux, & E. Pasquet. 2007. Complex biogeographic history of the cuckoo-shrikes and allies (Passeriformes: Campephagidae) revealed by mitochondrial and nuclear sequence data. *Molec. Phylog. Evol.* 44: 138–53.

Fuchs, J., J. Fjeldså, & E. Pasquet. 2006. An ancient African radiation of corvoid birds (Aves: Passeriformes) detected by mitochondrial and nuclear sequence data. *Zool. Scripta* 35: 375–385

Furness, R.W. 1996. Stercorariidae (skuas and jaegers). In J. del Hoyo, A. Elliott, & J. Sargatal, eds., *Handbook of the Birds of the World*, vol. 3: 556–71. Lynx Edicions, Barcelona.

Galbraith, I.C.J. 1956. Variation, relationships and evolution in the *Pachycephala pectoralis* superspecies (Aves, Muscicapidae). *Bull. Brit. Mus. Nat. Hist.* 4: 133–222.

Galbraith, I.C.J. 1967. The Black-tailed and Robust Whistlers *Pachycephala melanura* as a species distinct from the Golden Whistler *P. pectoralis. Emu* 66: 289–94.

Garcia-R., J.C., & S.A. Trewick. 2015. Dispersal and speciation in purple swamphens (Rallidae: *Porphyrio*). *Auk* 132: 140–55.

Gardner, J.L., J.W.H. Trueman, D. Ebert, L. Joseph, & R.D. Magrath. 2010. Phylogeny and evolution of the Meliphagoidea, the largest radiation of Australasian songbirds. *Molec. Phylog. Evol.* 55: 1087–1102.

Gelang, M., A. Cibois, E. Pasquet, U. Olsson, P. Alström, & P.G.P. Ericson. 2009. Phylogeny of babblers (Aves: Passeriformes): major lineages, family limits, and classification. *Zool. Scripta* 38: 225–36.

George, G.G. 1967. Bird Notes—Western Highlands, May–June 1967. *PNG Bird Soc. Newsl.* 23: 2–3.

George, G.G. 1973. Observations [Baiyer River Sanctuary]. *PNG Bird Soc. Newsl.* 92/93: 3.

George, G.G., & B.J. Coates. 1970. Preliminary list of Lake Kopiago birds. *PNG Bird Soc. Newsl.* 58: 2–4.

Germi, F., A. Salim, & A. Minganti. 2013. First records of Chinese Sparrowhawk *Accipiter soloensis* wintering in Papua (Indonesian New Guinea). *Forktail* 29: 43–47.

Gibbs, D. 1994. Undescribed taxa and new records from the Fakfak Mountains, Irian Jaya. *Bull. Brit. Orn. Club* 114: 4–11.

Gibbs, D. 1996. Mountain Eared Nightjar in Arfak Mountains, Irian Jaya: range extension and first description of nest and egg. *Dutch Birding* 18: 246–47.

Gibbs, D. 2009. West Papua (Irian Jaya) Indonesia, 12 January–12 March 1996. Site guide for bird watchers, fully revised. Unpublished.

Gibbs, D., E. Barnes, & J. Cox. 2001. *Pigeons and Doves*. Princeton Univ. Press, Princeton, NJ.

Gibson, R., & A. Baker. 2012. Multiple gene sequences resolve phylogenetic relationships in the shorebird suborder Scolopaci (Aves: Charadriiformes). *Molec. Phylog. Evol.* 64: 66–72.

Gill, B.J., & M.E. Hauber. Piecing together the epic transoceanic migration of the Long-tailed Cuckoo (*Eudynamys taitensis*): an analysis of museum and sight records. *Emu* 112: 326-332.

Gill, B.J., B.D. Bell, G.K. Chambers, D.G. Medway, R.L. Palma, R.P. Scofield, A.J.D. Tennyson, & T.H. Worthy. 2010. *Checklist of the Birds of New Zealand, Norfolk and Macquarie Islands, and the Ross Dependency, Antarctica.* 4th ed. Ornithological Society of New Zealand and Te Papa Press, Wellington, New Zealand.

Gill, F.B. 2014. Species taxonomy of birds: which null hypothesis? *Auk* 131: 150–61.

Gill, F.B., & D. Donsker, eds. 2015. *IOC World Bird Names* (vers. 5.1). doi:10.14344/IOC.ML.5.1.

Gill, F.B., M.T. Wright, S.B. Coyne, & R. Kirk. 2009. On hyphens and phylogeny. *Wilson J. Orn.* 121: 652–55.

Gilliard, E.T. 1950. Notes on the birds of south-eastern Papua. *Amer. Mus. Novit.* 1453: 1–40.

Gilliard, E. T. 1956. The systematics of the New Guinea Manucode *Manucodia ater. Amer. Mus. Novit.* 1770: 1–13.

Gilliard, E.T. 1959. The ecology of hybridization in New Guinea honeyeaters (Aves). *Amer. Mus. Novit.* 1937: 1–26.

Gilliard, E.T. 1961. Four new birds from the mountains of central New Guinea. *Amer. Mus. Novit.* 2031: 1–7.

Gilliard, E.T. 1969. *The Birds of Paradise and Bower Birds*. Weidenfeld and Nicolson, London.

Gilliard, E.T., & M. LeCroy. 1961. Birds of the Victor Emanuel and Hindenburg Mountains, New Guinea. *Bull. Amer. Mus. Nat. Hist.* 123: 1–86.

Gilliard, E.T., & M. LeCroy. 1966. Birds of the middle Sepik region, New Guinea. Results of the American Museum of Natural History expedition to New Guinea 1953–1954. *Bull. Amer. Mus. Nat. Hist.* 132: 247–75.

Gilliard, E.T., & M. LeCroy. 1967. Annotated list of birds of the Adelbert Mountains, New Guinea. Results of the 1959 Gilliard expedition. *Bull. Amer. Mus. Nat. Hist.* 138: 53–81.

Gilliard, E.T., & M. LeCroy. 1968. Birds of the Schrader Mountains region, New Guinea. Results of the American Museum of Natural History expedition to New Guinea in 1964. *Amer. Mus. Novit.* 2343: 1–41.

Gilliard, E.T., & M. LeCroy. 1970. Notes on birds from the Tamrau Mountains, New Guinea. *Amer. Mus. Novit.* 2420: 1–28.

van Gils, J., & P. Wiersma. 1996. Scolopacidae (sandpipers [species accounts]). In J. del Hoyo, A. Elliott, & J. Sargatal, eds., *Handbook of the Birds of the World*, vol. 3: 489–533. Lynx Edicions, Barcelona.

Gjershaug, J.O., H.R.L. Lerner, & O.H. Diserud. 2009. Taxonomy and distribution of the Pygmy Eagle *Aquila* (*Hieraaetus*) *weiskei* (Accipitriformes: Accipitridae). *Zootaxa* 2326: 24–38.

Glynn, W.E. 1995. Bat Hawk *Macheiramphus alcinus* sighting from Kobakma, Irian Jaya, Indonesia on 8th December 1990. *Muruk* 7: 122–23.

Gochfeld, M., & J. Burger. 1996. Sternidae (terns). In J. del Hoyo, A. Elliott, & J. Sargatal, eds., *Handbook of the Birds of the World*, vol. 3: 624–67. Lynx Edicions, Barcelona.

Gonzalez, J., H. Düttmann, & M. Wink. 2009. Phylogenetic relationships based on two mitochondrial genes and hybridization patterns in Anatidae. *J. Zoology, London* 279: 310–18.

Goodfellow, W. 1926. [Remarks on a recent field trip in British Papua collecting birds of paradise.] *Bull. Brit. Orn. Club* 46: 58–59.

Goodwin, D. 1982. *Estrildid Finches of the World*. Oxford Univ. Press, Oxford, UK.

Goodwin, D. 1983. *Pigeons and Doves of the World*. 2nd ed. Comstock Publishing, Cornell Univ. Press, Ithaca, NY.

Grant, C.H.B. 1959. The expedition of the British Ornithologists' Union to New Guinea, 1910–11. *Ibis* 101: 65–70.

Greensmith, A. 1975. Some notes on Melanesian seabirds. *Sunbird* 6: 77–89.

Greenway, J.C. 1934. Descriptions of four subspecies of birds from the Huon Gulf region, New Guinea. *Proc. New Engl. Zool. Club* 14: 1–3.

Greenway, J.C. 1935. Birds from the coastal range between the Markham and Waria Rivers, northeastern New Guinea. *Proc. New Engl. Zool. Club* 14: 15–106.

Greenway, J.C. 1942. A new manucode bird of paradise. *Proc. New Engl. Zool. Club* 19: 51–52.

Greenway, J.C. 1962. Oriolidae. In E. Mayr & J.C. Greenway, eds., *Check-list of Birds of the World*: vol. 15: 122–37. Mus. Comp. Zool., Harvard Univ., Cambridge, MA.

Greenway, J.C. 1966. Birds collected on Batanta, off western New Guinea, by E. Thomas Gilliard in 1964. *Amer. Mus. Novit.* 2258: 1–27.

Greenway, J.C. 1967. Daphoenosittinae; Climacteridae. In R.A. Paynter Jr., ed., *Check-list of Birds of the World*, vol. 12: 145–48; 162–66. Mus. Comp. Zool., Harvard Univ., Cambridge, MA.

Greenway, J.C. 1973. Type specimens of birds in the American Museum of Natural History. Part 1: Tinamidae–Rallidae. *Bull. Amer. Mus. Nat. Hist.* 150(3): 209–345.

Greenway, J.C. 1978. Type specimens of birds in the American Museum of Natural History. Part 2: Otididae–Picidae. *Bull. Amer. Mus. Nat. Hist.* 161: 1–303.

Gregory, P. 1993. Grey-streaked Flycatcher *Muscicapa griseisticta* at Ok Menga, Western Province. *Muruk* 6(1): 46.

Gregory, P. 1994a. Yellow-billed Spoonbill (*Platalea flavipes*) at Bensbach. *Muruk* 6(3): 45.

Gregory, P. 1994b. Great Knot (*Calidris tenuirostris*) at Tabubil. *Muruk* 6(3): 45.

Gregory, P. 1995a. *Birds of the Ok Tedi Area*. OTML, Port Moresby, Papua New Guinea.

Gregory, P. 1995b. Further studies of the birds of the Ok Tedi area, Western Province, Papua New Guinea. *Muruk* 7: 1–38.

Gregory, P. 1996. Notes on the Greater Melampitta (*Melampitta gigantea*) in the Tabubil area. *Muruk* 8: 36–37.

Gregory, P. 1997. Range extensions and unusual sightings from Western Province, Papua New Guinea. *Bull. Brit. Orn. Club* 117: 304–11.

Gregory, P. 2007a. Significant sightings from tour reports. *Muruk* 8: 99–128.

Gregory, P. 2007b. Acanthizidae (Australasian warblers). In J. del Hoyo, A. Elliott, & D. Christie, eds., *Handbook of the Birds of the World*, vol. 12: 544–611. Lynx Edicions, Barcelona.

Gregory, P. 2008a. *Birds of New Guinea and Its Offshore Islands*. Sicklebill Publications, Kuranda, Queensland.

Gregory, P. 2008b. Melanocharitidae (berrypeckers). In J. del Hoyo, A. Elliott, & D. Christie, eds., *Handbook of the Birds of the World*, vol. 13: 322–39. Lynx Edicions, Barcelona.

Gregory, P., ed. 2008c. Significant sightings from West Papuan tour reports. *Muruk* 9(1): 1–33.

Gregory, P. 2011. An overview of recent taxonomic changes to the avifauna of New Guinea. *Muruk* 10: 2–40.

Gregory, P., I. Burrows, R. Burrows, & G. Burrows. 1996a. Blue Rock-Thrush at Manokwari, a new record for Irian Jaya. *Kukila* 8: 154.

Gregory, P., I. Burrows, R. Burrows, & G. Burrows. 1996b. Red-rumped Swallow at Manokwari, a new species for Irian Jaya. *Kukila* 8: 153.

Gregory, P., S.A. Halse, R.P. Jaensch, W.R. Kay, P. Kulmoi, G.B. Pearson, & A.W. Storey. 1996c. The middle Fly waterbird survey 1994–95. *Muruk* 8: 1–7.

Gregory, P.A., & G.R. Johnston. 1993. Birds of the cold tropics: Dokfuma, Star Mountains, New Guinea. *Bull. Brit. Orn. Club* 113: 139–45.

Gregory-Smith, R., & J. Gregory-Smith. 1989. Bird notes from Wallai Island. *Muruk* 4: 21.

Greig-Smith, P.W. 1978. Social feeding of fruit-pigeons in New Guinea. *Emu* 78: 92–93.

Gressitt, J.L., ed. 1982. *Ecology and Biogeography of New Guinea*. W.R. Junk, the Hague.

Griffiths, C.S. 1999. Phylogeny of the Falconidae inferred from molecular and morphological data. *Auk* 116: 116–30.

Groth, J.G. 1998. Molecular phylogenetics of finches and sparrows: consequences of character state removal in cytochrome *b* sequences. *Molec. Phylog. Evol.* 10: 377–90.

Gyldenstolpe, N. 1955a. Notes on a collection of birds made in the Western Highlands, Central New Guinea, 1951. *Ark. Zool.* (2)8: 1–181.

Gyldenstolpe, N. 1955b. Birds collected by Dr. Sten Bergman during his expedition to Dutch New Guinea 1948–1949. *Ark. Zool.* (2)8: 183–397.

Hachisuka, M.U. 1952. Change of names among sunbirds and a woodpecker. *Bull. Brit. Orn. Club* 72: 22–23.

Hackett, S.J., *et mult.* 2008. A phylogenomic study of birds reveals their evolutionary history. *Science* 320: 1763–67.

Hadden, D. 1975a. Birds seen in the Tari area from 1 August to 14 August 1975. *PNG Bird Soc. Newsl.* 113: 8–9.

Hadden, D. 1975b. Yellow-bellied Mountain White-eye *Zosterops fuscicapilla* [new distributional record]. *PNG Bird Soc. Newsl.* 115: 16.

Hadden, D. 1976. At Bosavi Mission Station, on the lower slopes of Mt Bosavi, 9–15 December 1975. *PNG Bird Soc. Newsl.* 120: 6–7.

Hadden, D. 1981. *Birds of the North Solomons*. Handbook no. 6. Wau Ecology Institute, Wau, Papua New Guinea.

Hadden, D. 2004. *Birds and Bird-Lore of Bougainville and the North Solomons*. Dove Publications, Alderley, Queensland.

Halse, S.A., G.B. Pearson, R.P. Jaensch, P. Kulmoi, P. Gregory, W.R. Kay, & A.W Storey. 1996. Waterbird surveys of the Middle Fly River Floodplain. *Papua New Guinea Wildlife Research* 23: 557–69.

Hamilton, S.G., J. Erico, & M.K. Tarburton. 2001. Notes on the sixth specimen record of the Three-toed Swiftlet *Aerodramus papuensis* in Papua New Guinea. *Corella* 25: 12-14.

Han, K.L., M.B. Robbins, & M.J. Braun. 2010. A multi-gene estimate of phylogeny in the nightjars and nighthawks (Caprimulgidae). *Molec. Phylog. Evol.* 55: 443–53.

Hancock, J., & J. Kushlan. 1984. *The Herons Handbook*. Croom Helm, London.

Haring, E., B. Däubl, W. Pinsker, A. Kryukov, & A. Gamauf. 2012. Genetic divergences and intraspecific variation in corvids of the genus *Corvus* (Aves: Passeriformes: Corvidae)—a first survey based on museum specimens. *J. Zool. Syst. Evol. Research* 50: 230–46.

Harris, R.B., S.M. Birks, & A.D. Leache. 2014. Incubator birds: biogeographical origins and evolution of underground nesting in megapodes (Galliformes: Megapodiidae). *J. Biogeogr.* 41: 2045–56.

Harrison, C.J.O., & S.A. Parker. 1966. The taxonomic affinities of the New Guinea genera *Paramythia* and *Oreocharis*. *Bull. Brit. Orn. Club* 86: 15–20.

Harshman, J., *et mult.* 2008. Phylogenomic evidence for multiple losses of flight in ratite birds. *Proc. Nat. Acad. Sci. USA* 105: 13462–67.

Hartert, E. 1899a. On the birds collected by Mr. Meek on Rossel Island in the Louisiade Archipelago. *Novit. Zool.* 6: 76–84.

Hartert, E. 1899b. On the birds collected by Mr. Meek on St. Aignan [Misima] Island in the Louisiade Archipelago. *Novit. Zool.* 6: 206–16.

Hartert, E. 1929. On various forms of the genus *Tyto*. *Novit. Zool.* 35: 93–104.

Hartert, E. 1930. List of the birds collected by Ernst Mayr. *Novit. Zool.* 36: 27–128.

Hartert, E., K. Paludan, W. Rothschild, & E. Stresemann. 1936. Ornithologische Ergebnisse der Expedition Stein 1931–32. 4: Die vogel des Weyland-Gebirges und seines Vorlandes. *Mitt. Zool. Mus. Berl.* 21: 165–240.

Hartley, N.D.R. 1984. 1984 Ornithon: 20–23rd April, 1984 [Trobriand Islands]. *PNG Bird Soc. Newsl.* 209: 6–8.

Heads, M. 2001. Birds of paradise (Paradisaeidae) and bowerbirds (Ptilonorhynchidae): regional levels of biodiversity and terrane tectonics in New Guinea. *J. Zool.* London 255: 221–339.

Healey, C.J. 1976. Sympatry in *Parotia lawesii* and *P. carolae*. *Emu* 76: 85.

Heron, S.J. 1974. Further information on the birds of Papua New Guinea. *Emu* 74: 201–2.

Heron, S.J. 1975a. Grey Fantail *Rhipidura fuliginosa*. *PNG Bird Soc. Newsl.* 112: 14.

Heron, S.J. 1975b. The birds of the mangroves in Papua New Guinea. *Austral. Bird Watcher* 6: 69–92.

Heron, S.J. 1975c. A.S. Meek's journeys to the Aroa River in 1903 and 1904–05. *Emu* 75: 232–33.

Heron, S.J. 1976. Waders and terns seen at the mouth of the Angabunga River and nearby beach at Aviara, 6 km south of Bereina. *PNG Bird Soc. Newsl.* 125: 12–14.

Heron, S.J. 1977a. First New Guinea record of the Asian Dowitcher. *PNG Bird Soc. Newsl.* 129. 9.

Heron, S.J. 1977b. The Knot—second record for the New Guinea area. *PNG Bird Soc. Newsl.* 131: 8.

Heron, S.J. 1977c. Observations and bird notes [Red-necked Crake *Rallina tricolor*, Horsfield's Bronze Cuckoo *Chalcites basalis*]. *PNG Bird Soc. Newsl.* 134: 3.

Heron, S.J. 1977d. Birds observed in little known areas of the Goilala Sub-Province, Central Province. *PNG Bird Soc. Newsl.* 135: 6–10.

Heron, S.J. 1977f. The Black-tailed Whistler *Pachycephala melanura* in mainland Papua New Guinea. *PNG Bird Soc. Newsl.* 131: 11–12.

Heron, S.J. 1978a. Eurasian Wigeon *Anas penelope* near Bereina, Central Province. *PNG Bird Soc. Newsl.* 139/140: 5–6.

Heron, S.J 1978b. Waders of the Bereina District, Papua New Guinea. *PNG Bird Soc. Newsl.* 142: 5–7.

Heron, S.J. 1978c. Waders of the New Guinea region. *PNG Bird Soc. Newsl.* 149/150: 8–15.

Hicks, R. 1985a. A weekend at Bensbach 15–17 November 1985. *PNG Bird Soc. Newsl.* 218: 4–6.

Hicks, R. 1985b. Pelagic birding 14 December 1985. *PNG Bird Soc. Newsl.* 218: 7.

Hicks, R. 1986. Female House Sparrow *Passer domesticus* on Paga Hill, Port Moresby. A second record for PNG. *Muruk* 1(3): 20.

Hicks, R. 1987. An extension of altitude range for two mannikin species. *Muruk* 2: 60.

Hicks, R. 1988a. Recent observations [Short-tailed Shearwater *Ardenna tenuirostris*, Pied Cormorant *Phalacrocorax varius*]. *Muruk* 3: 69–76; 129–37.

Hicks, R. 1988b. White-throated Nightjar *Eurostopodus mystacalis* first record for the Port Moresby area. *Muruk* 3: 55–56.

Hicks, R. 1989a. House Sparrows *Passer domesticus* on Yule Island, Central Province. *Muruk* 4: 23.

Hicks, R. 1989b. Recent observations [House Sparrow *Passer domesticus*]. *Muruk* 4: 39–47.

Hicks, R. 1990. [Asian Dowitcher *Limnodromus semipalmatus* and Pin-tailed Snipe *Gallinago stenura* sightings.] *Muruk* 4: 90 105.

Hicks, R. 1991. Asian waterfowl census: Port Moresby area 1990. *Muruk* 5: 8–11.

Hicks, R. 1992 [Observation of Papuan Whipbird]. *Muruk* 5: 145–151.

Hicks, R., C.H.B. Eastwood, & W.F. Glynn. 1988. White Pygmy-Goose—a new species for the Port Moresby area. *Muruk* 3: 5.

Hicks, R., & B.W. Finch. 1987. Blue Rock Thrush on Paga Hill, Port Moresby. First record for the Australian region, east of the Moluccas. *Muruk* 2: 66–67.

Hicks, R., & R. Gregory-Smith. 1990. Welcome Swallow at Moitaka settling ponds. *Muruk* 4: 110.

Higgins, P.J., ed. 1999. *Handbook of Australian, New Zealand and Antarctic Birds.* Vol. 4, Parrots to Dollarbird. Oxford Univ. Press, Melbourne, Australia.

Higgins, P.J., L. Christidis, & H.A. Ford. 2008. Meliphagidae (honeyeaters). In J. del Hoyo, A. Elliott, & D. Christie, eds., *Handbook of the Birds of the World*, vol. 13: 498–691. Lynx Edicions, Barcelona.

Higgins, P.J., & S.J.J.F. Davies, eds. 1996. *Handbook of Australian, New Zealand and Antarctic Birds*. Vol. 3, Snipe to Pigeons. Oxford Univ. Press, Melbourne, Australia.

Higgins, P.J., & J.M. Peter, eds. 2002. *Handbook of Australian, New Zealand and Antarctic Birds*. Vol. 6, Pardalotes to Shrike-thrushes. Oxford Univ. Press, Melbourne, Australia.

Higgins, P.J., J.M. Peter, & S.J. Cowling, eds. 2006. *Handbook of Australian, New Zealand and Antarctic birds*. Vol. 7a–b, Boatbill to Starlings. Oxford Univ. Press, Melbourne, Australia.

Higgins, P.J., J.M. Peter, & W.K. Steele, eds. 2001. *Handbook of Australian, New Zealand and Antarctic birds*. Vol. 5, Tyrant-flycatchers to Chats. Oxford Univ. Press, Melbourne, Australia.

Hobcroft, D., & K.D. Bishop. ms. Birds of the Laughlan Islands [Budi Budi Is, SE Islands, PNG]. Unpublished ms.

Hockey, P.A.R. 1996. Haematopodidae (oystercatchers). In J. del Hoyo, A. Elliott, & J. Sargatal, eds., *Handbook of the Birds of the World*, vol. 3: 308–25. Lynx Edicions, Barcelona.

van den Hoek Ostende, L.W., R.W.R.J. Dekker, & G.O. Keijl. 1997. Type-specimens of birds in the National Museum of Natural History, Leiden. Part 1: Non-passerines. *NNM Tech. Bull.*, Leiden 1: 1–248.

Holmes, D.A. 1989. The Tree Sparrow reaches New Guinea. *Kukila* 4: 150–51.

Holmes, D.A. 1997. Records of House Sparrows in Irian Jaya. *Kukila* 9: 182–83.

Holyoak, D.T. 1970. The relation of the parrot genus *Opopsitta* to *Psittaculirostris*. *Emu* 70: 198.

Holyoak, D.T. 1973. Comments on taxonomy and relationships in the parrot subfamilies Nestorinae, Loriinae and Platycercinae. *Emu* 73: 157–76.

Holyoak, D.T. 1999a. Aegothelidae (owlet-nightjars). In J. del Hoyo, A. Elliott, & J. Sargatal, eds., *Handbook of the Birds of the World*, vol. 5: 252–65. Lynx Edicions, Barcelona.

Holyoak, D.T. 1999b. Podargidae (frogmouths). In J. del Hoyo, A. Elliott, & J. Sargatal, eds., *Handbook of the Birds of the World*, vol. 5: 266–87. Lynx Edicions, Barcelona.

Hoogerwerf, A. 1959. Enkele vorlopige mededelingen over de ekstereend, *Anseranas semipalmata*, in zuid Nieuw-Guinea. *Ardea* 47: 192–99.

Hoogerwerf, A. 1962. Enkele gegevens over het vogelschade-onderzoek bij het rijstproefbedrift "Koembe" te Koerik (nabije Merauke) zuid Nieuw-Guinea. *Meded. Agr. Proefst. Manokwari, Landbouwk. Serie*, 7: 1–110.

Hoogerwerf, A. 1964. On birds new for New Guinea or with a larger range than previously known. *Bull. Brit. Orn. Club* 84: 70–77; 94–96; 118–24; 142–48; 153–61.

Hoogerwerf, A. 1971. On a collection of birds from the Vogelkop near Manokwari, north-western New Guinea. *Emu* 71: 1–12; 73–83.

Hopkins, H.C.F. 1988. Grey Imperial Pigeon near Karawari Lodge, East Sepik Province. *Muruk* 3: 6.

Hornbuckle, J. 1991. Irian Jaya 1991. A report of a birding trip to western New Guinea with general advice and a guide to sites. Unpublished.

Hornbuckle, J., & I. Merrill. 2004. Eastern Indonesia: Sulawesi, Halmahera, west Irian Jaya, Ambon and Tanimbar, August–September 2004. http://jonathanhornbuckle.webs.com/easternindonesia2004.htm. Accessed 24-Sep-2015.

del Hoyo, J., and collaborators (eds.). 1992–2012. *Handbook of the Birds of the World*. 16 vols. Lynx Edicions, Barcelona.

del Hoyo, J., & N.J. Collar. 2014. *Illustrated Checklist of the Birds of the World*. Volume 1: non-passerines. Lynx Edicions, Barcelona.

Hudson, R., & J.W.H Conroy. 1975. Southern Giant Petrel in New Guinea. *Emu* 75: 43.

Hulme, D. 1976. The White-headed Kingfisher *Halcyon saurophaga* at Kupiano, Central Province. *PNG Bird Soc. Newsl.* 127: 10.

Hulme, D. 1977. The birds of Lumi, West Sepik Province. *PNG Bird Soc. Newsl.* 129: 5–6.

Hume, R.A. 1996. Burhinidae (thick-knees). In J. del Hoyo, A. Elliott, & J. Sargatal, eds., *Handbook of the Birds of the World*, vol. 3: 348–63. Lynx Edicions, Barcelona.

Hunt, A. 1970. Observations [White-bellied Sea-Eagle *Haliaeetus leucogaster* nest at Sirinumu Dam]. *PNG Bird Soc. Newsl.* 55: 1.

Ingram, G. 1976. Birds from some islands of the Torres Strait. *Sunbird* 7: 67–76.

International Commission on Zoological Nomenclature (ICZN). 1999. *International Code of Zoological Nomenclature*. 4th ed. Internat. Trust for Zool. Nomencl., London.

Iredale, T. 1950. *Birds of Paradise and Bower Birds*. Georgian House, Melbourne, Australia.

Iredale, T. 1956. *Birds of New Guinea*, 2 Vols. Georgian House, Melbourne, Australia.

Irestedt, M., P.-H. Fabre, H. Batahla-Filho, K.A. Jønsson, C.S. Roselaar, G. Sangster, & P.G.P. Ericson. 2013. The spatio-temporal colonization and diversification across the Indo-Pacific by a 'great speciator' (Aves: *Erythropitta erythrogaster*). *Proc. B Roy. Soc. Lond.* 280 (1759). doi:10.1098/rspb.2013.0309.

Irestedt, M., K.A. Jønsson, J. Fjeldså, L. Christidis, & P.G.P. Ericson. 2009. An unexpectedly long history of sexual selection in birds-of-paradise. *BMC Evol. Biol.* 9: 235. doi:10.1186/1471-2148-9-235.

Irestedt, M., & J.I. Ohlson. 2008. The division of the major songbird radiation into Passerida and 'core Corvoidea' (Aves: Passeriformes)—the species tree *vs.* gene trees. *Zool. Scripta* 37: 305–13.

Irestedt, M., J.I. Ohlson, D. Zuccon, M. Källersjö, & P.G.P. Ericson. 2006. Nuclear DNA from old collections of avian study skins reveals the evolutionary history of the Old World suboscine (Aves: Passeriformes). *Zool. Scripta* 35: 567–80.

Isles, A., & P. Menkhorst. 1976. Bird list, Finschhafen-Sattelberg area. *PNG Bird Soc. Newsl.* 121: 13–18.

Jaensch, R. 1995. Little Bitterns (*Ixobrychus minutus*) in the middle Fly wetlands. *Muruk* 7: 117–18.

Jaensch, R. 1996. A Little Bittern (*Ixobrychus minutus*) at Chambri Lake, middle Sepik wetlands. *Muruk* 8: 27–28.

Jarvis, E.D., *et mult.* 2014. Whole-genome analyses resolve early branches in the tree of life of modern birds. *Science* 346: 1320–31.

Jenni, D. 1996. Jacanidae (jacanas). In J. del Hoyo, A. Elliott, & J. Sargatal, eds., *Handbook of the Birds of the World*, vol. 3: 276–91. Lynx Edicions, Barcelona.

Johnsgard, P.A. 1979. Anseriformes. In E. Mayr & G.W. Cottrell, eds., *Check-list of Birds of the World*, vol. 1, 2nd ed.: 425–506. Mus. Comp. Zool., Harvard Univ., Cambridge, MA.

Johnson, G.R., & S.J. Richards. 1994. Notes on birds observed in the Western Province during July 1993. *Muruk* 6(3): 9.

Johnson, K.P., S. de Kort, K. Dinwoodey, A.C. Mateman, C. ten Cate, C.M. Lessells, & D.H. Clayton. 2001. A molecular phylogeny of the dove genera *Streptopelia* and *Columba*. *Auk* 118: 874-887.

Johnstone, R. 1981. Notes on the distribution, ecology, and taxonomy of the Red-crowned Pigeon (*Ptilinopus regina*) and Torres Strait Pigeon (*Ducula bicolor*) in Western Australia. *Rec. West. Austral. Mus.* 9: 7–22.

Johnstone, R. 2006. The birds of Gag Island, Western Papuan Islands, Indonesia. *Rec. West. Austral. Mus.* 23: 115–32.

Johnstone, R.E. 1982. Distribution, status, and variation of the Silver Gull *Larus novaehollandiae* Stephens, with notes on the *Larus cirrocephalus* species group. *Rec. West. Austral. Mus.* 10: 133–65.

Jones, D.N., R.W. Dekker, & C.S. Roselaar. 1995. *The Megapodes*. Oxford Univ. Press, Oxford, UK.

Jonkers, B., & H. Roersma 1990. New subspecies of *Lonchura spectabilis* from East Sepik Province, Papua New Guinea. *Dutch Birding* 12: 22–25.

Jønsson, K.A., R.C.K. Bowie, R.G. Moyle, L. Christidis, C.E. Filardi, J. Norman, & J. Fjeldså. 2008. Molecular phylogenetics and diversification within one of the most geographically variable bird species complexes *Pachycephala pectoralis/melanura*. *J. Avian Biol.* 39: 473–78.

Jønsson, K.A., R.C.K. Bowie, R.G. Moyle, L. Christidis, J. Norman, B.W. Benz, & J. Fjeldså. 2010a. Historical biogeography of an Indo-Pacific passerine bird family (Pachycephalidae): different colonization patterns in the Indonesian and Melanesian archipelagos. *J. Biogeogr.* 37: 245–57.

Jønsson, K.A., R.C.K. Bowie, R.G. Moyle, M. Irestedt, L. Christidis, J. Norman, & J. Fjeldså. 2010b. Phylogeny and biogeography of Oriolidae (Aves: Passeriformes). *Ecography* 33: 232–41.

Jønsson, K.A., R.C.K. Bowie, J.A.A. Nylander, L. Christidis, J. Norman, & J. Fjeldså. 2010c. Biogeographical history of the cuckoo-shrikes (Aves: Passeriformes): transoceanic colonization of Africa from Australo-Papua. *J. Biogeogr.* 37: 1767–81.

Jønsson, K.A., P.-H. Fabre, R.E. Ricklefs, & J. Fjeldså. 2011a. Major global radiation of corvoid birds originated in the proto-Papuan archipelago. *Proc. Nat. Acad. Sci. USA* 108: 2328–33.

Jønsson, K.A., P.-H. Fabre, R.E. Ricklefs, & M. Irestedt. 2012. Brains, tools, innovation and biogeography in crows and ravens. *BMC Evol. Biol.* 12: 72.

Jønsson, K.A., & J. Fjeldså. 2006. Determining biogeographic patterns of dispersal and diversification in oscine passerine birds in Australia, Southeast Asia, and Africa. *J. Biogeogr.* 33: 1155–65.

Jønsson, K.A., J Fjeldså, P.G.P. Ericson, & M. Irestedt. 2007. Systematic placement of an enigmatic Southeast Asian taxon *Eupetes macrocerus* and implications for the biogeography of a main songbird radiation, the Passerida. *Biol. Letters* 3(3): 323–26.

Jønsson, K.A., M. Irestedt, R.C.K. Bowie, L. Christidis, & J. Fjeldså. 2011b. Systematics and biogeography of Indo-Pacific ground doves. *Molec. Phylog. Evol.* 59: 538–43.

Jønsson, K.A., M. Irestedt, L. Christidis, S.M. Clegg, B.G. Holt, J. Fjeldså. 2014. Evidence of taxon cycles in an Indo-Pacific passerine bird radiation (Aves: *Pachycephala*). *Proc. B Roy. Soc. Lond.* 281. doi:10.1098/rspb.2013.1727.

Joseph, L. 2008. [Book review of] *Systematics and Taxonomy of Australian Birds*, by Les Christidis and Walter Boles. *Emu* 108: 365–72.

Joseph, L., A.S. Nyári, & M.J. Andersen. 2014a. Taxonomic consequences of cryptic speciation in the Golden Whistler *Pachycephala pectoralis* complex in mainland southern Australia. *Zootaxa* 3900(2): 294–300.

Joseph, L., B. Slikas, D. Alpers, & R. Schodde. 2001. Molecular systematics and phylogeography of New Guinean logrunners (Orthonychidae). *Emu* 101: 273–80.

Joseph, L., A. Toon, A.S. Nyári, N.W. Longmore, K.M.C. Rowe, T. Haryoko, J. Trueman, & J.L. Gardner. 2014b. A new synthesis of the molecular systematics and biogeography of honeyeaters (Passeriformes: Meliphagidae) highlights biogeographic and ecological complexity of a spectacular avian radiation. *Zool. Scripta* 43(3): 235–48.

Joseph, L., A. Toon, E.E. Schirtzinger, & T.F. Wright. 2011a. Molecular systematics of two enigmatic genera *Psittacella* and *Pezoporus* illuminate the ecological radiation of Australo-Papuan parrots (Aves: Psittaciformes). *Molec. Phylog. Evol.* 59: 675–84.

Joseph, L., A. Toon, E.E. Schirtzinger, T.F. Wright, & R. Schodde. 2012. A revised nomenclature and classification for family-group taxa of parrots (Psittaciformes). *Zootaxa* 3205: 26–40.

Joseph, L., T. Zeriga, G.J. Adcock, & N.E. Langmore. 2011b. Phylogeography and taxonomy of the Little Bronze-Cuckoo (*Chalcites minutillus*) in Australia's monsoon tropics. *Emu* 111: 113–19.

Jouanin, C., & J.-L. Mougin. 1979. Procellariiformes. In E. Mayr & G.W. Cottrell, eds., *Check-list of Birds of the World*, vol. 1, 2nd ed.: 48–121. Mus. Comp. Zool., Harvard Univ., Cambridge, MA.

de Juana, E., F. Suárez, & P.F. Donald. 2004. Alaudidae (larks). In J. del Hoyo, A. Elliott, & D. Christie, eds., *Handbook of the Birds of the World*, vol. 9: 496–601. Lynx Edicions, Barcelona.

Junge, G.C.A. 1937. The birds of south New Guinea. Part 1: Non-passerines. *Nova Guinea* 1: 125–88.

Junge, G.C.A. 1939. The birds of south New Guinea. Part 2: Passerines. *Nova Guinea* 3: 1–94.

Junge, G.C.A. 1952. New subspecies of birds from New Guinea. *Zool. Meded. Leiden* 31: 247–49.

Junge, G.C.A. 1953. Zoological results of the Dutch New Guinea expedition, 1939, no. 5. The birds. *Zool. Verh. Leiden* 20: 1–77.

Junge, G.C.A. 1956. New bird records from Biak Island. *Zool. Meded. Leiden* 34: 231–37.

Kahl, M.P. 1979. Ciconiiformes. In E. Mayr & G.W. Cottrell, eds. *Check-list of Birds of the World*, vol. 1, 2nd ed.: 245–52. Mus. Comp. Zool., Harvard Univ., Cambridge, MA.

Kear, J. 2005. *Ducks, Geese, and Swans*. Oxford Univ. Press, Oxford, UK.

Kearns, A.M., L. Joseph, & L.G. Cook. 2012. A multilocus coalescent analysis of the speciational history of the Australo-Papuan butcherbirds and their allies. *Molec. Phylog. Evol.* 66(3): 941–52.

Kearns, A.M., L. Joseph, K.E. Omland, & L.G. Cook. 2011. Testing the effect of transient Plio-Pleistocene barriers in monsoonal Australo-Papua: did mangrove habitats maintain genetic connectivity in the Black Butcherbird? *Molec. Ecol.* 20: 5042–59.

Keast, J.A. 1956. Variation in the Australian Oriolidae. *Proc. Roy. Zool. Soc. N.S.W.* (1954–55): 19–25.

Keast, J.A. 1957. Variation and speciation in the genus *Climacteris* Temminck (Aves: Sittidae). *Austral. J. Zool.* 5: 474–95.

Keast, J.A. 1977. Relationships of the New Guinean Red-backed 'Warbler' *Eugerygone rubra*. *Emu* 77: 228–29.

Kemp, A.C. 2001. Bucerotidae (hornbills). In J. del Hoyo, A. Elliott, & J. Sargatal, eds., *Handbook of the Birds of the World*, vol. 6: 436–523. Lynx Edicions, Barcelona.

Kennedy, M., R.D. Gray, & H.G. Spencer. 2000. The phylogenetic relationships of the shags and cormorants: can sequence data resolve a disagreement between behavior and morphology? *Molec. Phylog. Evol.* 17: 345–59.

Kennedy, R.S., T.H. Fisher, S.C.B. Harrap, A.C. Diesmos, & A.S. Manamtam. 2001. A new species of woodcock (Aves: Scolopacidae) from the Philippines and a re-evaluation of other Asian/Papuasian woodcock. *Forktail* 17: 1–12.

Kennerley, P.R., & K.D. Bishop. 2001. Bristle-thighed Curlew *Numenius tahitiensis* on Manus, Admiralty Islands, Papua New Guinea. *Stilt* 38: 53–54.

Kennerley, P.R., & D. Pearson. 2010. *Reed and Bush Warblers*. Helm Identification Guides. Christopher Helm, London.

Keysser, C. 1923. Einiges über das Vogelleben im Saruwaged-Gebirge (Deutch-Neuguinea) *Orn. Monatsber.* 31: 9–10.

Kimball, R.T., C.M. Saint Mary, & E.L. Braun. 2011. A macroevolutionary perspective on multiple sexual traits in the Phasianidae (Galliformes). *Internat. J. Evol. Biol.* 2011: article ID 423938, 16 pages.

King, B. 1979. New distributional records and field notes for some New Guinean birds. *Emu* 79: 146–48.

Kinnear, N.B. 1924. Note on *Myzomela obscura* with description of a new form *M. o. aruensis. Bull. Brit. Orn. Club.* 44: 68–69.

Kirchman, J.J. 2012. Speciation of flightless rails on islands: a DNA-based phylogeny of the typical rails of the Pacific. *Auk* 129: 56–69.

Kloet, R.S. de, & S.R. de Kloet. 2005. The evolution of the spindlin gene in birds: sequence analysis of an intron of the spindlin W and Z genes reveals four major divisions of the Psittaciformes. *Molec. Phylog. Evol.* 36: 706–21.

Kloska, C., & J. Nicolai. 1988. Fortpflanzungsverhalten des Kamm-Talegalla (*Aepypodius arfakianus* Salvadori). *J. Orn.* 129: 185–204.

König, C., & F. Weick. 2008. *Owls of the World.* 2nd ed. Christopher Helm, London.

Koopman, K.F. 1957. Evolution of the genus *Myzomela* (Aves: Meliphagidae). *Auk* 74: 49–72.

Kriegs, J.O., A. Matzke, G. Churakov, A. Kuritzin, G. Mayr, J. Brosius, & J. Schmitz. 2007. Waves of genomic hitchhikers shed light on the evolution of gamebirds (Aves: Galliformes). *BMC Evol. Biol.* 7: 190. doi:10.1186/1471-2148-7-190.

Kushlan, J., & J. Hancock. 2005. *The Herons.* Oxford Univ. Press, Oxford, UK.

Kusmierski, R., G. Borgia, A. Uy, & R.H. Crozier. 1997. Labile evolution of display traits in bowerbirds indicates reduced effects of phylogenetic constraint. *Proc. B Roy. Soc. Lond.* 264: 307–13.

Lamothe, L. 1993. Papuan Hawk-Owl *Uroglaux dimorpha* in the Lae-Bulolo area. *Muruk* 6(1): 14.

Larsen, C., M. Speed, N. Harvey, & H.A. Noyes. 2007. A molecular phylogeny of the nightjars (Aves: Caprimulgidae) suggests extensive conservation of primitive morphological traits across multiple lineages. *Molec. Phylog. Evol.* 42(3): 789–96.

Lavretsky, P., K.G. McCracken, & J.L. Peters. 2014. Phylogenetics of a recent radiation in the mallards and allies (Aves: *Anas*): inferences from a genomic transect and the multispecies coalescent. *Molec. Phylog. Evol.* 70: 402–11.

Layton, A. 1970. Observations [Common Starling *Sturnus vulgaris*]. *PNG Bird Soc. Newsl.* 61: 1.

LeCroy, M. 1969. *Acrocephalus arundinaceus orientalis*, first record in New Guinea. *Emu* 69: 119–20.

LeCroy, M. 1999. Type specimens of new forms of *Lonchura. Bull. Brit. Orn. Club* 119: 214–20.

LeCroy, M. 2010. Type specimens of birds in the American Museum of Natural History. Part 8: Passeriformes, Pachycephalidae to Nectariniidae. *Bull. Amer. Mus. Nat. Hist.* 333: 1–178.

LeCroy, M. 2011. Type specimens of birds in the American Museum of Natural History. Part 9: Passeriformes, Zosteropidae to Meliphagidae. *Bull. Amer. Mus. Nat. Hist.* 348: 1–193.

LeCroy, M. 2013. Type specimens of birds in the American Museum of Natural History. Part 11: Passeriformes, Parulidae to Viduinae. *Bull. Amer. Mus. Nat. Hist.* 381: 1–155.

LeCroy, M. 2014. Type specimens of birds in the American Museum of Natural History. Part 12: Passeriformes, Ploceidae to Corvidae. *Bull. Amer. Mus. Nat. Hist.* 393: 1–165.

LeCroy, M., & J. Diamond. 1995. Plumage variation in the Broad-billed Fairy-wren *Malurus grayi. Emu* 95: 185–93.

LeCroy, M., & W.S. Peckover. 1998. Misima's missing birds. *Bull. Brit. Orn. Club* 118: 217–37.

LeCroy, M., & W.S. Peckover. 1999. Plumages of the Red-collared Honeyeater *Myzomela rosenbergii* from Goodenough Island, D'Entrecasteux Islands, Papua New Guinea. *Bull. Brit. Orn. Club* 119: 62–65.

LeCroy, M., & W.S. Peckover. 2000. Birds observed on Goodenough and Wagifa Islands, Milne Bay Province. *Muruk* 8: 41–44.

LeCroy, M., W.S. Peckover, A. Kulupi, & J. Manseima. 1984. Bird observation on Normanby and Fergusson, D'Entrecasteaux Islands, Papua New Guinea. Wildlife in Papua New Guinea No. 83/1.

Lee, J.Y., L. Joseph, & S.V. Edwards. 2012. A species tree for the Austral-Papuan fairy wrens and allies (Aves: Maluridae). *Syst. Biol.* 61: 253–71.

Legge, S., S. Murphy, P. Igag, & A.L. Mack. 2004. Territoriality and density of an Australian migrant, the Buff-breasted Paradise Kingfisher, in the New Guinean non-breeding grounds. *Emu* 104: 15–20.

Lenz, N. 1999. Evolutionary ecology of the Regent Bowerbird *Sericulus chrysocephalus. Ökologie der Voegel* 22: 1–200.

Lerner, H.R.L., & D.P. Mindell. 2005. Phylogeny of eagles, Old World vultures, and other Accipitridae based on nuclear and mitochondrial DNA. *Molec. Phylog. Evol.* 37: 327–46.

Linck, E., S. Schaack, and J.P. Dumbacher. 2015. Genetic differentiation with a widespread "supertramp" taxon: Molecular phylogenetics of the Louisiade White-eye (*Zosterops griseotinctus*). *Mol. Phylog. Evol.* http://dx.doi.org/10.1016/j.ympev.2015.08.018.

Lindgren, E. 1971. Records of a new and uncommon species for the island of New Guinea. *Emu* 71: 134–36.

Lindgren, E. 1974. Another record of the Ruff *Philomachus pugnax* in Papua-New Guinea. *PNG Bird Soc. Newsl.* 104: 4–5.

Livezey, B.C. 1998. A phylogenetic analysis of the Gruiformes (Aves) based on morphological characters, with an emphasis on the rails (Rallidae). *Philos. Trans. Royal Soc. London (Biology)* 353: 2077–51.

Llimona, F., & J. del Hoyo. 1992. Podicipedidae (grebes). In J. del Hoyo, A. Elliott, & J. Sargatal, eds., *Handbook of the Birds of the World*, vol. 1: 174–97. Lynx Edicions, Barcelona.

Losos, J.B., D.M. Hillis, & H.W. Greene. 2012. Who speaks with a forked tongue? *Science* 388: 1428–29.

Loynes, K., L. Joseph, & J.S. Keogh. 2009. Multi-locus phylogeny clarifies the systematics of the Australo-Papuan robins (Family Petroicidae, Passeriformes). *Molec. Phylog. Evol.* 53: 212–19.

MacDonald, J.D. 1973. *Birds of Australia*. A.H. & A.W. Reed, Sydney.

Mack, A.L., ed. 1998. A biological assessment of the Lakekamu basin, Papua New Guinea. *RAP Working Papers* 9: 1–187. Conservation International, Washington, DC.

Mack, A.L., & L.E. Alonso, eds. 2000. A biological assessment of the Wapoga River area of northwestern Irian Jaya. *RAP Bull. Biol. Assess.* 14: 1–129. Conservation International, Washington, DC.

Mack, A.L., & J.P. Dumbacher. 2007. Birds of Papua. In A.J. Marshall & B.M. Beehler, eds., *The Ecology of Papua*, 654–88. Periplus, Singapore.

Mack, A.L., & D.D. Wright. 1993. Birds sightings from Lake Teberu. *Muruk* 6(1): 34–35.

Mack, A.L., & D.D. Wright. 1996. Notes on the occurrence and feeding of birds at Crater Mountain Biological Research Station, Papua New Guinea. *Emu* 96: 89–101.

Mackay, R.D. 1966. [Notes on observations of *Paradisaea apoda* at Nomad River.] *PNG Bird Soc. Newsl.* 12: 1.

Mackay, R.D. 1969. [Olsobip bird list.] *PNG Bird Soc. Newsl.* 46: 3.

Mackay, R.D. 1970. Observations [Sepik River, Ambunti to Angoram]. *PNG Bird Soc. Newsl.* 51: 1.

Mackay, R.D. 1973. Observations [Rainbow Bee-eater *Merops ornatus*]. *PNG Bird Soc. Newsl.* 86: 3.

Mackay, R.D. 1976. A new bird for PNG, Tawny Frogmouth *Podargus strigoides*. *PNG Bird Soc. Newsl.* 121: 10.

Mackay, R.D. 1978. Observations and bird notes [Bare-eyed Rail *Gymnocrex plumbeiventris*]. *PNG Bird Soc. Newsl.* 141: 5.

Mackay, R.D. 1980. A list of the birds of the Baiyer River Sanctuary and adjacent areas. *PNG Bird Soc. Newsl.* 167/168: 24–38.

Mackay, R.D. 1988. Gurney's Eagle *Aquila gurneyi* in the highlands. *Muruk* 3: 56.

Mackay, R.D. 1989. The bower of the Fire-maned Bowerbird *Sericulus bakeri*. *Austral. Bird Watcher* 13: 62–64.

Mackay, R.D. 1991. Additions to the avifauna of the Adelbert Range, PNG. *Muruk* 5: 12–15.

Mackay, R.D., & H.L. Bell. 1968. *Butorides striatus* Mangrove Heron (nest description). *PNG Bird Soc. Newls.* 30: 4.

Mackay, R.D., & M. Mackay. 1974a. Observations [Normanby Island]. *PNG Bird Soc. Newsl.* 103: 3.

Mackay, R.D., & M. Mackay. 1974b. Observations [Milne Bay District]. *PNG Bird Soc. Newsl.* 103: 2–3.

Maclean, G.L. 1996. Glareolidae (coursers and pratincoles). In J. del Hoyo, A. Elliott, & J. Sargatal, eds., *Handbook of the Birds of the World*, vol. 3: 364–83. Lynx Edicions, Barcelona.

Majnep, I.S., & R. Bulmer. 1977. *Birds of My Kalam Country*. Oxford Univ. Press, Auckland, New Zealand.

Marchant, S., & P.J. Higgins, eds. 1990. *Handbook of Australian, New Zealand and Antarctic Birds*. Vol. 1, Ratites to Ducks. Oxford Univ. Press, Melbourne, Australia.

Marchant, S., & P.J. Higgins, eds. 1993. *Handbook of Australian, New Zealand and Antarctic Birds*. Vol. 2, Raptors to Lapwings. Oxford Univ. Press, Melbourne, Australia.

Marien, D. 1950. Notes on some Asiatic Meropidae (birds). *J. Bombay Nat. Hist. Soc.* 49: 151–64.

Marks, J.S., R.J. Cannings, & H. Mikkola. 1999. Strigidae (typical owls). In J. del Hoyo, A. Elliott, & J. Sargatal, eds., *Handbook of the Birds of the World*, vol. 5: 76–243. Lynx Edicions, Barcelona.

Marshall, A.J., & B.M. Beehler, eds. 2007. *Ecology of Papua*. 2 volumes. Periplus Editions, Singapore.

Marshall, J.T. 1978. Systematics of smaller Asian night birds based on voice. *Orn. Monogr.* 25: 1–58.

Martínez-Vilalta, A., & A. Motis. 1992. Ardeidae (herons). In J. del Hoyo, A. Elliott, & J. Sargatal, eds., *Handbook of the Birds of the World*, vol. 1: 376–429. Lynx Edicions, Barcelona.

Mason, I.J. 1982. The identity of certain early Australian types referred to the Cuculidae. *Bull. Brit. Orn. Club* 102: 99–106.

Mason, I.J. 1983. A new subspecies of Masked Owl *Tyto novaehollandiae* (Stephens) from southern New Guinea. *Bull. Brit. Orn. Club* 103: 122–28.

Mason, I.J., & R.I. Forrester. 1996. Geographic differentiation in the Channel-billed Cuckoo *Scythrops novaehollandiae* Latham, with description of two new subspecies from Sulawesi and the Bismarck Archipelago. *Emu* 96: 217–33.

Mason, I.J., J.L. McKean, & M.L. Dudzinski. 1984. Geographical variation in the Pheasant Coucal *Centropus phasianinus* (Latham) and a description of a new subspecies from Timor. *Emu* 84: 1–15.

Mason, I.J., & R. Schodde. 1980. Subspeciation in the Rufous Owl *Ninox rufa* (Gould). *Emu* 80: 141–44.

Matheu, E., & J. del Hoyo. 1992. Threskionithidae (ibises and spoonbills). In J. del Hoyo, A. Elliott, & J. Sargatal, eds., *Handbook of the Birds of the World*, vol. 1: 472–506. Lynx Edicions, Barcelona.

Matheve, H., D. van den Schoor, E. Collaerts, & P. Collaerts. 2009. West Papua 18/07–15/08/2009. [Birding report.] Unpublished. http://users.ugent.be/~hmatheve/hm/PAPUA09.html. Accessed 24-Sep-2015

Mathews, G.M. 1927, 1930. *Systema Avium Australasianarum: A Systematic List of the Birds of the Australasian Region.* 2 vols. British Ornithologists' Union, London.

Matthew, J.S. 2007. Pomatostomidae (Australasian babblers). In J. del Hoyo, A. Elliott, & D. Christie, eds., *Handbook of the Birds of the World*, vol. 12: 322–37. Lynx Edicions, Barcelona.

Mauro, I. 2005. The field discovery, mound characteristics, bare parts, vocalizations, and behaviour of Bruijn's Brush-turkey, *Aepypodius bruijnii. Emu* 105: 273–81.

Mauro, I. 2006. Habitat, microdistribution and conservation status of the enigmatic Bruijn's Brush-turkey *Aepypodius bruijni. Bird Conserv. Internat.* 16(4): 279–92.

Mayr, E. 1931. Die Vogel des Saruwaged und Herzoggebirges (NO-Neuguinea). *Mitt. Zool. Mus. Berl.* 17: 639–723.

Mayr, E. 1936. New subspecies of birds from the New Guinea region. *Amer. Mus. Novit.* 869: 1–7.

Mayr, E. 1937a. Notes on the genus *Sericornis* Gould. *Amer. Mus. Novit.* 904: 1–25.

Mayr, E. 1937b. Birds collected during the Whitney South Sea Expedition. 35. Notes on New Guinea birds. 2. *Amer. Mus. Novit.* 939: 1–14.

Mayr, E. 1937c. Birds collected during the Whitney South Sea Expedition. 37. Notes on New Guinea birds. 3. *Amer. Mus. Novit.* 947: 1–11.

Mayr, E. 1938. Birds collected during the Whitney South Sea Expedition. 40. Notes on New Guinea birds. 4. *Amer. Mus. Novit.* 1006: 1–16.

Mayr, E. 1940. Birds collected during the Whitney South Sea Expedition. 41. Notes on New Guinea birds. 6. *Amer. Mus. Novit.* 1056: 1–12.

Mayr, E. 1941. *List of New Guinea Birds.* Amer. Mus. Nat. Hist., New York

Mayr, E. 1942. *Systematics and the Origin of Species.* Columbia University Press, New York.

Mayr, E. 1943. Notes on Australian birds. 2. *Emu* 43: 3–17.

Mayr, E. 1944. Birds collected during the Whitney South Sea Expedition. 54. Notes on some genera from the Southwest Pacific. *Amer. Mus. Novit.* 1269: 1–14.

Mayr, E. 1945. Birds collected during the Whitney South Sea Expedition. 55. Notes on the birds of northern Melanesia. 1. *Amer. Mus. Novit.* 1294: 1–12.

Mayr, E. 1948. Geographic variation in the Reed-Warbler. *Emu* 47: 205–10.

Mayr, E. 1949. Notes on the birds of northern Melanesia. 2. *Amer. Mus. Novit.* 1417: 1–38.

Mayr, E. 1955. Notes on the birds of northern Melanesia. 3. Passeres. *Amer. Mus. Novit.* 1707: 1–46.

Mayr, E. 1960. Southeast Asian Motacillidae. In E. Mayr & J.C. Greenway, eds., *Check-list of Birds of the World*, vol. 9: 129–67. Mus. Comp. Zool., Harvard Univ., Cambridge, MA.

Mayr, E. 1962. Grallinidae; Artamidae; Ptilonorhynchidae; Paradisaeidae. In E. Mayr & J.C. Greenway, eds., *Check-list of Birds of the World*, vol. 15: 159–60; 160–65; 172–80; 181–203. Mus. Comp. Zool., Harvard Univ., Cambridge, MA.

Mayr, E. 1963. *Animal Species and Evolution.* Belknap Press, Harvard Univ., Cambridge, MA.

Mayr, E. 1967. Pachycephalinae; Indo-Australian Zosteropidae. In R.A. Paynter Jr., ed., *Check-list of Birds of the World*, vol. 12: 3–51; 289–324. Mus. Comp. Zool., Harvard Univ., Cambridge, MA.

Mayr, E. 1968a. *Larius* Boddaert, 1783 (Aves): proposed suppression under the plenary powers Z. N. (S.) 1833. *Bull. Zool. Nomencl.* 25: 52–54.

Mayr, E. 1968b. Australo-Papuan Estrildidae. In R.A. Paynter Jr., ed., *Check-list of Birds of the World*, vol. 14: 306–97. Mus. Comp. Zool., Harvard Univ., Cambridge, MA.

Mayr, E. 1970. *Populations, Species, and Evolution.* Harvard Univ. Press, Cambridge, MA.

Mayr, E. 1979a. Pittidae. In M.A. Traylor, ed., *Check-list of Birds of the World*, vol. 8: 310–29. Mus. Comp. Zool., Harvard Univ., Cambridge, MA.

Mayr, E. 1979b. Struthioniformes. In E. Mayr & G.W. Cottrell, eds., *Check-list of Birds of the World*, vol. 1, 2nd ed.: 3–11. Mus. Comp. Zool., Harvard Univ., Cambridge, MA.

Mayr, E. 1986. Australasian Sylviinae; Maluridae; Acanthizidae; Monarchidae; Eopsaltriidae. In E. Mayr & G.W. Cottrell, eds., *Check-list of Birds of the World*, vol. 11: 3–294; 390–409; 409–64; 464–556; 556–83. Mus. Comp. Zool., Harvard Univ., Cambridge, MA.

Mayr, E., & S. Camras. 1938. Birds of the Crane Pacific Expedition. *Publ. Field Mus. Nat. Hist. Zool. Ser.* 20: 453–73.

Mayr, E., & H.M. Van Deusen. 1956. Results of the Archbold Expeditions. No. 74. The birds of Goodenough Island, Papua. *Amer. Mus. Novit.* 1792: 1–8.

Mayr, E., & J.M. Diamond. 2001. *The Birds of Northern Melanesia.* Oxford Univ. Press, Oxford, UK.

Mayr, E., & E.T. Gilliard. 1951. New species and subspecies of birds from the highlands of New Guinea. *Amer. Mus. Novit.* 1524: 1–15.

Mayr, E., & E.T. Gilliard. 1952. Altitudinal hybridization in New Guinea honeyeaters. *Condor* 54: 325–37.

Mayr, E., & E.T. Gilliard. 1954. Birds of Central New Guinea. *Bull. Amer. Mus. Nat. Hist.* 103: 311–74.

Mayr, E., & J.C. Greenway. 1956. Sequence of passerine families (Aves). *Breviora* Mus. Comp. Zool. Harvard 58: 1–11.

Mayr, E., & R. Meyer de Schauensee. 1939a. Zoological results of the Denison-Crockett South Pacific Expedition for the Academy of Natural Sciences of Philadelphia, 1937–1938. Part 1, The birds of the island of Biak. *Proc. Acad. Nat. Sci. Phila.* 91: 1–37.

Mayr, E., & R. Meyer de Schauensee. 1939b. Zoological results of the Denison-Crockett South Pacific Expedition for the Academy of Natural Sciences of Philadelphia, 1937–1938. Part 4, Birds from northwest New Guinea. *Proc. Acad. Nat. Sci. Phila.* 91: 97–144.

Mayr, E., & R. Meyer de Schauensee. 1939c. Zoological results of the Denison-Crockett South Pacific Expedition for the Academy of Natural Sciences of Philadelphia, 1937–1938. Part 5, Birds from western Papuan islands. *Proc. Acad. Nat. Sci. Phila.* 91: 145–63.

Mayr, E., & A.L. Rand. 1935. Results of the Archbold Expeditions. No. 6. Twenty-four apparently undescribed birds from New Guinea and the D'Entrecasteaux Archipelago. *Amer. Mus. Novit.* 814: 1–17.

Mayr, E., & A.L. Rand. 1936. Results of the Archbold Expeditions. No. 10. Two new subspecies of birds from New Guinea. *Amer. Mus. Novit.* 868: 1–3.

Mayr, E., & A.L. Rand. 1937. Birds of the 1933–1934 Papuan Expedition. *Bull. Amer. Mus. Nat. Hist.* 73: 1–248.

Mayr, E., & L.L. Short. 1970. *Species Taxa of North American Birds.* Publ. no. 9. Nuttall Orn. Club, Cambridge, MA.

Mayr, G. 2009. *Paleogene Fossil Birds.* Springer-Verlag, Berlin.

Mayr, G., A. Manegold, & U.S. Johansson. 2003. Monophyletic groups within 'higher land birds'—comparison of morphological and molecular data. *J. Zool. Syst. Evol. Research* 41: 233–48.

McCormack, J.E., M.G. Harvey, B.C. Faircloth, N.G. Crawford, T.C. Glenn, & R.T. Brumfield. 2012. A phylogeny of birds based on over 1,500 loci collected by target enrichment and high-throughput sequencing. *PLoS One* 8(1): e54848. doi:10.1371/journal.pone.0054848.

McCracken, K.G., & F.H. Sheldon. 1998. Molecular and osteological heron phylogenies: sources of incongruence. *Auk* 115: 27–141.

McDonald, P.G. 2003. Variable plumage and bare part coloration in the Brown Falcon *Falco berigora*: the influence of age and sex. *Emu* 103: 21–28.

McGowan, P.J.K. 1994. Phasianidae (pheasants and grouse). In J. del Hoyo, A. Elliott, & J. Sargatal, eds., *Handbook of the Birds of the World*, vol. 2: 434–553. Lynx Edicions, Barcelona.

McKean, J.L. 1969. The brush tongue of the Artamidae indicates more study is needed of relationships. *Bull. Brit. Orn. Club* 89: 129–30.

McKean, J.L. 1978. Some remarks on the taxonomy of Australasian oystercatchers. *Sunbird* 9: 3–6.

McKean, J.L., M.C. Bartlett, & C.M. Perrins. 1975. New records from the Northern Territory. *Austral. Bird Watcher* 6: 45–46.

Medway, L., & J.D. Pye. 1977. Echolocation and systematics of swiftlets. In B. Stonehouse & C.M. Perrins, eds., *Evolutionary Ecology*, 225–38. Univ. Park Press, Baltimore, MD.

Mees, G.F. 1953. The white-eyes of the Aroe Islands. *Zool. Meded. Leiden* 32: 25–30.

Mees, G.F. 1955. Description of a new race of *Zosterops novaeguineae* Salvadori (Aves, Zosteropidae). *Zool. Meded. Leiden* 34: 153–54.

Mees, G.F. 1958. Eenbastaardtussen *Lonchura tristissima* (Wallace) en *L. leucosticta* (D'Albertis & Salvadori). *Nova Guinea* (ns)9: 15–19.

Mees, G.F. 1961. A systematic revision of the Indo-Australian Zosteropidae. Part 2. *Zool. Verh. Leiden* 50: 1–168.

Mees, G.F. 1964a. A revision of the Australian owls (Strigidae and Tytonidae). *Zool. Verh. Leiden* 65: 1–62.

Mees, G.F. 1964b. Notes on two small collections of birds from New Guinea. *Zool. Verh. Leiden* 66: 1–37.

Mees, G.F. 1964c. Four new subspecies of birds from the Moluccas and New Guinea. *Zool. Meded. Leiden* 40(15): 125–30.

Mees, G.F. 1965. The avifauna of Misool. *Nova Guinea, Zool.* (ns)10(31): 139–203.

Mees, G.F. 1972. Der Vögel der insel Gebe. *Zool. Meded. Leiden* 46: 69–89.

Mees, G.F. 1977a. Geographic variation of *Caprimulgus macrurus* Horsfield (Aves, Caprimulgidae). *Zool. Verh. Leiden* 155: 1–47.

Mees, G.F. 1977b. The subspecies of *Chlidonias hybridus* (Pallas), their breeding, distribution, and migrations. *Zool. Verh. Leiden* 157: 1–64.

Mees, G.F. 1979. Die Nachweise van *Cuculus canorus* im Indo-Australischen. *Mitt. Zool. Mus. Berl.* 55 (Suppl.): 127–34.

Mees, G.F. 1980. Supplementary notes on the avifauna of Misool. *Zool. Meded. Leiden* 55: 1–10.

Mees, G.F. 1982a. [Book review of] *Nocturnal Birds of Australia*, by R. Schodde & I.J. Mason. *Emu* 82: 182–84.

Mees, G.F. 1982b. Birds from the lowlands of southern New Guinea (Merauke and Koembe). *Zool. Verh. Leiden* 191: 1–187.

Mees, G.F. 1982c. Bird records from the Moluccas. *Zool. Meded. Leiden* 56: 91–111.

Mees, G.F. 1994. Vogelkundig onderzoek op Nieuw Guinea in 1828: terugblik op de ornithologische resultaten van de reis van Zr. Ms. Korvet Triton naar de zuid-west kust van Nieuw Guinea. *Zool. Bijdr. Leiden* 40: 3–64.

Mees, G.F. 2006. The avifauna of Flores (Lesser Sunda Islands). *Zool. Meded. Leiden* 80(3): 1–261.

Meise, W. 1929. Verzeichnis der Typen des Staatlichen Museums für Tierkunde in Dresden. *Abh. Berichte Tierkunde und Völkerkunde Dresen* 17(4): 1–22.

Melville, D. 1979. Ornithological notes on a visit to Irian Jaya. *PNG Bird Soc. Newsl.* 161: 3–22.

Melville, D. 1980. Some observations on birds in Irian Jaya, New Guinea. *Emu* 80: 89–91.

Meyer de Schauensee, R. 1940. On a collection of birds from Waigeu. *Notul. Nat.* 45: 1–16.

Miller, A.H. 1964. A new species of warbler from New Guinea. *Auk* 81: 1–4.

Mitchell, K.J., B. Llamas, J. Soubrier, N.J. Rawlence, T.H. Worthy, J. Wood, M.S.Y. Lee, & A. Cooper. 2014. Ancient DNA reveals elephant birds and kiwi are sister taxa and clarifies ratite bird evolution. *Science* 344: 898–900.

Mittermeier, R.A., B.M. Beehler, G. Kula, & C.G. Mittermeier. 1997. Papua New Guinea. In R.A. Mittermeier, P. Robles Gil, & C.G. Mittermeier, eds., *Megadiversity*, 363–85. CEMEX, Mexico City.

Miyamoto, T. 1971. Note of birds collected at some places in the Territory of Papua and New Guinea. *Tori* 20: 191–203.

Mlíkovský, J. 1989. Note on the osteology and taxonomic position of Salvadori's Duck *Salvadorina waigiuensis* (Aves: Anseridae). *Bull. Brit. Orn. Club* 109: 22–25.

Mlíkovský, J. 2010. A preliminary review of the grebes, family Podicipedidae. *Brit. Orn. Club Occas. Publ.* 5: 125–31.

Monroe, B.L. 1989. The correct name of the Terek Sandpiper. *Bull. Brit. Orn. Club* 109: 106–7.

Mordue, T.A. 1981. Buff-breasted Sandpiper at Sangara, Oro Province. *PNG Bird Soc. Newsl.* 183/184: 36–37.

Mosey, H. 1956. Birds observed on a visit to New Guinea during June to August 1950. *Emu* 56: 357–66.

Moyle, R.G. 2006. A molecular phylogeny of the kingfishers (Alcedinidae) with insights into early biogeographic history. *Auk* 123: 487–99.

Moyle, R.G., C.E. Filardi, C.E. Smith, & J. Diamond. 2009. Explosive Pleistocene diversification and hemispheric expansion of a "great speciator." *Proc. Nat. Acad. Sci. USA* 106: 1863–68.

Moyle, R.G., J. Fuchs, E. Pasquet, & B.D. Marks. 2007. Feeding behavior, toe count, and the phylogenetic relationships among alcedinine kingfishers (Alcedinidae). *J. Avian Biol.* 38: 317–26.

Moyle, R.G., R.M. Jones, & M.J. Andersen. 2013. A reconsideration of *Gallicolumba* (Aves: Columbidae) relationships using fresh source material reveals pseodogenes, chimeras, and a novel phylogenetic hypothesis. *Molec. Phylog. Evol.* 66: 1060–66.

Murphy, S.A., M.C. Double, & S.M. Legge. 2007. The phylo-geography of palm cockatoos, *Probosciger aterrimus*, in the dynamic Australo-Papuan region. *J. Biogeogr.* 34: 1534–45.

Murray, A. 1986. Some notes on migratory species observed at Tabubil, Western Province, July 1985–June 1986. *PNG Bird Soc. Newsl.* 222: 10.

Murray, A. 1988a. An unidentified pygmy-parrot (*Micropsitta*). *Muruk* 3: 118.

Murray, A. 1988b. A study of the birds of Tabubil and Ok Tedi valley, Western Province, Papua New Guinea. *Muruk* 3: 89–93.

Murray, A., L.M. Murray, & R. Hicks. 1987. The Fairy Martin *Hirundo ariel* in Papua New Guinea. Three new sight records and a brief review. *Muruk* 2: 64–65

Naish, T., R. Powell, R. Levy, *et al.* 2009. Obliquity-paced Pliocene West Antarctic ice sheet ocillations. *Nature* 459: 322–28.

Norman, J.A., W.E. Boles, & L. Christidis. 2009a. Relation-ships of the New Guinean songbird genera *Amalocichla* and *Pachycare* based on mitochondrial and nuclear DNA sequences. *J. Avian Biol.* 40: 1–6.

Norman, J.A., L. Christidis, L. Joseph, B. Slikas, & D. Alpers. 2012. Unravelling a biogeographical knot: origin of the 'leapfrog' distribution pattern of Australo-Papuan sooty owls (Strigiformes) and logrunners (Passeriformes). *Proc. B Roy. Soc. Lond.* 269: 2127–33.

Norman, J.A., P.G.P. Ericson, K.A. Jønsson, J. Fjeldså, L. Christidis. 2009b. A multi-gene phylogeny reveals novel relationships for aberrant genera of Australo-Papuan core Corvoidea and polyphyly of the Pachycephalidae and Psophodidae (Aves: Passeriformes). *Molec. Phylog. Evol.* 52: 488–97.

Norman, J.A., F.E. Rheindt, D.L. Rowe, & L. Christidis. 2007. Speciation dynamics in the Australo-Papuan *Meliphaga* honeyeaters. *Molec. Phylog. Evol.* 42: 80–91.

Noske, R. 2007. Neosittidae (sittellas); Climacteridae (Aus-tralian treecreepers). In J. del Hoyo, A. Elliott, & D. Christie, eds., *Handbook of the Birds of the World*, vol. 12: 628–61. Lynx Edicions, Barcelona.

Nunn, G.B., & J. Cracraft. 1992. Phylogenetic relationships among the major lineages of the birds-of-paradise (Paradi-saeidae) using mitochondrial DNA gene sequences. *Molec. Phylog. Evol.* 5: 445–59.

Nunn, G.B., & S.E. Stanley. 1998. Body size effects and rates of cytochrome *b* evolution in tube-nosed seabirds. *Molec. Phylog. Evol.* 15(10): 1360–71.

Nyári, A.S., B.W. Benz, K.A. Jønsson, J. Fjeldså, & R.G. Moyle. 2009. Phylogenetic relationships of fantails (Aves: Rhipiduridae). *Zool. Scripta* 38: 553–61.

Nyári, A.S., & L. Joseph. 2011. Systematic dismantlement of *Lichenostomus* improves the basis for understanding relationships within the honeyeaters (Meliphagidae) and the historical development of Australo-Papuan bird communi-ties. *Emu* 111: 202–11.

Nyári, A.S., & L. Joseph. 2012. Evolution in Australasian mangrove forests: multilocus phylogenetic analysis of the *Gerygone* warblers (Aves: Acanthizidae). *PLoS One* 7(2): e31840. doi:10.1371/journal.pone.0031840.

Nyári, A.S., & L. Joseph. 2013. Comparative phylogeography of Australo-Papuan mangrove-restricted and mangrove-associated avifaunas. *Biol. J. Linn. Soc.* 109: 574–98.

Ogilvie-Grant, W.R. 1915. Report on the birds collected by the British Ornithologists' Union Expedition and the Wol-laston Expedition in Dutch New Guinea. *Ibis*, Jubilee Suppl. 2: xx; 1–336.

Oliver, J., & H.F. Hopkins. 1989. Brown-backed Honeyeater *Ramsayornis modestus* near Madang. *Muruk* 4: 24.

Olsen, P.D. 1999. Australian Strigidae (typical owls). In J. del Hoyo, A. Elliott, & J. Sargatal, eds., *Handbook of the Birds of the World*, vol. 5: 76–243. Lynx Edicions, Barcelona.

Olson, S.L. 1973. A classification of the Rallidae. *Wilson Bull.* 85: 381–416.

Olson, S.L. 1980. The significance of distribution of the Megapodiidae. *Emu* 80: 21–24.

Olson, S.L. 1987. The relationships of the New Guinean Ground-Robins *Amalocichla*. *Emu* 87: 247–48.

Olson, S.L., & K.I. Warheit. 1988. A new genus for *Sula abbotti*. *Bull. Brit. Orn. Club* 108: 9–12.

Olsson, U., P. Alström, P.G.P. Ericson, & P. Sundberg. 2005. Non-monophyletic taxa and cryptic species—evidence from a molecular phylogeny of leaf-warblers (*Phylloscopus*, Aves). *Molec. Phylog. Evol.* 36: 261–76.

Onley, D., & P. Scofield. 2007. *Albatrosses, Petrels and Shear-waters of the World*. Christopher Helm, London.

van Oort, E.D. 1909. Birds from southwestern and southern New Guinea. *Nova Guinea, Zool.* 9: 51–107.

Orenstein, R.I. 1975. Observations and comments on two stream-adapted birds of Papua New Guinea. *Bull. Brit. Orn. Club* 95: 161–65.

Orenstein, R.I. 1979. Wing-flashing in *Eugerygone rubra*. *Emu* 79: 43–44.

Orta, J. 1992a. Phaethontidae (tropicbirds). In J. del Hoyo, A. Elliott, & J. Sargatal, eds., *Handbook of the Birds of the World*, vol. 1: 280–89. Lynx Edicions, Barcelona.

Orta, J. 1992b. Phalacrocoracidae (cormorants and shags). In J. del Hoyo, A. Elliott, & J. Sargatal, eds., *Handbook of the Birds of the World*, vol. 1: 326–53. Lynx Edicions, Barcelona.

Orta, J. 1992c. Anhingidae (darters). In J. del Hoyo, A. Elliott, & J. Sargatal, eds., *Handbook of the Birds of the World*, vol. 1: 354–61. Lynx Edicions, Barcelona.

Orta, J. 1992d. Fregatidae (frigatebirds). In J. del Hoyo, A. Elliott, & J. Sargatal, eds., *Handbook of the Birds of the World*, vol. 1: 362–74. Lynx Edicions, Barcelona.

Ostende, L.W.H., R.W.R.J. Dekker, & G.O. Keijl. 1997. Type-specimens of birds in the National Museum of Natural History, Leiden. Part 1, Non-passerines. *NNM Tech. Bull.*, Leiden 1: 1–248.

Parker, S.A. 1981. Prolegomenon to further studies in the *Chrysococcyx malayanus* group (Aves: Cuculidae). *Zool. Verh. Leiden* 187: 1–56.

Parkes, K.C. 1949. A new button-quail from New Guinea. *Auk* 66: 84–86.

Parkes, K.C. 1978. A guide to forming and capitalizing compound names of birds in English. *Auk* 95: 324–26.

Parry, D.E. 1989. Black Swans at Merauke. *Kukila* 4: 64–65.

Pasquet, E., J.-M., Pons, & J. Fuchs. 2007. Evolutionary history and biogeography of the drongos (Dicruridae), a tropical Old World clade of corvoid passerines. *Molec. Phylog. Evol.* 45: 158–67.

Paton, T.A., & A.J. Baker. 2006. Sequences from 14 mitochondrial genes provide a well-supported phylogeny of the charadriiform birds congruent with the nuclear RAG-1 tree. *Molec. Phylog. Evol.* 39: 657–67.

Pavlova, A., R.M. Zink, S.V. Drovetski, Y. Red'kin, & S. Rohwer. 2003. Phylogenetic patterns in *Motacilla flava* and *Motacilla citreola*: species limits and population history. *Auk* 120: 744–58.

Payne, R.B. 1979. Ardeidae. In E. Mayr & G.W. Cottrell, eds., *Check-list of Birds of the World*, vol. 1, 2nd ed.: 193–244. Mus. Comp. Zool., Harvard Univ., Cambridge, MA.

Payne, R.B. 1997. Cuculidae (cuckoos). In J. del Hoyo, A. Elliott, & J. Sargatal, eds., *Handbook of the Birds of the World*, vol. 4: 508–607. Lynx Edicions, Barcelona.

Payne, R.B. 2005. *The Cuckoos*. Oxford Univ. Press, Oxford, UK.

Payne, R.B. 2010. Estrildidae (waxbills). In J. del Hoyo, A. Elliott, & D. Christie, eds., *Handbook of the Birds of the World*, vol. 15: 234–377. Lynx Edicions, Barcelona.

Pearson, D.L. 1975. Survey of the birds of a lowland-forest plot in the East Sepik District, Papua New Guinea. *Emu* 75: 175–77.

Peckover, W.S., & L.W.C. Filewood. 1976. *Birds of New Guinea and Tropical Australia*. Reed, Sydney.

Penhallurick, J., & M. Wink. 2004. Analysis of the taxonomy and nomenclature of the Procellariiformes based on complete nucleotide sequences of the mitochondrial cytochrome *b* gene. *Emu* 104: 125–47.

Pereira, S.L., & A.J. Baker. 2005. Multiple gene evidence for parallel evolution and retention of ancestral morphological states in the shanks (Charadriiformes: Scolopacidae). *Condor* 107 (3): 514–26.

Pereira, S.L., K.P. Johnson, D.H. Clayton, & A.J. Baker. 2007. Mitochondrial and nuclear DNA sequences support a Cretaceous origin of Columbiformes and a dispersal-driven radiation in the Paleogene. *Syst. Biol.* 56: 656–72.

Perron, R.M. 2011. The taxonomic status of *Casuarius bennetti papuanus* and *C. b. westermanni*. *Bull. Brit. Orn. Club* 131: 54–58.

Peters, J.L. 1931. *Check-list of Birds of the World*. Vol. 1 [Archaeornithes to falcons]. Harvard Univ. Press and Mus. Comp. Zool., Harvard Univ., Cambridge, MA.

Peters, J.L. 1934. *Check-list of Birds of the World*. Vol. 2 [megapodes to auks]. Harvard Univ. Press, Cambridge, MA.

Peters, J.L. 1937. *Check-list of Birds of the World*. Vol. 3 [pigeons and parrots]. Harvard Univ. Press, Cambridge, MA.

Peters, J.L. 1940a. A genus for *Eupetes caerulescens* Temminck. *Auk* 57: 94.

Peters, J.L. 1940b. *Check-list of Birds of the World*. Vol. 4 [cuckoos to crested swifts]. Harvard Univ. Press, Cambridge, MA.

Peters, J.L., E. Mayr, & H.G. Deignan. 1960. Alaudidae; Hirundinidae; Campephagidae. In E. Mayr & J.C. Greenway, eds., *Check-list of Birds of the World*, vol. 9: 3–80; 80–130; 167–221. Mus. Comp. Zool., Harvard Univ., Cambridge, MA.

Peterson, A.T. 2007. Geographic variation in size and coloration in the *Turdus poliocephalus* complex: a first review of species limits. *Sci. Pap. Nat. Hist. Mus. Kansas* 40: 1–17.

Peterson, A.T. 2014. Type specimens in modern ornithology are necessary and irreplaceable. *Auk* 131: 282–86.

Pierce, R. 2010. A Pycroft's Petrel (*Pterodroma pycrofti*) in Papua New Guinea. *Notornis* 56 (2009): 223–24.

Pierce, R.J. 1996. Recurvirostridae (stilts and avocets). In J. del Hoyo, A. Elliott, & J. Sargatal, eds., *Handbook of the Birds of the World*, vol. 3: 332–47. Lynx Edicions, Barcelona.

Piersma, W. 1996. Charadriidae (plovers); Scolopacidae (sandpipers) [family introductions]. In J. del Hoyo, A. Elliott, & J. Sargatal, eds., *Handbook of the Birds of the World*, vol. 3: 384–409; 444–87. Lynx Edicions, Barcelona.

Pizzey, G. 1980. *A Field Guide to the Birds of Australia*. Collins, Sydney.

Poggi, R. 1996. Use of archives for nomenclatorial purposes: clarifications and corrections of the dates of issue for volumes 1–8 (1870–1876) of the *Annali del Museo civico di Storia naturale di Genova*. *Arch. Nat. Hist.* 23(1): 99–105.

Polhemus, D.A. 2007. Tectonic geology of Papua. In A.J. Marshall & B.M. Beehler, eds., *The Ecology of Papua*, 137–64. Periplus, Singapore.

Pons, J.M., A. Hassanin, & P.-A. Crochet. 2005. Phylogenetic relationships with the Laridae (Charadriiformes: Aves) inferred from mitochondrial markers. *Molec. Phylog. Evol.* 37: 686–99.

Poole, A.F. 1994. Pandionidae (ospreys). In J. del Hoyo, A. Elliott, & J. Sargatal, eds., *Handbook of the Birds of the World*, vol. 2: 42–51. Lynx Edicions, Barcelona.

Portelli, D.J. 2013. Preliminary bird surveys of Lake Murray, Western Province, Papua New Guinea. Unpublished.

Pratt, T.K. 1982. Biogeography of birds in New Guinea. In J.L. Gressitt, ed., *Ecology and Biogeography of New Guinea*, 815–36. W.R. Junk, the Hague.

Pratt, T.K 1983. Additions to the avifauna of the Adelbert Range, Papua New Guinea. *Emu* 82 (1982): 117–25.

Pratt, T.K. 2000. Evidence for a previously unrecognized species of owlet-nightjar. *Auk* 117: 1–11.

Pratt, T.K., C.T. Atkinson, P.C. Banko, J.D. Jacobi, & B.L. Woodworth, eds. 2009. *Conservation Biology of Hawai`ian Forest Birds: Implications for Island Avifauna*. Yale Univ. Press, New Haven, CT.

Pratt, T.K., & B.M. Beehler. 2014. *Birds of New Guinea*. 2nd ed. Princeton Univ. Press, Princeton, NJ. [Issued in 2014 but dated 2015.]

Pratt, T.K., & M.P. Moore. 2004. Birds survey of Normanby and Tobweyama Islands, Papua New Guinea, 7–28 November, 2003. Unpublished.

Pratt, T.K., M.P. Moore, D. Mitchell, & M. Viula. 2005. A bird survey of the Louisiade Islands, Milne Bay Province, Papua New Guinea. 24 October to 23 November 2004. Report to the National Geographic Society. Unpublished.

Pratt, T.K., L.P. Morgan, A. Kulupi, & D. Mitchell. 2006. A bird survey of Woodlark and nearby islands, Milne Bay Province, Papua New Guinea. Report to the National Geographic Society. Unpublished.

Price, J.J., K.P. Johnson, S.E. Bush, & D.H. Clayton. 2005. Phylogenetic relationships of the Papuan Swiftlet *Aerodramus papuensis* and implications for the evolution of avian echolocation. *Ibis* 147: 790–96.

Price, J.J., K.P. Johnson, & D.H. Clayton. 2004. The evolution of echolocation in swiftlets. *J. Avian Biol.* 35: 135–43.

Pyle, P., A.J. Welch, & R.C. Fleischer. 2011. A new species of shearwater (*Puffinus*) recorded from Midway Atoll, northwestern Hawai`ian Islands. *Condor* 113: 518–27.

Quaisser, C., & B. Nicolai. 2006. Typusexemplare der Vogelsammlung im Museum Heineanum Halberstadt. *Abh. Ber. Mus. Hein.* 7(2): 1–105.

Ramirez, J.L., C.Y. Miyaki, & S.N. Del Lama. 2013. Molecular phylogeny of Threskiornithidae (Aves: Pelecaniformes) based on nuclear and mitochondrial DNA. *CMR Gen. Mol. Res.* 12(3): 2740-2750.

Rand, A.L. 1936a. Results of the Archbold Expeditions. No. 11. *Meliphaga analoga* and its allies. *Amer. Mus. Novit.* 872: 1–23.

Rand, A.L. 1936b. Results of the Archbold Expeditions. No. 12. Altitudinal variation in New Guinea birds. *Amer. Mus. Novit.* 890: 1–14.

Rand, A.L. 1938a. Results of the Archbold Expeditions. No. 19. On some nonpasserine New Guinea birds. *Amer. Mus. Novit.* 990: 1–15.

Rand, A.L. 1938b. Results of the Archbold Expeditions. No. 20. On some passerine New Guinea birds. *Amer. Mus. Novit.* 991: 1–20.

Rand, A.L. 1938c. Results of the Archbold Expeditions. No. 21. On some New Guinea birds. *Amer. Mus. Novit.* 992: 1–3.

Rand, A.L. 1940a. Results of the Archbold Expeditions. No. 25. New birds from the 1938–39 expedition. *Amer. Mus. Novit.* 1072: 1–14.

Rand, A.L. 1940b. Results of the Archbold Expeditions. No. 27. Ten new birds from New Guinea. *Amer. Mus. Novit.* 1074: 1–5.

Rand, A.L. 1941. Results of the Archbold Expeditions. No. 32. New and interesting birds from New Guinea. *Amer. Mus. Novit.* 1102: 1–15.

Rand, A.L. 1942a. Results of the Archbold Expeditions. No. 42. Birds of the 1936–1937 New Guinea Expedition. *Bull. Amer. Mus. Nat. Hist.* 79: 289–366.

Rand, A.L. 1942b. Results of the Archbold Expeditions. No. 43. Birds of the 1938–1939 New Guinea Expedition. *Bull. Amer. Mus. Nat. Hist.* 79: 425–515.

Rand, A.L. 1960. Laniidae. In E. Mayr, & J.C. Greenway, eds., *Check-list of Birds of the World*, vol. 9: 309–65. Mus. Comp. Zool., Harvard Univ., Cambridge, MA.

Rand, A.L. 1967. Nectariniidae. In R.A. Paynter Jr., ed., *Check-list of Birds of the World*, vol. 12: 208–89. Mus. Comp. Zool., Harvard Univ., Cambridge, MA.

Rand, A.L., & E.T. Gilliard. 1967. *Handbook of New Guinea Birds*. Weidenfeld & Nicolson, London/Natural History Press, Garden City, NY.

Rasmussen, P.C., & J. Anderton. 2005. *Birds of South Asia: The Ripley Guide*. Smithsonian Institution, Washington, DC/ Lynx Edicions, Barcelona.

Redman, N. 2011. First record of Baird's Sandpiper *Calidris bairdii* for Indonesia. *Kukila* 15: 122–23.

Restall, R. 1995. Proposed additions to the genus *Lonchura* (Estrildinae). *Bull. Brit. Orn. Club* 115: 140–57.

Restall, R. 1997. *Munias and Mannikins*. Yale Univ. Press, New Haven, CT.

Rheindt, F.E. 2012. New avian records from the little-explored Fakfak Mountains and the Onin Peninsula (West Papua). *Bull. Brit. Orn. Club* 132: 102–15.

Rheindt, F.E., & J.J. Austin. 2005. Major analytical and conceptual shortcomings in a recent taxonomic revision of the Procellariiformes—a reply to Penhallurick & Wink (2004). *Emu* 105: 181–86.

Rheindt, F.E., & R.O. Hutchinson. 2007. A photoshot odyssey through the confused avian taxonomy of Seram and Buru (southern Moluccas). *BirdingAsia* 7: 18–38.

Richards, A., & R. Rowland. 1995. List of birds recorded in Papua New Guinea during the period 16 October 1992 to 29 November 1992. *Muruk* 7: 75–96.

Ripley, S.D. 1942. *Trail of the Money Bird*. Longmans, Green, & Co., London.

Ripley, S.D. 1957. New birds from the Western Papuan Islands. *Postilla* Yale Peabody Mus. 31: 1–4.

Ripley, S.D. 1959. Comments on birds from the Western Papuan Islands. *Postilla* Yale Peabody Mus. 38: 1–17. [Mainly comments on birds of Kofiau Island.]

Ripley, S.D. 1960. Distribution and niche differentiation in species of megapodes in the Moluccas and Western Papuan area. *Proc. 12th Internat. Orn. Congr., Helsinki*: 631–40.

Ripley, S.D. 1964a. A systematic and ecological study of birds of New Guinea. *Yale Peabody Mus. Bull.* 19: 1–97.

Ripley, S.D. 1964b. Turdinae. In E. Mayr & R.A. Paynter Jr., eds., *Check-list of Birds of the World*, vol. 10: 13–227. Mus. Comp. Zool., Harvard Univ., Cambridge, MA.

Ripley, S.D. 1977. *Rails of the World: A Monograph of the Family Rallidae*. David R. Godine, Boston, MA.

Rocamora, G.J., & D. Yeatman-Berthelot. 2009. Dicruridae (drongos). In J. del Hoyo, A. Elliott, & D. Christie, eds., *Handbook of the Birds of the World*, vol. 14: 172–227. Lynx Edicions, Barcelona.

Roselaar, C.S. 1994. Systematic notes on Megapodiidae (Aves: Galliformes), including the description of five new subspecies. *Bull. Zoöl. Mus., Univ. Amsterdam* 14: 9–36.

Ross, J.D., & J.L. Bouzat. 2014. Genetic and morphometric diversity in the Lark Sparrow (*Chondestes grammicus*) suggest discontinuous clinal variation across major breeding regions associated with previously characterized subspecies. *Auk* 131: 298–313.

Rothschild, W. 1921. On some birds from the Weyland Mountains, Dutch New Guinea. *Novit. Zool.* 28: 280–94.

Rothschild, W. 1931. On a collection of birds made by Mr. F. Shaw Mayer in the Weyland Mountains, Dutch New Guinea, in 1930. *Novit. Zool.* 36: 250–76.

Rothschild, W., & E. Hartert. 1896. Contributions to the ornithology of the Papuan islands, 4. List of a collection made by Albert S. Meek on Fergusson, Trobriand, Egum, and Woodlark Islands. *Novit. Zool.* 3: 233–51.

Rothschild, W., & E. Hartert. 1901. Notes on Papuan birds. *Novit. Zool.* 8: 55–88; 102–62.

Rothschild, W., & E. Hartert. 1903. Notes on Papuan Birds. *Novit. Zool.* 10: 65–116; 196–231; 435–80.

Rothschild, W., & E. Hartert. 1907a. Notes on Papuan Birds. *Novit. Zool.* 14: 433–46.

Rothschild, W., & E. Hartert. 1907b. List of collections of birds made by A.S. Meek in the mountains on the upper Aroa River and on the Angabunga River, British New Guinea. *Novit. Zool.* 14: 447–83.

Rothschild, W., & E. Hartert. 1913. List of the collections of birds made by Albert S. Meek in the lower ranges of the Snow Mountains, on the Eilanden River, and on Mount Goliath, during the years 1910 and 1911. *Novit. Zool.* 20: 473–527.

Rothschild, W., & E. Hartert. 1914. On a collection of birds from Goodenough Island. *Novit. Zool.* 21: 1–9.

Rothschild, W., & E. Hartert. 1915a. The birds of Dampier [Karkar] Island. *Novit. Zool.* 22: 26–37.

Rothschild, W., & E. Hartert. 1915b. The birds of Vulcan [Manam] Island. *Novit. Zool.* 22: 38–45.

Rothschild, W., & E. Hartert. 1918a. A few additional notes on the birds of Rossel Island, Louisiade Group. *Novit. Zool.* 25: 311–12.

Rothschild, W., & E. Hartert. 1918b. Further notes on the birds of Sudest or Tagula Island in the Louisiade Group. *Novit. Zool.* 25: 311–12.

Rothschild, W., E. Stresemann, & K. Paludan. 1932. Ornithologische Ergebnisse der Expedition Stein 1931–1932: 1. Waigeu, 2. Numfor, 3. Japen. *Novit. Zool.* 38: 127–247.

Rowland, P. 1994a. A new altitude record and range extension for the Three-toed Swiftlet *Collocalia papuensis*. *Muruk* 6(3): 10.

Rowland, P. 1994b. Mountain Nightjar *Eurostopodus archboldi* breeding at Ambua. *Muruk* 6(3): 11.

Rowland, P. 1995. A specimen of Mountain Nightjar *Eurostopodus archboldi* from the Hindenburg Ranges. *Muruk* 7: 41.

Rowley, I. 1997. Cacatuidae (cockatoos). In J. del Hoyo, A. Elliott, & J. Sargatal, eds., *Handbook of the Birds of the World*, vol. 4: 246–79. Lynx Edicions, Barcelona.

Rowley, I., & E.M. Russell. 2007. Maluridae (fairy-wrens). In J. del Hoyo, A. Elliott, & D. Christie, eds., *Handbook of the Birds of the World*, vol. 12: 490–531. Lynx Edicions, Barcelona.

Rowley, I., & E.M. Russell. 2009. Artamidae (woodswallows). In J. del Hoyo, A. Elliott, & D. Christie, eds., *Handbook of the Birds of the World*, vol. 14: 286–307. Lynx Edicions, Barcelona.

Russell, E.M., & I. Rowley. 2009. Cracticidae (butcherbirds). In J. del Hoyo, A. Elliott, & D. Christie, eds., *Handbook of the Birds of the World*, vol. 12: 308–43. Lynx Edicions, Barcelona.

Ryan, P.G. 2006. Sylviidae (Old World warblers [various species accounts]). In J. del Hoyo, A. Elliott, & D. Christie, eds., *Handbook of the Birds of the World*, vol. 11: 577–709. Lynx Edicions, Barcelona.

Safford, R.J., & L.M. Smart. 1996. The continuing presence of Macgregor's Bird of Paradise *Macgregoria pulchra* on Mount Albert Edward, Papua New Guinea. *Bull. Brit. Orn. Club* 116: 186–88.

Saitoh, T., P. Alström, I. Nishiumi, Y. Shigeta, D. Williams, U. Olsson, & K. Ueda. 2010. Old divergences in a boreal bird supports long-term survival through the Ice Ages. *BMC Evol. Biol.* 10 (35): 1–13.

Saitoh, T., Y. Shigeta, & K. Ueda. 2008. Morphological differences among populations of the Arctic Warbler with some intraspecific taxonomic notes. *Ornith. Sci.* 7: 135–142.

Salomonsen, F. 1962. Whitehead's Swiftlet (*Collocalia whiteheadi* Ogilvie-Grant) in New Guinea and Melanesia. *Vidensk. Medd. Dansk Naturh. Foren.* 125: 509–12.

Salomonsen, F. 1966a. Preliminary descriptions of new honeyeaters (Aves, Meliphagidae). *Breviora* Mus. Comp. Zool. Harvard 254: 1–12.

Salomonsen, F. 1966b. Notes on the Green Heron (*Butorides striatus* Linnaeus) in Melanesia and Papua. *Vidensk. Medd. Dansk Naturh. Foren.* 129: 279–83.

Salomonsen, F. 1967. Meliphagidae; Dicaeidae. In R.A. Paynter Jr., ed. *Check-list of Birds of the World*, vol. 12: 166–208; 338–450. Mus. Comp. Zool., Harvard Univ., Cambridge, MA.

Salomonsen, F. 1983. Revision of the Melanesian swiftlets (Apodes, Aves) and their conspecific forms in the Indo-Australian and Polynesian region. *Kon. Danske Vidensk. Selskab Biolog. Skr.* 23(5): 1–108.

Salvadori, T. 1880–1882. *Ornitologia della Papuasia e delle Molucche*. G.B. Paravia di I. Vigliardi, Torino, Italy.

Sam, K., & B. Koane. 2014. New avian records along the elevational gradient of Mt. Wilhelm, Papua New Guinea. *Bull. Brit. Orn. Club* 134: 116–33.

Sangster, G. 2014. The application of species criteria in avian taxonomy and its implications for the debate over species concepts. *Biol. Rev.* 89: 199–214.

Sangster, G., P. Alström, E. Forsmark, & U. Olsson. 2010. Multi-locus phylogenetic analysis of Old World chats and flycatchers reveals extensive paraphyly at family, subfamily and genus level (Aves: Muscicapidae). *Molec. Phylog. Evol.* 57: 380–392.

Sangster, G., J.M. Collinson, P.-A. Crochet, A.G. Knox, D.T. Parkin, & S.C. Votier. 2013. Taxonomic recommendations for Western Palearctic birds: ninth report. *Ibis* 155: 898–907.

Schnitker, H. 2007. Revision der Systematik des Orangebrust-Zwergpapageis. *Papageien* 20: 100–105; 134–41.

Schodde, R. 1975. *Interim List of Australian Songbirds—Passerines*. Roy. Austral. Orn. Union, Melbourne, Australia.

Schodde, R. 1977. Contributions to Papuasian ornithology. 6. Survey of the birds of southern Bougainville Island, Papua New Guinea. *CSIRO Div. Wildl. Res. Tech. Paper* 34.

Schodde, R. 1978. The identity of five type-specimens of New Guinean birds. *Emu* 78: 1–6.

Schodde, R. 1984. First specimens of Campbell's Fairy-wren *Malurus campbelli*, from New Guinea. *Emu* 84: 249–50.

Schodde, R. 2006. Australasia's bird fauna today—origins and evolutionary development. In J.R. Merrick *et al.*, eds., *Evolution and Biogeography of Australasian Vertebrates*, 413–58. Auscipub, NSW, Australia.

Schodde, R. 2009. The identity of the type species of the cuckoo-dove genus *Macropygia* Swainson, 1837 (Columbidae). *Bull. Brit. Orn. Club* 129: 188–91.

Schodde, R. 2015. The identity of *Accipiter cirrocephalus rosselianus* Mayr, 1940. *Zootaxa* 3994(4): 597-599.

Schodde, R., & W.J. Bock. 2008. The valid name for the Grey Wagtail. *Bull. Brit. Orn. Club* 128: 132–33.

Schodde, R., & J.H. Calaby. 1972. The biogeography of Australo-Papuan bird and mammal faunas in relation to the Torres Strait. In D. Walker, ed., *Bridge and Barrier: The Natural and Cultural History of the Torres Strait*, 257–306. Publ. BG/3. Australian National Univ., Dept. of Biogeog. and Geomorph., Canberra, Australia.

Schodde, R., & L. Christidis. 2014. Relicts from Tertiary Australasia: undescribed families and subfamilies of songbirds (Passeriformes) and their zoologeographic signal. *Zootaxa* 3786(5): 501–22.

Schodde, R., E.C. Dickinson, F.D. Steinheimer, & W.J. Bock. 2010. The date of Latham's *Supplementum Indicis Ornithologici* 1801 or 1802? *South Austral. Ornithol.* 35(8): 231–35.

Schodde, R., & W.B. Hitchcock. 1968. Contributions to Papuasian ornithology. 1. Report on the birds of the Lake Kutubu area, Territory of Papua and New Guinea. *CSIRO Div. Wildl. Res. Tech. Paper* 13.

Schodde, R., G.M. Kirwan, & R. Porter. 2012. Morphological differentiation and speciation among darters (*Anhinga*). *Bull. Brit. Orn. Club* 132: 283–94.

Schodde, R., & I.J. Mason. 1974. Further observations on *Parotia wahnesi* and *P. lawesii* (Paradisaeidae). *Emu* 74: 200–201.

Schodde, R., & I.J. Mason. 1979. Revision of the Zitting Cisticola *Cisticola juncidis* (Rafinesque) in Australia, with description of a new subspecies. *Emu* 79: 49–53.

Schodde, R., & I.J. Mason. 1980. *Nocturnal Birds of Australia.* Lansdowne Editions, Melbourne, Australia.

Schodde, R., & I.J. Mason. 1997. Aves (Columbidae to Coraciidae). In W.W.K. Houston & A. Wells, eds., *Zoological Catalogue of Australia*, vol. 37.2. CSIRO Publishing, Collingwood, Victoria, Australia.

Schodde, R., & I.J. Mason. 1999. *The Directory of Australian Birds*. Passerines. CSIRO Publishing, Collingwood, Victoria, Australia.

Schodde, R., I.J. Mason, M.L. Dudzinski, & J.L. McKean. 1980. Variation in the Striated Heron *Butorides striatus* in Australasia. *Emu* 80: 203–12.

Schodde, R., & J.L. McKean. 1972. Distribution and taxonomic status of *Parotia lawesii helenae* De Vis. *Emu* 72: 113–14.

Schodde, R., & J.L. McKean. 1973a. The species of the genus *Parotia* (Paradisaeidae) and their relationships. *Emu* 73: 145–56.

Schodde, R., & J.L. McKean. 1973b. Distribution, taxonomy, and evolution of the gardener bowerbirds *Amblyornis* spp in eastern New Guinea, with descriptions of two new subspecies. *Emu* 73: 51–60.

Schodde, R., & J.L. McKean. 1976. The relationships of some monotypic genera of Australian oscines. In H.J. Frith & J.H. Calaby, eds., *Proc. 16th Internat. Orn. Congr., Canberra*: 530–41. Austral. Academy of Sci., Canberra, Australia.

Schodde, R., G.A. Smith, I.J. Mason, & R.G. Weatherly. 1979. Relationships and speciation in the Australian corellas (Psittacidae). *Bull. Brit. Orn. Club* 99: 128–37.

Schodde, R., G.F. van Tets, C.R. Champion, & G.S. Hope. 1975. Observations on birds at glacial altitudes on the Carstensz Massif, Western New Guinea. *Emu* 75: 65–72.

Schodde, R., & R.G. Weatherly. 1982. *The Fairy-wrens: A Monograph of the Maluridae*. Lansdowne Editions, Melbourne, Australia.

Schodde, R., & R.G. Weatherly. 1983. Campbell's Fairy-wren *Malurus campbelli*, a new species from New Guinea. *Emu* 82 (1982): 308–09.

Scholes, E., B.M. Beehler, D. Prawiradilaga, & T. Laman. ms. Taxonomic review and natural history of the bird-of-paradise *Parotia berlepschi* Kleinschmidt 1897 (Aves. Paradisaeidae) with photographic, audio, and video-based field observations. Unpublished ms.

Schweizer, M., T.F. Wright, J.V. Peñalba, E.E. Schirtzinger, & L. Joseph. 2015. Molecular phylogenetics suggests a New Guinean origin and frequent episodes of founder-event speciation in the nectarivorous lories and lorikeets (Aves: Psittaciformes). *Mol. Phylog. Evol.* 90: 34–48.

Setio, P., P.J. Kawatu, D. Kalo, D. Womsiwor, & B.M. Beehler. 2002. Birds of the Yongsu area, northern Cyclops Mountains, Papua, Indonesia. In S.J. Richards & S. Suryadi, eds., A biodiversity assessment of the Yongsu-Cyclops Mountains and the southern Mamberamo basin, Papua, Indonesia. *RAP Bull. Biol. Assess.* 25: 80–84. Conservation International, Washington, DC.

Shanahan, P.J. 1969. Birds of the Wau valley region, Morobe District. Unpublished.

Shany, N. 1995. Juvenile Papuan Hawk-Owl *Uroglaux dimorpha* near Vanimo. *Muruk* 7: 74.

Shapiro, B., D. Sibthorpe, A. Rambaut, J. Austin, G.M. Wragg, O.R.P. Bininda-Emonds, P.L.M. Lee, & A. Cooper. 2002. Flight of the Dodo. *Science* 295: 1683.

Sharpe, R.B. 1874–1898. *Catalogue of Birds in the British Museum*. British Museum of Natural History, London.

Sharpe, R.B. 1883. Contributions to the ornithology of New Guinea. Part 8. *Journ. Linn. Soc., Zool.* 16: 422–47.

Sheldon, F.H. 1987. Phylogeny of herons estimated from DNA-DNA hybridization studies. *Auk* 104: 97–108.

Sheldon, F.H., C.E. Jones, & K.G. McCracken. 2000. Relative patterns and rates of evolution in heron nuclear and mitochondrial DNA. *Molec. Biol. Evol.* 17: 437–50.

Sheldon, F.H., L.A. Whittingham, R.G. Moyle, B. Slikas, & D.W. Winkler. 2005. Phylogeny of swallows (Aves: Hirundinidae) estimated from nuclear and mitochondrial DNA sequences. *Molec. Phylog. Evol.* 35: 254–70

Shirihai, H. 2008. Rediscovery of Beck's Petrel *Pseudobulweria becki*, and other observations of tubenoses from the Bismarck Archipelago, Papua New Guinea. *Bull. Brit. Orn. Club* 128: 3–16.

Sibley, C.G., & J.E. Ahlquist. 1982a. The relationships of the Australo-Papuan scrub-robins *Drymodes* as indicated by DNA-DNA hybridization. *Emu* 82: 101–5.

Sibley, C.G., & J.E. Ahlquist. 1982b. The relationships of the Australo-Papuan sittellas *Daphoenositta* as indicated by DNA-DNA hybridization. *Emu* 82: 173–76.

Sibley, C.G., & J.E. Ahlquist. 1982c. The relationships of the Australasian whistlers *Pachycephala* as indicated by DNA-DNA hybridization. *Emu* 82: 199–202.

Sibley, C.G., & J.E. Ahlquist. 1982d. The relationships of the Australo-Papuan fairy-wrens as indicated by DNA-DNA hybridization. *Emu* 82: 251–55.

Sibley, C.G., & J.E. Ahlquist. 1984. The relationships of the Papuan genus *Peltops*. *Emu* 84: 181–83.

Sibley, C.G., & J.E. Ahlquist. 1985. The phylogeny and classification of passerine birds, based on comparisons of the genetic material, DNA. *Emu* 85: 1–14.

Sibley, C.G., & J.E. Ahlquist. 1987. The Lesser Melampitta is a bird of paradise. *Emu* 87: 66–68.

Sibley, C.G., & J.E. Ahlquist. 1990. *Phylogeny and Classification of Birds*. Yale Univ. Press, New Haven, CT.

Sibley, C.G., J.E. Ahlquist, & B.L. Monroe Jr. 1988. A classification of the living birds of the world based on DNA-DNA hybridisation studies. *Auk* 105: 409–23.

Sibley, C.G., & B.L. Monroe. 1990. *Distribution and Taxonomy of Birds of the World*. Yale Univ. Press, New Haven, CT.

Sibley, C.G., R. Schodde, & J.E. Ahlquist. 1984. The relationships of the treecreepers Climacteridae as indicated by DNA-DNA hybridization. *Emu* 84: 236–41.

Siegel-Causey, D. 1988. Phylogeny of the Phalacrocoracidae. *Condor* 90: 885–905.

Silvius, M.J., & A.W. Taufik. 1989. *Conservation and Land Use of Pulau Kimaam, Irian Jaya*. PHPA–AWB/Interwader, Bogor, Indonesia.

Silvius, M.J., A.W. Taufik, & F.R. Lambert. 1989. *Conservation and Land-use of the Wasur and Rawa Biru Reserves, Southeast Irian Jaya*. Report No. 10. Asian Wetland Bureau, Bogor, Indonesia.

Simpson, D.M. 1990. Paradise Lost? Birdwatching along the Fly River, Papua New Guinea. *Sea Swallow* 39: 53–57.

Simpson, K., & N. Day. 1996. *The Princeton Field Guide to the Birds of Australia*. Princeton Univ. Press, Princeton, NJ.

Sims, R.W. 1954. New race of button-quail (*Turnix maculosa*) from New Guinea. *Bull. Brit. Orn. Club* 74: 37–40.

Sims, R.W. 1956. Birds collected by Mr F. Shaw-Mayer in the Central Highlands of New Guinea 1950–1951. *Bull. Brit. Mus. Nat. Hist. Zool.* 3: 387–438.

Slikas, B.J., S.L. Olson, & R.C. Fleischer. 2002. Rapid, independent evolution of flightlessness in four species of Pacific Island rails (Rallidae): an analysis based on mitochondrial sequence data. *J. Avian Biol.* 33: 5–14.

Smith, G.A. 1977. Birds of Daru district, Western Province. *PNG Bird Soc. Newsl.* 129: 10–11.

Smith, J.M.B. 1976. Notes on birds breeding above 3,215 metres on Mt Wilhelm, Papua New Guinea. *Emu* 76: 220–21.

Smythe, F. 1975. *A Naturalist's Color Guide*. Amer. Mus. Nat. Hist., New York.

Snow, D.W. 1997. Proposed additions to the genus *Lonchura*: addenda and corrigenda. *Bull. Brit. Orn. Club* 117: 4.

Somadikarta, S. 1967. A recharacterization of *Collocalia papuensis* Rand, the Three-toed Swiftlet. *Proc. U.S. Nat. Mus.* 124: 1–8.

Somadikarta, S. 1975. An unrecorded specimen of *Collocalia papuensis* Rand. *Bull. Brit. Orn. Club* 95: 41–42.

Sorenson, M.D., E. O'Neal, J. Garcia-Moreno, & D.P. Mindell. 2003. More taxa, more characters: the Hoatzin problem is still unresolved. *Molec. Biol. Evol.* 20: 1484–99.

Sorenson, M.D, & R.B. Payne. 2005. A molecular genetic analysis of cuckoo phylogeny. In R.B. Payne, *The Cuckoos*, 68–94. Oxford Univ. Press, Oxford, UK.

Steadman, D.W. 2006. *Extinction and Biogeography of Tropical Pacific Birds*. University of Chicago Press, Chicago.

Stein, G. 1936. Ornithologische Ergebnisse der Expedition Stein, 1931–1932. 5. Beiträge zur Biologie papuanischer Vögel. *J. Orn.* 84: 21–57.

Steinbacher, J. 1979. Threskiornithidae. In E. Mayr & G.W. Cottrell, eds., *Check-list of Birds of the World*, vol. 1, 2nd ed.: 253–68. Mus. Comp. Zool., Harvard Univ., Cambridge, MA.

Stickney, E.H. 1943. Birds collected during the Whitney South Sea Expedition. 53. Northern shore birds in the Pacific. *Amer. Mus. Novit.* 1248: 1–9.

Stonor, C.R. 1937. On the systematic position of the Ptilonorhynchidae. *Proc. Zool. Soc. Lond.* 107: 475–90.

Storer, R.W. 1979. Podicipedidae. In E. Mayr & G.W. Cottrell, eds., *Check-list of Birds of the World*, vol. 1, 2nd ed.: 140–55. Mus. Comp. Zool., Harvard Univ., Cambridge, MA.

Storr, G.M. 1973. List of Queensland Birds. *Spec. Publ. West Austral. Mus.* 5.

Stresemann, E. 1922. Neue Formen aus dem papuanischen Gebiet. *J. Orn.* 70: 405–8.

Stresemann, E. 1923. Dr Bürger's ornithologische Ausbeute im Stromgebiet des Sepik. Ein Beitrag zur Kenntnis der Vogelwelt Neuguineas. *Arch. Naturgesch.* ser. A. 89 (7): 1–96; (8): 1–92.

Stresemann, E. 1934. Vier neue Unterarten von Paradiesvogeln. *Orn. Monatsber.* 42: 144–47.

Stresemann, E. 1941. *Calidris ferruginea* (Pontoppidan) statt *Calidris testacea* (Pallas). *Orn. Monatsber.* 49: 21.

Stresemann, E., & D. Amadon. 1979. Falconiformes. In E. Mayr & G.W. Cottrell, eds., *Check-list of Birds of the World*, vol. 1, 2nd ed.: 271–425. Mus. Comp. Zool., Harvard Univ., Cambridge, MA.

Stresemann, E., & J. Arnold. 1949. Speciation in the group of great reed-warblers. *J. Bombay Nat. Hist. Soc.* 48: 428–43.

Stresemann, E., & K. Paludan. 1932. Vorläufiges über die ornithologichen Ergebnisse der Expedition Stein 1931–32. 1. Zur Ornithologie der Insel Waigeu. *Orn. Monatsber.* 40: 13–18.

Stresemann, E., & K. Paludan. 1935. Ueber eine kleine Vogelsammlung aus dem Bezirk Merauke (Sud-Neu-guinea), angelegt von Dr. H. Nevermann. *Mitt. Zool. Mus. Berl.* 20: 447–63.

Stresemann, E., K. Paludan, E. Hartert, & W. Rothschild. 1934. Vorläufiges über die ornithologischen Ergebnisse der Expedition Stein 1931–32. 2. Zur Ornithologie des Weyland-Gebirges in Niederländisch-Neuguinea. *Orn. Monatsber.* 42: 43–46.

Stronach, N. 1980. Chestnut Teal *Anas castanea* at Bensbach (Western Province), a new species for the New Guinea region. *PNG Bird Soc. Newsl.* 173/174: 3.

Stronach, N. 1981a. Grey Falcon at Bensbach, Western Province, a new species for the New Guinea region. *PNG Bird Soc. Newsl.* 178/179: 5.

Stronach, N. 1981b. Notes on some birds of the Bensbach Region, Western Province. *PNG Bird Soc. Newsl.* 178/179: 6–12.

Stronach, N. 1981c. Pallid Cuckoo *Cuculus pallidus*, a new species for Papua New Guinea. *PNG Bird Soc. Newsl.* 181/182: 23.

Stronach, N. 1981d. Observations [White-throated Nightjar *Eurostopodus mystacalis*]. *PNG Bird Soc. Newsl.* 185/186: 39.

Stronach, N. 1981e. Observations to end of August, 1981 [Spectacled Monarch *Symposiachrus trivirgatus*]. *PNG Bird Soc. Newsl.* 181/182: 40.

Stronach, N. 1981f. The Black-faced Wood-swallow *Artamus cinereus* in the Western Province of Papua New Guinea. *PNG Bird Soc. Newsl.* 179/180: 11.

Stronach, N. 1990. The occurrence of the Grass Owl *Tyto capensis* and Richard's Pipit *Anthus novaeseelandiae* in the lowlands of New Guinea. *Bull. Brit. Orn. Club* 110: 181–84.

Summers-Smith, J.D. 2009. Passeridae (Old World sparrows). In J. del Hoyo, A. Elliott, & D. Christie, eds., *Handbook of the Birds of the World*, vol. 14: 760–813. Lynx Edicions, Barcelona.

Symons, F. 1992. Black-faced Wood-swallow *Artamus cinereus* at Wau (Morobe Province). *Muruk* 5: 90.

Tarburton, M. [undated.] Bird checklists for 672 Melanesian Islands. http://birdsofmelanesia.net/. Accessed 22-Nov-2015.

Taylor, P.B. 1996. Rallidae (rails). In J. del Hoyo, A. Elliott, & J. Sargatal, eds., *Handbook of the Birds of the World*, vol. 3: 108–209. Lynx Edicions, Barcelona.

Taylor, P.B. 1998. *Rails: A Guide to the Rails, Crakes, Gallinules and Coots of the World*. Yale Univ. Press, New Haven, CT.

Taylor, P.B. 2005. Campephagidae (cuckoo-shrikes). In J. del Hoyo, A. Elliott, & D. Christie, eds., *Handbook of the Birds of the World*, vol. 10: 40–123. Lynx Edicions, Barcelona.

Taylor, P.B. 2006. Muscicapidae (Old World flycatchers). Pp. 56-163 in J. del Hoyo, A. Elliott, & D. Christie, eds. *Handbook of Birds of the World*. Vol. 11. Lynx Edicions, Barcelona.

Tennyson, A.J.D., C.M. Miskelly, & M. LeCroy. 2014. Clarification of collection data for the type specimens of Hutton's Shearwater *Puffinus huttoni* Mathews, 1912, and implications for the accuracy of historic Subantarctic specimen data. *Bull. Brit. Orn. Club* 134: 242–46.

Thiollay, J.M. 1994. Accipitridae (hawks and eagles [various species accounts]). In J. del Hoyo, A. Elliott, & J. Sargatal, eds., *Handbook of the Birds of the World*, vol. 2: 52–205. Lynx Edicions, Barcelona.

Thomas, G., M.A. Willis, & T. Székely. 2004. A supertree approach to shorebird phylogeny. *BMC Evol. Biol.* 4: 28. http://www.biomedcentral.com/1471-2148/4/28. Accessed 24-Sep-2015.

Thomassen, H.A., R.J. den Tex, M.A.G. de Bakker, & G.D.E. Povel. 2005. Phylogenetic relationships amongst swifts and swiftlets: A multi locus approach. *Molec. Phylog. Evol.* 37: 264–77.

Thompson, D.W. 1900. On characteristic points in the cranial osteology of the parrots. *Proc. Zool. Soc. Lond.* 1899: 9–46.

Thompson, H.A.F. 1979. First Australasian record of Sand Martin *Riparia riparia*. *PNG Bird Soc. Newsl.* 151: 6–7.

Thompson, M.C. 1964. Two new distributional records of birds from the Southwest Pacific. *Ardea* 52: 121.

Tindige, K. 2003a. First record of Cinnamon Bittern *Ixobrychus cinnamomeus* for New Guinea. *Kukila* 12: 67.

Tindige, K. 2003b. First record of the Elegant Imperial Pigeon *Ducula concinna* on Biak Island, Irian Jaya (Papua). *Kukila* 12: 69. [Here reidentified as *Ducula geelvinkiana*.]

Tobias, J.A., N. Seddon, C.N. Spottiswoode, J.D. Pilgrim, L.D.C. Fishpool, & N.J. Collar. 2010. Quantitative criteria for species delimitation. *Ibis* 152: 724–46.

Tolhurst, L.P. 1988. Range extension of Red-headed Myzomela *Myzomela erythrocephala*. *Muruk* 3: 54.

Tolhurst, L.P. 1992. Extension to the range of the Lesser Black Coucal, *Centropus bernsteini*. *Muruk* 5: 92–93.

Tolhurst, L.P. 1996. Two new species for Misima Island, Milne Bay Province. *Muruk* 7: 34–35.

Toon, A., J.M. Hughes, & L. Joseph. 2010. Multilocus analysis of honeyeaters (Aves: Meliphagidae) highlights spatiotemporal heterogeneity in the influence of biogeographic barriers in the Australian monsoonal zone. *Mol. Ecol.* 19: 2980–94.

Tubb, J.A. 1945. Field notes on some New Guinea birds. *Emu* 44: 249–73.

Turner, A.K. 2004. Hirundinidae (swallows and martins). In J. del Hoyo, A. Elliott, & D. Christie, eds., *Handbook of the Birds of the World*, vol. 9: 602–85. Lynx Edicions, Barcelona.

Tyler, S.J. 2004. Motacillidae (pipits and wagtails). In J. del Hoyo, A. Elliott, & D. Christie, eds., *Handbook of the Birds of the World*, vol. 9: 686–787. Lynx Edicions, Barcelona.

Uy, A., & G. Borgia. 2000. Sexual selection drives rapid divergence in bowerbird display traits. *Evolution* 54: 273–278.

Vang, K. 1991. Woodlark Island. *Muruk* 5: 36–42.

Vaurie, C. 1962. Dicruridae; Old World Corvidae. In E. Mayr & J.C. Greenway, eds., *Check-list of Birds of the World*, vol. 15: 137–57; 204–84. Mus. Comp. Zool., Harvard Univ., Cambridge, MA.

Vaurie, C. 1965. *Birds of the Palearctic Fauna*. Non-Passeriformes. H.F. & G. Witherby, London.

Verbelen, P. 2014. First record of the Starry Owlet-nightjar *Aegotheles tatei* in Indonesian New Guinea (Papua). *Kukila* 18: 10–16.

Voison, C., & J.-F. Voison. 1997. A propos du Lori Papou. *Alauda* 65: 191-195.

Voous, K. 1963. Black-tailed Godwit (*Limosa limosa*) in New Guinea. *Ardea* 51: 253.

Wahlberg, N. 1987. Tufi. *PNG Bird Soc. Newsl.* 225: 4–5.

Wahlberg, N. 1990. Straw-necked Ibis *Threskiornis spinicollis* at Kanosia Lagoon, Central Province. *Muruk* 4: 109.

Walther, B.A., & P.J. Jones. 2008. Oriolidae (orioles). In J. del Hoyo, A. Elliott, & D. Christie, eds., *Handbook of the Birds of the World*, vol. 13: 692–731. Lynx Edicions, Barcelona.

Wardill, J.C., & T. Nando. 2000. Concentrations of wintering Streaked Shearwaters off the northern coast of West Papua. *Kukila* 11: 151.

Warham, J. 1961. The Birds of Raine Island, Pandora Cay, and Murray Island Sandbank, North Queensland. *Emu* 61: 77–93.

Warham, J. 1962. Bird Islands within the Barrier Reef and Torres Strait. *Emu* 62: 99–111.

Warne, J. 1989. Wilson's Storm Petrel *Oceanites oceanicus* in the Gulf of Papua. *Muruk* 4: 23.

Watson, G.W. 1986. Holarctic and Oriental Sylviidae. In E. Mayr & G.W. Cottrell, eds., *Check-list of Birds of the World*, vol. 11: 3–291. Mus. Comp. Zool., Harvard Univ., Cambridge, MA.

Watson, J.D., W.R. Wheeler, & E. Whitbourne. 1962. With the RAOU in Papua New Guinea, October 1960. *Emu* 62: 31–50; 67–98.

Wells, D.R. 1999. Hemiprocnidae (tree-swifts). In J. del Hoyo, A. Elliott, & J. Sargatal, eds., *Handbook of the Birds of the World*, vol. 5: 458–67. Lynx Edicions, Barcelona.

Weston, I. 1977. High mountain birds in the Enga Province. *PNG Bird Soc. Newsl.* 128: 14–19.

Wheeler, J.R. 1969. New bird species for PNG [Cotton Pygmy-Goose *Nettapus coromandelianus*]. *Geelong Nat.* 3: 125.

White, C.M.N., & M.D. Bruce. 1986. *Birds of Wallacea (Sulawesi, Moluccas, and Lesser Sunda Islands).* BOU checklist no. 7. Brit. Orn. Union, London.

White, N.E., M.J. Phillips, M.T. P. Gilbert, A. Alfaro-Núñez, E. Willerslev, P.R. Mawson, P.B.S. Spencer, M. Bunce. 2011. The evolutionary history of cockatoos (Aves: Psittaciformes: Cacatuidae) *Mol. Phylog. Evol.* 59: 615-623.

Whitney, B.M. 1987. The Pale-billed Sicklebill *Epimachus bruijnii* in Papua New Guinea. *Emu* 87: 244–46.

Wiersma, P. 1996. Charadriidae (plovers [species accounts]). In J. del Hoyo, A. Elliott, & J. Sargatal, eds., *Handbook of the Birds of the World*, vol. 3: 410–43. Lynx Edicions, Barcelona.

Wink, M., A. A. El-Sayed, H. Sauer-Gürth, & J. Gonzalez. 2009. Molecular phylogeny of owls (Strigiformes) inferred from DNA sequences of the mitochondrial cytochrome *b* and the nuclear RAG-1 gene. *Ardea* 97(4): 581–91.

Wink, M., H. Sauer-Gürth, & H.H. Witt. 2004. Phylogenetic differentiation of the Osprey *Pandion haliaetus* inferred from the nucleotide sequences of the mitochondrial cytochrome *b* gene. In R.D. Chancellor & B.-U. Meyburg, eds., *Raptors Worldwide*, 511–16. WWGMP & MME/Birdlife Hungary, Budapest.

Winker, K. 2010. Is it a species? *Ibis* 152: 679–82.

Winker, K., & S.M. Haig. 2010. Avian subspecies. *Orn. Monogr.* 67: 1–200.

Wolfe, A. 1968. *Dendrocygna eytoni* [at Waigani sewage ponds]. *PNG Bird Soc. Newsl.* 35: 2.

Wolfe, A. 1969. Observations [Little Black Cormorant *Phalacrocorax sulcirostris* and Darter *Anhinga novaehollandiae*]. *PNG Bird Soc. Newsl.* 45: 1.

Wood, V.J. 1970. The breeding and distribution of *Elanus caeruleus* in New Guinea. *Sunbird* 1: 48–55.

Woodall, P.F. 2001. Alcedinidae (kingfishers). In J. del Hoyo, A. Elliott, & J. Sargatal, eds., *Handbook of the Birds of the World*, vol. 6: 130–249. Lynx Edicions, Barcelona.

Worthy, T.H., & J.D. Scanlon. 2009. An Oligo-Miocene Magpie Goose (Aves: Anseranatidae) from Riversleigh, Northwestern Queensland, Australia. *Journ. Vert. Paleo.* 29: 205.

Woxvold, I., B. Ken, & K.P. Aplin. 2015. Birds. In S. Richards, & N. Whitmore, eds, A Rapid Biodiversity Assessment of Papua New Guinea's Hindenburg Wall Region, pages 103-129. Wildlife Conservation Society Papua New Guinea Program, Goroka, Papua New Guinea.

Yamamoto, T., A. Takahasi, N. Katsumata, K. Sato, & P.N. Trathan. 2010. At-sea distribution and behavior of Streaked Shearwaters (*Calonectris leucomelas*) during the nonbreeding period. *Auk* 127: 871–81.

Yosef, R. 2008. Laniidae (shrikes). In J. del Hoyo, A. Elliott, & D. Christie, eds., *Handbook of the Birds of the World*, vol. 13: 732–96. Lynx Edicions, Barcelona.

Zhou, X., Q. Lin, W. Fang, & X. Chen. 2014. The complete mitochondrial genomes of sixteen ardeid birds revealing the evolutionary process of the gene rearrangements. *BMC Genomics* 15: 573. doi:10.1186/1471-2164-15-573.

Ziswiler, V., H.R. Guttinger, & H. Bregulla. 1972. Monographie der Gattung *Erythrura* Swainson, 1837 (Aves, Passeres, Estrildidae). *Bon. Zool. Monogr.* 2.

Zuccon, D., A. Cibois, E. Pasquet, & P.G.P. Ericson. 2006. Nuclear and mitochondrial sequence data reveal the major lineages of starlings, mynas and related taxa. *Molec. Phylog. Evol.* 41: 333–344.

Zuccon, D., & P.G.P. Ericson. 2012. Molecular and morphological evidences place the extinct New Zealand endemic *Turnagra capensis* in the Oriolidae. *Molec. Phylog. Evol.* 62: 414–26.

Zwiers, P.B., G. Borgia, & R.C. Fleischer. 2008. Plumage based classification of the bowerbird genus *Sericulus* evaluated using a multi-gene, multi-genome analysis. *Molec. Phylog. Evol.* 46: 923–31.

Gazetteer of New Guinea Ornithology

Jennifer L. Mandeville *and* William S. Peckover

Entries marked with an asterisk (*) are considered alternative usage or spelling; that said, in many instances we are unable to determine preferred usage. All latitude and longitude coordinates are in decimal degrees, which facilitates use in digital mapping; see p. 36, convention no. 4 for explanation. The rightmost column includes alternative names, locational guidance, names of fieldworkers/dates of fieldwork, and published references (in parentheses).

NAME	LATITUDE	LONGITUDE	ELEV. (M)	BIRD REGION	ALTERNATIVE NAMES, RESEARCHERS, REFERENCES
Abalgamut Camp	-6.10	146.56		Huon Peninsula	Benz, B./2001
Abau	-10.17	148.76		SE Peninsula	south coast of SE Peninsula
Abid's Camp	-7.40	146.67		SE Peninsula	on the Bulldog Road above Edie Creek; Mirza, A.B.; Beehler, B.M.; Pratt, T.K.
Aculama River	-8.55	146.88		SE Peninsula	Akulama River
Adau River	-9.60	148.68		SE Peninsula	
Adelbert Mountains	-4.90	145.40	1,640	Sepik-Ramu	(Gilliard & LeCroy 1967, Pratt 1983)
Adi Island	-4.20	133.42		Bird's Neck	Pulu Adi*/Pulau Adi; (Gyldenstolpe 1955b)
Admosin Island	-5.08	145.80		Sepik-Ramu	Heinrich, B./1969
Adoa	-1.91	129.83		NW Islands	on Misool Island; (Mees 1965)
Adolphhafen	-7.75	147.42		SE Peninsula	
Aduale River	-8.52	146.93		SE Peninsula	Alabule/Arahure River
Aedo Camp	-6.69	145.11		S Lowlands	Mack, A./1996
Aeginimi Island	-8.33	142.92		Trans-Fly	
Agamuri village	-1.15	134.00	1,000	Bird's Head	Shaw Mayer, F.
Agats	-5.53	138.12		S Lowlands	
Agaun	-9.93	149.38	1,000	SE Peninsula	(Engilis & Cole 1997)
Agotu	-6.50	145.23	1,835	Eastern Ranges	Diamond, J.M., & Terborgh, J.W./1964
Agu River	-7.08	141.12		Trans-Fly	
Aibala River	-8.30	147.12		SE Peninsula	
Aibom	-4.27	143.20		Sepik-Ramu	
Aicora*/Aikora River	-8.28	147.63		SE Peninsula	Gira River*; Meek, A.S. and Eichhorn/1905
Aidoema*/Aiduma Island	-3.97	134.13		Bird's Neck	in Triton Bay of Bird's Neck
Aiema River	-7.08	141.99		Trans-Fly	
Aieme Creek or River	-9.77	147.73		SE Peninsula	
Aievi passage	-7.85	145.18		S Lowlands	in delta of the Purari River

NAME	LATITUDE	LONGITUDE	ELEV. (M)	BIRD REGION	ALTERNATIVE NAMES, RESEARCHERS, REFERENCES
Aifundi Islands	-0.40	135.25		Bay Islands	
Aikora/Aicora* River	-8.28	147.63		SE Peninsula	Gira River*
Aimaroe*/Aimaru Lakes	-1.28	132.27		Bird's Head	Lake Aimaru; (Gyldenstolpe 1955b)
Aiome	-5.14	144.72	110	Sepik-Ramu	Gilliard, E.T./1964
Aion Island	-9.30	152.83		SE Islands	off Woodlark Island (Pratt et al. 2006)
Aird River	-7.63	144.28		S Lowlands	
Aird River delta	-7.56	144.56		S Lowlands	
Aitape	-3.13	142.33		Sepik-Ramu	Eitape*
Aiyonka Mountain	-6.40	145.95	1,920	Eastern Ranges	
Aiyura	-6.32	145.90	1,554	Eastern Ranges	Smith, S./1977
Ajaura Hills	-9.72	148.70	400– 1,200	SE Peninsula	
Ajawi Island*	-0.20	135.00		Bay Islands	Ayawi Island; Mios Kairu
Ajibara River	-8.60	147.60		SE Peninsula	
Ajule Kajale Range	-8.75	147.83	600– 1,785	SE Peninsula	
Akaifu River	-8.40	146.37		SE Peninsula	
Akulama River	-8.55	146.88		SE Peninsula	Aculama River
Akura River	-6.25	143.80		Eastern Ranges	
Alabule River	-8.52	146.93		SE Peninsula	Aduale/Arahure River
Albert Mountains	-7.33	145.63	1,190	S Lowlands	
Alcester Island	-9.56	152.44		SE Islands	Nasikwabu Island; Hamlin, H./1928; (Pratt et al. 2006)
Alexishafen	-5.08	145.80		Sepik-Ramu	Dawson, W./1969
Aleya	-8.77	146.54		SE Peninsula	
Alice River	-6.17	141.12		S Lowlands	
Alkmaar	-4.68	138.73		S Lowlands	on the Lorentz River; Lorentz, H./1909
Alotau	-10.30	150.45		SE Peninsula	Alpha Helix New Guinea Expedition/1969
Ama	-4.10	141.67	30	Sepik-Ramu	
Amamapare	-4.88	136.80		S Lowlands	coastal port town for Freeport Mine
Amanab	-3.07	141.22	365	Sepik-Ramu	
Amawai River	-10.10	147.93		SE Peninsula	
Amazon Bay	-10.29	149.37		SE Peninsula	
Amazon Bay airstrip	-10.30	149.33	3	SE Peninsula	
Amazon Island	-10.17	149.32		SE Peninsula	
Ambeingawiok River	-6.90	145.68		Eastern Ranges	
Amberbaken	-0.67	133.25		Bird's Head	Amberbaki*; north coast of Bird's Head; Laglaize/1877
Amberbaki*	-0.67	133.25		Bird's Head	Amberbaken; north coast of Bird's Head; Laglaize/1877
Ambérno*/Amber-noh* River	-2.00	137.80		NW Lowlands	Mamberamo River

NAME	LATITUDE	LONGITUDE	ELEV. (M)	BIRD REGION	ALTERNATIVE NAMES, RESEARCHERS, REFERENCES
Amberpon*	-1.83	134.12		Bird's Head	Rumberpon Island
Ambiduru	-1.83	136.19		Bay Islands	Yapen Island; (Diamond 1985)
Amboin	-4.60	143.50		Sepik-Ramu	
Ambua Lodge	-5.97	143.05		Eastern Ranges	Frith, C. & D.
Ambua Peaks	-5.85	143.10		Eastern Ranges	south of Tari Gap
Ambulua airfield	-5.72	144.90	1,875	Eastern Ranges	
Ambulua village	-5.73	144.92		Eastern Ranges	
Ambunti	-4.22	142.82	50	Sepik-Ramu	Slater/1954
Amilibar	-5.75	144.52		Eastern Ranges	
Amin River	-5.05	141.88		Border Ranges	
Amogu River	-3.67	143.07		Sepik-Ramu	
Amphlett Islands	-9.27	150.83		SE Islands	
Amy Range	-6.60	144.37	1,715	Eastern Ranges	
Anagusa Island	-10.72	151.23		SE Islands	Dumbacher, J.P.
Anangalo airfield	-4.60	143.83	30	Sepik-Ramu	
Andai	-0.85	134.02		Bird's Head	in the Arfak Mts; Bruijn, A.A.; D'Albertis, L.; Beccari, O.
Andakombe airfield	-7.15	145.75	1,050	Eastern Ranges	
Angabanga/Ang-abunga* River	-8.60	147.03		SE Peninsula	St. Joseph's River*
Angakumia Ranges	-7.23	146.28	2,320	SE Peninsula	
Anggi Giji Lake	-1.37	133.92		Bird's Head	Lake Anggi Gigi*; *giji* is a small girl in Bahasa Indonesia
Anggi Gita Lake	-1.38	133.97		Bird's Head	*gita* is a small boy in Bahasa Indonesia
Anggura River	-6.30	143.65		S Lowlands	
Angoram	-4.07	144.07	12	Sepik-Ramu	
Angriffshafen	-2.75	141.45		Sepik-Ramu	
Anguganuk/ Anguganak	-3.60	142.25	low-lands	Sepik-Ramu	(Greig-Smith 1978)
Annanberg	-4.90	144.63		Sepik-Ramu	
Annie Inlet	-10.25	150.58		SE Peninsula	Hamlin, H./1929
Ansoes*	-1.73	135.80		Bay Islands	Ansus
Ansus	-1.73	135.80		Bay Islands	Ansoes*
Ansus Island	-1.77	135.78		Bay Islands	(Doherty, W. 1897, Meyer 1875, Powell 1883)
Antares Mountain or Peak	-4.89	140.90	4,170	Border Ranges	in the Star Mountains of Indonesia
Anumb River	-3.37	143.13		Sepik-Ramu	Sowan River
Aoeri Eilanden*	-2.17	134.75		Bay Islands	Auri Islands
Ape Hills	-2.85	141.13		Sepik-Ramu	
April River	-4.27	142.40		Sepik-Ramu	
April River airfield	-4.70	142.53	90	Sepik-Ramu	
Aptin River	-5.35	141.52		Border Ranges	

NAME	LATITUDE	LONGITUDE	ELEV. (M)	BIRD REGION	ALTERNATIVE NAMES, RESEARCHERS, REFERENCES
Ara River	-8.50	147.03		SE Peninsula	Aura River
Arahure River	-8.52	146.93		SE Peninsula	Aduale/Alabule River
Araltamu Island	-5.90	148.05		SE Peninsula	
Aramia River	-7.88	143.28		S Lowlands	
Arar Island	-0.35	130.24		NW Islands	off southwest coast of Waigeo Island
Arau district	-6.37	146.05	1,400	Eastern Ranges	
Arerr/Memunggal	-8.68	140.86		Trans-Fly	position approximate; (Bishop 2005a)
Arfak Mountains	-1.25	134.00		Bird's Head	Beccari, O.; D'Albertis, L.; Mayr, E.
Argau Creek	-8.18	146.72		SE Peninsula	
Arguni Bay	-3.10	133.70		Bird's Neck	
Arik River	-5.20	142.13		Border Ranges	
Arip valley	-5.38	141.29		S Lowlands	upper Fly
Armit Range	-7.22	145.76	2,805	Eastern Ranges	
Arndt Point	-6.48	147.85		Huon Peninsula	
Arnold River	-2.97	142.00		Sepik-Ramu	
Aroa lagoon	-9.02	146.80		SE Peninsula	Finch, B.
Aroa River	-9.07	146.80		SE Peninsula	
Aroana River	-9.02	146.82		SE Peninsula	
Aroe*/Arrou* Islands	-6.00	134.50		Aru Islands	Aru Islands
Aroma Coast	-10.13	148.12		SE Peninsula	
Aropokina	-8.77	146.65		SE Peninsula	
Arosele Camp	-6.60	145.19		Eastern Ranges	Benz, B./2002
Aru	-7.98	147.23		SE Peninsula	place in former German New Guinea
Aru Islands	-6.00	134.50		Aru Islands	Aroe*/Arrou* Islands
Arufi	-8.75	141.75		Trans-Fly	near Morehead
Aruma Apa	-9.28	147.62	1,888	SE Peninsula	Hufeisengebirge*/Horseshoe Mountain*/ Mount Maguli
Arun mission	-8.77	141.92		Trans-Fly	
Asapa airfield	-8.98	148.13	1,900	SE Peninsula	
Asaro River	-6.25	145.40		Eastern Ranges	
Aseki	-7.35	146.18	1,200	SE Peninsula	
Asimba	-8.60	147.58		SE Peninsula	
Asimba River	-8.60	147.60		SE Peninsula	
Asmat	-5.60	138.20		S Lowlands	
Astrolabe Bay	-5.35	145.92		Sepik-Ramu	Foerster; Wahnes; Beck, R./1928
Astrolabe Range or Mts	-9.50	147.50	1,074	SE Peninsula	
Atam*	-1.09	133.97		Bird's Head	Hatam/Hattam*; Beccari, O./1875
Atem River	-4.85	141.22		Border Ranges	
Atinju/Atinjoe* (Lakes)	-1.41	132.37		Bird's Head	(Gyldenstolpe 1955b)

NAME	LATITUDE	LONGITUDE	ELEV. (M)	BIRD REGION	ALTERNATIVE NAMES, RESEARCHERS, REFERENCES
Atkeri village	-1.75	130.09		NW Islands	on central-north coast of Misool Island; Ripley, S.D.; (Mees 1965)
Attack Island	-8.39	143.00		Trans-Fly	D'Albertis, L.
Auagum Range	-5.28	143.15	1,000–2,334	Eastern Ranges	
Auga River	-8.72	147.98		SE Peninsula	upper Angabanga River
Augu River	-8.52	146.93		SE Peninsula	
August River	-4.10	141.15		Sepik-Ramu	
Augustahafen	-	-		Sepik-Ramu	not located; apparently a coastal town north of Madang
Aukopmin	-5.57	141.32		Border Ranges	
Aura River	-8.50	147.03		SE Peninsula	Ara River
Aure	-7.02	145.55		Eastern Ranges	
Aure River	-7.07	145.32		S Lowlands	
Aure Scarp	-7.07	145.32	200–1,600	S Lowlands	
Auri Archipelago or Islands	-2.17	134.75		Bay Islands	Aoeri Eilanden*
Auwaiabaiwa	-9.32	147.53	500	SE Peninsula	
Avela	-8.80	147.18	1,500	SE Peninsula	
Avera/Aveve*	-8.63	147.08	1,640	SE Peninsula	upper Aroa River; Meek, A.S./1903
Avi Avi River	-7.80	146.45		SE Peninsula	Lakekamu basin
Aviara Beach	-8.64	146.50		SE Peninsula	near Bereina
Awaba	-8.02	142.75		S Lowlands	
Awaiama Bay	-10.22	150.52		SE Peninsula	Chad's Bay*
Awande	-6.56	145.57	1,920	Eastern Ranges	Diamond, J.M./1965; (Diamond 1972)
Awar	-4.15	144.83		Sepik-Ramu	
Aya River	-8.52	146.95		SE Peninsula	
Ayawi Island	-0.20	135.00		Bay Islands	Ajawi Island*; Mios Kairu
Ayu Islands	0.37	131.05		NW Islands	north of Waigeo Island
Aziana	-6.75	145.85		Eastern Ranges	
Aziana River	-6.73	145.65		Eastern Ranges	
Babauguina	-10.10	148.73		SE Peninsula	
Babooni*	-8.68	146.95	600	SE Peninsula	Bubuni; Meek, A.S./1905
Babu River	-7.90	147.17		SE Peninsula	
Badilu	-4.78	146.20		Huon Peninsula	
Bagabag Island	-4.80	146.22		Huon Peninsula	(Diamond & LeCroy 1979)
Bahrman Mountains*	-5.15	141.53		Border Ranges	Behrmann Hills
Baia River	-6.17	142.07		S Lowlands	foot of Karius Range (not Baiyer River); Woxvold, I.
Bailala River*	-7.95	145.42		S Lowlands	Vailala River
Baimuru	-7.50	144.82		S Lowlands	
Baitabag	-5.13	145.77		Sepik-Ramu	

NAME	LATITUDE	LONGITUDE	ELEV. (M)	BIRD REGION	ALTERNATIVE NAMES, RESEARCHERS, REFERENCES
Baiyer River	-5.40	144.07		Eastern Ranges	Sibley, C.G./1969
Baiyer River gorge	-5.83	144.18		Eastern Ranges	
Baiyer River Sanctuary	-5.50	144.15		Eastern Ranges	Mackay, R.; George, G.; Bishop, K.D.
Baliem River gorge	-4.50	139.30		Western Ranges	also known as Baliem gorge
Baliem/Balim* valley	-4.00	138.90		Western Ranges	Rand, A.L.
Balimo	-8.02	142.95		S Lowlands	
Bam Island	-3.62	144.82		Sepik-Ramu	(Cheshire 2011)
Bama	-0.52	132.33		Bird's Head	Tamrau Mountains
Bamoskabu	-0.75	132.42	701	Bird's Head	Ripley, S.D./1938; (Mayr & Meyer de Schauensee 1939b)
Bamu River	-8.02	143.53		S Lowlands	
Banahai	-	-		Huon Peninsula	(Mayr 1941: 224)
Bandaber	-8.93	141.28		Trans-Fly	
Baniara	-9.77	149.88		SE Peninsula	Hamlin, H./1929
Banz	-5.81	144.62	1,680	Eastern Ranges	
Baoen	-0.73	132.28	457	Bird's Head	Ripley, S.D./1938; (Mayr & Meyer de Schauensee 1939b)
Bapi River	-3.65	141.35		Sepik-Ramu	
Bariji River	-9.10	148.62		SE Peninsula	
Barkai Island	-6.68	134.67		Aru Islands	
Baroe	-1.83	132.25		Bird's Head	
Baroka	-8.72	146.63		SE Peninsula	near Hall Sound
Bartle Bay	-10.10	150.13		SE Peninsula	
Basilaki/Basilisk* Island	-10.62	151.00		SE Islands	Meek, A.S./1897
Basilisk Point	-10.25	150.67		SE Peninsula	
Batanta Island	-0.95	130.60		NW Islands	(Greenway 1966)
Batavia Bivak	-2.73	138.42		NW Lowlands	Mamberamo River; van Heurn/1920
Baubauguina	-10.10	148.73		SE Peninsula	Zimmer, J.T./1919
Bavo Island	-9.43	146.93		SE Peninsula	off Port Moresby
Bawe	2.98	134.72		Bird's Neck	
Bayern Bay	-7.07	147.07		SE Peninsula	
Beaufort River	-4.60	138.70		S Lowlands	Versteeg, G./1912
Bedidi River	-2.78	132.77		Bird's Neck	
Beehive Camp	-4.55	138.67	1,750	Western Ranges	Bijenkorf; Lorentz, H.
Behrmann Hills	-5.15	141.53	2,000–3,127	Border Ranges	Bahrmann Mountains*
Beipa	-8.53	146.55		SE Peninsula	Veipa
Belavista*/Bela Vista*/Bella Vista	-8.53	147.07		SE Peninsula	
Bele River Camp	-4.07	138.73		Western Ranges	on what is now the Ibele River; (Rand 1942b)
Bele River*	-3.98	138.87		Western Ranges	Ibele River; (Rand 1942b)

NAME	LATITUDE	LONGITUDE	ELEV. (M)	BIRD REGION	ALTERNATIVE NAMES, RESEARCHERS, REFERENCES
Belesana	-10.57	150.67		SE Islands	
Bena Bena River	-6.20	145.40		Eastern Ranges	
Bensbach River	-9.15	141.02		Trans-Fly	
Bensbach Wildlife Lodge	-8.87	141.25		Trans-Fly	
Benstead bluff	-5.03	141.13		Border Ranges	
Bentley Bay	-10.22	150.63		SE Peninsula	
Bentley Island	-10.72	151.23		SE Islands	
Beo	-0.07	130.70		NW Islands	
Beoga	-3.82	137.42	1,710	Western Ranges	
Bepondi Island	-0.42	135.27		Bay Islands	
Berau Peninsula*	-1.60	133.00		Bird's Head	Bird's Head/Doberai/Vogelkop* Peninsula
Bereina	-8.63	146.50		SE Peninsula	(Heron 1976)
Berlinhafen*	-3.13	142.48		Sepik-Ramu	Aitape
Bernhard Camp	-3.48	139.22	50	NW Lowlands	on the banks of the Taritatu/Idenburg* River; (Rand 1942b)
Beroro Pass	-5.05	141.12		Border Ranges	
Bertrand Island	-3.22	143.27		Sepik-Ramu	
Besir	-0.47	130.68		NW Islands	
Betaf	-2.12	139.27		NW Lowlands	
Bewani Mountains	-3.20	141.50		Sepik-Ramu	(Diamond 1969)
Bewani Patrol Post	-3.02	141.17		Sepik-Ramu	
Bewani River	-2.83	141.40		Sepik-Ramu	
Biagi*/Bihagi*/Biagge*	-8.07	148.02		SE Peninsula	near today's Kanga village; Meek, A.S./1906
Biak Island	-1.00	136.00		Bay Islands	Misori*/Misory*/Mysore* Island; sister island = Supiori
Biake No. 1 village	-4.08	141.10		Sepik-Ramu	
Biak-Utara Reserve	-0.42	135.83		Bay Islands	van Balen, S./2000
Bian River	-8.12	139.95		Trans-Fly	(Bishop 2005a)
Biaru River	-7.75	146.67		SE Peninsula	tributary of the Oreba River, in the Lakekamu basin
Bien River	-3.98	144.33		Sepik-Ramu	
Big Pig Island	-5.17	145.83		Sepik-Ramu	Diamond, J.M./1969
Big Wau Creek	-7.32	146.73		SE Peninsula	
Bigei River	-5.55	145.25		Sepik-Ramu	Peka River
Bijenkorf	-4.55	138.67	1,750	Western Ranges	Beehive Camp; Lorentz, H.
Bilbilokabip	-5.17	141.29		Border Ranges	Ok Tedi headwaters
Binaturi*/Benaturi* River	-9.15	142.95		Trans-Fly	Katau River
Binip River	-5.12	145.45		Sepik-Ramu	
Bintuni Bay	-2.35	133.70		Bird's Neck	MacCluer Gulf*/McCluer's Inlet*
Biolowat*/Bulowat	-7.05	146.59		SE Peninsula	Stevens, H./1932–33; (Greenway 1935)

NAME	LATITUDE	LONGITUDE	ELEV. (M)	BIRD REGION	ALTERNATIVE NAMES, RESEARCHERS, REFERENCES
Bioto	-8.73	146.62	<40	SE Peninsula	
Bioto Creek	-8.77	146.60		SE Peninsula	Zimmer, J.T./1920
Bioto Landing	-8.73	146.62	<40	SE Peninsula	
Bird's Head Peninsula	-1.50	133.00		Bird's Head	Berau*/Doberai/Vogelkop* Peninsula; name of bird region
Bird's Neck	-3.00	134.00		Bird's Neck	includes Onin and Bomberai Peninsulas; name of bird region
Biri	-2.82	137.83		NW Lowlands	
Biroe	-6.90	144.87		S Lowlands	
Bishop Museum field station	-7.37	146.67		SE Peninsula	transitioned to the Wau Ecology Institute in 1972
Bismarck Range	-5.80	145.00		Eastern Ranges	includes Mount Otto and Mount Wilhelm
Bitoi Creek	-7.25	146.92		SE Peninsula	
Bivak Eiland/Island Camp	-5.02	138.65		Western Ranges	Lorentz, H.
Bivak October	-1.43	133.99		Bird's Head	Arfak Mts; (Gyldenstolpe 1955b)
Black River	-5.73	141.83		S Lowlands	(Rand 1942a)
Black River Camp	-5.83	141.75	100	S Lowlands	(Rand 1942a)
Blucher Range	-5.68	141.97		Border Ranges	(Rand 1942a)
Blue Mountain	-4.42	141.28	1,955	Border Ranges	
Blup Blup Island	-3.50	144.60		Sepik-Ramu	Blupblup Island
Boana	-6.43	146.83		Huon Peninsula	southern slopes of Saruwaged Range
Bobairo/Bobare*/Bobiara*	-3.99	136.37		Western Ranges	(Junge 1953)
Bobo Island	-9.12	143.22		Trans-Fly	Bristow Island; near Daru Island
Boca de la Batalla	-10.62	150.75		SE Islands	
Boca de Tovar	-4.08	133.28		Bird's Neck	Nautilus Strait
Bodim	-2.28	138.83	90	NW Lowlands	(Ripley 1964a)
Boemi*/Bumi River	-3.38	135.42		NW Lowlands	
Bofu	-8.53	147.50		SE Peninsula	
Bog Camp	-2.58	138.59	1,650	NW Lowlands	Foja Mountains; (Beehler et al. 2012)
Bogadjim	-5.43	145.73		Huon Peninsula	(Gilliard & LeCroy 1967)
Bogia	-4.27	144.97		Sepik-Ramu	
Bogoya harbour*	-10.69	152.85		SE Islands	Bwagoaia harbour/Bugoiya* harbour
Boigu Island	-9.28	142.17		-	Torres Strait Islands, Australia
Boirave/Boilave*	-10.35	149.97		SE Peninsula	
Bokondini	-3.63	138.67	1,300	Western Ranges	
Boku	-9.75	148.02		SE Peninsula	
Bokukomana (East)	-9.75	148.02	400	SE Peninsula	
Bokukomana (West)	-9.68	147.87	160	SE Peninsula	
Bolobip	-5.37	141.65	1,520	Border Ranges	
Boma	-5.90	139.90		S Lowlands	

NAME	LATITUDE	LONGITUDE	ELEV. (M)	BIRD REGION	ALTERNATIVE NAMES, RESEARCHERS, REFERENCES
Bomai	-6.37	144.63	991	Eastern Ranges	(Diamond 1972)
Bomakia	-5.81	139.87		S Lowlands	
Bomana	-9.40	147.25		SE Peninsula	
Bomberai Peninsula	-3.20	133.20		Bird's Neck	(Diamond 1985)
Bomele	-4.83	139.90		S Lowlands	
Bon Kourangen (mountain)	-0.71	132.73		Bird's Head	Mount Kourange; (Mayr & Meyer de Schauensee 1939b)
Bona Bona Island	-10.50	149.83		SE Peninsula	
Bonahoi	-3.78	142.85		Sepik-Ramu	
Bondey	-	-		Bird's Head	Wandammen district; see p. 306: Mino anais orientalis (Schlegel 1871)
Bondobol	-9.00	141.49		Trans-Fly	northern edge of Bula plains; (Bishop 2005a)
Bonenau/Boneno*	-9.88	149.39		SE Peninsula	
Bongu	-5.48	145.80		Huon Peninsula	
Boni Island	-0.04	131.07		NW Islands	
Bonvouloir Islands	-10.25	151.87		SE Islands	
Booboonie*	-8.68	146.95	600	SE Peninsula	Bubuni/Baboonie*
Bootless Bay	-9.50	147.25		SE Peninsula	southeast of Port Moresby
Border Mountains	-3.77	141.17	1,080	Sepik-Ramu	
Border Ranges	-5.00	141.00		Border Ranges	bird region denoting the central ranges near the PNG-Indonesia border
Boroka	-8.82	146.60		SE Peninsula	near Hall Sound
Boroko	-9.47	147.20		SE Peninsula	a Port Moresby suburban neighborhood
Bosavi	-6.47	142.83	720	Eastern Ranges	on flank of Mount Bosavi
Bosim Point	-10.55	150.78		SE Islands	
Bosnek Island	-1.21	136.20		Bay Islands	
Boubouni*	-8.68	146.95	600	SE Peninsula	Babooni*/Booboonie*/Bubuni; Meek, A.S./1905
Bougainville Bay	-2.58	141.08		Sepik-Ramu	
Bougainville Mountains	-2.15	141.03		Sepik-Ramu	
Bougainville Strait	-0.12	130.22		NW Islands	
Bowutu Mountains	-7.72	147.17	1,000–2,756	SE Peninsula	
Bramble Haven atoll	-11.22	152.00		SE Islands	
Braunschweighafen	-7.48	147.30		SE Peninsula	Pajawa*/Paiewa; northwest of Morobe station
Bristow Island	-9.12	143.22		Trans-Fly	Bobo Island
Brown River	-9.20	147.23		SE Peninsula	various publications by H.L. Bell
Brown River Road	-9.02	146.85		SE Peninsula	various publications by H.L. Bell
Brown River Road 2.75-mile banding site	-9.18	147.20	<40	SE Peninsula	various publications by H.L. Bell
Brown River Road 3-mile banding site	-9.18	147.18	<40	SE Peninsula	various publications by H.L. Bell

NAME	LATITUDE	LONGITUDE	ELEV. (M)	BIRD REGION	ALTERNATIVE NAMES, RESEARCHERS, REFERENCES
Bubia	-6.72	146.88		SE Peninsula	on highway west of Lae; Alpha Helix New Guinea Expedition/1969
Bubu/Bubuu* River	-7.90	147.17		SE Peninsula	Shaw Mayer, F.
Bubui River	-6.58	147.82		Huon Peninsula	
Bubuni	-8.68	146.95	600	SE Peninsula	Babooni*/Booboonie*; Meek, A.S./1905
Buckel Mountains	-5.00	142.45		Border Ranges	
Budibudi Islands	-9.30	153.69		SE Islands	Budi Budi atoll; Laughlan or Lockland Is.
Bugi/Buji	-9.15	142.23		Trans-Fly	
Bugoiya harbour*	-10.69	152.85		SE Islands	Bogoya*/Bwagoaia harbour
Buiaimuba Point	-9.33	142.63		Trans-Fly	
Buisaval River	-7.27	146.98		SE Peninsula	near Salamaua
Bula plains	-9.13	141.33		Trans-Fly	Bulla plains*
Bulago River	-5.55	142.12		S Lowlands	
Bulla plains*	-9.13	141.33		Trans-Fly	Bula plains
Bulldog	-7.80	146.43		SE Peninsula	wartime road terminus of Bulldog Road on the Avi Avi River in the Lakekamu basin
Bulldog Road/Track*	-7.97	146.68		SE Peninsula	wartime truck road from Edie Creek to Lakekamu basin
Bulolo	-7.22	146.62	1,090	SE Peninsula	(Greenway 1935)
Bulolo Forestry College	-7.22	146.62		SE Peninsula	
Bulolo forestry station	-7.20	146.62		SE Peninsula	
Bulolo River	-7.25	146.67		SE Peninsula	
Bulolo-Watut Divide	-7.32	146.64	2,388	SE Peninsula	
Bulowat/Biolawat*	-7.12	146.62		SE Peninsula	confluence of the Bulolo and Watut Rivers; (Greenway 1935)
Bulowat/Biolowat*	-7.05	146.59		SE Peninsula	Stevens, H./1932–33; (Greenway 1935)
Bumi/Boemi* River	-3.38	135.42		NW Lowlands	
Buna/Buna Bay	-8.67	148.40		SE Peninsula	
Bundi	-5.75	145.23		Eastern Ranges	on northern flank of Mount Wilhelm
Buntibasa district	-6.45	146.12		Eastern Ranges	Shaw Mayer, F./1932
Bunum River	-6.25	146.48		Huon Peninsula	
Bunume River*	-2.92	141.13		Sepik-Ramu	Jabiri River
Burci River	-8.17	142.02		Trans-Fly	
Bürgers Mountain	-5.15	143.30	3,711	Eastern Ranges	Mount Bürgers/Yakopi Nalenk; (Stresemann 1923)
Burnett River	-5.78	142.18		S Lowlands	
Burnett River north arm	-5.75	142.22		S Lowlands	
Burrumtal*/Bulum valley	-6.51	147.48		Huon Peninsula	position approximate; (Mayr 1931)
Buso River	-6.73	147.20		Huon Peninsula	
Bussum	-6.50	147.83		Huon Peninsula	near Finschhafen

NAME	LATITUDE	LONGITUDE	ELEV. (M)	BIRD REGION	ALTERNATIVE NAMES, RESEARCHERS, REFERENCES
Busu River	-6.73	147.05		Huon Peninsula	
Busu River (alpine)	-6.37	147.18		Huon Peninsula	
But	-3.40	143.23		Sepik-Ramu	
Butaueng*/Butaweng	-6.60	147.82		Huon Peninsula	near Finschhafen
Button Islands	-9.62	149.62		SE Peninsula	
Buyawim River	-7.23	147.05		SE Peninsula	
Bwaga Bwaga Bay	-10.68	152.67		SE Islands	
Bwagabwaga	-10.68	152.67		SE Islands	
Bwagaoia	-10.68	152.83		SE Islands	Misima Island town
Bwagaoia harbour	-10.69	152.85		SE Islands	Bogoya harbour*/Bugoiya harbour*; main harbor of Misima Island
Bwaiiri Islands	-10.90	150.75		SE Islands	
Bwilei River	-7.13	146.87		SE Peninsula	
Bwin River	-2.87	141.13		Sepik-Ramu	
Cabo de San Lucas	-3.87	133.98		Bird's Neck	
Cabo Fresco	-10.53	150.80		SE Islands	
Calvados Islands	-11.13	152.75		SE Islands	
Cambaru	-8.42	137.62		Trans-Fly	False Cape
Cameron Range	-8.90	147.17		SE Peninsula	
Camp 6b, Utakwa River	-4.25	137.20	1,280	Western Ranges	Kloss, C.B.
Cannac Island	-9.28	153.50		SE Islands	near Budibudi Islands and Woodlark Island
Cape Awura	-4.03	134.32		Bird's Neck	
Cape Cretin	-6.67	147.84		Huon Peninsula	Becker, G./1944
Cape Cupola	-8.03	145.83		S Lowlands	
Cape Deliverance	-11.38	154.28		SE Islands	eastern point of Rossel Island
Cape Ducie	-10.22	150.60		SE Peninsula	
Cape D'Urville	-1.47	137.90		NW Lowlands	
Cape Ebora	-10.62	152.52		SE Islands	
Cape Franseski	-3.85	144.54		Sepik-Ramu	
Cape Frere	-10.08	150.17		SE Peninsula	
Cape Girgir	-3.85	144.54		Sepik-Ramu	
Cape Helen	-2.67	136.07		NW Lowlands	
Cape Kamdara	-2.32	140.12		NW Lowlands	Cape Korongwaah
Cape Killerton	-8.62	148.33		SE Peninsula	
Cape Korongwaah	-2.32	140.12		NW Lowlands	Cape Kamdara
Cape Manggua	-2.88	134.85		Bird's Neck	
Cape Maniburu/ Maniboeroe	-3.23	134.95		Bird's Neck	
Cape Manini	-1.75	137.25		NW Lowlands	
Cape Mansurbado	-1.72	137.32		NW Lowlands	
Cape Nambima	-4.03	134.52		Bird's Neck	

NAME	LATITUDE	LONGITUDE	ELEV. (M)	BIRD REGION	ALTERNATIVE NAMES, RESEARCHERS, REFERENCES
Cape Nelson	-9.00	149.25		SE Peninsula	
Cape of Good Hope	-0.33	132.42		Bird's Head	
Cape of St. Paul	-8.42	137.62		Trans-Fly	False Cape
Cape Oransbari	-1.35	134.28		Bird's Head	
Cape Ranbausawa	-2.20	136.52		NW Lowlands	
Cape Rigny	-5.47	145.98		Huon Peninsula	
Cape Rodney	-10.20	148.40		SE Peninsula	
Cape Rouwianui	-3.05	135.75		NW Lowlands	
Cape Sele	-1.43	130.93		Bird's Head	Tanjung Sele
Cape Sjeri	-1.65	134.13		Bird's Head	
Cape Steenboom	-4.92	136.83		S Lowlands	
Cape Suckling	-9.02	146.63		SE Peninsula	
Cape Tanahmerah	-2.40	140.33		NW Lowlands	Point Tanahmerah
Cape Tanggiri	-3.90	133.40		Bird's Neck	
Cape Vals/Cape Valsche	-8.42	137.62		Trans-Fly	False Cape
Cape Vogel	-9.70	150.00		SE Peninsula	
Cape Ward Hunt	-8.07	148.13		SE Peninsula	
Cape Winsop	-1.97	131.97		Bird's Head	
Cape Wituwai	-2.28	139.63		NW Lowlands	
Carrington River	-5.90	142.23		S Lowlands	upper Strickland basin
Carstensz Massif	-4.08	137.18		Western Ranges	Mount Carstensz/Puncak Jaya/Carstensz Toppen*
Carstensz meadow	-3.92	136.98		Western Ranges	position approximate
Casuarina River	-4.88	141.35		Border Ranges	
Catherine Island*	-10.23	150.60		SE Peninsula	Netuli Island
Cempedak Camp	-0.09	130.74	345	NW Islands	Mauro, I./2002
Cenderawasih Bay	-2.00	135.00		Bay Islands	Geelvink Bay/Teluk Cenderawasih
Chabrol Bay*	-0.35	130.93		NW Islands	Mayalabit/Majalabit* Bay; Guillemard/1883
Chad's Bay*	-10.22	150.52		SE Peninsula	Awaiama Bay; Meek, A.S./1899, 1901
Challis Head	-10.53	150.80		SE Islands	
Chambri Lake/Lakes	-4.30	143.13		Sepik-Ramu	usage varies from singular to plural; refers to a lake complex
Chapman Range	-7.92	146.75	3,588	SE Peninsula	
Charles Louis Mountains	-4.13	135.52	3,390	Western Ranges	
Chimbu Province	-6.00	144.90		Eastern Ranges	
China Strait	-10.62	150.67		SE Islands	
Chirima River	-8.65	147.65		SE Peninsula	
Chisholm River	-7.88	146.52		SE Peninsula	
Choqeri*/Choqueri*	-9.42	147.50		SE Peninsula	Sogeri; Goldie, A.
Chuave	-6.12	145.13		Eastern Ranges	

NAME	LATITUDE	LONGITUDE	ELEV. (M)	BIRD REGION	ALTERNATIVE NAMES, RESEARCHERS, REFERENCES
Clearwater Creek	-7.45	146.67		SE Peninsula	Alpha Helix New Guinea Expedition/1969
Cloudy Bay	-10.17	148.67		SE Peninsula	
Collingwood Bay	-9.43	149.20		SE Peninsula	
Conflict Islands	-10.77	151.82		SE Islands	
Coral Haven	-11.33	153.33		SE Islands	
Coutance Island	-10.22	148.10		SE Peninsula	Courtance Island*/Walaivele Island
Crater Lake	-5.68	143.65		Eastern Ranges	
Crater Mountain	-6.55	145.08	3,118	Eastern Ranges	Ebota Mountain
Crater Mountain research station	-6.72	145.09	900	Eastern Ranges	
Crater Mountain Wildlife Management Area	-6.73	145.10		Eastern Ranges	
Cromwell Mountains	-6.33	147.25		Huon Peninsula	
Crummer Peaks	-6.80	144.42	1,430	Eastern Ranges	
Crystal Rapids	-9.43	147.43	440	SE Peninsula	on Sogeri Plateau
Crystal River and gorge	-5.13	142.17		Border Ranges	
Cumbak Ridge	-7.40	146.70		SE Peninsula	Mount Kumbak; Alpha Helix New Guinea Expedition/1969
Cyclops Mountains	-2.50	140.60	0–2,160	NW Lowlands	
Czinyagi	-5.10	145.78		Sepik-Ramu	
Dablin Creek	-5.20	141.23		Border Ranges	near Tabubil
Dagie	-5.82	144.77		Eastern Ranges	in the Wahgi valley; (Gyldenstolpe 1955b)
Dagua	-3.42	143.32		Sepik-Ramu	
Dagwa*	-8.87	143.07		Trans-Fly	Dogwa; Archbold, R. and Rand, A.L./1934
D'Albertis Junction	-6.17	141.12		S Lowlands	
D'Albertis Point	-8.83	146.55		SE Peninsula	
Dallmannshafen*	-3.55	143.63		Sepik-Ramu	Wewak
Dama Damami River	-6.17	142.28		S Lowlands	
Damon Entrance	-3.90	144.55		Sepik-Ramu	
Dampier Island*	-4.65	145.97		Sepik-Ramu	Karkar Island; Meek, A.S./1914
Danau Bira	-2.47	138.02	427	NW Lowlands	Lake Holmes*; Diamond, J.M.
Danau Kurumoi*	-2.17	134.08		Bird's Neck	Lake Kurumoi
Danau Paniai	-3.89	136.32	1,735	Western Ranges	Paniai Lake; (Junge 1953)
Dap Range	-5.60	141.65	3,040	Border Ranges	
Darai Camp	-7.13	143.61		S Lowlands	
Darai Hills/Plateau	-7.13	143.59	300–800	S Lowlands	
Dark End Camp	-7.07	144.31		S Lowlands	Mack, A./2002
Darnley Island	-9.58	143.77	181	Trans-Fly	
Darowin	-8.15	140.47		Trans-Fly	
Daru	-9.07	143.20		Trans-Fly	town on Daru Island

NAME	LATITUDE	LONGITUDE	ELEV. (M)	BIRD REGION	ALTERNATIVE NAMES, RESEARCHERS, REFERENCES
Daru Island	-9.07	143.20		Trans-Fly	Yarru Island*; Archbold, R. and Rand, A.L./1934
Daru Point	-10.08	150.08		SE Peninsula	Archbold, R., Rand, A.L., and Tate, W./1936
Dauan Island	-9.43	142.53	296	Trans-Fly	
Daugo Island	-9.51	147.05		SE Peninsula	
Daumori Island	-8.30	142.87		Trans-Fly	
Dawari Island	-8.70	143.38		Trans-Fly	
Dawong	-6.87	146.77		SE Peninsula	Mayr, E./1929
Deboyne Island	-10.72	152.37		SE Islands	
Deboyne lagoon	-10.85	152.38		SE Islands	
Deception Bay	-7.75	144.67		S Lowlands	
Deikimdikin	-5.09	141.63		Border Ranges	in the Victor Emanuel Mountains; (Gilliard & LeCroy 1961)
Dejaoe River	-2.57	140.47		NW Lowlands	Smith, H./1945
Delena/Delana*	-8.86	146.57		SE Peninsula	Hamlin, H./1929
Demta	-2.35	140.15		NW Lowlands	
Dendawang Camp	-6.08	146.57		Huon Peninsula	Benz, B./2001
D'Entrecasteaux Archipelago	-9.50	150.60		SE Islands	
Depapre/Deprepare*	-2.47	140.37		NW Lowlands	"Deprepare" is apparently a misspelling
Deva Deva	-8.75	146.63	1,000	SE Peninsula	
Devil's Race	-5.63	142.17		S Lowlands	Hamlin, H./1929, Crandall, L./1928
Diap River	-5.40	142.10		S Lowlands	
Dibiri Island	-8.20	143.73		Trans-Fly	
Digimu River	-6.67	143.50		S Lowlands	
Digoel River*	-7.20	139.60		Trans-Fly	Digul/Uwimbu* River
Digul Mountains	-5.25	140.45		Border Ranges	
Digul River	-7.20	139.60		Trans-Fly	Uwimbu*/Digoel* River
Dilava	-8.54	146.97		SE Peninsula	
Dilava River	-9.02	146.82		SE Peninsula	
Dimisisi	-8.62	142.27		Trans-Fly	lower Fly River; Stronach, N.
Din River	-4.68	141.15		Sepik-Ramu	
Dinawa	-8.54	146.97		SE Peninsula	
Dinner Island*	-10.60	150.70		SE Islands	Samarai Island
Discovery Bay	-10.40	150.40		SE Peninsula	on the south side of Milne Bay proper
Ditschi	-1.50	134.00		Bird's Head	Mayr, E./1928
Djalan/Djalau River	-3.58	135.25		Bird's Neck	
Djokdjeroi	-0.38	131.78		Bird's Head	Ripley, S.D./1938
Dobbo*/Dobo	-5.77	134.22		Aru Islands	on Wamar Island of Aru Islands; Doherty, W./1897; Kuhn/1900; Cayley-Webster, H./1896
Doberai Peninsula	-1.60	133.00		Bird's Head	

NAME	LATITUDE	LONGITUDE	ELEV. (M)	BIRD REGION	ALTERNATIVE NAMES, RESEARCHERS, REFERENCES
Dobu Island	-9.75	150.83		SE Islands	between Fergusson and Normanby Islands; Hamlin, H./1928
Dobuwapa Bay	-10.40	150.40		SE Peninsula	
Dogura Bay	-10.00	149.85		SE Peninsula	
Dogura Inlet	-9.50	147.25		SE Peninsula	
Dogwa	-8.87	143.07		Trans-Fly	Dagwa*; on the Oriomo River
Dohunsehik	-1.47	133.83	1,400	Bird's Head	Mayr, E./1928
Doini Island	-10.70	150.72		SE Islands	Blanchard Island; Zimmer, J.T./1920
Dokfuma meadow	-5.17	141.10	3,080	Border Ranges	Star Mountains of PNG; Gregory, P.
Dolak/Dolok* Island	-7.90	138.60		Trans-Fly	Pulau Kimaam/Pulau Yos Sudarso/Frederik Hendrik Island
Dom Island	-0.95	131.23		NW Islands	Sorong Island
Doma Peaks	-5.90	143.13	3,568	Eastern Ranges	
Domara/Domara River/Domeri	-10.17	148.55		SE Peninsula	east of Cape Rodney
Dombo Island	-1.88	137.10		Bay Islands	
Domori Island	-8.30	142.87		Trans-Fly	
Donner Mountains	-5.05	141.58	2,458	Border Ranges	
Doormanpaad	-3.50	137.15	3,962	Border Ranges	Angemuk
Doormanpaad Bivak	-3.40	138.63	1,410– 1,450	Border Ranges	
Dore Bay*	-0.86	134.08		Bird's Head	Manokwari
Dorei Hum Bay	-0.77	131.53		Bird's Head	on the northwest coast of the Bird's Head
Dorei*	-0.86	134.08		Bird's Head	Manokwari; Doherty, W./1897; Kuhn/1900; Cayley-Webster, H./1896
Dorey harbour*	-0.86	134.08		Bird's Head	Manokwari
Doromena	-2.42	140.43		NW Lowlands	Jewett Jr, S.G./1945
Dosay	-	-		NW Lowlands	near Jayapura/Humboldt Bay?
Double Mountain Range	-4.67	142.70		Eastern Ranges	
Dourga River	-7.80	139.00		Trans-Fly	in Princess Marianne Strait
Dramai Island	-4.02	134.25		Bird's Neck	
Drei Zinnen Mountains	-4.73	141.20	2,621	Border Ranges	"three peaks" mountains
Dreikikir	-3.58	142.77		Sepik-Ramu	
Duchateau Islands	-11.28	152.38		SE Islands	
Duchess Island	-9.96	150.85		SE Islands	Olumwa Island; off w coast of Normanby Island. Pratt, T.K.; Dumbacher, J.P.
Dugan River	-5.10	141.65		Border Ranges	
Dugumenu Island	-8.80	151.92		SE Islands	in the Marshall Bennett Islands; Hamlin, H./1928
Dumae Creek	-9.80	149.00		SE Peninsula	near Agaun (Engilis & Cole 1997)
Dumoulin Islands	-10.90	150.75		SE Islands	
Dumpu	-5.90	145.73		Huon Peninsula	Berggy, J.

NAME	LATITUDE	LONGITUDE	ELEV. (M)	BIRD REGION	ALTERNATIVE NAMES, RESEARCHERS, REFERENCES
Duna Peaks	-5.85	142.53	2,830	Eastern Ranges	
Dunantina River	-6.25	145.43		Eastern Ranges	
Duperre Islets	-11.18	152.00		SE Islands	
Dupkla	-5.75	144.17		Eastern Ranges	Alpha Helix New Guinea Expedition/1969
Duru	-8.13	141.85		Border Ranges	
D'Urville Island	-3.35	143.55		Sepik-Ramu	
D'Urville Peak/Mount D'Urville	-9.48	147.38	1,048	SE Peninsula	
Dyke Ackland Bay	-9.00	148.75		SE Peninsula	
Eafa district	-9.20	147.71	1,520–1,840	SE Peninsula	Anthony, A.S./1895
East Cape	-10.28	150.72		SE Peninsula	
East Channel	-10.62	150.65		SE Islands	
East Island	-10.40	152.10		SE Islands	
Eastern Ranges	-6.40	145.60		Eastern Ranges	bird region encompassing the east-central sector of the Central Ranges
Ebota Mountain	-6.55	145.08		Eastern Ranges	Crater Mountain
Edere River	-4.22	135.92		Western Ranges	
Edie Creek village	-7.36	146.66		SE Peninsula	
Efman Island*	-0.93	131.12		NW Islands	Jefman Island
Efogi	-9.15	147.65		SE Peninsula	
Egum Atoll	-9.37	151.93		SE Islands	Meek, A.S./1895
Eiaus*/Eihaus	-10.70	152.77		SE Islands	
Eilanden River, eastern outflow	-5.85	138.23		S Lowlands	
Eilanden River, western outflow	-5.75	138.15		S Lowlands	Meek, A.S./1911
Eilogo plantation	-9.45	147.47		SE Peninsula	
Eitape*	-3.13	142.33		Sepik-Ramu	Aitape
Ekame Camp	-6.09	141.52		S Lowlands	Mack, A./2003
Ekeikei	-8.65	146.80	440	SE Peninsula	
Ekuti Divide	-7.43	146.58	3,278	SE Peninsula	
Elenagora	-7.78	146.05		S Lowlands	Pockley collected insects here/1922
Elevala/Elevara* River	-6.08	141.50		S Lowlands	
Elevala-Strickland Divide	-6.17	141.93	210	S Lowlands	
Elip River	-4.95	141.48		Border Ranges	
Eliptamin	-5.03	141.67	1,400	Border Ranges	
Ellangowan Island	-7.82	141.67		Trans-Fly	D'Albertis, L.
Eloa River	-7.72	146.50		SE Peninsula	Beehler, B.M.
Embukam	-	-		Trans-Fly	PNG Trans-Fly swamp; Bowe, M.
Emiapmin	-5.07	142.33		Border Ranges	

NAME	LATITUDE	LONGITUDE	ELEV. (M)	BIRD REGION	ALTERNATIVE NAMES, RESEARCHERS, REFERENCES
Empress Augusta River*	-4.30	143.30		Sepik-Ramu	Sepik River
Emuk Range	-5.47	141.58	2,300	Border Ranges	
Enarotali	-3.98	136.38		Western Ranges	Wilson, N./1962
Endrick River	-8.43	140.31		Trans-Fly	between the Kumbe and Merauke Rivers
Enga Province	-5.50	143.80		Eastern Ranges	
Engineer Islands	-10.58	151.33		SE Islands	
English Peaks	-8.75	147.47		SE Peninsula	near Mount Scratchley
Enu Island	-7.10	134.50		Aru Islands	Pulau Enu; near the Aru Islands
Eola	-8.55	146.98	960	SE Peninsula	
Eorila Creek	-9.30	147.49		SE Peninsula	
Epa	-8.72	146.72	40	SE Peninsula	Salvadori
Era Point	-9.48	147.14		SE Peninsula	
Eramboe*/Erembu	-8.01	140.96		Trans-Fly	(Mees 1964b)
Erave River	-6.65	144.55		Eastern Ranges	
Eriama Creek	-9.40	147.27		SE Peninsula	Bell, H.L.
Erima/Erima harbour	-5.42	145.73		Sepik-Ramu	in Astrolabe Bay
Erkwero	-1.57	132.35		Bird's Head	
Ero Creek	-8.57	147.27		SE Peninsula	
Ero Creek Camp	-8.55	147.28	1,690	SE Peninsula	below Murray Pass
Eroob Island	-9.58	143.77	181	Trans-Fly	
Esa'ala	-9.73	150.82		SE Islands	on Normanby Island
Etappenberg	-4.58	142.60		Sepik-Ramu	40 km southwest of Hunsteinspitze; Bürgers/1912
Ethel River	-8.82	146.60		SE Peninsula	Hall Sound; Zimmer, J.T./1920
Eti Gorua River	-7.75	146.72		SE Peninsula	
Etna Bay	-3.96	134.71		Bird's Neck	Cayley-Webster, H./1896
Evelyn River	-9.03	147.23		SE Peninsula	
Everill Junction	-7.58	141.38		Trans-Fly	Tambavwa Junction
Ewe Creek	-7.53	145.70		S Lowlands	
Ewe River	-6.72	141.38		S Lowlands	
Excellent Point	-10.20	150.55		SE Peninsula	
Fairfax harbour	-9.42	147.10		SE Peninsula	Port Moresby's harbor
Fakal	-1.88	129.91		NW Islands	village in western sector of Misool Island; (Mees 1965)
Fakfak Mountains	-3.00	132.50	150–1,620	Bird's Neck	
Fakfak/Fak-Fak*/Fak Fak*	-2.93	132.30		Bird's Neck	town on Onin Peninsula
Falls gorge	-5.73	142.15		S Lowlands	
False Cape	-8.42	137.62		Trans-Fly	
Fam Islands	-0.67	130.23		NW Islands	Islas de las Buenas Nuevas*

NAME	LATITUDE	LONGITUDE	ELEV. (M)	BIRD REGION	ALTERNATIVE NAMES, RESEARCHERS, REFERENCES
Fan	-1.33	132.35		Bird's Head	
Fane	-8.55	147.08	1,400	SE Peninsula	near Mafulu; Hamlin, H./1929
Faralulu district	-9.42	150.50		SE Peninsula	
Faringi/Dio River	-3.93	141.27		Sepik-Ramu	
Faur Island	-3.41	132.82		Bird's Neck	(Mayr 1941)
Favane	-8.47	147.13		SE Peninsula	
Fave	-8.55	147.08	1,400	SE Peninsula	
Feba	-1.48	137.90		NW Lowlands	
Fef	-0.80	132.42		Bird's Head	Ripley, S.D./1938
Feramin	-5.20	141.70		Border Ranges	
Fergusson Island	-9.60	150.75		SE Islands	
Fiamolu Mountains	-4.90	142.13	3,940	Border Ranges	
Fig Tree Point	-9.55	149.40		SE Peninsula	
Figi River	-4.97	141.92		Border Ranges	
Finisterre Range	-5.94	146.37	4,175	Huon Peninsula	
Finsch Coast	-3.00	142.00		Sepik-Ramu	
Finsch harbour	-6.55	147.85		Huon Peninsula	
Finschhafen	-6.52	147.48		Huon Peninsula	
Fio River*	-6.85	145.17		Eastern Ranges	Pio River
Fir Tree Creek	-9.55	149.40		SE Peninsula	Fig Tree Creek
Fir Tree Point	-9.55	149.40		SE Peninsula	Fig Tree Point
Fisherman's Island	-9.51	147.05		SE Peninsula	
Flamingo Bay	-5.55	138.00		S Lowlands	
Fly Islands	-7.45	147.33		SE Peninsula	
Fly River delta	-8.45	143.34		Trans-Fly	
Fly River, north entrance	-8.53	143.50		Trans-Fly	
Fly River, south entrance	-8.67	143.45		Trans-Fly	
Fofo Fofo	-8.60	146.90	840	SE Peninsula	
Foja Mountains	-2.58	138.75		NW Lowlands	Foya*/Gauttier* Mountains
Forbes Camp	-9.40	147.53		SE Peninsula	
Fore	-6.80	145.25		Eastern Ranges	
Forok	-3.62	143.75		Sepik-Ramu	
Foula	-8.57	147.08		SE Peninsula	
Foya Mountains*	-2.58	138.75		NW Lowlands	Foja/Gauttier* Mountains
Frak River	-4.72	141.53		Border Ranges	
Francisco River	-7.07	147.05		SE Peninsula	
Frederik Hendrik Island	-7.90	138.60		Trans-Fly	Dolak/Yos Sudarso/Kimaam Island
Freshwater Bay	-8.08	146.00		S Lowlands	
Frieda River	-4.32	142.00		Sepik-Ramu	

NAME	LATITUDE	LONGITUDE	ELEV. (M)	BIRD REGION	ALTERNATIVE NAMES, RESEARCHERS, REFERENCES
Fu River	-5.05	142.03		Border Ranges	
Gabagaba	-9.80	147.52		SE Peninsula	
Gabensis	-6.70	146.81		SE Peninsula	
Gabutau Point	-10.60	150.77		SE Islands	
Gadaisu	-10.37	149.78		SE Peninsula	
Gafutina River	-6.27	145.62		Eastern Ranges	
Gag/Gagi* Island	-0.44	129.88	349	NW Islands	
Gahom Camp	-4.62	142.75		Eastern Ranges	Mack, A./2003
Gaikarobi	-4.10	143.30		Sepik-Ramu	in the Sepik basin
Gaima Camp	-8.28	142.87		Trans-Fly	Archbold, R., Rand, A.L., Tate, W.W./1936
Galilu Point	-10.22	150.48		SE Peninsula	
Galley Reach	-9.12	146.88		SE Peninsula	
Galo Creek	-10.20	150.18		SE Peninsula	
Galuwala Creek	-9.28	149.33		SE Peninsula	
Galuwala village	-9.33	149.27		SE Peninsula	
Gam Island	-0.47	130.58		NW Islands	Gemien Island*
Gamoaoru Point	-10.10	148.18		SE Peninsula	
Gap, the	-9.13	147.70	2,000	SE Peninsula	
Garaina	-7.88	147.15		SE Peninsula	Alpha Helix New Guinea Expedition/1969
Gauttier Mountains*	-2.50	138.75		NW Lowlands	Foja Mountains
Gawa Island	-8.97	151.98		SE Islands	Marshall Bennett Islands; Brass & Petersons/1956
Gaza River	-	-		?	"northeastern New Guinea" (Rothschild 1925)
Gebe Island	-0.10	129.46		NW Islands	Waterstradt, J.
Gebroeders Mountains	-3.92	136.13		Western Ranges	within the Weyland Mountains; Shaw Mayer, F./1930
Gedik River	-3.62	141.50		Sepik-Ramu	
Geelvink Bay	-2.00	135.00		Bay Islands	Teluk Cenderawasih/Cenderawasih Bay
Geelvink East Point	-2.10	136.52		NW Lowlands	
Gemien Island*	-0.47	130.58		NW Islands	Gam Island
Gereka	-9.50	147.27		SE Peninsula	
Gerekanumu	-9.52	147.35		SE Peninsula	
Gerenda Camp	-8.43	147.38	3,800	SE Peninsula	
Gewoia	-9.20	148.48		SE Peninsula	
Gili Gili	-10.30	150.35		SE Peninsula	
Giluwe-Sugarloaf Mountains	-6.05	143.88		Eastern Ranges	
Gira River	-8.00	147.95		SE Peninsula	Aikora River
Giriwa River*	-8.67	148.38		SE Peninsula	Girua River
Girua airstrip	-8.80	148.32		SE Peninsula	
Girua River	-8.67	148.38		SE Peninsula	Giriwa River*

NAME	LATITUDE	LONGITUDE	ELEV. (M)	BIRD REGION	ALTERNATIVE NAMES, RESEARCHERS, REFERENCES
Goari hills*	-9.52	147.35	298	SE Peninsula	Guari hills
Goaribari Island	-7.78	144.23		S Lowlands	
Gobe	-6.83	143.42		S Lowlands	Kikori basin
Goedang Bivak	-3.40	138.63	1,410–1,450	Border Ranges	Mayr, E.
Goenong Tobi*	-0.83	133.25		Bird's Head	Gunung Tobi/Mount Tobi
Gogol River	-5.29	145.60		Sepik-Ramu	
Goladimitai	-6.60	145.17	1,600	Eastern Ranges	
Goldie River	-9.38	147.25		SE Peninsula	southeast branch of the Laloki River; Ramsay/1879
Gonema Oru	-10.08	148.17		SE Peninsula	Waipara/Ormond River
Gonema Oru River	-10.05	148.12		SE Peninsula	
Goodenough Bay	-9.90	150.40		SE Peninsula	
Goodenough Island	-9.35	150.25		SE Islands	Meek, A.S./1913
Goroka	-6.03	145.40		Eastern Ranges	
Goropu Mountains	-9.80	149.20	3,676	SE Peninsula	
Grand Valley	-4.00	138.90		Western Ranges	of the Baliem River
Grange Island	-10.32	148.90		SE Peninsula	Zimmer, J.T./1920
Gratzack Range	-4.88	141.10	1,000–2,785	Border Ranges	
Great Papuan Plateau	-6.62	142.83	200–1,000	Eastern Ranges	
Green River	-3.95	141.10		Sepik-Ramu	
Gu River	-5.70	141.43		Border Ranges	
Guabagusal	-7.25	146.98		SE Peninsula	
Guabe River	-5.50	145.90		Huon Peninsula	
Guam River	-4.55	144.62		Sepik-Ramu	Mete River
Guari	-8.03	146.87	1,800	SE Peninsula	
Guari hills	-9.52	147.35	298	SE Peninsula	Goari hills*
Guimu River	-8.07	147.38		SE Peninsula	
Gulewa	-10.63	152.73		SE Islands	
Gulf of Papua	-8.47	144.80		-	
Gulf Province	-7.30	144.88		S Lowlands	
Gum gorge	-5.45	141.52		Border Ranges	
Gumi Creek	-7.03	146.27		SE Peninsula	
Guna Isu	-10.70	150.20		SE Peninsula	Rugged Head
Gunung Bolema	-0.83	132.17		Bird's Head	Mount Bolema
Gunung Ngribou	-1.10	133.98	2,000	Bird's Head	Mount Ngribou
Gunung Yamin	-4.39	139.74		Border Ranges	Mount Goliath
Gurney airfield	-10.31	150.33	27	SE Peninsula	
Gusap Pass	-6.05	145.95		Eastern Ranges	
Gusap/Gusep*	-6.10	145.97		Eastern Ranges	Berggy, J.

NAME	LATITUDE	LONGITUDE	ELEV. (M)	BIRD REGION	ALTERNATIVE NAMES, RESEARCHERS, REFERENCES
Gusika	-6.56	147.84		Huon Peninsula	16 km north of Finschhafen
Gusyveti	-6.63	145.18	1,360	Eastern Ranges	
Guwasa	-6.47	145.10		Eastern Ranges	Diamond, J.M., Terborgh, J.W./1964
Guyebi	-5.73	145.23	2,000–2,134	Eastern Ranges	north slope Bismarck Range
Gwadi Creek	-9.67	147.27		SE Peninsula	
Gwenellif Creek	-3.99	141.87		Sepik-Ramu	
Habab Creek	-3.47	143.08		Sepik-Ramu	
Habbema Lake*	-4.17	138.67	3,225	Western Ranges	Lake Habbema
Habe Island	-	-		Trans-Fly	small island between Merauke and Dolak Island
Hagen Range	-5.77	144.00	3,791	Eastern Ranges	Mount Hagen
Hagen Road	-5.85	144.03		Eastern Ranges	Alpha Helix New Guinea Expedition/1969
Haia	-6.70	144.99	760	Eastern Ranges	Benz, B./2002
Haidana Island	-9.43	147.03		SE Peninsula	Collingwood Bay
Haitlal Island	-1.72	129.94		NW Islands	off north coast of Misool Island
Hak River	-4.92	141.47		Border Ranges	
Hakup River	-3.47	143.12		Sepik-Ramu	
Hall Bay	-8.77	146.54		SE Peninsula	D'Albertis/1875
Hall Sound	-8.83	146.60		SE Peninsula	Hamlin, H./1929
Hambitawuria	-2.88	132.33		Bird's Head	
Hamboa*	-5.58	145.75		Huon Peninsula	Hompua; Finisterre Range; position approximate
Hamo	-	-		SE Peninsula	near Aseki; Austral. Nat. Wildl. Collection
Hansemann Coast	-3.75	144.00		Sepik-Ramu	
Hapi Aifo River	-4.83	142.40		Border Ranges	
Hardy Point	-9.15	149.32		SE Peninsula	
Haro River	-4.75	145.37		Sepik-Ramu	
Harutami Range	-6.68	143.62		S Lowlands	
Hastings Island	-10.33	151.87		SE Islands	in the Bonvouloir Islands
Haszard Island	-10.59	151.37		SE Islands	in the Engineer Islands
Hatam	-1.09	133.97		Bird's Head	Atam*/Hattam*
Hathor gorge	-6.88	144.78		S Lowlands	Heifa gorge
Hattam*	-1.09	133.97		Bird's Head	Atam*/Hatam
Hatzfeldthafen	-4.40	145.20		Sepik-Ramu	
Hauser River	-3.97	141.03		Sepik-Ramu	
Haveri	-9.43	147.60		SE Peninsula	
Hawoi River	-7.30	143.50		S Lowlands	
Heath Island	-10.63	150.63		SE Islands	Rogeia Island*
Hegigio River	-6.83	143.65		S Lowlands	Aplin, E.K.
Heifa gorge	-6.88	144.78		S Lowlands	Hathor gorge
Heldsbach Coast	-6.50	147.83		Huon Peninsula	near Finschhafen

NAME	LATITUDE	LONGITUDE	ELEV. (M)	BIRD REGION	ALTERNATIVE NAMES, RESEARCHERS, REFERENCES
Hellwig Mountains	-4.53	138.68		Western Ranges	in Western Ranges
Henumai River	-4.77	141.83		Border Ranges	
Herbert River	-6.33	141.57		S Lowlands	
Heroana/Herowana*	-6.65	145.19	1,350	Eastern Ranges	
Herzog Mountains	-6.90	146.70	2,752	SE Peninsula	includes Mount Shungol and the Snake River Mountains
Hides Range	-6.47	143.18		Eastern Ranges	upper Kikori basin; position approximate
Higaturu oil palm plantation	-8.82	148.20	130	SE Peninsula	Mordue, T.
Hindenburg Mountains	-5.15	141.58	3,325	Border Ranges	
Hindenburg Wall	-5.18	141.33		Border Ranges	
Hiritano Highway	-9.02	146.85		SE Peninsula	
Hisiu	-9.03	146.72		SE Peninsula	Zimmer, J.T./1920
Hoiyevia/Hoiyeria*	-5.77	144.03		Eastern Ranges	Mount Hagen district; Blood, N.B./1945
Hol*/Holl*	-2.63	140.78		NW Lowlands	Holtekong; Mayr, E./1928
Hollandia*	-2.57	140.73		NW Lowlands	Jayapura/Sukarnapura*; Mayr, E./1931; Rand 1942b
Holnicote Bay	-8.58	148.33		SE Peninsula	
Holtekong/Holekang*	-2.63	140.78		NW Lowlands	Holl/*Hol*; coast of Humboldt Bay
Holuwon	-4.48	139.27	1,050	Border Ranges	Beehler, B.M./1980
Hombrom bluff	-9.38	147.33		SE Peninsula	
Hompua	5.58	145.75	914	Huon Peninsula	Hamboa*/Huambon*; Finisterre Range; position approximate
Hood Bay	-10.08	147.78		SE Peninsula	
Hood lagoon	-10.10	147.85		SE Peninsula	
Hood Point	-10.10	147.72		SE Peninsula	
Hori Hori	-1.62	132.30		Bird's Head	
Horseshoe Mountain*	9.32	147.53	1,680	SE Peninsula	Mount Maguli/Hufeisengebirge*
Hosken Island	-7.63	147.53		SE Peninsula	Meek, A.S./1913
Hsiduna area	-8.58	147.02	1,400	SE Peninsula	Bubuni
Huambon*	-6.35	147.72	914	Huon Peninsula	Hompua; Beck, R.H./1929
Hubrecht Mountains	-4.40	138.68	3,200	Western Ranges	
Hufeisengebirge*	-9.32	147.53		SE Peninsula	Mount Maguli/Horseshoe Mountain
Hula	-10.10	147.72		SE Peninsula	
Hum Island	-0.74	131.58		Bird's Head	off northwest coast of Bird's Head
Humboldt Bay	-2.63	140.78		NW Lowlands	
Hummock Island	-10.61	151.38		SE Islands	in the Engineer Islands
Hunstein Mountains	-4.51	142.67	1,520	Sepik-Ramu	
Hunstein Mountains	-4.52	142.67	1,520	Sepik-Ramu	
Hunstein River	-4.27	142.83		Sepik-Ramu	
Hunsteinspitze (Hunstein Peak)	-4.20	142.82		Sepik-Ramu	Mount Townsend; Bürgers/1912–13

NAME	LATITUDE	LONGITUDE	ELEV. (M)	BIRD REGION	ALTERNATIVE NAMES, RESEARCHERS, REFERENCES
Huon Gulf	-7.00	147.30		-	embayment south of the Huon Peninsula
Huon Peninsula	-6.30	147.00		Huon Peninsula	name of bird region
Huxley Peak	-8.90	147.87	3,800	SE Peninsula	of Mount Victoria
Hydrographer Mountains	-8.92	148.33	1,800	SE Peninsula	Eichhorn brothers/1918
Iahiluma Creek	-8.55	146.88		SE Peninsula	
Ianaba Island	-9.29	151.95		SE Islands	Yanaba Island; near Alcester Island
Iarago River	-8.92	146.98		SE Peninsula	
Iauarapi	-6.68	145.21		Eastern Ranges	Levitis, D.
Iave River*	-7.67	145.65		S Lowlands	Lohiki River
Ibarfornbek	-5.22	141.92	2,280	Border Ranges	
Ibele River	-3.75	139.15		Western Ranges	Bele River*
Ibibubari Island	-7.80	144.50		S Lowlands	
Idam River	-4.00	141.23		Sepik-Ramu	
Idenburg River*	-3.25	139.00		NW Lowlands	Taritatu River
Idenburg Top*	-4.03	137.05		Western Ranges	Ngga Pilimsit
Idihi Island	-9.43	146.88		SE Peninsula	
Idlers Bay	-9.47	147.09		SE Peninsula	
Idumava	-9.43	147.10		SE Peninsula	
Ifaar*/Ifar	-2.58	140.55		NW Lowlands	Mayr, E./1928
Ijapo Mountains	-2.73	141.03		Sepik-Ramu	
Ikeikei	-8.65	146.80		SE Peninsula	
Ikivari River	-8.93	146.87		SE Peninsula	
Ikwap Rier	-6.22	146.48		Huon Peninsula	
Ilaga	-4.00	137.63	2,320	Western Ranges	Ripley (1964)
Ilaga River	-4.00	137.62		Western Ranges	
Ilebaguma/Ilebeguma*	-4.93	145.44		Sepik-Ramu	
Ilimo/Ilimo Farm*	-9.40	147.12		SE Peninsula	suburb of Port Moresby
Ilkivip (Camp 4)	-5.18	141.62	2,195	Border Ranges	Gilliard, E.T./1954
Illolo	-9.44	147.36		SE Peninsula	
Imbai	-1.38	133.86		Bird's Head	Anggi Giji Lake
Imita Range/Imita Ridge	-9.33	147.55		SE Peninsula	
Imonda airfield	-3.33	141.15	292	Sepik-Ramu	
Inania	-8.88	146.95	280	SE Peninsula	
Inanwatan	-2.13	132.15		Bird's Head	
Inawa	-8.88	146.95		SE Peninsula	
Inawabui or Inauabui*	-8.70	146.63	< 40	SE Peninsula	Zimmer, J.T./1920
Inawaia	-8.63	146.57	< 40	SE Peninsula	
Inawi/Innawi*	-8.57	146.55	< 40	SE Peninsula	
Inim Range	-5.23	141.65	3,200	Border Ranges	

NAME	LATITUDE	LONGITUDE	ELEV. (M)	BIRD REGION	ALTERNATIVE NAMES, RESEARCHERS, REFERENCES
Iniok	-4.33	141.96		Sepik-Ramu	(I. Woxvold in litt.)
Iola	-8.55	146.98	960	SE Peninsula	Hamlin, H./1929
Ip River	-4.65	141.08		Sepik-Ramu	
Ipisi River	-9.12	146.90		SE Peninsula	
Irai Island	-10.77	151.69		SE Islands	Conflict Islands
Iris Strait	-4.03	134.28		Bird's Neck	
Irit River	-4.85	141.30		Border Ranges	
Irou River	-6.77	144.11	800–2,040	Eastern Ranges	
Irulu	-8.58	147.28		SE Peninsula	
Isatetu Creek	-7.20	143.82		S Lowlands	
Isla de los Ostiones	-4.07	134.62		Bird's Neck	
Isla de Santa Clara	-10.50	149.83		SE Peninsula	
Isla la Nabaja*	-4.20	133.42		Bird's Neck	Pulau Adi
Isla San Bartolome	-10.38	149.35		SE Peninsula	
Island River Camp	-4.83	139.90		S Lowlands	Eilanden River
Island River, eastern outflow	-5.85	138.23		S Lowlands	Eilanden River
Island River, western outflow	-5.75	138.15		S Lowlands	Eilanden River
Islas de las Buenas Nueva*	-0.67	130.23		NW Islands	Fam Islands
Isuno River	-6.97	146.35		SE Peninsula	
Isurava	-8.94	147.73		SE Peninsula	A village along Kokoda Track; the Isurava monument is a few kilometers to the southeast
Itamarina Island	-10.76	151.77		SE Islands	(Dumbacher et al. 2010)
Itoda	-4.04	136.09		Western Ranges	Wilson, N./1962
Ivani/Ivane* River	-8.45	147.95		SE Peninsula	
Ivimka Camp	-7.73	146.47	120	SE Peninsula	of the Lakekamu basin; Mack, A./2001; Beehler, B.M.; Atwood, F.
Iviri River	-7.72	147.02		SE Peninsula	
Ivori River	-7.43	145.48		S Lowlands	
Ivulu	-8.58	147.28		SE Peninsula	
Iwaka River	-4.45	136.77		S Lowlands	branch of the Kamura River; east of the Mimika River
Iwarum River	-4.53	144.98		Sepik-Ramu	
Jabiri River	-2.92	141.13		Sepik-Ramu	Bunume River*
Jabogema	-4.15	136.65		Western Ranges	
Jackson's field/airport	-9.44	147.22		SE Peninsula	the airport for Port Moresby
Jagei River*	-5.17	144.83		Sepik-Ramu	Yagei*/Ramu River; Tappenbeck/1896
Jais Aben	-5.13	146.80		Sepik-Ramu	coastal resort north of Madang
Jalan Korea	-2.62	140.27		NW Lowlands	

NAME	LATITUDE	LONGITUDE	ELEV. (M)	BIRD REGION	ALTERNATIVE NAMES, RESEARCHERS, REFERENCES
Jamna*/Yamna Island	-2.00	139.25		NW Lowlands	
Japen Island*	-1.80	136.30		Bay Islands	Yapen/Jobi* Island
Jar Island	-0.73	130.13		NW Islands	
Jassi River	-2.87	141.13		Sepik-Ramu	Nwin River
Jaur district*	-3.08	134.82		Bird's Neck	Yaur district
Jaur*	-3.00	134.75		Bird's Neck	Yaur
Javiwaitaiz River	-8.12	146.80		SE Peninsula	
Jayapura	-2.57	140.73		NW Lowlands	Hollandia*/Sukarnapura*
Jayawijaya Mountains	-4.50	139.50	4,700	Border Ranges	Oranje Mountains*
Jazira Doberai	-1.60	133.00		Bird's Head	Berau*/Doberai/Vogelkop* Peninsula
Jef Fa	-1.45	130.47		NW Islands	Ripley, S.D./1937
Jef Wa	-1.45	130.47		NW Islands	Ripley, S.D./1937
Jeflio	-1.13	131.23		NW Islands	west coast Bird's Head; (Gyldenstolpe 1955b)
Jefman Island	-0.93	131.12		NW Islands	Efman Island*
Jei River	-5.72	144.88		Eastern Ranges	
Jeimon/Jemon	-0.18	130.82		NW Islands	Kakiaij, J./1938
Jenkins Bay	-10.63	150.83		SE Islands	
Jimi/Jimmi* River	-5.40	144.25		Eastern Ranges	
Joanet/Joannet Island	-11.25	153.17		SE Islands	Pana Tinani Island
Jobi Island*	-1.80	136.30		Bay Islands	Japen*/Yapen Island; Meyer, A.B./1875
Jonanggee Mountain	-7.58	147.00		SE Peninsula	
Jop Island	-2.53	134.38		Bay Islands	
Josephstaal	-4.75	143.02	77	Sepik-Ramu	
June River	-6.68	141.23		S Lowlands	
Junzaing	-6.42	147.62		Huon Peninsula	Mayr, E./1929
Kabadi Plains	-9.03	146.88		SE Peninsula	Weiske, E./1899
Kaban Range	-5.55	141.68	1,200	Border Ranges	
Kabarei	-1.07	130.97		NW Islands	
Kabiufa mission	-6.08	145.37		Eastern Ranges	
Kabu River	-7.52	146.17		SE Peninsula	Kapu River
Kabui Bay	-0.45	130.70		NW Islands	
Kabuna*	-8.68	146.75		SE Peninsula	Kubuna; Hamlin, H./1929
Kabwum	-6.15	147.18		Huon Peninsula	
Kafiau*/Kavijaaw* Island	-1.18	129.83		NW Islands	Kofiau/Koffiao*/Poppa* Island
Kafu	-3.40	143.23		Sepik-Ramu	east of Aitape
Kagamuga	-5.84	144.30	1,600	Eastern Ranges	site of Mount Hagen airport
Kagi	-9.13	147.67	1,440	SE Peninsula	on Kokoda Track
Kaiaba	-10.66	152.80		SE Islands	
Kaiapit	-6.27	146.27		SE Peninsula	upper Markham valley
Kaiari district	-8.55	147.87	2,743	SE Peninsula	Anthony, A.S./1896

NAME	LATITUDE	LONGITUDE	ELEV. (M)	BIRD REGION	ALTERNATIVE NAMES, RESEARCHERS, REFERENCES
Kaibola	-8.40	151.07		SE Islands	
Kai-Ingri Camp	-5.57	143.03	3,315	Eastern Ranges	Kaijende Highlands; (Beehler et al. 2011)
Kaijende Highlands	-5.60	143.09	>2,400	Eastern Ranges	(Beehler et al. 2011)
Kaileuna Island	-8.53	150.95		SE Islands	Trobriand Islands; Hamlin, H./1928
Kailope River	-8.50	147.03		SE Peninsula	
Kaim River	-6.87	141.58		Trans-Fly	
Kaimana	-3.61	133.71		Bird's Neck	
Kaimari Island	-7.73	144.83		S Lowlands	
Kainantu	-6.30	145.87		Eastern Ranges	Rhodin, A./1977
Kaindan	-5.43	143.91	1,700	Eastern Ranges	Whiteside
Kaintiba	-7.50	146.00		SE Peninsula	
Kairiru Island	-3.35	143.55		Sepik-Ramu	
Kaironk Bush Camp	-5.22	144.50		Eastern Ranges	Benz, B./2007
Kaironk River valley	-5.25	144.47		Eastern Ranges	
Kairuku	-8.83	146.53		SE Peninsula	
Kais River	-2.02	132.05		Bird's Head	
Kaisa	-7.90	140.46		Trans-Fly	Kumbe River; (Mees 1982b)
Kaisenik	-7.37	146.73		SE Peninsula	in the upper Wau valley
Kaiser Friedrich-Wilhelmshafen*	-5.22	145.80		Sepik-Ramu	Madang
Kaiserin Augusta Fluss*	-3.83	144.53		Sepik-Ramu	Sepik River
Kaivaga	-9.47	147.15		SE Peninsula	
Kajan Mountains	-4.32	138.75		Western Ranges	Jayawijaya Mountains; Lorentz, H./1909–10
Kajumerah Bay	-3.97	134.38		Bird's Neck	San Juan del Prado*
Kakalo	-6.03	147.20	680	Huon Peninsula	
Kakoro	-7.84	146.53		SE Peninsula	in the Lakekamu basin
Kakoro airstrip	-7.83	146.53		SE Peninsula	in the Lakekamu basin
Kalamuk River	-1.48	131.58		Bird's Head	Kelamoek*/Karabra River
Kalawos	-0.92	132.18		Bird's Head	
Kali Sading (River)	-1.05	131.95		Bird's Head	Kelasading River*; Ripley, S.D./1938
Kalo Bay	-10.25	150.36		SE Peninsula	
Kam Wa	-1.45	130.47		NW Islands	Ripley, S.D./1937
Kamabun	-9.33	149.17		SE Peninsula	
Kamalio Baret	-4.23	143.28		Sepik-Ramu	
Kamarau Bay	-3.53	133.62		Bird's Neck	
Kambala	-3.85	133.45		Bird's Neck	on the Bomberai Peninsula
Kambrindo/Kambrindu	-4.13	143.83		Sepik-Ramu	
Kamfo village	-7.42	144.22		S Lowlands	near Kikori
Kamiali Bush Camp	-7.29	147.10		SE Peninsula	Benz, B./2004; Allison, A.
Kamindimbit	-4.27	143.33		Sepik-Ramu	

NAME	LATITUDE	LONGITUDE	ELEV. (M)	BIRD REGION	ALTERNATIVE NAMES, RESEARCHERS, REFERENCES
Kamoa Island*	-1.42	130.42		NW Islands	Kamual*/Kamuai Island; in the Schildpad Islands
Kamora*	-4.82	136.65		S Lowlands	Kamura River
Kampong [Kampung] Baru	-1.87	136.58		NW Lowlands	translates as "new village"; Stein, G./1931
Kampong Bahru	-1.83	132.25		Bird's Head	old spelling; translates as "new village"
Kampong Sururai	-1.40	133.92		Bird's Head	Thompson, M./1963
Kampong Wanggar	-3.40	135.33		NW Lowlands	Stein, G./1931
Kamtapia Range	-7.08	146.26	1,864	SE Peninsula	
Kamuai Island	-1.42	130.42		NW Islands	Kamual*/Kamoa* Island; in the Schildpad Islands
Kamual Island*	-1.42	130.42		NW Islands	Kamuai/Kamoa* Island; in the Schildpad Islands
Kana Kopi Point	-10.48	150.65		SE Islands	
Kandanam No. 2	-4.23	143.77		Sepik-Ramu	river village on stilts
Kandep	-5.85	143.52		Eastern Ranges	
Kandep Lakes	-5.88	143.60		Eastern Ranges	
Kanduwanam	-4.23	143.60		Sepik-Ramu	
Kanga	-8.79	147.65		SE Peninsula	village of the Biagge people
Kanganaman	-4.18	143.27		Sepik-Ramu	
Kanosia	-8.98	146.97		SE Peninsula	Zimmer, J.T./1920
Kanosia lagoon	-8.98	146.93		SE Peninsula	
Kanosia wharf	-8.99	146.95		SE Peninsula	
Kant River	-5.43	144.63		Eastern Ranges	
Kanudi	-9.43	147.13		SE Peninsula	Filewood, L./1973
Kanwa	-1.45	130.47		NW Islands	Ripley, S.D./1937
Kao River	-6.55	140.35		Trans-Fly	
Kapa Kapa	-9.80	147.52		SE Peninsula	
Kapaor*/Kapaur	-2.93	132.30		Bird's Neck	
Kapare River	-4.67	136.45		S Lowlands	western tributary of the Mimika River
Kapau River	-7.82	146.13		S Lowlands	
Kapaur	-2.93	132.18		Bird's Neck	on the Onin Peninsula; Doherty, W./1896
Kape Point	-10.23	150.83		SE Islands	East Cape
Kapu River	-7.52	146.17		SE Peninsula	Kabu River
Kara	-4.32	142.50		Border Ranges	
Karabra River	-1.48	131.58		Bird's Head	Kelamoek*/Kalamuk/Kladuk River
Karama River	-8.05	145.93		S Lowlands	
Karan	-3.30	142.50		Sepik-Ramu	
Kararau	-4.20	143.35		Sepik-Ramu	
Karau	-3.78	144.32		Sepik-Ramu	between Aitape and Sepik mouth
Karau lagoon	-3.78	144.18		Sepik-Ramu	
Karawari Lodge	-4.60	143.50		Sepik-Ramu	

NAME	LATITUDE	LONGITUDE	ELEV. (M)	BIRD REGION	ALTERNATIVE NAMES, RESEARCHERS, REFERENCES
Karawari River	-4.72	143.43		Sepik-Ramu	
Karawawi	-4.06	133.03	0	Bird's Neck	Thébaud & Mila/2014
Kardo*	-0.75	135.58		Bay Islands	Korido; Bruijn, A.A./1905
Karema	-9.18	147.20		SE Peninsula	Alpha Helix New Guinea Expedition/1969
Karimui	-6.32	144.80	1,113	Eastern Ranges	Diamond, J.M./1964–66
Karimui basin	-6.50	144.70	floor = 1,170	Eastern Ranges	
Karius Range	-5.67	142.50	2,830	Eastern Ranges	
Karkar Island	-4.65	145.97	1,280	Sepik-Ramu	
Karmanuntani River	-6.27	145.62		Eastern Ranges	
Karoefa*	-3.85	133.45		Bird's Neck	Karufa; on the Bomberai Peninsula
Karons	-0.50	133.50		Bird's Head	Laglaize, M./1877
Karoon country	-0.83	132.47		Bird's Head	Ripley, S.D./1938
Karu River	-5.13	142.28		Border Ranges	
Karufa	-3.85	133.45		Bird's Neck	Karoefa*; on the Bomberai Peninsula
Kasim	-1.32	131.03		NW Islands	village on northwest coast of Misool Island
Kasonaweja	-2.28	138.03		NW Lowlands	
Kasurai	-3.52	132.67		Bird's Head	
Kasuri River	-2.53	133.37		Bird's Neck	
Katau/Kataw* River	-9.15	142.95		Trans-Fly	Binaturi*/Benaturi* River; west of Daru
Katumbag	-6.05	144.78	1,585	Eastern Ranges	Gilliard, E.T./1950, 1952
Kaugel River	-6.33	144.53		Eastern Ranges	
Kaukas	-2.72	132.42		Bird's Neck	Kokas
Kaul Island	-4.67	146.00		Huon Peninsula	Alpha Helix New Guinea Expedition/1969
Kaviak	-4.67	146.00		Huon Peninsula	Alpha Helix New Guinea Expedition/1969
Kaviananga	7.60	141.30		Trans-Fly	
Kavuai Hills	-2.87	141.30		Sepik-Ramu	
Kaza River	-6.78	145.57		Eastern Ranges	Kusa River/Kuza River
Keagolo	-10.08	148.03		SE Peninsula	
Kebar valley	-0.82	133.03		Bird's Head	
Kegsugl	-5.83	145.10	2,530	Eastern Ranges	Gilliard, E.T./1950
Kekefi	-8.59	146.90	800	SE Peninsula	
Keki Lodge	-4.71	145.40		Sepik-Ramu	Beehler B.M./1998; Benz, B./2002
Keku	-5.40	145.63		Huon Peninsula	
Kelamoek* River	-1.48	131.58		Bird's Head	Kalamuk/Kladuk/Karabra River
Kelanoa	-6.01	147.50		SE Peninsula	
Kelasading River*	-1.05	131.95		Bird's Head	Kali Sading; Kakiaij, J./1938
Kelila	-3.77	138.67		Western Ranges	
Kemabu Plateau	-4.02	137.35		Western Ranges	
Keme Pass	-5.08	141.17		Border Ranges	
Kemp Welch River	-10.00	147.79		SE Peninsula	

NAME	LATITUDE	LONGITUDE	ELEV. (M)	BIRD REGION	ALTERNATIVE NAMES, RESEARCHERS, REFERENCES
Kena River	-5.38	142.43		S Lowlands	
Kepi	-6.52	139.32		S Lowlands	
Keram River	-4.12	144.08		Sepik-Ramu	Potter River*
Kerema	-7.97	145.46		S Lowlands	
Kerowagi	-5.88	144.84		Eastern Ranges	
Ketu River	-6.08	141.51		S Lowlands	(Bishop 2005b)
Keveri Hills	-9.92	148.75	200– 1,110	SE Peninsula	
Kewieng River	-5.95	146.62		Huon Peninsula	
Kikita village	-5.98	143.02		Eastern Ranges	
Kikori	-7.42	144.25		S Lowlands	
Kikori River	-7.67	144.27		S Lowlands	
Killerton Islands	-8.62	148.32		SE Peninsula	
Kilolo Creek	-7.47	146.67		SE Peninsula	Alpha Helix New Guinea Expedition/1969
Kimigomo village bush camp	-6.43	145.58		Eastern Ranges	Benz, B./2008
Kimil River	-5.75	144.52		Eastern Ranges	Wahgi valley
Kimita Island*	-10.85	152.97		SE Islands	Kimuta Island
Kimuta Island	-10.85	152.97		SE Islands	Kimita Island*; Meek, A.S./1897
Kiriwina Island	-8.50	151.08		SE Islands	Meek, A.S. /1895
Kiroro	-7.37	146.67		SE Peninsula	Alpha Helix New Guinea Expedition/1969
Kitava Island	-8.62	151.33		SE Islands	Trobriand Islands; Hamlin, H./1928
Kiterr	-7.92	140.90		Trans-Fly	In Wasur National Park; position approximate
Kiunga	-6.13	141.28	80–100	S Lowlands	
Kiwai Island	-8.58	143.42		Trans-Fly	
Kladuk/Kladoek* River	-1.55	131.67		Bird's Head	Kelamoek*/Kalamuk/Karabra River
Klagim	-0.92	132.17		Bird's Head	Ripley's local collectors, Jusup and Saban/1938
Klamono	-1.13	131.48		Bird's Head	
Kloofbivak	-4.42	138.75		Western Ranges	Versteeg, G./1912–13
Kobabaru	-4.07	141.18		Sepik-Ramu	
Kobakma	-3.68	139.00	899	Border Ranges	
Kobroor/Kobrur* Island	-6.14	134.55		Aru Islands	one of the major Aru Islands
Kodama Range	-7.60	146.65	2,728	SE Peninsula	
Kodige	-8.63	147.08	1,300	SE Peninsula	
Koemamba Islands*	-1.63	138.75		NW Lowlands	Kumamba Islands
Koemawa Mountains*	-3.98	133.00	1,490	Bird's Neck	Kumawa/Kumafa* Mountains
Koembe River*	-8.35	140.23		Trans-Fly	Kumbe River (Mees 1982b)
Koerik*	-8.25	140.28		Trans-Fly	Kurik

NAME	LATITUDE	LONGITUDE	ELEV. (M)	BIRD REGION	ALTERNATIVE NAMES, RESEARCHERS, REFERENCES
Koeroedoe Island*	-1.85	137.02		Bay Islands	Korido*/Korde*/Kordi*/Kurudu Island, just east of Yapen Island
Koffo*	-1.38	133.88		Bird's Head	Kofo, Anggi Giji Lake; Mayr, E./1928
Kofiau/Koffiao* Island	-1.18	129.83		NW Islands	Kavijaaw*/Kafiau*/Poppa* Island
Kofo	-1.38	133.88		Bird's Head	Koffo*
Kohari Hills	-2.83	141.03		Sepik-Ramu	
Koia River	-5.40	142.50		S Lowlands	
Koil Islands	-3.35	144.20		Sepik-Ramu	
Koitaki plantation	-9.42	147.43	480	SE Peninsula	
Koiyan	-6.08	147.02	960	Huon Peninsula	
Kokas	-2.72	132.42		Bird's Neck	Kaukas
Koki	-5.32	144.57		Eastern Ranges	
Kokoda station	-8.87	147.73		SE Peninsula	
Kokoda Trail*/Kokoda Track	-9.20	147.70		SE Peninsula	
Kol airfield	-5.73	144.83	1,570	Eastern Ranges	
Kom River	-5.03	141.67		Border Ranges	
Koma River	-6.85	145.17		Eastern Ranges	
Komimnung swamp	-4.02	144.67		Sepik-Ramu	lower Ramu
Komo	-6.07	142.87	1,600	Eastern Ranges	Campbell, R.
Komolom Island	-8.30	138.74		Trans-Fly	near Dolak Island
Kompiam	-5.41	143.93		Eastern Ranges	
Kon River	-5.67	144.80		Eastern Ranges	
Konda	-1.57	131.97		Bird's Head	
Kone district	-8.48	147.07		SE Peninsula	Anthony, A.S./1898
Konebada Point	-9.48	147.10		SE Peninsula	
Konstantinhafen	-5.48	145.77		Sepik-Ramu	head of Astrolabe Bay; German New Guinea; Meyer, A.B./1890; Kubary/1894
Koor River	-0.38	132.35		Bird's Head	Ripley, S.D./1938
Kopa/Kopar	-3.87	144.53		Sepik-Ramu	Jimi valley
Kopi	-7.36	144.21		S Lowlands	Kikori basin
Kopuana/Kupuana	-9.00	146.87		SE Peninsula	
Kordi*/Korde* Island	-1.85	137.02		Bay Islands	Kurudu Island, just east of Yapen Island
Kordo*	-0.81	135.56		Bay Islands	Korido; Bruijn, A.A./1905
Kore	-10.02	147.83		SE Peninsula	
Koreli Creek	-3.12	141.77		Sepik-Ramu	
Korido Island*	-1.85	137.02		Bay Islands	Kurudu Island, just east of Yapen Island
Korido/Korrido*	-0.81	135.56		Bay Islands	Kardo*/Kordo*, near Biak Island; Beccari, O.; Ripley, S.D.; Doherty, W.
Korimoro	-8.20	143.72		Trans-Fly	Maipani Point
Kori-sala	-5.52	142.76	780	Eastern Ranges	Archbold, R., Rand, A.L., and Tate, W.W./1936

NAME	LATITUDE	LONGITUDE	ELEV. (M)	BIRD REGION	ALTERNATIVE NAMES, RESEARCHERS, REFERENCES
Koroba	-5.68	142.73	1,720	Eastern Ranges	
Korobosea	-9.48	147.19		SE Peninsula	
Koronigl River	-5.95	144.83		Eastern Ranges	
Korosameri/ Korasameri* River	-4.30	143.40		Sepik-Ramu	
Korpera River	-7.75	146.67		SE Peninsula	
Kosipe River	-8.45	147.20		SE Peninsula	
Kosipi/Kosipe	-8.45	147.20	1,500	SE Peninsula	Bell, H.L./1970
Kotai	-7.82	146.13		SE Peninsula	Anthony, A.S./1898
Kotoi district	-9.08	147.33		SE Peninsula	Anthony, A.S./1898
Kotufa Creek	-5.10	142.28		Border Ranges	
Kowat	-4.83	145.35	1,160	Sepik-Ramu	
Kratke/Kraetke* Mountains	-6.90	145.93	3,587	Eastern Ranges	
Kri Island	-0.56	130.68		NW Islands	just south of Gam Island; site of Papua Diving resort
Krinjambi	-4.15	143.82		Sepik-Ramu	east bank of Sepik River
Krissa Trail/Camp	-2.76	141.34		Sepik-Ramu	Beehler, B.M. & C.H./1984
Kuabgen Range	-5.33	141.52	600–1,310	Border Ranges	
Kuama village	-8.50	147.42		SE Peninsula	
Kube Kube	-6.48	143.37	1,360	Eastern Ranges	in the Harutami Range
Kubor Mountains	-5.98	144.50	4,104	Eastern Ranges	
Kubuna	-8.68	146.75	40	SE Peninsula	Archbold Expedition/1933
Kubuna River	-8.80	146.77		SE Peninsula	
Kudjeru Gap	-7.50	146.75	1,880	SE Peninsula	in the upper Wau valley
Kuli mission/Kuli	-5.92	144.46		Eastern Ranges	base of the Kubor Mountains
Kulungtufu	-6.70	147.87		Huon Peninsula	
Kum River	-5.87	144.37		Eastern Ranges	
Kumafa Mountains*	-3.98	133.00	1,490	Bird's Neck	Koemawa*/Kumawa Mountains
Kumamba Islands	-1.63	138.75		NW Lowlands	Koemamba Islands*
Kumawa Mountains	-3.98	133.00	1,490	Bird's Neck	Koemawa*/Kumafa* Mountains
Kumbe	-8.35	140.23		Trans-Fly	
Kumbe River	-8.35	140.23		Trans-Fly	Koembe River*
Kumbu village	-4.70	145.40		Sepik-Ramu	Benz, B./2007
Kumburuf	-5.33	144.58		Eastern Ranges	
Kumdi	-5.70	144.15		Eastern Ranges	Blood, N.B./1944
Kumil River	-4.53	145.48		Sepik-Ramu	
Kumul Lodge	-5.77	143.95	2,600	Eastern Ranges	
Kumusi Point	-8.45	148.23		SE Peninsula	
Kumusi River	-8.53	147.95		SE Peninsula	Meek, A.S./1907
Kundiawa	-6.02	144.97		Eastern Ranges	

NAME	LATITUDE	LONGITUDE	ELEV. (M)	BIRD REGION	ALTERNATIVE NAMES, RESEARCHERS, REFERENCES
Kungi River	-4.07	142.99		Sepik-Ramu	
Kunimaipa River	-7.85	146.38		SE Peninsula	has been misspelled as Kunamaipa River*
Kunupi	-3.87	135.52		Western Ranges	on the middle Menoo River; Pratt brothers
Kup	-5.96	144.83		Eastern Ranges	Gilliard, E.T./1952
Kupbinga River	-7.07	145.55		S Lowlands	
Kuper Range	-7.22	146.81		SE Peninsula	
Kupiano	-10.07	148.18		SE Peninsula	
Kurai Hills	-7.96	146.42	2,120	SE Peninsula	
Kuri River	-2.37	133.98		Bird's Neck	
Kurik	-8.25	140.28		Trans-Fly	Koerik*
Kuriva A banding station	-9.05	147.10		SE Peninsula	near the Aroa River
Kuriva B banding station	-9.05	147.12		SE Peninsula	near the Aroa River
Kuriva River	-9.05	147.02		SE Peninsula	Veiya River
Kurudu Island	-1.85	137.02		Bay Islands	Korido*/Korde*/Kordi* Island, just east of Yapen Island
Kurulu	-3.92	138.85		Western Ranges	(Ripley 1964a)
Kusa/Kuza River	-6.78	145.57		Eastern Ranges	Kaza River
Kusing	-6.62	146.10	1,300	SE Peninsula	
Kuta Ridge	-5.90	144.22	1,880	Eastern Ranges	
Kutik River	-5.08	142.13		Border Ranges	
Kutubura River	-8.92	141.99		Trans-Fly	
Kuwarabai Range	-6.95	145.61	2,700	Eastern Ranges	
Kwau village	-1.03	133.91		Bird's Head	Arfak Mountains
Kwep River	-5.08	141.80		Border Ranges	
Kwerba village	-2.63	138.40		NW Lowlands	on the southwest verge of the Foja Mountains
Kwikila	-9.82	147.68		SE Peninsula	
Kyaka territory	-5.64	144.17	ca. 1,750	Eastern Ranges	(Diamond 1972)
Lababia Island	-7.23	147.17		SE Peninsula	
Labu	-6.75	146.95		SE Peninsula	
Labu swamps	-6.75	146.92		SE Peninsula	Schodde, R./1960
Lae/Lae airfield	-6.73	147.00		Huon Peninsula	
Lagaip River	-5.17	142.67		Eastern Ranges	
Lagare River	-3.47	135.83		Western Ranges	Shaw Mayer, F./1930
Lai River North	-5.22	144.25		Eastern Ranges	
Lai River South	-6.30	143.65		S Lowlands	
Laiagam	-5.50	143.48		Eastern Ranges	
Lakahia Island	-4.07	134.62		Bird's Neck	in Etna Bay
Lake Aimaroe*/ Aimaru	-1.28	132.27		Bird's Head	Aimaru/Aimaroe* Lakes

NAME	LATITUDE	LONGITUDE	ELEV. (M)	BIRD REGION	ALTERNATIVE NAMES, RESEARCHERS, REFERENCES
Lake Ambuve	-7.73	141.37		Trans-Fly	
Lake Anggi Gigi*	-1.37	133.92		Bird's Head	Anggi Giji Lake; west lake, Anggi Lakes
Lake Anggi Gita	-1.38	133.97		Bird's Head	east lake, Anggi Lakes
Lake Archbold	-3.68	138.87	1,220	Border Ranges	
Lake Atinjoe*/Atinju	-1.41	132.37		Bird's Head	
Lake Bitimi	-5.22	141.27		Border Ranges	
Lake Brown	-8.75	147.48	3,680	SE Peninsula	
Lake Chambri*	-4.30	143.13		Sepik-Ramu	Chambri Lake/Lakes
Lake Daviumbu	-7.58	141.30		Trans-Fly	Archbold, R., Rand, A.L., and Tate, W.W./1936
Lake Daviumbu Camp	-7.42	141.27		Trans-Fly	Archbold, R., Rand, A.L., and Tate, W.W./1936
Lake Gwam	-6.32	147.12	3,640	Huon Peninsula	
Lake Habbema	-4.15	138.65	3,225	Western Ranges	
Lake Herzog	-6.72	146.95		SE Peninsula	
Lake Holmes*	-2.47	138.02	427	NW Lowlands	Danau Bira
Lake Iaraguma	-9.28	147.05		SE Peninsula	
Lake Iviva	-5.43	143.53	ca. 2,500	Eastern Ranges	
Lake Jamoer*	-3.67	134.93		Bird's Neck	Lake Jamur*/Yamur
Lake Jamur*	-3.67	134.93		Bird's Neck	Lake Jamoer*/Yamur
Lake Kandep	-5.88	143.60		Eastern Ranges	
Lake Kopiago	-5.40	142.48		Eastern Ranges	
Lake Kurumoi	-2.17	134.08		Bird's Neck	Danau Kurumoi*; on the Bird's Neck
Lake Kutubu	-6.36	143.29	820	S Lowlands	Lake Marguerite*; (Schodde & Hitchcock 1968)
Lake Kutubu Patrol Post (abandoned)	-6.35	143.30		S Lowlands	
Lake Lifilikatabu	-9.42	147.37	560	SE Peninsula	in Varirata National Park
Lake Louise	-5.16	141.30		Border Ranges	in the Star Mountains of PNG
Lake Mackay	-8.74	147.47		SE Peninsula	
Lake Marguerite*	-6.36	143.29		S Lowlands	Lake Kutubu
Lake Murray	-7.00	141.13		Trans-Fly	
Lake Myola	-9.14	147.72		SE Peninsula	a dry lake bed and site of WWII airstrip
Lake Omha	-8.75	147.49	3,680	SE Peninsula	Beehler, B.M.; (Beehler 2008)
Lake Owa	-7.60	141.30		Trans-Fly	near Lake Murray
Lake Pangua	-7.73	141.38		Trans-Fly	near Lake Murray
Lake Paniai	-3.89	136.32		Western Ranges	Paniai Lake/Danau Paniai; (Junge 1953)
Lake Quarles	-4.32	138.65	3,585	Western Ranges	
Lake Rombebai	-1.87	137.92		NW Lowlands	Morris Lake
Lake Sentani	-2.62	140.50		NW Lowlands	Sentani Lake*; Mayr, E./1928
Lake Sibi	-4.92	141.78		Border Ranges	

NAME	LATITUDE	LONGITUDE	ELEV. (M)	BIRD REGION	ALTERNATIVE NAMES, RESEARCHERS, REFERENCES
Lake Sirunki	-5.43	143.53	ca. 2,500	Eastern Ranges	
Lake Sogolomik*	-5.24	141.26		Border Ranges	Lake Wangbin
Lake Tebera/Lake Teburu*	-6.78	144.65	600	Eastern Ranges	
Lake Tigi	-4.05	136.25		Western Ranges	
Lake Trist	-7.50	146.97	1,720	SE Peninsula	
Lake Vivien	-5.07	141.05	2,750	Border Ranges	
Lake Wamba	-6.00	146.55	2,520	Huon Peninsula	
Lake Wangbin	-5.24	141.26		Border Ranges	Lake Sogolomik*
Lake Wanum	-6.64	146.79		SE Peninsula	
Lake Wissel	-3.89	136.32		Western Ranges	Danau Paniai/Wissel Lakes/Paniai Lake
Lake Yamur	-3.67	134.93		Bird's Neck	Lake Jamoer*/Jamur*
Lakekamu basin	-7.91	146.49		SE Peninsula	Beehler, B.M., various years, 1978–90; (Beehler et al. 1994)
Lakekamu River	-7.87	146.47		SE Peninsula	
Lakeplain*	-3.00	138.40		NW Lowlands	Meervlatke*; Mamberamo basin
Laloki/Laroki* River	-9.39	147.31		SE Peninsula	Stone, O.C.
Lamari River	-6.90	145.42		Eastern Ranges	
Lamende Range	-5.92	143.83		Eastern Ranges	Shaw Mayer, F./1951
Lampusatu beach	-8.51	140.38		Trans-Fly	near Merauke; position approximate
Land Slip Range	-4.45	141.30	1,955	Border Ranges	
Langemak Bay	-6.58	147.83		Huon Peninsula	
Langimar River	-7.10	146.32		SE Peninsula	
Lasanga Island	-7.42	147.25		SE Peninsula	
Laughlan Islands	-9.27	153.62		SE Islands	Lockland Islands*/Budibudi atoll/Budi Budi Islands
Lea Lea	-9.78	146.98		SE Peninsula	
Left May River	-4.33	141.67		Sepik-Ramu	
Lehi Creek	-7.08	143.98		S Lowlands	
Lehuma	-1.48	134.05		Bird's Head	Arfak Mountains; (Hartert 1930)
Lelehudi village	-10.31	150.73		SE Peninsula	
Lemombui village (abandoned)	-3.92	141.88		Sepik-Ramu	
Lengguru River	-3.70	134.07	225	Bird's Neck	Thébaud & Mila/2014
Lepa Ridge	-5.47	144.22		Eastern Ranges	Trauna Ridge*/Mount Pugent*
Lepsius Point	-5.83	146.87		Huon Peninsula	
Leron River	-6.58	146.45		Huon Peninsula	
Lese Oalai	-8.28	146.28		SE Peninsula	
Libano Camp/Libano River	-6.40	142.98		Eastern Ranges	east of Mount Sisa
Liddle River	-5.80	142.18		S Lowlands	
Liki Island	-1.62	138.72		Western Ranges	

NAME	LATITUDE	LONGITUDE	ELEV. (M)	BIRD REGION	ALTERNATIVE NAMES, RESEARCHERS, REFERENCES
L'île de Soëk*	-1.00	136.00		Bay Islands	Biak Island
Liliki Bay	-1.63	150.83		NW Lowlands	in the Kumamba Islands, north of Sarmi
Lilius/Lisilus Island	-9.27	153.67		SE Islands	in the Budibudi Is; (K.D. Bishop in litt.)
Linsok	-0.10	130.72		NW Islands	Liussok; Kakiaij, J./1938
Little Pig Island	-5.17	145.84		Sepik-Ramu	Diamond, J.M./1969
Little Wau Creek	-7.33	146.73		SE Peninsula	
Liussok	-0.10	130.72		NW Islands	Linsok; Stein, G./1931
Lobo	-3.75	134.12		Bird's Neck	in Triton Bay; Temminck
Lockland Islands*	-9.27	153.62		SE Islands	Laughlan Islands/Budi Budi Islands
Loewelala	-0.97	132.15		Bird's Head	
Logaivu Mountain	-5.27	142.42	2,266	Border Ranges	
Logaiyu River	-5.42	142.68		Eastern Ranges	
Logari River Site	-3.01	136.56	275	NW Lowlands	Mack, A.
Lohiki	-7.67	145.65	40	S Lowlands	
Lohiki River	-7.67	145.65		S Lowlands	Iave River*
Lolebu	-6.50	147.80		Huon Peninsula	near Finschhafen; position approximate
Loloata Island	-9.54	147.29		SE Peninsula	southeast of Bootless Bay
Loloipa River	-8.45	146.95		SE Peninsula	
Lolorua	-8.95	146.93	< 40	SE Peninsula	Zimmer, J.T./1920
Lopon	-1.72	130.14		NW Islands	
Lordberg	-4.92	143.00	1,500	Eastern Ranges	30 km south of Hunsteinspitze; Bürgers/1912
Lorentz River	-5.50	138.03		S Lowlands	parent river of the Noord
Losavi cave	-6.65	145.17	1,350	Eastern Ranges	Clayton/2002
Louisiade Archipelago	-11.50	153.50		SE Islands	
Loupom Island	-10.38	149.36		SE Peninsula	
Lova/Lowa River	-8.39	146.99		SE Peninsula	
Lubuyavi Camp	-6.60	145.19	1,500	Eastern Ranges	near Heroana; Scholes, E./2000
Lufa	-6.33	145.30	1,920	Eastern Ranges	Diamond, J.M., & Terborgh, J.W./1964
Lumari-Vailala Divide	-6.90	145.38	2,990	Eastern Ranges	
Lumi	-3.48	142.03		Sepik-Ramu	
Lunn Island	-10.67	151.98		SE Islands	north of the Conflict Islands
Lusancay Islands	-8.42	150.17		SE Islands	
Lutur/Lutor* Island	-6.38	134.65		Aru Islands	
Luwak Island	-0.03	130.93		NW Islands	Rawak Island; off Waigeo Island
Mäanderberg*/ Maeanderberg	-4.05	141.68		Sepik-Ramu	
Mabadauan	-9.28	142.73		Trans-Fly	Archbold, R., Rand, A.L., and Tate, W.W./1936
MacCluer Gulf*	-2.50	132.50		Bird's Neck	McCluer's Inlet*/Bintuni Bay
MacGregor Junction	-5.75	141.72		S Lowlands	
Macleay*/Maclay Coast	-5.75	146.58		Huon Peninsula	

NAME	LATITUDE	LONGITUDE	ELEV. (M)	BIRD REGION	ALTERNATIVE NAMES, RESEARCHERS, REFERENCES
Mada Creek	-9.37	147.50		SE Peninsula	
Madang	-5.30	145.78		Sepik-Ramu	
Mademo Island	-2.03	139.23		NW Lowlands	Mapia Island*
Madi Creek	-8.92	146.98		SE Peninsula	
Madilogo Ridge	-9.20	147.53		SE Peninsula	
Madiri	-8.45	143.07		Trans-Fly	
Madiu	-8.59	146.90	800	SE Peninsula	
Maeanderberg/ Mäanderberg*	-4.07	141.68		Sepik-Ramu	Bürgers/1913 (Stresemann 1923)
Maeula district	-8.75	147.50		SE Peninsula	
Mafoor*/Mafor* Island	-1.03	134.87		Bay Islands	Mefoor*/Noemfoor*/Numfoor*/Numfor Island
Mafulu mission station	-8.52	147.03	1,140	SE Peninsula	
Mafulu/Mafooloo*	-8.52	147.02	1,250	SE Peninsula	Hamlin, Archbold Expeditions/1929, 1933
Maga	-2.06	130.43		NW Islands	
Magidobo	-6.18	142.77	1,450	Eastern Ranges	Aplin, K.
Magori village	-10.25	149.27		SE Peninsula	
Maguli Range	-9.28	147.62	1,888	SE Peninsula	Mount Maguli/Horsehoe Mountain*/ Hufeisengebirge*
Mai Kussa River	-9.15	142.20		Trans-Fly	
Mai Mai/Maimai	-3.73	142.38		Sepik-Ramu	Stalker, W./1908
Maiama River	-7.62	147.48		SE Peninsula	
Maijoebi	-1.48	134.13		Bird's Head	
Mailu district	-10.37	149.53		SE Peninsula	Orangerie Bay; Anthony, A.S./1895
Mailu Island	-10.39	149.35		SE Peninsula	
Maimafu Camp	-6.53	145.07		Eastern Ranges	Mack, A./1999
Maipani Point	-8.20	143.72		Trans-Fly	Korimoro
Maiwara	-5.00	145.50		Sepik-Ramu	Alpha Helix New Guinea Expedition/1969
Majepa	-4.09	136.26		Western Ranges	near Tigi Lake, in the Paniai Lake region; (Junge 1953)
Makimi Island	-3.07	135.77		Bay Islands	Shaw Mayer, F./1930
Malalaua	-8.08	146.17		SE Peninsula	
Malalo/Malalo mission	-7.02	146.98		SE Peninsula	near Salamaua
Malia/Nena River	-4.65	141.73		Sepik-Ramu	
Malingai	-4.18	143.23		Sepik-Ramu	
Malne River	-5.67	144.80		Eastern Ranges	
Malu	-4.22	142.87		Sepik-Ramu	Sepik River; Bürgers/1913
Mamama River	-8.97	148.00		SE Peninsula	
Mamasiwar/ Mimisiware	-2.87	134.45		Bay Islands	Ripley, S.D./1938
Mamba River	-8.87	147.75		SE Peninsula	

NAME	LATITUDE	LONGITUDE	ELEV. (M)	BIRD REGION	ALTERNATIVE NAMES, RESEARCHERS, REFERENCES
Mambare River	-8.10	148.00		SE Peninsula	Meek, A.S./1906
Mamberamo basin	-2.94	138.42		NW Lowlands	Lakeplain*/Meervlakte*
Mamberamo/Mam- berano*/Mamiramu* River	-2.00	137.80		NW Lowlands	
Mambrioe	-3.23	134.95		Bird's Neck	
Managuna Island	-10.68	152.88		SE Islands	
Manaifei River	-4.88	142.38		Border Ranges	
Manam Island	-4.83	145.03	1,807	Sepik-Ramu	Vulcan Island*
Manari	-9.20	147.62		SE Peninsula	
Mandum	-7.15	140.53		Trans-Fly	
Mangalim Creek	-5.28	141.30		Border Ranges	
Mangkawan Hills Camp	-0.11	130.79	140	NW Islands	Mauro, I./2002; (Waigeo Island)
Manien	-5.72	134.78		Aru Islands	Kühn, H.
Manim Island	-1.10	134.77		Bay Islands	
Manki logging area	-7.17	146.57		SE Peninsula	
Manki Range	-7.27	146.62	1,200– 1,587	SE Peninsula	
Manki village	-7.17	146.57	889	SE Peninsula	Bishop, K.D.
Mannasat	-6.33	147.33		Huon Peninsula	Grierson and Van Duesen/1964
Manoepi	-2.80	134.35		Bird's Neck	Shaw Mayer, F./1930
Manokwari	-0.86	134.08		Bird's Head	Dorey harbour*
Mansema	-1.05	134.03		Bird's Head	Arfak Mountains
Mansinam Island	-0.82	134.10		Bird's Head	Bruijn, A.A./1875
Manu Manu	-9.12	146.88		SE Peninsula	
Manuga Reefs	-11.01	153.35		SE Islands	between Rossel and Misima Islands
Manumbai	-6.03	134.30		Aru Islands	on Kobroor Island of Aru Islands; Cayley-Webster, H./1896
Maoke Range	-4.00	138.00		Western Ranges	central ranges of Indonesian NG
Mapi	-7.00	139.27		S Lowlands	
Mapia Island	0.82	134.33		Bay Islands	Pulau Pegun
Mapia Island	-2.03	139.23		NW Lowlands	south of Yamna Island
Mapia River	-4.27	136.07		S Lowlands	
Mappi River	-6.98	139.20		S Lowlands	
Maprik	-3.63	143.05		Sepik-Ramu	
Mapu River	-4.83	142.40		Border Ranges	
Marai Island	-1.75	135.82		Bay Islands	near Ansus, Yapen Island; Doherty, W./1897
Maransabadi Island	-2.03	134.57		Bird's Neck	
Maratambu	-4.96	145.39		Sepik-Ramu	Adelbert Mountains
Margarida	-10.30	149.33		SE Peninsula	
Margarima	-5.98	143.35		Eastern Ranges	

NAME	LATITUDE	LONGITUDE	ELEV. (M)	BIRD REGION	ALTERNATIVE NAMES, RESEARCHERS, REFERENCES
Mari Creek mine	-7.28	146.72	2,000	SE Peninsula	Meri Creek mine; Fenton, M.B.
Marienberg	-3.97	144.23		Sepik-Ramu	
Markham River/ Markham valley	-6.64	146.67		SE Peninsula	Smith, S./1977
Maro River	-8.47	140.35		Trans-Fly	
Marshall Bennett Islands	-8.97	151.97		SE Islands	
Marshall lagoon	-10.05	148.20		SE Peninsula	
Maruba River	-6.21	143.19		Eastern Ranges	(I. Woxvold in litt.)
Maruta passage	-10.50	149.92		SE Peninsula	
Masaweng River	-6.35	147.80		Huon Peninsula	
Masi Masi Island	-2.00	139.13		NW Lowlands	
Masian Point	-6.93	134.18		Aru Islands	Ngabordamlu Point
Masikeri Mountains	-2.00	134.00	1,114	Bird's Head	
Matapaila	-9.05	146.83	<40	SE Peninsula	
Matsika	-8.58	146.93	1,040	SE Peninsula	
Matterer Bay	-2.32	140.13		NW Lowlands	
Mauia Islet	-9.52	147.04		SE Peninsula	
Mave	-8.57	147.12	1,560	SE Peninsula	
May River	-4.25	141.88		Sepik-Ramu	
Mayalabit/Majalibit* Bay	-0.35	130.93		NW Islands	Chabrol* Bay
Mayri Bay	-10.48	149.45		SE Peninsula	
Mayum River	-5.27	143.20		Eastern Ranges	
Mbwei River	-7.25	145.52		S Lowlands	
McCluer's Inlet*	-2.50	132.50		Bird's Neck	MacCluer Gulf*/Bintuni Bay
McFarlane harbour	-10.08	149.17		SE Peninsula	
McNicholl Mountain	-5.50	142.83	3,240	Eastern Ranges	
Meervlakte*	-3.00	138.40		NW Lowlands	Lakeplain* or Mamberamo basin
Mefkadjem	-1.27	132.18		Bird's Head	
Mefoor Island*	-1.03	134.90		Bay Islands	Mafor*/Mafoor*/Noemfoor*/Numfoor*/ Numfor Island
Mega	-0.68	131.87		Bird's Head	
Mega River	-0.68	131.88		Bird's Head	
Meiro River	-5.20	145.78		Huon Peninsula	
Mekeo district	-8.50	146.60	<40	SE Peninsula	
Mekil biological field station	-4.81	141.65		Border Ranges	on Mount Mekil/Mount Stolle; Scholes, E.
Melokin Range	-5.42	141.60	1,555	Border Ranges	
Membe/Menebe	-5.63	144.20		Eastern Ranges	
Memeali Point	-10.63	150.78		SE Islands	
Memenga forest	-4.96	145.39		Sepik-Ramu	in the Adelbert Mountains

NAME	LATITUDE	LONGITUDE	ELEV. (M)	BIRD REGION	ALTERNATIVE NAMES, RESEARCHERS, REFERENCES
Memunggal/Arerr	-8.68	140.86		Trans-Fly	in Wasur National Park; position approximate
Mendi	-6.15	143.65	2,160	Eastern Ranges	
Mendi River	-6.25	143.70		Eastern Ranges	
Mendiptana	-5.87	140.68		S Lowlands	Mindip Tanah/Mindiptana
Mengino	-6.45	145.23	1,875	Eastern Ranges	Diamond, J.M.
Menoo River	-3.73	135.38		NW Lowlands	
Menyamya	-7.25	146.00	1,204	SE Peninsula	
Meos Kairu*	-0.20	135.00		Bay Islands	Mios Kairu; Meos and Mios mean island
Meos Korwar*	-0.40	135.27		Bay Islands	Mios Korwar; in the Aifundi Islands
Meos Noem*/Meos Num*	-1.50	135.25		Bay Islands	Mios Num Island
Meos War	-2.08	134.38		Bay Islands	
Meosuri	-2.17	134.75		Bay Islands	in the Auri Islands
Merauke	-8.50	140.35		Trans-Fly	Meek, A.S./1910
Merauke River	-8.47	140.35		Trans-Fly	Jackson, T./1924
Meren valley	-4.06	137.14		Western Ranges	
Meroka*	-9.37	147.51		SE Peninsula	Moroka; Anthony, A.S.
Mesan	-3.80	135.20		Bird's Neck	at the head of Geelvink Bay; position approximate
Meskor Bush Camp	-4.72	145.27		Sepik-Ramu	Benz, B./2007
Miak	-4.67	146.00		Huon Peninsula	Alpha Helix New Guinea Expedition/1969
Mianmin Divide Mountains	-4.67	141.65	2,813	Border Ranges	
Miarosa	-6.52	145.55		Eastern Ranges	Mirassa*; Diamond, J.M. & Terborgh, J.W./1964
Mibu Island	-8.73	143.45		Trans-Fly	
Middleburg Island	-0.37	132.20		Bird's Head	
Miei	-2.77	134.52		Bird's Neck	
Milali River	-4.75	142.05		Border Ranges	
Miliom	-3.50	142.05	1,510	Sepik-Ramu	Diamond, J.M./1966
Millport harbour	-10.35	149.47		SE Peninsula	
Milne Bay	-10.33	150.33		SE Peninsula	
Mimako	-4.65	136.47		S Lowlands	
Mimika River	-4.42	136.55		S Lowlands	Grant/1910
Mindabit	-4.28	143.90		Sepik-Ramu	Mindimbit
Mindik	-6.46	147.43	1,280	Huon Peninsula	Schodde, R.
Mindimbit	-4.28	143.90		Sepik-Ramu	Mindabit
Mindip Tanah	-5.87	140.68	100	S Lowlands	Mendiptana/Mindiptana
Mindiptana	-5.87	140.68		S Lowlands	Mindip Tanah/Mendiptana
Mingiman	-0.77	130.10		NW Islands	
Minj	-5.88	144.70		Eastern Ranges	

NAME	LATITUDE	LONGITUDE	ELEV. (M)	BIRD REGION	ALTERNATIVE NAMES, RESEARCHERS, REFERENCES
Minj-Nona Divide	-6.03	144.75	2,140	Eastern Ranges	
Minyip/Minyimp*	-5.80	144.15		Eastern Ranges	
Mios Kairu	-0.20	135.00		Bay Islands	Meos Kairu*
Mios Korwar	-0.40	135.25		Bay Islands	Meos Korwar*/Pondi*; in the Aifundi Islands
Mios Num/ Miosnom*/Miosnoum	-1.50	135.25		Bay Islands	Mios Num Island
Mipan	-6.75	141.15		S Lowlands	
Miptagin	-5.18	141.62	1,920	Border Ranges	Gilliard, E.T./1954
Mirassa*	-6.52	145.55		Eastern Ranges	Miarosa; Diamond, J.M., and Terborgh, J.W./1964
Mirip River	-5.13	141.57		Border Ranges	
Misima Island	-10.67	152.73		SE Islands	Saint Aignan Island*; Meek, A.S./1897; Hamlin/1930
Misool/Mysol*/Misol* Island	-1.90	130.00		NW Islands	Kühn, H./1900
Misori*/Misory* Island	-0.75	135.75		Bay Islands	Biak Island
Mittleres Sepikgebiet	-4.07	141.67		Sepik-Ramu	"middle Sepik district"
Miwa	-7.13	141.52		S Lowlands	
Mnier Hills	-0.17	131.13	870	NW Islands	
Moari	-1.37	134.12		Bird's Head	
Mobit Hills Camp	-0.13	130.79	285	NW Islands	Mauro, I./2002
Modijo*/Modio	-4.06	135.62		Western Ranges	Charles Louis Mountains
Moebrani*	-0.75	133.42		Bird's Head	Mubrani
Moeli*	-7.80	139.00		Trans-Fly	Muli/Princess Marianne Strait
Moeli* River or Strait*	-7.80	139.00		Trans-Fly	Muli River
Mogubu Point	-10.32	149.32		SE Peninsula	
Moitaka	-9.42	147.22		SE Peninsula	
Moitaka settling ponds	-9.38	147.18		SE Peninsula	Waigaini sewage ponds
Momi	-1.60	134.17		Bird's Head	Mayr, E./1928
Momos	-0.92	131.25		NW Islands	Guillemard/1884
Momsakten	-5.20	141.60	1,615	Border Ranges	Gilliard, E.T./1954
Mon	-8.82	146.53		SE Peninsula	Yule Island; (D'Albertis 1876)
Mondo	-8.55	147.10	1,560	SE Peninsula	
Mongi River	-6.72	147.60		Huon Peninsula	
Mongi-Busu	-6.37	147.20		Huon Peninsula	
Moni River	-9.57	148.68		SE Peninsula	Mayr, E./1928
Monkton River	-8.02	146.47		SE Peninsula	
Monoga River	-6.43	144.55		Eastern Ranges	
Montemont Islands	-11.30	152.30		SE Islands	
Mopa	-8.53	140.42		Trans-Fly	
Mopu Inlet	-8.10	146.07		SE Peninsula	

NAME	LATITUDE	LONGITUDE	ELEV. (M)	BIRD REGION	ALTERNATIVE NAMES, RESEARCHERS, REFERENCES
Mora airstrip	-6.37	143.23	852	Eastern Ranges	
Morabi	-9.12	146.88		SE Peninsula	
Morehead	-8.72	141.63		Trans-Fly	
Morehead River	-9.13	141.33		Trans-Fly	
Moresby Island	-10.62	151.00		SE Islands	
Mori	-3.25	141.73		Sepik-Ramu	
Moris	-	-		Bird's Head	Arfak Mountains; (Mayr 1941)
Moro	-6.36	143.23		Eastern Ranges	in the Kikori basin
Morobe Province/ district*	-7.10	146.50		SE and Huon Peninsula	
Morobe station/ Morobe harbour	-7.77	147.58		SE Peninsula	
Moroka/Moroke*	-9.37	147.51		SE Peninsula	Meroka*; Anthony, A.S.
Morris Lake	-1.87	137.92		NW Lowlands	Lake Rombebai
Moruka village	-9.25	147.08		SE Peninsula	Alpha Helix New Guinea Expedition/1969
Moso River	-2.63	140.93		NW Lowlands	
Mosquito Island	-7.43	147.32		SE Peninsula	in the Fly Islands
Mosquito Island	-9.77	149.88		SE Peninsula	near Baniara
Mosso	-2.67	140.92		NW Lowlands	
Mossy Ridge	-4.83	141.20		Border Ranges	
Motupore/Motupure* Island	-9.52	147.29		SE Peninsula	
Mou	-8.72	146.53		SE Peninsula	
Mount Abaipap	-0.10	130.52	700	NW Islands	
Mount Abilala	-5.82	146.17	3,648	Huon Peninsula	
Mount Aiome	-5.22	144.52	2,830	Eastern Ranges	
Mount Aiyang	-5.10	141.28	3,325	Border Ranges	
Mount Akrik	-5.13	141.15	2,574	Border Ranges	
Mount Albert Edward, east dome	-8.42	147.33	3,990	SE Peninsula	
Mount Albert Edward, west dome	-8.37	147.38	3,840	SE Peninsula	
Mount Albowagi	-3.48	143.23	1,238	Sepik-Ramu	
Mount Alexander	-9.33	147.72	1,640	SE Peninsula	
Mount Amung*/ Amungwiwa	-7.43	146.55	3,278	SE Peninsula	
Mount Andamunk	-5.22	143.35	3,725	Eastern Ranges	
Mount Angemuk	-3.52	138.58	3,810	Border Ranges	Doormanpaad
Mount Antares	-4.83	140.88	4,170	Border Ranges	in the Star Mountains of Indonesia
Mount Apatikaiogean	-10.68	152.68		SE Islands	on Misima Island
Mount Apmatari, east peak	-6.90	145.90	3,459	Eastern Ranges	
Mount Arfak	-1.15	133.98	2,950	Bird's Head	Bruijn, A.A./1905

NAME	LATITUDE	LONGITUDE	ELEV. (M)	BIRD REGION	ALTERNATIVE NAMES, RESEARCHERS, REFERENCES
Mount Astrolabe	-9.48	147.38	1,048	SE Peninsula	
Mount Auriga	-4.98	141.13	3,720	Border Ranges	
Mount Baik	-3.82	132.92	1,430	Bird's Neck	
Mount Bandji	-0.65	132.33	2,452	Bird's Head	Mount Bandjiet/Bantjiet
Mount Bangeta	-6.26	147.07	3,842	Huon Peninsula	high summit of Saruwaged Range
Mount Bantjiet	-0.65	132.33	2,452	Bird's Head	Mountain Bandjiet/Bandji; Gilliard, E.T./1964
Mount Barton	-8.98	148.32	1,800	SE Peninsula	
Mount Batchelor	-8.00	147.08	2,814	SE Peninsula	
Mount Bedego	-5.80	144.88	3,200	Eastern Ranges	
Mount Bellamy	-9.08	147.70	2,637	SE Peninsula	
Mount Besar	-0.85	130.63	1,070	NW Islands	highest mountain on Batanta Island; translates as "big mountain"
Mount Bewani	-3.22	141.17	1,515	Sepik-Ramu	
Mount Binnie Mining Camp	-5.20	141.13		Border Ranges	
Mount Bituba	-5.77	142.52	3,420	Eastern Ranges	
Mount Blucher	-5.57	141.82	1,365	Border Ranges	
Mount Bolema	-0.83	132.17		Bird's Head	Gunung Bolema
Mount Bon Kourangen	-0.63	132.90	2,582	Bird's Head	Mount Kourange/Bon Korangen
Mount Borebore	-9.07	147.23	403	SE Peninsula	Mount Frank Lawes
Mount Bosavi	-6.60	142.86	2,397	Eastern Ranges	
Mount Boue	-8.62	147.93	1,407	SE Peninsula	
Mount Bougainville	-2.65	141.03	1,220	Sepik-Ramu	
Mount Brown	-9.68	148.25	2,040	SE Peninsula	
Mount Bürgers	-5.15	143.30	3,711	Eastern Ranges	Bürgers Mountain/Yakopi Nalenk
Mount Buru	-4.23	134.95	1,393	Bird's Neck	
Mount Cameron	-8.88	147.19	ca. 1,900	SE Peninsula	Anthony, A.S./1896
Mount Cameron Range	-8.90	147.17	2,111	SE Peninsula	
Mount Capella	-4.99	141.08	4,015	Border Ranges	in the Star Mountains of PNG
Mount Capella, trig. station	-4.98	141.09	3,932	Border Ranges	
Mount Carstensz	-4.15	137.05	4,884	Western Ranges	Puncak Jaya/Carstensz Toppen/Carstensz Pyramid/Gunung Sukarno/Carstensz Massif
Mount Chalmers	-9.38	147.73	1,361	SE Peninsula	
Mount Chamberlin	-8.33	147.25	3,250	SE Peninsula	
Mount Chapman	-7.98	146.97	3,360	SE Peninsula	
Mount Clarence	-9.88	148.63	1,832	SE Peninsula	
Mount Dafonsoro	-2.50	140.33		NW Lowlands	highest of Cyclops Mountains; Jewett Jr, S.G./1945

NAME	LATITUDE	LONGITUDE	ELEV. (M)	BIRD REGION	ALTERNATIVE NAMES, RESEARCHERS, REFERENCES
Mount Danai	-0.20	131.01	950	NW Islands	
Mount Daum	-3.42	142.85	1,210	Sepik-Ramu	
Mount David	-4.77	139.88	2,290	Border Ranges	
Mount Dayman	-9.78	149.33	2,960	SE Peninsula	
Mount Derimapa	-3.70	135.92	1,700	Western Ranges	in the Weyland Mountains; Shaw Mayer, F./1930
Mount Diamond mission	-9.48	147.32		SE Peninsula	
Mount Diatau	-10.65	152.63	1,038	SE Islands	
Mount Dickson	-8.05	147.13	3,560	SE Peninsula	
Mount Donner	-5.08	141.63	2,458	Border Ranges	
Mount Douglas	-8.84	147.50	3,520	SE Peninsula	
Mount Duau	-6.98	144.57	1,560	Eastern Ranges	
Mount D'Urville/ D'Urville Peak	-9.48	147.38	1,048	SE Peninsula	
Mount Edagwaba*/ Edgwaba	-9.62	150.65	1,385	SE Islands	
Mount Emp	-8.75	146.75		SE Peninsula	
Mount Epa	-8.75	146.72	249	SE Peninsula	5 miles inland from Hall Sound; Salvadori. T.
Mount Essie	-8.12	146.98	3,655	SE Peninsula	
Mount Faim*/Faium	-5.35	142.17		S Lowlands	
Mount Favenc	-6.97	144.68	1,584	Eastern Ranges	
Mount Frank Lawes	-9.07	147.23	403	SE Peninsula	Mount Borebore
Mount Fubilan	-5.20	141.13	2,084	Border Ranges	
Mount Gahavisuka Park	-6.03	145.40		Eastern Ranges	
Mount Gaivara	-8.29	147.06	2,743	SE Peninsula	
Mount Garamambu	-4.28	143.00	512	Sepik-Ramu	
Mount Gayata	-9.09	147.47		SE Peninsula	Richardson Range, of the Owen Stanley Range; Weiske, E.
Mount Gebroeders	-3.67	135.95	2,780	Western Ranges	in the Weyland Mountains
Mount Gigira	-5.92	142.70		Eastern Ranges	
Mount Giluwe	-6.05	143.88	4,367	Eastern Ranges	
Mount Goliath	-4.69	139.83	3,340	Border Ranges	Gunung Yamin; Meek, A.S./1911
Mount Gong	-6.20	147.17	2,440	Huon Peninsula	
Mount Gonggia	-5.38	144.60	2,734	Eastern Ranges	
Mount Goodenough	-9.33	150.20		SE Islands	
Mount Gunduwai	-5.42	144.60	2,601	Eastern Ranges	
Mount Hagen	-5.74	144.04	3,791	Eastern Ranges	Gilliard, E.T./1950
Mount Hagen Base Camp	-5.80	144.05	2,560	Eastern Ranges	(Mayr & Gilliard 1954)
Mount Hagen Summit Camp	-5.27	144.05	3,414	Eastern Ranges	Gilliard, E.T./1950

NAME	LATITUDE	LONGITUDE	ELEV. (M)	BIRD REGION	ALTERNATIVE NAMES, RESEARCHERS, REFERENCES
Mount Hagen town	-5.86	144.22	1,729	Eastern Ranges	
Mount Haliago	-6.15	142.75	2,689	Eastern Ranges	
Mount Hides	-5.92	142.70	2,810	Eastern Ranges	
Mount Ialibu	-6.23	144.07	3,465	Eastern Ranges	
Mount Ifal	-4.98	141.72	3,350	Border Ranges	Gilliard, E.T./1954
Mount Iguntam	-5.68	145.00	3,960	Eastern Ranges	
Mount Imonda	-3.25	141.18	800	Sepik-Ramu	
Mount Jaya/Puncak Jaya	-4.15	137.05	5,030	Western Ranges	Mount Carstensz/Carstensz Toppen; Schodde et al./1975
Mount Jonanggee	-7.58	147.00	2,500	SE Peninsula	
Mount Juliana*	-4.72	140.28	4,700	Border Ranges	Mount Mandala
Mount Kainde*/ Kaindi	-7.57	146.67		SE Peninsula	Ziegler, A.C./1967; Alpha Helix New Guinea Expedition/1969
Mount Kanagioi	-4.67	145.97		Sepik/Ramu	Mount Kumbant/Mount Kunugui
Mount Karenmai	-5.38	144.45	2,080	Eastern Ranges	
Mount Karimui	-6.56	144.77	2,652	Eastern Ranges	
Mount Kebea	-8.57	146.97	1,080	SE Peninsula	
Mount Kemenagi	-6.47	143.38	1,360	Eastern Ranges	
Mount Kemp	-7.80	147.25	2,512	SE Peninsula	
Mount Kenevi	-9.12	147.68	3,416	SE Peninsula	
Mount Kilkerran	-9.47	150.77	1,720	SE Islands	once spelled Kilherran*
Mount Kinkain	-6.02	144.58		Eastern Ranges	Kubor Mountains; position approximate; Gilliard, E.T./1950
Mount Knutsford	-8.83	147.50	3,460	SE Peninsula	Anthony, A.S./1898
Mount Kofia	-8.57	146.97		SE Peninsula	
Mount Kominjim	-5.22	144.52	2,748	Eastern Ranges	in the Schrader Range; Gilliard, E.T./1964
Mount Kourange	-0.71	132.73		Bird's Head	Bon Kourangen
Mount Kraanvogel	-4.73	140.57	3,230	Border Ranges	
Mount Kubor	-6.10	144.77	3,969	Eastern Ranges	
Mount Kuia	-5.90	144.18	2,161	Eastern Ranges	
Mount Kumbak	-7.53	146.68	2,800	SE Peninsula	Bulldog Road, Ekuti Divide
Mount Kumbant	-4.67	145.97		Sepik/Ramu	Mount Kanagoi/Mount Kunugui
Mount Kunugui	-4.67	145.97	1,800	Sepik/Ramu	Mount Kanagoi/Mount Kumbant
Mount Kunupi	-3.83	135.92		Western Ranges	Pratt brothers
Mount Kwabu	-6.57	143.58	1,490	Eastern Ranges	
Mount Kweirok	-5.27	141.87	3,420	Border Ranges	
Mount Kwoka	-0.63	132.33	2,452	Bird's Head	
Mount Lamington	-8.93	148.17	1,689	SE Peninsula	Hamlin, H./1928
Mount Lawes	-9.33	147.23	479	SE Peninsula	
Mount Lawson	-7.68	146.62	2,722	SE Peninsula	
Mount Lehuma	-1.50	133.92	1,900	Bird's Head	Mayr, E./1928
Mount Leiwaro	-5.55	143.25	3,415	Eastern Ranges	Mount Liwaro

NAME	LATITUDE	LONGITUDE	ELEV. (M)	BIRD REGION	ALTERNATIVE NAMES, RESEARCHERS, REFERENCES
Mount Leonard Murray	-6.58	142.82		Eastern Ranges	
Mount Lilley	-8.82	147.18	2,111	SE Peninsula	
Mount Liwaro	-5.55	143.25	3,415	Eastern Ranges	Mount Leiwaro
Mount Logaivu	-5.27	142.42		Border Ranges	
Mount Maander*	-4.07	141.67		Sepik-Ramu	Maeanderberg
Mount Mabiom	-5.19	141.85	1,279	Border Ranges	position approximate; Archbold, R., Rand, A.L., and Tate, W.W./1936
Mount Maguli	-9.32	147.53		SE Peninsula	Horsehoe Mountain*/Hufeisengebirge*
Mount Mandala/Pun-cak Mandala*	-4.72	140.28	4,700	Border Ranges	Mount Juliana*
Mount Maneao	-9.68	149.63		SE Peninsula	
Mount Maneao, north peak	-9.78	149.34	2,680	SE Peninsula	
Mount Maneao, south peak	-9.82	149.33	2,546	SE Peninsula	
Mount Maori	-3.31	141.72	1,960	Sepik-Ramu	
Mount Maragubui	-5.60	144.33	2,585	Eastern Ranges	
Mount Maybole	-9.37	150.48	1,139	SE Peninsula	
Mount McIlwraith	-8.93	147.55	3,200	SE Peninsula	
Mount McNicholl	-5.50	142.83	3,240	Eastern Ranges	
Mount Mekil	-4.80	141.67	2,813	Border Ranges	Mount Stolle*
Mount Memenga	-4.97	145.54	1,412	Sepik-Ramu	(Gilliard & LeCroy 1967)
Mount Menawa	-3.20	141.72	1,885	Sepik-Ramu	Diamond, J.M./1966
Mount Mengam	-4.87	145.32	1,600	Sepik-Ramu	Adelbert Mountains
Mount Michael	-6.41	145.32	3,647	Eastern Ranges	Diamond, J.M.
Mount Mimye	-5.23	142.37	2,266	Border Ranges	
Mount Missim/Misim*/Mission*	-7.28	146.78	2,877	SE Peninsula	Stevens, H./1932; Pratt, T.K./1974–75, 1977–80; Beehler, B.M. 1975–76, 1978–1980; Pruett-Jones, S. & M.
Mount Moari	-1.37	134.08	1,998	Bird's Head	has been misspelled as Mount Maori
Mount Moiba	-9.91	149.63		SE Peninsula	north of Mount Simpson; position approximate
Mount Momoa	-8.68	147.55	2,320	SE Peninsula	
Mount Morait	-0.76	132.50		Bird's Head	eastern Tamrau Mountains; Beccari, O./1875
Mount Mura	-9.88	149.39	1,880	SE Peninsula	
Mount Murray	-6.75	144.02	2,254	Eastern Ranges	
Mount Musgrave	-6.87	144.57	1,110	Eastern Ranges	
Mount Musgrave	-8.93	147.48	3,040	SE Peninsula	Anthony, A.S./1898
Mount Namoa	-9.30	148.57	1,138	SE Peninsula	
Mount Nanson	-6.28	147.47		Huon Peninsula	Geisler
Mount Nebungesa	-8.68	147.78	3,440	SE Peninsula	
Mount Nelson	-8.13	147.22	3,520	SE Peninsula	

NAME	LATITUDE	LONGITUDE	ELEV. (M)	BIRD REGION	ALTERNATIVE NAMES, RESEARCHERS, REFERENCES
Mount Nibo	-3.33	141.65	1,510	Sepik-Ramu	Diamond, J.M./1966
Mount Nipa	-6.29	143.66		Eastern Ranges	
Mount Nisbet	-9.17	147.83	3,245	SE Peninsula	
Mount Nok/Mount Nok Base Camp	-0.08	130.75	880	NW Islands	on Waigeo Island; Mauro, I./2002
Mount Obree	-9.48	148.05	3,080	SE Peninsula	
Mount Oiamadawa'a	-9.30	150.20	2,200	SE Islands	easternmost summit on Goodenough Island
Mount Oiautukekea	-9.33	150.20	2,536	SE Islands	
Mount O-Mar	-6.02	144.58		Eastern Ranges	in the Kubor Mountains; (Mayr & Gilliard 1952: 5)
Mount Omeri*/Omeri Mountain	-8.15	146.62	1,680	SE Peninsula	
Mount Orata	-6.10	144.50		Eastern Ranges	Gilliard, E.T./1950
Mount Orian	-9.88	149.21		SE Peninsula	near Mount Suckling
Mount Otto	-5.98	145.48	3,456	Eastern Ranges	
Mount Peripatus	-4.78	141.17	1,599	Border Ranges	
Mount Piora	-6.72	145.83	3,557	Eastern Ranges	
Mount Popinausi	-8.82	147.38		SE Peninsula	
Mount Pryke	-7.20	146.90	1,945	SE Peninsula	
Mount Pugent	-5.47	144.22		Eastern Ranges	Lepa Ridge/Trauna Ridge*
Mount Riu	-11.52	153.43	806	SE Islands	Meek, A.S./1916
Mount Robinson	-5.23	141.18	1,645	Border Ranges	
Mount Rossel	-11.37	154.23	838	SE Islands	Meek, A.S./1915
Mount Saint Mary	-8.12	146.97	3,209	SE Peninsula	
Mount Sapau	-3.37	142.52	1,505	Sepik-Ramu	
Mount Sarawaket*/Saruwaged/Salawaket*	-6.32	147.08	4,121	Huon Peninsula	Saruwaged Range
Mount San Lai Camp	-0.04	130.85	780	NW Islands	Mauro, I./2002; Waigeo Island
Mount Schatteburg	-4.88	142.07	3,040	Border Ranges	
Mount Schrader	-4.98	144.05	ca. 2,080	Eastern Ranges	
Mount Scratchley Camp	-8.75	147.47	ca. 3,720	SE Peninsula	
Mount Scratchley repeater station	-8.73	147.43	3,572	SE Peninsula	
Mount Scratchley summit	-8.73	147.47	3,840	SE Peninsula	
Mount Semieng	-6.25	147.33	2,856	Huon Peninsula	Grierson and Van Duesen/1964
Mount Service	-8.92	147.55	3,960	SE Peninsula	
Mount Shoppe	-7.03	146.78	2,675	SE Peninsula	
Mount Shungol	-6.85	146.72	2,752	SE Peninsula	
Mount Sigul Mugal	-6.83	144.48	3,826	Eastern Ranges	in the Kubor Mountains; (Mayr & Gilliard 1952: 5)
Mount Silisigoda	-8.49	147.30	3,160	SE Peninsula	

NAME	LATITUDE	LONGITUDE	ELEV. (M)	BIRD REGION	ALTERNATIVE NAMES, RESEARCHERS, REFERENCES
Mount Simpson	-10.03	149.57	2,883	SE Peninsula	
Mount Simpson Bush Camp	-10.00	149.51		SE Peninsula	Benz, B./2008
Mount Sisa	-6.15	142.75	2,689	Eastern Ranges	includes O'Malley Peaks
Mount Sogolomik	-5.23	141.32	2,426	Border Ranges	
Mount Somoro/ Somero*	-3.48	142.19	1,408	Sepik-Ramu	Diamond, J.M./1966; Abid beg Mirza/ca. 1974
Mount Sorong	-3.65	135.93	1,524	Western Ranges	Shaw Mayer, F./1930
Mount Stolle*	-4.80	141.67	2,813	Border Ranges	Mount Mekil
Mount Stower	-7.12	146.53	1,825	SE Peninsula	
Mount Strong	-7.98	146.97	3,588	SE Peninsula	
Mount Suaru	-6.28	144.65	2,458	Eastern Ranges	
Mount Suckling	-9.72	148.97	3,676	SE Peninsula	
Mount Sugar Loaf/ The Sugar Loaf	-5.75	143.67	3,639	Eastern Ranges	
Mount Sullen	-3.42	142.18	1,650	Sepik-Ramu	
Mount Sumi	-6.55	143.57	1,350	Eastern Ranges	
Mount Sumuri	-3.90	135.50	2,500	Western Ranges	
Mount Tafa	-8.63	147.18	2,701	SE Peninsula	
Mount Tantam	-9.72	149.05	2,858	SE Peninsula	
Mount Tauwa	-6.75	144.32	1,430	Eastern Ranges	
Mount Thomson	-10.18	149.83	1,825	SE Peninsula	
Mount Thumb	-8.95	147.55	3,280	SE Peninsula	
Mount Thynne	-8.78	147.45	3,680	SE Peninsula	
Mount Totola	-9.09	147.47	1,600	SE Peninsula	
Mount Townsend	-4.20	142.82	468	Sepik-Ramu	
Mount Trafalgar	-9.13	149.17	1,644	SE Peninsula	
Mount Turu trig. station	-3.62	143.34	1,143	Sepik-Ramu	Diamond, J.M./1966
Mount Udon	-5.79	144.80	3,634	Eastern Ranges	
Mount Ulumam/ Uluman	-4.65	145.97		Huon Peninsula	
Mount Ulur	-6.25	147.33	2,856	Huon Peninsula	Grierson and Van Duesen/1964
Mount Ulus	-6.24	147.48		Huon Peninsula	Cromwell Mountains; position approximate
Mount Unbarn	-5.17	141.40	3,044	Border Ranges	
Mount Veriveri	-8.72	146.98	1,520	SE Peninsula	
Mount Victoria	-8.89	147.53	4,040	SE Peninsula	Huxley Peak denotes the high summit of Mount Victoria
Mount Wadimana	-10.03	149.67		SE Peninsula	
Mount Waimila	-0.04	131.19	710	NW Islands	
Mount Walawasi	-4.92	142.17	2,888	Border Ranges	
Mount Walker	-7.58	146.67	2,728	SE Peninsula	
Mount Wamburi	-5.75	145.58	3,520	Eastern Ranges	

NAME	LATITUDE	LONGITUDE	ELEV. (M)	BIRD REGION	ALTERNATIVE NAMES, RESEARCHERS, REFERENCES
Mount Wamtakin	-5.18	141.87	3,583	Border Ranges	
Mount Wilhelm	-5.78	145.03	4,509	Eastern Ranges	Gilliard, E.T./1950
Mount Wilhelmina	-4.25	138.75	4,730	Western Ranges	Puncak Trikora*
Mount Winter*	-8.80	147.47	3,520	SE Peninsula	Winter Height; Archbold, R., & Rand, A.L./1935
Mount Wondiwoi	-2.73	134.58	2,252	Bird's Neck	on the Wandammen Peninsula
Mount Yanakau	-5.65	143.38	3,327	Eastern Ranges	
Mount Yule	-8.20	146.78	3,276	SE Peninsula	
Mowi	-4.30	141.92	<40	Sepik-Ramu	
Mubi River	-6.80	143.62		S Lowlands	
Mubrani	-0.75	133.42		Bird's Head	Moebrani*
Mugu River	-5.50	144.15		Eastern Ranges	
Muli	-7.80	139.00		Trans-Fly	Moeli*
Muli Strait or River	-7.80	139.00		Trans-Fly	Moeli Strait or River*/Princess Marianne Strait
Mulik	-6.13	147.20	2,080	Huon Peninsula	
Muller Range	-5.40	142.30	3,623	Border Ranges	
Mullins harbour	-10.50	149.92		SE Peninsula	Meek, A.S./1900
Mumeng	-6.98	146.58		SE Peninsula	
Mumkui	-5.48	144.22		Eastern Ranges	
Mundamba	-4.62	143.83		Sepik-Ramu	
Mung River	-6.00	144.70		S Lowlands	
Murik Entrance	-3.78	144.18		Sepik-Ramu	
Murik Lakes	-3.82	144.23		Sepik-Ramu	
Murmur Pass	-5.96	143.86		Eastern Ranges	
Murray Island	-9.92	144.05	234	Trans-Fly	
Murray Pass	-8.51	147.35	2,760–3,196	SE Peninsula	
Murray Pass Camp	-8.52	147.33	2,840	SE Peninsula	
Murray Range	-6.57	144.00	2,254	Eastern Ranges	Wabau Range/Musgrave Mountains
Murray River	-5.85	142.15		S Lowlands	
Murua Island*	-9.17	152.75		SE Islands	Muyua Island/Woodlark Island; Meek, A.S./1897
Musa River	-9.05	148.92		SE Peninsula	
Muschu Island	-3.42	143.58		Sepik-Ramu	
Musgrave Mountains	-6.57	144.00		Eastern Ranges	Murray/Wabau Range
Musom	-6.57	146.99		SE Peninsula	
Muyua Island	-9.17	152.75		SE Islands	Murua Island*/Woodlark Island; Meek, A.S./1897
Myola	-9.14	147.72		SE Peninsula	Lake Myola. Along Kokoda Track there are two localities, Myola 1 nearest the track and Myola 2 to the east and formerly accessible by air allowing access for birders.

NAME	LATITUDE	LONGITUDE	ELEV. (M)	BIRD REGION	ALTERNATIVE NAMES, RESEARCHERS, REFERENCES
Mysore Island*	-1.00	136.00		Bay Islands	Biak Island
Nabire	-3.35	135.50		NW Lowlands	Richards, L./1962
Nabo Creek	-3.30	141.95		Sepik-Ramu	
Nade/Nadi*	-9.67	150.70		SE Islands	
Nadzab	-6.56	146.72		SE Peninsula	current airport servicing Lae city, based in Markham valley to west
Nagam River	-4.13	143.83		Sepik-Ramu	
Naiabui/Naia-bui*	-8.70	146.63		SE Peninsula	near Hall Sound; D'Albertis/1875, 1905
Namatota Island	-3.80	133.92		Bird's Neck	
Namau Province	-7.75	145.00		S Lowlands	
Namoai Bay	-10.62	150.75		SE Islands	
Nangamange	-6.92	141.12		Trans-Fly	
Nankina River	-6.00	146.77		Huon Peninsula	
Naoro	-9.27	147.62		SE Peninsula	
Naoro River	-9.12	147.53		SE Peninsula	
Napan/Nappan*	-2.90	134.85		Bird's Neck	
Napan-Wainami	-3.05	135.75		NW Lowlands	Napon; Shaw Mayer, F./1930
Napon	-3.05	135.75		NW Lowlands	Napan-Wainami; Shaw Mayer, F./1930
Nare Island	-10.75	151.32		SE Islands	
Narirogo Creek	-9.44	147.36		SE Peninsula	in Varirata National Park
Narrenuma	-9.68	150.92		SE Islands	Denton/1883
Naru Hills	-5.49	145.56		Huon Peninsula	Bailey, S.
Naru River	-5.50	145.70		Huon Peninsula	
Nasikwabu Island	-9.56	152.44		SE Islands	Alcester Island
Nassau Range*	-4.20	137.00	5,030	Western Ranges	Sudirman Range
Natural Tunnel	-5.57	143.07	3,200–3,600	Eastern Ranges	
Nautilus Strait	-4.08	133.28		Bird's Neck	Boca de Tovar
Navani Island	-10.78	152.38		SE Islands	
Nawawu	-4.88	145.32		Sepik-Ramu	
Nawen Hill	-3.58	143.62		Sepik-Ramu	
Ndalir	-8.51	140.39		Trans-Fly	near Merauke; position approximate
Nebilyer River	-6.13	144.20		Eastern Ranges	
Nefaar River	-2.58	140.55		NW Lowlands	
Nemayer River	-2.75	141.45		Sepik-Ramu	Paul River*
Nembi valley	-6.20	143.58		Eastern Ranges	
Neneba	-8.69	147.54	1,291	SE Peninsula	north slope of Mount Scratchley
Nengi River	-5.18	142.42		Border Ranges	
Neon basin*	-8.47	147.30		SE Peninsula	Neowa basin
Neowa basin	-8.47	147.30		SE Peninsula	Neon basin*
Netuli Island	-10.23	150.60		SE Islands	Catherine Island*

NAME	LATITUDE	LONGITUDE	ELEV. (M)	BIRD REGION	ALTERNATIVE NAMES, RESEARCHERS, REFERENCES
Neva Pass	-8.73	143.40		Trans-Fly	
New Year Creek	-7.27	145.63		S Lowlands	
Ngabordamlu Point	-6.93	134.18		Aru Islands	Masian Point
Niar River	-4.75	141.92		Border Ranges	
Nicora*/Nicura*/ Nikura/Nkora*	-8.80	146.62		SE Peninsula	D'Albertis, L.
Nigia River	-3.22	142.48		Sepik-Ramu	
Nimbokrang	-3.03	140.45		NW Lowlands	
Nimi River	-6.77	144.95		S Lowlands	
Nimoa Island	-11.30	153.22		SE Islands	
Nimode	-8.63	147.20	2,070	SE Peninsula	
Ninam River	-5.15	145.48		Sepik-Ramu	
Ninei	-1.45	134.03		Bird's Head	Mayr, E./1928
Ningerum	-5.67	141.13	91	S Lowlands	
Nipa	-6.17	143.47		Eastern Ranges	
Nivani Island	-10.78	152.38		SE Islands	
Noemfoor Island*	-1.03	134.90		Bay Islands	Mafor*/Mafoor*/Mefoor*/Numfoor*/Numfor Island
Nogales, Isla de	-10.28	149.35		SE Peninsula	
Nogola River	-2.90	138.45		NW Lowlands	
Noisaroe	-	-		Bird's Neck	"probably near head of Geelvink Bay" (Mayr 1941)
Nomad	-6.30	142.23	120	S Lowlands	
Nomad River	-6.30	142.20		S Lowlands	
Nondugl	-5.88	144.76	1,720	Eastern Ranges	Gilliard, E.T./1950
Nong River	-5.20	141.67		Border Ranges	Nunk River
Noord River	-5.50	138.03		S Lowlands	tributary of Lorentz River
Normanby Island	-9.90	150.90		SE Islands	Hamlin, H./1928
North Buckel Mountains	-5.00	142.55		Eastern Ranges	
North River	-3.97	141.20		Border Ranges	
North Wahgi swamp	-5.72	144.38	1,570	Eastern Ranges	
North-west Arm of the Ok Tedi	-5.12	141.20		Border Ranges	
Northwestern Islands	-0.62	130.60		NW Islands	bird region name for Raja Ampat Islands; also Western Papuan Islands
Northwestern Lowlands	-2.80	138.30		NW Lowlands	bird region of the Mamberamo basin and associated coastal ranges
Nosia village	-5.87	146.91		Huon Peninsula	Beehler, B.M./2012
Nowata	-9.98	149.72	608	SE Peninsula	
Nuguama River	-6.75	145.77		Eastern Ranges	
Num Island	-1.50	135.25		Bay Islands	Mios Num Island
Numanuma	-9.68	150.92		SE Islands	Denton/1883

NAME	LATITUDE	LONGITUDE	ELEV. (M)	BIRD REGION	ALTERNATIVE NAMES, RESEARCHERS, REFERENCES
Numfoor Island*	-1.03	134.90		Bay Islands	Mafor*/Mafoor*/Mefoor*/Noemfoor*/Numfor Island
Numfor Island	-1.03	134.90		Bay Islands	Mafor*/Mafoor*/Mefoor*/Noemfoor*/Numfoor* Island
Nunk River	-5.20	141.67		Border Ranges	Nong River
Nuru River	-5.50	145.70		Huon Peninsula	
Nusela Islands	-1.45	130.17		NW Islands	
Nustiga Island	-4.00	133.37		Bird's Neck	
Nwin River	-2.87	141.13		Sepik-Ramu	Jassi River
Oba Bay	-10.63	150.78		SE Islands	
Oba River	-6.85	139.27		S Lowlands	
Obana	-3.90	136.23		S Lowlands	
Obaoba/Boa Oba	-8.58	146.90	920	SE Peninsula	
Obo	-7.58	141.32		Trans-Fly	
Ocooma	-8.82	146.98		SE Peninsula	
October River	-4.07	141.10		Sepik-Ramu	
Odaki Hills	-2.78	141.33		Sepik-Ramu	
Odamun River	-6.85	138.80		S Lowlands	
Oenake Range	-2.07	141.07		Sepik-Ramu	
Oetakwa River*	-4.85	137.06		S Lowlands	Otakwa/Utakwa* River; Meek, A.S./1910
Ofi Creek	-9.27	147.52		SE Peninsula	
Ogeramnang/ Ogeranang*	-6.48	147.36	1,785	Huon Peninsula	
Oibada River	-10.35	149.62		SE Peninsula	Zimmer, J.T./1920
Oivi Creek	-8.85	147.98		SE Peninsula	
Oiwa	-7.31	146.26		SE Peninsula	near Aseki (CSIRO collection)
Ok Arip [River]	-5.32	141.32		Border Ranges	
Ok Birim [River]	-5.72	141.10		Border Ranges	
Ok Ma [River]	-5.35	141.22		Border Ranges	
Ok Mart [River]	-5.98	141.15		Border Ranges	
Ok Menga [River]	-5.37	141.30		Border Ranges	
Ok Om [River]	-5.00	142.00		Sepik-Ramu	
Ok Tarim [River]	-5.57	140.95		Border Ranges	
Ok Tedi [River]	-6.07	141.13		S Lowlands	
Okaba	-8.10	139.70		Trans-Fly	
Okapa	-6.56	145.60		Eastern Ranges	
Okasa	-6.57	145.65	1,280	Eastern Ranges	Diamond, J.M./1965
Okavai	-7.87	146.38		SE Peninsula	
Okbap	-4.60	140.43		Border Ranges	
Okoume	-8.83	147.05		SE Peninsula	
Oksapmin	-5.22	142.22	2,000	Border Ranges	

NAME	LATITUDE	LONGITUDE	ELEV. (M)	BIRD REGION	ALTERNATIVE NAMES, RESEARCHERS, REFERENCES
Oksibil	-4.90	140.62		Border Ranges	Border Ranges of Indonesian NG; "Ok" = river
Old Mawatta	-9.14	142.95		Trans-Fly	Archbold, R., Rand, A.L., and Tate, W.W./1936
Old Okavai	-7.92	146.43		SE Peninsula	
Olsobip	-5.38	141.52	467	Border Ranges	
Olumwa Island	-9.96	150.85		SE Islands	Duchess Island; off w coast of Normanby Island
O'Malley Peaks [of Mount Sisa]	-6.13	142.73	2,640	Eastern Ranges	
Omati River	-7.73	144.18		S Lowlands	
Omati River, outlet to Gulf of Papua 1	-7.80	144.18		S Lowlands	
Omati River, outlet to Gulf of Papua 2	-7.73	144.18		S Lowlands	
Omaura [mission]	-6.38	145.98	1,500	Eastern Ranges	
Omba River	-4.20	134.75		Bird's Neck	
Omeduku area	-8.60	147.08	ca. 2,000	SE Peninsula	
Omeri Mountain	-8.15	146.62	1,680	SE Peninsula	
Omong River	-5.97	144.80		Eastern Ranges	Gilliard, E.T./1950
Omula Hills	-2.88	141.32		Sepik-Ramu	
Omuru	-5.32	145.70		Sepik-Ramu	Alpha Helix New Guinea Expedition/1969
Omyaka Camp	-5.52	143.04	3,200	Eastern Ranges	(Beehler & Sine 2007)
Onin Peninsula	2.82	132.33		Bird's Neck	
Ononge	-8.67	147.25		SE Peninsula	
Oo Camp	-6.78	145.03		S Lowlands	Mack, A./1996
Oofafa River	-8.82	146.60		SE Peninsula	
Opanabu	-10.02	149.72		SE Peninsula	
Oranga	-5.10	145.47		Sepik-Ramu	
Orangerie Bay	-10.42	149.67		SE Peninsula	Zimmer, J.T./1920
Oranje Mountains*	-4.50	139.50	4,700	Border Ranges	Jayawijaya Mountains
Oransbari/Oranswari*	-1.35	134.28		Bird's Head	
Oreba River	7.92	146.43		SE Peninsula	also known as the Biaru River
Oreke River	-8.70	146.50		SE Peninsula	
Ori Ori/Oriori district	-8.75	147.33		SE Peninsula	Anthony, A.S./1896
Oriomo River	-9.03	143.18		Trans-Fly	
Oriopetana/ Oriropetana	-8.57	146.57	<40	SE Peninsula	Zimmer, J.T./1920
Ormond River	-10.08	148.17		SE Peninsula	Waipara/Gonema Oru
Ornapinka River	-6.30	145.83		Eastern Ranges	
Oro Bay	-8.90	148.48		SE Peninsula	Horton/1945
Orokolo	-7.85	145.32		S Lowlands	

NAME	LATITUDE	LONGITUDE	ELEV. (M)	BIRD REGION	ALTERNATIVE NAMES, RESEARCHERS, REFERENCES
Oronga	-5.10	145.47	80	Sepik-Ramu	in the Adelbert Mountains
Oroville Camp	-6.12	141.30		S Lowlands	Archbold, R., Rand, A.L., and Tate, W.W./1936
Otakwa River	-4.85	137.06		S Lowlands	Utakwa*/Oetakwa* River
Ovakumo	-8.82	146.98	560	SE Peninsula	
Owen Stanley Range	-8.90	147.50	4,000	SE Peninsula	
Ower's Corner	-9.42	147.42		SE Peninsula	
Owgarra	-8.50	147.03	1,200–2,450	SE Peninsula	upper Angabanga River; Meek, A.S./1905–6
Owi Island	-1.24	136.21		Bay Islands	
Paal Putih	-8.28	140.01		Trans-Fly	near the Maro River mouth
Padaido Islands	-1.25	136.60		Bay Islands	Traitors Islands
Paga Hill	-9.48	147.15		SE Peninsula	neighborhood of downtown Port Moresby
Paga Point	-9.48	147.14		SE Peninsula	neighborhood of downtown Port Moresby
Pagwi	-4.05	143.03		Sepik-Ramu	
Pahotouri*/Pahoturi River	-9.28	143.03		Trans-Fly	
Pai Creek	-7.23	146.08		SE Peninsula	
Paiewa River	-7.53	147.38		SE Peninsula	
Pajawa	-7.53	147.38		SE Peninsula	
Palimbai/Parembai	-4.18	143.23		Sepik-Ramu	
Palmer Junction	-5.90	141.55		S Lowlands	
Palmer Junction Camp	-5.92	141.50		S Lowlands	
Pamwi [=Pauwi?]	-1.78	137.89		NW Lowlands	lower Mamberamo; (Mayr 1941)
Panaeati Island	-10.68	152.37		SE Islands	
Panapompom/Pani-pom-pom* Island	-10.72	152.40		SE Islands	
Panasesa Island	-10.73	151.72		SE Islands	of the Conflict Islands; Hamlin, H./1929–30
Panatinane Island	-11.25	153.17		SE Islands	
Paniai Lake	-3.89	136.32	1,735	Western Ranges	Wissel Lakes*; (Junge 1953)
Papa	-9.30	146.99		SE Peninsula	
Param village bush camp	-9.99	149.49		SE Peninsula	Benz, B./2008
Parama Island	-9.00	143.42		Trans-Fly	
Pararoa Creek	-7.25	146.42		SE Peninsula	
Parimau	-4.33	136.58	30	S Lowlands	Mimika River
Parmana Point	-10.15	148.00		SE Peninsula	
Pasai	-3.60	135.83	2,130	Western Ranges	
Passam [forest]	-3.68	143.63		Sepik-Ramu	Alpha Helix New Guinea Expedition/1969
Passim	-1.92	134.08		Bird's Head	in the Arfak Mountains
Paul River*	-2.75	141.45		Sepik-Ramu	Nemayer River
Pauwi	-1.97	137.17		NW Lowlands	

NAME	LATITUDE	LONGITUDE	ELEV. (M)	BIRD REGION	ALTERNATIVE NAMES, RESEARCHERS, REFERENCES
Pawaia basin	-6.71	145.11		Eastern Ranges	southern sector of the Crater Mountain Wildlife Management Area
Pecan Island	-1.75	129.83		NW Islands	near Misool Island
Peka River	-5.55	145.25		Sepik-Ramu	Bigei River
Pelabuan/Biak	-1.19	136.08		Bay Islands	Biak Island; (Gibbs 2009)
Pelee Island	-2.20	130.26		NW Islands	Yetpelle Island; Beehler, B.M.
Penjara*/Penzaia*/ Penzara	-8.80	141.68		Trans-Fly	Archbold, R., Rand, A.L., and Tate, W.W./1936
Peripatusberg	-4.78	141.20		Border Ranges	in the Star Mountains of PNG
Piambil village	-6.04	144.01		Eastern Ranges	Benz, B./1998
Piau River	-6.85	144.77		S Lowlands	
Pig Island	-11.30	153.25		SE Islands	
Pindiu	-6.45	147.52		Huon Peninsula	
Pino River	-6.25	143.63		Eastern Ranges	
Pint River	-5.48	144.54		Eastern Ranges	
Pinu Creek	-9.05	146.82		SE Peninsula	
Pio River	-6.85	145.17		Eastern Ranges	Fio River*
Pioneer Bivak*/ Pionierbivak	-2.28	138.03		NW Lowlands	on the Mamberamo River
Pioneer Range	-7.17	146.90	2,490	SE Peninsula	
Piramid*	-3.89	138.73	1,680	Western Ranges	Pyramid
Pi-you-gona	-5.80	144.13	1,829	Eastern Ranges	Gilliard, E.T./1950
plains of Ulur Camp	-6.33	147.33		Huon Peninsula	Cromwell Mountains; Grierson and Van Duesen/1964
PNG North Coastal Ranges	-3.25	142.00		Sepik-Ramu	include the Bewani, Torricelli, and Prince Alexander Mountains
Poee	-2.70	140.58		NW Lowlands	Lake Sentani (Ripley 1964a)
Pogera*	-5.47	143.10		Eastern Ranges	Porgera
Point Masian	-6.93	134.18		Aru Islands	Masian/Ngabordamlu Point
Point Ngabordamlu	-6.93	134.18		Aru Islands	Masian/Ngabordamlu Point
Point Tanahmerah	-2.40	140.33		NW Lowlands	Cape Tanahmerah
Poire River	-3.08	141.93		Sepik-Ramu	
Pongani River	-8.90	148.48		SE Peninsula	
Popolé mission station	-8.50	147.03	1,100	SE Peninsula	
Popondetta	-8.75	148.25		SE Peninsula	
Poppa Island*	-1.18	129.83		NW Islands	Kofiau/Kafiau*/Kavijaaw*/Koffiao* Island
Pora River	-6.48	143.43		S Lowlands	
Porgera	-5.47	143.10		Eastern Ranges	Pogera*
Pori River	-5.23	142.55		Eastern Ranges	
Port Chalmers	-8.13	146.08		SE Peninsula	
Port Glasgow	-10.35	149.52		SE Peninsula	
Port Lerma	-10.63	150.78		SE Islands	

NAME	LATITUDE	LONGITUDE	ELEV. (M)	BIRD REGION	ALTERNATIVE NAMES, RESEARCHERS, REFERENCES
Port Moresby	-9.45	147.20		SE Peninsula	
Port Moresby harbour	-9.48	147.13		SE Peninsula	known as Fairfax Harbour
Port Romilly	-7.68	144.85		S Lowlands	
Potsdamhafen	-4.15	144.83		Sepik-Ramu	east of the mouth of the Ramu, opposite Manam Island
Potter River*	-4.12	144.08		Sepik-Ramu	Keram River
Poupe River	-8.88	147.38		SE Peninsula	
Powla	-8.58	147.17		SE Peninsula	
Prince Alexander Mountains	-3.62	143.33	1,178	Sepik-Ramu	easternmost of the PNG North Coastal Ranges
Princess Marianne Strait	-7.80	139.00		Trans-Fly	Muli Strait
Prinz Albrechthafen	-4.27	144.97		Sepik-Ramu	
Prinz Hendrik Peak	-4.13	138.20	3,943	Western Ranges	
Profi	-0.69	133.29		Bird's Head	in the Arfak Mountains
Puerto de San Francisco	-10.63	150.78		SE Islands	
Pugra	-8.63	147.20		SE Peninsula	
Pulan River	-2.92	141.10		Sepik-Ramu	
Pulau Adi	-4.20	133.42		Bird's Neck	Isla la Nabaja*/Pulu Adi*
Pulau Bivak	-5.02	138.65		Western Ranges	Lorentz, H.
Pulau Dolak [Dolak Island]	-7.90	138.60		Trans-Fly	Kimaam/Yos Sudarso/Dolok Island
Pulau Enu [Enu Island]	-7.10	134.50		Aru Islands	near the Aru Islands
Pulau Habe [Habe Island]	-8.15	139.13		Trans-Fly	
Pulau Karang [Karang Island]	-7.02	134.67		Aru Islands	
Pulau Kimaam [Dolak Island]	-7.90	138.60		Trans-Fly	Frederik Hendrik*/Yos Sudarso Island
Pulau Tsiof [Tsiof Island]	-0.88	131.20		NW Islands	
Pulau Yos Sudarso [Dolak Island]	-7.90	138.60		Trans-Fly	Kimaam/Yos Sudarso Island
Pulau/Pulu* Babi [Pig Island]	-5.92	134.15		Aru Islands	near the Aru Islands
Pulo Swangi* [Swangi Island]	-7.00	134.23		Aru Islands	Pulau Suangi
Pulu = old form of Pulau					means "island" in Malay and Bahasa Indonesia
Pulu Adi* [Adi Island]	-4.20	133.42		Bird's Neck	Pulau Adi/Isla la Nabaja*
Pulu Manim [Manim Island]	-1.12	134.78		Bay Islands	
Pulu Suangi [Swangi Island]	-7.00	134.23		Aru Islands	Pulau Suangi, Spirit Lighthouse

NAME	LATITUDE	LONGITUDE	ELEV. (M)	BIRD REGION	ALTERNATIVE NAMES, RESEARCHERS, REFERENCES
Puncak Jaya [Jaya Peak]	-4.08	137.18		Western Ranges	Mount Carstensz/Carstensz Massif
Puncak Trikora [Trikora Peak]	-4.26	138.68		Western Ranges	Mount Wilhelmina
Purari River	-7.23	145.32		S Lowlands	
Purari River, delta outflow	-7.85	145.18		S Lowlands	Aieve passage
Purari-Ramu Divide	-5.91	145.58	2,874	Eastern Ranges	
Pureni	-5.86	142.81	1,585	Eastern Ranges	
Puri River	-6.28	146.00		Eastern Ranges	
Puruya River	-7.07	145.55		S Lowlands	
Putei	-7.80	146.13		S Lowlands	
Puwani River	-3.90	141.15		Sepik-Ramu	
Pyramid	-3.89	138.73		Western Ranges	
Quarles meer	-4.32	138.65		Western Ranges	
Quarles valley bivak	-4.32	138.65	3,710	Western Ranges	
Raba Raba	-9.97	149.83		SE Peninsula	Rabaraba
Rabia Hills	-0.26	130.92	720	NW Islands	
Rai Coast	-5.75	146.50		Huon Peninsula	
Raja Ampat Islands	-0.62	130.60		NW Islands	here called the Northwestern Islands
Rakamanda mission	-5.52	143.60		Eastern Ranges	
Rakua River	-9.55	149.40		SE Peninsula	
Ramoi River	-0.95	131.27		Bird's Head	Beccari, O./1875
Ramu River	-4.20	144.70		Sepik-Ramu	
Randowaja	-1.85	136.53		NW Lowlands	
Rani Island	-0.92	135.50		Bay Islands	southwest of Supiori Island
Ransiki	-1.52	134.18		Bird's Head	
Rara Island	-10.81	152.49		SE Islands	(Dumbacher et al. 2011)
Rawa Biru	-8.68	140.89		Trans-Fly	Wasur National Park; position approximate
Rawak Island	-0.03	130.93		NW Islands	Luwak Island; off Waigeo island
Rawlinson Range	-6.53	147.28	3,200	Huon Peninsula	Wahnes, C./1905–6
Redscar Bay	-9.17	146.83		SE Peninsula	
Redscar Head	-9.27	146.90		SE Peninsula	
Renaia Island	-10.83	152.97		SE Islands	
Renard Islands	-10.85	153.00		SE Islands	
Rentoul River	-6.33	142.07		S Lowlands	tributary of the Nomad River
Rentoul River East Branch	-6.38	142.47		S Lowlands	
Reu Island	-9.25	152.95		SE Islands	off Woodlark Island (Pratt et al. 2006)
Richardson Range	-9.09	147.48		SE Peninsula	
Right May River	-4.57	141.57		Sepik-Ramu	
Rigo/Old Rigo	-9.73	147.67		SE Peninsula	

NAME	LATITUDE	LONGITUDE	ELEV. (M)	BIRD REGION	ALTERNATIVE NAMES, RESEARCHERS, REFERENCES
Rijak Bay	-10.63	152.75		SE Islands	
Rivo/Riwo	-5.15	145.80		Sepik-Ramu	
Rock Lookout	-7.20	146.32	2,320	SE Peninsula	
Rocky Pass	-10.62	150.75		SE Islands	
Roemberpon Island*	-1.83	134.20		Bird's Head	Rumberpon Island
Rogeia Island*	-10.63	150.63		SE Islands	Heath Island
Roguoia Creek	-9.10	147.55		SE Peninsula	
Roku	-9.47	147.08		SE Peninsula	
Rombebai Lake	-1.87	137.92		NW Lowlands	Morris Lake*
Ron Island*	-2.38	134.53		Bird's Neck	Roon Island; Doherty, W./1897
Rona*	-9.42	147.37		SE Peninsula	Rouna
Roon Island	-2.38	134.53		Bird's Neck	Ron Island*; Doherty, W./1897
Rorona	-9.03	146.88	<40	SE Peninsula	Zimmer, J.T./1920
Ross bluff	-5.03	141.07	3,727	Border Ranges	
Rossel Island	-11.37	154.13		SE Islands	Yela Island*; Eichhorn, Meek, Hamlin
Rouffaer River	-2.90	138.43		NW Lowlands	Tariku River
Rouna	-9.42	147.37		SE Peninsula	Rona*
Rouna Falls	-9.42	147.37		SE Peninsula	
Round Hill	-8.04	146.45	100	SE Peninsula	
Round Island	-11.33	153.22		SE Islands	near Tagula Island
Rubi	-3.17	134.97		Bird's Neck	
Rugged Head	-10.70	150.20		SE Peninsula	Guna Isu
Rugli	-5.67	144.20		Eastern Ranges	
Rumberpon Island	-1.83	134.12		Bird's Head	Amberpon*/Roemberpon* Island
Rumgenai	-5.91	141.30		S Lowlands	Gregory/1997
Rup Rup Island*	-3.50	144.60		Sepik-Ramu	Blupblup Island
Ruti airstrip	-5.30	144.25	488	Eastern Ranges	in the upper Jimi valley
Rutunabi	-6.50	145.17	2,000	Eastern Ranges	in Crater Mountain Wildlife Management Area
Sabang	-4.68	138.73		S Lowlands	
Saddle Mountain	-6.48	146.77		Huon Peninsula	Wahnes/1899; Nyman, E./1914
Safia airstrip	-9.60	148.63		SE Peninsula	
Safia-Pongani Gap	-9.50	148.50		SE Peninsula	
Sagara	-10.62	152.77		SE Islands	
Sagarai River	-10.42	149.88		SE Peninsula	
Sagewin Island	-0.95	130.63		NW Islands	
Sagewin Strait	-0.92	130.75		NW Islands	Saguien* Strait
Saguien* Strait	-0.92	130.75		NW Islands	Sagewin Strait
Saibai Island	-9.40	142.70	-		Torres Strait Islands, Australia
Saibutu	-9.47	150.55		SE Peninsula	
Saidor	-5.62	146.45		Huon Peninsula	

NAME	LATITUDE	LONGITUDE	ELEV. (M)	BIRD REGION	ALTERNATIVE NAMES, RESEARCHERS, REFERENCES
Sainkedoek*/ Sainkeduk	-0.78	132.22	61	Bird's Head	Ripley, S.D./1938
Saint Aignan Island*	-10.67	152.73		SE Islands	Misima Island; Meek, A.S./1897; Hamlin/1930
Saint Joseph River	-8.82	146.57		SE Peninsula	Angabanga River
Sakome	-1.45	134.07		Bird's Head	
Salamaua/Salamoa*	-7.05	147.05		SE Peninsula	
Salamo	-9.67	150.78		SE Islands	on Fergusson Island
Salawati Island	-0.80	130.92		NW Islands	Salawatty*/Salwatti* Island
Salem	-0.75	132.32		Bird's Head	
Saloga Island	-9.25	152.75		SE Islands	
Salumei River	-4.47	143.18		S Lowlands	
Salwatty Island*	-0.80	130.92		NW Islands	Salawati Island; Bruijn, A.A./1875
Samarai Island/town	-10.60	150.70		SE Islands	Dinner Island*; Beck/1929; Hamlin
Samati	-1.00	131.08		NW Islands	Guillemard/1885
Samboga River	-8.75	148.47		SE Peninsula	
Samia Creek/Samia River	-6.85	144.65		S Lowlands	
Samoa harbour	-7.03	147.03		SE Peninsula	
San Juan del Prado*	-3.97	134.38		Bird's Neck	Kajumerah Bay
San River	-4.87	141.53		Border Ranges	
San Roche passage	-10.47	149.85		SE Peninsula	
Sanchi River	-4.22	142.75		Sepik-Ramu	
Sandbank Bay	-10.18	148.57		SE Peninsula	
Sangara	-8.82	148.10		SE Peninsula	
Sangke River	-2.88	140.95		NW Lowlands	
Sankwep River	-6.62	147.00		Huon Peninsula	
Sansapor	-0.48	132.08		Bird's Head	Sausapor*
Saokorem	-0.50	133.17		Bird's Head	
Saonek	-0.42	130.82		NW Islands	
Saonek Island	-0.50	130.80		NW Islands	
Sapara/Saparo River	-6.03	147.52		Huon Peninsula	
Saparako	-	-		-	in the Huon Gulf area; (Reichenow 1891)
Sapoa/Sapoi Camp	-6.92	146.47		SE Peninsula	renamed Ivimka Camp in the Lakekamu basin/2002
Saporkren area	-0.44	130.73	140	NW Islands	Mauro, I./2002
Sareba Bay	-2.00	135.00		Bay Islands	
Sariba Island	-10.60	150.78		SE Islands	Meek, A.S.; Eichhorn
Sarmi	-1.87	138.75		NW Lowlands	Doherty, W./1905
Saruwaged/Sar-awaget*/Salawaket* Range	-6.33	146.83	4,121	Huon Peninsula	
Satop	-6.08	147.18		Huon Peninsula	

NAME	LATITUDE	LONGITUDE	ELEV. (M)	BIRD REGION	ALTERNATIVE NAMES, RESEARCHERS, REFERENCES
Sattelberg Mountains	-6.50	147.80		Huon Peninsula	
Sau valley	-5.36	144.03		Eastern Ranges	
Sauiti Creek	-7.20	143.83		S Lowlands	
Sauri Range	-6.55	143.77	1,666	Eastern Ranges	
Sausapor*	-0.48	132.08		Bird's Head	Sansapor
Sauwa River	-8.73	147.28		SE Peninsula	
Sauwa River East Arm	-8.68	147.32		SE Peninsula	
Sawa Sawaga Channel	-10.62	150.75		SE Islands	
Schatteburg Mountains	-4.90	142.13	3,940	Border Ranges	
Schildpad Islands	-1.45	130.47		NW Islands	Ripley, S.D./1937
Schouten Islands	-1.00	136.00		Bay Islands	the islands of Biak, Supiori, Numfor, etc.
Schouten Isles	-3.50	144.50		Sepik-Ramu	confusingly there are two distinct Schouten groups
Schrader Range	-5.30	144.40	2,320	Eastern Ranges	
Schraderberg	-4.88	144.22	2,070	Eastern Ranges	Bürgers/1913
Screw River	-4.12	142.92		Sepik-Ramu	
Sedjak	-0.75	132.75	503	Bird's Head	Ripley, S.D./1938
Sedjoi	-0.78	132.68	526	Bird's Head	Ripley, S.D./1938
Seerosensee	-4.25	142.97		Sepik-Ramu	lower Sepik River
Sek Island	-5.08	145.82		Sepik-Ramu	Diamond, J.M./1969
Sekak	-1.57	132.35		Bird's Head	
Sekru	-2.93	132.25		Bird's Head	Skru
Sele Strait	-1.34	130.92		NW Islands	
Seléh*/Tanjung Sele	-1.43	130.93		NW Islands	Sele Point or Cape; Bernstein, H.A.
Selekobo	-0.85	132.03		Bird's Head	southwestern flank of Tamrau Mountains
Selenek	-0.80	132.12		Bird's Head	
Seleo Island	-3.17	142.48		Sepik-Ramu	
Sempi	-5.01	145.79		Sepik-Ramu	Diamond, J.M./1969
Sena River	-6.40	144.82	1,372	Eastern Ranges	Diamond, J.M./1964
Sentani Lake*	-2.62	140.50		NW Lowlands	Lake Sentani; Mayr, E./1931
Sepik River	-4.20	144.00		Sepik-Ramu	
Sepik-Ramu	-4.50	143.40		Sepik-Ramu	bird region encompassing the Sepik and Ramu drainages
Sepik-Wahgi Divide	-5.70	144.30	3,634	Eastern Ranges	
Sereri/Duro Creek	-2.88	141.40		Sepik-Ramu	
Sergile	-0.88	131.25		NW Islands	
Serui/Seroei*	-1.88	136.23		Bay Islands	
Setekwa/Setakwa* River	-4.57	137.35		S Lowlands	Kloss, C.B./1912
Sevia	-6.30	147.65		Huon Peninsula	Beck, R./1929
Sewa Bay	-10.00	150.95		SE Islands	

NAME	LATITUDE	LONGITUDE	ELEV. (M)	BIRD REGION	ALTERNATIVE NAMES, RESEARCHERS, REFERENCES
Sewicki Lake	-3.34	133.83	123	Bird's Neck	Thébaud & Mila/2014
Sey River	-5.42	141.88		Border Ranges	
Siagara	-10.62	152.77		SE Islands	
Sialum	-6.08	147.58		Huon Peninsula	
Siassi Islands	-5.90	147.93		SE Peninsula	
Sibi River	-4.93	141.80		Border Ranges	
Sibil	-4.90	140.62		Border Ranges	village in the Border Ranges of Indonesian NG
Sibium Mountains	-9.32	148.42	2,230	SE Peninsula	
Sicklebill Camp	-6.48	145.20	2,134	Eastern Ranges	In Crater Mt WMA
Sideia Island	-10.60	150.68		SE Islands	
Sigufe	-8.78	147.23		SE Peninsula	
Sii River	-7.86	146.55		SE Peninsula	in the Lakekamu basin; Beehler, B.M. & C.H., and Pruett-Jones, S./1982
Sikuba/Sikubi	-8.78	147.23		SE Peninsula	
Sikwong	-7.07	146.10		SE Peninsula	
Silbattabatta	-6.02	134.37		Aru Islands	Aru Islands
Silisigoda	-8.49	147.30		SE Peninsula	
Simbai	-5.28	144.42	1,771	Eastern Ranges	Gilliard, E.T./1964
Simbana	-6.58	147.85		Huon Peninsula	Cotton, Nyman, E.
Simbang	-6.70	147.70		Huon Peninsula	Nyman, E./1899
Singarokai	-5.38	146.90		Huon Peninsula	Beehler, B.M./2012
Singgajebi Bay/Sing-gajeri Gulf	-2.92	134.83		Bird's Neck	
Sirebi River	-7.27	144.18		S Lowlands	
Sireru River	-7.00	144.30		S Lowlands	
Sirinumu Dam	-9.50	147.64		SE Peninsula	
Siriwo River	-2.98	135.87		Western Ranges	Shaw Mayer, F./1930
Sitipa River	-4.58	142.60		Sepik-Ramu	
Siwi	-1.50	134.00		Bird's Head	Mayr, E./1928
Siwi Utame Wildlife Management Area	-5.83	144.13	1,800–2,317	Eastern Ranges	
Skelton Island	-10.61	151.24		SE Islands	in the Engineer Islands
Skonga River	-4.58	141.18		Border Ranges	
Skru	-2.93	132.25		Bird's Head	Sekru
Slade Island	-10.59	151.20		SE Islands	in the Engineer Islands
Slate Creek	-7.22	146.53		SE Peninsula	
Snake Point	-6.17	141.12		S Lowlands	
Snake River	-7.05	146.58		SE Peninsula	
Snow Mountains*	-4.00	139.50	4,884	Western Ranges	Maoke Range; Meek, A.S./1910
Soabi Base airstrip	-6.06	142.14		S Lowlands	in the upper Strickland River; (I. Woxvold in litt.)

NAME	LATITUDE	LONGITUDE	ELEV. (M)	BIRD REGION	ALTERNATIVE NAMES, RESEARCHERS, REFERENCES
Soaru Range	-6.50	144.33		Eastern Ranges	
Sobger River	-3.72	140.30		NW Lowlands	
Soek Island*	-1.00	136.00		Bay Islands	Biak Island
Soemeet River	-0.80	132.17		Bird's Head	Ripley, S.D./1938
Soepiori Island*	-0.75	135.58		Bay Islands	Supiori Island; northern sister to Biak Island
Sogeram River	-4.80	144.72		Sepik-Ramu	
Sogeri district	-9.42	147.50		SE Peninsula	Sharpe
Sogeri Plateau	-9.43	147.42	450–500	SE Peninsula	
Sogeri/Sogere*	-9.42	147.42		SE Peninsula	Forbes, H.O./1885
Sogolomik	-5.23	141.23		Border Ranges	
Sol River	-5.17	141.65		Border Ranges	
Solal	-1.32	131.13		NW Islands	
Soliabeda/Soliabedo*	-6.70	144.83	610	Eastern Ranges	Diamond, J.M.
Somai	-6.65	145.22		Eastern Ranges	Benz, B./2002
Sopa River	-5.12	145.47		Sepik-Ramu	
Sopas hospital	-5.47	143.67		Eastern Ranges	
Soriabida*/Soliabeda	-6.70	144.83		Eastern Ranges	Diamond, J.M./1965
Sorido	-1.15	136.00		Bay Islands	Thompson, M./1963
Sorong district	-1.08	131.33		Bird's Head	
Sorong Island*	-0.88	131.25		Bird's Head	Dom Island; Ripley, S.D./1937
South Branch Right May River	-4.49	141.48		Sepik-Ramu	
South Buckel Mountains	-5.12	141.13	1,907	Border Ranges	
South Cape	-10.72	150.23		SE Peninsula	
South Naru	-5.50	145.75		Huon Peninsula	
South-East Cape	-10.72	150.23		SE Peninsula	
Southeastern Islands	-10.50	151.70		SE Islands	bird region encompassing the SE Islands of PNG
Southeastern Peninsula	-9.00	147.60		SE Peninsula	bird region: southeasternmost tail of New Guinea
Southern Highlands Province	-5.86	144.22		Eastern Ranges	
Southern Lowlands	-6.40	141.00		S Lowlands	bird region of the southern watershed lowlands of New Guinea
Sowan River	-3.37	143.13		Sepik-Ramu	Anumb River
Spear Island	-8.98	149.13		SE Peninsula	
Staniforth Range	-7.28	145.69	2,040	Eastern Ranges	
Stanton Point	-10.62	150.77		SE Islands	
Star Mountains (east)	-4.99	141.10		Border Ranges	in PNG
Star Mountains (west)	-4.90	140.91		Border Ranges	in Indonesian New Guinea
Stephansort	-5.42	145.72		Sepik-Ramu	near Madang; Nyman, E./1898
Stirling Range	-10.25	150.42	1,125	SE Peninsula	

NAME	LATITUDE	LONGITUDE	ELEV. (M)	BIRD REGION	ALTERNATIVE NAMES, RESEARCHERS, REFERENCES
Strickland River	-7.00	142.00		S Lowlands	
Strickland River gorge, northern entrance	-5.37	142.12		Eastern Ranges	also known as the Strickland gorge
Strickland River gorge, southern exit	-5.80	142.18		S Lowlands	also known as the Strickland gorge
Sturt Island	-8.17	142.25		Trans-Fly	Archbold, R., Rand, A.L., and Tate, W.W./1936
Suau Island	-10.70	150.25		SE Peninsula	
Subitana plantation	-9.42	147.53		SE Peninsula	
Sudest Island	-11.75	153.50		SE Islands	Tagula Island; Meek, A.S./1916
Sudirman Range	-4.20	137.00	5,030	Western Ranges	Nassau Range*
Sugar Loaf, the	-5.75	143.67	3,639	Eastern Ranges	
Sui	-8.82	143.40		Trans-Fly	
Suki	-8.05	141.72		Trans-Fly	
Suki lagoon	-8.08	141.77		Trans-Fly	
Suku	-8.92	147.33	400	SE Peninsula	upper Vanapa River; Anthony, A.S./1898
Suloga/Suloga harbour	-9.25	142.92		Trans-Fly	
Sumberbaba	-1.82	136.68		NW Lowlands	Richards, L./1962
Sumogi Island	-8.39	143.00		Trans-Fly	
Sunday Creek	-7.30	146.67		SE Peninsula	Alpha Helix New Guinea Expedition/1969
Sunshine	-7.07	146.59		SE Peninsula	in the upper Watut River
Superua River	-8.02	147.32		SE Peninsula	
Supiori Island	-0.77	135.58		Bay Islands	Soepiori Island*; northern sister to Biak Island
Suria	-9.08	147.48		SE Peninsula	Beehler, B.M., & Hare, A.
Surim Camp	-5.87	146.72	1,110	Huon Peninsula	Mack, A./2003
Surprise Creek	-7.27	146.53		SE Peninsula	Stevens, H./1932; (Greenway 1935)
Surprise Creek Camp	-7.25	146.60	884	SE Peninsula	Stevens, H./1932; (Greenway 1935)
Tab Island	-5.17	145.83		Sepik-Ramu	Diamond, J.M./1969
Tabai River	-3.03	135.83		Western Ranges	
Tabi*	-2.03	139.12		NW Lowlands	Takar; Doherty, W./1897; (Diamond 1981a)
Tabibuga	5.57	144.65	1,360	Eastern Ranges	
Tabu River	-4.62	141.15		Sepik-Ramu	
Tabubil	-5.37	141.23	600	Border Ranges	
Taburi district	-9.42	147.43		SE Peninsula	Adelbert Mountains
Tagari River	-6.08	142.93		Eastern Ranges	
Tagula (town)	-11.35	153.22		SE Islands	
Tagula Island	-11.75	153.50		SE Islands	Sudest Island; Hamlin, H./1929
Taibutu	-9.47	150.55		SE Peninsula	
Takar	-2.03	139.12		NW Lowlands	Tabi*; opposite Yamna Island; (Diamond 1981a)
Takin River	-5.22	141.82		Border Ranges	

NAME	LATITUDE	LONGITUDE	ELEV. (M)	BIRD REGION	ALTERNATIVE NAMES, RESEARCHERS, REFERENCES
Tamata	-8.35	147.87		SE Peninsula	Hamlin, H./1905
Tambanum	-4.20	143.60		Sepik-Ramu	
Tambavwa Junction	-7.58	141.38		Trans-Fly	Everill Junction
Tambul	-5.88	143.95		Eastern Ranges	Tumbul*
Tameo River	-10.17	150.27		SE Peninsula	
Tami Islands	-6.77	147.90		Huon Peninsula	
Tami River	-2.66	140.91		NW Lowlands	
Tamrau Mountains	-0.60	132.10	2,582	Bird's Head	
Tamulol	-1.95	130.28		NW Islands	on Misool Island; Ripley, S.D./1954
Tana	-1.48	137.90		NW Lowlands	Taua; mouth of the Mamberamo River
Tana Mera	-2.42	140.33		NW Lowlands	translates to "red earth"
Tana Merah/ Tanamerah	-6.08	140.28		Trans-Fly	translates to "red earth"
Tanahmerah Bay	-2.42	140.30		NW Lowlands	
Tantam	-9.72	149.05		SE Peninsula	
Tanubada ponds	-9.40	147.29		SE Peninsula	
Tapala	-8.08	146.10		SE Peninsula	near Malalaua and Kerema
Tapini	-8.35	146.98	960	SE Peninsula	
Tarara	-8.82	141.87		Trans-Fly	Archbold, R., Rand, A.L., and Tate, W.W./1936–37
Tarawai Island	-3.20	143.25		Sepik-Ramu	
Tarera Bay	-4.00	134.63		Bird's Neck	
Tarfia	-2.32	140.12		NW Lowlands	
Tari	-5.84	142.95	1,640	Eastern Ranges	
Tari Gap	-5.96	143.16		Eastern Ranges	Frith, C.B. and D.W./various years
Tariku River	-2.90	138.43		NW Lowlands	Rouffaer River*
Taritatu River	-3.25	139.00		NW Lowlands	Idenburg River*
Tarl River	-8.53	141.13		Trans-Fly	
Tarona Camp	-5.86	146.72		Huon Peninsula	Mack, A./2003
Taruve	-8.53	147.07		SE Peninsula	
Tau River	-4.75	142.17		Border Ranges	
Taua	-1.48	137.90		NW Lowlands	Tana; lower Mamberamo River
Taurama River, channel entrance	-7.78	144.00		S Lowlands	
Tauri River, outlet to Port Chalmers	-8.13	146.10		SE Peninsula	
Tau'telo Camp	-6.70	144.98	760	S Lowlands	Filardi, C., & Smith, C./2001
Teba	-1.48	137.90		NW Lowlands	Feba*; lower Mamberamo River
Tebi River	-6.69	144.10		S Lowlands	(Coates 1985)
Tekadu No. 1 aid post	-7.65	146.55		SE Peninsula	
Tekin	-5.20	142.20		Border Ranges	on the Arik River
Telefomin/Telefolmin*	-5.12	141.63	1,520	Border Ranges	Gilliard, E.T./1950

NAME	LATITUDE	LONGITUDE	ELEV. (M)	BIRD REGION	ALTERNATIVE NAMES, RESEARCHERS, REFERENCES
Teluk Cenderawasih	-2.00	135.00		Bay Islands	Geelvink/Cenderawasih Bay
Teluk Sebakor	-3.54	132.80		Bird's Neck	Sebakor Bay; on the Bomberai Peninsula
Tembagapura	-4.17	137.08		Western Ranges	"Copper Town" of the Freeport Mine above Timika
Tembroki	-1.37	133.97		Bird's Head	Tombrok
Teminaboean	-1.43	132.05		Bird's Head	
Tendanje Island	-3.22	143.25		Sepik-Ramu	
Teranmin River	-5.13	142.28		Border Ranges	
Teste Island	-10.95	151.07		SE Islands	40 km southeast of Samarai
Tetebedi	-9.17	148.07		SE Peninsula	
The Gap	-9.13	147.70	2,000	SE Peninsula	
Thompson Junction	-5.93	141.62		S Lowlands	junction of the Palmer and Tully Rivers
Three Pinnacles	-4.80	141.33	2,400	Border Ranges	
Thukari Creek	-6.90	144.78		S Lowlands	
Thurnwald Range	-4.83	141.43		Border Ranges	
Tifalmin	-5.12	141.42	1,480	Border Ranges	
Tigi Plantation	-5.58	144.30		Eastern Ranges	Alpha Helix New Guinea Expedition/1969
Tigibi	-5.93	143.05		Eastern Ranges	
Tigili River	-4.88	141.17		Border Ranges	
Tikiam Camp	-4.48	145.03		Sepik-Ramu	Mack, A./2003
Timalia River	-6.13	142.90		Eastern Ranges	position approximate; tributary of the Tagari River; Woxvold, I.
Timbe River	-5.93	147.07		Huon Peninsula	
Timbunke/Timbunki*	-4.18	143.52		Sepik-Ramu	
Timeepa/Timepa	-3.97	135.78	1,494	Western Ranges	
Timika	-4.05	136.90		S Lowlands	
Tip	-1.82	130.08		NW Islands	village in central-interior of Misool Island
Tirawiwa River	-3.03	136.37		Western Ranges	
Tiri River	-3.31	138.62		NW Lowlands	Mamberamo basin; (van Balen et al. 2003)
Tive Plateau	-6.37	144.63	800–1,200	Eastern Ranges	
Tiveri River	-7.85	146.38		SE Peninsula	
Tjobonda	-	-		Bird's Head	in the Arfak Mountains; Leglaize/1876
Tobweyama Island	-10.21	151.21		SE Islands	off Normanby Island. Pratt, T.K. and Moore, M. 2004
Toerey	-8.39	140.55		Trans-Fly	on the Merauke River; position approximate; (Mees 1982b)
Togoba	-5.90	144.15		Eastern Ranges	
Toiva Creek	-9.03	147.23		SE Peninsula	
Tol River	-5.07	147.67		Huon Peninsula	
Tomba	-5.78	143.97	2,530	Eastern Ranges	Gilliard, E.T./1950
Tomba Base Camp	-5.78	143.97	2,134–2,438	Eastern Ranges	(Mayr & Gilliard 1954)

NAME	LATITUDE	LONGITUDE	ELEV. (M)	BIRD REGION	ALTERNATIVE NAMES, RESEARCHERS, REFERENCES
Tombrok	-1.37	133.97		Bird's Head	Tembroki
Tomianumu	-9.42	147.53		SE Peninsula	
Tomu River	-6.62	142.13		Trans-Fly	
Tonda Wildlife Management Area	-9.02	141.30		Trans-Fly	
Tondon Range	-6.13	143.53		Eastern Ranges	
Topak River	-5.20	143.23		Eastern Ranges	
Tor River	-1.98	138.97		NW Lowlands	
Toro Pass, south entrance	-9.02	143.37		Trans-Fly	
Torres Strait	-10.00	143.00		-	
Torricelli Mountains	-3.37	142.17	1,560	Sepik-Ramu	
Townsville	-5.37	141.23		Border Ranges	
Traitors Islands	-1.25	136.60		Bay Islands	Padaido Islands
Trangan/Terangan* Island	-6.50	134.25		Aru Islands	in the Aru Islands; Kühn, H./1900
Trans-Fly	-8.50	141.50		Trans-Fly	bird region encompassing the south-central bulge of New Guinea
Trauna Farm	-5.50	144.20	1,200	Eastern Ranges	
Trauna Gap*/Trauna Ridge*	-5.49	144.20	ca. 1,220	Eastern Ranges	Lepa Ridge
Trauna River	-5.50	144.17		Eastern Ranges	
Trauna valley	-5.53	144.15		Eastern Ranges	Alpha Helix New Guinea Expedition/1969
Treub Mountains	-4.50	138.72	2,000–2,400	Western Ranges	
Treubbivak	-4.47	138.58		Western Ranges	Versteeg, G./1913
trig. station 7482	-4.99	141.23	3,218	Border Ranges	
trig. station NM/J/14	-6.00	146.49	3,917	Huon Peninsula	
Triton Bay	-3.83	134.13		Bird's Neck	Cayley-Webster, H.
Trobriand Islands	-8.53	150.95		SE Islands	Meek, A.S./1895
Tsau River	-5.66	144.60		Eastern Ranges	
Tsoma River	-6.84	145.17		Eastern Ranges	
Tsuewenkai*/ Tsuwenkai	-5.42	144.63		Eastern Ranges	Majnep, I.S.
Tua River	-6.65	144.55		Eastern Ranges	
Tubatuba or Tuba Tuba	-10.58	151.20		SE Islands	
Tubusereia/ Tupuselela*/ Tupuselia*	-9.55	147.32		SE Peninsula	
Tufi	-9.08	149.32		SE Peninsula	Hamlin, H./1929
Tuiu River	-9.05	146.82		SE Peninsula	
Tully River	-5.77	141.62		S Lowlands	
Tuman	-3.60	142.75		Sepik-Ramu	

NAME	LATITUDE	LONGITUDE	ELEV. (M)	BIRD REGION	ALTERNATIVE NAMES, RESEARCHERS, REFERENCES
Tuman River	-5.77	144.48		Eastern Ranges	
Tumbudu River	-5.32	142.32		Border Ranges	
Tumbul*	-5.88	143.95		Eastern Ranges	Tambul
Tuoa Creek	-7.12	145.52		S Lowlands	
Tupensomnai	-5.92	141.50		Border Ranges	
Turama River	-7.57	143.77		S Lowlands	
Ua-Ule Creek	-9.32	147.42		SE Peninsula	
Ubaigubi aid post	-6.49	145.17	1,600	Eastern Ranges	
Ubaigubi Lodge	-6.48	145.18	1,750	Eastern Ranges	
Ubaigubi Sicklebill Camp	-6.05	145.17	2,120	Eastern Ranges	Peckover, W.S. and Gillison, D.
Ubamurai	-6.60	145.20		Eastern Ranges	
Uberi	-9.36	147.50		SE Peninsula	on the Kokoda Track
Udabe River	-8.73	147.28		SE Peninsula	
Udava River	-8.90	147.37	360–690	SE Peninsula	
Ukarumpa	-6.33	145.88		Eastern Ranges	
Ukemupuko River	-4.58	136.08		S Lowlands	
Ukra swamp	-8.68	140.86		Trans-Fly	in Wasur National Park; position approximate
Ulap	-6.03	147.20	880	Huon Peninsula	
Ulya	-5.87	144.35	1,680	Eastern Ranges	
Umi River	-6.23	146.18		SE Peninsula	
Unchemchi	-5.22	141.58	1,783	Border Ranges	Gilliard, E.T./1950
Uni	-7.00	139.27		S Lowlands	
upper Angabunga* River	-8.50	147.03		SE Peninsula	Angabanga River
upper Setekwa River	-4.32	137.53	760	Western Ranges	
upper Vanapa valley	-8.58	147.25	960–2,000	SE Peninsula	Rand, A.L./1933
upper Watut River	-7.15	146.58		SE Peninsula	
Uranua Mountains	-7.67	147.24	400–1,800	SE Peninsula	
Uraru	-6.90	144.87		S Lowlands	Purari basin
Urawa River, mouth	-5.85	146.85		Huon Peninsula	
Urei River	-5.23	142.58		Eastern Ranges	
Uria River	-5.90	145.73		Huon Peninsula	
Urimo cattle station	-	-		Sepik-Ramu	60 km south-southwest of Wewak
Urisa	-3.21	133.77	0	Bird's Neck	Thébaud & Mila/2014
Urom River	-4.93	144.58		Sepik-Ramu	
Urumar Camp	-4.67	145.20		Sepik-Ramu	in the Adelbert Mts; Benz, B./2007
Urumuka River	-4.53	136.00		S Lowlands	
Uruna/Urunu	-8.58	147.28		SE Peninsula	

NAME	LATITUDE	LONGITUDE	ELEV. (M)	BIRD REGION	ALTERNATIVE NAMES, RESEARCHERS, REFERENCES
Usage/Usake River	-4.57	141.58		Sepik-Ramu	
Usino	-5.52	145.37		Sepik-Ramu	
Uta	-4.53	136.00		S Lowlands	
Uta River	-4.28	136.08		S Lowlands	
Utai	-3.42	141.58	210	Sepik-Ramu	Diamond, J.M.
Utakwa River*	-4.85	137.06		S Lowlands	Otakwa River; Meek, A.S./1910
Utanata River	-4.58	136.08		S Lowlands	
Utu	-5.17	145.53		Sepik-Ramu	
Uwimbu River*	-7.20	139.60		Trans-Fly	Digul River/Digoel* River
Vaifa	-8.53	146.87		SE Peninsula	
Vailala/Vaillala* River	-7.95	145.42		S Lowlands	Bailala River*
Vakena	-9.80	148.04		SE Peninsula	SE Peninsula
Vakena Mountains	-9.80	148.04		SE Peninsula	
Van Daalen River	-3.07	138.15		NW Lowlands	
Van Rees Mountains	-2.58	138.25		NW Lowlands	
Vanapa River	-9.08	146.97		SE Peninsula	Venapa River*
Vanimo	-2.07	141.03		Sepik-Ramu	
Vanumai/Vanamai*	-8.87	146.67	40	SE Peninsula	Hamlin, H./1929
Varirata National Park	-9.43	147.37	720	SE Peninsula	
Vasagabila	-9.10	147.12	<40	SE Peninsula	
Veimauri	-9.05	147.08		SE Peninsula	
Veimauri River	-9.07	146.97		SE Peninsula	Zimmer, J.T./1920–21
Veina Island*	-11.33	153.45		SE Islands	Yeina Island
Veipa	-8.53	146.55	<40	SE Peninsula	Beipa
Veiru Creek	-7.29	144.25		S Lowlands	Kikori basin; Woxvold, I.
Veiya River	-9.05	147.02		SE Peninsula	Kuriva River
Venapa River*	-9.08	146.97		SE Peninsula	Vanapa River
Verjus Dome	-8.45	146.83		SE Peninsula	
Vetapu River	-9.90	147.37		SE Peninsula	
Viai Island	-3.38	144.40		Sepik-Ramu	
Victor Emanuel Mountains	-5.25	141.07		Border Ranges	Gilliard, E.T./1954
Vikaiku	-8.53	146.87		SE Peninsula	
Viralieu Creek	-9.05	146.72		SE Peninsula	
Vitiaz Strait	-5.58	147.00		-	
Vogelkop Peninsula*	-1.60	133.00		Bird's Head	Berau*/Bird's Head/Doberai Peninsula
Vokeo Island	-3.23	144.10		Sepik-Ramu	Wogeo Island*
Vulcan Island*	-4.83	145.03		Sepik-Ramu	Manam Island; Meek, A.S./1913
Wa Samson/Wae* Samson River	-0.78	131.38		Bird's Head	Beccari, O./1875
Wabag	-5.05	143.72		Eastern Ranges	

NAME	LATITUDE	LONGITUDE	ELEV. (M)	BIRD REGION	ALTERNATIVE NAMES, RESEARCHERS, REFERENCES
Wabau Range	-6.57	144.00	2,254	Eastern Ranges	Murray Range/Musgrave Mountains
Wabo	-7.00	145.07		S Lowlands	
Wadapi	-1.87	136.42		Bay Islands	Wonapi; Yapen Island
Wafa	-6.75	146.10		SE Peninsula	
Wafa River	-6.68	146.17		SE Peninsula	
Waga River	-6.53	143.58		S Lowlands	
Wagifa Island	-9.50	149.37		SE Peninsula	
Wahgi Divide	-5.70	144.55		Eastern Ranges	Gilliard; Shaw Mayer, F.
Wahgi River	-6.37	143.18		Eastern Ranges	
Wahgi valley	-5.83	144.25		Eastern Ranges	
Wai Island	-0.70	130.71		NW Islands	north of Batanta; Beehler, B.M.
Wai Moi [River]	-5.77	141.62		S Lowlands	Wai Mio?; the word "Wai" means river
Wai Mungi [River]	-5.75	141.62		Border Ranges	the word "Wai" means river
Wai Pinyang [River]	-5.77	141.47		Border Ranges	the word "Wai" means river
Waibeem	-0.63	132.97		Bird's Head	in the Tamrau Mountains; Beehler, B.M., Marshall, A., & Safford, A./2006
Waifoi	-0.13	130.75		NW Islands	
Waifoi coastal forest	-0.10	130.71	0	NW Islands	Mauro, I./2002
Waigama/Waigamma*	-1.67	129.83		NW Islands	on the northwest coast of Misool Island; Guillemard/1886
Waigani sewage ponds*	-9.38	147.18		SE Peninsula	Moitaka settling ponds
Waigani swamp	-9.37	147.17		SE Peninsula	
Waigeo/Waigeu*/ Waigiou* Island	-0.23	131.00		NW Islands	
Waihali	-1.37	132.03		Bird's Head	
Waijeffi River	-0.42	131.80		Bird's Head	
Wailibit	-0.92	130.60		NW Islands	on the southern coast of Batanta Island; Gilliard, E.T./1964
Waima Beach	-8.64	146.46		SE Peninsula	
Waipara River	-10.08	148.17		SE Peninsula	Gonema Oru/Ormond River
Wair River	-1.78	137.38		NW Lowlands	
Wairemah	-0.13	130.73		NW Islands	
Waitapu	-8.55	147.25		SE Peninsula	
Wakatimi	-10.22	148.10		SE Peninsula	
Wakwak	-	-		S Lowlands	foot of the Western Ranges
Walckenaers Bay	-2.41	139.81		NW Lowlands	west of the Cyclops Mountains; (Diamond 1981a)
Walia	-5.33	143.23		Eastern Ranges	
Walis Island	-3.25	143.30		Sepik-Ramu	
Wall Mountain	-4.65	141.05	584	Border Ranges	
Walya village	-5.75	143.93		Eastern Ranges	Alpha Helix New Guinea Expedition/1969
Wamal	-8.12	139.10		Trans-Fly	Princess Marianne Strait

NAME	LATITUDE	LONGITUDE	ELEV. (M)	BIRD REGION	ALTERNATIVE NAMES, RESEARCHERS, REFERENCES
Wamar Island	-5.63	134.20		Aru Islands	Wammer*/Wamma* Island; near Wokam Island, in the Aru Islands
Wamea Island	-9.25	149.90		SE Peninsula	in the Amphlett Islands; Hamlin, H./1928
Wamena	-4.12	138.92	1,554	Western Ranges	Ripley, S.D./1960–61
Wamma Island*	-5.63	134.20		Aru Islands	Wammer*/Wamar Island; near Wokam Island, Aru Islands
Wammer Island*	-5.63	134.20		Aru Islands	Wamar/Wamma* Island; near Wokam Island, Aru Islands
Wanambai*	-6.03	134.30		Aru Islands	Manumbai; on Kobroor I of Aru Islands; Cayley-Webster, H./1896
Wanatonali Bay	-10.62	150.75		SE Islands	
Wandammen Bay or Inlet	-2.67	134.43		Bird's Neck	
Wandammen Mountains	-2.70	134.60		Bird's Neck	Wondiwoi Mountains/Mount Wondiwoi; Mayr, E.; Diamond, J.M
Wandammen Peninsula	-2.70	134.45		Bird's Neck	
Wando	-8.88	141.25		Trans-Fly	
Wanduli	-7.33	146.75		SE Peninsula	
Wanggar River	-3.40	135.50		NW Lowlands	(Hartert et al. 1931)
Wanggati	-6.85	139.19		S Lowlands	in Mappi district, Indonesian New Guinea; (Mees 1982b)
Wanggo River	-8.05	140.97		Trans-Fly	
Wangop River	-5.42	141.88		Border Ranges	
Wanigan Camp	-4.06	137.11		Western Ranges	position approximate
Wanigela	-9.33	149.17		SE Peninsula	
Wanigela River	-10.88	149.83		SE Peninsula	
Wanoem Bay*	-6.03	134.30		Aru Islands	Manumbai; on Kobroor Island of Aru Islands; Cayley-Webster, H./1896
Wanoembai*/ Wanumbai*	-6.03	134.30		Aru Islands	Manumbai; on Kobroor Island of Aru Islands; Cayley-Webster, H./1896
Wanti	-2.07	137.29		NW Lowlands	Waropen district
Wantoat airstrip	-6.13	146.47		Huon Peninsula	
Wanuma	-4.09	145.32	700	Sepik-Ramu	in the Adelbert Mountains
Wanumbai River	-6.00	134.71		Aru Islands	in the Aru Islands
Wapenamanda	-5.63	143.92		Eastern Ranges	
Wapenamanda gorge	-5.77	143.95		Eastern Ranges	
Wapoga River	-2.68	136.07		Western Ranges	
Wara River	-4.27	142.40		Sepik-Ramu	
Wara Sii	-7.86	146.55		SE Peninsula	in the Lakekamu basin; Beehler, B.M. & C.H., and Pruett-Jones, S./1982
Warafri	-1.08	136.24		Bay Islands	on Biak Island
Warbusi	-1.62	134.15		Bird's Head	
Waremag River Camp	-0.12	130.77	30	NW Islands	Mauro, I./2002
Wareo	-6.43	147.77		Huon Peninsula	Beck, R./1929

NAME	LATITUDE	LONGITUDE	ELEV. (M)	BIRD REGION	ALTERNATIVE NAMES, RESEARCHERS, REFERENCES
Wari Island	-10.95	151.07		SE Islands	
Waria River	-7.98	147.24		SE Peninsula	
Warmendi	-0.37	132.65		Bird's Head	
Waropen Bay/Waropen district	-2.03	137.19		NW Lowlands	Doherty, W./1896–98
Waropen Mountain	-2.38	137.03	1,059	NW Lowlands	
Warr Island	-2.01	134.75		Bay Islands	
Warsembo	-2.00	135.00		Bay Islands	Ruys, H.H./1905
Wasala River	-8.02	147.27		SE Peninsula	
Wasau Hills	-2.75	141.30		Sepik-Ramu	
Wasaunon Camp	-6.10	146.92	2,939	Huon Peninsula	in the YUS conservation area, Dabek, L.
Wasior	-2.72	134.50		Bird's Neck	Mayr, E./1928
Waskuk Hills	-4.18	142.73		Sepik-Ramu	
Wassi Kussa River	-9.15	142.05		Trans-Fly	Wassikussa River
Wasu	-5.97	147.20		Huon Peninsula	
Wasu Creek	-2.72	141.33		Sepik-Ramu	
Wasu Creek	-6.40	142.60		S Lowlands	
Wasua mission	-8.28	142.53		Trans-Fly	
Wasur National Park	-8.39	140.79		Trans-Fly	
Wataikwa River	-4.33	137.23		S Lowlands	Shortridge/1910
Watam	-3.92	144.20		Sepik-Ramu	
Water Tamak	-3.62	142.80		Sepik-Ramu	Lossin, R./1973
Waterstone	-2.72	141.33		Sepik-Ramu	
Water-val Bivak	-4.31	138.67		Western Ranges	Versteeg, G.
Watut River	-6.80	146.40		SE Peninsula	
Watut-Tauri Gap	-7.85	146.25		SE Peninsula	
Wau	-7.35	146.70		SE Peninsula	Stevens, H./1932; Shanahan, P.J.; Pratt, T.K.; Beehler, B.M.; (Greenway 1935, Beehler 1976)
Wau Ecology Institute	-7.35	146.72		SE Peninsula	Pratt, T.K.; Beehler, B.M.; Pruett-Jones, S. & M.
Waviai Island	-9.27	152.91		SE Islands	near Woodlark Island; (Pratt et al. 2006)
Wawin	-6.60	146.42		SE Peninsula	(Mees 1982b)
Wawoi River	-7.86	143.27		S Lowlands	(Bell 1967a)
Weam	-8.62	141.13		Trans-Fly	
Wedau	-10.13	150.06		SE Peninsula	
Wei Island	-3.38	144.40		Sepik-Ramu	
Weiga	-5.82	144.84	2,400	Eastern Ranges	Gyldenstolpe, N./1951
Wejos	-0.42	131.80		Bird's Head	
Wekabau Ridge Camp	-0.05	130.86	780	NW Islands	Mauro, I./2002
Welya	-5.77	143.95	2,440	Eastern Ranges	
Wendoe Mer River	-8.42	140.33		Trans-Fly	

NAME	LATITUDE	LONGITUDE	ELEV. (M)	BIRD REGION	ALTERNATIVE NAMES, RESEARCHERS, REFERENCES
Wengomanga	-7.37	146.28		SE Peninsula	position approximate; near Aseki; Australian National Wildlife Collection
Weni River	-3.73	141.83		Sepik-Ramu	
Wensudu	-2.07	139.18		NW Lowlands	Doherty, W.
Werar Hills	-0.19	131.19	760	NW Islands	
Wereave	-8.52	141.10	40	Trans-Fly	
Wesaoeni River	-0.50	133.03		Bird's Head	
West Branch Right May River	-4.48	141.47		Sepik-Ramu	
West Range	-4.35	141.33	1,750	Sepik-Ramu	
Western Papuan Islands*	-0.62	130.60		NW Islands	here called the Northwestern Islands; also Raja Ampat Islands
Western Ranges	-3.90	137.10		Western Ranges	bird region encompassing the western sector of the Central Ranges
Wewak	-3.57	143.06		Sepik-Ramu	
Weyland Mountains	-3.87	135.72	3,890	Western Ranges	Pratt brothers; Shaw Mayer, F.
Wharton Camp	-8.55	147.38	3,505	SE Peninsula	
Wharton Range	-8.42	147.42		SE Peninsula	Mount Albert Edward; First Archbold Expedition/1933–34
Wi River	-6.88	145.15		Eastern Ranges	position approximate; Woxvold, I.
Wichmann Range	-4.45	138.72	2,400–3,000	Western Ranges	Lorentz, H.
Windesi/Windehsi*	-2.43	134.23		Bird's Neck	northwestern Bird's Neck
Winter Height	-8.80	147.47	3,621	SE Peninsula	Mount Winter*
Wipim	-8.78	142.87		Trans-Fly	Alpha Helix New Guinea Expedition/1969
Wissel Lakes*	-3.89	136.32	1,735	Western Ranges	Paniai Lake; (Junge 1953)
Witu River	-2.28	139.63		NW Lowlands	
Wobo Hills	-2.73	141.25		Sepik-Ramu	
Woendi Atoll	-1.26	136.39		Bay Islands	southeast of Biak Island
Wogamush River	-4.30	142.38		Sepik-Ramu	
Wogeo Island*	-3.23	144.10		Sepik-Ramu	Vokeo Island
Wogupman River	-4.72	143.43		Sepik-Ramu	
Wogupmeri River	-4.70	143.43		Sepik-Ramu	
Woitape	-8.55	147.25		SE Peninsula	
Wok Agop	-5.38	141.10		Border Ranges	
Wok Bilak	-5.38	141.53		Border Ranges	
Wok Bilim	-5.30	141.52		Border Ranges	
Wok Bol	-5.37	141.60		Border Ranges	
Wok Diap	-5.50	141.80		Border Ranges	
Wok El	-5.32	141.82		Border Ranges	
Wok Feneng	-5.70	141.43		Border Ranges	
Wok Ilom	-5.33	141.52		Border Ranges	
Wok Isam	-5.32	141.52		Border Ranges	

NAME	LATITUDE	LONGITUDE	ELEV. (M)	BIRD REGION	ALTERNATIVE NAMES, RESEARCHERS, REFERENCES
Wok Isam (west arm)	-5.30	141.52		Border Ranges	
Wok Ium	-5.35	141.43		Border Ranges	
Wok Kup	-5.43	141.52		Border Ranges	
Wok Luap	-5.53	141.80		Border Ranges	
Wok Miriam	-5.33	141.53		Border Ranges	
Wok Narin	-5.55	141.78		Border Ranges	
Wok Sey	-5.43	141.92		Border Ranges	
Wok Wanik	-5.42	141.52		Border Ranges	
Wokam/Wokan* Island	-5.75	134.50		Aru Islands	in the Aru Islands; Kühn, H./1900
Wonapi	-1.87	136.42		Bay Islands	Wadapi; on Yapen Island
Wondiwoi Mountains	-2.68	134.62	2,252	Bird's Neck	on the Wandammen Peninsula; Mayr, E./1928
Wonenara	-6.80	145.88		Eastern Ranges	
Woodlark Island	-9.17	152.75		SE Islands	Muyua Island/Murua Island*; Meek, A.S./1897
Woroi*	-8.83	143.12		Trans-Fly	Wuroi
Wowolo	-10.28	149.37		SE Peninsula	
Wuroi	-8.83	143.12		Trans-Fly	Woroi*; Rand, A.L./1934
Wurup	-6.02	144.15		Eastern Ranges	
Wutung	-2.62	141.00		Sepik-Ramu	
Yagei River*	-5.17	144.83		Sepik-Ramu	Ramu River; Tappenbeck/1896
Yakaru Hills	-2.78	141.32		Sepik-Ramu	
Yakopi Nalenk Mountain	-5.15	143.30	3,711	Eastern Ranges	Mount Bürgers
Yakoulouma River	-8.55	146.88		SE Peninsula	
Yakwoi River	-7.22	146.02		SE Peninsula	
Yalu River	-6.60	146.92		Huon Peninsula	
Yalumet	-6.07	147.02		Huon Peninsula	
Yalumet River	-6.07	147.03		Huon Peninsula	
Yambon Gate	-4.27	142.75		Sepik-Ramu	
Yamna Island	-2.00	139.25		NW Lowlands	Jamna Island*; Doherty, W.
Yanaba Island	-9.29	151.95		SE Islands	Ianaba Island*; near Alcester I
Yandara	-5.75	145.00	1,829	Eastern Ranges	Shaw Mayer, F./1930
Yangaron Camp	-6.11	146.90		Huon Peninsula	Benz, B./2006
Yani River	-6.90	145.42		Eastern Ranges	
Yani River east arm	-6.75	145.38		Eastern Ranges	
Yaour Aolfe	-2.00	135.00		Bay Islands	
Yapen Island	-1.80	136.30		Bay Islands	Japen*/Jobi* Island; Meyer, A.B./1875
Yapsiei	-4.63	141.08		Sepik-Ramu	Flannery, T.
Yarru Island	-9.07	143.20		Trans-Fly	
Yati River	-6.23	146.18		SE Peninsula	
Yaur district	-3.08	134.82		Bird's Neck	Jaur district*

NAME	LATITUDE	LONGITUDE	ELEV. (M)	BIRD REGION	ALTERNATIVE NAMES, RESEARCHERS, REFERENCES
Yawan	-6.13	146.87		Huon Peninsula	Dabek, L., Jensen, R.
Yeflio	-1.13	131.23		NW Islands	west coast of the Bird's Head; (Gyldenstolpe 1955b)
Yeina Island	-11.33	153.45		SE Islands	Veina Island*
Yela Island*	-11.37	154.13		SE Islands	Rossel Island; Eichhorn, Meek, A.S., and Hamlin, H.
Yela River	-11.37	154.17		SE Islands	
Yellow River	-3.99	141.72		Sepik-Ramu	
Yellow valley	-4.04	137.18		Western Ranges	near Carstensz summit and Discovery valley; position approximate
Yetpelle Island	-2.20	130.26		NW Islands	Pelee Island
Yimas Lakes	-4.68	143.53		Sepik-Ramu	
Yimi/Yimmi	-4.05	142.83		Sepik-Ramu	
Yologe	-8.57	147.10		SE Peninsula	
Yonki town	-6.25	145.98		Eastern Ranges	
Yu Island	-0.03	129.60		NW Islands	Pulau Ju
Yuat River	-5.10	144.10		Sepik-Ramu	
Yule Island	-8.28	146.53		SE Peninsula	D'Albertis, L./1875
Yuma Range	-5.50	142.77	3,420	Eastern Ranges	
Yupna*/Yopno*/ Yupno River	-5.82	146.77		Huon Peninsula	in the YUS conservation area
YUS	-5.82	146.77		Huon Peninsula	a local region in north-central Huon Peninsula
YUS transect	-6.00	146.82		Huon Peninsula	Beehler, B.M., Mandeville, J.L., Freeman, B./2009–12
Zagahemi*/ Zagaheme*	-6.33	147.67		Huon Peninsula	Zakaheme; Beck, R./1929; Shaw Mayer, F./1930
Zakaheme	-6.33	147.67		Huon Peninsula	Zagahemi*/Zagaheme*; Beck, R./1929
Zenag	-6.95	146.62	1,000	SE Peninsula	A.H. Miller collected there, on the road from Wau to Lae
Zoogeographers' Gap	-3.90	135.00		Bird's Neck	eastern Bird's Neck

Index of English Bird Names and Topics

Note this includes the English names of the bird
(species and genera) as well as topical material from
the introductory sections.

abbreviations list 34
Aru Islands bird region 15
Astrapia, Arfak 431
    Black 431
    Huon 432
    Princess Stephanie's 432
    Ribbon-tailed 431
    Rothschild's 432
    Splendid 431
    Stephanie's 432
Avifauna, the 25
Avocet, Red-necked 166

Babbler, Australian 341
    Blue-capped 415
    Green-backed 354
    Grey-crowned 341
    Isidore's 340
    Isidore's Rufous 340
    New Guinea 340
    Northern 341
    Papuan 340
    Rufous 340
Barkcreeper, Papuan 283
Barn-Owl, Australian 214
Bay Islands bird region 14
Baza, Crested 202
    Pacific 201
Bee-eater, Blue-tailed 220
    Olive 220
    Rainbow 221
Bellbird, Piping 382
    Rufous-naped 381
Berryhunter, Mottled 368
Berrypecker, Arfak 346, 353
    Black 346
    Crested 354
    Eastern Crested 354

    Fan-tailed 347
    Green 348
    Lemon-breasted 347
    Long-tailed 347
    Mid-mountain 356
    New Guinea 353
    Obscure 345
    Painted 353
    Spotted 349
    Streaked 348
    Striated 348
    Thick-billed 349
    Tit 353
    Western Crested 353
    Yellow-bellied 347
Bird, Incubator 50
    New Guinea Friar 298
Bird of Paradise, Arfak 431
    Arfak Six-wired 421
    Black Sickle-billed 429
    Blue 435
    Blue-and-yellow Wattled 430
    Brown Sickle-billed 429
    Count Raggi's 438
    Crested 343
    Emperor 436
    Emperor of Germany 436
    Enamelled 421
    Goldie's 436
    Gray Sabre-tailed 429
    Great 437
    Greater 437
    Helen's 423
    Huon 432
    King 433
    King of Saxony 421
    Lawes's Six-wired 422
    Lesser 436
    Lesser Superb 427

    Little Emerald 437
    Little King 433
    Loria's 343
    Macgregor's 309
    Magnificent 434
    Multi-crested 343
    Queen Carola's Six-wired 423
    Prince Rudolph's 435
    Princess Stephanie's 432
    Ragglana 437
    Red 436
    Red-plumed 438
    Ribbontailed 432
    Shield-billed 344
    Sickle-crested 343
    Splendid 431
    Superb 426
    Twelve-wired 424
    Wahnes' Six-wired 422
    Waigeu 434
    Wattle-billed 344
    Wattled 344, 430
    White-plumed 436
    Wilson's 434
    Yellow-breasted 344
Bird's Head bird region 14
Bird's Neck bird region 14
bird regions for New Guinea 13
Bittern, Australian Little 109
    Black 110
    Black-backed 109
    Black-breasted Least 109
    Chinese Little 109
    Cinnamon 109
    Forest 108
    Little 109
    Mangrove 110
    Yellow 109
    Yellow-necked 110

Zebra 108
Black-bill 322
Blackbird, New Guinea 500
Blackwit, Lesser 439
Boatbill, Black-breasted 361
    Yellow-breasted 360
Boobook, Australian 217
    Barking 217
    Common 217
    Papuan 218, 219
    Southern 217
Booby, Abbott's 118
    Blue-faced 118
    Brown 119
    Masked 118
    Red-footed 119
    White 118
    White-bellied 119
Border Ranges bird region 15
Bowbill, Canary 352
    Grey-winged 352
Bowerbird, Adelbert 281
    Archbold's 279
    Beck's 281
    Black-faced Golden 280
    Buff-breasted 282
    Crested 278
    Fawn-breasted 282
    Fire-maned 281
    Flame 281
    Flamed 281
    Gardener 279
    Golden 281
    Golden-fronted 278
    Golden Regent 281
    Huon 278
    Lauterbach's 282
    MacGregor's 278
    Masked 280
    Plain 279
    Sanford's 280
    Streaked 277
    Striped 277
    Tomba 280
    Vogelkop 279
    Yellow-bellied 282
    Yellow-breasted 282
    Yellow-fronted 279
Brolga 135
Bronze-Cuckoo, Golden 143
    Malay 144
Bronzewing, New Guinea 68
Brush-Cuckoo, Grey-breasted 147
Brush-turkey, Black-billed 46
Brushturkey, Brown-collared 47
    Bruijn's 45
    Collared 47
    Maroon-throated 47
    Red-billed 45
    Red-legged 46

    Waigeo 45
    Wattled 44
    Yellow-legged 46
Bulbul, Sooty-headed 479
Bushchat, Pied 501
Bush-hen 132
Bush-hen, Common 132
    Pale-vented 132
    Rufous-tailed 131
Bushlark, Australasian 474
    Horsfield's 474
    Singing 474
Bustard 123
Bustard, Australian 123
Bustard Quail, Black-backed 185
Butcherbird, Black 363
    Black-backed 364
    Black-headed 364
    Hooded 364
    Louisiade 365
    Sudest 365
    Tagula 364
    White-rumped 365
    White-throated 364
Buttonquail, Red-backed 185
Buzzard, Bat-eating 203
    Grey-faced 207
    Long-tailed 202
    Long-tailed Honey 202

Cassowary, Australian 43
    Bennett's 42
    Double-wattled 43
    Dwarf 41
    Little 42
    Mountain 42
    Northern 43
    One-wattled 43
    Single-wattled 43
    Southern 43
    Two-wattled 43
Catbird, Black-eared 275
    Green 276
    Least 275
    Spotted 276
    White-eared 275
    White-throated 275
Chat, Pied 501
    Pied Bush 501
Christbird 171
Cicadabird 377
Cicadabird, Black 378
    Black-bellied 375
    Black-shouldered 376
    Common 376
    Grey-headed 377
    Papuan 375
    Sharpe's 376
Cisticola, Bright-capped 493

    Common 493
    Fan-tailed 493
    Golden 493
    Golden-headed 493
    Zitting 492
Cockatoo, Bare-eyed 240
    Goliath 239
    Great Black 239
    Greater Sulphur-crested 240
    Great Palm 239
    Palm 239
    Sulphur-crested 240
    Triton 240
    White 240
Coffinbird 153
conventions, in text 36
Coot 134
Coot, Black 134
    Common 134
    Eurasian 134
Corella, Bare-eyed 240
    Little 239
Cormorant 121
Cormorant, Australian Pied 121
    Big Black 121
    Black 121
    European 121
    Great 121
    Great Pied 121
    Large 121
    Large Black 121
    Little Black 121
    Little Pied 120
    Pied 121
Coucal, Bernstein's 137
    Biak 137
    Biak Island 137
    Black-billed 137
    Black Jungle 137
    Common 138
    Greater 137
    Greater Black 136
    Ivory-billed 137
    Lesser Black 137
    Menbek's 137
    Pheasant 137
Courser, Australian 186
Crake, Ashy 131
    Baillon's 130
    Grey-bellied 131
    Lesser Spotted 130
    Marsh 130
    Red-necked 126
    Sooty 130
    Spotless 130
    Tiny 130
    White-browed 131
Crane, Australian 135
    Sarus 135
Crow, Australian 454

Bare-eyed 453
Brown-capped 453
Brown-headed 453
Grey 453
King 415
Papuan 454
Torresian 454
Crowned-pigeon, Blue 73
  Common 73
  Grey 73
  Masked 73
  Scheepmaker's 74
  Sclater's 74
  White-tipped 74
Cuckoo, Ash-coloured 146
  Australian Bronze 143
  Black-eared 142
  Blyth's 148
  Brush 147
  Channel-billed 141
  Chestnut-breasted 146
  Chestnut-breasted Brush 146
  Fan-tailed 146
  Fan-tailed Brush 146
  Golden 143
  Golden Bronze- 143
  Grey-breasted Brush 147
  Grey-headed 147
  Himalayan 148
  Horsfield's 148
  Horsfield's Bronze 142
  Little Bronze 144
  Little Long-billed 142
  Long-billed 142
  Long-tailed 140
  Malay Bronze- 144
  Narrow-billed Bronze 143
  Oriental 148
  Pacific Long-tailed 140
  Pallid 145
  Rufous-tailed Bronze 143
  Rufous-throated Bronze 143
  Shining 143
  Shining Bronze 143
  Southern Muted 148
  Square-tailed 147
  White-crowned 145
  White-eared Bronze 143
Cuckoo-Dove, Amboina 67
  Bar-tailed 67
  Black-billed 67
  Brown 66
  Giant 66
  Great 65
  Long-tailed 66
  Mackinlay's 67
  Pink-breasted 67
  Rufous 68
  Rusty 67
  Slender billed 67

Spot-breasted 68
Cuckooshrike, Barred 369
  Black 378
  Black-faced 371
  Black-shouldered 376
  Blue-grey 369
  Boyer's 370
  Golden 372
  Gray's 378
  Hooded 369
  Large-billed 369
  Long-tailed 369
  New Guinea 379
  Orange 373
  Papuan Black 379
  Rufous-underwing 370
  Stout-billed 369
  White-bellied 371
  Yellow-eyed 370
Cuckoo-shrike, Little 372
  New Guinea 379
  Papuan 372
  White-breasted 372
  Yellow-eyed 370
Curlew, Australian 176
  Beach 164
  Bristle-thighed 175
  Bush 164
  Eastern 176
  Far Eastern 176
  Hudsonian 175
  Little 175

Dabchick 62
Dabchick, Australian 62
  Black-throated 62
  Eurasian Little 62
  Red-throated 62
Darter 122
Darter, Australasian 122
  Australian 122
  Oriental 122
dates of authority 31
Dollarbird 222
Dollarbird, Oriental 221
Dotterel, Asiatic 170
  Eastern 170
  Large 170
  Large-billed 170
  Large Sand 170
  Lesser Sand 170
  Little Ringed 168
  Mongolian 170
  Mongolian Sand 170
  Oriental 170
  Red-kneed 168
Dove, Bar-shouldered 71
  Bar-tailed Cuckoo- 67
  Beautiful Fruit- 84

Beccari's Ground 70
Black-billed Cuckoo- 67
Brown-backed Emerald 77
Brown Cuckoo- 66
Claret-breasted Fruit- 83
Common Emerald 75
Coroneted Fruit- 84
Diadem Fruit 85
Dwarf Fruit- 78
Emerald 76
Emerald Ground 76
Giant Cuckoo- 66
Goldenheart 69
Great Cuckoo- 66
Green-winged Ground 76
High Mountain Fruit 79
Knob-billed Fruit- 83
Least Fruit 78
Little Fruit 78
Mackinlay's Cuckoo- 67
Magnificent Fruit 78
Moluccan Fruit- 80
Mountain Fruit- 79
Orange-bellied Fruit- 82
Orange-fronted Fruit- 82
Ornate Fruit- 81
Pacific Emerald 76
Peaceful 71
Peaceful Ground 71
Pink-spotted Fruit- 81
Red-knobbed Fruit 83
Red-throated Ground 69
Rock 64
Rose-crowned Fruit- 85
Rufous-throated Ground 69
Rusty Cuckoo- 67
Slender-billed Cuckoo- 67
Small Green Fruit 78
Spot-breasted Cuckoo- 68
Spotted 65
Spotted-necked 65
Spotted Turtle- 65
Stephan's 77
Stephan's Emerald 76
Superb Fruit- 79
Wallace's Fruit- 81
White-bibbed Fruit- 79
White-breasted Fruit- 80
Wompoo Fruit- 77
Yellow-bibbed Fruit- 80
Zebra 71
Dowitcher, Asian 173
  Asiatic 173
  Long-billed 173
  Snipe-billed 173
Drongo, Mountain 407
  Papuan Mountain 407
  Pygmy 407
  Spangled 414
Duck, Australian Black 59

Australian White-eyed 61
Australian Wood- 57
Black 59
Burdekin 56
Diving Whistling 55
Grass Whistle- 54
Grey 59
Maned 57
New Zealand Grey 59
Pacific Black 59
Plumed Tree 54
Plumed Whistling 54
Salvadori's 57
Spot-billed 59
Spotted Tree 54
Spotted Whistling 54
Tufted 61
Wandering Whistling 54
Water Whistle- 54
White-eyed 61
Dwarf-Goose, Green 58

Eagle, Gurney's 204
Kapul 203
Little 204
New Guinea 203
New Guinea Harpy- 203
Papuan 203
Papuan Harpy 203
Pygmy 204
Red-backed Sea 206
Wedge-tailed 205
Whistling 206
White-bellied Sea- 206
White-breasted Fish- 207
White-breasted Sea 207
White-headed Sea 206
Eared-nightjar, Papuan 151
Spotted 151
White-throated 151
Eastern Ranges bird region 16
Egret, Cattle 112
Eastern Cattle 112
Eastern Great 114
Eastern Reef- 116
Great 113
Great White 114
Intermediate 114
Large 114
Lesser 114
Little 115
Pacific Reef 116
Pied 115
Plumed 114
Short-billed 114
Smaller 114
Snowy 115
Yellow-billed 114
English names treatment 29

Eupetes, High Mountain 356
Lowland 357
Mid-mountain 358

Fairy-warbler, Large-billed 339
Yellow-bellied 336
Fairywren, Broad-billed 286
Campbell's 286
Emperor 287
Orange-crowned 285
Wallace's 284
White-shouldered 287
Falcon, Bat 203
Brown 237
Grey 237
Little 237
Peregrine 237
False-sunbird, Canary 352
Grey-winged 352
False-thrush, Dubious 457
Sclater's 456
family accounts 29
Fantail, Arafura 410
Banded 413
Black 411
Black-and-white 407
Black Thicket- 407
Black-throated Thicket 408
Chestnut-backed 409
Chestnut-bellied 412
Dimorphic 409
Dimorphic Rufous 409
Drongo 407
Friendly 411
Grey-breasted Rufous 409
Mangrove 412
Mangrove Grey 412
Northern 412
Pheasant 412
Rufous 409
Rufous-backed 408
Sooty Thicket- 408
White-bellied Thicket- 407
White-breasted Thicket 408
White-browed 407
White-throated 413
Figbird 405
Figbird, Australasian 404
Green 405
Figparrot, Desmarest's 258
Fig-Parrot, Black-fronted 261
Blue-fronted 261
Creamy-breasted 261
Desmarest's 258
Double-eyed 261
Dusky-cheeked 261
Edward's 258
Golden-headed 258
Large 257

Orange-breasted 259
Red-faced 258
Salvadori's 259
Whiskered 259
Yellow-naped 258
Finch, Blood 512
Blue-faced 513
Chestnut 519
Crimson 512
Crimson-sided 512
Red-faced 512
Red-sided 512
Red-tailed 512
Spice 514
Firefinch, Australian 512
Firetail, Mountain 511
Fish-eagle, White-breasted 207
Fishtail 415
Flowerpecker, Australian 506
Louisiade 505
Mistletoe 506
Olive-crowned 503
Papuan 503, 504
Pectoral 503
Red-capped 504
Shining 503
Flycatcher, Australian Brown 463
Biak 445
Biak Black 445
Biak Monarch 449
Black and White 407, 468
Black and White Monarch 447
Black and Yellow Monarch 450
Black-breasted Flatbill 361
Black-faced 452
Black-fronted 448
Black Myiagra 445
Black-throated 467
Black-winged Monarch 452
Boat-billed 361
Broad-billed 444
Broad-billed Myiagra 445
Canary 461
Common Shining 444
Frilled 442
Frill-necked 442
Glossy 444
Gray-spotted 502
Grey-and-white 462
Grey-streaked 502
Grey-winged Monarch 452
Kofiau Monarch 448
Leaden 445
Leaden Myiagra 445
Lemon-bellied 463
Lemon-breasted Microeca 463
Little Yellow 461
Montane 461
New Guinea Frilled 442
Northern 413

Olive 462
Olive Microeca 462
Olive-yellow 462, 471
Paperbark 444
Papuan 461
Pearly-winged 452
Restless 444
River 462
Satin 446
Satin Myiagra 446
Shining 443
Shining Monarch 444
Spectacled 448
Spectacled Monarch 448
Spot-breasted 502
Spot-wing Monarch 447
Torrent 462
White-bellied 448
White-breasted 466
White-faced 473
White-lored 442
Yellow 461
Yellow-bellied 464
Yellow-breasted Flatbill 361
Yellow-legged 461
Yellow Microeca 461
Flycatcher-robin, Black 469
　Short-tailed 466
Flyeater, Brown-breasted 339
　Buff-breasted 339
　Yellow-bellied 336
Flyrobin, Black-sided 468
　Canary 461
　Lemon-bellied 463
　New Guinea 466
　Olive 462
　Orange-chinned 362
　Papuan 461
　Torrent 462
　White-winged 469
　Yellow-legged 461
Forest-Rail, Chestnut 124
　Forbes's 125
　Mayr's 125
　White-striped 125
Friarbird, Bald 298
　Brass's 296
　Dwarf 296
　Helmeted 297
　Little 297
　Meyer's 296
　New Guinea 298
　Noisy 298
　Yellow-throated 297
Frigatebird, Christmas Island 116
　Great 117
　Greater 117
　Least 117
　Lesser 117
　Pacific 117

Frogmouth, Giant 150
　Great Papuan 150
　Little Papuan 150
　Marbled 149
　Papuan 150
Fruitdove, Magnificent 78
Fruit-Dove, Beautiful 84
　Blue-spotted 85
　Claret-breasted 83
　Coronated 85
　Coroneted 84
　Crimson-capped 81
　Crimson-crowned 84
　Diadem 85
　Dwarf 78
　Gestroi's 82
　Golden-fronted 81
　Golden-shouldered 81
　Grey-breasted 84
　Grey-capped 85
　Knob-billed 83
　Lilac-capped 85
　Lilac-crowned 85
　Little Coroneted 85
　Moluccan 80
　Mountain 79
　Orange-bellied 82
　Orange-fronted 82
　Ornate 81
　Pink-capped 85
　Pink-spotted 81
　Purple-capped 79
　Purple-crowned 79
　Red-bibbed 84
　Red-breasted 84
　Red-throated 84
　Rose-crowned
　Splendid 81
　Superb 79
　Swainson's 85
　Wallace's 81
　Wallace's Green 81
　White-bibbed 79
　White-breasted 80
　Wompoo 77
　Yellow-bibbed 80
　Yellow-breasted 81
　Yellow-fronted 81
Fruitpigeon, Bar-breasted 90
　Black-belted 90
　Blackshouldered 90
　Grey 89
　Mountain Red-breasted 89
　Pacific 87
　Rufous-breasted 89
Fruit-pigeon, Black-knobbed 87
　Cinnamon 88
　Lilac-crowned 85
　Pink-spotted 81
　Purple-tailed 88

　Rufous-bellied 88
　Superb 79
　Wallace's 81
　White 91
Fulmar, Giant 97
future, the 27

Gallinule, Dusky 133
　Rufous-tailed 132
Gardener, Golden-maned 279
Gardenerbird, Brown 279
　Eastern 277
　Macgregor's 278
　Yellow-fronted 279
Garganey 60
genus accounts 29
geographic terms 35
Gerygone, Brown-breasted 339
　Fairy 337
　Green-backed 337
　Grey 335
　Grey-headed 337
　Large-billed 338
　Mangrove 339
　Mountain 335
　Treefern 339
　White-throated 336
　Yellow-bellied 336
Giant-Petrel, Antarctic 97
Godwit, Bar-tailed 174
　Black-tailed 174
Goldenface 325
Goldenheart 69
Goose, Cotton Pygmy 57
　Green Pygmy 58
　Magpie 53
　Pied 53
　Semipalmated 53
　White Pygmy- 58
Goshawk, Australasian 211
　Australian 211
　Black-mantled 211
　Brown 210
　Buergers's 213
　Chestnut-shouldered 213
　Doria's 208
　Grey 209, 210
　Grey-headed 211
　Meyer's 212
　New Guinea Grey-headed 211
　Papuan 212
　Rufous-breasted 210
　Variable 209
　Varied 210
　Vinous-chested 210
　White 210
Goura, Blue 73
　Common 73
　Great 74

Grey 73
Masked 73
Victoria 74
White-tipped 74
Grackle, Orange-faced 496
Grassbird, D'Albertis 491
  Fly River 491
  Little 491
  Papuan 490
  Tawny 490
Grasshopper-Warbler, Gray's 489
Grass-Owl, Eastern 215
Grebe, Australasian 62
  Australian Little 63
  Eurasian Little 62
  Red-throated 62
  Tricoloured 62
Greenshank 177
Greenshank, Common 177
  Little 176
Greybird, Australian 371
  Black 379
  Black-bellied 375
  Black-hooded 369
  Gray's 377
  Long-billed 377
  Müller's 376
  Papuan 372
  Rufous-underwing 370
  Stout-billed 369
  Yellow-eyed 370
Ground-Dove, Bronze 70
  Cinnamon 69
  Grey-bibbed 70
  Stephan's 77
  White-bibbed 69
  White-breasted 70
  White-throated 70
Ground-Pigeon, Grey 72
  Slaty 72
  Thick-billed 71
Ground-Robin, Greater 456
  Lesser 456
Gull, Black-headed 189
  Black-tailed 187
  Common Black-headed 189
  Red-billed 189
  Silver 189
Guria, Great 74

Hanging-Parrot, Papuan 262
Hardhead 61
Harpy-Eagle, New Guinea 203
Harrier, Australasian 209
  Papuan 209
  Pied 209
  Spotted-backed 209
  Swamp 208
Hawk, Australian Sparrow 212

Bat 202
Brown 237
Chestnut-shouldered 213
Crested 202
Crested Lizard 202
Doria's 207
Fish 199
Frog 207, 209
Grey Frog 209
Hawk-Owl, Barking 217
  Jungle 218
  Papuan 218
  Rufous 217
Heron, Buff-backed 113
  Dusky Grey 113
  Eastern Reef 116
  Giant 113
  Great-billed 113
  Little 112
  Little Pied 115
  Mangrove 111
  Nankeen Night- 110
  New Guinea Tiger 108
  New Guinea Zebra 108
  Pacific 113
  Pacific Reef 116
  Pied 115
  Reef 116
  Rufous Night- 110
  Striated 111
  Sumatran 113
  White 114
  White-faced 115
  White-necked 113
  Zebra 108
Historical biogeography of Region 25
Hobby, Australian 237
  Indian 236
  Oriental 236
Honeyeater, Allied 313
  Arfak 308
  Australian Brown 299
  Belford's 322
  Black-backed 304
  Black-backed Streaked 304
  Blackened 292
  Black-fronted 319
  Black-throated 319
  Blue-faced 300
  Brown 299, 305
  Brown-backed 307
  Brown-backed Streaked 304
  Carbon 292
  Cinnamon-browed 321
  Common Smoky 308
  Diamond 313
  Dusky 290
  Dwarf 351
  Eared 300
  Eastern Smoky 308

Elegant 316
Forest 315
Giant Wattled 309
Graceful 317
Gray 306
Gray-fronted 306
Grayish-brown 306
Gray's 296
Green-backed 301
Green-eyed 304
Grey-headed 301
Grey-streaked 303
Grey-throated Straightbilled 311
Guise's 304
Hill-forest 314
Huon 309
Huon Wattled 321
Large Spot-breasted 315
Large-tufted 312
Leaden 302
Leaden Streaked 302
Lemon-cheeked 319
Lesser Yellow-spotted 317
Long-bearded 324
Long-billed 310
Louisiades 313
Macgregor's 309
Mangrove Red-headed 292
Many-spotted 296
Many-streaked 303
Marbled 306
Mayr's 304
Mayr's Streaked 304
Meek's 303
Meek's Streaked 303
Mimic 313
Modest 307
Mottle-breasted 315
Mountain 314
Mountain Red-headed 293
Mountain Straight-billed 311
Nondescript 305
Obscure 318
Olive 299
Olive-brown 305
Olive Straight-billed 311
Olive-streaked 303
Orange-cheeked 319
Ornate 321
Pale-yellow 318
Plain 305
Plain Olive 299
Puff-backed 312
Pygmy 351
Red 291
Red-backed 304
Red-brown 290
Red-collared 289
Red-headed 292
Red-sided 303

Red-throated 291
Red-tinted 291
Rufous-backed 304
Rufous-backed Streaked 304
Rufous-banded 306
Rufous-breasted 307
Rufous-sided 303
Rufous-sided Streaked 303
Scarlet-bibbed 293
Scarlet-throated 293
Scrub 313
Short-bearded 324
Silver 299
Silver-eared 299
Silver-spangled 299
Smoky 308
Sooty 324
Spangle-crowned 306
Spangled 309
Spot-bellied 308
Spot-breasted 315
Spotted 295
Streak-capped 306
Streak-headed 306
Sub-bridled 319
Tagula 313
Tawny-breasted 294
Tawny Straight-billed 311
Varied 317
Vogelkop 321
Warbling 299
Wattled Smoky 308
Western Smoky 307
White-chinned 289, 301
White-eyed 301
White-fronted 321
White-marked 313
White-quilled 300
White-spangled 300
White-throated 289, 301
Yellow-browed 323
Yellow-gaped 316
Yellowish 318
Yellowish-streaked 303
Yellow-streaked 302, 318
Yellow-tinted 318
Hornbill, Blyth's 219
    Papuan 219
Huon Peninsula bird region 16

Ibis, Australian White 105
    Glossy 107
    Sacred 106
    Straw-necked 106
Ifrit 415
Ifrit, Blue-capped 415
Ifrita, Blue-capped 415
Imperial-pigeon, Bare-eyed 90
    Blue-tailed 86

Collared 86
Gold-eyed 86
Grey 89
Island 89
Pinon 90
Rufous-breasted 88
Shining 88
Yellow-eyed 86
Zoe 90

Jabiru 104
Jacana, Comb-crested 171
Jackywinter 463
Jaeger, Arctic 197
    Long-tailed 198
    Parasitic 198
    Pomarine 197
Jewel-babbler, Blue 356
    Brown-capped 357
    Brown-headed 357
    Chestnut-backed 357
    Dimorphic 357
    Lowland 357
    Mid-Mountain 358
    Spotted 355
Jungle-pigeon, Thick-billed 72

Kalanga 265
Kestrel, Australian 236
    Moluccan 236
    Nankeen 236
    Spotted 235
Kingfisher, Aru Giant 228
    Australian Paradise 224
    Azure 234
    Beach 231
    Biak Paradise- 223
    Black-sided 228
    Blue 229
    Blue-black 228
    Brown-backed Paradise 225
    Brown-headed Paradise- 225
    Buff-breasted Paradise- 224
    Collared 229
    Common 233
    Common Paradise- 223
    Dwarf 234
    Forest 228
    Gaudichaud's 228
    Greater Yellow-billed 232
    Hook-billed 226
    Islet 230
    Lesser Paradise 224
    Lesser Yellow-billed 232
    Little 234
    Kafiau Paradise 224
    Kofiau Paradise- 224
    Little Paradise- 224

Lowland Yellow-billed 232
Macleay's 229
Mangrove 230
Mountain 232
Mountain Yellow-billed 232
Numfor Paradise- 224
Papuan Dwarf 234
Pink-breasted Paradise 225
Red-breasted Paradise- 225
River 233
Rossel Paradise- 223
Rufous-bellied Giant 228
Sacred 231
Shovel-billed 227
the 233
Torresian 230
Variable Dwarf 234
White-headed 231
White-tailed Paradise 225
Yellow-billed 231
Kingparrot, Amboina 263
King-Parrot, Moluccan 263
    Papuan 263
Kissaba 421
Kite, Bat 203
    Black 205
    Black-breasted 202
    Black-eared 205
    Black-shouldered 200
    Black-winged 200
    Brahminy 206
    Common Black-shouldered 201
    Fork-tailed 205
    Pariah 205
    Red-backed 206
    Square-tailed 202
    Whistling 206
Knot 180
Knot, Common 180
    Eastern 180
    Great 180
    Red 180
Koel 140
Koel, Asian 140
    Black-capped 139
    Common 140
    Dwarf 139
    Eastern 139
    Indian 140
    Long-tailed 140
    Pacific 140
    Pacific Long-tailed 140
    White-crowned 145
Kokomo 219
Kookaburra, Blue-winged 227
    Gaudichaud's 228
    Hook-billed 226
    Rufous-bellied 228
    Shovel-billed 226
    Spangled 227

Landrail, Banded 128
Lapwing, Masked 167
Lark, Australasian 474
    Horsfield's 474
Latin terms 34
layout of the accounts 28
Leaf-Warbler, Biak 481
    Island 480
    Mountain 481
    New Guinea 481
    Numfor 481
Leatherhead 298
Leatherhead, Little 297
Lily-trotter 171
Logrunner 342
Logrunner, Australian 342
    Papuan 341
Longbill, Canary 352
    Dwarf 351
    Green-crowned 352
    Grey-bellied 351
    Grey-winged 352
    New Guinea 352
    Plumed 351
    Pygmy 351
    Slaty-chinned 352
    Slaty-headed 352
    Spectacled 350
    Yellow-bellied 351
Lorikeet, Bat 262
    Blue-eared 246
    Emerald 250
    Fairy 249
    Goldie's 252
    Green-naped 254
    Josephine's 247
    Little Red 249
    Lowland 246
    Musschenbroek's 250
    Orange-billed 250
    Papuan 248
    Plum-faced 244
    Pygmy 247
    Pygmy Streaked 247
    Rainbow 253
    Red-breasted 254
    Red-capped Streaked 253
    Red-chinned 247
    Red-collared 254
    Red-flanked 246
    Red-fronted 245
    Red-spotted 246
    Stella's 248
    Streaked 247
    Striated 247
    Whiskered 245
    Wilhelmina's 247
    Yellow-billed 250
    Yellow-fronted 246
Lory, Biak Red 255

    Black 256
    Black-capped 251
    Black-winged 255
    Brown 256
    Coconut 254
    Dusk-orange 255
    Dusky 255
    Duyvenbode's 256
    Eastern Black-capped 252
    Fairy 248
    Greater Streaked 257
    Josephine's 248
    Little Red 249
    Moluccan Red 255
    Moluccas Red 255
    Orange-billed Mountain 250
    Plum-faced Mountain 245
    Purple-bellied 252
    Rainbow 254
    Rajah 256
    Red-capped Streaked 253
    Red-chinned 247
    Red-fronted Blue-eared 246
    Streaked 247
    Violet-necked 254
    Western Black-capped 252
    White-rumped 255
    Yellow-billed Mountain 250
    Yellow-fronted Blue-eared 246
    Yellowish-streaked 257
    Yellow Streaked 247, 256
Lotusbird 171

Magpie, Australian 365
Magpielark 441
Magpie-lark, Australian 441
Mannikin, Alpine 520
    Arfak 516
    Black 519
    Black-breasted 519
    Black-faced 514
    Chestnut-breasted 518
    Eastern Alpine 520
    Grand 516
    Grand Valley 520
    Great-billed 516
    Grey-banded 516
    Grey-crowned 517
    Grey-headed 516
    Hooded 517
    Mayr's 517
    Moluccan 514
    New Britain 518
    Nutmeg 514
    Scaly-breasted 514
    Snow Mountain 520
    Spice 514
    Streak-headed 514
    Western Alpine 520

    White-crowned 517
    White-spotted 515
Manucode, Allied 420
    Black 421
    Crinkle-breasted 420
    Crinkle-collared 419
    Curl-breasted 419
    Curl-crested 419
    Glossy 420
    Glossy-mantled 421
    Green 420
    Green-breasted 420
    Jobi 420
    Trumpet 417
map of New Guinea bird regions 14
Marshbird, Striated 492
Marsh-Harrier, Eastern 209
    Pacific 209
    Papuan 209
    Spotted 209
Martin, Australian Tree 478
    Collared Sand 475
    Fairy 478
    Sand 475
    Tree 477
Megapode, Biak 48
    Dusky 50
    Melanesian 49
    Moluccan 47
    Orange-footed 49
Melampitta, Greater 439
    Lesser 439
Melidectes, Arfak 321
    Belford's 321
    Cinnamon-browed 321
    Foerster's 321
    Huon 321
    Long-bearded 324
    Mid-Mountain 321
    Ornamental 321
    Ornate 320
    Reichenow's 323
    Short-bearded 324
    Sooty 324
    Vogelkop 321
    White-fronted 321
    Yellow-browed 322
Meliphaga, Elegant 316
    Forest White-eared 315
    Graceful 316
    Louisiades 313
    Mimic 313
    Mottled 315
    Mountain 314
    Puff-backed 312
    Scrub 312
    Scrub White-eared 313
    Slender-billed 317
    Small Spot-breasted 314
    Southern White-eared 313

Spot-breasted 315
Sudest 313
Tagula 313
White-eared 314
White-eared Mountain 315
Yellow-gaped 315
Yellow-spotted 313
Melipotes, Arfak 308
Common 308
Huon 309
Wattled Smoky 308
Western 308
Microeca, Grey-headed 461
Mistletoebird 506
Monarch, Biak 449
Black 446
Black-and-gold 450
Black-backed 448
Black-faced 451
Black-winged 452
Broad-billed 445
Fantailed 446
Frilled 441
Golden 449
Hooded 447
Island 451
Island Grey-headed 451
Islet 450
Kafiau 448
Kofiau 447
Leaden 445
New Guinea Frilled 441
Ochre-collared 442
Pearly-winged 452
Rufous 447
Satin 446
Spectacled 448
Spot-winged 447
Monarch-flycatcher, Black 446
Rufous 447
Moorhen, Dusky 133
Pale-vented 132
Mountain-Pigeon, Bare-eyed 92
D'Albertis 92
Papuan 92
Mouse-babbler, Lowland 327
Mid-mountain 327
Mountain 329
Mouse-Warbler, Bicoloured 327
Black-backed 327
Chanting 327
High-mountain 329
Lowland 327
Mid-mountain 327
Mountain 328
Rusty 326
Sanford's 329
Mudlark 441
Munia, Black 519
Black-breasted 520

Black-faced 514
Chestnut-breasted 519
Eastern Alpine 520
Grand 516
Grey-banded 516
Grey-crowned 517
Grey-headed 517
Moluccan 514
New Guinea 518
Scaly-breasted 514
Streak-headed 515
Western Alpine 520
Muttonbird, King 100
Myiagra, Biak 445
Broad-billed 445
Leaden 445
Satin 446
Shining 444
Myna, Common 497
Golden 496
Golden-breasted 497
Indian 497
Orange-faced 496
Yellow-faced 496
Myzomela, Black 292
Black-and-Red 289
Dusky 290
Elfin 293
Mangrove Red-headed 292
Moluccan 290
Mountain 293
Mountain Red-headed 293
Papuan Black 291
Red 291
Red-collared 289
Red-headed 292
Red-headed Mountain 293
Red-throated 291
Red-tinted 291
Ruby-throated 290
Scarlet-bibbed 293
Sclater's 293
White-chinned 289
White-throated 289

nature conservation 26
Needletail, New Guinea 162
White-throated 162
New Guinea in context 12
New Guinea region definition 11
New Guinea's bird regions 13
Night-Heron, Nankeen 110
Rufous 111
Nightjar, Archbold's 151
Grey 152
Indian Jungle 152
Japanese 152
Jungle 152
Large-tailed 152

Long-tailed 153
Mountain 152
Papuan 151
Papuan Eared- 151
Spotted 151
Spotted Eared- 151
White-tailed 153
White-throated 151
White-throated Eared- 151
Noddy, Black 188
Brown 187
Common 188
White 189
White-capped 188
Northwestern Islands bird region 14
Northwestern Lowlands bird region 15
Nuthatch, Pink-faced 380

ordinal accounts 28
Oriole, Australian 405
Australian Yellow 406
Brown 405
Green 406
Green-backed 405
New Guinea 405
Olive-backed 405
Striated 405
White-bellied 405
Yellow 406
Yellow-bellied 406
Osprey 199
Osprey, Eastern 199
Owl, Australian Barn- 214
Australian Masked 214
Barking 217
Barn 215
Biak Scops- 216
Brown 218
Eastern Barn 215
Eastern Grass- 215
Grass 215
Jungle Hawk 218
Lesser Sooty 214
Masked 214
Moluccan Scops- 216
Papuan Boobook 218, 219
Papuan Hawk- 218
Pearly 215
Rufous 216
Sooty 213
Owlet-nightjar, Allied 156
Archbold's 155
Australian 156
Barred 155
Bennett's 156
Collared 156
Crested 156
Feline 153
Large 153

Mountain 154
Rand's 154
Spangled 154
Starry 153
Vogelkop 156
Wallace's 154
White-spotted 153
Oystercatcher, Australian Pied 165
Pied 165

Paradigalla, Long-tailed 430
Short-tailed 430
Paradise-Kingfisher, Aru 224
Australian 225
Biak 223
Brown-backed 225
Brown-headed 225
Buff-breasted 224
Common 222
Elliot's 224
Kofiau 224
Little 224
Numfor 224
Red-breasted 225
Rossel 223
White-tailed 225
Parotia, Arfak 421
Bronze 423
Carola's 423
Eastern 422
Foja 423
Helen's 423
Huon 422
Lawes's 422
Queen Carola's 423
Wahnes's 422
Western 421
Parrot, Bare-headed 241
Barred Little Tiger- 244
Black-cheeked Fig 261
Black-fronted Fig- 261
Blue-collared 266
Blue-faced Fig 262
Blue-fronted Fig- 261
Brehm's 242
Brehm's Tiger- 242
Buff-faced Pygmy 269
Buffy-faced Pygmy 269
Creamy-breasted Fig- 261
Desmarest's Fig 258
Double-eyed Fig- 261
Dusky-cheeked Fig- 261
Dwarf Fig 262
Eclectus 264
Edwards's Fig- 258
Geelvink Pygmy 268
Golden-headed Fig- 258
Great-billed 267
Green-winged King 264

Island 267
Large-billed 267
Large Fig- 257
Lilac-collared 267
Maradasz's 244
Madarasz's Tiger- 244
Modest 244
Modest Tiger- 243
Moluccan 267
Moluccan King- 263
Mountain Pygmy- 270
New Guinea Vulturine 241
Orange-breasted Fig- 259
Orange-fronted Hanging 262
Painted 243
Painted Tiger- 242
Papuan Hanging 262
Papuan King- 263
Pesquet's 241
Plain-breasted Little 244
Red-breasted Pygmy 269
Red-cheeked 265
Red-faced Fig 262
Red-sided 265
Red-sided Eclectus 265
Red-winged 264
Rose-breasted Pygmy- 270
Salvadori's Fig- 259
Simple 267
Timberline Tiger- 243
Two-eyed Fig 262
Vulturine 241
Whiskered Fig- 259
William's Fig 261
Yellow-capped Pygmy 268
Yellow-naped Fig- 258
Parrotfinch, Blue-faced 513
Blue-headed 513
Large-billed 513
Papuan 513
Parrot-Mannikin, Papuan 513
Pelican, Australian 104
Spectacled 104
Peltops, Highland 363
Lowland 362
Mountain 363
Singing 363
Peltops-flycatcher, Lowland 362
Singing 363
Petrel, Beck's 99
Black-bellied Storm- 95
Bulwer's 103
Collared 99
Cook's 98
Frigate 95
Gould's 99
Herald 98
Kermadec 98
Leach's Storm- 96
Matsudaira's Storm- 96

Phoenix 98
Providence 98
Pycroft's 99
Southern Giant 97
Stejneger's 99
Tahiti 99
Trinidade 99
White-bellied Storm- 95
White-faced Storm- 95
Wilson's Storm- 94
Phalarope, Northern 184
Red-necked 184
Pheasant-pigeon, Black-naped 73
Green-naped 73
Grey-naped 73
White-naped 73
Pigeon, Banded Imperial 90
Bare-eyed 92
Black-collared Fruit 90
Brown 67
Collared Imperial 86
Common 64
D'Albertis 92
Domestic 64
Eastern Nutmeg 91
Elegant Imperial 86
Feral 64
Floury Imperial 89
Geelvink Imperial 87
Green-collared 73
Green-winged 76
Grey-collared 73
Grey Imperial 89
Hackled 75
Island Imperial 89
Jungle Bronzewing 68
Magnificent Ground 73
Metallic 64
Mountain 92
Mueller's Imperial 90
Müller's Imperial 90
Nicobar 75
Noble Ground 73
Nutmeg 91
Pacific 87
Pacific Imperial 87
Papuan Mountain- 92
Pheasant 72
Pied Imperial 91
Pink-capped Fruit 85, 90
Pink-crowned 85
Pinon's Imperial 89
Purple-crowned 79
Purple-tailed Imperial 88
Red-breasted Imperial 89
Red-crowned 85
Rock 64
Rose-crowned 85
Rose-fronted 84
Rufescent Imperial 88

Southern Crowned 74
Spectacled Imperial 86
Spice 87
Spice Imperial 87
Stephan's Green-winged 77
Thick-billed Ground- 71
Torresian Imperial 91
Torres Strait 91
Torres Strait Imperial 91
Victoria Crowned 74
Vulturine 75
Western Crowned 73
White-capped Ground 68
White-eyed Imperial 86
White Nutmeg Imperial 91
White-tailed 75
White-throated 64
Wompoo 78
Wood 64
Zoe's Imperial 90
Pintail, Northern 60
Pipit, Alpine 523
    Australasian 523
    Australian 522
    New Guinea 523
    Richard's 523
Pitohui, Black 391
    Crested 382
    Hooded 404
    Mottle-breasted 390
    Northern Variable 401
    Raja Ampat 402
    Rusty 390
    Southern Variable 403
    Variable 402-404
    White-bellied 390
Pitta, Black-headed 273
    Blue-breasted 272
    Blue-winged 274
    Buff-breasted 274
    Hooded 273
    Lesser False 439
    Noisy 274
    Red-bellied 272
Ploughbill, Wattled 383
Plover, Black-bellied 167
    Eastern Golden 167
    Greater Oriental 170
    Greater Sand- 170
    Great Stone 164
    Grey 167
    Kentish 169
    Large Sand- 170
    Lesser Golden 167
    Lesser Sand- 169
    Little Ringed 168
    Masked 168
    Mongolian 170
    Oriental 170
    Pacific Golden 167

Red-capped 169
Rufous-capped 169
Silver 167
Spur-winged 168
Pratincole, Australian 186
    Collared 186
    Eastern 186
    Isabelline 186
    Large Indian 186
    Long-legged 186
    Oriental 187
Prion, Fairy 98
    Slender-billed 98
Pygmy-Goose, Asian 58
    Indian 58
    White 58
    White-quilled 58

Quail, Asian Blue 52
    Black-backed Bustard 185
    Blue-breasted 52
    Brown 51
    Chinese 52
    King 52
    Painted 52
    Red-backed 185
    Snow Mountain 50
    Snow Mountains 50
    Swamp 52
Quail-dove, Chestnut 69
Quail-thrush, Ajax 359
    False 359
    Painted 359

Racket-tail, Caroline 224
    Galatea 223
    Riedel's 223
Rail, Banded Land 128
    Bare-eyed 129
    Barred 127
    Buff-banded 128
    Chestnut 125, 129
    Chestnut-bellied 129
    Chestnut Forest- 124
    Forbes' 125
    Forbes's Chestnut 125
    Forbes's Forest- 125
    Grey-faced 132
    Land 128
    Lesser Spotted 130
    Lewin's 127
    Marsh 130
    Mayr's 126
    Mayr's Chestnut 126
    Mayr's Forest- 125
    New Guinea Chestnut 125
    New Guinea Flightless 132
    Papuan Flightless 132

Red-necked 126
Slate-breasted 127
Sooty 130
Tiny 130
Water 127
White-browed 131
White-striped 125
White-striped Chestnut 125
White-striped Forest- 125
Redshank, Common 176
    Spotted 176
Reed-Warbler, Australian 488
    Eastern Great 488
    Great 489
    Oriental 487
Reef-Egret, Eastern 116
    Pacific 116
Reeve 184
References and data sources 16
Riflebird, Albert 425
    Growling 425
    Magnificent 425
Robin, Alpine 460
    Archbold's New Guinea 460
    Ashy 465
    Banded Yellow 471
    Black-bibbed 467
    Black-capped 465
    Black-cheeked 465, 466
    Black-chinned 466
    Black-sided 467
    Black-throated 467
    Blue-grey 470
    Canary 461
    Cloud-forest 460
    Eastern White-eyed 459
    Garnet 460
    Greater Ground- 456
    Green-backed 457
    Grey 470
    Grey-headed 461
    Lesser Ground- 456
    Little 473
    Mangrove 469
    Mountain 459
    New Guinea 466
    Northern Scrub 464
    Olive-yellow 471
    Papuan Scrub- 464
    Pied 468
    Slaty 471
    Smoky 470
    Snow 460
    Snow Mountain 460
    Snow Mountains 460
    Subalpine 459
    Torrent 462
    White-eyed 458
    White-faced 472
    White-faced Yellow 473

White-rumped 468
White-throated 473
White-winged 469
Yellow-bellied 463
Yellow-legged 461
Robin-flycatcher, Mountain 460
Rock 460
Robin-whistler, Island 396
Rock-thrush, Blue 502
Roller, Broad-billed 222
Rubythroat 501
Rubythroat, Siberian 500
Ruff 183

Sanderling 180
Sandpiper, Asian Pectoral 182
Baird's 181
Broad-billed 182
Buff-breasted 183
Common 178
Curlew 183
Green 177
Marsh 176
Pectoral 182
Sharp-tailed 182
Siberian Pectoral 182
Terek 178
Western 179
Wood 177
Sand-Plover, Greater 170
Large 170
Lesser 169
Sandplover, Oriental 170
Satinbird, Crested 343
Loria's 342
Yellow-breasted 344
scientific names 30
Scops-Owl, Biak 216
Moluccan 216
Scrubfowl, Biak 47
Common 50
Dusky 49
Melanesian 49
Moluccan 47
New Guinea 49
Orange-footed 48
Wallace's 47
Scrub-Robin, Northern 464
Papuan 464
Scrubturkey, Black-billed 46
Bruijn's 45
Collared 47
Red-billed 46
Wattled 45
Scrubwren, Arfak 334
Beccari's 332
Buff-faced 334
Dusky 335

Grey-green 335
Large 332
Large Mountain 333
Mountain 333
Noisy 333
Olive 335
Pale-billed 329
Papuan 334
Perplexing 330–333
Tropical 331
Vogelkop 334
Sea-Eagle, White-bellied 206
White-breasted 207
Sepik-Ramu bird region 15
Sericornis, Arfak 334
Beccari's 332
Black and Green 334
Buff-faced 334
Grey-green 335
Large Mountain 333
Pale-billed 330
Papuan 335
Rufous 334
Shag, Little Black 121
Little Black and White 120
Shearwater, Audubon's 102
Black 102
Bonaparte's 101
Christmas 102
Christmas Island 102
Flesh-footed 101
Fleshy-footed 101
Heinroth's 103
Hutton's 102
Pink-footed 101
Short-tailed 101
Slender-billed 101
Sombre 100
Sooty 100
Streaked 101
Streak-headed 102
Tropical 102
Wedge-tailed 100
White-faced 102
Shelduck, Black-backed 56
Burdekin 56
Raja 56
White-headed 56
Shieldbill, Clicking 362
Lowland 362
Singing 363
Tinkling 363
Shrike, Black-capped 455
Black-headed 455
Brown 454
Long-tailed 455
Red-backed 455
Red-tailed 455
Schach 455
Shrike-flycatcher, Brown 388

Grey 389
Shrike-thrush, Rufous 388
Whistling 389
Shrikethrush, Black 390
Grey 388
Little 384
Rufous 388
Rusty 389
Sooty 388
Whistling 389
White-bellied 390
Shrike-tit, Wattled 383
Sicklebill, Black 429
Black-billed 427
Brown 429
Buff-tailed 428
Greater 429
Lowland 428
Meyer's 429
Pale-billed 428
White-billed 428
Silktail, Papuan 407
Sittella, Black 380
Papuan 379
Varied 380
Skua, Arctic 198
Long-tailed 198
Pomarine 197
South Polar 198
Snipe, Australian 172
Chinese 173
Japanese 172
Latham's 172
Marsh 173
Pin-tailed 173
Swinhoe's 173
Southeastern Islands bird region 16
Southeastern Peninsula bird region 16
Southern Lowlands bird region 15
Sparrow, English 510
House 510
Eurasian Tree 511
European Tree 511
Tree 511
Sparrowhawk, Australian 212
Black-mantled 211
Buergers's 213
Chinese 209
Collared 212
Horsfield's 209
species account format 29
species concepts 22
species status definitions 31
Spinetail, White-throated 162
Spoonbill, Black-billed 107
Royal 106
Yellow-billed 107
Starling 498
Starling, Colonial 494
Common 498

Eurasian 498
European 498
Island 496
Long-tailed 495
Metallic 494
Moluccan 496
Shining 494
Singing 495
Stilt, Black-winged 165
    White-headed 166
Stint, Eastern 181
    Little 181
    Long-toed 181
    Middendorf's 181
    Red-necked 181
    Rufous-necked 181
    Siberian Little 181
Stone-curlew, Beach 164
    Bush 164
    Reef 164
    Southern 164
Stork, Black-necked 104
    Green-necked 104
Storm-Petrel, Black-bellied 95
    Gould's 95
    Leach's 96
    Matsudaira's 96
    Sooty 96
    Swinhoe's 96
    Vieillot's 95
    White-bellied 95
    White-faced 95
    Wilson's 94
    Yellow-webbed 94
Straightbill, Green-backed 302
    Lowland 311
    Mountain 311
    Olive 310
    Tawny 311
subspecies treatment 23
Sunbird, Black 506
    Black-throated 509
    Olive-backed 508
    Rand's 509
    Yellow-bellied 509
    Yellow-breasted 509
Swallow, Bank 475
    Barn 475
    Common 476
    Hill 476
    House 476
    Pacific 476
    Red-rumped 477
    The 476
    Welcome 476
Swallow-shrike, White-breasted 366
Swamphen 133
Swamphen, Purple 133
Swan, Black 55
Swift, Fork-tailed 163

Moustached 157
New Guinea Spine-tailed 162
Pacific 163
Papuan Spine-tailed 161
White-rumped 163
Swiftlet, Bare-legged 159
    Glossy 158
    Lowland 160
    Mountain 160
    New Guinea 160
    Papuan 159
    Three-toed 159
    Uniform 160
    White-bellied 158
    Whitehead's 160
systematics 18

Talegalla, Brown-collared 47
Tattler, Grey-tailed 177
    Polynesian 178
    Siberian 177
    Wandering 178
taxonomy 21
Teal, Australian Grey 60
    Chestnut 60
    Cotton 58
    Garganey 60
    Grey 59
    Salvadori's 57
    Slender 60
    Sunda 60
terms, geographic 35
terms, table of 34
Tern, Atoll 189
    Black 196
    Black-naped 195
    Bridled 193
    Brown-winged 193
    Caspian 191
    Common 195
    Crested 191
    Eastern Common 195
    Fairy 189
    Great 191
    Great-crested 191
    Greater-crested 191
    Grey-backed 193
    Gull-billed 190
    Indian Lesser Crested 192
    Large 191
    Least 193
    Lesser Crested 191
    Little 192
    Marsh 196
    Roseate 194
    Sooty 194
    Spectacled 193
    Swift 191
    Whiskered 196

White 188
White-winged 196
White-winged Black 196
Thicket-Fantail, Black 407
    Sooty 408
    White-bellied 408
Thicket-flycatcher, Black-throated 467
    Green 458
    Grey 470
    Ground 465, 466
    Slaty 471
    White-rumped 469
    White-throated 459
    White-winged 469
    Yellow 471
Thicket-robin, White-winged 469
Thickhead, Brown 398
    Grey-headed 398
Thick-knee, Beach 164
    Bush 164
    Great 154
Thornbill, Grey 335
    Mountain 335
    New Guinea 335
    Papuan 335
Thrush, Greater New Guinea 456
    Ground 499
    Island 499
    Lesser New Guinea 457
    Russet-tailed 498
    Scaly 499
    White's 499
Tiger-heron, New Guinea 108
Tiger-Parrot, Barred Little 244
    Brehm's 242
    Madarasz's 244
    Modest 243
    Painted 242
Tit, Ploughshare 383
Torrentlark 440
Trans-Fly bird region 15
Treecreeper, Papuan 283
    White-throated 283
Tree-duck, Plumed 54
    Water 55
Treeswift, Moustached 157
    Whiskered 157
Triller, Biak 374
    Black-browed 373
    Pied 374
    Varied 374
    White-browed 374
    White-rumped 373
    White-winged 373
Tropicbird, Long-tailed 93
    Red-tailed 93
    White-tailed 93
    Yellow-billed 93
Trumpetbird 418
Turkey, Bruijn's Brush 45

Turnstone 179
Turnstone, Ruddy 179
Turtle-Dove, Spotted 65
type description 31
type locality 31

Velvet-thrush, Cloudforest 456
  Rusty 457

Wagtail, Eastern Yellow 521
  Green-headed 521
  Grey 522
  Willie 407
  Yellow 522
Warbler, Australian Reed- 488
  Biak Leaf- 481
  Bicoloured Mouse- 327
  Black and White Wren 288
  Black-headed Gerygone 338
  Black-throated 338
  Broad-billed Wren 285
  De Vis Tree 335
  Eastern Great Reed- 488
  Fairy 338
  Fan-tailed 493
  Fly River Grass 491
  Gray's 489
  Green-backed 337
  Great Reed- 488
  Grey Gerygone 335
  Island Leaf- 480
  Little Grass 492
  Mangrove 339
  Mountain Gerygone 335
  Mountain Leaf- 480
  Mountain Mouse- 328
  New Guinea Leaf- 481
  Numfor Leaf- 481
  Oriental Reed- 487
  Red-backed 460
  Rufous Wren 285
  Rusty Mouse- 326
  Streaked Fantail 493
  Treefern 339
  Treefern Gerygone 339
  Wallace's Wren 284
  White-throated 337
  White-throated Gerygone 337
Waterhen, Pale-vented 132
Wattlebird, Cinnamon-breasted 321

Short-tailed 430
  Yellow-browed 323
Western Ranges bird region 15
Wherrybill 361
Whimbrel 175
Whimbrel, Little 175
Whipbird, Papuan 354
Whistle-Duck, Grass 54
  Spotted 54
  Water 55
Whistler, Baliem 394
  Bismarck 393
  Black 391
  Black-backed 399
  Black-headed 399
  Black-tailed 395
  Brown 398
  Brown-backed 396
  Brownish 397
  Common Golden 394
  Dwarf 326
  Golden 393
  Golden-backed 395
  Golden-yellow 395
  Grey 398
  Grey-crowned 395
  Grey-headed 398
  Grey Mountain 396
  Hill Golden 393
  Island 396
  Lorentz's 396
  Louisiade 393
  Mangrove Golden 394
  Meyer's 395
  Modest 396
  Mottled 368
  Red-vented 368
  Regent 392
  Rufous 397
  Rufous-breasted 397
  Rufous-naped 382
  Rusty 397
  Rusty-breasted 397
  Schlegel's 392
  Sclater's 392
  Sooty 388
  Vogelkop 395
  White-bellied 399
  Yellow-backed 395
Whistling-duck, Grass 54

White-eye, Ashy-bellied 482
  Biak 484
  Black-fronted 483
  Capped 484
  Green-fronted 484
  Island 487
  Lemon-bellied 482
  Louisiade 486
  Louisiades 487
  Mangrove 482
  Moluccan 482
  Mountain 286
  New Guinea Mountain 486
  Oya Tabu 485
  Pale 482
  Papuan 486
  Soepiori 484
  Tagula 484
  Western Mountain 485
  White-throated 484
  Yellow 482
  Yellow-bellied Mountain 485
Wigeon 59
Wigeon, Eurasian 59
  European 59
Winter, Jacky 463
Woodcock, East Indian 172
  New Guinea 172
  Rufous 172
Wood-Duck, Australian 57
Woodswallow, Black-faced 367
  Gray 367
  Gray-breasted 367
  Great 366
  New Guinea 366
  White-bellied 367
  White-breasted 366
  White-rumped 366
Wood-swallow, Black-breasted 366
  Giant 366
  Greater 366
  New Guinea 366
Wren, Blue 287
  Blue-capped 284
  Broad-billed 285
  Imperial 287
  White-shouldered 288
Xanthotis, Brown 295
  Spotted 296
  Tawny-breasted 295

abbotti, Papasula 118
abbotti, Sula 118
aberrans, Trichoglossus 253
Acanthiza 335
Acanthizidae 325
Accipiter 209
Accipitridae 200
Accipitriformes 198
Acridotheres 497
Acrocephalidae 487
Acrocephalus 487
acrophila, Ptiloprora 304
Actitis 178
acuminata, Calidris 182
acuminata, Erolia 182
acuminatus, Totanus 182
acuta, Anas 60
adamsoni, Coracina 369
addenda, Gelochelidon 190
addenda, Lophorina 426
adelberti, Manucodia 418
adelberti, Sericornis 333
adiensis, Pitohui 401
adolphinae, Myzomela 293
aedificans, Amblyornis 278
Aegotheles 153
Aegothelidae 153
Aepypodius 44
Aerodramus 159
aeruginosus, Circus 209
aethiopicus, Threskiornis 106
aethiops, Poecilodryas 501
aethiops, Saxicola 501
affinis, Aegotheles 156
affinis, Colluricincla 385
affinis, Gelochelidon 190
affinis, Gerygone 338
affinis, Megapodius 48, 49
affinis, Melilestes 350

affinis, Milvus 205
affinis, Myiolestes 385
aicora, Meliphaga 315
aida, Malurus 287
aignani, Zosterops 487
Ailuroedus 275
ajax, Cinclosoma 359
ajax, Eupetes 359
alaris, Cinclosoma 359
alaris, Gallicolumba 69
alaris, Megaloprepia 77
Alaudidae 473
alba, Ardea 113
alba, Calidris 180
alba, Crocethia 180
alba, Daphoenositta 379
alba, Egretta 114
alba, Gygis 188
alba, Neositta 379
alba, Pterodroma 98
alba, Sterna 188
alba, Trynga 180
alba, Tyto 214
alberti, Hypotaenidia 127
alberti, Lewinia 127
alberti, Pachycephala 398
alberti, Pteridophora 421
alberti, Ptiloris 425
albertisi, Aegotheles 154
albertisi, Drepanephorus 427
albertisi, Drepanornis 427
albertisi, Epimachus 428
albertisii, Gymnophaps 92
albescens, Hirundo 476
albicans, Epimachus 38, 429
albicans, Falcinellus 429
albicauda, Melanocharis 348
albidior, Peneothello 470
albidior, Poecilodryas 470

albifacies, Poecilodryas 473
albifacies, Tregellasia 473
albifrons, Daphoenositta 380
albifrons, Henicophaps 68
albifrons, Machaerirhynchus 360
albifrons, Sittella 380
albifrons, Sterna 192
albifrons, Sternula 192
albigula, Machaerirhynchus 360
albigula, Myzomela 289
albigularis, Pachycephala 458
albigularis, Phylloscopus 480
albigularis, Seicercus 480
albipennis, Nettapus 58
albispecularis, Heteromyias 38, 465
albispecularis, Pachycephala 465
albispecularis, Poecilodryas 465
albiventris, Artamus 367
albiventris, Zosterops 482
alboauricularis, Lichmera 299
alboauricularis, Stigmatops 299
albogularis, Conopophila 306
albogularis, Entomophila 306
albogularis, Eurostopodus 151
albogularis, Melithreptus 301
albolimbata, Rhipidura 411
albolimbatus, Megalurus 491
albolimbatus, Poodytes 491
albolineata, Egretta 116
albonotata, Meliphaga 312
albonotata, Plesiodryas 467
albonotata, Poecilodryas 467
albonotata, Ptilotis 312
albonotatus, Megalestes 467
albopunctatum, Dicaeum 505
alboscapulatus, Malurus 287
albotaeniata, Amaurodryas 466
albotaeniata, Heteromyias 466
albus, Anous 189

albus, *Casmerodius* 114
Alcedinidae 233
Alcedo 233
alcinus, Machaerhamphus 202
alcinus, Macheiramphus 202
Aleadryas 381
alecto, Ara 239
alecto, Cracticus 363
alecto, Drymophila 443
alecto, Monarcha 444
alecto, Myiagra 443
alecto, Piezorhynchus 443
aliena, Mirafra 474
Alisterus 262
Alopecoenas 69
alpinum, Paramythia 353
alpinus, Megalurus 491
alpinus, Neopsittacus 250
alter, Manucodia 420
amabilis, Cyclopsitta 260
amabilis, Cyclopsittacus 260
amabilis, Eupetes 356
amabilis, Ptilorrhoa 356
Amalocichla 456
amati, Amblyornis 278
Amaurornis 131
Amblyornis 277
amboinensis, Alisterus 263
amboinensis, Columba 66
amboinensis, Macropygia 66
amboinensis, Psittacus 263
amethystina, Collocalia 158
amethystina, Loria 342
ampla, Paradisaea 435
anaethetus, Onychoprion 193
anaethetus, Sterna 193
anais, Mino 496
anaïs, Sericulus 496
analoga, Meliphaga 313
analoga, Ptilotis 313
Anas 58
Anatidae 54
andrewsi, Fregata 116
Androphobus 354
anggiensis, Fulica 134
angustifrons, Coracina 372
angustifrons, Graucalus 372
Anhinga 122
Anhingidae 122
Anous 187
Anseranas 53
Anseranatidae 53
Anseriformes 53
Anthus 522
antigone, Grus 135
Anurophasis 50
Aplonis 494
apoda, Paradisaea 437
Apodidae 157
approximans, Circus 208

approximans, Pachycephala 458
approximans, Pachycephalopsis 458
Aprosmictus 264
Apus 162
Aquila 204
archboldi, Aegotheles 154
archboldi, Dacelo 228
archboldi, Eurostopodus 151
archboldi, Lyncornis 151
archboldi, Petroica 460
archboldi, Sauromarptis 228
Archboldia 279
arctitorquis, Pachycephala 400
arcuata, Anas 54
arcuata, Dendrocygna 54
Ardea 112
Ardeidae 108
ardens, Sericulus 280
ardens, Xanthomelus 280
Ardeotis 123
arfaki, Oreocharis 353
arfaki, Oreopsittacus 244
arfaki, Parus 353
arfaki, Strix 214
arfaki, Trichoglossus 244
arfaki, Tyto 214
arfakiana, Gerygone 335
arfakiana, Melanocharis 345
arfakianum, Dicaeum 345
arfakianus, Aepypodius 44
arfakianus, Ailuroedus 276
arfakianus, Cacomantis 146
arfakianus, Sericornis 335
arfakianus, Tallegallus 44
argentauris, Lichmera 299
argentauris, Ptilotis 299
argus, Caprimulgus 151
argus, Eurostopodus 151
ariel, Atagen 117
ariel, Collocalia 478
ariel, Fregata 117
ariel, Hirundo 478
ariel, Petrochelidon 478
arminjoniana, Pterodroma 99
armiti, Heteromyias 38, 465
armiti, Poecilodryas 465
arossi, Macropygia 67
Arses 441
Artamidae 365
Artamus 366
arthuri, Pitohui 382
aru, Ducula 86
aruense, Edoliisoma 376
aruense, Edolisoma 376
aruensis, Arses 442
aruensis, Carterornis 449
aruensis, Casuarius 42
aruensis, Centropus 136
aruensis, Cyclopsitta 261

aruensis, Eclectus 266
aruensis, Geoffroyus 266
aruensis, Gerygone 337
aruensis, Manucodia 417
aruensis, Megapodius 48
aruensis, Meliphaga 312
aruensis, Monarcha 449
aruensis, Myiolestes 385
aruensis, Myzomela 290
aruensis, Nesocentor 136
aruensis, Noctua 216
aruensis, Otidiphaps 72
aruensis, Philemon 298
aruensis, Pitohui 403
aruensis, Pitta 272
aruensis, Psittacula 261
aruensis, Psittacus 266
aruensis, Ptilotis 312
aruensis, Rectes 403
aruensis, Sericornis 330
aruensis, Talegalla 46
aruensis, Tropidorhynchus 298
aruensis, Zosterops 485
arundinaceus, Acrocephalus 489
asiatica, Eupoda 170
asiatica, Mycteria 104
asiaticus, Ephippiorhyncus 104
asiaticus, Xenorhynchus 104
aspasia, Cinnyris 506
aspasia, Leptocoma 506
aspasia, Nectarinia 507
aspasioides, Leptocoma 507
aspasioides, Nectarinia 507
assimilis, Ninox 217
astigmaticus, Ailuroedus 276
Astrapia 430
astrolabae, Eurystopodus 151
ater, Manucodia 420
ater, Melipotes 309
ater, Phonygama 420
ater, Psittacus 256
aterrimus, Probosciger 239
aterrimus, Psittacus 239
atra, Chalcopsitta 256
atra, Colluricincla 38, 388
atra, Fulica 134
atra, Manucodia 421
atra, Myiagra 445
atra, Pachycephala 38, 388
atra, Rhipidura 411
atrata, Anas 55
atrata, Charmosyna 249
atratus, Cygnus 55
atratus, Falcinellus 429
atricapilla, Heteromyias 466
atricapilla, Poecilodryas 470
atrifrons, Zosterops 483
atrocaeruleus, Dicrurus 414
atrovirens, Campephaga 373
atrovirens, Lalage 373

attenua, Eos 254
atthis, Alcedo 233
atthis, Gracula 233
audax, Aquila 205
audax, Uroaetus 205
audax, Vultur 205
auga, Meliphaga 312
augustaevictoriae, Paradisaea 438
augustaevictoriae, Paradisea 438
aurantia, Carpophaga 90
aurantiacus, Carterornis 449
aurantiacus, Casuarius 43
aurantiacus, Monarcha 449
aurantiifrons, Loriculus 262
aurantiifrons, Ptilinopus 82
aurantiifrons, Ptilonopus 82
aurea, Coracias 280
aurea, Pachycephala 395
aureus, Sericulus 280
aureus, Xanthomelus 280
auriceps, Leptocoma 506
auriceps, Nectarinia 506
auricularis, Microeca 472
auricularis, Rhipidura 411
auricularis, Tregellasia 473
aurigaster, Pycnonotus 479
aurigaster, Turdus 479
auripennis, Seleucides 424
aurora, Loboparadisea 37, 344
austera, Notophoyx 115
austera, Xanthotis 294
australasiae, Ardea 111
australasiae, Nycticorax 111
australis, Acrocephalus 488
australis, Anthus 522
australis, Ardea 110
australis, Ardeotis 123
australis, Aythya 61
australis, Calamoherpe 488
australis, Choriotis 123
australis, Coturnix 52
australis, Dendrocygna 54
australis, Ephippiorhynchus 104
australis, Eupodotis 123
australis, Fulica 134
australis, Ixobrychus 110
australis, Mycteria 104
australis, Nyroca 61
australis, Otis 123
australis, Synoicus 52
Aviceda 201
axillaris, Coracina 369
axillaris, Elanus 200
axillaris, Graucalus 369
axillaris, Melanocharis 348
axillaris, Monarcha 446
axillaris, Neneba 348
axillaris, Pachycephalopsis 457
axillaris, Symposiachrus 446
Aythya 60

azurea, Alcedo 234
azurea, Alcyone 234
azureus, Ceyx 234

bailloni, Procellaria 102
bailloni, Puffinus 102
bairdii, Actodromas 181
bairdii, Calidris 181
bakeri, Sericulus 281
bakeri, Xanthomelus 281
baliem, Tyto 215
balim, Macropygia 67
balim, Malurus 287
balim, Pachycephala 394
balim, Pachycephalopsis 458
bangsi, Sterna 194
barnardi, Chalcites 144
barnardi, Chrysococcyx 144
barnesi, Astrachria 432
bartoni, Pachycephala 393
baru, Aerodramus 161
baru, Falco 236
barus, Collocalia 161
basalis, Chalcites 142
basalis, Chrysococcyx 143
basalis, Cuculus 142
basilica, Ducula 88
bastille, Crateroscelis 328
bastille, Meliphaga 315
batantae, Arses 441
batantae, Colluricincla 385
batantae, Edolisoma 378
batantae, Pinarolestes 385
batantae, Sericornis 330
batavorum, Loriculus 262
batesi, Melirrhophetes 321
baueri, Limosa 174
beccarii, Alopecoenas 70
beccarii, Chalcophaps 70
beccarii, Drymodes 464
beccarii, Drymoedus 464
beccarii, Gallicolumba 70
beccarii, Goura 74
beccarii, Micropsitta 269
beccarii, Nasiterna 269
beccarii, Otus 216
beccarii, Scops 216
beccarii, Sericornis 330, 331
becki, Meliphaga 314
becki, Pseudobulweria 99
belcheri, Heteroprion 98
belcheri, Pachyptila 37, 98
belensis, Saxicola 501
belfordi, Melidectes 321
belfordi, Melirrhophetes 321
bella, Charmosinopsis 249
bellus, Ptilinopus 79
bellus, Ptilonopus 79
bengalensis, Sterna 191

bengalensis, Thalasseus 191
bennetti, Casuarius 41
bennettii, Aegotheles 155
berauensis, Trichoglossus 253
bergii, Sterna 191
bergii, Thalasseus 191
bergmanii, Nectarinia 507
berigora, Falco 237
berigora, Ieracidea 237
berlepschi, Parotia 423
bernsteini, Centropus 137
bernsteini, Erythropitta 271
bernsteini, Pitta 271
bernsteinii, Chalcopsitta 256
bernsteinii, Monarcha 448
biaki, Larius 266
bicarunculatus, Casuarius 42
bicinia, Coracina 375
bicinia, Edolisoma 375
bicolor, Columba 91
bicolor, Ducula 91
bigilalae, Lonchura 515
bigilalei, Lonchura 515
bimaculata, Peneothello 468
bimaculatus, Myiolestes 468
bistriatus, Casuarius 42
bivitatta, Petroeca 459
bivittata, Petroica 459
blainvillii, Eurylaimus 362
blainvillii, Peltops 362
blissi, Crateroscelis 327
bloodi, Epimachus 38, 429
bloodi, Oreocharis 353
blythii, Cyclopsitta 257
blythii, Psittaculirostris 257
bonapartii, Todopsis 286
boobook, Ninox 217
borealis, Phylloscopus 479
boreonesioticus, Sericornis 333
boschmai, Lonchura 518
boyeri, Campephaga 370
boyeri, Coracina 370
brachyrhyncha, Rhipidura 409
brachyura, Poecilodryas 466
brachyurus, Accipiter 212
brachyurus, Heteromyias 466
brachyurus, Leucophantes 466
bracteatus, Dicrurus 414
brassi, Melidectes 322
brassi, Philemon 296
brehmii, Monarcha 449
brehmii, Psittacella 242
brehmii, Symposiachrus 449
brevicauda, Amalocichla 457
brevicauda, Aplonis 495
brevicauda, Drymoedus 457
brevicauda, Macruropsar 495
brevicauda, Paradigalla 430
brevicauda, Paramythia 354
brevipennis, Philemon 297

brevipennis, Pseudorectes 390
brevipennis, Rhectes 390
brevipes, Heteroscelis 177
brevipes, Procellaria 99
brevipes, Pterodroma 99
brevipes, Totanus 177
brevipes, Tringa 177
brevirostris, Drymaoedus 464
brevirostris, Syma 232
brevis, Reinwardtoena 65
brooki, Trichoglossus 254
bruijni, Pomareopsis 440
bruijnii, Aepypodius 45
bruijnii, Drepanornis 428
bruijnii, Epimachus 428
bruijnii, Grallina 440
bruijnii, Micropsitta 269
bruijnii, Nasiterna 269
bruijnii, Pomareopsis 440
bruijnii, Talegallus 45
brunnea, Ifrita 415
brunneicaudus, Pitohui 402
brunneicaudus, Rhectes 402
brunneiceps, Pitohui 403
brunneiceps, Rectes 403
brunneipectus, Gerygone 338
brunneipectus, Pseudogerygone 338
brunneivertix, Pitohui 402
brunneus, Melilestes 310
buccoides, Ailuroedus 275
buccoides, Kitta 275
buceroides, Philedon 297
buceroides, Philemon 297
Bucerotidae 219
Bucerotiformes 219
buergersi, Accipiter 213
buergersi, Erythrotriorchis 213
buergersi, Geoffroyus 266
buergersi, Mearnsia 162
buergersi, Psittacella 242
buergersi, Pteridophora 421
buergersi, Ptilorrhoa 358
buergersi, Sericornis 334
Bulweria 103
bulweria, Bulweria 103
bulweria, Procellaria 103
bürgersi, Astur 213
bürgersi, Chaetura 162
bürgersi, Eupetes 358
bürgersi, Geoffroyus 267
bürgersi, Pitohui 391
bürgersi, Psittacella 242
bürgersi, Sericornis 334
Burhinidae 163
Burhinus 163
buruanus, Ptilinopus 80
Butastur 207
Butorides 111

Cacatuidae 238
Cacomantis 145
caeruleiceps, Trichoglossus 254
caeruleogrisea, Campephaga 369
caeruleogrisea, Coracina 369
caerulescens, Eupetes 356
caerulescens, Ptilorrhoa 356
caeruleus, Elanus 200
caeruleus, Falco 200
cahni, Melidectes 320
calabyi, Tyto 214
calaminoros, Lonchura 514
calaminoros, Munia 514
caledonica, Ardea 110
caledonicus, Nycticorax 110
Calidris 179
calidus, Falco 238
calidus, Pomatorhinus 340
Caliechthrus 145
Caligavis 318
Calliope 500
calliope, Calliope 500
calliope, Erithacus 501
calliope, Luscinia 501
calliope, Motacilla 500
callopterus, Alisterus 263
callopterus, Aprosmictus 263
Caloenas 75
campbelli, Chenorhamphus 286
campbelli, Malurus 286
Campephagidae 368
Campochaera 372
candida, Gygis 188
candida, Sterna 188
candidior, Xanthotis 296
canescens, Monarcha 452
canescens, Turdus 500
caniceps, Lonchura 516
caniceps, Munia 516
canorus, Cuculus 147
cantans, Sericornis 332
cantoroides, Aplonis 495
cantoroides, Calornis 495
canutus, Calidris 180
canutus, Tringa 180
capensis, Tyto 215
capistratus, Trichoglossus 254
capitalis, Crateroscelis 326
caprata, Motacilla 501
caprata, Saxicola 501
Caprimulgidae 150
Caprimulgiformes 148
Caprimulgus 152
captata, Melanocharis 347
captus, Rallus 127
carbo, Pelecanus 121
carbo, Phalacrocorax 121
carbonarius, Dicrurus 414
carbonarius, Edolius 414
carbonarius, Turdus 499

carmichaeli, Rallicula 126
carneipes, Ardenna 101
carneipes, Puffinus 101
carolae, Melipotes 308
carolae, Parotia 423
carolinae, Macgregoria 309
carolinae, Pitohui 401
carolinae, Tanysiptera 224
Carterornis 449
carunculata, Paradigalla 430
caspia, Hydroprogne 191
caspia, Sterna 191
caspica, Motacilla 522
caspicus, Parus 522
cassicus, Cracticus 364
cassicus, Rhamphastos 364
castanea, Anas 60
castanea, Mareca 60
castaneiventris, Cacomantis 146
castaneiventris, Cuculus 146
castaneothorax, Amadina 518
castaneothorax, Lonchura 518
castaneothorax, Rhipidura 412
castaneoventris, Eulabeornis 129
castaneoventris, Gallirallus 129
castanonota, Ptilorrhoa 357
castanonotus, Eupetes 357
Casuariidae 40
Casuariiformes 40
Casuarius 40
casuarius, Casuarius 42
Catharacta 198
caudacuta, Chaetura 162
caudacuta, Hirundo 162
caudacutus, Hirundapus 162
caudata, Petroica 459
Cecropis 477
cenchroides, Falco 236
centrale, Dicaeum 504
centralis, Eupetes 355
centralis, Heteromyias 465
centralis, Ptilorrhoa 355
Centropodidae 136
Centropus 136
cervicalis, Cyclopsittacus 258
cervicalis, Otidiphaps 73
cervicalis, Psittaculirostris 258
cerviniventris, Chlamydera 282
cerviniventris, Drepanornis 428
cerviniventris, Pitohui 402
cerviniventris, Rectes 402
cervinus, Acrocephalus 311
Ceyx 233
Chaetorhynchus 406
Chalcites 142
chalconota, Carpophaga 88
chalconota, Ducula 88
chalconotus, Alopecoenas 70
Chalcophaps 75
Chalcopsitta 255

chalcothorax, Parotia 423
chalybata, Manucodia 420
chalybata, Paradisea 419
chalybatus, Manucodia 38, 419
chalybeocephala, Muscicapa 443
chalybeocephala, Myiagra 443
chalybeus, Centropus 137
chalybeus, Nesocentor 137
Charadriidae 166
Charadriiformes 163
Charadrius 168
Charmosyna 245
Chenonetta 57
Chenorhamphus 285
chinensis, Columba 65
chinensis, Coturnix 52
chinensis, Excalfactoria 52
chinensis, Spilopelia 65
chinensis, Streptopelia 65
chinensis, Synoicus 52
chinensis, Tetrao 52
chivae, Cacomantis 147
Chlamydera 281
Chlidonias 196
chlorcephala, Chalcostetha 507
chloris, Alcedo 229
chloris, Halcyon 229
chloris, Melanocharis 37, 346
chloris, Stigmatops 299
chloris, Todiramphus 229
chloris, Zosterops 482
chlorocephala, Chalcostetha 507
chloronota, Gerygone 337
chloronotus, Gerygone 337
chloroptera, Chalcopsitta 257
chloroptera, Melanocharis 346
chloropterus, Alisterus 263
chloropterus, Aprosmictus 263
chloropterus, Chalcopsittacus 257
chlororhynchus, Puffinus 100
chloroxantha, Micropsitta 268
christianae, Cinnyris 508
christianae, Leptocoma 508
Chroicocephalus 189
chrysenia, Parotia 423
chrysochlora, Chalcophaps 76
chrysochlora, Columba 76
chrysocome, Neneba 348
chrysogaster, Gerygone 336
chrysogenys, Lichenostomus 319
chrysogenys, Meliphaga 319
chrysogenys, Oreornis 319
chrysolaema, Zosterops 483
chrysolaemus, Zosterops 483
chrysomela, Carterornis 449
chrysomela, Monarcha 450
chrysomela, Muscicapa 449
chrysoptera, Daphoenositta 380
chrysoptera, Neositta 380
chrysopterus, Cicinnurus 434

chrysopterus, Diphyllodes 434
chrysotis, Meliphaga 295
chrysotis, Xanthotis 295
Ciconiidae 103
Ciconiiformes 103
Cinclosoma 359
Cinclosomatidae 355
cinctus, Charadrius 168
cinctus, Erythrogonys 168
cineracea, Myzomela 291
cinerascens, Drymophila 450
cinerascens, Gerygone 336
cinerascens, Monarcha 450
cinerascens, Oedistoma 350
cinerascens, Toxorhamphus 350
cinerea, Acanthiza 335
cinerea, Amaurornis 131
cinerea, Gerygone 335
cinerea, Motacilla 522
cinerea, Poliolimnas 131
cinerea, Porzana 131
cinerea, Ptilotis 306
cinerea, Scolpax 178
cinereiceps, Macropygia 66
cinereiceps, Pseudogerygone 337
cinereifrons, Heteromyias 38, 465
cinereifrons, Meliphaga 313
cinereifrons, Pycnopygius 305
cinereus, Artamus 367
cinereus, Poliolimnas 131
cinereus, Porphyrio 131
cinereus, Pycnopygius 306
cinereus, Xenus 178
cinnamomeus, Ailuroedus 275
cinnamomeus, Ardea 109
cinnamomeus, Ixobrychus 109
Cinnyris 508
Circus 208
cirrhocephalus, Accipiter 212
cirrocephalus, Accipiter 212
cirrocephalus, Sparvius 212
Cisticola 492
Cisticolidae 492
citreogaster, Pachycephala 393, 394
citreogularis, Philemon 297
citreogularis, Tropidorhynchus 297
citreola, Meliphaga 314
citrinella, Zosterops 482
citrinellus, Zosterops 482
clamosa, Rhipidura 408
clara, Eulacestoma 383
clarus, Pseudorectes 390
clarus, Rhectes 390
claudii, Casuarius 41
clelandi, Geopelia 71
clelandiae, Parotia 423
clelandiorum, Parotia 423
clementiae, Cinnyris 509
Climacteridae 282

Clytoceyx 226
Clytomyias 285
Cnemophilidae 342
Cnemophilus 342
cobana, Ethelornis 338
coccineifrons, Cicinnurus 433
coccineifrons, Cyclopsittacus 261
coccineopterus, Aprosmictus 264
coccineopterus, Ptistes 264
cochrani, Cinnyris 507
cochrani, Leptocoma 507
collaris, Pachycephala 393
collaris, Psittacella 243
collaris, Tachybaptus 62
Collocalia 158
Colluricincla 384
colonus, Halcyon 230
colonus, Todiramphus 230
Columba 64
Columbidae 63
Columbiformes 63
comrii, Manucodia 419
concinna, Carpophaga 86
concinna, Ducula 86
confirmata, Hemiprocne 157
connectens, Lophorina 426
connectens, Meliphaga 313
connectens, Rallus 127
connivens, Falco 217
connivens, Ninox 217
Conopophila 306
conspicillata, Gerygone 338
conspicillata, Microeca 338
conspicillatus, Pelecanus 104
constans, Ducula 91
cooki, Pterodroma 98
Coraciidae 221
Coraciiformes 220
Coracina 369
Cormobates 282
cornelia, Hermotimia 507
cornelia, Leptocoma 507
corniculatus, Merops 298
corniculatus, Philemon 298
coromanda, Ardea 112
coromanda, Ardeola 112
coromanda, Cancroma 112
coromanda, Egretta 112
coromandeliana, Anas 57
coromandelianus, Nettapus 57
coromandus, Bubulcus 113
coronatus, Todopsis 284
coronulatus, Ptilinopus 84
coronulatus, Ptilonopus 84
correcta, Plesiodryas 467
correctus, Poecilodryas 467
Corvidae 453
Corvus 453
Coturnix 51
Cracticidae 362

Cracticus 363
crassirostris, Ailuroedus 277
crassirostris, Larus 187
crassirostris, Rhamphocharis 38, 349
Crateroscelis 326
creatopus, Ardenna 101
crissalis, Zosterops 486
cristata, Columba 73
cristata, Goura 73
cristata, Rectes 382
cristata, Sterna 191
cristatus, Aegotheles 156
cristatus, Buteo 199
cristatus, Caprimulgus 156
cristatus, Lanius 454
cristatus, Ornorectes 382
cristatus, Pitohui 382
cristatus, Thalasseus 191
crockettorum, Meliphaga 316
crookshanki, Zosterops 485
cruentata, Myzomela 291
cryptoleuca, Peneothello 470
cryptoleucus, Poecilodryas 470
cryptorhynchus, Cicinnurus 433
Cuculidae 138
Cuculiformes 136
Cuculus 147
cuicui, Zosterops 462
cunctata, Macropygia 67
curonicus, Charadrius 169
cuvieri, Talegalla 45
cyanauchen, Lorius 251
cyanauchen, Psittacus 251
cyanicarpus, Geoffroyus 266
cyanocephala, Todopsis 287
cyanocephalus, Malurus 286
cyanocephalus, Todus 286
cyanogenia, Eos 255
cyanoleuca, Grallina 441
cyanoleuca, Myiagra 446
cyanoleucus, Corvus 441
cyanoleucus, Platyrhynchos 446
cyanotis, Entomyzon 300
cyanotis, Gracula 300
cyanus, Myiolestes 470
cyanus, Peneothello 470
Cyclopsitta 259
cyclopum, Charmosyna 248
cyclopum, Pachycephala 392
cyclopum, Phylloscopus 480
cyclopum, Seicercus 480
cyclopum, Sericornis 331
Cygnus 55

Dacelo 227
dactylatra, Sula 118
dahli, Pachycephala 395
dammermani, Ptiloprora 303

danae, Tanysiptera 225
Daphoenositta 379
darwini, Gerygone 337
dauma, Zoothera 499
daurica, Cecropis 477
daurica, Hirundo 477
decipiens, Pitohui 401
decipiens, Rectes 401
decollatus, Megapodius 49
decora, Paradisaea 436
decora, Paradisea 436
decorus, Ptilopus 84
deficiens, Crateroscelis 329
dejecta, Chibia 414
delicatula, Tyto 214
delicatulus, Strix 214
deliculata, Zosterops 483
deliculatus, Zosterops 483
Dendrocygna 54
desmarestii, Psittaculirostris 257
desmarestii, Psittacus 257
despecta, Colluricincla 386
despectus, Pinarolestes 386
destructa, Lonchura 516
destructa, Munia 516
Devioeca 461
devisi, Rhipidura 409
devittatus, Lorius 252
diamondi, Crateroscelis 328
diamondi, Phonygammus 417
Dicaeidae 503
Dicaeum 503
dichrous, Pitohui 404
dichrous, Rhectes 404
Dicruridae 413
Dicrurus 414
didimus, Accipiter 210
didimus, Coracina 371
digglesi, Erythropitta 272
digglesi, Pitta 272
diminuta, Cisticola 493
diminutus, Cisticola 493
dimorpha, Athene 218
dimorpha, Uroglaux 218
diophthalma, Cyclopsitta 261
diophthalma, Opopsitta 261
diophthalma, Psittacula 261
diophthalma, Psittaculirostris 262
discolor, Colluricincla 387
diversum, Dicaeum 504
dogwa, Accipiter 210
dogwa, Coturnix 51
dogwa, Malurus 287
dogwa, Synoicus 51
dohertyi, Gerygone 336
dohertyi, Malurus 286
dohertyi, Pitohui 401
domestica, Fringilla 510
domesticus, Passer 510
dominica, Pluvialis 166

doreya, Macropygia 66
doriae, Accipiter 208
doriae, Megatriorchis 207
dorotheae, Phaethon 93
dorsalis, Alisterus 263
dorsalis, Orthonyx 342
dorsalis, Pachycephala 399
dorsalis, Psittacus 263
dorsalis, Pycnopygius 306
dougallii, Sterna 194
Drepanornis 427
dryas, Rallicula 125
dryas, Rhipidura 410
Drymodes 464
dubia, Pachycephala 398
dubius, Charadrius 168
dubius, Ixobrychus 109
ducalis, Astrapia 432
Ducula 86
duivenbodei, Chalcopsitta 256
duivenbodei, Chalcopsittacus 256
dulciae, Pelagodroma 95
dumasi, Poecilodryas 466
dumontii, Mino 496
duperreyii, Megapodius 48

Eclectus 264
Edolisoma 375
edwardi, Porzana 130
edwardsii, Cyclopsittacus 258
edwardsii, Psittaculirostris 258
Egretta 114
egretta, Ardea 114
Elanus 200
eleonora, Cacatua 240
eleonora, Kakatoe 240
elisabeth, Cyanalcyon 229
ellioti, Philemon 298
ellioti, Tanysiptera 224
elliottsmithi, Astrapia 431
elliottsmithorum, Astrapia 431
emilii, Melidectes 320
Entomyzon 300
Eos 254
Ephippiorhynchus 104
Epimachus 428
episcopalis, Rhipidura 408
eques, Cinnyris 290
eques, Myzomela 290
erebus, Turdus 499
eremita, Megapodius 49
ernesti, Falco 238
ernesti, Munia 516
ernesti, Pachycephalopsis 458
erwini, Collocalia 158
erythrocephala, Myzomela 292
erythrogaster, Erythropitta 271
erythrogaster, Pitta 271
Erythrogonys 168

Erythropitta 271
erythropleura, Ptiloprora 303
erythropleura, Ptilotis 303
erythropterus, Aprosmictus 264
erythropterus, Psittacus 264
erythropus, Tringa 176
erythrothorax, Lorius 251
Erythrotriorchis 212
Erythrura 512
Esacus 164
esculenta, Collocalia 158
esculenta, Hirundo 158
Estrildidae 511
Eudynamys 139
Eugerygone 460
Eulabeornis 129
Eulacestoma 383
Eulacestomatidae 383
Eulipoa 47
Eurostopodus 150
Eurystomus 221
evangelinae, Neochmia 512
examinandus, Seicercus 479
excelsa, Collocalia 161
excelsa, Psittacella 243
excitus, Cacomantis 146
exhibita, Parotia 422
exiguus, Anthus 523
exilis, Cisticola 493
exilis, Malurus 493
extra, Diphyllodes 434
eytoni, Dendrocygna 54
eytoni, Leptotarsis 54

facialis, Ailuroedus 276
facialis, Meliphaga 314
falcinellus, Calidris 182
falcinellus, Limicola 182
falcinellus, Plegadis 107
falcinellus, Scolopax 182
falcinellus, Tantalus 107
Falco 235
Falconidae 235
Falconiformes 235
fallax, Glycichaera 301
fallax, Rhipidura 446
fallax, Symposiachrus 446
fallax, Timeliopsis 302
fasciatus, Accipiter 210
fasciatus, Astur 210
fascinans, Loxia 463
fascinans, Microeca 463
fasciolata, Locustella 489
fasciolatus, Acrocephalus 489
fastosus, Epimachus 429
fastuosus, Epimachus 429
fastuosus, Promerops 429
feminina, Astrapia 432
feminina, Lophorina 426

fergussonis, Melilestes 350
fergussonis, Oedistoma 350
fergussonis, Pachycephala 395
ferruginea, Calidris 183
ferruginea, Erolia 183
ferruginea, Sericornis 330
ferrugineus, Myiolestes 385
ferrugineus, Pitohui 390
ferrugineus, Pseudorectes 389
ferrugineus, Rhectes 390
ferrugineus, Sericornis 330
ferrugineus, Tringa 183
festetichi, Cyclopsittacus 261
filiger, Xanthotis 295
finschi, Paradisea 436
finschi, Ptilinopus 83
finschi, Ptilotis 305
finschi, Pycnopygius 305
finschii, Ducula 88
finschii, Erythropitta 272
finschii, Pitta 272
fitzroyi, Cacatua 240
flabelliformis, Cacomantis 146
flabelliformis, Cuculus 146
flava, Motacilla 522
flavescens, Lichenostomus 318
flavescens, Meliphaga 318
flavescens, Ptilotis 318
flavescens, Ptilotula 318
flaviceps, Campochaera 373
flavicipitis, Campochaera 373
flavicollis, Ardea 110
flavicollis, Dupetor 110
flavicollis, Ixobrychus 110
flavida, Meliphaga 313
flavifrons, Amblyornis 278
flavigaster, Microeca 463
flavipectus, Oedistoma 351
flavipes, Platalea 107
flavirictus, Meliphaga 315
flavirictus, Ptilotis 315
flavissima, Microeca 463
flaviventer, Machaerirhynchus 360
flaviventer, Meliphaga 295
flaviventer, Myzantha 294
flaviventer, Xanthotis 294
flaviventris, Melilestes 352
flaviventris, Toxorhamphus 352
flavocinctus, Mimetes 406
flavocinctus, Oriolus 406
flavogrisea, Pachycare 325
flavogrisea, Pachycephala 325
flavogriseum, Pachycare 37, 325
flavovirescens, Kempiella 462
flavovirescens, Microeca 462
flavus, Toxorhamphus 350
fluviatilis, Hydrochelidon 196
flyensis, Butorides 111
foersteri, Melidectes 321
foersteri, Melirrhophetes 321

forbesi, Myzomela 292
forbesi, Rallicula 125
forbesi, Rallina 125
fordi, Scythrops 141
forsteni, Trichoglossus 254
forsteri, Larus 189
fortior, Cacomantis 147
fortis, Colluricincla 387
fortis, Pachycephala 387
frater, Monarcha 452
fraterculus, Henicopernis 202
Fregata 116
Fregatidae 116
Fregetta 95
frenatus, Cyrtostomus 509
fretensis, Philemon 297
freycinet, Megapodius 49
frontalis, Daphoenositta 380
frontalis, Hirundo 476
frontata, Gallinula 133
fulgidus, Banksianus 241
fulgidus, Psittrichas 241
Fulica 134
fuliginosa, Oreospiza 511
fuliginosus, Haematopus 164
fuliginosus, Oreostruthus 511
fuligula, Aythya 61
fulva, Pluvialis 167
fulvigula, Euthyrhynchus 310
fulvigula, Timeliopsis 310
fulviventris, Plectorhyncha 311
fulviventris, Timeliopsis 311
fulvus, Charadrius 167
fumata, Myzomela 290
fumata, Ptilotis 290
fumigatus, Melipotes 308
fumosa, Crateroscelis 326
fumosa, Rhipidura 408
furva, Turnix 185
furvus, Turnix 185
fusca, Acanthochaera 324
fusca, Ptilotula 318
fuscata, Eos 255
fuscata, Pseudeos 255
fuscata, Sterna 194
fuscatus, Onychoprion 194
fuscescens, Monarcha 451
fuscicapilla, Timeliopsis 311
fuscicapilla, Zosterops 484
fuscicapillus, Corvus 453
fuscicapillus, Zosterops 485
fuscifrons, Cyclopsitta 260
fuscifrons, Cyclopsittacus 260
fuscior, Parotia 422
fuscirostris, Talegalla 46
fuscirostris, Talegallus 46
fusciventris, Xanthotis 294
fuscus, Melidectes 324
fuscus, Melionyx 324
fuscus, Pitohui 389

gagiensis, Pachycephala 398
gajduseki, Lonchura 518
galatea, Tanysiptera 222
galerita, Cacatua 240
galeritus, Psittacus 240
Gallicolumba 68
Galliformes 44
gallinacea, Irediparra 171
gallinacea, Parra 171
Gallinago 172
Gallinula 133
gamblei, Aleadryas 382
gamblei, Pachycephala 382
Garritornis 340
garzetta, Ardea 115
garzetta, Egretta 115
gaudichaud, Dacelo 228
gaudichaud, Sauromarptis 228
gaudichaudi, Dacelo 228
Gavicalis 317
geelvinianum, Dicaeum 504
geelvinkiana, Carpophaga 87
geelvinkiana, Ducula 87
geelvinkiana, Micropsitta 268
geelvinkiana, Nasiterna 268
geelvinkiana, Ptilopus 84
geelvinkianum, Dicaeum 504
geelvinkianus, Megapodius 47
geelvinkianus, Monarcha 451
geelvinkianus, Ptilinopus 84
geisleri, Drepanornis 428
geislerorum, Aeluroedus 275
geislerorum, Ailuroedus 275
geislerorum, Eupetes 357
geislerorum, Ptilorrhoa 357
Gelochelidon 190
geminus, Ptilinopus 85
geminus, Ptilonopus 85
Gennaeodryas 471
geoffroyi, Geoffroyus 265
geoffroyi, Psittacus 265
Geoffroyus 265
Geopelia 70
germana, Amblyornis 278
germana, Ptilotis 318
germanorum, Meliphaga 314
germanus, Amblyornis 278
Gerygone 335
gestroi, Ptilinopus 82
gestroi, Ptilonopus 82
gibberifrons, Anas 50
gigantea, Megalampitta 439
gigantea, Melampitta 440
gigantea, Mellopitta 439
gigantea, Procellaria 97
giganteus, Macronectes 97
gigas, Aegotheles 154
gilliardi, Eupetes 357
gilliardi, Melidectes 323, 324
gilliardi, Ptilorrhoa 357

giluwensis, Turnix 185
girrenera, Haliaetus 205
girrenera, Haliastur 205
giulianettii, Gerygone 480
giulianettii, Seicercus 480
giulianettii, Xanthotis 294
Glareola 186
glareola, Tringa 177
Glareolidae 186
Glycichaera 301
godmani, Cyclopsittacus 258
godmani, Psittaculirostris 258
goldiei, Cinclosoma 359
goldiei, Eupetes 359
goldiei, Macropygia 67
goldiei, Psitteuteles 252
goldiei, Trichoglossus 252
goldii, Ninox 218
goliath, Ardea 113
goliath, Probosciger 239
goliath, Psittacus 239
goliathi, Melipotes 308
goliathina, Charmosyna 248
goodfellowi, Casuarius 41
goodfellowi, Pitta 273
goodsoni, Edolisoma 378
goodsoni, Macropygia 67
goodsoni, Pinarolestes 385
gouldi, Circus 208
gouldi, Ixobrychus 110
gouldii, Monarcha 448
gouldii, Phonygammus 417
gouldii, Symposiachrus 448
Goura 73
gracilis, Anas 59
gracilis, Meliphaga 316
gracilis, Ptilotis 316
gracilis, Sterna 195
grallaria, Fregetta 95
grallaria, Procellaria 95
grallarius, Burhinus 164
grallarius, Charadrius 164
Grallina 440
gramineus, Megalurus 491
gramineus, Poodytes 491
gramineus, Sphenoeacus 491
grandis, Lonchura 516
grandis, Munia 516
grandis, Oreopsittacus 245
granti, Aethomyias 330
granti, Collocalia 160
granti, Meliphaga 315
granti, Paradisea 438
granti, Ptiloprora 302
granti, Sericornis 330
granti, Talegalla 45
grayi, Chenorhamphus 285
grayi, Malurus 285
grayi, Todopsis 285
gregalis, Geopelia 71

gregaria, Zosterops 483
gregarius, Zosterops 483
gretae, Meliphaga 312
grisea, Ardenna 100
grisea, Coracina 369
grisea, Procellaria 100
griseiceps, Pachycephala 398
griseigula, Euthyrhynchus 311
griseigula, Timeliopsis 311
griseigularis, Entomiza 300
griseigularis, Entomyzon 300
griseinucha, Macropygia 66
griseirostris, Melidectes 322
griseirostris, Melirrhophetes 322
griseisticta, Hemichelidon 502
griseisticta, Muscicapa 502
griseiventris, Plesiodryas 467
griseiventris, Poecilodryas 467
griseoceps, Kempiella 461
griseoceps, Microeca 461
griseogularis, Accipiter 209
griseogularis, Astur 209
griseonota, Pachycephala 400
griseotincta, Reinwardtoena 65
griseotincta, Zosterops 486
griseotinctus, Zosterops 486
grisescens, Microdynamis 139
griseus, Puffinus 100
Gruidae 135
Gruiformes 123
Grus 135
guilielmi, Paradisaea 436
guilielmi, Paradisea 436
guisei, Ptiloprora 304
guisei, Ptilotis 304
gularis, Rhipidura 413
gulielmi III, Opopsitta 260
gulielmi III, Psittacula 260
gulielmitertii, Cyclopsitta 260
gurneyi, Aquila 204
gurneyi, Heteropus 204
gurneyi, Spizaetus 204
guttata, Aethomyias 330
guttata, Dendrocygna 54
guttaticollis, Ailuroedus 276
guttatus, Caprimulgus 151
guttatus, Eurostopodus 151
guttatus, Sericornis 330
guttula, Monarcha 447
guttula, Muscicapa 447
guttula, Symposiachrus 447
guttulus, Monarcha 447
gutturalis, Anthus 523
gutturalis, Hirundo 475
Gymnocrex 129
Gymnophaps 92
gymnops, Melipotes 307
gymnorhynchus, Cicinnurus 433

habenichti, Erythropitta 272
habenichti, Pitta 272
haematodus, Psittacus 253
haematodus, Trichoglossus 253
Haematopodidae 164
Haematopus 164
hagenensis, Oreostruthus 512
hagenensis, Peneothello 469
Halcyon sancta 231
Halcyonidae 222
Haliaeetus 206
haliaetus, Falco 199
haliaetus, Pandion 199
Haliastur 206
halli, Macronectes 97
hallstromi, Psittacella 244
hallstromi, Pteridophora 421
halmaheira, Columba 64
halmaheira, Janthaenas 64
hamlini, Phylloscopus 480
hamlini, Seicercus 480
hanieli, Falco 237
hardwickii, Capella 172
hardwickii, Gallinago 172
hardwickii, Scolopax 172
harmonica, Colluricincla 388
harmonicus, Turdus 388
Harpyopsis 203
harterti, Arses 442
harterti, Machaerirhynchus 361
harterti, Megalurus 491
harterti, Melanorectes 391
harterti, Melanorhectes 391
harterti, Micropsitta 269
harterti, Psittacella 242
hattamensis, Pachycephala 457
hattamensis, Pachycephalopsis 457
hazarae, Caprimulgus 152
hebetior, Pitta 273
hecki, Casuarius 41
heinei, Oreocincla 498
heinei, Zoothera 498
heinrothi, Puffinus 103
heinrothi, Pufinns 103
helenae, Parotia 38, 422
helios, Astrapia 431
heliosyla, Ardea 108
heliosylus, Zonerodius 108
helviventris, Gallicolumba 69
helviventris, Ptilopus 69
Hemiprocne 157
Hemiprocnidae 157
Henicopernis 202
Henicophaps 68
henkei, Arses 442
henrici, Rhipidura 410
heraldica, Oestrelata 98
heraldica, Pterodroma 98
hercules, Cracticus 364
hermani, Poecilodryas 468

Heteromyias 465
Heteroscenes 145
heurni, Lonchura 516
heurni, Poecilodryas 472
heurni, Tregellasia 472
Hieraaetus 204
hilli, Nycticorax 111
Himantopus 165
himantopus, Charadrius 165
himantopus, Himantopus 165
hiogaster, Accipiter 209
hiogaster, Falco 209
Hirundapus 162
hirundinacea, Collocalia 160
hirundinacea, Motacilla 506
hirundinaceum, Dicaeum 506
hirundinaceus, Aerodramus 160
Hirundinidae 474
Hirundo 475
hirundo, Sterna 195
hispidoides, Alcedo 233
hoedtii, Noctua 218
hoeveni, Rallus 129
holerythra, Rectes 390
holerythrus, Pseudorectes 390
horsbrughi, Turnix 185
horsfieldi, Cuculus 148
horsfieldi, Mirafra 474
hottentottus, Dicrurus 415
humeralis, Athene 216
humeralis, Columba 71
humeralis, Geopelia 71
humeralis, Ninox 216
humeralis, Ptilinopus 82
humeralis, Ptilonopus 82
hunsteini, Cicinnurus 434
hunsteini, Diphyllodes 434
hunsteini, Pachycephala 458
hunsteini, Pachycephalopsis 458
hunsteini, Phonygama 418
hunsteini, Phonygammus 418
huonensis, Meliphaga 315
huonensis, Psittacella 244
huonensis, Ptilopus 85
huttoni, Puffinus 102
hybrida, Chlidonias 196
hybrida, Sterna 196
hybridus, Chlidonias 196
hybridus, Colluricincla 386
hybridus, Pinarolestes 386
Hydrobatidae 96
hydrocharis, Tanysiptera 224
Hydroprogne 190
hyperythra, Pachycephala 397
hyperythra, Rhipidura 412
hypoinochroa, Domicella 252
hypoinochrous, Lorius 252
hypoleuca, Coracina 372
hypoleuca, Pachycephala 396
hypoleuca, Petroica 467

hypoleuca, Poecilodryas 467
hypoleucos, Actitis 178
hypoleucos, Falco 237
hypoleucos, Phalacrocorax 121
hypoleucos, Tringa 178
hypoleucus, Artamus 367
hypoleucus, Elanus 200
hypoleucus, Graucalus 372
hypomelaena, Lonchura 514
hypopolia, Pachycephalopsis 459
Hypotaenidia 127
hypoxantha, Gerygone 338

ibis, Ardea 112
ibis, Ardeola 113
idenburgi, Butorides 111
idenburgi, Cinnyris 509
idenburgi, Myiolestes 386
idenburgi, Pachycephalopsis 458
idenburgi, Sericornis 332
Ieracidea 237
Ifrita 415
Ifritidae 415
ignicolle, Dicaeum 506
ignotum, Dicaeum 503
ijimae, Clivicola 475
ijimae, Riparia 475
iliolophus, Melilestes 350
iliolophus, Oedistoma 350
iliolophus, Toxorhamphus 351
imitator, Sericornis 332
impediens, Monarcha 451
imperator, Clytoceyx 226
incana, Scolopax 178
incana, Tringa 178
incanus, Heteroscelis 178
incerta, Amalocichla 456
incerta, Campephaga 375
incerta, Ptiloprora 303
incertum, Edolisoma 375
incertus, Eupetes 456
incertus, Pitohui 390
incertus, Pseudorectes 390
incinctus, Halcyon 229
incinctus, Todiramphus 229
incola, Podiceps 63
incola, Tachybaptus 63
incondita, Eos 255
inconspicua, Gerygone 337
indica, Chalcophaps 75
indica, Columba 75
indicus, Butastur 207
indicus, Caprimulgus 152
indicus, Falco 207
indistincta, Lichmera 299
indistincta, Meliphaga 299
indus, Falco 206
indus, Haliastur 206
inepta, Habroptila 132

inepta, Megacrex 132
ineptus, Amaurornis 132
inexpectata, Climacteris 283
inexpectata, Loria 343
inexpectata, Rigidipenna 149
infaustus, Cacomantis 147
infuscata, Myzomela 292
inornata, Amblyornis 278
inornata, Aplonis 494
inornata, Calornis 494
inornata, Muscicapa 450
inornata, Monarcha 450
inornatus, Ptilorhynchus 279
inquieta, Myiagra 444
inquietus, Turdus 444
inseparabilis, Cyclopsitta 261
inseparabilis, Cyclopsittacus 261
insignis, Aegotheles 153
insignis, Ardea 113
insignis, Artamus 366
insignis, Chalcopsitta 256
insignis, Clytomyias 285
insignis, Euaegotheles 153
insolitus, Ptilinopus 83
insolitus, Ptilopus 83
insperata, Gerygone 339
insularis, Arses 442
insularis, Halcyon 229
insularis, Monarcha 442
insularis, Otidiphaps 73
insularis, Pachycephalopsis 458
insulsus, Rallus 127
intensior, Charmosyna 246
intensior, Hypocharmosyna 246
intensitincta, Ducula 87
intercedens, Ptiloris 38, 425
intermedia, Ardea 114
intermedia, Cacatua 239
intermedia, Coracina 371
intermedia, Cyclopsitta 258
intermedia, Dacelo 227
intermedia, Egretta 114
intermedia, Meliphaga 317
intermedia, Mesophoyx 114
intermedia, Neositta 379
intermedia, Paradigalla 430
intermedia, Paradisaea 438
intermedia, Paradisea 438
intermedia, Rhipidura 410
intermedius, Diphyllodes 434
intermedius, Podargus 149
intermedius, Trichoglossus 253
intermixta, Psittacella 242
interposita, Megaloprepia 77
interposita, Rhamphocharis 349
interpres, Arenaria 179
interpres, Tringa 179
inversa, Drepanornis 428
iobiensis, Ptilinopus 83
iobiensis, Ptilopus 83

iozonus, Ptilinopus 82
iozonus, Ptilonopus 82
Irediparra 171
isabella, Glareola 186
isabella, Stiltia 186
isidorei, Garritornis 340
isidorei, Pomatorhinus 340
isidori, Pomatostomus 340
isura, Lophoictinia 202
Ixobrychus 109
ixoides, Ptilotis 305
ixoides, Pycnopygius 305

Jacanidae 172
jamesii, Phonygama 417
jamesii, Phonygammus 417
japonica, Cecropis 477
japonica, Hirundo 477
javanica, Mirafra 474
javanica, Sterna 196
javanicus, Chlidonias 196
jobiensis, Aeluroedus 276
jobiensis, Ailuroedus 276
jobiensis, Alopecoenas 69
jobiensis, Carpophaga 89
jobiensis, Centropus 136
jobiensis, Domicella 251
jobiensis, Ducula 89
jobiensis, Gallicolumba 70
jobiensis, Geoffroyus 265
jobiensis, Lorius 251
jobiensis, Manucodia 38, 420
jobiensis, Melidora 226
jobiensis, Pachycephala 398
jobiensis, Paradisaea 437
jobiensis, Paradisea 437
jobiensis, Philemon 298
jobiensis, Phlegoenas 69
jobiensis, Pionias 265
jobiensis, Pitohui 402
jobiensis, Rectes 402
jobiensis, Sericornis 332
jobiensis, Talegalla 46
jobiensis, Talegallus 46
jobiensis, Tropidorhynchus 298
johannae, Alopecoenas 70
johannae, Phlogoenas 70
joiceyi, Melidectes 322
joiceyi, Melirrhophetes 322
josefinae, Charmosyna 247
josefinae, Trichoglossus 247
jotaka, Caprimulgus 152
jubata, Anas 57
jubata, Chenonetta 57
jugularis, Certhia 508
jugularis, Cinnyris 508
jugularis, Nectarinia 509
julianae, Monarcha 447
julianae, Symposiachrus 447

juncidis, Cisticola 492
juncidis, Sylvia 492
jungei, Rhyticeros 219

kaporensis, Ptilinopus 82
karimuiensis, Myzomela 291
keiensis, Micropsitta 268
keiensis, Nasiterna 268
kempi, Cracticus 364
kempi, Kempiella 461
Kempiella 461
keraudrenii, Barita 417
keraudrenii, Manucodia 418
keraudrenii, Phonygammus 417
kerstingi, Macropygia 67
keysseri, Turdus 499
kinneari, Melidectes 322
kirhocephalus, Lanius 401
kirhocephalus, Pitohui 401
klossi, Pachycephala 392
klossi, Rallicula 124
kodonophonos, Pitohui 382
kombok, Amblyornis 278
kordensis, Carterornis 450
kordensis, Monarcha 450
kordensis, Rhipidura 413
kordoana, Charmosyna 245
kordoanus, Trichoglossus 245
kowaldi, Ifrita 415
kowaldi, Todopsis 415
krakari, Hypocharmosyna 247
krakari, Macropygia 67
kuboriensis, Cnemophilus 343
kuboriensis, Daphoenositta 380
kubuna, Rhipidura 409
kuehni, Pitta 272
kuehni, Xanthotis 295
kumawa, Melipotes 308
kumusi, Rhipidura 409
kumusii, Lonchura 517
kumusii, Munia 517
kumusii, Xanthotis 295
kunupi, Monarcha 452
kutubu, Malurus 287

lacustris, Hypotaenidia 128
lacustris, Rallus 128
laeta, Microeca 463
laetior, Alcyone 235
laetior, Ceyx 235
Lalage 373
lamonti, Synoicus 51
Laniidae 454
Lanius 454
lapponica, Limosa 174
lapponica, Scolopax 174
Laridae 187
lateralis, Casuarius 42

latipennis, Lophorina 426
lauterbachi, Arses 442
lauterbachi, Chlamydera 281
lauterbachi, Chlamydodera 281
laveryi, Cisticola 493
lawesii, Parotia 38, 422
leachii, Dacelo 227
lecroyae, Amblyornis 278
lecroyae, Pachycare 326
lecroyae, Pachycephalopsis 458
lepida, Alcedo 234
lepida, Coturnix 52
lepida, Excalfactoria 52
lepidus, Ceyx 234
lepidus, Cuculus 148
Leptocoma 506
lepturus, Phaethon 93
leschenaultii, Charadrius 170
lessonii, Alcyone 234
lessonii, Ceyx 234
leucocephalus, Himantopus 165
leucogaster, Falco 206
leucogaster, Haliaeetus 206
leucogaster, Pachycephala 400
leucogaster, Pelecanus 119
leucogaster, Sula 119
leucogastra, Pachycephala 399
leucolophus, Cacomantis 145
leucolophus, Caliechthrus 145
leucolophus, Cuculus 145
leucomela, Campephaga 374
leucomela, Lalage 374
leucomelas, Calonectris 101
leucomelas, Procellaria 101
leucomelas, Puffinus 102
leucopareia, Eutrygon 72
leucopareia, Trugon 72
leucophaea, Climacteris 283
leucophaea, Microeca 463
leucophrys, Porzana 131
leucophrys, Rhipidura 407
leucophrys, Turdus 407
leucops, Leucophantes 472
leucops, Tregellasia 472
leucoptera, Campephaga 374
leucoptera, Lalage 374
leucoptera, Pterodroma 99
leucoptera, Sterna 196
leucopterus, Chlidonias 196
leucopygialis, Artamus 366
leucorhoa, Oceanodroma 96
leucorhynchus, Pseudorectes 389
leucorhynchus, Rectes 389
leucorodia, Platalea 106
leucoryn, Artamus 366
leucorynchus, Artamus 366
leucorynchus, Lanius 366
leucosomus, Accipiter 209
leucosomus, Astur 209
leucospila, Corethrura 125

leucospila, Rallicula 125
leucospila, Rallina 125
leucostephes, Melidectes 321
leucostephes, Melirrhophetes 321
leucosticta, Lonchura 515
leucosticta, Munia 515
leucosticta, Ptilorrhoa 355
leucostictus, Eupetes 355
leucostigma, Pachycephala 368
leucostigma, Rhagologus 368
leucothorax, Gerygone 336
leucothorax, Rhipidura 408
leucura, Eopsaltria 470
leucura, Peneothello 470
levigaster, Gerygone 339
Lewinia 126
lherminieri, Puffinus 102, 103
Lichmera 299
limaria, Hypotaenidia 127
limarius, Rallus 128
Limnodromus 173
Limosa 174
limosa, Limosa 174
limosa, Scolopax 174
lineata, Coracina 369
lineatus, Ceblepyris 369
littleri, Butorides 112
livia, Columba 64
lobata, Tringa 184
lobatus, Phalaropus 184
Loboparadisea 344
lochmia, Pachycephala 382
Locustella 489
Locustellidae 489
Lonchura 513
longicauda, Coracina 369
longicauda, Falco 202
longicauda, Graucalus 369
longicauda, Henicopernis 202
longicauda, Melampitta 439
longicauda, Melanocharis 37, 346
longicauda, Talegalla 46
longicaudatus, Lanius 455
longicaudus, Stercorarius 198
longimembris, Strix 215
longimembris, Tyto 215
longipennis, Falco 237
longipennis, Sterna 195
longirostris, Chalcophaps 76
longirostris, Haematopus 165
longirostris, Myzomela 289
longirostris, Pterodroma 99
longirostris, Ptilotis 313
longirostris, Zosterops 486
Lophorina 426
lophotis, Xanthotis 296
lorentzi, Malurus 287
lorentzi, Pachycephala 396
lorentzi, Psittacella 242
lorentzi, Ptilotis 303

lorentzi, Rhipidura 411
loriae, Cnemophilus 342
loriae, Erythropitta 272
loriae, Eupetes 356
loriae, Loria 342
loriae, Pitta 272
loriae, Ptilorrhoa 356
Loriculus 262
Lorius 250
lory, Domicella 252
lory, Lorius 251
lory, Psittacus 251
louisiadensis, Coracina 372
louisiadensis, Cracticus 364
louisiadensis, Graucalus 372
louisiadensis, Myzomela 292
louisiadensis, Rhipidura 410
lucida, Myiagra 444
lucidus, Chalcites 143
lucidus, Chrysococcyx 143
lucidus, Cuculus 143
lucifer, Melidectes 321
lucionensis, Lanius 454
luctuosa, Ducula 91
lugubris, Fulica 134
lugubris, Melampitta 439
lugubris, Pachycephala 399
lunata, Sterna 193
lunatus, Onychoprion 193

maccormicki, Catharacta 198
maccormicki, Stercorarius 198
macgillivrayi, Megapodius 48
Macgregoria 309
macgregoriae, Amblyornis 277
macgregorii, Cnemophilus 37, 343
Machaerirhynchidae 360
Machaerirhynchus 360
Macheiramphus 202
mackinlayi, Macropygia 67
macklotii, Erythropitta 271
macklotii, Pitta 271
macleayii, Halcyon 229
macleayii, Todiramphus 229
macnicolli, Taeniaparadisea 431
Macronectes 97
macronyx, Budytes 521
macronyx, Motacilla 521
Macropygia 66
macrorrhina, Melidora 226
macrorrhinus, Dacelo 226
macrotarsa, Gelochelidon 190
macrotarsa, Sterna 190
macrurus, Caprimulgus 152
macrurus, Cincloramphus 491
macrurus, Megalurus 490
macrurus, Sphenoeacus 490
maculiceps, Melanocharis 347
maculiceps, Sarganura 347

maculipectus, Rhipidura 407
maculosa, Turnix 185
maculosus, Hemipodius 185
maculosus, Turnix 185
madagascariensis, Numenius 176
madagascariensis, Scolopax 176
madaraszi, Colluricincla 386
madaraszi, Pinarolestes 386
madaraszi, Psittacella 244
madaraszi, Ptilotis 295
madaraszi, Xanthotis 295
maeandrina, Colluricincla 387
maeandrinus, Pinarolestes 387
maforense, Dicaeum 504
maforensis, Campephaga 370
maforensis, Chalcostetha 507
maforensis, Coracina 370
maforensis, Eclectus 266
maforensis, Gerygone 481
maforensis, Leptocoma 507
maforensis, Macropygia 66
maforensis, Seicercus 481
mafulu, Coturnix 51
mafulu, Malurus 287
mafulu, Synoicus 51
magicus, Otus 216
magna, Aplonis 495
magnifica, Columba 77
magnifica, Craspedophora 425
magnifica, Lophorina 425
magnifica, Megaloprepia 77
magnifica, Paradisea 434
magnificus, Cicinnurus 434
magnificus, Diphyllodes 434
magnificus, Falcinellus 425
magnificus, Ptilinopus 78
magnificus, Ptiloris 38, 425
magnirostris, Burhinus 164
magnirostris, Esacus 164
magnirostris, Gerygone 338
magnirostris, Œdicnemus 164
magnirostris, Oriolus 405
magnirostris, Zosterops 486
magnus, Lamprotornis 495
major, Aegotheles 156
major, Lorius 251
major, Neopsittacus 250
major, Oreopsittacus 245
major, Psittacella 244
malaccensis, Passer 511
malayanus, Chrysococcyx 144
maldivarum, Glareola 187
Maluridae 284
Malurus 286
mamberana, Chaetura 162
manadensis, Monarcha 447
manadensis, Muscicapa 447
manadensis, Symposiachrus 447
manam, Centropus 137
manni, Aegotheles 154

Manucodia 419
manumudari, Monarcha 443
manumudari, Myiagra 443
maoriana, Pelagodroma 95
margaretae, Meliphaga 315
margaritae, Paradisaea 435
margaritae, Paradisornis 435
marginata, Ninox 217
mariae, Leptocoma 506
mariae, Lonchura 519
mariae, Nectarinia 506
marina, Pelagodroma 95
marina, Procellaria 95
marmorata, Ptilotis 306
marmoratus, Pycnopygius 306
massena, Trichoglossus 253
matsudairae, Halocyptena 96
matsudairae, Hydrobates 96
matsudairae, Oceanodroma 96
mauri, Calidris 179
maxima, Rallina 126
maximus, Artamus 366
maximus, Peneothello 470
maximus, Toxorhamphus 352
mayeri, Astrapia 431
mayri, Amblyornis 277
mayri, Eupetes 355
mayri, Lewinia 127
mayri, Lonchura 517
mayri, Megalurus 491
mayri, Munia 517
mayri, Phonygammus 418
mayri, Poecilodryas 472
mayri, Porzana 130
mayri, Ptiloprora 304
mayri, Ptilorrhoa 355
mayri, Rallicula 125
mayri, Rallina 126
mayri, Rallus 127
mayri, Tregellasia 472
mayri, Trugon 72
mayri, Turnix 185
mayri, Zapornia 130
Mearnsia 161
medius, Neopsittacus 250
meeki, Anthreptes 351
meeki, Caprimulgus 152
meeki, Edoliisoma 378
meeki, Erythropitta 272
meeki, Loriculus 262
meeki, Macropygia 67
meeki, Melanocharis 347
meeki, Melanorectes 391
meeki, Oedistoma 351
meeki, Pachycephala 400
meeki, Parotia 423
meeki, Pitohui 391
meeki, Pitta 272
meeki, Podargus 149
meeki, Pristorhamphus 347

meeki, Sericornis 334
meeki, Strix 214
meeki, Syma 232
meeki, Tyto 214
meeki, Zosterops 484
meekiana, Ptiloprora 302
meekiana, Ptilotis 302
mefoorana, Pitta 273
Megacrex 132
megala, Aviceda 201
megala, Baza 201
megala, Capella 173
megala, Gallinago 173
Megalampitta 439
Megaloprepia 77
megalorynchos, Psittacus 267
megalorynchos, Tanygnathus 267
Megalurus 490
Megapodiidae 44
Megapodius 47
megarhyncha, Colluricincla 384
megarhyncha, Halcyon 232
megarhyncha, Muscicapa 384
megarhyncha, Syma 232
megarhynchus, Chalcites 142
megarhynchus, Chrysococcyx 142
megarhynchus, Corvus 453
megarhynchus, Cuculus 142
megarhynchus, Epimachus 429
megarhynchus, Melilestes 310
megarhynchus, Myiolestes 386
megarhynchus, Ptilotis 310
megarhynchus, Rhamphomantis 142
Megatriorchis 207
melaena, Coracina 379
melaleuca, Muscipeta 407
melaleuca, Rhipidura 407
Melampitta 439
Melampittidae 438
melanocephalus, Aeluroedus 276
melanocephalus, Ailuroedus 276
Melanocharis 345
Melanocharitidae 345
melanochlamys, Accipiter 211
melanochlamys, Urospizias 211
melanogaster, Anhinga 122
melanogenia, Cyclopsitta 260
melanogenia, Psittacula 260
melanogenys, Poecilodryas 473
melanogenys, Tregellasia 473
melanolaema, Xanthotis 319
melanoleuca, Paradisea 424
melanoleuca, Seleucidis 424
melanoleucos, Halietor 120
melanoleucos, Hydrocorax 120
melanoleucos, Microcarbo 120
melanoleucos, Phalacrocorax 120
melanoleucus, Circus 209
melanoleucus, Seleucidis 424
melanolora, Campephaga 371

melanonotus, Carterornis 449
melanonotus, Monarcha 449
melanope, Motacilla 522
melanops, Coracina 371
melanops, Corvus 371
melanopsis, Monarcha 451
melanopsis, Muscicapa 451
melanopterus, Monarcha 448
melanopterus, Symposiachrus 448
Melanorectes 390
melanorhyncha, Colluricincla 386
melanorhynchus, Ardea 114
melanorhynchus, Myiolestes 386
melanosternon, Hamirostra 202
melanotis, Ailuroedus 275
melanotis, Ptilonorhycnhus 275
melanotos, Calidris 182
melanotos, Erolia 182
melanotos, Tringa 182
melanotus, Porphyrio 133
melanura, Carpophaga 91
melanura, Ducula 91
melanura, Pachycephala 394
melanuroides, Limosa 174
melas, Coracina 379
melas, Edolisoma 378
melas, Lanius 378
Melidectes 320
Melidora 226
Melilestes 310
Melionyx 323
Meliphaga 312
Meliphagidae 288
Melipotes 307
Melithreptus 301
mellori, Eulabeornis 128
mellori, Hypotaenidia 128
melvillensis, Myristicivora 91
melvillensis, Pandion 199
melvillensis, Ptilotula 318
menawa, Ptilorrhoa 356
menbeki, Centropus 136
mentalis, Cracticus 364
meridionalis, Cormobates 283
meridionalis, Pitohui 404
meridionalis, Rectes 404
Meropidae 220
Merops 220
metallica, Aplonis 494
metallicus, Lamprotornis 494
meyeri, Acrocephalus 488
meyeri, Chrysococcyx 144
meyeri, Epimachus 38, 429
meyeri, Euthyrhynchus 311
meyeri, Myzomela 292
meyeri, Pachycephala 395
meyeri, Philemon 296
meyeri, Pitohui 402
meyeri, Talegalla 46
meyeri, Tanysiptera 223

meyeri, Timeliopsis 311
meyeri, Xanthotis 294
meyerianus, Accipiter 212
meyerianus, Astur 212
meyerii, Chalcites 143
meyerii, Chrysococcyx 143
meyerii, Edoliosoma 376
meyerii, Edolisoma 376
Microcarbo 120
microcera, Ducula 87
Microdynamis 139
Microeca 462
Micropsitta 268
micropteryx, Trichoglossus 253
migrans, Falco 205
migrans, Milvus 205
miles, Hoplopterus 168
miles, Lobibyx 168
miles, Tringa 167
miles, Vanellus 167
Milvus 205
mimikae, Conopophila 306
mimikae, Meliphaga 315
mimikae, Myiagra 444
mimikae, Pseudogerygone 338
mimikae, Ptilotis 315
minima, Chalcophaps 76
minima, Eudynamis 140
minimus, Henicopernis 202
minimus, Ptilinopus 78
Mino 496
minor, Fregata 117
minor, Geoffroyus 266
minor, Goura 73
minor, Lophorina 427
minor, Paradisaea 436
minor, Paradisea 436
minor, Pelecanus 117
minor, Tanysiptera 223
minor, Zosterops 484
minus, Edoliisoma 375
minuta, Calidris 181
minuta, Erolia 181
minuta, Tringa 181
minutillus, Chalcites 144
minutillus, Chrysococcyx 144
minutus, Anous 188
minutus, Ixobrychus 109
minutus, Numenius 175
miosnomensis, Pachycephala 398
miquelii, Ptilinopus 79
miquelii, Ptilopus 79
Mirafra 474
miranda, Daphoenositta 380
misimae, Collocalia 158
misimae, Pachycephala 393
misoliensis, Aepypodius 45
misoliensis, Ailuroedus 276
misoliensis, Pinarolestes 385
misoriense, Dicaeum 504

misoriensis, Accipiter 210
misoriensis, Chalcites 144
misoriensis, Lamprococcyx 144
misoriensis, Micropsitta 268
misoriensis, Nasiterna 268
misoriensis, Phylloscopus 481
misoriensis, Seicercus 481
misoriensis, Urospizias 210
misulae, Accipiter 210
mixtus, Melidectes 320
modesta, Ardea 113
modesta, Glyciphila 307
modesta, Pachycephala 396
modesta, Poecilodryas 396
modesta, Psittacella 243
modestus, Ramsayornis 307
molucca, Ibis 105
molucca, Lonchura 514
molucca, Loxia 514
molucca, Threskiornis 105
moluccana, Amaurornis 131
moluccana, Porzana 131
moluccanus, Trichoglossus 254
moluccarum, Butorides 111
moluccensis, Falco 235
moluccensis, Tinnunculus 235
moluccus, Threskiornis 106
monacha, Alcippe 327
monacha, Crateroscelis 327
monacha, Pachycephala 399
Monachella 462
Monarcha 450
Monarchidae 440
mongolus, Charadrius 169
monorthonyx, Anurophasis 50
monorthonyx, Coturnix 50
montana, Coracina 375
montana, Fringilla 511
montana, Lonchura 520
montana, Meliphaga 314
montana, Ptilotis 314
montana, Timeliopsis 310
montanum, Edolisoma 375
montanus, Megalurus 491
montanus, Passer 511
montanus, Peltops 363
Monticola 501
monticola, Coturnix 52
monticola, Lonchura 520
monticola, Munia 520
monticola, Pitohui 404
monticola, Sericornis 333
monticola, Synoicus 52
montium, Paramythia 354
montona, Campephaga 375
moresbyae, Lonchura 515
moretoni, Malurus 287
morinella, Arenaria 179
morio, Coracina 376
moroka, Pachycephala 396

morphnoides, Hieraaetus 204
moszkowskii, Alisterus 263
moszkowskii, Aprosmictus 263
moszkowskii, Edoliisoma 378
Motacilla 521
Motacillidae 520
muelleriana, Monachella 462
muellerii, Ducula 90
mülleri, Mimeta 406
mülleri, Rhipidura 412
mülleriana, Muscicapa 462
mullerii, Columba 90
mullerii, Ducula 90
müllerii, Edoliisoma 376
multipunctata, Tyto 214
multistriata, Charmosyna 247
multistriata, Charmosynopsis 247
murchisonianus, Falco 237
murina, Acanthiza 335
murina, Crateroscelis 326
murina, Gerygone 335
murinus, Brachypteryx 326
muscalis, Cinclosoma 359
muscalis, Megalurus 490
Muscicapa 502
Muscicapidae 500
musschenbroekii, Nanodes 250
musschenbroekii, Neopsittacus 250
Myiagra 443
myolae, Lonchura 520
myristicivora, Columba 87
myristicivora, Ducula 87
mysolensis, Aplonis 496
mysolensis, Calornis 496
mysorensis, Chalcostetha 507
mysorensis, Geoffroyus 266
mysorensis, Leptocoma 507
mysorensis, Malurus 286
mysorensis, Pionias 266
mysorensis, Todopsis 286
mysorensis, Zosterops 484
mystacalis, Caprimulgus 151
mystacalis, Eurostopodus 151
mystacea, Aplonis 495
mystacea, Calornis 495
mystacea, Hemiprocne 157
mystaceus, Cypselus 157
Myzomela 288

naimii, Malurus 287
naina, Columba 78
nainus, Ptilinopus 78
nana, Myiagra 444
nana, Seisura 444
nanus, Ptilinopus 78
nativitatis, Puffinus 102
nebularia, Scolopax 177
nebularia, Tringa 177
Nectariniidae 506

neglecta, Carpophaga 86
neglecta, Ducula 86
neglecta, Gerygone 336
neglecta, Grallina 441
neglecta, Petrochelidon 478
neglecta, Pterodroma 98
neglectus, Burhinus 164
neglectus, Esacus 164
nehrkorni, Edoliisoma 376
nehrkorni, Edolisoma 376
Neochmia 512
Neopsittacus 250
neos, Myiolestes 386
Neosittidae 379
neoxena, Hirundo 476
Nettapus 57
neumanni, Eupetes 357
neumanni, Gallinula 134
neumanni, Ptilorrhoa 357
neumanni, Phonygammus 418
nevermanni, Lonchura 517
nicobarica, Caloenas 75
nicobarica, Columba 75
niedda, Lophorina 426
niger, Chlidonias 196
niger, Dicaeum 346
nigra, Astrapia 431
nigra, Melanocharis 346
nigra, Paradisea 431
nigra, Sterna 196
nigrescens, Melanorectes 390
nigrescens, Pitohui 391
nigrescens, Rectes 390
nigricans, Centropus 138
nigricans, Hirundo 478
nigricans, Petrochelidon 477
nigricans, Polophilus 138
nigriceps, Drymodes 464
nigriceps, Poecilodryas 472
nigriceps, Tanysiptera 225
nigriceps, Tregellasia 472
nigricissus, Eupetes 356
nigricissus, Ptilorrhoa 356
nigrifrons, Cyclopsitta 260
nigrifrons, Cyclopsittacus 260
nigripectus, Machaerirhynchus 361
nigripectus, Pitohui 403
nigripes, Ardea 115
nigripes, Egretta 115
nigrirostris, Macropygia 67
nigrirostris, Monarcha 451
nigriscapularis, Hermotimia 507
nigriscapularis, Leptocoma 507
nigrita, Myzomela 291
nigrocyanea, Halcyon 228
nigrocyaneus, Todiramphus 228
nigrogularis, Trichoglossus 254
nigromentalis, Rhipidura 413
nigroorbitalis, Poecilodryas 472
nigroorbitalis, Tregellasia 472

nigropectus, Eulacestoma 383
nigrorufa, Crateroscelis 327
nigrorufa, Sericornis 327
nigroviridis, Sericornis 334
nilotica, Gelochelidon 190
nilotica, Sterna 190
Ninox 216
nitens, Collocalia 158
nitidum, Dicaeum 505
nitidus, Carterornis 450
nitidus, Poecilodryas 450
niveifrons, Aleadryas 381
niveifrons, Pachycephala 381
nobilis, Otidiphaps 72
normani, Artamus 367
normani, Austartamus 367
normantoni, Cacatua 240
notata, Gerygone 336
nouhuysi, Melidectes 324
nouhuysi, Melionyx 324
nouhuysi, Melirrhophetes 324
nouhuysi, Sericornis 330, 332
novaeguineae, Chaetura 161
novaeguineae, Cinnyris 351
novaeguineae, Coturnix 52
novaeguineae, Falco 237
novaeguineae, Fulica 134
novaeguineae, Harpyopsis 203
novaeguineae, Hieracidea 237
novaeguineae, Mearnsia 161
novaeguineae, Orthonyx 341
novaeguineae, Paradisea 437
novaeguineae, Parra 171
novaeguineae, Philemon 297
novaeguineae, Pitta 273
novaeguineae, Toxorhamphus 351
novaeguineae, Tropidorhynchus 297
novaeguineae, Zosterops 485
novaeguineensis, Piezorhynchus 443
novaehollandiae, Accipiter 210
novaehollandiae, Accipiter 210
novaehollandiae, Anhinga 122
novaehollandiae, Ardea 115
novaehollandiae, Chroicocephalus 189
novaehollandiae, Coracina 371
novaehollandiae, Egretta 115
novaehollandiae, Hydralector 171
novaehollandiae, Larus 189
novaehollandiae, Notophoyx 115
novaehollandiae, Phalacrocorax 121
novaehollandiae, Plotus 122
novaehollandiae, Podiceps 62
novaehollandiae, Recurvirostra 166
novaehollandiae, Scythrops 141
novaehollandiae, Strix 214
novaehollandiae, Tachybaptus 62
novaehollandiae, Turdus 371
novaehollandiae, Tyto 214
novaehollandiae, Vanellus 168
novaeseelandiae, Anthus 523

novaeseelandiae, Ninox 217
novaeseelandiae, Strix 217
novaezealandiae, Limosa 174
novus, Machaerirhynchus 360
novus, Rhagologus 368
nubicola, Amblyornis 278
nuchalis, Melidectes 320
nuditarsus, Aerodramus 159
nuditarsus, Collocalia 159
Numenius 175
numforana, Coracina 376
numforanum, Edolisoma 376
numforensis, Collocalia 158
nupta, Lichmera 299
nupta, Stigmatops 299
Nycticorax 110
nymani, Myzomela 290
nymanni, Myzomela 290
nympha, Tanysiptera 225

oblita, Pitta 272
obscura, Aviceda 201
obscura, Caligavis 318
obscura, Colluricincla 386
obscura, Meliphaga 319
obscura, Myzomela 290
obscura, Ptilotis 318
obscura, Rectes 386
obscuratus, Cacomantis 147
obscuratus, Centropus 138
obscurifrons, Dicaeum 504
obscurior, Lalage 374
obscurior, Pachycephala 392
obscurus, Lichenostomus 319
obscurus, Oreornis 319
obscurus, Rhagologus 368
occasa, Gerygone 338
occidentalis, Amalocichla 456
occidentalis, Cyclopsittacus 258
occidentalis, Melidectes 324
occidentalis, Microeca 461
occidentalis, Psittaculirostris 258
occidentalis, Ptiloprora 302
occidentis, Paramythia 353
occidentis, Talegalla 46
occipitalis, Casuarius 43
oceanica, Procellaria 94
oceanicus, Oceanites 94
Oceanites 94
Oceanitidae 94
Oceanodroma 96
ocellatus, Podargus 149
ochracea, Syma 232
ochrogaster, Alcyone 234
ochrogaster, Ceyx 234
ochromelas, Melidectes 321
ochromelas, Melirrhophetes 321
ochropus, Tringa 177
ocrophus, Tringa 177

octogenarii, Pachycephala 392
ocularis, Glyciphila 299
ocularis, Lichmera 299
Oedistoma 350
olivacea, Gallinula 132
olivacea, Gerygone 336
olivacea, Lichmera 300
olivacea, Paramythia 353
olivacea, Sericornis 335
olivacentior, Amalocichla 457
olivaceum, Paramythia 353
olivaceus, Amaurornis 132
olivaceus, Psilopus 336
olivascens, Oedistoma 351
onerosa, Gerygone 339
Onychoprion 193
oorti, Ailuroedus 275
oorti, Clytomyias 285
oorti, Sericornis 333
ophthalmica, Cacatua 240
optatus, Cuculus 148
Oreocharis 353
Oreoicidae 381
oreophila, Zosterops 486
oreophilus, Cacomantis 147
oreophilus, Zosterops 486
Oreopsittacus 244
Oreornis 319
Oreostruthus 511
orientalis, Acrocephalus 487
orientalis, Coracias 221
orientalis, Cuculus 139
orientalis, Eudynamys 139
orientalis, Eurystomus 221
orientalis, Gallicolumba 69
orientalis, Geoffroyus 266
orientalis, Gracula 496
orientalis, Melanocharis 347
orientalis, Meliphaga 314
orientalis, Mino 496
orientalis, Ptilotis 314
orientalis, Salicaria 487
Oriolidae 400
Oriolus 405
oriomo, Coracina 372
ornata, Charmosyna 246
ornatus, Merops 220
ornatus, Ptilinopus 81
ornatus, Ptilopus 81
Ornorectes 382
orru, Corvus 454
Orthonychidae 341
Orthonyx 341
osculans, Chalcites 142
osculans, Chrysococcyx 142
osculans, Misocalius 142
ostralegus, Haematopus 164
Otididae 123
Otidiformes 123
Otidiphaps 72

Otus 216
oustaleti, Megapodius 49

Pachycare 325
Pachycephala 391
Pachycephalidae 384
Pachycephalopsis 457
Pachyptila 98
pacifica, Ardea 113
pacifica, Ardenna 100
pacifica, Columba 87
pacifica, Coracias 221
pacifica, Ducula 87
pacifica, Hirundo 163
pacifica, Procellaria 100
pacificus, Apus 163
pacificus, Eurystomus 222
pacificus, Puffinus 100
pallescens, Lalage 374
pallescens, Stercorarius 198
pallida, Columba 145
pallida, Crateroscelis 327
pallida, Ducula 145
pallida, Gerygone 339
pallida, Glycichaera 302
pallida, Megacrex 132
pallida, Melanocharis 346
pallida, Psittacella 242
pallidimas, Accipiter 210
pallidior, Charmosyna 246
pallidior, Charmosynopsis 246
pallidior, Myzomela 289
pallidipes, Zosterops 487
pallidogularis, Zosterops 483
pallidus, Cacomantis 145
pallidus, Cuculus 145
pallidus, Heteroscenes 145
pallidus, Oreostruthus 511
pallidus, Pitohui 402
palmeri, Myiolestes 385
palmerstoni, Pelecanus 117
palpebrosa, Gerygone 337
palustris, Porzana 130
palustris, Zapornia 130
Pandion 199
Pandionidae 199
paniaiae, Phylloscopus 480
Papasula 118
papou, Charmosyna 248
papou, Psittacus 248
papua, Anhinga 122
papua, Aprosmictus 264
papua, Geopelia 71
papuae, Meliphaga 313
papuana, Aythya 61
papuana, Devioeca 461
papuana, Erythrura 513
papuana, Gymnorhina 365
papuana, Microeca 461

papuana, *Myiagra* 445
papuanus, Accipiter 212
papuanus, Astur 212
papuanus, Casuarius 41
papuanus, Charadrius 169
papuanus, Cracticus 365
papuanus, Falco 236
papuanus, Machaerhamphus 203
papuanus, Macheiramphus 203
papuensis, Acanthiza 334
papuensis, Aerodramus 159
papuensis, Archboldia 279
papuensis, Butorides 111
papuensis, Caprimulgus 151
papuensis, Chaetorhynchus 407
papuensis, Collocalia 159
papuensis, Coracina 371
papuensis, Corvus 371
papuensis, Daphoenositta 379
papuensis, Eurostopodus 151
papuensis, Excalfactoria 52
papuensis, Geocichla 499
papuensis, Megalurus 492
papuensis, Merula 499
papuensis, Neosita 379
papuensis, Philemon 297
papuensis, Podargus 150
papuensis, Sericornis 334
papuensis, Sitta 379
papuensis, Turdus 499
papuensis, Tyto 215
papuensis, Zoothera 499
par, Eupetes 358
par, Ptilorrhoa 358
Paradigalla 430
Paradisaea 435
Paradisaeidae 416
Paradisornis 434
Paramythia 353
Paramythiidae 352
parasiticus, Larus 197
parasiticus, Stercorarius 197
Parotia 421
parva, Eudynamis 139
parva, Microdynamis 139
parva, Rallicula 125
Passer 510
Passeridae 510
Passeriformes 271
pastinator, Cacatua 240
pectorale, Dicaeum 503
pectoralis, Columba 83
pectoralis, Lewinia 127
pectoralis, Pachycephala 394
pectoralis, Psittacus 265
pectoralis, Ptilinopus 83
pectoralis, Rallus 127
Pelagodroma 94
Pelecanidae 104
Pelecanus 104

pelewensis, Anas 59
pelewensis, Caloenas 75
Peltops 362
penelope, Anas 59
penelope, Mareca 59
Peneothello 468
peninsularis, Crateroscelis 329
peregrinus, Falco 237
periophthalmicus, Monarcha 452
perlata, Columba 81
perlatus, Ptilinopus 81
perneglecta, Pachycephala 398
personata, Sula 118
perspicillata, Columba 86
perspicillata, Ducula 86
perspicillata, Sericornis 334
perspicillatus, Sericornis 334
perstriata, Ptiloprora 303
perstriata, Ptilotis 303
Petrochelidon 477
Petroica 459
Petroicidae 455
phaeopus, Numenius 175
phaeopus, Scolopax 175
Phaethon 93
Phaethontidae 92
Phaethontiformes 92
phaeton, Fringilla 512
phaeton, Neochmia 512
phaionota, Pachycephala 396
phaionotum, Pachycephala 396
phaionotus, Myiolestes 396
phaionotus, Pachycephala 396
Phalacrocoracidae 120
Phalacrocorax 120
Phalaropus 184
phasiana, Rhipidura 412
phasianella, Macropygia 67
Phasianidae 50
phasianinus, Centropus 137
phasianinus, Cuculus 137
Philemon 296
philemon, Xanthotis 294
philipi, Casuarius 43
philippensis, Gallirallus 128
philippensis, Hypotaenidia 128
philippensis, Monticola 502
philippensis, Rallus 128
philippensis, Turdus 502
philippinus, Merops 220
Philomachus 183
Phoenicopteriformes 61
Phonygammus 416
Phylloscopidae 479
Phylloscopus 479
picata, Ardea 114
picata, Egretta 114
picata, Hydranassa 115
picata, Notophoyx 115
picta, Psittacella 243

picticollis, Casuarius 41
piersmai, Calidris 180
pileata, Sterna 188
pileatus, Anous 188
piliatus, Chenorhamphus 285
pinon, Columba 89
pinon, Ducula 89
piperata, Ptilotis 349
piperata, Rhamphocharis 38, 349
pistrinaria, Ducula 89
Pitohui 400
Pitta 273
Pittidae 271
placens, Climacteris 283
placens, Cormobates 283
placens, Eopsaltria 471
placens, Gennaeodryas 471
placens, Poecilodryas 471
placens, Sternula 192
placentis, Charmosyna 246
placentis, Psittacus 246
placida, Geopelia 71
plagosus, Chalcites 143
plagosus, Cuculus 143
Platalea 106
Plegadis 107
Plesiodryas 467
plicatus, Aceros 219
plicatus, Buceros 219
plicatus, Rhyticeros 219
plotus, Pelecanus 119
plotus, Sula 119
plumbea, Coturnix 51
plumbea, Ptiloprora 302
plumbea, Ptilotis 302
plumbeicollis, Ptilinopus 81
plumbeicollis, Ptilopus 81
plumbeiventris, Eulabeornis 129
plumbeiventris, Gymnocrex 129
plumbeiventris, Rallus 129
plumbeus, Synoecus 51
plumeriferus, Herodias 114
plumifer, Aegotheles 155
plumifera, Aegotheles 155
plumifera, Ardea 114
plumiferus, Herodias 114
pluto, Myzomela 292
Pluvialis 166
Podargidae 149
Podargus 149
Podicipedidae 61
Poecilodryas 467
poecilorhyncha, Anas 59
poecilurus, Chalcites 144
poecilurus, Chrysococcyx 144
poikilosternos, Xanthotis 295
poliocephala, Gerygone 480
poliocephala, Glycichaera 301
poliocephala, Microeca 461
poliocephalus, Accipiter 211

poliocephalus, Phylloscopus 480
poliocephalus, Seicercus 480
poliocephalus, Turdus 499
poliopsa, Edoliidoma 377
poliopsa, Edolisoma 377
poliopterus, Melilestes 352
poliopterus, Toxorhamphus 352
poliosoma, Pachycephalopsis 458
poliura, Megaloprepia 77
polychloros, Eclectus 265
polycryptus, Accipiter 211
polygramma, Meliphaga 296
polygramma, Ptilotis 295
polygrammica, Campephaga 374
polygrammica, Lalage 374
polygrammus, Xanthotis 295
polyphonus, Melidectes 320
pomarinus, Lestris 197
pomarinus, Stercorarius 197
Pomatostomidae 340
Pomatostomus 340
pontifex, Sericornis 333
Porphyrio 132
porphyrio, Fulica 133
porphyrio, Porphyrio 133
Porzana 130, 131
postrema, Ducula 89
praecipua, Ptilotis 304
praedicta, Ptiloprora 303
praerepta, Monarcha 450
prasina, Neneba 348
prasinonota, Pachycephala 382
prasinorrhous, Ptilinopus 80
prasinorrhous, Ptilonopus 80
pratti, Crateroscelis 328
primitiva, Myzomela 290
princeps, Accipiter 211
princeps, Melidectes 324
princeps, Melionyx 324
prionurus, Cuculus 146
Probosciger 238
Procellariidae 97
Procellariiformes 93
propinqua, Chibia 414
propinquus, Centropus 138
proteus, Pitohui 402
proxima, Gerygone 339
proxima, Ptilotis 305
proximus, Pycnopygius 305
Pseudeos 255
Pseudobulweria 99
pseudogeelvinkianus, Ptilinopus 84
pseudohumeralis, Ptilinopus 82
Pseudorectes 389
pseutes, Syma 232
Psittacella 241
Psittaciformes 238
Psittaculidae 241
Psittaculirostris 257
Psitteuteles 252

Psittrichas 241
Psittrichasidae 240
Psophodidae 354
Pteridophora 421
Pterodroma 98
Ptilinopus 78
Ptilonorhynchidae 274
Ptiloprora 302
Ptiloris 424
Ptilorrhoa 355
Ptilotula 318
pucherani, Geoffroyus 265
puella, Columba 77
puella, Megaloprepia 77
Puffinus 102
pugnax, Calidris 184
pugnax, Philomachus 183
pugnax, Tringa 183
pulchella, Charmosyna 249
pulchella, Columba 84
pulchellus, Nettapus 58
pulchellus, Ptilinopus 84
pulcher, Aegotheles 153
pulcher, Eupetes 358
pulchra, Macgregoria 309
pulchra, Paradisaea 436
pulchra, Ptilorrhoa 358
pullicauda, Neopsittacus, 37, 250
pulverulenta, Eopsaltria 470
pulverulenta, Peneonanthe 470
pulverulenta, Peneothello 469
pulverulenta, Poecilodryas 470
pulverulentus, Myiolestes 469
punctulata, Lonchura 514
punctulata, Loxia 514
purpuratus, Ptilinopus 85
purpureoviolaceus, Phonygama 418
purpureoviolaceus, Phonygammus 418
pusilla, Alcedo 235
pusilla, Alcyone 235
pusilla, Ceyx 234
pusilla, Ninox 217
pusilla, Porzana 130
pusilla, Zapornia 130
pusillus, Ceyx 234
pusillus, Rallus 130
pusio, Micropsitta 269
pusio, Nasiterna 269
Pycnonotidae 478
Pycnonotus 478
Pycnopygius 305
pycrofti, Pterodroma 99
pygmaea, Dendrocygna 54
pygmaea, Goura 73
pygmaeum, Oedistoma 351
pyrrhophanus, Cacomantis 146
pyrrhophanus, Cuculus 146

quadricolor, Cyanalcyon 228

quadricolor, Todiramphus 228
quadrigemmus, Ptilopus 85
quadrimaculata, Peneothello 469
quadrimaculatus, Poecilodryas 469
queenslandica, Ninox 217
querquedula, Anas 60
querquedula, Spatula 60
quoyi, Barita 363
quoyi, Cracticus 363
quoyii, Megapodius 49

radjah, Anas 56
radjah, Radjah 56
radjah, Tadorna 56
raggiana, Paradisaea 437
raggiana, Paradisea 437
Rallicula 124
Rallidae 124
Rallina 126
ramsayi, Lonchura 519
Ramsayornis 307
ramuensis, Opopsitta 260
randi, Malurus 287
randi, Pachycare 325
randi, Rallus 128
randi, Sericornis 331
rara, Meliphaga 313
recondita, Astrapia 431
Recurvirostra 166
Recurvirostridae 165
reductus, Rallus 128
regia, Paradisea 433
regia, Platalea 106
regina, Ptilinopus 85
regius, Cicinnurus 433
reichenowi, Edoliisoma 377
reichenowi, Edolisoma 377
reichenowi, Pachycephala 397
reinwardt, Megapodius 48
reinwardti, Reinwardtoena 65
reinwardtii, Reinwardtoena 65
Reinwardtoena 65
reinwardtsi, Columba 65
reinwardtsi, Reinwardtoena 65
remota, Pachycephala 393
respublica, Cicinnurus 434
respublica, Diphyllodes 434
respublica, Lophorina 434
rex, Clytoceyx 226
rex, Dacelo 226
rex, Paradisea 433
Rhagologidae 367
Rhagologus 367
Rhamphocharis 348
Rhamphomantis 141
Rhipidura 407
Rhipiduridae 406
rhodinolaema, Carpophaga 89
rhodinolaema, Ducula 89

rhododendri, Anthus 523
Rhyticeros 219
richardsi, Anthus 523
richardsoni, Porzana 130
ridibundus, Chroicocephalus 189
ridibundus, Larus 189
riedelii, Tanysiptera 223
Riparia 475
riparia, Hirundo 475
riparia, Riparia 475
ripleyi, Crateroscelis 329
rivoli, Columba 79
rivoli, Ptilinopus 79
robertsoni, Mino 497
robusta, Crateroscelis 328
robusta, Gerygone 328
robusta, Pallenura 522
rogersi, Anas 59
rogersi, Calidris 180
rogersi, Chalcophaps 76
roratus, Eclectus 264
roratus, Larius 264
roratus, Psittacus 264
rosenbergii, Myzomela 289
rosenbergii, Pitta 273
rosenbergii, Scolopax 172
rosenbergii, Trichoglossus 253
rosseli, Dicaeum 504
rosseliana, Gerygone 339
rosseliana, Ninox 218
rosseliana, Pachycephala 393
rosseliana, Tanysiptera 223
rosselianus, Accipiter 211
rosselianus, Lorius 252
rosselianus, Monarcha 451
rostrata, Mellopitta 439
rostrata, Procellaria 99
rostrata, Pseudobulweria 99
rostrata, Pterodroma 100
rostratum, Edoliosoma 377
rostratum, Edolisoma 377
rothschildi, Astrapia 432
rothschildi, Charmosyna 249
rothschildi, Charmosynopsis 249
rothschildi, Heteromyias 465
rothschildi, Zosterops 484
rubecula, Myiagra 445
rubecula, Todus 445
rubicunda, Antigone 135
rubicunda, Ardea 135
rubicunda, Grus 135
rubiensis, Carpophaga 89
rubiensis, Lorius 251
rubiensis, Manucodia 420
rubiensis, Monarcha 447
rubiensis, Pitohui 401
rubiensis, Rhectes 401
rubiensis, Symposiachrus 447
rubiensis, Tchitrea 447
rubiensis, Xanthotis 294

rubra, Eugerygone 460
rubra, Paradisaea 436
rubra, Paradisea 436
rubra, Pseudogerygone 460
rubra, Rallicula 124
rubra, Rallina 124
rubra, Uranornis 436
rubricauda, Phaethon 93
rubricauda, Phaeton 93
rubrifrons, Chalcopsitta 257
rubrigulare, Dicaeum 504
rubrigularis, Charmosyna 247
rubrigularis, Trichoglossus 247
rubripes, Sula 119
rubritorquis, Trichoglossus 254
rubrobrunnea, Myzomela 290
rubrocoronatum, Dicaeum 504
rubronotata, Charmosyna 245
rubronotatus, Coriphilus 245
rudolphi, Paradisaea 435
rudolphi, Paradisornis 435
rufa, Anhinga 122
rufa, Athene 216
rufa, Ninox 216
rufescens, Gerygone 334
rufescens, Sericornis 334
ruficapillus, Charadrius 169
ruficollis, Buceros 219
ruficollis, Calidris 181
ruficollis, Chalcites 143
ruficollis, Chrysococcyx 143
ruficollis, Erolia 181
ruficollis, Gerygone 339
ruficollis, Lamprococcyx 143
ruficollis, Myiagra 444
ruficollis, Platyrhynchos 444
ruficollis, Podiceps 62
ruficollis, Tachybaptus 62
ruficollis, Trynga 181
ruficrissa, Gallinula 131
rufidorsa, Rhipidura 408
rufifrons, Muscicapa 409
rufifrons, Rhipidura 409
rufigaster, Columba 88
rufigaster, Ducula 88
rufigula, Gallicolumba 69
rufigula, Peristera 69
rufinucha, Aleadryas 381
rufinucha, Pachycephala 381
rufitergum, Tadorna 56
rufiventer, Cuculus 139
rufiventer, Eydynamys 139
rufiventris, Pachycephala 399
rufiventris, Platyrhynchos 412
rufiventris, Rhipidura 412
rufocrissalis, Melidectes 322
rufocrissalis, Melirrhophetes 322
rufolateralis, Myiagra 443
rufolateralis, Piezorhynchus 443
russatus, Chalcites 142, 144

rustica, Hirundo 475

sacra, Ardea 116
sacra, Demigretta 116
sacra, Egretta 116
sagittata, Coracias 405
sagittatus, Oriolus 405
salvadoriana, Tanysiptera 225
salvadorii, Aegotheles 154
salvadorii, Carpophaga 90
salvadorii, Cinnyris 507
salvadorii, Corvus 454
salvadorii, Cyclopsittacus 259
salvadorii, Ducula 90
salvadorii, Lorius 251
salvadorii, Merops 220
salvadorii, Pachycephala 397
salvadorii, Paradisaea 437
salvadorii, Pitohui 401
salvadorii, Psittaculirostris 259
salvadorii, Ptilinopus 84
salvadorii, Ptilopus 84
salvadorii, Ptilotis 319
salvadorii, Sphecotheres 404
Salvadorina 56
sancta, Halcyon 231
sanctus, Todiramphus 231
sanfordi, Archboldia 280
sanfordi, Crateroscelis 328
sanfordi, Rhamphomantis 142
sanguinea, Cacatua 239
sanguineus, Cnemophilus 343
saruwagedi, Peneothello 469
saruwagedi, Poecilodryas 469
saturata, Ptilorrhoa 358
saturata, Scolopax 172
saturatior, Coturnix 51
saturatior, Eugerygone 460
saturatior, Synoicus 51
saturatior, Xanthotis 294
saturatus, Cuculus 148
saturatus, Eupetes 358
saturatus, Machaerirhynchus 361
saurophaga, Halcyon 231
saurophagus, Todiramphus 231
Saxicola 501
schach, Lanius 455
scheepmakeri, Goura 74
schillmolleri, Caprimulgus 152
schistacea, Meliornis 302
schistaceus, Melanorectes 391
schistaceus, Rectes 391
schisticeps, Campephaga 377
schisticeps, Coracina 378
schisticeps, Edolisoma 377
schisticinus, Astur 211
schlegeli, Henicophaps 68
schlegeli, Rynchaenas 68
schlegelii, Caprimulgus 152

schlegelii, Pachycephala 392
schoddei, Scythrops 141
schraderensis, Melidectes 322
scintillata, Chalcopsitta 256
sciurorum, Myiagra 445
sclateri, Myzomela 293
sclateriana, Amalocichla 456
sclaterii, Casuarius 42
sclaterii, Goura 74
scolopacea, Eudynamys 140
scolopacea, Limosa 173
scolopaceus, Eydynamys 140
scolopaceus, Limnodromus 173
Scolopacidae 171
Scolopax 172
scratchleyana, Lonchura 517
scratchleyana, Munia 517
Scythrops 141
sefilata, Paradisea 421
sefilata, Parotia 421
Seicercus 479
sejuncta, Ducula 87
Seleucidis 424
sellamontis, Syma 232
semipalmata, Anas 53
semipalmata, Anseranas 53
semipalmatus, Limnodromus 173
semipalmatus, Macrorhamphus 173
senex, Pitohui 402
separata, Carpophaga 86
sepik, Meliphaga 315
sepikensis, Lonchura 518
sepikiana, Charmosyna 248
sepikiana, Charmosyne 248
sepikiana, Mirafra 474
sepikiana, Pachycephala 397
septentrionalis, Clytoceyx 226
septentrionalis, Gallicolumba 69
septentrionalis, Megaloprepia 77
septentrionalis, Toxorhamphus 352
septentrionalis, Xanthotis 296
sericea, Leptocoma 508
sericea, Loboparadisea 37, 344
sericea, Nectarinia 508
sericeus, Cinnyris 506
Sericornis 329
Sericulus 280
serrata, Sterna 194
serratus, Onychoprion 194
setekwa, Dicaeum 504
setekwa, Meliphaga 312
severus, Falco 236
sharpei, Eulabeornis 129
sharpei, Ptilotis 312
sharpei, Zosterops 483
sharpii, Donacicola 519
sharpii, Lonchura 519
shawmayeri, Casuarius 41
shawmayeri, Zosterops 486
sibilans, Eupetes 356

sibilans, Ptilorrhoa 356
sibirica, Calidris 182
sibirica, Limicola 182
sigillata, Peneothello 469
sigillata, Poecilodryas 469
sigillifer, Erythrura 513
sigillifer, Lobospingus 513
similis, Cicinnurus 433
simillima, Motacilla 521
simillima, Pitta 274
simplex, Geoffroyus 266
simplex, Pachycephala 398
simplex, Pionias 266
simplex, Ptilotis 305
sinensis, Ardea 109
sinensis, Ixobrychus 109
sinensis, Sterna 192
sinensis, Sternula 192
sintillatus, Psittacus 256
Sipodotus 284
sloetii, Campephaga 372
sloetii, Campochaera 372
smaragdina, Ducula 88
sociabilis, Micropsitta 268
socialis, Neopsittacus 250
solandri, Pterodroma 98
solitaria, Ceyx 234
solitarius, Ceyx 234
solitarius, Monticola 502
solitarius, Turdus 502
soloensis, Accipiter 209
soloensis, Falco 209
solomonensis, Butorides 111
solomonensis, Ptilinopus 80
solomonensis, Ptilonopus 80
somu, Lorius 251
sonoroides, Gavicalis 317
sonoroides, Ptilotis 317
sordida, Pitta 273
sordidus, Halcyon 230
sordidus, Todiramphus 230
sordidus, Turdus 273
soror, Pachycephala 392
speciosus, Ptilinopus 81
speciosus, Ptilopus 81
spectabilis, Chalcopsitta 256
spectabilis, Donacicola 517
spectabilis, Lonchura 517
Sphecotheres 404
sphenurus, Haliastur 206
sphenurus, Milvus 206
sphinx, Lophorina 427
spilodera, Entomophila 329
spilodera, Sericornis 329
spilogaster, Ptilotis 295
spilogaster, Xanthotis 295
spilonotus, Circus 209
spilorrhoa, Carpophaga 91
spilorrhoa, Ducula 91
spilothorax, Circus 208

spinicaudus, Pachycephala 394
spinicaudus, Pteruthius 394
spinicollis, Ibis 106
spinicollis, Threskiornis 106
splendidissima, Astrapia 431
spodiopygius, Aerodramus 159
squamata, Eos 254
squamata, Rhipidura 410
squamatus, Psittacus 254
squatarola, Charadrius 167
squatarola, Pluvialis 167
squatarola, Squatarola 167
squatarola, Tringa 167
stagnatilis, Scolopax 176
stagnatilis, Tringa 176
stegmanni, Charadrius 170
steini, Climacteris 283
steini, Collocalia 160
steini, Meliphaga 315
steini, Monarcha 451
steini, Myzomela 292
steini, Poecilodryas 468
steini, Rallicula 125
stellae, Charmosyna 248
stenolophus, Microglossus 239
stenozona, Aviceda 201
stenozona, Baza 201
stentoreus, Acrocephalus 489
stenura, Gallinago 173
stenura, Scolopax 173
stephani, Chalcophaps 76
stephaniae, Astrapia 432
stephaniae, Astrarchia 432
Stercorariidae 197
Stercorarius 197
Sterna 194
Sternula 192
stevensi, Meliphaga 316
stictocephalus, Pycnonotus 306
stictocephalus, Pycnopygius 306
stictolaema, Cyanalcyon 228
stictolaemus, Todiramphus 228
Stiltia 186
stolida, Sterna 187
stolidus, Anous 187
stonii, Aeluraedus 275
stonii, Ailuroedus 275
stramineipectus, Pitohui 401
strenua, Campephaga 369
strepitans, Pomatorhinus 341
Streptopelia 64
streptophora, Rhipidura 410
stresemanni, Epimachus 429
stresemanni, Lanius 455
stresemanni, Megalurus 490
stresemanni, Melidectes 323
stresemanni, Melilestes 310
stresemanni, Micropsitta 269
stresemanni, Pachycephala 396
stresemanni, Sericornis 333

striata, Ardea 111
striata, Ardeola 111
striata, Butorides 111
striata, Geopelia 71
striativentris, Melanocharis 348
striativentris, Neneba 348
striatus, Butorides 111
strictipennis, Ibis 105
strictipennis, Threskiornis 105
Strigidae 215
Strigiformes 213
striolata, Cecropis 477
strophium, Ptilinopus 79
Sturnidae 494
Sturnus 497
stygia, Lonchura 519
suavissima, Cyclopsitta 260
subalaris, Amblyornis 277
subalaris, Coracina 370
subalaris, Graucalus 370
subalter, Manucodia 420
subaurantium, Pachycare 325
subcollaris, Psittacella 243
subcristata, Aviceda 201
subcristatus, Lepidogenys 201
subcyanea, Peneothello 471
subcyanea, Poecilodryas 471
subcyanocephalus, Eudynamis 140
subcyanocephalus, Eudynamys 140
subflavescens, Ducula 91
subfrenata, Caligavis 319
subfrenata, Meliphaga 319
subfrenata, Ptilotis 319
subfrenatus, Lichenostomus 319
subfrenatus, Oreornis 319
subminuta, Calidris 181
subminuta, Erolia 181
subminuta, Tringa 181
subpallida, Pachycare 326
subpallidum, Pachycare 326
subplacens, Charmosyna 246
subplacens, Trichoglossus 246
subrubra, Rallicula 124
subruficollis, Calidris 183
subruficollis, Tringa 183
subruficollis, Tryngites 183
subtuberosus, Philemon 298
sudestensis, Eopsaltria 398
sudestensis, Pachycephala 398
sudestiensis, Geoffroyus 266
sueurii, Lalage 373
Sula 118
sula, Pelecanus 119
sula, Sula 119
sulcirostris, Carbo 121
sulcirostris, Phalacrocorax 121
Sulidae 117
sumatrana, Ardea 113
sumatrana, Sterna 195
sumbae, Acrocephalus 488

superba, Columba 79
superba, Lophorina 426
superba, Paradisea 426
superbus, Ptilinopus 79
superciliaris, Drymodes 464
superciliosa, Anas 59
superciliosa, Colluricincla 389
superciliosus, Merops 220
superflua, Colluricincla 386
superflua, Dacelo 227
superfluus, Pinarolestes 386
sylvia, Sericornis 301
sylvia, Tanysiptera 224
Syma 231
Symposiachrus 446
Synoicus 51
syringanuchalis, Chalcopsittacus 256
szalayi, Mimeta 405
szalayi, Oriolus 405

tabuensis, Porzana 130
tabuensis, Rallus 130
tabuensis, Zapornia 130
Tachybaptus 62
tachycrypta, Colluricincla 389
Tadorna 56
tagulae, Aerodramus 160
tagulae, Collocalia 160
tagulana, Gerygone 339
tagulanum, Edoliosoma 377
tagulanum, Edolisoma 377
tagulanus, Philemon 298
tahitica, Hirundo 476
tahitiensis, Numenius 175
taitensis, Cuculus 140
taitensis, Eudynamys 141
taitensis, Urodynamis 140
taivana, Motacilla 521
Talegalla 45
Tanygnathus 267
Tanysiptera 222
tappenbecki, Colluricincla 386
tappenbecki, Musciparus 287
tarara, Ducula 91
tarara, Gerygone 337
tarara, Microeca 463
tararae, Xanthotis 294
tarrali, Globicera 87
tatei, Aegotheles 153
teerinki, Lonchura 519
telefolminensis, Pachycephala 396
telefolminensis, Rallicula 124
telescophthalmus, Arses 442
telescopthalmus, Arses 441
telescopthalmus, Muscicapa 441
temmincki, Orthonyx 342
temmincki, Ptilinopus 79
temporalis, Pomatorhinus 341
temporalis, Pomatostomus 341

tenebricosa, Tyto 213
tenebricosus, Strix 213
tenebrosa, Colluricincla 38, 388
tenebrosa, Gallinula 133
tenebrosa, Pachycephala 388
tenebrosus, Rhectes 388
tener, Loriculus 262
tentelare, Syma 232
tenuifrons, Zosterops 483
tenuirostre, Edolisoma 376
tenuirostris, Ardenna 101
tenuirostris, Cacatua 240
tenuirostris, Calidris 180
tenuirostris, Coracina 377
tenuirostris, Gracaulus 376
tenuirostris, Macropygia 67
tenuirostris, Procellaria 100
tenuirostris, Puffinus 100
tenuirostris, Totanus 180
terborghi, Aegotheles 155
terek, Tringa 178
terrestris, Trugon 72
testacea, Calidris 183
teysmanni, Cinnyris 509
Thalasseus 191
theomacha, Ninox 218
theomacha, Spiloglaux 218
thierfelderi, Centropus 137
thomasi, Melidectes 323
threnothorax, Rhipidura 408
Threskiornis 105
Threskiornithidae 105
tibialis, Rectes 403
tibicen, Coracias 365
tibicen, Cracticus 365
tibicen, Gymnorhina 365
tigrina, Columba 65
tigrina, Streptopelia 65
Timeliopsis 310
timoriensis, Cincloramphus 490
timoriensis, Megalurus 490
timoriensis, Mirafra 474
tobata, Tringa 184
Todiramphus 228
tommasonis, Edoliisoma 378
tommasonis, Edolisoma 378
torotoro, Halcyon 232
torotoro, Syma 231
torquata, Hypotaenidia 127
torquatus, Gallirallus 128
torquatus, Melidectes 320
torquatus, Rallus 127
torresii, Thalasseus 192
totanus, Scolopax 176
totanus, Tringa 176
toxopeusi, Daphoenositta 379
toxopeusi, Neositta 379
Toxorhamphus 351
transfreta, Cacatua 239
Tregellasia 472

tricarunculatus, Casuarius 42
Trichoglossus 253
trichroa, Erythrura 513
trichroa, Fringilla 513
tricolor, Ceblepyris 373
tricolor, Lalage 373
tricolor, Podiceps 62
tricolor, Rallina 126
tricolor, Tachybaptus 62
trigeminus, Ptilonopus 85
Tringa 176
tristis, Acridotheres 497
tristis, Corvus 453
tristis, Gymnocorvus 453
tristis, Paradisea 497
tristissima, Lonchura 514
tristissima, Munia 514
triton, Cacatua 240
trivialis, Philemon 297
trivirgata, Drymophila 448
trivirgatus, Monarcha 448
trivirgatus, Phylloscopus 481
trivirgatus, Symposiachrus 448
trobriandi, Lalage 374
trobriandi, Manucodia 419
trobriandi, Pachycephala 387
tropica, Fregetta 95
tropica, Thalassidroma 95
trouessarti, Pseudobulweria 99
Trugon 71
Tryngites 183
tschegrava, Hydroprogne 191
tschutschensis, Motacilla 521
Turdidae 498
Turdus 499
Turnicidae 184
Turnix 185
turtur, Pachyptila 98
tymbonomus, Cuculus 147
tyro, Dacelo 227
tyro, Sauromarptis 227
Tyto 213
Tytonidae 213

ultimus, Epimachus 429
ultramontanus, Dicrurus 414
umbrina, Colluricincla 388
umbrosa, Melanocharis 346
umbrosa, Ptiloprora 304
unappendiculatus, Casuarius 43
unicolor, Melanocharis 346
unicus, Pycnopygius 305
uniformis, Chlamydera 281
Urodynamis 140
Uroglaux 218
uropygialis, Ducula 88
uropygialis, Eupetes 358
uropygialis, Lonchura 518
uropygialis, Pitohui 403

uropygialis, Ptilorrhoa 358
uropygialis, Rectes 403
utakwensis, Ptilotis 319

vagans, Arachnothera 310
vagans, Melilestes 310
valia, Cyrtostomus 508
vana, Lonchura 516
vana, Munia 516
Vanellus 167
vanikorensis, Aerodramus 160
vanikorensis, Collocalia 160
vanikorensis, Hirundo 160
variegatus, Numenius 175
variegatus, Tantalus 175
variolosus, Cacomantis 147
variolosus, Cuculus 147
varius, Pelecanus 121
varius, Phalacrocorax 121
velox, Thalasseus 191
veredus, Charadrius 170
veredus, Eupoda 170
veredus, Eupodia 170
veronica, Leptocoma 507
versicolor, Gavicalis 317
versicolor, Lichenostomus 318
versicolor, Meliphaga 318
versicolor, Pitta 274
versicolor, Ptilotis 317
versteegi, Turdus 499
versteri, Melanocharis 347
versteri, Pristorhamphus 347
vicaria, Paecilodryas 468
vicaria, Peneothello 468
vicina, Cinnyris 507
vicina, Meliphaga 313
vicina, Ptilotis 313
vicinus, Ptilinopus 84
vicinus, Ptilopus 84
victoria, Goura 74
victoria, Lophyrus 74
victoriana, Orthonyx 342
victorianus, Orthonyx 342
vidua, Rhipidura 413
vieilloti, Sphecotera 404
vieilloti, Sphecotheres 404
violaceum, Dicaeum 504
violaceus, Mino 496
virago, Cyclopsitta 261
virago, Cyclopsittacus 261
virago, Pristorhamphus 347
virgata, Crateroscelis 333
virgatus, Sericornis 330, 333
viridescens, Rhamphocharis 349
viridicrissalis, Lorius 251
viridifrons, Meliphaga 318
viridipectus, Nasiterna 268
viridipectus, Pachycephala 392
viridis, Androphilus 354

viridis, Androphobus 354
viridis, Columba 83
viridis, Ptilinopus 83
viridis, Sphecotheres 405
visi, Ptilotis 294
visi, Xanthotis 294
vitiensis, Columba 64
vittata, Pachyptila 98
vittatum, Edolisoma 378
vulcani, Tanysiptera 223
vulgaris, Meliphaga 317
vulgaris, Sturnus 498
vulpes, Rhipidura 411

wadai, Goura 74
wahgiensis, Artamus 366
wahgiensis, Elanus 200
wahgiensis, Lonchura 518
wahgiensis, Megalurus 491
wahgiensis, Myzomela 289
wahgiensis, Neositta 379
wahgiensis, Rallus 128
wahgiensis, Saxicola 501
wahgiensis, Tregellasia 473
wahgiensis, Zosterops 485
wahnesi, Charmosyna 248
wahnesi, Gerygone 337
wahnesi, Parotia 422
wahnesi, Pseudogerygone 337
waigeuense, Edolisoma 378
waigeuense, Oedistoma 351
waigeuensis, Aviceda 201
waigeuensis, Collocalia 160
waigeuensis, Pachycephala 398
waigiouensis, Eurystomus 221
waigiuensis, Anas 57
waigiuensis, Melidora 226
waigiuensis, Salvadorina 57
wallaceana, Alcyone 234
wallacei, Eulipoa 47
wallacei, Megapodius 47
wallacii, Aegotheles 154
wallacii, Ptilinopus 81
wallacii, Ptilonopus 81
wallacii, Sipodotus 284
wallacii, Todopsis 284
wandamensis, Melanorectes 391
wandamensis, Pitohui 391
wardelli, Myiagra 443
wardelli, Piezorhynchus 443
weberi, Trichoglossus 254
websteri, Ceyx 234
weiskei, Cacomantis 146
weiskei, Eutolmaetus 204
weiskei, Hieraaetus 204
wellsi, Syma 232
weylandi, Sericornis 331
whiteheadi, Collocalia 160

wiedenfeldi, Aegotheles 155
wilhelminae, Aprosmictus 263
wilhelminae, Charmosyna 247
wilhelminae, Trichoglossus 247
wollastoni, Anthus 523
wondiwoi, Aegotheles 154
wondiwoi, Sericornis 331
wuroi, Myiolestes 385
wuroi, Sericornis 330
wuroi, Zosterops 485

xanthogenys, Machaerirhynchus 360
xanthorhynchus, Cuculus 144
Xanthotis 293
Xenus 178

yorki, Aerodramus 160
yorki, Caprimulgus 152
yorki, Collocalia 160
yorki, Eulabeornis 128
ypsilophora, Coturnix 51
ypsilophorus, Coturnix 51

Zapornia 130
zimmeri, Microeca 463
zoeae, Columba 90
zoeae, Ducula 90
Zonerodius 108
zonurus, Ptilinopus 81
zonurus, Ptilopus 81
Zoothera 498
Zosteropidae 481
Zosterops 482

140° east

of

ria

A

the island of
New Guinea

projection: Sinusoidal *center 136° east*
© 2012 Conservation International
cartography: M.Denil, K.Koenig
data: NASA SRTM
ESRI

elevation *meters*
2000
1000
500
100
50